Special Triangles

Name	Characteristic	Examples
Right Triangle	Triangle has a right angle.	
Isosceles Triangle	Triangle has two equal sides.	$AB = BC$
Equilateral Triangle	Triangle has three equal sides.	$AB = BC = CA$
Similar Triangles	Corresponding angles are equal; corresponding sides are proportional.	$A = D, B = E, C = F$ $$\frac{AB}{DE} = \frac{AC}{DF} = \frac{BC}{EF}$$

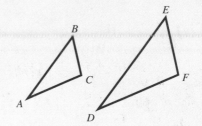

Beginning and Intermediate Algebra

Algebra

SECOND EDITION

BEGINNING AND INTERMEDIATE ALGEBRA

SECOND EDITION

Margaret L. Lial
American River College

John Hornsby
University of New Orleans

 **ADDISON-WESLEY**

An Imprint of Addison Wesley Longman, Inc.

Reading, Massachusetts • Menlo Park, California • New York • Harlow, England
Don Mills, Ontario • Sydney • Mexico City • Madrid • Amsterdam

Publisher: Jason Jordan

Acquisitions Editor: Jennifer Crum

Project Manager: Kari Heen

Developmental Editor: Terry McGinnis

Managing Editor: Ron Hampton

Production Supervisor: Kathleen A. Manley

Production Services: Elm Street Publishing Services, Inc.

Compositor: Typo-Graphics

Artists: Precision Graphics, Jim Bryant, Bob Giuliani, Darwin and Vally Hennings, and Gary Torrisi

Marketing Manager: Craig Bleyer

Prepress Buyer: Caroline Fell

Manufacturing Coordinator: Evelyn Beaton

Text and Cover Designer: Susan Carsten

Cover Illustration: © Peter Siu/SIS

Library of Congress Cataloging-in-Publication Data
Lial, Margaret L.
    Beginning and intermediate algebra.—2nd ed./Margaret L. Lial, John Hornsby.
      p.   cm.
    Includes index.
    ISBN 0-321-04133-X (student edition)
    ISBN 0-321-04130-5 (annot. inst. ed.)
    1. Algebra.   I. Hornsby, John.   II. Title.
QA152.2.L513   2000
512.9—dc21   99-22008
                                     CIP

Reprinted with corrections, May 2000

Printed in the U.S.A.

23456789-VH-02 01 00

Contents

CHAPTER 9   Linear Systems   509

CHAPTER 10   Roots and Radicals   594

CHAPTER 14 **Sequences and Series** **862**

APPENDIX A **Review of Decimals and Percents** **905**

APPENDIX B **Sets** **909**

APPENDIX C **Joint and Combined Variation** **914**

APPENDIX D **Synthetic Division** **919**

APPENDIX E **Review of Exponents, Polynomials, and Factoring (Transition from Beginning to Intermediate Algebra)** **924**

Preface

The second edition of *Beginning and Intermediate Algebra* is designed to accommodate instructors and students who wish to eliminate the topic overlap between typical beginning and intermediate algebra texts. This text integrates the appropriate topics from the successful Lial/Hornsby parent texts, *Beginning Algebra,* Eighth Edition, and *Intermediate Algebra,* Eighth Edition, in one book and features a flexible Table of Contents that can be easily adapted for a variety of course structures.

This revision of *Beginning and Intermediate Algebra* reflects our ongoing commitment to creating the best possible text and supplements package using the most up-to-date strategies for helping students succeed. One of these strategies, evident in our new Table of Contents and consistent with current teaching practices, involves the early introduction of functions and graphing lines in a rectangular coordinate system. We believe that this pedagogy has a great deal of merit as it provides students with the important "input-output" concept that will be an integral part of later mathematics courses. This organization also allows an early treatment of interesting interpretations of data in the form of line and bar graphs—two pictorial representations that students already see on a daily basis in magazines and newspapers. Chapter 4 introduces three sections on linear equations in two variables, ordered pairs, graphing, and slope, with a gentle introduction to the function concept in the form of input-output relationships. This allows students to read graphs in the chapters that immediately follow and to slowly develop an understanding of the basic idea of a function. Chapter 8 reviews lines and graphing presented in Chapter 4, then introduces the more involved concepts of forms of equations of lines and inequalities in two variables. The function concept is addressed again here, this time with a thorough discussion of domain, range, function notation, and other function-related topics. Also consistent with this approach, graphs of quadratic equations are included earlier in the text when quadratic equations are solved (Chapter 11) rather than with the material on conic sections as in the previous edition.

If you choose not to cover graphing linear equations and functions earlier as the new edition suggests, you can defer Chapter 4 and combine it with Chapter 8 as in the previous edition. Material on graphing in Section 5.1 can easily be omitted. Applied problems in Chapters 5–7 that refer to the function concept can be used without actually working through Chapter 4. Alternatively, the material in Chapter 8 can be covered immediately after Chapter 4, if desired.

Other up-to-date pedagogical strategies to foster student success include a strong emphasis on vocabulary and problem solving, an increased number of real world applications in both examples and exercises, and a focus on relevant industry themes throughout the text.

Another strategy for student success, an exciting new CD-ROM called "Pass the Test," debuts with this edition of *Beginning and Intermediate Algebra.* Directly correlated to the text's content, "Pass the Test" helps students master concepts by providing interactive pretests, chapter tests, section reviews, and InterAct Math tutorial exercises.

To support an increased emphasis on graphical manipulation, the CD-ROM also includes a graphing tool that can be used for open-ended, student-directed exploration of number lines and coordinate graphs, as well as for exercises relevant to the graphing content throughout the book.

The *Student's Study Guide and Journal,* redesigned and enhanced with an optional journal feature for those who would like to incorporate more writing in their mathematics curriculum, provides an additional strategy for student success.

Although *Beginning and Intermediate Algebra,* Second Edition, integrates many new elements, it also retains the time-tested features of the Lial/Hornsby series: learning objectives for each section, careful exposition, fully developed examples, Cautions and Notes, and design features that highlight important definitions, rules, and procedures. Since the hallmark of any mathematics text is the quality of its exercise sets, we have carefully developed exercise sets that provide ample opportunity for drill and, at the same time, test conceptual understanding. In preparing this edition we have also addressed the standards of the National Council of Teachers of Mathematics and the American Mathematical Association of Two-Year Colleges, incorporating many new exercises focusing on concepts, writing, graph interpretation, technology use, collaborative work, and analysis of data from a wide variety of sources in the world around us.

CONTENT CHANGES

We have fine-tuned and polished presentations of topics throughout the text based on user and reviewer feedback. Some of the content changes you may notice include the following:

- Operations with real numbers are consolidated from four sections to two sections in Chapter 1.

- We consistently emphasize problem solving using a six-step problem-solving strategy, first introduced in Section 2.3 and continually reinforced in examples and the exercise sets throughout the text. Section 2.6 contains a comprehensive discussion of problem solving.

- New Chapter 4 introduces graphing and slope, along with an intuitive introduction to functions using input-output. Equations with two variables are presented earlier so that the applications used in later chapters (for example, with polynomials and rational expressions) can be more realistic and relevant. Both graphing and functions are continued in later chapters.

- New material is included on graphing parabolas in Section 5.1.

- Division of polynomials is consolidated in Section 5.6.

- For increased flexibility, Chapter 8 begins with a review section on graphs and slopes of lines. The presentation of functions in Section 8.4 now includes new examples with more visual representations of function concepts.

- Material on determinants and Cramer's rule is consolidated in Section 9.7.

- Quadratic functions and graphs are presented earlier in the text when quadratic equations are solved (Chapter 11) rather than with the material on conic sections, as in the previous edition.

- The material on inverse, exponential, and logarithmic functions (Chapter 12) is now placed before conic sections (Chapter 13) instead of after it.

• For additional flexibility, three new appendixes have been included. Because of the many requests for "transitional" material, Appendix E, in conjunction with the Summary exercise group in Chapter 2, form an excellent package of review topics between beginning and intermediate algebra.

NEW FEATURES

We believe students and instructors will welcome the following new features:

Industry Themes To help motivate the material, each chapter features a particular industry that is presented in the chapter opener and revisited in examples and exercises in the chapter. Identified by special icons, these examples and exercises incorporate sourced data, often in the form of graphs and tables. Featured industries include business, health care, entertainment, sports, transportation, and others. (See pages 1, 88, and 199.)

New Examples and Exercises We have added 25% more real application problems with data sources. These examples and exercises often relate to the industry themes. They are designed to show students how algebra is used to describe and interpret data in everyday life. (See pages 9, 178, and 235.)

The Olympic Committee has come to rely more and more on television rights and major corporate sponsors to finance the games. The pie charts show the funding plans for the first Olympics in Athens and the 1996 Olympics in Atlanta. Use proportions and the figures to answer the questions in Exercises 35 and 36.

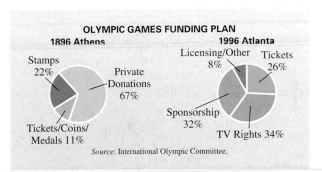

35. In the 1996 Olympics, total revenue of $350 million was raised. There were 10 major sponsors.
   (a) Write a proportion to find the amount of revenue provided by tickets. Solve it.
   (b) What amount was provided by sponsors? Assuming the sponsors contributed equally, how much was provided per sponsor?
   (c) What amount was raised by TV rights?

36. Suppose the amount of revenue raised in the 1896 Olympics was equivalent to the $350 million in 1996.
   (a) Write a proportion for the amount of revenue provided by stamps and solve it.
   (b) What amount (in dollars) would have been provided by private donations?
   (c) In the 1988 Olympics, there were 9 major sponsors, and the total revenue was $95 million. What is the ratio of major sponsors in 1988 to those in 1996? What is the ratio of revenue in 1988 to revenue in 1996?

Technology Insights Exercises   Technology is part of our lives, and we assume that all students of this text have access to scientific calculators. *While graphing calculators are not required for this text,* it is likely that students will go on to courses that use them. For this reason, we have included Technology Insights exercises in selected exercise sets. These exercises illustrate the power of graphing calculators and provide an opportunity for students to interpret typical results seen on graphing calculator screens. (See pages 221, 236, and 297.)

Mathematical Journal Exercises   While we continue to include conceptual and writing exercises that require short written answers, new journal exercises have been added that ask students to fully explain terminology, procedures, and methods, document their understanding using examples, or make connections between concepts. Instructors who wish to incorporate a journal component in their classes will find these exercises especially useful. For the greatest possible flexibility, both writing exercises and journal exercises are indicated with icons in the Annotated Instructor's Edition, but not in the Student Edition. (See pages 35, 103, and 271.)

Group Activities   Appearing at the end of each chapter, these activities allow students to apply the industry theme of the chapter to its mathematical content in a collaborative setting. (See pages 149, 190, and 300.)

Test Your Word Power   To help students understand and master mathematical vocabulary, this new feature has been incorporated at the end of each chapter. Key terms from the chapter are presented with four possible definitions in multiple-choice format. Answers and examples illustrating each term are provided at the bottom of the appropriate page. (See pages 76, 238, and 301.)

Transitional Material   Appendix E and the Summary exercises in Chapter 2 provide material for instructors who want to review beginning algebra topics before starting intermediate algebra. (See pages 190 and 924.)

HALLMARK FEATURES

We have retained the popular features of the previous edition of the text. Some of these features are as follows:

Learning Objectives   Each section begins with clearly stated numbered objectives, and material in the section is keyed to these objectives. In this way students know exactly what is being covered in each section.

OBJECTIVES

1   Learn the definition of *factor*.

2   Write fractions in lowest terms.

3   Multiply and divide fractions.

4   Add and subtract fractions.

5   Solve applied problems that involve fractions.

6   Interpret data in a circle graph.

Cautions and Notes We often give students warnings of common errors and emphasize important ideas in Cautions and Notes that appear throughout the exposition.

Connections Retained from the previous edition, Connections boxes have been streamlined and now often appear at the beginning or the end of the exposition in selected sections. They continue to provide connections to the real world or to other mathematical concepts, historical background, and thought-provoking questions for writing or class discussion. (See pages 105, 225, and 250.)

Problem Solving Increased emphasis has been given to our six-step problem-solving method to aid students in solving application problems. This method is continually reinforced in examples and exercises throughout the text. (See pages 105, 107, and 354.)

Ample and Varied Exercise Sets Algebra students require a large number and variety of practice exercises to master the material. This text contains approximately 7700 exercises, including over 2100 review exercises, plus numerous conceptual and writing exercises, journal exercises, and challenging exercises that go beyond the examples. More illustrations, diagrams, graphs, and tables now accompany exercises. Multiple-choice, matching, true/false, and completion exercises help to provide variety. Exercises suitable for calculator use are marked with a calculator icon ▦ in both the Student Edition and the Annotated Instructor's Edition. (See pages 34, 143, and 170.)

Relating Concepts Previously titled Mathematical Connections, these sets of exercises often appear near the end of selected sections. They tie together topics and highlight the relationships among various concepts and skills. For example, they may show how algebra and geometry are related, or how a graph of a linear equation in two variables is related to the solution of the corresponding linear equation in one variable. Instructors have told us that these sets of exercises make great collaborative activities for small groups of students. (See pages 69, 234, and 256.)

Ample Opportunity for Review Each chapter concludes with a Chapter Summary that features Key Terms and Symbols, Test Your Word Power, and a Quick Review of each section's content. Chapter Review Exercises keyed to individual sections are included as well as mixed review exercises and a Chapter Test. Following every chapter after Chapter 1, there is a set of Cumulative Review Exercises that covers material going back to the first chapter. Students always have an opportunity to review material that appears earlier in the text, and this provides an excellent way to prepare for the final examination in the course. (See pages 192–198 and 238–246.)

SUPPLEMENTS

Our extensive supplements package includes the Annotated Instructor's Edition, testing materials, study guides, solutions manuals, CD-ROM software, videotapes, and a Web site. For more information on these and other helpful supplements, contact your Addison Wesley Longman sales representative.

Annotated Instructor's Edition (ISBN 0-321-04130-5)

For immediate access, the Annotated Instructor's Edition provides answers to all text exercises and Group Activities in color in the margin or next to the corresponding exercise, as well as Chalkboard Examples and Teaching Tips. To assist instructors in assigning homework, additional icons not shown in the Student Edition indicate journal exercises ▤ , writing exercises ✎ , and challenging exercises ▲ .

Exercises designed for calculator use ▦ , ▦ are indicated in both the Student Edition and the Annotated Instructor's Edition.

Instructor's Solutions Manual (ISBN 0-321-06215-9)

The *Instructor's Solutions Manual* provides solutions to all exercises, including answer art, and lists of all writing, journal, challenging, Relating Concepts, and calculator exercises.

Answer Book (ISBN 0-321-06216-7)

The *Answer Book* contains answers to all exercises and lists of all writing, journal, challenging, Relating Concepts, and calculator exercises. Instructors may ask the bookstore to order multiple copies of the *Answer Book* for students to purchase.

Printed Test Bank (ISBN 0-321-06214-0)

The *Printed Test Bank* contains short answer and multiple-choice versions of a placement test and final exam; six forms of chapter tests for each chapter, including four open-response (short answer) and two multiple-choice forms; 10 to 20 additional exercises per objective for instructors to use for extra practice, quizzes, or tests; answer keys to all of the above listed tests and exercises; and lists of all writing, journal, challenging, Relating Concepts, and calculator exercises.

 TestGen-EQ with QuizMaster EQ (ISBN 0-321-06217-5)

This fully networkable software presents a friendly graphical interface which enables professors to build, edit, view, print and administer tests. Tests can be printed or easily exported to HTML so they can be posted to the Web for student practice.

FOR THE STUDENT

Student's Study Guide and Journal (ISBN 0-321-06220-5)
The *Student's Study Guide and Journal* contains a "Chart Your Progress" feature for students to track their scores on homework assignments, quizzes, and tests, additional practice for each learning objective, section summary outlines that give students additional writing opportunities and help with test preparation, and self-tests with answers at the end of each chapter. A manual icon at the beginning of each section in the Student Edition identifies section coverage.

Student's Solutions Manual (ISBN 0-321-06219-1)
The *Student's Solutions Manual* provides solutions to all odd-numbered exercises (journal and writing exercises excepted). A manual icon at the beginning of each section in the Student Edition identifies section coverage.

InterAct Math Tutorial Software (ISBN 0-321-06218-3 (Student Version))
This tutorial software correlates with every odd-numbered exercise in the text. The program is highly interactive with sample problems and interactive guided solutions accompanying every exercise. The program recognizes common student errors and provides customized feedback with sophisticated answer recognition capabilities. The management system (InterAct Math Plus) allows instructors to create, administer, and track tests, and to monitor student performance during practice sessions.

"Real to Reel" Videotapes (0-321-05662-0)
This videotape series provides separate lessons for each section in the book. A videotape icon at the beginning of each section identifies section coverage. All objectives, topics, and problem-solving techniques are covered and content is specific to *Beginning and Intermediate Algebra,* Second Edition.

"Pass the Test" Interactive CD-ROM (ISBN 0-321-06206-X)
This CD helps students to master the course content by providing interactive pre-tests, chapter tests, section reviews, and InterAct tutorial exercises. After studying a chapter in class, students take a pre-test to determine what areas in that chapter need additional work. They are then directed to section reviews and tutorial exercises for continued practice. Students continue to take chapter tests and practice their skills until they have mastered the chapter. A unique graphing tool is provided for exploring the relationship between graphs and their algebraic representation.

World Wide Web Supplement (www.LialAlgebra.com)
Students can visit the Web site to explore additional real world applications related to the chapter themes, look up words in a complete glossary, and work through graphing calculator tutorials. The tutorials consist of step-by-step procedures as well as practice exercises for mastering basic graphing calculator skills.

MathXL (http://www.mathxl.com)
Available on-line with a pre-assigned ID and password by ordering a new copy of *Beginning and Intermediate Algebra,* Second Edition, with ISBN 0-201-68367-9, MathXL helps students prepare for tests by allowing them to take practice tests that are similar to the chapter tests in their text. Students also get a personalized study plan that identifies strengths and pinpoints topics where more review is needed. For more information on subscriptions, contact your Addison Wesley Longman sales representative.

Math Tutor Center

The Addison Wesley Longman Math Tutor Center is staffed by qualified mathematics instructors who provide students with tutoring on text examples, exercises, and problems. Tutoring assistance is provided by telephone, fax, and e-mail and is available five days a week, seven hours a day. A registration number for the tutoring service may be obtained by calling our toll-free customer service number, 1-800-922-0579 and requesting ISBN 0-201-44461-5. Registration for the service is active for one or more of the following time periods depending on the course duration: Fall (8/31–1/31), Spring (1/1–6/30), or Summer (5/1–8/31). The Math Tutor Center service is also available for other Addison Wesley Longman textbooks in developmental math, precalculus math, liberal arts math, applied math, applied calculus, calculus, and introductory statistics. For more information, please contact your Addison Wesley Longman sales representative.

Spanish Glossary (ISBN 0-321-01647-5)

This book includes math terms that would be encountered in Basic Math through College Algebra.

ACKNOWLEDGMENTS

For a series to last through eight editions, it is necessary for the authors to rely on comments, criticisms, and suggestions of users, nonusers, instructors, and students. We are grateful for the many responses that we have received over the years. We wish to thank the following individuals who reviewed this edition of the text:

Josette Ahlering, *Central Missouri State University*

Vickie Aldrich, *Dona Ana Branch Community College*

Jose Alvarado, *University of Texas–Pan American*

Lisa Anderson, *Ventura College*

Robert B. Baer, *Miami University–Hamilton Campus*

Judy Barclay, *Cuesta College*

Peter Blaskiewicz, *McClennan Community College*

Julie R. Bonds, *Sonoma State University*

Beverly R. Broomell, *Suffolk Community College*

Marc D. Campbell, *Daytona Beach Community College*

Cheryl V. Cantwell, *Seminole Community College*

Stanley Carter, *Central Missouri State University*

Don C. Chandler, *El Paso Community College*

Terry Cheng, *Irvine Valley College*

Jeff Clark, *Santa Rosa Junior College*

Ted Corley, *Glendale Community College*

Charles N. Curtis, *Missouri Southern State College*

Lisa Delong Cuneo, *Pennsylvania State University–Dubois Campus*

Marlene Demerjian, *College of the Canyons*

Glenn DiStefano, *Louisiana State University at Alexandria*

Richard N. Dodge, *Jackson Community College*

Warrene C. Ferry, *Jones County Junior College*

Jeanne Fitzgerald, *Phoenix College*

Linda Franko, *Cuyahoga Community College*

Linda L. Galloway, *Macon State College*

Theresa A. Geiger, *St. Petersburg Junior College*

Steve Grosteffon, *Santa Fe Community College*

Martha Haehl, *Maple Woods Community College*

Adam Hall, *Belleville Area College*

Melissa Harper, *Embry-Riddle Aeronautical University*

Margret Hathaway, *Kansas City Kansas Community College*

W. Hildebrand, *Montgomery College*

Matthew Hudock, *St. Philips College*

Dale W. Hughes, *Johnson County Community College*

Linda Hurst, *Central Texas College*

Nancy Johnson, *Broward Community College–North*

Robert Kaiden, *Lorain County Community College*

Michael Karelius, *American River College*

Judy Kasabian, *El Camino College*

Margaret Kimbell, *Texas State Technical College*

Linda Kodama, *Kapiolani Community College*

Jeff A. Koleno, *Lorain County Community College*

Inessa Levi, *University of Louisville*

Mitchel Levy, *Broward Community College–Central*

William R. Livingston, *Missouri Southern State College*

Ellen Lund, *San Jacinto College*

Linda Marable, *Nashville State Technical Institute*

Doug Martin, *Mt. San Antonio College*

Jim Matovina, *Community College of Southern Nevada*

Eric Matsuoka, *Leeward Community College*

Philip Meyer, *Skyline College*

Larry Mills, *Johnson County Community College*

Mary Ann (Molly) Misko, *Gadsden State Community College*

Michael Montano, *Riverside Community College*

Lionel Mordecai, *Southwestern College*

Elsie Newman, *Owens Community College*

Kathy Nickell, *College of DuPage*

Joanne V. Peeples, *El Paso Community College*

Martin Peres, *Broward Community College–North*

Robert J. Rapalje, *Seminole Community College*

Janice Rech, *University of Nebraska–Omaha*

Dale Rice, *Belleville Area College*

Dale Rohm, *University of Wisconsin–Stevens Point*

Joyce Saxon, *Morehead University*

Richard Semmler, *Northern Virginia Community College*

Steve Shattuck, *Central Missouri State University*

Kathleen A. Smith, *University of Central Arkansas*

LeeAnn Spahr, *Durham Technical Community College*

Elizbeth Suco, *Miami Dade Community College–Wolfson*

Bettie A. Truit, *Black Hawk Community College*

Gail Wiltse, *St. John's River Community College*

Alice Woolam, *Pensacola Junior College*

No author can complete a project of this magnitude without the help of many other individuals. Our sincere thanks go to Jenny Crum of Addison Wesley Longman who coordinated the package of texts of which this book is a part. Other dedicated staff at Addison Wesley Longman who worked long and hard to make this revision a success include Jason Jordan, Kari Heen, Susan Carsten, Meredith Nightingale, and Kathy Manley.

While Terry McGinnis has assisted us for many years "behind the scenes" in producing our texts, she has contributed far more to these revisions than ever. There is no question that these books are improved because of her attention to detail and consistency, and we are most grateful for her work above and beyond the call of duty. Kitty Pellissier continues to do an outstanding job in checking the answers to exercises.

Steven Pusztai of Elm Street Publishing Services provided excellent production work. As usual, Paul Van Erden created an accurate, useful index. Becky Troutman prepared the Index of Applications. We are also grateful to Tommy Thompson who made suggestions for the feature "For the Student: 10 Ways to Succeed with Algebra," to Vickie Aldrich, Lucy Gurrola, and Sharon North who wrote the Group Activity features, and to Janis Cimperman of St. Cloud University and Steve C. Ouellette of the Walpole Massachusetts State Public Schools who provided additional answer-checking.

We wish to add a special thank you to Harold Hiken of the University of Wisconsin–Milwaukee, who has been a supporter of the Lial family of textbooks for many years. His input, suggestions, and comments have been most helpful, and we are grateful for his dedication to our books.

To these individuals and all the others who have worked on this series for 30 years, remember that we could not have done it without you. We hope that you share with us our pride in these books.

Margaret L. Lial
John Hornsby

An Introduction to Calculators

There is little doubt that the appearance of handheld calculators nearly three decades ago and the later development of scientific and graphing calculators have changed the methods of learning and studying mathematics forever. Where the study of computations with tables of logarithms and slide rules made up an important part of mathematics courses prior to 1970, today the widespread availability of calculators make their study a topic only of historical significance.

Most consumer models of calculators are inexpensive. At first, however, they were costly. One of the first consumer models available was the Texas Instruments SR-10, which sold for about $150 in 1973. It could perform the four operations of arithmetic and take square roots, but could do very little more.

Today calculators come in a large array of different types, sizes, and prices. *For the course for which this textbook is intended, the most appropriate type is the scientific calculator,* which costs $10–$20.

In this introduction, we explain some of the features of scientific and graphing calculators. However, remember that calculators vary among manufacturers and models, and that while the methods explained here apply to many of them, they may not apply to your specific calculator. For this reason, it is important to remember that *this introduction is only a guide, and is not intended to take the place of your owner's manual.* Always refer to the manual in the event you need an explanation of how to perform a particular operation.

SCIENTIFIC CALCULATORS

Scientific calculators are capable of much more than the typical four-function calculator that you might use for balancing your checkbook. Most scientific calculators use *algebraic logic.* (Models sold by Texas Instruments, Sharp, Casio, and Radio Shack, for example, use algebraic logic.) A notable exception is Hewlett Packard, a company whose calculators use *Reverse Polish Notation* (RPN). In this introduction, we explain the use of calculators with algebraic logic.

ARITHMETIC OPERATIONS

To perform an operation of arithmetic, simply enter the first number, press the operation key ($+$, $-$, $\times$, or $\div$), enter the second number, and then press the $=$ key. For example, to add 4 and 3, use the following keystrokes.

$$\boxed{4}\ \boxed{+}\ \boxed{3}\ \boxed{=}\ \boxed{\qquad\qquad 7}$$

CHANGE SIGN KEY

The key marked $\boxed{\pm}$ allows you to change the sign of a display. This is particularly useful when you wish to enter a negative number. For example, to enter -3, use the following keystrokes.

$$\boxed{3}\ \boxed{\pm}\ \boxed{\qquad\quad -3}$$

MEMORY KEY

Scientific calculators can hold a number in memory for later use. The label of the memory key varies among models; two of these are $\boxed{M}$ and $\boxed{STO}$. $\boxed{M+}$ and $\boxed{M-}$ allow you to add to or subtract from the value currently in memory. The memory recall key, labeled $\boxed{\text{MR}}$, $\boxed{\text{RM}}$, or $\boxed{\text{RCL}}$, allows you to retrieve the value stored in memory.

Suppose that you wish to store the number 5 in memory. Enter 5, then press the key for memory. You can then perform other calculations. When you need to retrieve the 5, press the key for memory recall.

If a calculator has a constant memory feature, the value in memory will be retained even after the power is turned off. Some advanced calculators have more than one memory. It is best to read the owner's manual for your model to see exactly how memory is activated.

CLEARING/CLEAR ENTRY KEYS

These keys allow you to clear the display or clear the last entry entered into the display. They are usually marked $\boxed{\text{C}}$ and $\boxed{\text{CE}}$. In some models, pressing the $\boxed{\text{C}}$ key once will clear the last entry, while pressing it twice will clear the entire operation in progress.

SECOND FUNCTION KEY

This key is used in conjunction with another key to activate a function that is printed *above* an operation key (and not on the key itself). It is usually marked $\boxed{\text{2nd}}$. For example, suppose you wish to find the square of a number, and the squaring function (explained in more detail later) is printed above another key. You would need to press $\boxed{\text{2nd}}$ before the desired squaring function can be activated.

SQUARE ROOT KEY

Pressing the square root key, $\boxed{\sqrt{x}}$, will give the square root (or an approximation of the square root) of the number in the display. For example, to find the square root of 36, use the following keystrokes.

$$\boxed{3} \ \boxed{6} \ \boxed{\sqrt{x}} \qquad \boxed{ 6}$$

The square root of 2 is an example of an irrational number (Chapter 10). The calculator will give an approximation of its value, since the decimal for $\sqrt{2}$ never terminates and never repeats. The number of digits shown will vary among models. To find an approximation of $\sqrt{2}$, use the following keystrokes.

$$\boxed{2} \ \boxed{\sqrt{x}} \qquad \boxed{1.4142136} \qquad \text{An approximation}$$

SQUARING KEY

This key, $\boxed{x^2}$, allows you to square the entry in the display. For example, to square 35.7, use the following keystrokes.

$$\boxed{3} \ \boxed{5} \ \boxed{.} \ \boxed{7} \ \boxed{x^2} \qquad \boxed{1274.49}$$

The squaring key and the square root key are often found on the same key, with one of them being a second function (that is, activated by the second function key, described above).

RECIPROCAL KEY

The key marked $\boxed{1/x}$ is the reciprocal key. (When two numbers have a product of 1, they are called *reciprocals*. See Chapter 1.) Suppose that you wish to find the reciprocal of 5. Use the following keystrokes.

$$\boxed{5} \ \boxed{1/x} \qquad \boxed{0.2}$$

INVERSE KEY

Some calculators have an inverse key, marked $\boxed{\text{INV}}$. Inverse operations are operations that "undo" each other. For example, the operations of squaring and taking the square root are inverse operations. The use of the $\boxed{\text{INV}}$ key varies among different models of calculators, so read your owner's manual carefully.

EXPONENTIAL KEY

The key marked $\boxed{x^y}$ or $\boxed{y^x}$ allows you to raise a number to a power. For example, if you wish to raise 4 to the fifth power (that is, find 4^5, as explained in Chapter 5), use the following keystrokes.

ROOT KEY

Some calculators have this key specifically marked $\boxed{\sqrt[x]{}}$ or $\boxed{\sqrt[y]{}}$; with others, the operation of taking roots is accomplished by using the inverse key in conjunction with the exponential key. Suppose, for example, your calculator is of the latter type and you wish to find the fifth root of 1024. Use the following keystrokes.

Notice how this "undoes" the operation explained in the exponential key discussion above.

PI KEY

The number π is an important number in mathematics. It occurs, for example, in the area and circumference formulas for a circle. By pressing the $\boxed{\pi}$ key, you can display the first few digits of π. (Because π is irrational, the display shows only an approximation.) One popular model gives the following display when the $\boxed{\pi}$ key is pressed:

$$\boxed{\qquad 3.1415927\qquad}.$$

METHODS OF DISPLAY

When decimal approximations are shown on scientific calculators, they are either *truncated* or *rounded*. To see how a particular model is programmed, evaluate 1/18 as an example. If the display shows .0555555 (last digit 5), it truncates the display. If it shows .0555556 (last digit 6), it rounds off the display.

When very large or very small numbers are obtained as answers, scientific calculators often express these numbers in scientific notation (Chapter 5). For example, if you multiply 6,265,804 by 8,980,591, the display might look like this:

$$\boxed{5.6270623 \qquad 13}.$$

The "13" at the far right means that the number on the left is multiplied by 10^{13}. This means that the decimal point must be moved 13 places to the right if the answer is to be expressed in its usual form. Even then, the value obtained will only be an approximation: 56,270,623,000,000.

GRAPHING CALCULATORS

Graphing calculators are becoming increasingly popular in mathematics classrooms. While you are not expected to have a graphing calculator to study from this book, we do include a feature in many exercise sets called *Technology Insights* that asks you to interpret typical graphing calculator screens. These exercises can help to prepare you for future courses where graphing calculators may be recommended or even required.

BASIC FEATURES

Graphing calculators provide many features beyond those found on scientific calculators. In addition to the typical keys found on scientific calculators, they have keys that can be used to create graphs, make tables, analyze data, and change settings. One of the major differences between graphing and scientific calculators is that a graphing calculator has a larger viewing screen with graphing capabilities. The screens below illustrate the graphs of $y = x$ and $y = x^2$.

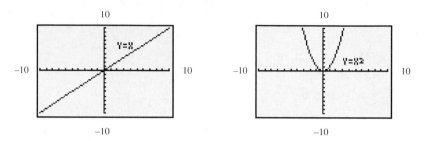

If you look closely at the screens, you will see that the graphs appear to be "jagged" rather than smooth, as they should be. The reason for this is that graphing calculators have much lower resolution than a computer screen. Because of this, graphs generated by graphing calculators must be interpreted carefully.

EDITING INPUT

The screen of a graphing calculator can display several lines of text at a time. This feature allows you to view both previous and current expressions. If an incorrect expression is entered, an error message is displayed. The erroneous expression can be viewed and corrected by using various editing keys, much like a word-processing program. You do not need to enter the entire expression again. Many graphing calculators can also recall past expressions for editing or updating. The screen on the left below shows how two expressions are evaluated. The final line is entered incorrectly, and the resulting error message is shown in the screen on the right.

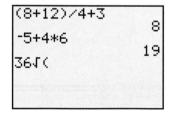   

ORDER OF OPERATIONS

Arithmetic operations on graphing calculators are usually entered as they are written in mathematical equations. For example, to evaluate $\sqrt{36}$ on a typical scientific calculator, you would first enter 36 and then press the square root key. As seen above, this is not the correct syntax for a graphing calculator. To find this root, you would first press the square root key, and then enter 36. See the screen on the left at the top of the next page. The order of operations on a graphing calculator is also important, and current models assist the user by inserting parentheses when typical errors might occur. The open parenthesis that follows the square root symbol is automatically entered by the calculator, so that an expression such as $\sqrt{2 \times 8}$ will not be calculated incorrectly as $\sqrt{2} \times 8$. Compare the two entries and their results in the screen on the right.

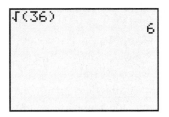

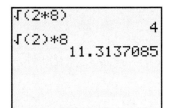

VIEWING WINDOWS

The viewing window for a graphing calculator is similar to the viewfinder in a camera. A camera usually cannot take a photograph of an entire view of a scene. The camera must be centered on some object and can only capture a portion of the available scenery. A camera with a zoom lens can photograph different views of the same scene by zooming in and out. Graphing calculators have similar capabilities. The xy-coordinate plane is infinite. The calculator screen can only show a finite, rectangular region in the plane, and it must be specified before the graph can be drawn. This is done by setting both minimum and maximum values for the x- and y-axes. The scale (distance between tick marks) is usually specified as well. Determining an appropriate viewing window for a graph is often a challenge, and many times it will take a few attempts before a satisfactory window is found.

The screen on the left shows a "standard" viewing window, and the graph of $y = 2x + 1$ is shown on the right. Using a different window would give a different view of the line.

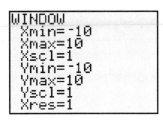

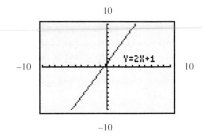

LOCATING POINTS ON A GRAPH: TRACING AND TABLES

Graphing calculators allow you to trace along the graph of an equation, and, while doing this, display the coordinates of points on the graph. See the screen on the left at the top of the next page, which indicates that the point (2, 5) lies on the graph of $y = 2x + 1$. Tables for equations can also be displayed. The screen on the right shows a partial table for this same equation. Note the middle of the screen, which indicates that when $x = 2$, $y = 5$.

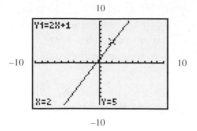

ADDITIONAL FEATURES

There are many features of graphing calculators that go far beyond the scope of this book. These calculators can be programmed, much like computers. Many of them can solve equations at the stroke of a key, analyze statistical data, and perform symbolic algebraic manipulations. Mathematicians from the past would have been amazed by today's calculators. Many important equations in mathematics cannot be solved by hand. However, their solutions can often be approximated using a calculator. Calculators also provide the opportunity to ask "What if . . . ?" more easily. Values in algebraic expressions can be altered and conjectures tested quickly.

FINAL COMMENTS

Despite the power of today's calculators, they cannot replace human thought. **In the entire problem-solving process, your brain is the most important component.** Calculators are only tools, and like any tool, they must be used appropriately in order to enhance our ability to understand mathematics. Mathematical insight may often be the quickest and easiest way to solve a problem; a calculator may neither be needed nor appropriate. By applying mathematical concepts, you can make the decision whether or not to use a calculator.

Beginning

and Intermediate

Algebra

SECOND EDITION

The Real Number System

Growth in productivity—the amount of goods and services produced for each hour of work—is the most important factor in improving living standards of our country's population. In the latter part of the 1990s, the U.S. construction industry

finally began to revive after a long period of decline. Having survived many years of downsizing and layoffs, manufacturing-related industries are again contributing to economic growth. This bar graph shows the increase in construction spending in the latter half of 1996 and all of 1997. During which month and year was construction spending highest? Several exercises in Section 1.2 use this graph. Throughout this chapter, we will see other specific examples and exercises that relate to construction and manufacturing.

Construction/ Manufacturing

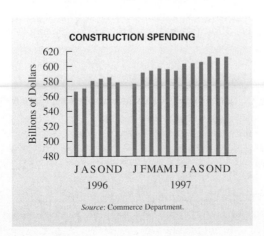

CONSTRUCTION SPENDING

Billions of Dollars

620
600
580
560
540
520
500
480

J A S O N D   J F M A M J J A S O N D
1996   1997

Source: Commerce Department.

1.1 Fractions

OBJECTIVES

 Learn the definition of *factor*.

2 Write fractions in lowest terms.

3 Multiply and divide fractions.

4 Add and subtract fractions.

5 Solve applied problems that involve fractions.

6 Interpret data in a circle graph.

FOR EXTRA HELP

SSG Sec. 1.1
SSM Sec. 1.1

Pass the Test Software

InterAct Math
   Tutorial Software

Video 1

As preparation for the study of algebra, this section begins with a brief review of arithmetic. In everyday life the numbers seen most often are the **natural numbers,**

$$1, 2, 3, 4, \ldots ,$$

the **whole numbers,**

$$0, 1, 2, 3, 4, \ldots ,$$

and **fractions,** such as

$$\frac{1}{2}, \quad \frac{2}{3}, \quad \text{and} \quad \frac{15}{7}.$$

The parts of a fraction are named as follows.

$$\text{Fraction bar} \rightarrow \frac{4}{7} \quad \left(\frac{a}{b} = a \div b \right) \quad \begin{array}{l} \leftarrow \text{Numerator} \\ \leftarrow \text{Denominator} \end{array}$$

As we will see later, the fraction bar represents division $\left(\dfrac{a}{b} = a \div b \right)$ and also serves as a grouping symbol.

CONNECTIONS

A common use of fractions is to measure dimensions of tools, amounts of building materials, and so on. We need to add, subtract, multiply, and divide fractions in order to solve many types of measurement problems.

FOR DISCUSSION OR WRITING
Discuss some situations in your experience where you have needed to perform the operations of addition, subtraction, multiplication, or division on fractions. (*Hint:* To get you started, think of art projects, carpentry projects, adjusting recipes, and working on cars.)

OBJECTIVE 1 **Learn the definition of** *factor.* In the statement $2 \times 9 = 18$, the numbers 2 and 9 are called **factors** of 18. Other factors of 18 include 1, 3, 6, and 18. The result of the multiplication, 18, is called the **product.**

The number 18 is **factored** by writing it as the product of two or more numbers. For example, 18 can be factored in several ways, as $6 \cdot 3$, or $18 \cdot 1$, or $9 \cdot 2$, or $3 \cdot 3 \cdot 2$. In algebra, a raised dot $\cdot$ is often used instead of the $\times$ symbol to indicate multiplication.

A natural number (except 1) is **prime** if it has only itself and 1 as factors. "Factors" are understood here to mean natural number factors. (By agreement, the number 1 is not a prime number.) The first dozen primes are

$$2, 3, 5, 7, 11, 13, 17, 19, 23, 29, 31, 37.$$

A natural number (except 1) that is not prime is a **composite** number.

It is often useful to find all the **prime factors** of a number—those factors that are prime numbers. For example, the only prime factors of 18 are 2 and 3.

E X A M P L E 1   Factoring Numbers

Write the number as the product of prime factors.

(a) 35

Write 35 as the product of the prime factors 5 and 7, or as

$$35 = 5 \cdot 7.$$

(b) 24

One way to begin is to divide by the smallest prime, 2, to get

$$24 = 2 \cdot 12.$$

Now divide 12 by 2 to find factors of 12.

$$24 = 2 \cdot 2 \cdot 6$$

Since 6 can be written as $2 \cdot 3$,

$$24 = 2 \cdot 2 \cdot 2 \cdot 3,$$

where all factors are prime.

 It is not necessary to start with the smallest prime factor, as shown in Example 1(b). In fact, no matter which prime factor we start with, we will *always* obtain the same prime factorization.

O B J E C T I V E 2 Write fractions in lowest terms. We use prime numbers to write fractions in *lowest terms.* A fraction is in **lowest terms** when the numerator and denominator have no factors in common (other than 1). By the **basic principle of fractions,** if the numerator and denominator of a fraction are multiplied or divided by the *same* nonzero number, the value of the fraction is unchanged. To write a fraction in lowest terms, use these steps.

Writing a Fraction in Lowest Terms

Step 1   Write the numerator and the denominator as the product of prime factors.

Step 2   Divide the numerator and the denominator by the **greatest common factor,** the product of all factors common to both.

E X A M P L E 2   Writing Fractions in Lowest Terms

Write the fraction in lowest terms.

(a) $\dfrac{10}{15} = \dfrac{2 \cdot 5}{3 \cdot 5} = \dfrac{2 \cdot 1}{3 \cdot 1} = \dfrac{2}{3}$

Since 5 is the greatest common factor of 10 and 15, dividing both numerator and denominator by 5 gives the fraction in lowest terms.

(b) $\dfrac{15}{45} = \dfrac{3 \cdot 5}{3 \cdot 3 \cdot 5} = \dfrac{1 \cdot 1}{3 \cdot 1 \cdot 1} = \dfrac{1}{3}$

The factored form shows that 3 and 5 are the common factors of both 15 and 45. Dividing both 15 and 45 by $3 \cdot 5 = 15$ gives $\frac{15}{45}$ in lowest terms as $\frac{1}{3}$.

We can simplify this process by finding the greatest common factor in the numerator and denominator by inspection. For instance, in Example 2(b), we can use 15 rather than $3 \cdot 5$.

$$\frac{15}{45} = \frac{15}{3 \cdot 15} = \frac{1}{3 \cdot 1} = \frac{1}{3}$$

Errors may occur when writing fractions in lowest terms if the factor 1 is not included. To see this, refer to Example 2(b). In the equation

$$\frac{3 \cdot 5}{3 \cdot 3 \cdot 5} = \frac{?}{3},$$

if 1 is not written in the numerator when dividing common factors, you may make an error. The ? should be replaced by 1.

OBJECTIVE **3** Multiply and divide fractions. The basic operations on whole numbers, addition, subtraction, multiplication, and division, also apply to fractions. We multiply two fractions by first multiplying their numerators and then multiplying their denominators. This rule is written in symbols as follows.

Multiplying Fractions

If $\dfrac{a}{b}$ and $\dfrac{c}{d}$ are fractions, then $\dfrac{a}{b} \cdot \dfrac{c}{d} = \dfrac{a \cdot c}{b \cdot d}$.

EXAMPLE 3 Multiplying Fractions

Find the product of $\frac{3}{8}$ and $\frac{4}{9}$, and write it in lowest terms.

First, multiply $\frac{3}{8}$ and $\frac{4}{9}$.

$$\frac{3}{8} \cdot \frac{4}{9} = \frac{3 \cdot 4}{8 \cdot 9} \qquad \text{Multiply numerators; multiply denominators.}$$

It is easiest to write a fraction in lowest terms while the product is in factored form. Factor 8 and 9 and then divide out common factors in the numerator and denominator.

$$\frac{3 \cdot 4}{8 \cdot 9} = \frac{3 \cdot 4}{2 \cdot 4 \cdot 3 \cdot 3} = \frac{1 \cdot 3 \cdot 4}{2 \cdot 4 \cdot 3 \cdot 3} \qquad \text{Factor. Introduce a factor of 1.}$$

$$= \frac{1}{2 \cdot 3} \qquad \text{3 and 4 are common factors.}$$

$$= \frac{1}{6} \qquad \text{Lowest terms}$$

Two fractions are **reciprocals** of each other if their product is 1. For example, $\frac{3}{4}$ and $\frac{4}{3}$ are reciprocals since

$$\frac{3}{4} \cdot \frac{4}{3} = \frac{12}{12} = 1.$$

Also, $\frac{7}{11}$ and $\frac{11}{7}$ are reciprocals of each other. We use the reciprocal to divide fractions. To *divide* two fractions, multiply the first fraction by the reciprocal of the second fraction.

Dividing Fractions

For the fractions $\dfrac{a}{b}$ and $\dfrac{c}{d}$, $\dfrac{a}{b} \div \dfrac{c}{d} = \dfrac{a}{b} \cdot \dfrac{d}{c}$.

(To divide by a fraction, multiply by its reciprocal.)

The reason this method works will be explained in Chapter 7. The answer to a division problem is called a **quotient.** For example, the quotient of 20 and 10 is 2, since $20 \div 10 = 2$.

┌ **E X A M P L E   4**   Dividing Fractions

Find the following quotients, and write them in lowest terms.

(a) $\dfrac{3}{4} \div \dfrac{8}{5} = \dfrac{3}{4} \cdot \dfrac{5}{8} = \dfrac{3 \cdot 5}{4 \cdot 8} = \dfrac{15}{32}$ Multiply by the reciprocal of $\frac{8}{5}$.

(b) $\dfrac{3}{4} \div \dfrac{5}{8} = \dfrac{3}{4} \cdot \dfrac{8}{5} = \dfrac{3 \cdot 8}{4 \cdot 5} = \dfrac{3 \cdot 4 \cdot 2}{4 \cdot 5} = \dfrac{6}{5}$

O B J E C T I V E   4   Add and subtract fractions. To find the **sum** of two fractions having the same denominator, add the numerators and keep the same denominator.

Adding Fractions

If $\dfrac{a}{b}$ and $\dfrac{c}{b}$ are fractions, then $\dfrac{a}{b} + \dfrac{c}{b} = \dfrac{a + c}{b}$.

┌ **E X A M P L E   5**   Adding Fractions with the Same Denominator

Add.

(a) $\dfrac{3}{7} + \dfrac{2}{7} = \dfrac{3 + 2}{7} = \dfrac{5}{7}$ Add numerators and keep the same denominator.

(b) $\dfrac{2}{10} + \dfrac{3}{10} = \dfrac{2 + 3}{10} = \dfrac{5}{10} = \dfrac{1}{2}$

If the fractions to be added do not have the same denominators, the procedure above can still be used, but only *after* the fractions are rewritten with a common denominator. For example, to rewrite $\frac{3}{4}$ as a fraction with a denominator of 32,

$$\frac{3}{4} = \frac{?}{32},$$

find the number that can be multiplied by 4 to give 32. Since $4 \cdot 8 = 32$, use the number 8. By the basic principle, we can multiply the numerator and the denominator by 8.

$$\frac{3}{4} = \frac{3 \cdot 8}{4 \cdot 8} = \frac{24}{32}$$

> **Finding the Least Common Denominator**
>
> To add or subtract fractions with different denominators, find the **least common denominator (LCD)** as follows.
>
> *Step 1* Factor both denominators.
>
> *Step 2* For the LCD, use every factor that appears in any factored form. If a factor is repeated, use the largest number of repeats in the LCD.

The next example shows this procedure.

E X A M P L E 6   Adding Fractions with Different Denominators

Add the following fractions.

(a) $\dfrac{4}{15} + \dfrac{5}{9}$

To find the least common denominator, first factor both denominators.

$$15 = 5 \cdot 3 \qquad \text{and} \qquad 9 = 3 \cdot 3$$

Since 5 and 3 appear as factors, and 3 is a factor of 9 twice, the LCD is

$$5 \cdot 3 \cdot 3 \qquad \text{or} \qquad 45.$$

Write each fraction with 45 as denominator.

$$\frac{4}{15} = \frac{4 \cdot 3}{15 \cdot 3} = \frac{12}{45} \qquad \text{and} \qquad \frac{5}{9} = \frac{5 \cdot 5}{9 \cdot 5} = \frac{25}{45}$$

Now add the two equivalent fractions.

$$\frac{4}{15} + \frac{5}{9} = \frac{12}{45} + \frac{25}{45} = \frac{37}{45}$$

(b) $3\dfrac{1}{2} + 2\dfrac{3}{4}$

These numbers are called mixed numbers. A **mixed number** is understood to be the sum of a whole number and a fraction. We can add mixed numbers using either of two methods.

Method 1

Rewrite both numbers as follows.

$$3\frac{1}{2} = 3 + \frac{1}{2} = \frac{3}{1} + \frac{1}{2} = \frac{6}{2} + \frac{1}{2} = \frac{6+1}{2} = \frac{7}{2}$$

$$2\frac{3}{4} = 2 + \frac{3}{4} = \frac{8}{4} + \frac{3}{4} = \frac{8+3}{4} = \frac{11}{4}$$

Now add. The common denominator is 4.

$$3\frac{1}{2} + 2\frac{3}{4} = \frac{7}{2} + \frac{11}{4} = \frac{14}{4} + \frac{11}{4} = \frac{25}{4} \quad \text{or} \quad 6\frac{1}{4}$$

Method 2

Write $3\frac{1}{2}$ as $3\frac{2}{4}$. Then add vertically.

$$3\frac{1}{2} \quad \rightarrow \quad 3\frac{2}{4}$$
$$+\,2\frac{3}{4} \qquad\qquad +\,2\frac{3}{4}$$
$$\rule{2cm}{0.4pt} \qquad\qquad \rule{2cm}{0.4pt}$$
$$\qquad\qquad\qquad 5\frac{5}{4}$$

Since $\frac{5}{4} = 1\frac{1}{4}$,

$$5\frac{5}{4} = 5 + 1\frac{1}{4} = 6\frac{1}{4}, \quad \text{or} \quad \frac{25}{4}.$$

NOTE To multiply and divide mixed numbers, follow the same general procedure shown in Example 6(b), Method 1. First change to fractions, then perform the operation, and then convert back to a mixed number if desired. For example,

$$3\frac{1}{2} \cdot 2\frac{3}{4} = \frac{7}{2} \cdot \frac{11}{4} = \frac{77}{8} \quad \text{or} \quad 9\frac{5}{8}$$

$$3\frac{1}{2} \div 2\frac{3}{4} = \frac{7}{2} \div \frac{11}{4} = \frac{7}{2} \cdot \frac{4}{11} = \frac{14}{11} \quad \text{or} \quad 1\frac{3}{11}.$$

The **difference** between two numbers is found by subtraction. For example, $9 - 5 = 4$ so the difference between 9 and 5 is 4. Subtraction of fractions is similar to addition. Just subtract the numerators instead of adding them; again, keep the same denominator.

Subtracting Fractions

$$\frac{a}{b} - \frac{c}{b} = \frac{a - c}{b}$$

EXAMPLE 7 Subtracting Fractions

Subtract. Write the differences in lowest terms.

(a) $\dfrac{15}{8} - \dfrac{3}{8} = \dfrac{15 - 3}{8}$ Subtract numerators; keep the same denominator.

$\qquad\qquad = \dfrac{12}{8} = \dfrac{3}{2}$ Lowest terms

(b) $\dfrac{7}{18} - \dfrac{4}{15}$

Here, $18 = 2 \cdot 3 \cdot 3$ and $15 = 3 \cdot 5$, so the LCD is $2 \cdot 3 \cdot 3 \cdot 5 = 90$.

$$\frac{7}{18} - \frac{4}{15} = \frac{7 \cdot 5}{2 \cdot 3 \cdot 3 \cdot 5} - \frac{4 \cdot 2 \cdot 3}{2 \cdot 3 \cdot 3 \cdot 5} = \frac{35}{90} - \frac{24}{90} = \frac{11}{90}$$

(c) $\dfrac{15}{32} - \dfrac{11}{45}$

Since $32 = 2 \cdot 2 \cdot 2 \cdot 2 \cdot 2$ and $45 = 3 \cdot 3 \cdot 5$, there are no common factors, and the LCD is $32 \cdot 45 = 1440$.

$$\dfrac{15}{32} - \dfrac{11}{45} = \dfrac{15 \cdot \mathbf{45}}{32 \cdot \mathbf{45}} - \dfrac{11 \cdot \mathbf{32}}{45 \cdot \mathbf{32}} \qquad \text{Get a common denominator.}$$

$$= \dfrac{675}{1440} - \dfrac{352}{1440}$$

$$= \dfrac{323}{1440} \qquad \text{Subtract.}$$

OBJECTIVE **5** Solve applied problems that involve fractions. Applied problems often require work with fractions. For example, when a carpenter reads diagrams and plans, he or she often must work with fractions whose denominators are 2, 4, 8, 16, or 32, as shown in the next example.

EXAMPLE 8 Adding Fractions to Solve a Manufacturing (Woodworking) Problem

The diagram in Figure 1 appears in the book *Woodworker's 39 Sure-Fire Projects*. It is the front view of a corner bookcase/desk. Add the fractions shown in the diagram to find the height of the bookcase/desk.

We must add the following measures (in inches):

$$\dfrac{3}{4}, \quad 4\dfrac{1}{2}, \quad 9\dfrac{1}{2}, \quad \dfrac{3}{4}, \quad 9\dfrac{1}{2}, \quad \dfrac{3}{4}, \quad 4\dfrac{1}{2}.$$

Begin by changing $4\frac{1}{2}$ to $4\frac{2}{4}$ and $9\frac{1}{2}$ to $9\frac{2}{4}$, since the common denominator is 4. Then, use Method 2 from Example 6(b).

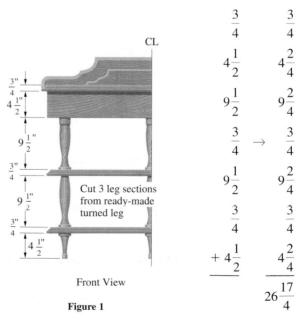

Front View

Figure 1

$$\begin{array}{cc}
\dfrac{3}{4} & \dfrac{3}{4} \\[6pt]
4\dfrac{1}{2} & 4\dfrac{2}{4} \\[6pt]
9\dfrac{1}{2} & 9\dfrac{2}{4} \\[6pt]
\dfrac{3}{4} \quad \rightarrow & \dfrac{3}{4} \\[6pt]
9\dfrac{1}{2} & 9\dfrac{2}{4} \\[6pt]
\dfrac{3}{4} & \dfrac{3}{4} \\[6pt]
+ 4\dfrac{1}{2} & 4\dfrac{2}{4} \\ \hline
& 26\dfrac{17}{4}
\end{array}$$

Since $\frac{17}{4} = 4\frac{1}{4}$, $26\frac{17}{4} = 26 + 4\frac{1}{4} = 30\frac{1}{4}$. The height is $30\frac{1}{4}$ inches. It is best to give answers as mixed numbers in applications like this.

OBJECTIVE **6** Interpret data in a circle graph. A **circle graph** or **pie chart** is often used to give a pictorial representation of data. A circle is used to indicate the total of all the categories represented. The circle is divided into sectors, or wedges (like pieces of pie) whose sizes show the relative magnitudes of the categories. The sum of all the fractional parts must be 1 (for 1 whole circle).

 EXAMPLE 9 Using a Pie Chart to Interpret Information

The pie chart in Figure 2 shows the job categories of African Americans employed in 1993 (age 16 or older).

JOB CATEGORIES FOR AFRICAN-AMERICANS

Source: U.S. Department of Labor.

Figure 2

(a) In a group of 150,000 such employees, about how many would we expect to be employed in precision production/repair?

To find the answer, we multiply the fraction indicated in the chart for the category $\left(\frac{2}{25}\right)$ by the number of people in the group (150,000):

$$\frac{2}{25} \cdot 150,000 = \frac{300,000}{25} = 12,000.$$

About 12,000 people in the group are employed in precision production/repair.

(b) *Estimate* the number employed in service.

The fraction $\frac{23}{100}$ is approximately $\frac{1}{4}$. Therefore, a good estimate for this number is

$$\frac{1}{4} \cdot 150,000 = 37,500.$$

1.1 EXERCISES

Decide whether each statement is true or false. If it is false, say why.

1. In the fraction $\frac{3}{7}$, 3 is the numerator and 7 is the denominator.

2. The mixed number equivalent of $\frac{41}{5}$ is $8\frac{1}{5}$.

3. The fraction $\frac{17}{51}$ is in lowest terms.

4. The reciprocal of $\dfrac{8}{2}$ is $\dfrac{4}{1}$.

5. The product of 8 and 2 is 10.

6. The difference between 12 and 2 is 6.

Identify each number as prime, composite, or neither. If the number is composite, write it as the product of prime factors. See Example 1.

7. 19	**8.** 31	**9.** 64	**10.** 99
11. 3458	**12.** 1025	**13.** 1	**14.** 0
15. 30	**16.** 40	**17.** 500	**18.** 700
19. 124	**20.** 120	**21.** 29	**22.** 83

Write each fraction in lowest terms. See Example 2.

23. $\dfrac{8}{16}$ **24.** $\dfrac{4}{12}$ **25.** $\dfrac{15}{18}$ **26.** $\dfrac{16}{20}$

27. $\dfrac{15}{45}$ **28.** $\dfrac{16}{64}$ **29.** $\dfrac{144}{120}$ **30.** $\dfrac{132}{77}$

31. One of the following is the correct way to write $\dfrac{16}{24}$ in lowest terms. Which one is it?

(a) $\dfrac{16}{24} = \dfrac{8+8}{8+16} = \dfrac{8}{16} = \dfrac{1}{2}$ (b) $\dfrac{16}{24} = \dfrac{4 \cdot 4}{4 \cdot 6} = \dfrac{4}{6}$

(c) $\dfrac{16}{24} = \dfrac{8 \cdot 2}{8 \cdot 3} = \dfrac{2}{3}$ (d) $\dfrac{16}{24} = \dfrac{14+2}{21+3} = \dfrac{2}{3}$

32. For the fractions $\dfrac{p}{q}$ and $\dfrac{r}{s}$, which one of the following can serve as a common denominator?

(a) $q \cdot s$ (b) $q + s$ (c) $p \cdot r$ (d) $p + r$

Find each product or quotient, and write it in lowest terms. See Examples 3 and 4.

33. $\dfrac{4}{5} \cdot \dfrac{6}{7}$ **34.** $\dfrac{5}{9} \cdot \dfrac{10}{7}$ **35.** $\dfrac{1}{10} \cdot \dfrac{12}{5}$ **36.** $\dfrac{6}{11} \cdot \dfrac{2}{3}$

37. $\dfrac{15}{4} \cdot \dfrac{8}{25}$ **38.** $\dfrac{4}{7} \cdot \dfrac{21}{8}$ **39.** $2\dfrac{2}{3} \cdot 5\dfrac{4}{5}$ **40.** $3\dfrac{3}{5} \cdot 7\dfrac{1}{6}$

41. $\dfrac{5}{4} \div \dfrac{3}{8}$ **42.** $\dfrac{7}{6} \div \dfrac{9}{10}$ **43.** $\dfrac{32}{5} \div \dfrac{8}{15}$ **44.** $\dfrac{24}{7} \div \dfrac{6}{21}$

45. $\dfrac{3}{4} \div 12$ **46.** $\dfrac{2}{5} \div 30$ **47.** $2\dfrac{5}{8} \div 1\dfrac{15}{32}$ **48.** $2\dfrac{3}{10} \div 7\dfrac{4}{5}$

49. Write a summary explaining how to multiply and divide two fractions. Give examples.

50. Write a summary explaining how to add and subtract two fractions. Give examples.

Find each sum or difference, and write it in lowest terms. See Examples 5–7.

51. $\dfrac{7}{12} + \dfrac{1}{12}$ **52.** $\dfrac{3}{16} + \dfrac{5}{16}$ **53.** $\dfrac{5}{9} + \dfrac{1}{3}$ **54.** $\dfrac{4}{15} + \dfrac{1}{5}$

55. $3\dfrac{1}{8} + \dfrac{1}{4}$ **56.** $5\dfrac{3}{4} + \dfrac{2}{3}$ **57.** $\dfrac{7}{12} - \dfrac{1}{9}$ **58.** $\dfrac{11}{16} - \dfrac{1}{12}$

59. $6\dfrac{1}{4} - 5\dfrac{1}{3}$ **60.** $8\dfrac{4}{5} - 7\dfrac{4}{9}$ **61.** $\dfrac{5}{3} + \dfrac{1}{6} - \dfrac{1}{2}$ **62.** $\dfrac{7}{15} + \dfrac{1}{6} - \dfrac{1}{10}$

The following chart appears on a package of Quaker Quick Grits.

	Microwave		Stove Top	
Servings	1	1	4	6
Water	$\frac{3}{4}$ cup	1 cup	3 cups	4 cups
Grits	3 Tbsp	3 Tbsp	$\frac{3}{4}$ cup	1 cup
Salt (optional)	dash	dash	$\frac{1}{4}$ tsp	$\frac{1}{2}$ tsp

Use the chart to answer the questions in Exercises 63 and 64.

63. How many cups of water would be needed for 8 microwave servings?

64. How many tsp of salt would be needed for 5 stove top servings? (*Hint:* 5 is halfway between 4 and 6.)

Work each problem. See Example 8.

65. On Tuesday, February 10, 1998, Earthlink stock on the NASDAQ exchange closed the day at $4\frac{5}{8}$ (dollars) ahead of where it had opened. It closed at $38\frac{5}{8}$ (dollars). What was its opening price?

66. A report in *USA Today* on February 10, 1998, stated that Teva Pharmaceutical skidded $9\frac{9}{16}$ (dollars) to $37\frac{1}{2}$ (dollars) after the Israeli drugmaker said fourth-quarter net income was likely to be below expectations. What was its price before the skid?

67. A hardware store sells a 40-piece socket wrench set. The measure of the largest socket is $\frac{3}{4}$ inch, while the measure of the smallest socket is $\frac{3}{16}$ inch. What is the difference between these measures?

68. Two sockets in a socket wrench set have measures of $\frac{9}{16}$ inch and $\frac{3}{8}$ inch. What is the difference between these two measures?

69. A motel owner has decided to expand his business by buying a piece of property next to the motel. The property has an irregular shape, with five sides as shown in the figure. Find the total distance around the piece of property. (This is called the *perimeter* of the figure.)

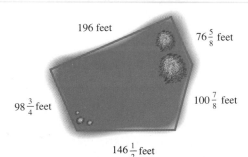

196 feet

$76\frac{5}{8}$ feet

$98\frac{3}{4}$ feet

$100\frac{7}{8}$ feet

$146\frac{1}{2}$ feet

70. Find the perimeter of the triangle in the figure.

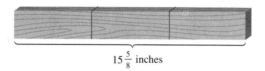

$5\frac{1}{4}$ feet $7\frac{1}{2}$ feet

$10\frac{1}{8}$ feet

71. A piece of board is $15\frac{5}{8}$ inches long. If it must be divided into 3 pieces of equal length, how long must each piece be?

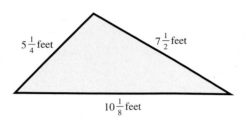

$15\frac{5}{8}$ inches

72. If one serving of a macaroni and cheese meal requires $\frac{1}{8}$ cup of chopped onions, how many cups of onions will $7\frac{1}{2}$ servings require?

73. Tex's favorite recipe for barbecue sauce calls for $2\frac{1}{3}$ cups of tomato sauce. The recipe makes enough barbecue sauce to serve 7 people. How much tomato sauce is needed for 1 serving?

74. A cake recipe calls for $1\frac{3}{4}$ cups of sugar. A caterer has $15\frac{1}{2}$ cups of sugar on hand. How many cakes can he make?

The pie chart gives fractional job categories of white employees in the workforce in 1993 (age 16 or older). Use the chart to answer the questions in Exercises 75 and 76. See Example 9.

75. In a random sample of 5000 such employees, about how many would we expect to be employed in service occupations?

76. In a random sample of 7500 such employees, about how many would we expect to be employed in farming/forestry/fishing occupations?

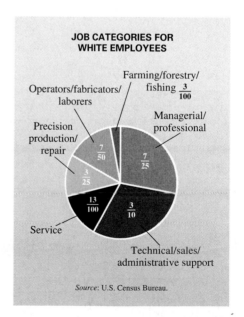

JOB CATEGORIES FOR WHITE EMPLOYEES

Farming/forestry/fishing $\frac{3}{100}$

Operators/fabricators/laborers

Managerial/professional

Precision production/repair

$\frac{7}{50}$

$\frac{7}{25}$

$\frac{3}{25}$

$\frac{13}{100}$

$\frac{3}{10}$

Service

Technical/sales/administrative support

Source: U.S. Census Bureau.

77. At the conclusion of the AWL softball league season, batting statistics for five players were as follows.

Player	At-bats	Hits	Home Runs
Stephanie Baldock	40	9	2
Jennifer Crum	36	12	3
Jason Jordan	11	5	1
Greg Tobin	16	8	0
David Perry	20	10	2

Answer each of the following, using estimation skills as necessary.

(a) Which player got a hit in exactly $\frac{1}{3}$ of his or her at-bats?

(b) Which player got a hit in just less than $\frac{1}{2}$ of his or her at-bats?

(c) Which player got a home run in just less than $\frac{1}{10}$ of his or her at-bats?

(d) Which player got a hit in just less than $\frac{1}{4}$ of his or her at-bats?

(e) Which two players got hits in exactly the same fractional parts of their at-bats? What was the fractional part, expressed in lowest terms?

78. For each of the following, write a fraction in lowest terms that represents the region described.

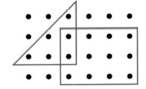

(a) the dots in the rectangle as a part of the dots in the entire figure

(b) the dots in the triangle as a part of the dots in the entire figure

(c) the dots in the overlapping region of the triangle and the rectangle as a part of the dots in the triangle alone

(d) the dots in the overlapping region of the triangle and the rectangle as a part of the dots in the rectangle alone

79. Estimate the best approximation for the following sum:

$$\frac{14}{26} + \frac{98}{99} + \frac{100}{51} + \frac{90}{31} + \frac{13}{27}.$$

(a) 6   **(b)** 7   **(c)** 5   **(d)** 8

80. In the local softball league, the first five games produced the following results: David Horwitz got 8 hits in 20 at-bats, and Chalon Bridges got 12 hits in 30 at-bats. David claims that he did just as well as Chalon. Is he correct? Why or why not?

1.2 Exponents, Order of Operations, and Inequality

OBJECTIVES

1 Use exponents.

2 Use the order of operations rules.

3 Use more than one grouping symbol.

4 Know the meanings of $\neq$, $<$, $>$, $\leq$, and $\geq$.

5 Translate word statements to symbols.

6 Write statements that change the direction of inequality symbols.

7 Interpret data in a bar graph.

FOR EXTRA HELP

📖 **SSG** Sec. 1.2
SSM Sec. 1.2

💿 **Pass the Test Software**

💿 **InterAct Math Tutorial Software**

📼 **Video I**

OBJECTIVE 1 Use exponents. In a multiplication problem, the same factor may appear several times. For example, in the product

$$3 \cdot 3 \cdot 3 \cdot 3 = 81,$$

the factor 3 appears four times. In algebra, repeated factors are written with an *exponent*. For example, in $3 \cdot 3 \cdot 3 \cdot 3$, the number 3 appears as a factor four times, so the product is written as 3^4, and is read "3 to the **fourth power**."

$$3 \cdot 3 \cdot 3 \cdot 3 = 3^4$$

The number 4 is the **exponent** or **power** and 3 is the **base** in the **exponential expression** 3^4. A natural number exponent, then, tells how many times the base is used as a factor. A number raised to the first power is simply that number. For example, $5^1 = 5$ and $\left(\frac{1}{2}\right)^1 = \frac{1}{2}$.

EXAMPLE 1 Evaluating an Exponential Expression

Find the values of the following.

(a) 5^2 $\mathbf{5 \cdot 5} = 25$
            5 is used as a factor 2 times.

Read 5^2 as "5 squared."

(b) 6^3 $\mathbf{6 \cdot 6 \cdot 6} = 216$
            6 is used as a factor 3 times.

Read 6^3 as "6 cubed."

(c) $2^5 = 2 \cdot 2 \cdot 2 \cdot 2 \cdot 2 = 32$ 2 is used as a factor 5 times.
Read 2^5 as "2 to the fifth power."

(d) $\left(\dfrac{2}{3}\right)^3 = \dfrac{2}{3} \cdot \dfrac{2}{3} \cdot \dfrac{2}{3} = \dfrac{8}{27}$ $\frac{2}{3}$ is used as a factor 3 times.

OBJECTIVE 2 Use the order of operations rules. Many problems involve more than one operation. To indicate the order in which the operations should be performed, we often use *grouping symbols*. If no grouping symbols are used, we apply the order of operations rules discussed below.

Consider the expression $5 + 2 \cdot 3$. To show that the multiplication should be performed before the addition, parentheses can be used to write

$$5 + (2 \cdot 3) = 5 + 6 = 11.$$

If addition is to be performed first, the parentheses should group $5 + 2$ as follows.

$$(5 + 2) \cdot 3 = 7 \cdot 3 = 21$$

Other grouping symbols used in more complicated expressions are brackets [ ], braces { }, and fraction bars. (For example, in $\frac{8-2}{3}$, the expression $8 - 2$ is considered to be grouped in the numerator.)

To work problems with more than one operation, use the following **order of operations.** This order is used by most calculators and computers.

Order of Operations

If grouping symbols are present, simplify within them, innermost first (and above and below fraction bars separately), in the following order.

Step 1   Apply all exponents.

Step 2   Do any multiplications or divisions in the order in which they occur, working from left to right.

Step 3   Do any additions or subtractions in the order in which they occur, working from left to right.

If no grouping symbols are present, start with Step 1.

A dot has been used to show multiplication; another way to show multiplication is with parentheses. For example, 3(7), (3)7, and (3)(7) each mean $3 \cdot 7$ or 21. The next example shows the use of parentheses for multiplication.

E X A M P L E   2   **Using the Order of Operations**

Find the values of the following.

(a) $9(6 + 11)$

Using the order of operations given above, work first inside the parentheses.

$$9(\mathbf{6 + 11}) = 9(\mathbf{17}) \qquad \text{Work inside parentheses.}$$
$$= 153 \qquad \text{Multiply.}$$

(b) $6 \cdot 8 + 5 \cdot 2$

Do any multiplications, working from left to right, and then add.

$$\mathbf{6 \cdot 8 + 5 \cdot 2} = 48 + 10 \qquad \text{Multiply.}$$
$$= 58 \qquad \text{Add.}$$

(c) $2(\mathbf{5 + 6}) + 7 \cdot 3 = 2(\mathbf{11}) + 7 \cdot 3 \qquad \text{Work inside parentheses.}$
$$= 22 + 21 \qquad \text{Multiply.}$$
$$= 43 \qquad \text{Add.}$$

(d) $9 - 2^3 + 5$

Find 2^3 first.

$$9 - 2^3 + 5 = 9 - 2 \cdot 2 \cdot 2 + 5 \qquad \text{Use the exponent.}$$
$$= 9 - 8 + 5 \qquad \text{Multiply.}$$
$$= 1 + 5 \qquad \text{Subtract.}$$
$$= 6 \qquad \text{Add.}$$

(e) $72 \div 2 \cdot 3 + 4 \cdot 2^3$

$$72 \div 2 \cdot 3 + 4 \cdot 2^3 = 72 \div 2 \cdot 3 + 4 \cdot 8 \qquad \text{Use the exponent.}$$
$$= 36 \cdot 3 + 4 \cdot 8 \qquad \text{Perform the division.}$$
$$= 108 + 32 \qquad \text{Perform the multiplications.}$$
$$= 140 \qquad \text{Add.}$$

Notice that the multiplications and divisions are performed from left to right *as they appear;* then the additions and subtractions should be done from left to right, *as they appear.*

OBJECTIVE **3** Use more than one grouping symbol. An expression with double (or *nested*) parentheses, such as $2(8 + 3(6 + 5))$, can be confusing. For clarity, square brackets, [ ], often are used in place of one pair of parentheses. Fraction bars also act as grouping symbols. The next example explains these situations.

EXAMPLE 3 Using Brackets and Fraction Bars as Grouping Symbols

Simplify each expression.

(a) $2[8 + 3(6 + 5)]$

Work first within the parentheses, and then simplify inside the brackets until a single number remains.

$$
\begin{aligned}
2[8 + 3(\mathbf{6 + 5})] &= 2[8 + 3(\mathbf{11})] \\
&= 2[8 + \mathbf{33}] \\
&= 2[\mathbf{41}] \\
&= 82
\end{aligned}
$$

(b) $\dfrac{4(5 + 3) + 3}{2(3) - 1}$

The expression can be written as the quotient

$$[4(5 + 3) + 3] \div [2(3) - 1],$$

which shows that the fraction bar groups the numerator and denominator separately. Simplify both numerator and denominator, then divide, if possible.

$$
\begin{aligned}
\frac{4(5 + 3) + 3}{2(3) - 1} &= \frac{4(\mathbf{8}) + 3}{2(3) - 1} \qquad \text{Work inside parentheses.} \\
&= \frac{32 + 3}{6 - 1} \qquad \text{Multiply.} \\
&= \frac{35}{5} \qquad \text{Add and subtract.} \\
&= 7 \qquad \text{Divide.}
\end{aligned}
$$

 Parentheses and fraction bars are used as grouping symbols to indicate an expression that represents a single number. That is why we must first simplify within parentheses and above and below fraction bars.

OBJECTIVE **4** Know the meanings of $\neq, <, >, \leq,$ and $\geq$. So far, we have used the symbols for the operations of arithmetic and the symbol for equality ($=$). The equality symbol with a slash through it, $\neq$, means "is not equal to." For example,

$$7 \neq 8$$

indicates that 7 is not equal to 8.

If two numbers are not equal, then one of the numbers must be less than the other. The symbol $<$ represents "is less than," so that "7 is less than 8" is written

$$7 < 8.$$

Also, write "6 is less than 9" as $6 < 9$.

The symbol $>$ means "is greater than." Write "8 is greater than 2" as

$$8 > 2.$$

The statement "17 is greater than 11" becomes $17 > 11$.

Keep the meanings of the symbols $<$ and $>$ clear by remembering that the symbol always *points to the smaller number.* For example, write "8 is less than 15" by pointing the symbol toward the 8:

Smaller number $\rightarrow$ $8 < 15.$

Two other symbols, $\leq$ and $\geq$, also represent the idea of inequality. The symbol $\leq$ means "is less than or equal to," so that

$$5 \leq 9$$

means "5 is less than or equal to 9." This statement is true, since $5 < 9$ is true. **If either the $<$ part or the $=$ part is true, then the inequality $\leq$ is true.**

The symbol $\geq$ means "is greater than or equal to." Again,

$$9 \geq 5$$

is true because $9 > 5$ is true. Also, $8 \leq 8$ is true since $8 = 8$ is true. But it is not true that $13 \leq 9$ because neither $13 < 9$ nor $13 = 9$ is true.

EXAMPLE 4   Using Inequality Symbols

Determine whether each statement is true or false.

(a) $6 \neq 6$

The statement is false because 6 *is equal to* 6.

(b) $5 < 19$

Since 5 represents a number that is indeed less than 19, this statement is true.

(c) $15 \leq 20$

The statement $15 \leq 20$ is true, since $15 < 20$.

(d) $25 \geq 30$

Both $25 > 30$ and $25 = 30$ are false. Because of this, $25 \geq 30$ is false.

(e) $12 \geq 12$

Since $12 = 12$, this statement is true.

OBJECTIVE 5   Translate word statements to symbols. An important part of algebra deals with translating words into algebraic notation.

PROBLEM SOLVING

As we will see throughout this book, the ability to solve problems using mathematics is based on translating the words of the problem into symbols. The next example is the first of many that illustrate such translations.

┌─ **E X A M P L E 5**   Translating from Words to Symbols

Write each word statement in symbols.

(a) Twelve **equals** ten **plus** two.

$$12 = 10 + 2$$

(b) Nine **is less than** ten.

$$9 < 10$$

(c) Fifteen **is not equal to** eighteen.

$$15 \neq 18$$

(d) Seven **is greater than** four.

$$7 > 4$$

(e) Thirteen **is less than or equal to** forty.

$$13 \leq 40$$

(f) Eleven **is greater than or equal to** eleven.

$$11 \geq 11$$

■

O B J E C T I V E ▐ 6 ▐ Write statements that change the direction of inequality symbols. Any statement with $<$ can be converted to one with $>$, and any statement with $>$ can be converted to one with $<$. We do this by reversing the order of the numbers and the direction of the symbol. For example, the statement $6 < 10$ can be written with $>$ as $10 > 6$. Similarly, the statement $4 \leq 10$ can be changed to $10 \geq 4$.

┌─ **E X A M P L E 6**   Converting between Inequality Symbols

The following examples show the same statement written in two equally correct ways.

(a) $9 < 16$ $16 > 9$

(b) $5 > 2$ $2 < 5$

(c) $3 \leq 8$ $8 \geq 3$

(d) $12 \geq 5$ $5 \leq 12$

Note that in each pair of inequalities, the point of the inequality symbol points toward the smaller number.

■

Here is a summary of the symbols discussed in this section.

Symbols of Equality and Inequality

$=$	is equal to	$\neq$	is not equal to
$<$	is less than	$>$	is greater than
$\leq$	is less than or equal to	$\geq$	is greater than or equal to

CAUTION The equality and inequality symbols are used to write mathematical *sentences* that describe the relationship between two numbers. On the other hand, the symbols for operations ($+$, $-$, $\times$, $\div$) are used to write mathematical *expressions* that represent a single number. For example, compare the sentence $4 < 10$ with the expression $4 + 10$, which represents the number 14.

O B J E C T I V E ▐ 7 ▐ Interpret data in a bar graph. **Bar graphs** are often used to summarize data in a concise manner.

 **E X A M P L E 7**   Interpreting Inequality Concepts Using a Bar Graph

Figure 3 shows a bar graph that depicts the percentage growth of the ten fastest-growing U.S. manufacturing industries from 1993 to 1994.

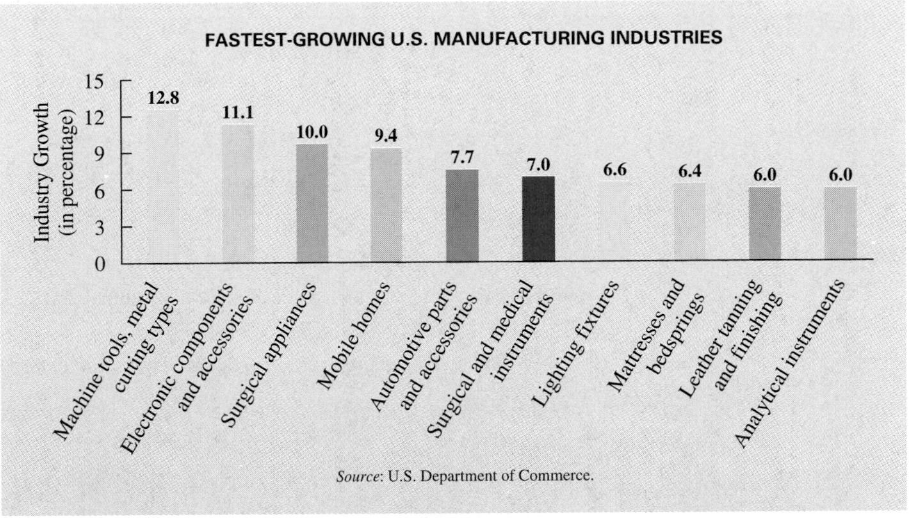

FASTEST-GROWING U.S. MANUFACTURING INDUSTRIES

Source: U.S. Department of Commerce.

Figure 3

(a) Which industries showed growth greater than 9.4%?

Machine tools, metal cutting types (at 12.8%), electronic components and accessories (at 11.1%), and surgical appliances (at 10.0%) all showed growth greater than 9.4%.

(b) Which industries showed growth greater than or equal to 9.4%?

The three industries listed in part (a) and the mobile home industry, at 9.4%, showed growth greater than or equal to 9.4%. (These are industries that showed *at least* 9.4% growth.)

1.2 EXERCISES

Decide whether each statement is true or false. If it is false, explain why.

1. When evaluated, $4 + 3(8 - 2)$ is equal to 42.

2. $3^3 = 9$

3. The statement "4 is 12 less than 16" is interpreted $4 = 12 - 16$.

4. The statement "6 is 4 less than 10" is interpreted $6 < 10 - 4$.

Find the value of each exponential expression. See Example 1.

5. 7^2 **6.** 4^2 **7.** 12^2 **8.** 14^2

9. 4^3 **10.** 5^3 **11.** 10^3 **12.** 11^3

13. 3^4 **14.** 6^4 **15.** 4^5 **16.** 3^5

17. $\left(\dfrac{2}{3}\right)^4$ **18.** $\left(\dfrac{3}{4}\right)^3$ **19.** $(.04)^3$ **20.** $(.05)^4$

21. Explain in your own words how to evaluate a power of a number, such as 6^3.

22. Explain why any power of 1 must be equal to 1.

Find the value of each expression. See Examples 2 and 3.

23. $9 \cdot 5 - 13$ **24.** $7 \cdot 6 - 11$ **25.** $\dfrac{1}{4} \cdot \dfrac{2}{3} + \dfrac{2}{5} \cdot \dfrac{11}{3}$

study

26. $\dfrac{9}{4} \cdot \dfrac{2}{3} + \dfrac{4}{5} \cdot \dfrac{5}{3}$

27. $9 \cdot 4 - 8 \cdot 3$

28. $11 \cdot 4 + 10 \cdot 3$

29. $(4.3)(1.2) + (2.1)(8.5)$

30. $(2.5)(1.9) + (4.3)(7.3)$

31. $5[3 + 4(2^2)]$

32. $6[2 + 8(3^3)]$

33. $3^2[(11 + 3) - 4]$

34. $4^2[(13 + 4) - 8]$

35. $\dfrac{6(3^2 - 1) + 8}{3 \cdot 2 - 2}$

36. $\dfrac{2(8^2 - 4) + 8}{4 \cdot 3 - 10}$

37. $\dfrac{4(6 + 2) + 8(8 - 3)}{6(4 - 2) - 2^2}$

38. $\dfrac{6(5 + 1) - 9(1 + 1)}{5(8 - 6) - 2^3}$

39. Write an explanation of how you would use the order of operations to simplify $4 + 3(2^2 - 1)^3$.

40. When evaluating $(4^2 + 3^3)^4$, what is the *last* exponent that would be applied?

Tell whether each statement is true or false. In Exercises 45–54, first simplify the expression involving an operation. See Examples 2–4.

41. $5 < 6$

42. $3 < 7$

43. $8 \geq 17$

44. $10 \geq 41$

45. $17 \leq 18 - 1$

46. $12 \geq 10 + 2$

47. $6 \cdot 8 + 6 \cdot 6 \geq 0$

48. $4 \cdot 20 - 16 \cdot 5 \geq 0$

49. $6[5 + 3(4 + 2)] \leq 70$

50. $6[2 + 3(2 + 5)] \leq 135$

51. $\dfrac{9(7 - 1) - 8 \cdot 2}{4(6 - 1)} > 3$

52. $\dfrac{2(5 + 3) + 2 \cdot 2}{2(4 - 1)} > 1$

53. $8 \leq 4^2 - 2^2$

54. $10^2 - 8^2 > 6^2$

Write each word statement in symbols. See Example 5.

55. Fifteen is equal to five plus ten.

56. Twelve is equal to twenty minus eight.

57. Nine is greater than five minus four.

58. Ten is greater than six plus one.

59. Sixteen is not equal to nineteen.

60. Three is not equal to four.

61. Two is less than or equal to three.

62. Five is less than or equal to nine.

Write each statement in words and decide whether it is true or false.

63. $7 < 19$

64. $9 < 10$

65. $3 \neq 6$

66. $9 \neq 13$

67. $8 \geq 11$

68. $4 \leq 2$

69. Construct a true statement that involves an addition on the left side, the symbol $\geq$, and a multiplication on the right side.

70. Construct a false statement that involves subtraction on the left side, the symbol $\leq$, and a division on the right side. Then tell why the statement is false and how it could be changed to become true.

Write each statement with the inequality symbol reversed while keeping the same meaning. See Example 6.

71. $5 < 30$

72. $8 > 4$

73. $12 \geq 3$

74. $25 \leq 41$

75. What English-language phrase is used to express the fact that one person's age *is less than* another person's age?

76. What English-language phrase is used to express the fact that one person's height *is greater than* another person's height?

Skip

77. $12 \geq 12$ is a true statement. Suppose that someone tells you the following: "$12 \geq 12$ is false, because even though 12 is equal to 12, 12 is not greater than 12." How would you respond to this?

78. The symbol $\neq$ means "is not equal to." How do you think we read the symbol $\not>$? Is this equivalent to $<$? How do you think we read $\not<$?

The bar graph shows total construction spending in billions of dollars at seasonally adjusted rates. Use the graph to answer the questions in Exercises 79 and 80. See Example 7.

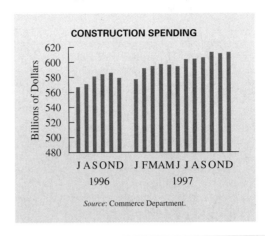

CONSTRUCTION SPENDING

Source: Commerce Department.

79. In five of the months depicted, construction spending decreased from the previous month. What were these five months?

80. What was the first month shown in which construction spending was greater than $600 billion?

81. According to the Tuesday, February 3, 1998, issue of *The New York Times,* the National Association of Purchasing Management's index declined to 52.4 in January, 1998, from its level of 53.1 the previous month.
 (a) By how much did the index decline?
 (b) To find the percent of decline, divide the amount from part (a) by 53.1 and convert to a percent. What was the percent of decline?

82. According to the same article cited in Exercise 81, construction spending rose .1% in 1997. If construction spending the previous year was $611.2 billion, what was the amount spent in 1997? (*Hint:* Find .1% of 611.2 and add it to 611.2 to get the amount in billions of dollars.)

1.3 Variables, Expressions, and Equations

OBJECTIVES

1 Define *variable,* and find the value of an algebraic expression, given the values of the variables.

2 Convert phrases from words to algebraic expressions.

3 Identify solutions of equations.

OBJECTIVE **1** Define *variable,* and find the value of an algebraic expression, given the values of the variables. A **variable** is a symbol, usually a letter, such as x, y, or z, used to represent any unknown number. An **algebraic expression** is a collection of numbers, variables, operation symbols, and grouping symbols (such as parentheses). For example,

$$6(x + 5), \qquad 2m - 9, \qquad \text{and} \qquad 8p^2 + 6p + 2$$

are all algebraic expressions. In the algebraic expression $2m - 9$, the expression $2m$ indicates the product of 2 and m, just as $8p^2$ shows the product of 8 and p^2. Also, $6(x + 5)$ means the product of 6 and $x + 5$. An algebraic expression has different numerical values for different values of the variable.

4 Identify solutions of equations from a set of numbers.

5 Distinguish between an *expression* and an *equation*.

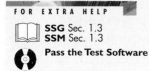

FOR EXTRA HELP

SSG Sec. 1.3
SSM Sec. 1.3

Pass the Test Software

InterAct Math
  Tutorial Software

Video 1

EXAMPLE 1   Evaluating Expressions

Find the numerical values of the following algebraic expressions when $m = 5$.

(a) $8m$
Replace m with 5, to get

$$8m = 8 \cdot 5 = 40.$$

(b) $3m^2$
For $m = 5$,

$$3m^2 = 3 \cdot 5^2 = 3 \cdot 25 = 75.$$

CAUTION In Example 1(b), notice that $3m^2$ means $3 \cdot m^2$; it *does not* mean $3m \cdot 3m$. The product $3m \cdot 3m$ is indicated by $(3m)^2$.

EXAMPLE 2   Evaluating Expressions

Find the value of each expression when $x = 5$ and $y = 3$.

(a) $2x + 7y$
Replace x with 5 and y with 3. Follow the order of operations; multiply first, then add.

$$\begin{aligned}
2x + 7y &= 2 \cdot 5 + 7 \cdot 3 &&\text{Let } x = 5 \text{ and } y = 3.\\
&= 10 + 21 &&\text{Multiply.}\\
&= 31 &&\text{Add.}
\end{aligned}$$

(b) $\dfrac{9x - 8y}{2x - y}$
Replace x with 5 and y with 3.

$$\begin{aligned}
\frac{9x - 8y}{2x - y} &= \frac{9 \cdot 5 - 8 \cdot 3}{2 \cdot 5 - 3} &&\text{Let } x = 5 \text{ and } y = 3.\\[2mm]
&= \frac{45 - 24}{10 - 3} &&\text{Multiply.}\\[2mm]
&= \frac{21}{7} &&\text{Subtract.}\\[2mm]
&= 3 &&\text{Divide.}
\end{aligned}$$

(c)
$$\begin{aligned}
x^2 - 2y^2 &= 5^2 - 2 \cdot 3^2 &&\text{Let } x = 5 \text{ and } y = 3.\\
&= 25 - 2 \cdot 9 &&\text{Use the exponents.}\\
&= 25 - 18 &&\text{Multiply.}\\
&= 7 &&\text{Subtract.}
\end{aligned}$$

OBJECTIVE 2 Convert phrases from words to algebraic expressions. In Section 1.2 we saw how to translate from words to symbols.

PROBLEM SOLVING

Sometimes variables must be used to change word phrases into algebraic expressions. The next example illustrates this. Such translations are used in problem solving.

E X A M P L E   3   Changing Word Phrases to Algebraic Expressions

Change the following word phrases to algebraic expressions. Use x as the variable.

(a) The **sum** of a number and 9

"**Sum**" is the answer to an addition problem. This phrase translates as

$$x + 9 \qquad \text{or} \qquad 9 + x.$$

(b) 7 **minus** a number

"**Minus**" indicates subtraction, so the answer is $7 - x$.

    Here $x - 7$ would *not* be correct; this statement translates as "a number minus 7," not "7 minus a number." The expressions $7 - x$ and $x - 7$ are rarely equal. For example, if $x = 10$, $10 - 7 \neq 7 - 10$. ($7 - 10$ is a *negative number*, discussed in Section 1.4.)

(c) 7 **less than** a number

Write 7 **less than** a number as $x - 7$. In this case $7 - x$ would not be correct, because "less than" means "subtracted from."

(d) The **product** of 11 and a number

$$11 \cdot x \qquad \text{or} \qquad 11x$$

As mentioned earlier, $11x$ means 11 times x. No symbol is needed to indicate the product of a number and a variable.

(e) 5 **divided by** a number $x\overline{)5} \ \ \text{or} \ \ 5 \div x$

This translates as $\frac{5}{x}$. The expression $\frac{x}{5}$ would *not* be correct here.

(f) The **product** of 2, and the **sum** of a number and 8

We are multiplying 2 times another number. This number is the sum of x and 8, written $x + 8$. Using parentheses for this sum, the final expression is

$$2(x + 8).$$

O B J E C T I V E   3   Identify solutions of equations.   An **equation** is a statement that two algebraic expressions are equal. Therefore, an equation *always* includes the equality symbol, $=$. Examples of equations are

$$x + 4 = 11, \qquad 2y = 16, \qquad \text{and} \qquad 4p + 1 = 25 - p.$$

Solving an Equation

To **solve** an equation means to find the values of the variable that make the equation true. Such values of the variable are called the **solutions** of the equation.

E X A M P L E   4   Deciding whether a Number Is a Solution

Decide whether the given number is a solution of the equation.

(a) $5p + 1 = 36; \quad 7$

Replace p with 7.

$$5p + 1 = 36$$
$$5 \cdot 7 + 1 = 36 \qquad ? \qquad \text{Let } p = 7.$$
$$35 + 1 = 36 \qquad ?$$
$$36 = 36 \qquad \text{True}$$

The number 7 is a solution of the equation.

(b) $9m - 6 = 32;$ 4

$$9m - 6 = 32$$
$$9 \cdot 4 - 6 = 32 \qquad ? \qquad \text{Let } m = 4.$$
$$36 - 6 = 32 \qquad ?$$
$$30 = 32 \qquad \text{False}$$

The number 4 is not a solution of the equation.

OBJECTIVE 4   Identify solutions of equations from a set of numbers. A **set** is a collection of objects. In mathematics, these objects are most often numbers. The objects that belong to the set, called **elements** of the set, are written between **set braces.** For example, the set containing the numbers 1, 2, 3, 4, and 5 is written as

$$\{1, 2, 3, 4, 5\}.$$

For more information about sets, see Appendix B at the back of this book.

In some cases, the set of numbers from which the solutions of an equation must be chosen is specifically stated. One way of determining solutions is direct substitution of all possible replacements. The ones that lead to a true statement are solutions.

EXAMPLE 5   Finding a Solution from a Given Set

Change each word statement to an equation. Use x as the variable. Then find all solutions of the equation from the set

$$\{0, 2, 4, 6, 8, 10\}.$$

(a) The sum of a number and four is six.

The word "is" suggests "equals." If x represents the unknown number, then translate as follows.

The sum of a number and four   is   six.
$$x + 4 = 6$$

Try each number from the given set $\{0, 2, 4, 6, 8, 10\}$, in turn, to see that 2 is the only solution of $x + 4 = 6$.

(b) 9 more than five times a number is 49.

Use x to represent the unknown number. We start with $5x$ and then add 9 to it. The word "is" translates as =.

$$5x + 9 = 49$$

Try each number from $\{0, 2, 4, 6, 8, 10\}$. The solution is **8**, since $5 \cdot 8 + 9 = 49$.

(c) The sum of a number and 12 is equal to four times the number.

If x represents the number, "the sum of a number and 12" is represented by $x + 12$. The translation is

$$x + 12 = 4x.$$

Trying each replacement leads to a true statement when $x = 4$, since $4 + 12 = 4(4) = 16$.

O B J E C T I V E 5 Distinguish between an *expression* and an *equation.* Students often have trouble distinguishing between equations and expressions. Remember that an equation is a sentence; an expression is a phrase.

$$4x + 5 = 9 \qquad\qquad 4x + 5$$
$$\uparrow \qquad\qquad\qquad \uparrow$$

Equation   Expression
(to solve)   (to simplify or evaluate)

E X A M P L E 6 Distinguishing between Equations and Expressions

Decide whether each of the following is an equation or an expression.

(a) $2x - 5y$

There is no equals sign, so this is an expression.

(b) $2x = 5y$

Because of the equals sign, this is an equation.

1.3 EXERCISES

Fill in each blank with the correct response.

1. If $x = 3$, then the value of $x + 7$ is _____.

2. If $x = 1$ and $y = 2$, then the value of $4xy$ is _____.

3. The sum of 12 and x is represented by the expression _____. If $x = 9$, the value of that expression is _____.

4. If x can be chosen from the set $\{0, 1, 2, 3, 4, 5\}$, the only solution of $x + 5 = 9$ is _____.

5. Will the equation $x = x + 4$ ever have a solution? _____.

6. $2x + 3$ is a(n) _____, while $2x + 3 = 8$ is a(n) _____.
   (equation/expression)   (equation/expression)

Exercises 7–12 cover some of the concepts introduced in this section. Give a short explanation for each.

7. Why is $2x^3$ not the same as $2x \cdot 2x \cdot 2x$? Explain.

8. Why are "5 less than a number" and "5 is less than a number" translated differently?

9. Explain in your own words why, when evaluating the expression $4x^2$ for $x = 3$, 3 must be squared *before* multiplying by 4.

10. What value of x would cause the expression $2x + 3$ to equal 9? Explain your reasoning.

11. There are many pairs of values of x and y for which $2x + y$ will equal 6. Name two such pairs and describe how you determined them.

12. Suppose that for the equation $3x - y = 9$, the value of x is given as 4. What would be the corresponding value of y? How do you know this?

Find the numerical value **(a)** *if x = 4 and* **(b)** *if x = 6. See Example 1.*

13. $x + 9$ **14.** $x - 1$ **15.** $5x$ **16.** $7x$ **17.** $4x^2$

18. $5x^2$ **19.** $\dfrac{x + 1}{3}$ **20.** $\dfrac{x - 2}{5}$ **21.** $\dfrac{3x - 5}{2x}$ **22.** $\dfrac{4x - 1}{3x}$

23. $3x^2 + x$ **24.** $2x + x^2$ ▦ **25.** $6.459x$ ▦ **26.** $.74x^2$

Find the numerical value if **(a)** *x = 2 and y = 1 and* **(b)** *x = 1 and y = 5. See Example 2.*

27. $8x + 3y + 5$ **28.** $4x + 2y + 7$ **29.** $3(x + 2y)$ **30.** $2(2x + y)$

31. $x + \dfrac{4}{y}$ **32.** $y + \dfrac{8}{x}$ **33.** $\dfrac{x}{2} + \dfrac{y}{3}$ **34.** $\dfrac{x}{5} + \dfrac{y}{4}$

35. $\dfrac{2x + 4y - 6}{5y + 2}$ **36.** $\dfrac{4x + 3y - 1}{x}$ **37.** $2y^2 + 5x$ **38.** $6x^2 + 4y$

39. $\dfrac{3x + y^2}{2x + 3y}$ **40.** $\dfrac{x^2 + 1}{4x + 5y}$ ▦ **41.** $.841x^2 + .32y^2$ ▦ **42.** $.941x^2 + .2y^2$

Change each word phrase to an algebraic expression. Use x as the variable to represent the number. See Example 3.

43. Twelve times a number   **44.** Nine times a number

45. Seven added to a number   **46.** Thirteen added to a number

47. Two subtracted from a number   **48.** Eight subtracted from a number

49. A number subtracted from seven   **50.** A number subtracted from fourteen

51. The difference between a number and 6   **52.** The difference between 6 and a number

53. 12 divided by a number   **54.** A number divided by 12

55. The product of 6 and four less than a number   **56.** The product of 9 and five more than a number

57. In the phrase "Four more than the product of a number and 6," does the word *and* signify the operation of addition? Explain.

58. Suppose that the directions on a test read "Solve the following expressions." How would you politely correct the person who wrote these directions? What alternative directions might you suggest?

Decide whether each given number is a solution of the equation. See Example 4.

59. $5m + 2 = 7$;   1   **60.** $3r + 5 = 8$;   1

61. $2y + 3(y - 2) = 14$;   3   **62.** $6a + 2(a + 3) = 14$;   2

63. $6p + 4p + 9 = 11$; $\dfrac{1}{5}$ **64.** $2x + 3x + 8 = 20$; $\dfrac{12}{5}$

65. $3r^2 - 2 = 46$;   4   **66.** $2x^2 + 1 = 19$;   3

67. $\dfrac{z + 4}{2 - z} = \dfrac{13}{5}$; $\dfrac{1}{3}$ **68.** $\dfrac{x + 6}{x - 2} = \dfrac{37}{5}$; $\dfrac{13}{4}$

Change each word statement to an equation. Use x as the variable. Find all solutions from the set {2, 4, 6, 8, 10}. *See Example 5.*

69. The sum of a number and 8 is 18.   **70.** A number minus three equals 1.

71. Sixteen minus three-fourths of a number is 13.   **72.** The sum of six-fifths of a number and 2 is 14.

73. One more than twice a number is 5.   **74.** The product of a number and 3 is 6.

75. Three times a number is equal to 8 more than twice the number.   **76.** Twelve divided by a number equals $\dfrac{1}{3}$ times that number.

Identify as an expression *or an* equation. *See Example 6.*

77. $3x + 2(x - 4)$ **78.** $5y - (3y + 6)$ **79.** $7t + 2(t + 1) = 4$
80. $9r + 3(r - 4) = 2$ **81.** $x + y = 3$ **82.** $x + y - 3$

 Mathematicians who study statistics have developed methods of determining mathematical models. Loosely speaking, a **mathematical model** is an equation that can be used to determine unknown quantities. We cannot always expect a model to give us an accurate answer, but at least we can obtain a rough estimate. For example, based on data from the U.S. Bureau of Labor Statistics, average hourly earnings of production workers in manufacturing industries in the United States during the period from 1990 through 1996 can be approximated by the equation $y = .319x - 624.31$, where x represents the year and y represents the hourly earnings in dollars.

 Use this equation to approximate the average hourly earnings during the following years. Do this by replacing x in the equation by the year, and then simplifying the right side of the equation to find y, the hourly earnings. Then compare your answer to the actual earnings (given in parentheses) and tell whether the approximation is greater than or less than the actual earnings, and by how much.

skip

83. 1990 ($10.83)
84. 1994 ($12.07)
85. 1995 ($12.37)
86. 1996 ($12.78)

1.4 Real Numbers and the Number Line

OBJECTIVES

1 Set up number lines.

2 Classify numbers.

3 Tell which of two different real numbers is smaller.

4 Find additive inverses and absolute values of real numbers.

5 Interpret real number meanings from a table of data.

FOR EXTRA HELP

 SSG Sec. 1.4
SSM Sec. 1.4

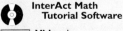 **Pass the Test Software**

InterAct Math
  Tutorial Software

 Video I

OBJECTIVE **1** Set up number lines. In Section 1.1 we introduced two important sets of numbers, the *natural numbers* and the *whole numbers*.

Natural Numbers

$\{1, 2, 3, 4, \ldots\}$ is the set of **natural numbers.**

Whole Numbers

$\{0, 1, 2, 3, \ldots\}$ is the set of **whole numbers.**

 The three dots show that the list of numbers continues in the same way indefinitely.

These numbers, along with many others, can be represented on **number lines** like the one pictured in Figure 4. We draw a number line by choosing any point on the line and calling it 0. Choose any point to the right of 0 and call it 1. The distance between 0 and 1 gives a unit of measure used to locate other points, as shown in Figure 4. The points labeled in Figure 4 and those continuing in the same way to the right correspond to the set of whole numbers.

Figure 4

All the whole numbers starting with 1 are located to the right of 0 on the number line. But numbers may also be placed to the left of 0. These numbers, written -1, -2, -3, and so on, shown in Figure 5, are called **negative numbers.** (The minus sign is used to show that these numbers are located to the *left* of 0.) The numbers to the *right* of 0 are **positive numbers.** The number 0 itself is neither positive nor negative. Positive numbers and negative numbers are called **signed numbers.**

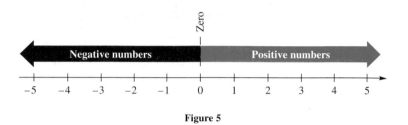

Figure 5

CONNECTIONS

There are many practical applications of negative numbers. For example, temperatures sometimes fall below zero. The lowest temperature ever recorded in meteorological records was $-128.6°F$ at Vostok, Antarctica, on July 22, 1983. A business that spends more than it takes in has a negative "profit." Altitudes below sea level can be represented by negative numbers. The shore surrounding the Dead Sea is 1312 feet below sea level; this can be represented as -1312 feet. (*Source: The World Almanac and Book of Facts,* 1995.)

OBJECTIVE 2 Classify numbers. The set of numbers marked on the number line in Figure 5, including positive and negative numbers and zero, is part of the set of *integers.*

Integers

$\{. . . , -3, -2, -1, 0, 1, 2, 3, . . .\}$ is the set of **integers.**

Not all numbers are integers. For example, $\frac{1}{2}$ is not; it is a number halfway between the integers 0 and 1. Also, $3\frac{1}{4}$ is not an integer. Several numbers that are not integers are *graphed* in Figure 6. The **graph** of a number is a point on the number line. The number is called the **coordinate** of the point. Think of the graph of a set of numbers as a picture of the set. All the numbers in Figure 6 can be written as quotients of integers. These numbers are examples of *rational numbers.*

Figure 6

Rational Numbers

$\{x \mid x$ is a quotient of two integers, with denominator not 0$\}$ is the set of **rational numbers.**

    (Read the part in the braces as "the set of all numbers x such that x is a quotient of two integers, with denominator not 0.")

The set symbolism used in the definition of rational numbers,

$$\{x \mid x \text{ has a certain property}\},$$

is called **set-builder notation.** This notation is convenient to use when it is not possible to list all the elements of the set.

Since any integer can be written as the quotient of itself and 1, all integers also are rational numbers.

    Although many numbers are rational, not all are. For example, a square that measures one unit on a side has a diagonal whose length is the square root of 2, written $\sqrt{2}$. See Figure 7. It can be shown that $\sqrt{2}$ cannot be written as a quotient of integers. Because of this, $\sqrt{2}$ is not rational; it is *irrational*. Other examples of irrational numbers are $\sqrt{3}$, $\sqrt{7}$, $-\sqrt{10}$, and π (the ratio of the *circumference* of a circle to its diameter).

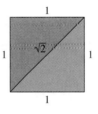

Figure 7

    All numbers, both rational and irrational, can be represented by points on the number line and are called *real numbers*.

Real Numbers

$\{x \mid x$ is a rational or an irrational number$\}$ is the set of **real numbers.**

Real numbers can be written as decimals. Any rational number will have a decimal that either comes to an end (terminates) or repeats in a fixed "block" of digits. For example, $\frac{2}{5} = .4$ and $\frac{27}{100} = .27$ are rational numbers with terminating decimals; $\frac{1}{3} = .3333. . .$ and $\frac{3}{11} = .27272727 . . .$ are repeating decimals. The decimal representation of an irrational number will neither terminate nor repeat. (A review of operations with decimals can be found in Appendix A.)

    An example of a number that is not a real number is the square root of a negative number like $\sqrt{-5}$. These numbers are discussed in a later chapter.

    Two ways to represent the relationships among the various types of numbers are shown in Figure 8. Part (a) also gives some examples. Notice that every real number is either a rational number or an irrational number.

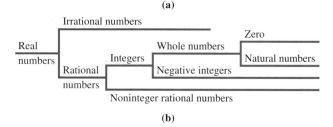

Rational numbers	Irrational numbers
$\frac{4}{9}, -\frac{5}{8}, \frac{11}{7}$	$-\sqrt{8}$
Integers $-11, -6, -4$	$\sqrt{15}$
Whole numbers 0	$\sqrt{23}$
Natural numbers 1, 2, 3, 4, 5, 37, 40	π $\frac{\pi}{4}$

All numbers shown are real numbers.

(a)

Real numbers
- Irrational numbers
- Rational numbers
  - Integers
    - Whole numbers
      - Zero
      - Natural numbers
    - Negative integers
  - Noninteger rational numbers

(b)

Figure 8

EXAMPLE 1 Determining whether a Number Belongs to a Set

List the numbers in the set

$$\left\{ -5, \quad -\frac{2}{3}, \quad 0, \quad \sqrt{2}, \quad 3\frac{1}{4}, \quad 5, \quad 5.8 \right\}$$

that belong to each of the following sets of numbers.

(a) Natural numbers
The only natural number in the set is 5.

(b) Whole numbers
The whole numbers consist of the natural numbers and 0. So the elements of the set that are whole numbers are 0 and 5.

(c) Integers
The integers in the set are -5, 0, and 5.

(d) Rational numbers
The rational numbers are -5, $-\frac{2}{3}$, 0, $3\frac{1}{4}$, 5, 5.8, since each of these numbers *can* be written as the quotient of two integers. For example, $5.8 = \frac{58}{10}$.

(e) Irrational numbers
The only irrational number in the set is $\sqrt{2}$.

(f) Real numbers
All the numbers in the set are real numbers.

OBJECTIVE **3** Tell which of two different real numbers is smaller. Given any two whole numbers, you probably can tell which number is smaller. But what happens with negative numbers, as in the set of integers? Positive numbers decrease as the corresponding points on the number line go to the left. For example, $8 < 12$, and 8 is to the left of 12 on the number line. This ordering is extended to all real numbers by definition.

The Ordering of Real Numbers

For any two real numbers a and b, a **is less than** b if a is to the left of b on the number line.

This means that any negative number is smaller than 0, and any negative number is smaller than any positive number. Also, 0 is smaller than any positive number.

EXAMPLE 2 Determining the Order of Real Numbers

Is it true that $-3 < -1$?

To decide whether the statement is true, locate both numbers, -3 and -1, on a number line, as shown in Figure 9. Since -3 is to the left of -1 on the number line, -3 is smaller than -1. The statement $-3 < -1$ is true.

Figure 9

 In Section 1.2 we saw how it is possible to rewrite a statement involving $<$ as an equivalent statement involving $>$. The question in Example 2 can also be worded as follows: Is it true that $-1 > -3$? This is, of course, also a true statement.

We can say that for any two real numbers a and b, a **is greater than** b if a is to the right of b on the number line.

OBJECTIVE **4** Find additive inverses and absolute values of real numbers. By a property of the real numbers, for any real number x (except 0), there is exactly one number on the number line the same distance from 0 as x but on the opposite side of 0. For example, Figure 10 shows that the numbers 3 and -3 are each the same distance from 0 but are on opposite sides of 0. The numbers 3 and -3 are called **additive inverses,** or **opposites,** of each other.

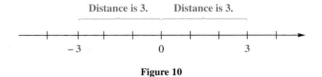

Figure 10

Additive Inverse

The **additive inverse** of a number x is the number that is the same distance from 0 on the number line as x, but on the opposite side of 0.

The additive inverse of the number 0 is 0 itself. In fact, 0 is the only real number that is its own additive inverse. Other additive inverses occur in pairs. For example, 4 and -4, and 5 and -5, are additive inverses of each other. Several pairs of additive inverses are shown in Figure 11.

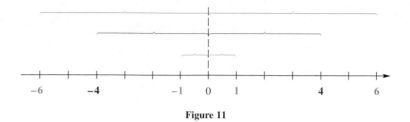

Figure 11

The additive inverse of a number can be indicated by writing the symbol $-$ in front of the number. With this symbol, the additive inverse of 7 is written -7. The additive inverse of -4 is written $-(-4)$, and can also be read "the opposite of -4" or "the negative of -4." Figure 11 suggests that 4 is an additive inverse of -4. Since a number can have only one additive inverse, the symbols 4 and $-(-4)$ must represent the same number, which means that

$$-(-4) = 4.$$

This idea can be generalized as follows.

Double Negative Rule

For any real number x,

$$-(-x) = x.$$

The following chart shows several numbers and their additive inverses.

Number	Additive Inverse
-4	$-(-4)$ or 4
0	0
19	-19
$-\dfrac{2}{3}$	$\dfrac{2}{3}$

 The chart suggests that the additive inverse of a number is found by changing the sign of the number.

An important property of additive inverses will be studied later in this chapter: $a + (-a) = (-a) + a = 0$ for all real numbers a.

As mentioned above, additive inverses are numbers that are the same distance from 0 on the number line. See Figure 11. This idea can also be expressed by saying that a number and its additive inverse have the same absolute value. The **absolute value** of a real number can be defined as the distance between 0 and the number on the number line. The symbol for the absolute value of the number x is $|x|$, read "the absolute value of x." For example, the distance between 2 and 0 on the number line is 2 units, so that

$$|2| = 2.$$

Because the distance between -2 and 0 on the number line is also 2 units,

$$|-2| = 2.$$

Since distance is a physical measurement, which is never negative, **the absolute value of a number is never negative.** For example, $|12| = 12$ and $|-12| = 12$, since both 12 and -12 lie at a distance of 12 units from 0 on the number line. Also, since 0 is a distance of 0 units from 0, $|0| = 0$.

In symbols, the absolute value of x is defined as follows.

Formal Definition of Absolute Value

$$|x| = \begin{cases} x & \text{if } x \geq 0 \\ -x & \text{if } x < 0 \end{cases}$$

By this definition, if x is a positive number or 0, then its absolute value is x itself. For example, since 8 is a positive number, $|8| = 8$. However, if x is a negative number, then its absolute value is the additive inverse of x. This means that if $x = -9$, then $|-9| = -(-9) = 9$, since the additive inverse of -9 is 9.

CAUTION The formal definition of absolute value can be confusing if it is not read carefully. The "$-x$" in the second part of the definition *does not* represent a negative number. Since x is negative in the second part, $-x$ represents the opposite of a negative number, that is, a positive number. *The absolute value of a number is never negative.*

EXAMPLE 3   Finding Absolute Value

Simplify by finding the absolute value.

(a) $|5| = 5$ **(b)** $|-5| = -(-5) = 5$

(c) $-|5| = -(5) = -5$ **(d)** $-|-14| = -(14) = -14$

(e) $|8 - 2| = |6| = 6$ **(f)** $-|8 - 2| = -|6| = -6$

Part (e) of Example 3 shows that absolute value bars are also grouping symbols. You must perform any operations that appear inside absolute value symbols before finding the absolute value.

OBJECTIVE 5 Interpret real number meanings from a table of data. A table of data provides a concise way of relating information. The final example gives an illustration of this.

 EXAMPLE 4   Interpreting Data

The table shows the annual percent change from the previous year in the Consumer Price Index for 1986 and 1987.

	Percent Change	
Category	1986	1987
Food	3.2	4.1
Shelter	5.5	4.7
Rent, residential	5.8	4.1
Fuel and other utilities	−2.3	−1.1
Apparel and upkeep	.9	4.4
Private transportation	−4.7	−3.0
New cars	4.2	3.6
Gasoline	−21.9	−4.0
Public transportation	5.9	3.5
Medical care	7.5	6.6
Entertainment	3.4	3.3
Commodities	−.9	3.2

Source: Bureau of Labor Statistics, U.S. Dept. of Labor.

(a) What category in which year represents the greatest drop?

To find the greatest drop, we must find the negative entry with the largest absolute value. The entry for gasoline in 1986 is −21.9, so this is the category and year with the greatest drop.

(b) Which percent change is represented by a larger drop: 1986 fuel and other utilities or 1987 gasoline?

The two entries are −2.3 and −4.0, respectively. The larger drop is −4.0, so 1987 gasoline is the answer.

(c) What is the smallest percent change (positive or negative) in these two years?

The smallest percent change (positive or negative) is .9 and −.9, both found in 1986.

1.4 EXERCISES

In Exercises 1–6, give a number that satisfies the given condition.

1. An integer between 3.5 and 4.5

2. A rational number between 3.8 and 3.9

3. A whole number that is not positive and is less than 1

4. A whole number greater than 4.5

5. An irrational number that is between $\sqrt{11}$ and $\sqrt{13}$

6. A real number that is neither negative nor positive

In Exercises 7–10, decide whether each statement is true or false.

7. Every natural number is positive.

8. Every whole number is positive.

9. Every integer is a rational number.

10. Every rational number is a real number.

For Exercises 11 and 12, see Example 1.

11. List all numbers from the set

$$\left\{-9, -\sqrt{7}, -1\frac{1}{4}, -\frac{3}{5}, 0, \sqrt{5}, 3, 5.9, 7\right\}$$

that are

(**a**) natural numbers; (**b**) whole numbers; (**c**) integers;
(**d**) rational numbers; (**e**) irrational numbers; (**f**) real numbers.

12. List all numbers from the set

$$\left\{-5.3, -5, -\sqrt{3}, -1, -\frac{1}{9}, 0, 1.2, 1.8, 3, \sqrt{11}\right\}$$

that are

(**a**) natural numbers; (**b**) whole numbers; (**c**) integers;
(**d**) rational numbers; (**e**) irrational numbers; (**f**) real numbers.

13. Explain in your own words the different sets of numbers introduced in this section, and give an example of each kind.

14. What two possible situations exist for the decimal representation of a rational number?

Use an integer to express each number representing a change in the following applications.

15. In February 1998, the number of housing starts in the United States increased from the previous month by 93,000 units. (*Source: The Wall Street Journal.*)

16. The Wolfsburg, Germany, Volkswagen plant turns out 1550 fewer cars per day than it did in 1991. (*Source:* Paul Klebnikov, "Bringing Back the Beetle," *Forbes,* April 7, 1997.)

17. The boiling point of chlorine is approximately 30° below 0° Fahrenheit.

18. The height of Mt. Arenal, an active volcano in Costa Rica, is 5436 feet above sea level. (*Source: The Universal Almanac,* 1997, John W. Wright, General Editor.)

19. Between 1980 and 1990, the population of the District of Columbia decreased by 31,532. (*Source:* U.S. Bureau of the Census.)

20. In 1994, the country of Taiwan produced 159,376 more passenger cars than commercial vehicles. (*Source:* American Automobile Manufacturers Association.)

21. The city of New Orleans lies 8 feet below sea level. (*Source:* U.S. Geological Survey, *Elevations and Distances in the United States,* 1990.)

22. When the wind speed is 20 miles per hour and the actual temperature is 25° Fahrenheit, the wind-chill factor is 3° below 0° Fahrenheit. (Give three responses.)

Graph each group of numbers on a number line. See Figure 6.

23. $0, 3, -5, -6$

24. $2, 6, -2, -1$

25. $-2, -6, -4, 3, 4$

26. $-5, -3, -2, 0, 4$

27. $\frac{1}{4}, 2\frac{1}{2}, -3\frac{4}{5}, -4, -1\frac{5}{8}$

28. $5\frac{1}{4}, 4\frac{5}{9}, -2\frac{1}{3}, 0, -3\frac{2}{5}$

29. Match each expression in Column I with its value in Column II. Some choices in Column II may not be used.

I	II
(**a**) $\lvert -7 \rvert$	**A.** 7
(**b**) $-(-7)$	**B.** -7
(**c**) $-\lvert -7 \rvert$	**C.** neither A nor B
(**d**) $-\lvert -(-7) \rvert$	**D.** both A and B

30. Fill in the blanks with the correct values: The opposite of -2 is _____, while the absolute value of -2 is _____. The additive inverse of -2 is _____, while the additive inverse of the absolute value of -2 is _____.

*Find **(a)** the opposite (or additive inverse) of each number and **(b)** the absolute value of each number.*

31. -2 **32.** -8 **33.** 6 **34.** 11

35. $7 - 4$ **36.** $8 - 3$ **37.** $7 - 7$ **38.** $3 - 3$

39. Look at Exercises 35 and 36 and use the results to complete the following: If $a - b > 0$, then the absolute value of $a - b$ in terms of a and b is _____.

40. Look at Exercises 37 and 38 and use the results to complete the following: If $a - b = 0$, then the absolute value of $a - b$ is _____.

Select the smaller of the two given numbers. See Examples 2 and 3.

41. $-12, -4$ **42.** $-9, -14$ **43.** $-8, -1$

44. $-15, -16$ **45.** $3, |-4|$ **46.** $5, |-2|$

47. $|-3|, |-4|$ **48.** $|-8|, |-9|$ **49.** $-|-6|, -|-4|$

50. $-|-2|, -|-3|$ **51.** $|5 - 3|, |6 - 2|$ **52.** $|7 - 2|, |8 - 1|$

Decide whether each statement is true or false. See Examples 2 and 3.

53. $6 > -(-2)$ **54.** $-8 > -(-2)$ **55.** $-4 \leq -(-5)$

56. $-6 \leq -(-3)$ **57.** $|-6| < |-9|$ **58.** $|-12| < |-20|$

59. $-|8| > |-9|$ **60.** $-|12| > |-15|$ **61.** $-|-5| \geq -|-9|$

62. $-|-12| \leq -|-15|$ **63.** $|6 - 5| \geq |6 - 2|$ **64.** $|13 - 8| \leq |7 - 4|$

The table shows the percent change in the Producer Price Index for selected commodities from 1994 to 1995 and from 1995 to 1996. Use the table to answer the questions in Exercises 65–68. See Example 4.

Commodity	Change from 1994 to 1995	Change from 1995 to 1996*
Interior solvent-based paint	16.4	11.1
Plastic construction products	10.9	−3.1
Softwood plywood	11.3	−14.1
Bright nails	1.8	−.3
Gypsum products	18.4	−.3

*1996 data is preliminary.
Source: U.S. Bureau of Labor Statistics.

65. Which category in what year represents the greatest drop?

66. Which of these is the greater number: the change in plastic construction products from 1995 to 1996 or the change in bright nails from 1994 to 1995?

67. True or false? The absolute value of the data in the first column for softwood plywood is less than the absolute value of the data in the second column for the same commodity.

68. Which category represents the greatest increase?

For each statement give a pair of values for a and b that make it true, and then give a pair of values that make it false.

69. $|a + b| = |a - b|$

70. $|a - b| = |b - a|$

71. $|a + b| = -|a + b|$

72. $|-(a + b)| = -(a + b)$

Give three numbers between −6 and 6 that satisfy each given condition.

73. Positive real numbers but not integers

74. Real numbers but not positive numbers

75. Real numbers but not whole numbers

76. Rational numbers but not integers

77. Real numbers but not rational numbers

78. Rational numbers but not negative numbers

79. Students often say "Absolute value is always positive." Is this true? If not, explain why.

80. True or false: If a is negative, $|a| = -a$.

1.5   Addition and Subtraction of Real Numbers

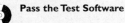

OBJECTIVES

1 Add two numbers with the same sign.

2 Add positive and negative numbers.

3 Use the definition of subtraction.

4 Use the order of operations with real numbers.

5 Interpret words and phrases involving addition and subtraction.

6 Use signed numbers to interpret data.

FOR EXTRA HELP

 SSG Sec. 1.5
SSM Sec. 1.5

 Pass the Test Software

**InterAct Math
  Tutorial Software**

Video 2

In this section and the next section, we extend the rules for operations with positive numbers to the negative numbers, beginning with addition and subtraction.

OBJECTIVE **1** Add two numbers with the same sign. A number line can be used to explain addition of real numbers.

EXAMPLE 1   Adding Positive Numbers on a Number Line

Use a number line to find the sum $2 + 3$.

Add the positive numbers 2 and 3 on the number line by starting at 0 and drawing an arrow two units to the *right,* as shown in Figure 12. This arrow represents the number 2 in the sum $2 + 3$. Then, from the right end of this arrow draw another arrow three units to the right. The number below the end of this second arrow is 5, so $2 + 3 = 5$.

Figure 12

EXAMPLE 2   Adding Negative Numbers on a Number Line

Use a number line to find the sum $-2 + (-4)$. (Parentheses are placed around the -4 to avoid the confusing use of $+$ and $-$ next to each other.)

Add the negative numbers -2 and -4 on the number line by starting at 0 and drawing an arrow two units to the *left,* as shown in Figure 13. The arrow is drawn to the left to represent the addition of a *negative* number. From the left end of this first arrow, draw a second arrow four units to the left. The number below the end of this second arrow is -6, so $-2 + (-4) = -6$.

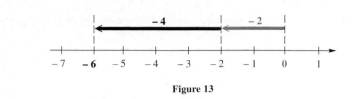

Figure 13

In Example 2, the sum of the two negative numbers -2 and -4 is a negative number whose distance from 0 is the sum of the distance of -2 from 0 and the distance of -4 from 0. That is, *the sum of two negative numbers is the negative of the sum of their absolute values.*

$$-2 + (-4) = -(|-2| + |-4|) = -(2 + 4) = -6$$

Adding Numbers with the Same Signs

To add two numbers with the *same* signs, add the absolute values of the numbers. The sum has the same sign as the numbers being added.

E X A M P L E   3   Adding Two Negative Numbers

Find the sums.

(a) $-2 + (-9) = -(|-2| + |-9|) = -(2 + 9) = -11$

(b) $-8 + (-12) = -20$

(c) $-15 + (-3) = -18$

O B J E C T I V E   2   Add positive and negative numbers. We can use a number line to explain the sum of a positive number and a negative number.

E X A M P L E   4   Adding Numbers with Different Signs

Use a number line to find the sum $-2 + 5$.

Find the sum $-2 + 5$ on the number line by starting at 0 and drawing an arrow two units to the left. From the left end of this arrow, draw a second arrow five units to the right, as shown in Figure 14. The number below the end of the second arrow is 3, so $-2 + 5 = 3$.

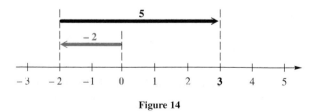

Figure 14

Adding Numbers with Different Signs

To add two numbers with *different* signs, subtract the smaller absolute value from the larger absolute value. The answer has the sign of the number with the larger absolute value.

For example, to add -12 and 5, find their absolute values: $|-12| = 12$ and $|5| = 5$. Then find the difference between these absolute values: $12 - 5 = 7$. Since $|-12| > |5|$, the sum will be negative, so that the final answer is $-12 + 5 = -7$.

While a number line is useful in showing the rules for addition, it is important to be able to do the problems quickly, "in your head."

EXAMPLE 5 Adding Mentally

Check each answer, trying to work the addition mentally. If you get stuck, use a number line.

(a) $7 + (-4) = 3$ **(b)** $-8 + 12 = 4$

(c) $-\dfrac{1}{2} + \dfrac{1}{8} = -\dfrac{4}{8} + \dfrac{1}{8} = -\dfrac{3}{8}$ Remember to get a common denominator first.

(d) $\dfrac{5}{6} + \left(-\dfrac{4}{3}\right) = -\dfrac{1}{2}$ **(e)** $-4.6 + 8.1 = 3.5$

(f) $-16 + 16 = 0$ **(g)** $42 + (-42) = 0$

Parts (f) and (g) in Example 5 suggest that the sum of a number and its additive inverse is 0. This is always true, and this property is discussed further in Section 1.7.

The rules for adding signed numbers are summarized as follows.

Adding Signed Numbers

Like signs Add the absolute values of the numbers. The sum has the same sign as the given numbers.

Unlike signs Find the difference between the larger absolute value and the smaller. The sum has the sign of the number with the larger absolute value.

OBJECTIVE **3** Use the definition of subtraction. We can illustrate subtraction of 4 from 7, written $7 - 4$, using a number line. As seen in Figure 15, we begin at 0 and draw an arrow seven units to the right. From the right end of this arrow, draw an arrow four units to the left. The number at the end of the second arrow shows that $7 - 4 = 3$.

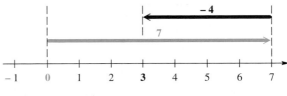

Figure 15

The procedure used to find $7 - 4$ is exactly the same procedure that would be used to find $7 + (-4)$, so that

$$7 - 4 = 7 + (-4).$$

This suggests that *subtraction* of a positive number from a larger positive number is the same as *adding* the additive inverse of the smaller number to the larger. This result leads to the definition of subtraction for all real numbers.

Definition of Subtraction

For any real numbers x and y,

$$x - y = x + (-y).$$

That is, to *subtract y* from *x, add the additive inverse* (or *opposite*) of *y* to *x*. This definition gives the following procedure for subtracting signed numbers.

Subtracting Signed Numbers

Step 1 Change the subtraction symbol to the addition symbol.

Step 2 Change the sign of the number being subtracted.

Step 3 Add.

EXAMPLE 6 **Using the Definition of Subtraction**

Subtract.

Change $-$ to $+$.

No change

Additive inverse of 3

(a) $12 - 3 = 12 + (-3) = 9$

(b) $5 - 7 = 5 + (-7) = -2$

Change $-$ to $+$.

No change

Additive inverse of -5

(c) $-3 - (-5) = -3 + (5) = 2$

(d) $-6 - 9 = -6 + (-9) = -15$

(e) $8 - (-4) = 8 + (4) = 12$

OBJECTIVE 4 **Use the order of operations with real numbers.** As before, with problems that have grouping symbols, first do any operations inside the parentheses and brackets. Work from the inside out.

EXAMPLE 7 **Adding and Subtracting with Grouping Symbols**

Perform the indicated operations.

(a)

$$
\begin{aligned}
-6 - [2 - (8 + 3)] &= -6 - [2 - 11] & \text{Add.}\\
&= -6 - [2 + (-11)] & \text{Use the definition of subtraction.}\\
&= -6 - (-9) & \text{Add.}\\
&= -6 + (9) & \text{Use the definition of subtraction.}\\
&= 3 & \text{Add.}
\end{aligned}
$$

(b)

$$
\begin{aligned}
5 + [(-3 - 2) - (4 - 1)] &= 5 + [(-3 + (-2)) - 3]\\
&= 5 + [(-5) - 3]\\
&= 5 + [(-5) + (-3)]\\
&= 5 + (-8)\\
&= -3
\end{aligned}
$$

(c) $\dfrac{2}{3} - \left[\dfrac{1}{12} - \left(-\dfrac{1}{4}\right)\right] = \dfrac{8}{12} - \left[\dfrac{1}{12} - \left(-\dfrac{3}{12}\right)\right]$ Get a common denominator.

$= \dfrac{8}{12} - \left[\dfrac{1}{12} + \dfrac{3}{12}\right]$ Use the definition of subtraction.

$= \dfrac{8}{12} - \dfrac{4}{12}$ Add.

$= \dfrac{4}{12}$ Subtract.

$= \dfrac{1}{3}$ Lowest terms

(d) $|4 - 7| + 2|6 - 3| = |-3| + 2|3|$ Work within absolute value symbols.

$= 3 + 2 \cdot 3$ Evaluate the absolute values.

$= 3 + 6$ Multiply.

$= 9$ Add.

OBJECTIVE 5 Interpret words and phrases involving addition and subtraction. We begin by working with addition words and phrases.

PROBLEM SOLVING

As we mentioned earlier, problem solving often requires translating words and phrases into symbols. The word *sum* is one of the words that indicates addition. The chart below lists some of the words and phrases that also signify addition.

Word or Phrase	Example	Numerical Expression and Simplification
Sum of	The *sum of* −3 and 4	−3 + 4 = 1
Added to	5 *added to* −8	−8 + 5 = −3
More than	12 *more than* −5	−5 + 12 = 7
Increased by	−6 *increased by* 13	−6 + 13 = 7
Plus	3 *plus* 14	3 + 14 = 17

EXAMPLE 8 Interpreting Words and Phrases Involving Addition

Write a numerical expression for each phrase, and simplify the expression.

(a) The sum of −8 and 4 and 6

$$-8 + 4 + 6 = -4 + 6 = 2$$

Add in order from left to right.

(b) 3 more than −5, increased by 12

$$-5 + 3 + 12 = -2 + 12 = 10$$

We now look at how we interpret words and phrases that involve subtraction.

PROBLEM SOLVING

In order to solve problems that involve subtraction, we must be able to interpret key words and phrases that indicate subtraction. *Difference* is one of them. Some of these are given in the chart below.

Word or Phrase	Example	Numerical Expression and Simplification
Difference between	The *difference between* -3 and -8	$-3 - (-8) = -3 + 8 = 5$
Subtracted from	12 *subtracted from* 18	$18 - 12 = 6$
Less	6 *less* 5	$6 - 5 = 1$
Less than	6 *less than* 5	$5 - 6 = 5 + (-6) = -1$
Decreased by	9 *decreased by* -4	$9 - (-4) = 9 + 4 = 13$
Minus	8 *minus* 5	$8 - 5 = 3$

 When you are subtracting two numbers, it is important that you write them in the correct order, because, in general, $a - b \neq b - a$. For example, $5 - 3 \neq 3 - 5$. For this reason, *think carefully before interpreting an expression involving subtraction.* (This difficulty did not arise for addition.)

┌ **E X A M P L E 9**   Interpreting Words and Phrases Involving Subtraction

Write a numerical expression for each phrase, and simplify the expression.

(a) The difference between -8 and 5

It is conventional to write the numbers in the order they are given when "difference between" is used.

$$-8 - 5 = -8 + (-5) = -13$$

(b) 4 subtracted from the sum of 8 and -3

Here addition is also used, as indicated by the word *sum*. First, add 8 and -3. Next, subtract 4 *from* this sum.

$$[8 + (-3)] - 4 = 5 - 4 = 1$$

(c) 4 less than -6

Be careful with order here. 4 must be taken *from* -6.

$$-6 - 4 = -6 + (-4) = -10$$

Notice that "4 less than -6" differs from "4 *is less than* -6." The second of these is symbolized as $4 < -6$ (which is a false statement).

(d) 8, decreased by 5 **less than** 12

First, write "5 less than 12" as $12 - 5$. Next, subtract $12 - 5$ from 8.

$$8 - (12 - 5) = 8 - 7 = 1$$

OBJECTIVE **6** Use signed numbers to interpret data. The Producer Price Index is the oldest continuous statistical series published by the Bureau of Labor Statistics. It measures the average changes in prices received by producers of all commodities produced in the United States. The next example shows how signed numbers can be used to interpret such data.

EXAMPLE 10   Using a Signed Number in Data Interpretation

The bar graph in Figure 16 gives the Producer Price Index (PPI) for construction materials between 1985 and 1990.

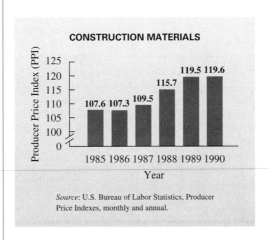

CONSTRUCTION MATERIALS

Source: U.S. Bureau of Labor Statistics, Producer Price Indexes, monthly and annual.

Figure 16

(a) Use a signed number to represent the change in the PPI from 1987 to 1988.

To find this change, we start with the index number from 1988 and subtract from it the index number from 1987.

$$115.7 \quad - \quad 109.5 \quad = \quad 6.2$$

The 1988 index   The 1987 index   A positive number indicates an increase.

(b) Use a signed number to represent the change in the PPI from 1985 to 1986.

Use the same procedure as in part (a).

$$107.3 \quad - \quad 107.6 \quad = \quad 107.3 + (-107.6) = -.3$$

The 1986 index   The 1985 index   A negative number indicates a decrease.

EXAMPLE 11   Solving a Problem Involving Subtraction

The record high temperature in the United States was 134° Fahrenheit, recorded at Death Valley, California, in 1913. The record low was −80°F, at Prospect Creek, Alaska, in 1971. See Figure 17. What is the difference between these highest and lowest temperatures? (*Source: The World Almanac and Book of Facts, 1998.*)

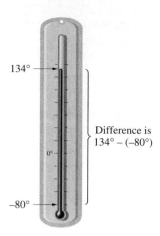

Figure 17

We must subtract the lowest temperature from the highest temperature.

$$134 - (-80) = 134 + 80 \qquad \text{Use the definition of subtraction.}$$
$$= 214 \qquad \text{Add.}$$

The difference between the two temperatures is 214°F.

1.5 EXERCISES

Fill in each blank with the correct response.

1. The sum of two negative numbers will always be a _____ number.
   (positive/negative)

2. The sum of a number and its opposite will always be _____ .

3. To simplify the expression $8 + [-2 + (-3 + 5)]$, I should begin by adding _____ and _____, according to the rule for order of operations.

4. If I am adding a positive number and a negative number, and the negative number has the larger absolute value, the sum will be a _____ number.
   (positive/negative)

5. Explain in words how to add signed numbers. Consider the various cases and give examples.

6. Explain in words how to subtract signed numbers.

Find each sum or difference. See Examples 1–7.

7. $6 + (-4)$

8. $12 + (-9)$

9. $7 + (-10)$

10. $4 + (-8)$

11. $-7 + (-3)$

12. $-11 + (-4)$

13. $-10 + (-3)$

14. $-16 + (-7)$

15. $-12.4 + (-3.5)$

16. $-21.3 + (-2.5)$

17. $-8 + 7$

18. $-12 + 10$

19. $5 + [14 + (-6)]$

20. $7 + [3 + (-14)]$

21. $4 - 7$

22. $8 - 13$

23. $6 - 10$

24. $9 - 14$

25. $-7 - 3$

26. $-12 - 5$

27. $-10 - 6$

28. $-13 - 16$

29. $7 - (-4)$

30. $9 - (-6)$

31. $6 - (-13)$

32. $13 - (-3)$

33. $-7 - (-3)$

34. $-8 - (-6)$

35. $3 - (4 - 6)$

36. $6 - (7 - 14)$

37. $-3 - (6 - 9)$

38. $-4 - (5 - 12)$

39. $\dfrac{1}{2} - \left(-\dfrac{1}{4}\right)$

40. $\dfrac{1}{3} - \left(-\dfrac{4}{3}\right)$

41. $-8 + [3 + (-1) + (-2)]$

42. $-7 + [5 + (-8) + 3]$

43. $\dfrac{5}{8} - \left(-\dfrac{1}{2} - \dfrac{3}{4}\right)$

44. $\dfrac{9}{10} - \left(\dfrac{1}{8} - \dfrac{3}{10}\right)$

45. $[(-3.1) - 4.5] - (.8 - 2.1)$

46. $[(-7.8) - 9.3] - (.6 - 3.5)$

47. $[-5 + (-7)] + [-4 + (-9)] + [13 + (-12)]$

48. $[-3 + (-11)] + [13 + (-3)] + [19 + (-7)]$

49. $-4 + [(-6 - 9) - (-7 + 4)]$

50. $-8 + [(-3 - 10) - (-4 + 1)]$

51. Is it possible to add a negative number to another negative number and get a positive number? If so, give an example.

52. Under what conditions will the sum of a positive number and a negative number be a number which is neither negative nor positive?

53. Make up a subtraction problem so that the difference between two negative numbers is a negative number.

54. Make up a subtraction problem so that the difference between two negative numbers is a positive number.

Simplify each expression involving absolute value. Remember that absolute value bars serve as grouping symbols. See Example 7(d).

55. $|3 - 8| - |-2 + 8|$

56. $|-2 - 7| - |9 - (-3)|$

57. $\left|\dfrac{2}{3} - \dfrac{7}{3}\right| + 2\left|-\dfrac{4}{9} + \dfrac{5}{9}\right|$

58. $\left|-\dfrac{1}{5} + \dfrac{3}{5}\right| + 3\left|\dfrac{3}{10} - \dfrac{8}{10}\right|$

Write a numerical expression for each phrase and simplify. See Example 8.

59. The sum of -5 and 12 and 6

60. The sum of -3 and 5 and -12

61. 14 added to the sum of -19 and -4

62. -2 added to the sum of -18 and 11

63. The sum of -4 and -10, increased by 12

64. The sum of -7 and -13, increased by 14

65. 4 more than the sum of 8 and -18

66. 10 more than the sum of -4 and -6

Write a numerical expression for each phrase and simplify. See Example 9.

67. The difference between 4 and -8

68. The difference between 7 and -14

69. 8 less than −2

70. 9 less than −13

71. The sum of 9 and −4, decreased by 7

72. The sum of 12 and −7, decreased by 14

73. 12 less than the difference between 8 and −5

74. 19 less than the difference between 9 and −2

The bar graph shows the federal budget outlays for national defense for the years 1985 through 1994. Use a signed number to represent the change in outlay for each of the following time periods. See Example 10.

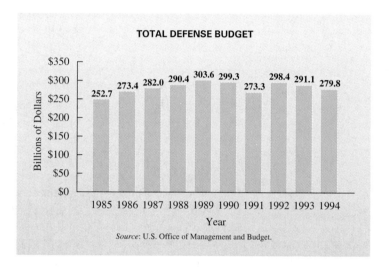

Source: U.S. Office of Management and Budget.

75. 1988 to 1989   **76.** 1986 to 1987

77. 1990 to 1991   **78.** 1993 to 1994

The two charts show the heights of some selected mountains and the depths of some selected trenches. Use the information given to find the answers in Exercises 79–82.

Mountain	Height (in feet)	Trench	Depth (in feet, as a negative number)
Foraker	17,400	Philippine	−32,995
Wilson	14,246	Cayman	−24,721
Pikes Peak	14,110	Java	−23,376

Source: The World Almanac and Book of Facts, 1998.

79. What is the difference between the height of Mt. Foraker and the depth of the Philippine Trench?

80. What is the difference between the height of Pikes Peak and the depth of the Java Trench?

81. How much deeper is the Cayman Trench than the Java Trench?

82. How much deeper is the Philippine Trench than the Cayman Trench?

Use this idea when working Exercises 83 and 84: If a player steals 65 bases in 1990 and 58 in 1991, the change in stolen bases is represented by $58 - 65 = -7$. *His number went* down *by 7.*

83. The bar graph shows the number of bases stolen by baseball player Rickey Henderson. Use a signed number to represent the change in the number of stolen bases from one year to the next. See Example 10. (*Source:* David S. Neft and Richard M. Cohen, *The Sports Encyclopedia: Baseball 1997.*)

     (a) From 1991 to 1992   **(b)** From 1992 to 1993

     (c) From 1989 to 1990   **(d)** From 1993 to 1994

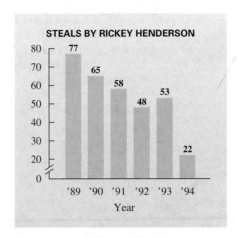

84. The bar graph shows the number of bases stolen by baseball player Kenny Lofton. Use a signed number to represent the change in the number of stolen bases from one year to the next. See Example 10. (*Source:* David S. Neft and Richard M. Cohen, *The Sports Encyclopedia: Baseball 1997.*)

     (a) From 1993 to 1994   **(b)** From 1994 to 1995

     (c) From 1995 to 1996   **(d)** From 1996 to 1997

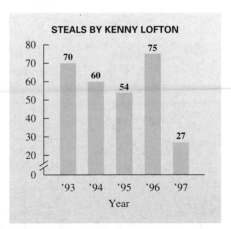

Solve each problem. (Refer to Figure 17 for Exercises 85–87.) See Example 11.

85. On January 23, 1943, the temperature rose 49°F in two minutes in Spearfish, South Dakota. If the starting temperature was −4°F, what was the temperature two minutes later? (*Source: The Guinness Book of World Records.*)

86. The lowest temperature ever recorded in Little Rock, Arkansas, was −5°F. The highest temperature ever recorded there was 117°F more than the lowest. What was this highest temperature? (*Source: The World Almanac and Book of Facts,* 1998.)

87. The coldest temperature recorded in Chicago, Illinois, was −27°F in 1985. The record low in Huron, South Dakota, was set in 1994 and was 14°F lower than −27°F. What was the record low in Huron? (*Source: The World Almanac and Book of Facts, 1998.*)

88. No one knows just why humpback whales love to heave their 45-ton bodies out of the water, but leap they do. (This activity is called *breaching.*) Mark and Debbie, two researchers based on the island of Maui, noticed that one of their favorite whales, "Pineapple," leaped 15 feet above the surface of the ocean while her mate cruised 12 feet below the surface. What is the difference between these two distances?

89. The top of Mt. Whitney, visible from Death Valley, has an altitude of 14,494 feet above sea level. The bottom of Death Valley is 282 feet below sea level. Using zero as sea level, find the difference between these two elevations. (*Source: The World Almanac and Book of Facts, 1998.*)

90. The highest point in Louisiana is Driskill Mountain, at an altitude of 535 feet. The lowest point is at Spanish Fort, 8 feet below sea level. Using zero as sea level, find the difference between these two elevations. (*Source: The World Almanac and Book of Facts, 1998.*)

91. A female polar bear weighed 660 pounds when she entered her winter den. She lost 45 pounds during each of the first two months of hibernation, and another 205 pounds before leaving the den with her two cubs in March. How much did she weigh when she left the den?

92. On three consecutive passes, Troy Aikman of the Dallas Cowboys passed for a gain of 6 yards, was sacked for a loss of 12 yards, and passed for a gain of 43 yards. What positive or negative number represents the total net yardage for the plays?

93. Kim Falgout owes $870.00 on her Master Card account. She returns two items costing $35.90 and $150.00 and receives credits for these on the account. Next, she makes a purchase of $82.50, and then two more purchases of $10.00 each. She makes a payment of $500.00. She then incurs a finance charge of $37.23. How much does she still owe?

94. A welder working with stainless steel must use precise measurements. Suppose a welder attaches two pieces of steel that are each 3.60 inches in length, and then attaches an additional three pieces that are each 9.10 inches long. She finally cuts off a piece that is 7.60 inches long. Find the length of the welded piece of steel.

1.6 Multiplication and Division of Real Numbers

OBJECTIVES

1   Find the product of a positive number and a negative number.

2   Find the product of two negative numbers.

3   Identify factors of integers.

4   Use the reciprocal of a number to apply the definition of division.

5   Use the order of operations when multiplying and dividing signed numbers.

6   Evaluate expressions involving variables.

7   Interpret words and phrases involving multiplication and division.

8   Translate simple sentences into equations.

FOR EXTRA HELP

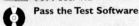
SSG Sec. 1.6
SSM Sec. 1.6

Pass the Test Software

InterAct Math Tutorial Software

Video 2

In this section we learn how to multiply with positive and negative numbers. We already know how to multiply positive numbers and that the product of two positive numbers is positive. We also know that the product of 0 and any positive number is 0, so we extend that property to all real numbers.

Multiplication by Zero

For any real number x, $x \cdot 0 = 0$.

OBJECTIVE 1   Find the product of a positive number and a negative number.   In order to define the product of a positive and a negative number so that the result is consistent with the multiplication of two positive numbers, look at the following pattern.

$$3 \cdot 5 = 15$$
$$3 \cdot 4 = 12$$
$$3 \cdot 3 = 9$$
$$3 \cdot 2 = 6$$
$$3 \cdot 1 = 3$$
$$3 \cdot 0 = 0$$
$$3 \cdot (-1) = ?$$

Numbers decrease by 3.

What should $3(-1)$ equal? The product $3(-1)$ represents the sum

$$-1 + (-1) + (-1) = -3,$$

so the product should be -3. Also,

$$3(-2) = -2 + (-2) + (-2) = -6$$

and

$$3(-3) = -3 + (-3) + (-3) = -9.$$

These results maintain the pattern in the list, which suggests the following rule.

Multiplying Numbers with Different Signs

For any positive real numbers x and y,

$$x(-y) = -(xy) \quad \text{and} \quad (-x)y = -(xy).$$

That is, the product of two numbers with opposite signs is negative.

EXAMPLE 1   Multiplying a Positive Number and a Negative Number

Find the products using the multiplication rule given above.

(a) $8(-5) = -(8 \cdot 5) = -40$ (b) $5(-4) = -(5 \cdot 4) = -20$

(c) $(-7)(2) = -(7 \cdot 2) = -14$ (d) $(-9)(3) = -(9 \cdot 3) = -27$

OBJECTIVE **2** Find the product of two negative numbers. The product of two positive numbers is positive, and the product of a positive and a negative number is negative. What about the product of two negative numbers? Look at another pattern.

$$(-5)(4) = -20 \quad \text{| Numbers increase by 5.}$$
$$(-5)(3) = -15$$
$$(-5)(2) = -10$$
$$(-5)(1) = -5$$
$$(-5)(0) = 0$$
$$(-5)(-1) = ?$$

The numbers on the left of the equals sign (in color) decrease by 1 for each step down the list. The products on the right increase by 5 for each step down the list. To maintain this pattern, $(-5)(-1)$ should be 5 more than $(-5)(0)$, or 5 more than 0, so

$$(-5)(-1) = 5.$$

The pattern continues with

$$(-5)(-2) = 10$$
$$(-5)(-3) = 15$$
$$(-5)(-4) = 20$$
$$(-5)(-5) = 25,$$

and so on. This pattern suggests the next rule.

Multiplying Two Negative Numbers

For any positive real numbers x and y,

$$(-x)(-y) = xy.$$

The product of two negative numbers is positive.

EXAMPLE 2 Multiplying Two Negative Numbers

Find the products using the multiplication rule given above.

(a) $(-9)(-2) = 9 \cdot 2 = 18$ **(b)** $(-6)(-12) = 6 \cdot 12 = 72$

(c) $(-8)(-1) = 8 \cdot 1 = 8$ **(d)** $(-15)(-2) = 15 \cdot 2 = 30$

A summary of multiplying signed numbers is given here.

Multiplying Signed Numbers

The product of two numbers having the *same* sign is *positive*, and the product of two numbers having *different* signs is *negative*.

OBJECTIVE **3** Identify factors of integers. In Section 1.1 the definition of a *factor* was given for whole numbers. (For example, since $9 \cdot 5 = 45$, both 9 and 5 are factors of 45.) The definition can now be extended to integers.

If the product of two integers is a third integer, then each of the two integers is a *factor* of the third. For example, $(-3)(-4) = 12$, so -3 and -4 are both factors of 12. The factors of 12 are $-12, -6, -4, -3, -2, -1, 1, 2, 3, 4, 6,$ and 12.

The following chart shows several integers and the factors of those integers.

Integer	Factors
18	$-18, -9, -6, -3, -2, -1, 1, 2, 3, 6, 9, 18$
20	$-20, -10, -5, -4, -2, -1, 1, 2, 4, 5, 10, 20$
15	$-15, -5, -3, -1, 1, 3, 5, 15$
7	$-7, -1, 1, 7$
1	$-1, 1$

OBJECTIVE 4 Use the reciprocal of a number to apply the definition of division. In Section 1.5 we saw that the difference between two numbers is found by adding the additive inverse of the second number to the first. Similarly, the *quotient* of two numbers is found by *multiplying* by the *reciprocal*, or *multiplicative inverse*. By definition, since

$$8 \cdot \frac{1}{8} = \frac{8}{8} = 1 \qquad \text{and} \qquad \frac{5}{4} \cdot \frac{4}{5} = \frac{20}{20} = 1,$$

the reciprocal or multiplicative inverse of 8 is $\frac{1}{8}$, and of $\frac{5}{4}$ is $\frac{4}{5}$.

Reciprocal or Multiplicative Inverse

Pairs of numbers whose product is 1 are **reciprocals,** or **multiplicative inverses,** of each other.

The following chart shows several numbers and their multiplicative inverses.

Number	Multiplicative Inverse (Reciprocal)
4	$\frac{1}{4}$
$.3 = \frac{3}{10}$	$\frac{10}{3}$
-5	$\frac{1}{-5}$ or $-\frac{1}{5}$
$-\frac{5}{8}$	$-\frac{8}{5}$
0	None
1	1
-1	-1

Why is there no multiplicative inverse for the number 0? Suppose that k is to be the multiplicative inverse of 0. Then $k \cdot 0$ should equal 1. But $k \cdot 0 = 0$ for any number k. Since there is no value of k that is a solution of the equation $k \cdot 0 = 1$, the following statement can be made.

0 has no multiplicative inverse.

By definition, the quotient of x and y is the product of x and the multiplicative inverse of y.

Definition of Division

For any real numbers x and y, with $y \neq 0$, $\quad \dfrac{x}{y} = x \cdot \dfrac{1}{y}.$

The definition of division indicates that y, the number to divide by, cannot be 0. The reason is that 0 has no multiplicative inverse, so that $\frac{1}{0}$ is not a number. **Because 0 has no multiplicative inverse,** *division by 0 is undefined.* **If a division problem turns out to involve division by 0, write "undefined."**

While division by zero $\left(\frac{a}{0}\right)$ is undefined, we may divide 0 by any nonzero number. In fact, if $a \neq 0$,

$$\frac{0}{a} = 0.$$

Since division is defined in terms of multiplication, all the rules for multiplying signed numbers also apply to dividing them.

EXAMPLE 3 Using the Definition of Division

Find the quotients using the definition of division.

(a) $\dfrac{12}{3} = 12 \cdot \dfrac{1}{3} = 4$

(b) $\dfrac{-10}{2} = -10 \cdot \dfrac{1}{2} = -5$

(c) $\dfrac{-14}{-7} = -14\left(\dfrac{1}{-7}\right) = 2$

(d) $-\dfrac{2}{3} \div \left(-\dfrac{4}{5}\right) = -\dfrac{2}{3} \cdot \left(-\dfrac{5}{4}\right) = \dfrac{5}{6}$

(e) $\dfrac{0}{13} = 0\left(\dfrac{1}{13}\right) = 0 \qquad \dfrac{0}{a} = 0 \quad (a \neq 0)$

(f) $\dfrac{-10}{0} \qquad$ Undefined

When dividing fractions, multiplying by the reciprocal works well. However, using the definition of division directly with integers is awkward. It is easier to divide in the usual way, then determine the sign of the answer. The following rule for division can be used instead of multiplying by the reciprocal.

Dividing Signed Numbers

The quotient of two numbers having the same sign is positive; the quotient of two numbers having *different* signs is *negative.*

Note that these are the same as the rules for multiplication.

E X A M P L E   4   Dividing Signed Numbers

Find the quotients.

(a) $\dfrac{8}{-2} = -4$

(b) $\dfrac{-4.5}{-.09} = 50$

(c) $-\dfrac{1}{8} \div \left(-\dfrac{3}{4}\right) = -\dfrac{1}{8} \cdot \left(-\dfrac{4}{3}\right) = \dfrac{1}{6}$

From the definitions of multiplication and division of real numbers,

$$\frac{-40}{8} = -40 \cdot \frac{1}{8} = -5, \text{ and } \frac{40}{-8} = 40\left(\frac{1}{-8}\right) = -5, \text{ so that}$$

$$\frac{-40}{8} = \frac{40}{-8}.$$

Based on this example, the quotient of a positive and a negative number can be expressed in any of the following three forms.

For any positive real numbers x and y, $\dfrac{-x}{y} = \dfrac{x}{-y} = -\dfrac{x}{y}.$

The quotient of two negative numbers can be expressed as a quotient of two positive numbers.

For any positive real numbers x and y, $\dfrac{-x}{-y} = \dfrac{x}{y}.$

OBJECTIVE 5   Use the order of operations when multiplying and dividing signed numbers.

E X A M P L E   5   Using the Order of Operations

Perform the indicated operations.

(a) $(-9)(2) - (-3)(2)$

First find all products, working from left to right.

$$(-9)(2) - (-3)(2) = -18 - (-6)$$

Now perform the subtraction.

$$-18 - (-6) = -18 + 6 = -12$$

(b) $(-6)(-2) - (3)(-4) = 12 - (-12) = 12 + 12 = 24$

(c) $-5(-2 - 3) = -5(-5) = 25$

(d) $\dfrac{5(-2) - (3)(4)}{2(1 - 6)}$

Simplify the numerator and denominator separately. Then write in lowest terms.

$$\frac{5(-2) - (3)(4)}{2(1 - 6)} = \frac{-10 - 12}{2(-5)} = \frac{-22}{-10} = \frac{11}{5}$$

We now summarize the rules for operations with signed numbers.

Operations with Signed Numbers

Addition
Like signs   Add the absolute values of the numbers. The sum has the same sign as the numbers.
Unlike signs   Subtract the number with the smaller absolute value from the one with the larger. Give the sum the sign of the number having the larger absolute value.

Subtraction
Add the additive inverse, or opposite, of the second number.

Multiplication and Division
Like signs   The product or quotient of two numbers with like signs is positive.
Unlike signs   The product or quotient of two numbers with unlike signs is negative.
Division by 0 is undefined.
0 divided by a nonzero number equals 0.

OBJECTIVE 6   Evaluate expressions involving variables.   The next examples show numbers substituted for variables where the rules for multiplying and dividing signed numbers must be used.

E X A M P L E 6   Evaluating an Expression for Numerical Values

Evaluate each expression, given that $x = -1$, $y = -2$, and $m = -3$.

(a) $(3x + 4y)(-2m)$

First substitute the given values for the variables. Then use the order of operations to find the value of the expression.

$(3x + 4y)(-2m) = [3(-1) + 4(-2)][-2(-3)]$ Put parentheses around the number for each variable.

$= [-3 + (-8)][6]$ Find the products.

$= (-11)(6)$ Add inside the parentheses.

$= -66$ Multiply.

(b) $2x^2 - 3y^2$

Use parentheses as shown.

$2(-1)^2 - 3(-2)^2 = 2(1) - 3(4)$ Apply the exponents.

$= 2 - 12$ Multiply.

$= -10$ Subtract.

(c) $\dfrac{4y^2 + x}{m}$

$$\frac{4(-2)^2+(-1)}{-3} = \frac{4(4) + (-1)}{-3}$$ Apply the exponent.

$$= \frac{16 + (-1)}{-3}$$ Multiply.

$$= \frac{15}{-3}$$ Add.

$$= -5$$ Divide.

Notice how the fraction bar was used as a grouping symbol.

OBJECTIVE 7 Interpret words and phrases involving multiplication and division. Just as there are words and phrases that indicate addition and subtraction, certain ones also indicate multiplication and division.

PROBLEM SOLVING

The word *product* refers to multiplication. The chart below gives other key words and phrases that indicate multiplication.

Word or Phrase	Example	Numerical Expression and Simplification
Product of	The *product of* -5 and -2	$(-5)(-2) = 10$
Times	13 *times* -4	$13(-4) = -52$
Twice (meaning "2 times")	*Twice* 6	$2(6) = 12$
Of (used with fractions)	$\frac{1}{2}$ *of* 10	$\frac{1}{2}(10) = 5$
Percent of	12% *of* -16	$.12(-16) = -1.92$

EXAMPLE 7 Interpreting Words and Phrases Involving Multiplication

Write a numerical expression for each phrase and simplify. Use the order of operations.

(a) The product of 12 and the sum of 3 and -6
Here 12 is multiplied by "the sum of 3 and -6."
$$12[3 + (-6)] = 12(-3) = -36$$

(b) Twice the difference between 8 and -4
$$2[8 - (-4)] = 2[8 + 4] = 2(12) = 24$$

(c) Two-thirds of the sum of -5 and -3
$$\frac{2}{3}[-5 + (-3)] = \frac{2}{3}[-8] = -\frac{16}{3}$$

(d) 15% of the difference between 14 and -2
Remember that $15\% = .15$.
$$.15[14 - (-2)] = .15(14 + 2) = .15(16) = 2.4$$

PROBLEM SOLVING

The word *quotient* refers to the answer in a division problem. In algebra, quotients are usually represented with a fraction bar; the symbol ÷ is seldom used. When translating applied problems involving division, use a fraction bar. The chart gives some key phrases associated with division.

Phrase	Example	Numerical Expression and Simplification
Quotient of	The *quotient of* −24 and 3	$\frac{-24}{3} = -8$
Divided by	−16 *divided by* −4	$\frac{-16}{-4} = 4$
Ratio of	The *ratio of* 2 to 3	$\frac{2}{3}$

It is customary to write the first number named as the numerator and the second as the denominator when interpreting a phrase involving division, as shown in the next example.

EXAMPLE 8 Interpreting Words and Phrases Involving Division

Write a numerical expression for each phrase, and simplify the expression.

(a) The **quotient of** 14 and the sum of −9 and 2

"Quotient" indicates division. The number 14 is the numerator and "the sum of −9 and 2" is the denominator.

$$\frac{14}{-9+2} = \frac{14}{-7} = -2$$

(b) The product of 5 and −6, **divided by** the difference between −7 and 8

The numerator of the fraction representing the division is obtained by multiplying 5 and −6. The denominator is found by subtracting −7 and 8.

$$\frac{5(-6)}{-7-8} = \frac{-30}{-15} = 2$$

OBJECTIVE 8 Translate simple sentences into equations. In this section and the previous one, important words and phrases involving the four operations of arithmetic have been introduced. We can use these words and phrases to interpret sentences that translate into equations.

EXAMPLE 9 Translating Words Into an Equation

Write the following in symbols, using x as the variable, and guess or use trial and error to find the solution. All solutions come from the list of integers between −12 and 12, inclusive.

(a) Three **times** a number **is** −18.

The word *times* indicates multiplication, and the word *is* translates as the equals sign (=).

$$3x = -18$$

Since the integer between -12 and 12, inclusive, that makes this statement true is -6, the solution of the equation is -6.

(b) The **sum** of a number and 9 **is** 12.

$$x + 9 = 12$$

Since $3 + 9 = 12$, the solution of this equation is 3.

(c) The **difference between** a number and 5 **is** 0.

$$x - 5 = 0$$

Since $5 - 5 = 0$, the solution of this equation is 5.

(d) The **quotient of** 24 and a number **is** -2.

$$\frac{24}{x} = -2$$

Here, x must be a negative number, since the numerator is positive and the quotient is negative. Since $\frac{24}{-12} = -2$, the solution is -12.

It is important to recognize the distinction between the types of problems found in Examples 7 and 8 and Example 9. In Examples 7 and 8, the phrases translate as *expressions,* while in Example 9, the sentences translate as *equations.* Remember that an equation is a sentence with an $=$ sign, while an expression is a phrase.

$$\frac{5(-6)}{-7 - 8} \quad \text{is an \textbf{expression,}}$$

$$3x = -18 \quad \text{is an \textbf{equation.}}$$

1.6 EXERCISES

Fill in each blank with one of the following: greater than 0, less than 0, equal to 0.

1. A positive number is _____.

2. A negative number is _____.

3. The product or the quotient of two numbers with the same sign is _____.

4. The product or the quotient of two numbers with different signs is _____.

5. If three negative numbers are multiplied together, the product is _____.

6. If two negative numbers are multiplied and then their product is divided by a negative number, the result is _____.

7. If a negative number is squared and the result is added to a positive number, the final answer is _____.

8. The reciprocal of a negative number is _____.

9. If three positive numbers, five negative numbers, and zero are multiplied, the product is _____.

10. The fifth power of a negative number is _____.

Find each product. See Examples 1 and 2.

11. $(-4)(-5)$ 12. $(-4)(-6)$ 13. $(-7)(4)$

14. $(-8)(5)$ 15. $(-4)(-20)$ 16. $(-8)(-30)$

17. $(-8)(0)$ **18.** $(-12)(0)$ **19.** $\left(-\dfrac{3}{8}\right)\left(-\dfrac{20}{9}\right)$

20. $\left(-\dfrac{5}{4}\right)\left(-\dfrac{6}{25}\right)$ **21.** $(-6)\left(-\dfrac{1}{4}\right)$ **22.** $(-8)\left(-\dfrac{1}{2}\right)$

Find all integer factors of each given number.

23. 32   **24.** 36   **25.** 40   **26.** 50   **27.** 31   **28.** 17

Find each quotient. See Examples 3 and 4.

29. $\dfrac{-15}{5}$ **30.** $\dfrac{-18}{6}$ **31.** $\dfrac{20}{-10}$ **32.** $\dfrac{28}{-4}$

33. $\dfrac{-160}{-10}$ **34.** $\dfrac{-260}{-20}$ **35.** $\dfrac{0}{-3}$ **36.** $\dfrac{0}{-6}$

37. $\dfrac{-10.252}{-.4}$ **38.** $\dfrac{-29.584}{-.8}$ **39.** $\left(-\dfrac{3}{4}\right) \div \left(-\dfrac{1}{2}\right)$ **40.** $\left(-\dfrac{3}{16}\right) \div \left(-\dfrac{5}{8}\right)$

41. Explain in words how to multiply signed numbers. Consider the various cases and give examples.

42. Explain in words how to divide signed numbers. Give examples.

Perform the indicated operations. See Examples 5(a), (b), and (c).

43. $7 - 3 \cdot 6$ **44.** $8 - 2 \cdot 5$

45. $-10 - (-4)(2)$ **46.** $-11 - (-3)(6)$

47. $-7(3 - 8)$ **48.** $-5(4 - 7)$

49. $(12 - 14)(1 - 4)$ **50.** $(8 - 9)(4 - 12)$

51. $(7 - 10)(10 - 4)$ **52.** $(5 - 12)(19 - 4)$

53. $(-2 - 8)(-6) + 7$ **54.** $(-9 - 4)(-2) + 10$

55. $3(-5) + |\,3 - 10\,|$ **56.** $4(-8) + |\,4 - 15\,|$

Perform each indicated operation. See Example 5(d).

57. $\dfrac{-5(-6)}{9 - (-1)}$ **58.** $\dfrac{-12(-5)}{7 - (-5)}$ **59.** $\dfrac{-21(3)}{-3 - 6}$

60. $\dfrac{-40(3)}{-2 - 3}$ **61.** $\dfrac{-10(2) + 6(2)}{-3 - (-1)}$ **62.** $\dfrac{8(-1) + 6(-2)}{-6 - (-1)}$

63. $\dfrac{-27(-2) - (-12)(-2)}{-2(3) - 2(2)}$ **64.** $\dfrac{-13(-4) - (-8)(-2)}{(-10)(2) - 4(-2)}$

65. Explain the method you would use to evaluate $3x + 2y$ if $x = -3$ and $y = 4$.

66. If x and y are both replaced by negative numbers, is the value of $4x + 8y$ positive or negative?

Evaluate each expression if $x = 6$, $y = -4$, and $a = 3$. See Example 6.

67. $5x - 2y + 3a$ **68.** $6x - 5y + 4a$ **69.** $(2x + y)(3a)$

70. $(5x - 2y)(-2a)$ **71.** $\left(\dfrac{1}{3}x - \dfrac{4}{5}y\right)\left(-\dfrac{1}{5}a\right)$ **72.** $\left(\dfrac{5}{6}x + \dfrac{3}{2}y\right)\left(-\dfrac{1}{3}a\right)$

73. $(-5 + x)(-3 + y)(3 - a)$ **74.** $(6 - x)(5 + y)(3 + a)$ **75.** $-2y^2 + 3a$

76. $5x - 4a^2$ **77.** $\dfrac{2y^2 - x}{a + 10}$ **78.** $\dfrac{xy + 8a}{x - y}$

Write a numerical expression for each phrase and simplify. See Examples 7 and 8.

79. The product of -9 and 2, added to 9

80. The product of 4 and -7, added to -12

81. Twice the product of -1 and 6, subtracted from -4

82. Twice the product of -8 and 2, subtracted from -1

83. Nine subtracted from the product of 1.5 and -3.2

84. Three subtracted from the product of 4.2 and -8.5

85. The product of 12 and the difference between 9 and -8

86. The product of -3 and the difference between 3 and -7

87. The quotient of -12 and the sum of -5 and -1

88. The quotient of -20 and the sum of -8 and -2

89. The sum of 15 and -3, divided by the product of 4 and -3

90. The sum of -18 and -6, divided by the product of 2 and -4

91. The product of $-\dfrac{1}{2}$ and $\dfrac{3}{4}$, divided by $-\dfrac{2}{3}$

92. The product of $-\dfrac{2}{3}$ and $-\dfrac{1}{5}$, divided by $\dfrac{1}{7}$

Write each statement in symbols, using x as the variable, and find the solution by guessing or by using trial and error. All solutions come from the set of integers between -12 and 12, inclusive. See Example 9.

93. Six times a number is -42.

94. Four times a number is -36.

95. The quotient of a number and 3 is -3.

96. The quotient of a number and 4 is -1.

97. 6 less than a number is 4.

98. 7 less than a number is 2.

99. When 5 is added to a number, the result is -5.

100. When 6 is added to a number, the result is -3.

To find the average of a group of numbers, we add the numbers and then divide the sum by the number of terms added. For example, to find the average of 14, 8, 3, 9, *and* 1, *we add them and then divide by* 5:

$$\frac{14 + 8 + 3 + 9 + 1}{5} = \frac{35}{5} = 7.$$

The average of these numbers is 7.

Exercises 101–106 involve finding the average of a group of numbers.

101. Find the average of 23, 18, 13, -4, and -8.

102. Find the average of 18, 12, 0, -4, and -10.

103. What is the average of all integers between -10 and 14, inclusive of both?

104. What is the average of all even integers between -18 and 4, inclusive of both?

105. The table shows average hourly earnings of various fields in private industry for 1995. Find the average of these amounts.

Private Industry Group	1995 Hourly Earnings
Mining	$15.30
Construction	15.08
Manufacturing	12.37
Transportation, public utilities	14.23
Wholesale trade	12.43
Retail trade	7.69
Finance, insurance, real estate	12.33
Service	11.39

Source: U.S. Bureau of Labor Statistics.

106. During the 1990s, the number of labor union or employee association members has remained fairly constant. The chart shows the number of members, in thousands, for the years 1990–1996.

Year	Number (in thousands)
1990	16,740
1991	16,568
1992	16,390
1993	16,598
1994	16,748
1995	16,360
1996	16,269

Source: U.S. Bureau of Labor Statistics.

What is the average number, in thousands, for this seven-year period?

107. If the average of a group of numbers is 0, what is the sum of all the numbers?

108. Suppose there is a group of numbers with some positive and some negative. Under what conditions will the average be a positive number? Under what conditions will the average be negative?

The operation of division is used in divisibility tests. *A divisibility test allows us to determine whether a given number is divisible (without remainder) by another number. For example, a number is divisible by 2 if its last digit is divisible by 2, and not otherwise.*

109. Tell why **(a)** 3,473,986 is divisible by 2 and   **(b)** 4,336,879 is not divisible by 2.

110. An integer is divisible by 3 if the sum of its digits is divisible by 3, and not otherwise. Show that **(a)** 4,799,232 is divisible by 3 and   **(b)** 2,443,871 is not divisible by 3.

111. An integer is divisible by 4 if its last two digits form a number divisible by 4, and not otherwise. Show that **(a)** 6,221,464 is divisible by 4 and **(b)** 2,876,335 is not divisible by 4.

112. An integer is divisible by 5 if its last digit is divisible by 5, and not otherwise. Show that **(a)** 3,774,595 is divisible by 5 and **(b)** 9,332,123 is not divisible by 5.

113. An integer is divisible by 6 if it is divisible by both 2 and 3, and not otherwise. Show that **(a)** 1,524,822 is divisible by 6 and **(b)** 2,873,590 is not divisible by 6.

114. An integer is divisible by 8 if its last three digits form a number divisible by 8, and not otherwise. Show that **(a)** 2,923,296 is divisible by 8 and **(b)** 7,291,623 is not divisible by 8.

115. An integer is divisible by 9 if the sum of its digits is divisible by 9, and not otherwise. Show that **(a)** 4,114,107 is divisible by 9 and **(b)** 2,287,321 is not divisible by 9.

116. An integer is divisible by 12 if it is divisible by both 3 and 4, and not otherwise. Show that **(a)** 4,253,520 is divisible by 12 and **(b)** 4,249,474 is not divisible by 12.

1.7　Properties of Real Numbers

OBJECTIVES

1 Use the commutative properties.

2 Use the associative properties.

3 Use the identity properties.

4 Use the inverse properties.

5 Use the distributive property.

FOR EXTRA HELP

SSG Sec. 1.7
SSM Sec. 1.7

Pass the Test Software

InterAct Math
　Tutorial Software

Video 2

If you were asked to find the sum $3 + 89 + 97$, you might mentally add $3 + 97$ to get 100, and then add $100 + 89$ to get 189. While the rule for order of operations says to add from left to right, we may change the order of the terms and group them in any way we choose without affecting the sum. These are examples of shortcuts that we use in everyday mathematics. These shortcuts are justified by the basic properties of addition and multiplication, discussed in this section. In these properties, a, b, and c represent real numbers.

OBJECTIVE 1 **Use the commutative properties.** The word *commute* means to go back and forth. Many people commute to work or to school. If you travel from home to work and follow the same route from work to home, you travel the same distance each time. The **commutative properties** say that if two numbers are added or multiplied in any order, the result is the same.

Commutative Properties

$$a + b = b + a$$
$$ab = ba$$

EXAMPLE 1 **Using the Commutative Properties**

Use a commutative property to complete each statement.

(a) $-8 + 5 = 5 +$ _____

By the commutative property for addition, the missing number is -8, since $-8 + 5 = 5 + (-8)$.

(b) $(-2)(7) =$ _____ (-2)

By the commutative property for multiplication, the missing number is 7, since $(-2)(7) = (7)(-2)$.

OBJECTIVE ▸2◂ Use the associative properties. When we *associate* one object with another, we tend to think of those objects as being grouped together. The **associative properties** say that when we add or multiply three numbers, we can group the first two together or the last two together and get the same answer.

Associative Properties

$$(a + b) + c = a + (b + c)$$
$$(ab)c = a(bc)$$

EXAMPLE 2   Using the Associative Properties

Use an associative property to complete each statement.

(a) $8 + (-1 + 4) = (8 + \underline{\quad}) + 4$
The missing number is -1.

(b) $[2 \cdot (-7)] \cdot 6 = 2 \cdot \underline{\quad}$
The completed expression on the right should be $2 \cdot [(-7) \cdot 6]$.

By the associative property of addition, the sum of three numbers will be the same no matter how the numbers are "associated" in groups. For this reason, parentheses can be left out in many addition problems. For example, both

$$(-1 + 2) + 3 \qquad \text{and} \qquad -1 + (2 + 3)$$

can be written as

$$-1 + 2 + 3.$$

In the same way, parentheses also can be left out of many multiplication problems.

EXAMPLE 3   Distinguishing between the Associative and Commutative Properties

(a) Is $(2 + 4) + 5 = 2 + (4 + 5)$ an example of the associative property or the commutative property?
The order of the three numbers is the same on both sides of the equals sign. The only change is in the grouping, or association, of the numbers. Therefore, this is an example of the associative property.

(b) Is $6(3 \cdot 10) = 6(10 \cdot 3)$ an example of the associative property or the commutative property?
The same numbers, 3 and 10, are grouped on each side. On the left, the 3 appears first, but on the right, the 10 appears first. Since the only change involves the order of the numbers, this statement is an example of the commutative property.

(c) Is $(8 + 1) + 7 = 8 + (7 + 1)$ an example of the associative property or the commutative property?
In the statement, both the order and the grouping are changed. On the left the order of the three numbers is 8, 1, and 7. On the right it is 8, 7, and 1. On the left the 8 and 1 are grouped, and on the right the 7 and 1 are grouped. Therefore, both the associative and the commutative properties are used.

EXAMPLE 4 Using the Commutative and Associative Properties

Find the sum: $23 + 41 + 2 + 9 + 25$.

The commutative and associative properties make it possible to choose pairs of numbers whose sums are easy to add.

$$23 + 41 + 2 + 9 + 25 = (41 + 9) + (23 + 2) + 25$$
$$= 50 + 25 + 25$$
$$= 100$$

OBJECTIVE 3 Use the identity properties. If a child wears a costume on Halloween, the child's appearance is changed, but his or her *identity* is unchanged. The identity of a real number is left unchanged when identity properties are applied. The **identity properties** say that the sum of 0 and any number equals that number, and the product of 1 and any number equals that number.

Identity Properties

$$a + 0 = a \quad \text{and} \quad 0 + a = a$$
$$a \cdot 1 = a \quad \text{and} \quad 1 \cdot a = a$$

The number 0 leaves the identity, or value, of any real number unchanged by addition. For this reason, 0 is called the **identity element for addition.** Since multiplication by 1 leaves any real number unchanged, 1 is the **identity element for multiplication.**

EXAMPLE 5 Using the Identity Properties

These statements are examples of the identity properties.

(a) $-3 + 0 = -3$

(b) $1 \cdot \dfrac{1}{2} = \dfrac{1}{2}$

We use the identity property for multiplication to write fractions in lowest terms and to get common denominators.

EXAMPLE 6 Using the Identity Property to Simplify Expressions

Simplify the following expressions.

(a) $\dfrac{49}{35}$

$$\dfrac{49}{35} = \dfrac{7 \cdot 7}{5 \cdot 7} \qquad \text{Factor.}$$

$$= \dfrac{7}{5} \cdot \dfrac{7}{7} \qquad \text{Write as a product.}$$

$$= \dfrac{7}{5} \cdot 1 \qquad \text{Divide.}$$

$$= \dfrac{7}{5} \qquad \text{Identity property}$$

(b) $\dfrac{3}{4} + \dfrac{5}{24}$

$$\dfrac{3}{4} + \dfrac{5}{24} = \dfrac{3}{4} \cdot 1 + \dfrac{5}{24} \qquad \text{Identity property}$$

$$= \dfrac{3}{4} \cdot \dfrac{6}{6} + \dfrac{5}{24} \qquad \text{Get a common denominator.}$$

$$= \dfrac{18}{24} + \dfrac{5}{24} \qquad \text{Multiply.}$$

$$= \dfrac{23}{24} \qquad \text{Add.}$$

OBJECTIVE 4   Use the inverse properties. Each day before you go to work or school, you probably put on your shoes before you leave. Before you go to sleep at night, you probably take them off, and this leads to the same situation that existed before you put them on. These operations from everyday life are examples of inverse operations. The **inverse properties** of addition and multiplication lead to the additive and multiplicative identities, respectively. Recall that $-a$ is the **additive inverse** of a and $\frac{1}{a}$ is the **multiplicative inverse,** or **reciprocal,** of the nonzero number a. The sum of the numbers a and $-a$ is 0, and the product of the nonzero numbers a and $\frac{1}{a}$ is 1.

Inverse Properties

$$a + (-a) = 0 \qquad \text{and} \qquad -a + a = 0$$
$$a \cdot \dfrac{1}{a} = 1 \qquad \text{and} \qquad \dfrac{1}{a} \cdot a = 1 \quad (a \neq 0)$$

EXAMPLE 7   Using the Inverse Properties

The following statements are examples of the inverse properties.

(a) $\dfrac{2}{3} \cdot \dfrac{3}{2} = 1$ **(b)** $(-5)\left(-\dfrac{1}{5}\right) = 1$

(c) $-\dfrac{1}{2} + \dfrac{1}{2} = 0$ **(d)** $4 + (-4) = 0$

In the next example, we show how the various properties are used to simplify an expression.

EXAMPLE 8   Using Properties to Simplify an Expression

Simplify $-2x + 10 + 2x$, using the properties discussed in this section.

$$-2x + 10 + 2x = (-2x + 10) + 2x \qquad \text{Order of operations}$$
$$= [10 + (-2x)] + 2x \qquad \text{Commutative property}$$
$$= 10 + [(-2x) + 2x] \qquad \text{Associative property}$$
$$= 10 + 0 \qquad \text{Inverse property}$$
$$= 10 \qquad \text{Identity property}$$

Note that for *any* value of *x*, $-2x$ and $2x$ are additive inverses; this is why we can use the inverse property in this simplification.

 The detailed procedure shown in Example 8 is seldom, if ever, used in practice. We include the example to show how the properties of this section apply, even though steps may be skipped when actually doing the simplification.

OBJECTIVE 5 Use the distributive property. The everyday meaning of the word *distribute* is "to give out from one to several." An important property of real number operations involves this idea.

Look at the following statements.

$$2(5 + 8) = 2(13) = 26$$

$$2(5) + 2(8) = 10 + 16 = 26$$

Since both expressions equal 26,

$$2(5 + 8) = 2(5) + 2(8).$$

This result is an example of the **distributive property,** the only property involving *both* addition and multiplication. With this property, a product can be changed to a sum or difference.

The distributive property says that multiplying a number *a* by a sum of numbers $b + c$ gives the same result as multiplying *a* by *b* and *a* by *c* and then adding the two products.

Distributive Property

$$a(b + c) = ab + ac \qquad \text{and} \qquad (b + c)a = ba + ca$$

As the arrows show, the *a* outside the parentheses is "distributed" over the *b* and *c* inside. Another form of the distributive property is valid for subtraction.

$$a(b - c) = ab - ac \qquad \text{and} \qquad (b - c)a = ba - ca$$

The distributive property also can be extended to more than two numbers.

$$a(b + c + d) = ab + ac + ad$$

 The distributive property can be used "in reverse." For example, we can write

$$ac + bc = (a + b)c.$$

EXAMPLE 9 Using the Distributive Property

Use the distributive property to rewrite each expression.

(a) $5(9 + 6) = 5 \cdot 9 + 5 \cdot 6$ Distributive property

$\qquad\qquad\quad = 45 + 30$ Multiply.

$\qquad\qquad\quad = 75$ Add.

(b) $4(x + 5 + y) = 4x + 4 \cdot 5 + 4y$ Distributive property

$\qquad\qquad\qquad\quad = 4x + 20 + 4y$ Multiply.

(c) $-2(x + 3) = -2x + (-2)(3)$ Distributive property
$$= -2x - 6 \qquad\text{Multiply.}$$

(d) $3(k - 9) = 3k - 3 \cdot 9$ Distributive property
$$= 3k - 27 \qquad\text{Multiply.}$$

(e) $8(3r + 11t + 5z) = 8(3r) + 8(11t) + 8(5z)$ Distributive property
$$= (8 \cdot 3)r + (8 \cdot 11)t + (8 \cdot 5)z \qquad\text{Associative property}$$
$$= 24r + 88t + 40z \qquad\text{Multiply.}$$

(f) $6 \cdot 8 + 6 \cdot 2 = 6(8 + 2)$ Distributive property
$$= 6(10) = 60 \qquad\text{Add, then multiply.}$$

(g) $4x - 4m = 4(x - m)$ Distributive property

(h) $6x - 12 = 6 \cdot x - 6 \cdot 2 = 6(x - 2)$ Distributive property

The symbol $-a$ may be interpreted as $-1 \cdot a$. Similarly, when a negative sign precedes an expression within parentheses, it may also be interpreted as a factor of -1. The distributive property is used to remove parentheses from expressions such as $-(2y + 3)$. We do this by first writing $-(2y + 3)$ as $-1 \cdot (2y + 3)$.

$$-(2y + 3) = -1 \cdot (2y + 3)$$
$$= -1 \cdot (2y) + (-1) \cdot (3) \qquad\text{Distributive property}$$
$$= -2y - 3 \qquad\text{Multiply.}$$

EXAMPLE 10   Using the Distributive Property to Remove Parentheses

Write without parentheses.

(a) $-(7r - 8) = -1(7r) + (-1)(-8)$ Distributive property
$$= -7r + 8 \qquad\text{Multiply.}$$

(b) $-(-9w + 2) = 9w - 2$

The properties discussed here are the basic properties that justify how we do algebra. You should know them by name because we will be referring to them frequently. Here is a summary of these properties.

Properties of Addition and Multiplication

For any real numbers a, b, and c, the following properties hold.

Commutative Properties $a + b = b + a \qquad ab = ba$

Associative Properties $(a + b) + c = a + (b + c)$
$$(ab)c = a(bc)$$

Identity Properties   There is a real number 0 such that
$$a + 0 = a \qquad\text{and}\qquad 0 + a = a.$$
There is a real number 1 such that
$$a \cdot 1 = a \qquad\text{and}\qquad 1 \cdot a = a.$$

Properties of Addition and Multiplication (continued)

Inverse Properties	For each real number a, there is a single real number $-a$ such that

$$a + (-a) = 0 \quad \text{and} \quad (-a) + a = 0.$$

For each nonzero real number a, there is a single real number $\frac{1}{a}$ such that

$$a \cdot \frac{1}{a} = 1 \quad \text{and} \quad \frac{1}{a} \cdot a = 1.$$

Distributive Property $\qquad a(b + c) = ab + ac \qquad (b + c)a = ba + ca$

1.7 EXERCISES

Match each item in Column I with the correct choice(s) from Column II. Choices may be used once, more than once, or not at all.

I

1. Identity element for addition
2. Identity element for multiplication
3. Additive inverse of a
4. Multiplicative inverse, or reciprocal, of the nonzero number a
5. The number that is its own additive inverse
6. The two numbers that are their own multiplicative inverses
7. The only number that has no multiplicative inverse
8. An example of the associative property
9. An example of the commutative property
10. An example of the distributive property

II

A. $(5 \cdot 4) \cdot 3 = 5 \cdot (4 \cdot 3)$

B. 0

C. $-a$

D. -1

E. $5 \cdot 4 \cdot 3 - 60$

F. 1

G. $(5 \cdot 4) \cdot 3 = 3 \cdot (5 \cdot 4)$

H. $5(4 + 3) = 5 \cdot 4 + 5 \cdot 3$

I. $\dfrac{1}{a}$

Decide whether each statement is an example of the commutative, associative, identity, inverse, or distributive property. See Examples 1, 2, 3, 5, 6, 7, and 9.

11. $7 + 18 - 18 + 7$

12. $13 + 12 = 12 + 13$

13. $5(13 \cdot 7) = (5 \cdot 13) \cdot 7$

14. $-4(2 \cdot 6) = (-4 \cdot 2) \cdot 6$

15. $-6 + (12 + 7) = (-6 + 12) + 7$

16. $(-8 + 13) + 2 = -8 + (13 + 2)$

17. $-6 + 6 = 0$

18. $12 + (-12) = 0$

19. $\left(\dfrac{2}{3}\right)\left(\dfrac{3}{2}\right) = 1$

20. $\left(\dfrac{5}{8}\right)\left(\dfrac{8}{5}\right) = 1$

21. $2.34 + 0 = 2.34$

22. $-8.456 + 0 = -8.456$

23. $(4 + 17) + 3 = 3 + (4 + 17)$

24. $(-8 + 4) + (-12) = -12 + (-8 + 4)$

25. $6(x + y) = 6x + 6y$

26. $14(t + s) = 14t + 14s$

27. $-\dfrac{5}{9} = -\dfrac{5}{9} \cdot \dfrac{3}{3} = -\dfrac{15}{27}$

28. $\dfrac{13}{12} = \dfrac{13}{12} \cdot \dfrac{7}{7} = \dfrac{91}{84}$

29. $5(2x) + 5(3y) = 5(2x + 3y)$

30. $3(5t) - 3(7r) = 3(5t - 7r)$

31. The following conversation actually took place between one of the authors of this book and his son, Jack, when Jack was four years old:

DADDY: "Jack, what is 3 + 0?"

JACK: "3."

DADDY: "Jack, what is 4 + 0?"

JACK: "4. And Daddy, *string* plus zero equals *string*!"

What property of addition did Jack recognize?

32. The distributive property holds for multiplication with respect to addition. Is there a distributive property for addition with respect to multiplication? If not, give an example to show why.

33. Write a paragraph explaining in your own words the following properties of addition and multiplication: commutative, associative, identity, inverse.

34. Write a paragraph explaining in your own words the distributive property of multiplication with respect to addition. Give examples.

Use the indicated property to write a new expression that is equal to the given expression. Then simplify the new expression if possible. See Examples 1, 2, 5, 7, and 9.

35. $r + 7$; commutative

36. $t + 9$; commutative

37. $s + 0$; identity

38. $w + 0$; identity

39. $-6(x + 7)$; distributive

40. $-5(y + 2)$; distributive

41. $(w + 5) + (-3)$; associative

42. $(b + 8) + (-10)$; associative

Use the properties of this section to simplify each expression. See Examples 7 and 8.

43. $6t + 8 - 6t + 3$

44. $9r + 12 - 9r + 1$

45. $\frac{2}{3}x - 11 + 11 - \frac{2}{3}x$

46. $\frac{1}{5}y + 4 - 4 - \frac{1}{5}y$

47. $\left(\frac{9}{7}\right)(-.38)\left(\frac{7}{9}\right)$

48. $\left(\frac{4}{5}\right)(-.73)\left(\frac{5}{4}\right)$

49. $t + (-t) + \frac{1}{2}(2)$

50. $w + (-w) + \frac{1}{4}(4)$

51. Evaluate $25 - (6 - 2)$ and evaluate $(25 - 6) - 2$. Do you think subtraction is associative?

52. Evaluate $180 \div (15 \div 3)$ and evaluate $(180 \div 15) \div 3$. Do you think division is associative?

53. Suppose that a student shows you the following work.

$$-3(4 - 6) = -3(4) - 3(6) = -12 - 18 = -30$$

The student has made a very common error. Explain the student's mistake, and work the problem correctly.

54. Explain how the procedure of changing $\frac{3}{4}$ to $\frac{9}{12}$ requires the use of the multiplicative identity element, 1.

Use the distributive property to rewrite each expression. Simplify if possible. See Example 9.

55. $5x + x$

56. $6q + q$

57. $4(t + 3)$

58. $5(w + 4)$

59. $-8(r + 3)$

60. $-11(x + 4)$

61. $-5(y - 4)$

62. $-9(g - 4)$

63. $-\frac{4}{3}(12y + 15z)$

64. $-\frac{2}{5}(10b + 20a)$

65. $8 \cdot z + 8 \cdot w$

66. $4 \cdot s + 4 \cdot r$

67. $7(2v) + 7(5r)$

68. $13(5w) + 13(4p)$

69. $8(3r + 4s - 5y)$

70. $2(5u - 3v + 7w)$ **71.** $q + q + q$ **72.** $m + m + m + m$

73. $-5x + x$ **74.** $-9p + p$

Use the distributive property to write each expression without parentheses. See Example 10.

75. $-(4t + 3m)$ **76.** $-(9x + 12y)$ **77.** $-(-5c - 4d)$

78. $-(-13x - 15y)$ **79.** $-(-3q + 5r - 8s)$ **80.** $-(-4z + 5w - 9y)$

81. The operations of "getting out of bed" and "taking a shower" are not commutative. Give an example of another pair of everyday operations that are not commutative.

82. The phrase "dog biting man" has two different meanings, depending on how the words are associated:

(dog biting) man   dog (biting man)

Give another example of a three-word phrase that has different meanings depending on how the words are associated.

RELATING CONCEPTS (EXERCISES 83–86)

In Section 1.6 we used a pattern to see that the product of two negative numbers is a positive number. In the group of exercises that follows, we show another justification for determining the sign of the product of two negative numbers.

Work Exercises 83–86 in order.

83. Evaluate the expression $-3[5 + (-5)]$ by using the rules for order of operations.

84. Write the expression in Exercise 83 using the distributive property. Do not simplify the products.

85. The product -3×5 should be one of the terms you wrote when answering Exercise 84. Based on the results in Section 1.6, what is this product?

86. In Exercise 83, you should have obtained 0 as an answer. Now, consider the following, using the results of Exercises 83 and 85.

$$-3[5 + (-5)] = -3(5) + (-3)(-5)$$
$$0 = -15 + ?$$

The question mark represents the product $(-3)(-5)$. When added to -15, it must give a sum of 0. Therefore, how must we interpret $(-3)(-5)$?

Did you make the connection that a rule can be obtained in more than one way, with consistent results from each method?

1.8 Simplifying Expressions

OBJECTIVES

1. Simplify expressions.
2. Identify terms and numerical coefficients.
3. Identify like terms.

OBJECTIVE 1 Simplify expressions. In this section we show how to simplify expressions using the properties of addition and multiplication introduced in the previous section.

EXAMPLE 1 Simplifying Expressions

Simplify the following expressions.

(a) $4x + 8 + 9$

Since $8 + 9 = 17$,

$$4x + 8 + 9 = 4x + 17.$$

4 Combine like terms.

5 Simplify expressions from word phrases.

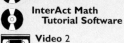

FOR EXTRA HELP

📖 **SSG** Sec. 1.8
SSM Sec. 1.8

💿 **Pass the Test Software**

💿 **InterAct Math Tutorial Software**

📼 **Video 2**

(b) $4(3m - 2n)$
Use the distributive property first.

$$4(3m - 2n) = 4(3m) - 4(2n) \qquad \text{Arrows denote distributive property.}$$
$$= (4 \cdot 3)m - (4 \cdot 2)n \qquad \text{Associative property}$$
$$= 12m - 8n$$

(c) $6 + 3(4k + 5) = 6 + 3(4k) + 3(5) \qquad \text{Distributive property}$
$$= 6 + (3 \cdot 4)k + 3(5) \qquad \text{Associative property}$$
$$= 6 + 12k + 15$$
$$= 6 + 15 + 12k \qquad \text{Commutative property}$$
$$= 21 + 12k$$

(d) $5 - (2y - 8) = 5 - 1 \cdot (2y - 8) \qquad \text{Replace } - \text{ with } -1.$
$$= 5 - 2y + 8 \qquad \text{Distributive property}$$
$$= 5 + 8 - 2y \qquad \text{Commutative property}$$
$$= 13 - 2y$$

NOTE In Example 1, parts (c) and (d), a different use of the commutative property would have resulted in answers of $12k + 21$ and $-2y + 13$. These answers also would be acceptable.

The steps using the commutative and associative properties will not be shown in the rest of the examples, but you should be aware that they are usually involved.

OBJECTIVE 2 Identify terms and numerical coefficients. A **term** is a number, a variable, or a product or quotient of numbers and variables raised to powers.* Examples of terms include

$$-9x^2, \quad 15y, \quad -3, \quad 8m^2n, \quad \frac{2}{p}, \quad \text{and} \quad k.$$

The **numerical coefficient** of the term $9m$ is 9, the numerical coefficient of $-15x^3y^2$ is -15, the numerical coefficient of x is 1, and the numerical coefficient of 8 is 8. In the expression $\frac{x}{3}$, the numerical coefficient of x is $\frac{1}{3}$. Do you see why?

CAUTION It is important to be able to distinguish between *terms* and *factors*. For example, in the expression $8x^3 + 12x^2$, there are two *terms*, $8x^3$ and $12x^2$. On the other hand, in the one-term expression $(8x^3)(12x^2)$, $8x^3$ and $12x^2$ are *factors*.

Several examples of terms and their numerical coefficients follow.

_____

*Another name for certain terms, **monomial,** is introduced in Chapter 4.

Term	Numerical Coefficient
$-7y$	-7
$8p$	8
$34r^3$	34
$-26x^5yz^4$	-26
$-k$	-1
$\dfrac{x}{7}$	$\dfrac{1}{7}$

OBJECTIVE **3** Identify like terms. Terms with exactly the same variables that have the same exponents are **like terms.** For example, $9m$ and $4m$ have the same variable and are like terms. Also, $6x^3$ and $-5x^3$ are like terms. The terms $-4y^3$ and $4y^2$ have different exponents and are **unlike terms.**

Here are some additional examples.

$5x$ and $-12x$	$3x^2y$ and $5x^2y$	Like terms
$4xy^2$ and $5xy$	$-7w^3z^3$ and $2xz^3$	Unlike terms

OBJECTIVE **4** Combine like terms. Recall the distributive property:

$$x(y + z) = xy + xz.$$

As seen in the previous section, this statement can also be written "backward" as

$$xy + xz = x(y + z).$$

This form of the distributive property may be used to find the sum or difference of like terms. For example,

$$3x + 5x = (3 + 5)x = 8x.$$

This process is called **combining like terms.**

 NOTE Remember that *only like terms may be combined.* For example, $5x^2 + 2x \neq 7x^3$.

EXAMPLE 2 Combining Like Terms

Combine like terms in the following expressions.

(a) $9m + 5m$

Use the distributive property as given above.

$$9m + 5m = (9 + 5)m = 14m$$

(b) $6r + 3r + 2r = (6 + 3 + 2)r = 11r$ Distributive property

(c) $4x + x = 4x + 1x = (4 + 1)x = 5x$ (Note: $x = 1x$.)

(d) $16y^2 - 9y^2 = (16 - 9)y^2 = 7y^2$

(e) $32y + 10y^2$ cannot be combined because $32y$ and $10y^2$ are unlike terms. The distributive property cannot be used here to combine coefficients.

When an expression involves parentheses, the distributive property is used both "forward" and "backward" to combine like terms, as shown in the following example.

E X A M P L E   3 Simplifying Expressions Involving Like Terms

Combine like terms in the following expressions.

(a) $14y + 2(6 + 3y) = 14y + 2(6) + 2(3y)$ Distributive property

$\qquad\qquad\qquad\qquad = 14y + 12 + 6y$ Multiply.

$\qquad\qquad\qquad\qquad = 20y + 12$ Combine like terms.

(b) $9k - 6 - 3(2 - 5k) = 9k - 6 - 3(2) - 3(-5k)$ Distributive property

$\qquad\qquad\qquad\qquad\quad = 9k - 6 - 6 + 15k$ Multiply.

$\qquad\qquad\qquad\qquad\quad = 24k - 12$ Combine like terms.

(c) $-(2 - r) + 10r = -1(2 - r) + 10r$ Replace $-$ with -1.

$\qquad\qquad\qquad\quad = -1(2) - 1(-r) + 10r$ Distributive property

$\qquad\qquad\qquad\quad = -2 + 1r + 10r$ Multiply.

$\qquad\qquad\qquad\quad = -2 + 11r$ Combine like terms.

(d) $5(2a - 6) - 3(4a - 9) = 10a - 30 - 12a + 27$ Distributive property

$\qquad\qquad\qquad\qquad\qquad = -2a - 3$ Combine like terms.

Example 3(d) shows that the commutative property can be used with subtraction by treating the subtracted terms as the addition of their additive inverses.

 Examples 2 and 3 suggest that like terms may be combined by adding or subtracting the coefficients of the terms and keeping the same variable factors.

O·B·J·E·C·T·I·V·E 5 Simplify expressions from word phrases. Earlier we saw how to translate words, phrases, and statements into expressions and equations. Now we can simplify translated expressions by combining like terms.

E X A M P L E   4 Converting Words to a Mathematical Expression

Convert to a mathematical expression, and simplify: The sum of 9, five times a number, four times the number, and six times the number.

The word "sum" indicates that the terms should be added. Use x to represent the number. Then the phrase translates as follows.

$\qquad\qquad 9 + 5x + 4x + 6x$ Write as a mathematical expression.

$\qquad\qquad\quad = 9 + 15x$ Combine like terms.

 In Example 4, we are dealing with an expression to be simplified, *not* an equation to be solved.

1.8 EXERCISES

Decide whether each statement is true or false.

1. $6t + 5t^2 = 11t^3$

2. $9xy^2 - 3x^2y = 6xy$

3. $8r^2 + 3r - 12r^2 + 4r = -4r^2 + 7r$

4. $4 + 3t^3 = 7t^3$

In Exercises 5–8, choose the letter of the correct response.

5. Which one of the following is true for all real numbers x?
  (a) $6 + 2x = 8x$ **(b)** $6 - 2x = 4x$
  (c) $6x - 2x = 4x$ **(d)** $3 + 8(4x - 6) = 11(4x - 6)$

6. Which one of the following is an example of a pair of like terms?
  (a) $6t, 6w$ **(b)** $-8x^2y, 9xy^2$ **(c)** $5ry, 6yr$ **(d)** $-5x^2, 2x^3$

7. Which one of the following is an example of a term with numerical coefficient 5?
  (a) $5x^3y^7$ **(b)** x^5 **(c)** $\dfrac{x}{5}$ **(d)** 5^2xy^3

8. Which one of the following is a correct translation for "six times a number, subtracted from the product of eleven and the number" (if x represents the number)?
  (a) $6x - 11x$ **(b)** $11x - 6x$ **(c)** $(11 + x) - 6x$ **(d)** $6x - (11 + x)$

Simplify each expression. See Example 1.

9. $4r + 19 - 8$

10. $7t + 18 - 4$

11. $5 + 2(x - 3y)$

12. $8 + 3(s - 6t)$

13. $-2 - (5 - 3p)$

14. $-10 - (7 - 14r)$

Give the numerical coefficient of each term.

15. $-12k$ **16.** $-23y$ **17.** $5m^2$ **18.** $-3n^6$ **19.** xw

20. pq **21.** $-x$ **22.** $-t$ **23.** 74 **24.** 98

25. Give an example of a pair of like terms in the variable x, such that one of them has a negative numerical coefficient, one has a positive numerical coefficient, and their sum has a positive numerical coefficient.

26. Give an example of a pair of unlike terms such that each term has x as the only variable factor.

Identify each group of terms as like or unlike.

27. $8r, -13r$ **28.** $-7a, 12a$ **29.** $5z^4, 9z^3$ **30.** $8x^5, -10x^3$

31. $4, 9, -24$ **32.** $7, 17, -83$ **33.** x, y **34.** t, s

35. There is an old saying, "You can't add apples and oranges." Explain how this saying can be applied to the goal of Objective 4 in this section.

36. Explain how the distributive property is used in combining $6t + 5t$ to get $11t$.

Simplify each expression by combining like terms. See Examples 1–3.

37. $4k + 3 - 2k + 8 + 7k - 16$

38. $9x + 7 - 13x + 12 + 8x - 15$

39. $-\dfrac{4}{3} + 2t + \dfrac{1}{3}t - 8 - \dfrac{8}{3}t$

40. $-\dfrac{5}{6} + 8x + \dfrac{1}{6}x - 7 - \dfrac{7}{6}$

41. $-5.3r + 4.9 - 2r + .7 + 3.2r$

42. $2.7b + 5.8 - 3b + .5 - 4.4b$

43. $2y^2 - 7y^3 - 4y^2 + 10y^3$

44. $9x^4 - 7x^6 + 12x^4 + 14x^6$

45. $13p + 4(4 - 8p)$

46. $5x + 3(7 - 2x)$

47. $-4(y - 7) - 6$

48. $-5(t - 13) - 4$

49. $-5(5y - 9) + 3(3y + 6)$

50. $-3(2t + 4) + 8(2t - 4)$

51. $-4(-3k + 3) - (6k - 4) - 2k + 1$ **52.** $-5(8j + 2) - (5j - 3) - 3j + 17$

53. $-7.5(2y + 4) - 2.9(3y - 6)$ **54.** $8.4(6t - 6) + 2.4(9 - 3t)$

Convert each phrase into a mathematical expression. Use x as the variable. Combine like terms when possible. See Example 4.

55. Five times a number, added to the sum of the number and three

56. Six times a number, added to the sum of the number and six

57. A number multiplied by -7, subtracted from the sum of 13 and six times the number

58. A number multiplied by 5, subtracted from the sum of 14 and eight times the number

59. Six times a number added to -4, subtracted from twice the sum of three times the number and 4 (*Hint: Twice* means two times.)

60. Nine times a number added to 6, subtracted from triple the sum of 12 and 8 times the number (*Hint: Triple* means three times.)

61. Write the expression $9x - (x + 2)$ using words, as in Exercises 55–60.

62. Write the expression $2(3x + 5) - 2(x + 4)$ using words, as in Exercises 55–60.

RELATING CONCEPTS (EXERCISES 63-70)

Work Exercises 63–70 in order. They will help prepare you for graphing later in the text.

63. Evaluate the expression $x + 2$ for the values of x shown in the chart.

x	x + 2
0	
1	
2	
3	

64. Based on your results from Exercise 63, complete the following statement: For every increase of 1 unit for x, the value of $x + 2$ increases by _____ unit(s).

65. Repeat Exercise 63 for these expressions:
(a) $x + 1$ (b) $x + 3$ (c) $x + 4$

66. Based on your results from Exercises 63 and 65, make a conjecture (an educated guess) about what happens to the value of an expression of the form $x + b$ for any value of b, as x increases by 1 unit.

67. Repeat Exercise 63 for these expressions:
(a) $2x + 2$ (b) $3x + 2$ (c) $4x + 2$

68. Based on your results from Exercise 67, complete the following statement: For every increase of 1 unit for x, the value of $mx + 2$ increases by _____ units.

69. Repeat Exercise 63 and compare your results to those in Exercise 67 for these expressions:
(a) $2x + 7$ (b) $3x + 5$ (c) $4x + 1$

70. Based on your results from Exercises 63–69, complete the following statement: For every increase of 1 unit for x, the value of $mx + b$ increases by _____ units.

Did you make the connection that an increase of 1 for x yields an increase of m for $mx + b$?

CHAPTER 1 GROUP ACTIVITY

Comparing Shapes of Houses

Objective: Use arithmetic skills to make comparisons.

People throughout the world live in different shaped homes. As the population of the earth continues to grow, issues of housing and heating become increasingly important. This activity will explore perimeters and areas of some of these different shaped homes. Floor plans for three different homes are given below.

A. As a group, look at the dimensions of the given floor plans. Considering only the dimensions, which plan do you think has the greatest area?

B. Now have each student in your group pick one floor plan. For each plan, find the following. (Round all answers to the nearest whole number. In Plan one, let $\pi = 3.14$.)
   1. The area of the plan
   2. The perimeter or circumference (the distance around the outside) of the plan

C. Share your findings with the group and answer the following questions.
   1. What did you determine about the areas of the three floor plans?
   2. Which plan has the smallest perimeter? Which has the largest perimeter?
   3. Why do you think houses with round floor plans might be more energy efficient?
   4. What advantages do you think houses with square or rectangular floor plans have? Why do you think floor plans with these shapes are most common for homes today?

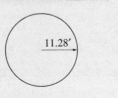 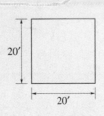 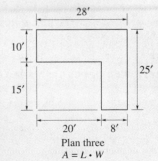

Plan one
$A = \pi r^2$
$C = 2\pi r$

Plan two
$A = s^2$

Plan three
$A = L \cdot W$

CHAPTER 1 SUMMARY

KEY TERMS

1.1			
natural numbers	mixed number	signed numbers	identity property
whole numbers	difference	integers	identity element for
numerator	**1.2** exponent (power)	graph	addition
denominator	base	coordinate	identity element for
factor	exponential	rational numbers	multiplication
product	expression	set-builder notation	inverse property
factored	grouping symbols	irrational numbers	distributive property
prime	**1.3** variable	real numbers	**1.8** term
composite	algebraic expression	additive inverse	numerical coefficient
greatest common	equation	(opposite)	like and unlike terms
factor	solution	absolute value	combining like terms
lowest terms	set	**1.6** multiplicative	
reciprocal	element	inverse	
quotient	**1.4** number line	(reciprocal)	
sum	negative numbers	**1.7** commutative property	
least common	positive numbers	associative property	
denominator (LCD)			

NEW SYMBOLS

a^n	n factors of a	$\leq$	is less than or equal to (read from left to right)	$\lvert x \rvert$	absolute value of x
[ ]	square brackets (used as grouping symbols)	$\geq$	is greater than or equal to (read from left to right)	$\dfrac{1}{x}$ or $1/x$	the multiplicative inverse, or reciprocal, of the nonzero number x
$=$	is equal to	{ }	set braces		
$\neq$	is not equal to	$\{x \mid x$ **has a certain property**$\}$	set-builder notation	$a(b), (a)(b), a \cdot b,$ **or** ab	a times b
$<$	is less than (read from left to right)	$-x$	the additive inverse, or opposite, of x	$\dfrac{a}{b}$ or a/b	a divided by b
$>$	is greater than (read from left to right)				

TEST YOUR WORD POWER

See how well you have learned the vocabulary in this chapter. Answers, with examples, are given at the bottom of the next page.

1. A **factor** is
(a) the answer in an addition problem
(b) the answer in a multiplication problem
(c) one of two or more numbers that are added to get another number
(d) one of two or more numbers that are multiplied to get another number.

2. A number is **prime** if
(a) it cannot be factored
(b) it has just one factor
(c) it has only itself and 1 as factors
(d) it has at least two different factors.

3. An **exponent** is
(a) a symbol that tells how many numbers are being multiplied

(b) a number raised to a power
(c) a number that tells how many times a factor is repeated
(d) one of two or more numbers that are multiplied.

4. A **variable** is
(a) a symbol used to represent an unknown number
(b) a value that makes an equation true
(c) a solution of an equation
(d) the answer in a division problem.

5. An **integer** is
(a) a positive or negative number
(b) a natural number, its opposite, or zero
(c) any number that can be graphed on a number line
(d) the quotient of two numbers.

6. A **coordinate** is
(a) the number that corresponds to a point on a number line
(b) the graph of a number
(c) any point on a number line
(d) the distance from 0 on a number line.

7. The **absolute value** of a number is
(a) the graph of the number
(b) the reciprocal of the number
(c) the opposite of the number
(d) the distance between 0 and the number on a number line.

8. A **term** is
(a) a numerical factor
(b) a number or a product or quotient of numbers and variables raised to powers
(c) one of several variables with the same exponents
(d) a sum of numbers and variables raised to powers.

9. A **numerical coefficient** is
(a) the numerical factor in a term
(b) the number of terms in an expression
(c) a variable raised to a power
(d) the variable factor in a term.

QUICK REVIEW

CONCEPTS	EXAMPLES

1.1 FRACTIONS

Operations with Fractions

Addition/Subtraction:

1. To add/subtract fractions with the same denominator, add/subtract the numerators and keep the same denominator.

2. To add/subtract fractions with different denominators, find the LCD and write each fraction with this LCD. Then follow the procedure above.

Multiplication: Multiply numerators and multiply denominators.

Division: Multiply the first fraction by the reciprocal of the second fraction.

Perform the operations.

$$\frac{2}{5}+\frac{7}{5}=\frac{2+7}{5}=\frac{9}{5}$$

$$\frac{2}{3}-\frac{1}{2}=\frac{4}{6}-\frac{3}{6} \quad 6 \text{ is the LCD.}$$
$$=\frac{4-3}{6}=\frac{1}{6}$$

$$\frac{4}{3}\cdot\frac{5}{6}=\frac{20}{18}=\frac{10}{9}$$

$$\frac{6}{5}\div\frac{1}{4}=\frac{6}{5}\cdot\frac{4}{1}=\frac{24}{5}$$

1.2 EXPONENTS, ORDER OF OPERATIONS, AND INEQUALITY

Order of Operations
Simplify within parentheses or above and below fraction bars first, in the following order.

1. Apply all exponents.

2. Do any multiplications or divisions from left to right.

3. Do any additions or subtractions from left to right. If no grouping symbols are present, start with Step 1.

Simplify $36-4(2^2+3)$.

$$36-4(2^2+3)=36-4(4+3)$$
$$=36-4(7)$$
$$=36-28$$
$$=8$$

CONCEPTS	EXAMPLES

1.3 VARIABLES, EXPRESSIONS, AND EQUATIONS

Evaluate an expression with a variable by substituting a given number for the variable.

Evaluate $2x + y^2$ if $x = 3$ and $y = -4$.

$$2x + y^2 = 2(3) + (-4)^2$$
$$= 6 + 16$$
$$= 22$$

Values of a variable that make an equation true are solutions of the equation.

Is 2 a solution of $5x + 3 = 18$?

$$5(2) + 3 = 18 \quad ?$$
$$13 = 18 \qquad \text{False}$$

2 is not a solution.

1.4 REAL NUMBERS AND THE NUMBER LINE

The Ordering of Real Numbers
a is less than b if a is to the left of b on the number line.

Graph -2, 0, and 3.

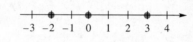

$$-2 < 3 \qquad\qquad 3 > 0 \qquad\qquad 0 < 3$$

The additive inverse of x is $-x$.

The absolute value of x, $|x|$, is the distance between x and 0 on the number line.

$$-(5) = -5 \qquad -(-7) = 7 \qquad -0 = 0$$
$$|13| = 13 \qquad |0| = 0 \qquad |-5| = 5$$

1.5 ADDITION AND SUBTRACTION OF REAL NUMBERS

Rules for Addition
To add two numbers with the same sign, add their absolute values. The sum has that same sign.

To add two numbers with different signs, subtract their absolute values. The sum has the sign of the number with larger absolute value.

Add.

$$9 + 4 = 13$$
$$-8 + (-5) = -13$$
$$7 + (-12) = -5$$
$$-5 + 13 = 8$$

Definition of Subtraction
$$x - y = x + (-y)$$

Rules for Subtraction
1. Change the subtraction symbol to the addition symbol.
2. Change the sign of the number being subtracted.
3. Add, using the rules for addition.

Subtract.

$$5 - (-2) = 5 + 2 = 7$$

$$-3 - 4 = -3 + (-4) = -7$$

$$-2 - (-6) = -2 + 6 = 4$$
$$13 - (-8) = 13 + 8 = 21$$

1.6 MULTIPLICATION AND DIVISION OF REAL NUMBERS

Rules for Multiplying and Dividing Signed Numbers
The product (or quotient) of two numbers having the *same sign* is *positive;* the product (or quotient) of two numbers having *different signs* is *negative.*

Multiply or divide.

$$6 \cdot 5 = 30 \qquad (-7)(-8) = 56 \qquad \frac{20}{4} = 5$$

$$\frac{-24}{-6} = 4 \qquad (-6)(5) = -30 \qquad (6)(-5) = -30$$

$$\frac{-18}{9} = -2 \qquad\qquad \frac{49}{-7} = -7$$

CONCEPTS	EXAMPLES

Definition of Division

$$\frac{x}{y} = x \cdot \frac{1}{y}, \quad y \neq 0$$

$$\frac{10}{2} = 10 \cdot \frac{1}{2} = 5$$

Division by 0 is undefined.

$\frac{5}{0}$ is undefined.

0 divided by a nonzero number equals 0.

$\frac{0}{5} = 0$

1.7   PROPERTIES OF REAL NUMBERS

Commutative

$$a + b = b + a$$
$$ab = ba$$

$$7 + (-1) = -1 + 7$$
$$5(-3) = (-3)5$$

Associative

$$(a + b) + c = a + (b + c)$$
$$(ab)c = a(bc)$$

$$(3 + 4) + 8 = 3 + (4 + 8)$$
$$[(-2)(6)](4) = (-2)[(6)(4)]$$

Identity

$$a + 0 = a \qquad 0 + a = a$$
$$a \cdot 1 = a \qquad 1 \cdot a = a$$

$$-7 + 0 = -7 \qquad 0 + (-7) = -7$$
$$9 \cdot 1 = 9 \qquad 1 \cdot 9 = 9$$

Inverse

$$a + (-a) = 0 \qquad -a + a = 0$$
$$a \cdot \frac{1}{a} = 1 \qquad \frac{1}{a} \cdot a = 1 \quad (a \neq 0)$$

$$7 + (-7) = 0 \qquad -7 + 7 = 0$$
$$2\left(\frac{1}{2}\right) = 1 \qquad -\frac{1}{2}(-2) = 1$$

Distributive

$$a(b + c) = ab + ac$$
$$(b + c)a = ba + ca$$
$$a(b - c) = ab - ac$$

$$5(4 + 2) = 5(4) + 5(2)$$
$$(4 + 2)5 = 4(5) + 2(5)$$
$$9(5 - 4) = 9(5) - 9(4)$$

1.8   SIMPLIFYING EXPRESSIONS

Only like terms may be combined.

Simplify $-3y^2 + 6y^2 + 14y^2 = 17y^2$.

$$4(3 + 2x) - 6(5 - x)$$
$$= 12 + 8x - 30 + 6x \qquad \text{Distributive property}$$
$$= 14x - 18$$

CHAPTER 1 REVIEW EXERCISES

[1.1]   *Perform each operation.**

1. $\frac{8}{5} \div \frac{32}{15}$

2. $\frac{3}{8} + 3\frac{1}{2} - \frac{3}{16}$

*For help with any of these exercises, refer to the section given in brackets.

3. The pie chart illustrates how 800 people responded to a survey that asked "Do you believe that there was a conspiracy to assassinate John F. Kennedy?" What fractional part of the group did not have an opinion?

4. Based on the chart in Exercise 3, how many people responded "yes"?

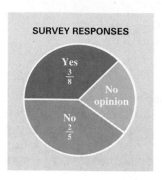

SURVEY RESPONSES

Yes $\frac{3}{8}$

No opinion

No $\frac{2}{5}$

[1.2] *Find the value of each exponential expression.*

5. 5^4

6. $\left(\frac{3}{5}\right)^3$

7. $(.02)^5$

8. $(.001)^3$

Find the value of each expression.

9. $8 \cdot 5 - 13$

10. $7[3 + 6(3^2)]$

11. $\dfrac{9(4^2 - 3)}{4 \cdot 5 - 17}$

12. $\dfrac{6(5 - 4) + 2(4 - 2)}{3^2 - (4 + 3)}$

Tell whether each statement is true or false.

13. $12 \cdot 3 - 6 \cdot 6 \le 0$

14. $3[5(2) - 3] > 20$

15. $9 \le 4^2 - 8$

Write each word statement in symbols.

16. Thirteen is less than seventeen.

17. Five plus two is not equal to ten.

18. Americans are literally bombarded by mail-order catalogs on a daily basis. The bar graph shows the estimated number of catalogs mailed to consumers and businesses during the years 1983–1996.

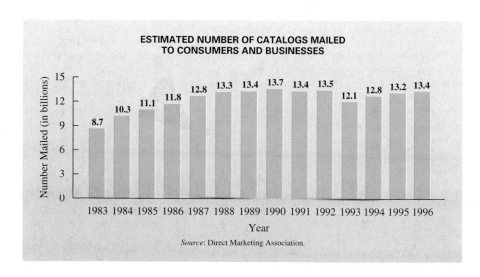

ESTIMATED NUMBER OF CATALOGS MAILED TO CONSUMERS AND BUSINESSES

Number Mailed (in billions)

1983: 8.7, 1984: 10.3, 1985: 11.1, 1986: 11.8, 1987: 12.8, 1988: 13.3, 1989: 13.4, 1990: 13.7, 1991: 13.4, 1992: 13.5, 1993: 12.1, 1994: 12.8, 1995: 13.2, 1996: 13.4

Year

Source: Direct Marketing Association.

(a) In which years were *fewer than* 13.4 billion catalogs mailed?

(b) In which years were *at least* 12.8 billion mailed?

(c) How many *total* catalogs were mailed in the five years having the largest numbers of mailings?

[1.3]   *Find the numerical value of each expression if x = 6 and y = 3.*

19. $2x + 6y$ **20.** $4(3x - y)$ **21.** $\dfrac{x}{3} + 4y$ **22.** $\dfrac{x^2 + 3}{3y - x}$

Change each word phrase to an algebraic expression. Use x as the variable to represent the number.

23. Six added to a number

24. A number subtracted from eight

25. Nine subtracted from six times a number

26. Three-fifths of a number added to 12

Decide whether the given number is a solution of the given equation.

27. $5x + 3(x + 2) = 22;$ 2

28. $\dfrac{t + 5}{3t} = 1;$ 6

Change each word statement to an equation. Use x as the variable. Then find the solution from the set {0, 2, 4, 6, 8, 10}.

29. Six less than twice a number is 10.

30. The product of a number and 4 is 8.

[1.4]   *Graph each group of numbers on a number line.*

31. $-4, -\dfrac{1}{2}, 0, 2.5, 5$

32. $-2, |-3|, -3, |-1|$

Classify each number, using the sets natural numbers, whole numbers, integers, rational numbers, irrational numbers, real numbers.

33. $\dfrac{4}{3}$

34. $\sqrt{6}$

Select the smaller number in each pair.

35. $-10, 5$ **36.** $-8, -9$ **37.** $-\dfrac{2}{3}, -\dfrac{3}{4}$ **38.** $0, -|23|$

Decide whether each statement is true or false.

39. $12 > -13$ **40.** $0 > -5$ **41.** $-9 < -7$ **42.** $-13 \geq -13$

*For the following, **(a)** find the opposite of the number and **(b)** find the absolute value of the number.*

43. -9 **44.** 0 **45.** 6 **46.** $-\dfrac{5}{7}$

Simplify each number by removing absolute value symbols.

47. $|-12|$ **48.** $-|3|$ **49.** $-|-19|$ **50.** $-|9 - 2|$

[1.5]   *Perform the indicated operations.*

51. $-10 + 4$ **52.** $14 + (-18)$ **53.** $-8 + (-9)$

54. $\dfrac{4}{9} + \left(-\dfrac{5}{4}\right)$ **55.** $-13.5 + (-8.3)$ **56.** $(-10 + 7) + (-11)$

57. $[-6 + (-8) + 8] + [9 + (-13)]$ **58.** $(-4 + 7) + (-11 + 3) + (-15 + 1)$

59. $-7 - 4$ **60.** $-12 - (-11)$

61. $5 - (-2)$ **62.** $-\dfrac{3}{7} - \dfrac{4}{5}$

63. $2.56 - (-7.75)$ **64.** $(-10 - 4) - (-2)$

65. $(-3 + 4) - (-1)$ **66.** $-(-5 + 6) - 2$

Write a numerical expression for each phrase, and simplify the expression.

67. 19 added to the sum of -31 and 12

68. 13 more than the sum of -4 and -8

69. The difference between -4 and -6

70. Five less than the sum of 4 and -8

Find the solution of the equation from the set $\{-3, -2, -1, 0, 1, 2, 3\}$ by guessing or by trial and error.

71. $x + (-2) = -4$

72. $12 + x = 11$

Solve each problem.

73. Like many people, Kareem Dunlap neglects to keep up his checkbook balance. When he finally balanced his account, he found the balance was $-\$23.75$, so he deposited $\$50.00$. What is his new balance?

74. The low temperature in Yellowknife, in the Canadian Northwest Territories, one January day was $-26°$F. It rose $16°$ that day. What was the high temperature?

75. Eric owed his brother $28. He repaid $13 but then borrowed another $14. What positive or negative amount represents his present financial status?

76. If the temperature drops $7°$ below its previous level of $-3°$, what is the new temperature?

77. A football team gained 3 yards on the first play from scrimmage, lost 12 yards on the second play, and then gained 13 yards on the third play. How many yards did the team gain or lose altogether?

78. In 1985, the construction industry had 4,480,000 employees. By 1990, this number had increased by 759,000 employees, but it experienced a decrease of 530,000 employees by 1994. How many employees were there in 1994? (*Source:* U.S. Bureau of the Census.)

[1.6] *Perform the indicated operations.*

79. $(-12)(-3)$

80. $15(-7)$

81. $\left(-\dfrac{4}{3}\right)\left(-\dfrac{3}{8}\right)$

82. $(-4.8)(-2.1)$

83. $5(8 - 12)$

84. $(5 - 7)(8 - 3)$

85. $2(-6) - (-4)(-3)$

86. $3(-10) - 5$

87. $\dfrac{-36}{-9}$

88. $\dfrac{220}{-11}$

89. $-\dfrac{1}{2} \div \dfrac{2}{3}$

90. $-33.9 \div (-3)$

91. $\dfrac{-5(3) - 1}{8 - 4(-2)}$

92. $\dfrac{5(-2) - 3(4)}{-2[3 - (-2)] - 1}$

93. $\dfrac{10^2 - 5^2}{8^2 + 3^2 - (-2)}$

94. $\dfrac{(.6)^2 + (.8)^2}{(-1.2)^2 - (-.56)}$

Evaluate each expression if $x = -5$, $y = 4$, and $z = -3$.

95. $6x - 4z$

96. $5x + y - z$

97. $5x^2$

98. $z^2(3x - 8y)$

Write a numerical expression for each phrase, and simplify the expression.

99. Nine less than the product of -4 and 5

100. Five-sixths of the sum of 12 and -6

101. The quotient of 12 and the sum of 8 and -4

102. The product of -20 and 12, divided by the difference between 15 and -15

Write each sentence in symbols, using x as the variable, and find the solution by guessing or by trial and error. All solutions come from the list of integers between -12 and 12.

103. 8 times a number is -24.

104. The quotient of a number and 3 is -2.

105. The payrolls and average salaries of 5 of the 28 major league baseball teams on opening day of 1998 are listed here.

Team	Payroll	Average Salary
Baltimore Orioles	$68,988,134	$2,555,116
New York Yankees	63,460,567	2,440,791
Cleveland Indians	59,583,500	2,127,982
Atlanta Braves	59,536,000	2,126,286
Texas Rangers	55,304,595	1,975,164

Source: The Associated Press.

(a) What is the average of the five payrolls? Round to the nearest dollar.
(b) What is the average of the five average salaries? Round to the nearest dollar.

106. The bar graph shows the 1993 sales in millions of dollars of four of the largest brands in the United States. What was the average of these sales?

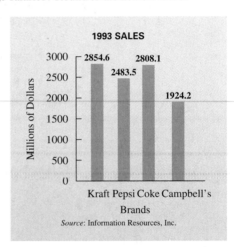

[1.7] *Decide whether each statement is an example of the commutative, associative, identity, inverse, or distributive property.*

107. $6 + 0 = 6$

108. $5 \cdot 1 = 5$

109. $-\dfrac{2}{3}\left(-\dfrac{3}{2}\right) = 1$

110. $17 + (-17) = 0$

111. $5 + (-9 + 2) = [5 + (-9)] + 2$

112. $w(xy) = (wx)y$

113. $3x + 3y = 3(x + y)$

114. $(1 + 2) + 3 = 3 + (1 + 2)$

Use the distributive property to rewrite each expression. Simplify if possible.

115. $7y + y$ **116.** $-12(4 - t)$ **117.** $3(2s) + 3(5y)$ **118.** $-(-4r + 5s)$

119. Evaluate $25 - (5 - 2)$ and $(25 - 5) - 2$. Use this example to explain why subtraction is not associative.

120. Evaluate $180 \div (15 \div 5)$ and $(180 \div 15) \div 5$. Use this example to explain why division is not associative.

[1.8] *Combine terms whenever possible.*

121. $2m + 9m$

122. $15p^2 - 7p^2 + 8p^2$

123. $5p^2 - 4p + 6p + 11p^2$

124. $-2(3k - 5) + 2(k + 1)$

125. $7(2m + 3) - 2(8m - 4)$

126. $-(2k + 8) - (3k - 7)$

MIXED REVIEW EXERCISES*

Perform the indicated operations.

127. $[(-2) + 7 - (-5)] + [-4 - (-10)]$

128. $\left(-\dfrac{5}{6}\right)^2$

129. $\dfrac{6(-4) + 2(-12)}{5(-3) + (-3)}$

130. $\dfrac{3}{8} - \dfrac{5}{12}$

131. $\dfrac{8^2 + 6^2}{7^2 + 1^2}$

132. $-16(-3.5) - 7.2(-3)$

133. $2\dfrac{5}{6} - 4\dfrac{1}{3}$

134. $-8 + [(-4 + 17) - (-3 - 3)]$

135. $-\dfrac{12}{5} \div \dfrac{9}{7}$

136. $(-8 - 3) - 5(2 - 9)$

137. $5x^2 - 12y^2 + 3x^2 - 9y^2$

138. $-4(2t + 1) - 8(-3t + 4)$

139. Write a sentence or two explaining the special considerations involving zero when dividing.

140. "Two negatives give a positive" is often heard from students. Is this correct? Use more precise language in explaining what this means.

141. Use x as the variable and write an expression for "the product of 5 and the sum of a number and 7." Then use the distributive property to rewrite the expression.

142. The highest temperature ever recorded in Albany, New York, was 99°F, while the lowest was 112° less than the highest. What was the lowest temperature ever recorded in Albany? (*Source: The World Almanac and Book of Facts, 1998.*)

The year-to-year percentage change in net corporate income for companies in the DJ-Global U.S. Index is given in the accompanying graph for the years 1994 through the second quarter of 1997.

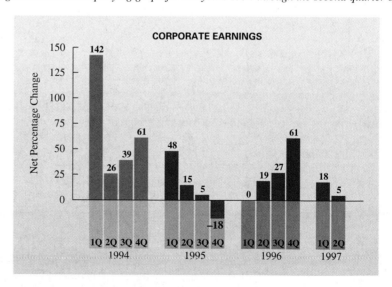

143. What signed number represents the change from the fourth quarter in 1994 to the fourth quarter in 1995?

144. What signed number represents the change from the third quarter in 1995 to the third quarter in 1996?

---

*The order of exercises in this final group does not correspond to the order in which topics occur in the chapter. This random ordering should help you prepare for the chapter test in yet another way.

CHAPTER 1 TEST

1. Write $\dfrac{63}{99}$ in lowest terms.   **2.** Add: $\dfrac{5}{8} + \dfrac{11}{12} + \dfrac{7}{15}$.   **3.** Divide: $\dfrac{19}{15} \div \dfrac{6}{5}$.

4. The pie chart indicates the market share of different means of intercity transportation, based on 1,230 million passengers carried.
  (a) How many of these passengers used air travel?
  (b) How many of these passengers did not use the bus?

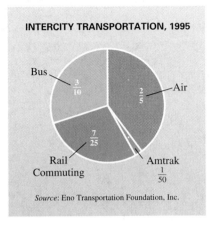

INTERCITY TRANSPORTATION, 1995

Bus $\dfrac{3}{10}$

Air $\dfrac{2}{5}$

Rail Commuting $\dfrac{7}{25}$

Amtrak $\dfrac{1}{50}$

Source: Eno Transportation Foundation, Inc.

5. Decide whether $4[-20 + 7(-2)] \le 135$ is true or false.

6. Graph the group of numbers $-1, -3, \mid -4 \mid, \mid -1 \mid$ on a number line.

7. To which of the following sets does $-\dfrac{2}{3}$ belong: natural numbers, whole numbers, integers, rational numbers, irrational numbers, real numbers?

8. Explain how a number line can be used to show that -8 is less than -1.

9. Write in symbols: The quotient of -6 and the sum of 2 and -8. Simplify the expression.

10. According to the bar graph shown here, which subgroups of Asian and Pacific Islander-owned businesses had revenues that were
  (a) less than $13 billion?   **(b)** greater than or equal to $16.2 billion?

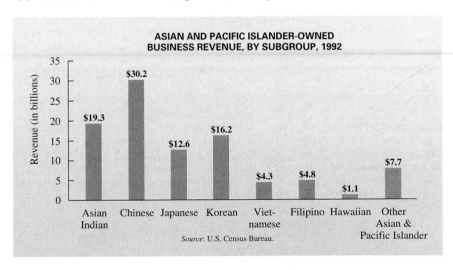

ASIAN AND PACIFIC ISLANDER-OWNED BUSINESS REVENUE, BY SUBGROUP, 1992

Revenue (in billions)

Asian Indian $19.3
Chinese $30.2
Japanese $12.6
Korean $16.2
Vietnamese $4.3
Filipino $4.8
Hawaiian $1.1
Other Asian & Pacific Islander $7.7

Source: U.S. Census Bureau.

Perform the indicated operations.

11. $-2 - (5 - 17) + (-6)$ **12.** $-5\dfrac{1}{2} + 2\dfrac{2}{3}$

13. $-6 - [-7 + (2 - 3)]$

14. $4^2 + (-8) - (2^3 - 6)$

15. $(-5)(-12) + 4(-4) + (-8)^2$

16. $\dfrac{-7 - (-6 + 2)}{-5 - (-4)}$

17. $\dfrac{30(-1 - 2)}{-9[3 - (-2)] - 12(-2)}$

Find the solution for each equation from the set $\{-6, -4, -2, 0, 2, 4, 6\}$ by guessing or by trial and error.

18. $-x + 3 = -3$

19. $-3x = -12$

Evaluate each expression, given $x = -2$ and $y = 4$.

20. $3x - 4y^2$

21. $\dfrac{5x + 7y}{3(x + y)}$

Solve each problem.

22. The average sales prices of new single-family homes in the United States for the years 1990 through 1995 are shown in the chart. Determine the change from one year to the next by subtraction.

	Year	Average Sales Price	Change from Previous Year
	1990	$149,800	
	1991	$147,200	$-$2600
(a)	1992	$144,100	_____
(b)	1993	$147,700	_____
(c)	1994	$154,500	_____
(d)	1995	$158,700	_____

Source: U.S. Bureau of the Census.

23. For a certain system of rating major league baseball relief pitchers, 3 points are awarded for a save, 3 points are awarded for a win, 2 points are subtracted for a loss, and 2 points are subtracted for a blown save. If Dennis Eckersley of the Boston Red Sox has 4 saves, 3 wins, 2 losses, and 1 blown save, how many points does he have?

24. The bar graph shows the number of corporate name changes that occurred during the years 1990–1996. These changes were mostly brought on by mergers and acquisitions. Use a signed number to represent each of the following.
(a) the change from 1990 to 1991
(b) the change from 1991 to 1996
(c) the change from 1992 to 1996

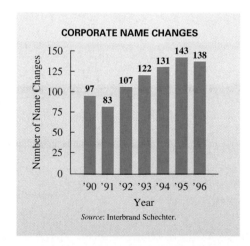

Match each property in Column I with the example of it in Column II.

I	II

25. Commutative   **A.** $3x + 0 = 3x$

26. Associative   **B.** $(5 + 2) + 8 = 8 + (5 + 2)$

27. Inverse   **C.** $-3(x + y) = -3x + (-3y)$

28. Identity   **D.** $-5 + (3 + 2) = (-5 + 3) + 2$

29. Distributive

   E. $-\dfrac{5}{3}\left(-\dfrac{3}{5}\right) = 1$

30. What property is used to show that $3(x + 1) = 3x + 3$?

31. Consider the expression $-6[5 + (-2)]$.
   (a) Evaluate it by first working within the brackets.
   (b) Evaluate it by using the distributive property.
   (c) Why must the answers in items (a) and (b) be the same?

Simplify by combining like terms.

32. $8x + 4x - 6x + x + 14x$ **33.** $5(2x - 1) - (x - 12) + 2(3x - 5)$

Linear Equations and Applications

The first organized worldwide sporting event was the Summer Olympics I, held in 1896 in Athens, Greece. Three hundred eleven men from 13 nations gathered to compete. One hundred years later, the 1996 Summer Olympics, held in Atlanta, attracted 10,341 competitors—6562 men and 3779 women, from 197 nations. The results of an Olympic women's swimming event from 1968 to 1996 are given in the table. (Winning times are given in minutes and seconds to the nearest hundredth.)

200-Meter Freestyle

1968 Deborah Meyer, U.S.	2:10.50
1972 Shane Gould, Australia	2:03.56
1976 Kornelia Ender, East Germany	1:59.26
1980 Barbara Krause, East Germany	1:58.33
1984 Mary Wayte, U.S.	1:59.23
1988 Heike Friederich, East Germany	1:57.65
1992 Nicole Haislett, U.S.	1:57.90
1996 Claudia Poll, Costa Rica	1:58.16

Which year had the fastest winning time? In Section 2.6, we'll see how to determine the winning speed from this information.

In the United States, the first organized professional sport was baseball. The first baseball league was founded in 1876, and the first World Series playoff was held in 1882.* Both football and basketball worked their way from college campus to professional league status.

And, as you'll discover in this chapter, sports involves mathematics; the vast quantity of statistics kept by professional sports leagues fills numerous books.

*All data from *The Universal Almanac*, 1997, John W. Wright, General Editor.

Visit our Web site at www.LialAlgebra.com

2.1 The Addition and Multiplication Properties of Equality

OBJECTIVES

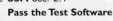

1 Identify linear equations.

2 Use the addition property of equality.

3 Use the multiplication property of equality.

4 Simplify equations, and then use the properties of equality.

FOR EXTRA HELP

📖 **SSG** Sec. 2.1
SSM Sec. 2.1

 Pass the Test Software

InterAct Math
   Tutorial Software

Video 3

Quite possibly the most important topic in the study of beginning algebra is the solution of equations. We will investigate many types of equations in this book, and the properties introduced in this section are essential in solving them.

OBJECTIVE 1 Identify linear equations. The simplest type of equation is a *linear equation.* Before we can solve a linear equation we must be able to recognize one.

Linear Equation

A **linear equation in one variable** can be written in the form

$$Ax + B = 0$$

for real numbers A and B, with $A \neq 0$.

For example,

$$4x + 9 = 0, \qquad 2x - 3 = 5, \qquad \text{and} \qquad x = 7$$

are linear equations; the last two can be written in the specified form using the properties to be developed in this section. However,

$$x^2 + 2x = 5, \qquad \frac{1}{x} = 6 \qquad \text{and} \qquad |\, 2x + 6 \,| = 0$$

are *not* linear equations.

Methods of solving linear equations are introduced in this section and in Section 2.2. As discussed in Section 1.3, a *solution* of an equation is a number that when substituted for the variable makes the equation true; that is, *satisfies* the equation. An equation is solved by finding its **solution set,** the set of all solutions. Equations that have exactly the same solution sets are **equivalent equations.** Linear equations are solved by using a series of steps to produce equivalent equations until an equation of the form

$$x = \text{a number}$$

is obtained.

OBJECTIVE 2 Use the addition property of equality. Consider the equation

$$x = 4.$$

Suppose we add 7 to both sides of the equation, getting

$$x + 7 = 11.$$

Now, adding $2x$ to both sides gives

$$3x + 7 = 2x + 11.$$

Verify that $x = 4$ satisfies each of these three equations, so they are thus equivalent equations. To solve the third equation, $3x + 7 = 2x + 11$, all we have to do is reverse the steps shown earlier by adding the negative in each case. The **addition property of equality** justifies this idea.

Addition Property of Equality

If A, B, and C are mathematical expressions that represent real numbers, then the equations

$$A = B \qquad \text{and} \qquad A + C = B + C$$

have exactly the same solution. In words, the same expression may be added to both sides of an equation without changing the solution.

Note that, in the addition property, C represents a mathematical expression. This means that numbers or terms with variables, or even sums of terms that represent real numbers, can be added to both sides of an equation.

E X A M P L E   1   Using the Addition Property of Equality

Solve $x - 16 = 7$.

    If the left side of this equation were just x, the solution would be found. Get x alone by using the addition property of equality and adding 16 on both sides.

$$x - 16 = 7$$
$$x - 16 + 16 = 7 + 16 \qquad \text{Add 16 on both sides.}$$
$$x = 23$$

Note that we combined the steps that change $x - 16 + 16$ to $x + 0$ and $x + 0$ to x. We will combine these steps from now on. Check by substituting 23 for x in the *original* equation.

$$x - 16 = 7 \qquad \text{Original equation}$$
$$23 - 16 = 7 \quad ? \qquad \text{Let } x = 23.$$
$$7 = 7 \qquad \text{True}$$

Since the check results in a true statement, $\{23\}$ is the solution set.

In this example, why was 16 added to both sides of the equation $x - 16 = 7$? The equation would be solved if it could be rewritten so that one side contained only the variable and the other side contained only a number. Since $x - 16 + 16 = x + 0 = x$, adding 16 on the left side simplifies that side to just x, the variable, as desired.

The addition property of equality says that the same expression may be *added* to both sides of an equation. As was shown in Chapter 1, subtraction is defined in terms of addition. Therefore, the addition property also permits *subtracting* the same expression from both sides of an equation.

┌──

E X A M P L E   2   Subtracting a Variable Expression to Solve an Equation

Solve the equation $3k - 12 + k + 2 = 5 + 3k + 2$.

Begin by combining like terms on each side of the equation to get

$$4k - 10 = 7 + 3k.$$

Next, get all terms that contain variables on the same side of the equation and all terms without variables on the other side. One way to start is to subtract $3k$ from both sides.

$$4k - 10 - 3k = 7 + 3k - 3k \qquad \text{Subtract } 3k \text{ from both sides.}$$
$$k - 10 = 7 \qquad \text{Combine terms.}$$
$$k - 10 + 10 = 7 + 10 \qquad \text{Add 10 to both sides.}$$
$$k = 17$$

Check by substituting 17 for k in the original equation.

$$3k - 12 + k + 2 = 5 + 3k + 2 \qquad \text{Original equation}$$
$$3(17) - 12 + 17 + 2 = 5 + 3(17) + 2 \qquad ? \qquad \text{Let } k = 17.$$
$$51 - 12 + 17 + 2 = 5 + 51 + 2 \qquad ? \qquad \text{Multiply.}$$
$$58 = 58 \qquad \text{True}$$

The check results in a true statement, so the solution set is $\{17\}$.

└──

 Subtracting $3k$ in Example 2 is the same as adding $-3k$, since by the definition of subtraction, $a - b = a + (-b)$.

OBJECTIVE **3**   Use the multiplication property of equality. The addition property of equality by itself is not enough to solve an equation like $3x + 2 = 17$.

$$3x + 2 = 17$$
$$3x + 2 - 2 = 17 - 2 \qquad \text{Subtract 2 from both sides.}$$
$$3x = 15$$

The variable x is not alone on one side of the equation; the equation has $3x$ instead. Another property is needed to change $3x = 15$ to $x = $ a number.

If $3x = 15$, then $3x$ and 15 both represent the same number. Multiplying both $3x$ and 15 by the same number will also result in an equality. The **multiplication property of equality** states that both sides of an equation can be multiplied by the same nonzero expression.

Multiplication Property of Equality

If A, B, and C are mathematical expressions that represent real numbers, then the equations

$$A = B \quad \text{and} \quad AC = BC$$

have exactly the same solution. (Assume that $C \neq 0$.) In words, both sides of an equation may be multiplied by the same nonzero expression without changing the solution.

This property can be used to solve $3x = 15$. The $3x$ on the left must be changed to $1x$, or x, instead of $3x$. To get x, multiply both sides of the equation by $\frac{1}{3}$. Use $\frac{1}{3}$ since it is the reciprocal of the coefficient of x. This works because $3 \cdot \frac{1}{3} = \frac{3}{3} = 1$.

$$3x = 15$$

$$\frac{1}{3}(3x) = \frac{1}{3} \cdot 15 \qquad \text{Multiply both sides by } \tfrac{1}{3}.$$

$$\left(\frac{1}{3} \cdot 3\right)x = \frac{1}{3} \cdot 15 \qquad \text{Associative property}$$

$$1x = 5 \qquad \text{Multiplicative inverse property}$$

$$x = 5 \qquad \text{Identity property}$$

The solution set of the equation is {5}. Check by substituting 5 for x in the original equation. From now on we shall combine the last two steps shown in this example.

Just as the addition property of equality permits subtracting the same number from both sides of an equation, the multiplication property of equality permits dividing both sides of an equation by the same nonzero number. For example, the equation $3x = 15$, solved above by multiplication, could also be solved by dividing both sides by 3, as follows.

$$3x = 15$$

$$\frac{3x}{3} = \frac{15}{3} \qquad \text{Divide by 3.}$$

$$x = 5 \qquad \text{Simplify.}$$

In practice, it is usually easier to multiply on each side if the coefficient of the variable is a fraction, and divide on each side if the coefficient is an integer. For example, to solve

$$-\frac{3}{4}x = 12$$

it is easier to multiply by $-\frac{4}{3}$ than to divide by $-\frac{3}{4}$. On the other hand, to solve

$$-5x = -20$$

it is easier to divide by -5 than to multiply by $-\frac{1}{5}$.

EXAMPLE 3   Dividing Each Side of an Equation by a Nonzero Number

Solve $2.1x = 6.09$.

Divide both sides of the equation by 2.1.

$$\frac{2.1x}{2.1} = \frac{6.09}{2.1}$$

$$x = 2.9 \qquad \text{Divide.}$$

You may wish to use a calculator to find the quotient $6.09 \div 2.1$. Check that the solution is 2.9.

$$2.1(2.9) = 6.09 \qquad ? \qquad \text{Let } x = 2.9.$$

$$6.09 = 6.09 \qquad \qquad \text{True}$$

The solution set is $\{2.9\}$.

In the next two examples, multiplication produces the solution more quickly than division.

EXAMPLE 4   Using the Multiplication Property of Equality with a Fraction

Solve $\frac{a}{4} = 3$.

Replace $\frac{a}{4}$ by $\frac{1}{4}a$ since division by 4 is the same as multiplication by $\frac{1}{4}$. Get a alone by multiplying both sides by 4, the reciprocal of the coefficient of a.

$$\frac{a}{4} = 3$$

$$\frac{1}{4}a = 3 \qquad \text{Change } \frac{a}{4} \text{ to } \frac{1}{4}a.$$

$$4 \cdot \frac{1}{4}a = 4 \cdot 3 \qquad \text{Multiply by 4.}$$

$$1a = 12 \qquad \text{Inverse property}$$

$$a = 12 \qquad \text{Identity property}$$

Check the answer.

$$\frac{a}{4} = 3 \qquad \text{Original equation}$$

$$\frac{12}{4} = 3 \qquad ? \qquad \text{Let } a = 12.$$

$$3 = 3 \qquad \text{True}$$

Since 12 satisfies the equation, the solution set is $\{12\}$.

EXAMPLE 5   Using the Multiplication Property of Equality in Two Ways

Solve $\frac{3}{4}h = 6$.

We will show two ways to solve this equation.

Method 1

To get h alone, multiply both sides of the equation by $\frac{4}{3}$, the reciprocal of $\frac{3}{4}$.

$$\frac{3}{4}h = 6$$

$$\frac{4}{3}\left(\frac{3}{4}h\right) = \frac{4}{3} \cdot 6 \qquad \text{Multiply by } \tfrac{4}{3}.$$

$$1 \cdot h = \frac{4}{3} \cdot \frac{6}{1} \qquad \text{Inverse property}$$

$$h = 8 \qquad \text{Identity property}$$

Method 2

Begin by multiplying both sides of the equation by 4 to eliminate the denominator.

$$\frac{3}{4}h = 6$$

$$4\left(\frac{3}{4}h\right) = 4 \cdot 6 \qquad \text{Multiply by 4.}$$

$$3h = 24$$

$$\frac{3h}{3} = \frac{24}{3} \qquad \text{Divide by 3.}$$

$$h = 8$$

Using either method, the solution set is {8}. Check the answer by substitution in the original equation.

OBJECTIVE 4 **Simplify equations, and then use the properties of equality.** The next example uses the distributive property to first simplify an equation.

┌ **E X A M P L E 6** **Simplifying an Equation before Solving**

Solve $2(3 + 7x) - (1 + 15x) = 2$.

Use the distributive property to first simplify the equation.

$$2(3 + 7x) - 1(1 + 15x) = 2 \qquad \text{Replace the } - \text{ sign with } -1.$$

$$6 + 14x - 1 - 15x = 2 \qquad \text{Distributive property}$$

$$5 - x = 2 \qquad \text{Combine like terms.}$$

$$5 - x - 5 = 2 - 5 \qquad \text{Subtract 5.}$$

$$-x = -3$$

The variable is alone on the left, but its coefficient is -1, since $-x = -1 \cdot x$. When this occurs, simply multiply both sides by -1 (the reciprocal of -1).

$$-x = -3$$

$$-1 \cdot x = -3 \qquad\qquad {\scriptstyle -x \,=\, -1\,\cdot\, x}$$

$$-1(-1 \cdot x) = -1(-3) \qquad \text{Multiply by } -1.$$

$$x = 3$$

The solution set is {3}. Check by substituting into the original equation. (Incidentally, *dividing* by -1 would also allow us to solve an equation of the form $-x = a$.)

NOTE From the final steps in Example 6, we can see that the following is true.

$$\text{If } -x = a, \text{ then } x = -a.$$

CONNECTIONS

The use of algebra to solve equations and applied problems is very old. The 3600-year-old Rhind Papyrus includes the following "word problem." "Aha, its whole, its seventh, it makes 19." This brief sentence describes the equation

$$x + \frac{x}{7} = 19.$$

The word *aha* was used as we would use the words "Let x equal . . . ." The solution of this equation is $16\frac{5}{8}$. The word *algebra* is from the work *Risab al-jabr m'al muquabalah,* written in the ninth century by Muhammed ibn Musa Al-Khowarizmi. The title means "the science of transposition and cancellation." These ideas are used in equation solving. From Latin versions of Khowarizmi's text, "al-jabr" became the broad term covering the art of equation solving.

2.1 EXERCISES

1. Which of the pairs of equations are equivalent equations?
   (a) $x + 2 = 6$ and $x = 4$ **(b)** $10 - x = 5$ and $x = -5$
   (c) $x + 3 = 9$ and $x = 6$ **(d)** $4 + x = 8$ and $x = -4$

2. Decide whether each of the following is an expression or an equation. If it is an expression, simplify it. If it is an equation, solve it.
   (a) $5x + 8 - 4x + 7$ **(b)** $-6y + 12 + 7y - 5$
   (c) $5x + 8 - 4x = 7$ **(d)** $-6y + 12 + 7y = -5$

3. In your own words, state the addition and multiplication properties of equality and give examples.

4. State how you would find the solution of a linear equation if your next-to-last step reads "$-x = 5$."

Solve each equation using the addition property of equality. Check the solution. See Examples 1 and 2.

5. $x + 6 = 12$ 6. $x - 4 = 9$ 7. $x - 8.4 = -2.1$

8. $y + 15.5 = -5.1$ 9. $\frac{2}{5}w - 6 = \frac{7}{5}w$ 10. $-\frac{2}{7}z + 2 = \frac{5}{7}z$

11. $5.6x + 2 = 4.6x$ 12. $9.1x - 5 = 8.1x$ 13. $3p + 6 = 10 + 2p$

14. $8b - 4 = -6 + 7b$ 15. $1.2y - 4 = .2y - 4$ 16. $7.7r + 6 = 6.7r + 6$

17. $\frac{1}{2}x + 2 = -\frac{1}{2}x$ 18. $\frac{1}{5}x - 7 = -\frac{4}{5}x$ 19. $3x + 7 - 2x = 0$

20. $5x + 4 - 4x = 0$

21. Which of the following are not linear equations in one variable?
    (a) $x^2 - 5x + 6 = 0$ **(b)** $x^3 = x$
    (c) $3x - 4 = 0$ **(d)** $7x - 6x = 3 + 9x$

    Define a linear equation in one variable in words.

22. Refer to the definition of linear equation in one variable given in this section. Why is the restriction $A \neq 0$ necessary?

Solve each equation. First simplify both sides of the equation as much as possible. Check each solution. See Examples 2 and 6.

23. $5t + 3 + 2t - 6t = 4 + 12$

24. $4x + 3x - 6 - 6x = 10 + 3$

25. $6x + 5 + 7x + 3 = 12x + 4$

26. $4x - 3 - 8x + 1 = -5x + 9$

27. $5.2q - 4.6 - 7.1q = -.9q - 4.6$

28. $-4.0x + 2.7 - 1.6x = -4.6x + 2.7$

29. $\frac{5}{7}x + \frac{1}{3} = \frac{2}{5} - \frac{2}{7}x + \frac{2}{5}$

30. $\frac{6}{7}s - \frac{3}{4} = \frac{4}{5} - \frac{1}{7}s + \frac{1}{6}$

31. $(5y + 6) - (3 + 4y) = 10$

32. $(8r - 3) - (7r + 1) = -6$

33. $2(p + 5) - (9 + p) = -3$

34. $4(k - 6) - (3k + 2) = -5$

35. $-6(2b + 1) + (13b - 7) = 0$

36. $-5(3w - 3) + (1 + 16w) = 0$

37. $10(-2x + 1) = -19(x + 1)$

38. $2(2 - 3r) = -5(r - 3)$

39. $-2(8p + 2) - 3(2 - 7p) = 2(4 + 2p)$

40. $-5(1 - 2z) + 4(3 - z) = 7(3 + z)$

41. In the statement of the multiplication property of equality in this section, there is a restriction that $C \neq 0$. What would happen if you should multiply both sides of an equation by 0?

42. Which one of the equations that follow does not require the use of the multiplication property of equality to solve it?

   (a) $3x - 5x = 6$ **(b)** $-\frac{1}{4}x = 12$ **(c)** $5x - 4x = 7$ **(d)** $\frac{x}{3} = -2$

Solve each equation and check the solution. See Examples 1–6.

43. $5x = 30$

44. $7x = 56$

45. $3a = -15$

46. $5k = -70$

47. $10t = -36$

48. $4s = -34$

49. $-6x = -72$

50. $-8x = -64$

51. $2r = 0$

52. $5x = 0$

53. $\frac{1}{4}y = -12$

54. $\frac{1}{5}p = -3$

55. $-y = 12$

56. $-t = 14$

57. $-x = -\frac{4}{7}$

58. $-m = -\frac{9}{5}$

59. $.2t = 8$

60. $.9x = 18$

61. $4x + 3x = 21$

62. $9x + 2x = 121$

63. $5m + 6m - 2m = 63$

64. $11r - 5r + 6r = 168$

65. $\frac{x}{7} = -5$

66. $\frac{k}{8} = -3$

67. $-\frac{2}{7}p = -5$

68. $-\frac{3}{8}y = -2$

69. $-\frac{7}{9}c = \frac{3}{5}$

70. $-\frac{5}{6}d = \frac{4}{9}$

71. $-2.1m = 25.62$

72. $-3.9a = -31.2$

73. Write an equation that requires the use of the multiplication property of equality, where both sides must be multiplied by $\frac{2}{3}$, and the solution is a negative number.

74. Write an equation that requires the use of the multiplication property of equality, where both sides must be divided by 100, and the solution is not an integer.

Write an equation using the information given in the problem. Use x as the variable. Then solve the equation.

75. Three times a number is 17 more than twice the number. Find the number.

76. If six times a number is subtracted from seven times the number, the result is -9. Find the number.

77. If five times a number is added to three times the number, the result is the sum of seven times the number and 9. Find the number.

78. When a number is multiplied by 4, the result is 6. Find the number.

79. When a number is divided by -5, the result is 2. Find the number.

80. If twice a number is divided by 5, the result is 4. Find the number.

2.2 More on Solving Linear Equations

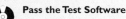
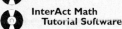
OBJECTIVE **1** **Learn and use the four steps for solving a linear equation.** To solve linear equations in general, follow these steps.

Solving a Linear Equation

Step 1 **Simplify each side separately.** Use the distributive property to clear parentheses and combine terms, as needed.

Step 2 **Isolate the variable terms on one side.** If necessary, use the addition property to get all variable terms on one side of the equation and all numbers on the other.

Step 3 **Isolate the variable.** Use the multiplication property, if necessary, to get the equation in the form $x = $ a number.

Step 4 **Check.** Check the solution by substituting into the *original* equation.

E X A M P L E   1   Using the Four Steps to Solve an Equation

Solve $3r + 4 - 2r - 7 = 4r + 3$.

Step 1 $3r + 4 - 2r - 7 = 4r + 3$

$r - 3 = 4r + 3$ Combine like terms.

Step 2 $r - 3 - r = 4r + 3 - r$ Use the addition property of equality. Subtract r.

$-3 = 3r + 3$

$-3 - 3 = -3 + 3r + 3$ Add -3.

$-6 = 3r$

Step 3 $\dfrac{-6}{3} = \dfrac{3r}{3}$ Use the multiplication property of equality. Divide by 3.

$-2 = r$ or $r = -2$

Step 4   Substitute -2 for r in the original equation.

$3r + 4 - 2r - 7 = 4r + 3$

$3(-2) + 4 - 2(-2) - 7 = 4(-2) + 3$?   Let $r = -2$.

$-6 + 4 + 4 - 7 = -8 + 3$?   Multiply.

$-5 = -5$ True

The solution set of the equation is $\{-2\}$.

In Step 2 of Example 1, the terms were added and subtracted so that the variable term ended up on the right. Choosing differently would lead to the variable term being on the left side of the equation. Usually there is no advantage either way.

EXAMPLE 2 Using the Four Steps to Solve an Equation

Solve $4(k - 3) - k = k - 6$.

Step 1 Before combining like terms, use the distributive property to simplify $4(k - 3)$.

$$4(k - 3) - k = k - 6$$
$$4 \cdot k - 4 \cdot 3 - k = k - 6 \qquad \text{Distributive property}$$
$$4k - 12 - k = k - 6$$
$$3k - 12 = k - 6 \qquad \text{Combine like terms.}$$

Step 2
$$-k + 3k - 12 = -k + k - 6 \qquad \text{Add } -k.$$
$$2k - 12 = -6$$
$$2k - 12 + 12 = -6 + 12 \qquad \text{Add 12.}$$
$$2k = 6$$

Step 3
$$\frac{2k}{2} = \frac{6}{2} \qquad \text{Divide by 2.}$$
$$k = 3$$

Step 4 Check your answer by substituting 3 for k in the original equation. Remember to do the work inside the parentheses first.

$$4(k - 3) - k = k - 6$$
$$4(3 - 3) - 3 = 3 - 6 \qquad ? \qquad \text{Let } k = 3.$$
$$4(0) - 3 = 3 - 6 \qquad ?$$
$$0 - 3 = 3 - 6 \qquad ?$$
$$-3 = -3 \qquad \text{True}$$

The solution set of the equation is $\{3\}$.

EXAMPLE 3 Using the Four Steps to Solve an Equation

Solve $8a - (3 + 2a) = 3a + 1$.

Step 1 Simplify.

$$8a - (3 + 2a) = 3a + 1$$
$$8a - 3 - 2a = 3a + 1 \qquad \text{Distributive property}$$
$$6a - 3 = 3a + 1 \qquad \text{Combine terms.}$$

Step 2
$$-3a + 6a - 3 = -3a + 3a + 1 \qquad \text{Add } -3a.$$
$$3a \quad 3 = 1$$
$$3a - 3 + 3 = 1 + 3 \qquad \text{Add 3.}$$
$$3a = 4$$

Step 3
$$\frac{3a}{3} = \frac{4}{3} \qquad \text{Divide by 3.}$$
$$a = \frac{4}{3}$$

Step 4 Check the solution in the original equation.

$$8a - (3 + 2a) = 3a + 1$$

$$8\left(\frac{4}{3}\right) - \left[3 + 2\left(\frac{4}{3}\right)\right] = 3\left(\frac{4}{3}\right) + 1 \qquad ? \qquad \text{Let } a = \tfrac{4}{3}.$$

$$\frac{32}{3} - \left[3 + \frac{8}{3}\right] = 4 + 1 \qquad ?$$

$$\frac{32}{3} - \left[\frac{9}{3} + \frac{8}{3}\right] = 5 \qquad ?$$

$$\frac{32}{3} - \frac{17}{3} = 5 \qquad ?$$

$$5 = 5 \qquad \qquad \text{True}$$

The check shows that $\left\{\frac{4}{3}\right\}$ is the solution set.

Be very careful with signs when solving equations like the one in Example 3. When a subtraction sign appears immediately in front of a quantity in parentheses, such as in the expression

$$8 - (3 + 2a),$$

 remember that the $-$ sign acts like a factor of -1 and affects the sign of *every* term within the parentheses. Thus,

$$8 - (3 + 2a) = 8 + (-1)(3 + 2a) = 8 - 3 - 2a.$$
$$\qquad\qquad\qquad\qquad\qquad\qquad\quad \uparrow \quad \uparrow$$

Change to $-$ in *both* terms.

E X A M P L E 4 **Using the Four Steps to Solve an Equation**

Solve $4(8 - 3t) = 32 - 8(t + 2)$.

Step 1 Use the distributive property.

$$4(8 - 3t) = 32 - 8(t + 2) \qquad \text{Given equation}$$
$$32 - 12t = 32 - 8t - 16 \qquad \text{Distributive property}$$
$$32 - 12t = 16 - 8t$$

Step 2
$$32 - 12t + \mathbf{12t} = 16 - 8t + \mathbf{12t} \qquad \text{Add } 12t.$$
$$32 = 16 + 4t$$
$$32 - \mathbf{16} = 16 + 4t - \mathbf{16} \qquad \text{Subtract } 16.$$
$$16 = 4t$$

Step 3
$$\frac{16}{4} = \frac{4t}{4} \qquad \text{Divide by } 4.$$
$$4 = t \quad \text{or} \quad t = 4$$

Step 4   Check the solution.

$$4(8 - 3t) = 32 - 8(t + 2)$$
$$4(8 - 3 \cdot 4) = 32 - 8(4 + 2) \qquad ? \qquad \text{Let } t = 4.$$
$$4(8 - 12) = 32 - 8(6) \qquad ?$$
$$4(-4) = 32 - 48 \qquad ?$$
$$-16 = -16 \qquad \text{True}$$

Since 4 satisfies the equation, the solution set is {4}.

OBJECTIVE **2** Solve equations with fractions or decimals as coefficients. We can clear an equation of fractions by multiplying both sides by the least common denominator (LCD) of all denominators in the equation. It is a good idea to do this immediately after using the distributive property to clear parentheses. Most students make fewer errors working with integer coefficients.

EXAMPLE 5   Solving an Equation with Fractions as Coefficients

Solve $\frac{2}{3}x - \frac{1}{2}x = -\frac{1}{6}x - 2$.

The LCD of all the fractions in the equation is 6. Start by multiplying both sides of the equation by 6.

$$\frac{2}{3}x - \frac{1}{2}x = -\frac{1}{6}x - 2$$
$$6\left(\frac{2}{3}x - \frac{1}{2}x\right) = 6\left(-\frac{1}{6}x - 2\right) \qquad \text{Multiply by 6.}$$
$$6\left(\frac{2}{3}x\right) + 6\left(-\frac{1}{2}x\right) = 6\left(-\frac{1}{6}x\right) + 6(-2) \qquad \text{Distributive property}$$
$$4x - 3x = -x - 12$$

Now use the four steps to solve this simpler equivalent equation.

Step 1 $\qquad\qquad x = -x - 12 \qquad$ Combine like terms.

Step 2 $\qquad\qquad x + x = x - x - 12 \qquad$ Add x.
$$2x = -12$$

Step 3 $\qquad\qquad \frac{2x}{2} = \frac{-12}{2} \qquad$ Divide by 2.
$$x = -6$$

Step 4   Check the answer.

$$\frac{2}{3}(-6) - \frac{1}{2}(-6) = -\frac{1}{6}(-6) - 2 \qquad ? \qquad \text{Let } x = -6.$$
$$-4 + 3 = 1 - 2 \qquad ?$$
$$-1 = -1 \qquad \text{True}$$

The solution set of the equation is {−6}.

When clearing equations of fractions or decimals be sure to multiply *every* term on both sides of the equation by the least common denominator.

The multiplication property can also be used to clear an equation of decimals.

┌ **E X A M P L E   6**   Solving an Equation with Decimal Coefficients

Solve $.20t + .10(20 - t) = .18(20)$.

Step 1   Begin by clearing parentheses.

$$.20t + .10(20 - t) = .18(20)$$
$$.20t + .10(20) - .10t = .18(20) \qquad \text{Distributive property}$$

To clear decimals, multiply both sides of the equation by 100, since the decimals are all hundredths. A number can be multiplied by 100 by moving the decimal point two places to the right.

$$.20t + .10(20) - .10t = .18(20) \qquad \text{Multiply by 100.}$$
$$20t + 10(20) - 10t = 18(20)$$
$$20t + 200 - 10t = 360$$
$$10t + 200 = 360 \qquad \text{Combine like terms.}$$

Step 2
$$10t + 200 - \mathbf{200} = 360 - \mathbf{200} \qquad \text{Subtract 200.}$$
$$10t = 160$$

Step 3
$$\frac{10t}{\mathbf{10}} - \frac{160}{\mathbf{10}} \qquad \text{Divide by 10.}$$
$$t = 16$$

Step 4   Check that {16} is the solution set by substituting into the original equation.

 To avoid errors when multiplying a term like $.10(20 - t)$ by 100, use the distributive property to clear parentheses first, then multiply each term by 100 to eliminate decimals.

O B J E C T I V E   3   Recognize equations with no solutions or infinitely many solutions.   Each equation that we have solved so far has had exactly one solution. As the next examples show, linear equations may also have no solutions or infinitely many solutions. (The four steps are not identified in these examples. See if you can identify them.)

┌ **E X A M P L E   7**   Solving an Equation That Has Infinitely Many Solutions

Solve $5x - 15 = 5(x - 3)$.

$$5x - 15 = 5(x - 3)$$
$$5x - 15 = 5x - 15 \qquad \text{Distributive property}$$
$$-5x + 5x - 15 = -5x + 5x - 15 \qquad \text{Add } -5x.$$
$$-15 = -15$$
$$\mathbf{15} - 15 = \mathbf{15} - 15 \qquad \text{Add 15.}$$
$$0 = 0$$

The variable has "disappeared." Since the last statement (0 = 0) is *true, any* real number is a solution. (We could have predicted this from the line in the solution that says $5x - 15 = 5x - 15$, which is certainly true for *any* value of x.) Indicate the solution set as {all real numbers}.

 When you are solving an equation like the one in Example 7, do not write {0} as the solution set. While 0 is a solution, there are infinitely many other solutions.

E X A M P L E   8   Solving an Equation That Has No Solution

Solve $2x + 3(x + 1) = 5x + 4$.

$$2x + 3(x + 1) = 5x + 4$$
$$2x + 3x + 3 = 5x + 4 \qquad \text{Distributive property}$$
$$5x + 3 = 5x + 4 \qquad \text{Combine terms.}$$
$$5x + 3 - 5x = 5x + 4 - 5x \qquad \text{Subtract } 5x \text{ from each side.}$$
$$3 = 4 \qquad \text{Combine terms.}$$

Again, the variable has disappeared, but this time a *false* statement (3 = 4) results. When this happens, the equation has no solution. Its solution set is the **empty set,** or **null set,** symbolized $\emptyset$.

The chart below summarizes the solution sets of the three types of linear equations.

Final Equation	Number of Solutions	Solution Set
x = a number	One	{a number}
A true statement with no variable, such as 0 = 0	Infinite	{all real numbers}
A false statement with no variable, such as 3 = 4	None	$\emptyset$

OBJECTIVE 4   Write expressions for two related unknown quantities. We continue our work with translating from words to symbols.

PROBLEM SOLVING

Often we are given a problem in which the sum of two quantities is a particular number, and we are asked to find the values of the two quantities. Example 9 shows how to express the unknown quantities in terms of a single variable.

E X A M P L E   9   Translating Phrases into an Algebraic Expression

Two numbers have a sum of 23. If one of the numbers is represented by k, find an expression for the other number.

First, suppose that the sum of two numbers is 23, and one of the numbers is **10**. How would you find the other number? You would subtract **10** from 23 to get 13: $23 - 10 = 13$. So instead of using **10** as one of the numbers, use k as stated in the problem. The other number would be obtained in the same way. You must subtract k from 23. Therefore, an expression for the other number is $23 - k$.

The approach used in Example 9, first writing an expression with a trial number, can also be useful when translating applied problems to equations.

2.2 EXERCISES

1. In your own words, give the four steps used to solve a linear equation. Use an example to demonstrate the steps.

Solve each equation and check your solution. See Examples 1–4, 7, and 8.

2. $5k + 4 = 14$

3. $2m + 8 = 16$

4. $12h - 5 = 10h + 5$

5. $-4x - 1 = -2x + 1$

6. $-2p + 4 = -(5p + 1)$

7. $4x + 6 = -(x - 2)$

8. $3(4x + 2) = -6(x - 5)$

9. $5(2m + 3) = 3(4m + 9)$

10. $6(3w + 5) = 2(10w + 10)$

11. $6(4x - 1) = 12(2x + 3)$

12. $6(2x + 8) = 4(3x - 6)$

13. $3(2x - 4) = 6(x - 2)$

14. $3(6 - 4x) = 2(-6x + 9)$

15. $7r - 5r + 2 = 5r - r$

16. $9p - 4p + 6 = 7p - 3p$

17. $11x - 5(x + 3) - 6x = 0$

18. $6x - 4(x + 2) - 2x = 0$

19. After working correctly through several steps of a linear equation, a student obtains the equation $7x = 3x$. Then the student divides both sides by x to get $7 = 3$, and gives $\emptyset$ as the answer. Is this correct? If not, explain why.

20. Which one of the following linear equations does not have all real numbers as its solution?
    (a) $5x = 4x + x$ (b) $2(x + 6) = 2x + 12$
    (c) $\frac{1}{2}x = .5x$ (d) $3x = 2x$

21. Based on the discussion in this section, if an equation has decimals or common fractions as coefficients, what additional step will make the work easier?

22. Explain why an equation involving $\frac{1}{x - 2}$ cannot have 2 as a solution.

Solve each equation by clearing fractions or decimals. See Examples 5 and 6.

23. $\frac{3}{5}t - \frac{1}{10}t = t - \frac{5}{2}$

24. $-\frac{2}{7}r + 2r = \frac{1}{2}r + \frac{17}{2}$

25. $-\frac{1}{4}(x - 12) + \frac{1}{2}(x + 2) = x + 4$

26. $\frac{1}{9}(y + 18) + \frac{1}{3}(2y + 3) = y + 3$

27. $\frac{2}{3}k - \left(k + \frac{1}{4}\right) = \frac{1}{12}(k + 4)$

28. $-\frac{5}{6}q - \left(q - \frac{1}{2}\right) = \frac{1}{4}(q + 1)$

29. $.20(60) + .05x = .10(60 + x)$

30. $.30(30) + .15x = .20(30 + x)$

31. $1.00x + .05(12 - x) = .10(63)$

32. $.92x + .98(12 - x) = .96(12)$

33. $.06(10,000) + .08x = .072(10,000 + x)$

34. $.02(5000) + .03x = .025(5000 + x)$

RELATING CONCEPTS (EXERCISES 35–40)

Work Exercises 35–40 in order.

35. Consider the term 100*ab*. Evaluate it for $a = 2$ and $b = 4$.

36. Based on your study of Section 1.7, will you get the same answer for Exercise 35 if you evaluate (100*a*)*b* for $a = 2$ and $b = 4$? Why or why not?

37. Is the term (100*a*)(100*b*) equivalent to 100*ab*? Why or why not?

38. If your answer to Exercise 37 is *no,* explain why the distributive property is not involved.

39. Consider the equation

$$.05(x + 2) + .10x = 2.00.$$

To simplify our work, we could begin by multiplying both sides by 100. When applied to the first term on the left, our expression would be

$$100 \cdot .05(x + 2).$$

Is this expression equivalent to $[100 \cdot .05](x + 2)$? (*Hint:* Compare to the expressions in Exercises 35 and 36, letting $a = .05$ and $b = x + 2$.)

40. Students often want to "distribute" the 100 to both .05 and $(x + 2)$ in the expression $100 \cdot .05(x + 2)$. Is this correct? (*Hint:* Compare your answer to those in Exercises 37 and 38.)

Did you make the connection that the distributive property is not used when the product of two quantities is multiplied by a third quantity?

Solve each equation and check the solution. See Examples 1–8.

41. $10(2x - 1) = 8(2x + 1) + 14$

42. $9(3k - 5) = 12(3k - 1) - 51$

43. $-(4y + 2) - (-3y - 5) = 3$

44. $-(6k - 5) - (-5k + 8) = -3$

45. $\frac{1}{2}(x + 2) + \frac{3}{4}(x + 4) = x + 5$

46. $\frac{1}{3}(x + 3) + \frac{1}{6}(x - 6) = x + 3$

47. $.10(x + 80) + .20x = 14$

48. $.30(x + 15) + .40(x + 25) = 25$

49. $4(x + 8) = 2(2x + 6) + 20$

50. $4(x + 3) = 2(2x + 8) - 4$

51. $9(v + 1) - 3v = 2(3v + 1) - 8$

52. $8(t - 3) + 4t = 6(2t + 1) - 10$

Write the answer to each problem as an algebraic expression. See Example 9.

53. Two numbers have a sum of 11. One of the numbers is *q*. Find the other number.

54. The product of two numbers is 9. One of the numbers is *k*. What is the other number?

55. A football player gained *x* yards rushing. On the next down he gained 7 yards. How many yards did he gain altogether?

56. A baseball player got 65 hits one season. He got *h* of the hits in one game. How many hits did he get in the rest of the games?

57. Mary is *a* years old. How old will she be in 12 years? How old was she 5 years ago?

58. Tom has *r* quarters. Find the value of the quarters in cents.

59. A bank teller has *t* dollars, all in five-dollar bills. How many five-dollar bills does the teller have?

60. A plane ticket costs *b* dollars for an adult and *d* dollars for a child. Find the total cost for 3 adults and 2 children.

2.3 An Introduction to Applications of Linear Equations

OBJECTIVES

 1 Learn the six steps for solving applied problems.

2 Solve problems involving unknown numbers.

3 Solve problems involving sums of quantities.

4 Solve problems involving supplementary and complementary angles.

5 Solve problems involving consecutive integers.

FOR EXTRA HELP

📖 **SSG** Sec. 2.3
SSM Sec. 2.3

💿 **Pass the Test Software**

💿 **InterAct Math Tutorial Software**

📼 **Video** 3

CONNECTIONS

The purpose of algebra is to solve real problems. Since such problems are stated in words, not mathematical symbols, the first step in solving them is to translate the problem into one or more mathematical statements. This is the hardest step for most people. George Polya (1888–1985), a native of Budapest, Hungary, wrote the modern classic *How to Solve It*. In this book he proposed a four-step process for problem solving:

1. Understand the problem.

2. Devise a plan.

3. Carry out the plan.

4. Look back and check.

FOR DISCUSSION OR WRITING

Compare Polya's four-step process with the six steps given in this section. Identify which of the steps in our list match Polya's four steps. Trial and error is also a useful problem-solving tool. Where does this method fit into Polya's steps?

OBJECTIVE **1** Learn the six steps for solving applied problems. We now begin to look at how algebra is used to solve applied problems. It must be emphasized that many *meaningful* applications of mathematics require concepts that are beyond the level of this book. Some of the problems you will encounter will seem "contrived," and to some extent they are. But the skills you will develop in solving simple problems will help you in solving more realistic problems in chemistry, physics, biology, business, and other fields.

PROBLEM SOLVING

In earlier sections we learned how to translate words, phrases, and sentences into mathematical expressions and equations. Now we will use these translations to solve applied problems using algebra. While there is no specific method that enables you to solve all kinds of applied problems, the following six-step method is suggested.

Solving an Applied Problem

Step 1 **Decide what you are asked to find.**   Read the problem carefully. Choose a variable to represent the number you are asked to find. *Write down* what the variable represents.

Step 2 **Write down any other pertinent information.**   If there are other unknown quantities, express them using the variable. Draw figures or diagrams and use charts, if they apply.

Step 3 **Write an equation.**   Translate the problem into an equation.

Step 4 **Solve the equation.**   Use the properties to solve the equation.

Step 5 **Answer the question(s) posed.**   Make sure you answer the question in the problem. In some cases, you need more than the solution of the equation.

Step 6 **Check.**   Check your solution using the original words of the problem. Be sure your answer makes sense.

PROBLEM SOLVING

The third step is often the hardest. To translate the problem into an equation, write the given phrases as mathematical expressions. Since equal mathematical expressions are names for the same number, translate any words that mean *equal* or *same* as =. The = sign leads to an equation to be solved.

OBJECTIVE **2** Solve problems involving unknown numbers. Some of the simplest applied problems involve unknown numbers.

EXAMPLE 1 Finding the Value of an Unknown Number

The product of 4, and a number decreased by 7, is 100. Find the number.

Step 1   Read the problem carefully. Decide what you are being told to find, and then choose a variable to represent the unknown quantity. In this problem, we are told to find a number, so we write

$$\text{Let } x = \text{the number.}$$

Step 2   There are no other unknown quantities to find.

Step 3   Translate the problem into an equation.

The product of 4,	and	a number	decreased by	7,	is	100.
↓		↓	↓	↓	↓	↓
4 ·	(	x	−	7)	=	100

Because of the commas in the given problem, writing the equation as $4x - 7 = 100$ is incorrect. The equation $4x - 7 = 100$ corresponds to the statement "The product of 4 and a number, decreased by 7, is 100."

Step 4   Solve the equation.

$$4(x - 7) = 100$$
$$4x - 28 = 100 \qquad \text{Distributive property}$$
$$4x = 128 \qquad \text{Add 28 to both sides.}$$
$$x = 32 \qquad \text{Divide by 4.}$$

Step 5   The number is 32.

Step 6   Check the solution by using the original words of the problem. When 32 is decreased by 7, we get $32 - 7 = 25$. If 4 is multiplied by 25, we get 100, as the problem required. The answer, 32, is correct.

 The commas in the statement of the problem in Example 1 are important in translating correctly.

OBJECTIVE **3** Solve problems involving sums of quantities. A common type of problem in elementary algebra involves finding two quantities when the sum of the quantities is known. In Example 9 of the previous section, we prepared for this type of problem by writing mathematical expressions for two related unknown quantities.

PROBLEM SOLVING

In general, to solve such problems, choose a variable to represent one of the unknowns and then represent the other quantity in terms of the same variable, using information obtained in the problem. Then write an equation based on the words of the problem. The next example illustrates these ideas.

E X A M P L E   2   Finding the Numbers of Olympic Medals Won by the U.S.

In the 1996 Olympics, U.S. contestants won 12 more gold than silver medals. They won a total of 76 gold and silver medals. Find the number of each type of medal won. (*Source:* United States Olympic Committee.)

Step 1   Let x = the number of silver medals.

Step 2   Let $x + 12$ = the number of gold medals.

Step 3   Now write an equation.

The total	is	the number of silver	plus	the number of gold.
↓	↓	↓	↓	↓
76	=	x	+	$(x + 12)$

Step 4   Solve the equation.

$$76 = 2x + 12 \qquad \text{Combine terms.}$$
$$76 - 12 = 2x + 12 - 12 \qquad \text{Subtract 12.}$$
$$64 = 2x \qquad \text{Combine terms.}$$
$$32 = x \qquad \text{Divide by 2.}$$

Step 5   Because x represents the number of silver medals, the U.S. won 32 silver medals. Because $x + 12$ represents the number of gold medals, the U.S. won $32 + 12 = 44$ gold medals.

Step 6   Since there were 44 gold and 32 silver medals, the total number of medals was $44 + 32 = 76$. Because $44 - 32 = 12$, there were 12 more gold medals than silver medals. This information agrees with what is given in the problem, so the answers check.

The problem in Example 2 could also have been solved by letting x represent the number of gold medals. Then $x - 12$ would represent the number of silver medals. The equation would then be

$$76 = x + (x - 12).$$

The solution of this equation is 44, which is the number of gold medals. The number of silver medals would then be $44 - 12 = 32$. The answers are the same, whichever approach is used.

EXAMPLE 3 Finding the Number of Orders for Tea

The owner of P. J.'s Coffeehouse found that on one day the number of orders for tea was $\frac{1}{3}$ the number of orders for coffee. If the total number of orders for the two drinks was 76, how many orders were placed for tea?

Step 1   Let x = the number of orders for coffee.

Step 2   Let $\frac{1}{3}x$ = the number of orders for tea.

Step 3   Use the fact that the total number of orders was 76 to write an equation.

The total	is	orders for coffee	plus	orders for tea.
↓	↓	↓	↓	↓
76	=	x	+	$\frac{1}{3}x$

Step 4   Now solve the equation.

$$76 = \frac{4}{3}x \qquad \text{Combine like terms.}$$

$$\frac{3}{4}(76) = \frac{3}{4}\left(\frac{4}{3}x\right) \qquad \text{Multiply by } \tfrac{3}{4}.$$

$$57 = x$$

Step 5   In this problem, x *does not represent the quantity that we are asked to find.* The number of orders placed for tea was $\frac{1}{3}x$. So $\frac{1}{3}(57) = 19$ is the number of orders for tea.

Step 6   The number of coffee orders (x) was 57 and the number of tea orders was 19; 19 is one-third of 57, and $19 + 57 = 76$. Since this agrees with the information given in the problem, the answer is correct.

PROBLEM SOLVING

In Example 3, it was easier to let the variable represent the quantity that was *not* asked for. This required an extra step in Step 5 to find the number of orders for tea. In some cases, this approach is easier than letting the variable represent the quantity that we are asked to find. Experience in solving problems will indicate when this approach is useful, and experience comes only from solving many problems!

EXAMPLE 4 Analyzing a Gasoline/Oil Mixture

A lawn trimmer uses a mixture of gasoline and oil. For each ounce of oil the mixture contains 16 ounces of gasoline. If the tank holds 68 ounces of the mixture, how many ounces of oil and how many ounces of gasoline does it require when it is full?

Step 1   Let x = the number of ounces of oil required when full.

Step 2   Let $16x$ = the number of ounces of gasoline required when full.

Amount of gasoline		Amount of oil		Total amount in tank
↓		↓		↓

Step 3

$16x$	+	x	=	68

Step 4 $17x = 68$ Combine terms.

                                                    $x = 4$ Divide by 17.

Step 5   The trimmer requires 4 ounces of oil and $16(4) = 64$ ounces of gasoline when full.

Step 6   Since $4 + 64 = 68$, and 64 is 16 times 4, the answers check.

PROBLEM SOLVING

Sometimes it is necessary to find three unknown quantities in an applied problem. Frequently the three unknowns are compared in *pairs*. When this happens, it is usually easiest to let the variable represent the unknown found in both pairs. The next example illustrates this.

EXAMPLE 5   Dividing a Board into Pieces

The instructions for a woodworking project require three pieces of wood. The longest piece must be twice the length of the middle-sized piece, and the shortest piece must be 10 inches shorter than the middle-sized piece. Maria Gonzales has a board 70 inches long that she wishes to use. How long can each piece be?

Steps 1 and 2   Since the middle-sized piece appears in both pairs of comparisons, let x represent the length of the middle-sized piece. We have

$$x = \text{the length of the middle-sized piece}$$
$$2x = \text{the length of the longest piece}$$
$$x - 10 = \text{the length of the shortest piece.}$$

A sketch is helpful here. See Figure 1.

| | $2x$ | x | $x - 10$ |

Figure 1

	Longest	Middle-sized	Shortest	Total length
	↓	↓	↓	↓
Step 3	$2x$	$+$ x	$+$ $(x - 10)$	$=$ 70

Step 4 $4x - 10 = 70$ Combine terms.

                      $4x - 10 + 10 = 70 + 10$ Add 10 to each side.

                                $4x = 80$ Combine terms.

                                 $x = 20$ Divide by 4 on each side.

Step 5   The middle-sized piece is 20 inches long, the longest piece is $2(20) = 40$ inches long, and the shortest piece is $20 - 10 = 10$ inches long.

Step 6   Check to see that the sum of the lengths is 70 inches, and that all conditions of the problem are satisfied.

OBJECTIVE **4** Solve problems involving supplementary and complementary angles. The next example deals with concepts from geometry. An angle can be measured by a unit called the **degree** (°). Two angles whose sum is 90° are said to be **complementary,** or complements of each other. Two angles whose sum is 180° are said to be **supplementary,** or supplements of each other. See Figure 2. If x represents the degree measure of an angle, then

$90 - x$ represents the degree measure of its complement, and

$180 - x$ represents the degree measure of its supplement.

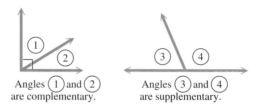

Angles ① and ② are complementary.　　Angles ③ and ④ are supplementary.

Figure 2

E X A M P L E　6　Finding the Measure of an Angle

Find the measure of an angle whose supplement is 10° more than twice its complement.

Step 1　Let　　　　　x = the degree measure of the angle.

Step 2　Then　　$90 - x$ = the degree measure of its complement;
　　　　　　　$180 - x$ = the degree measure of its supplement.

Step 3

Supplement	is	10	more than	twice	its complement.
↓	↓	↓	↓	↓	↓
$180 - x$	=	10	+	2 ·	$(90 - x)$

Step 4　Solve the equation.

$$180 - x = 10 + 180 - 2x \qquad \text{Distributive property}$$
$$180 - x = 190 - 2x \qquad \text{Combine terms.}$$
$$180 - x + 2x = 190 - 2x + 2x \qquad \text{Add } 2x.$$
$$180 + x = 190 \qquad \text{Combine terms.}$$
$$180 + x - 180 = 190 - 180 \qquad \text{Subtract 180.}$$
$$x = 10$$

Step 5　The measure of the angle is 10°.

Step 6　The complement of 10° is 80° and the supplement of 10° is 170°. 170° is equal to 10° more than twice 80° (170 = 10 + 2(80) is true); therefore, the answer is correct.

OBJECTIVE **5** Solve problems involving consecutive integers. Two integers that differ by 1 are called **consecutive integers.** For example, 3 and 4, 6 and 7, and −2 and −1 are pairs of consecutive integers. In general, if x represents an integer, $x + 1$ represents the next larger consecutive integer.

　　Consecutive even integers, such as 8 and 10, differ by 2. Similarly, consecutive odd integers, such as 9 and 11, also differ by two. In general, if x represents an even integer, $x + 2$ represents the next larger consecutive even integer. The same holds true for odd integers; that is, if x is an odd integer, $x + 2$ is the next larger odd integer.

┌ **E X A M P L E 7** Finding Consecutive Integers

Two pages that face each other in this book have 569 as the sum of their page numbers. What are the page numbers?

Because the two pages face each other, they must have page numbers that are consecutive integers.

Step 1 Let $x =$ the smaller page number.

Step 2 Then $x + 1 =$ the larger page number.

Step 3 Because the sum of the page numbers is 569, the equation is

$$x + (x + 1) = 569.$$

Step 4 Solve the equation.

$$
\begin{aligned}
x + (x + 1) &= 569 \\
2x + 1 &= 569 \qquad \text{Combine like terms.} \\
2x &= 568 \qquad \text{Subtract 1.} \\
x &= 284 \qquad \text{Divide by 2.}
\end{aligned}
$$

Step 5 The smaller page number is 284 and the larger page number is $284 + 1 = 285$.

Step 6 The sum of 284 and 285 is 569. Our answer is correct.

In the final example we do not number the steps. See if you can identify them.

┌ **E X A M P L E 8** Finding Consecutive Odd Integers

If the smaller of two consecutive odd integers is doubled, the result is 7 more than the larger of the two integers. Find the two integers.

Let x be the smaller integer. Since the two numbers are consecutive *odd* integers, then $x + 2$ is the larger. Now write an equation.

If the smaller is doubled	the result is	7	more than	the larger.
↓	↓	↓	↓	↓
$2x$	$=$	7	$+$	$x + 2$

Solve the equation.

$$
\begin{aligned}
2x &= 7 + x + 2 \\
2x &= 9 + x \\
2x + (-x) &= 9 + x + (-x) \\
x &= 9
\end{aligned}
$$

The first integer is 9 and the second is $9 + 2 = 11$. To check our answer we see that when 9 is doubled, we get 18, which is 7 more than the larger odd integer, 11. Our answer is correct.

2.3 EXERCISES

1. Which one of the following would not be a reasonable answer in an applied problem that requires finding the number of coins in a jar?

   (a) 7   **(b)** 0   **(c)** $6\frac{2}{3}$ **(d)** 80

2. Which one of the following would not be a reasonable answer in an applied problem that requires finding someone's age (in years)?

   (a) $5\frac{1}{2}$ **(b)** 7   **(c)** 12.25   **(d)** -4

3. Explain in your own words the six-step method for solving applied problems.

4. List some words that will translate as "=" in an applied problem.

Solve each problem. Use the six-step method. See Example 1.

5. If 1 is added to a number and this sum is doubled, the result is 5 more than the number. Find the number.

6. If 2 is subtracted from a number and this difference is tripled, the result is 4 more than the number. Find the number.

7. If 3 is added to twice a number and this sum is multiplied by 4, the result is the same as if the number is multiplied by 7 and 8 is added to the product. What is the number?

8. The sum of three times a number and 12 more than the number is the same as the difference between -6 and twice the number. What is the number?

Solve each problem. Use the six-step method. See Examples 2–5.

9. The U.S. Senate has 100 members. After the 1996 election, during the second session, there were 7 more Republicans than Democrats. One seat was vacant. How many Democrats and Republicans were there in the Senate? (*Source:* U.S. Congress, Joint Committee on Printing, *Congressional Directory,* biennial.)

10. The total number of Democrats and Republicans in the U.S. House of Representatives in 1996 was 433. There were 39 fewer Democrats than Republicans. How many members of each party were there? (*Source:* U.S. Congress, Joint Committee on Printing, *Congressional Directory,* biennial.)

11. There were 2783 more men than women competing in the 1996 Olympic Games in Atlanta. The total number of competitors was 10,341. How many men and how many women competed? (*Source: The Universal Almanac,* 1997, John W. Wright, General Editor.)

12. In the first Super Bowl, played in 1967, Green Bay and Kansas City scored a total of 45 points. Green Bay won by 25 points. What was the score of the first Super Bowl? (*Source: The Universal Almanac,* 1997, John W. Wright, General Editor.)

13. In the 1998 U.S. Senior Open, Hale Irwin finished with 1 stroke more than the winner, Vicente Fernandez. The sum of their scores was 571. Find their scores. (*Source:* Television coverage.)

14. Nagaraj Nanjappa has a strip of paper 39 inches long. He wants to cut it into two pieces so that one piece will be 9 inches shorter than the other. How long should the two pieces be?

15. In one day Akilah Cadet received 13 packages. Federal Express delivered three times as many as Airborne Express, while United Parcel Service delivered 2 less than Airborne Express. How many packages did each service deliver to Akilah?

16. In her job at the post office, Janie Quintana works a 6.5-hour day. She sorts mail, sells stamps, and does supervisory work. One day she sold stamps twice as long as she sorted mail, and she supervised .5 hour longer than she sorted mail. How many hours did she spend at each task?

17. Venus is 31.2 million miles farther from the sun than Mercury, while Earth is 57 million miles farther from the sun than Mercury. If the total of the distances from these three planets to the sun is 196.2 million miles, how far away from the sun is Mercury? (All distances given here are *mean (average)* distances.) (*Source: The Universal Almanac,* 1997, John W. Wright, General Editor.)

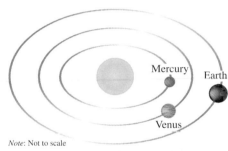

Note: Not to scale

18. Together, Saturn, Jupiter, and Mars have a total of 41 satellites (moons). Jupiter has 5 fewer satellites than Saturn, and Mars has 20 fewer satellites than Saturn. How many satellites does Mars have? (*Source: The Universal Almanac,* 1997, John W. Wright, General Editor.)

19. During the 1996 baseball season, Atlanta Braves pitchers John Smoltz, Greg Maddux, and Mark Wohlers pitched a total of 576 innings. Maddux pitched $8\frac{2}{3}$ fewer innings than Smoltz and $167\frac{2}{3}$ more innings than Wohlers. How many innings did each player pitch? (*Source:* Street and Smith's *Baseball Yearbook,* 1997.)

20. During their World Championship year of 1996, New York Yankees pitchers Dwight Gooden, David Cone, and Jeff Nelson pitched a total of 317 innings. Gooden pitched $96\frac{1}{3}$ innings more than Nelson, and Cone pitched $2\frac{1}{3}$ fewer innings than Nelson. How many innings did each player pitch? (*Source:* Street and Smith's *Baseball Yearbook,* 1997.)

21. The sum of the measures of the angles of any triangle is 180°. In triangle *ABC,* angles *A* and *B* have the same measure, while the measure of angle *C* is 60° larger than each of *A* and *B*. What are the measures of the three angles?

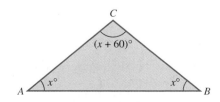

22. (See Exercise 21.) In triangle *ABC,* the measure of angle *A* is 141° more than the measure of angle *B.* The measure of angle *B* is the same as the measure of angle *C.* Find the measure of each angle.

23. A husky running the Iditarod (a thousand-mile race between Anchorage and Nome, Alaska) burns $5\frac{3}{8}$ calories in exertion for every 1 calorie burned in thermoregulation in extreme cold. According to one scientific study, a husky in top condition burns an amazing total of 11,200 calories per day. How many calories are burned for exertion and how many are burned for regulation of body temperature?

24. The 1997 edition of *A Guide Book of United States Coins* lists the value of a Mint State-65 (uncirculated) 1950 Jefferson nickel minted at Denver as $\frac{4}{3}$ the value of a similar condition 1944 nickel minted at Philadelphia. Together the total value of the two coins is $14.00. What is the value of each coin?

25. A pharmacist found that at the end of the day she had $\frac{4}{3}$ as many prescriptions for antibiotics as she did for tranquilizers. She had 42 prescriptions altogether for these two types of drugs. How many did she have for tranquilizers?

26. In a mixture of concrete, there are 3 pounds of cement mix for every 1 pound of gravel. If the mixture contains a total of 140 pounds of these two ingredients, how many pounds of gravel are there?

27. A mixture of nuts contains only peanuts and cashews. For every ounce of cashews there are 5 ounces of peanuts. If the mixture contains a total of 27 ounces, how many ounces of each type of nut does the mixture contain?

28. An insecticide contains 95 centigrams of inert ingredient for every 1 centigram of active ingredient. If a quantity of the insecticide weighs 336 centigrams, how much of each type of ingredient does it contain?

Use the concepts of this section to answer each question.

29. If the sum of two numbers is *k,* and one of the numbers is *m,* how can you express the other number?

30. If the product of two numbers is *r,* and one of the numbers is *s* ($s \neq 0$), how can you express the other number?

31. Is there an angle whose supplement is equal to its complement? If so, what is the measure of the angle?

32. Is there an angle that is equal to its supplement? Is there an angle that is equal to its complement? If the answer is yes to either question, give the measure of the angle.

33. If *x* represents an integer, how can you express the next smaller consecutive integer in terms of *x*?

34. Express three consecutive even integers, all in terms of *x,* if *x* represents the middle of the three.

Solve each problem. See Example 6.

35. Find the measure of an angle whose supplement measures 10 times the measure of its complement.

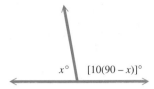

$x°$ $[10(90 - x)]°$

36. Find the measure of an angle whose supplement measures 4 times the measure of its complement.

37. Find the measure of an angle, if its supplement measures 38° less than three times its complement.

38. Find the measure of an angle, if its supplement measures 39° more than twice its complement.

39. Find the measure of an angle such that the sum of the measures of its complement and its supplement is 160°.

40. Find the measure of an angle such that the difference between the measures of its supplement and three times its complement is 10°.

Solve each problem. See Examples 7 and 8.

41. The sum of two consecutive integers is 137. Find the integers.

42. The sum of two consecutive integers is −357. Find the integers.

43. Find two consecutive even integers such that the smaller added to three times the larger gives a sum of 46.

44. Find two consecutive odd integers such that twice the larger is 17 more than the smaller.

45. Two pages that are back-to-back in this book have 203 as the sum of their page numbers. What are the page numbers?

46. If the sum of three consecutive even integers is 60, what is the smallest even integer?

47. When the smaller of two consecutive integers is added to three times the larger, the result is 43. Find the integers.

48. If five times the smaller of two consecutive integers is added to three times the larger, the result is 59. Find the integers.

49. If 6 is subtracted from the largest of three consecutive odd integers, with this result multiplied by 2, the answer is 23 less than the sum of the first and twice the second of the integers. Find the integers.

50. If the first and third of three consecutive even integers are added, the result is 22 less than three times the second integer. Find the integers.

Apply the ideas of this section to solve each problem based on the graph.

51. In 1994, the funding for Head Start programs increased by .55 billion dollars from the funding in 1993. In 1995, the increase was .20 billion dollars over the funding in 1994. For those three years the total funding was 9.64 billion dollars. How much was funded in each of these years? (*Source:* U.S. Department of Health and Human Services.)

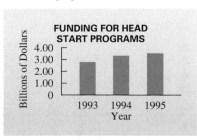

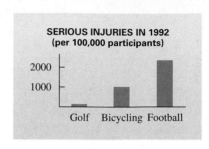

52. According to data provided by the National Safety Council, in 1992 the number of serious injuries per 100,000 participants in football, bicycling, and golf is illustrated in the graph. There were 800 more in bicycling than in golf, and there were 1267 more in football than in bicycling. Altogether there were 3179 serious injuries per 100,000 participants. How many such serious injuries were there in each sport?

SERIOUS INJURIES IN 1992
(per 100,000 participants)

2000

1000

Golf   Bicycling   Football

2.4 Formulas and Applications from Geometry

OBJECTIVES

1 Solve a formula for one variable given the values of the other variables.

2 Use a formula to solve a geometric application.

3 Solve problems about angle measures.

4 Solve a formula for a specified variable.

FOR EXTRA HELP

 SSG Sec. 2.4
SSM Sec. 2.4

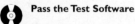 **Pass the Test Software**

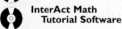 **InterAct Math Tutorial Software**

Video 3

Many applied problems can be solved with formulas. There are formulas for geometric figures such as squares and circles, for distance, for money earned on bank savings, and for converting English measurements to metric measurements, for example. For reference, the formulas used in this book are given on the inside covers.

OBJECTIVE **1** Solve a formula for one variable given the values of the other variables. Given the values of all but one of the variables in a formula, the value of the remaining variable can be found by using the methods introduced in this chapter for solving equations.

EXAMPLE 1 Using a Formula to Evaluate a Variable

Find the value of the remaining variable in each of the following.

(a) $A = LW$; $A = 64, L = 10$

As shown in Figure 3, this formula gives the area of a rectangle with length L and width W. Substitute the given values into the formula and then solve for W.

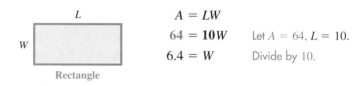

L

W

Rectangle

$A = LW$

Figure 3

$$A = LW$$
$$\mathbf{64 = 10}W \qquad \text{Let } A = 64, L = 10.$$
$$6.4 = W \qquad \text{Divide by 10.}$$

Check that the width of the rectangle is 6.4.

(b) $A = \dfrac{1}{2}(b + B)h$; $A = 210, B = 27, h = 10$

This formula gives the area of a trapezoid with parallel sides of lengths b and B and distance h between the parallel sides. See Figure 4. Again, begin by substituting the given values into the formula.

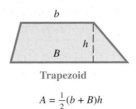

Trapezoid

$$A = \frac{1}{2}(b + B)h$$

Figure 4

$$A = \frac{1}{2}(b + B)h$$

$$210 = \frac{1}{2}(b + \mathbf{27})(\mathbf{10}) \qquad A = 210, B = 27, h = 10$$

Now solve for b.

$$210 = \frac{1}{2}(10)(b + 27) \qquad \text{Commutative property}$$

$$210 = 5(b + 27)$$

$$210 = 5b + 135 \qquad \text{Distributive property}$$

$$75 = 5b \qquad \text{Subtract 135.}$$

$$15 = b \qquad \text{Divide by 5.}$$

Check that the length of the shorter parallel side, *b,* is 15.

OBJECTIVE 2 Use a formula to solve a geometric application. For applications involving geometric figures, a drawing or diagram is useful in Step 2 of the six-step method.

Recall that the *perimeter* of a two-dimensional figure is the distance around the figure, or the sum of the lengths of its sides. Similarly, the **circumference** of a ball (or sphere) is the greatest distance around the ball. (See Figure 5.)

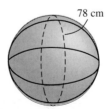 **EXAMPLE 2** Finding the Radius of a Basketball

A professional basketball has a circumference of at most 78 centimeters. See Figure 5. What is the radius of the circular cross section through the center of the ball?

78 cm

Figure 5

Step 1   We want to find the radius of the circular cross section shown, so let $r =$ the radius in centimeters.

Step 2   See Figure 5.

Step 3   The formula for the circumference of a circle is

$$C = 2\pi r.$$

To find the radius r, we substitute 78 for C. We will use the approximate value 3.14 for π.

$$78 = 2(3.14)r \qquad C = 78,\ \pi \approx 3.14*$$

Step 4   Solve the equation.

$$78 = 6.28r \qquad \text{Multiply.}$$
$$r = 12.42 \qquad \text{Divide by 6.28; round.}$$

Step 5   The radius of the indicated cross section is 12.42 centimeters.

Step 6   Check. If $r = 12.42$ and $\pi = 3.14$, the circumference is $2(3.14)(12.42)$ or about 78, as required.

The **area** of a geometric figure is a measure of the surface covered by the figure. Example 3 shows an application of area.

E X A M P L E   3   Finding the Height of a Triangular Sail

The area of a triangular sail of a sailboat is 126 square feet. The base of the sail is 12 feet. Find the height of the sail.

Step 1   Since we must find the height of the triangular sail,

let $h =$ the height of the sail in feet.

Step 2   See Figure 6.

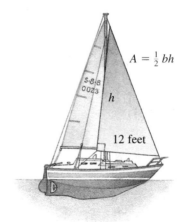

$$A = \tfrac{1}{2}bh$$

h

12 feet

Figure 6

Step 3   The formula for the area of a triangle is $A = \frac{1}{2}bh$, where A is the area, b is the base, and h is the height. Using the information given in the problem, substitute 126 for A and 12 for b in the formula.

$$A = \frac{1}{2}bh$$

$$\mathbf{126} = \frac{1}{2}(\mathbf{12})h \qquad A = 126,\ b = 12$$

---

*The number π is *not exactly* equal to 3.14.

Step 4   Solve the equation.

$$126 = \frac{12}{2}h$$

$$126 = 6h$$

$$21 = h \qquad \text{Divide by 6.}$$

Step 5   The height of the sail is 21 feet.

Step 6   Check to see that the values $A = 126$, $b = 12$, and $h = 21$ satisfy the formula for the area of a triangle.

OBJECTIVE ▣3 Solve problems about angle measures. Angle measures are vital to the study of *geodesy*, the measurement of the earth's surface, and other fields. Methods of calculating angle measures are discussed in the next example.

Refer to Figure 7, which shows two intersecting lines forming angles that are numbered ①, ②, ③, and ④. Angles ① and ③ lie "opposite" each other. They are called **vertical angles.** Another pair of vertical angles are ② and ④. In geometry, it is shown that vertical angles have equal measures.

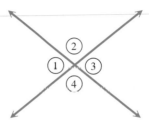

Figure 7

Now look at angles ① and ②. When their measures are added, we get 180°, the measure of a **straight angle,** so the angles are supplements. There are three other such pairs of angles: ② and ③, ③ and ④, ① and ④.

EXAMPLE 4   Finding Angle Measures

Refer to the appropriate figures in each part.

(a) Find the measure of each marked angle in Figure 8.

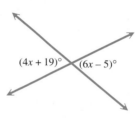

Figure 8

Since the marked angles are vertical angles, they have the same measures. Set $4x + 19$ equal to $6x - 5$ and solve.

$$4x + 19 = 6x - 5$$
$$-4x + 4x + 19 = -4x + 6x - 5 \qquad \text{Add } -4x.$$
$$19 = 2x - 5$$
$$19 + 5 = 2x - 5 + 5 \qquad \text{Add } 5.$$
$$24 = 2x$$
$$12 = x \qquad \text{Divide by 2.}$$

Now we must find the required angle measures. Since $x = 12$, one angle has measure $4(12) + 19 = 67$ degrees. The other has the same measure, since $6(12) - 5 = 67$ as well. Each angle measures $67°$.

(b) Find the measure of each marked angle in Figure 9.

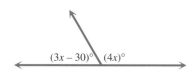

$(3x - 30)°$ $(4x)°$

Figure 9

The measures of the marked angles must add to $180°$ since together they form a straight angle. The equation to solve is

$$(3x - 30) + 4x = 180.$$
$$7x - 30 = 180 \qquad \text{Combine like terms.}$$
$$7x - 30 + 30 = 180 + 30 \qquad \text{Add 30.}$$
$$7x = 210$$
$$x = 30 \qquad \text{Divide by 7.}$$

To find the measures of the angles, substitute 30 for x in the two expressions.

$$3x - 30 = 3(30) - 30 = 90 - 30 = 60$$
$$4x = 4(30) = 120$$

The two angle measures are $60°$ and $120°$, which are supplementary.

 In Example 4, the answer is not the value of x. Remember to substitute the value of the variable into the expression for each angle you are asked to find.

OBJECTIVE **4** Solve a formula for a specified variable. Sometimes it is necessary to solve a number of problems that use the same formula. For example, a surveying class might need to solve several problems that involve the formula for the area of a rectangle, $A = LW$. Suppose that in each problem the area (A) and the length (L) of a rectangle are given and the width (W) must be found. Rather than solving for W each time the formula is used, it would be simpler to *rewrite the formula* so that it is solved for W. This process is called **solving for a specified variable.** Solving a formula for a specified variable requires the same steps used earlier to solve equations with just one variable.

The formula for converting temperatures given in degrees Celsius to degrees Fahrenheit is

$$F = \frac{9}{5}C + 32.$$

The next example shows how to solve this formula for C.

EXAMPLE 5 Solving for a Specified Variable

Solve $F = \frac{9}{5}C + 32$ for C.

To clear the fraction, multiply on both sides by 5.

$$F = \frac{9}{5}C + 32 \qquad \text{Solve for } C.$$

$$5F = 5\left(\frac{9}{5}C + 32\right) \qquad \text{Multiply by 5.}$$

$$5F = 9C + 160$$

Now, to get $9C$ alone on one side, subtract 160 from both sides. Then divide both sides by 9.

$$5F - 160 = 9C + 160 - 160 \qquad \text{Subtract 160.}$$

$$5F - 160 = 9C$$

$$\frac{5F - 160}{9} = C \qquad \text{Divide by 9}$$

This result can be written in a different form using the distributive property to rewrite the numerator as $5(F - 32)$. Then we have

$$C = \frac{5(F - 32)}{9} = \frac{5}{9}(F - 32).$$

 When solving a formula for a specified variable, treat that variable as if it were the *only* variable in the equation, and treat all others as if they were numbers. Use the method of solving equations described in Sections 2.1 and 2.2 to solve for the specified variable.

EXAMPLE 6 Solving for a Specified Variable

Solve $A = \frac{1}{2}(b + B)h$ for B.

This is the formula for the area of a trapezoid. To get B alone, begin by multiplying both sides by 2 to clear the fraction.

$$A = \frac{1}{2}(b + B)h$$

$$2A = 2 \cdot \frac{1}{2}(b + B)h \qquad \text{Multiply by 2. (Do not distribute the 2 on the right side.)}$$

$$2A = (b + B)h \qquad 2 \cdot \frac{1}{2} = \frac{2}{2} = 1$$

$$2A = bh + Bh \qquad \text{Distributive property}$$

Undo what was done to B by first subtracting bh on both sides. Then divide both sides by h.

$$2A - bh = Bh \qquad\qquad \text{Subtract } bh.$$

$$\frac{2A - bh}{h} = B \quad \text{or} \quad B = \frac{2A - bh}{h} \qquad \text{Divide by } h.$$

The result can be written in a different form as follows.

$$B = \frac{2A - bh}{h} = \frac{2A}{h} - \frac{bh}{h} = \frac{2A}{h} - b$$

Either form is correct.

2.4 EXERCISES

1. In your own words, explain what is meant by the *perimeter* of a geometric figure.

2. In your own words, explain what is meant by the *area* of a geometric figure.

Decide whether perimeter or area would be used to solve a problem concerning the measure of the quantity.

3. Carpeting for a bedroom
4. Sod for a lawn
5. Fencing for a yard
6. Baseboards for a living room
7. Tile for a bathroom
8. Fertilizer for a garden
9. Determining the cost for replacing a linoleum floor with a wood floor
10. Determining the cost for planting rye grass in a lawn for the winter

In the following exercises a formula is given along with the values of all but one of the variables in the formula. Find the value of the variable that is not given. See Example 1.

11. $P = 2L + 2W$ (perimeter of a rectangle); $L = 6, W = 4$

12. $P = 2L + 2W$; $L = 8, W = 5$

13. $P = 4s$ (perimeter of a square); $s = 6$

14. $P = 4s$; $s = 12$

15. $A = \dfrac{1}{2}bh$ (area of a triangle); $b = 10, h = 14$

16. $A = \dfrac{1}{2}bh$; $b = 8, h = 16$

17. $d = rt$ (distance formula); $d = 100, t = 2.5$

18. $d = rt$; $d = 252, r = 45$

19. $I = prt$ (simple interest); $p = 5000, r = .025, t = 7$

20. $I = prt$; $p = 7500, r = .035, t = 6$

21. $A = \dfrac{1}{2}h(b + B)$ (area of a trapezoid); $h = 7, b = 12, B = 14$

22. $A = \dfrac{1}{2}h(b + B)$; $h = 3, b = 19, B = 31$

 23. $C = 2\pi r$ (circumference of a circle); $C = 8.164, \pi = 3.14$

 24. $C = 2\pi r$; $C = 16.328, \pi = 3.14$

 25. $A = \pi r^2$ (area of a circle); $r = 12, \pi = 3.14$

26. $A = \pi r^2$; $r = 4, \pi = 3.14$

27. If a formula contains exactly five variables, how many values would you need to be given in order to find the value of any one variable?

28. The formula for changing Celsius to Fahrenheit is given in Example 5 as $F = \dfrac{9}{5}C + 32$. Sometimes it is seen as $F = \dfrac{9C}{5} + 32$. These are both correct. Why is it true that $\dfrac{9}{5}C$ is equal to $\dfrac{9C}{5}$?

*The **volume** of a three-dimensional object is a measure of the space occupied by the object. For example, we would need to know the volume of a gasoline tank in order to know how many gallons of gasoline it would take to completely fill the tank. In each of the following exercises, a formula for the volume, V, of a three-dimensional object is given, along with values for the other variables. Solve for V. See Example 1.*

29. $V = LWH$ (volume of a rectangular-sided box); $L = 12, W = 8, H = 4$

30. $V = LWH$; $L = 10, W = 5, H = 3$

31. $V = \dfrac{1}{3}Bh$ (volume of a pyramid); $B = 36, h = 4$

32. $V = \dfrac{1}{3}Bh$; $B = 12, h = 13$

33. $V = \dfrac{4}{3}\pi r^3$ (volume of a sphere); $r = 6, \pi = 3.14$

34. $V = \dfrac{4}{3}\pi r^3$; $r = 12, \pi = 3.14$

Use a formula to write an equation for each application, and use the six-step problem-solving method of Section 2.3 to solve it. Formulas may be found on the inside covers of this book. See Examples 2 and 3. Use 3.14 for π, if applicable.

35. Recently, a prehistoric ceremonial site dating to about 3000 B.C. was discovered at Stanton Drew in southwestern England. The site, which is larger than Stonehenge, is a near perfect circle, consisting of nine concentric rings that probably held upright wooden posts. Around this timber temple is a wide, encircling ditch enclosing an area with a diameter of 443 feet. Find this enclosed area. (*Source: Archaeology,* vol. 51, no. 1, Jan./Feb. 1998.)

36. The U.S. Postal Service requires that any box sent through the mail have length plus girth (distance around) totaling no more than 108 inches. The maximum volume that meets this condition is contained by a box with a square end 18 inches on each side. What is the length of the box? What is the maximum volume?

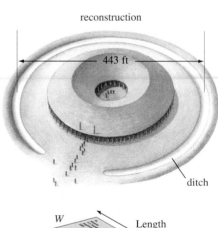

37. The Skydome in Toronto, Canada, is the first stadium with a hard-shell, retractable roof. The steel dome is 630 feet in diameter. To the nearest foot, what is the circumference of this dome?

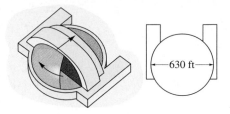

38. The largest drum ever constructed was played at the Royal Festival Hall in London in 1987. It had a diameter of 13 feet. What was the area of the circular face of the drum? (*Source: The Guinness Book of World Records.*)

39. The newspaper, *The Constellation,* printed in 1859 in New York City as part of the Fourth of July celebration, had length 51 inches and width 35 inches. What was the perimeter? What was the area? (*Source: The Guinness Book of World Records.*)

40. The *Daily Banner,* published in Roseberg, Oregon, in the nineteenth century, had page size 3 inches by 3.5 inches. What was the perimeter? What was the area? (*Source: The Guinness Book of World Records.*)

41. A color television set with a liquid crystal display was manufactured by Epson in 1985 and had dimensions 3 inches by $6\frac{3}{4}$ inches by $1\frac{1}{8}$ inches. What was the volume of this set?

(*Source: The Guinness Book of World Records.*)

42. A sea lock at Zeebrugge, Belgium, measures 1640 feet by 187 feet by 75.4 feet. What is the volume of the lock? (It is a rectangular solid, so use the formula for the volume of such a figure.) (*Source: The Guinness Book of World Records.*)

43. The survey plat (a two-dimensional map plotted from a plane survey) in the figure shows two lots that form a trapezoid. The measures of the parallel sides are 115.80 feet and 171.00 feet. The height of the trapezoid is 165.97 feet. Find the combined area of the two lots. Round your answer to the nearest hundredth of a square foot.

44. Lot A in the figure is in the shape of a trapezoid. The parallel sides measure 26.84 feet and 82.05 feet. The height of the trapezoid is 165.97 feet. Find the area of Lot A. Round your answer to the nearest hundredth of a square foot.

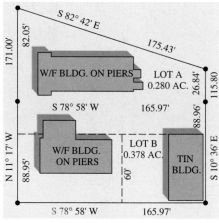

Property survey of lot in New Roads, Louisiana.

Find the measure of each marked angle. See Example 4.

45.

$(10x+7)°$ $(7x+3)°$

46.

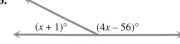
$(x+1)°$ $(4x-56)°$

47.

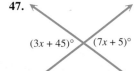
$(3x+45)°$ $(7x+5)°$

48.

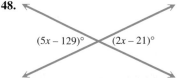
$(5x-129)°$ $(2x-21)°$

49.
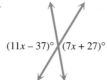

$(11x - 37)°$ $(7x + 27)°$

50.
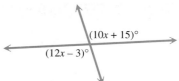

$(10x + 15)°$

$(12x - 3)°$

Solve each formula for the specified variable. See Examples 5 and 6.

51. $d = rt$ for r

52. $I = prt$ for r

53. $I = prt$ for p

54. $V = LWH$ for H

55. $P = a + b + c$ for a

56. $P = a + b + c$ for b

57. $A = \dfrac{1}{2}bh$ for b

58. $A = \dfrac{1}{2}bh$ for h

59. $A = p + prt$ for r

60. $P = 2L + 2W$ for W

61. $V = \pi r^2 h$ for h

62. $V = \dfrac{1}{3}\pi r^2 h$ for h

63. $y = mx + b$ for m

64. $Ax + By = C$ for y

RELATING CONCEPTS (EXERCISES 65–68)

Students in beginning algebra are often faced with the following problem: "I know I did the work correctly, but my answer doesn't look like the one given in the book. What's wrong?" In many cases, there are equally acceptable *equivalent* answers for a problem.

Work Exercises 65–68 in order, *to see how two seemingly different answers to a problem can both be correct.*

65. Solve the formula $P = 2L + 2W$ for W using the following steps.

    (a) Subtract $2L$ from both sides.

    (b) Divide both sides by 2 to get $\dfrac{P - 2L}{2} = W$.

66. Solve the formula $P = 2L + 2W$ for W using the following steps.

    (a) Divide each term on both sides by 2.

    (b) Subtract L from both sides to get $\dfrac{P}{2} - L = W$.

67. Compare the results in Exercises 65(b) and 66(b). They are both acceptable answers, but they are written in different forms. To show they are equivalent, follow the steps below and provide the justification for each step.

    (a) $\dfrac{P}{2} - L = \dfrac{P}{2} - \dfrac{2}{2} \cdot L$

    (b) $\phantom{\dfrac{P}{2} - L} = \dfrac{P}{2} - \dfrac{2}{2} \cdot \dfrac{L}{1}$

    (c) $\phantom{\dfrac{P}{2} - L} = \dfrac{P}{2} - \dfrac{2L}{2}$

    (d) $\phantom{\dfrac{P}{2} - L} = \dfrac{P - 2L}{2}$

68. Suppose that the expression $\dfrac{5T + 4}{4}$ is obtained in a similar problem. What is another valid answer?

Did you make the connection that answers for a problem may be written in different, yet equivalent forms?

2.5 Ratios and Proportions

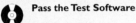

OBJECTIVE 1 Write ratios. Ratios provide a way of comparing two numbers or quantities using division. A **ratio** is a quotient of two quantities with the same units.

The ratio of the number a to the number b is written

$$a \text{ to } b, \qquad a:b, \qquad \text{or} \qquad \frac{a}{b}.$$

This last way of writing a ratio is most common in algebra. Note that the order must be $\frac{a}{b}$, not $\frac{b}{a}$.

Percents are ratios where the second number is always 100. For example, 50% represents the ratio of 50 to 100, 27% represents the ratio of 27 to 100, and so on.

When ratios are used in comparing units of measure, the units should be the same. This is shown in Example 1.

EXAMPLE 1 Writing a Ratio

Write a ratio for each word phrase.

(a) The ratio of 5 hours to 3 hours
This ratio can be written as $\frac{5}{3}$.

(b) The ratio of 5 hours to 3 days
First convert 3 days to hours: 3 days $= 3 \cdot 24 = 72$ hours. The ratio of 5 hours to 3 days is thus

$$\frac{5 \text{ hours}}{72 \text{ hours}} \quad \text{or} \quad \frac{5}{72}.$$

 Because the units of the two quantities in a ratio must be the same, ratios do not have units in their final forms.

OBJECTIVE 2 Decide whether proportions are true. A ratio is used to compare two numbers or amounts. A **proportion** is a statement that two ratios are equal. For example,

$$\frac{3}{4} = \frac{15}{20}$$

is a proportion that says that the ratios $\frac{3}{4}$ and $\frac{15}{20}$ are equal. In the proportion

$$\frac{a}{b} = \frac{c}{d},$$

a, b, c, and d are the **terms** of the proportion. Beginning with the proportion

$$\frac{a}{b} = \frac{c}{d}$$

and multiplying both sides by the common denominator, bd, gives

$$bd \cdot \frac{a}{b} = bd \cdot \frac{c}{d}$$

$$ad = bc.$$

The same products *ad* and *bc* can be found by multiplying diagonally.

$$\frac{a}{b} = \frac{c}{d}$$

bc

ad

This is called **cross multiplication,** and *ad* and *bc* are called **cross products.**

Cross Products

If $\dfrac{a}{b} = \dfrac{c}{d}$, then the cross products *ad* and *bc* are equal.

Also, if $ad = bc$, then $\dfrac{a}{b} = \dfrac{c}{d}$ $(b, d \neq 0)$.

From the rule given above,

$$\text{if} \quad \frac{a}{b} = \frac{c}{d} \quad \text{then} \quad ad = bc.$$

However, if $\frac{a}{c} = \frac{b}{d}$, then $ad = cb$, or $ad = bc$. This means that the two proportions are equivalent and

$$\text{the proportion } \frac{a}{b} = \frac{c}{d} \text{ can also be written as } \frac{a}{c} = \frac{b}{d}.$$

Sometimes one form is more convenient to work with than the other.

EXAMPLE 2 Deciding Whether a Proportion Is True

Decide whether the following proportions are true or false.

(a) $\dfrac{3}{4} = \dfrac{15}{20}$

Check to see whether the cross products are equal.

$$\frac{3}{4} = \frac{15}{20}$$

$4 \cdot 15 = 60$

$3 \cdot 20 = 60$

The cross products are equal, so the proportion is true.

(b) $\dfrac{6}{7} = \dfrac{30}{32}$

The cross products are $6 \cdot 32 = 192$ and $7 \cdot 30 = 210$. The cross products are different, so the proportion is false.

The cross product method cannot be used directly if there is more than one term on either side of the equals sign. For example, you cannot use the method directly to solve the equation

$$\frac{4}{x} + 3 = \frac{1}{9},$$

because there are two terms on the left side.

OBJECTIVE **3** Solve proportions. Four numbers are used in a proportion. If any three of these numbers are known, the fourth can be found.

EXAMPLE 3 Solving an Equation Using Cross Products

(a) Find x in the proportion

$$\frac{63}{x} = \frac{9}{5}.$$

The cross products must be equal, so

$$63 \cdot 5 = 9x$$
$$315 = 9x.$$

Divide both sides by 9 to get

$$35 = x.$$

The solution set is $\{35\}$.

(b) Solve the equation

$$\frac{m - 2}{5} = \frac{m + 1}{3}.$$

Find the cross products, and set them equal to each other.

$$3(m - 2) = 5(m + 1) \qquad \text{Be sure to use parentheses.}$$
$$3m - 6 = 5m + 5 \qquad \text{Distributive property}$$
$$3m = 5m + 11 \qquad \text{Add 6.}$$
$$-2m = 11 \qquad \text{Subtract } 5m.$$
$$m = -\frac{11}{2} \qquad \text{Divide by } -2.$$

The solution set is $\left\{ -\frac{11}{2} \right\}$.

OBJECTIVE **4** Solve applied problems using proportions. Proportions occur in many practical applications.

EXAMPLE 4 Applying Proportions

A local store is offering 3 packs of toothpicks for $.87. How much would it charge for 10 packs?

Let x = the cost of 10 packs of toothpicks.

Set up a proportion. One ratio in the proportion can involve the number of packs, and the other can involve the costs. Make sure that the corresponding numbers appear in the numerator and the denominator.

$$\frac{\text{Cost of 3}}{\text{Cost of 10}} = \frac{3}{10}$$

$$\frac{.87}{x} = \frac{3}{10}$$

$$3x = .87(10) \qquad \text{Cross products}$$

$$3x = 8.7$$

$$x = 2.90 \qquad \text{Divide by 3.}$$

The 10 packs should cost $2.90. As shown earlier, the proportion could also be written as $\frac{3}{.87} = \frac{10}{x}$, which would give the same cross products.

 Many people would solve the problem in Example 4 mentally as follows: Three packs cost $.87, so one pack costs $.87/3 = $.29. Then ten packs will cost $10($.29) = $2.90. If you do the problem this way, you are using proportions and probably not even realizing it!

An important application that uses proportions is *unit pricing*—deciding which size of an item offered in different sizes produces the best price per unit. For example, suppose you can buy 36 ounces of pancake syrup for $3.89. To find the price per unit, set up the proportion

$$\frac{36 \text{ ounces}}{1 \text{ ounce}} = \frac{3.89}{x}$$

and solve for *x*.

$$36x = 3.89 \qquad \text{Cross products}$$

$$x = \frac{3.89}{36} \qquad \text{Divide by 36.}$$

$$x \approx .108 \qquad \text{Use a calculator.}$$

Thus, the price for 1 ounce is $.108, or about 11 cents. Notice that the unit price is the ratio of the cost for 36 ounces, $3.89, to the number of ounces, 36, which means that the unit price for an item is found by dividing the cost by the number of units.

EXAMPLE 5 Determining Unit Price to Obtain the Best Buy

Besides the 36-ounce size discussed above, the local supermarket carries two other sizes of a popular brand of pancake syrup, priced as follows.

Size	Price
36-ounce	$3.89
24-ounce	$2.79
12-ounce	$1.89

Which size is the best buy? That is, which size has the lowest unit price?

To find the best buy, divide the price by the number of units to get the price per ounce. Each result in the following table was found by using a calculator and rounding the answer to three decimal places.

Size	Unit Cost (dollars per ounce)	
36-ounce	$\frac{\$3.89}{36} = \$.108$	← The best buy
24-ounce	$\frac{\$2.79}{24} = \$.116$	
12-ounce	$\frac{\$1.89}{12} = \$.158$	

Since the 36-ounce size produces the lowest price per unit, it would be the best buy. (*Be careful:* Sometimes the largest container *does not* produce the lowest price per unit.)

OBJECTIVE 5 Solve direct variation problems. Suppose that gasoline costs $1.50 per gallon. Then 1 gallon costs $1.50, 2 gallons cost 2($1.50) = $3.00, 3 gallons cost 3($1.50) = $4.50, and so on. Using a proportion,

$$\frac{\text{Cost of 1 gallon}}{\text{Total cost}} = \frac{1 \text{ gallon}}{x \text{ gallons}}$$

$$x \cdot \text{Cost of 1 gallon} = 1 \cdot \text{Total cost}.$$

Thus, the total cost is obtained by multiplying the number of gallons by the price per gallon. In general, if k equals the price per gallon and x equals the number of gallons, then the total cost y is equal to kx. Notice that as number of gallons increases, total cost increases.

The preceding discussion is an example of variation. As in the gasoline example, two variables *vary directly* if one is a multiple of the other.

Direct Variation

y **varies directly** as x if there exists a number k such that

$$y = kx.$$

Another way to say this is that y **is directly proportional to** x.

EXAMPLE 6 Using Direct Variation to Find the Cost of Gasoline

If 6 gallons of gasoline cost $8.52, find the cost of 20 gallons.

Here, the total cost of the gasoline, y, varies directly as (or is directly proportional to) the number of gallons purchased, x. This means there is a number k such that $y = kx$. To find y when $x = 20$, use the fact that 6 gallons cost $8.52.

$$y = kx$$
$$8.52 = k(6) \qquad \text{Let } x = 6, y = 8.52.$$
$$1.42 = k \qquad \text{Divide by 6.}$$

Since $k = 1.42$,

$$y = 1.42x.$$

When $x = 20$,

$$y = 1.42(20) = 28.4,$$

so the cost to purchase 20 gallons of gasoline is $28.40.

2.5 EXERCISES

Determine the ratio and write it in lowest terms. See Example 1.

1. 25 feet to 40 feet

2. 16 miles to 48 miles

3. 18 dollars to 72 dollars

4. 300 people to 250 people

5. 144 inches to 6 feet

6. 60 inches to 2 yards

7. 5 days to 40 hours

8. 75 minutes to 2 hours

9. Which one of the following ratios is not the same as the ratio 2 to 5?
(a) .4   **(b)** 4 to 10   **(c)** 20 to 50   **(d)** 5 to 2

10. Give three ratios that are equivalent to the ratio 4 to 3.

11. Explain the distinction between *ratio* and *proportion*. Give examples.

12. Suppose that someone told you to use cross products in order to multiply fractions. How would you explain to the person what is wrong with his or her thinking?

Decide whether each proportion is true or false. See Example 2.

13. $\dfrac{5}{35} = \dfrac{8}{56}$

14. $\dfrac{4}{12} = \dfrac{7}{21}$

15. $\dfrac{120}{82} = \dfrac{7}{10}$

16. $\dfrac{27}{160} = \dfrac{18}{110}$

17. $\dfrac{\frac{1}{2}}{5} = \dfrac{1}{10}$

18. $\dfrac{\frac{1}{3}}{6} = \dfrac{1}{18}$

Solve each equation. See Example 3.

19. $\dfrac{k}{4} = \dfrac{175}{20}$

20. $\dfrac{49}{56} = \dfrac{z}{8}$

21. $\dfrac{x}{6} = \dfrac{18}{4}$

22. $\dfrac{z}{80} = \dfrac{20}{100}$

23. $\dfrac{3y - 2}{5} = \dfrac{6y - 5}{11}$

24. $\dfrac{2p + 7}{3} = \dfrac{p - 1}{4}$

Solve each problem by setting up and solving a proportion. See Example 4.

25. A chain saw requires a mixture of 2-cycle engine oil and gasoline. According to the directions on a bottle of Oregon 2-cycle Engine Oil, for a 50 to 1 ratio requirement, approximately 2.5 fluid ounces of oil are required for 1 gallon of gasoline. For 2.75 gallons, how many fluid ounces of oil are required?

26. The directions on the bottle mentioned in Exercise 25 indicate that if the ratio requirement is 24 to 1, approximately 5.5 ounces of oil are required for 1 gallon of gasoline. If gasoline is to be mixed with 22 ounces of oil, how much gasoline is to be used?

 27. In 1998, the average exchange rate between U.S. dollars and United Kingdom pounds was 1 pound to $1.6762. Margaret went to London and exchanged her U.S. currency for U.K. pounds, and received 400 pounds. How much in U.S. money did Margaret exchange?

28. If 3 U.S. dollars can be exchanged for 4.5204 Swiss francs, how many Swiss francs can be obtained for $49.20? (Round to the nearest hundredth.)

29. If 6 gallons of premium unleaded gasoline cost $3.72, how much would it cost to completely fill a 15-gallon tank?

30. If sales tax on a $16.00 compact disc is $1.32, how much would the sales tax be on a $120.00 compact disc player?

31. The distance between Kansas City, Missouri, and Denver is 600 miles. On a certain wall map, this is represented by a length of 2.4 feet. On the map, how many feet would there be between Memphis and Philadelphia, two cities that are actually 1000 miles apart?

32. The distance between Singapore and Tokyo is 3300 miles. On a certain wall map, this distance is represented by 11 inches. The actual distance between Mexico City and Cairo is 7700 miles. How far apart are they on the same map?

33. Biologists tagged 250 fish in Willow Lake on October 5. On a later date they found 7 tagged fish in a sample of 350. Estimate the total number of fish in Willow Lake to the nearest hundred.

34. On May 13 researchers at Argyle Lake tagged 420 fish. When they returned a few weeks later, their sample of 500 fish contained 9 that were tagged. Give an approximation of the fish population in Argyle Lake to the nearest hundred.

 The Olympic Committee has come to rely more and more on television rights and major corporate sponsors to finance the games. The pie charts show the funding plans for the first Olympics in Athens and the 1996 Olympics in Atlanta. Use proportions and the figures to answer the questions in Exercises 35 and 36.

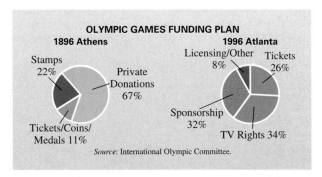

35. In the 1996 Olympics, total revenue of $350 million was raised. There were 10 major sponsors.
  (a) Write a proportion to find the amount of revenue provided by tickets. Solve it.
  (b) What amount was provided by sponsors? Assuming the sponsors contributed equally, how much was provided per sponsor?
  (c) What amount was raised by TV rights?

36. Suppose the amount of revenue raised in the 1896 Olympics was equivalent to the $350 million in 1996.
  (a) Write a proportion for the amount of revenue provided by stamps and solve it.
  (b) What amount (in dollars) would have been provided by private donations?
  (c) In the 1988 Olympics, there were 9 major sponsors, and the total revenue was $95 million. What is the ratio of major sponsors in 1988 to those in 1996? What is the ratio of revenue in 1988 to revenue in 1996?

A supermarket was surveyed to find the prices charged for items in various sizes. Find the best buy (based on price per unit) for each particular item. See Example 5.

37. Trash bags
  20 count: $3.09
  30 count: $4.59

38. Black pepper
  1-ounce size: $ .99
  2-ounce size: $1.65
  4-ounce size: $4.39

39. Breakfast cereal
   15-ounce size: $2.99
   25-ounce size: $4.49
   31-ounce size: $5.49

40. Cocoa mix
   8-ounce size: $1.39
   16-ounce size: $2.19
   32-ounce size: $2.99

41. Tomato ketchup
   14-ounce size: $ .89
   32-ounce size: $1.19
   64-ounce size: $2.95

42. Cut green beans
   8-ounce size: $ .45
   16-ounce size: $ .49
   50-ounce size: $1.59

Two triangles are **similar** if they have the same shape (but not necessarily the same size). Similar triangles have sides that are proportional. The figure shows two similar triangles. Notice that the ratios of the corresponding sides are all equal to $\frac{3}{2}$:

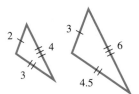

$$\frac{3}{2} = \frac{3}{2} \qquad \frac{4.5}{3} = \frac{3}{2} \qquad \frac{6}{4} = \frac{3}{2}.$$

If we know that two triangles are similar, we can set up a proportion to solve for the length of an unknown side.

Use a proportion to find the length x, given that the pair of triangles are similar.

43.

44.

45.

46.

*For the problems in Exercises 47 and 48, (**a**) draw a sketch consisting of two right triangles, depicting the situation described, and (**b**) solve the problem. (Source: The Guinness Book of World Records.)*

47. An enlarged version of the chair used by George Washington at the Constitutional Convention casts a shadow 18 feet long at the same time a vertical pole 12 feet high casts a shadow 4 feet long. How tall is the chair?

48. One of the tallest candles ever constructed was exhibited at the 1897 Stockholm Exhibition. If it cast a shadow 5 feet long at the same time a vertical pole 32 feet high cast a shadow 2 feet long, how tall was the candle?

The Consumer Price Index, issued by the U.S. Bureau of Labor Statistics, provides a means of determining the purchasing power of the U.S. dollar from one year to the next. Using the period from 1982 to 1984 as a measure of 100.0, the Consumer Price Index from 1990 to 1995 is shown here.

Year	Consumer Price Index
1990	130.7
1991	136.2
1992	140.3
1993	144.5
1994	148.2
1995	152.4

Source: Bureau of Labor Statistics.

To use the Consumer Price Index to predict a price in a particular year, we can set up a proportion and compare it with a known price in another year, as follows:

$$\frac{\text{Price in year } A}{\text{Index in year } A} = \frac{\text{Price in year } B}{\text{Index in year } B}.$$

Use the Consumer Price Index figures above to find the amount that would be charged for the use of the same amount of electricity that cost $225 in 1990. Give your answer to the nearest dollar.

49. in 1992   **50.** in 1993   **51.** in 1994   **52.** in 1995

53. The Consumer Price Index figures for shelter for the years 1981 and 1991 are 90.5 and 146.3. If shelter for a particular family cost $3000 in 1981, what would be the comparable cost in 1991? Give your answer to the nearest dollar.

54. Due to a volatile fuel oil market in the early 1980s, the price of fuel decreased during the first three quarters of the decade. The Consumer Price Index figures for 1982 and 1986 were 105.0 and 74.1. If it cost you $21.50 to fill your tank with fuel oil in 1982, how much would it have cost to fill the same tank in 1986? Give your answer to the nearest cent.

▌RELATING CONCEPTS (EXERCISES 55-58)

In Section 2.2 we learned that to make the solution process easier, if an equation involves fractions, we can multiply both sides of the equation by the least common denominator of all the fractions in the equation. A proportion consists of two fractions equal to each other, so a proportion is a special case of this kind of equation.

Work Exercises 55–58 in order *to see how the process of solving by cross products is justified.*

55. In the equation $\frac{x}{6} = \frac{2}{5}$, what is the least common denominator of the two fractions?

56. Solve the equation in Exercise 55 as follows:
   (a) Multiply both sides by the LCD. What is the equation that you obtain?
   (b) Solve for x by dividing both sides by the coefficient of x. What is the solution?

57. Solve the equation in Exercise 55 as follows:
   (a) Set the cross products equal. What is the equation you obtain?
   (b) Repeat part (b) of Exercise 56.

58. Compare your results from Exercises 56(b) and 57(b). What do you notice?

Did you make the connection that solving a proportion by cross products is justified by the general method of solving an equation with fractions?

Solve each variation problem. See Example 6.

  59. The interest on an investment varies directly as the rate of simple interest. If the interest is $48 when the interest rate is 5%, find the interest when the rate is 4.2%.

60. For a given base, the area of a triangle varies directly as its height. Find the area of a triangle with a height of 6 inches, if the area is 10 square inches when the height is 4 inches.

61. The distance a spring stretches is directly proportional to the force applied. If a force of 30 pounds stretches a spring 16 inches, how far will a force of 50 pounds stretch the spring?

62. The perimeter of a square is directly proportional to the length of its side. What is the perimeter of a square with a 5-centimeter side, if a square with a 12-centimeter side has a perimeter of 48 centimeters?

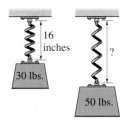

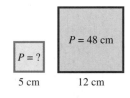

63. According to *The Guinness Book of World Records,* the longest recorded voyage in a paddleboat is 2226 miles in 103 days; the boat was propelled down the Mississippi River by the foot power of two boaters. Assuming a constant rate, how far would they have gone in 120 days? (*Hint:* Distance varies directly as time.)

2.6 More about Problem Solving

OBJECTIVES

1 Use percent in problems involving rates.

2 Solve problems involving mixtures.

3 Solve problems involving simple interest.

4 Solve problems involving denominations of money.

5 Solve problems involving distance, rate, and time.

FOR EXTRA HELP

 **SSG** Sec. 2.6
SSM Sec. 2.6

 Pass the Test Software

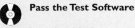 **InterAct Math Tutorial Software**

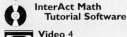 **Video 4**

OBJECTIVE 1 Use percent in problems involving rates. Recall that percent means "per hundred." Thus, percents are ratios where the second number is always 100. For example, 50% represents the ratio of 50 to 100 and 27% represents the ratio of 27 to 100.

PROBLEM SOLVING

Percents are often used in problems involving mixing different concentrations of a substance or different interest rates. In each case, to get the amount of pure substance or the interest, we multiply.

Mixture Problems	Interest Problems (annual)
base × rate (%) = percentage	principal × rate (%) = interest
$b \times r = p$	$p \times r = I$

In an equation, the percent always appears as a decimal. For example, 35% is written as .35, not 35.

┌ **E X A M P L E   1**   Using Percent to Find a Percentage

(a) If a chemist has 40 liters of a 35% acid solution, then the amount of pure acid in the solution is

$$40 \quad \times \quad .35 \quad = \quad 14 \text{ liters.}$$

Amount of solution   Rate of concentration   Amount of pure acid

(b) If $1300 is invested for one year at 7% simple interest, the amount of interest earned in the year is

$$\$1300 \quad \times \quad .07 \quad = \quad \$91.$$

Principal   Interest rate   Interest earned

PROBLEM SOLVING

In the examples that follow, we use charts to organize the information in the problems. A chart enables us to more easily set up the equation for the problem, which is usually the most difficult step. The six steps described in Section 2.3 are used but are not always specifically numbered.

OBJECTIVE 2 Solve problems involving mixtures. In the next example, we use percent to solve a mixture problem.

┌ **E X A M P L E   2**   Solving a Mixture Problem

A chemist needs to mix 20 liters of 40% acid solution with some 70% solution to get a mixture that is 50% acid. How many liters of the 70% solution should be used?

Step 1   Let x = the number of liters of 70% solution that are needed.

Step 2   Recall from part (a) of Example 1 that the amount of pure acid in this solution will be given by the product of the percent of strength and the number of liters of solution, or

liters of pure acid in x liters of 70% solution = $.70x$.

The amount of pure acid in the 20 liters of 40% solution is

liters of pure acid in the 40% solution = $.40(20) = 8$.

The new solution will contain $20 + x$ liters of 50% solution. The amount of pure acid in this solution is

liters of pure acid in the 50% solution = $.50(20 + x)$.

The given information is summarized in the chart below.

Liters of Mixture	Rate (as a decimal)	Liters of Pure Acid
x	.70	.70(x)
20	.40	.40(20)
20 + x	.50	.50(20 + x)

Step 3   The number of liters of pure acid in the 70% solution added to the number of liters of pure acid in the 40% solution will equal the number of liters of pure acid in the final mixture, so the equation is

Pure acid in 70%	plus	pure acid in 40%	is	pure acid in 50%.
$\downarrow$	$\downarrow$	$\downarrow$	$\downarrow$	$\downarrow$
$.70x$	$+$	$.40(20)$	$=$	$.50(20 + x).$

Step 4   Clear parentheses, then multiply by 100 to clear decimals.

$$.70x + .40(20) = .50(20) + .50x \qquad \text{Distributive property}$$
$$70x + 40(20) = 50(20) + 50x \qquad \text{Multiply by 100.}$$

Solve for *x*.

$$70x + 800 = 1000 + 50x$$
$$20x + 800 = 1000 \qquad \text{Subtract } 50x.$$
$$20x = 200 \qquad \text{Subtract } 800.$$
$$x = 10 \qquad \text{Divide by } 20.$$

Steps 5 and 6   Check this solution to see that the chemist needs to use 10 liters of 70% solution.

OBJECTIVE 3   Solve problems involving simple interest.   The next example uses the formula for simple interest, $I = prt$. Remember that when $t = 1$, the formula becomes $I = pr$, as shown in the Problem-Solving box at the beginning of this section. Once again the idea of multiplying the total amount (principal) by the rate (rate of interest) gives the percentage (amount of interest).

┌ **E X A M P L E   3**   Solving a Simple Interest Problem

Elizabeth Thornton receives an inheritance. She plans to invest part of it at 9% and $2000 more than this amount in a less secure investment at 10%. To earn $1150 per year in interest, how much should she invest at each rate?

Let $x =$ the amount invested at 9% (in dollars);

$x + 2000 =$ the amount invested at 10% (in dollars).

Use a chart to arrange the information given in the problem.

Amount Invested in Dollars	Rate of Interest	Interest for One Year
x	.09	$.09x$
$x + 2000$	.10	$.10(x + 2000)$

We multiply amount by rate to get the interest earned. Since the total interest is to be $1150, the equation is

Interest at 9%	plus	interest at 10%	is	total interest.
$\downarrow$	$\downarrow$	$\downarrow$	$\downarrow$	$\downarrow$
$.09x$	$+$	$.10(x + 2000)$	$=$	$1150.$

Clear parentheses; then clear decimals.

$$.09x + .10x + .10(2000) = 1150 \qquad \text{Distributive property}$$
$$9x + 10x + 10(2000) = 115{,}000 \qquad \text{Multiply by 100.}$$

Now solve for x.

$$9x + 10x + 20{,}000 = 115{,}000$$
$$19x + 20{,}000 = 115{,}000 \qquad \text{Combine terms.}$$
$$19x = 95{,}000 \qquad \text{Subtract 20,000.}$$
$$x = 5000 \qquad \text{Divide by 19.}$$

She should invest \$5000 at 9% and \$5000 + \$2000 = \$7000 at 10%.

 Although decimals were cleared in Examples 2 and 3, the equations also can be solved without doing so.

OBJECTIVE 4 Solve problems involving denominations of money. If a cash drawer contains 37 quarters, the total value of the coins is

$$\underset{\substack{\uparrow \\ \text{Number of coins}}}{37} \quad \times \quad \underset{\substack{\uparrow \\ \text{Denomination}}}{\$.25} \quad = \quad \underset{\substack{\uparrow \\ \text{Total value}}}{\$9.25.}$$

PROBLEM SOLVING

Problems that involve different denominations of money or items with different monetary values are very similar to mixture and interest problems. The basic relationship is the same.

Money Problems

Number × Value of one = Total value

A chart is helpful for these problems, too.

EXAMPLE 4 Solving a Problem about Money

A bank teller has 25 more five-dollar bills than ten-dollar bills. The total value of the money is \$200. How many of each denomination of bill does he have?

We must find the number of each denomination of bill that the teller has.

$$\text{Let} \qquad x = \text{the number of ten-dollar bills;}$$
$$x + 25 = \text{the number of five-dollar bills.}$$

Organize the given information in a chart.

Number of Bills	Denomination in Dollars	Dollar Value
x	10	$10x$
$x + 25$	5	$5(x + 25)$

Multiplying the number of bills by the denomination gives the monetary value. The value of the tens added to the value of the fives must be $200:

Value of fives	plus	value of tens	is	$200.
↓	↓	↓	↓	↓
$5(x + 25)$	$+$	$10x$	$=$	$200.$

Solve this equation.

$$5x + 125 + 10x = 200 \qquad \text{Distributive property}$$
$$15x + 125 = 200 \qquad \text{Combine terms.}$$
$$15x = 75 \qquad \text{Subtract 125.}$$
$$x = 5 \qquad \text{Divide by 15.}$$

Since x represents the number of tens, the teller has 5 tens and $5 + 25 = 30$ fives. Check that the value of this money is $5(\$10) + 30(\$5) = \$200$.

EXAMPLE 5   Solving a Problem about Ticket Prices

Lindsay Davenport, the winner of the 1998 U.S. Open women's tennis championship, plays World Team Tennis for the Sacramento Capitals. Tickets are $35 for box seats and $25 for all other seats. There are 5 times as many other seats as box seats. How many of each kind of ticket must the promoter sell for ticket sales revenue to reach $16,000?

Choose a variable and use it to write the revenue from each kind of ticket.

$$\text{Let} \quad x = \text{the number of box seats};$$
$$5x = \text{the number of other seats}.$$

Again, organize the information in a chart.

Number of Seats	Ticket Price	Revenue
x	35	$35x$
$5x$	25	$125x$

Total sales revenue is

Revenue from box seats	plus	revenue from other seats	is	total revenue.
↓	↓	↓	↓	↓
$35x$	$+$	$125x$	$=$	$16,000.$

To solve the equation, combine the terms on the left side of the equation, then divide.

$$160x = 16,000$$
$$x = 100 \qquad \text{Divide by 160.}$$

Thus, there are 100 box seats and $5(100) = 500$ other seats. Check that this produces revenue of $100(\$35) + 500(\$25) = \$16,000$, as required.

OBJECTIVE 5   Solve problems involving distance, rate, and time.   If an automobile travels at an average rate of 50 miles per hour for two hours, then it travels $50 \times 2 = 100$ miles. This is an example of the basic relationship between distance, rate, and time:

$$\text{distance} = \text{rate} \times \text{time},$$

given by the formula $d = rt$. By solving, in turn, for r and t in the formula, we obtain two other equivalent forms of the formula. The three forms are given below.

Distance, Rate, Time Relationship

$$d = rt \qquad r = \frac{d}{t} \qquad t = \frac{d}{r}$$

The following examples illustrate the uses of these formulas.

 E X A M P L E 6   Finding Distance, Rate, or Time

(a) The speed of sound is 1088 feet per second at sea level at 32°F. In 5 seconds under these conditions, sound travels

$$\underset{\text{Rate}}{1088} \quad \times \quad \underset{\text{Time}}{5} \quad = \quad \underset{\text{Distance}}{5440 \text{ feet.}}$$

Here, we found distance given rate and time, using $d = rt$.

(b) The winner of the first Indianapolis 500 race (in 1911) was Ray Harroun, driving a Marmon Wasp at an average speed of 74.59 miles per hour.* To complete the 500 miles, it took him

$$\underset{\text{Rate}}{\overset{\text{Distance} \rightarrow}{\frac{500}{74.59}}} = 6.70 \text{ hours} \quad \text{(rounded).} \quad \leftarrow \text{Time}$$

Here, we found time given rate and distance, using $t = \dfrac{d}{r}$. To convert .7 hour to minutes, multiply by 60: .7(60) = 42 minutes, so the race took Harroun 6 hours, 42 minutes to complete.

(c) In the 1996 Olympic Games in Atlanta, Claudia Poll of Costa Rica won the women's 200-meter freestyle swimming event in 1 minute, 58.16 seconds or 60 + 58.16 = 118.16 seconds.* Her rate was

$$\underset{\text{Time}}{\overset{\text{Distance} \rightarrow}{\frac{200}{118.16}}} = 1.69 \text{ meters per second} \quad \text{(rounded).} \quad \leftarrow \text{Rate}$$

Here, we found rate given distance and time, using $r = \dfrac{d}{t}$.

*Source: The Universal Almanac, 1997, John W. Wright, General Editor.

▌PROBLEM SOLVING

The next example shows how to solve a typical application of the formula $d = rt$. We continue to use the six-step method, but now we also include a sketch along with a chart in Step 2. The sketch shows what is happening in the problem and will help you set up the equation.

┌ **E X A M P L E 7** Solving a Motion Problem

Two cars leave Baton Rouge, Louisiana, at the same time and travel east on Interstate 10. One travels at a constant speed of 55 miles per hour and the other travels at a constant speed of 63 miles per hour. In how many hours will the distance between them be 24 miles?

Step 1 We are looking for time.

Let t = the number of hours until the distance between them is 24 miles.

Step 2 The sketch in Figure 10 shows what is happening in the problem.

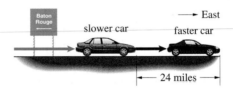

Figure 10

Now, construct a chart like the one below. Fill in the information given in the problem, and use t for the time traveled by each car. Multiply rate by time to get the expressions for distances traveled.

	Rate	Time	Distance
Faster car	63	t	$63t$
Slower car	55	t	$55t$

Difference is 24 miles.

Steps 3 and 4 Refer to Figure 10 and notice that the *difference* between the larger distance and the smaller distance is 24 miles. Now write the equation and solve it.

$$63t - 55t = 24$$
$$8t = 24 \qquad \text{Combine terms.}$$
$$t = 3 \qquad \text{Divide by 8.}$$

Steps 5 and 6 After 3 hours the faster car will have traveled $63 \times 3 = 189$ miles, and the slower car will have traveled $55 \times 3 = 165$ miles. Since $189 - 165 = 24$, the conditions of the problem are satisfied. It will take 3 hours for the distance between them to be 24 miles.

EXAMPLE 8 Solving a Motion Problem

Two planes leave Los Angeles at the same time. One heads south to San Diego; the other heads north to San Francisco. The San Francisco plane flies 50 miles per hour faster. In $\frac{1}{2}$ hour, the planes are 275 miles apart. What are their speeds?

Let r represent the speed of the slower plane. Then $r + 50$ will represent the speed of the faster plane. Fill in a chart.

	Rate	Time	Distance
Slower plane	r	$\frac{1}{2}$	$\frac{1}{2}r$
Faster plane	$r + 50$	$\frac{1}{2}$	$\frac{1}{2}(r + 50)$

Sum is 275 miles.

As Figure 11 shows, the planes are headed in *opposite* directions.

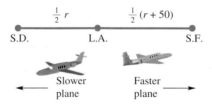

Figure 11

The *sum* of their distances equals 275 miles. We must solve the equation

$$\frac{1}{2}r + \frac{1}{2}(r + 50) = 275.$$

$$r + (r + 50) = 550 \qquad \text{Multiply by 2.}$$
$$2r + 50 = 550 \qquad \text{Combine terms.}$$
$$2r = 500 \qquad \text{Subtract 50.}$$
$$r = 250 \qquad \text{Divide by 2.}$$

The slower plane (headed south) has a speed of 250 miles per hour. The speed of the faster plane is $250 + 50 = 300$ miles per hour. To check, verify that

$$\frac{1}{2}(250) + \frac{1}{2}(300) = 275 \text{ miles.}$$

In motion problems like the ones in Examples 7 and 8, once you have filled in two pieces of information in each row of the chart, you can automatically fill in the third piece of information, using the appropriate formula relating distance, rate, and time. Set up the equation based on your sketch and the information in the chart.

Another way to solve the problems in this section is given in Chapter 9.

Most difficulties in solving problems of the types discussed here occur because students do not take enough time to read the problem carefully and organize the information given in the problem. Organizing the information in a chart greatly increases the chances of solving the problem correctly.

2.6 EXERCISES

Solve each problem. See Example 1 and the Problem-Solving box before Example 4.

1. How much pure acid is in 250 milliliters of a 14% acid solution?

2. How much pure alcohol is in 150 liters of a 30% alcohol solution?

3. If $10,000 is invested for one year at 3.5% simple interest, how much interest is earned?

4. If $25,000 is invested at 3% simple interest for 2 years, how much interest is earned?

5. What is the monetary amount of 283 nickels?

6. What is the monetary amount of 35 half-dollars?

 In Exercises 7–8, find the time based on the information provided. Use a calculator and round your answer to the nearest thousandth. See Example 6.

	Event and Year	Participant	Distance	Rate
7.	Indianapolis 500, 1996	Buddy Lazier	500 miles	149.956 mph
8.	Daytona 500, 1996	Dale Jarrett	500 miles	154.308 mph

Source: Indianapolis Motor Speedway Hall of Fame and Museum.

 In Exercises 9–10, find the rate based on the information provided. Use a calculator and round your answer to the nearest hundredth. See Example 6.

	Event and Year	Participant	Distance	Time
9.	Summer Olympics, 100-meter Hurdles, Women, 1996	Ludmila Engquist	100 meters	12.58 seconds
10.	Summer Olympics, 400-meter Relay, Women, 1996	United States	400 meters	41.95 seconds

Source: The Universal Almanac, 1997, John W. Wright, General Editor.

 Use your knowledge of percent and the six-step method to solve these problems.

11. According to a Knight-Ridder Newspapers report, as of May 31, 1992, the nation's "consumer-debt burden" was 16.4%. This means that the average American had consumer debts, such as credit card bills, auto loans, and so on, totaling 16.4% of his or her take-home pay. Suppose that Paul Eldersveld has take-home pay of $3250 per month. What is 16.4% of his monthly take-home pay?

12. Several years ago, General Motors announced that it would raise prices on the following year's vehicles by an average of 1.6%. If a certain vehicle had an original price of $10,526 and this price was raised 1.6%, to the nearest dollar, what would the new price be?

13. The 1916 dime minted in Denver is quite rare. The 1979 edition of *A Guide Book of United States Coins* listed its value in extremely fine condition as $625. The 1997 value had increased to $2400. What was the percent increase in the value of this coin?

14. In 1963, the value of a 1903 Morgan dollar minted in New Orleans in uncirculated condition was $1500. Due to a discovery of a large hoard of these dollars late that year, the value plummeted. Its value as listed in the 1997 edition of *A Guide Book of United States Coins* was $175. What percent of its 1963 value was its 1997 value?

 15. The pie chart shows the breakdown, by approximate percents, of age groups buying team sports equipment in 1995. According to the National Sporting Goods Association, about $15,691 million was spent that year for team equipment. How much was spent by each age group?
 (a) 18–24
 (b) 25–34
 (c) 45–64

 16. The pie chart shows the breakdown for the year 1995 of purchases of aerobic shoes. Approximately $372 million was spent on aerobic shoes that year. Determine the amount spent by each group.
 (a) Under 14
 (b) 35–44
 (c) 45–64

PURCHASES OF TEAM SPORTS EQUIPMENT

Source: National Sporting Goods Association.

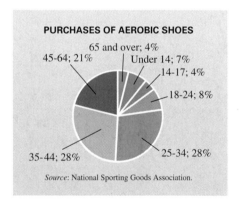

PURCHASES OF AEROBIC SHOES

Source: National Sporting Goods Association.

17. Express the amount of alcohol in *r* liters of pure water.

18. Express the amount of alcohol in *k* liters of pure alcohol.

Work each mixture problem. Use the six-step method. See Example 2.

19. How many gallons of 50% antifreeze must be mixed with 80 gallons of 20% antifreeze to get a mixture that is 40% antifreeze?

Rate	Gallons of Mixture	Gallons of Antifreeze
.50	x	$.50x$
.20	80	$.20(80)$
.40	$x + 80$	$.40(x + 80)$

20. How many liters of 25% acid solution must be added to 80 liters of 40% solution to get a solution that is 30% acid?

Rate	Liters of Mixture	Liters of Acid
.25	x	$.25x$
.40	80	$.40(80)$
.30	$x + 80$	$.30(x + 80)$

21. A certain metal is 20% tin. How many kilograms of this metal must be mixed with 80 kilograms of a metal that is 70% tin to get a metal that is 50% tin?

Rate	Kilograms of Metal	Kilograms of Pure Tin
.20	x	
.70		
.50		

22. A pharmacist has 20 liters of a 10% drug solution. How many liters of 5% solution must be added to get a mixture that is 8%?

Rate	Liters of Solution	Liters of Pure Drug
	20	20(.10)
.05		
.08		

23. How many liters of a 60% acid solution must be mixed with a 75% acid solution to get 20 liters of a 72% solution?

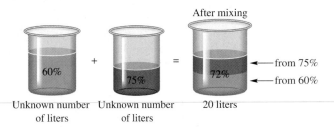

24. How many gallons of a 12% indicator solution must be mixed with a 20% indicator solution to get 10 gallons of a 14% solution?

25. Minoxidil is a drug that has recently proven to be effective in treating male pattern baldness. A pharmacist wishes to mix a solution that is 2% minoxidil. She has on hand 50 milliliters of a 1% solution, and she wishes to add some 4% solution to it to obtain the desired 2% solution. How much 4% solution should she add?

26. Water must be added to 20 milliliters of a 4% minoxidil solution to dilute it to a 2% solution. How many milliliters of water should be used?

Use the concepts of this section to answer each question.

27. Suppose that a chemist is mixing two acid solutions, one of 20% concentration and the other of 30% concentration. Which one of the following concentrations could *not* be obtained?
  (a) 22%   **(b)** 24%
  (c) 28%   **(d)** 32%

28. Suppose that pure alcohol is added to a 24% alcohol mixture. Which one of the following concentrations could *not* be obtained?
  (a) 22%   **(b)** 26%
  (c) 28%   **(d)** 30%

Work each investment problem using simple interest. See Example 3.

29. Li Nguyen invested some money at 3% and $4000 less than that amount at 5%. The two investments produced a total of $200 interest in one year. How much was invested at each rate?

30. LaShondra Williams inherited some money from her uncle. She deposited part of the money in a savings account paying 2%, and $3000 more than that amount in a different account paying 3%. Her annual interest income was $690. How much did she deposit at each rate?

31. With income earned by selling the rights to his life story, an actor invests some of the money at 3% and $30,000 more than twice as much at 4%. The total annual interest earned from the investments is $5600. How much is invested at each rate?

32. An artist invests her earnings in two ways. Some goes into a tax-free bond paying 6%, and $6000 more than three times as much goes into mutual funds paying 5%. Her total annual interest income from the investments is $825. How much does she invest at each rate?

Work Exercises 33–38 involving different monetary rates. See Example 4.

33. A bank teller has some five-dollar bills and some twenty-dollar bills. The teller has 5 more twenties than fives. The total value of the money is $725. Find the number of five-dollar bills that the teller has.

Number of Bills	Denomination	Value
x	5	
$x + 5$	20	

34. A coin collector has $1.70 in dimes and nickels. She has 2 more dimes than nickels. How many nickels does she have?

Number of Coins	Denomination	Value
x	.05	.05x
	.10	

35. A cashier has a total of 126 bills, made up of fives and tens. The total value of the money is $840. How many of each kind does he have?

36. A convention manager finds that she has $1290, made up of twenties and fifties. She has a total of 42 bills. How many of each kind does she have?

37. A merchant wishes to mix candy worth $5 per pound with 40 pounds of candy worth $2 per pound to get a mixture that can be sold for $3 per pound. How many pounds of $5 candy should be used?

38. At Vern's Grill, hamburgers cost 90 cents each, and a bag of french fries costs 40 cents. How many hamburgers and how many bags of french fries can a customer buy with $8.80 if he wants twice as many hamburgers as bags of french fries?

Vern's Grill	
Hamburgers	90¢ each
French fries	40¢ a bag
Soda pop	25¢ each
Shakes	90¢ each
Cherry pie	90¢ a slice
Ice cream	40¢ a scoop

39. Read Example 2. Can a problem of this type have a fraction as an answer? Now read Example 4. Can a problem of this type have a fraction as an answer? Explain.

RELATING CONCEPTS (EXERCISES 40-43)

Sometimes applied problems that may seem different actually apply the same concepts.

Work Exercises 40–43 in order, to see how this happens.

40. Consider the following problem: A hoard of coins consists of only nickels and dimes. There are 3400 coins, and the value of the money is $290. How many of each denomination are there?
   (a) Write an equation you would use to solve the problem. Let *x* represent the number of nickels in the hoard.
   (b) Solve the problem.

41. Consider the following problem: An investor deposits $3400 in two accounts. One account is a passbook savings account that pays 5% interest, and the other is a money market account that pays 10% interest. After one year, the total interest earned is $290. How much did she invest at each rate?
   (a) Write an equation you would use to solve the problem. Let *x* represent the amount invested in the passbook savings account.
   (b) Solve the problem.

42. Compare the equations you wrote in Exercises 40(a) and 41(a). What do you notice about them?

43. If, in either of the problems in Exercises 40 and 41, you let *x* represent the other unknown quantity, will you get the same *solution* to the equation? Will you get the same *answers* to the problem?

Did you make the connection that applications that appear different may be solved in the same way?

44. How do you think the pie charts in Exercises 15 and 16 were constructed so that the sections represented the appropriate percentages accurately?

Solve each problem. See Example 6.

45. A driver averaged 53 miles per hour and took 10 hours to travel from Memphis to Chicago. What is the distance between Memphis and Chicago?

46. A small plane traveled from Warsaw to Rome, averaging 164 miles per hour. The trip took 2 hours. What is the distance from Warsaw to Rome?

47. Suppose that an automobile averages 45 miles per hour, and travels for 30 minutes. Is the distance traveled $45 \times 30 = 1350$ miles? If not, explain why not, and give the correct distance.

48. Which of the following choices is the best *estimate* for the average speed of a trip of 405 miles that lasted 8.2 hours?
   (a) 50 miles per hour   **(b)** 30 miles per hour
   (c) 60 miles per hour   **(d)** 40 miles per hour

Solve each problem. See Examples 7 and 8.

49. St. Louis and Portland are 2060 miles apart. A small plane leaves Portland, traveling toward St. Louis at an average speed of 90 miles per hour. Another plane leaves St. Louis at the same time, traveling toward Portland, averaging 116 miles per hour. How long will it take them to meet?

	r	t	d
Plane leaving Portland	90	t	90t
Plane leaving St. Louis	116	t	116t

50. Atlanta and Cincinnati are 440 miles apart. John leaves Cincinnati, driving toward Atlanta at an average speed of 60 miles per hour. Pat leaves Atlanta at the same time, driving toward Cincinnati in her antique auto, averaging 28 miles per hour. How long will it take them to meet?

	r	t	d
John	60	t	60t
Pat	28	t	28t

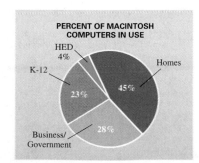

51. Two trains leave a city at the same time. One travels north and the other travels south at 20 miles per hour faster. In 2 hours the trains are 280 miles apart. Find their speeds.

	r	t	d
Northbound	x	2	
Southbound	x + 20	2	

52. Two planes leave an airport at the same time, one flying east, the other flying west. The eastbound plane travels 150 miles per hour slower. They are 2250 miles apart after 3 hours. Find the speed of each plane.

	r	t	d
Eastbound	x − 150	3	
Westbound	x	3	

53. At a given hour two steamboats leave a city in the same direction on a straight canal. One travels at 18 miles per hour and the other travels at 25 miles per hour. In how many hours will the boats be 35 miles apart?

54. From a point on a straight road, Lupe and Maria ride bicycles in the same direction. Lupe rides 10 miles per hour and Maria rides 12 miles per hour. In how many hours will they be 5 miles apart?

The remaining applications in this exercise set are not organized by type of problem. Use the problem-solving techniques described in this chapter to solve them.

55. According to a survey by Information Resources, Inc., single men spent an average of $10.87 more than single women when buying frozen dinners during the 52 weeks that ended August 22, 1993. Together, a single man and a single woman would spend a total of $92.29, according to the survey. What were the average expenditures for a single man and for a single woman?

56. Team Marketing Report, a sports-business newsletter, computes a Fan Cost Index for a trip to a major league baseball park. The index consists of the cost of four average-priced tickets, two small beers, four small sodas, four hot dogs, parking for one car, two game programs, and two twill baseball caps. For the 1994 season, the New York Yankees had the highest Fan Cost Index, while the Cincinnati Reds had the lowest. The Yankees' index was $43.37 less than twice that of the Reds. What was the Fan Cost Index for each of these teams if the *total* of the two was $194.56?

57. The pie chart shows the percents for the different categories of Macintosh computer units in use at the end of 1997. At that time there were approximately 21,782,000 units in use. Find the approximate number of units used in each category. (*Source:* Apple Computer, Inc., Cupertino, California.)

PERCENT OF MACINTOSH COMPUTERS IN USE

HED 4%
K-12
23%
Homes
45%
28%
Business/ Government

58. At the start of play on September 22, 1997, the standings of the Central Division of the American League were as shown. "Winning percentage" is commonly expressed as a decimal rounded to the nearest thousandth. To find the winning percentage of a team, divide the number of wins by the total number of games played. Find the winning percentage of each of the following teams.

	Won	Lost
Cleveland	83	71
Chicago	77	78
Milwaukee	76	78
Kansas City	64	90
Minnesota	63	91

    (a) Cleveland   **(b)** Chicago
    (c) Milwaukee

59. In an automobile race, a driver was 240 miles from the finish line after 3 hours. Another driver, who was in a later race, traveled at the same speed as the first driver. After 2.5 hours the second driver was 300 miles from the finish. Find the speed of each driver.

60. Fran Liberto sold a painting that she found at a garage sale to an art dealer. With the $12,000 she got, she invested part at 4% in a certificate of deposit and the rest at 5% in a municipal bond. Her total annual income from the two investments was $515. How much did she invest at each rate? (*Hint:* Let x represent the amount invested at 4%. Then $12,000 - x$ represents the amount invested at 5%.)

CHAPTER 2 GROUP ACTIVITY

Are You a Race-Walker?

Objective: Use proportions to calculate walking speeds.

Materials: Students will need a large area for walking. Each group will need a stopwatch.

Race-walking at speeds exceeding 8 miles per hour is a high fitness, long-distance competitive sport. The table below contains the gold medal winners of the 1996 Olympic race-walking competition. Complete the table by applying the proportion given below to find the race-walker's steps per minute.

    Use 10 km ≈ 6.21 miles. (Round all answers except those for steps per minute to the nearest thousandths. Round steps per minute to the nearest whole number.)

$$\frac{70 \text{ steps per minute}}{2 \text{ miles per hour}} = \frac{x \text{ steps per minute}}{y \text{ miles per hour}}$$

Event	Gold Medal Winner	Country	Time in Hours: Minutes: Seconds	Time in Minutes	Time in Hours	y Miles per Hour	x Steps per Minute
10 km Walk, Women	Yelena Nikolayeva	Russia	0:41:49				
20 km Walk, Men	Jefferson Perez	Ecuador	1:20:07				
50 km Walk, Men	Robert Korzeniowski	Poland	3:43:30				

Source: 1999 World Almanac.

A. Using a stopwatch, take turns counting how many steps each member of the group takes in one minute while walking at a normal pace. Record the results in the chart below. Then do it again at a fast pace. Record these results.

Name	Normal Pace		Fast Pace	
	x Steps per Minute	*y* Miles per Hour	*x* Steps per Minute	*y* Miles per Hour

B. Use the proportion above to convert the numbers from part A to miles per hour and complete the chart.

    1. Find the average speed for the group at a normal pace and at a fast pace.

    2. What is the minimum number of steps per minute you would have to take to be a race-walker?

    3. At a fast pace did anyone in the group walk fast enough to be a race-walker? Explain how you decided.

CHAPTER 2 SUMMARY

KEY TERMS

2.1 linear equation in one variable solution set equivalent equations	**2.2** empty (null) set **2.3** degree complementary angles supplementary angles consecutive integers	**2.4** perimeter circumference area vertical angles straight angle	**2.5** ratio proportion cross products vary directly as (is proportional to)

NEW SYMBOLS

∅ empty set	1° one degree	a to b, $a:b$, or $\dfrac{a}{b}$ the ratio of a to b

TEST YOUR WORD POWER

See how well you have learned the vocabulary in this chapter. Answers, with examples, are given at the bottom of the next page.

1. A **solution set** is the set of numbers that
(a) make an expression undefined
(b) make an equation false
(c) make an equation true
(d) make an expression equal to 0.

2. The **empty set** is a set
(a) with 0 as its only element
(b) with an infinite number of elements
(c) with no elements
(d) of ideas.

3. **Complementary angles** are angles
(a) formed by two parallel lines
(b) whose sum is 90°
(c) whose sum is 180°
(d) formed by perpendicular lines.

4. Supplementary angles are angles
(a) formed by two parallel lines
(b) whose sum is 90°
(c) whose sum is 180°
(d) formed by perpendicular lines.

5. A ratio
(a) compares two quantities using a quotient
(b) says that two quotients are equal
(c) is a product of two quantities
(d) is a difference between two quantities.

6. A proportion
(a) compares two quantities using a quotient
(b) says that two quotients are equal
(c) is a product of two quantities
(d) is a difference between two quantities.

QUICK REVIEW

CONCEPTS	EXAMPLES

2.1 THE ADDITION AND MULTIPLICATION PROPERTIES OF EQUALITY

The same expression may be added to (or subtracted from) each side of an equation without changing the solution.

Solve $x - 6 = 12$.

$$x - 6 + 6 = 12 + 6 \qquad \text{Add 6.}$$
$$x = 18 \qquad \text{Combine terms.}$$

Solution set: $\{18\}$

Each side of an equation may be multiplied (or divided) by the same nonzero expression without changing the solution.

Solve $\dfrac{3}{4}x = -9$.

$$\frac{4}{3} \cdot \frac{3}{4}x = \frac{4}{3}(-9) \qquad \text{Multiply by } \tfrac{4}{3}.$$
$$x = -12$$

Solution set: $\{-12\}$

2.2 MORE ON SOLVING LINEAR EQUATIONS

Solving a Linear Equation

1. Clear parentheses and combine like terms to simplify each side.

2. Get the variable term on one side, a number on the other.

3. Get the equation into the form $x =$ a number.

4. Check by substituting the result into the original equation.

Solve the equation $2x + 3(x + 1) = 38$.

$$2x + 3x + 3 = 38 \qquad \text{Clear parentheses.}$$
$$5x + 3 = 38 \qquad \text{Combine like terms.}$$

$$5x + 3 - 3 = 38 - 3 \qquad \text{Subtract 3.}$$
$$5x = 35 \qquad \text{Combine terms.}$$

$$\frac{5x}{5} = \frac{35}{5} \qquad \text{Divide by 5.}$$
$$x = 7$$

$$2x + 3(x + 1) = 38 \qquad \text{Check.}$$
$$2(7) + 3(7 + 1) = 38 \quad ? \qquad \text{Let } x = 7.$$
$$14 + 24 = 38 \quad ? \qquad \text{Multiply.}$$
$$38 = 38 \qquad \text{True}$$

Solution set: $\{7\}$

Answers to Test Your Word Power

1. (c) *Example:* $\{8\}$ is the solution set of $2x + 5 = 21$.   **2.** (c) *Example:* The empty set $\emptyset$ is the solution set of $5x + 3 = 5x + 4$.

3. (b) *Example:* Angles with measures 35° and 55° are complementary angles.   **4.** (c) *Example:* Angles with measures 112° and 68° are supplementary angles.   **5.** (a) *Example:* $\dfrac{7 \text{ inches}}{12 \text{ inches}} = \dfrac{7}{12}$ **6.** (b) *Example:* $\dfrac{2}{3} = \dfrac{8}{12}$

CONCEPTS	EXAMPLES

2.3 AN INTRODUCTION TO APPLICATIONS OF LINEAR EQUATIONS

Solving an Applied Problem Using the Six-Step Method

One number is 5 more than another. Their sum is 21. Find both numbers.

Step 1   Choose a variable to represent the unknown.

Let x be the smaller number.

Step 2   Determine expressions for any other unknown quantities, using the variable. Draw figures or diagrams and use charts if they apply.

Let $x + 5$ be the larger number.

Step 3   Write an equation.

Step 4   Solve the equation.

$$x + (x + 5) = 21$$

$2x + 5 = 21$	Combine terms.
$2x + 5 - 5 = 21 - 5$	Subtract 5.
$2x = 16$	Combine terms.
$x = 8$	Divide by 2.

Step 5   Answer the question(s) asked in the problem.

The numbers are 8 and 13.

Step 6   Check your solution by using the original words of the problem. Be sure that the answer is appropriate and makes sense.

13 is 5 more than 8, and $8 + 13 = 21$. It checks.

2.4 FORMULAS AND APPLICATIONS FROM GEOMETRY

To find the values of one of the variables in a formula given values for the others, substitute the known values into the formula.

Find L if $A = LW$, given that $A = 24$ and $W = 3$.

$24 = L \cdot 3$	$A = 24, W = 3$
$\dfrac{24}{3} = \dfrac{L \cdot 3}{3}$	Divide by 3.
$8 = L$	

To solve a formula for one of the variables, isolate that variable by treating the other variables as numbers and using the steps for solving equations.

Solve $A = \dfrac{1}{2} bh$ for b.

$2A = 2\left(\dfrac{1}{2}\boldsymbol{bh}\right)$	Multiply by 2.
$2A = \boldsymbol{bh}$	
$\dfrac{2A}{h} = \boldsymbol{b}$	Divide by h.

2.5 RATIOS AND PROPORTIONS

To write a ratio, express quantities in the same units.

Express as a ratio: 4 feet to 8 inches.

$$4 \text{ feet to } 8 \text{ inches} = 48 \text{ inches to } 8 \text{ inches}$$

$$= \frac{48}{8} = \frac{6}{1} \quad \text{or} \quad 6 \text{ to } 1 \quad \text{or} \quad 6:1$$

To solve a proportion, use cross products.

Solve $\dfrac{x}{12} = \dfrac{35}{60}$.

$60x = 12 \cdot 35$	Cross products
$60x = 420$	Multiply.
$\dfrac{60x}{60} = \dfrac{420}{60}$	Divide by 60.
$x = 7$	

Solution set: $\{7\}$

(continued)

CONCEPTS	EXAMPLES

Solving Direct Variation Problems

1. Write the variation equation using $y = kx$.

2. Find k by substituting the given values of x and y into the equation.

3. Write the equation with the value of k from Step 2 and the given value of x or y. Solve for the remaining variable.

At a constant speed (or rate) distance varies directly as time. Find the time to travel 85 miles at a constant speed if it takes 1.5 hours to go 67.5 miles at the same constant speed.

Let t represent the unknown time. The direct variation equation is $d = kt$. Substitute $d = 67.5$ and $t = 1.5$ to find k.

$$67.5 = k(1.5)$$
$$45 = k \qquad \text{Divide by 1.5.}$$

Now substitute $k = 45$ and $d = 85$ in the equation $d = kt$.

$$85 = 45t$$
$$1.89 = t \qquad \text{Divide by 45; round up.}$$

It will take approximately 1.89 hours to travel 85 miles.

2.6 MORE ABOUT PROBLEM SOLVING

Problems involving applications of percent can be solved using charts.

A sum of money is invested at simple interest in two ways. Part is invested at 12%, and $20,000 less than that amount is invested at 10%. If the total interest for one year is $9000, find the amount invested at each rate.

Let $\qquad x = $ amount invested at 12%;

$x - 20,000 = $ amount invested at 10%.

Dollars Invested	Rate of Interest	Interest for One Year
x	.12	.12x
$x - 20,000$	.10	.10($x - 20,000$)

$$.12x + .10(x - 20,000) = 9000$$
$$.12x + .10x + .10(-20,000) = 9000 \qquad \text{Distributive property}$$
$$12x + 10x + 10(-20,000) = 900,000 \qquad \text{Multiply by 100.}$$
$$12x + 10x - 200,000 = 900,000$$
$$22x - 200,000 = 900,000 \qquad \text{Combine terms.}$$
$$22x = 1,100,000 \qquad \text{Add 200,000.}$$
$$x = 50,000 \qquad \text{Divide by 22.}$$

$50,000 is invested at 12% and $30,000 is invested at 10%.

The three forms of the formula relating distance, rate, and time, are $d = rt$, $r = \dfrac{d}{t}$, and $t = \dfrac{d}{r}$.

To solve a problem about distance, set up a sketch showing what is happening in the problem.

Two cars leave from the same point, traveling in opposite directions. One travels at 45 miles per hour and the other at 60 miles per hour. How long will it take them to be 210 miles apart?

Let $t = $ time it takes for them to be 210 miles apart.

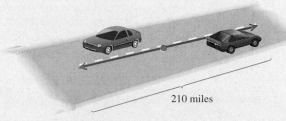

210 miles

(continued)

CONCEPTS	EXAMPLES
Make a chart using the information given in the problem, along with the unknown quantities.	The chart gives the information from the problem, with expressions for distance obtained by using $d = rt$.

	Rate	Time	Distance
One car	45	t	$45t$
Other car	60	t	$60t$

The sum of the distances, $45t$ and $60t$, must be 210 miles.

$$45t + 60t = 210$$
$$105t = 210 \quad \text{Combine like terms.}$$
$$t = 2 \quad \text{Divide by 2.}$$

It will take them 2 hours to be 210 miles apart.

CHAPTER 2 REVIEW EXERCISES

[2.1–2.2] *Solve each equation.*

1. $m - 5 = 1$ **2.** $y + 8 = -4$ **3.** $3k + 1 = 2k + 8$

4. $5k = 4k + \dfrac{2}{3}$ **5.** $(4r - 2) - (3r + 1) = 8$ **6.** $3(2y - 5) = 2 + 5y$

7. $7k = 35$ **8.** $12r = -48$ **9.** $2p - 7p + 8p = 15$

10. $\dfrac{m}{12} = -1$ **11.** $\dfrac{5}{8}k = 8$ **12.** $12m + 11 = 59$

13. $3(2x + 6) - 5(x + 8) = x - 22$ **14.** $5x + 9 - (2x - 3) = 2x - 7$

15. $\dfrac{1}{2}r - \dfrac{r}{3} = \dfrac{r}{6}$ **16.** $.10(x + 80) + .20x = 14$

17. $3x - (-2x + 6) = 4(x - 4) + x$ **18.** $2(y - 3) - 4(y + 12) = -2(y + 27)$

 [2.3] *Use the six-step method to solve each problem.*

19. Soccer is the world's most popular sport, with over 20 million participants. The number of participating nations in the 1990 World Cup was 6 less than in the 1986 World Cup. In 1994, 28 more nations participated than in 1986. How many nations participated in 1986 if there was a total of 358 participating nations in those three years? (*Source:* FIFA.)

20. In a recent year, the state of Florida had a total of 120 members in its House of Representatives, consisting of only Democrats and Republicans. There were 30 more Democrats than Republicans. How many representatives from each party were there?

21. The land area of Hawaii is 5213 square miles greater than the area of Rhode Island. Together, the areas total 7637 square miles. What is the area of each of the two states?

22. The height of Seven Falls in Colorado is $\dfrac{5}{2}$ the height of Twin Falls in Idaho. The sum of the heights is 420 feet. Find the height of each. (*Source: World Almanac and Book of Facts.*)

23. The supplement of an angle measures 10 times the measure of its complement. What is the measure of the angle?

[2.4]   *A formula is given along with the values for all but one of the variables. Find the value of the variable that is not given.*

24. $A = \frac{1}{2}bh$; $A = 44, b = 8$

25. $A = \frac{1}{2}h(b + B)$; $b = 3, B = 4, h = 8$

26. $C = 2\pi r$; $C = 29.83, \pi = 3.14$

27. $V = \frac{4}{3}\pi r^3$; $r = 6, \pi = 3.14$

Solve the formula for each specified variable.

28. $A = LW$;   for L

29. $A = \frac{1}{2}h(b + B)$;   for h

Find the measure of each marked angle.

30.

31.

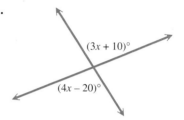

Solve each problem.

32. The Ziegfield Room in Reno, Nevada, has a circular turntable on which its showgirls dance. The circumference of the table is 62.5 feet. What is the diameter? What is the radius? What is the area? (Use $\pi = 3.14$.) (*Source: The Guinness Book of World Records, 1997.*)

33. A baseball diamond is a square with a side of 90 feet. The pitcher's rubber is located 60.5 feet from home plate as shown in the figure. Find the measures of the angles marked in the figure.

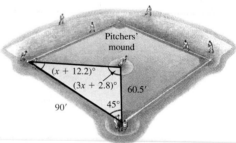

34. What is wrong with the following problem? "The formula for the area of a trapezoid is $A = \frac{1}{2}h(b + B)$. If $h = 12$ and $b = 14$, find the value of A."

[2.5]   *Give a ratio for each word phrase, writing fractions in lowest terms.*

35. 60 centimeters to 40 centimeters

36. 5 days to 2 weeks

37. 90 inches to 10 feet

38. 3 months to 3 years

Solve each equation.

39. $\dfrac{p}{21} = \dfrac{5}{30}$

40. $\dfrac{5 + x}{3} = \dfrac{2 - x}{6}$

41. $\dfrac{y}{5} = \dfrac{6y - 5}{11}$

Use proportions to solve each problem.

42. If 2 pounds of fertilizer will cover 150 square feet of lawn, how many pounds would be needed to cover 500 square feet?

43. In a camera, the ratio of the length of the image on the film to the object length equals the ratio of the distance from the lens to the image compared to the distance from the lens to the object. See the figure. How far from the lens should a child 1 meter tall be to fit on 35 mm film (35 mm by 35 mm), if the distance from the lens to the film is 7 mm?

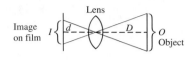

44. The tax on a \$24.00 item is \$2.04. How much tax would be paid on a \$36.00 item?

45. The distance between two cities on a road map is 32 centimeters. The two cities are actually 150 kilometers apart. The distance on the map between two other cities is 80 centimeters. How far apart are these cities?

 46. In the 1996 Olympic Games held in Atlanta, Romania earned a total of 20 medals. Two of every ten medals were gold. How many gold medals were earned? (*Source: The Universal Almanac,* 1997, John W. Wright, General Editor.)

 47. Which is the best buy for a popular breakfast cereal?
15-ounce size: \$2.69
20-ounce size: \$3.29
25.5-ounce size: \$3.49

 [2.6] *Solve each problem.*

48. The San Francisco Giants baseball team has borrowed \$160 million of the \$290 million to build their new Pacific Bell Park, scheduled to open in 2002. Corporate sponsors will provide the rest of the cost. What percent of the cost of the new park is borrowed? (*Source: Michael K. Ozanian, "Fields of Debt," Forbes,* December 15, 1997, vol. 160, no. 13, p. 174.)

 49. (Refer to Exercise 48.) It has been estimated that Bell Park's annual cash flow will be 1.07 times the cost of the interest on the loan of \$160 million. Suppose the money is borrowed at a 9% annual interest rate. How much will the annual cash flow amount to? (*Source: Michael K. Ozanian, "Fields of Debt," Forbes,* December 15, 1997, vol. 160, no. 13, p. 174.)

50. A nurse must mix 15 liters of a 10% solution of a drug with some 60% solution to get a 20% mixture. How many liters of the 60% solution will be needed?

51. Todd Cardella invested \$10,000 from which he earns an annual income of \$550 per year. He invested part of it at 5% annual interest and the remainder in bonds paying 6% interest. How much did he invest at each rate?

 52. In 1846, the vessel *Yorkshire* traveled from Liverpool to New York, a distance of 3150 miles, in 384 hours. What was the *Yorkshire*'s average speed? Round your answer to the nearest tenth.

53. Sue Fredine drove from Louisville to Dallas, a distance of 819 miles, averaging 63 miles per hour. What was her driving time?

54. Two planes leave St. Louis at the same time. One flies north at 350 miles per hour and the other flies south at 420 miles per hour. In how many hours will they be 1925 miles apart?

 55. Jim leaves his house on his bicycle and averages 5 miles per hour. His wife, Annie, leaves $\frac{1}{2}$ hour later, following the same path and averaging 8 miles per hour. How long will it take for Annie to catch up to Jim?

56. The circumference of a circle varies directly as its radius. A circle with a circumference of 5 inches has a radius of approximately .796 inches. Find the radius of a circle with a circumference of 17.5 inches.

57. Typically, the supply of an item varies directly as the price. When a computer game was priced at $50, a merchant had a supply of 40 games. After the game went on sale for $35, what was the supply?

58. The pressure exerted by a liquid at a given point is proportional to the depth of the point beneath the surface of the liquid. The pressure at 10 meters is $\frac{80}{3}$ newtons per square centimeter. What pressure is exerted at 30 meters?

59. Write your own applied problem, and use the six-step method to solve it.

60. Make a list of the different types of applied problems presented in this chapter. Which type is your favorite? Which type is your least favorite? Explain.

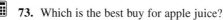

MIXED REVIEW EXERCISES*

Solve.

61. $\dfrac{y}{7} = \dfrac{y-5}{2}$

62. $I = prt$ for r

63. $-2x = -4 - 2(2 - x)$

64. $2k - 5 = 4k + 13$

65. $.05x + .02x = 4.9$

66. $2 - 3(y - 5) = 4 + y$

67. $9x - (7x + 2) = 3x + (2 - x)$

68. $\dfrac{1}{3}s + \dfrac{1}{2}s + 7 = \dfrac{5}{6}s + 5 + 2$

69. A student solved $3 - (8 + 4x) = 2x + 7$ and gave the answer as {6}. Verify that this answer is incorrect by checking it in the equation. Then explain the error and give the correct solution set. (*Hint:* The error involves the subtraction sign.)

70. A recipe for biscuit tortoni calls for $\frac{2}{3}$ cup of macaroon cookie crumbs. The recipe is for 8 servings. How many cups of macaroon cookie crumbs would be needed for 30 servings?

71. The Golden Gate Bridge in San Francisco is 2605 feet longer than the Brooklyn Bridge. Together, their spans total 5795 feet. How long is each bridge? (*Source: World Almanac and Book of Facts.*)

72. A student solves $\dfrac{3}{x} = \dfrac{5}{x}$ by cross multiplying and gets 0 as an answer. What is wrong with this answer?

73. Which is the best buy for apple juice?
32-ounce size: $1.19
48-ounce size: $1.79
64-ounce size: $1.99

74. If 1 quart of oil must be mixed with 24 quarts of gasoline, how much oil would be needed for 192 quarts of gasoline?

75. Two trains are 390 miles apart. They start at the same time and travel toward one another, meeting 3 hours later. If the speed of one train is 30 miles per hour more than the speed of the other train, find the speed of each train.

76. One side of a triangle is 3 centimeters longer than the shortest side. The third side is twice as long as the shortest side. If the perimeter of the triangle is 39 centimeters, find the length of the shortest side.

---

*The order of the exercises in this final group does not correspond to the order in which topics occur in the chapter. This random ordering should help you in your preparation for the chapter test.

77. The shorter base of a trapezoid is 42 centimeters long, and the longer base is 48 centimeters long. The area of the trapezoid is 360 square centimeters. Find the height of the trapezoid.

42 cm

h

48 cm

$A = 360$ square centimeters

78. The perimeter of a square is 200 meters. Find the length of a side.

79. Can $4 + \dfrac{5}{x} = \dfrac{2}{3}$ be solved directly by cross multiplication? Explain your answer.

CHAPTER 2 TEST

Solve each equation.

1. $5x + 9 = 7x + 21$

2. $-\dfrac{4}{7}x = -12$

3. $7 - (m - 4) = -3m + 2(m + 1)$

4. $.06(x + 20) + .08(x - 10) = 4.6$

5. $-8(2x + 4) = -4(4x + 8)$

Solve each problem.

6. The three largest islands in the Hawaiian island chain are Hawaii (the Big Island), Maui, and Kauai. Together, their areas total 5300 square miles. The island of Hawaii is 3293 square miles larger than the island of Maui, and Maui is 177 square miles larger than Kauai. What is the area of each island?

7. The formula for the perimeter of a rectangle is $P = 2L + 2W$.
(a) Solve for W.
(b) If $P = 116$ and $L = 40$, find the value of W.

Find the measure of each marked angle.

8.

$(3x + 15)°$ $(4x - 5)°$

9. Find the measure of an angle if its supplement measures 10° more than three times its complement.

Solve the equation.

10. $\dfrac{y + 5}{3} = \dfrac{y - 3}{4}$

Solve each problem.

11. The circumference of a circle varies directly as the radius. A circle with a radius of 6 centimeters has a circumference of 37.68 centimeters. What is the circumference of a circle with a radius of 10 centimeters?

12. The distance between Milwaukee and Boston is 1050 miles. On a certain map this distance is represented by 21 inches. On the same map Seattle and Cincinnati are 46 inches apart. What is the actual distance between Seattle and Cincinnati?

13. Cathy Wacaser invested some money at 3% simple interest and $6000 more than that amount at 4.5% simple interest. After one year her total interest from the two accounts was $870. How much did she invest at each rate?

14. The Tampa Bay Lightning hockey team has debt of about $177 million, with a franchise value of only $75 million. What percent of the franchise value is the debt? (*Source:* Michael K. Ozanian, "Fields of Debt," *Forbes,* December 15, 1997, vol. 160, no. 13, p. 174(2).)

15. Two cars leave from the same point, traveling in opposite directions. One travels at a constant rate of 50 miles per hour while the other travels at a constant rate of 65 miles per hour. How long will it take for them to be 460 miles apart?

Exercises 16 and 17 refer to the 1997 All-Star basketball game held February 9 in Cleveland. (Source: *The ESPN 1998 Information Please Sports Almanac.*)

16. The East won the game by 12 points. The total of the two team scores was 252. What was the final score of the game?

17. The high scorer of the game was Glenn Rice, playing for the Eastern Conference. He scored 7 more points than Latrell Sprewell, the high scorer on the Western Conference team. The total number of points scored by both players was 45. How many points did Rice score?

CUMULATIVE REVIEW EXERCISES CHAPTERS 1–2

Beginning with this chapter, each chapter in the text will conclude with a set of cumulative review exercises designed to cover the major topics from the beginning of the course. This feature allows students to constantly review topics that have been introduced up to that point. Section references for these exercises are given with the answers at the back of the book.

1. Write $\dfrac{108}{144}$ in lowest terms.

Perform the indicated operations.

2. $\dfrac{5}{6} + \dfrac{1}{4} - \dfrac{7}{15}$

3. $\dfrac{9}{8} \cdot \dfrac{16}{3} \div \dfrac{5}{8}$

Translate from words to symbols. Use x as the variable, if necessary.

4. The difference between half a number and 18.

5. The quotient of 6 and 12 more than a number is 2.

6. True or false? $\dfrac{8(7) - 5(6 + 2)}{3 \cdot 5 + 1} \geq 1$

Perform the indicated operations.

7. $9 - (-4) + (-2)$

8. $\dfrac{-4(9)(-2)}{-3^2}$

9. $(-7 - 1)(-4) + (-4)$

10. Find the value of $\dfrac{3x^2 - y^3}{-4z}$ when $x = -2$, $y = -4$, and $z = 3$.

Name the property illustrated.

11. $7(k + m) = 7k + 7m$

12. $3 + (5 + 2) = 3 + (2 + 5)$

13. Simplify $-4(k + 2) + 3(2k - 1)$ by combining terms.

Solve each equation and check the solution.

14. $2r - 6 = 8r$

15. $4 - 5(a + 2) = 3(a + 1) - 1$

16. $\dfrac{2}{3}y + \dfrac{3}{4}y = -17$

17. $-(m - 1) = 3 - 2m$

18. $\dfrac{2x + 3}{5} = \dfrac{x - 4}{2}$

19. $\dfrac{y - 2}{3} = \dfrac{2y + 1}{5}$

Solve each formula for the indicated variable.

20. $3x + 4y = 24$ for y

21. $A = P(1 + ni)$ for n

Solve each problem.

22. For a woven hanging, Miguel Hidalgo needs three pieces of yarn, which he will cut from a 40-centimeter piece. The longest piece is to be 3 times as long as the middle-sized piece, and the shortest piece is to be 5 centimeters shorter than the middle-sized piece. What lengths should he cut?

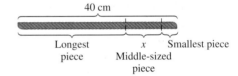

23. A radio telescope at the Max Planck Institute in Germany has a 328-foot diameter. What is the circumference of this telescope? (Use $\pi = 3.14$.) (*Source: The Guinness Book of World Records.*)

24. A cook wants to increase a recipe that serves 6 to make enough for 20 people. The recipe calls for $1\dfrac{1}{4}$ cups of grated cheese. How much cheese will be needed to serve 20?

25. Two cars are 400 miles apart. Both start at the same time and travel toward one another. They meet 4 hours later. If the speed of one car is 20 miles per hour faster than the other, what is the speed of each car?

Linear Inequalities and Absolute Value

3

The concepts of inequality, union and intersection of sets, and absolute value are often used when interpreting data from tables and graphs and solving applications. In this chapter, the theme is entertainment and leisure, so many of the applications will come from these fields.

 Entertainment

3.1 Linear Inequalities in One Variable

3.2 Set Operations and Compound Inequalities

3.3 Absolute Value Equations and Inequalities

Summary: Exercises on Solving Linear and Absolute Value Equations and Inequalities

Americans are spending more time and money on entertainment and leisure than ever. One beneficiary of this entertainment boom is Broadway. A recent study by the League of American Theaters and Producers gave the increase, shown in the graph, in the number of theater tickets sold (in millions) from the 1990–1991 season through the 1996–1997 season.

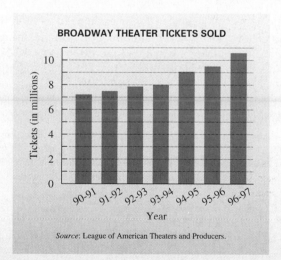

BROADWAY THEATER TICKETS SOLD

Source: League of American Theaters and Producers.

As the graph indicates, theater attendance has increased from approximately 7.3 million in 1990–1991 to approximately 10.6 million in 1996–1997. Based on the graph, in which seasons were ticket sales less than 8 million *or* greater than 10 million? We will use this graph again in the exercises for Section 3.1.

 Visit our Web site at www.LialAlgebra.com

3.1 Linear Inequalities in One Variable

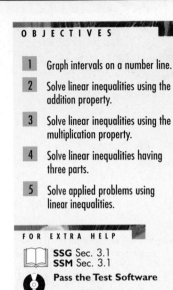

OBJECTIVES

1 Graph intervals on a number line.

2 Solve linear inequalities using the addition property.

3 Solve linear inequalities using the multiplication property.

4 Solve linear inequalities having three parts.

5 Solve applied problems using linear inequalities.

FOR EXTRA HELP

📖 **SSG** Sec. 3.1
   SSM Sec. 3.1

💿 **Pass the Test Software**

💿 **InterAct Math**
    Tutorial Software

📼 **Video 5**

Solving inequalities is closely related to the methods of solving equations. In this section we introduce properties that are essential for solving inequalities.

Inequalities are algebraic expressions related by

$<$ "is less than,"

$\leq$ "is less than or equal to,"

$>$ "is greater than,"

$\geq$ "is greater than or equal to."

We solve an inequality by finding all real number solutions for it. For example, the solution set of $x \leq 2$ includes all real numbers that are less than or equal to 2, and not just the integers less than or equal to 2. For example, -2.5, -1.7, -1, $\frac{7}{4}$, $\frac{1}{2}$, $\sqrt{2}$, and 2 are all real numbers less than or equal to 2, and are therefore solutions of $x \leq 2$.

OBJECTIVE **1** Graph intervals on a number line. A good way to show the solution set of an inequality is by graphing. We graph all the real numbers satisfying $x \leq 2$ by placing a square bracket at 2 on a number line and drawing an arrow extending from the bracket to the left (to represent the fact that all numbers less than 2 are also part of the graph). The graph is shown in Figure 1.

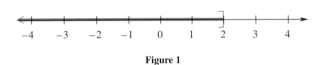

Figure 1

The set of numbers less than or equal to 2 is an example of an **interval** on the number line. To write intervals, we use **interval notation.** For example, using this notation, the interval of all numbers less than or equal to 2 is written as $(-\infty, 2]$. The negative infinity symbol $-\infty$ does not indicate a number. It is used to show that the interval includes all real numbers less than 2. As on the number line, the square bracket indicates that 2 is part of the solution. A parenthesis is always used next to the infinity symbol. The set of real numbers is written in interval notation as $(-\infty, \infty)$.

EXAMPLE 1   Graphing an Interval Written in Interval Notation on a Number Line

Write each inequality in interval notation and graph it.

(a) $x > -5$

The statement $x > -5$ says that x can represent any number greater than -5, but x cannot equal -5. The interval is written as $(-5, \infty)$. We show this on a graph by placing a parenthesis at -5 and drawing an arrow to the right, as in Figure 2. The parenthesis at -5 shows that -5 is not part of the graph.

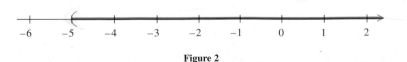

Figure 2

(b) $-1 \le x < 3$

This statement is read "-1 is less than or equal to x *and* x is less than 3." Thus, we want the set of numbers that are *between* -1 and 3, with -1 included and 3 excluded. In interval notation, we write this as $[-1, 3)$, using a square bracket at -1 because it is part of the graph, and a parenthesis at 3, because 3 is not part of the graph. The graph is shown in Figure 3.

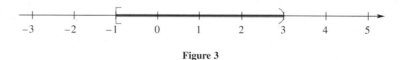

Figure 3

We now summarize the various types of intervals.

Type of Interval	Set	Interval Notation	Graph
Open interval	$\{x \mid a < x\}$	(a, ∞)	
	$\{x \mid a < x < b\}$	(a, b)	
	$\{x \mid x < b\}$	$(-\infty, b)$	
	$\{x \mid x \text{ is a real number}\}$	$(-\infty, \infty)$	
Half-open interval	$\{x \mid a \le x\}$	$[a, \infty)$	
	$\{x \mid a < x \le b\}$	$(a, b]$	
	$\{x \mid a \le x < b\}$	$[a, b)$	
	$\{x \mid x \le b\}$	$(-\infty, b]$	
Closed interval	$\{x \mid a \le x \le b\}$	$[a, b]$	

An **inequality** says that two expressions are *not* equal. Solving inequalities is similar to solving equations.

Linear Inequality

A **linear inequality in one variable** can be written in the form

$$ax < b,$$

where a and b are real numbers, with $a \ne 0$.

Examples of linear inequalities include

$$x + 5 < 2, \qquad y - 3 \geq 5, \qquad \text{or} \qquad 2k + 5 \leq 10.$$

(Throughout this section we give the definitions and rules only for $<$, but they are also valid for $>$, $\leq$, and $\geq$.)

OBJECTIVE **2** **Solve linear inequalities using the addition property.** An inequality is solved by finding all numbers that make the inequality true. Usually, an inequality has an infinite number of solutions. These solutions, like the solutions of equations, are found by producing a series of simpler equivalent inequalities. **Equivalent inequalities** are inequalities with the same solution set. We use the addition and multiplication properties of inequality to produce equivalent inequalities.

Addition Property of Inequality

For all real numbers a, b, and c, the inequalities

$$a < b \qquad \text{and} \qquad a + c < b + c$$

are equivalent. (The same number may be added to both sides of an inequality without changing the solution set.)

As with equations, the addition property can be used to *subtract* the same number from both sides of an inequality.

EXAMPLE 2 Using the Addition Property of Inequality

Solve $x - 7 < -12$.

Add 7 to both sides.

$$x - 7 + 7 < -12 + 7$$
$$x < -5$$

The solution set, $(-\infty, -5)$, is graphed in Figure 4.

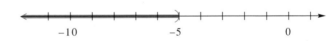

Figure 4

EXAMPLE 3 Using the Addition Property of Inequality

Solve the inequality $14 + 2m \leq 3m$, and graph the solution set.

First, subtract $2m$ from both sides.

$$14 + 2m \leq 3m$$
$$14 + 2m - 2m \leq 3m - 2m \qquad \text{Subtract } 2m.$$
$$14 \leq m \qquad \text{Combine like terms.}$$

The inequality $14 \leq m$ (14 is less than or equal to m) can also be written $m \geq 14$ (m is greater than or equal to 14). Notice that in each case, the inequality symbol

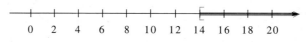

points to the smaller expression, 14. The solution set in interval notation is $[14, \infty)$. The graph is shown in Figure 5.

Figure 5

Errors often occur in graphing inequalities when the variable term is on the right side. (This is probably due to the fact that we read from left to right.) To guard against such errors, it is a good idea to rewrite these inequalities so that the variable is on the left, as discussed in Example 3.

OBJECTIVE **3**   Solve linear inequalities using the multiplication property.   An inequality such as $3x \leq 15$ can be solved by dividing both sides by 3. This is done with the multiplication property of inequality, which is a little more involved than the corresponding property for equations. To see how this property works, start with the true statement

$$-2 < 5.$$

Multiply both sides by, say, 8.

$$-2(8) < 5(8)$$
$$-16 < 40 \qquad \text{True}$$

This gives a true statement. Start again with $-2 < 5$, and this time multiply both sides by -8.

$$-2(-8) < 5(-8)$$
$$16 < -40 \qquad \text{False}$$

The result, $16 < -40$, is false. To make it true, change the direction of the inequality symbol to get

$$16 > -40.$$

As these examples suggest, multiplying both sides of an inequality by a *negative* number reverses the direction of the inequality symbol. The same is true for dividing by a negative number, since division is defined in terms of multiplication.

Multiplication Property of Inequality

For all real numbers a, b, and c, with $c \neq 0$,

(a) the inequalities

$$a < b \qquad \text{and} \qquad ac < bc$$

are equivalent if $c > 0$;

(b) the inequalities

$$a < b \qquad \text{and} \qquad ac > bc$$

are equivalent if $c < 0$. (Both sides of an inequality may be multiplied by a *positive* number without changing the direction of the inequality symbol. Multiplying or dividing by a *negative* number reverses the inequality symbol.)

> CAUTION Remember to reverse the direction of the inequality symbol when multiplying or dividing by a negative number.

E X A M P L E   4   Using the Multiplication Property of Inequality

Solve each inequality, then graph its solution set.

(a) $5m \leq -30$

Use the multiplication property to divide both sides by 5. Since $5 > 0$, do *not* reverse the inequality symbol.

$$5m \leq -30$$

$$\frac{5m}{5} \leq \frac{-30}{5} \qquad \text{Divide by 5.}$$

$$m \leq -6$$

The solution set, graphed in Figure 6, is the interval $(-\infty, -6]$.

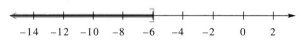

Figure 6

(b) $-4k \leq 32$

Divide both sides by -4. Since $-4 < 0$, the inequality symbol must be reversed.

$$-4k \leq 32$$

$$\frac{-4k}{-4} \geq \frac{32}{-4} \qquad \text{Divide by } -4 \text{ and reverse symbol.}$$

$$k \geq -8$$

Figure 7 shows the graph of the solution set, $[-8, \infty)$.

Figure 7

The steps used in solving a linear inequality are given below.

Solving a Linear Inequality

Step 1   **Simplify each side separately.**   Simplify each side of the inequality as much as possible by using the distributive property to clear parentheses and by combining like terms as needed.

Step 2   **Isolate the variable terms on one side.**   Use the addition property of inequality to get all terms with variables on one side of the inequality and all numbers on the other side.

Step 3   **Isolate the variable.**   Use the multiplication property to change the inequality to the form $x < k$ or $x > k$.

Remember: Reverse the direction of the inequality symbol **only** when **multiplying or dividing both sides of an inequality by a negative number.**

┌─ **E X A M P L E   5**   Solving a Linear Inequality Using the Distributive Property

Solve $-3(x + 4) + 2 \geq 8 - x$. Graph the solution set.

Step 1	$-3x - 12 + 2 \geq 8 - x$	Clear parentheses.
	$-3x - 10 \geq 8 - x$	Combine like terms.
Step 2	$-3x - 10 + x \geq 8 - x + x$	Add x on both sides.
	$-2x - 10 \geq 8$	
	$-2x - 10 + 10 \geq 8 + 10$	Add 10 on both sides.
	$-2x \geq 18$	
Step 3	$\dfrac{-2x}{-2} \leq \dfrac{18}{-2}$	Divide by -2, change $\geq$ to $\leq$.
	$x \leq -9$	

Figure 8 shows the graph of the solution set, $(-\infty, -9]$.

Figure 8

┌─ **E X A M P L E   6**   Solving a Linear Inequality with Fractions

Solve $-\dfrac{2}{3}(r - 3) - \dfrac{1}{2} < \dfrac{1}{2}(5 - r)$, and graph the solution set.

    To clear fractions, multiply both sides by the common denominator, 6. Then use the distributive property, and combine terms.

	$-\dfrac{2}{3}(r - 3) - \dfrac{1}{2} < \dfrac{1}{2}(5 - r)$	
	$-4(r - 3) - 3 < 3(5 - r)$	Multiply by 6, the LCD.
Step 1	$-4r + 12 - 3 < 15 - 3r$	Distributive property
	$-4r + 9 < 15 - 3r$	Combine like terms.
Step 2	$3r - 4r + 9 < 3r + 15 - 3r$	Add $3r$ to both sides.
	$-r + 9 < 15$	
	$-r + 9 - 9 < 15 - 9$	Subtract 9.
	$-r < 6$	

Step 3   To solve for r, multiply both sides of the inequality by -1. Since -1 is negative, change the direction of the inequality symbol.

$$(-1)(-r) > (-1)(6) \qquad \text{Multiply by } -1, \text{ change } < \text{ to } >.$$
$$r > -6$$

The solution set, $(-6, \infty)$, is graphed in Figure 9.

Figure 9

OBJECTIVE **4** Solve linear inequalities having three parts. For some applications, it is necessary to work with an inequality such as

$$3 < x + 2 < 8,$$

where $x + 2$ is *between* 3 and 8. To solve this inequality, we subtract 2 from each of the three parts of the inequality, giving

$$3 - 2 < x + 2 - 2 < 8 - 2$$
$$1 < x < 6.$$

The solution set, $(1, 6)$, is graphed in Figure 10.

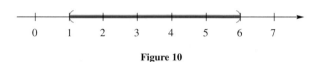

Figure 10

CAUTION When using inequalities with three parts like the one above, it is important to have the numbers in the correct positions. It would be *wrong* to write an inequality as $8 < x + 2 < 3$, since this would imply that $8 < 3$, a false statement. In general, three-part inequalities are written so that the symbols point in the same direction, and they both point toward the smaller number.

EXAMPLE 7 Solving a Three-Part Inequality

Solve the inequality $-2 \le 3k - 1 \le 5$, and graph the solution set.

To begin, we add 1 to each of the three parts.

$$-2 + 1 \le 3k - 1 + 1 \le 5 + 1 \qquad \text{Add 1.}$$
$$-1 \le 3k \le 6$$
$$\frac{-1}{3} \le \frac{3k}{3} \le \frac{6}{3} \qquad\qquad \text{Divide by 3.}$$
$$-\frac{1}{3} \le k \le 2$$

A graph of the solution set, $\left[-\frac{1}{3}, 2\right]$, is shown in Figure 11.

Figure 11

The types of solutions to linear equations or linear inequalities are shown in the box that follows.

Solution Sets of Linear Equations and Inequalities

Equation or Inequality	Typical Solution Set	Graph of Solution Set
Linear equation $ax = b$	$\{p\}$	
Linear inequality $ax < b$	$(-\infty, p)$ or (p, ∞)	
Three-part inequality $b < ax < c$	(p, q)	

OBJECTIVE **5** Solve applied problems using linear inequalities. In addition to the familiar "is less than" and "is greater than," the expressions "is no more than," and "is at least" also denote inequalities.

PROBLEM SOLVING

Expressions for inequalities sometimes appear in applied problems. The chart below shows how to interpret these expressions.

Word Expression	Interpretation	Word Expression	Interpretation
a is at least b	$a \geq b$	a is at most b	$a \leq b$
a is no less than b	$a \geq b$	a is no more than b	$a \leq b$

The final examples show how applied problems are solved using linear inequalities.

EXAMPLE 8   Solving an Application Using a Linear Inequality

A rental company charges $15 to rent a chain saw, plus $2 per hour. Rusty Brauner can spend no more than $35 to clear some logs from his yard. What is the maximum amount of time he can keep the rented saw?

Let h = the number of hours he can rent the saw. He must pay $15, plus $2h$, to rent the saw for h hours, and this amount must be *no more than* $35.

$$
\begin{array}{ccc}
\underbrace{\text{Cost of renting}} & \underbrace{\text{is no more than}} & \underbrace{\text{35 dollars.}} \\
15 + 2h & \leq & 35 \\
15 + 2h - 15 & \leq & 35 - 15 \qquad \text{Subtract 15.} \\
2h & \leq & 20 \\
h & \leq & 10 \qquad \text{Divide by 2.}
\end{array}
$$

He can keep the saw for a maximum of 10 hours. (Of course, he may keep it for less time, as indicated by the inequality $h \leq 10$.)

┌ **E X A M P L E   9**   Finding an Average Test Score

Andrew has test grades of 86, 88, and 78 on his first three tests in geometry. If he wants an average of at least 80 after his fourth test, what are the possible scores he can make on his fourth test?

Let x = Andrew's score on his fourth test. To find his average after 4 tests, add the test scores and divide by 4.

$$\underbrace{\text{Average}}_{} \quad \underbrace{\substack{\text{is at}\\\text{least}}}_{} \quad \underbrace{80.}_{}$$

$$\frac{86 + 88 + 78 + x}{4} \geq 80$$

$$\frac{252 + x}{4} \geq 80 \qquad \text{Add the known scores.}$$

$$4\left(\frac{252 + x}{4}\right) \geq 4(80) \qquad \text{Multiply by 4.}$$

$$252 + x \geq 320$$

$$252 - 252 + x \geq 320 - 252 \qquad \text{Subtract 252.}$$

$$x \geq 68 \qquad \text{Combine terms.}$$

He must score 68 or more on the fourth test to have an average of *at least* 80.

3.1 EXERCISES

Match each inequality with the correct graph or interval notation. See Example 1.

1. $x \leq 3$ **A.** [graph with bracket at 3, 0 and 3 labeled]

2. $x > 3$

3. $x < 3$ **B.** [graph with parenthesis at 3, 0 and 3 labeled]

4. $x \geq 3$

5. $-3 \leq x \leq 3$

6. $-3 < x < 3$ **C.** $(3, \infty)$

D. $(-\infty, 3]$

E. $(-3, 3)$

F. $[-3, 3]$

7. Explain how to determine whether to use parentheses or brackets when graphing the solution set of an inequality.

8. Describe the steps used to solve a linear inequality. Explain when it is necessary to reverse the inequality symbol.

Solve each inequality. Give the solution set in both interval and graph forms. See Examples 2–6.

9. $4x + 1 \geq 21$ **10.** $5t + 2 \geq 52$ **11.** $\dfrac{3k - 1}{4} > 5$ **12.** $\dfrac{5z - 6}{8} < 8$

13. $-4x < 16$ **14.** $-2m > 10$ **15.** $-\dfrac{3}{4}r \geq 30$ **16.** $-1.5y \leq -\dfrac{9}{2}$

17. $-1.3m \geq -5.2$ **18.** $-2.5y \leq -1.25$ **19.** $\dfrac{2k - 5}{-4} > 5$ **20.** $\dfrac{3z - 2}{-5} < 6$

21. $y + 4(2y - 1) \geq y$

22. $m - 2(m - 4) \leq 3m$

23. $-(4 + r) + 2 - 3r < -14$

24. $-(9 + k) - 5 + 4k \geq 4$

25. $-3(z - 6) > 2z - 2$

26. $-2(y + 4) \leq 6y + 16$

27. $\frac{2}{3}(3k - 1) \geq \frac{3}{2}(2k - 3)$

28. $\frac{7}{5}(10m - 1) < \frac{2}{3}(6m + 5)$

29. $-\frac{1}{4}(p + 6) + \frac{3}{2}(2p - 5) < 10$

30. $\frac{3}{5}(k - 2) - \frac{1}{4}(2k - 7) \leq 3$

31. $3(2x - 4) - 4x < 2x + 3$

32. $7(4 - x) + 5x < 2(16 - x)$

33. $8\left(\frac{1}{2}x + 3\right) < 8\left(\frac{1}{2}x - 1\right)$

34. $10x + 2(x - 4) < 12x - 10$

35. A student solved the inequality $5x < -20$ by dividing both sides by 5 and reversing the direction of the inequality symbol. His reasoning was that since -20 is a negative number, reversing the direction of the symbol was required. Is this correct? Explain why or why not.

*No because # being ÷'d must be neg. →
not # being ÷'d into.*

36. Which is the graph of $-2 < x$?

(a)

(b)

(c)

(d)

Solve the inequality. Give the solution set in both interval and graph forms. See Example 7.

37. $-4 < x - 5 < 6$

38. $-1 < x + 1 < 8$

39. $-9 \leq k + 5 \leq 15$

40. $-4 \leq m + 3 \leq 10$

41. $-6 \leq 2z + 4 \leq 16$

42. $-15 < 3p + 6 < -12$

43. $-19 \leq 3x - 5 \leq 1$

44. $-16 < 3t + 2 < -10$

45. $-1 < \frac{2x - 5}{6} \leq 5$

46. $-3 \leq \frac{3m + 1}{4} \leq 3$

47. $4 \leq 5 - 9x < 8$

48. $4 \leq 3 - 2x < 8$

Stop

Answer the questions in Exercises 49–52 based on the given graph.

49. In which months did the percent of tornadoes exceed 7.7%?

50. In which months was the percent of tornadoes at least 12.9%?

51. The data used to determine the graph were based on the number of tornadoes sighted in the United States during the last twenty years. A total of 17,252 tornadoes were reported. In which months were fewer than 1500 reported?

52. How many more tornadoes occurred during March than October? (Use the total given in Exercise 51.)

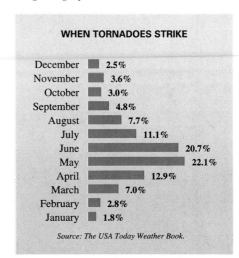

WHEN TORNADOES STRIKE

December 2.5%
November 3.6%
October 3.0%
September 4.8%
August 7.7%
July 11.1%
June 20.7%
May 22.1%
April 12.9%
March 7.0%
February 2.8%
January 1.8%

Source: The USA Today Weather Book.

This bar graph (first given in the chapter introduction) shows the increase in the number of theater tickets sold (in millions) from the 1990–1991 season through the 1996–1997 season. Use the graph to answer the questions in Exercises 53–56.

53. During which seasons were ticket sales less than or equal to 8 million?

54. During which seasons were ticket sales greater than 8 million but less than 10 million?

55. Suppose ticket sales were as follows: 7.3 million during the 1990–1991 season, 7.5 million during the 1991–1992 season, and 7.9 million during the 1992–1993 season. How many million tickets would need to be sold during the 1993–1994 season to average sales of at least 7.75 million tickets each season over this period of time? (*Hint:* See Example 9.)

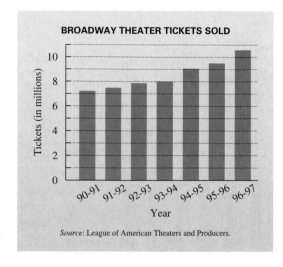

56. Refer to Exercise 55. Were enough tickets sold during the 1993–1994 season to average sales of at least 7.75 million tickets each season from 1990–1994? Explain your answer.

Solve each problem. See Examples 8 and 9.

57. In a midwestern city, taxicabs charge $1.50 for the first $\frac{1}{5}$ mile and $.25 for each additional $\frac{1}{5}$ mile. Dantrell Davis has only $3.75 in his pocket. What is the maximum distance he can travel (not including a tip for the cabbie)?

58. Eleven years ago taxicab fares in the city in Exercise 57 were $.90 for the first $\frac{1}{7}$ mile and $.10 for each additional $\frac{1}{7}$ mile. Based on the information given there and the answer you found, how much farther could Dantrell Davis have traveled at that time?

59. Margaret Westmoreland earned scores of 90 and 82 on her first two tests in English Literature. What score must she make on her third test to keep an average of 84 or greater?

60. Susan Carsten scored 92 and 96 on her first two tests in Methods in Teaching Art. What score must she make on her third test to keep an average of 90 or greater?

61. A couple wishes to rent a car for one day while on vacation. Ford Automobile Rental wants $35.00 per day and 14¢ per mile, while Chevrolet-For-A-Day wants $34.00 per day and 16¢ per mile. After how many miles would the price to rent the Chevrolet exceed the price to rent a Ford?

62. Jane and Terry Brandsma vacationed in Long Island for a week. They needed to rent a car, so they checked out two rental firms. Avis wanted $28 per day, with no mileage fee. Downtown Toyota wanted $108 per week and 14¢ per mile. How many miles would they have to drive before the Avis price is less than the Toyota price?

A product will produce a profit only when the revenue R from selling the product exceeds the cost C of producing it. In Exercises 63 and 64 find the smallest whole number of units x that must be sold for the business to show a profit for the item described.

63. Peripheral Visions, Inc. finds that the cost to produce x studio quality videotapes is $C = 20x + 100$, while the revenue produced from them is $R = 24x$ (C and R in dollars).

64. Speedy Delivery finds that the cost to make x deliveries is $C = 3x + 2300$, while the revenue produced from them is $R = 5.50x$ (C and R in dollars).

RELATING CONCEPTS (EXERCISES 65–69)

Work Exercises 65–69 in order.

65. Solve the linear equation $5(x + 3) - 2(x - 4) = 2(x + 7)$, and graph the solution set on a number line.

66. Solve the linear inequality $5(x + 3) - 2(x - 4) > 2(x + 7)$, and graph the solution set on a number line.

67. Solve the linear inequality $5(x + 3) - 2(x - 4) < 2(x + 7)$, and graph the solution set on a number line.

68. Graph all the solution sets of the equation and inequalities in Exercises 65–67 on the same number line. What set do you obtain?

69. Based on the results of Exercises 65–67, complete the following using a conjecture (educated guess): The solution set of $-3(x + 2) = 3x + 12$ is $\{-3\}$, and the solution set of $-3(x + 2) < 3x + 12$ is $(-3, \infty)$. Therefore the solution set of $-3(x + 2) > 3x + 12$ is _____.

Did you make the connection that the solution sets of the $<$ and $>$ inequalities that correspond to a linear equation are separated by the solution of the equation?

70. Suppose that $-1 < x < 5$. Complete this inequality: _____ $< -x <$ _____.

71. Suppose that $4 < y < 1$. What can you say about y?

72. What is wrong with writing a statement of inequality as $10 > 5 < 8$? What three-part inequality would avoid this problem?

3.2 Set Operations and Compound Inequalities

OBJECTIVES

1. Find the intersection of two sets.
2. Solve compound inequalities with the word *and*.
3. Find the union of two sets.
4. Solve compound inequalities with the word *or*.

In this section we discuss the use of the words *and* and *or* as they relate to sets and inequalities.

OBJECTIVE **1** Find the intersection of two sets. The intersection of two sets is defined using the word *and*.

Intersection of Sets

For any two sets A and B, the **intersection** of A and B, symbolized $A \cap B$, is defined as follows:

$$A \cap B = \{x \mid x \text{ is an element of } A \textbf{ and } x \text{ is an element of } B\}.$$

EXAMPLE 1   Finding the Intersection of Two Sets

Let $A = \{1, 2, 3, 4\}$ and $B = \{2, 4, 6\}$. Find $A \cap B$.

The set $A \cap B$ contains those elements that belong to both A *and* B at the same time: the numbers 2 and 4. Therefore,

$$A \cap B = \{1, 2, 3, 4\} \cap \{2, 4, 6\}$$
$$= \{2, 4\}.$$

A **compound inequality** consists of two inequalities linked by a connective word such as *and* or *or*. Examples of compound inequalities are

$$x + 1 \leq 9 \quad \text{and} \quad x - 2 > 3$$

and

$$2x \geq 4 \quad \text{or} \quad 3x - 6 < 5.$$

OBJECTIVE 2 Solve compound inequalities with the word *and*. Use the following steps.

Solving Inequalities with *and*

Step 1   Solve each inequality in the compound inequality individually.

Step 2   Since the inequalities are joined with *and*, the solution will include all numbers that satisfy both solutions in Step 1 at the same time (the intersection of the solution sets).

The next examples show how to solve a compound inequality with *and*.

EXAMPLE 2   Solving a Compound Inequality with *and*

Solve the compound inequality

$$x + 1 \leq 9 \quad \text{and} \quad x - 2 \geq 3.$$

Step 1   Solve each inequality in the compound inequality individually.

$$x + 1 \leq 9 \quad \text{and} \quad x - 2 \geq 3$$
$$x \leq 8 \quad \text{and} \quad x \geq 5$$

Step 2   Since the inequalities are joined with the word *and*, the solution set will include all numbers that satisfy both solutions in Step 1 at the same time. Thus, the compound inequality is true whenever $x \leq 8$ and $x \geq 5$ are both true. The top graph in Figure 12 shows $x \leq 8$ and the bottom graph shows $x \geq 5$.

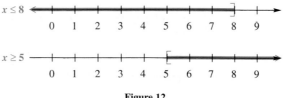

Figure 12

Next, we find the intersection of the two graphs in Figure 12 to get the solution set of the compound inequality. The solution set consists of all numbers between 5 and 8, including both 5 and 8. This is the intersection of the two graphs, written in interval notation as [5, 8]. See Figure 13.

Figure 13

E X A M P L E   3   Solving a Compound Inequality with *and*

Solve the compound inequality

$$-3x - 2 > 4 \quad \text{and} \quad 5x - 1 \le -21.$$

Step 1   Solve each inequality separately.

$$-3x - 2 > 4 \quad \text{and} \quad 5x - 1 \le -21$$
$$-3x > 6 \quad \text{and} \quad 5x \le -20$$
$$x < -2 \quad \text{and} \quad x \le -4$$

The graphs of $x < -2$ and $x \le -4$ are shown in Figure 14.

Figure 14

Step 2   Now find all values of x that satisfy both conditions; that is, the real numbers that are less than -2 and also less than or equal to -4. As shown by the graph in Figure 15, the solution set is $(-\infty, -4]$.

Figure 15

E X A M P L E   4   Solving a Compound Inequality with *and*

Solve $x + 2 < 5$ and $x - 10 > 2$.

First solve each inequality separately.

$$x + 2 < 5 \quad \text{and} \quad x - 10 > 2$$
$$x < 3 \quad \text{and} \quad x > 12$$

The graphs of $x < 3$ and $x > 12$ are shown in Figure 16.

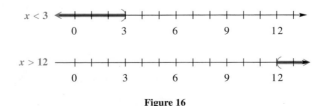

Figure 16

There is no number that is both less than 3 *and* greater than 12, so the given compound inequality has no solution. The solution set is ∅. See Figure 17.

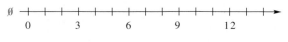

Figure 17

OBJECTIVE 3 **Find the union of two sets.** The union of two sets is defined using the word *or*.

Union of Sets

For any two sets *A* and *B*, the **union** of *A* and *B*, symbolized *A* ∪ *B*, is defined as follows:

$$A \cup B = \{x \mid x \text{ is an element of } A \textbf{ or } x \text{ is an element of } B\}.$$

EXAMPLE 5 Finding the Union of Two Sets

Find the union of the sets $A = \{1, 2, 3, 4\}$ and $B = \{2, 4, 6\}$.

We begin by listing all the elements of set *A:* 1, 2, 3, 4. Then we list any additional elements from set *B*. In this case the elements 2 and 4 are already listed, so the only additional element is 6. Therefore,

$$A \cup B = \{1, 2, 3, 4\} \cup \{2, 4, 6\} = \{1, 2, 3, 4, 6\}.$$

The union consists of all elements in either *A or B* (or both).

Notice in Example 5, that even though the elements 2 and 4 appeared in both sets *A* and *B*, they are only written once in *A* ∪ *B*.

OBJECTIVE 4 **Solve compound inequalities with the word *or*.** Use the following steps.

Solving Inequalities with *or*

Step 1 Solve each inequality in the compound inequality individually.

Step 2 Since the inequalities are joined with *or*, the solution will include all numbers that satisfy either one or both of the solutions in Step 1 (the union of the solution sets).

The next examples show how to solve a compound inequality with *or*.

EXAMPLE 6 Solving a Compound Inequality with *or*

Solve $6x - 4 < 2x$ or $-3x \le -9$.

Solve each inequality separately (Step 1).

$$
\begin{aligned}
6x - 4 &< 2x & \text{or} & & -3x &\le -9 \\
4x &< 4 \\
x &< 1 & \text{or} & & x &\ge 3
\end{aligned}
$$

The graphs of these two inequalities are shown in Figure 18.

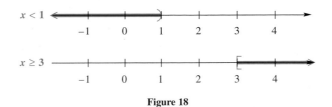

Figure 18

Since the inequalities are joined with *or*, find the union of the two sets (Step 2). The union is shown in Figure 19, and is written $(-\infty, 1) \cup [3, \infty)$.

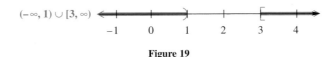

Figure 19

When inequalities are used to write the solution set in Example 6, it *should* be written as

$$x < 1 \text{ or } x \geq 3,$$

which keeps the numbers 1 and 3 in their order on the number line. Writing $3 \leq x < 1$ would imply that $3 \leq 1$, which is **FALSE**. Remember, there is no short-cut way to write the solution of such a union.

EXAMPLE 7   Solving a Compound Inequality with *or*

Solve $-4x + 1 \geq 9$ or $5x + 3 \geq -12$.

Solve each inequality separately.

$$
\begin{array}{rcl}
-4x + 1 \geq 9 & \text{or} & 5x + 3 \geq -12 \\
-4x \geq 8 & \text{or} & 5x \geq -15 \\
x \leq -2 & \text{or} & x \geq -3
\end{array}
$$

The graphs of these two inequalities are shown in Figure 20.

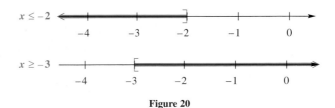

Figure 20

By taking the union, we obtain every real number as a solution, since every real number satisfies at least one of the two inequalities. The set of all real numbers is written in interval notation as $(-\infty, \infty)$ and graphed in Figure 21.

Figure 21

E X A M P L E 8   Applying Intersection and Union

The five highest-grossing films of all time (adjusted for inflation) are listed below. Gross income is given through May 4, 1997.

Five All-Time Highest-Grossing Films

Film	Admissions	Gross Income
Gone with the Wind	197,548,731	$869,214,425
Star Wars	176,063,374	$774,992,216
E.T.	135,987,938	$598,346,929
The Ten Commandments	131,000,000	$576,400,000
The Sound of Music	130,571,938	$574,514,287

Source: Exhibitor Relations Co.

List the elements of the following sets.

(a) the set of top five films with admissions greater than 170,000,000 *and* gross greater than $800,000,000

The only film that satisfies both conditions is *Gone with the Wind,* so the set is

$$\{Gone\ with\ the\ Wind\}.$$

(b) the set of top five films with admissions less than 170,000,000 *or* gross greater than $700,000,000

Here, a film that satisfies at least one of the conditions is in the set. This set includes all five films:

$$\{Gone\ with\ the\ Wind,\ Star\ Wars,\ E.T.,\ The\ Ten\ Commandments,$$
$$The\ Sound\ of\ Music\}.$$

3.2 EXERCISES

Decide whether the statement is true or false. If it is false, explain why.

1. The union of the solution sets of $x + 1 = 5$, $x + 1 < 5$, and $x + 1 > 5$ is $(-\infty, \infty)$.

2. The intersection of the sets $\{x \mid x \geq 7\}$ and $\{x \mid x \leq 7\}$ is $\emptyset$.

3. The union of the sets $(-\infty, 8)$ and $(8, \infty)$ is $\{8\}$.

4. The intersection of the sets $(-\infty, 8]$ and $[8, \infty)$ is $\{8\}$.

5. The intersection of the set of rational numbers and the set of irrational numbers is $\{0\}$.

6. The union of the set of rational numbers and the set of irrational numbers is the set of real numbers.

Let $A = \{1, 2, 3, 4, 5, 6\}$, $B = \{1, 3, 5\}$, $C = \{1, 6\}$, and $D = \{4\}$. Specify each of the following sets. See Examples 1 and 5.

7. $B \cap A$ **8.** $A \cap B$ **9.** $A \cap D$ **10.** $B \cap C$

11. $B \cap \emptyset$ **12.** $A \cap \emptyset$ **13.** $A \cup B$ **14.** $B \cup D$

15. $B \cup C$ **16.** $C \cup B$ **17.** $C \cup D$ **18.** $D \cup C$

19. Use the sets *A*, *B*, and *C* for Exercises 7–18 to show that *A* ∩ (*B* ∩ *C*) is equal to (*A* ∩ *B*) ∩ *C*. This is true for any choices of sets. What property does this illustrate? (*Hint:* See Section 1.7.)

20. Repeat Exercise 19, showing that *A* ∪ (*B* ∪ *C*) is equal to (*A* ∪ *B*) ∪ *C*.

21. Write a few sentences showing how the concept of intersection can be applied to a real-life situation.

22. A compound inequality uses one of the words *and* or *or*. Explain how you will determine whether to use *intersection* or *union* when graphing the solution set.

Two sets are specified by graphs. Graph the intersection of the two sets.

23.   **24.**
25.   **26.**

For each compound inequality, give the solution set in both interval and graph forms. See Examples 2–4.

27. $x - 3 \leq 6$ and $x + 2 \geq 7$ **28.** $x + 5 \leq 11$ and $x - 3 \geq -1$

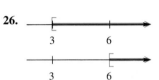

29. $-3x > 3$ and $x + 3 > 0$ **30.** $-3x < 3$ and $x + 2 < 6$

31. $3x - 4 \leq 8$ and $-4x + 1 \geq -15$ **32.** $7x + 6 \leq 48$ and $-4x \geq -24$

Two sets are specified by graphs. Graph the union of the two sets.

33.   **34.**
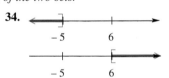
35.   **36.**
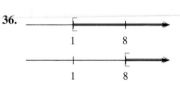

For each compound inequality, give the solution set in both interval and graph forms. See Examples 6 and 7.

37. $x + 2 > 7$ or $1 - x > 6$ **38.** $x + 1 > 3$ or $x + 4 < 2$

39. $x + 1 > 3$ or $-4x + 1 > 5$ **40.** $3x < x + 12$ or $x + 1 > 10$

Express each set in the simplest interval form.

41. $(-\infty, -1] \cap [-4, \infty)$ **42.** $[-1, \infty) \cap (-\infty, 9]$

43. $(-\infty, -6] \cap [-9, \infty)$ **44.** $(5, 11] \cap [6, \infty)$

45. $(-\infty, 3) \cup (-\infty, -2)$ **46.** $[-9, 1] \cup (-\infty, -3)$

47. $[3, 6] \cup (4, 9)$ **48.** $[-1, 2] \cup (0, 5)$

Stop

For each compound inequality, decide whether intersection or union should be used. Then give the solution set in both interval and graph forms. See Examples 2, 3, 4, 6, and 7.

49. $x < -1$ and $x > -5$

50. $x > -1$ and $x < 7$

51. $x < 4$ or $x < -2$

52. $x < 5$ or $x < -3$

53. $-3x \leq -6$ or $-3x \geq 0$

54. $2x - 6 \leq -18$ and $2x \geq -18$

 55. $x + 1 \geq 5$ and $x - 2 \leq 10$

56. $-8x \leq -24$ or $-5x \geq 15$

Use the graphs to answer the questions in Exercises 57 and 58.

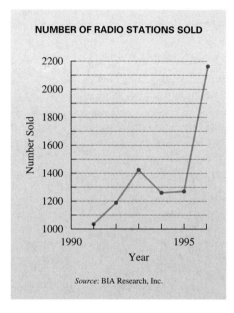

NUMBER OF RADIO STATIONS SOLD

Source: BIA Research, Inc.

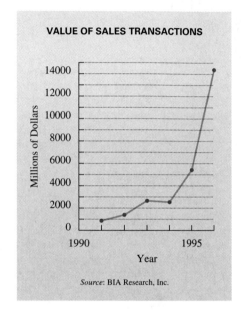

VALUE OF SALES TRANSACTIONS

Source: BIA Research, Inc.

57. In which years did the number of stations sold exceed 1200 *and* the value of transactions exceed 3000 (million dollars)?

58. In which years was the number of stations sold greater than 1300 *or* the value of transactions less than 3000 (million dollars)?

RELATING CONCEPTS (EXERCISES 59–64)

The figures represent the backyards of neighbors Luigi, Maria, Than, and Joe. Find the area and the perimeter of each yard. Suppose that each resident has 150 feet of fencing and enough sod to cover 1400 square feet of lawn. Give the name or names of the residents whose yards satisfy the following descriptions.

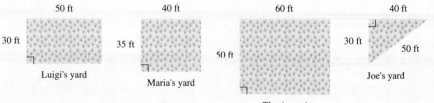

59. The yard can be fenced *and* the yard can be sodded.

60. The yard can be fenced *and* the yard cannot be sodded.

61. The yard cannot be fenced *and* the yard can be sodded.

62. The yard cannot be fenced *and* the yard cannot be sodded.

63. The yard can be fenced *or* the yard can be sodded.

64. The yard cannot be fenced *or* the yard can be sodded.

Did you make the connection that intersections and unions are useful in applied as well as theoretical situations?

3.3 Absolute Value Equations and Inequalities

OBJECTIVES

1 Use the distance definition of absolute value.

2 Solve equations of the form $|ax + b| = k$, for $k > 0$.

3 Solve inequalities of the form $|ax + b| < k$ and of the form $|ax + b| > k$, for $k > 0$.

4 Solve absolute value equations that involve rewriting.

5 Solve equations of the form $|ax + b| = |cx + d|$.

6 Solve special cases of absolute value equations and inequalities.

FOR EXTRA HELP

SSG Sec. 3.3
SSM Sec. 3.3

Pass the Test Software

InterAct Math Tutorial Software

Video 5

Sometimes a model requires that a linear expression must be positive. We use absolute value to ensure that this will happen. Such situations are expressed as absolute value equations or inequalities.

OBJECTIVE 1 Use the distance definition of absolute value. In Chapter 1 we saw that the absolute value of a number x, written $|x|$, represents the distance from x to 0 on the number line. For example, the solutions of $|x| = 4$ are 4 and -4, as shown in Figure 22. Since absolute value represents distance from 0, it is reasonable to interpret the solutions of $|x| > 4$ to be all numbers that are *more* than 4 units from 0. The set $(-\infty, -4) \cup (4, \infty)$ fits this description. Figure 23 shows the graph of the solution set of $|x| > 4$. The solution set of $|x| < 4$ consists of all numbers that are *less* than 4 units from 0 on the number line. Another way of thinking of this is to think of all numbers between -4 and 4. This set of numbers is given by $(-4, 4)$, as shown in Figure 24.

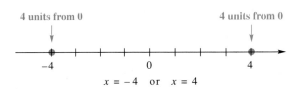

$x = -4$ or $x = 4$

Figure 22

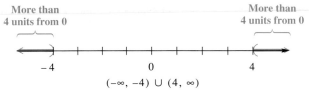

$(-\infty, -4) \cup (4, \infty)$

Figure 23

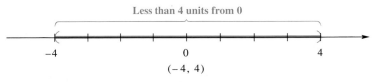

$(-4, 4)$

Figure 24

The equation and inequalities just described are examples of **absolute value equations and inequalities.** They involve the absolute value of a variable expression, and generally take the form

$$|ax + b| = k, \qquad |ax + b| > k, \qquad \text{or} \qquad |ax + b| < k,$$

where k is a positive number. We solve them by rewriting them as compound equations or inequalities.

O B J E C T I V E **2**   Solve equations of the form $|ax + b| = k$, for $k > 0$.   The first example shows how to solve a typical absolute value equation. Remember that since absolute value refers to distance from the origin, most absolute value equations will have two cases.

E X A M P L E   1   Solving an Absolute Value Equation

Solve $|2x + 1| = 7$.

For $|2x + 1|$ to equal 7, $2x + 1$ must be 7 units from 0 on the number line. This can happen only when $2x + 1 = 7$ or $2x + 1 = -7$. We solve this compound equation as follows.

$$
\begin{array}{rcl}
2x + 1 = 7 & \text{or} & 2x + 1 = -7 \\
2x = 6 & \text{or} & 2x = -8 \\
x = 3 & \text{or} & x = -4
\end{array}
$$

The solution set is $\{-4, 3\}$. Its graph is shown in Figure 25.

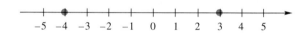

Figure 25

O B J E C T I V E **3**   Solve inequalities of the form $|ax + b| < k$ and of the form $|ax + b| > k$, for $k > 0$.

E X A M P L E   2   Solving an Absolute Value Inequality with $>$

Solve $|2x + 1| > 7$.

This absolute value inequality must be rewritten as

$$2x + 1 > 7 \qquad \text{or} \qquad 2x + 1 < -7,$$

because $2x + 1$ must represent a number that is *more* than 7 units from 0 on either side of the number line. Now we solve the compound inequality.

$$
\begin{array}{rcl}
2x + 1 > 7 & \text{or} & 2x + 1 < -7 \\
2x > 6 & \text{or} & 2x < -8 \\
x > 3 & \text{or} & x < -4
\end{array}
$$

The solution set, $(-\infty, -4) \cup (3, \infty)$, is graphed in Figure 26. Notice that the graph consists of two intervals.

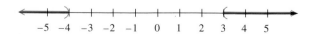

Figure 26

E X A M P L E   3   Solving an Absolute Value Inequality with $<$

Solve $|2x + 1| < 7$.

Here the expression $2x + 1$ must represent a number that is less than 7 units from 0 on the number line. Another way of thinking of this is to realize that $2x + 1$ must be between -7 and 7. This is written as the three-part inequality

$$-7 < 2x + 1 < 7.$$

We solved such inequalities in Section 3.1 by working with all three parts at the same time.

$$-7 < 2x + 1 < 7$$
$$-8 < 2x < 6 \qquad \text{Subtract 1 from each part.}$$
$$-4 < x < 3 \qquad \text{Divide each part by 2.}$$

The solution set is $(-4, 3)$, and the graph consists of a single interval, as shown in Figure 27.

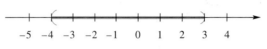

Figure 27

Look back at Figures 25, 26, and 27. These are the graphs of $|2x + 1| = 7$, $|2x + 1| > 7$, and $|2x + 1| < 7$. If we find the union of the three sets, we get the set of all real numbers. This is because for any value of x, $|2x + 1|$ will satisfy one and only one of the following: It is equal to 7, greater than 7, or less than 7.

When solving absolute value equations and inequalities of the types in Examples 1, 2, and 3, remember the following.

1. The methods described apply when the constant is alone on one side of the equation or inequality and is *positive*.
2. Absolute value inequalities written in the form $|ax + b| > k$ and absolute value equations translate into "or" compound statements.
3. Absolute value inequalities in the form $|ax + b| < k$ translate into "and" compound statements, which may be written as three-part inequalities.
4. An "or" statement *cannot* be written in three parts. It would be *incorrect* to use

$$-7 > 2x + 1 > 7$$

in Example 2, since this would imply that $-7 > 7$, which is false.

OBJECTIVE 4 Solve absolute value equations that involve rewriting. Sometimes an absolute value equation or inequality requires some rewriting before it can be set up as a compound statement, as shown in the next example.

EXAMPLE 4 Solving an Absolute Value Equation Requiring Rewriting

Solve the equation $|x + 3| + 5 = 12$.

First get the absolute value expression alone on one side of the equals sign. Do this by subtracting 5 on each side.

$$|x + 3| + 5 - 5 = 12 - 5 \qquad \text{Subtract 5.}$$
$$|x + 3| = 7$$

Then use the method shown in Example 1.

$$x + 3 = 7 \qquad \text{or} \qquad x + 3 = -7$$
$$x = 4 \qquad \text{or} \qquad x = -10$$

Check that the solution set is $\{4, -10\}$ by substituting in the original equation. ∎

Use a similar method to solve an absolute value *inequality* that requires rewriting. The methods used in Examples 1, 2, and 3 are summarized here.

Solving Absolute Value Equations and Inequalities

Let k be a positive number, and p and q be two numbers.

1. To solve $|ax + b| = k$, solve the compound equation

$$ax + b = k \qquad \text{or} \qquad ax + b = -k.$$

The solution set is of the form $\{p, q\}$, with two numbers.

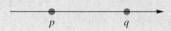

2. To solve $|ax + b| > k$, solve the compound inequality

$$ax + b > k \qquad \text{or} \qquad ax + b < -k.$$

The solution set is of the form $(-\infty, p) \cup (q, \infty)$, which consists of two separate intervals.

3. To solve $|ax + b| < k$, solve the compound inequality

$$-k < ax + b < k.$$

The solution set is of the form (p, q), a single interval.

Some people prefer to write the compound statements in parts 1 and 2 of the summary as

$$ax + b = k \qquad \text{or} \qquad -(ax + b) = k$$

and

$$ax + b > k \qquad \text{or} \qquad -(ax + b) > k.$$

These forms are equivalent to those we give in the summary and produce the same results.

OBJECTIVE 5   Solve equations of the form $|ax + b| = |cx + d|$.   For two expressions to have the same absolute value, they must either be equal or be negatives of each other.

Solving $|ax + b| = |cx + d|$

To solve an absolute value equation of the form

$$|ax + b| = |cx + d|,$$

solve the compound equation

$$ax + b = cx + d \qquad \text{or} \qquad ax + b = -(cx + d).$$

EXAMPLE 5   Solving an Equation with Two Absolute Values

Solve the equation $|z + 6| = |2z - 3|$.

This equation is satisfied either if $z + 6$ and $2z - 3$ are equal to each other, or if $z + 6$ and $2z - 3$ are negatives of each other. Thus, we have

$$z + 6 = 2z - 3 \qquad \text{or} \qquad z + 6 = -(2z - 3).$$

Solve each equation.

$$6 + 3 = 2z - z \qquad \text{or} \qquad z + 6 = -2z + 3$$
$$\qquad\qquad\qquad\qquad\qquad\qquad 3z = -3$$
$$9 = z \qquad\qquad \text{or} \qquad\qquad z = -1$$

The solution set is $\{9, -1\}$.

OBJECTIVE 6   Solve special cases of absolute value equations and inequalities.   When a typical absolute value equation or inequality involves a *negative* constant or *zero* alone on one side, use the properties of absolute value to solve. Keep in mind the following.

1. The absolute value of an expression can never be negative: $|a| \geq 0$ for all real numbers a.

2. The absolute value of an expression equals 0 only when the expression is equal to 0.

The next two examples illustrate these special cases.

EXAMPLE 6   Solving Special Cases of Absolute Value Equations

Solve each equation.

(a) $|5r - 3| = -4$

Since the absolute value of an expression can never be negative, there are no solutions for this equation. The solution set is $\emptyset$.

(b) $|7x - 3| = 0$

The expression $7x - 3$ will equal 0 *only* for the solution of the equation

$$7x - 3 = 0.$$

The solution of this equation is $\frac{3}{7}$. The solution set is $\left\{\frac{3}{7}\right\}$. It consists of only one element.

E X A M P L E 7   Solving Special Cases of Absolute Value Inequalities

Solve each of the following inequalities.

(a) $|x| \geq -4$

The absolute value of a number is never negative. For this reason, $|x| \geq -4$ is true for *all* real numbers. The solution set is $(-\infty, \infty)$.

(b) $|k + 6| - 3 < -5$

Add 3 to both sides to get the absolute value expression alone on one side.

$$|k + 6| < -2$$

There is no number whose absolute value is less than -2, so this inequality has no solution. The solution set is $\emptyset$.

(c) $|m - 7| + 4 \leq 4$

Adding -4 to both sides gives

$$|m - 7| \leq 0.$$

The value of $|m - 7|$ will never be less than 0. However, $|m - 7|$ will equal 0 when $m = 7$. Therefore, the solution set is $\{7\}$.

CONNECTIONS

Absolute value is used to find the relative error of a measurement in science, engineering, manufacturing, and other fields. If x_t represents the true value of a measurement and x represents the measured value, then the *relative error in x* equals the absolute value of the difference between x_t and x divided by x_t. That is,

$$\text{relative error in } x = \left|\frac{x_t - x}{x_t}\right|.$$

In many situations in the work world, the relative error must be less than some predetermined amount. For example, suppose a machine filling *quart* milk cartons is set for a relative error no greater than .05. Here $x_t = 32$ ounces, the relative error $= .05$ ounce, and we must find x, given

$$\left|\frac{32 - x}{32}\right| = \left|1 - \frac{x}{32}\right| \leq .05.$$

FOR DISCUSSION OR WRITING
With this tolerance level, how many ounces may a carton contain?

3.3 EXERCISES

Keeping in mind that the absolute value of a number can be interpreted as the distance between the graph of the number and 0 on the number line, match each absolute value equation or inequality with the graph of its solution set.

Choices

1. (a) $|x| = 5$ **A.**
    (b) $|x| < 5$
    (c) $|x| > 5$
    (d) $|x| \le 5$ **B.**
    (e) $|x| \ge 5$

      C.

      D.

      E.

2. (a) $|x| = 9$ **A.**
    (b) $|x| > 9$
    (c) $|x| \ge 9$
    (d) $|x| < 9$ **B.**
    (e) $|x| < 9$

      C.

      D.

      E.

3. How many solutions will $|ax + b| = k$, have if (a) $k = 0$; (b) $k > 0$; (c) $k < 0$?

4. The graph of the solution set of $|2x + 1| = 9$ is given here.

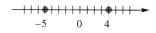

Without actually doing the algebraic work, graph the solution set of each inequality, referring to the graph above.
   (a) $|2x + 1| < 9$ (b) $|2x + 1| > 9$

Solve each equation. See Example 1.

5. $|x| = 12$ **6.** $|k| = 14$ **7.** $|4x| = 20$ **8.** $|5x| = 30$

9. $|y - 3| = 9$ **10.** $|p - 5| = 13$ **11.** $|2x + 1| = 7$ **12.** $|2y + 3| = 19$

13. $|4r - 5| = 17$ **14.** $|5t - 1| = 21$ **15.** $|2y + 5| = 14$ **16.** $|2x - 9| = 18$

17. $\left|\dfrac{1}{2}x + 3\right| = 2$ **18.** $\left|\dfrac{2}{3}q - 1\right| = 5$ **19.** $\left|1 - \dfrac{3}{4}k\right| = 7$ **20.** $\left|2 - \dfrac{5}{2}m\right| = 14$

21. Decide when to use *and* and when to use *or* to solve each of the following. Assume k is a positive number.

(a) $|ax + b| = k$ **(b)** $|ax + b| < k$ **(c)** $|ax + b| > k$

22. The graph of the solution set of $|3y - 4| < 5$ is given here.

Without actually doing the algebraic work, graph the solution set of each of the following, referring to the graph above.

(a) $|3y - 4| = 5$ **(b)** $|3y - 4| > 5$

Solve each inequality and graph the solution set. See Example 2.

23. $|x| > 3$ **24.** $|y| > 5$ **25.** $|k| \geq 4$ **26.** $|r| \geq 6$

27. $|t + 2| > 10$ **28.** $|4x + 1| \geq 21$ **29.** $|3 - x| > 5$ **30.** $|5 - x| > 3$

Solve each inequality and graph the solution set. See Example 3. (Hint: Compare your answers to those in Exercises 23–30.)

31. $|x| \leq 3$ **32.** $|y| \leq 5$ **33.** $|k| < 4$ **34.** $|r| < 6$

35. $|t + 2| \leq 10$ **36.** $|4x + 1| < 21$ **37.** $|3 - x| \leq 5$ **38.** $|5 - x| \leq 3$

Decide which method you should use to solve each absolute value equation or inequality. Find the solution set and graph. See Examples 1–3.

39. $|-4 + k| > 9$ **40.** $|-3 + t| > 8$ **41.** $|r + 5| > 20$ **42.** $|3x - 1| < 8$

43. $|7 + 2z| = 5$ **44.** $|9 - 3p| = 3$ **45.** $|3r - 1| \leq 11$ **46.** $|2s - 6| \leq 6$

47. $|-6x - 6| \leq 1$ **48.** $|-2x - 6| \leq 5$ **49.** $|3x - 1| \geq 8$ **50.** $|r + 5| \leq 20$

51. Write an absolute value equation in the variable x that states that the distance between x and 4 is equal to 9.

52. Write an absolute value inequality in the variable x that states that the distance between $2x$ and -3 is **(a)** greater than 4 **(b)** less than 4.

Solve each equation or inequality. Give the solution set in set notation for equations and in interval notation for inequalities. See Example 4.

53. $|x + 4| + 1 = 2$ **54.** $|y + 5| - 2 = 12$ **55.** $|2x + 1| + 3 > 8$

56. $|6x - 1| - 2 > 6$ **57.** $|x + 5| - 6 \leq -1$ **58.** $|r - 2| - 3 \leq 4$

Solve each equation. See Example 5.

59. $|3x + 1| = |2x + 4|$ **60.** $|7x + 12| = |x - 8|$

61. $\left| m - \dfrac{1}{2} \right| = \left| \dfrac{1}{2}m - 2 \right|$ **62.** $\left| \dfrac{2}{3}r - 2 \right| = \left| \dfrac{1}{3}r + 3 \right|$

63. $|6x| = |9x + 1|$ **64.** $|13y| = |2y + 1|$

65. $|2p - 6| = |2p + 11|$ **66.** $|3x - 1| - |3x + 9|$

Solve each equation or inequality. See Examples 6 and 7.

67. $|12t - 3| = -8$ **68.** $|13w + 1| = -3$ **69.** $|4x + 1| = 0$

70. $|6r - 2| = 0$ **71.** $|2q - 1| < -6$ **72.** $|8n + 4| < -4$

73. $|x + 5| > -9$ **74.** $|x + 9| > -3$ **75.** $|7x + 3| \leq 0$

76. $|4x - 1| \leq 0$ **77.** $|5x - 2| \geq 0$ **78.** $|4 + 7x| \geq 0$

79. $|10z + 7| > 0$ **80.** $|4x + 1| > 0$

skip

RELATING CONCEPTS (EXERCISES 81-84)

The ten tallest buildings in Kansas City, Missouri, are listed along with their heights.

Building	Height (in feet)
One Kansas City Place	626
AT&T Town Pavilion	590
Hyatt Regency	504
Kansas City Power and Light	476
City Hall	443
Federal Office Building	413
Commerce Tower	402
City Center Square	402
Southwest Bell Telephone	394
Pershing Road Associates	352

Source: The World Almanac and Book of Facts.

Use the information in the chart to **work Exercises 81–84 in order.**

 81. To find the average of a group of numbers, we add the numbers and then divide by the number of items added. Use a calculator to find the average of the heights.

82. Let k represent your answer for Exercise 81. Then for a height x, the expression $|x - k| < 50$ says that the height is within 50 feet of the average height k. Using your results from Exercise 81, list the buildings that are within 50 feet of the average.

83. Repeat Exercise 82, but find the buildings that are within 75 feet of the average.

84. (a) Write an absolute value inequality that describes the height of a building that is *not* within 75 feet of the average.

(b) Solve the inequality you wrote in part (a).

(c) Use the result of part (b) to find the buildings that are not within 75 feet of the average.

(d) Confirm that your answer to part (c) makes sense by comparing it with your answer to Exercise 83.

Did you make the connection that absolute value statements are useful when comparing quantities with an average?

To determine when a manufacturing process is out of control, sample measurements x are compared to the required measurement x_1. Whenever $|x - x_1| \geq k$, for a predetermined value k, the process is out of control. See the Connections box at the end of this section.

85. A box of oatmeal must contain 16 ounces. The machine that fills the boxes is set to fill them within .5 ounce of 16 ounces.
   (a) Write an absolute value inequality to describe this situation.
   (b) Solve the inequality for x.
   (c) For what values of x is the machine out of control?

86. A machine produces .25-inch bolts with an error of not more than .002 inch. Answer parts (a)–(c) of Exercise 85 for this machine.

S U M M A R Y Exercises on Solving Linear and Absolute Value Equations
and Inequalities

Students often have difficulty distinguishing between the various types of equations and inequalities introduced in Chapters 2 and 3. This section of miscellaneous equations and inequalities provides practice in solving all such types. You might wish to refer to the boxes in these chapters that summarize the various methods of solution.

Solve each equation or inequality.

1. $4z + 1 = 49$

2. $|m - 1| = 6$

3. $6q - 9 = 12 + 3q$

4. $3p + 7 = 9 + 8p$

5. $|a + 3| = -4$

6. $2m + 1 \leq m$

7. $8r + 2 \geq 5r$

8. $4(a - 11) + 3a = 20a - 31$

9. $2q - 1 = -7$

10. $|3q - 7| - 4 = 0$

11. $6z - 5 \leq 3z + 10$

12. $|5z - 8| + 9 \geq 7$

13. $9y - 3(y + 1) = 8y - 7$

14. $|y| \geq 8$

15. $9y - 5 \geq 9y + 3$

16. $13p - 5 > 13p - 8$

17. $|q| < 5.5$

18. $4z - 1 = 12 + z$

19. $\frac{2}{3}y + 8 = \frac{1}{4}y$

20. $-\frac{5}{8}y \geq -20$

21. $\frac{1}{4}p < -6$

22. $7z - 3 + 2z = 9z - 8z$

23. $\frac{3}{5}q - \frac{1}{10} = 2$

24. $|r - 1| < 7$

25. $r + 9 + 7r = 4(3 + 2r) - 3$

26. $6 - 3(2 - p) < 2(1 + p) + 3$

27. $|2p - 3| > 11$

28. $\frac{x}{4} - \frac{2x}{3} = -10$

29. $|5a + 1| \leq 0$

30. $5z - (3 + z) \geq 2(3z + 1)$

31. $-2 \leq 3x - 1 \leq 8$

32. $-1 \leq 6 - x \leq 5$

33. $|7z - 1| = |5z + 3|$

34. $|p + 2| = |p + 4|$

35. $|1 - 3x| \geq 4$

36. $\frac{1}{2} \leq \frac{2}{3}r \leq \frac{5}{4}$

37. $-(m + 4) + 2 = 3m + 8$

38. $\frac{p}{6} - \frac{3p}{5} = p - 86$

39. $-6 \leq \frac{3}{2} - x \leq 6$

40. $|5 - y| < 4$

41. $|y - 1| \geq -6$

42. $|2r - 5| = |r + 4|$

43. $8q - (1 - q) = 3(1 + 3q) - 4$

44. $8y - (y + 3) = -(2y + 1) - 12$

45. $|r - 5| = |r + 9|$

46. $|r + 2| < -3$

47. $2x + 1 > 5$ or $3x + 4 < 1$

48. $1 - 2x \geq 5$ and $7 + 3x \geq -2$

C H A P T E R   3   G R O U P   A C T I V I T Y

Comparing Vacation Options

Objective: Use tables, linear equations, and compound inequalities to compare admission costs to Universal Studios theme parks.

The North family, which includes parents Wayne and Sharon and daughters Marissa age 7 and Ce-Ce age 5, plans to visit Universal Studios Escape on their vacation to Orlando this summer. The Universal Studios complex consists of two theme parks: Universal Studios Florida and the new Islands of Adventure theme park (which opened in May 1999 and is made up of five "islands" situated around a manmade lake). The theme parks are connected by an entertainment district called Universal Studios Citywalk, a 30-acre complex offering live music, dance clubs, themed restaurants, bars, cafes, theatres and shops. Estimated costs of the new theme parks range from $2.6 billion to $2.8 billion.*

A. A one-day pass to the Islands of Adventure is the same cost as a pass to Universal Studios Florida: $46.64 for adults, $37.10 for children ages 3–9, and free for children under age 3.

   1. Complete the following table.

Number of Days	Ticket Cost		Total Cost (in dollars)
	Adults	Children	
1			
2			
x			

   2. Create a linear equation that gives the total cost of the family's admission based on the number of days spent at the parks.

   3. Assume the North family plans to visit the parks as many days as possible but wishes to keep the total admission cost between $400 and $500. Use the results found in Exercise A2 to set up a compound inequality. Solve this inequality to find the number of days the family can visit the parks based on their budget.

B. Universal Studios also offers two package admission plans.

   Plan 1 A 2-day Escape pass, which allows guests to visit both parks in the same day, costs $84.75 for adults and $68.85 for children ages 3–9.

   Plan 2 A 3-day Escape pass costs $105.95 for adults and $84.75 for children ages 3–9.

   1. How much can the North family save on a two-day visit by purchasing 2-day Escape passes? Express this range of admission costs using a compound inequality.

   2. How much can the North family save on a three-day visit by purchasing 3-day Escape passes? Express this range of admission costs using a compound inequality.

   3. If the family decides to visit the parks for four days, what is the most economical way to do so? What is the least economical way to do so? Express this range of admission costs using a compound inequality.

Sources: St. Louis Post Dispatch.

   Universal Studios Escape Web site: www.uescape.com.

CHAPTER 3 SUMMARY

KEY TERMS

3.1 interval interval notation inequality	**linear inequality in one variable** equivalent inequalities	**3.2** intersection compound inequality union	**3.3** absolute value equation absolute value inequality

NEW SYMBOLS

∞	infinity	$(-\infty, \infty)$	the set of real numbers	$\cup$	set union
$-\infty$	negative infinity	$\cap$	set intersection		

TEST YOUR WORD POWER

See how well you have learned the vocabulary in this chapter. Answers, with examples, are given at the bottom of the page.

1. An **inequality** is
(a) a statement that two algebraic expressions are equal
(b) a point on a number line
(c) an equation with no solutions
(d) a statement with algebraic expressions related by $<$, $\leq$, $>$, or $\geq$.

2. Interval notation is
(a) a portion of a number line

(b) a special notation for describing a point on a number line
(c) a way to use symbols to describe an interval on a number line
(d) a notation to describe unequal quantities.

3. The **intersection** of two sets A and B is the set of elements that belong
(a) to both A and B
(b) to either A or B, or both

(c) to either A or B, but not both
(d) to just A.

4. The **union** of two sets A and B is the set of elements that belong
(a) to both A and B
(b) to either A or B, or both
(c) to either A or B, but not both
(d) to just B.

QUICK REVIEW

CONCEPTS	EXAMPLES
3.1 LINEAR INEQUALITIES IN ONE VARIABLE	

Solving Linear Inequalities in One Variable	Solve $3(x + 2) - 5x \leq 12$.
Step 1 Simplify each side of the inequality by clearing parentheses and combining like terms.	$3x + 6 - 5x \leq 12$
Step 2 Get all terms with variables on one side and all terms without variables on the other side.	$-2x + 6 \leq 12$
Step 3 Write the inequality in the form $x < k$ or $x > k$.	$-2x \leq 6$
	$\dfrac{-2x}{-2} \geq \dfrac{6}{-2}$
	$x \geq -3$
If an inequality is multiplied or divided by a *negative* number, the inequality symbol *must be reversed.*	The solution set is $[-3, \infty)$ and is graphed below.

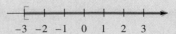

$$-3 \quad -2 \quad -1 \quad 0 \quad 1 \quad 2 \quad 3$$

CONCEPTS	EXAMPLES

3.2 SET OPERATIONS AND COMPOUND INEQUALITIES

Solving a Compound Inequality
Step 1 Solve each inequality in the compound inequality individually.
Step 2 If the inequalities are joined with *and*, the solution set is the intersection of the two individual solution sets. If the inequalities are joined with *or*, the solution set is the union of the two individual solution sets.

Solve $x + 1 > 2$ and $2x < 6$.

$$x + 1 > 2 \quad \text{and} \quad 2x < 6$$
$$x > 1 \quad \text{and} \quad x < 3$$

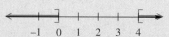

The solution set is $(1, 3)$.

Solve $x \geq 4$ or $x \leq 0$.

The solution set is $(-\infty, 0] \cup [4, \infty)$.

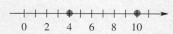

3.3 ABSOLUTE VALUE EQUATIONS AND INEQUALITIES

Solving Absolute Value Equations and Inequalities
Let k be a positive number.
To solve $|ax + b| = k$, solve the compound equation
$$ax + b = k \quad \text{or} \quad ax + b = -k.$$

Solve $|x - 7| = 3$.

$$x - 7 = 3 \quad \text{or} \quad x - 7 = -3$$
$$x = 10 \quad \text{or} \quad x = 4$$

The solution set is $\{4, 10\}$.

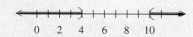

To solve $|ax + b| > k$, solve the compound inequality
$$ax + b > k \quad \text{or} \quad ax + b < -k.$$

Solve $|x - 7| > 3$.

$$x - 7 > 3 \quad \text{or} \quad x - 7 < -3$$
$$x > 10 \quad \text{or} \quad x < 4$$

The solution set is $(-\infty, 4) \cup (10, \infty)$.

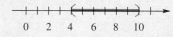

To solve $|ax + b| < k$, solve the compound inequality
$$-k < ax + b < k.$$

Solve $|x - 7| < 3$.

$$-3 < x - 7 < 3$$
$$4 < x < 10 \qquad \text{Add 7.}$$

The solution set is $(4, 10)$.

To solve an absolute value equation of the form
$$|ax + b| = |cx + d|$$
solve the compound equation
$$ax + b = cx + d \quad \text{or} \quad ax + b = -(cx + d).$$

Solve $|x + 2| = |2x - 6|$.

$$x + 2 = 2x - 6 \quad \text{or} \quad x + 2 = -(2x - 6)$$
$$x = 8 \quad \text{or} \quad x = \frac{4}{3}$$

The solution set is $\left\{\frac{4}{3}, 8\right\}$.

CHAPTER 3 REVIEW EXERCISES

[3.1] *Solve each inequality. Express the solution set in interval form.*

1. $-\dfrac{2}{3}k < 6$

2. $-5x - 4 \geq 11$

3. $\dfrac{6a + 3}{-4} < -3$

4. $5 - (6 - 4k) \geq 2k - 7$

5. $8 \leq 3y - 1 < 14$

6. $\dfrac{5}{3}(m - 2) + \dfrac{2}{5}(m + 1) > 1$

Solve each problem.

7. To pass algebra, a student must have an average of at least 70% on five tests. On the first four tests, a student has grades of 75%, 79%, 64%, and 71%. What possible grades on the fifth test would guarantee a passing grade in the class?

8. The perimeter of a rectangular playground must be no greater than 120 meters. The width of the playground must be 22 meters. Find the possible lengths of the playground.

9. The hit movie *Titanic* earned more in Europe than in the United States. The average movie ticket in London, for example, costs the equivalent of $10.59. (*Source: Parade Magazine,* September 13, 1998, p. 25.) A student group from the United States is touring London and wishes to see the movie there. If $1000 is available to purchase tickets and the group receives a $50 discount from the tour company, how many tickets can be purchased?

10. While solving the inequality $10x + 2(x - 4) < 12x - 13$, a student did all the work correctly and obtained the statement $-8 < -13$. The student did not know what to do at this point, because the variable "disappeared." How would you help the student interpret this result?

[3.2] *Let $A = \{a, b, c, d\}$, $B = \{a, c, e, f\}$, and $C = \{a, e, f, g\}$. Find each set.*

11. $A \cap B$

12. $A \cap C$

13. $B \cup C$

14. $A \cup C$

Solve each compound inequality. Graph the solution set.

15. $x \leq 4$ and $x < 3$

16. $x + 4 > 12$ and $x - 2 < 1$

17. $x > 5$ or $x \leq -1$

18. $x - 4 > 6$ or $x + 3 \leq 18$

Express each union or intersection in simplest interval form.

19. $(-3, \infty) \cap (-\infty, 4)$

20. $(-\infty, 6) \cap (-\infty, 2)$

21. $(4, \infty) \cup (9, \infty)$

22. $(1, 2) \cup (1, \infty)$

23. According to the Bureau of Labor Statistics, the following are the median weekly earnings of full-time workers by occupation for men and women.

Occupation	Men	Women
Managerial and professional specialty	$ 852	$616
Mathematical and computer scientists	$1005	$754
Waiters and waitresses	$ 300	$264
Bus drivers	$ 482	$354

Give the occupation that satisfies each description.
(a) The median earnings for men are less than $900 *and* for women are greater than $500.
(b) The median earnings for men are greater than $900 *or* for women are greater than $600.

[3.3] *Solve each absolute value equation.*

24. $|x| = 7$

25. $|3k - 7| = 8$

26. $|z - 4| = -12$

27. $|4a + 2| - 7 = -3$

28. $|3p + 1| = |p + 2|$

29. $|2m - 1| = |2m + 3|$

Solve each absolute value inequality. Give the solution set in interval form.

30. $|-y + 6| \le 7$

31. $|2p + 5| \le 1$

32. $|x + 1| \ge -3$

33. $|5r - 1| > 9$

34. $|11x - 3| \le -2$

35. $|11x - 3| \le 0$

36. Write an inequality that states that the distance between x and 14 is greater than 12.

MIXED REVIEW EXERCISES

Solve.

37. $5 - (6 - 4k) > 2k - 5$

38. $x < 3$ and $x \ge -2$

39. $|3k + 6| \ge 0$

40. $|2k - 7| + 4 = 11$

41. In order to qualify for a company pension plan, an employee must average at least $1000 per month in earnings. During the first four months of the year, an employee made $900, $1200, $1040, and $760. What possible amounts earned during the fifth month will qualify the employee?

42. $|p| < 14$

43. $\dfrac{3}{4}(a - 2) - \dfrac{1}{3}(5 - 2a) < -2$

44. $-4 < 3 - 2k < 9$

45. $-.3x + 2.1(x - 4) \le -6.6$

46. $|5r - 1| > 14$

47. $x \ge -2$ or $x < 4$

48. $|m - 1| = |2m + 3|$

49. $|m + 3| \le 1$

In Exercises 50 and 51, sketch the graph of each solution set.

50. $x > 6$ and $x < 8$

51. $-5x + 1 \ge 11$ or $3x + 5 \ge 26$

52. The solution set of $|3x + 4| = 7$ is shown on the number line.

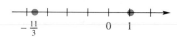

(a) What is the solution set of $|3x + 4| \ge 7$?

(b) What is the solution set of $|3x + 4| \le 7$?

CHAPTER 3 TEST

1. What is the special rule that must be remembered when multiplying or dividing both sides of an inequality by a negative number?

Solve each inequality. Give the solution set in both interval and graph forms.

2. $4 - 6(x + 3) \le -2 - 3(x + 6) + 3x$

3. $-\dfrac{4}{7}x > -16$

4. $-6 \le \dfrac{4}{3}x - 2 \le 2$

5. Which one of the following inequalities is equivalent to $x < -3$?

(a) $-3x < 9$ (b) $-3x > -9$ (c) $-3x > 9$ (d) $-3x < -9$

6. The graph shows the number (in millions) of U.S. citizen departures to Europe. For the period shown in the graph, between which two years did the number of departures increase the most? For which two years did they stay the same?

DEPARTURES TO EUROPE

Source: U.S. Office of Tourism Industries.

Solve each problem.

7. A student must have an average grade of at least 80% on the four tests in a course to get a grade of B. The student had grades of 83%, 76%, and 79% on the first three tests. What minimum grade on the fourth test would guarantee a B in the course?

8. A product will break even or produce a profit only if the revenue R from selling the product is at least the cost C of producing it. Suppose that the cost to produce x units of carpet is $C = 50x + 5000$, while the revenue is $R = 60x$. For what values of x is R at least equal to C?

9. Let $A = \{1, 2, 5, 7\}$ and $B = \{1, 5, 9, 12\}$. Find

(a) $A \cap B$ (b) $A \cup B$.

10. Solve each compound inequality.

(a) $3k \geq 6$ and $k - 4 < 5$ (b) $-4x \leq -24$ or $4x - 2 < 10$

Solve each absolute value equation or inequality.

11. $|4x - 3| = 7$ **12.** $|4x - 3| > 7$ **13.** $|4x - 3| < 7$

14. $|3 - 5x| = |2x + 8|$ **15.** $|-3x + 4| - 4 < -1$

16. If $k < 0$, what is the solution set of

(a) $|5x + 3| < k$ (b) $|5x + 3| > k$ (c) $|5x + 3| = k$?

CUMULATIVE REVIEW EXERCISES CHAPTERS 1–3

Let $A = \left\{-8, -\dfrac{2}{3}, -\sqrt{6}, 0, \dfrac{4}{5}, 9, \sqrt{36}\right\}$. *Simplify the elements of A as necessary and then list the elements that belong to the set.*

1. Natural numbers   **2.** Whole numbers   **3.** Integers

4. Rational numbers   **5.** Irrational numbers   **6.** Real numbers

Add or subtract, as indicated.

7. $-\dfrac{4}{3} - \left(-\dfrac{2}{7}\right)$ **8.** $|-4| - |2| + |-6|$

Evaluate each expression.

9. $(-3)^5$

10. $\left(\dfrac{6}{7}\right)^3$

11. Which one of the following is not a real number: $-\sqrt{36}$ or $\sqrt{-36}$?

12. Which one of the following is undefined: $\dfrac{4-4}{4+4}$ or $\dfrac{4+4}{4-4}$?

Evaluate if $a = 2$, $b = -3$, and $c = 4$.

13. $-3a + 2b - c$

14. $-2b^2 - 4c$

15. $-8(a^2 + b^3)$

16. $\dfrac{3a^3 - b}{4 + 3c}$

Use the properties of real numbers to simplify each expression.

17. $-7r + 5 - 13r + 12$

18. $-(3k + 8) - 2(4k - 7) + 3(8k + 12)$

Identify the property of real numbers illustrated by each equation.

19. $(a + b) + 4 = 4 + (a + b)$

20. $4x + 12x = (4 + 12)x$

21. $-9 + 9 = 0$

22. What is the reciprocal, or multiplicative inverse, of $-\dfrac{2}{3}$?

Solve each equation.

23. $-4x + 7(2x + 3) = 7x + 36$

24. $-\dfrac{3}{5}x + \dfrac{2}{3}x = 2$

25. $.06x + .03(100 + x) = 4.35$

26. $P = a + b + c$; for b

Solve each inequality. Give the solution set in both interval and graph forms.

27. $3 - 2(x + 7) \leq -x + 3$

28. $-4 < 5 - 3x \leq 0$

29. $2x + 1 > 5$ or $2 - x > 2$

30. $\left| -7k + 3 \right| \geq 4$

Solve each problem.

31. Kathy Manley invested some money at 7% interest and the same amount at 10%. Her total interest for the year was $150 less than one-tenth of the total amount she invested. How much did she invest at each rate?

32. A dietician must use three foods, A, B, and C, in a diet. He must include twice as many grams of food A as food C, and 5 grams of food B. The three foods must total at most 24 grams. What is the largest amount of food C that the dietician can use?

33. Laurie Reilly got scores of 88 and 78 on her first two tests. What score must she make on her third test to keep an average of 80 or greater?

34. Jack and Jill are running in the Fresh Water Fun Run. Jack runs at 7 miles per hour and Jill runs at 5 miles per hour. If they start at the same time, how long will it be before Jack is $\dfrac{1}{4}$ mile ahead of Jill?

35. How much pure alcohol should be added to 7 liters of 10% alcohol to increase the concentration to 30% alcohol?

36. A coin collection contains 29 coins. It consists of pennies, nickels, and quarters. The number of quarters is 4 less than the number of nickels, and the face value of the collection is $2.69. How many of each denomination are there in the collection?

Clark's rule is a formula used in reducing drug dosage according to weight from the recommended adult dosage to a child dosage. It is as follows.

$$\frac{\text{Weight of child in pounds}}{150} \times \text{adult dose} = \text{child's dose}$$

37. Find a child's dosage if the child weighs 55 pounds and the recommended adult dosage is 120 milligrams.

38. Find a child's dosage if the child weighs 75 pounds and the recommended adult dosage is 40 drops.

39. Since 1975, the number of daily newspapers has steadily declined. According to the graph,
   (a) by how much did the number of daily newspapers decrease between 1990 and 1995?
   (b) by what *percent* did the number of daily newspapers decrease from 1990 to 1995?

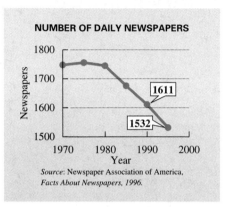

NUMBER OF DAILY NEWSPAPERS

Source: Newspaper Association of America, *Facts About Newspapers, 1996.*

40. The body-mass index, or BMI, of a person is given by the formula

$$\text{BMI} = \frac{704 \times (\text{weight in pounds})}{(\text{height in inches})^2}.$$

Ken Griffey, Jr., is listed as being 6 feet, 3 inches tall and weighing 205 pounds. What is his BMI? (*Source: Readers Digest,* October 1993.)

Linear Equations in Two Variables

Calvin Coolidge, 30th president of the United States, once said "The chief business of America is business." Small businesses account for 99 percent of the 19 million nonfarm businesses in the United States today. Small businesses employ 55 percent of the private workforce, make 44 percent of all sales in America, and produce 38 percent of the nation's gross national product.*

Business

4.1 Linear Equations in Two Variables

4.2 Graphing Linear Equations in Two Variables

4.3 The Slope of a Line

The number of new consumer packaged-goods products introduced in the United States from 1991–1996 is illustrated in the line graph. The graph shows that, with the exception of 1994, this number has increased from year to year at a steady pace. What might account for the one-year decrease? We will return to this graph in the exercises for Section 4.1.

NEW CONSUMER PRODUCTS

Source: Marketing Intelligence Service, Ltd.

Mathematics plays an important role in the business world. Preparing a business plan, pricing, borrowing money, determining market share, and tracking costs, revenue, and profit all require a good understanding of mathematics. Throughout this chapter many of the examples and exercises will further illustrate the use of mathematics in business management.

---

*Data from *The Universal Almanac,* 1997, John W. Wright, General Editor.

Visit our Web site at www.LialAlgebra.com

OBJECTIVES

1 Interpret graphs.

2 Write a solution as an ordered pair.

3 Decide whether a given ordered pair is a solution of a given equation.

4 Complete ordered pairs for a given equation.

5 Plot ordered pairs.

FOR EXTRA HELP

SSG Sec. 4.1
SSM Sec. 4.1

Pass the Test Software

InterAct Math
   Tutorial Software

Video 6

Graphs are prevalent in our society. Pie charts (circle graphs) and bar graphs were introduced in Chapter 1, and we have seen many examples of them in the early chapters of this book. It is important to be able to interpret graphs correctly.

OBJECTIVE 1 Interpret graphs. We begin with a bar graph where we must estimate the heights of the bars.

EXAMPLE 1 Interpreting Bar Graphs

Venture capital is money invested in new, often speculative, business enterprises. The amount of venture capital invested in companies has risen during the decade of the 1990s, as shown in the graph in Figure 1. Use the graph to determine the following.

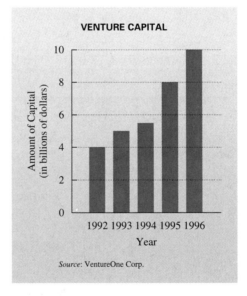

VENTURE CAPITAL

Source: VentureOne Corp.

Figure 1

(a) What amount of venture capital was provided in 1992?

Move horizontally from the top of the bar for 1992 to the scale on the left to see that about $4 billion was provided.

(b) In what year was that amount doubled?

Follow the line for $8 billion across to the right. The bar for 1995 just touches that line, so the amount provided in 1992 was doubled in 1995.

EXAMPLE 2 Interpreting Line Graphs

The line graph in Figure 2 shows the total number of deals that involved venture capital in the 1990s. Use the graph to estimate the number of deals in 1993 and in 1995.

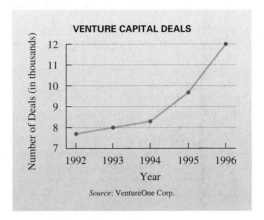

Figure 2

In 1993 the point lies on the line that corresponds to 8000, so 8000 venture capital deals were closed in 1993. The point for 1995 is about $\frac{2}{3}$, or .7, of the way between the lines for 9 and 10. Thus, we estimate 9700 deals were made in 1995.

We solved linear equations in one variable and explored their applications in Chapter 2. Now we want to extend those ideas to *linear equations in two variables*.

Linear Equation

A **linear equation in two variables** is an equation that can be put in the form

$$Ax + By = C,$$

where *A, B,* and *C* are real numbers and *A* and *B* are not both 0.

OBJECTIVE 2 Write a solution as an ordered pair. A solution of a linear equation in *two* variables requires *two* numbers, one for each variable. For example, the equation $y = 4x + 5$ is satisfied if x is replaced with 2 and y is replaced with 13, since

$$13 = 4(2) + 5. \qquad \text{Let } x = 2; y = 13.$$

The pair of numbers $x = 2$ and $y = 13$ gives a solution of the equation $y = 4x + 5$. The phrase "$x = 2$ and $y = 13$" is abbreviated

x-value ⌝ ⌐ y-value
↓   ↓
(2, 13)
⎵
Ordered pair

with the *x*-value, 2, and the *y*-value, 13, given as a pair of numbers written inside parentheses. *The x-value is always given first.* A pair of numbers such as (2, 13) is called an

ordered pair. As the name indicates, the order in which the numbers are written is important. The ordered pairs (**2, 13**) and (**13, 2**) are not the same. The second pair indicates that $x = 13$ and $y = 2$. (Of course, letters other than x and y may be used in the equation with the numbers.)

OBJECTIVE [3] Decide whether a given ordered pair is a solution of a given equation. An ordered pair that is a solution of an equation is said to *satisfy* the equation.

EXAMPLE 3   Deciding Whether an Ordered Pair Satisfies an Equation

Decide whether the given ordered pair is a solution of the given equation.

(a) $(3, 2)$; $2x + 3y = 12$

To see whether $(3, 2)$ is a solution of the equation $2x + 3y = 12$, we substitute 3 for x and 2 for y in the given equation.

$$2x + 3y = 12$$
$$2(3) + 3(2) = 12 \qquad ? \qquad \text{Let } x = 3; \text{ let } y = 2.$$
$$6 + 6 = 12 \qquad ?$$
$$12 = 12 \qquad\qquad \text{True}$$

This result is true, so $(3, 2)$ satisfies $2x + 3y = 12$.

(b) $(-2, -7)$; $m + 5n = 33$

$$(-2) + 5(-7) = 33 \qquad ? \qquad \text{Let } m = -2; \text{ let } n = -7.$$
$$-2 + (-35) = 33 \qquad ?$$
$$-37 = 33 \qquad\qquad \text{False}$$

This result is false, so $(-2, -7)$ is *not* a solution of $m + 5n = 33$.

OBJECTIVE [4] Complete ordered pairs for a given equation. Choosing a number for one variable in a linear equation makes it possible to find the value of the other variable, as shown in the next example.

EXAMPLE 4   Completing an Ordered Pair

Complete the ordered pair $(7, \quad)$ for the equation $y = 4x + 5$.

In this ordered pair, $x = 7$. (Remember that x always comes first.) To find the corresponding value of y, replace x with 7 in the equation $y = 4x + 5$.

$$y = 4(7) + 5 = 28 + 5 = 33$$

The ordered pair is $(7, 33)$.

Ordered pairs often are displayed in a **table of values** as in the next example. The table may be written either vertically or horizontally.

┌ **E X A M P L E   5**   Completing a Table of Values

Complete the given table of values for each equation. Then write the results as ordered pairs.

(a) $x - 2y = 8$

x	y
2	
10	
	0
	-2

To complete the first two ordered pairs, let $x = 2$ and $x = 10$, respectively.

If	$x = 2,$		If	$x = 10,$	
then	$x - 2y = 8$		then	$x - 2y = 8$	
becomes	$\mathbf{2} - 2y = 8$		becomes	$\mathbf{10} - 2y = 8$	
	$-2y = 6$			$-2y = -2$	
	$y = -3.$			$y = 1.$	

Now complete the last two ordered pairs by letting $y = 0$ and $y = -2$, respectively.

If	$y = 0,$		If	$y = -2,$	
then	$x - 2y = 8$		then	$x - 2y = 8$	
becomes	$x - 2(\mathbf{0}) = 8$		becomes	$x - 2(\mathbf{-2}) = 8$	
	$x - 0 = 8$			$x + 4 = 8$	
	$x = 8.$			$x = 4.$	

The completed table of values is as follows.

x	y
2	-3
10	1
8	0
4	-2

The corresponding ordered pairs are $(2, -3)$, $(10, 1)$, $(8, 0)$, and $(4, -2)$.

(b) $x = 5$

x	y
	-2
	6
	3

The given equation is $x = 5$. No matter which value of y might be chosen, the value of x is always the same, 5.

x	y
5	-2
5	6
5	3

The ordered pairs are $(5, -2)$, $(5, 6)$, and $(5, 3)$.

We can think of $x = 5$ in Example 5(b) as an equation in two variables by rewriting $x = 5$ as $x + 0y = 5$. This form of the equation shows that for any value of y, the value of x is 5. Similarly, $y = -2$ is the same as $0x + y = -2$.

Earlier in this book, we saw that linear equations in *one* variable had either one, zero, or an infinite number of real number solutions. Every linear equation in *two* variables has an infinite number of ordered pairs as solutions. Each choice of a number for one variable leads to a particular real number for the other variable.

To graph these solutions, represented as the ordered pairs (x, y), we need *two* number lines, one for each variable. These two number lines are drawn as shown in Figure 3. The horizontal number line is called the **x-axis.** The vertical line is called the **y-axis.** Together, the x-axis and y-axis form a **rectangular coordinate system.** It is also called the **Cartesian coordinate system,** in honor of René Descartes.

The coordinate system is divided into four regions, called **quadrants.** These quadrants are numbered counterclockwise, as shown in Figure 3. Points on the axes themselves are not in any quadrant. The point at which the x-axis and y-axis meet is called the **origin.** The origin, labeled 0 in Figure 3, is the point corresponding to $(0, 0)$.

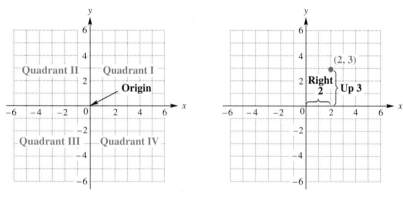

Figure 3   Figure 4

CONNECTIONS

The coordinate system that we use to plot points is credited to René Descartes (1596–1650), a French mathematician of the seventeenth century. It has been said that he developed the coordinate system while lying in bed, watching an insect move across the ceiling. He realized that he could locate the position of the insect at any given time by finding its distance from each of two perpendicular walls.

OBJECTIVE ⑤ **Plot ordered pairs.** By referring to the two axes, every point on the plane can be associated with an ordered pair. The numbers in the ordered pair are called the **coordinates** of the point. For example, locate the point associated with the ordered pair $(2, 3)$ by starting at the origin. Since the x-coordinate is 2, go 2 units to the right along the x-axis. Then, since the y-coordinate is 3, turn and go up 3 units on a line parallel to the y-axis. This is called **plotting** the point $(2, 3)$. (See Figure 4.) From now on we refer to the point with x-coordinate 2 and y-coordinate 3 as the point $(2, 3)$.

> **NOTE** When we graphed on a number line, one number corresponded to each point. On a plane, however, both numbers in the ordered pair are needed to locate a point. The ordered pair is a name for the point.

E X A M P L E   6   Plotting Ordered Pairs

Plot the given points on a coordinate system.

(a) $(1, 5)$ **(b)** $(-2, 3)$ **(c)** $(-1, -4)$

(d) $(7, -2)$ **(e)** $\left(\dfrac{3}{2}, 2\right)$ **(f)** $(5, 0)$

Locate the point $(-1, -4)$, for example, by first going 1 unit to the left along the x-axis. Then turn and go 4 units down, parallel to the y-axis. Plot the point $\left(\dfrac{3}{2}, 2\right)$, by going $\dfrac{3}{2}$ (or $1\dfrac{1}{2}$) units to the right along the x-axis. Then turn and go 2 units up, parallel to the y-axis. Figure 5 shows the graphs of the points in this example.

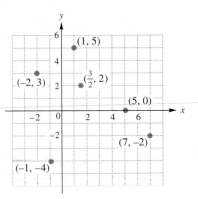

Figure 5

E X A M P L E   7   Completing Ordered Pairs to Estimate the Number of Business Incorporations

New business incorporations have increased steadily from 1990–1996. Their number can be closely approximated by the equation

$$y = 25.59x - 50{,}280,$$

where x is the year, and y is the number of incorporations in thousands.* This approximation is only valid for the years under examination, 1990–1996, since it was based on data for those years.

(a) Complete the table of ordered pairs for this linear equation. Round the y-values to the nearest thousand.

x	1990	1994	1996
y			

To find y when $x = 1990$, substitute into the equation.

$$y = 25.59(\mathbf{1990}) - 50{,}280 \qquad \text{Let } x = 1990.$$
$$= 644 \text{ (rounded)}$$

---

*Based on data from Dun & Bradstreet.

This means that, in 1990, there were 644,000 new incorporations. Find the y-values for 1994 and 1996 similarly. The completed table follows.

x	1990	1994	1996
y	644	746	798

(b) Interpret the ordered pair with $x = 1996$; with $x = 1994.5$.

The ordered pair (1996, 798) means that in 1996, 798,000 businesses were incorporated. We can interpret 1994.5 to mean halfway through 1994.

$$y = 25.59(1994.5) - 50,280 \qquad \text{Let } x = 1994.5.$$
$$= 759$$

There were approximately 759,000 incorporations.

(c) Graph the ordered pairs found in part (a).

The ordered pairs are graphed in Figure 6. Notice how the axes are labeled. In this application, x represents the year, and y represents the number of corporations in thousands. Different scales are used on the two axes because the two sets of numbers differ so much in size. Here, each square represents one unit in the horizontal direction and 50 units in the vertical direction. Because the numbers in the first ordered pair are quite large, we show a break in both axes near the origin.

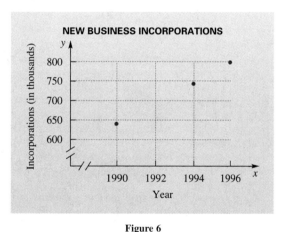

Figure 6

We can also think of ordered pairs as representing an input value x and an output value y. If we input x into the equation, the output is y. For instance, in Example 7, if we input the year into the equation

$$y = 25.59x - 50,280,$$

we get the number of incorporations (in thousands) in that year as the output. We encounter many examples of this type of relationship every day.

- The cost to fill the tank with gasoline depends on how many gallons are needed; the number of gallons is the input, and the cost is the output.

- The distance traveled depends on the traveling time; input a time, and the output is a distance.

- The growth of a plant depends on the amount of sun it gets; the input is the amount of sun, and the output is the growth.

This idea is illustrated in Figure 7 with an input-output "machine."

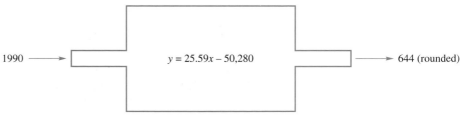

$$1990 \longrightarrow \quad y = 25.59x - 50{,}280 \quad \longrightarrow 644 \text{ (rounded)}$$

An input-output machine

Figure 7

4.1 EXERCISES

Fill in each blank with the correct response.

1. The symbol (x, y) _____ represent an ordered pair, while the symbols $[x, y]$ and
   (does/does not)
   $\{x, y\}$ _____ represent ordered pairs.
   (do/do not)

2. The ordered pair $(3, 2)$ is a solution of the equation $2x - 5y =$ _____ .

3. The point whose graph has coordinates $(-4, 2)$ is in quadrant _____ .

4. The point whose graph has coordinates $(0, 5)$ lies along the _____ -axis.

5. The ordered pair $(4,$ _____ $)$ is a solution of the equation $y = 3$.

6. The ordered pair $($ _____ $, -2)$ is a solution of the equation $x = 6$.

Use the bar graph to respond to each statement or question. See Example 1.

7. Between which pairs of consecutive years did the winnings increase? How much was the increase in each case?

8. How much did winnings decline between 1991 and 1992?

9. Which year had the greatest winnings?

10. Which year had the least winnings?

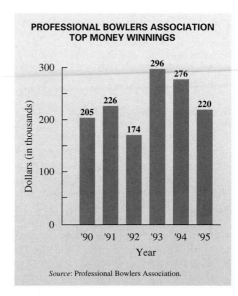

PROFESSIONAL BOWLERS ASSOCIATION
TOP MONEY WINNINGS

Source: Professional Bowlers Association.

Use the line graph to respond to each statement or question. See Example 2.

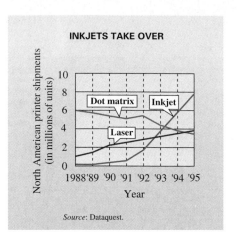

INKJETS TAKE OVER

11. Which one of the following would be the best estimate for the total number of units of printer shipments in North America in 1989?
   (a) 5.8 million   (b) 1.5 million
   (c) .2 million   (d) 7.5 million

12. In what year did the number of inkjet printer shipments first exceed the number of laser printer shipments?

13. What type of printer has shown the most rapid increase in shipments since 1992?

14. Describe how the trend in shipments of dot matrix printers has differed from the shipments of inkjet and laser printers since 1992.

This line graph (first given in the chapter introduction) shows the number of new consumer packaged-goods products for the years 1991–1996. Use this graph for Exercises 15–17. See Example 2.

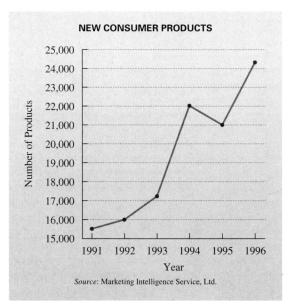

NEW CONSUMER PRODUCTS

Source: Marketing Intelligence Service, Ltd.

15. Between which two years did the number of new products decrease? Between which two years did the greatest increase occur?

16. Estimate the number of new products in 1993, 1994, 1995, and 1996.

17. Estimate the increase in new products from 1993 to 1994 and from 1995 to 1996. Does your answer confirm your answer to Exercise 15?

18. Define a linear equation in one variable and a linear equation in two variables, and give examples of each.

Decide whether the given ordered pair is a solution of the given equation. See Example 3.

19. $x + y = 9$; $(0, 9)$ 20. $x + y = 8$; $(0, 8)$ 21. $2p - q = 6$; $(4, 2)$

22. $2v + w = 5$; $(3, -1)$ 23. $4x - 3y = 6$; $(2, 1)$ 24. $5x - 3y = 15$; $(5, 2)$

25. $y = 3x$; $(2, 6)$ **26.** $x = -4y$; $(-8, 2)$ **27.** $x = -6$; $(-6, 5)$

28. $y = 2$; $(4, 2)$ **29.** $x + 4 = 0$; $(-6, 2)$ **30.** $x - 6 = 0$; $(4, 2)$

31. Explain why there are an infinite number of solutions for a linear equation in two variables. How does this differ from the number of solutions for a linear equation in one variable?

32. Give the ordered pairs that correspond to the points labeled in the figure. (All coordinates are integers.)

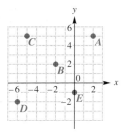

33. Do $(4, -1)$ and $(-1, 4)$ represent the same ordered pair? Explain why.

34. Do the ordered pairs $(3, 4)$ and $(4, 3)$ correspond to the same point on the plane? Explain why.

Complete each ordered pair for the equation $y = 2x + 7$. See Example 4.

35. $(2, \quad)$ **36.** $(0, \quad)$ **37.** $(\quad , 0)$ **38.** $(\quad , -3)$

Complete each ordered pair for the equation $y = -4x - 4$. See Example 4.

39. $(0, \quad)$ **40.** $(\quad , 0)$ **41.** $(\quad , 16)$ **42.** $(\quad , 24)$

43. Explain why it would be easier to find the corresponding y-value for $x = \dfrac{1}{3}$ than for $x = \dfrac{1}{7}$ in the equation $y = 6x + 2$.

44. For the equation $y = mx + b$, what is the y-value corresponding to $x = 0$ for *any* value of m?

Complete each table of values. See Example 5.

45. $2x + 3y = 12$ **46.** $4x + 3y = 24$ **47.** $3x - 5y = -15$ **48.** $4x - 9y = -36$

x	y
0	
	0
	8

x	y
0	
	0
	4

x	y
0	
	0
	-6

x	y
	0
0	
	8

49. $x = -9$ **50.** $x = 12$ **51.** $y = -6$ **52.** $y = -10$

x	y
	6
	2
	-3

x	y
	3
	8
	0

x	y
8	
4	
-2	

x	y
4	
0	
-4	

Plot each ordered pair in a rectangular coordinate system. See Example 6.

53. $(6, 2)$ **54.** $(5, 3)$ **55.** $(-4, 2)$ **56.** $(-3, 5)$

57. $\left(-\dfrac{4}{5}, -1\right)$ **58.** $\left(-\dfrac{3}{2}, -4\right)$ **59.** $(0, 4)$ **60.** $(-3, 0)$

Fill in each blank with the word positive *or the word* negative.

The point with coordinates (x, y) is in

61. quadrant III if *x* is _____ and *y* is _____.

62. quadrant II if *x* is _____ and *y* is _____.

63. quadrant IV if *x* is _____ and *y* is _____.

64. quadrant I if *x* is _____ and *y* is _____.

Complete each table of values and then plot the ordered pairs. See Examples 5 and 6.

65. $x - 2y = 6$

x	y
0	
	0
2	
	-1

66. $2x - y = 4$

x	y
0	
	0
1	
	-6

67. $3x - 4y = 12$

x	y
0	
	0
-4	
	-4

68. $2x - 5y = 10$

x	y
0	
	0
-5	
	-3

69. $y + 4 = 0$

x	y
0	
5	
-2	
-3	

70. $x - 5 = 0$

x	y
	1
	0
	6
	-4

Solve each problem. See Example 7.

71. Suppose that it costs $5000 to start up a business selling snow cones. Furthermore, it costs $.50 per cone in labor, ice, syrup, and overhead. Then the cost to make *x* snow cones is given by *y* dollars, where $y = .50x + 5000$. Express as an ordered pair each of the following.

   (a) When 100 snow cones are made, the cost is $5050. (*Hint:* What does *x* represent? What does *y* represent?)

   (b) When the cost is $6000, the number of snow cones made is 2000.

72. It costs a flat fee of $20 plus $5 per day to rent a pressure washer. Therefore, the cost to rent the pressure washer for *x* days is given by $y = 5x + 20$, where *y* is in dollars. Express as an ordered pair each of the following.

   (a) When the washer is rented for 5 days, the cost is $45.

   (b) I paid $50 when I returned the washer, so I must have rented it for 6 days.

73. In statistics, ordered pairs are used to decide whether two quantities are related in such a way that one can be predicted from the other. These ordered pairs are plotted on a graph, called a *scatter diagram.*

   Some major league baseball fans are concerned by the increase in time to complete a game. Make a scatter diagram by plotting the following ordered pairs of years since 1980 and minutes beyond two hours to complete a game: (5, 40), (7, 48), (9, 46), (11, 49), (13, 48), (15, 57). As shown in the figure, the horizontal axis is used

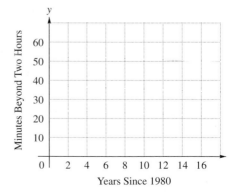

LENGTHS OF BASEBALL GAMES

to represent the years since 1980, and the vertical axis represents the minutes beyond two hours.

  (a) Graph these points on a similar grid. Do the points lie in an approximately linear pattern?

  (b) Could the number of years since 1980 be used to predict the length of a game?

  (c) What is the input? What is the output?

  (*Source: The New York Times,* May 30, 1995, p. B9.)

74. The maximum benefit for the heart from exercising occurs if the heart rate is in the target heart rate zone. The line graph shows the upper limit of the zone (in beats per minute) for various ages.

  (a) Use the graph to estimate the ordered pairs in the form (*x, y*) that describe the upper limit of the target heart rate zone for ages 20, 40, 60, and 80.

  (b) What is the input here? What is the output?

  (*Source:* Robert V. Hockey, *Physical Fitness: The Pathway to Healthy Living,* Times Mirror/Mosby College Publishing, 1989, pp. 85–87.)

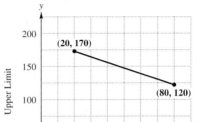

TARGET HEART RATE ZONE

75. The lower limit of the target heart rate zone (see Exercise 74) is given in the accompanying graph of the equation $y = .7(220 - x)$, where the horizontal axis represents age and the vertical axis represents heart rate.

  (a) Write ordered pairs (*x, y*) for ages 30, 40, 60, and 70 by estimating the *y*-values from the graph.

  (b) What is the input here? What is the output?

TARGET HEART RATE ZONE

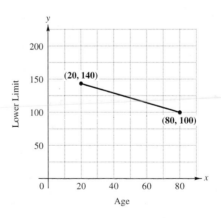

76. What is the heart rate zone for age 20? For age 40? (Refer to Exercises 74 and 75.)

77. Should the graph in Exercise 75 be used to estimate the lower limit of the target heart rate zone for ages below 20 or above 80? Why or why not?

78. The five corporations receiving the most patents for inventions in 1996 are listed below. Write a set of ordered pairs (company, number of patents) for these data. Give the inputs and outputs.

Corporation	Number of Patents
General Electric	819
IBM	1867
Hitachi	963
Canon	1541
Motorola	1064

Source: U.S. Patent and Trademark Office.

4.2 Graphing Linear Equations in Two Variables

OBJECTIVES

1 Graph linear equations.

2 Find intercepts.

3 Graph linear equations of the form $Ax + By = 0$.

4 Graph linear equations of the form $y = k$ or $x = k$.

5 Define a function.

FOR EXTRA HELP

SSG Sec. 4.2
SSM Sec. 4.2

Pass the Test Software

InterAct Math Tutorial Software

Video 6

In this section we will use a few ordered pairs that satisfy a linear equation to graph the equation.

OBJECTIVE 1 **Graph linear equations.** We know that infinitely many ordered pairs satisfy a linear equation. Some ordered pairs that are solutions of $x + 2y = 7$ are graphed in Figure 8.

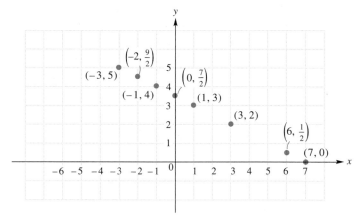

Figure 8

Notice that the points plotted in this figure all appear on a straight line, as shown in Figure 9. In fact, all ordered pairs satisfying the equation $x + 2y = 7$ correspond to points that lie on this same straight line, and the coordinates of any point on the line give a solution of the equation. This line is a "picture" of all the solutions of the equation $x + 2y = 7$. Only a portion of the line is shown here, but it extends indefinitely in both directions, as suggested by the arrowhead on each end of the line. The line is called the **graph** of the equation and the process of plotting the ordered pairs and drawing the line through the corresponding points is called **graphing.**

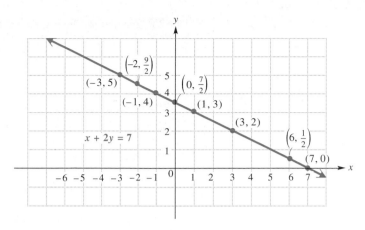

Figure 9

The preceding discussion can be generalized.

Graph of a Linear Equation

The graph of any linear equation in two variables is a straight line.

Notice that the word *line* appears in the name "*linear* equation."

Since two distinct points determine a line, we can graph a straight line by finding any two different points on the line. However, it is a good idea to plot a third point as a check.

EXAMPLE 1 Graphing a Linear Equation

Graph the linear equation $2y = -3x + 6$.

Although this equation is not in the form $Ax + By = C$, it *could* be put in that form, and so is a linear equation.

For most linear equations, two different points on the graph can be found by first letting $x = 0$, and then letting $y = 0$. Doing this gives the ordered pairs $(0, 3)$ and $(2, 0)$. We get a third ordered pair (as a check) by letting x or y equal some other number. For example, if $x = -2$, we find that $y = 6$, giving the ordered pair $(-2, 6)$. These three ordered pairs are shown in the table of values with Figure 10. Plot the corresponding points, then draw a line through them. This line, shown in Figure 10, is the graph of $2y = -3x + 6$.

x	y
0	3
2	0
-2	6

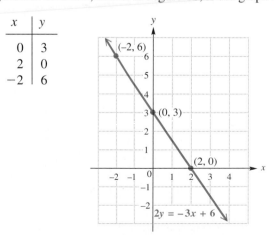

Figure 10

OBJECTIVE [2] Find intercepts. In Figure 10 the graph intersects (crosses) the *y*-axis at (0, 3) and the *x*-axis at (2, 0). For this reason (0, 3) is called the **y-intercept** and (2, 0) is called the **x-intercept** of the graph. The intercepts are particularly useful for graphing linear equations, as in Example 1.

Finding Intercepts

We find the *x*-intercept by letting $y = 0$ in the given equation and solving for *x*.

We find the *y*-intercept by letting $x = 0$ in the given equation and solving for *y*.

EXAMPLE 2   Finding Intercepts

Find the intercepts for the graph of $2x + y = 4$. Draw the graph.

Find the *y*-intercept by letting $x = 0$; find the *x*-intercept by letting $y = 0$.

$$2x + y = 4 \qquad\qquad 2x + y = 4$$
$$2(\mathbf{0}) + y = 4 \qquad\qquad 2x + \mathbf{0} = 4$$
$$0 + y = 4 \qquad\qquad 2x = 4$$
$$y = 4 \qquad\qquad x = 2$$

The *y*-intercept is (0, 4). The *x*-intercept is (2, 0). The graph with the two intercepts shown in color is given in Figure 11. We get a third point as a check. For example, choosing $x = 1$ gives $y = 2$. These three ordered pairs are shown in the table with Figure 11. Plot (0, 4), (2, 0), and (1, 2) and draw a line through them. This line, shown in Figure 11, is the graph of $2x + y = 4$.

x	*y*
0	4
2	0
1	2

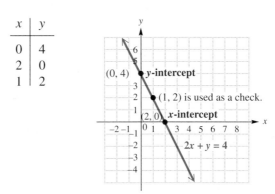

Figure 11

 When choosing *x*- or *y*-values to find ordered pairs to plot, be careful to choose so that the resulting points are not too close together. For example, using $(-1, -1)$, $(0, 0)$, and $(1, 1)$ may result in an inaccurate line. It is better to choose points where the *x*-values differ by at least 2.

OBJECTIVE [3] Graph linear equations of the form $Ax + By = 0$. In earlier examples, the *x*- and *y*-intercepts were used to help draw the graphs. This is not always possible, as the following examples show. Example 3 shows what to do when the *x*- and *y*-intercepts are the same point.

┌ **EXAMPLE 3**   Graphing an Equation of the Form $Ax + By = 0$

Graph the linear equation $x - 3y = 0$.

If we let $x = 0$, then $y = 0$, giving the ordered pair $(0, 0)$. Letting $y = 0$ also gives $(0, 0)$. This is the same ordered pair, so choose two other values for x or y. Choosing 2 for y gives $x - 3 \cdot 2 = 0$, giving the ordered pair $(6, 2)$. For a check point, we choose -6 for x getting -2 for y. This ordered pair, $(-6, -2)$, along with $(0, 0)$ and $(6, 2)$, was used to get the graph shown in Figure 12.

x	y
0	0
6	2
-6	-2

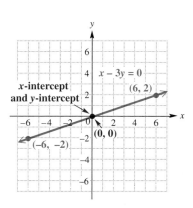

Figure 12

Example 3 can be generalized as follows.

Line through the Origin

If A and B are real numbers, the graph of a linear equation of the form

$$Ax + By = 0$$

goes through the origin $(0, 0)$.

OBJECTIVE **4** **Graph linear equations of the form $y = k$ or $x = k$.** The equation $y = -4$ is a linear equation in which the coefficient of x is 0. (Write $y = -4$ as $0x + y = -4$ to see this.) Also, $x = 3$ is a linear equation in which the coefficient of y is 0. These equations lead to horizontal or vertical straight lines, as the next examples show.

┌ **EXAMPLE 4**   Graphing an Equation of the Form $y = k$

Graph the linear equation $y = -4$.

As the equation states, for any value of x, y is always equal to -4. To get ordered pairs that are solutions of this equation, we choose any numbers for x, always using -4 for y. Three ordered pairs that satisfy the equation are shown in the table of values with Figure 13 on the next page. Drawing a line through these points gives the horizontal line shown in Figure 13.

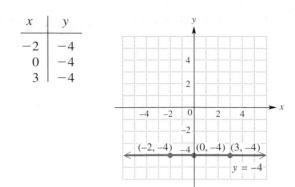

x	y
-2	-4
0	-4
3	-4

Figure 13

Horizontal Line

The graph of the linear equation $y = k$, where k is a real number, is the horizontal line going through the point $(0, k)$.

E X A M P L E   5   Graphing an Equation of the Form $x = k$

Graph the linear equation $x - 3 = 0$.

First add 3 to each side of the equation $x - 3 = 0$ to get $x = 3$. All the ordered pairs that are solutions of this equation have an x-value of 3. Any number can be used for y. We show three ordered pairs that satisfy the equation in the table of values with Figure 14. Drawing a line through these points gives the vertical line shown in Figure 14.

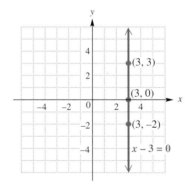

x	y
3	3
3	0
3	-2

Figure 14

Vertical Line

The graph of the linear equation $x = k$, where k is a real number, is the vertical line going through the point $(k, 0)$.

In particular, notice that the horizontal line $y = 0$ is the x-axis and the vertical line $x = 0$ is the y-axis.

CONNECTIONS

Beginning in this chapter we include information on the basic features of graphing calculators. The most obvious feature is their ability to graph equations. We must solve the equation for y in order to input it into the calculator. Also, we must select an appropriate "window" for the graph. The window is determined by the minimum and maximum values of x and y. Graphing calculators have a standard window, often from $x = -10$ to $x = 10$ and from $y = -10$ to $y = 10$. We indicate this as $[-10, 10]$, $[-10, 10]$, with the x-interval shown first.

For example, to graph the equation $2x + y = 4$, discussed in Example 2, we first solve for y.

$$2x + y = 4$$
$$y = -2x + 4 \qquad \text{Subtract } 2x.$$

If we input this equation as $y = -2x + 4$ and choose the standard window, the calculator shows the following graph.

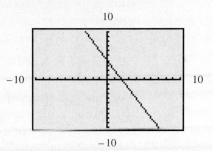

The x-value of the point on the graph where $y = 0$ (the x-intercept) gives the solution of the equation

$$y = 0,$$

or $\qquad\qquad -2x + 4 = 0.$ $\qquad$ Substitute $-2x + 4$ for y.

Since each tick mark on the x-axis represents 1, the graph shows that $(2, 0)$ is the x-intercept.

Working backwards, we use this idea to graphically find the solution of an equation in *one* variable as follows.

1. Rewrite the equation with 0 on one side.

2. Replace 0 with y and graph the equation in two variables.

3. Use the graph to find the x-value of the point where $y = 0$.

For example, we use this method to solve $6(3x + 5) = 2(10x + 10)$ as follows. First, rewrite the equation with 0 on one side.

$$6(3x + 5) = 2(10x + 10)$$
$$6(3x + 5) - 2(10x + 10) = 0 \qquad\qquad \text{Subtract } 2(10x + 10) \text{ on both sides.}$$

Second, replace 0 with y, and graph the equation in two variables.

$$6(3x + 5) - 2(10x + 10) = y$$

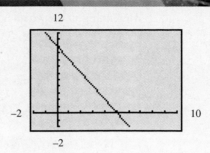

Third, use the graph to find the x-value of the point on the line where $y = 0$, the x-intercept. As shown in the figure, this occurs at $x = 5$. Verify this solution by substitution in the original equation.

FOR DISCUSSION OR WRITING
How would you rewrite these equations, with one side equal to 0, to input them into a graphing calculator for solution? (It is not necessary to clear parentheses or combine terms.)

1. $3x + 4 - 2x - 7 = 4x + 3$
2. $5x - 15 = 3(x - 2)$

The different forms of straight-line equations and the methods of graphing them are given in the following summary.

Graphing Straight Lines

Equation	To Graph	Example
$y = k$	Draw a horizontal line, through $(0, k)$.	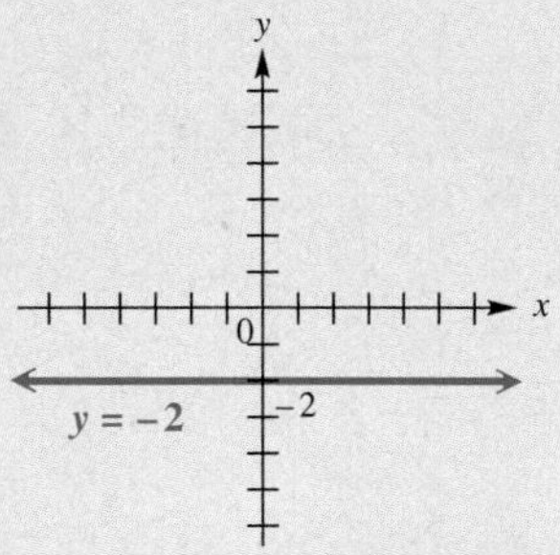 (graph showing $y = -2$, horizontal line through -2 on y-axis)
$x = k$	Draw a vertical line, through $(k, 0)$.	(graph showing $x = 4$, vertical line through 4 on x-axis)

(handwritten annotations):
ex { y = 2 / y = -4

ex { x = 4 / x = -5

slope on vert line is no slope

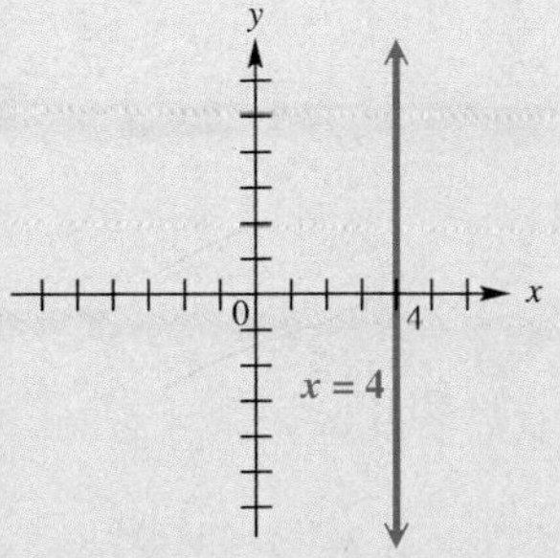

Graphing Straight Lines (continued)

Equation	To Graph	Example
$Ax + By = 0$	Graph goes through (0, 0). Get additional points that lie on the graph by choosing any value of x or y, except 0.	
$Ax + By = C$ but not of the types above	Find any two points the line goes through. A good choice is to find the intercepts: let $x = 0$, and find the corresponding value of y; then let $y = 0$, and find x. As a check, get a third point by choosing a value of x or y that has not yet been used.	

OBJECTIVE 5 Define a function. In the previous section, we saw several examples of relationships between two variables where each input value x produced an output value y. If each input x produces just *one* output y, we call the relationship a **function.** When a set of input values and a set of output values are related by a linear equation, $Ax + By = C$, each x-value corresponds to just one y-value, so linear equations define functions, unless $B = 0$. If $B = 0$, the equation becomes $Ax = C$ or $x = \frac{C}{A}$, where $\frac{C}{A}$ represents a real number. In this case, as we saw in Example 5, one input value x corresponds to an infinite number of output values y and the equation does not define a function.

┌ **EXAMPLE 6** Deciding Whether a Set of Ordered Pairs Defines a Function

Which of the following sets defines a function?

(a) $\{(1, -2), (2, 3), (0, 4), (4, 7)\}$

The input 1 is paired with the output -2, the input 2 is paired with the output 3, and so on. Every input value corresponds to exactly one output value, so the set defines a function.

(b) $\{(2, -4), (1, -1), (0, 0), (1, 1), (2, 4)\}$

Here, the input value 2 corresponds to the output values -4 and 4, so this set does not define a function. (What two output values correspond to the input value 1?)

(c) The ordered pairs that satisfy the equation $2x + 3y = 12$

Because this is a linear equation with $B \neq 0$, it defines a function.

(d) The ordered pairs that satisfy the equation $5x = 10$

In the context of two variables, when the coefficient of y is zero, as here, any value can be used for y. The only restriction imposed by the equation is that x must equal

2. Thus, $(2, 0)$, $(2, 5)$, $(2, -1)$, and any ordered pair with $x = 2$ satisfies this relationship. This is not a function, however, because $x = 2$ is paired with more than one real number.

(e) The ordered pairs that satisfy the equation $4y = 16$

Every ordered pair in this set has a y-value of 4. Since there is no restriction here on x, every real number is paired with $y = 4$. For example, $(-1, 4)$, $(2.5, 4)$, $(100, 4)$, and so on belong to this set, which defines a function, because every x corresponds to exactly one y, namely 4.

(f) The relationship between number of gallons needed to fill the gas tank and cost for a fill-up

This defines a function with ordered pairs of the form (number of gallons, cost). For a specific number of gallons, there is exactly one cost.

(g) The relationship between time spent traveling at a constant speed and number of miles traveled

Here the ordered pairs have the form (time, number of miles). A specific amount of time will correspond to exactly one number of miles, so this, too, defines a function.

Notice that in those sets that define functions, *no x-value appears more than once* (with a different *y*-value). The sets that were not functions had the same x as the input in at least two ordered pairs.

4.2 EXERCISES

In Exercises 1–6, match the information about the graphs with the linear equations in (a)–(e).

(a) $x = 5$ **(b)** $y = -3$ **(c)** $2x - 5y = 8$ **(d)** $x + 4y = 0$ **(e)** $3x + y = -4$

1. The graph of the equation has x-intercept $(4, 0)$.

2. The graph of the equation has y-intercept $(0, -4)$.

3. The graph of the equation goes through the origin.

4. The graph of the equation is a vertical line.

5. The graph of the equation is a horizontal line.

6. The graph of the equation goes through $(9, 2)$.

Complete the given ordered pairs using the given equation. Then graph each equation by plotting the points and drawing a line through them. See Examples 1 and 2.

7. $y = -x + 5$
$(0, \quad)$, $(\quad, 0)$, $(2, \quad)$

8. $y = x - 2$
$(0, \quad)$, $(\quad, 0)$, $(5, \quad)$

9. $y = \dfrac{2}{3}x + 1$
$(0, \quad)$, $(3, \quad)$, $(-3, \quad)$

10. $y = -\dfrac{3}{4}x + 2$
$(0, \quad)$, $(4, \quad)$, $(-4, \quad)$

11. $3x = -y - 6$
$(0, \quad)$, $(\quad, 0)$, $\left(-\dfrac{1}{3}, \quad\right)$

12. $x = 2y + 3$
$(\quad, 0)$, $(0, \quad)$, $\left(\quad, \dfrac{1}{2}\right)$

Find the x-intercept and the y-intercept for the graph of each equation. See Example 2.

13. $2x - 3y = 24$ **14.** $-3x + 8y = 48$ **15.** $x + 6y = 0$ **16.** $3x - y = 0$

17. What is the equation of the x-axis?

18. What is the equation of the y-axis?

19. A student attempted to graph $4x + 5y = 0$ by finding intercepts. She first let $x = 0$ and found y; then she let $y = 0$ and found x. In both cases, the resulting point was $(0, 0)$. She knew that she needed at least two different points to graph the line, but was unsure what to do next since finding intercepts gave her only one point. How would you explain to her what to do next?

20. Write a paragraph summarizing how to graph a linear equation in two variables.

Graph each linear equation. See Examples 1–5.

21. $x = y + 2$ **22.** $x = -y + 6$ **23.** $x - y = 4$ **24.** $x - y = 5$

25. $2x + y = 6$ **26.** $-3x + y = -6$ **27.** $3x + 7y = 14$ **28.** $6x - 5y = 18$

29. $y - 2x = 0$ **30.** $y + 3x = 0$ **31.** $y = -6x$ **32.** $y = 4x$

33. $y + 1 = 0$ **34.** $y - 3 = 0$ **35.** $x = -2$ **36.** $x = 4$

TECHNOLOGY INSIGHTS (EXERCISES 37-42)

In each exercise below, a calculator-generated graph of a linear equation in one variable with one side equal to 0 is shown. Accompanying the graph is the equation itself, where y is expressed in terms of x on the left side. Solve the equation using the methods of Section 2.2, and show that the solution you get is the same as the x-intercept (labeled "zero") on the calculator screen.

37. $8 - 2(3x - 4) - 2x = 0$ **38.** $5(2x - 1) - 4(2x + 1) - 7 = 0$

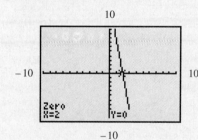   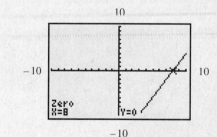

39. $.6x - .1x - x + 2.5 = 0$ **40.** $-\dfrac{2}{7}x + 2x - \dfrac{1}{2}x - \dfrac{17}{2} = 0$

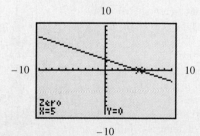   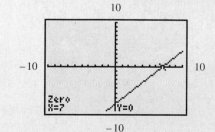

41. Use the results of Exercises 37–40 to explain how the x-intercept of the graph of an equation in two variables corresponds to the solution of an equation in one variable.

42. A horizontal line has no x-intercept. If you try to solve $5x - (3x + 2x) + 4 = 0$, you get no solution. What would the graph of $y = 5x - (3x + 2x) + 4$ look like on a graphing calculator?

Solve each problem.

43. The number of master's degrees earned in business management increased during the years 1971–1994 as shown in the figure. If $x = 0$ represents 1970, $x = 5$ represents 1975, $x = 10$ represents 1980, and so on, the number of master's degrees can be approximated by

$$y = 2.81x + 24.1,$$

where y is in thousands. (This is a *linear model* for the data.)

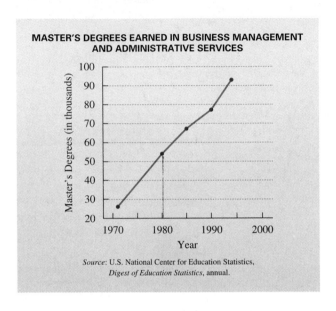

MASTER'S DEGREES EARNED IN BUSINESS MANAGEMENT AND ADMINISTRATIVE SERVICES

Source: U.S. National Center for Education Statistics, *Digest of Education Statistics*, annual.

(a) Use the equation to approximate the number of such degrees in the years 1980, 1985, 1990, and 1994.

(b) Estimate the y-values from the graph for the same years.

(c) Are the approximations using the equation close to the values you read from the graph?

44. Sporting goods sales (in billions of dollars) from 1988–1996 are approximated by the equation

$$y = 1.625x + 40.75,$$

where $x = 0$ corresponds to 1986, $x = 2$ corresponds to 1988, and so on. Sales for even-numbered years in that time period are plotted in the accompanying figure, which also shows the graph of the linear equation. As the graph indicates, the actual sales in 1992 (when the U.S. economy was depressed) of about \$47 billion were less than the sales approximated by the equation.

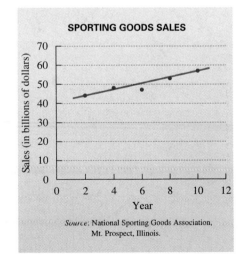

SPORTING GOODS SALES

Source: National Sporting Goods Association, Mt. Prospect, Illinois.

(a) Use the *equation* to approximate the sales in each of the even-numbered years.

(b) Does this equation define a function? Why or why not?

45. The height y of a woman (in centimeters) is a function of the length of her radius bone x (from the wrist to the elbow) and is defined by $y = 73.5 + 3.9x$. Estimate the heights of women with radius bones of the following lengths.

(a) 23 centimeters

(b) 25 centimeters

(c) 20 centimeters

(d) Graph $y = 73.5 + 3.9x$.

46. As a rough estimate, the weight of a man taller than about 60 inches is a linear function approximated by $y = 5.5x - 220$, where x is the height of the person in inches, and y is the weight in pounds. Estimate the weights of men whose heights are as follows.

(a) 62 inches   (b) 64 inches

(c) 68 inches   (d) 72 inches

(e) Graph $y = 5.5x - 220$.

47. The graph shows that the value of a certain automobile over its first five years is a function of the year. Use the graph to estimate the depreciation (loss in value) during the following years.

Automobile Value

(a) First   (b) Second   (c) Fifth

(d) What is the total depreciation over the 5-year period?

48. The demand for an item is a function of its price. As price goes up, demand goes down. On the other hand, when price goes down, demand goes up. Suppose the demand for a certain Beanie Baby is 1000 when its price is $30 and 8000 when it costs $15.

(a) Let x be the price and y be the demand for the Beanie Baby. Graph the two given pairs of prices and demands.

(b) Assume the relationship is linear. Draw a line through the two points from part (a). From your graph estimate the demand if the price drops to $10.

(c) Use the graph to estimate the price if the demand is 4000.

Decide whether each set of ordered pairs defines a function. See Example 6.

49. $\{(0, 5), (2, 3), (4, 1), (6, -1), (8, -3)\}$

50. $\{(1, 3), (2, 3), (3, 3), (4, 3)\}$

51. The ordered pairs that satisfy $-x + 2y = 9$

52. The ordered pairs that satisfy $y = 4$

53. The ordered pairs that satisfy $x = 8$

54. The ordered pairs that satisfy $5x + y = 7$

55. The relationship in Exercise 43 between year and number of master's degrees

56. The relationship in Exercise 44 between year and sporting goods sales

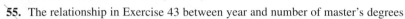

4.3 The Slope of a Line

OBJECTIVES

1 Find the slope of a line given two points.

2 Find the slope from the equation of a line.

3 Use the slope to determine whether two lines are parallel, perpendicular, or neither.

FOR EXTRA HELP

SSG Sec. 4.3
SSM Sec. 4.3

Pass the Test Software

InterAct Math Tutorial Software

Video 6

When two variables are related in such a way that the value of one depends on the value of the other, we can form ordered pairs of corresponding numbers. For example, if $x + y = 7$, then when $x = 2$, $y = 5$, and when $x = -3$, $y = 10$. We write these pairs as $(2, 5)$ and $(-3, 10)$, respectively, with the understanding that the first number in the ordered pair represents x and the second number represents y. To indicate two nonspecific ordered pairs that satisfy a particular equation relating x and y, we use *subscript notation*. We write the pairs as (x_1, y_1) and (x_2, y_2). (Read x_1 as "x-sub-one" and x_2 as "x-sub-two.")

We can graph a straight line if at least two different points on the line are known. A line also can be graphed by using just one point on the line if the "steepness" of the line is known.

OBJECTIVE 1 **Find the slope of a line given two points.** One way to measure the steepness of a line is to compare the vertical change in the line (the rise) to the horizontal change (the run) while moving along the line from one fixed point to another. This measure of steepness is called the *slope* of the line.

Figure 15 shows a line with the points (x_1, y_1) and (x_2, y_2). As we move along the line from the point (x_1, y_1) to the point (x_2, y_2), y changes by $y_2 - y_1$ units. This is the vertical change. Similarly, x changes by $x_2 - x_1$ units, the horizontal change. The ratio of the change in y to the change in x gives the slope of the line. We usually denote slope with the letter m. The slope of a line is defined as follows.

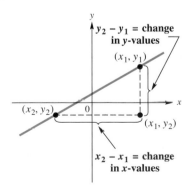

Figure 15

Slope Formula

The **slope** of the line through the points (x_1, y_1) and (x_2, y_2) is

$$m = \frac{\text{change in } y}{\text{change in } x} = \frac{y_2 - y_1}{x_2 - x_1} \quad \text{if } x_1 \neq x_2.$$

The slope of a line tells how fast y changes for each unit of change in x; that is, the slope gives the ratio of the change in y to the change in x. The change in y is called the **rise,** and the change in x is called the **run.**

CONNECTIONS

The idea of slope is used in many everyday situations. For example, because $10\% = \frac{1}{10}$, a highway with a 10% grade (or slope) rises one meter for every 10 horizontal meters. The highway sign shown below is used to warn of a downgrade ahead that may be long or steep. Architects specify the pitch of a roof using slope; a $\frac{5}{12}$ roof means that the roof rises 5 feet for every 12 feet in the horizontal direction. The slope of a stairwell also indicates the ratio of the vertical rise to the horizontal run. The slope of the stairs in the figure is $\frac{8}{14}$.

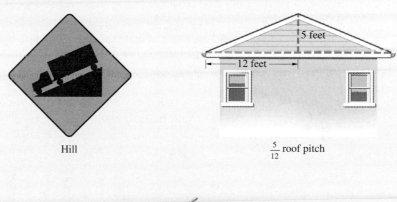

Hill

$\frac{5}{12}$ roof pitch

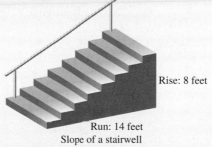

Rise: 8 feet

Run: 14 feet

Slope of a stairwell

FOR DISCUSSION OR WRITING
Describe some other everyday examples of slope.

> **E X A M P L E 1**   Finding the Slope of a Line
>
> Find the slope of each of the following lines.
>
> **(a)** The line through $(-4, 7)$ and $(1, -2)$
>
> Use the definition of slope. Let $(-4, 7) = (x_2, y_2)$ and $(1, -2) = (x_1, y_1)$. Then
>
> $$\text{slope} = \frac{\text{change in } y}{\text{change in } x}$$
>
> $$m = \frac{y_2 - y_1}{x_2 - x_1}$$
>
> $$= \frac{7 - (-2)}{-4 - 1} = \frac{9}{-5} = -\frac{9}{5}.$$
>
> As Figure 16 shows, this line has a slope of $-\dfrac{9}{5}$ (which can also be written as $\dfrac{-9}{5}$ or $\dfrac{9}{-5}$). One way of interpreting this is that the line drops vertically 9 units for a horizontal change of 5 units to the right.
>
>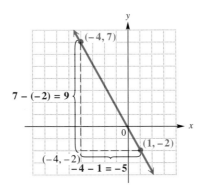
>
> **Figure 16**
>
> **(b)** The line through $(12, -5)$ and $(-9, -2)$
>
> $$m = \frac{-5 - (-2)}{12 - (-9)} = \frac{-3}{21} = -\frac{1}{7}$$
>
> The same slope is found by subtracting in reverse order.
>
> $$\frac{-2 - (-5)}{-9 - 12} = \frac{3}{-21} = -\frac{1}{7}$$

> **CAUTION** It makes no difference which point is (x_1, y_1) or (x_2, y_2); however, it is important to be consistent. Start with the x- and y-values of one point (either one) and subtract the corresponding values of the other point.

In Example 1(a) the slope is negative and the corresponding line in Figure 16 falls from left to right. As Figure 17(a) shows, this is generally true of lines with negative slopes. Lines with positive slopes go up (rise) from left to right, as shown in Figure 17(b).

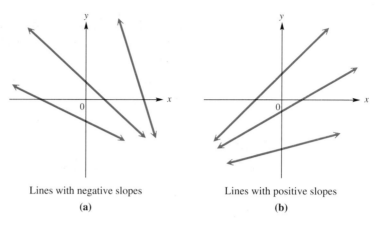

Lines with negative slopes	Lines with positive slopes
(a)	**(b)**

Figure 17

Positive and Negative Slopes

A line with positive slope rises from left to right.

A line with negative slope falls from left to right.

E X A M P L E 2   Finding the Slope of a Horizontal Line

Find the slope of the line through $(-8, 4)$ and $(2, 4)$.

Use the definition of slope.

$$m = \frac{4 - 4}{-8 - 2} = \frac{0}{-10} = 0$$

As shown in Figure 18, the line through these two points is horizontal, with equation $y = 4$. *All horizontal lines have a slope of 0,* since the difference in y-values is always 0.

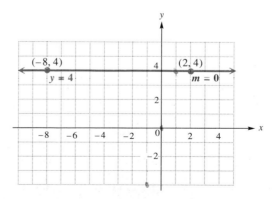

Figure 18

E X A M P L E 3   Finding the Slope of a Vertical Line

Find the slope of the line through (6, 2) and (6, −9).

$$m = \frac{2 - (-9)}{6 - 6} = \frac{11}{0} \qquad \text{Undefined}$$

Since division by 0 is undefined, the slope is undefined. The graph in Figure 19 shows that the line through these two points is vertical, with equation $x = 6$. All points on a vertical line have the same x-value, so *the slope of any vertical line is undefined.*

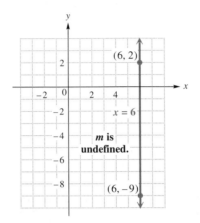

Figure 19

Slopes of Horizontal and Vertical Lines

Horizontal lines, with equations of the form $y = k$, **have slope 0.**

Vertical lines, with equations of the form $x = k$, **have undefined slope.**

O B J E C T I V E 2   Find the slope from the equation of a line.   The slope of a line also can be found directly from its equation. For example, the slope of the line

$$y = -3x + 5$$

can be found using any two points on the line. Get two points by choosing two differ-ent values of x, say −2 and 4, and finding the corresponding y-values.

If $x = -2$:	If $x = 4$:
$y = -3(-2) + 5$	$y = -3(4) + 5$
$y = 6 + 5$	$y = -12 + 5$
$y = 11.$	$y = -7.$

The ordered pairs are (−2, 11) and (4, −7). Now use the slope formula to find the slope.

$$m = \frac{11 - (-7)}{-2 - 4} = \frac{18}{-6} = -3$$

The slope, −3, is the same number as the coefficient of x in the equation $y = -3x + 5$. It can be shown that this always happens, *as long as the equation is solved for y.* This fact is used to find the slope of a line from its equation.

Finding the Slope of a Line from its Equation

Step 1   Solve the equation for y.

Step 2   The slope is given by the coefficient of x.

E X A M P L E 4   Finding Slope from an Equation

Find the slope of each of the following lines.

(a) $2x - 5y = 4$

Solve the equation for y.

$$2x - 5y = 4$$
$$-5y = -2x + 4 \qquad \text{Subtract } 2x \text{ from each side.}$$
$$y = \frac{2}{5}x - \frac{4}{5} \qquad \text{Divide each side by } -5.$$

The slope is given by the coefficient of x, so the slope is $m = \dfrac{2}{5}$.

(b) $8x + 4y = 1$

Solve the equation for y.

$$8x + 4y = 1$$
$$4y = -8x + 1 \qquad \text{Subtract } 8x \text{ from each side.}$$
$$y = -2x + \frac{1}{4} \qquad \text{Divide each side by } 4.$$

The slope of this line is given by the coefficient of x, -2.

OBJECTIVE 3   Use the slope to determine whether two lines are parallel, perpendicular, or neither. Two lines in a plane that never intersect are **parallel.** We use slopes to tell whether two lines are parallel. For example, Figure 20 shows the graph of $x + 2y = 4$ and the graph of $x + 2y = -6$. These lines appear to be parallel. Solve for y to find that both $x + 2y = 4$ and $x + 2y = -6$ have a slope of $-\frac{1}{2}$. Nonvertical parallel lines always have equal slopes.

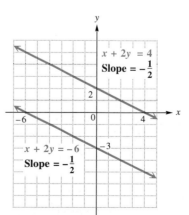

Figure 20

Figure 21 shows the graph of $x + 2y = 4$ and the graph of $2x - y = 6$. These lines appear to be **perpendicular** (meet at a 90° angle). Solving for y shows that the slope of $x + 2y = 4$ is $-\frac{1}{2}$, while the slope of $2x - y = 6$ is 2. The product of $-\frac{1}{2}$ and 2 is

$$-\frac{1}{2}(2) = -1.$$

This is true in general; the product of the slopes of two perpendicular lines, neither of which is vertical, is always -1. This means that the slopes of perpendicular lines are negative reciprocals: if one slope is the nonzero number a, the other is $-\frac{1}{a}$.

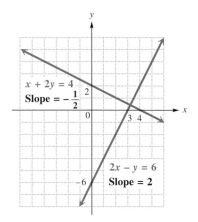

Figure 21

Parallel and Perpendicular Lines

Two nonvertical lines with the same slope are parallel; two perpendicular lines, neither of which is vertical, have slopes that are negative reciprocals of each other.

EXAMPLE 5 Deciding Whether Two Lines Are Parallel or Perpendicular

Decide whether the lines are *parallel, perpendicular,* or *neither.*

(a) $x + 2y = 7$
        $-2x + y = 3$

Find the slope of each line by first solving each equation for y.

$$x + 2y = 7 \qquad\qquad -2x + y = 3$$
$$2y -= -x + 7 \qquad\qquad y -= 2x + 3$$
$$y = -\frac{1}{2}x + \frac{7}{2}$$

$$\text{Slope: } -\frac{1}{2} \qquad\qquad\qquad \text{Slope: } 2$$

Since the slopes are not equal, the lines are not parallel. Check the product of the slopes: $-\frac{1}{2}(2) = -1$. The two lines are perpendicular because the product of their slopes is -1, indicating that the slopes are negative reciprocals.

(b) $3x - y = 4$
$6x - 2y = 9$

Find the slopes. Both lines have a slope of 3, so the lines are parallel.

(c) $4x + 3y = 6$
$2x - y = 5$

Here the slopes are $-\frac{4}{3}$ and 2. These two straight lines are neither parallel nor perpendicular.

CONNECTIONS

Because the viewing window of a graphing calculator is a rectangle, the graphs of perpendicular lines will not appear perpendicular unless appropriate intervals are used for x and y. Graphing calculators usually have a key to select a "square" window automatically. In a square window, the x-interval is about 1.5 times the y-interval. The equations used in Figure 21 are graphed with the standard (nonsquare) window and then with a square window in the screens below.

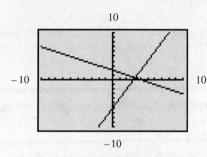

A standard (nonsquare) window
[−10, 10] by [−10, 10]
Lines do not appear perpendicular.

(a)

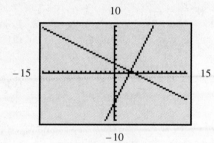

A square window
[−15, 15] by [−10, 10]
Lines appear perpendicular.

(b)

Figure 21

4.3 EXERCISES

1. What is meant by "rise"? What is meant by "run"?

Use the coordinates of the indicated points to find the slope of each line. See Example 1.

2.

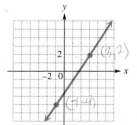

3.

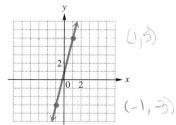

4.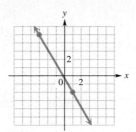

5.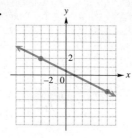

6.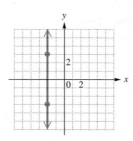

7.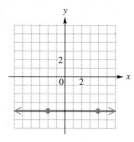

8. Look at the graph in Exercise 2 and answer the following.
   (a) Start at the point $(-1, -4)$ and count vertically up to the horizontal line that goes through the other plotted point. What is this vertical change? (Remember: "up" means positive, "down" means negative.)
   (b) From this new position, count horizontally to the other plotted point. What is this horizontal change? (Remember: "right" means positive, "left" means negative.)
   (c) What is the quotient of the numbers found in parts (a) and (b)? What do we call this number?

9. Refer to Exercise 8. If we were to *start* at the point $(3, 2)$ and *end* at the point $(-1, -4)$, do you think that the answer to part (c) would be the same? Explain why or why not.

On a pair of axes similar to the one shown, sketch the graph of a straight line having the indicated slope.

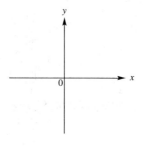

10. negative

11. positive

12. undefined

13. zero

14. Explain in your own words what is meant by *slope* of a line.

15. A student was asked to find the slope of the line through the points $(2, 5)$ and $(-1, 3)$. His answer, $-\dfrac{2}{3}$, was incorrect. He showed his work as

$$\frac{3 - 5}{2 - (-1)} = \frac{-2}{3} = -\frac{2}{3}.$$

What was his error? Give the correct slope.

Find the slope of the line going through each pair of points. See Examples 1–3.

16. $(4, -1)$ and $(-2, -8)$

17. $(1, -2)$ and $(-3, -7)$

18. $(-8, 0)$ and $(0, -5)$

19. $(0, 3)$ and $(-2, 0)$

20. $(6, -5)$ and $(-12, -5)$ **21.** $(4, 3)$ and $(-6, 3)$

22. $(-8, 6)$ and $(-8, -4)$ **23.** $(-12, 3)$ and $(-12, -7)$

24. $(3.1, 2.6)$ and $(1.6, 2.1)$ **25.** $\left(-\dfrac{7}{5}, \dfrac{3}{10}\right)$ and $\left(\dfrac{1}{5}, -\dfrac{1}{2}\right)$

Find the slope of each line. See Example 4.

26. $y = 2x - 3$ **27.** $y = 5x + 12$ **28.** $2y = -x + 4$ **29.** $4y = x + 1$

30. $-6x + 4y = 4$ **31.** $3x - 2y = 3$ **32.** $y = 4$ **33.** $x = 6$

The figure shows a line that has a positive slope (because it rises from left to right) and a positive y-value for the y-intercept (because it intersects the y-axis above the origin).

*For each line shown, decide whether (**a**) the slope is positive, negative, or zero and (**b**) the y-value of the y-intercept is positive, negative, or zero.*

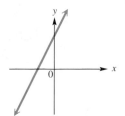

34.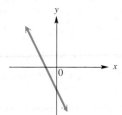

35.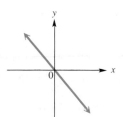

36.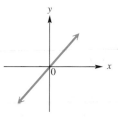

37.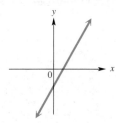

38.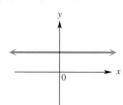

39.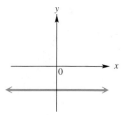

40. If two nonvertical lines are parallel, what do we know about their slopes? If two lines are perpendicular and neither is parallel to an axis, what do we know about their slopes? Why must the lines be nonvertical?

41. If two lines are both vertical or both horizontal, which of the following are they?
    (a) parallel   **(b)** perpendicular   **(c)** neither parallel nor perpendicular

42. If a line is vertical, what is true of any line that is perpendicular to it?

For each pair of equations, give the slopes of the lines and then determine whether the two lines are parallel, perpendicular, or neither parallel nor perpendicular. See Example 5.

43. $2x + 5y = 4$
    $4x + 10y = 1$

44. $-4x + 3y = 4$
    $-8x + 6y = 0$

45. $8x - 9y = 6$
    $8x + 6y = -5$

46. $5x - 3y = -2$
    $3x - 5y = -8$

47. $3x - 2y = 6$
    $2x + 3y = 3$

48. $3x - 5y = -1$
    $5x + 3y = 2$

For Exercises 49 and 50, you may wish to refer to the Connections box preceding Example 1.

49. What is the slope (or pitch) of this roof?

50. What is the slope (or grade) of this hill?

RELATING CONCEPTS   (EXERCISES 51–56)

Figure A depicts public school enrollment (in thousands) in grades 9–12 in the United States. Figure B gives the (average) number of public school students per computer.

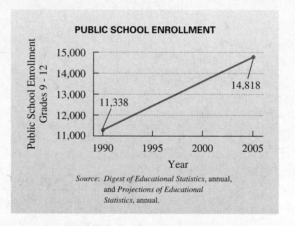

Figure A

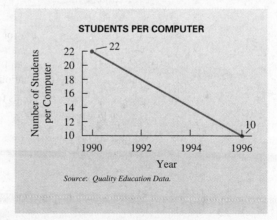

Figure B

Work Exercises 51–56 in order.

51. Use the ordered pairs (1990, 11,338) and (2005, 14,818) to find the slope of the line in Figure A.

RELATING CONCEPTS (EXERCISES 51–56) (CONTINUED)

52. The slope of the line in Figure A is _____. This means that
                                                                  (positive/negative)
during the period represented, enrollment _____.
                                                          (increased/decreased)

53. The slope of a line represents its *rate of change.* Based on Figure A, what was the increase in students *per year* during the period shown?

54. Use the given ordered pairs to find the slope of the line in Figure B.

55. The slope of the line in Figure B is _____. This means that
                                                                  (positive/negative)
during the period represented, the number of students per computer

_____.
   (increased/decreased)

56. Based on Figure B, what was the decrease in students per computer *per year* during the period shown?

Did you make the connection between the sign of the slope of the line and the increase or decrease in the quantity represented by *y*?

Work each business-related problem.

57. The table gives the number of shopping centers in the years 1991–1996. The equation

$$y = .797x - 1549,$$

where *x* represents the year and *y* is the number of shopping centers in thousands, approximates these data quite well.

Number of Shopping Centers (in thousands)

Year	1991	1992	1993	1994	1995	1996
Number of Centers	38.0	39.0	39.6	40.4	41.2	42.1

Source: International Council of Shopping Centers.

(a) Give three ordered pairs for this *equation* and indicate the input and output numbers for each.

(b) Compare the *y*-values you find with the values given for the number of centers in the table. Are the approximations close?

58. The line graph shows the points for the ordered pairs from the table in Exercise 57. A straight line graph of the equation in Exercise 57 is also shown. Except for one point, the straight line is almost the same as the line graph. Which point is farthest from the straight line?

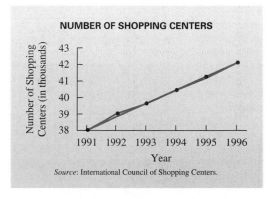

NUMBER OF SHOPPING CENTERS

Source: International Council of Shopping Centers.

59. The growth in retail square footage, in billions, is shown in the line graph. This graph looks like a straight line. If the change in square footage each year is the same, then it is a straight line. Find the change in square footage for the years shown in the graph. (*Hint:* To find the change in square footage from 1991 to 1992, subtract the *y*-value for 1991 from the *y*-value for 1992.) Is the graph a straight line?

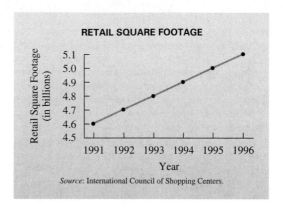

60. Find the slope of the line in Exercise 59 by using any two of the points shown on the line. How does the slope compare with the yearly change in square footage?

TECHNOLOGY INSIGHTS (EXERCISES 61–66)

61. Two views of the same line are shown in the accompanying calculator screen, along with coordinates of two points displayed at the bottoms. What is the slope of this line?

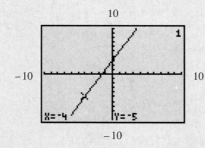 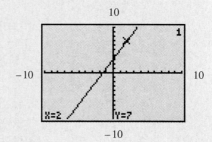

62. Repeat Exercise 61 for the line shown here.

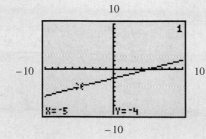 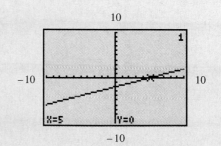

TECHNOLOGY INSIGHTS (EXERCISES 61-66) (CONTINUED)

Some graphing calculators have the capa-
bility of displaying a table of points for a
graph. The table shown here gives several
points that lie on a line designated Y_1.

X	Y₁	
-12	-.8	
-10	0	
-8	.8	
-6	1.6	
-4	2.4	
-2	3.2	
0	4	

X=-12

63. Use any pair of points displayed to find the slope of the line.
64. What is the *x*-intercept of the line?
65. What is the *y*-intercept of the line?
66. Which one of the two lines shown is the graph of Y_1?

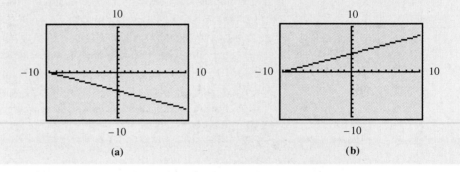

(a)   (b)

CHAPTER 4 GROUP ACTIVITY

▦ Determining Business Profit

Objective: Use equations, tables, and graphs to make business decisions.

Graphs are used by businesses to analyze and make decisions. This activity will explore such a process.

The Parent Teacher Organization of a school decides to sell bags of popcorn as a fund-raiser. The parents want to estimate their profit. Costs include $14 for popcorn (enough for 680 bags) and $7 for bags to hold the popcorn.

A. Write a profit formula.

1. From the information above, determine the total costs. (Assume the organization is buying enough supplies for 680 bags of popcorn.)

2. If the bags sell for $.25 each, write an expression for the total sales of *n* bags of popcorn.

3. Since Profit = Sales − Cost, write an equation that represents the profit *P* for *n* bags of popcorn sold.

4. Work in pairs. One person should use the profit equation from above to complete the table on the next page. The other person should decide on an appropriate scale and graph the profit equation.

(continued)

n	P
0	
	0
100	

B. Choose a different price for a bag of popcorn (between $.20 and $.75).

1. Write the profit equation for this cost.

2. Switch roles, that is, if you drew the graph in part A, now make a table of values for this equation. Have your partner graph the profit equation on the same coordinate system you used in part A.

C. Compare your findings and answer the following questions.

1. What is the break-even point (that is, when profits are 0 or sales equal costs) for each equation?

2. What does it mean if you end up with a negative value for P?

3. If you sell all 680 bags, what will your profits be for the two different prices? Explain how you would estimate this from your graph.

4. What are the advantages and/or disadvantages of charging a higher price?

5. Together, decide on the price you would charge for a bag of popcorn. Explain why you chose this price.

CHAPTER 4 SUMMARY

KEY TERMS

4.1 linear equation	rectangular	plot	slope
ordered pair	(Cartesian)	**4.2** graph, graphing	rise
table of values	coordinate system	y-intercept	run
x-axis	quadrants	x-intercept	parallel lines
y-axis	origin	function	perpendicular lines
	coordinates	**4.3** subscript notation	

NEW SYMBOLS

(a, b) an ordered pair (x_1, y_1) x-sub-one, y-sub-one m slope

TEST YOUR WORD POWER

See how well you have learned the vocabulary in this chapter. Answers, with examples, are given at the bottom of the next page.

1. An **ordered pair** is a pair of numbers written
(a) in numerical order between brackets
(b) between parentheses or brackets
(c) between parentheses in which order is important
(d) between parentheses in which order does not matter.

TEST YOUR WORD POWER (CONTINUED)

2. The **coordinates** of a point are
(a) the numbers in the corresponding ordered pair
(b) the solution of an equation
(c) the values of the x- and y-intercepts
(d) the graph of the point.

3. An **intercept** is
(a) the point where the x-axis and y-axis intersect
(b) a pair of numbers written in parentheses in which order matters
(c) one of the four regions determined by a rectangular coordinate system
(d) the point where a graph intersects the x-axis or the y-axis.

4. A **function** is
(a) the numbers in an ordered pair
(b) a set of ordered pairs in which each x-value corresponds to exactly one y-value
(c) a pair of numbers written between parentheses in which order matters
(d) the set of all ordered pairs that satisfy an equation.

5. The **slope** of a line is
(a) the measure of the run over the rise of the line
(b) the distance between two points on the line
(c) the ratio of the change in y to the change in x along the line

(d) the horizontal change compared to the vertical change of two points on the line.

6. Two lines in a plane are **parallel** if
(a) they represent the same line
(b) they never intersect
(c) they intersect at a 90° angle
(d) one has a positive slope and one has a negative slope.

7. Two lines in a plane are **perpendicular** if
(a) they represent the same line
(b) they never intersect
(c) they intersect at a 90° angle
(d) one has a positive slope and one has a negative slope.

QUICK REVIEW

CONCEPTS	EXAMPLES

4.1 LINEAR EQUATIONS IN TWO VARIABLES

An ordered pair is a solution of an equation if it satisfies the equation.

Is $(2, -5)$ or $(0, -6)$ a solution of $4x - 3y = 18$?

$$4(2) - 3(-5) = 23 \neq 18 \qquad 4(0) - 3(-6) = 18$$
$(2, -5)$ is not a solution. $(0, -6)$ is a solution.

If a value of either variable in an equation is given, the other variable can be found by substitution.

Complete the ordered pair $(0, \quad)$ for $3x = y + 4$.

$$3(0) = y + 4$$
$$0 = y + 4$$
$$-4 = y$$

The ordered pair is $(0, -4)$.

Plot the ordered pair $(-2, 4)$ by starting at the origin, going 2 units to the left, then going 4 units up.

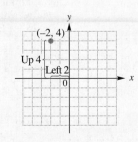

CONCEPTS	EXAMPLES

4.2 GRAPHING LINEAR EQUATIONS IN TWO VARIABLES

The graph of $y = k$ is a horizontal line through $(0, k)$.

The graph of $x = k$ is a vertical line through $(k, 0)$.

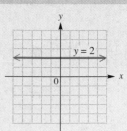 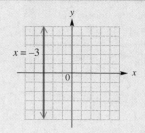

The graph of $Ax + By = 0$ goes through the origin.
Find and plot another point that satisfies the equation.
Then draw the line through the two points.

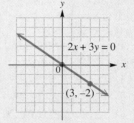

To graph a linear equation:
1. Find at least two ordered pairs that satisfy the equation.
2. Plot the corresponding points.
3. Draw a straight line through the points.

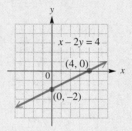

A set of ordered pairs defines a function if each input x corresponds to exactly one output y; that is, if no x-value occurs in more than one ordered pair.

The set $\{(4, 3), (8, 6), (12, 9)\}$ defines a function, but the set $\{(-1, 2), (0, 2), (-1, -2)\}$ does not because -1 is the x-value in two ordered pairs.

4.3 THE SLOPE OF A LINE

The slope of the line through (x_1, y_1) and (x_2, y_2) is

$$m = \frac{y_2 - y_1}{x_2 - x_1} \ (x_1 \neq x_2).$$

The line through $(-2, 3)$ and $(4, -5)$ has slope

$$m = \frac{-5 - 3}{4 - (-2)} = \frac{-8}{6} = -\frac{4}{3}.$$

Horizontal lines have slope 0.

The line $y = -2$ has slope 0.

Vertical lines have undefined slope.

The line $x = 4$ has **undefined slope**.

To find the slope of a line from its equation, solve for y. The slope is the coefficient of x.

Find the slope of $3x - 4y = 12$.

$$-4y = -3x + 12$$

$$y = \frac{3}{4}x - 3$$

The slope is $\frac{3}{4}$.

Parallel lines have the same slope.

The lines $y = 3x - 1$ and $y = 3x + 4$ are parallel because both have slope 3.

The slopes of perpendicular lines are negative reciprocals (that is, their product is -1).

The lines $y = -3x - 1$ and $y = \frac{1}{3}x + 4$ are perpendicular because their slopes are -3 and $\frac{1}{3}$ and $-3\left(\frac{1}{3}\right) = -1$.

CHAPTER 4 REVIEW EXERCISES

[4.1] *Complete the given ordered pairs for each equation.*

1. $y = 3x + 2$; $(-1, \quad)$, $(0, \quad)$, $(\quad , 5)$

2. $4x + 3y = 6$; $(0, \quad)$, $(\quad , 0)$, $(-2, \quad)$

3. $x - 7 = 0$; $(\quad , -3)$, $(\quad , 0)$, $(\quad , 5)$

Decide whether the given ordered pair is a solution of the given equation.

4. $x + y = 7$; $(2, 5)$

5. $2x + y = 5$; $(-1, 3)$

6. $3x - y = 4$; $\left(\dfrac{1}{3}, -3\right)$

Name the quadrant in which each pair lies. Then plot each ordered pair in a rectangular coordinate system.

7. $(2, 3)$ **8.** $(-4, 2)$

9. $(3, 0)$ **10.** $(0, -6)$

11. If $xy > 0$, in what quadrant or quadrants must (x, y) lie?

12. On what axis does the point $(k, 0)$ lie for any real value of k?

[4.2] *Find the x- and y-intercepts for the line that is the graph of each equation, and graph the line.*

13. $y = 2x + 5$ **14.** $3x + 2y = 8$

15. $x + 2y = -4$ **16.** $x + y = 0$

Decide whether each set defines a function.

17. $\{(5, -3), (-4, -3), (-1, 8), (2, 8)\}$

18. The ordered pairs that satisfy the equation $3x - 9 = 0$

19. The following relationship between years and sponsorship spending: Sponsorship spending in North America (in billions of dollars) is closely approximated by $y = .425x - 1.225$, where x is the number of years since 1980. For example, in 1985, $x = 5$ and $y = .425(5) - 1.225 = .9$ billion dollars. (*Source: IEG Sponsorship Report.*)

20. Describe the input values and the output values for the relationship in Exercise 19.

[4.3] *Find the slope of each line.*

21. Through $(2, 3)$ and $(-4, 6)$ **22.** Through $(0, 6)$ and $(1, 6)$

23. Through $(2, 5)$ and $(2, 8)$ **24.** $y = 3x - 4$

25. $y = \dfrac{2}{3}x + 1$ **26.**

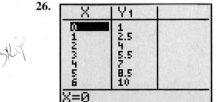

27.

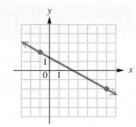

28. $y = 5$

29. A line perpendicular to the graph of $y = -3x + 3$

30. Explain why the signs of the slopes of perpendicular lines (neither of which is vertical) cannot be the same.

Decide whether the lines in each pair are parallel, perpendicular, or neither.

31. $3x + 2y = 6$
   $6x + 4y = 8$

32. $x - 3y = 1$
   $3x + y = 4$

33. $x - 2y = 8$
   $x + 2y = 8$

MIXED REVIEW EXERCISES

34. Use the bar graph to estimate the growth rates for motor vehicles and parts sales for the years 1992–1994. (The rate for 1992, for example, represents the percent increase in sales from 1991 to 1992.) Sales in 1993 were approximately $456 billion. Use the growth rate in 1994 to estimate sales in that year.

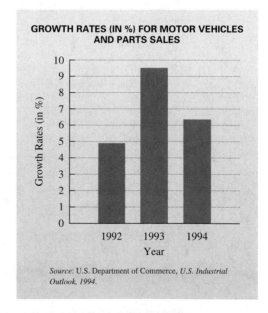

Source: U.S. Department of Commerce, *U.S. Industrial Outlook, 1994.*

35. Complete these ordered pairs for $x = 3y$:   (0,   ),   (8,   ),   (, −3).

In Exercises 36–38, find the x- and y-intercepts and slope.

36. $x = -2$

37. $y = 2x + 3$

38. $11x - 3y = 4$

39. Find the slope of the line through $(4, -1)$ and $(-2, -3)$.

40. What is the slope of a line perpendicular to a line with undefined slope?

41. Is $\left(0, -\dfrac{16}{3}\right)$ a solution of $5x - 3y = 16$?

Graph each equation given in Exercises 42–44.

42. $x + 3y = 9$ **43.** $x - 5 = 0$ **44.** $2x - y = 3$

45. Two points determine a line. Explain why it is a good idea to plot three points before drawing the line.

46. The set of ordered pairs gives the five U.S. cities with the largest Hispanic populations in 1990. Data is in millions. (*Source:* U.S. Census Bureau.)

  {(New York, 1.78), (Los Angeles, 1.39), (Chicago, .55), (San Antonio, .52), (Houston, .45)}

Does this set define a function? What are the inputs and the outputs?

CHAPTER 4 TEST

The graph shows the cost of 30 seconds of advertising time on the annual Super Bowl telecast. Use the graph to respond to Exercises 1 and 2.

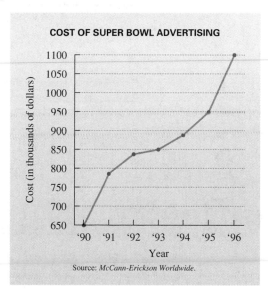

COST OF SUPER BOWL ADVERTISING

Source: McCann-Erickson Worldwide.

 1. Which one of the following is the best estimate for the advertising cost in 1993?
     (a) $820,000   **(b)** $848,000   **(c)** $910,000

2. To the nearest percent, what was the percent increase in the advertising cost from 1995 to 1996?

Complete the ordered pairs for the given equation.

3. $3x + 5y = -30$;   (0,   ),   (, 0),   (, 3)

4. $y + 12 = 0$;   (0,   ),   (−4,   ), $\left(\dfrac{5}{2}, \right)$

5. Is $(4, -1)$ a solution of $4x - 7y = 9$?

6. How do you find the *x*-intercept and the *y*-intercept for a linear equation in two variables?

Graph each linear equation. Give the x- and y-intercepts.

7. $3x + y = 6$ **8.** $y + 3 = 0$

9. Does the set of ordered pairs {(90, 650), (91, 780), (92, 830), (93, 850), (94, 880), (95, 950), (96, 1100)}, shown as points on the graph for Exercises 1 and 2, define a function? Explain why or why not.

10. Describe the inputs and outputs for the set of ordered pairs in Exercise 9.

Find the slope of each line in Exercises 11–16.

11. Through $(-4, 6)$ and $(-1, -2)$

12. $2x + y = 10$

13.

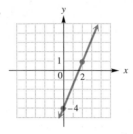

14. (These are two different views of the same line.)

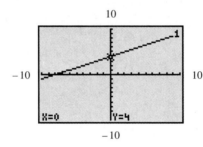
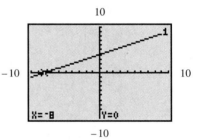

15. A line parallel to the graph of $y = -4x + 6$

16. A line perpendicular to the graph of $y = -4x + 6$

CUMULATIVE REVIEW EXERCISES CHAPTERS 1–4

Perform the indicated operations.

1. $16\frac{7}{8} - 3\frac{1}{10}$

2. $\frac{3}{4} \div \frac{5}{8}$

3. $-11 + 20 + (-2)$

4. $\dfrac{(-3)^2 - (-4)(2^4)}{5 \cdot 2 - (-2)^3}$

5. Rafael and Elda Muñoz are painting a bedroom for their new baby. They painted $\frac{1}{4}$ of the room on Saturday and $\frac{1}{3}$ of the room on Sunday. How much of the room is still unpainted?

6. True or false? $\dfrac{4(3 - 9)}{2 - 6} \geq 6$

7. Find the value of $xz^3 - 5y^2$ when $x = -2$, $y = -4$, and $z = 3$.

8. What property does $3(-2 + x) = -6 + 3x$ illustrate?

9. Simplify $-4p - 6 + 3p + 8$ by combining terms.

Solve.

10. $V = \dfrac{1}{3}\pi r^2 h$ for h

11. $6 - 3(1 + a) = 2(a + 5) - 2$

12. $-(m - 1) = 3 - 2m$

13. $\dfrac{y - 2}{3} = \dfrac{2y + 1}{5}$

14. $-5z \geq 4z - 18$

15. $2 < -6(z + 1) < 10$

Solve each problem.

16. Kimshana Lavoris earned $200 working part time during July, $375 during August, and $325 during September. If her average income for the four months from July through October must be at least $300, what possible amounts could she earn in October?

17. Mount Mayon in the Philippines is the most perfectly shaped conical volcano in the world. Its base is a perfect circle with a 39-mile circumference and it has a height of 8000 feet. (One mile is 5280 feet.) Find the radius of the circular base to the nearest mile. (*Hint:* This problem has some unneeded information.)

18. How much of a 20% chemical solution must be mixed with 30 liters of a 60% solution to get a 50% mixture?

19. The winning times in seconds for the women's 1000-meter speed skating event in the Winter Olympics for the years from 1960 to 1998 can be closely approximated by the equation

$$y = -.4685x + 95.07,$$

where x is the number of years since 1960. That is, $x = 5$ represents 1965, $x = 10$ represents 1970, and so on. Complete the table of ordered pairs for this linear equation. Round the y-values to the nearest hundredth of a second. (*Source: The Universal Almanac,* 1997, John W. Wright, General Editor.)

x	y
12	
28	
36	

20. Baby boomers are expected to inherit $10.4 trillion from their parents over the next 45 years, an average of $50,000 each. The pie chart shows how they plan to spend their inheritance. How much of the $50,000 is expected to go toward the purchase of a home? How much to retirement?

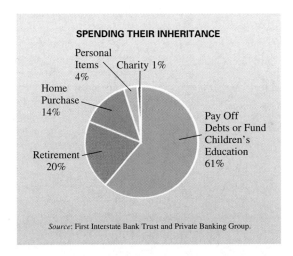

SPENDING THEIR INHERITANCE

Personal Items 4%
Charity 1%
Home Purchase 14%
Pay Off Debts or Fund Children's Education 61%
Retirement 20%

Source: First Interstate Bank Trust and Private Banking Group.

Consider the linear equation $3x + 2y = 12$. Find the following.

21. The x- and y-intercepts   **22.** The graph   **23.** The slope

24. Are the lines with equations $x + 5y = -6$ and $y = 5x - 8$ parallel, perpendicular, or neither?

25. California's exports to Pacific Rim nations, with the exception of China, fell during the first half of 1998, as shown in the bar graph. Is the set whose ordered pairs represent the data from the graph as (importer, % change) a function? Give the ordered pairs with the smallest change and the largest change. How is absolute value used in your answer?

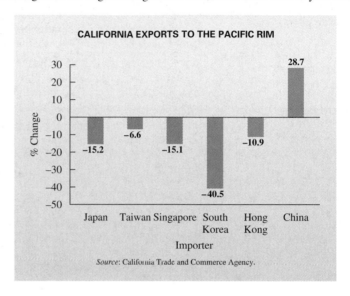

CALIFORNIA EXPORTS TO THE PACIFIC RIM

Source: California Trade and Commerce Agency.

Polynomials and Exponents

The number of passengers traveling by air has increased rapidly in the last decade. (See Exercise 46 in Section 5.1.) From 1985 to 1995, scheduled air carriers enjoyed an increase in net profits of $1,514,000,000. Surprisingly, in spite of more passengers traveling in planes that are consistently full, consumer complaints against U.S. airlines have generally decreased, as shown in the figure. There is one year, however, when all three graphs indicate an increase in each category from one year to the next, rather than a decrease. In which year did this occur?

Aeronautics

5.1 Addition and Subtraction of Polynomials; Graphing Simple Polynomials

5.2 The Product Rule and Power Rules for Exponents

5.3 Multiplication of Polynomials

5.4 Special Products

5.5 Integer Exponents and the Quotient Rule

5.6 Division of Polynomials

5.7 An Application of Exponents: Scientific Notation

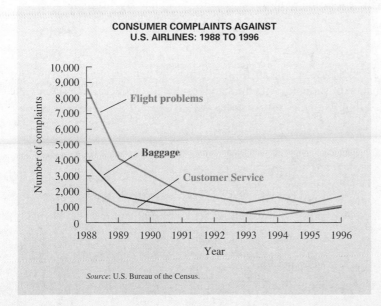

The graphs in the figure can each be approximated by a *polynomial function*. Polynomials are one of the topics studied in this chapter. Throughout the chapter we will see other examples of polynomials that describe information about the aeronautics industry.

5.1 Addition and Subtraction of Polynomials; Graphing Simple Polynomials

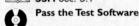

OBJECTIVES

1. Identify terms and coefficients.
2. Add like terms.
3. Know the vocabulary for polynomials.
4. Evaluate polynomials.
5. Add and subtract polynomials.
6. Graph equations defined by polynomials with degree 2.

FOR EXTRA HELP

SSG Sec. 5.1
SSM Sec. 5.1

Pass the Test Software

InterAct Math
Tutorial Software

Video 7

OBJECTIVE 1 Identify terms and coefficients. In Chapter 1 we saw that in an expression such as

$$4x^3 + 6x^2 + 5x + 8,$$

the quantities $4x^3$, $6x^2$, $5x$, and 8 are called *terms*. As mentioned earlier, in the term $4x^3$, the number 4 is called the *numerical coefficient*, or simply the *coefficient*, of x^3. In the same way, 6 is the coefficient of x^2 in the term $6x^2$, 5 is the coefficient of x in the term $5x$, and 8 is the coefficient in the term 8. A constant term, like 8 in the expression above, can be thought of as $8x^0$, where

$$x^0 \text{ is defined to equal } 1.$$

We explain the reason for this definition later in this chapter.

EXAMPLE 1 Identifying Coefficients

Name the (numerical) coefficient of each term in these expressions.

(a) $4x^3$
The coefficient is 4.

(b) $x - 6x^4$
The coefficient of x is 1 because $x = 1 \cdot x$. The coefficient of x^4 is -6 since $x - 6x^4$ can be written as the sum $x + (-6x^4)$.

(c) $5 - v^3$
The coefficient of the term 5 is 5 because $5 = 5v^0$. By writing $5 - v^3$ as a sum, $5 + (-v^3)$, or $5 + (-1v^3)$, the coefficient of v^3 can be identified as -1.

OBJECTIVE 2 Add like terms. Recall from Section 1.8 that *like terms* have exactly the same combination of variables with the same exponents on the variables. Only the coefficients may differ. Examples of like terms are

$$19m^5 \quad \text{and} \quad 14m^5,$$
$$6y^9, \quad -37y^9, \quad \text{and} \quad y^9,$$
$$3pq \quad \text{and} \quad -2pq,$$
$$2xy^2 \quad \text{and} \quad -xy^2.$$

Using the distributive property, we add like terms by adding their coefficients.

EXAMPLE 2 Adding Like Terms

Simplify each expression by adding like terms.

(a) $-4x^3 + 6x^3 = (-4 + 6)x^3 = 2x^3$ Distributive property

(b) $9x^6 - 14x^6 + x^6 = (9 - 14 + 1)x^6 = -4x^6$

(c) $12m^2 + 5m + 4m^2 = (12 + 4)m^2 + 5m = 16m^2 + 5m$

(d) $3x^2y + 4x^2y - x^2y = (3 + 4 - 1)x^2y = 6x^2y$

In Example 2(c), we cannot combine $16m^2$ and $5m$. These two terms are unlike because the exponents on the variables are different. *Unlike terms* have different variables or different exponents on the same variables.

OBJECTIVE **3** Know the vocabulary for polynomials. A **polynomial in** x is a term or the sum of a finite number of terms of the form ax^n, for any real number a and any whole number n. For example,

$$16x^8 - 7x^6 + 5x^4 - 3x^2 + 4$$

is a polynomial in x (the 4 can be written as $4x^0$). This polynomial is written in **descending powers** of the variable, since the exponents on x decrease from left to right. On the other hand,

$$2x^3 - x^2 + \frac{4}{x}$$

is not a polynomial in x, since a variable appears in a denominator. Of course, we could define *polynomial* using any variable and not just x, as in Example 2(c). In fact, polynomials may have terms with *more* than one variable, as in Example 2(d).

The **degree of a term** is the sum of the exponents on the variables. For example, $3x^4$ has degree 4, while $6x^{17}$ has degree 17. The term $5x$ has degree 1, -7 has degree 0 (since -7 can be written as $-7x^0$), and $2x^2y$ has degree $2 + 1 = 3$ (y has an exponent of 1.) The **degree of a polynomial** is the highest degree of any nonzero term of the polynomial. For example, $3x^4 - 5x^2 + 6$ is of degree 4, the polynomial $5x + 7$ is of degree 1, 3 (or $3x^0$) is of degree 0, and $x^2y + xy - 5xy^2$ is of degree 3.

Three types of polynomials are very common and are given special names. A polynomial with exactly three terms is called a **trinomial.** (*Tri-* means "three," as in *tri*angle.) Examples are

$$9m^3 - 4m^2 + 6, \qquad 19y^2 + 8y + 5, \qquad \text{and} \qquad -3m^5n^2 + 2n^3 - m^4.$$

A polynomial with exactly two terms is called a **binomial.** (*Bi-* means "two," as in *bi*cycle.) Examples are

$$-9x^4 + 9x^3, \qquad 8m^2 + 6m, \qquad \text{and} \qquad 3m^5n^2 - 9m^2n^4.$$

A polynomial with only one term is called a **monomial.** (*Mon(o)-* means "one," as in *mono*rail.) Examples are

$$9m, \qquad -6y^5, \qquad a^2b^2, \qquad \text{and} \qquad 6.$$

EXAMPLE 3 Classifying Polynomials

For each polynomial, first simplify if possible by combining like terms. Then give the degree and tell whether it is a monomial, a binomial, a trinomial, or none of these.

(a) $2x^3 + 5$

The polynomial cannot be simplified. The degree is 3. The polynomial is a binomial.

(b) $4xy - 5xy + 2xy$

Add like terms to simplify: $4xy - 5xy + 2xy = xy$, which is a monomial of degree 2.

OBJECTIVE **4** Evaluate polynomials. A polynomial usually represents different numbers for different values of the variable, as shown in the next example.

┌───

EXAMPLE 4 Evaluating a Polynomial

Find the value of $3x^4 + 5x^3 - 4x - 4$ when $x = -2$ and when $x = 3$.
First, substitute -2 for x.

$$3x^4 + 5x^3 - 4x - 4 = 3(-2)^4 + 5(-2)^3 - 4(-2) - 4$$
$$= 3 \cdot 16 + 5 \cdot (-8) + 8 - 4$$
$$= 48 - 40 + 8 - 4$$
$$= 12$$

Next, replace x with 3.

$$3x^4 + 5x^3 - 4x - 4 = 3(3)^4 + 5(3)^3 - 4(3) - 4$$
$$= 3 \cdot 81 + 5 \cdot 27 - 12 - 4$$
$$= 362$$

───

 CAUTION

Notice the use of parentheses around the numbers that are substituted for the variable in Example 4. This is particularly important when substituting a negative number for a variable that is raised to a power, so that the sign of the product is correct.

───

CONNECTIONS

In Section 4.2 we introduced the idea of a function: for every input x, there is one output y. Polynomials often provide a way of defining functions that approximate data collected over a period of time. For example, according to the U.S. National Aeronautics and Space Administration (NASA), the budget in millions of dollars for space station research for 1996–2001 can be approximated by the polynomial equation

$$y = -10.25x^2 - 126.04x + 5730.21,$$

where $x = 0$ represents 1996, $x = 1$ represents 1997, and so on, up to $x = 5$ representing 2001. The actual budget for 1998 was 5327 million dollars; an input of $x = 2$ (for 1998) gives approximately $y = 5437$. Considering the magnitude of the numbers, this is a very good approximation.

FOR DISCUSSION OR WRITING
Use the given polynomial equation to approximate the budget in other years between 1996 and 2001. Compare to the actual figures given here.

Year	Budget (in millions of dollars)
1996	5710
1997	5675
1998	5327
1999	5306
2000	5077
2001	4832

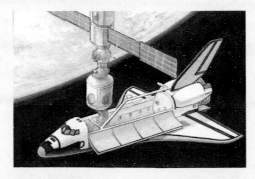

OBJECTIVE 5 Add and subtract polynomials. Polynomials may be added, subtracted, multiplied, and divided.

Adding Polynomials

To add two polynomials, add like terms.

E X A M P L E   5   Adding Polynomials Vertically

Add $6x^3 - 4x^2 + 3$ and $-2x^3 + 7x^2 - 5$.

Write like terms in columns.

$$
\begin{array}{r}
6x^3 - 4x^2 + 3 \\
-2x^3 + 7x^2 - 5 \\
\hline
\end{array}
$$

Now add, column by column.

$$
\begin{array}{ccc}
6x^3 & -4x^2 & 3 \\
-2x^3 & 7x^2 & -5 \\
\hline
4x^3 & 3x^2 & -2
\end{array}
$$

Add the three sums together.

$$4x^3 + 3x^2 + (-2) = 4x^3 + 3x^2 - 2$$

The polynomials in Example 5 also can be added horizontally, as shown in the next example.

E X A M P L E   6   Adding Polynomials Horizontally

Add $6x^3 - 4x^2 + 3$ and $-2x^3 + 7x^2 - 5$.

Write the sum as

$$(6x^3 - 4x^2 + 3) + (-2x^3 + 7x^2 - 5).$$

Use the associative and commutative properties to rewrite this sum with the parentheses removed and with the subtractions changed to additions of inverses.

$$6x^3 + (-4x^2) + 3 + (-2x^3) + 7x^2 + (-5)$$

Place like terms together.

$$6x^3 + (-2x^3) + (-4x^2) + 7x^2 + 3 + (-5)$$

Combine like terms to get

$$4x^3 + 3x^2 + (-2), \qquad \text{or simply} \qquad 4x^3 + 3x^2 - 2,$$

the same answer found in Example 5.

Earlier, we defined the difference $x - y$ as $x + (-y)$. (We find the difference $x - y$ by adding x and the opposite of y.) For example,

$$7 - 2 = 7 + (-2) = 5 \qquad \text{and} \qquad -8 - (-2) = -8 + 2 = -6.$$

A similar method is used to subtract polynomials.

Subtracting Polynomials

To subtract two polynomials, change all the signs on the second polynomial and add the result to the first polynomial.

E X A M P L E 7 Subtracting Polynomials

(a) Perform the subtraction $(5x - 2) - (3x - 8)$.
By the definition of subtraction,

$$(5x - 2) - (3x - 8) = (5x - 2) + [-(3x - 8)].$$

As shown in Chapter 1, the distributive property gives

$$-(3x - 8) = -1(3x - 8) = -3x + 8,$$

so

$$(5x - 2) - (3x - 8) = (5x - 2) + (-3x + 8) = 2x + 6.$$

(b) Subtract $6x^3 - 4x^2 + 2$ from $11x^3 + 2x^2 - 8$.
Write the problem.

$$(11x^3 + 2x^2 - 8) - (6x^3 - 4x^2 + 2)$$

Change all the signs in the second polynomial and add the two polynomials.

$$(11x^3 + 2x^2 - 8) + (-6x^3 + 4x^2 - 2) = 5x^3 + 6x^2 - 10$$

To check a subtraction problem, use the fact that if $a - b = c$, then $a = b + c$. For example, $6 - 2 = 4$, so we check by writing $6 = 2 + 4$, which is correct. Check the polynomial subtraction above by adding $6x^3 - 4x^2 + 2$ and $5x^3 + 6x^2 - 10$. Since the sum is $11x^3 + 2x^2 - 8$, the subtraction was performed correctly.

Subtraction also can be done in columns (vertically). We will use vertical subtraction in Section 5.6 when we study polynomial division.

E X A M P L E 8 Subtracting Polynomials Vertically

Use the method of subtracting by columns to find

$$(14y^3 - 6y^2 + 2y - 5) - (2y^3 - 7y^2 - 4y + 6).$$

Arrange like terms in columns.

$$\begin{array}{r} 14y^3 - 6y^2 + 2y - 5 \\ 2y^3 - 7y^2 - 4y + 6 \\ \hline \end{array}$$

Change all signs in the second row, and then add.

$$\begin{array}{r} 14y^3 - 6y^2 + 2y - 5 \\ -2y^3 + 7y^2 + 4y - 6 \\ \hline 12y^3 + y^2 + 6y - 11 \end{array}$$

Change all signs.

Add.

Either the horizontal or the vertical method may be used to add and subtract polynomials.

Polynomials in more than one variable are added and subtracted by combining like terms, just as with single variable polynomials.

E X A M P L E   9   Adding and Subtracting Polynomials with More Than One Variable

Add or subtract as indicated.

(a) $(4a + 2ab - b) + (3a - ab + b)$

$$(4a + 2ab - b) + (3a - ab + b) = 4a + 2ab - b + 3a - ab + b$$
$$= 7a + ab$$

(b) $(2x^2y + 3xy + y^2) - (3x^2y - xy - 2y^2)$

$$(2x^2y + 3xy + y^2) - (3x^2y - xy - 2y^2)$$
$$= 2x^2y + 3xy + y^2 - 3x^2y + xy + 2y^2$$
$$= -x^2y + 4xy + 3y^2$$

O B J E C T I V E   6   Graph equations defined by polynomials with degree 2. In Chapter 4 we introduced graphs of straight lines. These graphs were defined by linear equations (which are actually polynomial equations of degree 1). By selective point-plotting, we can find the graphs of polynomial equations of degree 2.

E X A M P L E   1 0   Graphing Equations Defined by Polynomials with Degree 2

(a) Graph $y = x^2$.

Select several values for x; then find the corresponding y-values. For example, selecting $x = 2$ gives

$$y = 2^2 = 4,$$

and so the point (2, 4) is on the graph of $y = x^2$. (Recall that in an ordered pair such as (2, 4), the x-value comes first and the y-value second.) We show some ordered pairs that satisfy $y = x^2$ in a table next to Figure 1. If the ordered pairs from the table are plotted on a coordinate system and a smooth curve drawn through them, the graph is as shown in Figure 1.

x	y
3	9
2	4
1	1
0	0
-1	1
-2	4
-3	9

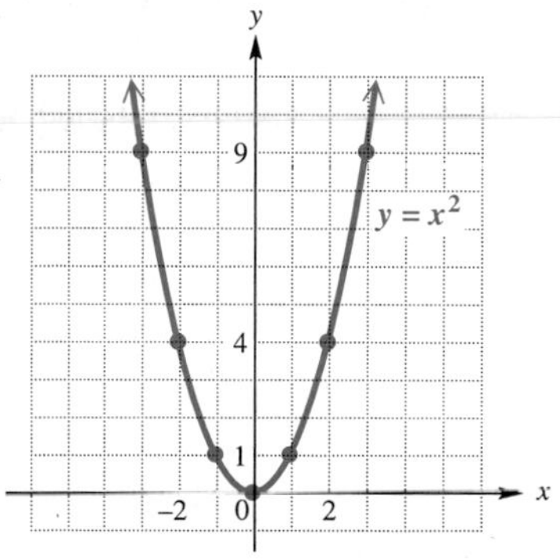

Figure 1

The graph of $y = x^2$ is the graph of a function, since each input x is related to just one output y. The curve in Figure 1 is called a **parabola.** The point (0, 0), the lowest point on this graph, is called the **vertex** of the parabola. The vertical line through the vertex (the y-axis here) is called the **axis** of the parabola. The axis of a parabola is a **line of symmetry** for the graph. If the graph is folded on this line, the two halves will match.

(b) Graph $y = -x^2 + 3$.

Once again plot points to obtain the graph. For example, if $x = -2$,

$$y = -(-2)^2 + 3 = -4 + 3 = -1.$$

This point and several others are shown in the table that accompanies the graph in Figure 2. The vertex of this parabola is $(0, 3)$. This time the vertex is the *highest* point of the graph. The graph opens downward because x^2 has a negative coefficient.

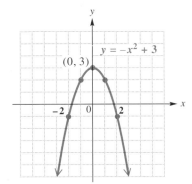

x	y
-2	-1
-1	2
0	3
1	2
2	-1

Figure 2

 All polynomials of degree 2 have parabolas as their graphs. When graphing by plotting points, it is necessary to continue finding points until the vertex and points on either side of it are located. (In this section, all parabolas have their vertices on the *x*-axis or the *y*-axis.)

5.1 EXERCISES

Fill in each blank with the correct response.

1. In the term $7x^5$, the coefficient is _____ and the exponent is _____.

2. The expression $5x^3 - 4x^2$ has _____ term(s).
   (how many?)

3. The degree of the term $-4x^8$ is _____.

4. The polynomial $4x^2 - y^2$ _____ an example of a trinomial.
   (is/is not)

5. When $x^2 + 10$ is evaluated for $x = 4$, the result is _____.

6. $5x$ —— $+ 3x^3 - 7x$ is a trinomial of degree 4.

7. $3xy + 2xy - 5xy =$ _____.

8. _____ is an example of a monomial with coefficient 5, in the variable x, having degree 9.

For each polynomial, determine the number of terms and name the coefficients of the terms. See Example 1.

9. $6x^4$	**10.** $-9y^5$	**11.** t^4	**12.** s^7
13. $-19r^2 - r$	**14.** $2y^3 - y$	**15.** $x + 8x^2 + 5x^3$	**16.** $v - 2v^3 - v^7$

In each polynomial add like terms whenever possible. Write the result in descending powers of the variable. See Example 2.

17. $-3m^5 + 5m^5$ **18.** $-4y^3 + 3y^3$ **19.** $2r^5 + (-3r^5)$

20. $-19y^2 + 9y^2$ **21.** $.2m^5 - .5m^2$ **22.** $-.9y + .9y^2$

23. $-3x^5 + 2x^5 - 4x^5$ **24.** $6x^3 - 8x^3 + 9x^3$ **25.** $-4p^7 + 8p^7 + 5p^9$

26. $-3a^8 + 4a^8 - 3a^2$ **27.** $-4y^2 + 3y^2 - 2y^2 + y^2$ **28.** $3r^5 - 8r^5 + r^5 + 2r^5$

For each polynomial first simplify, if possible, and write it in descending powers of the variable. Then give the degree of the resulting polynomial and tell whether it is a monomial, a binomial, a trinomial, or none of these. See Example 3.

29. $6x^4 - 9x$ **30.** $7t^3 - 3t$

31. $5m^4 - 3m^2 + 6m^4 - 7m^3$ **32.** $6p^5 + 4p^3 - 8p^5 + 10p^2$

33. $\dfrac{5}{3}x^4 - \dfrac{2}{3}x^4$ **34.** $\dfrac{4}{5}r^6 + \dfrac{1}{5}r^6$

35. $.8x^4 - .3x^4 - .5x^4 + 7$ **36.** $1.2t^3 - .9t^3 - .3t^3 + 9$

*Find the value of each polynomial when (**a**) $x = 2$ and when (**b**) $x = -1$. See Example 4.*

37. $2x^5 - 4x^4 + 5x^3 - x^2$ **38.** $2x^2 + 5x + 1$

39. $-3x^2 + 14x - 2$ **40.** $-2x^2 + 3$

TECHNOLOGY INSIGHTS (EXERCISES 41–42)

The graphing calculator screen shown here indicates that 1 has been stored into the memory location designated X, and then the polynomial $X^2 - 3X + 6$ has been evaluated. The result is 4. (This can be verified by hand using direct substitution, as shown in Example 4.)

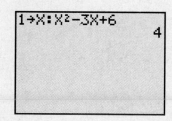

Predict the result the calculator will give for the following screens.

41. $2 \rightarrow X : 2X^2 - 3X - 5$ **42.** $3 \rightarrow X : X^2 + 5X - 10$

RELATING CONCEPTS   (EXERCISES 43–46)

As explained earlier in this section, the polynomial equation

$$y = -10.25x^2 - 126.04x + 5730.21$$

gives a good approximation of NASA's budget for space station research, in millions of dollars, for 1996–2001, where $x = 0$ represents 1996, $x = 1$ represents 1997, and so on. If we evaluate the polynomial for a specific input value x, we will get one and only one output value y as a result. This idea is basic to the study of functions (first introduced in Section 4.2), one of the most important concepts in mathematics.

Work Exercises 43–46 in order.

43. If gasoline costs \$1.25 per gallon, then the monomial $1.25x$ gives the cost of x gallons. Evaluate this monomial for 4, and then use the result to fill in the blanks: If _____ gallons are purchased, the cost is _____.

44. If it costs \$15 to rent a chain saw plus \$2 per day, the binomial $2x + 15$ gives the cost to rent the chain saw for x days. Evaluate this polynomial for 6 and then use the result to fill in the blanks: If the saw is rented for _____ days, the cost is _____.

45. If an object is thrown upward under certain conditions its height in feet is given by the trinomial $-16x^2 + 60x + 80$, where x is in seconds. Evaluate this polynomial for 2.5 and then use the result to fill in the blanks: If _____ seconds have elapsed, the height of the object is _____ feet.

46. The polynomial $2.69x^2 + 4.75x + 452.43$ gives a good approximation for the number of revenue passenger miles, in billions, for the U.S. airline industry during the period from 1990 to 1995, where $x = 0$ represents 1990. Use this polynomial to approximate the number of revenue passenger miles in 1991. (*Hint:* Any power of 1 is equal to 1, so simply add the coefficients and the constant.) (*Source:* Air Transportation Association of America.)

Did you make the connection that a polynomial can be related to data, using a particular data value for input to get another value as output?

Add or subtract as indicated. See Examples 5 and 8.

47. Add.
$$3m^2 + 5m$$
$$2m^2 - 2m$$

48. Add.
$$4a^3 - 4a^2$$
$$6a^3 + 5a^2$$

49. Subtract.
$$12x^4 - \ \ x^2$$
$$8x^4 + 3x^2$$

50. Subtract.
$$13y^5 - \ \ y^3$$
$$7y^5 + 5y^3$$

51. Add.

$$\frac{2}{3}x^2 + \frac{1}{5}x + \frac{1}{6}$$
$$\frac{1}{2}x^2 - \frac{1}{3}x + \frac{2}{3}$$

52. Add.

$$\frac{4}{7}y^2 - \frac{1}{5}y + \frac{7}{9}$$
$$\frac{1}{3}y^2 - \frac{1}{3}y + \frac{2}{5}$$

53. Add.

$$9m^3 - 5m^2 + 4m - 8$$
$$\underline{-3m^3 + 6m^2 + 8m - 6}$$

54. Add.

$$12r^5 + 11r^4 - 7r^3 - 2r^2$$
$$\underline{-8r^5 + 10r^4 + 3r^3 + 2r^2}$$

55. Subtract.

$$12m^3 - 8m^2 + 6m + 7$$
$$\underline{-3m^3 + 5m^2 - 2m - 4}$$

56. Subtract.

$$5a^4 - 3a^3 + 2a^2 - a + 6$$
$$\underline{-6a^4 + \ a^3 - \ a^2 + a - 1}$$

57. After reading Examples 5–8, explain whether you have a preference regarding horizontal or vertical addition and subtraction of polynomials.

58. Write a paragraph explaining how to add and subtract polynomials. Give an example using addition.

Perform the indicated operations. See Examples 6 and 7.

59. $(8m^2 - 7m) - (3m^2 + 7m - 6)$

60. $(x^2 + x) - (3x^2 + 2x - 1)$

61. $(16x^3 - x^2 + 3x) + (-12x^3 + 3x^2 + 2x)$

62. $(-2b^6 + 3b^4 - b^2) + (b^6 + 2b^4 + 2b^2)$

63. $(7y^4 + 3y^2 + 2y) - (18y^4 - 5y^2 + y)$

64. $(8t^5 + 3t^3 + 5t) - (19t^5 - 6t^3 + t)$

65. $(9a^4 - 3a^2 + 2) + (4a^4 - 4a^2 + 2) + (-12a^4 + 6a^2 - 3)$

66. $(4m^2 - 3m + 2) + (5m^2 + 13m - 4) - (16m^2 + 4m - 3)$

67. $[(8m^2 + 4m - 7) - (2m^2 - 5m + 2)] - (m^2 + m + 1)$

68. $[(9b^3 - 4b^2 + 3b + 2) - (-2b^3 - 3b^2 + b)] - (8b^3 + 6b + 4)$

69. $[(3x^2 - 2x + 7) - (4x^2 + 2x - 3)] - [(9x^2 + 4x - 6) + (-4x^2 + 4x + 4)]$

70. $[(6t^2 - 3t + 1) - (12t^2 + 2t - 6)] - [(4t^2 - 3t - 8) + (-6t^2 + 10t - 12)]$

71. Without actually performing the operations, determine mentally the coefficient of x^2 in the simplified form of $(-4x^2 + 2x - 3) - (-2x^2 + x - 1) + (-8x^2 + 3x - 4)$.

72. Without actually performing the operations, determine mentally the coefficient of x in the simplified form of $(-8x^2 - 3x + 2) - (4x^2 - 3x + 8) - (-2x^2 - x + 7)$.

Add or subtract as indicated. See Example 9.

73. $(6b + 3c) + (-2b - 8c)$

74. $(-5t + 13s) + (8t - 3s)$

75. $(4x + 2xy - 3) - (-2x + 3xy + 4)$

76. $(8ab + 2a - 3b) - (6ab - 2a + 3b)$

77. $(5x^2y - 2xy + 9xy^2) - (8x^2y + 13xy + 12xy^2)$

78. $(16t^3s^2 + 8t^2s^3 + 9ts^4) - (-24t^3s^2 + 3t^2s^3 - 18ts^4)$

For Exercises 79–82, use the formulas found on the inside covers.

Find the perimeter of each rectangle.

79.

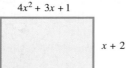

$4x^2 + 3x + 1$

$x + 2$

80.

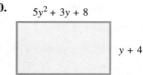

$5y^2 + 3y + 8$

$y + 4$

*Find **(a)** a polynomial representing the perimeter of each triangle and **(b)** the measures of the angles of the triangle.*

81.

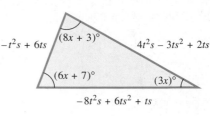

$2y - 3t$
$(10x + 3)°$
$5y + 3t$
$(5x - 1)°$
$(8x + 2)°$
$16y + 5t$

82.

$(8x + 3)°$
$-t^2s + 6ts$
$4t^2s - 3ts^2 + 2ts$
$(6x + 7)°$
$(3x)°$
$-8t^2s + 6ts^2 + ts$

The concepts required to work Exercises 83–86 have been covered, but the usual wording of the problem has been changed. Perform the indicated operations.

83. Subtract $9x^2 - 6x + 5$ from $3x^2 - 2$.

84. Find the difference when $9x^4 + 3x^2 + 5$ is subtracted from $8x^4 - 2x^3 + x - 1$.

85. Find the difference between the sum of $5x^2 + 2x - 3$ and $x^2 - 8x + 2$ and the sum of $7x^2 - 3x + 6$ and $-x^2 + 4x - 6$.

86. Subtract the sum of $9t^3 - 3t + 8$ and $t^2 - 8t + 4$ from the sum of $12t + 8$ and $t^2 - 10t + 3$.

Graph each of the following by completing the table of values. See Example 10.

87. $y = x^2 - 4$

x	y
-2	
-1	
0	
1	
2	

88. $y = x^2 - 6$

x	y
-2	
-1	
0	
1	
2	

89. $y = 2x^2 - 1$

x	y
-2	
-1	
0	
1	
2	

90. $y = 2x^2 + 2$

x	y
-2	
-1	
0	
1	
2	

91. $y = 4 - x^2$

x	y
-2	
-1	
0	
1	
2	

92. $y = 2 - x^2$

x	y
-2	
-1	
0	
1	
2	

93. $y = (x + 3)^2$

x	-5	-4	-3	-2	-1
y					

94. $y = (x - 4)^2$

x	2	3	4	5	6
y					

5.2 The Product Rule and Power Rules for Exponents

OBJECTIVE 1 Identify bases and exponents. Recall from Section 1.2 that in the expression 5^2, the number 5 is the *base* and 2 is the *exponent*. The expression 5^2 is called an *exponential expression*. Usually we do not write the exponent when it is 1; however, sometimes it is convenient to do so. In general, for any quantity a, $a^1 = a$.

E X A M P L E 1   Determining the Base and Exponent in an Exponential Expression

Evaluate each exponential expression. Name the base and the exponent.

		Base	Exponent
(a) $5^4 = 5 \cdot 5 \cdot 5 \cdot 5 = 625$		5	4
(b) $-5^4 = -1 \cdot 5^4 = -1 \cdot (5 \cdot 5 \cdot 5 \cdot 5) = -625$		5	4
(c) $(-5)^4 = (-5)(-5)(-5)(-5) = 625$		-5	4

Note the differences between parts (b) and (c) of Example 1. In -5^4 the lack of parentheses shows that the exponent 4 refers only to the base 5, and not -5; in $(-5)^4$ the parentheses show that the exponent 4 refers to the base -5. In summary, $-a^n$ and $(-a)^n$ are not necessarily the same.

Expression	Base	Exponent	Example
$-a^n$	a	n	$-3^2 = -(3 \cdot 3) = -9$
$(-a)^n$	$-a$	n	$(-3)^2 = (-3)(-3) = 9$

O B J E C T I V E 2 Use the product rule for exponents. By the definition of exponents,

$$2^4 \cdot 2^3 = \overbrace{(2 \cdot 2 \cdot 2 \cdot 2)}^{4 \text{ factors}}\overbrace{(2 \cdot 2 \cdot 2)}^{3 \text{ factors}}$$
$$= \underbrace{2 \cdot 2 \cdot 2 \cdot 2 \cdot 2 \cdot 2 \cdot 2}_{4 + 3 = 7 \text{ factors}}$$
$$= 2^7.$$

Also,

$$6^2 \cdot 6^3 = (6 \cdot 6)(6 \cdot 6 \cdot 6)$$
$$= 6 \cdot 6 \cdot 6 \cdot 6 \cdot 6$$
$$= 6^5.$$

Generalizing from these examples, $2^4 \cdot 2^3 = 2^{4+3} = 2^7$ and $6^2 \cdot 6^3 = 6^{2+3} = 6^5$, suggests the **product rule for exponents.**

Product Rule for Exponents

For any positive integers m and n, $a^m \cdot a^n = a^{m+n}$.
(Keep the same base; add the exponents.)
Example: $6^2 \cdot 6^5 = 6^{2+5} = 6^7$

E X A M P L E 2   Using the Product Rule

Use the product rule for exponents to find each result when possible.

(a) $6^3 \cdot 6^5 = 6^{3+5} = 6^8$ by the product rule.

(b) $(-4)^7(-4)^2 = (-4)^{7+2} = (-4)^9$

(c) $x^2 \cdot x = x^2 \cdot x^1 = x^{2+1} = x^3$

(d) $m^4 m^3 m^5 = m^{4+3+5} = m^{12}$

(e) $2^3 \cdot 3^2$

The product rule does not apply to the product $2^3 \cdot 3^2$, since the bases are different.

$$2^3 \cdot \mathbf{3^2} = 8 \cdot 9 = 72$$

(f) $2^3 + 2^4$

The product rule does not apply to $2^3 + 2^4$, since this is a *sum,* not a *product.*

$$2^3 + 2^4 = 8 + 16 = 24$$

(g) $(2x^3)(3x^7)$

Since $2x^3$ means $2 \cdot x^3$ and $3x^7$ means $3 \cdot x^7$, we can use the associative and commutative properties to get

$$(2x^3)(3x^7) = (2 \cdot 3) \cdot (x^3 \cdot x^7) = 6x^{10}.$$

Be sure that you understand the difference between *adding* and *multiplying* exponential expressions. For example,

$$8x^3 + 5x^3 = 13x^3, \qquad \text{but} \qquad (8x^3)(5x^3) = 8 \cdot 5x^{3+3} = 40x^6.$$

OBJECTIVE ▮**3** Use the rule $(a^m)^n = a^{mn}$. We simplify an expression such as $(8^3)^2$ with the product rule for exponents.

$$(8^3)^2 = (8^3)(8^3) = 8^{3+3} = 8^6$$

The exponents in $(8^3)^2$ are multiplied to give the exponent in 8^6. As another example,

$$(5^2)^3 = 5^2 \cdot 5^2 \cdot 5^2 = 5^{2+2+2} = 5^6,$$

and $2 \cdot 3 = 6$. These examples suggest **power rule (a) for exponents.**

Power Rule (a) for Exponents

For any positive integers m and n, $\qquad (a^m)^n = a^{mn}$.
(Raise a power to a power by multiplying exponents.)
Example: $(3^2)^4 = 3^{2 \cdot 4} = 3^8$

EXAMPLE 3 Using Power Rule (a)

Use power rule (a) for exponents to simplify each expression.

(a) $(2^5)^3 = 2^{5 \cdot 3} = 2^{15}$ **(b)** $(5^7)^2 = 5^{7(2)} = 5^{14}$

(c) $(x^2)^5 = x^{2(5)} = x^{10}$ **(d)** $(n^3)^2 = n^{3(2)} = n^6$

OBJECTIVE ▮**4** Use the rule $(ab)^m = a^m b^m$. The properties studied in Chapter 1 can be used to develop two more rules for exponents. Using the definition of an exponential expression and the commutative and associative properties, we can rewrite the expression $(4 \cdot 8)^3$ as follows.

$$(4 \cdot 8)^3 = (4 \cdot 8)(4 \cdot 8)(4 \cdot 8) \qquad \text{Definition of exponent}$$
$$= (4 \cdot 4 \cdot 4) \cdot (8 \cdot 8 \cdot 8) \qquad \text{Commutative and associative properties}$$
$$= 4^3 \cdot 8^3 \qquad \text{Definition of exponent}$$

This example suggests **power rule (b) for exponents.**

Power Rule (b) for Exponents

For any positive integer m, $(ab)^m = a^m b^m$.
(Raise a product to a power by raising each factor to the power.)
Example: $(2p)^5 = 2^5 p^5$

EXAMPLE 4   Using Power Rule (b)

Use power rule (b) for exponents to simplify each expression.

(a) $(3xy)^2 = 3^2 x^2 y^2 = 9x^2 y^2$

(b) $5(pq)^2 = 5(p^2 q^2) \qquad$ Power rule (b)
$\qquad\qquad = 5p^2 q^2 \qquad$ Multiply.

(c) $3(2m^2 p^3)^4 = 3[2^4 (m^2)^4 (p^3)^4] \qquad$ Power rule (b)
$\qquad\qquad\qquad = 3 \cdot 2^4 m^8 p^{12} \qquad$ Power rule (a)
$\qquad\qquad\qquad = 48 m^8 p^{12} \qquad$ Multiply.

(d) $(-5^6)^3 = (-1 \cdot 5^6)^3 = (-1)^3 \cdot (5^6)^3 = -1 \cdot 5^{18} = -5^{18}$

OBJECTIVE 5 Use the rule $\left(\dfrac{a}{b}\right)^m = \dfrac{a^m}{b^m}$. Since the quotient $\frac{a}{b}$ can be written as $a\left(\frac{1}{b}\right)$, we can use power rule (b), together with some of the properties of real numbers, to get **power rule (c) for exponents.**

Power Rule (c) for Exponents

For any positive integer m,

$$\left(\frac{a}{b}\right)^m = \frac{a^m}{b^m} \quad (b \neq 0).$$

(Raise a quotient to a power by raising both numerator and denominator to the power.)
Example: $\left(\dfrac{5}{3}\right)^2 = \dfrac{5^2}{3^2}$

EXAMPLE 5   Using Power Rule (c)

Use power rule (c) for exponents to simplify each expression.

(a) $\left(\dfrac{2}{3}\right)^5 = \dfrac{2^5}{3^5}$
   (b) $\left(\dfrac{m}{n}\right)^3 = \dfrac{m^3}{n^3} \quad (n \neq 0)$

We list the rules for exponents discussed in this section below. These rules are basic to the study of algebra and should be *memorized*.

Rules for Exponents

For positive integers m and n:

		Examples
Product rule	$a^m \cdot a^n = a^{m+n}$	$6^2 \cdot 6^5 = 6^{2+5} = 6^7$
Power rules (a)	$(a^m)^n = a^{mn}$	$(3^2)^4 = 3^{2 \cdot 4} = 3^8$
(b)	$(ab)^m = a^m b^m$	$(2p)^5 = 2^5 p^5$
(c)	$\left(\dfrac{a}{b}\right)^m = \dfrac{a^m}{b^m}$ $(b \neq 0)$	$\left(\dfrac{5}{3}\right)^2 = \dfrac{5^2}{3^2}$

As shown in the next example, more than one rule may be needed to simplify an expression with exponents.

EXAMPLE 6   Using Combinations of Rules

Use the rules for exponents to simplify each expression.

(a) $\left(\dfrac{2}{3}\right)^2 \cdot 2^3 = \dfrac{2^2}{3^2} \cdot \dfrac{2^3}{1}$ Power rule (c)

$= \dfrac{2^2 \cdot 2^3}{3^2 \cdot 1}$ Multiply the fractions.

$= \dfrac{2^5}{3^2}$ Product rule

(b) $(5x)^3 (5x)^4 = (5x)^7$ Product rule

$= 5^7 x^7$ Power rule (b)

(c) $(2x^2 y^3)^4 (3xy^2)^3 = 2^4 (x^2)^4 (y^3)^4 \cdot 3^3 x^3 (y^2)^3$ Power rule (b)

$= 2^4 \cdot 3^3 x^8 y^{12} x^3 y^6$ Power rule (a)

$= 16 \cdot 27 x^{11} y^{18}$ Product rule

$= 432 x^{11} y^{18}$

(d) $(-x^3 y)^2 (-x^5 y^4)^3$

Think of the negative sign in each factor as -1.

$(-1 \cdot x^3 y)^2 (-1 \cdot x^5 y^4)^3$

$= (-1)^2 (x^3)^2 (y)^2 \cdot (-1)^3 (x^5)^3 (y^4)^3$ Power rule (b)

$= (-1)^2 (x^6)(y^2)(-1)^3 (x^{15})(y^{12})$ Power rule (a)

$= (-1)^5 (x^{21})(y^{14})$ Product rule

$= -x^{21} y^{14}$

5.2 EXERCISES

Decide whether each statement is true or false.

1. $3^3 = 9$

2. $(-2)^4 = 2^4$

3. $(a^2)^3 = a^5$

4. $\left(\dfrac{1}{4}\right)^2 = \dfrac{1}{4^2}$

Write each expression using exponents.

5. $w \cdot w \cdot w \cdot w \cdot w \cdot w$

6. $t \cdot t \cdot t \cdot t \cdot t \cdot t \cdot t$

7. $\dfrac{1}{4 \cdot 4 \cdot 4 \cdot 4}$

8. $\dfrac{1}{3 \cdot 3 \cdot 3}$

9. $(-7x)(-7x)(-7x)(-7x)$

10. $(-8p)(-8p)$

11. $\left(\dfrac{1}{2}\right)\left(\dfrac{1}{2}\right)\left(\dfrac{1}{2}\right)\left(\dfrac{1}{2}\right)\left(\dfrac{1}{2}\right)\left(\dfrac{1}{2}\right)$

12. $\left(-\dfrac{1}{4}\right)\left(-\dfrac{1}{4}\right)\left(-\dfrac{1}{4}\right)\left(-\dfrac{1}{4}\right)\left(-\dfrac{1}{4}\right)$

13. Explain how the expressions $(-3)^4$ and -3^4 are different.

14. Explain how the expressions $(5x)^3$ and $5x^3$ are different.

Identify the base and the exponent for each exponential expression. In Exercises 15–18, also evaluate each expression. See Example 1.

15. 3^5

16. 2^7

17. $(-3)^5$

18. $(-2)^7$

19. $(-6x)^4$

20. $(-8x)^4$

21. $-6x^4$

22. $-8x^4$

23. Explain why the product rule does not apply to the expression $5^2 + 5^3$. Then evaluate the expression by finding the individual powers and adding the results.

24. Repeat Exercise 23 for the expression $(-4)^3 + (-4)^4$.

Use the product rule to simplify each expression. Write the answer in exponential form. See Example 2.

25. $5^2 \cdot 5^6$

26. $3^6 \cdot 3^7$

27. $4^2 \cdot 4^7 \cdot 4^3$

28. $5^3 \cdot 5^8 \cdot 5^2$

29. $(-7)^3(-7)^6$

30. $(-9)^8(-9)^5$

31. $t^3 \cdot t^8 \cdot t^{13}$

32. $n^5 \cdot n^6 \cdot n^9$

33. $(-8r^4)(7r^3)$

34. $(10a^7)(-4a^3)$

35. $(-6p^5)(-7p^5)$

36. $(-5w^8)(-9w^8)$

37. Explain why the product rule does not apply to the expression $3^2 \cdot 4^3$. Then evaluate the expression by finding the individual powers and multiplying the results.

38. Repeat Exercise 37 for the expression $(-3)^3 \cdot (-2)^5$.

TECHNOLOGY INSIGHTS (EXERCISES 39-42)

The graphing calculator screen shown here on the left reinforces the fact that $(-3)^4$ and -3^4 are not equal, since the first is 81 and the second is -81. The screen on the right shows how a calculator displays whether a statement is true or false: If the calculator returns a 1, the statement is true; it returns a 0 for a false statement.

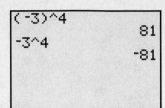

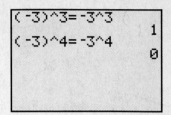

Evaluate each expression in the statement, and then tell whether the calculator would return a 1 or a 0.

39.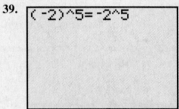

40.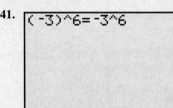

41.

42.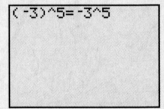

Use the power rules for exponents to simplify each expression. Write the answer in exponential form. See Examples 3–5.

43. $(4^3)^2$

44. $(8^3)^6$

45. $(t^4)^5$

46. $(y^6)^5$

47. $(7r)^3$

48. $(11x)^4$

49. $(5xy)^5$

50. $(9pq)^6$

51. $(-5^2)^6$

52. $(-9^4)^8$

53. $(-8^3)^5$

54. $(-7^5)^7$

55. $8(qr)^3$

56. $4(vw)^5$

57. $\left(\dfrac{1}{2}\right)^3$

58. $\left(\dfrac{1}{3}\right)^3$

59. $\left(\dfrac{a}{b}\right)^3$ $(b \neq 0)$

60. $\left(\dfrac{r}{t}\right)^4$ $(t \neq 0)$

61. $\left(\dfrac{9}{5}\right)^8$

62. $\left(\dfrac{12}{7}\right)^3$

Use a combination of the rules of exponents introduced in this section to simplify each expression. See Example 6.

63. $\left(\dfrac{5}{2}\right)^{3} \cdot \left(\dfrac{5}{2}\right)^{2}$

64. $\left(\dfrac{3}{4}\right)^{5} \cdot \left(\dfrac{3}{4}\right)^{6}$

65. $\left(\dfrac{9}{8}\right)^{3} \cdot 9^{2}$

66. $\left(\dfrac{8}{5}\right)^{4} \cdot 8^{3}$

67. $(2x)^{9}(2x)^{3}$

68. $(6y)^{5}(6y)^{8}$

69. $(-6p)^{4}(-6p)$

70. $(-13q)^{3}(-13q)$

71. $(6x^{2}y^{3})^{5}$

72. $(5r^{5}t^{6})^{7}$

73. $(x^{2})^{3}(x^{3})^{5}$

74. $(y^{4})^{5}(y^{3})^{5}$

75. $(2w^{2}x^{3}y)^{2}(x^{4}y)^{5}$

76. $(3x^{4}y^{2}z)^{3}(yz^{4})^{5}$

77. $(-r^{4}s)^{2}(-r^{2}s^{3})^{5}$

78. $(-ts^{6})^{4}(-t^{3}s^{5})^{3}$

79. $\left(\dfrac{5a^{2}b^{5}}{c^{6}}\right)^{3} \quad (c \neq 0)$

80. $\left(\dfrac{6x^{3}y^{9}}{z^{5}}\right)^{4} \quad (z \neq 0)$

81. A student tried to simplify $(10^{2})^{3}$ as 1000^{6}. Is this correct? If not, how is it simplified using the product rule for exponents?

82. Explain why $(3x^{2}y^{3})^{4}$ is *not* equivalent to $(3 \cdot 4)x^{8}y^{12}$.

Find the area of each figure.* (Leave π in the answer for Exercise 86.) Use the formulas found on the inside covers.

83.
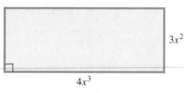
$3x^{2}$

$4x^{3}$

84.
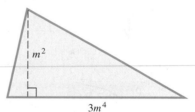
m^{2}

$3m^{4}$

85.
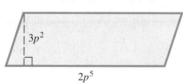
$3p^{2}$

$2p^{5}$

86.
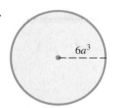
$6a^{3}$

Find the volume of each figure. Use the formulas found on the inside covers.

87.
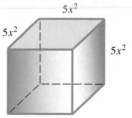
$5x^{2}$

$5x^{2}$

$5x^{2}$

88.
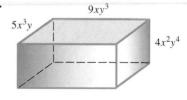
$9xy^{3}$

$5x^{3}y$

$4x^{2}y^{4}$

89. Assume a is a positive number greater than 1. Arrange the following terms in order from smallest to largest: $-(-a)^{3}, -a^{3}, (-a)^{4}, -a^{4}$. Explain how you decided on the order.

90. Describe a rule to tell whether an exponential expression with a negative base is positive or negative.

*The small square in the figures for Exercises 83–85 indicates a right angle (90°).

In Chapter 2 we used the formula for simple interest, $I = prt$, which deals with interest paid only on the principal. With **compound interest,** interest is paid on the principal and the interest earned earlier. The formula for compound interest, which involves an exponential expression, is

$$A = P(1 + r)^n.$$

Here A is the amount accumulated from a principal of P dollars left untouched for n years with an annual interest rate r (expressed as a decimal).

Use this formula and a calculator to find A to the nearest cent in Exercises 91–94.

91. $P = \$250, r = .04, n = 5$ **92.** $P = \$400, r = .04, n = 3$

93. $P = \$1500, r = .035, n = 6$ **94.** $P = \$2000, r = .025, n = 4$

5.3 Multiplication of Polynomials

OBJECTIVES

1 Multiply a monomial and a polynomial.

2 Multiply two polynomials.

3 Multiply binomials by the FOIL method.

FOR EXTRA HELP

SSG Sec. 5.3
SSM Sec. 5.3

Pass the Test Software

InterAct Math
Tutorial Software

Video 7

OBJECTIVE 1 Multiply a monomial and a polynomial. As shown in the previous section, the product of two monomials is found by using the rules for exponents and the commutative and associative properties. For example,

$$(-8m^6)(-9m^4) = -8(-9)(m^6)(m^4) = 72m^{6+4} = 72m^{10}.$$

CAUTION Do not confuse *addition* of terms with *multiplication* of terms. For example,

$$7q^5 + 2q^5 = 9q^5, \qquad \text{but} \qquad (7q^5)(2q^5) = 7 \cdot 2q^{5+5} = 14q^{10}.$$

To find the product of a monomial and a polynomial with more than one term, we use the distributive property with the method just shown.

EXAMPLE 1   Multiplying a Monomial and a Polynomial

Use the distributive property to find each product.

(a) $4x^2(3x + 5)$

$$4x^2(3x + 5) = (4x^2)(3x) + (4x^2)(5) \qquad \text{Distributive property}$$
$$= 12x^3 + 20x^2 \qquad \text{Multiply monomials.}$$

(b) $-8m^3(4m^3 + 3m^2 + 2m - 1)$

$$-8m^3(4m^3 + 3m^2 + 2m - 1)$$
$$= (-8m^3)(4m^3) + (-8m^3)(3m^2) \qquad \text{Distributive property}$$
$$+ (-8m^3)(2m) + (-8m^3)(-1)$$
$$= -32m^6 - 24m^5 - 16m^4 + 8m^3 \qquad \text{Multiply monomials.}$$

OBJECTIVE 2 Multiply two polynomials. We can use the distributive property repeatedly to find the product of any two polynomials. For example, to find the product of the polynomials $x + 1$ and $x - 4$, think of $x - 4$ as a single quantity and use the distributive property as follows.

$$(x + 1)(x - 4) = x(x - 4) + 1(x - 4)$$

Now use the distributive property twice to find $x(x - 4)$ and $1(x - 4)$.

$$x(x - 4) + \mathbf{1}(x - 4) = x(x) + x(-4) + \mathbf{1}(x) + \mathbf{1}(-4)$$
$$= x^2 - 4x + x - 4$$
$$= x^2 - 3x - 4$$

(We could have treated $x + 1$ as the single quantity instead to get $(x + 1)x + (x + 1)(-4)$ in the first step. Verify that this approach gives the same final result, $x^2 - 3x - 4$.)

We give a rule for multiplying any two polynomials below.

Multiplying Polynomials

To multiply two polynomials, multiply each term of the second polynomial by each term of the first polynomial and add the products.

E X A M P L E 2   Multiplying Two Polynomials

Multiply $(m^2 + 5)(4m^3 - 2m^2 + 4m)$.

Multiply each term of the second polynomial by each term of the first.

$$(m^2 + 5)(4m^3 - 2m^2 + 4m)$$
$$= m^2(4m^3) - m^2(2m^2) + m^2(4m) + \mathbf{5}(4m^3) - \mathbf{5}(2m^2) + \mathbf{5}(4m)$$
$$= 4m^5 - 2m^4 + 4m^3 + 20m^3 - 10m^2 + 20m$$

Now combine like terms.

$$- 4m^5 - 2m^4 + 24m^3 - 10m^2 + 20m$$

When at least one of the factors in a product of polynomials has three or more terms, the multiplication can be simplified by writing one polynomial above the other vertically.

E X A M P L E 3   Multiplying Vertically

Multiply $(x^3 + 2x^2 + 4x + 1)(3x + 5)$ using the vertical method.

Write the polynomials as follows.

$$\begin{array}{r} x^3 + 2x^2 + 4x + 1 \\ 3x + 5 \\ \hline \end{array}$$

It is not necessary to line up terms in columns, because any terms may be multiplied (not just like terms). Begin by multiplying each of the terms in the top row by 5.

Step 1

$$\begin{array}{r} x^3 + \ \ 2x^2 + \ \ 4x + 1 \\ 3x + 5 \\ \hline 5x^3 + 10x^2 + 20x + 5 \end{array} \qquad 5(x^3 + 2x^2 + 4x + 1)$$

Notice how this process is similar to multiplication of whole numbers. Now multiply each term in the top row by $3x$. Be careful to place like terms in columns, since the final step will involve addition (as in multiplying two whole numbers).

Step 2

$$\begin{array}{r} x^3 + \ \ 2x^2 + \ \ 4x + 1 \\ 3x + 5 \\ \hline 5x^3 + 10x^2 + 20x + 5 \\ 3x^4 + 6x^3 + 12x^2 + \ \ 3x \end{array} \qquad 3x(x^3 + 2x^2 + 4x + 1)$$

Step 3   Add like terms.

$$x^3 + 2x^2 + 4x + 1$$
$$3x + 5$$
$$\overline{5x^3 + 10x^2 + 20x + 5}$$
$$3x^4 + 6x^3 + 12x^2 + 3x$$
$$\overline{3x^4 + 11x^3 + 22x^2 + 23x + 5}$$

The product is $3x^4 + 11x^3 + 22x^2 + 23x + 5$.

OBJECTIVE **3** **Multiply binomials by the FOIL method.** In algebra, many of the polynomials to be multiplied are both binomials (with just two terms). For these products a shortcut that eliminates the need to write out all the steps is used. To develop this shortcut, multiply $x + 3$ and $x + 5$ using the distributive property.

$$(x + 3)(x + 5) = x(x + 5) + 3(x + 5)$$
$$= x(x) + x(5) + 3(x) + 3(5)$$
$$= x^2 + 5x + 3x + 15$$
$$= x^2 + 8x + 15$$

The first term in the second line, $x(x)$, is the product of the first terms of the two binomials.

$$(x + 3)(x + 5) \qquad \text{Multiply the first terms: } x(x).$$

The term $x(5)$ is the product of the first term of the first binomial and the last term of the second binomial. This is the **outer product.**

$$(x + 3)(x + 5) \qquad \text{Multiply the outer terms: } x(5).$$

The term $3(x)$ is the product of the last term of the first binomial and the first term of the second binomial. The product of these middle terms is the **inner product.**

$$(x + 3)(x + 5) \qquad \text{Multiply the inner terms: } 3(x).$$

Finally, $3(5)$ is the product of the last terms of the two binomials.

$$(x + 3)(x + 5) \qquad \text{Multiply the last terms: } 3(5).$$

The inner product and the outer product should be added mentally, so that the three terms of the answer can be written without extra steps as

$$(x + 3)(x + 5) = x^2 + 8x + 15.$$

A summary of these steps is given below. This procedure is sometimes called the **FOIL method,** which comes from the abbreviation for *First, Outer, Inner, Last.*

Multiplying Binomials by the FOIL Method

Step 1 **Multiply the first terms.** Multiply the two first terms of the binomials to get the first term of the answer.

Step 2 **Find the outer and inner products.** Find the outer product and the inner product and add them (mentally if possible) to get the middle term of the answer.

Step 3 **Multiply the last terms.** Multiply the two last terms of the binomials to get the last term of the answer.

E X A M P L E 4 Using the FOIL Method

Find the product $(x + 8)(x - 6)$ by the FOIL method.

Step 1 F Multiply the *first* terms: $x(x) = x^2$.

Step 2 O Find the product of the *outer* terms: $x(-6) = -6x$.

   I Find the product of the *inner* terms: $8(x) = 8x$.

   Add the outer and inner products mentally: $-6x + 8x = 2x$.

Step 3 L Multiply the *last* terms: $8(-6) = -48$.

The product $(x + 8)(x - 6)$ is the sum of the terms found in the three steps above, so

$$(x + 8)(x - 6) = x^2 - 6x + 8x - 48 = x^2 + 2x - 48.$$

As a shortcut, this product can be found in the following manner.

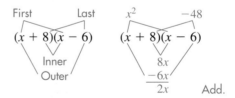

It is not possible to add the inner and outer products of the FOIL method if unlike terms result, as shown in the next example.

E X A M P L E 5 Using the FOIL Method

Multiply $(9x - 2)(3y + 1)$.

First	$(9x - 2)(3y + 1)$	$27xy$
Outer	$(9x - 2)(3y + 1)$	$9x$
Inner	$(9x - 2)(3y + 1)$	$-6y$
Last	$(9x - 2)(3y + 1)$	-2

Unlike terms

$$\underset{\text{F} \quad \text{O} \quad \text{I} \quad \text{L}}{(9x - 2)(3y + 1) = 27xy + 9x - 6y - 2}$$

┌───

E X A M P L E   6   Using the FOIL Method

Find the following products.

$$\begin{array}{cccc} \text{F} & \text{O} & \text{I} & \text{L} \end{array}$$

(a) $(2k + 5y)(k + 3y) = (2k)(k) + (2k)(3y) + (5y)(k) + (5y)(3y)$

$$= 2k^2 + 6ky + 5ky + 15y^2$$

$$= 2k^2 + 11ky + 15y^2 \qquad \text{Combine like terms.}$$

(b) $(7p + 2q)(3p - q) = 21p^2 - pq - 2q^2$ FOIL

(c) $2x^2(x - 3)(3x + 4) = 2x^2(3x^2 - 5x - 12)$ FOIL

$$= 6x^4 - 10x^3 - 24x^2 \qquad \text{Distributive property}$$

───

5.3 EXERCISES

1. Match each product in Column I with the correct monomial in Column II.

I	II
(a) $(5x^3)(6x^5)$	**A.** $125x^{15}$
(b) $(-5x^5)(6x^3)$	**B.** $30x^8$
(c) $(5x^5)^3$	**C.** $-216x^9$
(d) $(-6x^3)^3$	**D.** $-30x^8$

2. Match each product in Column I with the correct polynomial in Column II.

I	II
(a) $(x - 5)(x + 3)$	**A.** $x^2 + 8x + 15$
(b) $(x + 5)(x + 3)$	**B.** $x^2 - 8x + 15$
(c) $(x - 5)(x - 3)$	**C.** $x^2 - 2x - 15$
(d) $(x + 5)(x - 3)$	**D.** $x^2 + 2x - 15$

Find each product. See Example 1.

3. $(-5a^9)(-8a^5)$ **4.** $(-3m^6)(-5m^4)$ **5.** $-2m(3m + 2)$

6. $-5p(6 + 3p)$ **7.** $3p(8 - 6p + 12p^3)$ **8.** $4x(3 + 2x + 5x^3)$

9. $-8z(2z + 3z^2 + 3z^3)$ **10.** $-7y(3 + 5y^2 - 2y^3)$

11. $7x^2y(2x^3y^2 + 3xy - 4y)$ **12.** $9xy^3(-3x^2y^4 + 6xy - 2x)$

Find each product. See Examples 2 and 3.

13. $(6x + 1)(2x^2 + 4x + 1)$ **14.** $(9y - 2)(8y^2 - 6y + 1)$

15. $(4m + 3)(5m^3 - 4m^2 + m - 5)$ **16.** $(y + 4)(3y^3 - 2y^2 + y + 3)$

17. $(2x - 1)(3x^5 - 2x^3 + x^2 - 2x + 3)$ **18.** $(2a + 3)(a^4 - a^3 + a^2 - a + 1)$

19. $(5x^2 + 2x + 1)(x^2 - 3x + 5)$ **20.** $(2m^2 + m - 3)(m^2 - 4m + 5)$

Find each product. Use the FOIL method. See Examples 4–6.

21. $(n - 2)(n + 3)$ **22.** $(r - 6)(r + 8)$ **23.** $(x + 6)(x - 6)$

24. $(y + 9)(y - 9)$ **25.** $(4r + 1)(2r - 3)$ **26.** $(5x + 2)(2x - 7)$

27. $(3x + 2)(3x - 2)$ **28.** $(7x + 3)(7x - 3)$ **29.** $(3q + 1)(3q + 1)$

30. $(4w + 7)(4w + 7)$ **31.** $(3x + y)(x - 2y)$ **32.** $(5p + m)(2p - 3m)$

33. $(-3t + 4)(t + 6)$ **34.** $(-5x + 9)(x - 2)$

35. $3y^3(2y + 3)(y - 5)$ **36.** $5t^4(t + 3)(3t - 1)$

37. Find a polynomial that represents the area of this square.

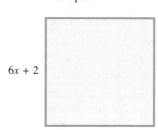

$6x + 2$

38. Find a polynomial that represents the area of this rectangle.

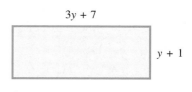

$3y + 7$

$y + 1$

39. Perform the following multiplications:

$$(x + 4)(x - 4); \quad (y + 2)(y - 2); \quad (r + 7)(r - 7).$$

Observe your answers, and explain the pattern that can be found.

40. Repeat Exercise 39 for the following:

$$(x + 4)(x + 4); \quad (y - 2)(y - 2); \quad (r + 7)(r + 7).$$

Find each product.

41. $\left(3p + \dfrac{5}{4}q\right)\left(2p - \dfrac{5}{3}q\right)$

42. $\left(-x + \dfrac{2}{3}y\right)\left(3x - \dfrac{3}{4}y\right)$

43. $(m^3 - 4)(2m^3 + 3)$

44. $(4a^2 + b^2)(a^2 - 2b^2)$

45. $(2k^3 + h^2)(k^2 - 3h^2)$

46. $(4x^3 - 5y^4)(x^2 + y)$

47. $3p^3(2p^2 + 5p)(p^3 + 2p + 1)$

48. $5k^2(k^2 - k + 4)(k^3 - 3)$

49. $-2x^5(3x^2 + 2x \quad 5)(4x + 2)$

50. $-4x^3(3x^4 + 2x^2 - x)(-2x + 1)$

Find a polynomial that represents the area of each shaded region. In Exercises 53 and 54 leave π in your answer. Use the formulas found on the inside covers.

51.

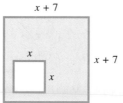

$x + 7$

x

x

$x + 7$

52.

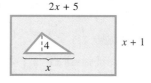

$2x + 5$

4

x

$x + 1$

53.

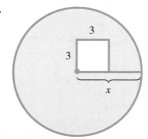

3

3

x

54.

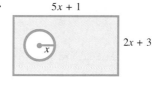

$5x + 1$

x

$2x + 3$

RELATING CONCEPTS (EXERCISES 55–62)

Work Exercises 55–62 in order. Refer to the figure as necessary.

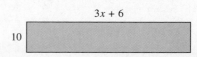

$3x + 6$

10

55. Find a polynomial that represents the area of the rectangle.

RELATING CONCEPTS (EXERCISES 55–62) (CONTINUED)

56. Suppose you know that the area of the rectangle is 600 square yards. Use this information and the polynomial from Exercise 55 to write an equation that allows you to solve for x.

57. Solve for x.

58. What are the dimensions of the rectangle (assume units are all in yards)?

59. Suppose the rectangle represents a lawn and it costs $3.50 per square yard to lay sod on the lawn. How much will it cost to sod the entire lawn?

60. Use the result of Exercise 58 to find the perimeter of the lawn.

61. Again, suppose the rectangle represents a lawn and it costs $9.00 per yard to fence the lawn. How much will it cost to fence the lawn?

62. (a) Suppose that it costs k dollars per square yard to sod the lawn. Determine a polynomial in the variables x and k that represents the cost to sod the entire lawn.

   (b) Suppose that it costs r dollars per yard to fence the lawn. Determine a polynomial in the variables x and r that represents the cost to fence the lawn.

Did you make the connection that sodding requires knowing the area, while fencing requires knowing the perimeter?

63. Explain the FOIL method of multiplying two binomials. Give an example.

64. Why does the FOIL method not apply to the product of a binomial and a trinomial? Give an example.

5.4 Special Products

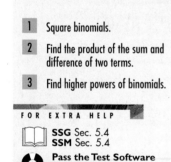

OBJECTIVES

1. Square binomials.

2. Find the product of the sum and difference of two terms.

3. Find higher powers of binomials.

FOR EXTRA HELP

SSG Sec. 5.4
SSM Sec. 5.4

Pass the Test Software

InterAct Math Tutorial Software

Video 7

In this section, we develop patterns for certain binomial products that occur frequently.

OBJECTIVE 1 Square binomials. The square of a binomial can be found quickly by using the method shown in Example 1.

EXAMPLE 1   Squaring a Binomial

Find $(m + 3)^2$.

Squaring $m + 3$ by the FOIL method gives

$$(m + 3)(m + 3) = m^2 + 3m + 3m + 9 = m^2 + 6m + 9.$$

The result has the square of both the first and the last terms of the binomial:

$$m^2 = m^2 \quad \text{and} \quad 3^2 = 9.$$

The middle term is twice the product of the two terms of the binomial, since both the outer and inner products are $(m)(3)$ and

$$(m)(3) + (m)(3) = 2(m)(3) = 6m.$$

This example suggests the following rule.

Square of a Binomial

The square of a binomial is a trinomial consisting of the square of the first term, plus twice the product of the two terms, plus the square of the last term of the binomial. For x and y,

$$(x + y)^2 = x^2 + 2xy + y^2$$

and

$$(x - y)^2 = x^2 - 2xy + y^2.$$

E X A M P L E 2   Squaring Binomials

Use the rule to find each product.

(a) $(5z - 1)^2 = (5z)^2 - 2(5z)(1) + 1^2 = 25z^2 - 10z + 1$
Recall that $(5z)^2 = 5^2z^2 = 25z^2$.

(b) $(3b + 5r)^2 = (3b)^2 + 2(3b)(5r) + (5r)^2 = 9b^2 + 30br + 25r^2$

(c) $(2a - 9x)^2 = 4a^2 - 36ax + 81x^2$

(d) $\left(4m + \dfrac{1}{2}\right)^2 = (4m)^2 + 2(4m)\left(\dfrac{1}{2}\right) + \left(\dfrac{1}{2}\right)^2 = 16m^2 + 4m + \dfrac{1}{4}$

(e) $t(a + 2b)^2 = t(a^2 + 4ab + 4b^2)$ Square the binomial.
           $= a^2t + 4abt + 4b^2t$ Distributive property

 A common error when squaring a binomial is forgetting the middle term of the product. In general, $(x + y)^2 \neq x^2 + y^2$ and $(x - y)^2 \neq x^2 - y^2$.

O B J E C T I V E 2 Find the product of the sum and difference of two terms. Binomial products of the form $(x + y)(x - y)$ also occur frequently. In these products, one binomial is the sum of two terms, and the other is the difference of the same two terms. As an example, the product of $a + 2$ and $a - 2$ is

$$(a + 2)(a - 2) = a^2 - 2a + 2a - 4 = a^2 - 4.$$

Using the FOIL method, the product of $x + y$ and $x - y$ is the difference of two squares.

Product of the Sum and Difference of Two Terms

The product of the sum and difference of the two terms x and y is

$$(x + y)(x - y) = x^2 - y^2.$$

 The product $(x + y)(x - y)$ cannot be written as $(x + y)^2$ or as $(x - y)^2$, since one factor involves addition and the other involves subtraction.

┌─ **E X A M P L E   3**   Finding the Product of the Sum and Difference of Two Terms

Find each product.

(a) $(x + 4)(x - 4)$

Use the pattern for the sum and difference of two terms.

$$(x + 4)(x - 4) = x^2 - 4^2 = x^2 - 16$$

(b) $(3 - w)(3 + w)$

By the commutative property, this product is the same as $(3 + w)(3 - w)$.

$$(3 - w)(3 + w) = (3 + w)(3 - w) = 3^2 - w^2 = 9 - w^2$$

(c) $(a - b)(a + b) = a^2 - b^2$

┌─ **E X A M P L E   4**   Finding the Product of the Sum and Difference of Two Terms

Find each product.

(a) $(5m + 3)(5m - 3)$

Use the rule for the product of the sum and difference of two terms.

$$(5m + 3)(5m - 3) = (5m)^2 - 3^2 = 25m^2 - 9$$

(b) $(4x + y)(4x - y) = (4x)^2 - y^2 = 16x^2 - y^2$

(c) $\left(z - \dfrac{1}{4}\right)\left(z + \dfrac{1}{4}\right) = z^2 - \dfrac{1}{16}$

(d) $r(x^2 + y)(x^2 - y) = r(x^4 - y^2)$
$$= rx^4 - ry^2$$

The product formulas of this section will be important later, particularly in Chapters 6 and 7. Therefore, it is important to memorize these formulas and practice using them.

O B J E C T I V E 3 Find higher powers of binomials. The methods used in the previous section and this section can be combined to find higher powers of binomials.

┌─ **E X A M P L E   5**   Finding Higher Powers of Binomials

Find each product.

(a) $(x + 5)^3$

$(x + 5)^3 = (x + 5)^2(x + 5)$ $a^3 = a^2 \cdot a$

$\qquad = (x^2 + 10x + 25)(x + 5)$ Square the binomial.

$\qquad = x^3 + 10x^2 + 25x + 5x^2 + 50x + 125$ Multiply polynomials.

$\qquad = x^3 + 15x^2 + 75x + 125$ Combine like terms.

(b) $(2y - 3)^4$

$(2y - 3)^4 = (2y - 3)^2(2y - 3)^2$ $a^4 = a^2 \cdot a^2$

$\qquad = (4y^2 - 12y + 9)(4y^2 - 12y + 9)$ Square each binomial.

$\qquad = 16y^4 - 48y^3 + 36y^2 - 48y^3 + 144y^2$ Multiply polynomials.

$\qquad\quad - 108y + 36y^2 - 108y + 81$

$\qquad = 16y^4 - 96y^3 + 216y^2 - 216y + 81$ Combine like terms.

5.4 EXERCISES

1. Consider the square $(2x + 3)^2$.
   (a) What is the square of the first term, $(2x)^2$?
   (b) What is twice the product of the two terms, $2(2x)(3)$?
   (c) What is the square of the last term, 3^2?
   (d) Write the final product, which is a trinomial, using your results in parts (a)–(c).

2. Explain in your own words how to square a binomial. Give an example.

Find each square. See Examples 1 and 2.

3. $(p + 2)^2$ 4. $(r + 5)^2$ 5. $(a - c)^2$

6. $(p - y)^2$ 7. $(4x - 3)^2$ 8. $(5y + 2)^2$

9. $(8t + 7s)^2$ 10. $(7z - 3w)^2$ 11. $\left(5x + \dfrac{2}{5}y\right)^2$

12. $\left(6m - \dfrac{4}{5}n\right)^2$ 13. $x(2x + 5)^2$ 14. $t(3t - 1)^2$

15. $-(4r - 2)^2$ 16. $-(3y - 8)^2$

17. Consider the product $(7x + 3y)(7x - 3y)$.
   (a) What is the product of the first terms, $(7x)(7x)$?
   (b) Multiply the outer terms, $(7x)(-3y)$. Then multiply the inner terms, $(3y)(7x)$. Add the results. What is this sum?
   (c) What is the product of the last terms, $(3y)(-3y)$?
   (d) Write the complete product using your answers in parts (a) and (c). Why is the sum found in part (b) omitted here?

18. Explain in your own words how to find the product of the sum and the difference of two terms. Give an example.

Find each product. See Examples 3 and 4.

19. $(q + 2)(q - 2)$ 20. $(x + 8)(x - 8)$

21. $(2w + 5)(2w - 5)$ 22. $(3z + 8)(3z - 8)$

23. $(10x + 3y)(10x - 3y)$ 24. $(13r + 2z)(13r - 2z)$

25. $(2x^2 - 5)(2x^2 + 5)$ 26. $(9y^2 - 2)(9y^2 + 2)$

27. $\left(7x + \dfrac{3}{7}\right)\left(7x - \dfrac{3}{7}\right)$ 28. $\left(9y + \dfrac{2}{3}\right)\left(9y - \dfrac{2}{3}\right)$

29. $p(3p + 7)(3p - 7)$ 30. $q(5q - 1)(5q + 1)$

RELATING CONCEPTS (EXERCISES 31–40)

Special products can be illustrated by using areas of rectangles. Use the figure, and **work Exercises 31–36 in order** *to justify the special product* $(a + b)^2 = a^2 + 2ab + b^2$.

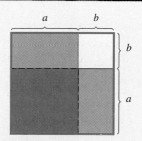

31. Express the area of the large square as the square of a binomial.

32. Give the monomial that represents the area of the red square.

33. Give the monomial that represents the sum of the areas of the blue rectangles.

34. Give the monomial that represents the area of the yellow square.

35. What is the sum of the monomials you obtained in Exercises 32–34?

36. Explain why the binomial square you found in Exercise 31 must equal the polynomial you found in Exercise 35.

To understand how the special product $(a + b)^2 = a^2 + 2ab + b^2$ can be applied to a purely numerical problem, work Exercises 37–40 in order.

37. Evaluate 35^2 using either traditional paper-and-pencil methods or a calculator.

38. The number 35 can be written as $30 + 5$. Therefore, $35^2 = (30 + 5)^2$. Use the special product for squaring a binomial with $a = 30$ and $b = 5$ to write an expression for $(30 + 5)^2$. Do not simplify at this time.

39. Use the rules for order of operations to simplify the expression you found in Exercise 38.

40. How do the answers in Exercises 37 and 39 compare?

Did you make the connections among geometry, algebra, and arithmetic in these exercises?

The special product

$$(a + b)(a - b) = a^2 - b^2$$

can be used to perform some multiplication problems. For example,

$$
\begin{aligned}
51 \times 49 &= (50 + 1)(50 - 1) & 102 \times 98 &= (100 + 2)(100 - 2) \\
&= 50^2 - 1^2 = 2500 - 1 & &= 100^2 - 2^2 \\
&= 2499 & &= 10{,}000 - 4 \\
& & &= 9996.
\end{aligned}
$$

Once these patterns are recognized, multiplications of this type can be done mentally.

Use this method to calculate each product mentally.

41. 101×99 **42.** 103×97

43. 201×199 **44.** 301×299

45. $20\frac{1}{2} \times 19\frac{1}{2}$ **46.** $30\frac{1}{3} \times 29\frac{2}{3}$

Find each product. See Example 5.

47. $(m - 5)^3$ **48.** $(p + 3)^3$ **49.** $(2a + 1)^3$

50. $(3m - 1)^3$ **51.** $(3r - 2t)^4$ **52.** $(2z + 5y)^4$

53. Explain how the expressions $x^2 + y^2$ and $(x + y)^2$ differ.

54. Does $a^3 + b^3$ equal $(a + b)^3$? Explain your answer.

Determine a polynomial that represents the area of each figure. Use the formulas found on the inside covers.

55.

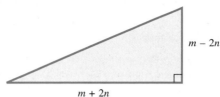

$m - 2n$

$m + 2n$

56.

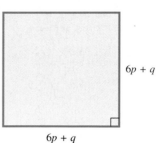

$6p + q$

$6p + q$

57.

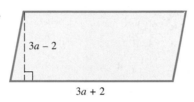

$3a - 2$

$3a + 2$

58.

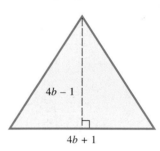

$4b - 1$

$4b + 1$

59.

$x + 2$

60.

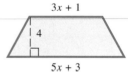

$3x + 1$

4

$5x + 3$

In Exercises 61 and 62, refer to the figure shown here.

61. Find a polynomial that represents the volume of the cube.

62. If the value of x is 6, what is the volume of the cube?

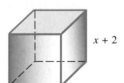

$x + 2$

5.5 Integer Exponents and the Quotient Rule

In our work so far, all exponents have been positive integers. To develop meanings for exponents other than positive integers (such as 0 and negative integers), we want to define them in such a way that the same rules for exponents apply.

OBJECTIVE 1   Use zero as an exponent.   Suppose we want to find a meaning for an expression such as

$$6^0,$$

where 0 is used as an exponent. If we were to multiply this by 6^2, for example, we would still want the product rule to be valid. Therefore, we would have

$$6^0 \cdot 6^2 = 6^{0+2} = 6^2.$$

FOR EXTRA HELP

📖 **SSG** Sec. 5.5
SSM Sec. 5.5

💿 **Pass the Test Software**

💿 **InterAct Math**
   Tutorial Software

📼 **Video** 8

So multiplying 6^2 by 6^0 should give 6^2. Since 6^0 is acting as if it were 1 here, we should define 6^0 to equal 1. This is the definition for 0 used as an exponent with any nonzero base.

Definition of Zero Exponent

For any nonzero real number a, $\qquad a^0 = 1$.
Example: $17^0 = 1$

EXAMPLE 1   Using Zero Exponents

Evaluate each exponential expression.

(a) $60^0 = 1$

(b) $(-60)^0 = 1$

(c) $-60^0 = -(1) = -1$

(d) $y^0 = 1$, if $y \neq 0$.

(e) $6y^0 = 6(1) = 6$, if $y \neq 0$.

(f) $(6y)^0 = 1$, if $y \neq 0$.

Notice the difference between parts (b) and (c) of Example 1. In Example 1(b), the base is -60 and the exponent is 0. Any nonzero base raised to a zero exponent is 1. But in Example 1(c), the base is 60. Then $60^0 = 1$, and $-60^0 = -1$.

OBJECTIVE 2   Use negative numbers as exponents.   Now suppose that we want to give a meaning to

$$6^{-2}$$

so that the product rule is still valid. If we multiply 6^{-2} by 6^2, we get

$$6^{-2} \cdot 6^2 = 6^{-2+2} = 6^0 = 1.$$

The expression 6^{-2} is acting as if it were the reciprocal of 6^2 since their product is 1. The reciprocal of 6^2 may be written $\dfrac{1}{6^2}$, leading us to define 6^{-2} as $\dfrac{1}{6^2}$. This is a particular case of the definition of negative exponents.

Definition of a Negative Exponent

For any nonzero real number a and any integer n, $\qquad a^{-n} = \dfrac{1}{a^n}$.

Example: $3^{-2} = \dfrac{1}{3^2}$

By definition, a^{-n} and a^n are reciprocals, since

$$a^n \cdot a^{-n} = a^n \cdot \frac{1}{a^n} = 1.$$

The definition of a^{-n} also can be written as

$$a^{-n} = \frac{1}{a^n} = \left(\frac{1}{a}\right)^n.$$

For example,

$$6^{-3} = \left(\frac{1}{6}\right)^3 \quad \text{and} \quad \left(\frac{1}{3}\right)^{-2} = 3^2.$$

E X A M P L E   2   Using Negative Exponents

Simplify by using the definition of negative exponents.

(a) $3^{-2} = \dfrac{1}{3^2} = \dfrac{1}{9}$

(b) $5^{-3} = \dfrac{1}{5^3} = \dfrac{1}{125}$

(c) $\left(\dfrac{1}{2}\right)^{-3} = 2^3 = 8$

(d) $\left(\dfrac{2}{5}\right)^{-4} = \left(\dfrac{5}{2}\right)^4 = \dfrac{5^4}{2^4} = \dfrac{625}{16}$ $\frac{2}{5}$ and $\frac{5}{2}$ are reciprocals.

(e) $4^{-1} - 2^{-1} = \dfrac{1}{4} - \dfrac{1}{2} = \dfrac{1}{4} - \dfrac{2}{4} = -\dfrac{1}{4}$

(f) $p^{-2} = \dfrac{1}{p^2}, \quad p \neq 0$

(g) $\dfrac{3}{3^{-1}} = \dfrac{3}{\frac{1}{3}} = 3 \div \dfrac{1}{3} = 3 \cdot 3 = 9$

(h) $\dfrac{2}{x^{-4}}, \quad x \neq 0$

$$\frac{2}{x^{-4}} = 2 \cdot \frac{1}{x^{-4}} = 2 \cdot x^4 \quad \text{or} \quad 2x^4$$

A negative exponent does not indicate a negative number; negative exponents lead to reciprocals.

Expression	Example	
a^{-n}	$3^{-2} = \dfrac{1}{3^2} = \dfrac{1}{9}$	Not negative
$-a^{-n}$	$-3^{-2} = -\dfrac{1}{3^2} = -\dfrac{1}{9}$	Negative

Examples 2(f) and 2(h) suggest that the definition of a negative exponent allows us to move factors in a fraction between the numerator and denominator if we also change the signs of the exponents. For example,

$$\frac{2^{-3}}{3^{-4}} = \frac{\dfrac{1}{2^3}}{\dfrac{1}{3^4}}$$ Definition of negative exponent

$$= \frac{1}{2^3} \cdot \frac{3^4}{1}$$ Division by a fraction

$$\frac{2^{-3}}{3^{-4}} = \frac{3^4}{2^3}.$$ Multiplication of fractions

This fact is generalized below.

Changing from Negative to Positive Exponents

For any nonzero numbers a and b, and any integers m and n,

$$\frac{a^{-m}}{b^{-n}} = \frac{b^n}{a^m}.$$

Example: $\dfrac{3^{-5}}{2^{-4}} = \dfrac{2^4}{3^5}$

EXAMPLE 3   Changing from Negative to Positive Exponents

Write with only positive exponents. Assume all variables represent nonzero real numbers.

(a) $\dfrac{4^{-2}}{5^{-3}} = \dfrac{5^3}{4^2}$

(b) $\dfrac{m^{-5}}{p^{-1}} = \dfrac{p^1}{m^5} = \dfrac{p}{m^5}$

(c) $\dfrac{a^{-2}b}{3d^{-3}} = \dfrac{bd^3}{3a^2}$

(d) $x^3y^{-4} = \dfrac{x^3y^{-4}}{1} = \dfrac{x^3}{y^4}$

Be careful. We cannot change negative exponents to positive exponents using this rule if the exponents occur in a sum of terms. For example,

$$\frac{5^{-2} + 3^{-1}}{7 - 2^{-3}}$$

CAUTION cannot be written with positive exponents using the rule given here. We would have to use the definition of a negative exponent to rewrite this expression with positive exponents, as

$$\frac{\dfrac{1}{5^2} + \dfrac{1}{3}}{7 - \dfrac{1}{2^3}}.$$

OBJECTIVE **3** Use the quotient rule for exponents. How do we simplify the quotient of two exponential expressions with the same base? We know that

$$\frac{6^5}{6^3} = \frac{6 \cdot 6 \cdot 6 \cdot 6 \cdot 6}{6 \cdot 6 \cdot 6} = 6^2.$$

Notice that $5 - 3 = 2$. Also,

$$\frac{6^2}{6^4} = \frac{6 \cdot 6}{6 \cdot 6 \cdot 6 \cdot 6} = \frac{1}{6^2} = 6^{-2}.$$

Here, $2 - 4 = -2$. These examples suggest the quotient rule for exponents.

Quotient Rule for Exponents

For any nonzero real number a and any integers m and n,

$$\frac{a^m}{a^n} = a^{m-n}.$$

(Keep the base; subtract the exponents.)

Example: $\dfrac{5^8}{5^4} = 5^{8-4} = 5^4$

┌─
EXAMPLE 4 Using the Quotient Rule for Exponents

Simplify by using the quotient rule for exponents. Write answers with positive exponents.

(a) $\dfrac{5^8}{5^6} = 5^{8-6} = 5^2$ **(b)** $\dfrac{4^2}{4^9} = 4^{2-9} = 4^{-7} = \dfrac{1}{4^7}$

(c) $\dfrac{5^{-3}}{5^{-7}} = \dfrac{5^7}{5^3} = 5^{7-3} = 5^4$ **(d)** $\dfrac{q^5}{q^{-3}} = q^5 q^3 = q^8, \quad q \neq 0$

(e) $\dfrac{3^2 x^5}{3^4 x^3} = \dfrac{3^2}{3^4} \cdot \dfrac{x^5}{x^3} = 3^{2-4} \cdot x^{5-3} = 3^{-2} x^2 = \dfrac{x^2}{3^2} = \dfrac{x^2}{9}, \quad x \neq 0$

(f) $\dfrac{(m+n)^{-2}}{(m+n)^{-4}} = \dfrac{(m+n)^4}{(m+n)^2} = (m+n)^{4-2} = (m+n)^2, \quad m \neq -n$

(g) $\dfrac{7x^{-3}y^2}{2^{-1}x^2y^{-5}} = \dfrac{7 \cdot 2^1 y^2 y^5}{x^2 x^3} = \dfrac{14y^7}{x^5}$
─┘

Working as shown in Examples 4(c) and 4(d), first changing negative exponents to positive exponents, avoids the potential for errors.

We list all the definitions and rules for exponents given in this section and Section 5.2 in the following summary.

Definitions and Rules for Exponents

If no denominators are zero, for any integers m and n:

		Examples
Product rule	$a^m \cdot a^n = a^{m+n}$	$7^4 \cdot 7^3 = 7^7$
Zero exponent	$a^0 = 1$	$(-3)^0 = 1$
Negative exponent	$a^{-n} = \dfrac{1}{a^n}$	$5^{-3} = \dfrac{1}{5^3}$
Quotient rule	$\dfrac{a^m}{a^n} = a^{m-n}$	$\dfrac{2^2}{2^5} = 2^{-3} = \dfrac{1}{2^3}$
Power rules (a)	$(a^m)^n = a^{mn}$	$(4^2)^3 = 4^6$
(b)	$(ab)^m = a^m b^m$	$(3k)^4 = 3^4 k^4$
(c)	$\left(\dfrac{a}{b}\right)^m = \dfrac{a^m}{b^m}$	$\left(\dfrac{2}{3}\right)^{10} = \dfrac{2^{10}}{3^{10}}$
(d)	$\dfrac{a^{-m}}{b^{-n}} = \dfrac{b^n}{a^m}$	$\dfrac{5^{-3}}{3^{-5}} = \dfrac{3^5}{5^3}$
(e)	$\left(\dfrac{a}{b}\right)^{-m} = \left(\dfrac{b}{a}\right)^m$	$\left(\dfrac{4}{7}\right)^{-2} = \left(\dfrac{7}{4}\right)^2$

OBJECTIVE **4** Use combinations of rules. As shown in the next example, sometimes we may need to use more than one rule to simplify an expression.

EXAMPLE 5 Using a Combination of Rules

Simplify each expression.

(a) $\dfrac{(4^2)^3}{4^5}$

Use power rule (a) and then the quotient rule.

$$\frac{(4^2)^3}{4^5} = \frac{4^6}{4^5} = 4^{6-5} = 4^1 = 4$$

(b) $(2x)^3 (2x)^2$

Use the product rule first. Then use power rule (b).

$$(2x)^3 (2x)^2 = (2x)^5 = 2^5 x^5 \quad \text{or} \quad 32x^5$$

(c) $\left(\dfrac{2x^3}{5}\right)^{-4}$

Use power rules (a), (b), (c), and (e).

$$\left(\frac{2x^3}{5}\right)^{-4} = \left(\frac{5}{2x^3}\right)^4 \qquad \text{Power rule (e)}$$

$$= \frac{5^4}{2^4 (x^3)^4} \qquad \text{Power rules (b) and (c)}$$

$$= \frac{5^4}{16x^{12}} \qquad \text{Power rule (a)}$$

(d) $\left(\dfrac{3x^{-2}}{4^{-1}y^3}\right)^{-3} = \dfrac{3^{-3}x^6}{4^3y^{-9}}$ Power rules

$= \dfrac{x^6y^9}{3^3 \cdot 4^3}$ Power rule (d)

$= \dfrac{x^6y^9}{27(64)}$ Definition of exponent

$= \dfrac{x^6y^9}{1728}$

 Since the steps can be done in different orders, there are many equally good ways to simplify a problem like Example 5(d).

CONNECTIONS

 In Exercises 91–94 of Section 5.2, we gave an example of an exponential expression that modeled the growth of money left at compound interest. Exponential expressions can also model the way in which quantities grow or decay. (These are examples of functions.) For instance, U.S. commercial space revenues for 1990–1995 can be modeled by the exponential equation

$$y = 3501.09(1.17)^x,$$

where y is in millions of dollars, and $x = 0$ corresponds to 1990, $x = 1$ corresponds to 1991, and so on, up to $x = 5$ for 1995. Using a calculator, we can find that in 1991 the revenues were approximately $3501.09(1.17)^1 = 4096.28$ million dollars. (This is actually a bit lower than the true amount of 4370 million dollars.) (*Source:* U.S. Department of Commerce, International Trade Administration, *U.S. Industrial Outlook 1994*; and unpublished data.)

FOR DISCUSSION OR WRITING
Use the model and a calculator to find the approximate revenues for 1990, 1992, 1993, 1994, and 1995.

5.5 EXERCISES

Decide whether each statement is true or false.

1. $(-2)^{-4} < 0$ **2.** $-2^{-4} < 0$ **3.** $(-5)^0 > 0$

4. $-5^0 < 0$ **5.** $1 - 12^0 = 1$ **6.** $-(-3)^0 = -1$

The given expression is either equal to 0, 1, or -1. Decide which is correct. See Example 1.

7. 9^0 **8.** 5^0 **9.** $(-4)^0$ **10.** $(-10)^0$

11. -9^0 **12.** -5^0 **13.** $(-2)^0 - 2^0$ **14.** $(-8)^0 - 8^0$

15. $\dfrac{0^{10}}{10^0}$ **16.** $\dfrac{0^5}{5^0}$

Evaluate each expression. See Examples 1 and 2.

17. $7^0 + 9^0$ **18.** $8^0 + 6^0$ **19.** 4^{-3} **20.** 5^{-4}

21. $\left(\dfrac{1}{2}\right)^{-4}$ **22.** $\left(\dfrac{1}{3}\right)^{-3}$ **23.** $\left(\dfrac{6}{7}\right)^{-2}$ **24.** $\left(\dfrac{2}{3}\right)^{-3}$

25. $(-3)^{-4}$ **26.** $(-4)^{-3}$ **27.** $5^{-1} + 3^{-1}$ **28.** $6^{-1} + 2^{-1}$

Use the quotient rule to simplify each expression. Write the expression with positive exponents. Assume that all variables represent nonzero real numbers. See Examples 2–4.

29. $\dfrac{5^8}{5^5}$ **30.** $\dfrac{11^6}{11^3}$ **31.** $\dfrac{9^4}{9^5}$ **32.** $\dfrac{7^3}{7^4}$ **33.** $\dfrac{5}{5^{-1}}$

34. $\dfrac{6}{6^{-2}}$ **35.** $\dfrac{x^{12}}{x^{-3}}$ **36.** $\dfrac{y^4}{y^{-6}}$ **37.** $\dfrac{1}{6^{-3}}$ **38.** $\dfrac{1}{5^{-2}}$

39. $\dfrac{2}{r^{-4}}$ **40.** $\dfrac{3}{s^{-8}}$ **41.** $\dfrac{4^{-3}}{5^{-2}}$ **42.** $\dfrac{6^{-2}}{5^{-4}}$ **43.** $p^5 q^{-8}$

44. $x^{-8} y^4$ **45.** $\dfrac{r^5}{r^{-4}}$ **46.** $\dfrac{a^6}{a^{-4}}$ **47.** $\dfrac{6^4 x^8}{6^5 x^3}$ **48.** $\dfrac{3^8 y^5}{3^{10} y^2}$

49. $\dfrac{(a+b)^{-3}}{(a+b)^{-4}}$ **50.** $\dfrac{(x+y)^{-8}}{(x+y)^{-9}}$ **51.** $\dfrac{(x+2y)^{-3}}{(x+2y)^{-5}}$ **52.** $\dfrac{(p-3q)^{-2}}{(p-3q)^{-4}}$

RELATING CONCEPTS (EXERCISES 53–56)

In Objective 1, we showed how 6^0 acts as 1 when it is applied to the product rule, thus motivating the definition for 0 as an exponent. We can also use the quotient rule to motivate this definition.

Work Exercises 53–56 in order.

53. Consider the expression $\dfrac{25}{25}$. What is its simplest form?

54. Because $25 = 5^2$, the expression $\dfrac{25}{25}$ can be written as the quotient of powers of 5. Write the expression in this way.

55. Apply the quotient rule for exponents to the expression you wrote in Exercise 54. Give the answer as a power of 5.

56. Your answers in Exercises 53 and 55 must be equal because they both represent $\dfrac{25}{25}$.

Write this equality. What definition does this result support?

Did you make the connection that the definition of 0 as an exponent can be justified using division rules as well as multiplication rules?

Use a combination of the rules for exponents to simplify each expression. Write answers with only positive exponents. Assume that all variables represent nonzero real numbers. See Example 5.

57. $\dfrac{(7^4)^3}{7^9}$ **58.** $\dfrac{(5^3)^2}{5^2}$ **59.** $x^{-3} \cdot x^5 \cdot x^{-4}$ **60.** $y^{-8} \cdot y^5 \cdot y^{-2}$

61. $\dfrac{(3x)^{-2}}{(4x)^{-3}}$ **62.** $\dfrac{(2y)^{-3}}{(5y)^{-4}}$ **63.** $\left(\dfrac{x^{-1}y}{z^2}\right)^{-2}$ **64.** $\left(\dfrac{p^{-4}q}{r^{-3}}\right)^{-3}$

65. $(6x)^4 (6x)^{-3}$ **66.** $(10y)^9 (10y)^{-8}$ **67.** $\dfrac{(m^7 n)^{-2}}{m^{-4} n^3}$ **68.** $\dfrac{(m^8 n^{-4})^2}{m^{-2} n^5}$

69. $\dfrac{(x^{-1}y^2z)^{-2}}{(x^{-3}y^3z)^{-1}}$ **70.** $\dfrac{(a^{-2}b^{-3}c^{-4})^{-5}}{(a^2b^3c^4)^5}$ **71.** $\left(\dfrac{xy^{-2}}{x^2y}\right)^{-3}$ **72.** $\left(\dfrac{wz^{-5}}{w^{-3}z}\right)^{-2}$

73. Consider the following typical student **error**:

$$\frac{16^3}{2^2} = \left(\frac{16}{2}\right)^{3-2} = 8^1 = 8.$$

Explain what the student did incorrectly, and then give the correct answer.

74. Consider the following typical student **error**:

$$-5^4 = (-5)^4 = 625.$$

Explain what the student did incorrectly, and then give the correct answer.

5.6 Division of Polynomials

OBJECTIVES

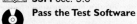

1 Divide a polynomial by a monomial.

2 Divide a polynomial by a polynomial.

FOR EXTRA HELP

📖 **SSG** Sec. 5.6
   SSM Sec. 5.6

Pass the Test Software

**InterAct Math
   Tutorial Software**

▣ **Video 8**

OBJECTIVE 1 Divide a polynomial by a monomial. We add two fractions with a common denominator as follows.

$$\frac{a}{c} + \frac{b}{c} = \frac{a+b}{c}.$$

Looking at this statement in reverse gives us a rule for dividing a polynomial by a monomial.

Dividing a Polynomial by a Monomial

To divide a polynomial by a monomial, divide each term of the polynomial by the monomial:

$$\frac{a+b}{c} = \frac{a}{c} + \frac{b}{c} \quad (c \neq 0).$$

The parts of a division problem are named in the diagram.

$$\text{Dividend} \rightarrow \quad \frac{12x^2 + 6x}{6x} = 2x + 1 \quad \leftarrow \text{Quotient}$$
$$\text{Divisor} \rightarrow$$

EXAMPLE 1 Dividing a Polynomial by a Monomial

Divide $5m^5 - 10m^3$ by $5m^2$.

Use the rule above, with $+$ replaced by $-$. Then use the quotient rule for exponents.

$$\frac{5m^5 - 10m^3}{5m^2} = \frac{5m^5}{5m^2} - \frac{10m^3}{5m^2} = m^3 - 2m$$

Recall from arithmetic that division problems can be checked by multiplication:

$$\frac{63}{7} = 9 \quad \text{because} \quad 7 \cdot 9 = 63.$$

To check the polynomial quotient, multiply $m^3 - 2m$ by $5m^2$. Because

$$5m^2(m^3 - 2m) = 5m^5 - 10m^3,$$

the quotient is correct.

Since division by 0 is undefined, the quotient

$$\frac{5m^5 - 10m^3}{5m^2}$$

is undefined if $m = 0$. From now on we assume that no denominators are 0.

E X A M P L E   2   Dividing a Polynomial by a Monomial

Divide $\dfrac{16a^5 - 12a^4 + 8a^2}{4a^3}$.

Divide each term of $16a^5 - 12a^4 + 8a^2$ by $4a^3$.

$$\frac{16a^5 - 12a^4 + 8a^2}{4a^3} = \frac{16a^5}{4a^3} - \frac{12a^4}{4a^3} + \frac{8a^2}{4a^3}$$

$$= 4a^2 - 3a + \frac{2}{a}$$

The result is not a polynomial because of the expression $\frac{2}{a}$, which has a variable in the denominator. While the sum, difference, and product of two polynomials are always polynomials, the quotient of two polynomials may not be.

Again, check by multiplying.

$$4a^3\left(4a^2 - 3a + \frac{2}{a}\right) = 4a^3(4a^2) - 4a^3(3a) + 4a^3\left(\frac{2}{a}\right)$$

$$= 16a^5 - 12a^4 + 8a^2$$

E X A M P L E   3   Dividing a Polynomial by a Monomial with a Negative Coefficient

Divide $-7x^3 + 12x^4 - 4x$ by $-4x$.

The polynomial should be written in descending powers before dividing. Write it as $12x^4 - 7x^3 - 4x$; then divide by $-4x$.

$$\frac{12x^4 - 7x^3 - 4x}{-4x} = \frac{12x^4}{-4x} - \frac{7x^3}{-4x} + \frac{-4x}{-4x}$$

$$= -3x^3 + \frac{7x^2}{4} + 1 = -3x^3 + \frac{7}{4}x^2 + 1$$

Check by multiplying.

In Example 3, notice the quotient $\dfrac{-4x}{-4x} = 1$. It is a common error to leave **CAUTION** this term out of the answer. Checking by multiplication will show that the answer $-3x^3 + \dfrac{7}{4}x^2$ is not correct.

OBJECTIVE 2 Divide a polynomial by a polynomial. We use a method of "long division" to divide a polynomial by a polynomial (other than a monomial). This method is similar to the method of long division used for two whole numbers. For comparison, the division of whole numbers is shown alongside the division of polynomials. Both polynomials must be written with descending powers before beginning the division process.

Step 1   Divide 27 into 6696.

$$27\overline{)6696}$$

Divide $2x + 3$ into $8x^3 - 4x^2 - 14x + 15$.

$$2x + 3\overline{)8x^3 - 4x^2 - 14x + 15}$$

Step 2   27 divides into 66 **2** times; $2 \cdot 27 = \boxed{54}$.

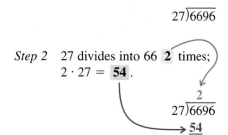

$$\begin{array}{r} 2 \\ 27\overline{)6696} \\ 54 \end{array}$$

$2x$ divides into $8x^3$ **$4x^2$** times; $4x^2(2x + 3) = \boxed{8x^3 + 12x^2}$.

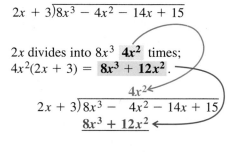

$$\begin{array}{r} 4x^2 \\ 2x + 3\overline{)8x^3 - 4x^2 - 14x + 15} \\ 8x^3 + 12x^2 \end{array}$$

Step 3   Subtract: $66 - 54 = 12$; then bring down the next digit.

$$\begin{array}{r} 2 \\ 27\overline{)6696} \\ 54\downarrow \\ 129 \end{array}$$

Subtract: $-4x^2 - 12x^2 = -16x^2$; then bring down the next term.

$$\begin{array}{r} 4x^2 \\ 2x + 3\overline{)8x^3 - 4x^2 - 14x + 15} \\ 8x^3 + 12x^2 \quad \downarrow \\ -16x^2 - 14x \end{array}$$

(To subtract two polynomials, change the sign of the second and then add.)

Step 4   27 divides into 129 **4** times; $4 \cdot 27 = \boxed{108}$.

$$\begin{array}{r} 24 \\ 27\overline{)6696} \\ 54 \\ 129 \\ 108 \end{array}$$

$2x$ divides into $-16x^2$ **$-8x$** times; $-8x(2x + 3) = \boxed{-16x^2 - 24x}$.

$$\begin{array}{r} 4x^2 - 8x \\ 2x + 3\overline{)8x^3 - 4x^2 - 14x + 15} \\ 8x^3 + 12x^2 \\ -16x^2 - 14x \\ -16x^2 - 24x \end{array}$$

Step 5   Subtract: $129 - 108 = 21$; then bring down the next digit.

$$\begin{array}{r} 24 \\ 27\overline{)6696} \\ 54 \\ 129 \\ 108 \\ 216 \end{array}$$

Subtract: $-14x - (-24x) = 10x$; then bring down the next term.

$$\begin{array}{r} 4x^2 - 8x \\ 2x + 3\overline{)8x^3 - 4x^2 - 14x + 15} \\ 8x^3 + 12x^2 \\ -16x^2 - 14x \\ -16x^2 - 24x \\ 10x + 15 \end{array}$$

Step 6   27 divides into 216 **8** times; $2x$ divides into $10x$ **5** times;
8 · 27 = **216** . $5(2x + 3) =$ **$10x + 15$** .

$$\begin{array}{r} 248 \\ 27\overline{)6696} \\ 54 \\ \overline{129} \\ 108 \\ \overline{216} \\ 216 \end{array}$$

$$\begin{array}{r} 4x^2 - 8x + 5 \\ 2x + 3\overline{)8x^3 - 4x^2 - 14x + 15} \\ \underline{8x^3 + 12x^2} \\ -16x^2 - 14x \\ \underline{-16x^2 - 24x} \\ 10x + 15 \\ \underline{10x + 15} \end{array}$$

6696 divided by 27 is 248. $8x^3 - 4x^2 - 14x + 15$ divided by
There is no remainder. $2x + 3$ is $4x^2 - 8x + 5$. There is no
remainder.

Step 7   Check by multiplying.   Check by multiplying.
27 · 248 = 6696 $(2x + 3)(4x^2 - 8x + 5)$
    $= 8x^3 - 4x^2 - 14x + 15$

Notice that at each step in the polynomial division process, the *first* term was divided into the *first* term.

E X A M P L E   4   Dividing a Polynomial by a Polynomial

Divide $5x + 4x^3 - 8 - 4x^2$ by $2x - 1$.

   Both polynomials must be written in descending powers of x. Rewrite the first polynomial as $4x^3 - 4x^2 + 5x - 8$. Then begin the division process.

$$\begin{array}{r} 2x^2 - x + 2 \\ 2x - 1\overline{)4x^3 - 4x^2 + 5x - 8} \\ \underline{4x^3 - 2x^2} \\ -2x^2 + 5x \\ \underline{-2x^2 + x} \\ 4x - 8 \\ \underline{4x - 2} \\ -6 \end{array}$$

Step 1 $2x$ divides into $4x^3$ **($2x^2$)** times;
$2x^2(2x - 1) = 4x^3 - 2x^2$.

Step 2   Subtract; bring down the next term.

Step 3 $2x$ divides into $-2x^2$ **($-x$)** times;
$-x(2x - 1) = -2x^2 + x$.

Step 4   Subtract; bring down the next term.

Step 5 $2x$ divides into $4x$ **2** times;
$2(2x - 1) = 4x - 2$.

Step 6   Subtract. The remainder is -6.

   In a division problem like this one, the division process stops when the degree of the remainder is less than the degree of the divisor. Thus, $2x - 1$ divides into $4x^3 - 4x^2 + 5x - 8$ with a quotient of $2x^2 - x + 2$ and a remainder of -6. Write the remainder as a fraction with $2x - 1$ as the denominator. The result is not a polynomial because of the remainder.

$$\frac{4x^3 - 4x^2 + 5x - 8}{2x - 1} = 2x^2 - x + 2 + \frac{-6}{2x - 1}$$

Step 7 Check by multiplying.

$$(2x - 1)\left(2x^2 - x + 2 + \frac{-6}{2x - 1}\right)$$

$$= (2x - 1)(2x^2) + (2x - 1)(-x) + (2x - 1)(2) + (2x - 1)\left(\frac{-6}{2x - 1}\right)$$

$$= 4x^3 - 2x^2 - 2x^2 + x + 4x - 2 - 6$$

$$= 4x^3 - 4x^2 + 5x - 8$$

E X A M P L E 5   Dividing into a Polynomial with Missing Terms

Divide $x^3 + 2x - 3$ by $x - 1$.

Here the polynomial $x^3 + 2x - 3$ is missing the x^2 term. When terms are missing, use 0 as the coefficient for the missing terms; in this case, $0x^2$ is a "placeholder."

$$x^3 + 2x - 3 = x^3 + 0x^2 + 2x - 3$$

Now divide.

$$
\begin{array}{r}
x^2 + x + 3 \\
x - 1 \overline{\smash{\big)}\ x^3 + 0x^2 + 2x - 3} \\
\underline{x^3 - x^2} \\
x^2 + 2x \\
\underline{x^2 - x} \\
3x - 3 \\
\underline{3x - 3}
\end{array}
$$

The remainder is 0. The quotient is $x^2 + x + 3$. Check by multiplying.

$$(x^2 + x + 3)(x - 1) = x^3 + 2x - 3$$

E X A M P L E 6   Dividing by a Polynomial with Missing Terms

Divide $x^4 + 2x^3 + 2x^2 - x - 1$ by $x^2 + 1$.

Since $x^2 + 1$ has a missing x term, write it as $x^2 + 0x + 1$. Then proceed through the division process.

$$
\begin{array}{r}
x^2 + 2x + 1 \\
x^2 + 0x + 1 \overline{\smash{\big)}\ x^4 + 2x^3 + 2x^2 - x - 1} \\
\underline{x^4 + 0x^3 + x^2} \\
2x^3 + x^2 - x \\
\underline{2x^3 + 0x^2 + 2x} \\
x^2 - 3x - 1 \\
\underline{x^2 + 0x + 1} \\
-3x - 2 \qquad \leftarrow \text{Remainder}
\end{array}
$$

When the result of subtracting ($-3x - 2$, in this case) is a polynomial of smaller degree than the divisor ($x^2 + 0x + 1$), that polynomial is the remainder. Write the result as

$$x^2 + 2x + 1 + \frac{-3x - 2}{x^2 + 1}.$$

CONNECTIONS

In Section 5.1, we found the value of a polynomial in x for a given value of x by substituting that number for x. Surprisingly, we can accomplish the same thing by division. Suppose we want to find the value of $2x^3 - 4x^2 + 3x - 5$ for $x = -3$. Instead of substituting -3 for x in the polynomial, we divide the polynomial by $x - (-3) = x + 3$. The remainder will give the value of the polynomial for $x = -3$. In general, when a polynomial P is divided by $x - r$, the remainder is equal to P evaluated at $x = r$.

FOR DISCUSSION OR WRITING

1. Evaluate $2x^3 - 4x^2 + 3x - 5$ for $x = -3$.

2. Divide $2x^3 - 4x^2 + 3x - 5$ by $x + 3$. Give the remainder.

3. Compare the answers to Exercises 1 and 2. What do you notice?

4. Choose another polynomial and evaluate it both ways at some value of the variable. Do the answers agree?

5.6 EXERCISES

Fill in each blank with the correct response.

1. In the statement $\dfrac{6x^2 + 8}{2} = 3x^2 + 4$, _____ is the dividend, _____ is the divisor, and _____ is the quotient.

2. The expression $\dfrac{3x + 12}{x}$ is undefined if $x =$ _____ .

3. To check the division shown in Exercise 1, multiply _____ by _____ and show that the product is _____ .

4. The expression $5x^2 - 3x + 6 + \dfrac{2}{x}$ _____ a polynomial.
   (is/is not)

5. Explain why the division problem $\dfrac{16m^3 - 12m^2}{4m}$ can be performed using the methods of this section, while the division problem $\dfrac{4m}{16m^3 - 12m^2}$ cannot.

6. Suppose that a polynomial in the variable x has degree 5 and it is divided by a monomial in the variable x having degree 3. What is the degree of the quotient?

Perform each division. See Examples 1–3.

7. $\dfrac{60x^4 - 20x^2 + 10x}{2x}$

8. $\dfrac{120x^6 - 60x^3 + 80x^2}{2x}$

9. $\dfrac{20m^5 - 10m^4 + 5m^2}{5m^2}$

10. $\dfrac{12t^5 - 6t^3 + 6t^2}{6t^2}$

11. $\dfrac{8t^5 - 4t^3 + 4t^2}{2t}$

12. $\dfrac{8r^4 - 4r^3 + 6r^2}{2r}$

13. $\dfrac{4a^5 - 4a^2 + 8}{4a}$

14. $\dfrac{5t^8 + 5t^7 + 15}{5t}$

Divide each polynomial by $3x^2$. See Examples 1–3.

15. $12x^5 - 9x^4 + 6x^3$

16. $24x^6 - 12x^5 + 30x^4$

17. $3x^2 + 15x^3 - 27x^4$

18. $3x^2 - 18x^4 + 30x^5$

19. $36x + 24x^2 + 6x^3$

20. $9x - 12x^2 + 9x^3$

21. $4x^4 + 3x^3 + 2x$

22. $5x^4 - 6x^3 + 8x$

Perform each division. See Examples 1–3.

23. $\dfrac{-27r^4 + 36r^3 - 6r^2 - 26r + 2}{-3r}$

24. $\dfrac{-8k^4 + 12k^3 + 2k^2 - 7k + 3}{-2k}$

25. $\dfrac{2m^5 - 6m^4 + 8m^2}{-2m^3}$

26. $\dfrac{6r^5 - 8r^4 + 10r^2}{-2r^4}$

27. $(20a^4 - 15a^5 + 25a^3) \div (5a^4)$

28. $(16y^5 - 8y^2 + 12y) \div (4y^2)$

29. $(120x^{11} - 60x^{10} + 140x^9 - 100x^8) \div (10x^{12})$

30. $(120x^{12} - 84x^9 + 60x^8 - 36x^7) \div (12x^9)$

31. The quotient in Exercise 21 is $\dfrac{4x^2}{3} + x + \dfrac{2}{3x}$. Notice how the third term is written with x in the denominator. Would $\dfrac{2}{3}x$ be an acceptable form for this term? Why or why not?

32. Refer to the quotient given in Example 2 in this section. Write it as an equivalent expression using negative exponents as necessary.

Use the appropriate formula, found on the inside covers, to answer each question.

33. The area of the rectangle is given by the polynomial $15x^3 + 12x^2 - 9x + 3$. What is the polynomial that expresses the length?

34. The area of the triangle is given by the polynomial $24m^3 + 48m^2 + 12m$. What is the polynomial that expresses the length of the base?

35. The quotient of a certain polynomial and $-7m^2$ is $9m^2 + 3m + 5 - \dfrac{2}{m}$. Find the polynomial.

36. Suppose that a polynomial of degree n is divided by a monomial of degree m to get a *polynomial* quotient.
  (a) How do m and n compare in value?
  (b) What is the expression that gives the degree of the quotient?

RELATING CONCEPTS (EXERCISES 37–40)

Our system of numeration is called a decimal system. It is based on powers of ten. In a whole number such as 2846, each digit is understood to represent the number of powers of ten for its place value. The 2 represents two thousands (2×10^3), the 8 represents eight hundreds (8×10^2), the 4 represents four tens (4×10^1), and the 6 represents six ones (or units) (6×10^0). In expanded form we write

$$2846 = (2 \times 10^3) + (8 \times 10^2) + (4 \times 10^1) + (6 \times 10^0).$$

*Keeping this information in mind, **work Exercises 37–40 in order.***

37. Divide 2846 by 2, using paper-and-pencil methods: $2\overline{)2846}$.

38. Write your answer in Exercise 37 in expanded form.

39. Use the methods of this section to divide the polynomial $2x^3 + 8x^2 + 4x + 6$ by 2.

40. Compare your answers in Exercises 38 and 39. How are they similar? How are they different? For what value of x does the answer in Exercise 39 equal the answer in Exercise 38?

Did you make the connection between division of whole numbers and division of polynomials?

Perform each division. See Example 4.

41. $\dfrac{x^2 - x - 6}{x - 3}$

42. $\dfrac{m^2 - 2m - 24}{m - 6}$

43. $\dfrac{2y^2 + 9y - 35}{y + 7}$

44. $\dfrac{2y^2 + 9y + 7}{y + 1}$

45. $\dfrac{p^2 + 2p + 20}{p + 6}$

46. $\dfrac{x^2 + 11x + 16}{x + 8}$

47. $(r^2 - 8r + 15) \div (r - 3)$

48. $(t^2 + 2t - 35) \div (t - 5)$

49. $\dfrac{12m^2 - 20m + 3}{2m - 3}$

50. $\dfrac{12y^2 + 20y + 7}{2y + 1}$

51. $\dfrac{4a^2 - 22a + 32}{2a + 3}$

52. $\dfrac{9w^2 + 6w + 10}{3w - 2}$

53. $\dfrac{8x^3 - 10x^2 - x + 3}{2x + 1}$

54. $\dfrac{12t^3 - 11t^2 + 9t + 18}{4t + 3}$

55. $\dfrac{8k^4 - 12k^3 - 2k^2 + 7k - 6}{2k - 3}$

56. $\dfrac{27r^4 - 36r^3 - 6r^2 + 26r - 24}{3r - 4}$

57. $\dfrac{5y^4 + 5y^3 + 2y^2 - y - 3}{y + 1}$

58. $\dfrac{2r^3 - 5r^2 - 6r + 15}{r - 3}$

59. $\dfrac{3k^3 - 4k^2 - 6k + 10}{k - 2}$

60. $\dfrac{5z^3 - z^2 + 10z + 2}{z + 2}$

61. $\dfrac{6p^4 - 16p^3 + 15p^2 - 5p + 10}{3p + 1}$

62. $\dfrac{6r^4 - 11r^3 - r^2 + 16r - 8}{2r - 3}$

Perform each division. See Examples 5 and 6.

63. $\dfrac{5 - 2r^2 + r^4}{r^2 - 1}$

64. $\dfrac{4t^2 + t^4 + 7}{t^2 + 1}$

65. $\dfrac{y^3 + 1}{y + 1}$

66. $\dfrac{y^3 - 1}{y - 1}$

67. $\dfrac{a^4 - 1}{a^2 - 1}$

68. $\dfrac{a^4 - 1}{a^2 + 1}$

69. $\dfrac{x^4 - 4x^3 + 5x^2 - 3x + 2}{x^2 + 3}$

70. $\dfrac{3t^4 + 5t^3 - 8t^2 - 13t + 2}{t^2 - 5}$

71. $\dfrac{2x^5 + 9x^4 + 8x^3 + 10x^2 + 14x + 5}{2x^2 + 3x + 1}$

72. $\dfrac{4t^5 - 11t^4 - 6t^3 + 5t^2 - t + 3}{4t^2 + t - 3}$

73. $(3a^2 - 11a + 17) \div (2a + 6)$

74. $(4x^2 + 11x - 8) \div (3x + 6)$

75. Suppose that one of your classmates asks you the following question: "How do I know when to stop the division process in a problem like the one in Exercise 69?" How would you respond?

76. Suppose that someone asks you if the following division problem is correct:

$$(6x^3 + 4x^2 - 3x + 9) \div (2x - 3) = 4x^2 + 9x - 3.$$

Tell how, by looking only at the *first term* of the quotient, you immediately know that the problem has been worked incorrectly.

Find a polynomial that describes each quantity required. Use the formulas found on the inside covers.

77. Give the length of the rectangle.

5x + 2

The area is $5x^3 + 7x^2 - 13x - 6$ square units.

78. Find the measure of the base of the parallelogram.

x − 1

The area is $2x^3 + 2x^2 - 3x - 1$ square units.

79. If the distance traveled is $5x^3 - 6x^2 + 3x + 14$ miles and the rate is $x + 1$ miles per hour, what is the time traveled?

80. If it costs $4x^5 + 3x^4 + 2x^3 + 9x^2 - 29x + 2$ dollars to fertilize a garden, and fertilizer costs $x + 2$ dollars per square yard, what is the area of the garden?

RELATING CONCEPTS (EXERCISES 81–84)

Students often would like to know whether the quotient obtained in a polynomial division problem is actually correct or whether an error was made. While the method described here is not 100% foolproof, it is quick and will at least give a fairly good idea as to the accuracy of the result. To illustrate, suppose that $4x^4 + 2x^3 - 14x^2 + 19x + 10$ is divided by $2x + 5$. Lakeisha and Stan obtain the following answers:

Lakeisha   *Stan*

$2x^3 - 4x^2 + 3x + 2$ $2x^3 - 4x^2 - 3x + 2.$

As a "quick check" we can evaluate Lakeisha's answer for $x = 1$ and Stan's answer for $x = 1$:

Lakeisha: When $x = 1$, her answer gives $2(1)^3 - 4(1)^2 + 3(1) + 2 = 3$.

Stan: When $x = 1$, his answer gives $2(1)^3 - 4(1)^2 - 3(1) + 2 = -3$.

Now, if the original quotient of the two polynomials is evaluated for $x = 1$, we get

$$\frac{4x^4 + 2x^3 - 14x^2 + 19x + 10}{2x + 5} = \frac{4(1)^4 + 2(1)^3 - 14(1)^2 + 19(1) + 10}{2(1) + 5}$$

$$= \frac{21}{7}$$

$$= 3.$$

Because Stan's answer, -3, is different from the quotient 3 just obtained, Stan can conclude his answer is incorrect. Lakeisha's answer, 3, agrees with the quotient just obtained, and while this does not *guarantee* that she is correct, at least she can feel better and go on to the next problem.

In Exercises 81–84, a division problem is given, along with two possible answers. One is correct and one is incorrect. Use the method just described to determine which one is correct and which one is not.

81. Problem

$$\frac{2x^2 + 3x - 14}{x - 2}$$

Possible answers

(a) $2x + 7$ **(b)** $2x - 7$

82. Problem

$$\frac{x^4 + 4x^3 - 5x^2 - 12x + 6}{x^2 - 3}$$

Possible answers

(a) $x^2 - 4x - 2$ **(b)** $x^2 + 4x - 2$

83. Problem

$$\frac{2y^3 + 17y^2 + 37y + 7}{2y + 7}$$

Possible answers

(a) $y^2 + 5y + 1$ **(b)** $y^2 - 5y + 1$

84. In the explanation preceding Exercise 81 we used 1 for the value of x to check our work. This is because a polynomial is easy to evaluate for 1. Why is this so? Why would we not be able to use 1 if the divisor is $x - 1$?

Did you make the connections between division of polynomials and evaluating polynomials?

5.7 An Application of Exponents: Scientific Notation

OBJECTIVES

 Express numbers in scientific notation.

 Convert numbers in scientific notation to numbers without exponents.

 Use scientific notation in calculations.

FOR EXTRA HELP

SSG Sec. 5.7
SSM Sec. 5.7

Pass the Test Software

InterAct Math
Tutorial Software

Video 8

CONNECTIONS

According to the Aerospace Industries Association of America, in 1996 aerospace industry profits were 7326 million dollars. Another way to express this number is 7,326,000,000. Isaac Asimov wrote in *Asimov on Numbers* that the radius of the nucleus of a carbon atom has been calculated as .00000000000038 centimeter. These examples use numbers from real life that are very large or very small, thus requiring many zeros when written in decimal form. Because of difficulties working with many zeros, these numbers are often expressed as powers of 10, as discussed in this section.

FOR DISCUSSION OR WRITING

1. Based on your experience with multiplication of whole numbers, how can you easily multiply a number by 100? By 10,000?

2. How can you easily divide a number by 100? By 10,000?

3. Discuss the role of the digit 0 in the following numbers: 32,000 and .00032.

OBJECTIVE Express numbers in scientific notation. In **scientific notation,** a number is written in the form $a \times 10^n$, where n is an integer and $1 \leq |a| < 10$. When a number is multiplied by a power of 10, such as 10^1, $10^2 = 100$, $10^3 = 1000$, $10^{-1} = .1$, $10^{-2} = .01$, $10^{-3} = .001$, and so on, the net effect is to move the decimal point to the right if it is a positive power and to the left if it is a negative power. This is shown in the examples below. (In work with scientific notation, the times symbol, $\times$, is commonly used.)

23.19×10^1	$= 23.19 \times 10 = 231.9$	Decimal point moves 1 place to the right.
23.19×10^2	$= 23.19 \times 100 = 2319.$	Decimal point moves 2 places to the right.
23.19×10^3	$= 23.19 \times 1000 = 23190.$	Decimal point moves 3 places to the right.
23.19×10^{-1}	$= 23.19 \times .1 = 2.319$	Decimal point moves 1 place to the left.
23.19×10^{-2}	$= 23.19 \times .01 = .2319$	Decimal point moves 2 places to the left.
23.19×10^{-3}	$= 23.19 \times .001 = .02319$	Decimal point moves 3 places to the left.

A number in scientific notation is always written with the decimal point after the first nonzero digit and then multiplied by the appropriate power of 10. For example, 35 is written 3.5×10^1, or 3.5×10; 56,200 is written 5.62×10^4, since

$$56,200 = 5.62 \times \mathbf{10,000} = 5.62 \times \mathbf{10^4}.$$

To write a number in scientific notation, follow these steps.

Writing a Number in Scientific Notation

Step 1 Move the decimal point to the right of the first nonzero digit.

Step 2 Count the number of places you moved the decimal point.

Step 3 The number of places in Step 2 is the absolute value of the exponent on 10.

Step 4 The exponent on 10 is positive if you made the number smaller in Step 1. The exponent is negative if you made the number larger in Step 1. If the decimal point is not moved, the exponent is 0.

EXAMPLE 1 Using Scientific Notation

Write each number in scientific notation.

(a) 93,000,000

The number will be written in scientific notation as 9.3×10^n. To find the value of *n,* first compare 9.3 with the original number, 93,000,000. Is 9.3 larger or smaller? Here, 9.3 is *smaller* than 93,000,000. Therefore, we must multiply by a *positive* power of 10 so the product 9.3×10^n will equal the larger number.

Move the decimal point to follow the first nonzero digit (the 9). Count the number of places the decimal point was moved.

$$93,000,000 \quad \text{7 places}$$

Since the decimal point was moved 7 places, and since *n* is positive, $93,000,000 = 9.3 \times 10^7$.

(b) $63,200,000,000 = 6.3200000000 = 6.32 \times 10^{10}$

10 places

(c) .00462

Move the decimal point to the right of the first nonzero digit, and count the number of places the decimal point was moved.

$$.00462 \quad \text{3 places}$$

Since 4.62 is *larger* than .00462, the exponent must be *negative.*

$$.00462 = 4.62 \times 10^{-3}$$

(d) $.0000762 = 7.62 \times 10^{-5}$

Another way to choose the exponent when writing a number in scientific notation is this: If the original number is "large," like 93,000,000, use a *positive* exponent on 10, since positive is larger than negative. However, if the original number is "small," like .00462, use a *negative* exponent on 10, since negative is smaller than positive.

OBJECTIVE 2 Convert numbers in scientific notation to numbers without exponents. To convert a number written in scientific notation to a number without exponents, work in reverse. Multiplying by a positive power of 10 will make the number larger; multiplying by a negative power of 10 will make the number smaller.

EXAMPLE 2 Writing Numbers without Exponents

Write each number without exponents.

(a) 6.2×10^3

Since the exponent is positive, make 6.2 larger by moving the decimal point 3 places to the right.

$$6.2 \times 10^3 = 6.200 = 6200.$$

(b) $4.283 \times 10^5 = 4.28300 = 428,300$ Move 5 places to the right.

(c) $7.04 \times 10^{-3} = .00704$ Move 3 places to the left.

As these examples show, the exponent tells the number of places and the direction that the decimal point is moved.

OBJECTIVE **3** Use scientific notation in calculations. The next example uses scientific notation with products and quotients.

EXAMPLE 3 Multiplying and Dividing with Scientific Notation

Write each product or quotient without exponents.

(a) $(6 \times 10^3)(5 \times 10^{-4})$

$(6 \times 10^3)(5 \times 10^{-4}) = (6 \times 5)(10^3 \times 10^{-4})$ Commutative and associative properties

$= 30 \times 10^{-1}$ Product rule for exponents

$= 3$ Write without exponents.

(b) $\dfrac{4 \times 10^{-5}}{2 \times 10^3} = \dfrac{4}{2} \times \dfrac{10^{-5}}{10^3} = 2 \times 10^{-8} = .00000002$

 Multiplying or dividing numbers written in scientific notation may produce an answer in the form $a \times 10^0$. Since $10^0 = 1$, $a \times 10^0 = a$. For example,

$$(8 \times 10^{-4})(5 \times 10^4) = 40 \times 10^0 = 40.$$

The final example comes from Isaac Asimov's *Asimov on Numbers*.

EXAMPLE 4 Calculating Using Scientific Notation

Fill in the blank in this sentence: "We next suppose that the Greatest Possible Volume is packed perfectly tightly (leaving no empty spaces) with objects of the Smallest Possible Volume. If we divided 7.8×10^{84} by 1.7×10^{-40}, we find that the number of protons it takes to fill the Observable Universe is _____."

$$\frac{7.8 \times 10^{84}}{1.7 \times 10^{-40}} = \frac{7.8}{1.7} \times 10^{84-(-40)}$$ Divide.

$$= \frac{7.8}{1.7} \times 10^{124}$$ Subtract exponents.

$$\approx 4.6 \times 10^{124}$$ Use a calculator.

5.7 EXERCISES

Match each number written in scientific notation in Column I with the correct choice from Column II.

I	II
1. 4.6×10^{-4}	**A.** .00046
2. 4.6×10^4	**B.** 46,000
3. 4.6×10^5	**C.** 460,000
4. 4.6×10^{-5}	**D.** .000046

Determine whether or not each number is written in scientific notation as defined in Objective 1. If it is not, write it as such.

5. 4.56×10^3 **6.** 7.34×10^5 **7.** 5,600,000   **8.** 34,000

9. $.8 \times 10^2$ **10.** $.9 \times 10^3$ **11.** .004   **12.** .0007

13. Explain in your own words what it means for a number to be written in scientific notation. Give examples.

14. Explain how to multiply a number by a positive power of ten. Then explain how to multiply a number by a negative power of ten.

Write each number in scientific notation. See Example 1.

15. 5,876,000,000

16. 9,994,000,000

17. 82,350

18. 78,330

19. .000007

20. .0000004

21. .00203

22. .0000578

Write each number without exponents. See Example 2.

23. 7.5×10^5

24. 8.8×10^6

25. 5.677×10^{12}

26. 8.766×10^9

27. 6.21×10^0

28. 8.56×10^0

29. 7.8×10^{-4}

30. 8.9×10^{-5}

31. 5.134×10^{-9}

32. 7.123×10^{-10}

Use properties and rules for exponents to perform the indicated operations, and write each answer without exponents. See Example 3.

33. $(2 \times 10^8) \times (3 \times 10^3)$

34. $(4 \times 10^7) \times (3 \times 10^3)$

35. $(5 \times 10^4) \times (3 \times 10^2)$

36. $(8 \times 10^5) \times (2 \times 10^3)$

37. $(3 \times 10^{-4}) \times (2 \times 10^8)$

38. $(4 \times 10^{-3}) \times (2 \times 10^7)$

39. $\dfrac{9 \times 10^{-5}}{3 \times 10^{-1}}$

40. $\dfrac{12 \times 10^{-4}}{4 \times 10^{-3}}$

41. $\dfrac{8 \times 10^3}{2 \times 10^2}$

42. $\dfrac{5 \times 10^4}{1 \times 10^3}$

43. $\dfrac{2.6 \times 10^{-3}}{2 \times 10^2}$

44. $\dfrac{9.5 \times 10^{-1}}{5 \times 10^3}$

▌TECHNOLOGY INSIGHTS (EXERCISES 45-50)

Graphing calculators such as the TI-83 can display numbers in scientific notation (when in scientific mode), using the format shown in the screen on the left. For example, the calculator displays 5.4E3 to represent 5.4×10^3, the scientific notation form for 5400. 5.4E−4 means 5.4×10^{-4}. It will also perform operations with numbers entered in scientific notation, as shown in the screen on the right. Notice how the rules for exponents are applied.

```
5400
                5.4E3
5.4*10^(-4)
               5.4E-4
```

```
(2E5)*(4E-2)
                 8E3
(2E5)/(4E-2)
                 5E6
(2E5)³
                8E15
```

TECHNOLOGY INSIGHTS (EXERCISES 45–50) (CONTINUED)

Predict the display the calculator would give for the expression shown in each screen.

45.
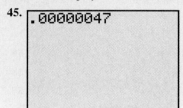

46.

47.
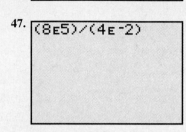

48.
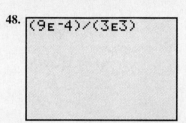

49.
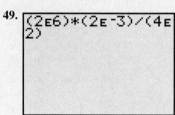

50.

Each of the following statements comes from Astronomy! A Brief Edition *by James B. Kaler (Addison-Wesley, 1997). If the number in the statement is in scientific notation, write it in standard form without using exponents. If the number is in standard form, write it in scientific notation.*

51. Multiplying this view over the whole sky yields a galaxy count of more than 10 *billion.* (page 496)

52. The circumference of the solar orbit is . . . about 4.7 million km . . . . (in reference to the orbit of Jupiter, page 395)

53. The solar luminosity requires that 2×10^9 kg of mass be converted into energy every second. (page 327)

54. At maximum, a cosmic ray particle—a mere atomic nucleus of only 10^{-13} cm across—can carry the energy of a professionally pitched baseball. (page 445)

 Each of the following statements is based on data from the aerospace industry. Write each number in scientific notation.

55. In 1996, aerospace industry profits were $7,326,000,000. (*Source:* Aerospace Industries Association of America.)

56. In 1995, the total receipts for international transportation transactions of the United States were $46,537,000,000. (*Source:* U.S. Bureau of Economic Analysis.)

57. In 1994, the total value of U.S. aircraft shipments was \$30,262,000,000. (*Source:* U.S. Department of Commerce, International Trade Administration.)

58. In 1992, the total value of net orders for U.S. civil jet transport aircraft was \$16,640,000,000. (*Source:* Aerospace Industries Association of America.)

Solve each problem.

59. The distance to Earth from the planet Pluto is 4.58×10^9 kilometers. In April 1983, Pioneer 10 transmitted radio signals from Pluto to Earth at the speed of light, 3.00×10^5 kilometers per second. How long (in seconds) did it take for the signals to reach Earth?

60. In Exercise 59, how many hours did it take for the signals to reach Earth?

 61. The accompanying graph depicts aerospace industry sales in current dollars. The figures at the tops of the bars represent billions of dollars.

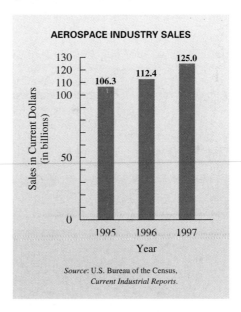

AEROSPACE INDUSTRY SALES

Source: U.S. Bureau of the Census, *Current Industrial Reports*.

(a) For each of the years, write the figure represented in scientific notation. The figure for 1996 is preliminary, and the figure for 1997 is an estimate.

(b) If a line segment is drawn between the tops of the bars for 1995 and 1997, what is its slope?

62. In a recent year, the state of Texas had about 1.3×10^6 farms with an average of 7.1×10^2 acres per farm. What was the total number of acres devoted to farmland in Texas that year? (*Source:* National Agricultural Statistics Service, U.S. Dept. of Agriculture.)

CHAPTER 5 GROUP ACTIVITY

Measuring the Flight of a Rocket

Objective: Graph quadratic equations to solve an application problem.

A physics class observes the launch of two rockets that are slightly different. One has an initial velocity of 48 feet per second. The second one has an initial velocity of 64 feet per second. Follow the steps below to determine the maximum height each rocket will achieve and how long it will take it to reach that height.

This formula is used to find the height of a projectile when shot vertically into the air.

$$H = -16t^2 + Vt$$

H = height above the launch pad in feet, t = time in seconds, and V = initial velocity.

A. One student should complete the table for the first rocket. The other student should do the same for the second rocket. Use the formula for the height of a projectile and the values provided.

Initial Velocity = 48 Feet per Second		Initial Velocity = 64 Feet per Second	
Time (seconds)	Height (feet)	Time (seconds)	Height (feet)
0		0	
.5		.5	
1		1	
1.5		1.5	
2		2	
2.5		2.5	
3		3	
		3.5	
		4	

B. Graph the results. You must determine the scales for the two axes on the coordinate system. Clearly label each graph and its scale.

C. Answer the following questions for each rocket.

   1. Find the time in seconds when the rocket is at height 0 feet. Why does this happen twice?

   2. What is the maximum height the rocket reached?

   3. At what time (in seconds) did the rocket reach its maximum height?

D. Compare the data from both rockets and answer the following questions.

   1. Which rocket had the greater maximum height? Explain how you decided.

   2. Which rocket reached its maximum height first? Explain how you decided.

   3. Consider the answers for Exercises D1 and D2 and discuss why this occurred.

   4. Do both rockets reach a height of 48 feet? If yes, at what times? What observations about projectiles might this lead to?

   5. What are some other possible uses for the formula of the height of a projectile?

CHAPTER 5 SUMMARY

KEY TERMS

5.1 polynomial	trinomial	vertex	inner product
descending powers	binomial	axis	FOIL
degree of a term	monomial	line of symmetry	**5.7** scientific notation
degree of a polynomial	parabola	**5.3** outer product	

NEW SYMBOLS

x^{-n} x to the negative n power

TEST YOUR WORD POWER

See how well you have learned the vocabulary in this chapter. Answers, with examples, are given at the bottom of the page.

1. A **polynomial** is an algebraic expression made up of
(a) a term or a finite product of terms with positive coefficients and exponents
(b) a term or a finite sum of terms with real coefficients and whole number exponents
(c) the product of two or more terms with positive exponents
(d) the sum of two or more terms with whole number coefficients and exponents.

2. The **degree of a term** is
(a) the number of variables in the term
(b) the product of the exponents on the variables

(c) the smallest exponent on the variables
(d) the sum of the exponents on the variables.

3. A **trinomial** is a polynomial with
(a) only one term
(b) exactly two terms
(c) exactly three terms
(d) more than three terms.

4. A **binomial** is a polynomial with
(a) only one term
(b) exactly two terms
(c) exactly three terms
(d) more than three terms.

5. A **monomial** is a polynomial with
(a) only one term
(b) exactly two terms
(c) exactly three terms
(d) more than three terms.

6. **FOIL** is a method for
(a) adding two binomials
(b) adding two trinomials
(c) multiplying two binomials
(d) multiplying two trinomials.

Answers to Test Your Word Power

1. (b) *Example:* $5x^3 + 2x^2 - 7$ **2.** (d) *Examples:* The term 6 has degree 0, $3x$ has degree 1, $-2x^8$ has degree 8, and $5x^3y^4$ has degree 6.
3. (c) *Example:* $2a^2 - 3ab + b^2$ **4.** (b) *Example:* $3t^3 + 5t$ **5.** (a) *Examples:* -5 and $4xy^5$
6. (c) *Example:* $(m + 4)(m - 3) = m(m) - 3m + 4m + 4(-3) = m^2 + m - 12$

CONCEPTS	EXAMPLES

5.1 ADDITION AND SUBTRACTION OF POLYNOMIALS; GRAPHING SIMPLE POLYNOMIALS

Addition

Add like terms.

Add.

$$2x^2 + 5x - 3$$
$$\underline{5x^2 - 2x + 7}$$
$$7x^2 + 3x + 4$$

Subtraction

Change the signs of the terms in the second polynomial and add to the first polynomial.

Subtract.

$$(2x^2 + 5x - 3) - (5x^2 - 2x + 7)$$
$$= (2x^2 + 5x - 3) + (-5x^2 + 2x - 7)$$
$$= -3x^2 + 7x - 10$$

Graphing Simple Polynomials

To graph a simple polynomial equation such as $y = x^2 - 2$, plot points near the vertex. (In this chapter, all parabolas have a vertex on the x-axis or the y-axis.)

Graph $y = x^2 - 2$.

x	y
-2	2
-1	-1
0	-2
1	-1
2	2

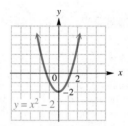

5.2 THE PRODUCT RULE AND POWER RULES FOR EXPONENTS

For any integers m and n with no denominators zero:

Perform the operations by using rules for exponents.

Product Rule

$a^m \cdot a^n = a^{m+n}$

$$2^4 \cdot 2^5 = 2^9$$

Power Rules

(a) $(a^m)^n = a^{mn}$

$$(3^4)^2 = 3^8$$

(b) $(ab)^m = a^m b^m$

$$(6a)^5 = 6^5 a^5$$

(c) $\left(\dfrac{a}{b}\right)^m = \dfrac{a^m}{b^m} \quad (b \neq 0)$

$$\left(\frac{2}{3}\right)^4 = \frac{2^4}{3^4}$$

5.3 MULTIPLICATION OF POLYNOMIALS

Multiplying Polynomials

Multiply each term of the first polynomial by each term of the second polynomial. Then add like terms.

Multiply.

$$3x^3 - 4x^2 + 2x - 7$$
$$\underline{\qquad\qquad 4x + 3}$$
$$9x^3 - 12x^2 + 6x - 21$$
$$\underline{12x^4 - 16x^3 + 8x^2 - 28x\qquad}$$
$$12x^4 - 7x^3 - 4x^2 - 22x - 21$$

FOIL Method for Multiplying Binomials

Step 1 Multiply the two first terms to get the first term of the answer.

Step 2 Find the outer product and the inner product and mentally add them, when possible, to get the middle term of the answer.

Step 3 Multiply the two last terms to get the last term of the answer.

Multiply. $(2x + 3)(5x - 4)$

$$2x(5x) = 10x^2$$

$$2x(-4) + 3(5x) = 7x$$

$$3(-4) = -12$$

The product of $(2x + 3)$ and $(5x - 4)$ is $10x^2 + 7x - 12$.

CONCEPTS	EXAMPLES

5.4   SPECIAL PRODUCTS

Square of a Binomial

$(a + b)^2 = a^2 + 2ab + b^2$

$(a - b)^2 = a^2 - 2ab + b^2$

Multiply.

$$(3x + 1)^2 = 9x^2 + 6x + 1$$
$$(2m - 5n)^2 = 4m^2 - 20mn + 25n^2$$

Product of the Sum and Difference of Two Terms

$(a + b)(a - b) = a^2 - b^2$

$$(4a + 3)(4a - 3) = 16a^2 - 9$$

5.5   INTEGER EXPONENTS AND THE QUOTIENT RULE

If $a \neq 0$, for integers m and n:

Simplify by using the rules for exponents.

Zero Exponent

$$a^0 = 1$$

$$15^0 = 1$$

Negative Exponent

$$a^{-n} = \frac{1}{a^n}$$

$$5^{-2} = \frac{1}{5^2} = \frac{1}{25}$$

Quotient Rule

$$\frac{a^m}{a^n} = a^{m-n}$$

$$\frac{a^{-m}}{b^{-n}} = \frac{b^n}{a^m} \quad (b \neq 0)$$

$$\left(\frac{a}{b}\right)^{-m} = \left(\frac{b}{a}\right)^m \quad (b \neq 0)$$

$$\frac{4^8}{4^3} = 4^5$$

$$\frac{4^{-2}}{3^{-5}} = \frac{3^5}{4^2}$$

$$\left(\frac{6}{5}\right)^{-3} = \left(\frac{5}{6}\right)^3$$

5.6   DIVISION OF POLYNOMIALS

Dividing a Polynomial by a Monomial

Divide each term of the polynomial by the monomial.

$$\frac{a + b}{c} = \frac{a}{c} + \frac{b}{c}$$

Divide.

$$\frac{4x^3 - 2x^2 + 6x - 8}{2x} = 2x^2 - x + 3 - \frac{4}{x}$$

Dividing a Polynomial by a Polynomial

Use "long division."

Divide.

$$
\begin{array}{r}
2x - 5 \\
3x + 4{\overline{\smash{\big)}\,6x^2 - 7x - 21}} \\
\underline{6x^2 + 8x} \\
-15x - 21 \\
\underline{-15x - 20} \\
-1 \quad \leftarrow \text{Remainder}
\end{array}
$$

The final quotient is $2x - 5 + \dfrac{-1}{3x + 4}$.

5.7   AN APPLICATION OF EXPONENTS: SCIENTIFIC NOTATION

To write a number in scientific notation (as $a \times 10^n$), move the decimal point to follow the first nonzero digit. The number of places the decimal point is moved is the absolute value of n. If moving the decimal point makes the number smaller, n is positive. If it makes the number larger, n is negative. If the decimal point is not moved, n is 0.

Write in scientific notation.

$$247 = 2.47 \times 10^2$$
$$.0051 = 5.1 \times 10^{-3}$$
$$4.8 = 4.8 \times 10^0$$

Write without exponents.

$$3.25 \times 10^5 = 325,000$$
$$8.44 \times 10^{-6} = .00000844$$

CHAPTER 5 REVIEW EXERCISES

[5.1] *Combine terms where possible in each polynomial. Write the answer in descending powers of the variable. Give the degree of the answer. Identify the polynomial as a monomial, binomial, trinomial, or none of these.*

1. $9m^2 + 11m^2 + 2m^2$

2. $-4p + p^3 - p^2 + 8p + 2$

3. $12a^5 - 9a^4 + 8a^3 + 2a^2 - a + 3$

4. $-7y^5 - 8y^4 - y^5 + y^4 + 9y$

5. $(12r^4 - 7r^3 + 2r^2) - (5r^4 - 3r^3 + 2r^2 - 1)$

6. Simplify $(5x^3y^2 - 3xy^5 + 12x^2) - (-9x^2 - 8x^3y^2 + 2xy^5)$.

Add or subtract as indicated.

7. Add.

$\begin{array}{r} -2a^3 + 5a^2 \\ \underline{3a^3 - a^2} \end{array}$

8. Subtract.

$\begin{array}{r} 6y^2 - 8y + 2 \\ \underline{5y^2 + 2y - 7} \end{array}$

9. Subtract.

$\begin{array}{r} -12k^4 - 8k^2 + 7k \\ \underline{k^4 + 7k^2 - 11k} \end{array}$

Graph each equation by completing the table of values.

10. $y = -x^2 + 5$

x	-2	-1	0	1	2
y					

11. $y = 3x^2 - 2$

x	-2	-1	0	1	2
y					

[5.2] **12.** Why does the product rule for exponents not apply to the expression $7^2 + 7^4$?

Use the product rule, power rules, or both to simplify each expression. Write the answers in exponential form.

13. $4^3 \cdot 4^8$

14. $(-5)^6(-5)^5$

15. $(-8x^4)(9x^3)$

16. $(2x^2)(5x^3)(x^9)$

17. $(19x)^5$

18. $(-4y)^7$

19. $5(pt)^4$

20. $\left(\dfrac{7}{5}\right)^6$

21. $(6x^2z^4)^2(x^3yz^2)^4$

22. $\left(\dfrac{2m^3n}{p^2}\right)^3$

[5.3] *Find each product.*

23. $(a + 2)(a^2 - 4a + 1)$

24. $(3r - 2)(2r^2 + 4r - 3)$

25. $(5p^2 + 3p)(p^3 - p^2 + 5)$

26. $(m - 9)(m + 2)$

27. $(3k - 6)(2k + 1)$

28. $(a + 3b)(2a - b)$

29. $(6k + 5q)(2k - 7q)$

30. $(s - 1)^3$

31. Find a polynomial that represents the area of the rectangle shown.

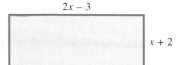

$2x - 3$

$x + 2$

32. If the side of a square has a measure represented by $5x^4 + 2x^2$, what polynomial represents its area?

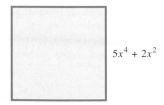

$5x^4 + 2x^2$

[5.4] *Find each product.*

33. $(a + 4)^2$

34. $(2r + 5t)^2$

35. $(6m - 5)(6m + 5)$

36. $(5a + 6b)(5a - 6b)$

37. $(r + 2)^3$

38. $t(5t - 3)^2$

39. Choose values for x and y to show that in general, the following hold true.

    (a) $(x + y)^2 \neq x^2 + y^2$ **(b)** $(x + y)^3 \neq x^3 + y^3$

40. Write an explanation on how to raise a binomial to the third power. Give an example.

41. Refer to Exercise 39. Suppose that you happened to let $x = 0$ and $y = 1$. Would your results be sufficient to illustrate the truth, in general, of the inequalities shown? If not, what would you need to do as your next step in working the exercise?

42. What is the volume of a cube with one side having length $x^2 + 2$ centimeters?

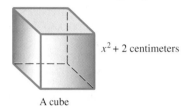

$x^2 + 2$ centimeters

A cube

43. What is the volume of a sphere with radius $x + 1$ inches?

$x + 1$ inches

A sphere

[5.5] *Evaluate each expression.*

44. $6^0 + (-6)^0$

45. $(-23)^0 - (-23)^0$

46. -10^0

Write each expression using only positive exponents.

47. -7^{-2}

48. $\left(\dfrac{5}{8}\right)^{-2}$

49. $(5^{-2})^{-4}$

50. $9^3 \cdot 9^{-5}$

51. $2^{-1} + 4^{-1}$

52. $\dfrac{6^{-5}}{6^{-3}}$

Simplify. Write each answer with only positive exponents. Assume that all variables represent nonzero real numbers.

53. $\dfrac{x^{-7}}{x^{-9}}$

54. $\dfrac{y^4 \cdot y^{-2}}{y^{-5}}$

55. $(6r^{-2})^{-1}$

56. $(3p)^4(3p^{-7})$

57. $\dfrac{ab^{-3}}{a^4 b^2}$

58. $\dfrac{(6r^{-1})^2(2r^{-4})}{r^{-5}(r^2)^{-3}}$

[5.6] *Perform each division.*

59. $\dfrac{-15y^4}{9y^2}$

60. $\dfrac{6y^4 - 12y^2 + 18y}{6y}$

61. $(-10m^4n^2 + 5m^3n^2 + 6m^2n^4) \div (5m^2n)$

62. What polynomial, when multiplied by $6m^2n$, gives the product $12m^3n^2 + 18m^6n^3 - 24m^2n^2$?

63. One of your friends in class simplified $\dfrac{6x^2 - 12x}{6}$ as $x^2 - 12x$. Is this correct? If not, what error did your friend make and how would you explain the correct method of performing the division?

Perform each division.

64. $\dfrac{2r^2 + 3r - 14}{r - 2}$

65. $\dfrac{10a^3 + 9a^2 - 14a + 9}{5a - 3}$

66. $\dfrac{x^4 + 3x^3 - 5x^2 - 3x + 4}{x^2 - 1}$

67. $\dfrac{m^4 + 4m^3 - 5m^2 - 12m + 6}{m^2 - 3}$

[5.7]   *Write each number in scientific notation.*

68. 48,000,000

69. 28,988,000,000

70. .0000000824

Write each number without exponents.

71. 2.4×10^4

72. 7.83×10^7

73. 8.97×10^{-7}

Perform each indicated operation and write the answer without exponents.

74. $(2 \times 10^{-3}) \times (4 \times 10^5)$

75. $\dfrac{8 \times 10^4}{2 \times 10^{-2}}$

76. $\dfrac{12 \times 10^{-5} \times 5 \times 10^4}{4 \times 10^3 \times 6 \times 10^{-2}}$

The quote is taken from the source cited. Write each number in scientific notation in the quote without exponents.

77. The muon, a close relative of the electron produced by the bombardment of cosmic rays against the upper atmosphere, has a half-life of 2 millionths of a second (2×10^{-6}s). (Excerpt from *Conceptual Physics,* 6th edition, by Paul G. Hewitt. Copyright © by Paul G. Hewitt. Published by HarperCollins College Publishers.)

78. There are 13 red balls and 39 black balls in a box. Mix them up and draw 13 out one at a time without returning any ball . . . the probability that the 13 drawings each will produce a red ball is . . . 1.6×10^{-12}. (Warren Weaver, *Lady Luck,* New York: Doubleday & Company, Inc., 1963, pp. 298–299.)

The quote is taken from the source cited. Write each number in scientific notation.

79. An electron and a positron attract each other in two ways: the electromagnetic attraction of their opposite electric charges, and the gravitational attraction of their two masses. The electromagnetic attraction is

$$4,200,000,000,000,000,000,000,000,000,000,000,000,000$$

times as strong as the gravitational. (Isaac Asimov, *Isaac Asimov's Book of Facts,* New York: Bell Publishing Company, 1981, p. 106.)

80. A Boeing 747 would not do well in a dogfight; it is too big to maneuver with the lively agility required of an Air Force fighter. It performs best when it flies straight and level, and this is true of the worldwide airline industry as a whole. That industry features annual revenues that approach $200 billion. (T. A. Heppenheimer, *Turbulent Skies: The History of Commercial Aviation,* New York: John Wiley & Sons, 1995, p. 345.)

81. According to *The Wall Street Journal Almanac 1998,* the U.S. airline industry earned record profits in 1996 of $2.82 billion and set records for both the number of passengers and the amount of cargo carried. Passenger traffic increased 7% to 578.4 billion revenue passenger miles, while cargo traffic rose 4.6% to 17.7 billion revenue ton miles.

Write each of the following numbers from the above paragraph using scientific notation.

(a) 2.82 billion   (b) 578.4 billion   (c) 17.7 billion

82. According to Campbell, Mitchell, and Reece in *Biology Concepts and Connections* (Benjamin Cummings, 1994, p. 230), "(t)he amount of DNA in a human cell is about 1000 times greater than the DNA in *E. coli.* Does this mean humans have 1000 times as many genes as the 2000 in *E. coli*? The answer is probably no; the human genome is thought to carry between 50,000 and 100,000 genes, which code for various proteins (as well as for tRNA and rRNA)."

Write each of the following numbers from the above quote using scientific notation.

(a) 1000   (b) 2000   (c) 50,000   (d) 100,000

RELATING CONCEPTS (EXERCISES 83-88)

In Exercises 83–88 we use the letters P and Q to denote the following polynomials:

$$P:\quad x - 3$$
$$Q:\quad x^3 + x^2 - 11x - 3.$$

Work these exercises in order.

83. (a) Evaluate P for $x = 5$.
　(b) Evaluate Q for $x = 5$.
　(c) Find the polynomial represented by $P + Q$.
　(d) Evaluate the polynomial found in part (c) for $x = 5$ and verify that it is equal to the sum of the values you found in parts (a) and (b).

84. (a) Evaluate P for $x = 4$.
　(b) Evaluate Q for $x = 4$.
　(c) Find the polynomial represented by $P - Q$.
　(d) Evaluate the polynomial found in part (c) for $x = 4$ and verify that it is equal to the difference between the values you found in parts (a) and (b).

85. (a) Evaluate P for $x = -3$.
　(b) Evaluate Q for $x = -3$.
　(c) Find the polynomial represented by $P \cdot Q$.
　(d) Evaluate the polynomial found in part (c) for $x = -3$ and verify that it is equal to the product of the values you found in parts (a) and (b).

86. (a) Evaluate P for $x = 6$.
　(b) Evaluate Q for $x = 6$.
　(c) Find the polynomial represented by $\dfrac{Q}{P}$.
　(d) Evaluate the polynomial found in part (c) for $x = 6$ and verify that it is equal to the quotient of the values you found in parts (a) and (b).

87. The concept illustrated in these exercises is this: If we evaluate P for a particular value of x and evaluate Q for the same value of x, and then perform an operation on P and Q, the resulting polynomial will give the appropriate value when evaluated for x. The values used in Exercises 83–86 were chosen arbitrarily. Repeat each exercise for a different value of x.

88. Refer to Exercise 86. Why could we *not* choose 3 as a replacement for x? Is there any other replacement for x that would not be valid?

Did you make the connection between operations with polynomials and operations with those same polynomials evaluated for particular values of the variables?

MIXED REVIEW EXERCISES

Perform the indicated operations. Write with positive exponents only. Assume all variables represent nonzero real numbers.

89. $5^0 + 7^0$

90. $\left(\dfrac{6r^2p}{5}\right)^3$

91. $(12a + 1)(12a - 1)$

92. 2^{-4}

93. $(8^{-3})^4$

94. $\dfrac{2p^3 - 6p^2 + 5p}{2p^2}$

95. $\dfrac{(2m^{-5})(3m^2)^{-1}}{m^{-2}(m^{-1})^2}$

96. $(3k - 6)(2k^2 + 4k + 1)$

97. $\dfrac{r^9 \cdot r^{-5}}{r^{-2} \cdot r^{-7}}$

98. $(2r + 5s)^2$

99. $(-5y^2 + 3y - 11) + (4y^2 - 7y + 15)$

100. $(2r + 5)(5r - 2)$

101. $\dfrac{2y^3 + 17y^2 + 37y + 7}{2y + 7}$

102. $(25x^2y^3 - 8xy^2 + 15x^3y) \div (10x^2y^3)$

103. $(6p^2 - p - 8) - (-4p^2 + 2p - 3)$

104. $\dfrac{5^8}{5^{19}}$

105. $(-7 + 2k)^2$

106. $\left(\dfrac{x}{y^{-3}}\right)^{-4}$

CHAPTER 5 TEST

For each polynomial, combine terms when possible and write the polynomial in descending powers of the variable. Give the degree of the simplified polynomial. Decide whether the simplified polynomial is a monomial, binomial, trinomial, or none of these.

1. $5x^2 + 8x - 12x^2$

2. $13n^3 - n^2 + n^4 + 3n^4 - 9n^2$

3. Use the table to complete a set of ordered pairs that lie on the graph of $y = 2x^2 - 4$. Then graph the equation.

x	y
-2	
-1	
0	
1	
2	

Perform the indicated operations.

4. $(2y^2 - 8y + 8) + (-3y^2 + 2y + 3) - (y^2 + 3y - 6)$

5. $(-9a^3b^2 + 13ab^5 + 5a^2b^2) - (6ab^5 + 12a^3b^2 + 10a^2b^2)$

6. Subtract.

$9t^3 - 4t^2 + 2t + 2$
$\underline{9t^3 + 8t^2 - 3t - 6}$

7. $3x^2(-9x^3 + 6x^2 - 2x + 1)$

8. $(t - 8)(t + 3)$

9. $(4x + 3y)(2x - y)$

10. $(5x - 2y)^2$

11. $(10v + 3w)(10v - 3w)$

12. $(2r - 3)(r^2 + 2r - 5)$

13. What polynomial expression represents the area of this square?

$3x + 9$

Evaluate each expression.

14. 5^{-4}

15. $(-3)^0 + 4^0$

16. $4^{-1} + 3^{-1}$

17. Use the rules for exponents to simplify $\dfrac{(3x^2y)^2(xy^3)^2}{(xy)^3}$. Assume x and y are nonzero.

Simplify, and write the answer using only positive exponents. Assume that variables represent nonzero numbers.

18. $\dfrac{8^{-1} \cdot 8^4}{8^{-2}}$

19. $\dfrac{(x^{-3})^{-2}(x^{-1}y)^2}{(xy^{-2})^2}$

20. Do you agree or disagree with the following statement?

$$3^{-4} \text{ represents a negative number.}$$

Justify your answer.

Perform each division.

21. $\dfrac{8y^3 - 6y^2 + 4y + 10}{2y}$

22. $(-9x^2y^3 + 6x^4y^3 + 12xy^3) \div (3xy)$

23. $(3x^3 - x + 4) \div (x - 2)$

24. (a) Write 45,000,000,000 using scientific notation.
   (b) Write 3.6×10^{-6} without using exponents.

   (c) Write the quotient without using scientific notation: $\dfrac{9.5 \times 10^{-1}}{5 \times 10^3}$.

 25. According to an article in *Aviation Week and Space Technology* (March 23, 1998), the MD-17 jet has the capacity to carry more than a 170,000-pound payload. Suppose we consider a group of 1000 such jets.
   (a) Write the numbers 170,000 and 1000 using scientific notation.
   (b) Multiply these two numbers to find the total payload that this group can carry. Express your answer using scientific notation.

CUMULATIVE REVIEW EXERCISES CHAPTERS 1–5

Write each fraction in lowest terms.

1. $\dfrac{28}{16}$

2. $\dfrac{55}{11}$

Perform each operation.

3. $\dfrac{2}{3} + \dfrac{1}{8}$

4. $\dfrac{7}{4} - \dfrac{9}{5}$

5. A contractor installs toolsheds. Each requires $1\dfrac{1}{4}$ cubic yards of concrete. How much concrete would be needed for 25 sheds?

6. A retailer has $34,000 invested in her business. She finds that last year she earned 5.4% on this investment. How much did she earn?

7. List all positive integer factors of 45.

8. If $a < 0$ and $b > 0$, what is the sign of $b - a$?

Find the value of each expression if $x = -2$ and $y = 4$.

9. $\dfrac{4x - 2y}{x + y}$

10. $x^3 - 4xy$

Perform the indicated operations.

11. $\dfrac{(-13 + 15) - (3 + 2)}{6 - 12}$

12. $-7 - 3[2 + (5 - 8)]$

Decide which property justifies each statement.

13. $(9 + 2) + 3 = 9 + (2 + 3)$

14. $6(4 + 2) = 6(4) + 6(2)$

15. Use the bar graph to find a signed number that represents the change in the number of airline passenger deaths
  (a) from 1994 to 1995
  (b) from 1995 to 1996.

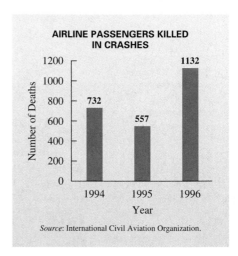

AIRLINE PASSENGERS KILLED IN CRASHES

Source: International Civil Aviation Organization.

16. Simplify the expression $-3(2x^2 - 8x + 9) - (4x^2 + 3x + 2)$.

Solve each equation.

17. $2 - 3(t - 5) = 4 + t$

18. $2(5h + 1) = 10h + 4$

19. $d = rt$ for r

20. $\dfrac{x}{5} = \dfrac{x - 2}{7}$

21. $\dfrac{1}{3}p - \dfrac{1}{6}p = -2$

22. $.05x + .15(50 - x) = 5.50$

23. $4 - (3x + 12) = (2x - 9) - (5x - 1)$

Solve each problem.

24. A 1-ounce mouse takes about 16 times as many breaths as does a 3-ton elephant. If the two animals take a combined total of 170 breaths per minute, how many breaths does each take during that time period? (*Source:* Christopher McGowan, *Dinosaurs, Spitfires, and Sea Dragons,* Harvard University Press, 1991.)

25. If a number is subtracted from 8 and this difference is tripled, the result is 3 times the number. Find this number, and you will learn how many times a dolphin rests during a 24-hour period.

26. In 1995 and 1996, there were the same number of aircraft accidents involving passenger fatalities throughout worldwide scheduled air services. In 1994, there were 2 more than in each of the other years. How many accidents were there in each of these years if there was a total of 68 altogether? (*Source:* International Civil Aviation Organization.)

Solve each inequality.

27. $-8x \leq -80$ **28.** $-2(x + 4) > 3x + 6$ **29.** $-3 \leq 2x + 5 < 9$

Solve the problem.

30. One side of a triangle is twice as long as a second side. The third side of the triangle is 17 feet long. The perimeter of the triangle cannot be more than 50 feet. Find the longest possible values for the other two sides of the triangle.

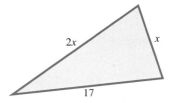

31. Complete the table of values for $-2x + 4y = 8$.

x	0		2		4
y		0		1	

32. Graph $y = -3x + 6$.

33. Graph $y = (x + 4)^2$, using the x-values $-6, -5, -4, -3$, and -2 to obtain a set of points.

Perform the indicated operations.

34. $(7x^3 - 12x^2 - 3x + 8) + (6x^2 + 4) - (-4x^3 + 8x^2 - 2x - 2)$

35. $6x^5(3x^2 - 9x + 10)$ **36.** $(7x + 4)(9x + 3)$

37. $(5x + 8)^2$ **38.** $\dfrac{14x^3 - 21x^2 + 7x}{7x}$

39. $\dfrac{y^3 - 3y^2 + 8y - 6}{y - 1}$

Evaluate each expression.

40. $4^{-1} + 3^0$ **41.** $2^{-4} \cdot 2^5$ **42.** $\dfrac{8^{-5} \cdot 8^7}{8^2}$

43. Write with positive exponents only: $\dfrac{(a^{-3}b^2)^2}{(2a^{-4}b^{-3})^{-1}}$.

44. Write in scientific notation: 34,500.

45. Write without exponents: 5.36×10^{-7}.

6 Factoring and Applications

In the early days of the American frontier, it took four horses 75 days to haul one wagonload of goods a thousand miles. By the late 1800s, roads, canals, and railways connected "island" communities, greatly expediting the movement of people and goods. Whereas a stagecoach traveled 50 miles in a day, trains now covered 50 miles in an *hour*—over 700 miles per day. The vehicle that really changed twentieth-century American travel, however, was the automobile. Today emerging technologies are further transforming modern transportation systems. Just one example is the global network of computers moving millions of bits of data per second.

Much of this emerging technology depends on mathematics. Understanding mathematics makes it possible to model data using functions. In Chapter 5, we used polynomials of degree 2 to model data. Now, in this chapter, we introduce a method for solving polynomial equations of degree 2 called *quadratic equations,* and we give further examples of them as models. For instance, average fuel economy trends in miles per gallon for the automotive industry are closely approximated by the quadratic equation

$$y = -.04x^2 + .93x + 21,$$

where x represents the year. The years are coded so that $x = 0$ represents 1978, $x = 2$ represents 1980, $x = 4$ represents 1982, and so on. This equation was developed from the data in the table.

Fuel Economy

Year	1978	1980	1982	1984	1986	1988	1990	1992	1994	1996
Miles per Gallon	19.9	23.1	25.1	25.0	25.9	26.0	25.4	25.1	24.7	24.9

Source: National Highway Traffic Safety Administration.

Substituting 0 for x in the equation gives $y = 21$, the approximate average miles per gallon in 1978. For 1996, substituting $1996 - 1978 = 18$ for x in the equation gives $y = 24.78$ as the approximate average miles per gallon. Which of these is a closer approximation to the data in the table? We will return to this transportation model in Section 6.6, Example 4.

Visit our Web site at www.LialAlgebra.com

6.1 The Greatest Common Factor; Factoring by Grouping

Recall from Chapter 1 that to **factor** means to write a quantity as a product. That is, factoring is the opposite of multiplying. For example,

Multiplying	*Factoring*
$6 \cdot 2 = 12,$	$12 = 6 \cdot 2.$
Factors Product	Product Factors

Other factored forms of 12 are

$$(-6)(-2), \quad 3 \cdot 4, \quad (-3)(-4), \quad 12 \cdot 1, \quad \text{and} \quad (-12)(-1).$$

More than two factors may be used, so another factored form of 12 is $2 \cdot 2 \cdot 3$. The positive integer factors of 12 are

$$1, 2, 3, 4, 6, 12.$$

OBJECTIVE 1 Find the greatest common factor of a list of terms. An integer that is a factor of two or more integers is called a **common factor** of those integers. For example, 6 is a common factor of 18 and 24 since 6 is a factor of both 18 and 24. Other common factors of 18 and 24 are 1, 2, and 3. The **greatest common factor** of a list of integers is the largest common factor of those integers. Thus, 6 is the greatest common factor of 18 and 24, since it is the largest of the common factors of these numbers.

 Factors of a number are also divisors of the number. The greatest common factor is actually the same as the greatest common divisor.

CONNECTIONS

There are many rules for deciding what numbers divide into a given number. Here are some especially useful divisibility rules for small numbers. It is surprising how many people do not know them.

A Whole Number Divisible by:	Must Have the Following Property:
2	Ends in 0, 2, 4, 6, or 8
3	Sum of its digits is divisible by 3
4	Last two digits form a number divisible by 4
5	Ends in 0 or 5
6	Divisible by both 2 and 3
8	Last three digits form a number divisible by 8
9	Sum of its digits is divisible by 9
10	Ends in 0

Recall from Chapter 1 that a prime number has only itself and 1 as factors. In Section 1.1 we factored numbers into prime factors. This is the first step in finding the greatest common factor of a list of numbers. We find the greatest common factor (GCF) of a list of numbers as follows.

Finding the Greatest Common Factor (GCF)

Step 1 **Factor.** Write each number in prime factored form.

Step 2 **List common factors.** List each prime number that is a factor of every number in the list.

Step 3 **Choose smallest exponents.** Use as exponents on the common prime factors the *smallest* exponent from the prime factored forms. (If a prime does not appear in one of the prime factored forms, it cannot appear in the greatest common factor.)

Step 4 **Multiply.** Multiply the primes from Step 3. If there are no primes left after Step 3, the greatest common factor is 1.

E X A M P L E 1 Finding the Greatest Common Factor for Numbers

Find the greatest common factor for each list of numbers.

(a) 30, 45

First write each number in prime factored form.

$$30 = 2 \cdot 3 \cdot \mathbf{5}$$
$$45 = 3 \cdot 3 \cdot \mathbf{5}$$

Now, take each prime the *least* number of times it appears in all the factored forms. There is no 2 in the prime factored form of 45, so there will be no 2 in the greatest common factor. The least number of times 3 appears in all the factored forms is 1, and the least number of times 5 appears is also 1. From this, the GCF is

$$\mathbf{3^1} \cdot \mathbf{5^1} = 15.$$

(b) 72, 120, 432

Find the prime factored form of each number.

$$72 = 2 \cdot 2 \cdot 2 \cdot \mathbf{3} \cdot 3$$
$$120 = 2 \cdot 2 \cdot 2 \cdot \mathbf{3} \cdot 5$$
$$432 = 2 \cdot 2 \cdot 2 \cdot 2 \cdot \mathbf{3} \cdot 3 \cdot 3$$

The least number of times 2 appears in all the factored forms is 3, and the least number of times 3 appears is 1. There is no 5 in the prime factored form of either 72 or 432, so the GCF is

$$2^3 \cdot 3 = 24.$$

(c) 10, 11, 14

Write the prime factored form of each number.

$$10 = 2 \cdot 5$$
$$11 = 11$$
$$14 = 2 \cdot 7$$

There are no primes common to all three numbers, so the GCF is 1.

The greatest common factor can also be found for a list of variable terms. For example, the terms x^4, x^5, x^6, and x^7 have x^4 as the greatest common factor because each of these terms can be written with x^4 as a factor.

$$x^4 = 1 \cdot x^4, \qquad x^5 = x \cdot x^4, \qquad x^6 = x^2 \cdot x^4, \qquad x^7 = x^3 \cdot x^4$$

 The exponent on a variable in the GCF is the *smallest* exponent that appears in the factors.

E X A M P L E   2   Finding the Greatest Common Factor for Variable Terms

Find the greatest common factor for each list of terms.

(a) $21m^7$, $-18m^6$, $45m^8$, $-24m^5$

$$21m^7 = 3 \cdot 7 \cdot \boldsymbol{m^7}$$
$$-18m^6 = -1 \cdot 2 \cdot 3^2 \cdot \boldsymbol{m^6}$$
$$45m^8 = 3^2 \cdot 5 \cdot \boldsymbol{m^8}$$
$$-24m^5 = -1 \cdot 2^3 \cdot 3 \cdot \boldsymbol{m^5}$$

First, 3 is the greatest common factor of the coefficients 21, -18, 45, and -24. The smallest exponent on m is 5, so the GCF of the terms is $3m^5$.

(b) x^4y^2, x^7y^5, x^3y^7, y^{15}

$$x^4y^2 = x^4 \cdot \boldsymbol{y^2}$$
$$x^7y^5 = x^7 \cdot \boldsymbol{y^5}$$
$$x^3y^7 = x^3 \cdot \boldsymbol{y^7}$$
$$y^{15} = \boldsymbol{y^{15}}$$

There is no x in the last term, y^{15}, so x will not appear in the greatest common factor. There is a y in each term, however, and 2 is the smallest exponent on y. The GCF is y^2.

O B J E C T I V E   2   Factor out the greatest common factor. We use the idea of a greatest common factor to write a polynomial (a sum) in factored form as a product. For example, the polynomial

$$3m + 12$$

has two terms, $3m$ and 12. The greatest common factor for these two terms is 3. We can write $3m + 12$ so that each term is a product with 3 as one factor.

$$3m + 12 = 3 \cdot m + 3 \cdot 4$$

Now use the distributive property.

$$3m + 12 = 3 \cdot m + 3 \cdot 4 = 3(m + 4)$$

The factored form of $3m + 12$ is $3(m + 4)$. This process is called **factoring out the greatest common factor.**

The polynomial $3m + 12$ is *not* in factored form when written as

$$3 \cdot m + 3 \cdot 4.$$

 The *terms* are factored, but the polynomial is not. The factored form of $3m + 12$ is the *product*

$$3(m + 4).$$

EXAMPLE 3   Factoring Out the Greatest Common Factor

Factor out the greatest common factor.

(a) $20m^5 + 10m^4 + 15m^3$

The GCF for the terms of this polynomial is $5m^3$.

$$20m^5 + 10m^4 + 15m^3 = (5m^3)(4m^2) + (5m^3)(2m) + (5m^3)3$$
$$= 5m^3(4m^2 + 2m + 3)$$

Check by multiplying $5m^3$ and $4m^2 + 2m + 3$. You should get the original polynomial.

(b) $x^5 + x^3 = (x^3)x^2 + (x^3)1 = x^3(x^2 + 1)$

(c) $20m^7p^2 - 36m^3p^4 = 4m^3p^2(5m^4 - 9p^2)$

(d) $a(a + 3) + 4(a + 3)$

The binomial $a + 3$ is the greatest common factor here.

$$a(a + 3) + 4(a + 3) = (a + 3)(a + 4)$$

 Be sure to include the 1 in a problem like Example 3(b). Always check that the factored form can be multiplied out to give the original polynomial.

OBJECTIVE 3   Factor by grouping. Common factors are used in **factoring by grouping,** as explained in the next example.

EXAMPLE 4   Factoring by Grouping

Factor by grouping.

(a) $2x + 6 + ax + 3a$

The first two terms have a common factor of 2, and the last two terms have a common factor of a.

$$2x + 6 + ax + 3a = 2(x + 3) + a(x + 3)$$

The expression is still not in factored form because it is the *sum* of two terms. Now, however, $x + 3$ is a common factor and can be factored out.

$$2x + 6 + ax + 3a = 2(x + 3) + a(x + 3) = (x + 3)(2 + a)$$

The final result is in factored form because it is a *product*. Note that the goal in factoring by grouping is to get a common factor, $x + 3$ here, so that the last step is possible.

Same

(b) $m^2 + 6m + 2m + 12 = m(m + 6) + 2(m + 6)$

$$= (m + 6)(m + 2)$$

(c) $6xy - 21x - 8y + 28 = 3x(2y - 7) - 4(2y - 7) = (2y - 7)(3x - 4)$

Must be same

Since the quantities in parentheses in the second step must be the same, it was necessary here to factor out -4 rather than 4.

 Use negative signs carefully when grouping, as in Example 4(c). Otherwise, sign errors may result.

Use these steps when factoring four terms by grouping.

Factoring by Grouping

Step 1   **Group terms.**   Collect the terms into two groups so that each group has a common factor.

Step 2   **Factor within groups.**   Factor out the greatest common factor from each group.

Step 3   **Factor the entire polynomial.**   Factor a common binomial factor from the results of Step 2.

Step 4   **If necessary, rearrange terms.**   If Step 2 does not result in a common binomial factor, try a different grouping.

E X A M P L E   5   Rearranging Terms Before Factoring by Grouping

Factor by grouping.

(a) $10x^2 - 12y^2 + 15xy - 8xy$

Factoring out the common factor of 2 from the first two terms and the common factor of xy terms from the last two terms gives

$$10x^2 - 12y^2 + 15xy - 8xy = 2(5x^2 - 6y^2) + xy(15 - 8).$$

This did not lead to a common factor, so we try rearranging the terms. There is usually more than one way to do this. Let's try

$$10x^2 - 8xy - 12y^2 + 15xy,$$

grouping the first two terms and the last two terms as follows.

$$10x^2 - 8xy - 12y^2 + 15xy = 2x(5x - 4y) + 3y(-4y + 5x)$$

$$= 2x(5x - 4y) + 3y(5x - 4y)$$

$$= (5x - 4y)(2x + 3y)$$

(b) $2xy + 12 - 3y - 8x$

We need to rearrange these terms to get two groups that each have a common factor. Trial and error suggests the following grouping.

$$2xy + 12 - 3y - 8x = (2xy - 3y) + (-8x + 12)$$
$$= y(2x - 3) - 4(2x - 3) \qquad \text{Factor each group.}$$
$$= (2x - 3)(y - 4) \qquad \text{Factor out the common binomial factor.}$$

6.1 EXERCISES

1. Is 3 the greatest common factor of 18, 24, and 42? If not, what is?

2. Is pq the greatest common factor of pq^2, p^2, and p^2q^2? If not, what is?

3. Factoring is the opposite of what operation?

4. How can you check your answer when you factor a polynomial?

5. Give an example of three numbers whose greatest common factor is 5.

6. Explain how to find the greatest common factor of a list of terms. Use examples.

Find the greatest common factor for each list of terms. See Examples 1 and 2.

7. $16y$, 24
8. $18w$, 27
9. $30x^3$, $40x^6$, $50x^7$
10. $60z^4$, $70z^8$, $90z^9$
11. $12m^3n^2$, $18m^5n^4$, $36m^8n^3$
12. $25p^5r^7$, $30p^7r^8$, $50p^5r^3$
13. $-x^4y^3$, $-xy^2$
14. $-a^4b^5$, $-a^3b$
15. $42ab^3$, $-36a$, $90b$, $-48ab$
16. $45c^3d$, $75c$, $90d$, $-105cd$

An expression is factored when it is written as a product, not a sum. Which of the following are not factored?

17. $2k^2(5k)$
18. $2k^2(5k + 1)$
19. $2k^2 + (5k + 1)$
20. $(2k^2 + 1)(5k + 1)$

21. Is $-xy$ a common factor of $-x^4y^3$ and $-xy^2$? If so, what is the other factor that when multiplied by $-xy$ gives $-x^4y^3$?

22. Is $-a^5b^2$ a common factor of $-a^4b^5$ and $-a^3b$?

Complete each factoring.

23. $12 = 6($ $)$
24. $18 = 9($ $)$
25. $3x^2 = 3x($ $)$
26. $8x^3 = 8x($ $)$
27. $9m^4 = 3m^2($ $)$
28. $12p^5 = 6p^3($ $)$
29. $-8z^9 = -4z^5($ $)$
30. $-15k^{11} = -5k^8($ $)$
31. $6m^4n^5 = 3m^3n($ $)$
32. $27a^3b^2 = 9a^2b($ $)$
33. $-14x^4y^3 = 2xy($ $)$
34. $-16m^3n^3 = 4mn^2($ $)$

Factor out the greatest common factor. See Example 3.

35. $12y - 24$
36. $18p + 36$
37. $10a^2 - 20a$
38. $15x^3 - 30x^2$

39. $65y^{10} + 35y^6$

40. $100a^5 + 16a^3$

41. $11w^3 - 100$

42. $13z^5 - 80$

43. $9m^2n^3 + 24m^2n^2$

44. $19p^2y - 38p^2y^3$

45. $13y^8 + 26y^4 - 39y^2$

46. $5x^5 + 25x^4 - 20x^3$

47. $45q^4p^5 + 36qp^6 + 81q^2p^3$

48. $125a^3z^5 + 60a^4z^4 - 85a^5z^2$

49. $a^5 + 2a^3b^2 - 3a^5b^2 + 4a^4b^3$

50. $x^6 + 5x^4y^3 - 6xy^4 + 10xy$

51. $c(x + 2) - d(x + 2)$

52. $r(5 - x) + t(5 - x)$

53. $m(m + 2n) + n(m + 2n)$

54. $3p(1 - 4p) - 2q(1 - 4p)$

Students often have difficulty when factoring by grouping because they are not able to tell when the polynomial is indeed factored. For example,

$$5y(2x - 3) + 8t(2x - 3)$$

is not in factored form, because it is the *sum* of two terms, $5y(2x - 3)$ and $8t(2x - 3)$. However, because $2x - 3$ is a common factor of these two terms, the expression can now be factored as

$$(2x - 3)(5y + 8t).$$

The factored form is a *product* of two factors, $2x - 3$ and $5y + 8t$.

Determine whether each expression is in factored form or is not in factored form. If it is not in factored form, factor it if possible.

55. $8(7t + 4) + x(7t + 4)$

56. $3r(5x - 1) + 7(5x - 1)$

57. $(8 + x)(7t + 4)$

58. $(3r + 7)(5x - 1)$

59. $18x^2(y + 4) + 7(y + 4)$

60. $12k^3(s - 3) + 7(s + 3)$

61. Tell why it is not possible to factor the expression in Exercise 60.

62. Summarize the method of factoring a polynomial with four terms by grouping. Give an example.

Factor by grouping. See Examples 4 and 5.

63. $p^2 + 4p + 3p + 12$

64. $m^2 + 2m + 5m + 10$

65. $a^2 - 2a + 5a - 10$

66. $y^2 - 6y + 4y - 24$

67. $7z^2 + 14z - az - 2a$

68. $5m^2 + 15mp - 2mp - 6p^2$

69. $18r^2 + 12ry - 3xr - 2xy$

70. $8s^2 - 4st + 6sy - 3yt$

71. $3a^3 + 3ab^2 + 2a^2b + 2b^3$

72. $4x^3 + 3x^2y + 4xy^2 + 3y^3$

73. $1 - a + ab - b$

74. $6 - 3x - 2y + xy$

75. $16m^3 - 4m^2p^2 - 4mp + p^3$

76. $10t^3 - 2t^2s^2 - 5ts + s^3$

77. $5m + 15 - 2mp - 6p$

78. $y^2 - 3y - xy + 3x$

79. $18r^2 + 12ry - 3ry - 2y^2$

80. $3a^3 + 3ab^2 + 2a^2b + 2b^3$

81. $a^5 + 2a^5b - 3 - 6b$

82. $a^2b - 4a - ab^4 + 4b^3$

▌ RELATING CONCEPTS   (EXERCISES 83-88)

In most cases, the choice of which pairs of terms to group when factoring by grouping can be made in several ways.

Work Exercises 83–88 in order *to see how this applies to the polynomial in Example 5(b).*

83. Start with the polynomial in Example 5(b), $2xy + 12 - 3y - 8x$, and rearrange the terms as follows: $2xy - 8x + (-3y) + 12$. What properties from Section 1.7 allow us to do this?

RELATING CONCEPTS (EXERCISES 83-88)

84. Group the first pair of terms in the rearranged polynomial. What is the greatest common factor of this pair?

85. Now group the second pair of terms in the rearranged polynomial. Is -3 a common factor of this second pair?

86. Factor the greatest common factor from the first pair and -3 from the second pair.

87. Is your result from Exercise 86 in factored form? If not, why?

88. If your answer to Exercise 87 is *no,* factor the polynomial. Is it the same result as the one shown in Example 5(b) in this section?

Did you make the connection that although there is more than one way to factor a polynomial by grouping, you should get the same answer?

89. Refer to Exercise 73. The answer given in the back of the book is $(1 - a)(1 - b)$. A student factored this same polynomial and got the result $(a - 1)(b - 1)$.
(a) Is this student's answer correct?
(b) If your answer to part (a) is *yes,* explain why these two seemingly different answers are both acceptable.

90. A student factored $18x^3y^2 + 9xy$ as $9xy(2x^2y)$. Is this correct? If not, explain the error and factor correctly.

6.2 Factoring Trinomials

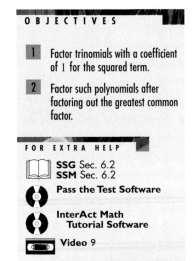

OBJECTIVES

1 Factor trinomials with a coefficient of 1 for the squared term.

2 Factor such polynomials after factoring out the greatest common factor.

FOR EXTRA HELP

📖 **SSG** Sec. 6.2
SSM Sec. 6.2

💿 **Pass the Test Software**

💿 **InterAct Math Tutorial Software**

📼 **Video 9**

Using FOIL, the product of the polynomials $k - 3$ and $k + 1$ is

$$(k - 3)(k + 1) = k^2 - 2k - 3.$$

Now suppose we are given the polynomial $k^2 - 2k - 3$ and want to rewrite it as the product $(k - 3)(k + 1)$. This product is called the **factored form** of $k^2 - 2k - 3$, and the process of finding the factored form is called **factoring.**

OBJECTIVE 1 Factor trinomials with a coefficient of 1 for the squared term. When factoring polynomials with integer coefficients, we use only integers for the numerical factors. For example, $x^2 + 5x + 6$ can be factored by finding integers a and b such that

$$x^2 + 5x + 6 = (x + a)(x + b).$$

To find these integers a and b, first multiply the two factors on the right-hand side of the equation:

$$(x + a)(x + b) = x^2 + ax + bx + ab.$$

By the distributive property,

$$x^2 + ax + bx + ab = x^2 + (a + b)x + ab.$$

Comparing this result with $x^2 + 5x + 6$ shows that we must find integers a and b having a sum of 5 and a product of 6.

$$x^2 + \mathbf{5}x + \mathbf{6} = x^2 + (a + b)x + \boldsymbol{ab}$$

Sum of a and b is 5. ⟶ Product of a and b is 6.

Since many pairs of integers have a sum of 5, it is best to begin by listing those pairs of integers whose product is 6. Both 5 and 6 are positive, so we consider only pairs in which both integers are positive.

Product	*Sum*
$1 \cdot 6 = 6$	$1 + 6 = 7$
$2 \cdot 3 = \mathbf{6}$	$2 + 3 = \mathbf{5}$ Sum is 5.

Both pairs have a product of 6, but only the pair 2 and 3 has a sum of 5. So 2 and 3 are the required integers, and

$$x^2 + 5x + 6 = (x + \mathbf{2})(x + \mathbf{3}).$$

Check by multiplying the binomials using FOIL. Make sure that the sum of the outer and inner products produces the correct middle term.

$$(x + 2)(x + 3) = x^2 + \mathbf{5}x + 6$$

$$\begin{array}{c} 2x \\ \underline{3x} \\ 5x \end{array} \quad \text{Add.}$$

This method of factoring can be used only for trinomials that have 1 as the coefficient of the squared term. Methods for factoring other trinomials are given in the next section.

┌ **E X A M P L E   1**   Factoring a Trinomial with All Terms Positive

Factor $m^2 + 9m + 14$.

Look for two integers whose product is 14 and whose sum is 9. List the pairs of integers whose product is 14. Then examine the sums. Again, only positive integers are needed because all signs are positive.

14, 1	$14 + 1 = 15$
7, 2	$7 + 2 = \mathbf{9}$ Sum is 9.

From the list, 7 and 2 are the required integers, since $7 \cdot 2 = 14$ and $7 + 2 = 9$. Thus $m^2 + 9m + 14 = (m + 2)(m + 7)$.

 In Example 1, the answer also could have been written $(m + 7)(m + 2)$. Because of the commutative property of multiplication, the order of the factors does not matter.

The trinomials in the examples so far had all positive terms. If a trinomial has one or more negative terms, both positive and negative factors must be considered.

┌─
E X A M P L E 2 Factoring a Trinomial with Two Negative Terms

Factor $p^2 - 2p - 15$.

Find two integers whose product is -15 and whose sum is -2. If these numbers do not come to mind right away, we can find them (if they exist) by listing all the pairs of integers whose product is -15. Because the last term, -15, is negative, we need pairs of integers with different signs.

$$15, -1 \qquad 15 + (-1) = 14$$
$$5, -3 \qquad 5 + (-3) = 2$$
$$-15, 1 \qquad -15 + 1 = -14$$
$$-5, 3 \qquad -5 + 3 = -2 \qquad \text{Sum is } -2.$$

The necessary integers are -5 and 3, and

$$p^2 - 2p - 15 = (p - 5)(p + 3).$$
─ ■

┌─
E X A M P L E 3 Deciding Whether a Polynomial Is Prime

(a) Factor $x^2 - 5x + 12$.

List all pairs of integers whose product is 12. Since the middle term is negative and the last term is positive, we need pairs with both numbers negative. Then examine the sums.

$$-12, -1 \qquad -12 + (-1) = -13$$
$$-6, -2 \qquad -6 + (-2) = -8$$
$$-3, -4 \qquad -3 + (-4) = -7$$

None of the pairs of integers has a sum of -5. Because of this, the trinomial $x^2 - 5x + 12$ *cannot be factored using only integer factors.* A polynomial that cannot be factored using only integer factors is called a **prime polynomial.**

(b) $k^2 - 8k + 11$

There is no pair of integers whose product is 11 and whose sum is -8, so $k^2 - 8k + 11$ is a prime polynomial.
─ ■

The procedure for factoring a trinomial of the form $x^2 + bx + c$ is summarized here.

Factoring $x^2 + bx + c$

Find two integers whose product is c and whose sum is b.

1. Both integers must be positive if b and c are positive.
2. Both integers must be negative if c is positive and b is negative.
3. One integer must be positive and one must be negative if c is negative.

┌─
E X A M P L E 4 Factoring a Trinomial with Two Variables

Factor $z^2 - 2bz - 3b^2$.

To factor $z^2 - 2bz - 3b^2$, look for two expressions whose product is $-3b^2$ and whose sum is $-2b$. The expressions are $-3b$ and b, with

$$z^2 - 2bz - 3b^2 = (z - 3b)(z + b).$$
─ ■

OBJECTIVE Factor such polynomials after factoring out the greatest common factor. The trinomial in the next example does not have a coefficient of 1 for the squared term. (In fact, there is no squared term.) However, there may be a common factor.

E X A M P L E 5   Factoring a Trinomial with a Common Factor

Factor $4x^5 - 28x^4 + 40x^3$.

First, factor out the greatest common factor, $4x^3$.

$$4x^5 - 28x^4 + 40x^3 = 4x^3(x^2 - 7x + 10)$$

Now factor $x^2 - 7x + 10$. The integers -5 and -2 have a product of 10 and a sum of -7. The complete factored form is

$$4x^5 - 28x^4 + 40x^3 = 4x^3(x - 5)(x - 2).$$

 When factoring, always look for a common factor first. Remember to include the common factor as part of the answer. As a check, multiplying out the factored form should always give the original polynomial.

6.2 EXERCISES

1. In factoring a trinomial in x as $(x + a)(x + b)$, what must be true of a and b if the coefficient of the last term of the trinomial is negative?

2. In Exercise 1, what must be true of a and b if the coefficient of the last term is positive?

3. In your own words, explain the meaning of a *prime polynomial*.

4. A teacher asked a class to factor $m^3 + 3m^2 + 2m$. One student did not understand why her answer, $(m + 1)(m + 2)$ was incorrect. Explain her mistake, and give the correct answer.

In Exercises 5–8, list all pairs of integers with the given product. Then find the pair whose sum is given.

5. Product: 48   Sum: -19

6. Product: 48   Sum: 14

7. Product: -24 Sum: -5

8. Product: -36 Sum: -16

9. Which one of the following is the correct factored form of $x^2 - 12x + 32$?
   (a) $(x - 8)(x + 4)$ (b) $(x + 8)(x - 4)$
   (c) $(x - 8)(x - 4)$ (d) $(x + 8)(x + 4)$

10. Explain the steps you would use to factor $2x^3 + 8x^2 - 10x$.

Complete the following. See Examples 1–3.

11. $p^2 + 11p + 30 = (p + 5)(\quad)$ 12. $x^2 + 10x + 21 = (x + 7)(\quad)$

13. $x^2 + 15x + 44 = (x + 4)(\quad)$ 14. $r^2 + 15r + 56 = (r + 7)(\quad)$

15. $x^2 - 9x + 8 = (x - 1)(\quad)$ 16. $t^2 - 14t + 24 = (t - 2)(\quad)$

17. $y^2 - 2y - 15 = (y + 3)(\quad)$ 18. $t^2 - t - 42 = (t + 6)(\quad)$

19. $x^2 + 9x - 22 = (x - 2)(\quad)$ 20. $x^2 + 6x - 27 = (x - 3)(\quad)$

21. $y^2 - 7y - 18 = (y + 2)(\quad)$ 22. $y^2 - 2y - 24 = (y + 4)(\quad)$

RELATING CONCEPTS (EXERCISES 23-30)

To check your factoring of a trinomial, remember that when the factors are multiplied, the product must include *every* term of the original trinomial. In a trinomial such as $x^2 + x - 12$, students often factor so that the first and third terms are correct, but the middle term is incorrect, differing only in the sign.

Work Exercises 23–30 in order to see how this problem can be avoided.

23. Add the pair of numbers: -3 and 4.

24. Add the pair of numbers: 3 and -4.

25. In Exercise 24, we added the *opposites* of the numbers in Exercise 23. How does the sum in Exercise 24 compare with the sum in Exercise 23?

26. Consider this factored form, given for $x^2 + x - 12$: $(x + 3)(x - 4)$. Multiply the two binomials. Is this the correct factored form? Why or why not?

27. Consider this factored form, given for $x^2 + x - 12$: $(x - 3)(x + 4)$. Multiply the two binomials. Is this the correct factored form? Why or why not?

28. In Exercises 26 and 27, one factored form is correct and one is not. For the one that is not, how does the sign of the middle term of the product compare with the signs of the second terms of the correct factored form?

29. Compare the two factored forms shown in Exercises 26 and 27, and use the result of Exercise 28 to complete the following statement: When I factor a trinomial into a product of binomials, and the middle term of the product is different only in sign, I should _____ in order to obtain the correct factored form.

30. Given that the factored form $(x + 5)(x - 3)$ is incorrect only in the sign of the middle term of the product for a particular trinomial, what would be the correct factored form?

Did you make the connection that if only the sign of the middle term is incorrect when you check the product of the binomial factors, then only the signs of the second terms in the binomials need to be changed?

Factor completely. If the polynomial cannot be factored, write prime. *See Examples 1–3.*

31. $y^2 + 9y + 8$ **32.** $a^2 + 9a + 20$ **33.** $b^2 + 8b + 15$

34. $x^2 + 6x + 8$ **35.** $m^2 + m - 20$ **36.** $p^2 + 4p - 5$

37. $y^2 - 8y + 15$ **38.** $y^2 - 6y + 8$ **39.** $x^2 + 4x + 5$

40. $t^2 + 11t + 12$ **41.** $t^2 - 8t + 16$ **42.** $s^2 - 10s + 25$

43. $r^2 - r - 30$ **44.** $q^2 - q - 42$ **45.** $n^2 - 12n - 35$

46. $x^2 - 4x + 12$

Factor completely. See Examples 4 and 5.

47. $r^2 + 3ra + 2a^2$ **48.** $x^2 + 5xa + 4a^2$

49. $t^2 - tz - 6z^2$ **50.** $a^2 - ab - 12b^2$

51. $x^2 + 4xy + 3y^2$ **52.** $p^2 + 9pq + 8q^2$

53. $v^2 - 11vw + 30w^2$ **54.** $v^2 - 11vx + 24x^2$

55. $4x^2 + 12x - 40$ **56.** $5y^2 - 5y - 30$

57. $2t^3 + 8t^2 + 6t$ **58.** $3t^3 + 27t^2 + 24t$

59. $2x^6 + 8x^5 - 42x^4$ **60.** $4y^5 + 12y^4 - 40y^3$

61. $m^3n - 10m^2n^2 + 24mn^3$ **62.** $y^3z + 3y^2z^2 - 54yz^3$

63. Use the FOIL method from Section 5.3 to show that $(2x + 4)(x - 3) = 2x^2 - 2x - 12$. If you are asked to completely factor $2x^2 - 2x - 12$, why would it be incorrect to give $(2x + 4)(x - 3)$ as your answer?

64. If you are asked to completely factor the polynomial $3x^2 + 9x - 12$, why would it be incorrect to give $(x - 1)(3x + 12)$ as your answer?

Use a combination of the factoring methods discussed in this section to factor each polynomial.

65. $a^5 + 3a^4b - 4a^3b^2$ **66.** $m^3n - 2m^2n^2 - 3mn^3$ **67.** $y^3z + y^2z^2 - 6yz^3$

68. $k^7 - 2k^6m - 15k^5m^2$ **69.** $z^{10} - 4z^9y - 21z^8y^2$ **70.** $x^9 + 5x^8w - 24x^7w^2$

71. $(a + b)x^2 + (a + b)x - 12(a + b)$

72. $(x + y)n^2 + (x + y)n + 16(x + y)$

73. $(2p + q)r^2 - 12(2p + q)r + 27(2p + q)$

74. $(3m - n)k^2 - 13(3m - n)k + 40(3m - n)$

75. What polynomial can be factored as $(a + 9)(a + 4)$?

76. What polynomial can be factored as $(y - 7)(y + 3)$?

6.3 More on Factoring Trinomials

OBJECTIVES

1 Factor trinomials by grouping when the coefficient of the squared term is not 1.

2 Factor trinomials using FOIL.

FOR EXTRA HELP

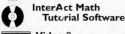

SSG Sec. 6.3
SSM Sec. 6.3

Pass the Test Software

**InterAct Math
Tutorial Software**

Video 9

Trinomials such as $2x^2 + 7x + 6$, in which the coefficient of the squared term is *not* 1, can be factored with an extension of the method presented in the last section.

OBJECTIVE 1 **Factor trinomials by grouping when the coefficient of the squared term is not 1.** Recall that a trinomial such as $m^2 + 3m + 2$ is factored by finding two numbers whose product is 2 and whose sum is 3. To factor $2x^2 + 7x + 6$, we look for two integers whose product is $2 \cdot 6 = 12$ and whose sum is 7.

$$
\begin{array}{c}
\text{Sum is 7.} \\
2x^2 + 7x + 6 \\
\text{Product is } 2 \cdot 6 = 12.
\end{array}
$$

By considering the pairs of positive integers whose product is 12, the necessary integers are found to be 3 and 4. We use these integers to write the middle term, $7x$, as $7x = 3x + 4x$. With this, the trinomial $2x^2 + 7x + 6$ becomes

$$2x^2 + 7x + 6 = 2x^2 + \underbrace{3x + 4x}_{7x = 3x + 4x} + 6.$$

Factor the new polynomial by grouping as in Section 6.1.

$$2x^2 + 3x + 4x + 6 = x(2x + 3) + 2(2x + 3)$$
$$= (2x + 3)(x + 2)$$

The common factor of $2x + 3$ was factored out to get

$$2x^2 + 7x + 6 = (2x + 3)(x + 2).$$

Check by finding the product of $2x + 3$ and $x + 2$.

We could have written the middle term in the polynomial $2x^2 + 7x + 6$ as $7x = 4x + 3x$ to get

$$2x^2 + 7x + 6 = 2x^2 + 4x + 3x + 6$$
$$= 2x(x + 2) + 3(x + 2)$$
$$= (x + 2)(2x + 3).$$

Either result is correct.

EXAMPLE 1 Factoring Trinomials by Grouping

Factor the trinomial.

(a) $6r^2 + r - 1$

We must find two integers with a product of $6(-1) = -6$ and a sum of 1.

Sum is 1.
↓
$$6r^2 + r - 1 = 6r^2 + 1r - 1$$

Product is $6(-1) = -6$.

The integers are -2 and 3. We write the middle term, $+r$, as $-2r + 3r$, so that

$$6r^2 + r - 1 = 6r^2 - 2r + 3r - 1.$$

Factor by grouping on the right-hand side.

$$6r^2 + r - 1 = 6r^2 - 2r + 3r - 1$$
$$= 2r(3r - 1) + 1(3r - 1) \qquad \text{The binomials must be the same.}$$
$$= (3r - 1)(2r + 1)$$

(b) $12z^2 - 5z - 2$

Look for two integers whose product is $12(-2) = -24$ and whose sum is -5. The required integers are 3 and -8, and

$$12z^2 - 5z - 2 = 12z^2 + 3z - 8z - 2 \qquad -5z = 3z - 8z$$
$$= 3z(4z + 1) - 2(4z + 1) \qquad \text{Group terms and factor each group.}$$
$$= (4z + 1)(3z - 2). \qquad \text{Factor out } 4z + 1.$$

(c) $10m^2 + mn - 3n^2$

Two integers whose product is $10(-3) = -30$ and whose sum is 1 are -5 and 6. Rewrite the trinomial with four terms.

$$10m^2 + mn - 3n^2 = 10m^2 - 5mn + 6mn - 3n^2 \qquad mn = -5mn + 6mn$$
$$= 5m(2m - n) + 3n(2m - n) \qquad \text{Group terms and factor each group.}$$
$$= (2m - n)(5m + 3n) \qquad \text{Factor out the common factor.}$$

OBJECTIVE 2 Factor trinomials using FOIL. The rest of this section shows an alternative method of factoring trinomials in which the coefficient of the squared term is not 1. This method uses trial and error.

To factor $2x^2 + 7x + 6$ (the trinomial factored at the beginning of this section) by trial and error, we must use FOIL backwards. We want to write $2x^2 + 7x + 6$ as the product of two binomials.

$$2x^2 + 7x + 6 = (\qquad)(\qquad)$$

The product of the two first terms of the binomials is $2x^2$. The possible factors of $2x^2$ are $2x$ and x or $-2x$ and $-x$. Since all terms of the trinomial are positive, we consider only positive factors. Thus, we have

$$2x^2 + 7x + 6 = (2x \qquad)(x \qquad).$$

The product of the two last terms, 6, can be factored as $1 \cdot 6$, $6 \cdot 1$, $2 \cdot 3$, or $3 \cdot 2$. Try each pair to find the pair that gives the correct middle term.

Since $2x + 6 = 2(x + 3)$, the binomial $2x + 6$ has a common factor of 2, while $2x^2 + 7x + 6$ has no common factor other than 1. The product $(2x + 6)(x + 1)$ cannot be correct.

 If the original polynomial has no common factor, then none of its binomial factors will either.

Now try the numbers 2 and 3 as factors of 6. Because of the common factor of 2 in $2x + 2$, $(2x + 2)(x + 3)$ will not work. Try $(2x + 3)(x + 2)$.

$$(2x + 3)(x + 2) = 2x^2 + 7x + 6 \qquad \text{Correct}$$

$$
\begin{array}{c}
3x \\
\underline{4x} \\
7x \qquad \text{Add.}
\end{array}
$$

Finally, we see that $2x^2 + 7x + 6$ factors as

$$2x^2 + 7x + 6 = (2x + 3)(x + 2).$$

Check by multiplying $2x + 3$ and $x + 2$.

┌ **E X A M P L E   2**   Factoring a Trinomial with All Terms Positive Using FOIL
Factor $8p^2 + 14p + 5$.
    The number 8 has several possible pairs of factors, but 5 has only 1 and 5 or -1 and -5. For this reason, it is easier to begin by considering the factors of 5. Ignore the negative factors since all coefficients in the trinomial are positive. If $8p^2 + 14p + 5$ can be factored, the factors will have the form

$$(\qquad + 5)(\qquad + 1).$$

The possible pairs of factors of $8p^2$ are $8p$ and p, or $4p$ and $2p$. Try various combinations, checking the middle term in each case.

Since the sum $14p$ is the correct middle term, the trinomial $8p^2 + 14p + 5$ factors as $(4p + 5)(2p + 1)$.

E X A M P L E 3 Factoring a Trinomial with a Negative Middle Term Using FOIL

Factor $6x^2 - 11x + 3$.

Since 3 has only 1 and 3 or -1 and -3 as factors, it is better here to begin by factoring 3. The last term of the trinomial $6x^2 - 11x + 3$ is positive and the middle term has a negative coefficient, so consider only negative factors. We need negative factors because the *product* of two negative factors is positive and their *sum* is negative, as required. Use -3 and -1 as factors of 3:

$$(\quad - 3)(\quad - 1).$$

The factors of $6x^2$ may be either $6x$ and x, or $2x$ and $3x$. Try $2x$ and $3x$.

$$(2x - 3)(3x - 1) \qquad \text{Correct}$$
$$\begin{array}{c} -9x \\ -2x \\ \hline -11x \quad \text{Add.} \end{array}$$

These factors give the correct middle term, so

$$6x^2 - 11x + 3 = (2x - 3)(3x - 1).$$

E X A M P L E 4 Factoring a Trinomial with a Negative Last Term Using FOIL

Factor $8x^2 + 6x - 9$.

The integer 8 has several possible pairs of factors, as does -9. Since the last term is negative, one positive factor and one negative factor of -9 are needed. Since the coefficient of the middle term is small, it is wise to avoid large factors such as 8 or 9. Let us try 4 and 2 as factors of 8, and 3 and -3 as factors of -9, and check the middle term.

$$(4x + 3)(2x - 3) \qquad \text{Incorrect}$$
$$\begin{array}{c} 6x \\ -12x \\ \hline -6x \quad \text{Add.} \end{array}$$

Now, try exchanging 3 and -3, since only the sign of the middle term is incorrect.

$$(4x - 3)(2x + 3) \qquad \text{Correct}$$
$$\begin{array}{r} -6x \\ 12x \\ \hline 6x \end{array} \quad \text{Add.}$$

This time we got the correct middle term, so

$$8x^2 + 6x - 9 = (4x - 3)(2x + 3).$$

E X A M P L E 5 Factoring a Trinomial with Two Variables

Factor $12a^2 - ab - 20b^2$.

There are several pairs of factors of $12a^2$, including $12a$ and a, $6a$ and $2a$, and $3a$ and $4a$, just as there are many pairs of factors of $-20b^2$, including $-20b$ and b, $10b$ and $-2b$, $-10b$ and $2b$, $4b$ and $-5b$, and $-4b$ and $5b$. Once again, since the desired middle term is small, avoid the larger factors. Try the factors $6a$ and $2a$ and $4b$ and $-5b$.

$$(6a + 4b)(2a - 5b)$$

This cannot be correct, as mentioned before, since $6a + 4b$ has a common factor while the given trinomial has none. Try $3a$ and $4a$ with $4b$ and $-5b$.

$$(3a + 4b)(4a - 5b) = 12a^2 + ab - 20b^2 \qquad \text{Incorrect}$$

Here the middle term has the wrong sign, so change the signs in the factors.

$$(3a - 4b)(4a + 5b) - 12a^2 - ab - 20b^2 \qquad \text{Correct}$$

E X A M P L E 6 Factoring a Trinomial with a Common Factor

Factor $28x^5 - 58x^4 - 30x^3$.

First factor out the greatest common factor, $2x^3$.

$$28x^5 - 58x^4 - 30x^3 = 2x^3(14x^2 - 29x - 15)$$

Now try to factor $14x^2 - 29x - 15$. Try $7x$ and $2x$ as factors of $14x^2$ and -3 and 5 as factors of -15.

$$(7x - 3)(2x + 5) = 14x^2 + 29x - 15 \qquad \text{Incorrect}$$

The middle term differs only in sign, so change the signs in the two factors.

$$(7x + 3)(2x - 5) = 14x^2 - 29x - 15 \qquad \text{Correct}$$

Finally, the factored form of $28x^5 - 58x^4 - 30x^3$ is

$$28x^5 - 58x^4 - 30x^3 = 2x^3(7x + 3)(2x - 5).$$

[CAUTION] Remember to include the common factor in the final result.

┌─
E X A M P L E   7   Factoring a Trinomial with a Negative Common Factor

Factor $-24a^3 - 42a^2 + 45a$.

   The common factor could be $3a$ or $-3a$. If we factor out $-3a$, the first term of the trinomial factor will be positive, which makes it easier to factor.

$$-24a^3 - 42a^2 + 45a = -3a(8a^2 + 14a - 15)$$ Factor out the greatest common factor.

$$= -3a(4a - 3)(2a + 5)$$ Use trial and error.
└─

6.3 EXERCISES

In Exercises 1–6 complete the steps in factoring the trinomial $2x^2 + x - 21$ by grouping.

   1. Find the product of _____ and _____ .

   2. Find factors of _____ that have a sum of _____ .

   3. Write the middle term x as _____ + _____ .

   4. Factor the polynomial _____ by grouping as (_____) + (_____).

   5. Factor out the common factor of _____ .

   6. The factored form is _____ .

Fill in the blanks in Exercises 7–10.

   7. $6a^2 + 7ab - 20b^2 = (3a$ _____$)($ _____ $+ 5b)$

   8. $9m^2 - 3mn - 2n^2 = (3m$ _____$)($ _____ $- 2n)$

   9. $3x^2 - 9x - 30 = 3(x^2$ _____$) = 3(x$ _____$)(x$ _____$)$

   10. $4z^3 - 10z^2 - 6z =$ _____$(2z^2$ _____$) = 2z($ _____ $- 3)(2z$ _____$)$

Decide which is the correct factored form of the given polynomial.

   11. $4y^2 + 17y - 15$
      (a) $(y + 5)(4y - 3)$
      (b) $(2y - 5)(2y + 3)$
   13. $4k^2 + 13mk + 3m^2$
      (a) $(4k + m)(k + 3m)$
      (b) $(4k + 3m)(k + m)$

   12. $12c^2 - 7c - 12$
      (a) $(6c - 2)(2c + 6)$
      (b) $(4c + 3)(3c - 4)$
   14. $2x^2 + 11x + 12$
      (a) $(2x + 3)(x + 4)$
      (b) $(2x + 4)(x + 3)$

   15. For the polynomial $12x^2 + 7x - 12$, 2 is not a common factor. Explain why the binomial $2x - 6$, then, cannot be a factor of the polynomial.

   16. Explain how the signs of the last terms of the two binomial factors of a trinomial are determined.

Factor completely. Use either method described in this section. See Examples 1–7.

   17. $3a^2 + 10a + 7$

   19. $4r^2 + r - 3$

   21. $15m^2 + m - 2$

   23. $8m^2 - 10m - 3$

   25. $20x^2 + 11x - 3$

   27. $21m^2 + 13m + 2$

   29. $20y^2 + 39y - 11$

   31. $6b^2 + 7b + 2$

   18. $7r^2 + 8r + 1$

   20. $4r^2 + 3r - 10$

   22. $6x^2 + x - 1$

   24. $12s^2 + 11s - 5$

   26. $20x^2 - 28x - 3$

   28. $38x^2 + 23x + 2$

   30. $10x^2 + 11x - 6$

   32. $6w^2 + 19w + 10$

33. $24x^2 - 42x + 9$

34. $48b^2 - 74b - 10$

35. $40m^2q + mq - 6q$

36. $15a^2b + 22ab + 8b$

37. $2m^3 + 2m^2 - 40m$

38. $3x^3 + 12x^2 - 36x$

39. $15n^4 - 39n^3 + 18n^2$

40. $24a^4 + 10a^3 - 4a^2$

41. $18x^5 + 15x^4 - 75x^3$

42. $32z^5 - 20z^4 - 12z^3$

43. $15x^2y^2 - 7xy^2 - 4y^2$

44. $14a^2b^3 + 15ab^3 - 9b^3$

45. $12p^2 + 7pq - 12q^2$

46. $6m^2 - 5mn - 6n^2$

47. $25a^2 + 25ab + 6b^2$

48. $6x^2 - 5xy - y^2$

49. $6a^2 - 7ab - 5b^2$

50. $25g^2 - 5gh - 2h^2$

51. $6m^6n + 7m^5n^2 + 2m^4n^3$

52. $12k^3q^4 - 4k^2q^5 - kq^6$

53. $5 - 6x + x^2$

54. $7 + 8x + x^2$

55. $16 + 16x + 3x^2$

56. $18 + 65x + 7x^2$

57. $-10x^3 + 5x^2 + 140x$

58. $-18k^3 - 48k^2 + 66k$

If a trinomial has a negative coefficient for the squared term, such as $-2x^2 + 11x - 12$, it is usually easier to factor by first factoring out the common factor -1:

$$-2x^2 + 11x - 12 = -1(2x^2 - 11x + 12)$$
$$= -1(2x - 3)(x - 4).$$

Use this method to factor each trinomial. See Example 7.

59. $-x^2 - 4x + 21$

60. $-x^2 + x + 72$

61. $-3x^2 - x + 4$

62. $-5x^2 + 2x + 16$

63. $2a^2 \quad 5ab \quad 2b^2$

64. $-3p^2 + 13pq - 4q^2$

65. The answer given in the back of the book for Exercise 59 is $1(x + 7)(x - 3)$. Is $(x + 7)(3 - x)$ also a correct answer? Explain why or why not.

66. One answer for Exercise 60 is $-1(x + 8)(x - 9)$. Is $(-x - 8)(-x + 9)$ also a correct answer? Explain.

Factor each polynomial. Remember to factor out the greatest common factor as the first step.

67. $25q^2(m + 1)^3 - 5q(m + 1)^3 - 2(m + 1)^3$

68. $18x^2(y - 3)^2 - 21x(y - 3)^2 - 4(y - 3)^2$

69. $15x^2(r + 3)^3 - 34xy(r + 3)^3 - 16y^2(r + 3)^3$

70. $4t^2(k + 9)^7 + 20ts(k + 9)^7 + 25s^2(k + 9)^7$

Find all integers k so that the trinomial can be factored using the methods of this section. (Hint: Try all possible factored forms with the given first and last terms. The coefficient of x will give the values of k.)

71. $5x^2 + kx - 1$

72. $2c^2 + kc - 3$

73. $2m^2 + km + 5$

74. $3y^2 + ky + 3$

RELATING CONCEPTS (EXERCISES 75-82)

One of the most common questions that beginning algebra students ask is this: "If my answer doesn't look exactly like the one given in the back of the book, is it necessarily incorrect?" Often there are several different equivalent forms of an answer that are all correct.

Work Exercises 75–82 in order to see how and why this is possible for factoring problems.

75. Factor the integer 35 as the product of two prime numbers.

76. Factor the integer 35 as the product of the negatives of two prime numbers.

77. Verify the following factored form: $6x^2 - 11x + 4 = (3x - 4)(2x - 1)$.

78. Verify the following factored form: $6x^2 - 11x + 4 = (4 - 3x)(1 - 2x)$.

79. Compare the two valid factored forms in Exercises 77 and 78. How do the factors in each case compare?

80. Suppose you know that the correct factored form of a particular trinomial is $(7t - 3)(2t - 5)$. Based on your observations in Exercises 77–79, what is another valid factored form?

81. Look at your results in Exercises 75 and 76, and fill in the blanks: If an integer factors as the product of a and b, then it also factors as the product of _____ and _____.

82. Look at your results in Exercises 77 and 78 and fill in the blanks: If a trinomial factors as the product of the binomials P and Q, then it also factors as the product of the binomials _____ and _____.

Did you make the connection that the product of the negatives of two factors is the same as the product of the two factors?

6.4 Special Factoring Rules

OBJECTIVES

1 Factor the difference of two squares.

2 Factor a perfect square trinomial.

3 Factor the difference of two cubes.

4 Factor the sum of two cubes.

FOR EXTRA HELP

SSG Sec. 6.4
SSM Sec. 6.4

Pass the Test Software

InterAct Math Tutorial Software

Video 10

By reversing the rules for multiplying binomials that we learned in the last chapter, we get rules for factoring polynomials in certain forms.

OBJECTIVE 1 Factor the difference of two squares. Recall from the last chapter that

$$(x + y)(x - y) = x^2 - y^2.$$

Based on this product, we factor a **difference of two squares** as follows.

Difference of Two Squares

$$x^2 - y^2 = (x + y)(x - y)*$$

_____

*A pair of expressions like $x + y$ and $x - y$ are called *conjugates*.

┌ **E X A M P L E   1** Factoring a Difference of Squares

Factor each difference of two squares.

(a) $x^2 - 49 = x^2 - 7^2 = (x + 7)(x - 7)$

(b) $y^2 - m^2 = (y + m)(y - m)$

(c) $z^2 - \dfrac{9}{16} = z^2 - \left(\dfrac{3}{4}\right)^2 = \left(z + \dfrac{3}{4}\right)\left(z - \dfrac{3}{4}\right)$

(d) $x^2 - 8$

Because 8 is not the square of an integer, this is a prime polynomial.

(e) $p^2 + 16$

The polynomial is not the *difference* of two squares. Using FOIL,

$$(p + 4)(p - 4) = p^2 - 16$$
$$(p - 4)(p - 4) = p^2 - 8p + 16$$
and $$(p + 4)(p + 4) = p^2 + 8p + 16$$

so $p^2 + 16$ is a prime polynomial.

 As Example 1(e) suggests, after any common factor is removed, the sum of two squares cannot be factored.

┌ **E X A M P L E   2** Factoring More Complex Differences of Squares

Factor completely.

(a) $9a^2 - 4b^2$

This is a difference of two squares because

$$9a^2 - 4b^2 = (3a)^2 - (2b)^2,$$

so $9a^2 - 4b^2 = (3a + 2b)(3a - 2b)$.

(b) $81y^2 - 36$

First factor out the common factor of 9.

$$81y^2 - 36 = 9(9y^2 - 4)$$
$$= 9(3y + 2)(3y - 2)$$

(c) $p^4 - 36 = (p^2)^2 - 6^2 - (p^2 + 6)(p^2 - 6)$

Neither $p^2 + 6$ nor $p^2 - 6$ can be factored further.

(d) $m^4 - 16 = (m^2)^2 - 4^2$

$$= (m^2 + 4)(m^2 - 4) \qquad \text{Difference of squares}$$
$$= (m^2 + 4)(m + 2)(m - 2) \qquad \text{Difference of squares}$$

CAUTION Remember to factor again when any of the *factors* is a difference of squares, as in Example 2(d).

OBJECTIVE 2 Factor a perfect square trinomial. The expressions 144, $4x^2$, and $81m^6$ are called *perfect squares,* since

$$144 = 12^2, \qquad 4x^2 = (2x)^2, \qquad \text{and} \qquad 81m^6 = (9m^3)^2.$$

A **perfect square trinomial** is a trinomial that is the square of a binomial. For example, $x^2 + 8x + 16$ is a perfect square trinomial since it is the square of the binomial $x + 4$:

$$x^2 + 8x + 16 = (x + 4)^2.$$

For a trinomial to be a perfect square, two of its terms must be perfect squares. For this reason, $16x^2 + 4x + 15$ cannot be a perfect square trinomial since only the term $16x^2$ is a perfect square.

On the other hand, even though two of the terms are perfect squares, the trinomial may not be a perfect square trinomial. For example, $x^2 + 6x + 36$ has two perfect square terms, but it is not a perfect square trinomial. (Try to find a binomial that can be squared to give $x^2 + 6x + 36$; it cannot be done.)

We can multiply to see that the square of a binomial gives the following perfect square trinomials.

Perfect Square Trinomial

$$x^2 + 2xy + y^2 = (x + y)^2$$
$$x^2 - 2xy + y^2 = (x - y)^2$$

The middle term of a perfect square trinomial is always twice the product of the two terms in the squared binomial. (This was shown in Section 5.4.) Use this to check any attempt to factor a trinomial that appears to be a perfect square.

While perfect square trinomials can be factored using the procedures of Sections 6.2 and 6.3, it is usually more efficient to recognize the pattern and factor accordingly. The following example illustrates this.

EXAMPLE 3 Factoring a Perfect Square Trinomial

Factor the perfect square trinomial.

(a) $x^2 + 10x + 25$

The term x^2 is a perfect square and so is 25. Try to factor the trinomial as

$$x^2 + 10x + 25 = (x + 5)^2.$$

To check, take twice the product of the two terms in the squared binomial.

$$\text{Twice} \rightarrow 2 \cdot x \cdot 5 = 10x$$

First term ⎯↑ ↑⎯ Last term
of binomial   of binomial

Since $10x$ is the middle term of the trinomial, the trinomial is a perfect square and can be factored as $(x + 5)^2$.

(b) $x^2 - 22xz + 121z^2$

The first and last terms are perfect squares ($121 = 11^2$). Check to see whether the middle term of $x^2 - 22xz + 121z^2$ is twice the product of the first and last terms of the binomial $x - 11z$.

$$\text{Twice} \rightarrow 2 \cdot \boldsymbol{x} \cdot \boldsymbol{11z} = 22xz$$

First term ⎤↑   ↑⎡ Last term
of binomial   of binomial

Since twice the product of the first and last terms of the binomial is the middle term, $x^2 - 22xz + 121z^2$ is a perfect square trinomial and

$$x^2 - 22xz + 121z^2 = (x - 11z)^2.$$

(c) $9m^2 - 24m + 16 = (3m)^2 - 2(3m)(4) + 4^2 = (3m - 4)^2$

Twice ⎤↑ ↑ ↑⎡ Last term
First
term

(d) $25y^2 + 20y + 16$

The first and last terms are perfect squares.

$$25y^2 = (5y)^2 \qquad \text{and} \qquad 16 = 4^2$$

Twice the product of the first and last terms of the binomial $5y + 4$ is

$$2 \cdot 5y \cdot 4 = 40y,$$

which is not the middle term of $25y^2 + 20y + 16$. This polynomial is not a perfect square. In fact, the polynomial cannot be factored even with the methods of Section 6.3; it is a prime polynomial.

The sign of the second term in the squared binomial is always the same as the sign of the middle term in the trinomial. Also, the first and last terms of a perfect square trinomial must be positive, since they are squares. For example, the polynomial $x^2 - 2x - 1$ cannot be a perfect square because the last term is negative.

OBJECTIVE 3 Factor the difference of two cubes. The difference of two squares was factored above; we can also factor the **difference of two cubes.** Use the following pattern.

Difference of Two Cubes

$$x^3 - y^3 = (x - y)(x^2 + xy + y^2)$$

This pattern *should be memorized.* Multiply on the right to see that the pattern gives the correct factors.

$$
\begin{array}{r}
x^2 + xy + y^2 \\
x - y \\
\hline
-x^2y - xy^2 - y^3 \\
x^3 + x^2y + xy^2 \\
\hline
x^3 \qquad\qquad - y^3
\end{array}
$$

Notice the pattern of the terms in the factored form of $x^3 - y^3$.

- $x^3 - y^3 =$ (a binomial factor)(a trinomial factor)
- The binomial factor has the difference of the cube roots of the given terms.
- The terms in the trinomial factor are all positive.
- What you write in the binomial factor determines the trinomial factor:

$$x^3 - y^3 = (x - y)(\ \underset{\text{First term squared}}{x^2}\ +\ \underset{\substack{\text{positive} \\ \text{product of} \\ \text{the terms}}}{xy}\ +\ \underset{\substack{\text{second term} \\ \text{squared}}}{y^2}\ ).$$

The polynomial $x^3 - y^3$ is not equivalent to $(x - y)^3$, because $(x - y)^3$ can also be written as

$$(x - y)^3 = (x - y)(x - y)(x - y)$$
$$= (x - y)(x^2 - 2xy + y^2)$$

but

$$x^3 - y^3 = (x - y)(x^2 + xy + y^2).$$

E X A M P L E   4   **Factoring Differences of Cubes**

Factor the following.

(a) $m^3 - 125$

Let $x = m$ and $y = 5$ in the pattern for the difference of two cubes.

$$x^3 - y^3 = (x - y)(x^2 + xy + y^2)$$
$$m^3 - 125 = m^3 - 5^3 = (m - 5)(m^2 + 5m + 5^2) \qquad \text{Let } x = m,\ y = 5.$$
$$= (m - 5)(m^2 + 5m + 25)$$

(b) $8p^3 - 27$

Since $8p^3 = (2p)^3$ and $27 = 3^3$, substitute into the rule using $2p$ for x and 3 for y.

$$8p^3 - 27 = (2p)^3 - 3^3$$
$$= (2p - 3)[(2p)^2 + (2p)3 + 3^2]$$
$$= (2p - 3)(4p^2 + 6p + 9)$$

(c) $4m^3 - 32 = 4(m^3 - 8)$
$$= 4(m^3 - 2^3)$$
$$= 4(m - 2)(m^2 + 2m + 4)$$

(d) $125t^3 - 216s^6 = (5t)^3 - (6s^2)^3$
$$= (5t - 6s^2)[(5t)^2 + (5t)(6s^2) + (6s^2)^2]$$
$$= (5t - 6s^2)(25t^2 + 30ts^2 + 36s^4)$$

CAUTION A common error in factoring the difference of two cubes, such as $x^3 - y^3 = (x - y)(x^2 + xy + y^2)$, is to try to factor $x^2 + xy + y^2$. It is easy to confuse this factor with a perfect square trinomial, $x^2 + 2xy + y^2$. Because there is no 2 in $x^2 + xy + y^2$, it is very unusual to be able to further factor an expression of the form $x^2 + xy + y^2$.

OBJECTIVE **4** Factor the sum of two cubes. A sum of two squares, such as $m^2 + 25$, cannot be factored using real numbers, but the **sum of two cubes** can be factored by the following pattern, *which should be memorized.*

Sum of Two Cubes

$$x^3 + y^3 = (x + y)(x^2 - xy + y^2)$$

Compare the pattern for the *sum* of two cubes with the pattern for the *difference* of two cubes. The only difference between them is the positive and negative signs.

$$x^3 - y^3 = (x - y)(x^2 + xy + y^2)$$

Same sign · Opposite sign · Positive

$$x^3 + y^3 = (x + y)(x^2 - xy + y^2)$$

Same sign · Opposite sign · Positive

Observing these relationships should help you to remember these patterns.

EXAMPLE 5 Factoring Sums of Cubes

Factor.

(a) $k^3 + 27 = k^3 + 3^3$
$$= (k + 3)(k^2 - 3k + 3^2)$$
$$= (k + 3)(k^2 - 3k + 9)$$

(b) $8m^3 + 125 = (2m)^3 + 5^3$
$$= (2m + 5)[(2m)^2 - (2m)(5) + 5^2]$$
$$= (2m + 5)(4m^2 - 10m + 25)$$

(c) $1000a^6 + 27b^3 = (10a^2)^3 + (3b)^3$
$$= (10a^2 + 3b)[(10a^2)^2 - (10a^2)(3b) + (3b)^2]$$
$$= (10a^2 + 3b)(100a^4 - 30a^2b + 9b^2)$$

The methods of factoring discussed in this section are summarized here. All these rules should be memorized.

Special Factorizations

Difference of two squares	$x^2 - y^2 = (x + y)(x - y)$
Perfect square trinomials	$x^2 + 2xy + y^2 = (x + y)^2$
	$x^2 - 2xy + y^2 = (x - y)^2$
Difference of two cubes	$x^3 - y^3 = (x - y)(x^2 + xy + y^2)$
Sum of two cubes	$x^3 + y^3 = (x + y)(x^2 - xy + y^2)$

Remember the *sum* of two *squares* can be factored only if the terms have a common factor.

6.4 EXERCISES

1. To help you factor the difference of squares, complete the following list of squares.

   $1^2 =$ _____ $2^2 =$ _____ $3^2 =$ _____ $4^2 =$ _____ $5^2 =$ _____

   $6^2 =$ _____ $7^2 =$ _____ $8^2 =$ _____ $9^2 =$ _____ $10^2 =$ _____

   $11^2 =$ _____ $12^2 =$ _____ $13^2 =$ _____ $14^2 =$ _____ $15^2 =$ _____

   $16^2 =$ _____ $17^2 =$ _____ $18^2 =$ _____ $19^2 =$ _____ $20^2 =$ _____

2. The following powers of x are all perfect squares: $x^2, x^4, x^6, x^8, x^{10}$. Based on this observation, we may make a conjecture (an educated guess) that if the power of a variable is divisible by _____ (with 0 remainder), then we have a perfect square.

3. To help you factor the sum or difference of cubes, complete the following list of cubes.

   $1^3 =$ _____ $2^3 =$ _____ $3^3 =$ _____ $4^3 =$ _____ $5^3 =$ _____

   $6^3 =$ _____ $7^3 =$ _____ $8^3 =$ _____ $9^3 =$ _____ $10^3 =$ _____

4. The following powers of x are all perfect cubes: $x^3, x^6, x^9, x^{12}, x^{15}$. Based on this observation, we may make a conjecture that if the power of a variable is divisible by _____ (with 0 remainder), then we have a perfect cube.

5. Identify the monomial as a perfect square, a perfect cube, both of these, or neither of these.
   (a) $64x^6y^{12}$ **(b)** $125t^6$ **(c)** $49x^{12}$ **(d)** $81r^{10}$

6. What must be true for x^n to be both a perfect square and a perfect cube?

Factor each binomial completely. Use your answers in Exercises 1 and 2 as necessary. See Examples 1 and 2.

7. $y^2 - 25$ **8.** $t^2 - 16$ **9.** $9r^2 - 4$ **10.** $4x^2 - 9$

11. $36m^2 - \dfrac{16}{25}$ **12.** $100b^2 - \dfrac{4}{49}$ **13.** $36x^2 - 16$ **14.** $32u^2 - 8$

15. $196p^2 - 225$ **16.** $361q^2 - 400$ **17.** $16r^2 - 25a^2$ **18.** $49m^2 - 100p^2$

19. $100x^2 + 49$ **20.** $81w^2 + 16$ **21.** $p^4 - 49$ **22.** $r^4 - 25$

23. $x^4 - 1$ **24.** $y^4 - 16$ **25.** $p^4 - 256$ **26.** $16k^4 - 1$

27. When a student was directed to factor $x^4 - 81$ completely, his teacher did not give him full credit for the answer $(x^2 + 9)(x^2 - 9)$. The student argued that since his answer does indeed give $x^4 - 81$ when multiplied out, he should be given full credit. Was the teacher justified in her grading of this item? Why or why not?

28. The binomial $4x^2 + 16$ is a sum of two squares that *can* be factored. How is this binomial factored? When can the sum of two squares be factored?

Factor each trinomial completely. It may be necessary to factor out the greatest common factor first. See Example 3.

29. $w^2 + 2w + 1$

30. $p^2 + 4p + 4$

31. $x^2 - 8x + 16$

32. $x^2 - 10x + 25$

33. $t^2 + t + \dfrac{1}{4}$

34. $m^2 + \dfrac{2}{3}m + \dfrac{1}{9}$

35. $x^2 - 1.0x + .25$

36. $y^2 - 1.4y + .49$

37. $2x^2 + 24x + 72$

38. $3y^2 - 48y + 192$

39. $16x^2 - 40x + 25$

40. $36y^2 - 60y + 25$

41. $49x^2 - 28xy + 4y^2$

42. $4z^2 - 12zw + 9w^2$

43. $64x^2 + 48xy + 9y^2$

44. $9t^2 + 24tr + 16r^2$

45. $-50h^2 + 40hy - 8y^2$

46. $-18x^2 - 48xy - 32y^2$

Factor each binomial completely. Use your answers in Exercises 3 and 4 as necessary. See Examples 4 and 5.

47. $a^3 + 1$

48. $m^3 + 8$

49. $a^3 - 1$

50. $m^3 - 8$

51. $p^3 + q^3$

52. $x^3 + z^3$

53. $27x^3 - 1$

54. $64y^3 - 27$

55. $8p^3 + 729q^3$

56. $64x^3 + 125y^3$

57. $y^3 - 8x^3$

58. $w^3 - 216z^3$

59. $27a^3 - 64b^3$

60. $125m^3 - 8p^3$

61. $125t^3 + 8s^3$

62. $27r^3 + 1000s^3$

63. State and give the name of each of the five special factoring rules. For each rule, give the numbers of three exercises from this exercise set that are examples.

▌ RELATING CONCEPTS (EXERCISES 64–67)

We have seen that multiplication and factoring are reverse processes. We know that multiplication and division are also related: To check a division problem, we multiply the quotient by the divisor to get the dividend.

*To see how factoring and division are related, **work Exercises 64–67 in order.***

64. Factor $10x^2 + 11x - 6$.

65. Use long division to divide $10x^2 + 11x - 6$ by $2x + 3$.

66. Could we have predicted the result in Exercise 65 from the result in Exercise 64? Explain.

67. Divide $x^3 - 1$ by $x - 1$. Use your answer to factor $x^3 - 1$.

Did you make the connection that the quotient in a division problem is a factor of the dividend?

Extend the methods of factoring presented so far in this chapter to factor each polynomial completely.

68. $(m + n)^2 - (m - n)^2$

69. $(a - b)^3 - (a + b)^3$

70. $m^2 - p^2 + 2m + 2p$

71. $3r - 3k + 3r^2 - 3k^2$

RELATING CONCEPTS (EXERCISES 72-77)

*It is possible to use two different methods of factoring a particular polynomial and get the same answer. To see this, **work Exercises 72–77 in order,** referring to the polynomial*

$$(x^2 + 2x + 1) - 4.$$

72. Notice that the first three terms of the polynomial are grouped. Factor that trinomial, which is a perfect square trinomial, to obtain a difference of squares.

73. Factor the difference of squares obtained in Exercise 72.

74. Combine the constant terms within the factors from Exercise 73 to obtain the simplest forms of the factors.

75. Now start over and write the original polynomial $(x^2 + 2x + 1) - 4$ as a trinomial, by combining the constant terms.

76. Factor the trinomial obtained in Exercise 75 using the methods of Sections 6.2 and 6.3.

77. How do your results compare in Exercises 74 and 76? Which method do you prefer?

Did you make the connection that factoring a polynomial by two different methods should give the same answer?

78. In the polynomial $9y^2 + 14y + 25$, the first and last terms are perfect squares. Can the polynomial be factored? If it can, factor it. If it cannot, explain why it is not a perfect square trinomial.

Find the value of the indicated variable. (Hint: Perform the squaring on the right side, then solve the equation using the steps given in Chapter 2.)

79. Find a value of b so that $x^2 + bx + 25 = (x + 5)^2$.

80. For what value of c is $4m^2 - 12m + c = (2m - 3)^2$?

81. Find a so that $ay^2 - 12y + 4 = (3y - 2)^2$.

82. Find b so that $100a^2 + ba + 9 = (10a + 3)^2$.

SUMMARY Exercises on Factoring

These mixed exercises are included to give you practice in selecting an appropriate method for factoring a particular polynomial. As you factor a polynomial, ask yourself these questions to decide on a suitable factoring technique.

Factoring a Polynomial

Step 1   Is there a common factor?

Step 2   How many terms are in the polynomial?
Four terms   Can the polynomial be factored by grouping?
Three terms   Is it a perfect square trinomial? If the trinomial is not a perfect square, check to see whether the coefficient of the squared term is 1. If so, use the method of Section 6.2. If the coefficient of the squared term is not 1, use the general factoring methods of Section 6.3.
Two terms   Check to see whether it is either the difference of two squares or the sum or difference of two cubes.

Step 3   Can any factors be factored further?

SUMMARY EXERCISES

Factor each polynomial completely.

1. $a^2 - 4a - 12$
2. $a^2 + 17a + 72$
3. $6y^2 - 6y - 12$
4. $7y^6 + 14y^5 - 168y^4$
5. $6a + 12b + 18c$
6. $m^2 - 3mn - 4n^2$
7. $p^2 - 17p + 66$
8. $z^2 - 6z + 7z - 42$
9. $10z^2 - 7z - 6$
10. $2m^2 - 10m - 48$
11. $m^2 - n^2 + 5m - 5n$
12. $15y + 5$
13. $8a^5 - 8a^4 - 48a^3$
14. $8k^2 - 10k - 3$
15. $z^2 - 3za - 10a^2$
16. $50z^2 - 100$
17. $x^2 - 4x - 5x + 20$
18. $100n^2r^2 + 30nr^3 - 50n^2r$
19. $6n^2 - 19n + 10$
20. $9y^2 + 12y - 5$
21. $16x + 20$
22. $m^2 + 2m - 15$
23. $6y^2 - 5y - 4$
24. $m^2 - 81$
25. $6z^2 + 31z + 5$
26. $5z^2 + 24z - 5 + 3z + 15$
27. $4k^2 - 12k + 9$
28. $8p^2 + 23p - 3$
29. $54m^2 - 24z^2$
30. $8m^2 - 2m - 3$
31. $3k^2 + 4k - 4$
32. $45a^3b^5 - 60a^4b^2 + 75a^6b^4$
33. $14k^3 + 7k^2 - 70k$
34. $5 + r - 5s - rs$
35. $y^4 - 16$
36. $20y^5 - 30y^4$
37. $8m - 16m^2$
38. $k^2 - 16$
39. $z^3 - 8$
40. $y^2 - y - 56$
41. $k^2 + 9$
42. $27p^{10} - 45p^9 - 252p^8$
43. $32m^9 + 16m^5 + 24m^3$
44. $8m^3 + 125$
45. $16r^2 + 24rm + 9m^2$
46. $z^2 - 12z + 36$
47. $15h^2 + 11hg - 14g^2$
48. $5z^3 - 45z^2 + 70z$
49. $k^2 - 11k + 30$
50. $64p^2 - 100m^2$
51. $3k^3 - 12k^2 - 15k$
52. $y^2 - 4yk - 12k^2$
53. $1000p^3 + 27$
54. $64r^3 - 343$
55. $6 + 3m + 2p + mp$
56. $2m^2 + 7mn - 15n^2$
57. $16z^2 - 8z + 1$
58. $125m^4 - 400m^3n + 195m^2n^2$
59. $108m^2 - 36m + 3$
60. $100a^2 - 81y^2$
61. $64m^2 - 40mn + 25n^2$
62. $4y^2 - 25$
63. $32z^3 + 56z^2 - 16z$
64. $10m^2 + 25m - 60$
65. $20 + 5m + 12n + 3mn$
66. $4 - 2q - 6p + 3pq$
67. $6a^2 + 10a - 4$
68. $36y^6 - 42y^5 - 120y^4$
69. $a^3 - b^3 + 2a - 2b$
70. $16k^2 - 48k + 36$
71. $64m^2 - 80mn + 25n^2$
72. $72y^3z^2 + 12y^2 - 24y^4z^2$
73. $8k^2 - 2kh - 3h^2$
74. $2a^2 - 7a - 30$
75. $(m + 1)^3 + 1$
76. $8a^3 - 27$
77. $10y^2 - 7yz - 6z^2$
78. $m^2 - 4m + 4$
79. $8a^2 + 23ab - 3b^2$
80. $a^4 - 625$

RELATING CONCEPTS (EXERCISES 81-88)

A binomial may be *both* a difference of squares *and* a difference of cubes. One example of such a binomial is $x^6 - 1$. Using the techniques of Section 6.4, one factoring method will give the complete factored form, while the other will not.

Work Exercises 81–88 in order to determine the method to use if you have to make such a decision.

81. Factor $x^6 - 1$ as the difference of two squares.

82. The factored form obtained in Exercise 81 consists of a difference of cubes multiplied by a sum of cubes. Factor each binomial further.

83. Now start over and factor $x^6 - 1$ as the difference of two cubes.

84. The factored form obtained in Exercise 83 consists of a binomial which is a difference of squares and a trinomial. Factor the binomial further.

85. Compare your results in Exercises 82 and 84. Which one of these is the complete factored form?

86. Verify that the trinomial in the factored form in Exercise 84 is the product of the two trinomials in the factored form in Exercise 82.

87. Use the results of Exercises 81–86 to complete the following statement: In general, if I must choose between factoring first using the method for difference of squares or the method for difference of cubes, I should choose the _____ method to eventually obtain the complete factored form.

88. Find the *complete* factored form of $x^6 - 729$ using the knowledge you have gained in Exercises 81–87.

Did you make the connection that when it is possible to factor a polynomial as the difference of squares or the difference of cubes, only one method gives a complete factored form?

6.5 Solving Quadratic Equations by Factoring

OBJECTIVES

1 Solve quadratic equations by factoring.

2 Solve other equations by factoring.

FOR EXTRA HELP

SSG Sec. 6.5
SSM Sec. 6.5

Pass the Test Software

InterAct Math Tutorial Software

Video 10

In this section we introduce *quadratic equations,* equations that contain a squared term and no terms of higher degree.

Quadratic Equation

An equation that can be written in the form

$$ax^2 + bx + c = 0,$$

where a, b, and c are real numbers, with $a \neq 0$, is a **quadratic equation.**

The parabolas graphed in Section 5.1 are graphs of quadratic equations with $b = 0$, or with $b = 0$ and $c = 0$. The form $ax^2 + bx + c = 0$ is the **standard form** of a quadratic equation. For example,

$$x^2 + 5x + 6 = 0, \qquad 2x^2 - 5x = 3, \qquad \text{and} \qquad y^2 = 4$$

are all quadratic equations but only $x^2 + 5x + 6 = 0$ is in standard form.

OBJECTIVE 1 Solve quadratic equations by factoring. We use the **zero-factor property** to solve a quadratic equation by factoring.

Zero-Factor Property

If a and b are real numbers and   **if $ab = 0$, then $a = 0$ or $b = 0$.**

In other words, if the product of two numbers is 0, then at least one of the numbers must be 0. This means that one number *must* be 0, but both *may be* 0.

E X A M P L E   1   Using the Zero-Factor Property

Solve the equation $(x + 3)(2x - 1) = 0$.

The product $(x + 3)(2x - 1)$ is equal to 0. By the zero-factor property, the only way that the product of these two factors can be 0 is if at least one of the factors is 0. Therefore, either $x + 3 = 0$ or $2x - 1 = 0$. Solve each of these two linear equations as in Chapter 2.

$$x + 3 = 0 \qquad \text{or} \qquad 2x - 1 = 0$$
$$x = -3 \qquad \text{or} \qquad 2x = 1 \qquad \text{Add 1 to both sides.}$$
$$x = \frac{1}{2} \qquad \text{Divide by 2.}$$

Since both of these equations have a solution, the equation $(x + 3)(2x - 1) = 0$ has two solutions, -3 and $\frac{1}{2}$. Check these answers by substituting -3 for x in the original equation. Then start over and substitute $\frac{1}{2}$ for x.

If $x = -3$, then

$$(-3 + 3)[2(-3) - 1] = 0 \quad ?$$
$$0(-7) = 0 \quad ?$$
$$0 = 0. \qquad \text{True}$$

If $x = \frac{1}{2}$, then

$$\left(\frac{1}{2} + 3\right)\left(2 \cdot \frac{1}{2} - 1\right) = 0 \quad ?$$
$$\frac{7}{2}(1 - 1) = 0 \quad ?$$
$$\frac{7}{2} \cdot 0 = 0 \quad ?$$
$$0 = 0. \qquad \text{True}$$

Both -3 and $\frac{1}{2}$ produce true statements, so the solution set is $\left\{-3, \frac{1}{2}\right\}$.

 The word "or" as used in Example 1 means "one or the other or both."

In Example 1 the equation to be solved was presented with the polynomial in factored form. If the polynomial in an equation is not already factored, first make sure that the equation is in standard form. Then factor the polynomial.

E X A M P L E   2   Solving a Quadratic Equation Not in Standard Form

Solve the equation $x^2 - 5x = -6$.

First, rewrite the equation with all terms on one side by adding 6 to both sides.

$$x^2 - 5x = -6$$
$$x^2 - 5x + 6 = 0 \qquad \text{Add 6.}$$

Now factor $x^2 - 5x + 6$. Find two numbers whose product is 6 and whose sum is -5. These two numbers are -2 and -3, so the equation becomes

$$(x - 2)(x - 3) = 0. \qquad \text{Factor.}$$
$$x - 2 = 0 \quad \text{or} \quad x - 3 = 0 \qquad \text{Zero-factor property}$$
$$x = 2 \quad \text{or} \quad x = 3 \qquad \text{Solve each equation.}$$

Check both solutions by substituting first 2 and then 3 for x in the original equation. The solution set is $\{2, 3\}$.

In summary, follow these steps to solve quadratic equations by factoring.

Solving a Quadratic Equation by Factoring

Step 1 **Write in standard form.** Write the equation in standard form: all terms on one side of the equals sign, with 0 on the other side.

Step 2 **Factor.** Factor completely.

Step 3 **Use the zero-factor property.** Set each factor with a variable equal to 0, and solve the resulting equations.

Step 4 **Check.** Check each solution in the original equation.

 Not all quadratic equations can be solved by factoring. A more general method for solving such equations is given in Chapter 11.

E X A M P L E   3   Solving a Quadratic Equation with a Common Factor

Solve $4p^2 + 40 = 26p$.

Subtract $26p$ from each side and write in descending powers to get

$$4p^2 - 26p + 40 = 0.$$
$$2(2p^2 - 13p + 20) = 0 \qquad \text{Factor out 2.}$$
$$2(2p - 5)(p - 4) = 0 \qquad \text{Factor the trinomial.}$$
$$2 = 0 \quad \text{or} \quad 2p - 5 = 0 \quad \text{or} \quad p - 4 = 0 \qquad \text{Zero-factor property}$$

The equation $2 = 0$ has no solution. Solve the equation in the middle by first adding 5 on both sides of the equation. Then divide both sides by 2. Solve the equation on the right by adding 4 to both sides.

$$2p - 5 = 0 \quad \text{or} \quad p - 4 = 0$$
$$2p = 5 \quad \text{or} \quad p = 4$$
$$p = \frac{5}{2}$$

The solutions of $4p^2 + 40 = 26p$ are $\frac{5}{2}$ and 4; check them by substituting in the original equation to see that the solution set is $\left\{\frac{5}{2}, 4\right\}$.

 A common error is to include 2 as a solution in Example 3. Only *variable* factors lead to solutions.

┌ **E X A M P L E   4**   Solving Special Quadratic Equations

Solve the equations.

(a) $16m^2 - 25 = 0$

Factor the left-hand side of the equation as the difference of two squares.

$$(4m + 5)(4m - 5) = 0$$

$4m + 5 = 0$	or	$4m - 5 = 0$	Zero-factor property
$4m = -5$	or	$4m = 5$	Solve each equation.
$m = -\dfrac{5}{4}$	or	$m = \dfrac{5}{4}$	

Check the two solutions, $-\frac{5}{4}$ and $\frac{5}{4}$, in the original equation. The solution set is $\left\{-\frac{5}{4}, \frac{5}{4}\right\}$.

(b) $y^2 = 2y$

First write the equation in standard form.

$y^2 - 2y = 0$		Standard form
$y(y - 2) = 0$		Factor.
$y = 0$ or $y - 2 = 0$		Zero-factor property
$y = 2$		Solve.

The solution set is $\{0, 2\}$.

(c) $k(2k + 1) = 3$

$k(2k + 1) = 3$		
$2k^2 + k = 3$		Distributive property
$2k^2 + k - 3 = 0$		Subtract 3.
$(k - 1)(2k + 3) = 0$		Factor.
$k - 1 = 0$ or $2k + 3 = 0$		Zero-factor property
$k = 1$ or $2k = -3$		
$k = -\dfrac{3}{2}$		

The solution set is $\left\{1, -\frac{3}{2}\right\}$.

In Example 4(b) it is tempting to begin by dividing both sides of the equation by y to get $y = 2$. Note, however, that the other solution, 0, is not found by this method.

CAUTION   In Example 4(c) we could not use the zero-factor property to solve the equation in its given form because of the 3 on the right. Remember that the zero-factor property applies only to a product that equals 0.

OBJECTIVE **2** Solve other equations by factoring. We can also use the zero-factor property to solve equations that involve more than two factors with variables, as shown in Example 5. (These equations are *not* quadratic equations. Why not?)

EXAMPLE 5 Solving Equations with More Than Two Variable Factors

Solve the equations.

(a) $6z^3 - 6z = 0$

$$6z^3 - 6z = 0$$
$$6z(z^2 - 1) = 0 \qquad \text{Factor out } 6z.$$
$$6z(z + 1)(z - 1) = 0 \qquad \text{Factor } z^2 - 1.$$

By an extension of the zero-factor property this product can equal 0 only if at least one of the factors is 0. Write and solve three equations, one for each factor with a variable.

$$6z = 0 \qquad \text{or} \qquad z + 1 = 0 \qquad \text{or} \qquad z - 1 = 0$$
$$z = 0 \qquad \text{or} \qquad z = -1 \qquad \text{or} \qquad z = 1$$

Check by substituting, in turn, 0, -1, and 1 in the original equation. The solution set is $\{-1, 0, 1\}$.

(b) $(3x - 1)(x^2 - 9x + 20) = 0$

$$(3x - 1)(x^2 - 9x + 20) = 0$$
$$(3x - 1)(x - 5)(x - 4) = 0 \qquad \text{Factor } x^2 - 9x + 20.$$
$$3x - 1 = 0 \qquad \text{or} \qquad x - 5 = 0 \qquad \text{or} \qquad x - 4 = 0 \qquad \text{Zero-factor property}$$
$$x = \frac{1}{3} \qquad \text{or} \qquad x = 5 \qquad \text{or} \qquad x = 4$$

The solutions of the original equation are $\frac{1}{3}$, 4, and 5. Check each solution to verify that the solution set is $\left\{\frac{1}{3}, 4, 5\right\}$.

In Example 5(b), it would be unproductive to begin by multiplying the two factors together. Keep in mind that the zero-factor property requires the product of two or more factors; this product must equal 0. Always consider first whether an equation is given in an appropriate form for the zero-factor property.

CONNECTIONS

Galileo Galilei (1564–1642) developed theories to explain physical phenomena and set up experiments to test his ideas. According to legend, Galileo dropped objects of different weights from the tower of Pisa to disprove the Aristotelian view that heavier objects fall faster than lighter objects. He developed the formula for freely falling objects described by $d = 16t^2$, where d is the distance in feet that an object falls (disregarding air resistance) in t seconds, regardless of weight.*

---

*From *Mathematical Ideas,* 8th edition, by Charles D. Miller, Vern E. Heeren, and E. John Hornsby, Jr., Addison-Wesley, 1997, page 442.

CONNECTIONS (CONTINUED)

The equation $d = 16t^2$ is a quadratic equation in the variable t. To find the number of seconds it would take an object to fall 256 feet, for example, substitute 256 for d and solve for t. On the other hand, to find how far the object would fall in 2 seconds, substitute 2 for t and calculate d. This idea of distance traveled depending on the time elapsed is an example of an important mathematical concept, the *function*. Functions were introduced in Chapter 4 and are discussed in more detail in Chapter 8.

FOR DISCUSSION OR WRITING

1. How long would it take an object to fall 256 feet?

2. How far would an object fall in 2 seconds?

6.5 EXERCISES

Fill in the blanks.

1. A quadratic equation is an equation that can be put in the form _____ $= 0$.

2. The form $ax^2 + bx + c = 0$ is called _____ form.

3. If a quadratic equation is in standard form, to solve the equation we should begin by _____ the polynomial.

4. Once the polynomial in a quadratic equation is factored, what is the next step to solve the equation?

Solve each equation and check your solutions. See Example 1.

5. $(3k + 8)(k + 7) = 0$ **6.** $(5r + 9)(r + 6) = 0$

7. $t(t + 4) = 0$ **8.** $x(x - 10) = 0$

9. Students often become confused as to how to handle a constant, such as 2 in the equation $2x(3x - 4) = 0$. How would you explain to someone how to solve this equation and how to handle the constant 2?

10. As shown in Example 5, the zero-factor property can be extended to more than two factors. For example, to solve $(x - 4)(x + 3)(2x - 7) = 0$, we would set each factor equal to 0 and solve three equations. Find the solution set of this equation.

11. Why do you think that 9 is called a *double solution* of the equation $(x - 9)^2 = 0$?

12. Define *quadratic equation* in your own words, without using an algebraic expression. Then give examples.

Solve each equation and check your solutions. See Examples 2–4.

13. $y^2 + 3y + 2 = 0$ **14.** $p^2 + 8p + 7 = 0$ **15.** $y^2 - 3y + 2 = 0$

16. $r^2 - 4r + 3 = 0$ **17.** $x^2 = 24 - 5x$ **18.** $t^2 = 2t + 15$

19. $5x^2 = 15 + 10x$ **20.** $2m^2 = 8 + 6m$ **21.** $z^2 = -2 - 3z$

22. $p^2 = 2p + 3$ **23.** $m^2 + 8m + 16 = 0$ **24.** $b^2 - 6b + 9 = 0$

25. $3x^2 + 5x - 2 = 0$ **26.** $6r^2 - r - 2 = 0$ **27.** $6p^2 = 4 - 5p$

28. $6x^2 = 4 + 5x$ **29.** $9s^2 + 12s = -4$ **30.** $36x^2 + 60x = -25$

31. $2y^2 - 18 = 0$ **32.** $3m^2 - 300 = 0$ **33.** $16k^2 - 49 = 0$

34. $4w^2 - 9 = 0$ **35.** $n^2 = 121$ **36.** $x^2 = 400$

37. What is wrong with this reasoning for solving the equation in Exercise 35? "To solve $n^2 = 121$, I must find a number that, when multiplied by itself, gives 121. Because $11^2 = 121$, the solution of the equation is 11."

38. What is wrong with this reasoning in solving $x^2 = 7x$? "To solve $x^2 = 7x$, first divide both sides by x to get $x = 7$. Therefore, the solution is 7."

Solve each equation and check your solutions. See Examples 4 and 5.

39. $x^2 = 7x$ **40.** $t^2 = 9t$ **41.** $6r^2 = 3r$

42. $10y^2 = -5y$ **43.** $g(g - 7) = -10$ **44.** $r(r - 5) = -6$

45. $3z(2z + 7) = 12$ **46.** $4b(2b + 3) = 36$

47. $2(y^2 - 66) = -13y$ **48.** $3(t^2 + 4) = 20t$

49. $3x(x + 1) = (2x + 3)(x + 1)$ **50.** $2k(k + 3) = (3k + 1)(k + 3)$

51. $(2r + 5)(3r^2 - 16r + 5) = 0$ **52.** $(3m + 4)(6m^2 + m - 2) = 0$

53. $(2x + 7)(x^2 + 2x - 3) = 0$ **54.** $(x + 1)(6x^2 + x - 12) = 0$

55. $9y^3 - 49y = 0$ **56.** $16r^3 - 9r = 0$

57. $r^3 - 2r^2 - 8r = 0$ **58.** $x^3 - x^2 - 6x = 0$

59. $a^3 + a^2 - 20a = 0$ **60.** $y^3 - 6y^2 + 8y = 0$

61. $r^4 = 2r^3 + 15r^2$ **62.** $x^3 = 3x + 2x^2$

63. $6p^2(p + 1) = 4(p + 1) - 5p(p + 1)$ **64.** $6x^2(2x + 3) - 5x(2x + 3) = 4(2x + 3)$

65. $(k + 3)^2 - (2k - 1)^2 = 0$ **66.** $(4y - 3)^3 - 9(4y - 3) = 0$

67. Verify that the solutions of $(x - 3)(x + 2) = 1$ are not found from the two equations

$$x - 3 = 1 \quad \text{and} \quad x + 2 = 1.$$

Explain why.

68. What is wrong with the following solution?

$$4x^2 = 4x$$
$$x = 1 \qquad \text{Divide both sides by } 4x.$$

Explain how to solve the equation correctly.

◢ TECHNOLOGY INSIGHTS (EXERCISES 69-72)

In Section 4.2 we showed how an equation in one variable can be solved by getting 0 on one side, then replacing 0 with y to get a corresponding equation in two variables. The x-values of the x-intercepts of the graph of the two-variable equation then give the solutions of the original equation.

Use the calculator screens to determine the solution set of each quadratic equation. Verify your answers by substitution.

69. $x^2 + .4x - .05 = 0$

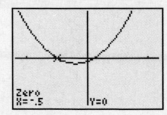 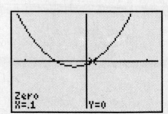

TECHNOLOGY INSIGHTS (EXERCISES 69–72) (CONTINUED)

70. $2x^2 - 7.2x + 6.3 = 0$

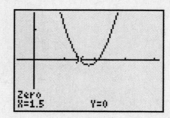 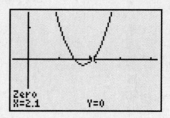

71. $2x^2 + 7.2x + 5.5 = 0$

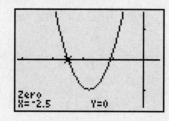 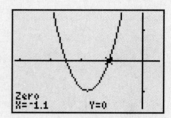

72. $4x^2 - x - 33 = 0$

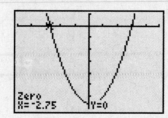 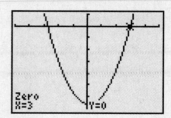

6.6 Applications of Quadratic Equations

We can now use factoring to solve quadratic equations that arise from applied problems.

OBJECTIVE 1 Solve problems about geometric figures. Most problems in this section require one of the formulas given on the inside covers. We still follow the six-step problem-solving method from Section 2.3 and continue the work with formulas and geometric problems begun in Section 2.4.

EXAMPLE 1 Solving an Area Problem

The Goldsteins are planning to add a rectangular porch to their house. The width of the porch will be 4 feet less than its length, and they want it to have an area of 96 square feet. Find the length and width of the porch.

Step 1 Let x = the length of the porch;

Step 2 $x - 4$ = the width (the width is 4 less than the length). See Figure 1.

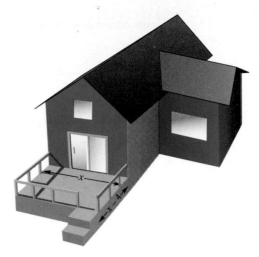

Figure 1

Step 3   The area of a rectangle is given by the formula

$$\text{area} = \text{length} \times \text{width}.$$

$$A = LW$$

$$\mathbf{96 = x(x - 4)} \qquad \text{Let } A = 96, L = x, W = x - 4.$$

Step 4 $$96 = x^2 - 4x \qquad \text{Distributive property}$$

$$0 = x^2 - 4x - 96 \qquad \text{Subtract 96 from both sides.}$$

$$0 = (x - 12)(x + 8) \qquad \text{Factor.}$$

$$x - 12 = 0 \qquad \text{or} \qquad x + 8 = 0 \qquad \text{Zero-factor property}$$

$$x = 12 \qquad \text{or} \qquad x = -8 \qquad \text{Solve.}$$

Step 5   The solutions of the equations are 12 and -8. Since a rectangle cannot have a negative length, discard the solution -8. Then 12 feet is the length of the porch and $12 - 4 = 8$ feet is the width.

Step 6   As a check, note that the width is 4 less than the length and the area is $8 \cdot 12 = 96$ square feet, as required.

 When solving an applied problem, remember Step 6. Always check solutions against physical facts.

The next application involves *perimeter,* the distance around a figure, as well as area.

┌ **E X A M P L E   2**   **Solving an Area and Perimeter Problem**

The length of a rectangular rug is 4 feet more than the width. The area of the rug is numerically 1 more than the perimeter. See Figure 2. Find the length and width of the rug.

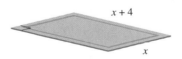

Figure 2

Let x = the width of the rug.

Then $x + 4$ = the length of the rug.

The area is the product of the length and width, so

$$A = LW.$$

Substituting $x + 4$ for the length and x for the width gives

$$A = (x + 4)x.$$

Now substitute into the formula for perimeter.

$$P = 2L + 2W$$
$$P = 2(x + 4) + 2x$$

According to the information given in the problem, the area is numerically 1 more than the perimeter. Write this as an equation.

The area   is   1   more than   the perimeter.

$$(x + 4)x = 1 + 2(x + 4) + 2x$$

Simplify and solve this equation.

$$x^2 + 4x = 1 + 2x + 8 + 2x$$ Distributive property

$$x^2 + 4x = 9 + 4x$$ Combine terms.

$$x^2 = 9$$ Subtract $4x$ from both sides.

$$x^2 - 9 = 0$$ Subtract 9 from both sides.

$$(x + 3)(x - 3) = 0$$ Factor.

$$x + 3 = 0 \quad \text{or} \quad x - 3 = 0$$ Zero-factor property

$$x = -3 \quad \text{or} \quad x = 3$$

A rectangle cannot have a negative width, so ignore -3. The only valid solution is 3, so the width is 3 feet and the length is $3 + 4 = 7$ feet. Check to see that the area is numerically 1 more than the perimeter. The rug is 3 feet wide and 7 feet long.

OBJECTIVE 2 Solve problems using the Pythagorean formula. The next example requires the **Pythagorean formula** from geometry.

Pythagorean Formula

If a right triangle (a triangle with a 90° angle) has longest side of length c and two other sides of lengths a and b, then

$$a^2 + b^2 = c^2.$$

(See the figure.) The longest side, the **hypotenuse,** is opposite the right angle. The two shorter sides are the **legs** of the triangle.

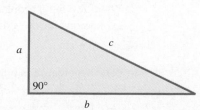

E X A M P L E 3   Using the Pythagorean Formula

Ed and Mark leave their office with Ed traveling north and Mark traveling east. When Mark is 1 mile farther than Ed from the office, the distance between them is 2 miles more than Ed's distance from the office. Find their distances from the office and the distance between them.

$$\text{Let} \qquad x \text{ represent Ed's distance from the office.}$$

$$\text{Then} \quad x + 1 \text{ represents Mark's distance from the office,}$$

$$\text{and} \quad x + 2 \text{ represents the distance between them.}$$

Label a right triangle as in Figure 3.

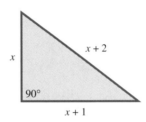

Figure 3

Substitute into the Pythagorean formula,

$$a^2 + b^2 = c^2$$
$$x^2 + (x + 1)^2 = (x + 2)^2.$$

Since $(x + 1)^2 = x^2 + 2x + 1$, and since $(x + 2)^2 = x^2 + 4x + 4$, the equation becomes

$$x^2 + x^2 + 2x + 1 = x^2 + 4x + 4.$$

$x^2 - 2x - 3 = 0$	Standard form
$(x - 3)(x + 1) = 0$	Factor.
$x - 3 = 0 \quad \text{or} \quad x + 1 = 0$	Zero-factor property
$x = 3 \quad \text{or} \quad x = -1$	Solve.

Since -1 cannot be the length of a side of a triangle, 3 is the only possible answer. Therefore, Ed is 3 miles north of the office, Mark is $3 + 1 = 4$ miles east of the office, and they are $3 + 2 = 5$ miles apart. Check to see that $3^2 + 4^2 = 5^2$ is true.

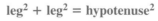

When solving a problem involving the Pythagorean formula, be sure that the expressions for the sides are properly placed.

$$\textbf{leg}^2 + \textbf{leg}^2 = \textbf{hypotenuse}^2$$

(The hypotenuse is opposite the right angle.)

CONNECTIONS

When a carpenter builds a floor for a rectangular room, it is essential that the corners of the floor are at right angles; otherwise, problems will occur when the walls are constructed, when flooring is laid, and so on. To check that the floor is "squared off," the carpenter can use the *converse* of the Pythagorean formula: If $a^2 + b^2 = c^2$, then the angle opposite side c is a right angle.

FOR DISCUSSION OR WRITING

Suppose a carpenter is building an 8-foot by 12-foot room. After the floor is built, the carpenter finds that the length of the diagonal of the floor is 14 feet, 8 inches. Is the floor "squared off" properly? If not, what should the diagonal measure?

OBJECTIVE **3** Solve problems using quadratic models. In the next example, we use the quadratic model for fuel economy from the beginning of this chapter.

EXAMPLE 4 Using a Quadratic Equation That Models Gasoline Mileage

Earlier we gave the quadratic expression $-.04x^2 + .93x + 21$, that models the automotive industry fuel economy trend. If we set the expression equal to y, then y is the average miles per gallon for the industry in year x. Recall, x is coded so that $x = 0$ represents 1978, $x = 2$ represents 1980, and so on. The equation

$$y = -.04x^2 + .93x + 21$$

was developed from the following data.

Fuel Economy

Year	Miles per Gallon
1978	19.9
1980	23.1
1982	25.1
1984	25.0
1986	25.9
1988	26.0
1990	25.4
1992	25.1
1994	24.7
1996	24.9

Source: National Highway Traffic Safety Administration.

(a) From the table, in what year did the number of miles per gallon appear to peak?

Look down the second column for the largest number, 26.0 miles per gallon. Reading across to the first column, we see that this mileage corresponds to 1988.

(b) Use the equation to find the miles per gallon in 1992.

The miles per gallon in year x is given by y. In 1992, $x = 1992 - 1978 = 14$. Substitute 14 for x in the equation $y = -.04x^2 + .93x + 21$.

$$y = -.04(14)^2 + .93(14) + 21 = 26.18 \qquad \text{Use a calculator.}$$

From the table, the actual data for 1992 is 25.1, so this approximation is a little high.

(c) Repeat part (b) for 1986.

For 1986, $x = 1986 - 1978 = 8$, and

$$y = -.04(8)^2 + .93(8) + 21 = 25.88. \qquad \text{Let } x = 8.$$

How does this approximation compare to the actual data in the table?

(d) Which of the results in parts (b) and (c) is the better approximation of the data in the table?

The approximation for 1986 is much closer than the approximation for 1992.

6.6 EXERCISES

Complete these statements which review the six-step method for solving applied problems first introduced in Chapter 2.

1. Read the problem carefully, choose _____ to represent _____ and write it down.

2. Write down a mathematical _____ for any other unknown quantities. Draw a _____ if it would be helpful.

3. Translate the problem into _____.

4. Solve the _____ and answer the _____ in the problem. _____ your answer.

In Exercises 5–8, a figure and a corresponding geometric formula are given. Complete each problem using the problem-solving steps from Chapter 2.

(a) *Write an equation using the formula and the given information. (Step 3)*

(b) *Solve the equation, giving only the solution(s) that make sense in the problem. (Step 4)*

(c) *Use the solution(s) to find the indicated dimensions of the figure. (Step 5)*

(d) *Check your solution. (Step 6)*

5. Area of a rectangle: $A = LW$
The area of this rectangle is 80 square units. Find its length and its width.

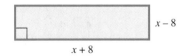

6. Area of a parallelogram: $A = bh$
The area of this parallelogram is 45 square units. Find its base and its height.

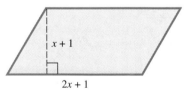

7. Area of a triangle: $A = \frac{1}{2}bh$
The area of this triangle is 60 square units. Find its base and its height.

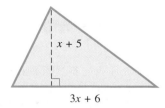

8. Volume of a rectangular Chinese box: $V - LWH$
The volume of this box is 192 cubic units. Find its length and its width.

Solve each problem. Check your answer to be sure that it is reasonable. Refer to the formulas found on the inside covers. See Examples 1 and 2.

9. The length of a VHS videocassette shell is 3 inches more than its width. The area of the rectangular top side of the shell is 28 square inches. Find the length and the width of the videocassette shell.

10. A plastic box that holds a standard audiocassette has a length 4 centimeters longer than its width. The area of the rectangular top of the box is 77 square centimeters. Find the length and the width of the box.

11. The dimensions of a certain IBM computer monitor screen are such that its length is 3 inches more than its width. If the length is increased by 1 inch while the width remains the same, the area is increased by 8 square inches. What are the dimensions of the screen?

12. The keyboard of the computer mentioned in Exercise 11 is 11 inches longer than it is wide. If both its length and width are increased by 2 inches, the area of the top of the keyboard is increased by 58 square inches. What are the length and the width of the keyboard?

13. The area of a triangle is 30 square inches. The base of the triangle measures 2 inches more than twice the height of the triangle. Find the measures of the base and the height.

14. A certain triangle has its base equal in measure to its height. The area of the triangle is 72 square meters. Find the equal base and height measure.

15. A ten-gallon aquarium holding African cichlids is 3 inches higher than it is wide. Its length is 21 inches, and its volume is 2730 cubic inches. What are the height and width of the aquarium?

16. Nana Nantambu wishes to build a box to hold her tools. It is to be 2 feet high, and the width is to be 3 feet less than its length. If its volume is to be 80 cubic feet, find the length and the width of the box.

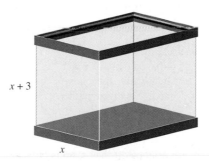

Use the Pythagorean formula to solve each problem. See Example 3.

17. The hypotenuse of a right triangle is 1 centimeter longer than the longer leg. The shorter leg is 7 centimeters shorter than the longer leg. Find the length of the longer leg of the triangle.

18. The longer leg of a right triangle is 1 meter longer than the shorter leg. The hypotenuse is 1 meter shorter than twice the shorter leg. Find the length of the shorter leg of the triangle.

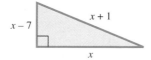

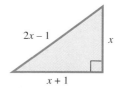

19. A ladder is resting against a wall. The top of the ladder touches the wall at a height of 15 feet. Find the distance from the wall to the bottom of the ladder if the length of the ladder is one foot more than twice its distance from the wall.

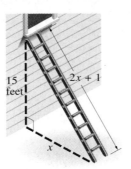

15 feet

$2x + 1$

x

20. Two cars leave an intersection. One car travels north; the other travels east. When the car traveling north had gone 24 miles, the distance between the cars was four miles more than three times the distance traveled by the car heading east. Find the distance between the cars at that time.

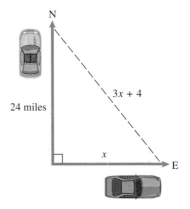

N

24 miles

$3x + 4$

x

E

21. A garden has the shape of a right triangle with one leg 2 meters longer than the other. The hypotenuse is two meters less than twice the length of the shorter leg. Find the length of the shorter leg.

22. The hypotenuse of a right triangle is 1 foot more than twice the length of the shorter leg. The longer leg is 1 foot less than twice the length of the shorter leg. Find the length of the shorter leg.

If an object is propelled upward from a height of s feet at an initial velocity of v feet per second, then its height h after t seconds (disregarding air resistance*) is given by the equation

48 feet

$$h = -16t^2 + vt + s,$$

where h is in feet. For example, if the object is propelled from a height of 48 feet with an initial velocity of 32 feet per second, its height h is given by the equation $h = -16t^2 + 32t + 48$.

Use this information in Exercises 23–26.

23. After how many seconds is the height 64 feet? (*Hint:* Let $h = 64$ and solve for t.)

24. After how many seconds is the height 60 feet?

25. After how many seconds does the object hit the ground? (*Hint:* When the object hits the ground, $h = 0$.)

26. The quadratic equation from Exercise 25 has two solutions, yet only one of them is appropriate for answering the question. Why is this so?

_____

*From now on, in all problems of this sort involving an object propelled upward or dropped from a height, we give formulas that disregard air resistance.

If an object is propelled upward from ground level with an initial velocity of 64 *feet per second, its height h in feet t seconds later is given by the equation* $h = -16t^2 + 64t$. *Use this information in Exercises 27–30.*

27. After how many seconds is the height 48 feet? (*Hint:* Let $h = 48$ and solve for t.)

28. It can be shown using concepts developed in other courses that the object reaches its maximum height 2 seconds after it is propelled. What is this maximum height?

29. After how many seconds does the object hit the ground? (*Hint:* Let $h = 0$.)

30. The quadratic equation from Exercise 29 has two solutions, yet only one of them is appropriate for answering the question. Why is this so?

Exercises 31 and 32 require the formula for the volume of a pyramid,

$$V = \frac{1}{3}Bh,$$

where B is the area of the base.

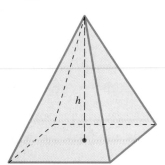

31. Suppose a pyramid has a rectangular base whose width is 3 centimeters less than the length. If the height is 8 centimeters and the volume is 144 cubic centimeters, find the length of the base.

32. The volume of a pyramid is 32 cubic meters. Suppose the numerical value of the height is 10 less than the numerical value of the area of the base. Find the area of the base.

If an object is dropped, the distance d in feet it falls in t seconds is given by

$$d = \frac{1}{2}gt^2,$$

where g is approximately 32 *feet per second*^{2}. *Find the distance an object would fall in the following times.*

33. 4 seconds

34. 8 seconds

How long would it take an object to fall from the top of the building described? Use a calculator as necessary, and round your answer to the nearest tenth of a second.

35. Navarre Building, New York City   512 feet

36. One Canada Square, London   800 feet

37. Central Plaza, Hong Kong   1028 feet

38. Las Vegas Stratosphere Tower, Las Vegas   1149 feet

Solve each problem. Refer to the discussion of consecutive integers and Examples 7 and 8 in Section 2.3.

39. The product of two consecutive integers is 11 more than their sum. Find the integers.

40. The product of two consecutive integers is 4 less than 4 times their sum. Find the integers.

41. Find three consecutive odd integers such that 3 times the sum of all three is 18 more than the product of the smaller two.

42. Find three consecutive odd integers such that the sum of all three is 42 less than the product of the larger two.

43. Find three consecutive even integers such that the sum of the squares of the smaller two is equal to the square of the largest.

44. Find three consecutive even integers such that the square of the sum of the smaller two is equal to twice the largest.

As we saw in Example 4, quadratic expressions are often used to approximate data. Exercises 45–47 further illustrate this kind of application.

45. The table shows the trend in transit ridership (in billions) from 1975 to 1995. The data is modeled fairly well by the quadratic expression $-.012x^2 + .381x + 5.96$. If y represents the number of passengers (in billions), then

$$y = -.012x^2 + .381x + 5.96.$$

Again, the years are coded, with $x = 5$ corresponding to 1975, $x = 10$ corresponding to 1980, and so on.

Transit Ridership Trend

Year	Number of Passengers (in billions)
1975	7.5
1980	8.5
1985	8.8
1990	9.0
1995	8.0

Source: American Public Transit Association.

(a) In what year shown in the table did the ridership peak?

(b) Use the equation to approximate the ridership y in 1990. (*Hint:* Let $x = 20$.) Does the equation give a good approximation of the value shown in the table?

(c) To what value does x correspond in 1995?

(d) Repeat part (b) for 1995.

(e) Which of the results in parts (b) and (d) is the better approximation of the actual data in the table?

gkp

46. Although motor vehicle accidents declined in New York City from 1990 to 1996, taxi accidents increased during that period, as indicated in the bar graph. The data in the graph is approximated by the quadratic equation

$$y = -143.3x^2 + 1823.3x + 6820.$$

Here, y represents the number of accidents in year x, where $x = 0$ corresponds to 1990, $x = 1$ corresponds to 1991, and so on.

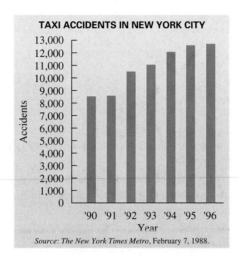

47. (See Exercise 46.) The number of people injured in taxi accidents also increased over the years 1990 to 1996. The equation

$$y = -416.72x^2 + 4416.7x + 8500,$$

where y represents the number injured in year x, with $x = 0$ corresponding to 1990, $x = 1$ corresponding to 1991, and so on, approximates the data quite well.

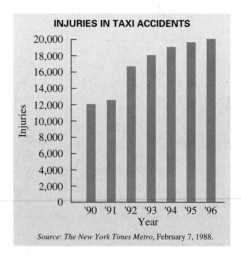

(a) What value of x corresponds to 1998?

(b) According to the equation, if this trend continued, how many taxi accidents occurred in 1998?

(a) Use the equation to approximate the number injured in 1998.

(b) Find the ratio for 1998 (see Section 2.5) of the number of accidents (answer to Exercise 46(b)) to the number of people injured (answer to part (a) of this exercise). What does this tell us about the average number of people injured per accident?

48. A square piece of cardboard is to be formed into an open-topped box by cutting 3-inch squares from the corners and folding up the sides. See the figure. The volume V of the resulting box is given by the formula $V = 3(x - 6)^2$, where x is the original length of each side of the piece of cardboard. What is the volume of the box if the original length of each side of the piece of cardboard is 14 inches?

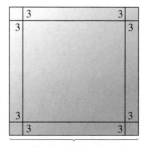

x inches

Original Piece of Cardboard

The Resulting Box

RELATING CONCEPTS (EXERCISES 49-52)

One of the many known proofs of the Pythagorean formula is based on the figures shown.

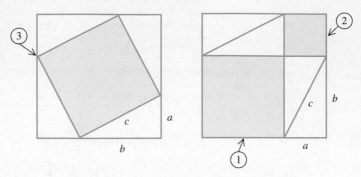

Figure A   Figure B

Refer to the appropriate figure and ***answer Exercises 49–52 in order.***

49. What is an expression for the area of the dark square labeled ③ in Figure A?

50. The five regions in Figure A are equal in area to the six regions in Figure B. What is an expression for the area of the square labeled ① in Figure B?

51. What is an expression for the area of the square labeled ② in Figure B?

52. Represent this statement using algebraic expressions: The sum of the areas of the dark regions in Figure B is equal to the area of the dark region in Figure A. What does this equation represent?

Did you make the connection that the Pythagorean formula is the result of the fact that the dark areas in the two figures are equal?

CHAPTER 6 GROUP ACTIVITY

 Transportation for Tomorrow

Objective: Use quadratic equations to find vehicle speed and stopping distance given acceleration rate.

In 1998 major automotive manufacturers began to produce electric cars. In this activity the speed and stopping distances of three different electric vehicles with given acceleration rates will be examined. Acceleration is the rate of change of speed with respect to time and is given in feet per second per second. Complete the table below for Cars 1, 2, and 3 to find the time, speed, and distance for each vehicle. (Round to the nearest whole number.)

Car	Acceleration Time	Speed	Stopping Distance
1			
2			
3			

A. Car 1: The Toyota Rav4EV has an acceleration constant of $a = 4.89$ feet per second per second.*

   1. Determine how long it would take for this vehicle to go $\frac{3}{10}$ mile (1584 feet) by solving the formula $d = at^2$, where d represents distance in feet, a is the acceleration constant, and t is time in seconds.

   2. Now find the speed r in miles per hour the car has achieved using the distance formula $d = rt$. (Convert time to hours, distance to miles.)

   3. Next find how many feet it will take the car to stop on a dry road at that speed (in miles per hour) using the following quadratic equation: $d = .045r^2 + 1.1r$.

B. Determine the same information for two other vehicles with the following acceleration constants.
Car 2: $a = 3.96$ feet per second per second
Car 3: $a = 5.48$ feet per second per second

Source: Ecomall.com.

CHAPTER 6 SUMMARY

KEY TERMS

6.1 factor
  common factor
  greatest common
    factor
	factoring by grouping
	6.2 factored form
	  factoring
	  prime polynomial
		6.4 perfect square
		    trinomial
		6.5 quadratic equation
		  standard form
			6.6 hypotenuse
			  legs

TEST YOUR WORD POWER

See how well you have learned the vocabulary in this chapter. Answers, with examples, are given at the bottom of the page.

1. Factoring is
(a) a method of multiplying polynomials
(b) the process of writing a polynomial as a product
(c) the answer in a multiplication problem
(d) a way to add the terms of a polynomial.

2. A polynomial is in **factored form** when
(a) it is prime
(b) it is written as a sum

(c) the squared term has a coefficient of 1
(d) it is written as a product.

3. A **perfect square trinomial** is a trinomial
(a) that can be factored as the square of a binomial
(b) that cannot be factored
(c) that is multiplied by a binomial
(d) where all terms are perfect squares.

4. A **quadratic equation** is a polynomial equation of
(a) degree one
(b) degree two
(c) degree three
(d) degree four.

5. A **hypotenuse** is
(a) either of the two shorter sides of a triangle
(b) the shortest side of a triangle
(c) the side opposite the right angle in a triangle
(d) the longest side in any triangle.

Answers to Test Your Word Power

1. (b) *Example:* $x^2 - 5x - 14 = (x - 7)(x + 2)$ 2. (d) *Example:* The factored form of $x^2 - 5x - 14$ is $(x - 7)(x + 2)$.   3. (a) *Example:* $a^2 + 2a + 1$ is a perfect square trinomial; its factored form is $(a + 1)^2$.   4. (b) *Examples:* $y^2 - 3y + 2 = 0$, $x^2 - 9 = 0$, $2m^2 = 6m + 8$ 5. (c) *Example:* In Figure 3 of Section 6.6, the hypotenuse is the side labeled $x + 2$.

CHAPTER 6   Factoring and Applications

QUICK REVIEW

CONCEPTS	EXAMPLES

6.1 THE GREATEST COMMON FACTOR; FACTORING BY GROUPING

Finding the Greatest Common Factor (GCF)

1. Include the largest numerical factor of every term.

2. Include each variable that is a factor of every term raised to the smallest exponent that appears in a term.

Find the greatest common factor of

$$4x^2y, \qquad -6x^2y^3, \qquad 2xy^2.$$

$$4x^2y = 2^2 \cdot x^2 \cdot y$$

$$-6x^2y^3 = -1 \cdot 2 \cdot 3 \cdot x^2 \cdot y^3$$

$$2xy^2 = 2 \cdot x \cdot y^2$$

The greatest common factor is $2xy$.

Factoring by Grouping

Step 1 Group the terms so that each group has a common factor.

Step 2 Factor out the greatest common factor in each group.

Step 3 Factor a common binomial factor from the result of Step 2.

Step 4 If Step 3 cannot be performed, try a different grouping.

Factor $3x^2 + 5x - 24xy - 40y$.

$$(3x^2 + 5x) + (-24xy - 40y)$$

$$x(3x + 5) - 8y(3x + 5)$$

$$(3x + 5)(x - 8y)$$

6.2 FACTORING TRINOMIALS

To factor $x^2 + bx + c$, find m and n such that $mn = c$ and $m + n = b$.

$$mn = c$$
$$\downarrow$$
$$x^2 + bx + c$$
$$\uparrow$$
$$m + n = b$$

Then

$$x^2 + bx + c = (x + m)(x + n).$$

Factor $x^2 + 6x + 8$.

$$mn = 8$$
$$\downarrow$$
$$x^2 + 6x + 8$$
$$\uparrow$$
$$m + n = 6$$

$m = 2$ and $n = 4$

$$x^2 + 6x + 8 = (x + 2)(x + 4)$$

6.3 MORE ON FACTORING TRINOMIALS

To factor $ax^2 + bx + c$:

By Grouping

Find m and n:

$$mn = ac$$
$$\downarrow \qquad\qquad \downarrow$$
$$ax^2 + bx + c$$
$$\uparrow$$
$$m + n = b$$

By Trial and Error

Use FOIL backwards.

Factor $3x^2 + 14x - 5$.

$$\uparrow\!\!-\!15\!-\!\!\uparrow$$

$mn = -15$, $m + n = 14$

By trial and error or by grouping.

$$3x^2 + 14x - 5 = (3x - 1)(x + 5)$$

6.4 SPECIAL FACTORING RULES

Difference of Squares

$x^2 - y^2 = (x + y)(x - y)$

Perfect Square Trinomial

$x^2 + 2xy + y^2 = (x + y)^2$

$x^2 - 2xy + y^2 = (x - y)^2$

Factor.

$$4x^2 - 9 = (2x + 3)(2x - 3)$$

$$9x^2 + 6x + 1 = (3x + 1)^2$$

$$4x^2 - 20x + 25 = (2x - 5)^2$$

CONCEPTS	EXAMPLES

Difference of Cubes
$x^3 - y^3 = (x - y)(x^2 + xy + y^2)$

$m^3 - 8 = m^3 - 2^3 = (m - 2)(m^2 + 2m + 4)$

Sum of Cubes
$x^3 + y^3 = (x + y)(x^2 - xy + y^2)$

$27 + z^3 = 3^3 + z^3 = (3 + z)(9 - 3z + z^2)$

6.5 SOLVING QUADRATIC EQUATIONS BY FACTORING

Zero-Factor Property
If a and b are real numbers and if $ab = 0$, then $a = 0$ or $b = 0$.

If $(x - 2)(x + 3) = 0$, then $x - 2 = 0$ or $x + 3 = 0$.

Solving a Quadratic Equation by Factoring

Solve $2x^2 = 7x + 15$.

Step 1 Write in standard form.

$$2x^2 - 7x - 15 = 0$$

Step 2 Factor.

$$(2x + 3)(x - 5) = 0$$

Step 3 Use the zero-factor property.

$$2x + 3 = 0 \quad \text{or} \quad x - 5 = 0$$
$$2x = -3 \qquad\qquad x = 5$$
$$x = -\frac{3}{2}$$

Step 4 Check.

Both solutions satisfy the original equation; the solution set is $\left\{-\dfrac{3}{2}, 5\right\}$.

6.6 APPLICATIONS OF QUADRATIC EQUATIONS

Pythagorean Formula
In a right triangle, the square of the hypotenuse equals the sum of the squares of the legs.

$$a^2 + b^2 = c^2$$

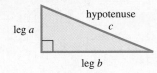

In a right triangle, one leg measures 2 feet longer than the other. The hypotenuse measures 4 feet longer than the shorter leg. Find the lengths of the three sides of the triangle.

Let $x =$ the length of the shorter leg. Then

$$x^2 + (x + 2)^2 = (x + 4)^2.$$

Verify that the solutions of this equation are -2 and 6. Discard -2 as a solution. Check that the sides are 6, $6 + 2 = 8$, and $6 + 4 = 10$ feet in length.

CHAPTER 6 REVIEW EXERCISES

[6.1] *Factor out the greatest common factor or factor by grouping.*

1. $7t + 14$

2. $60z^3 + 30z$

3. $2xy - 8y + 3x - 12$

4. $6y^2 + 9y + 4y + 6$

[6.2] *Factor completely.*

5. $x^2 + 5x + 6$

6. $y^2 - 13y + 40$

7. $q^2 + 6q - 27$

8. $r^2 - r - 56$

9. $r^2 - 4rs - 96s^2$

10. $p^2 + 2pq - 120q^2$

11. $8p^3 - 24p^2 - 80p$

12. $3x^4 + 30x^3 + 48x^2$

13. $p^7 - p^6q - 2p^5q^2$

14. $3r^5 - 6r^4s - 45r^3s^2$

[6.3]

15. To begin factoring $6r^2 - 5r - 6$, what are the possible first terms of the two binomial factors if we consider only positive integer coefficients?

16. What is the first step you would use to factor $2z^3 + 9z^2 - 5z$?

Factor completely.

17. $2k^2 - 5k + 2$

18. $3r^2 + 11r - 4$

19. $6r^2 - 5r - 6$

20. $10z^2 - 3z - 1$

21. $8v^2 + 17v - 21$

22. $24x^5 - 20x^4 + 4x^3$

23. $-6x^2 + 3x + 30$

24. $10r^3s + 17r^2s^2 + 6rs^3$

[6.4]

25. Which one of the following is the difference of two squares?
    (a) $32x^2 - 1$ (b) $4x^2y^2 - 25z^2$ (c) $x^2 + 36$ (d) $25y^3 - 1$

26. Which one of the following is a perfect square trinomial?
    (a) $x^2 + x + 1$ (b) $y^2 - 4y + 9$ (c) $4x^2 + 10x + 25$ (d) $x^2 - 20x + 100$

Factor completely.

27. $n^2 - 49$

28. $25b^2 - 121$

29. $49y^2 - 25w^2$

30. $144p^2 - 36q^2$

31. $x^2 + 100$

32. $r^2 - 12r + 36$

33. $9t^2 - 42t + 49$

34. $m^3 + 1000$

35. $125k^3 + 64x^3$

36. $343x^3 - 64$

[6.5] *Solve each equation and check your solutions.*

37. $z^2 + 4z + 3 = 0$

38. $2m^2 - 10m + 8 = 0$

39. $x(x - 8) = -15$

40. $3z^2 - 11z - 20 = 0$

41. $81t^2 - 64 = 0$

42. $y^2 = 8y$

43. $3n(n - 5) = 18$

44. $t^2 - 14t + 49 = 0$

45. $t^2 = 12(t - 3)$

46. $(5z + 2)(z^2 + 3z + 2) = 0$

[6.6] *Solve each problem.*

47. The length of a rectangle is 6 meters more than the width. The area is 40 square meters. Find the length and width of the rectangle.

48. The length of a rectangle is three times the width. If the width were increased by 3 meters while the length remained the same, the new rectangle would have an area of 30 square meters. Find the length and width of the original rectangle.

49. The volume of a box is to be 120 cubic meters. The width of the box is to be 4 meters, and the height 1 meter less than the length. Find the length and height of the box.

50. The sides of a right triangle have lengths (in feet) that are consecutive integers. What are the lengths of the sides?

If an object is propelled straight up from ground level with an initial velocity of 128 feet per second, its height h in feet after t seconds is $h = 128t - 16t^2$.

Find the height of the object after the following periods of time.

51. 1 second

52. 2 seconds

53. 4 seconds

54. When does the object described above return to the ground?

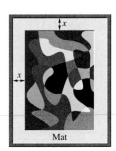

55. A 9-inch by 12-inch picture is to be placed on a cardboard mat so that there is an equal border around the picture. The area of the finished mat and picture is to be 208 square inches. How wide will the border be?

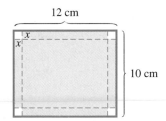

Mat

56. A box is made from a 12-centimeter by 10-centimeter piece of cardboard by cutting equal-sized squares from each corner and folding up the sides. The area of the bottom of the box is to be 48 square centimeters. Find the length of a side of the cutout squares.

12 cm

x

x

10 cm

57. In 1994, Greyhound Lines, Inc. was near its second bankruptcy in three years. Since then, a new CEO, Craig R. Lentzsch, appears to have turned the company around. The bar graph and table of values show the operating income for 1994, when Lentzsch took over, 1995, and 1996. (*Source:* Company Reports, Rothschild Inc.)

Year	Operating Income (in millions of dollars)
1994	−65
1995	9.4
1996	37

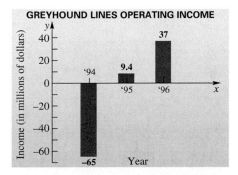

GREYHOUND LINES OPERATING INCOME

Using the data, we constructed the quadratic equation

$$y = -23.4x^2 + 285x - 831,$$

which gives the operating income y (in millions of dollars) in year x. We use $x = 4$ to represent 1994, $x = 5$ to represent 1995, and so on.

(a) Use the equation to predict the operating income in 1997.

(b) Use the equation to predict the operating income in 1998.

(c) Comment on the validity of the answers for parts (a) and (b).

MIXED REVIEW EXERCISES

Factor completely.

58. $z^2 - 11zx + 10x^2$ **59.** $3k^2 + 11k + 10$

60. $15m^2 + 20mp - 12mp - 16p^2$ **61.** $y^4 - 625$

62. $6m^3 - 21m^2 - 45m$ **63.** $24ab^3c^2 - 56a^2bc^3 + 72a^2b^2c$

64. $25a^2 + 15ab + 9b^2$ **65.** $12x^2yz^3 + 12xy^2z - 30x^3y^2z^4$

66. $2a^5 - 8a^4 - 24a^3$ **67.** $12r^2 + 18rq - 10rq - 15q^2$

68. $1000a^3 + 27$ **69.** $49t^2 + 56t + 16$

Solve.

70. $t(t - 7) = 0$ **71.** $x(x + 3) = 10$ **72.** $4x^2 - x - 3 = 0$

73. The numbers of alternative-fueled vehicles, in thousands, in use for the years 1995–1997 are given in the table.

Alternative-Fueled Vehicles

Year	Number (in thousands)
1995	333
1996	357
1997	386

Source: Energy Information Administration, Alternatives to Traditional Fuels, 1993.

Using statistical methods, we constructed the quadratic equation

$$y = 2.5x^2 - 453.5x + 20,850$$

to model the number of vehicles y in year x. Here we used $x = 95$ for 1995, $x = 96$ for 1996, and so on. Because only three years of data were used to determine the model, we must be particularly careful about using it to estimate for years before 1995 or after 1997.

(a) What prediction for 1998 is given by the equation?

(b) Why might the prediction for 1998 be unreliable?

74. The sum of two consecutive even integers is 34 less than their product. Find the integers.

75. The floor plan for a house is a rectangle with length 7 meters more than its width. The area is 170 square meters. Find the width and length of the house.

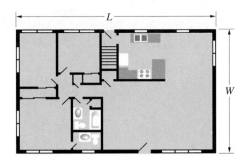

76. The triangular sail of a schooner has an area of 30 square meters. The height of the sail is 4 meters more than the base. Find the base of the sail.

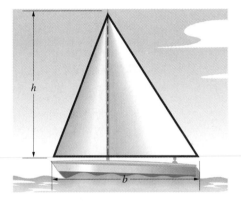

77. Two cars left an intersection at the same time. One traveled north. The other traveled 14 miles farther, but to the east. How far apart were they then, if the distance between them was 4 miles more than the distance traveled east?

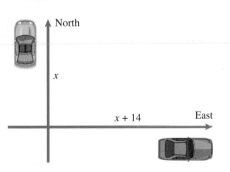

78. A ladder is leaning against a building. The distance from the bottom of the ladder to the building is 4 feet less than the length of the ladder. How high up the side of the building is the top of the ladder if that distance is 2 feet less than the length of the ladder?

79. A bicyclist heading east and a motorist traveling south left an intersection at the same time. When the motorist had gone 17 miles farther than the bicyclist, the distance between them was 1 mile more than the distance traveled by the motorist. How far apart were they then? (*Hint:* Draw a sketch.)

80. Although $(2x + 8)(3x - 4) = 6x^2 + 16x - 32$ is a true statement, the polynomial is not factored completely. Explain why and give the completely factored form.

CHAPTER 6 TEST

1. Which one of the following is the correct, completely factored form of $2x^2 - 2x - 24$?
   (a) $(2x + 6)(x - 4)$ (b) $(x + 3)(2x - 8)$
   (c) $2(x + 4)(x - 3)$ (d) $2(x + 3)(x - 4)$

Factor each polynomial completely. If it cannot be factored, write prime.

2. $2m^3n^2 + 3m^3n - 5m^2n^2$ 3. $x^2 - 5x - 24$

4. $2x^2 + x - 3$ 5. $10z^2 - 17z + 3$

6. $t^2 + 2t + 3$ 7. $x^2 + 36$

8. $12 - 6a + 2b - ab$ 9. $9y^2 - 64$

10. $4x^2 - 28xy + 49y^2$ 11. $-2x^2 - 4x - 2$

12. $6t^4 + 3t^3 - 108t^2$ 13. $r^3 - 125$

14. $8k^3 + 64$

15. Why is $(p + 3)(p + 3)$ not the correct factored form of $p^2 + 9$?

Solve each equation.

16. $2r^2 - 13r + 6 = 0$ 17. $25x^2 - 4 = 0$

18. $x(x - 20) = -100$ 19. $t^3 = 9t$

Solve each problem.

20. If an object is propelled from ground level at an initial velocity of 96 feet per second, after t seconds its height h in feet is given by the formula $h = -16t^2 + 96t$. After how many seconds will its height h be 108 feet?

21. A carpenter needs to cut a brace to support a wall stud. See the figure. The brace should be 7 feet less than three times the length of the stud. The brace will be fastened on the floor 1 foot less than twice the length of the stud away from the stud.
    (a) Let x represent the length of the stud. Write an expression for the length of the brace.
    (b) Write an expression for the distance from the wall to where the brace is fastened.
    (c) How long should the brace be? (*Hint:* Use the Pythagorean formula.)

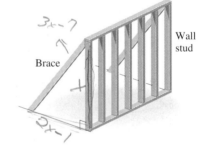

22. Assume the data given in Exercise 73 of the Chapter 6 Review Exercises represent the number of alternative-fueled vehicles at the beginning of each of the years shown. Then 95.5 would represent the number of such vehicles halfway through 1995 (July 1). Use the equation that models the data,

$$y = 2.5x^2 - 453.5x + 20{,}850,$$

to approximate the numbers of alternative-fueled vehicles on July 1 of 1995 and 1996. (Round your answers to the nearest whole number.)

23. Why isn't "$x = \dfrac{2}{3}$" the correct response to "Solve the equation $x^2 = \dfrac{4}{9}$"?

CUMULATIVE REVIEW EXERCISES CHAPTERS 1-6

Solve each equation.

1. $3x + 2(x - 4) = 4(x - 2)$

2. $.3x + .9x = .06$

3. $\frac{2}{3}y - \frac{1}{2}(y - 4) = 3$

4. Solve for P: $A = P + Prt$.

Solve each problem.

5. Find the measures of the marked angles.

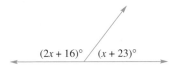

$(2x + 16)°$ $(x + 23)°$

6. In a recent year, the United States exported to the Bahamas $315 million more in goods than it imported. Together, the two amounts totaled $1167 million. How much were the exports and how much were the imports?

7. In a mixture of concrete, there are 3 pounds of cement mix for every 1 pound of gravel. If the mixture contains a total of 140 pounds of these two ingredients, how many pounds of gravel are there?

8. Fill in each blank with *positive* or *negative*. The point with coordinates (a, b) is in
   (a) quadrant II if a is _____ and b is _____.
   (b) quadrant III if a is _____ and b is _____.

9. The sales of a small company are a function of the number of years x it has been in business. Sales (in thousands) are given by $y = 12x + 3$. Write ordered pairs for this function for the second and fifth years it has been in business.

10. What are the x- and y-intercepts of the graph of the sales equation $y = 12x + 3$?

11. Graph the linear equation $y = 12x + 3$ for $x \geq 0$.

12. The points on the graph show the number of U.S. radio stations in the years 1990–1995, along with the graph of a linear equation that models the data. Use the ordered pairs shown on the graph to find the slope of the line. Interpret the slope.

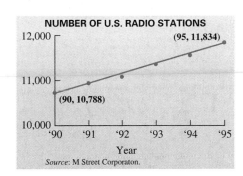

NUMBER OF U.S. RADIO STATIONS

(95, 11,834)

(90, 10,788)

Year

Source: M Street Corporaton.

Evaluate each expression.

13. $2^{-3} \cdot 2^5$

14. $\left(\frac{3}{4}\right)^{-2}$

15. $\left(\frac{4^{-3} \cdot 4^4}{4^5}\right)^{-1}$

16. Simplify $\dfrac{(p^2)^3 p^{-4}}{(p^{-3})^{-1} p}$ and write the answer using only positive exponents. Assume $p \neq 0$.

Perform the indicated operations.

17. $(2k^2 + 4k) - (5k^2 - 2) - (k^2 + 8k - 6)$ **18.** $3m^3(2m^5 - 5m^3 + m)$

19. $(y^2 + 3y + 5)(3y - 1)$ **20.** $(2p + 3q)(2p - 3q)$

21. $\dfrac{8x^4 + 12x^3 - 6x^2 + 20x}{2x}$ **22.** $(12p^3 + 2p^2 - 12p + 5) \div (2p - 2)$

Factor completely.

23. $2a^2 + 7a - 4$ **24.** $10m^2 + 19m + 6$

25. $15x^2 - xy - 6y^2$ **26.** $9x^2 + 6x + 1$

27. $-32t^2 - 112tz - 98z^2$ **28.** $25r^2 - 81t^2$

29. $100x^2 + 25$ **30.** $2pq + 6p^3q + 8p^2q$

31. $2ax - 2bx + ay - by$

Solve each equation.

32. $(2p - 3)(p + 2)(p - 6) = 0$ **33.** $6m^2 + m - 2 = 0$

Solve each problem.

34. The difference between the squares of two consecutive even integers is 28 less than the square of the smaller integer. Find the two integers.

35. The length of the hypotenuse of a right triangle is twice the length of the shorter leg, plus 3 meters. The longer leg is 7 meters longer than the shorter leg. Find the lengths of the sides.

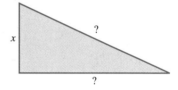

Rational Expressions

One of the reasons that people are living longer today than in the past is the increased efficiency of health care. Organ transplants allow at-risk patients to survive longer than they would have a decade ago. Heart, liver, kidney, lung, pancreas, and cornea transplants are now available. The table shows the number of heart transplants in the United States for each year from 1990 through 1994.

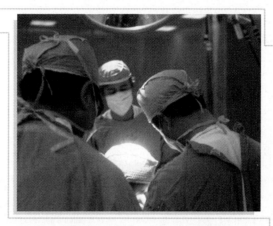

Health Care

7.1 The Fundamental Property of Rational Expressions

7.2 Multiplication and Division of Rational Expressions

7.3 The Least Common Denominator

7.4 Addition and Subtraction of Rational Expressions

7.5 Complex Fractions

7.6 Solving Equations Involving Rational Expressions

Summary: Exercises on Operations and Equations with Rational Expressions

7.7 Applications of Rational Expressions

Year	1990	1991	1992	1993	1994
Number of Heart Transplants	2108	2125	2171	2297	2340

Source: U.S. Department of Health and Human Services, Public Health Service, Division of Organ Transplantation, and United Network for Organ Sharing.

In 1995, the number of heart transplants was still on the rise, and compared to ten years before, in 1985, the increase was drastic. In those two years, the total number of transplants was 3080, and the ratio of the number in 1995 to the number in 1985 was approximately 33 to 10. How many transplants were performed in each of these two years? To answer this question, write an equation. If we let x represent the number of transplants in 1995, then $3080 - x$ represents the number of transplants in 1985. Writing this information as the ratio

$$\frac{\text{Number of transplants in 1995}}{\text{Number of transplants in 1985}},$$

we have

$$\frac{x}{3080 - x} = \frac{33}{10}.$$

The fractions on each side of this equation are *rational expressions,* the topic of this chapter. The equation is an example of a *rational equation.* In Section 7.6, we show how to solve this type of equation. You are asked to solve this problem in Exercise 71 of Section 7.7.

▬▬▬ 7.1 The Fundamental Property of Rational Expressions

OBJECTIVES

1. Find the values for which a rational expression is undefined.

2. Find the numerical value of a rational expression.

3. Write rational expressions in lowest terms.

4. Recognize equivalent forms of rational expressions.

FOR EXTRA HELP

📖 **SSG** Sec. 7.1
SSM Sec. 7.1

💿 **Pass the Test Software**

💿 **InterAct Math Tutorial Software**

📼 **Video 11**

The quotient of two integers (with divisor not zero) is called a rational number. In the same way, the quotient of two polynomials with divisor not equal to zero is called a *rational expression*. The techniques of factoring, studied in Chapter 6, are essential in working with rational expressions.

Rational Expression

A **rational expression** is an expression of the form

$$\frac{P}{Q}$$

where P and Q are polynomials, with $Q \neq 0$.

Examples of rational expressions include $\dfrac{-6x}{x^3 + 8}$, $\dfrac{9x}{y + 3}$, and $\dfrac{2m^3}{9}$.

OBJECTIVE 1 Find the values for which a rational expression is undefined. A fraction with denominator 0 is *not* a rational expression, since division by 0 is not possible. For this reason, be careful when substituting a number for a variable in the denominator of a rational expression. For example, in

$$\frac{8x^2}{x - 3}$$

x can take on any value except 3. When $x = 3$, the denominator becomes $3 - 3 = 0$, making the expression undefined.

 NOTE The numerator of a rational expression may be *any* number.

To determine the values for which a rational expression is undefined, use the following procedure.

Determining When a Rational Expression Is Undefined

Step 1   Set the denominator of the rational expression equal to 0.

Step 2   Solve this equation.

Step 3   The solutions of the equation are the values that make the rational expression undefined.

This procedure is illustrated in Example 1.

EXAMPLE 1   Finding Values That Make Rational Expressions Undefined

Find any values for which the following rational expressions are undefined.

(a) $\dfrac{p + 5}{3p + 2}$

Remember that the *numerator* may be any number; we must find any value of p that makes the *denominator* equal to 0 since division by 0 is undefined.

Step 1   Set the denominator equal to 0.

$$3p + 2 = 0$$

Step 2   Solve this equation.

$$3p = -2$$

$$p = -\frac{2}{3}$$

Step 3   Since $p = -\frac{2}{3}$ will make the denominator 0, the given expression is undefined for $-\frac{2}{3}$.

(b) $\dfrac{9m^2}{m^2 - 5m + 6}$

Find the numbers that make the denominator 0 by solving the equation

$$m^2 - 5m + 6 = 0.$$

$$(m - 2)(m - 3) = 0 \qquad \text{Factor.}$$

$$m - 2 = 0 \quad \text{or} \quad m - 3 = 0 \qquad \text{Zero-factor property}$$

$$m = 2 \quad \text{or} \quad m = 3 \qquad \text{Solve.}$$

The original expression is undefined for 2 and for 3.

(c) $\dfrac{2r}{r^2 + 1}$

This denominator cannot equal 0 for any value of r, since r^2 is always greater than or equal to 0 and adding 1 makes the sum greater than 0. Thus, there are no values for which this rational expression is undefined.

OBJECTIVE 2   Find the numerical value of a rational expression.

EXAMPLE 2   Evaluating a Rational Expression

Find the numerical value of $\dfrac{3x + 6}{2x - 4}$ for the given values of x.

(a) $x = 1$

Find the value of the rational expression by substituting 1 for x.

$$\frac{3x + 6}{2x - 4} = \frac{3(1) + 6}{2(1) - 4} \qquad \text{Let } x = 1.$$

$$= \frac{9}{-2}$$

$$= -\frac{9}{2}$$

(b) $x = 2$

Substituting 2 for x makes the denominator 0, so the rational expression is undefined when $x = 2$.

OBJECTIVE **3** Write rational expressions in lowest terms. A fraction such as $\frac{2}{3}$ is said to be in lowest terms. How can "lowest terms" be defined? We use the idea of greatest common factor to give this definition, which applies to all rational expressions.

Lowest Terms

A rational expression $\frac{P}{Q}$ ($Q \neq 0$) is in **lowest terms** if the greatest common factor of its numerator and denominator is 1.

The properties of rational numbers also apply to rational expressions. We use the fundamental property of rational expressions to write a rational expression in lowest terms.

Fundamental Property of Rational Expressions

If $\frac{P}{Q}$ is a rational expression and if K represents any polynomial, where $K \neq 0$, then

$$\frac{PK}{QK} = \frac{P}{Q}.$$

This property is based on the identity property of multiplication, since

$$\frac{PK}{QK} = \frac{P}{Q} \cdot \frac{K}{K} = \frac{P}{Q} \cdot 1 = \frac{P}{Q}.$$

The fundamental property suggests the method used in the next example to write a fraction in lowest terms. We show the procedure with both a common fraction and a rational expression involving variables. Notice the similarity.

EXAMPLE 3 Writing in Lowest Terms

Write each expression in lowest terms.

(a) $\dfrac{30}{72}$ **(b)** $\dfrac{14k^2}{2k^3}$

Begin by factoring.

$$\frac{30}{72} = \frac{2 \cdot 3 \cdot 5}{2 \cdot 2 \cdot 2 \cdot 3 \cdot 3} \qquad\qquad \frac{14k^2}{2k^3} = \frac{2 \cdot 7 \cdot k \cdot k}{2 \cdot k \cdot k \cdot k}$$

Group any factors common to the numerator and denominator.

$$\frac{30}{72} = \frac{5(\mathbf{2 \cdot 3})}{2 \cdot 2 \cdot 3(\mathbf{2 \cdot 3})} \qquad\qquad \frac{14k^2}{2k^3} = \frac{7(\mathbf{2 \cdot k \cdot k})}{k(\mathbf{2 \cdot k \cdot k})}$$

Use the fundamental property.

$$\frac{30}{72} = \frac{5}{2 \cdot 2 \cdot 3} = \frac{5}{12} \qquad\qquad \frac{14k^2}{2k^3} = \frac{7}{k}$$

E X A M P L E   4   Writing in Lowest Terms

Write each rational expression in lowest terms.

(a) $\dfrac{3x - 12}{5x - 20}$

   Begin by factoring both numerator and denominator. Then use the fundamental property.

$$\frac{3x - 12}{5x - 20} = \frac{3(x - 4)}{5(x - 4)}$$

$$= \frac{3}{5}$$

The rational expression $\dfrac{3x - 12}{5x - 20}$ equals $\dfrac{3}{5}$ for all values of x, where $x \neq 4$ (since the denominator of the original rational expression is 0 when x is 4).

(b) $\dfrac{m^2 + 2m - 8}{2m^2 - m - 6}$

$$\frac{m^2 + 2m - 8}{2m^2 - m - 6} = \frac{(m + 4)(m - 2)}{(2m + 3)(m - 2)} \qquad \text{Factor.}$$

$$= \frac{m + 4}{2m + 3} \qquad \text{Fundamental property}$$

Thus, $\dfrac{m^2 + 2m - 8}{2m^2 - m - 6} = \dfrac{m + 4}{2m + 3}$ for $m \neq -\dfrac{3}{2}$ and $m \neq 2$, since the denominator of the original expression is 0 for these values of m.

   From now on, we will write statements of equality of rational expressions with the understanding that they apply only to those real numbers that make neither denominator equal to 0.

One of the most common errors in algebra occurs when students attempt to write rational expressions in lowest terms *before factoring*. The fundamental property is applied only *after* the numerator and denominator are expressed in factored form. For example, although x appears in both the numerator and denominator in Example 4(a), and 12 and 20 have a common factor of 4, the fundamental property cannot be used before factoring because $3x$, $5x$, 12, and 20 are *terms,* not *factors.* Terms are *added* or *subtracted;* factors are *multiplied* or *divided.* For example,

$$\frac{6 + 2}{3 + 2} = \frac{8}{5}, \quad \textbf{not} \quad \frac{6}{3} + \frac{2}{2} = 2 + 1 = 3.$$

Also, $\dfrac{2x + 3}{4x + 6} = \dfrac{2x + 3}{2(2x + 3)} = \dfrac{1}{2},$ but $\dfrac{x^2 + 6}{x + 3} \neq x + 2.$

E X A M P L E 5   Writing in Lowest Terms (Factors Are Opposites)

Write $\dfrac{x - y}{y - x}$ in lowest terms.

At first glance, there does not seem to be any way in which $x - y$ and $y - x$ can be factored to get a common factor. This is not the case, however, since the numerator can be factored as

$$x - y = -1(-x + y) = -1(y - x).$$

Now use the fundamental property to simplify.

$$\frac{x - y}{y - x} = \frac{-1(y - x)}{1(y - x)} = \frac{-1}{1} = -1$$

Either the numerator or the denominator could have been factored in the first step.

In Example 5, the binomials in the numerator and denominator are opposites. A general rule for this situation follows.

A fraction in which the numerator and denominator show subtraction of the same terms but in opposite order is equal to -1. More generally, if a factor of the numerator is the opposite of a factor in the denominator, the quotient of those two factors is -1.

E X A M P L E 6   Writing in Lowest Terms (Factors Are Opposites)

Write each rational expression in lowest terms.

(a) $\dfrac{2 - m}{m - 2}$

Since $2 - m$ and $m - 2$ (or $-2 + m$) are opposites, we use the rule above.

$$\frac{2 - m}{m - 2} = -1$$

(b) $\dfrac{4x^2 - 9}{6 - 4x}$

Factor the numerator and denominator and use the rule above.

$$\frac{4x^2 - 9}{6 - 4x} = \frac{(2x + 3)(2x - 3)}{2(3 - 2x)}$$

$$= \frac{2x + 3}{2}(-1)$$

$$= -\frac{2x + 3}{2}$$

(c) $\dfrac{3 + r}{3 - r}$

The quantity $3 - r$ *is not* the opposite of $3 + r$. This rational expression cannot be simplified.

OBJECTIVE **4**   Recognize equivalent forms of rational expressions. When working with rational expressions, it is important to be able to recognize equivalent forms of an expression. For example, the common fraction $-\frac{5}{6}$ can also be written as $\frac{-5}{6}$ and as $\frac{5}{-6}$. Look again at Example 6(b). The form of the answer given there is only one of several acceptable forms. The $-$ sign representing the -1 factor is in front of the fraction, on the same line as the fraction bar. The -1 factor may be placed in front of the fraction, in the numerator, or in the denominator. Some other acceptable forms of the answer are

$$\frac{-(2x + 3)}{2}, \qquad \frac{-2x - 3}{2}, \qquad \text{and} \qquad \frac{2x + 3}{-2}.$$

However, $\dfrac{-2x + 3}{2}$ is *not* an acceptable form, because the sign preceding 3 in the numerator should be $-$ rather than $+$. (The form $\dfrac{2x + 3}{-2}$, listed above, is seldom used.)

E X A M P L E   7   Writing Equivalent Forms of a Rational Expression

Write four equivalent forms of the rational expression $-\dfrac{3x + 2}{x - 6}$.

If we let the negative sign preceding the fraction apply to the numerator, we have the equivalent form

$$\frac{-(3x + 2)}{x - 6}.$$

By distributing the negative sign in this expression, we have another equivalent form,

$$\frac{-3x - 2}{x - 6}.$$

If we let the negative sign apply to the denominator of the fraction, we get

$$\frac{3x + 2}{-(x - 6)}$$

or, distributing once again,

$$\frac{3x + 2}{-x + 6}.$$

CAUTION   In Example 7, it would be incorrect to distribute the negative sign to *both* the numerator *and* the denominator. This would lead to the *opposite* of the original expression.

CONNECTIONS

In Chapter 5 we used long division to find the quotient of two polynomials. For example, we found $(2x^2 + 5x - 12) \div (2x - 3)$ as follows:

$$
\begin{array}{r}
x + 4 \\
2x - 3\overline{)2x^2 + 5x - 12} \\
\underline{2x^2 - 3x} \\
8x - 12 \\
\underline{8x - 12} \\
0
\end{array}
$$

The quotient is $x + 4$. We also get the same quotient by expressing the division problem as a rational expression (fraction) and writing this rational expression in lowest terms.

$$\frac{2x^2 + 5x - 12}{2x - 3} = \frac{(2x - 3)(x + 4)}{2x - 3} = x + 4$$

FOR DISCUSSION OR WRITING

What kind of division problem has a quotient that cannot be found by reducing a fraction to lowest terms? Try using rational expressions to solve the following division problems. Then use long division to compare.

1. $(3x^2 + 11x + 8) \div (x + 2)$

2. $(x^3 - 8) \div (x^2 + 2x + 4)$

7.1 EXERCISES

1. Fill in the blanks with the correct responses.

(a) The rational expression $\dfrac{x + 5}{x - 3}$ is undefined when $x =$ _____, and is equal to 0 when

$x =$ _____.

(b) The rational expression $\dfrac{p - q}{q - p}$ is undefined when $p =$ _____, and in all other cases

its simplified form is _____.

2. Make the correct choice for the blank.

(a) $\dfrac{4 - r^2}{4 + r^2}$ _____ equal to -1.
(is/is not)

(b) $\dfrac{5 + 2x}{3 - x}$ and $\dfrac{-5 - 2x}{x - 3}$ _____ equivalent rational expressions.
(are/are not)

3. Define *rational expression* in your own words, and give an example.

4. Give an example of a rational expression that is not in lowest terms, and then show the steps required to write it in lowest terms.

Find any values for which each rational expression is undefined. See Example 1.

5. $\dfrac{12}{5y}$ **6.** $\dfrac{-7}{3z}$ **7.** $\dfrac{4x^2}{3x + 5}$ **8.** $\dfrac{2x^3}{3x + 4}$

9. $\dfrac{5m + 2}{m^2 + m - 6}$ **10.** $\dfrac{2r - 5}{r^2 - 5r + 4}$ **11.** $\dfrac{3x - 1}{x^2 + 2}$ **12.** $\dfrac{4q + 2}{q^2 + 9}$

*Find the numerical value of each rational expression when **(a)** $x = 2$ and **(b)** $x = -3$. See Example 2.*

13. $\dfrac{5x - 2}{4x}$

14. $\dfrac{3x + 1}{5x}$

15. $\dfrac{2x^2 - 4x}{3x}$

16. $\dfrac{4x^2 - 1}{5x}$

17. $\dfrac{(-3x)^2}{4x + 12}$

18. $\dfrac{(-2x)^3}{3x + 9}$

19. $\dfrac{5x + 2}{2x^2 + 11x + 12}$

20. $\dfrac{7 - 3x}{3x^2 - 7x + 2}$

21. If 2 is substituted for x in the rational expression $\dfrac{x - 2}{x^2 - 4}$, the result is $\dfrac{0}{0}$. An often-heard statement is "Any number divided by itself is 1." Does this mean that this expression is equal to 1 for $x = 2$? If not, explain.

22. For $x \neq 2$, the rational expression $\dfrac{2(x - 2)}{x - 2}$ is equal to 2. Can the same be said for $\dfrac{2x - 2}{x - 2}$? Explain.

Write each rational expression in lowest terms. See Examples 3 and 4.

23. $\dfrac{18r^3}{6r}$

24. $\dfrac{27p^2}{3p}$

25. $\dfrac{4(y - 2)}{10(y - 2)}$

26. $\dfrac{15(m - 1)}{9(m - 1)}$

27. $\dfrac{(x + 1)(x - 1)}{(x + 1)^2}$

28. $\dfrac{(t + 5)(t - 3)}{(t - 1)(t + 5)}$

29. $\dfrac{7m + 14}{5m + 10}$

30. $\dfrac{8z - 24}{4z - 12}$

31. $\dfrac{m^2 - n^2}{m + n}$

32. $\dfrac{a^2 - b^2}{a - b}$

33. $\dfrac{12m^2 - 3}{8m - 4}$

34. $\dfrac{20p^2 - 45}{6p - 9}$

35. $\dfrac{3m^2 - 3m}{5m - 5}$

36. $\dfrac{6t^2 - 6t}{2t - 2}$

37. $\dfrac{9r^2 - 4s^2}{9r + 6s}$

38. $\dfrac{16x^2 - 9y^2}{12x - 9y}$

39. $\dfrac{zw + 4z - 3w - 12}{zw + 4z + 5w + 20}$

40. $\dfrac{km + 4k + 4m + 16}{km + 4k + 5m + 20}$

41. $\dfrac{5k^2 - 13k - 6}{5k + 2}$

42. $\dfrac{7t^2 - 31t - 20}{7t + 4}$

43. $\dfrac{2x^2 - 3x - 5}{2x^2 - 7x + 5}$

44. $\dfrac{3x^2 + 8x + 4}{3x^2 - 4x - 4}$

Write each expression in lowest terms. See Examples 5 and 6.

45. $\dfrac{6 - t}{t - 6}$

46. $\dfrac{2 - k}{k - 2}$

47. $\dfrac{m^2 - 1}{1 - m}$

48. $\dfrac{a^2 - b^2}{b - a}$

49. $\dfrac{q^2 - 4q}{4q - q^2}$

50. $\dfrac{z^2 - 5z}{5z - z^2}$

51. $\dfrac{p + 6}{p - 6}$

52. $\dfrac{5 - x}{5 + x}$

53. The area of the rectangle is represented by $x^4 + 10x^2 + 21$. What is the width?

$\left(Hint: \text{Use } W = \dfrac{A}{L}.\right)$

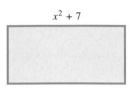

$x^2 + 7$

54. The volume of the box is represented by

$$(x^2 + 8x + 15)(x + 4).$$

Find the polynomial that represents the area of the bottom of the box.

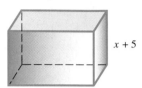

$x + 5$

Write four equivalent expressions for each of the following. See Example 7.

55. $-\dfrac{x + 4}{x - 3}$

56. $-\dfrac{x + 6}{x - 1}$

57. $-\dfrac{2x - 3}{x + 3}$

58. $-\dfrac{5x - 6}{x + 4}$

59. $\dfrac{-3x + 1}{5x - 6}$

60. $\dfrac{-2x - 9}{3x + 1}$

Write each expression in lowest terms.

61. $\dfrac{m^2 - n^2 - 4m - 4n}{2m - 2n - 8}$

62. $\dfrac{x^2y + y + x^2z + z}{xy + xz}$

63. $\dfrac{b^3 - a^3}{a^2 - b^2}$

64. $\dfrac{k^3 + 8}{k^2 - 4}$

65. $\dfrac{z^3 + 27}{z^3 - 3z^2 + 9z}$

66. $\dfrac{1 - 8r^3}{8r^2 + 4r + 2}$

TECHNOLOGY INSIGHTS (EXERCISES 67–72)

In the tables shown in Exercises 67–72, the expressions for Y_1 and Y_2 are equivalent except for the value for which Y_1 shows an error. In the table on the right, $Y_1 = \dfrac{x^2 - 4}{x + 2}$, and so, an error message appears at $x = -2$. In each case, Y_1 is defined by a rational expression. Predict the denominator of Y_1 in each case.

X	Y₁	Y₂
-5	-7	-7
-4	-6	-6
-3	-5	-5
-2	ERROR	-4
-1	-3	-3
0	-2	-2
1	-1	-1

Y₂⊟X−2

67.

X	Y₁	Y₂
-6	-9	-9
-5	-8	-8
-4	-7	-7
-3	ERROR	-6
-2	-5	-5
-1	-4	-4
0	-3	-3

Y₂⊟X−3

68.

X	Y₁	Y₂
-7	-11	-11
-6	-10	-10
-5	-9	-9
-4	ERROR	-8
-3	-7	-7
-2	-6	-6
-1	-5	-5

Y₂⊟X−4

69.

X	Y₁	Y₂
-8	13	13
-7	12	12
-6	11	11
-5	ERROR	10
-4	9	9
-3	8	8
-2	7	7

Y₂⊟5−X

70.

X	Y₁	Y₂
-9	15	15
-8	14	14
-7	13	13
-6	ERROR	12
-5	11	11
-4	10	10
-3	9	9

Y₂⊟6−X

TECHNOLOGY INSIGHTS (EXERCISES 67-72) (CONTINUED)

71.

X	Y₁	Y₂
0	9	9
1	13	13
2	19	19
3	ERROR	27
4	37	37
5	49	49
6	63	63

Y₂☐X²+3X+9

72.

X	Y₁	Y₂
1	21	21
2	28	28
3	37	37
4	ERROR	48
5	61	61
6	76	76
7	93	93

Y₂☐X²+4X+16

7.2 Multiplication and Division of Rational Expressions

OBJECTIVES

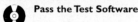

1 Multiply rational expressions.

2 Divide rational expressions.

FOR EXTRA HELP

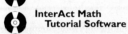

📖 **SSG** Sec. 7.2
SSM Sec. 7.2

💿 **Pass the Test Software**

💿 **InterAct Math**
   Tutorial Software

📼 **Video** 11

OBJECTIVE 1 Multiply rational expressions. The product of two fractions is found by multiplying the numerators and multiplying the denominators. Rational expressions are multiplied in the same way.

Multiplying Rational Expressions

The product of the rational expressions $\frac{P}{Q}$ and $\frac{R}{S}$ is

$$\frac{P}{Q} \cdot \frac{R}{S} = \frac{PR}{QS}.$$

The next example shows the multiplication of two common fractions and the multiplication of two rational expressions with variables so that you can compare the steps.

EXAMPLE 1 Multiplying Rational Expressions

Multiply. Write answers in lowest terms.

(a) $\dfrac{3}{10} \cdot \dfrac{5}{9}$ **(b)** $\dfrac{6}{x} \cdot \dfrac{x^2}{12}$

Find the product of the numerators and the product of the denominators.

$$\frac{3}{10} \cdot \frac{5}{9} = \frac{3 \cdot 5}{10 \cdot 9} \qquad\qquad \frac{6}{x} \cdot \frac{x^2}{12} = \frac{6 \cdot x^2}{x \cdot 12}$$

Use the fundamental property to write each product in lowest terms.

$$\frac{3}{10} \cdot \frac{5}{9} = \frac{1 \cdot 3 \cdot 5}{2 \cdot 5 \cdot 3 \cdot 3} = \frac{1}{6} \qquad\qquad \frac{6}{x} \cdot \frac{x^2}{12} = \frac{6 \cdot x \cdot x}{2 \cdot 6 \cdot x} = \frac{x}{2}$$

Notice in the second step above that the products were left in factored form since common factors must be identified to write the product in lowest terms.

It is also possible to divide out common factors in the numerator and denominator *before* multiplying the rational expressions. For example,

$$\frac{6}{5} \cdot \frac{35}{22} = \frac{2 \cdot 3}{\mathbf{5}} \cdot \frac{\mathbf{5} \cdot 7}{2 \cdot 11} \qquad \text{Identify common factors.}$$

$$= \frac{3 \cdot 7}{11} \qquad \text{Lowest terms}$$

$$= \frac{21}{11}. \qquad \text{Multiply in numerator.}$$

E X A M P L E 2   Multiplying Rational Expressions

Multiply. Write the product in lowest terms.

$$\frac{x + y}{2x} \cdot \frac{x^2}{(x + y)^2}$$

Use the definition of multiplication.

$$\frac{x + y}{2x} \cdot \frac{x^2}{(x + y)^2} = \frac{(x + y)x^2}{2x(x + y)^2} \qquad \begin{array}{l}\text{Multiply numerators; multiply} \\ \text{denominators.}\end{array}$$

$$= \frac{(\mathbf{x + y})x \cdot x}{2x(\mathbf{x + y})(x + y)} \qquad \text{Factor; identify common factors.}$$

$$= \frac{x}{2(x + y)} \cdot \frac{x(\mathbf{x + y})}{x(\mathbf{x + y})} \qquad \text{Definition of multiplication}$$

$$= \frac{x}{2(x + y)} \qquad \text{Lowest terms}$$

Notice the factor $\dfrac{x(x + y)}{x(x + y)}$ in the third line. Since it equals 1, the final product is $\dfrac{x}{2(x + y)}$.

E X A M P L E 3   Multiplying Rational Expressions

Multiply. Write the product in lowest terms.

$$\frac{x^2 + 3x}{x^2 - 3x - 4} \cdot \frac{x^2 - 5x + 4}{x^2 + 2x - 3}$$

First factor the numerators and denominators. Then use the fundamental property to write the product in lowest terms.

$$\frac{x^2 + 3x}{x^2 - 3x - 4} \cdot \frac{x^2 - 5x + 4}{x^2 + 2x - 3}$$

$$= \frac{x(x + 3)}{(x - 4)(x + 1)} \cdot \frac{(x - 4)(x - 1)}{(x + 3)(x - 1)} \qquad \text{Factor.}$$

$$= \frac{x(x + 3)(\mathbf{x - 4})(x - 1)}{(x - 4)(x + 1)(x + 3)(x - 1)} \qquad \text{Multiply numerators; multiply denominators.}$$

$$= \frac{x}{x + 1} \qquad \text{Lowest terms}$$

The quotients $\dfrac{x+3}{x+3}$, $\dfrac{x-4}{x-4}$, and $\dfrac{x-1}{x-1}$ are all equal to 1, justifying the final product

$\dfrac{x}{x+1}$.

OBJECTIVE 2 Divide rational expressions. To develop a method for dividing rational numbers and rational expressions, consider the following problem. Suppose that you have $\frac{7}{8}$ gallon of milk and you wish to find how many quarts you have. Since a quart is $\frac{1}{4}$ gallon, you must ask yourself, "How many $\frac{1}{4}$s are there in $\frac{7}{8}$?" This would be interpreted as

$$\frac{7}{8} \div \frac{1}{4} \qquad \text{or} \qquad \frac{\frac{7}{8}}{\frac{1}{4}}$$

since the fraction bar means division.

The fundamental property of rational expressions discussed earlier can be applied to rational number values of P, Q, and K. With $P = \frac{7}{8}$, $Q = \frac{1}{4}$, and $K = 4$,

$$\frac{P}{Q} = \frac{P \cdot K}{Q \cdot K} = \frac{\frac{7}{8} \cdot 4}{\frac{1}{4} \cdot 4} = \frac{\frac{7}{8} \cdot 4}{1} = \frac{7}{8} \cdot \frac{4}{1}.$$

So, to divide $\frac{7}{8}$ by $\frac{1}{4}$, we must multiply $\frac{7}{8}$ by the reciprocal of $\frac{1}{4}$, namely 4. Since $\left(\frac{7}{8}\right)(4) = \frac{7}{2}$, there are $\frac{7}{2}$ or $3\frac{1}{2}$ quarts in $\frac{7}{8}$ gallon.

The discussion above illustrates the rule for dividing common fractions. To divide $\frac{a}{b}$ by $\frac{c}{d}$, multiply $\frac{a}{b}$ by the reciprocal of $\frac{c}{d}$. Division of rational expressions is defined in the same way.

Dividing Rational Expressions

If $\frac{P}{Q}$ and $\frac{R}{S}$ are any two rational expressions, with $\frac{R}{S} \neq 0$, then

$$\frac{P}{Q} \div \frac{R}{S} = \frac{P}{Q} \cdot \frac{S}{R} = \frac{PS}{QR}.$$

The next example shows the division of two common fractions and the division of two rational expressions involving variables.

EXAMPLE 4 Dividing Rational Expressions

Divide. Write answers in lowest terms.

(a) $\dfrac{5}{8} \div \dfrac{7}{16}$

(b) $\dfrac{y}{y+3} \div \dfrac{4y}{y+5}$

Multiply the first expression and the reciprocal of the second.

$$\frac{5}{8} \div \frac{7}{16} = \frac{5}{8} \cdot \frac{\mathbf{16}}{\mathbf{7}} \qquad \text{Reciprocal of } \tfrac{7}{16} \qquad\qquad \frac{y}{y+3} \div \frac{4y}{y+5}$$

$$= \frac{5 \cdot 16}{8 \cdot 7} \qquad\qquad\qquad\qquad = \frac{y}{y+3} \cdot \frac{\mathbf{y+5}}{\mathbf{4y}} \qquad \text{Reciprocal of } \tfrac{4y}{y+5}$$

$$= \frac{5 \cdot 8 \cdot 2}{8 \cdot 7} \qquad\qquad\qquad\qquad = \frac{y(y+5)}{(y+3)(4y)}$$

$$= \frac{5 \cdot 2}{7} \qquad \text{Fundamental} \qquad\qquad = \frac{y+5}{4(y+3)} \qquad \text{Fundamental}$$
$$\qquad\qquad \text{property} \qquad\qquad\qquad\qquad\qquad\qquad\qquad \text{property}$$

$$= \frac{10}{7}$$

E X A M P L E 5 Dividing Rational Expressions

Divide. Write the quotient in lowest terms.

$$\frac{(3m)^2}{(2p)^3} \div \frac{6m^3}{16p^2}$$

Use the properties of exponents as necessary.

$$\frac{(3m)^2}{(2p)^3} \div \frac{6m^3}{16p^2} = \frac{(3m)^2}{(2p)^3} \cdot \frac{16p^2}{6m^3} \qquad \text{Multiply by reciprocal.}$$

$$= \frac{(3m)(3m)}{(2p)(2p)(2p)} \cdot \frac{16p^2}{6m^3} \qquad \text{Meaning of exponent}$$

$$= \frac{9 \cdot 16m^2p^2}{8 \cdot 6p^3m^3} \qquad\qquad \text{Multiply numerators.}$$
$$\qquad\qquad\qquad\qquad\qquad \text{Multiply denominators.}$$

$$= \frac{3}{mp} \qquad\qquad\qquad\qquad \text{Lowest terms}$$

E X A M P L E 6 Dividing Rational Expressions

Divide. Write the quotient in lowest terms.

$$\frac{x^2 - 4}{(x+3)(x-2)} \div \frac{(x+2)(x+3)}{2x}$$

$$= \frac{x^2 - 4}{(x+3)(x-2)} \cdot \frac{2x}{(x+2)(x+3)} \qquad \text{Use the definition of division.}$$

$$= \frac{(x+2)(x-2)}{(x+3)(x-2)} \cdot \frac{2x}{(x+2)(x+3)} \qquad \begin{array}{l}\text{Be sure numerators and}\\ \text{denominators are factored.}\end{array}$$

$$= \frac{(x+2)(x-2)(2x)}{(x+3)(x-2)(x+2)(x+3)} \qquad \begin{array}{l}\text{Multiply numerators and}\\ \text{multiply denominators.}\end{array}$$

$$= \frac{2x}{(x+3)^2} \qquad\qquad\qquad\qquad \begin{array}{l}\text{Use the fundamental property}\\ \text{to write in lowest terms.}\end{array}$$

In Example 6, only the numerator had to be factored. Remember that *all* numerators and denominators must be factored before the fundamental property can be applied.

EXAMPLE 7   Dividing Rational Expressions (Factors Are Opposites)

Divide. Write the quotient in lowest terms.

$$\frac{m^2 - 4}{m^2 - 1} \div \frac{2m^2 + 4m}{1 - m}$$

$$= \frac{m^2 - 4}{m^2 - 1} \cdot \frac{1 - m}{2m^2 + 4m} \qquad \text{Use the definition of division.}$$

$$= \frac{(m + 2)(m - 2)}{(m + 1)(m - 1)} \cdot \frac{1 - m}{2m(m + 2)} \qquad \begin{array}{l}\text{Factor; } 1 - m \text{ and } m - 1 \\ \text{differ only in sign.}\end{array}$$

$$= \frac{-1(m - 2)}{2m(m + 1)} \qquad \text{From Section 7.1, } \frac{1 - m}{m - 1} = -1.$$

$$= \frac{2 - m}{2m(m + 1)} \qquad \begin{array}{l}\text{Use the distributive property} \\ \text{in the numerator.}\end{array}$$

In summary, follow these steps to multiply and divide rational expressions.

Multiplying or Dividing Rational Expressions

Step 1   **Note the operation.**   If the operation is division, use the definition of division to rewrite as multiplication.

Step 2   **Factor.**   Factor all numerators and denominators completely.

Step 3   **Multiply.**   Multiply numerators and multiply denominators.

Step 4   **Write in lowest terms.**   Use the fundamental property to write the answer in lowest terms.

7.2 EXERCISES

1. Match each multiplication problem in Column I with the correct product in Column II.

I	II

(a) $\dfrac{5x^3}{10x^4} \cdot \dfrac{10x^7}{2x}$ **A.** $\dfrac{2}{5x^5}$

(b) $\dfrac{10x^4}{5x^3} \cdot \dfrac{10x^7}{2x}$ **B.** $\dfrac{5x^5}{2}$

(c) $\dfrac{5x^3}{10x^4} \cdot \dfrac{2x}{10x^7}$ **C.** $\dfrac{1}{10x^7}$

(d) $\dfrac{10x^4}{5x^3} \cdot \dfrac{2x}{10x^7}$ **D.** $10x^7$

2. Match each division problem in Column I with the correct quotient in Column II.

I

(a) $\dfrac{5x^3}{10x^4} \div \dfrac{10x^7}{2x}$

(b) $\dfrac{10x^4}{5x^3} \div \dfrac{10x^7}{2x}$

(c) $\dfrac{5x^3}{10x^4} \div \dfrac{2x}{10x^7}$

(d) $\dfrac{10x^4}{5x^3} \div \dfrac{2x}{10x^7}$

II

A. $\dfrac{5x^5}{2}$

B. $10x^7$

C. $\dfrac{2}{5x^5}$

D. $\dfrac{1}{10x^7}$

Multiply. Write each answer in lowest terms. See Examples 1 and 2.

3. $\dfrac{15a^2}{14} \cdot \dfrac{7}{5a}$

4. $\dfrac{27k^3}{9k} \cdot \dfrac{24}{9k^2}$

5. $\dfrac{12x^4}{18x^3} \cdot \dfrac{-8x^5}{4x^2}$

6. $\dfrac{12m^5}{-2m^2} \cdot \dfrac{6m^6}{28m^3}$

7. $\dfrac{2(c+d)}{3} \cdot \dfrac{18}{6(c+d)^2}$

8. $\dfrac{4(y-2)}{x} \cdot \dfrac{3x}{6(y-2)^2}$

Divide. Write each answer in lowest terms. See Examples 4 and 5.

9. $\dfrac{9z^4}{3z^5} \div \dfrac{3z^2}{5z^3}$

10. $\dfrac{35q^8}{9q^5} \div \dfrac{25q^6}{10q^5}$

11. $\dfrac{4t^4}{2t^5} \div \dfrac{(2t)^3}{-6}$

12. $\dfrac{-12a^6}{3a^2} \div \dfrac{(2a)^3}{27a}$

13. $\dfrac{3}{2y-6} \div \dfrac{6}{y-3}$

14. $\dfrac{4m+16}{10} \div \dfrac{3m+12}{18}$

RELATING CONCEPTS (EXERCISES 15–18)

We know that division by 0 is undefined. For example, $5 \div 0$ cannot be a real number because there is no real number that when multiplied by 0 gives 5 as a product. When dividing rational expressions, we know that no denominator can be 0 in either of the expressions in the problem, and that furthermore, the numerator of the divisor cannot be 0.

Work Exercises 15–18 in order, *referring to the division problem*

$$\dfrac{x-6}{x+4} \div \dfrac{x+7}{x+5}.$$

15. Why must we have the restriction $x \neq -4$?

16. Why must we have the restriction $x \neq -5$?

17. Why must we have the restriction $x \neq -7$?

18. Why is 6 allowed as a replacement for x even though 6 causes the numerator in the first fraction to be 0?

Did you make the connection that division by 0 is not allowed in both the individual fractions and the problem as a whole?

19. Explain in your own words how to multiply rational expressions. Illustrate with an example.

20. Explain in your own words how to divide rational expressions. Illustrate with an example.

Multiply or divide. Write each answer in lowest terms. See Examples 3, 6, and 7.

21. $\dfrac{5x - 15}{3x + 9} \cdot \dfrac{4x + 12}{6x - 18}$

22. $\dfrac{8r + 16}{24r - 24} \cdot \dfrac{6r - 6}{3r + 6}$

23. $\dfrac{2 - t}{8} \div \dfrac{t - 2}{6}$

24. $\dfrac{4}{m - 2} \div \dfrac{16}{2 - m}$

25. $\dfrac{27 - 3z}{4} \cdot \dfrac{12}{2z - 18}$

26. $\dfrac{5 - x}{5 + x} \cdot \dfrac{x + 5}{x - 5}$

27. $\dfrac{p^2 + 4p - 5}{p^2 + 7p + 10} \div \dfrac{p - 1}{p + 4}$

28. $\dfrac{z^2 - 3z + 2}{z^2 + 4z + 3} \div \dfrac{z - 1}{z + 1}$

29. $\dfrac{2k^2 - k - 1}{2k^2 + 5k + 3} \div \dfrac{4k^2 - 1}{2k^2 + k - 3}$

30. $\dfrac{2m^2 - 5m - 12}{m^2 + m - 20} \div \dfrac{4m^2 - 9}{m^2 + 4m - 5}$

31. $\dfrac{2k^2 + 3k - 2}{6k^2 - 7k + 2} \cdot \dfrac{4k^2 - 5k + 1}{k^2 + k - 2}$

32. $\dfrac{2m^2 - 5m - 12}{m^2 - 10m + 24} \div \dfrac{4m^2 - 9}{m^2 - 9m + 18}$

33. $\dfrac{m^2 + 2mp - 3p^2}{m^2 - 3mp + 2p^2} \div \dfrac{m^2 + 4mp + 3p^2}{m^2 + 2mp - 8p^2}$

34. $\dfrac{r^2 + rs - 12s^2}{r^2 - rs - 20s^2} \div \dfrac{r^2 - 2rs - 3s^2}{r^2 + rs - 30s^2}$

35. $\dfrac{m^2 + 3m + 2}{m^2 + 5m + 4} \cdot \dfrac{m^2 + 10m + 24}{m^2 + 5m + 6}$

36. $\dfrac{z^2 - z - 6}{z^2 - 2z - 8} \cdot \dfrac{z^2 + 7z + 12}{z^2 - 9}$

37. $\dfrac{y^2 + y - 2}{y^2 + 3y - 4} \div \dfrac{y + 2}{y + 3}$

38. $\dfrac{r^2 + r - 6}{r^2 + 4r - 12} \div \dfrac{r + 3}{r - 1}$

39. $\dfrac{2m^2 + 7m + 3}{m^2 - 9} \cdot \dfrac{m^2 - 3m}{2m^2 + 11m + 5}$

40. $\dfrac{m^2 + 2mp - 3p^2}{m^2 - 3mp + 2p^2} \div \dfrac{m^2 + 4mp + 3p^2}{m^2 + 2mp - 8p^2}$

41. $\dfrac{r^2 + rs - 12s^2}{r^2 - rs - 20s^2} \div \dfrac{r^2 - 2rs - 3s^2}{r^2 + rs - 30s^2}$

42. $\dfrac{(x + 1)^3(x + 4)}{x^2 + 5x + 4} \div \dfrac{x^2 + 2x + 1}{x^2 + 3x + 2}$

43. $\dfrac{(q - 3)^4(q + 2)}{q^2 + 3q + 2} \div \dfrac{q^2 - 6q + 9}{q^2 + 4q + 4}$

44. $\dfrac{(x + 4)^3(x - 3)}{x^2 - 9} \div \dfrac{x^2 + 8x + 16}{x^2 + 6x + 9}$

In working each exercise, remember how grouping symbols are used (Section 1.2), how to factor sums and differences of cubes (Section 6.4), and how to factor by grouping (Section 6.1).

45. $\dfrac{x + 5}{x + 10} \div \left(\dfrac{x^2 + 10x + 25}{x^2 + 10x} \cdot \dfrac{10x}{x^2 + 15x + 50} \right)$

46. $\dfrac{m - 8}{m - 4} \div \left(\dfrac{m^2 - 12m + 32}{8m} \cdot \dfrac{m^2 - 8m}{m^2 - 8m + 16} \right)$

47. $\dfrac{3a - 3b - a^2 + b^2}{4a^2 - 4ab + b^2} \cdot \dfrac{4a^2 - b^2}{2a^2 - ab - b^2}$

48. $\dfrac{4r^2 - t^2 + 10r - 5t}{2r^2 + rt + 5r} \cdot \dfrac{4r^3 + 4r^2t + rt^2}{2r + t}$

49. $\dfrac{-x^3 - y^3}{x^2 - 2xy + y^2} \div \dfrac{3y^2 - 3xy}{x^2 - y^2}$

50. $\dfrac{b^3 - 8a^3}{4a^3 + 4a^2b + ab^2} \div \dfrac{4a^2 + 2ab + b^2}{-a^3 - ab^3}$

51. If the rational expression $\dfrac{5x^2y^3}{2pq}$ represents the area of a rectangle and $\dfrac{2xy}{p}$ represents the length, what rational expression represents the width?

[shaded rectangle] Width

Length $= \dfrac{2xy}{p}$

The area is $\dfrac{5x^2y^3}{2pq}$.

52. If you are given the problem $\dfrac{4y + 12}{2y - 10} \div \dfrac{?}{y^2 - y - 20} = \dfrac{2(y + 4)}{y - 3}$, what must be the polynomial that is represented by the question mark?

7.3 The Least Common Denominator

OBJECTIVES

1 Find the least common denominator for a group of fractions.

2 Rewrite rational expressions with given denominators.

FOR EXTRA HELP

📖 **SSG** Sec. 7.3
SSM Sec. 7.3

💿 **Pass the Test Software**

💿 **InterAct Math Tutorial Software**

📼 **Video** 11

OBJECTIVE **1** Find the least common denominator for a group of fractions. Just as with common fractions, adding or subtracting rational expressions (to be discussed in the next section) often requires finding a **least common denominator (LCD),** the least expression that all denominators divide into without a remainder. For example, the least common denominator for $\frac{2}{9}$ and $\frac{5}{12}$ is 36, since 36 is the smallest positive number that both 9 and 12 divide into evenly.

Least common denominators often can be found by inspection. For example, the LCD for $\frac{1}{6}$ and $\frac{2}{3m}$ is $6m$. In other cases, the LCD can be found by a procedure similar to that used in Chapter 6 for finding the greatest common factor.

Finding the Least Common Denominator

Step 1 **Factor.** Factor each denominator into prime factors.

Step 2 **List the factors.** List each different denominator factor the *greatest* number of times it appears in any of the denominators.

Step 3 **Multiply.** Multiply the denominator factors from Step 2 to get the LCD.

When each denominator is factored into prime factors, every prime factor must divide evenly into the LCD.

In Example 1, the LCD is found both for numerical denominators and algebraic denominators.

EXAMPLE 1 Finding the Least Common Denominator

Find the LCD for each pair of fractions.

(a) $\dfrac{1}{24}, \dfrac{7}{15}$ **(b)** $\dfrac{1}{8x}, \dfrac{3}{10x}$

Step 1 Write each denominator in factored form with numerical coefficients in prime factored form.

$$24 = 2 \cdot 2 \cdot 2 \cdot 3 \qquad\qquad 8x = 2 \cdot 2 \cdot 2 \cdot x$$
$$= 2^3 \cdot 3 \qquad\qquad\qquad = 2^3 \cdot x$$
$$15 = 3 \cdot 5 \qquad\qquad 10x = 2 \cdot 5 \cdot x$$

Step 2 We find the LCD by taking each different factor the *greatest* number of times it appears as a factor in any of the denominators.

The factor 2 appears three times in one product and not at all in the other, so the greatest number of times 2 appears is three. The greatest number of times both 3 and 5 appear is one.

Here 2 appears three times in one product and once in the other, so the greatest number of times 2 appears is three. The greatest number of times 5 appears is one, and the greatest number of times x appears in either product is one.

Step 3 $\text{LCD} = 2 \cdot 2 \cdot 2 \cdot 3 \cdot 5$
$\qquad\quad = 2^3 \cdot 3 \cdot 5$
$\qquad\quad = 120$

$\text{LCD} = 2 \cdot 2 \cdot 2 \cdot 5 \cdot x$
$\qquad\quad = 2^3 \cdot 5 \cdot x$
$\qquad\quad = 40x$

┌ **E X A M P L E 2** Finding the LCD

Find the LCD for $\dfrac{5}{6r^2}$ and $\dfrac{3}{4r^3}$.

Step 1   Factor each denominator.

$$6r^2 = 2 \cdot 3 \cdot r^2$$
$$4r^3 = 2 \cdot 2 \cdot r^3 = 2^2 \cdot r^3$$

Step 2   The greatest number of times 2 appears is two, the greatest number of times 3 appears is one, and the greatest number of times r appears is three; therefore,

Step 3 $$\text{LCD} = 2^2 \cdot 3 \cdot r^3 = 12r^3.$$

┌ **E X A M P L E 3** Finding the LCD

Find the LCD.

(a) $\dfrac{6}{5m}, \dfrac{4}{m^2 - 3m}$

Factor each denominator.

$$5m = 5 \cdot m$$
$$m^2 - 3m = m(m - 3)$$

Take each different factor the greatest number of times it appears as a factor.

$$\text{LCD} - 5 \cdot m \cdot (m - 3) = 5m(m - 3)$$

Since m is not a *factor* of $m - 3$, both factors, m and $m - 3$, must appear in the LCD.

(b) $\dfrac{1}{r^2 - 4r - 5}, \dfrac{1}{r^2 - r - 20}, \dfrac{1}{r^2 - 10r + 25}$

Factor each denominator.

$$r^2 - 4r - 5 = (r - 5)(r + 1)$$
$$r^2 - r - 20 = (r - 5)(r + 4)$$
$$r^2 - 10r + 25 = (r - 5)^2$$

The LCD is $(r - 5)^2(r + 1)(r + 4)$.

(c) $\dfrac{1}{q - 5}, \dfrac{3}{5 - q}$

The expression $5 - q$ can be written as $-1(q - 5)$, since

$$-1(q - 5) = -q + 5 = 5 - q.$$

Because of this, either $q - 5$ or $5 - q$ can be used as the LCD.

O B J E C T I V E 2 Rewrite rational expressions with given denominators. Once we find the LCD, we can use the fundamental property to write equivalent rational expressions with this LCD. The next example shows how to do this with both numerical and algebraic fractions.

E X A M P L E 4   Writing a Fraction with a Given Denominator

Rewrite each expression with the indicated denominator.

(a) $\dfrac{3}{8} = \dfrac{}{40}$

(b) $\dfrac{9k}{25} = \dfrac{}{50k}$

For each example, first factor the denominator on the right. Then compare the denominator on the left with the one on the right to decide what factors are missing.

$$\dfrac{3}{8} = \dfrac{}{5 \cdot 8}$$

A factor of 5 is missing. Multiply by $\frac{5}{5}$ to get a denominator of 40.

$$\dfrac{3}{8} = \dfrac{3}{8} \cdot \dfrac{5}{5} = \dfrac{15}{40}$$
$$\downarrow$$
$$\tfrac{5}{5} = 1$$

$$\dfrac{9k}{25} = \dfrac{}{25 \cdot 2k}$$

Factors of 2 and k are missing. Get a denominator of $50k$ by multiplying by $\frac{2k}{2k}$.

$$\dfrac{9k}{25} = \dfrac{9k}{25} \cdot \dfrac{2k}{2k} = \dfrac{18k^2}{50k}$$
$$\downarrow$$
$$\tfrac{2k}{2k} = 1$$

Notice the use of the multiplicative identity property in each part of this example.

E X A M P L E 5   Writing a Fraction with a Given Denominator

Rewrite the following rational expression with the indicated denominator.

$$\dfrac{12p}{p^2 + 8p} = \dfrac{}{p^3 + 4p^2 - 32p}$$

Factor $p^2 + 8p$ as $p(p + 8)$. Compare with the denominator on the right which factors as $p(p + 8)(p - 4)$. The factor $p - 4$ is missing, so multiply $\dfrac{12p}{p(p + 8)}$ by $\dfrac{p - 4}{p - 4}$.

$$\dfrac{12p}{p^2 + 8p} = \dfrac{12p}{p(p + 8)} \cdot \dfrac{p - 4}{p - 4} \quad \text{Multiplicative identity property}$$
$$= \dfrac{12p(p - 4)}{p(p + 8)(p - 4)} \quad \text{Multiplication of rational expressions}$$
$$= \dfrac{12p^2 - 48p}{p^3 + 4p^2 - 32p} \quad \text{Multiply the factors.}$$

In the next section we add and subtract rational expressions, which sometimes requires the steps illustrated in Examples 4 and 5. While it is beneficial to leave the denominator in factored form, we multiplied the factors in the denominator in Example 5 to give the answer in the same form as the original problem.

7.3 EXERCISES

Choose the correct response in Exercises 1–4.

1. Suppose that the greatest common factor of a and b is 1. Then the least common denominator for $\dfrac{1}{a}$ and $\dfrac{1}{b}$ is

   (a) *a.*   (b) *b.*   (c) *ab.*   (d) 1.

2. If a is a factor of b, then the least common denominator for $\dfrac{1}{a}$ and $\dfrac{1}{b}$ is

   (a) *a.*   (b) *b.*   (c) *ab.*   (d) 1.

3. The least common denominator for $\dfrac{11}{20}$ and $\dfrac{1}{2}$ is

   (a) 40.   (b) 2.   (c) 20.   (d) none of these.

4. Suppose that we wish to write the fraction $\dfrac{1}{(x-4)^2(y-3)}$ with denominator $(x-4)^3(y-3)^2$. We must multiply both the numerator and the denominator by

   (a) $(x-4)(y-3)$.   (b) $(x-4)^2$.   (c) $x-4$.   (d) $(x-4)^2(y-3)$.

Find the LCD for the fractions in each list. See Examples 1–3.

5. $\dfrac{-7}{15}, \dfrac{21}{20}$

6. $\dfrac{9}{10}, \dfrac{12}{25}$

7. $\dfrac{17}{100}, \dfrac{23}{120}, \dfrac{43}{180}$

8. $\dfrac{17}{250}, \dfrac{-21}{300}, \dfrac{127}{360}$

9. $\dfrac{9}{x^2}, \dfrac{8}{x^5}$

10. $\dfrac{12}{m^7}, \dfrac{13}{m^8}$

11. $\dfrac{-2}{5p}, \dfrac{15}{6p}$

12. $\dfrac{14}{15k}, \dfrac{9}{4k}$

13. $\dfrac{17}{15y^2}, \dfrac{55}{36y^4}$

14. $\dfrac{4}{25m^3}, \dfrac{-9}{10m^4}$

15. $\dfrac{13}{5a^2b^3}, \dfrac{29}{15a^5b}$

16. $\dfrac{-7}{3r^4s^5}, \dfrac{-22}{9r^6s^8}$

17. $\dfrac{7}{6p}, \dfrac{15}{4p-8}$

18. $\dfrac{7}{8k}, \dfrac{-23}{12k-24}$

▌RELATING CONCEPTS (EXERCISES 19–22)

Suppose we want to find the LCD for the two common fractions

$$\frac{1}{24} \quad \text{and} \quad \frac{1}{20}.$$

In their prime factored forms, the denominators are

$$24 = 2^3 \cdot 3$$

and

$$20 = 2^2 \cdot 5.$$

*Refer to this information as necessary and **work Exercises 19–22 in order.***

19. What is the prime factored form of the LCD of the two fractions?

20. Suppose that two algebraic fractions have denominators $(t+4)^3(t-3)$ and $(t+4)^2(t+8)$. What is the factored form of the LCD of these?

21. What is the similarity between your answers in Exercises 19 and 20?

22. Comment on the following statement: The method for finding the LCD for two algebraic fractions is the same as the method for finding the LCD for two common fractions.

Did you make the connection between finding the LCD for common fractions and for algebraic fractions?

Find the LCD for the fractions in each list. See Examples 1–3.

23. $\dfrac{37}{6r-12}, \dfrac{25}{9r-18}$

24. $\dfrac{-14}{5p-30}, \dfrac{5}{6p-36}$

25. $\dfrac{5}{12p+60}, \dfrac{17}{p^2+5p}, \dfrac{16}{p^2+10p+25}$

26. $\dfrac{13}{r^2+7r}, \dfrac{-3}{5r+35}, \dfrac{-7}{r^2+14r+49}$

27. $\dfrac{3}{8y+16}, \dfrac{22}{y^2+3y+2}$

28. $\dfrac{-2}{9m-18}, \dfrac{-9}{m^2-7m+10}$

29. $\dfrac{12}{m-3}, \dfrac{-4}{3-m}$

30. $\dfrac{-17}{8-a}, \dfrac{2}{a-8}$

31. $\dfrac{29}{p-q}, \dfrac{18}{q-p}$

32. $\dfrac{16}{z-x}, \dfrac{8}{x-z}$

33. $\dfrac{6}{a^2+6a}, \dfrac{-5}{a^2+3a-18}$

34. $\dfrac{8}{y^2-5y}, \dfrac{-2}{y^2-2y-15}$

35. $\dfrac{-5}{k^2+2k-35}, \dfrac{-8}{k^2+3k-40}, \dfrac{9}{k^2-2k-15}$

36. $\dfrac{19}{z^2+4z-12}, \dfrac{16}{z^2+z-30}, \dfrac{6}{z^2+2z-24}$

37. Suppose that $(2x-5)^2$ is the LCD for two fractions. Is $(5-2x)^2$ also acceptable as an LCD? Why?

38. Suppose that $(4t-3)(5t-6)$ is the LCD for two fractions. Is $(3-4t)(6-5t)$ also acceptable as an LCD? Why?

■ RELATING CONCEPTS (EXERCISES 39-44)

Work Exercises 39–44 in order.

39. Suppose that you want to write $\dfrac{3}{4}$ as an equivalent fraction with denominator 28. By what number must you multiply both the numerator and the denominator?

40. If you write $\dfrac{3}{4}$ as an equivalent fraction with denominator 28, by what number are you actually multiplying the fraction?

41. What property of multiplication is being used when we write a common fraction as an equivalent one with a larger denominator? (See Section 1.7.)

42. Suppose that you want to write $\dfrac{2x+5}{x-4}$ as an equivalent fraction with denominator $7x-28$. By what number must you multiply both the numerator and the denominator?

43. If you write $\dfrac{2x+5}{x-4}$ as an equivalent fraction with denominator $7x-28$, by what number are you actually multiplying the fraction?

44. Repeat Exercise 41, changing "a common" to "an algebraic."

Did you make the connection between writing common fractions with larger denominators and writing algebraic fractions with more denominator factors?

Write each rational expression on the left with the indicated denominator. See Examples 4 and 5.

45. $\dfrac{15m^2}{8k} = \dfrac{}{32k^4}$

46. $\dfrac{5t^2}{3y} = \dfrac{}{9y^2}$

47. $\dfrac{19z}{2z-6}=\dfrac{}{6z-18}$

48. $\dfrac{2r}{5r-5}=\dfrac{}{15r-15}$

49. $\dfrac{-2a}{9a-18}=\dfrac{}{18a-36}$

50. $\dfrac{-5y}{6y+18}=\dfrac{}{24y+72}$

51. $\dfrac{6}{k^2-4k}=\dfrac{}{k(k-4)(k+1)}$

52. $\dfrac{15}{m^2-9m}=\dfrac{}{m(m-9)(m+8)}$

53. $\dfrac{36r}{r^2-r-6}=\dfrac{}{(r-3)(r+2)(r+1)}$

54. $\dfrac{4m}{m^2-8m+15}=\dfrac{}{(m-5)(m-3)(m+2)}$

55. $\dfrac{a+2b}{2a^2+ab-b^2}=\dfrac{}{2a^3b+a^2b^2-ab^3}$

56. $\dfrac{m-4}{6m^2+7m-3}=\dfrac{}{12m^3+14m^2-6m}$

57. $\dfrac{4r-t}{r^2+rt+t^2}=\dfrac{}{t^3-r^3}$

58. $\dfrac{3x-1}{x^2+2x+4}=\dfrac{}{x^3-8}$

59. $\dfrac{2(z-y)}{y^2+yz+z^2}=\dfrac{}{y^4-z^3y}$

60. $\dfrac{2p+3q}{p^2+2pq+q^2}=\dfrac{}{(p+q)(p^3+q^3)}$

61. Write an explanation of how to find the least common denominator for a group of denominators. Give an example.

62. Write an explanation of how to write a rational expression as an equivalent rational expression with a given denominator. Give an example.

7.4   Addition and Subtraction of Rational Expressions

OBJECTIVES

1   Add rational expressions having the same denominator.

2   Add rational expressions having different denominators.

3   Subtract rational expressions.

FOR EXTRA HELP

 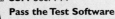 **SSG** Sec. 7.4
SSM Sec. 7.4

Pass the Test Software

InterAct Math
   Tutorial Software

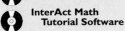   **Video 11**

To add and subtract rational expressions, we use our previous work on finding least common denominators and writing fractions with the LCD.

OBJECTIVE 1   Add rational expressions having the same denominator.   We find the sum of two rational expressions with a procedure similar to the one used for adding two fractions.

> **Adding Rational Expressions**
>
> If $\dfrac{P}{Q}$ and $\dfrac{R}{Q}$ are rational expressions, then
>
> $$\frac{P}{Q}+\frac{R}{Q}=\frac{P+R}{Q}.$$

Again, the first example shows how addition of rational expressions compares with that of rational numbers.

E X A M P L E 1   Adding Rational Expressions with the Same Denominator

Add.

(a) $\dfrac{4}{7} + \dfrac{2}{7}$

(b) $\dfrac{3x}{x+1} + \dfrac{2x}{x+1}$

The denominators are the same, so the sum is found by adding the two numerators and keeping the same (common) denominator.

$$\frac{4}{7} + \frac{2}{7} = \frac{4+2}{7}$$

$$= \frac{6}{7}$$

$$\frac{3x}{x+1} + \frac{2x}{x+1} = \frac{3x+2x}{x+1}$$

$$= \frac{5x}{x+1}$$

O B J E C T I V E 2   Add rational expressions having different denominators. Use the steps given below to add two rational expressions with different denominators. These are the same steps used to add fractions with different denominators.

Adding with Different Denominators

Step 1   **Find the LCD.**   Find the least common denominator (LCD).

Step 2   **Rewrite fractions.**   Rewrite each rational expression as an equivalent fraction with the LCD as the denominator.

Step 3   **Add.**   Add the numerators to get the numerator of the sum. The LCD is the denominator of the sum.

Step 4   **Write in lowest terms.**   Use the fundamental property to write the answer in lowest terms.

E X A M P L E 2   Adding Rational Expressions with Different Denominators

Add.

(a) $\dfrac{1}{12} + \dfrac{7}{15}$

(b) $\dfrac{2}{3y} + \dfrac{1}{4y}$

Step 1   First find the LCD using the methods of the previous section.

$$\text{LCD} = 2^2 \cdot 3 \cdot 5 = 60 \qquad \qquad \text{LCD} = 2^2 \cdot 3 \cdot y = 12y$$

Step 2   Now rewrite each rational expression as a fraction with the LCD (either 60 or 12y) as the denominator.

$$\frac{1}{12} + \frac{7}{15} = \frac{1(5)}{12(5)} + \frac{7(4)}{15(4)}$$

$$= \frac{5}{60} + \frac{28}{60}$$

$$\frac{2}{3y} + \frac{1}{4y} = \frac{2(4)}{3y(4)} + \frac{1(3)}{4y(3)}$$

$$= \frac{8}{12y} + \frac{3}{12y}$$

Step 3 Since the fractions now have common denominators, add the numerators.

Step 4 Write in lowest terms if necessary.

$$\frac{5}{60} + \frac{28}{60} = \frac{5 + 28}{60} \qquad\qquad \frac{8}{12y} + \frac{3}{12y} = \frac{8 + 3}{12y}$$

$$= \frac{33}{60} = \frac{11}{20} \qquad\qquad\qquad = \frac{11}{12y}$$

E X A M P L E 3 Adding Rational Expressions

Add and write the sum in lowest terms.

$$\frac{2x}{x^2 - 1} + \frac{-1}{x + 1}$$

Step 1 Since the denominators are different, find the LCD.

$$x^2 - 1 = (x + 1)(x - 1)$$

$$x + 1 \text{ is prime.}$$

The LCD is $(x + 1)(x - 1)$.

Step 2 Rewrite each rational expression as a fraction with common denominator $(x + 1)(x - 1)$.

$$\frac{2x}{x^2 - 1} + \frac{-1}{x + 1} = \frac{2x}{(x + 1)(x - 1)} + \frac{-1(x - 1)}{(x + 1)(x - 1)} \qquad \text{Multiply second fraction by } \frac{x - 1}{x - 1}.$$

$$= \frac{2x}{(x + 1)(x - 1)} + \frac{-x + 1}{(x + 1)(x - 1)} \qquad \text{Distributive property}$$

Step 3
$$= \frac{2x - x + 1}{(x + 1)(x - 1)} \qquad \text{Add numerators; keep the same denominator.}$$

$$= \frac{x + 1}{(x + 1)(x - 1)} \qquad \text{Combine like terms in the numerator.}$$

Step 4
$$= \frac{\mathbf{1}(x + 1)}{(x + 1)(x - 1)} \qquad \text{Identity property for multiplication}$$

$$= \frac{1}{x - 1} \qquad \text{Fundamental property of rational expressions}$$

E X A M P L E 4 Adding Rational Expressions

Add and write the sum in lowest terms.

$$\frac{2x}{x^2 + 5x + 6} + \frac{x + 1}{x^2 + 2x - 3}$$

Begin by factoring the denominators completely.

$$\frac{2x}{(x + 2)(x + 3)} + \frac{x + 1}{(x + 3)(x - 1)}$$

The LCD is $(x + 2)(x + 3)(x - 1)$. Use the fundamental property of rational expressions to rewrite each fraction with the LCD.

$$\frac{2x}{(x + 2)(x + 3)} + \frac{x + 1}{(x + 3)(x - 1)}$$

$$= \frac{2x(x - 1)}{(x + 2)(x + 3)(x - 1)} + \frac{(x + 1)(x + 2)}{(x + 3)(x - 1)(x + 2)}$$

$$= \frac{2x(x - 1) + (x + 1)(x + 2)}{(x + 2)(x + 3)(x - 1)} \quad \text{Add numerators; keep the same denominator.}$$

$$= \frac{2x^2 - 2x + x^2 + 3x + 2}{(x + 2)(x + 3)(x - 1)} \quad \text{Distributive property}$$

$$= \frac{3x^2 + x + 2}{(x + 2)(x + 3)(x - 1)} \quad \text{Combine terms.}$$

Since $3x^2 + x + 2$ cannot be factored, the rational expression cannot be simplified further. It is usually best to leave the denominator in factored form since it is then easier to identify common factors in the numerator and denominator.

Rational expressions to be added or subtracted may have denominators that are opposites of each other. The next example illustrates this.

E X A M P L E 5 Adding Rational Expressions with Denominators That Are Opposites

Add and write the sum in lowest terms.

$$\frac{y}{y - 2} + \frac{8}{2 - y}$$

To get a common denominator of $y - 2$, multiply the second expression by -1 in both the numerator and the denominator.

$$\frac{y}{y - 2} + \frac{8}{2 - y} = \frac{y}{y - 2} + \frac{8(-1)}{(2 - y)(-1)} \quad \text{Fundamental property}$$

$$= \frac{y}{y - 2} + \frac{-8}{y - 2} \quad \text{Distributive property}$$

$$= \frac{y - 8}{y - 2} \quad \text{Add numerators; keep the same denominator.}$$

If we had chosen to use $2 - y$ as the common denominator, the final answer would be in the form $\frac{8 - y}{2 - y}$, which is equivalent to $\frac{y - 8}{y - 2}$.

O B J E C T I V E 3 Subtract rational expressions. To subtract rational expressions, use the following rule.

Subtracting Rational Expressions

If $\frac{P}{Q}$ and $\frac{R}{Q}$ are rational expressions, then

$$\frac{P}{Q} - \frac{R}{Q} = \frac{P - R}{Q}.$$

E X A M P L E   6   Subtracting Rational Expressions

Subtract and write the difference in lowest terms.

$$\frac{2m}{m-1} - \frac{2}{m-1}$$

By the definition of subtraction,

$$\frac{2m}{m-1} - \frac{2}{m-1} = \frac{2m-2}{m-1}$$ Subtract numerators; keep the same denominator.

$$= \frac{2(m-1)}{m-1}$$ Factor the numerator.

$$= 2.$$ Write in lowest terms.

E X A M P L E   7   Subtracting Rational Expressions with Different Denominators

Subtract and write the difference in lowest terms.

$$\frac{9}{x-2} - \frac{3}{x} = \frac{9x}{x(x-2)} - \frac{3(x-2)}{x(x-2)}$$ The LCD is $x(x-2)$.

$$= \frac{9x - 3(x-2)}{x(x-2)}$$ Subtract numerators; keep the same denominator.

$$= \frac{9x - 3x + 6}{x(x-2)}$$ Distributive property

$$= \frac{6x+6}{x(x-2)}$$ Combine like terms in the numerator.

We could factor the final numerator in Example 7 to get an answer in the form $\frac{6(x+1)}{x(x-2)}$; however, the fundamental property would not apply, since there are no common factors that would allow us to write the answer in lower terms. When a rational expression cannot be simplified further, we will leave numerators in polynomial form and denominators in factored form.

E X A M P L E   8   Subtracting Rational Expressions with Denominators That Are Opposites

Subtract and write the difference in lowest terms.

$$\frac{3x}{x-5} - \frac{2x-25}{5-x}$$

The denominators are opposites, so either may be used as the common denominator. We choose $x - 5$.

$$\frac{3x}{x - 5} - \frac{2x - 25}{5 - x} = \frac{3x}{x - 5} - \frac{(2x - 25)}{5 - x} \cdot \frac{-1}{-1} \qquad \text{Fundamental property}$$

$$= \frac{3x}{x - 5} - \frac{(-2x + 25)}{x - 5} \qquad \text{Multiply.}$$

$$= \frac{3x - (-2x + 25)}{x - 5} \qquad \text{Subtract numerators.}$$

$$= \frac{3x + 2x - 25}{x - 5} \qquad \text{Distributive property}$$

$$= \frac{5x - 25}{x - 5} \qquad \text{Combine terms.}$$

$$= \frac{5(x - 5)}{x - 5} \qquad \text{Factor.}$$

$$= 5 \qquad \text{Lowest terms}$$

 Sign errors often occur in subtraction problems like the ones in Examples 7 and 8. Use parentheses after the subtraction sign to avoid this common error. Remember that the numerator of the fraction being subtracted must be treated as a single quantity.

EXAMPLE 9 Subtracting Rational Expressions

Subtract and write the difference in lowest terms.

$$\frac{6x}{x^2 - 2x + 1} - \frac{1}{x^2 - 1}$$

Begin by factoring.

$$\frac{6x}{x^2 - 2x + 1} - \frac{1}{x^2 - 1} = \frac{6x}{(x - 1)(x - 1)} - \frac{1}{(x - 1)(x + 1)}$$

From the factored denominators, we can now identify the common denominator, $(x - 1)(x - 1)(x + 1)$. Use the factor $x - 1$ twice, since it appears twice in the first denominator.

$$\frac{6x}{(x - 1)(x - 1)} - \frac{1}{(x - 1)(x + 1)}$$

$$= \frac{6x(x + 1)}{(x - 1)(x - 1)(x + 1)} - \frac{1(x - 1)}{(x - 1)(x - 1)(x + 1)} \qquad \text{Fundamental property}$$

$$= \frac{6x(x + 1) - 1(x - 1)}{(x - 1)(x - 1)(x + 1)} \qquad \text{Subtract.}$$

$$= \frac{6x^2 + 6x - x + 1}{(x - 1)(x - 1)(x + 1)} \qquad \text{Distributive property}$$

$$= \frac{6x^2 + 5x + 1}{(x - 1)(x - 1)(x + 1)} \qquad \text{Combine like terms.}$$

The result may be written as $\dfrac{6x^2 + 5x + 1}{(x - 1)^2(x + 1)}$.

When adding and subtracting rational expressions, several different equivalent forms of the answer often exist. If your answer does not look exactly like the one given in the back of the book, check to see if you have written an equivalent form.

7.4 EXERCISES

RELATING CONCEPTS (EXERCISES 1-8)

Work Exercises 1–8 in order. This will help you later in determining whether your answer is equivalent to the one given in the answer section if it doesn't match the form shown.

Jill worked a problem involving the sum of two rational expressions correctly. Her answer was given as $\dfrac{5}{x-3}$. Jack worked the same problem and gave his answer as $\dfrac{-5}{3-x}$. We want to decide whether Jack's answer was also correct.

1. Evaluate the fraction $\dfrac{5}{7-3}$ using the rule for order of operations.

2. Multiply the fraction given in Exercise 1 by -1 in both the numerator and the denominator, using the distributive property as necessary. Leave it in unsimplified form.

3. In Exercise 2, what number did you actually multiply the *fraction* by?

4. Simplify the result you found in Exercise 2 and compare it to the one found in Exercise 1. How do they compare?

5. Now look at the answers given by Jill and Jack above. Based on what you learned in Exercises 1–4, determine whether Jack's answer was also correct. Why or why not?

6. Jason Jordan, a perceptive algebra student, made the following comment and was praised by his teacher for his insight: "I can see that if I change the sign of each term in a fraction, the result I get is equivalent to the fraction that I started with." Explain why Jason's observation is correct.

7. Jennifer Crum, another perceptive algebra student, followed Jason's comment with one of her own: "I can see that if I put a negative sign in front of a fraction and change the signs of all terms in either the numerator or the denominator, but not both, then the result I get is equivalent to the fraction that I started with." The teacher knew that some real education was happening in the classroom. Explain why Jennifer's observation is also correct.

8. Use the concepts of Exercises 1–7 to determine whether each rational expression is equivalent or not equivalent to $\dfrac{y-4}{3-y}$.

   (a) $\dfrac{y-4}{y-3}$ (b) $\dfrac{-y+4}{-3+y}$ (c) $-\dfrac{y-4}{y-3}$

   (d) $-\dfrac{y-4}{3-y}$ (e) $-\dfrac{4-y}{y-3}$ (f) $\dfrac{y+4}{3+y}$

Did you make the connection that rational expressions can be written in equivalent forms?

Add or subtract. Write the answer in lowest terms. See Examples 1 and 6.

9. $\dfrac{4}{m} + \dfrac{7}{m}$

10. $\dfrac{5}{p} + \dfrac{11}{p}$

11. $\dfrac{a+b}{2} - \dfrac{a-b}{2}$

12. $\dfrac{x-y}{2} - \dfrac{x+y}{2}$

13. $\dfrac{x^2}{x+5} + \dfrac{5x}{x+5}$

14. $\dfrac{t^2}{t-3} + \dfrac{-3t}{t-3}$

15. $\dfrac{y^2-3y}{y+3} + \dfrac{-18}{y+3}$

16. $\dfrac{r^2-8r}{r-5} + \dfrac{15}{r-5}$

17. Explain with an example how to add or subtract rational expressions with the same denominators.

18. Explain with an example how to add or subtract rational expressions with different denominators.

Add or subtract. Write the answer in lowest terms. See Examples 2, 3, 4, and 7.

19. $\dfrac{z}{5} + \dfrac{1}{3}$

20. $\dfrac{p}{8} + \dfrac{3}{5}$

21. $\dfrac{5}{7} - \dfrac{r}{2}$

22. $\dfrac{10}{9} - \dfrac{z}{3}$

23. $-\dfrac{3}{4} - \dfrac{1}{2x}$

24. $-\dfrac{5}{8} - \dfrac{3}{2a}$

25. $\dfrac{x+1}{6} + \dfrac{3x+3}{9}$

26. $\dfrac{2x-6}{4} + \dfrac{x+5}{6}$

27. $\dfrac{x+3}{3x} + \dfrac{2x+2}{4x}$

28. $\dfrac{x+2}{5x} + \dfrac{6x+3}{3x}$

29. $\dfrac{7}{3p^2} - \dfrac{2}{p}$

30. $\dfrac{12}{5m^2} - \dfrac{5}{m}$

31. $\dfrac{x}{x-2} + \dfrac{4}{x+2} - \dfrac{8}{x^2-4}$

32. $\dfrac{2x}{x-1} + \dfrac{3}{x+1} - \dfrac{4}{x^2-1}$

33. $\dfrac{t}{t+2} + \dfrac{5-t}{t} - \dfrac{4}{t^2+2t}$

34. $\dfrac{2p}{p-3} + \dfrac{2+p}{p} - \dfrac{-6}{p^2-3p}$

35. What are the two possible LCDs that could be used for the sum

$$\dfrac{10}{m-2} + \dfrac{5}{2-m}?$$

36. If one form of the correct answer to a sum or difference of rational expressions is $\dfrac{4}{k-3}$, what would an alternate form of the answer be if the denominator is $3-k$?

Add or subtract. Write the answer in lowest terms. See Examples 5 and 8.

37. $\dfrac{4}{x-5} + \dfrac{6}{5-x}$

38. $\dfrac{10}{m-2} + \dfrac{5}{2-m}$

39. $\dfrac{-1}{1-y} + \dfrac{3-4y}{y-1}$

40. $\dfrac{-4}{p-3} - \dfrac{p+1}{3-p}$

41. $\dfrac{2}{x-y^2} + \dfrac{7}{y^2-x}$

42. $\dfrac{-8}{p-q^2} + \dfrac{3}{q^2-p}$

43. $\dfrac{x}{5x-3y} - \dfrac{y}{3y-5x}$

44. $\dfrac{t}{8t-9s} - \dfrac{s}{9s-8t}$

45. $\dfrac{3}{4p-5} + \dfrac{9}{5-4p}$

46. $\dfrac{8}{3-7y} - \dfrac{2}{7y-3}$

In these subtraction problems, the rational expression that follows the subtraction sign has a numerator with more than one term. Be very careful with signs and find each difference.

47. $\dfrac{2m}{m-n} - \dfrac{5m+n}{2m-2n}$

48. $\dfrac{5p}{p-q} - \dfrac{3p+1}{4p-4q}$

49. $\dfrac{5}{x^2-9} - \dfrac{x+2}{x^2+4x+3}$

50. $\dfrac{1}{a^2-1} - \dfrac{a-1}{a^2+3a-4}$

51. $\dfrac{2q + 1}{3q^2 + 10q - 8} - \dfrac{3q + 5}{2q^2 + 5q - 12}$

52. $\dfrac{4y - 1}{2y^2 + 5y - 3} - \dfrac{y + 3}{6y^2 + y - 2}$

Perform the indicated operations. See Examples 1–9.

53. $\dfrac{4}{r^2 - r} + \dfrac{6}{r^2 + 2r} - \dfrac{1}{r^2 + r - 2}$

54. $\dfrac{6}{k^2 + 3k} - \dfrac{1}{k^2 - k} + \dfrac{2}{k^2 + 2k - 3}$

55. $\dfrac{x + 3y}{x^2 + 2xy + y^2} + \dfrac{x - y}{x^2 + 4xy + 3y^2}$

56. $\dfrac{m}{m^2 - 1} + \dfrac{m - 1}{m^2 + 2m + 1}$

57. $\dfrac{r + y}{18r^2 + 12ry - 3ry - 2y^2} + \dfrac{3r - y}{36r^2 - y^2}$

58. $\dfrac{2x - z}{2x^2 - 4xz + 5xz - 10z^2} - \dfrac{x + z}{x^2 - 4z^2}$

Perform the indicated operations. Remember the order of operations.

59. $\left(\dfrac{-k}{2k^2 - 5k - 3} + \dfrac{3k - 2}{2k^2 - k - 1}\right)\dfrac{2k + 1}{k - 1}$

60. $\left(\dfrac{3p + 1}{2p^2 + p - 6} - \dfrac{5p}{3p^2 - p}\right)\dfrac{2p - 3}{p + 2}$

61. $\dfrac{k^2 + 4k + 16}{k + 4}\left(\dfrac{-5}{16 - k^2} + \dfrac{2k + 3}{k^3 - 64}\right)$

62. $\dfrac{m - 5}{2m + 5}\left(\dfrac{-3m}{m^2 - 25} - \dfrac{m + 4}{125 - m^3}\right)$

63. Refer to the rectangle in the figure.
(a) Find an expression that represents its perimeter. Give the simplified form.
(b) Find an expression that represents its area. Give the simplified form.

$\dfrac{3k + 1}{10}$

$\dfrac{5}{6k + 2}$

64. Refer to the triangle in the figure. Find an expression that represents its perimeter.

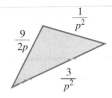

$\dfrac{1}{p^2}$

$\dfrac{9}{2p}$

$\dfrac{3}{p^2}$

7.5 Complex Fractions

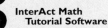

The quotient of two mixed numbers in arithmetic, such as $2\frac{1}{2} \div 3\frac{1}{4}$, can be written as a fraction:

$$2\frac{1}{2} \div 3\frac{1}{4} = \dfrac{2\frac{1}{2}}{3\frac{1}{4}} = \dfrac{2 + \frac{1}{2}}{3 + \frac{1}{4}}.$$

The last expression is the quotient of expressions that involve fractions. In algebra, some rational expressions also have fractions in the numerator, or denominator, or both.

Complex Fraction

A rational expression with fractions in the numerator, denominator, or both, is called a **complex fraction**.

Examples of complex fractions include

$$\frac{2 + \dfrac{1}{2}}{3 + \dfrac{1}{4}}, \qquad \frac{\dfrac{3x^2 - 5x}{6x^2}}{2x - \dfrac{1}{x}}, \qquad \text{and} \qquad \frac{3 + x}{5 - \dfrac{2}{x}}.$$

The parts of a complex fraction are named as follows.

$$\frac{\dfrac{2}{p} - \dfrac{1}{q}}{\dfrac{3}{p} + \dfrac{5}{q}}$$

← Numerator of complex fraction
← Main fraction bar
← Denominator of complex fraction

OBJECTIVE 1 Simplify a complex fraction by writing it as a division problem (Method 1). Since the main fraction bar represents division in a complex fraction, one method of simplifying a complex fraction involves division.

Method 1

To simplify a complex fraction:

Step 1   Write both the numerator and denominator as single fractions.

Step 2   Change the complex fraction to a division problem.

Step 3   Perform the indicated division.

Once again, in this section the first example shows complex fractions from both arithmetic and algebra.

EXAMPLE 1   Simplifying Complex Fractions by Method 1

Simplify each complex fraction.

(a) $\dfrac{\dfrac{2}{3} + \dfrac{5}{9}}{\dfrac{1}{4} + \dfrac{1}{12}}$

(b) $\dfrac{6 + \dfrac{3}{x}}{\dfrac{x}{4} + \dfrac{1}{8}}$

Step 1   First, write each numerator as a single fraction.

$$\frac{2}{3} + \frac{5}{9} = \frac{2(3)}{3(3)} + \frac{5}{9}$$

$$= \frac{6}{9} + \frac{5}{9} = \frac{11}{9}$$

$$6 + \frac{3}{x} = \frac{6}{1} + \frac{3}{x}$$

$$= \frac{6x}{x} + \frac{3}{x} = \frac{6x + 3}{x}$$

Do the same thing with each denominator.

$$\frac{1}{4} + \frac{1}{12} = \frac{1(3)}{4(3)} + \frac{1}{12}$$

$$= \frac{3}{12} + \frac{1}{12} = \frac{4}{12}$$

$$\frac{x}{4} + \frac{1}{8} = \frac{x(2)}{4(2)} + \frac{1}{8}$$

$$= \frac{2x}{8} + \frac{1}{8} = \frac{2x + 1}{8}$$

Step 2   The original complex fraction can now be written as follows.

$$\frac{\dfrac{11}{9}}{\dfrac{4}{12}} \qquad\qquad \frac{\dfrac{6x+3}{x}}{\dfrac{2x+1}{8}}$$

Step 3   Now use the rule for division and the fundamental property.

$$\frac{11}{9} \div \frac{4}{12} = \frac{11}{9} \cdot \frac{12}{4} \qquad\qquad \frac{6x+3}{x} \div \frac{2x+1}{8} = \frac{6x+3}{x} \cdot \frac{8}{2x+1}$$

$$= \frac{11 \cdot 3 \cdot 4}{3 \cdot 3 \cdot 4} \qquad\qquad\qquad\quad = \frac{3(2x+1)}{x} \cdot \frac{8}{2x+1}$$

$$= \frac{11}{3} \qquad\qquad\qquad\qquad\qquad\quad = \frac{24}{x}$$

EXAMPLE 2   Simplifying a Complex Fraction by Method 1

Simplify the complex fraction.

$$\frac{\dfrac{xp}{q^3}}{\dfrac{p^2}{qx^2}}$$

Here the numerator and denominator are already single fractions, so use the division rule and then the fundamental property.

$$\frac{xp}{q^3} \div \frac{p^2}{qx^2} = \frac{xp}{q^3} \cdot \frac{qx^2}{p^2} = \frac{x^3}{q^2 p}$$

EXAMPLE 3   Simplifying a Complex Fraction by Method 1

Simplify the complex fraction.

$$\frac{\dfrac{3}{x+2} - 4}{\dfrac{2}{x+2} + 1} = \frac{\dfrac{3}{x+2} - \dfrac{4(x+2)}{x+2}}{\dfrac{2}{x+2} + \dfrac{1(x+2)}{x+2}} \qquad \text{Write both second terms with a denominator of } x+2.$$

$$= \frac{\dfrac{3 - 4(x+2)}{x+2}}{\dfrac{2 + 1(x+2)}{x+2}} \qquad \text{Subtract in the numerator. Add in the denominator.}$$

$$= \frac{\dfrac{3 - 4x - 8}{x+2}}{\dfrac{2 + x + 2}{x+2}} \qquad \text{Distributive property}$$

$$\frac{\dfrac{3 - 4x - 8}{x + 2}}{\dfrac{2 + x + 2}{x + 2}} = \frac{\dfrac{-5 - 4x}{x + 2}}{\dfrac{4 + x}{x + 2}}$$ Combine terms.

$$= \frac{-5 - 4x}{x + 2} \cdot \frac{x + 2}{4 + x}$$ Multiply by the reciprocal.

$$= \frac{-5 - 4x}{4 + x}$$ Lowest terms

OBJECTIVE 2 Simplify a complex fraction by multiplying by the least common denominator (Method 2). As an alternative method, a complex fraction may be simplified by a method that uses the fundamental property of rational expressions. Since any expression can be multiplied by a form of 1 to get an equivalent expression, we may multiply both the numerator and the denominator of a complex fraction by the same nonzero expression to get an equivalent complex fraction. If we choose the expression to be the LCD of all the fractions within the complex fraction, the complex fraction will be simplified. This is Method 2.

Method 2

To simplify a complex fraction:

Step 1   Find the LCD of all fractions within the complex fraction.

Step 2   Multiply both the numerator and the denominator of the complex fraction by this LCD using the distributive property as necessary. Write in lowest terms.

In the next example, Method 2 is used to simplify the complex fractions from Example 1.

EXAMPLE 4   Simplifying Complex Fractions by Method 2

Simplify each complex fraction.

(a) $\dfrac{\dfrac{2}{3} + \dfrac{5}{9}}{\dfrac{1}{4} + \dfrac{1}{12}}$

(b) $\dfrac{6 + \dfrac{3}{x}}{\dfrac{x}{4} + \dfrac{1}{8}}$

Step 1   Find the LCD for all denominators in the complex fraction.

The LCD for 3, 9, 4, and 12 is 36. | The LCD for x, 4, and 8 is $8x$.

Step 2   Multiply numerator and denominator of the complex fraction by the LCD.

$$\frac{\dfrac{2}{3} + \dfrac{5}{9}}{\dfrac{1}{4} + \dfrac{1}{12}} = \frac{36\left(\dfrac{2}{3} + \dfrac{5}{9}\right)}{36\left(\dfrac{1}{4} + \dfrac{1}{12}\right)}$$ | $$\frac{6 + \dfrac{3}{x}}{\dfrac{x}{4} + \dfrac{1}{8}} = \frac{8x\left(6 + \dfrac{3}{x}\right)}{8x\left(\dfrac{x}{4} + \dfrac{1}{8}\right)}$$

$$= \frac{36\left(\frac{2}{3}\right) + 36\left(\frac{5}{9}\right)}{36\left(\frac{1}{4}\right) + 36\left(\frac{1}{12}\right)} \qquad\qquad = \frac{8x(6) + 8x\left(\frac{3}{x}\right)}{8x\left(\frac{x}{4}\right) + 8x\left(\frac{1}{8}\right)} \qquad \text{Distributive property}$$

$$= \frac{24 + 20}{9 + 3} \qquad\qquad\qquad = \frac{48x + 24}{2x^2 + x}$$

$$= \frac{44}{12} = \frac{4 \cdot 11}{4 \cdot 3} \qquad\qquad = \frac{24(2x + 1)}{x(2x + 1)} \qquad \text{Factor.}$$

$$= \frac{11}{3} \qquad\qquad\qquad\quad = \frac{24}{x} \qquad\qquad \text{Lowest terms}$$

E X A M P L E   5   Simplifying a Complex Fraction by Method 2

Simplify the complex fraction.

$$\frac{\dfrac{3}{5m} - \dfrac{2}{m^2}}{\dfrac{9}{2m} + \dfrac{3}{4m^2}}$$

The LCD for $5m$, m^2, $2m$, and $4m^2$ is $20m^2$. Multiply numerator and denominator by $20m^2$.

$$\frac{\dfrac{3}{5m} - \dfrac{2}{m^2}}{\dfrac{9}{2m} + \dfrac{3}{4m^2}} = \frac{20m^2\left(\dfrac{3}{5m} - \dfrac{2}{m^2}\right)}{20m^2\left(\dfrac{9}{2m} + \dfrac{3}{4m^2}\right)}$$

$$= \frac{20m^2\left(\dfrac{3}{5m}\right) - 20m^2\left(\dfrac{2}{m^2}\right)}{20m^2\left(\dfrac{9}{2m}\right) + 20m^2\left(\dfrac{3}{4m^2}\right)} \qquad \text{Distributive property}$$

$$= \frac{12m - 40}{90m + 15}$$

Either of the two methods shown in this section can be used to simplify a complex fraction. You may want to choose one method and stick with it to eliminate confusion. However, some students prefer to use Method 1 for problems like Example 2, which is the quotient of two fractions. They prefer Method 2 for problems like Examples 1, 3, 4, and 5, which have sums or differences in the numerators or denominators or both.

CONNECTIONS

Some numbers can be expressed as *continued fractions,* which are infinite complex fractions. For example, the irrational number $\sqrt{2}$ can be expressed as follows.

$$\sqrt{2} = 1 + \cfrac{1}{2 + \cfrac{1}{2 + \cfrac{1}{2 + \cfrac{1}{2 + \cdots}}}}$$

Better and better approximations of $\sqrt{2}$ can be found by using more and more terms of the fraction. We give the first three approximations here.

$$1 + \frac{1}{2} = 1.5$$

$$1 + \cfrac{1}{2 + \cfrac{1}{2}} = 1 + \cfrac{1}{\frac{5}{2}} = 1 + \frac{2}{5} = \frac{7}{5} = 1.4$$

$$1 + \cfrac{1}{2 + \cfrac{1}{2 + \cfrac{1}{2}}} = 1 + \cfrac{1}{2 + \cfrac{1}{\frac{5}{2}}} = 1 + \cfrac{1}{2 + \frac{2}{5}} = 1 + \cfrac{1}{\frac{12}{5}}$$

$$= 1 + \frac{5}{12} = \frac{17}{12} = 1.41\overline{6}$$

A calculator gives $\sqrt{2} \approx 1.414213562$ to nine decimal places.

FOR DISCUSSION OR WRITING
Give the next approximation of $\sqrt{2}$ and compare it to the calculator value shown above. How many places after the decimal agree?

7.5 EXERCISES

Note: In many problems involving complex fractions, several different equivalent forms of the answer exist. If your answer does not look exactly like the one given in the back of the book, check to see if your answer is an equivalent form.

1. Consider the complex fraction $\cfrac{\dfrac{1}{2} - \dfrac{1}{3}}{\dfrac{5}{6} - \dfrac{1}{12}}$. Answer each of the following, outlining Method 1 for simplifying this complex fraction.

   (a) To combine the terms in the numerator, we must find the LCD of $\frac{1}{2}$ and $\frac{1}{3}$. What is this LCD? Determine the simplified form of the numerator of the complex fraction.

   (b) To combine the terms in the denominator, we must find the LCD of $\frac{5}{6}$ and $\frac{1}{12}$. What is this LCD? Determine the simplified form of the denominator of the complex fraction.

(c) Now use the results from parts (a) and (b) to write the complex fraction as a division problem using the symbol $\div$.

(d) Perform the operation from part (c) to obtain the final simplification.

2. Consider the same complex fraction given in Exercise 1, $\dfrac{\dfrac{1}{2} - \dfrac{1}{3}}{\dfrac{5}{6} - \dfrac{1}{12}}$. Answer each of the

following, outlining Method 2 for simplifying this complex fraction.

(a) We must determine the LCD of all the fractions within the complex fraction. What is this LCD?

(b) Multiply every term in the complex fraction by the LCD found in part (a), but do not combine the terms in the numerator and the denominator yet.

(c) Combine the terms from part (b) to obtain the simplified form of the complex fraction.

3. Which one of the following complex fractions is equivalent to $\dfrac{3 - \dfrac{1}{2}}{2 - \dfrac{1}{4}}$? Answer this

question without showing any work, and explain the reason for the equivalence.

(a) $\dfrac{3 + \dfrac{1}{2}}{2 + \dfrac{1}{4}}$ (b) $\dfrac{-3 + \dfrac{1}{2}}{2 - \dfrac{1}{4}}$ (c) $\dfrac{-3 - \dfrac{1}{2}}{-2 - \dfrac{1}{4}}$ (d) $\dfrac{-3 + \dfrac{1}{2}}{-2 + \dfrac{1}{4}}$

4. Only one of the choices below is equal to $\dfrac{\dfrac{1}{2} + \dfrac{1}{4}}{\dfrac{1}{3} + \dfrac{1}{12}}$. Which one is it?

Answer this question without showing any work, and explain the reason for the equivalence.

(a) $\dfrac{9}{5}$ (b) $-\dfrac{9}{5}$ (c) $-\dfrac{5}{9}$ (d) -12

5. In your own words, describe Method 1 for simplifying complex fractions. Illustrate with the example $\dfrac{\dfrac{1}{2}}{\dfrac{2}{3}}$.

6. In your own words, describe Method 2 for simplifying complex fractions. Illustrate with the example $\dfrac{\dfrac{1}{2}}{\dfrac{2}{3}}$.

Simplify each complex fraction. Use either method. See Examples 1–5.

7. $\dfrac{-\dfrac{4}{3}}{\dfrac{2}{9}}$ 8. $\dfrac{-\dfrac{5}{6}}{\dfrac{5}{4}}$ 9. $\dfrac{\dfrac{x}{y^2}}{\dfrac{x^2}{y}}$ 10. $\dfrac{\dfrac{p^4}{r}}{\dfrac{p^2}{r^2}}$

11. $\dfrac{\dfrac{4a^4b^3}{3a}}{\dfrac{2ab^4}{b^2}}$ 12. $\dfrac{\dfrac{2r^4t^2}{3t}}{\dfrac{5r^2t^5}{3r}}$ 13. $\dfrac{\dfrac{m+2}{3}}{\dfrac{m-4}{m}}$ 14. $\dfrac{\dfrac{q-5}{q}}{\dfrac{q+5}{3}}$

15. $\dfrac{\dfrac{2}{x} - 3}{\dfrac{2 - 3x}{2}}$

16. $\dfrac{6 + \dfrac{2}{r}}{\dfrac{3r + 1}{4}}$

17. $\dfrac{\dfrac{1}{x} + x}{\dfrac{x^2 + 1}{8}}$

18. $\dfrac{\dfrac{3}{m} - m}{\dfrac{3 - m^2}{4}}$

19. $\dfrac{a - \dfrac{5}{a}}{a + \dfrac{1}{a}}$

20. $\dfrac{q + \dfrac{1}{q}}{q + \dfrac{4}{q}}$

21. $\dfrac{\dfrac{5}{8} + \dfrac{2}{3}}{\dfrac{7}{3} - \dfrac{1}{4}}$

22. $\dfrac{\dfrac{6}{5} - \dfrac{1}{9}}{\dfrac{2}{5} + \dfrac{5}{3}}$

23. $\dfrac{\dfrac{1}{x^2} + \dfrac{1}{y^2}}{\dfrac{1}{x} - \dfrac{1}{y}}$

24. $\dfrac{\dfrac{1}{a^2} - \dfrac{1}{b^2}}{\dfrac{1}{a} - \dfrac{1}{b}}$

25. $\dfrac{\dfrac{2}{p^2} - \dfrac{3}{5p}}{\dfrac{4}{p} + \dfrac{1}{4p}}$

26. $\dfrac{\dfrac{2}{m^2} - \dfrac{3}{m}}{\dfrac{2}{5m^2} + \dfrac{1}{3m}}$

27. $\dfrac{\dfrac{5}{x^2y} - \dfrac{2}{xy^2}}{\dfrac{3}{x^2y^2} + \dfrac{4}{xy}}$

28. $\dfrac{\dfrac{1}{m^3p} + \dfrac{2}{mp^2}}{\dfrac{4}{mp} + \dfrac{1}{m^2p}}$

29. $\dfrac{\dfrac{1}{4} - \dfrac{1}{a^2}}{\dfrac{1}{2} + \dfrac{1}{a}}$

30. $\dfrac{\dfrac{1}{9} - \dfrac{1}{m^2}}{\dfrac{1}{3} + \dfrac{1}{m}}$

31. $\dfrac{\dfrac{1}{z + 5}}{\dfrac{4}{z^2 - 25}}$

32. $\dfrac{\dfrac{1}{a + 1}}{\dfrac{2}{a^2 - 1}}$

33. $\dfrac{\dfrac{1}{m + 1} - 1}{\dfrac{1}{m + 1} + 1}$

34. $\dfrac{\dfrac{2}{x - 1} + 2}{\dfrac{2}{x - 1} - 2}$

35. $\dfrac{\dfrac{1}{m - 1} + \dfrac{2}{m + 2}}{\dfrac{2}{m + 2} - \dfrac{1}{m - 3}}$

36. $\dfrac{\dfrac{5}{r + 3} - \dfrac{1}{r - 1}}{\dfrac{2}{r + 2} + \dfrac{3}{r + 3}}$

37. In a fraction, what operation does the fraction bar represent?

38. What property of real numbers justifies Method 2 of simplifying complex fractions?

RELATING CONCEPTS (EXERCISES 39–42)

In order to find the average of two numbers, we add them and divide by 2. Suppose that we wish to find the average of $\frac{3}{8}$ and $\frac{5}{6}$.

Work Exercises 39–42 in order, to see how a complex fraction occurs in a problem like this.

39. Write in symbols: the sum of $\dfrac{3}{8}$ and $\dfrac{5}{6}$, divided by 2. Your result should be a complex fraction.

40. Simplify the complex fraction from Exercise 39 using Method 1.

41. Simplify the complex fraction from Exercise 39 using Method 2.

42. Your answers in Exercises 40 and 41 should be the same. Which method did you prefer? Why?

Did you make the connection between finding the average of two fractions and simplifying a complex fraction?

The expressions in Exercises 43–48 are called continued fractions. *Simplify these continued fractions by starting at "the bottom" and working upward.*

43. $1 + \dfrac{1}{1 + \dfrac{1}{1 + 1}}$

44. $5 + \dfrac{5}{5 + \dfrac{5}{5 + 5}}$

45. $7 - \dfrac{3}{5 + \dfrac{2}{4 - 2}}$

46. $3 - \dfrac{2}{4 + \dfrac{2}{4 - 2}}$

47. $r + \dfrac{r}{4 - \dfrac{2}{6 + 2}}$

48. $\dfrac{2q}{7} - \dfrac{q}{6 + \dfrac{8}{4 + 4}}$

7.6 Solving Equations Involving Rational Expressions

OBJECTIVES

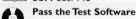

1 Distinguish between expressions with rational coefficients and equations with terms that are rational expressions.

2 Solve equations with rational expressions.

3 Solve a formula for a specified variable.

FOR EXTRA HELP

📖 **SSG** Sec. 7.6
    SSM Sec. 7.6

💿 **Pass the Test Software**

💿 **InterAct Math**
    Tutorial Software

📼 **Video** 12

In Section 2.2 we solved equations with fractions as coefficients. By using the multiplication property of equality, we cleared the fractions by multiplying by the LCD. We continue this work here.

OBJECTIVE 1 Distinguish between expressions with rational coefficients and equations with terms that are rational expressions. Before solving equations with rational expressions, you must understand the difference between *sums* and *differences* of terms with rational coefficients, and *equations* with terms that are rational expressions. Sums and differences are *simplified,* while equations are *solved.*

┌ **EXAMPLE 1**   Distinguishing between Expressions and Equations

Identify each of the following as an expression or an equation. If it is an expression, simplify it. If it is an equation, solve it.

(a) $\dfrac{3}{4}x - \dfrac{2}{3}x$

This is a difference of two terms, so it is an expression. (There is no equals sign.) Simplify by finding the LCD, writing each coefficient with this LCD, and combining like terms.

$$\frac{3}{4}x - \frac{2}{3}x = \frac{9}{12}x - \frac{8}{12}x \qquad \text{Get a common denominator.}$$

$$= \frac{1}{12}x \qquad \text{Combine like terms.}$$

(b) $\dfrac{3}{4}x - \dfrac{2}{3}x = \dfrac{1}{2}$

Because of the equals sign, this is an equation to be solved. We proceed as in Section 2.2, using the multiplication property of equality to clear fractions. The LCD is 12.

$$\frac{3}{4}x - \frac{2}{3}x = \frac{1}{2}$$

$$12\left(\frac{3}{4}x - \frac{2}{3}x\right) = 12\left(\frac{1}{2}\right) \qquad \text{Multiply by 12.}$$

$$12\left(\frac{3}{4}x\right) - 12\left(\frac{2}{3}x\right) = 12\left(\frac{1}{2}\right) \qquad \text{Distributive property}$$

$$9x - 8x = 6 \qquad \text{Multiply.}$$
$$x = 6 \qquad \text{Combine like terms.}$$

The solution, 6, should be checked in the original equation. Because it leads to a true statement, the solution set is {6}.

The ideas of Example 1 can be summarized as follows.

When adding or subtracting, the LCD must be kept throughout the simplification. When solving an equation, the LCD is used to multiply both sides so that denominators are eliminated.

OBJECTIVE 2 Solve equations with rational expressions. The next few examples illustrate clearing an equation of fractions in order to solve it.

EXAMPLE 2 Solving an Equation Involving Rational Expressions

Solve $\dfrac{p}{2} - \dfrac{p-1}{3} = 1$.

Multiply both sides by the LCD, 6.

$$6\left(\frac{p}{2} - \frac{p-1}{3}\right) = 6 \cdot 1$$

$$6\left(\frac{p}{2}\right) - 6\left(\frac{p-1}{3}\right) = 6 \qquad \text{Distributive property}$$

$$3p - 2(p-1) = 6$$

Be careful to put parentheses around $p - 1$; otherwise you may get an incorrect solution. Continue simplifying and solve the equation.

$$3p - 2p + 2 = 6 \qquad \text{Distributive property}$$
$$p + 2 = 6 \qquad \text{Combine like terms.}$$
$$p = 4 \qquad \text{Subtract 2.}$$

Check to see that {4} is the solution set by replacing p with 4 in the original equation.

 The most common error in equations like the one found in Example 2 occurs when parentheses are not used for the numerator $p - 1$. If parentheses are not used, a sign error may result.

The equations in Examples 1(b) and 2 did not have variables in denominators. When solving equations that have a variable in a denominator, remember that the number 0 cannot be a denominator. Therefore, the solution cannot be a number that will make the denominator equal 0. Example 3 illustrates this.

E X A M P L E   3   Solving an Equation Involving Rational Expressions with No Solution

Solve $\dfrac{x}{x-2} = \dfrac{2}{x-2} + 2$.

Multiply both sides by the LCD, $x - 2$.

$$(x-2)\left(\frac{x}{x-2}\right) = (x-2)\left(\frac{2}{x-2}\right) + (x-2)(2)$$

$$x = 2 + 2x - 4 \qquad \text{Distributive property}$$

$$x = -2 + 2x \qquad \text{Combine terms.}$$

$$-x = -2 \qquad \text{Subtract } 2x.$$

$$x = 2 \qquad \text{Multiply by } -1.$$

The proposed solution is 2. If we substitute 2 into the original equation, we get

$$\frac{2}{2-2} = \frac{2}{2-2} + 2 \qquad ?$$

$$\frac{2}{0} = \frac{2}{0} + 2. \qquad ?$$

Notice that 2 makes both denominators equal 0. Because 0 cannot be the denominator of a fraction, the solution set is $\emptyset$.

While it is always a good idea to check solutions to guard against arithmetic and algebraic errors, it is *essential* to check proposed solutions when variables appear in denominators in the original equation. Some students like to determine which numbers cannot be solutions *before* solving the equation.

Solving Equations with Rational Expressions

Step 1   **Multiply by the LCD.**   Multiply both sides of the equation by the least common denominator. (This clears the equation of fractions.)

Step 2   **Solve.**   Solve the resulting equation.

Step 3   **Check.**   Check each proposed solution by substituting it in the original equation. Reject any that cause a denominator to equal 0.

E X A M P L E   4   Solving an Equation Involving Rational Expressions

Solve $\dfrac{2}{x^2-x} = \dfrac{1}{x^2-1}$.

Step 1   Begin by finding the LCD. Since $x^2 - x$ can be factored as $x(x-1)$, and $x^2 - 1$ can be factored as $(x+1)(x-1)$, the LCD is $x(x+1)(x-1)$.

$$\frac{2}{x(x-1)} = \frac{1}{(x+1)(x-1)} \qquad \text{Factor the denominators.}$$

Step 2   Notice that 0, −1, and 1 cannot be solutions of this equation. Multiply both sides of the equation by $x(x + 1)(x − 1)$.

$$x(x + 1)(x − 1)\frac{2}{x(x − 1)} = x(x + 1)(x − 1)\frac{1}{(x + 1)(x − 1)}$$

$$2(x + 1) = x$$

$$2x + 2 = x \qquad \text{Distributive property}$$

$$2 = −x \qquad \text{Subtract } 2x.$$

$$x = −2 \qquad \text{Multiply by } −1.$$

Step 3   The proposed solution is −2, which does not make any denominator equal 0. A check will verify that no arithmetic or algebraic errors have been made. Thus, the solution set is $\{−2\}$.

E X A M P L E   5   Solving an Equation Involving Rational Expressions

Solve $\dfrac{1}{x − 1} + \dfrac{1}{2} = \dfrac{2}{x^2 − 1}$.

Factor the denominator on the right.

$$\frac{1}{x − 1} + \frac{1}{2} = \frac{2}{(x + 1)(x − 1)}$$

Notice that 1 and −1 cannot be solutions of this equation. Multiply both sides of the equation by the LCD, $2(x + 1)(x − 1)$.

$$2(x + 1)(x − 1)\left(\frac{1}{x − 1} + \frac{1}{2}\right) = 2(x + 1)(x − 1)\frac{2}{(x + 1)(x − 1)}$$

$$2(x + 1)(x − 1)\frac{1}{x − 1} + \mathbf{2}(x + 1)(x − 1)\frac{1}{\mathbf{2}} = 2(x + 1)(x − 1)\frac{2}{(x + 1)(x − 1)}$$

$$2(x + 1) + (x + 1)(x − 1) = 4$$

$$2x + 2 + x^2 − 1 = 4 \qquad \text{Distributive property}$$

$$x^2 + 2x + 1 = 4$$

$$x^2 + 2x − 3 = 0 \qquad \text{Get 0 on the right side.}$$

Factoring gives

$$(x + 3)(x − 1) = 0.$$

$$x + 3 = 0 \quad \text{or} \quad x − 1 = 0 \qquad \text{Zero-factor property}$$

$$x = −3 \quad \text{or} \qquad x = 1$$

−3 and 1 are proposed solutions. However, as noted above, 1 makes an original denominator equal 0, so 1 is not a solution. Substituting −3 for x gives a true statement, so $\{−3\}$ is the solution set.

E X A M P L E   6   Solving an Equation Involving Rational Expressions

Solve $\dfrac{1}{k^2 + 4k + 3} + \dfrac{1}{2k + 2} = \dfrac{3}{4k + 12}$.

Factoring each denominator gives the equation

$$\frac{1}{(k+1)(k+3)} + \frac{1}{2(k+1)} = \frac{3}{4(k+3)}.$$

The LCD is $4(k+1)(k+3)$, indicating that -1 and -3 cannot be solutions of the equation. Multiply both sides by this LCD.

$$4(k+1)(k+3)\left(\frac{1}{(k+1)(k+3)} + \frac{1}{2(k+1)}\right)$$
$$= 4(k+1)(k+3)\frac{3}{4(k+3)}$$

$$4(k+1)(k+3)\frac{1}{(k+1)(k+3)} + 2\cdot 2(k+1)(k+3)\frac{1}{2(k+1)}$$
$$= 4(k+1)(k+3)\frac{3}{4(k+3)}$$

$$4 + 2(k+3) = 3(k+1)$$
$$4 + 2k + 6 = 3k + 3 \qquad \text{Distributive property}$$
$$2k + 10 = 3k + 3$$
$$10 - 3 = 3k - 2k$$
$$7 = k$$

The proposed solution, 7, does not make an original denominator equal 0. A check shows that the algebra is correct, so $\{7\}$ is the solution set.

OBJECTIVE 3   Solve a formula for a specified variable. Solving a formula for a specified variable was discussed in Chapter 2. In the next example, this procedure is applied to a formula involving fractions.

EXAMPLE 7   Solving for a Specified Variable

Solve the formula $\frac{1}{a} = \frac{1}{b} + \frac{1}{c}$ for c.

The LCD of all the fractions in the equation is abc, so multiply both sides by abc.

$$abc\left(\frac{1}{a}\right) = abc\left(\frac{1}{b} + \frac{1}{c}\right)$$
$$abc\left(\frac{1}{a}\right) = abc\left(\frac{1}{b}\right) + abc\left(\frac{1}{c}\right) \qquad \text{Distributive property}$$
$$bc = ac + ab$$

Since we are solving for c, get all terms with c on one side of the equation. Do this by subtracting ac from both sides.

$$bc - ac = ab \qquad \text{Subtract } ac.$$

Factor out the common factor c on the left.

$$c(b-a) = ab \qquad \text{Factor out } c.$$

Finally, divide both sides by the coefficient of c, which is $b-a$.

$$c = \frac{ab}{b-a}$$

Students often have trouble in the step that involves factoring out the variable for which they are solving. In Example 7, we had to factor out c on the left side so that we could divide both sides by $b - a$.

When solving an equation for a specified variable, be sure that the specified variable appears alone on only one side of the equals sign in the final equation.

7.6 EXERCISES

Identify as an expression or an equation. If an expression appears, simplify it. If an equation appears, solve it. See Example 1.

1. $\dfrac{7}{8}x + \dfrac{1}{5}x$

2. $\dfrac{4}{7}x + \dfrac{3}{5}x$

3. $\dfrac{7}{8}x + \dfrac{1}{5}x = 1$

4. $\dfrac{4}{7}x + \dfrac{3}{5}x = 1$

5. $\dfrac{3}{5}y - \dfrac{7}{10}y$

6. $\dfrac{2}{3}y - \dfrac{7}{4}y$

7. $\dfrac{3}{5}y - \dfrac{7}{10}y = 1$

8. $\dfrac{2}{3}y - \dfrac{7}{4}y = -13$

When solving an equation with variables in denominators, we must determine the values that cause these denominators to equal 0, so we can reject these values if they appear as possible solutions. Find all values for which at least one denominator is equal to 0. Do not solve. See Examples 3–6.

9. $\dfrac{3}{x + 2} - \dfrac{5}{x} = 1$

10. $\dfrac{7}{x} + \dfrac{9}{x - 3} = 5$

11. $\dfrac{-1}{(x + 3)(x - 4)} = \dfrac{1}{2x + 1}$

12. $\dfrac{8}{(x - 8)(x + 2)} = \dfrac{7}{3x - 10}$

13. $\dfrac{4}{x^2 + 8x - 9} + \dfrac{1}{x^2 - 4} = 0$

14. $\dfrac{-3}{x^2 + 9x - 10} - \dfrac{12}{x^2 - 16} = 0$

15. Explain how the LCD is used in a different way when adding and subtracting rational expressions compared to solving equations with rational expressions.

16. If we multiply both sides of the equation $\dfrac{6}{x + 5} = \dfrac{6}{x + 5}$ by $x + 5$, we get $6 = 6$. Are all real numbers solutions of this equation? Explain.

Solve each equation and check. See Examples 1(b), 2, and 3.

17. $\dfrac{5}{m} - \dfrac{3}{m} = 8$

18. $\dfrac{4}{y} + \dfrac{1}{y} = 2$

19. $\dfrac{5}{y} + 4 = \dfrac{2}{y}$

20. $\dfrac{11}{q} = 3 - \dfrac{1}{q}$

21. $\dfrac{3x}{5} - 6 = x$

22. $\dfrac{5t}{4} + t = 9$

23. $\dfrac{4m}{7} + m = 11$

24. $a - \dfrac{3a}{2} = 1$

25. $\dfrac{z - 1}{4} = \dfrac{z + 3}{3}$

26. $\dfrac{r - 5}{2} = \dfrac{r + 2}{3}$

27. $\dfrac{3p + 6}{8} = \dfrac{3p - 3}{16}$

28. $\dfrac{2z + 1}{5} = \dfrac{7z + 5}{15}$

29. $\dfrac{2x + 3}{x} = \dfrac{3}{2}$

30. $\dfrac{5 - 2y}{y} = \dfrac{1}{4}$

31. $\dfrac{k}{k - 4} - 5 = \dfrac{4}{k - 4}$

32. $\dfrac{-5}{a + 5} = \dfrac{a}{a + 5} + 2$

33. $\dfrac{q + 2}{3} + \dfrac{q - 5}{5} = \dfrac{7}{3}$

34. $\dfrac{t}{6} + \dfrac{4}{3} = \dfrac{t - 2}{3}$

35. $\dfrac{x}{2} = \dfrac{5}{4} + \dfrac{x - 1}{4}$

36. $\dfrac{8p}{5} = \dfrac{3p - 4}{2} + \dfrac{5}{2}$

Solve each equation and check. Be very careful with signs. See Example 2.

37. $\dfrac{a+7}{8} - \dfrac{a-2}{3} = \dfrac{4}{3}$

38. $\dfrac{x+3}{7} - \dfrac{x+2}{6} = \dfrac{1}{6}$

39. $\dfrac{p}{2} - \dfrac{p-1}{4} = \dfrac{5}{4}$

40. $\dfrac{r}{6} - \dfrac{r-2}{3} = -\dfrac{4}{3}$

41. $\dfrac{3x}{5} - \dfrac{x-5}{7} = 3$

42. $\dfrac{8k}{5} - \dfrac{3k-4}{2} = \dfrac{5}{2}$

Solve each equation and check your answer. See Examples 3–6.

43. $\dfrac{2}{m} = \dfrac{m}{5m+12}$

44. $\dfrac{x}{4-x} = \dfrac{2}{x}$

45. $\dfrac{-2}{z+5} + \dfrac{3}{z-5} = \dfrac{20}{z^2-25}$

46. $\dfrac{3}{r+3} - \dfrac{2}{r-3} = \dfrac{-12}{r^2-9}$

47. $\dfrac{3}{x-1} + \dfrac{2}{4x-4} = \dfrac{7}{4}$

48. $\dfrac{2}{p+3} + \dfrac{3}{8} = \dfrac{5}{4p+12}$

49. $\dfrac{y}{3y+3} = \dfrac{2y-3}{y+1} - \dfrac{2y}{3y+3}$

50. $\dfrac{2k+3}{k+1} - \dfrac{3k}{2k+2} = \dfrac{-2k}{2k+2}$

51. $\dfrac{5x}{14x+3} = \dfrac{1}{x}$

52. $\dfrac{m}{8m+3} = \dfrac{1}{3m}$

53. $\dfrac{2}{x-1} - \dfrac{2}{3} = \dfrac{-1}{x+1}$

54. $\dfrac{5}{p-2} = 7 - \dfrac{10}{p+2}$

55. $\dfrac{x}{2x+2} = \dfrac{-2x}{4x+4} + \dfrac{2x-3}{x+1}$

56. $\dfrac{5t+1}{3t+3} = \dfrac{5t-5}{5t+5} + \dfrac{3t-1}{t+1}$

57. $\dfrac{8x+3}{x} = 3x$

58. $\dfrac{2}{y} = \dfrac{y}{5y-12}$

59. $\dfrac{3y}{y^2+5y+6} = \dfrac{5y}{y^2+2y-3} - \dfrac{2}{y^2+y-2}$

60. $\dfrac{m}{m^2+m-2} + \dfrac{m}{m^2-1} = \dfrac{m}{m^2+3m+2}$

61. $\dfrac{x+4}{x^2-3x+2} - \dfrac{5}{x^2-4x+3} = \dfrac{x-4}{x^2-5x+6}$

62. $\dfrac{3}{r^2+r-2} - \dfrac{1}{r^2-1} = \dfrac{7}{2(r^2+3r+2)}$

63. If you are solving a formula for the letter k, and your steps lead to the equation $kr - mr = km$, what would be your next step?

64. If you are solving a formula for the letter k, and your steps lead to the equation $kr - km = mr$, what would be your next step?

Solve each formula for the specified variable. See Example 7.

65. $m = \dfrac{kF}{a}$ for F

66. $I = \dfrac{kE}{R}$ for E

67. $m = \dfrac{kF}{a}$ for a

68. $I = \dfrac{kE}{R}$ for R

69. $I = \dfrac{E}{R+r}$ for R

70. $I = \dfrac{E}{R+r}$ for r

71. $h = \dfrac{2A}{B+b}$ for A

72. $d = \dfrac{2S}{n(a+L)}$ for S

73. $d = \dfrac{2S}{n(a+L)}$ for a

74. $h = \dfrac{2A}{B+b}$ for B

75. $\dfrac{1}{x} = \dfrac{1}{y} - \dfrac{1}{z}$ for y

76. $\dfrac{3}{k} = \dfrac{1}{p} + \dfrac{1}{q}$ for q

77. $9x + \dfrac{3}{z} = \dfrac{5}{y}$ for z

78. $\dfrac{1}{a} = \dfrac{1}{b} + \dfrac{1}{c}$ for a

RELATING CONCEPTS (EXERCISES 79-84)

In Section 7.4 we saw how, after adding or subtracting two rational expressions, the result can often be simplified to lowest terms. In this section we learned that multiplying both sides of an equation by the LCD may lead to a solution that must be rejected because it causes a denominator to equal 0.

Work Exercises 79–84 in order, *to see a relationship between these two concepts.*

79. Solve the equation $\dfrac{x^2}{x-3} + \dfrac{2x-15}{x-3} = 0$ by multiplying both sides by $x-3$. What is the solution? What is the number that must be rejected as a solution?

80. Combine the two rational expressions on the left side of the given equation in Exercise 79 and reduce to lowest terms.

 (a) If you set the simplified form equal to 0 and solve, how does your solution compare to the actual solution in Exercise 79?

 (b) If you set the common factor that you divided out equal to 0 and solve, how does your solution compare to the rejected solution in Exercise 79?

81. Consider the equation $\dfrac{1}{x-1} + \dfrac{1}{2} - \dfrac{2}{x^2-1} = 0$, which is equivalent to the equation solved in Example 5. According to Example 5, what is the solution? What is the number that must be rejected as a solution?

82. Combine the three rational expressions on the left side of the given equation in Exercise 81 and reduce to lowest terms. Repeat parts (a) and (b) of Exercise 80, comparing this time to the equation in Exercise 81.

83. Write a short paragraph summarizing what you have learned in working Exercises 79–82.

84. Devise a procedure whereby you could solve an equation involving rational expressions which would not lead to values that must be rejected.

Did you make the connection between rational expressions that can be simplified and rational equations with rejected solutions?

TECHNOLOGY INSIGHTS (EXERCISES 85-90)

The first line in the calculator screen on the left shows that 6 has been stored into memory location X. The expression $\dfrac{1}{X+4}$ is then calculated to be $\dfrac{1}{10}$, as expected.

However, if the calculator is directed to evaluate $\dfrac{1}{X-6}$, an error message occurs, as shown in the screen on the right. The value 6 leads to division by 0, which is not allowed.

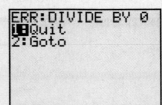

TECHNOLOGY INSIGHTS (EXERCISES 85-90) (CONTINUED)

For the following screens, either evaluate the expression or tell whether an error message would occur for the expression entered.

85.
```
5→X
              5
-3/(X+5)▶Frac
```

86.
```
8→X
              8
-4/(X+8)▶Frac
```

87.
```
12→X
             12
-3/(12-X)
```

88.
```
16→X
             16
-8/(16-X)
```

89.
```
3→X
              3
1/((X+4)*(X-3))▶
Frac
```

90.
```
5→X
              5
1/((X-5)*(X-3))▶
Frac
```

SUMMARY Exercises on Operations and Equations with Rational Expressions

We have performed the four operations of arithmetic with rational expressions and solved equations with rational expressions. The exercises in this summary include a mixed variety of problems of these types. To work them, recall the procedures explained in the earlier sections of this chapter. They are summarized here.

Multiplication of Rational Expressions	Multiply numerators and multiply denominators. Use the fundamental property to express in lowest terms.
Division of Rational Expressions	First, change the second fraction to its reciprocal; then multiply as described above.
Addition of Rational Expressions	Find the least common denominator (LCD) if necessary. Write all rational expressions with this LCD. Add numerators, and keep the same denominator. Express in lowest terms.

Subtraction of Rational Expressions	Find the LCD if necessary. Write all rational expressions with this LCD. Subtract numerators (use parentheses as required), and keep the same denominator. Express in lowest terms.
Solving Equations with Rational Expressions	Multiply both sides of the equation by the LCD of all the rational expressions in the equation. Solve, using methods described in earlier chapters. Be sure to check all proposed solutions and reject any that cause a denominator to equal 0.

A common error that was mentioned in Section 7.6 bears repeating here. Students often confuse *operations* on rational expressions with the *solution of equations* with rational expressions. For example, the four possible operations on the rational expressions $\frac{1}{x}$ and $\frac{1}{x-2}$ can be performed as follows.

Add:

$$\frac{1}{x} + \frac{1}{x-2} = \frac{x-2}{x(x-2)} + \frac{x}{x(x-2)}$$ Write with a common denominator.

$$= \frac{x-2+x}{x(x-2)}$$ Add numerators; keep the same denominator.

$$= \frac{2x-2}{x(x-2)}$$ Combine like terms.

Subtract:

$$\frac{1}{x} - \frac{1}{x-2} = \frac{x-2}{x(x-2)} - \frac{x}{x(x-2)}$$ Write with a common denominator.

$$= \frac{x-2-x}{x(x-2)}$$ Subtract numerators; keep the same denominator.

$$= \frac{-2}{x(x-2)}$$ Combine like terms.

Multiply:

$$\frac{1}{x} \cdot \frac{1}{x-2} = \frac{1}{x(x-2)}$$ Multiply numerators and multiply denominators.

Divide:

$$\frac{1}{x} \div \frac{1}{x-2} = \frac{1}{x} \cdot \frac{x-2}{1} = \frac{x-2}{x}$$ Change to multiplication by the reciprocal of the second fraction.

On the other hand, consider the *equation*

$$\frac{1}{x} + \frac{1}{x-2} = \frac{3}{4}.$$

Neither 0 nor 2 can be solutions of this equation, since each will cause a denominator to equal 0. We use the multiplication property of equality to multiply both sides by the LCD, $4x(x - 2)$, giving an equation with no denominators.

$$4x(x - 2)\frac{1}{x} + 4x(x - 2)\frac{1}{x - 2} = 4x(x - 2)\frac{3}{4}$$

$$4x - 8 + 4x = 3x^2 - 6x \qquad \text{Distributive property}$$

$$0 = 3x^2 - 14x + 8 \qquad \text{Get 0 on one side.}$$

$$0 = (3x - 2)(x - 4) \qquad \text{Factor.}$$

$$3x - 2 = 0 \quad \text{or} \quad x - 4 = 0 \qquad \text{Zero-factor property}$$

$$x = \frac{2}{3} \quad \text{or} \quad x = 4$$

Both $\frac{2}{3}$ and 4 are solutions since neither makes a denominator equal 0; the solution set is $\left\{\frac{2}{3}, 4\right\}$.

In conclusion, remember the following points when working exercises involving rational expressions.

Points to Remember When Working with Rational Expressions

1. The fundamental property is applied only after numerators and denominators have been *factored*.

2. When adding and subtracting rational expressions, the common denominator must be kept throughout the problem and in the final result.

3. Always look to see if the answer is in lowest terms; if it is not, use the fundamental property.

4. When solving equations with rational expressions, reject any proposed solution that causes an original denominator to equal 0.

SUMMARY EXERCISES

For each exercise, indicate "operation" if an operation is to be performed or "equation" if an equation is to be solved. Then perform the operation or solve the equation as the case may be.

1. $\frac{4}{p} + \frac{6}{p}$

2. $\frac{x^3y^2}{x^2y^4} \cdot \frac{y^5}{x^4}$

3. $\frac{1}{x^2 + x - 2} \div \frac{4x^2}{2x - 2}$

4. $\frac{8}{m - 5} = 2$

5. $\frac{2y^2 + y - 6}{2y^2 - 9y + 9} \cdot \frac{y^2 - 2y - 3}{y^2 - 1}$

6. $\frac{2}{k^2 - 4k} + \frac{3}{k^2 - 16}$

7. $\frac{x - 4}{5} = \frac{x + 3}{6}$

8. $\frac{3t^2 - t}{6t^2 + 15t} \div \frac{6t^2 + t - 1}{2t^2 - 5t - 25}$

9. $\frac{4}{p + 2} + \frac{1}{3p + 6}$

10. $\frac{1}{y} + \frac{1}{y - 3} = -\frac{5}{4}$

11. $\frac{3}{t - 1} + \frac{1}{t} = \frac{7}{2}$

12. $\frac{6}{y} - \frac{2}{3y}$

13. $\frac{5}{4z} - \frac{2}{3z}$

14. $\frac{k + 2}{3} = \frac{2k - 1}{5}$

15. $\dfrac{1}{m^2 + 5m + 6} + \dfrac{2}{m^2 + 4m + 3}$

16. $\dfrac{2k^2 - 3k}{20k^2 - 5k} \div \dfrac{2k^2 - 5k + 3}{4k^2 + 11k - 3}$

17. $\dfrac{2}{x + 1} + \dfrac{5}{x - 1} = \dfrac{10}{x^2 - 1}$

18. $\dfrac{3}{x + 3} + \dfrac{4}{x + 6} = \dfrac{9}{x^2 + 9x + 18}$

19. $\dfrac{4t^2 - t}{6t^2 + 10t} \div \dfrac{8t^2 + 2t - 1}{3t^2 + 11t + 10}$

20. $\dfrac{x}{x - 2} + \dfrac{3}{x + 2} = \dfrac{8}{x^2 - 4}$

7.7 Applications of Rational Expressions

OBJECTIVES

1 Solve problems about numbers and data.

2 Solve problems about distance.

3 Solve problems about work.

4 Solve problems about variation.

FOR EXTRA HELP

📖 **SSG** Sec. 7.7
SSM Sec. 7.7

💿 **Pass the Test Software**

💿 **InterAct Math Tutorial Software**

📼 **Video 12**

Every time we learn to solve a new type of equation, we are able to apply our knowledge to solving new types of applications. In Section 7.6 we solved equations involving rational expressions; now we can solve applications that involve this type of equation. The six-step problem solving method of Chapter 2 still applies.

OBJECTIVE 1 Solve problems about numbers and data. We begin with an example about an unknown number.

EXAMPLE 1 Solving a Problem about an Unknown Number

If the same number is added to both the numerator and the denominator of the fraction $\frac{2}{5}$, the result is $\frac{2}{3}$. Find the number.

Step 1 Let $x = $ the number added to the numerator and the denominator.

Step 2 Then

$$\frac{2 + x}{5 + x}$$

represents the result of adding the same number to both the numerator and denominator.

Step 3 Since this result is $\frac{2}{3}$, the equation is

$$\frac{2 + x}{5 + x} = \frac{2}{3}.$$

Step 4 Solve this equation by multiplying both sides by the LCD, $3(5 + x)$.

$$3(5 + x)\frac{2 + x}{5 + x} = 3(5 + x)\frac{2}{3}$$

$$3(2 + x) = 2(5 + x)$$

$$6 + 3x = 10 + 2x \qquad \text{Distributive property}$$

$$x = 4 \qquad \text{Subtract } 2x; \text{ subtract } 6.$$

Step 5 The number is 4.

Step 6 Check the solution in the words of the original problem. If 4 is added to both the numerator and denominator of $\frac{2}{5}$, the result is $\frac{6}{9} = \frac{2}{3}$, as required.

EXAMPLE 2 Examining the Leading Causes of Death in the United States

In 1996, approximately 734,000 Americans died of heart disease, the leading cause of death in the United States. Cancer was not far behind. However, there was a large drop between cancer and the third leading cause, stroke. Approximately 380,000 more deaths were caused by cancer than stroke, and the ratio of cancer deaths to stroke deaths was 3.4 to 1. How many deaths were caused by cancer, and how many were caused by stroke?

Let x represent the number of deaths caused by stroke. Then $x + 380,000$ deaths were caused by cancer. Since the ratio of cancer deaths to stroke deaths was 3.4 to 1 (or simply 3.4), we can write the following equation.

$$\frac{x + 380,000}{x} = 3.4$$

Now solve the equation.

$$x + 380,000 = 3.4x \qquad \text{Multiply by } x.$$
$$380,000 = 2.4x \qquad \text{Subtract } x.$$
$$x \approx 158,000 \qquad \text{Divide by 2.4.}$$

Since x represents the number of stroke deaths, there were about 158,000 deaths due to stroke, and about $x + 380,000 = 538,000$ deaths due to cancer. A check shows that these numbers satisfy the conditions of the problem. (*Note:* These are only approximations based on actual data from the Centers for Disease Control and Prevention and the National Center for Health Statistics.)

OBJECTIVE 2 Solve problems about distance. Recall from Chapter 2 the following formulas relating distance, rate, and time.

Distance, Rate, and Time Relationship

$$d = rt \qquad r = \frac{d}{t} \qquad t = \frac{d}{r}$$

You may wish to refer to Example 6 in Section 2.6, which illustrates the basic use of these formulas.

PROBLEM SOLVING

In Section 2.6 we solved applications involving distance, rate, and time. Recall the importance of setting up a chart to organize the information given in these problems.

EXAMPLE 3 Solving a Problem about Distance, Rate, and Time

The Big Muddy River has a current of 3 miles per hour. A motorboat takes the same amount of time to go 12 miles downstream as it takes to go 8 miles upstream. What is the speed of the boat in still water?

This problem requires the distance formula, $d = rt$. Let $x =$ the speed of the boat in still water. Since the current pushes the boat when the boat is going downstream, the speed of the boat downstream will be the sum of the speed of the boat and the speed of the current, or $x + 3$ miles per hour. Also, the boat's speed going upstream is $x - 3$ miles per hour. See Figure 1 on the next page.

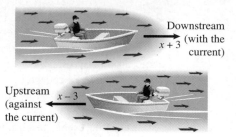

Figure 1

Summarize all of the information in a chart.

	d	r	t
Downstream	12	x + 3	
Upstream	8	x − 3	

Fill in the last column, representing time, by using the formula $t = \dfrac{d}{r}$.

For the time upstream,

$$\frac{d}{r} = \frac{8}{x - 3},$$

and for the time downstream,

$$\frac{d}{r} = \frac{12}{x + 3}.$$

Now complete the chart.

	d	r	t
Downstream	12	x + 3	$\dfrac{12}{x + 3}$
Upstream	8	x − 3	$\dfrac{8}{x - 3}$

Times are equal.

According to the original problem, the time upstream equals the time downstream. The two times from the chart must therefore be equal, giving the equation

$$\frac{12}{x + 3} = \frac{8}{x - 3}.$$

Solve this equation by multiplying both sides by the LCD, $(x + 3)(x - 3)$.

$$(x + 3)(x - 3)\frac{12}{x + 3} = (x + 3)(x - 3)\frac{8}{x - 3}$$

$$12(x - 3) = 8(x + 3)$$

$$12x - 36 = 8x + 24 \qquad \text{Distributive property}$$

$$4x = 60 \qquad\qquad \text{Subtract } 8x; \text{ add } 36.$$

$$x = 15 \qquad\qquad \text{Divide by 4.}$$

The speed of the boat in still water is 15 miles per hour. Check this solution by first finding the speed of the boat downstream, which is $15 + 3 = 18$ miles per hour. Traveling 12 miles would take

$$d = rt$$
$$12 = 18t$$
$$t = \frac{2}{3} \text{ hour.}$$

On the other hand, the speed of the boat upstream is $15 - 3 = 12$ miles per hour, and traveling 8 miles would take

$$d = rt$$
$$8 = 12t$$
$$t = \frac{2}{3} \text{ hour.}$$

The time upstream equals the time downstream, as required.

CONNECTIONS

Sometimes what seems to be the obvious answer is not correct. Consider the following problem.

A car travels from A to B at 40 miles per hour and returns at 60 miles per hour. What is its average rate for the entire trip?

The correct answer is not 50 miles per hour! Remembering the distance-rate-time relationship and letting x = the distance between A and B, we can simplify a complex fraction to find the correct answer.

$$\text{Average rate for entire trip} = \frac{\text{Total distance}}{\text{Total time}}$$

$$= \frac{x + x}{\dfrac{x}{40} + \dfrac{x}{60}}$$

$$= \frac{2x}{\dfrac{3x}{120} + \dfrac{2x}{120}}$$

$$= \frac{2x}{\dfrac{5x}{120}}$$

$$= 2x \cdot \frac{120}{5x}$$

$$= 48$$

The average rate for the entire trip is 48 miles per hour.

FOR DISCUSSION OR WRITING
Find the average rate for the entire trip if the speed from A to B is 50 miles per hour. Repeat for 55 miles per hour. Will the result always be less than $\frac{1}{2}$ the sum of the two speeds? What might account for this?

OBJECTIVE **3** Solve problems about work. Suppose that you can mow your lawn in 4 hours. After 1 hour, you will have mowed $\frac{1}{4}$ of the lawn. After 2 hours, you will have mowed $\frac{2}{4}$ or $\frac{1}{2}$ of the lawn, and so on. This idea is generalized as follows.

Rate of Work

If a job can be completed in t units of time, then the rate of work is

$$\frac{1}{t} \text{ job per unit of time.}$$

PROBLEM SOLVING

The relationship between problems involving work and problems involving distance is a very close one. Recall that the formula $d = rt$ says that distance traveled is equal to rate of travel multiplied by time traveled. Similarly, the fractional part of a job accomplished is equal to the rate of the work multiplied by the time worked. In the lawn-mowing example, after 3 hours, the fractional part of the job done is

$$\underbrace{\frac{1}{4}}_{\substack{\text{Rate of}\\\text{work}}} \cdot \underbrace{3}_{\substack{\text{Time}\\\text{worked}}} = \underbrace{\frac{3}{4}}_{\substack{\text{Fractional part}\\\text{of job done}}}.$$

After 4 hours, $\left(\frac{1}{4}\right)(4) = 1$ whole job has been done.

These ideas are used in solving problems about the length of time needed to do a job.

EXAMPLE 4 Solving a Problem about Work

With a riding lawn mower, Mateo, the groundskeeper in a large park, can cut the lawn in 8 hours. With a small mower, his assistant Chet needs 14 hours to cut the same lawn. If both Mateo and Chet work on the lawn, how long will it take to cut it?

Let x = the number of hours it will take for Mateo and Chet to mow the lawn, working together.

Certainly, x will be less than 8, since Mateo alone can mow the lawn in 8 hours. Begin by making a chart. Based on the previous discussion, Mateo's rate alone is $\frac{1}{8}$ job per hour, and Chet's rate is $\frac{1}{14}$ job per hour.

	Rate	Time Working Together	Fractional Part of the Job Done When Working Together
Mateo	$\frac{1}{8}$	x	$\frac{1}{8}x$
Chet	$\frac{1}{14}$	x	$\frac{1}{14}x$

Sum is 1 whole job.

Since together Mateo and Chet complete 1 whole job, we must add their individual fractional parts and set the sum equal to 1.

$$\underset{\substack{\text{done by Mateo}}}{\text{Fractional part}} + \underset{\substack{\text{done by Chet}}}{\text{Fractional part}} = 1 \text{ whole job}$$

$$\frac{1}{8}x \quad + \quad \frac{1}{14}x \quad = \quad 1$$

$$56\left(\frac{1}{8}x + \frac{1}{14}x\right) = 56(1) \qquad \text{Multiply by LCD, 56.}$$

$$56\left(\frac{1}{8}x\right) + 56\left(\frac{1}{14}x\right) = 56(1) \qquad \text{Distributive property}$$

$$7x + 4x = 56$$

$$11x = 56 \qquad \text{Combine like terms.}$$

$$x = \frac{56}{11} \qquad \text{Divide by 11.}$$

Working together, Mateo and Chet can mow the lawn in $\frac{56}{11}$ hours, or $5\frac{1}{11}$ hours. (Is this answer reasonable?)

An alternative approach in work problems is to consider the part of the job that can be done in 1 hour. For instance, in Example 4 Mateo can do the entire job in 8 hours, and Chet can do it in 14. Thus, their work rates, as we saw in Example 4, are $\frac{1}{8}$ and $\frac{1}{14}$, respectively. Since it takes them x hours to complete the job when working together, in one hour they can mow $\frac{1}{x}$ of the lawn. The amount mowed by Mateo in one hour plus the amount mowed by Chet in one hour must equal the amount they can do together. This leads to the equation

$$\text{Amount by Mateo} \rightarrow \quad \underset{\mathbf{8}}{\mathbf{1}} + \overset{\overset{\text{Amount by Chet}}{\downarrow}}{\underset{\mathbf{14}}{\mathbf{1}}} = \underset{x}{\mathbf{1}}. \quad \leftarrow \text{Amount together}$$

Compare this with the equation in Example 4. Multiplying both sides by $56x$ leads to

$$7x + 4x = 56,$$

the same equation found in the third step in the example. The same solution results.

OBJECTIVE **4** **Solve problems about variation.** In Section 2.5 we discussed direct variation. (See Example 6 in that section.) Recall that y *varies directly* as x if there exists a constant (a numerical value) k such that $y = kx$. This value is called the **constant of variation.** For example, the circumference C of a circle varies directly as its diameter d, because $C = \pi d$. Here, the constant of variation is π.

Some types of variation involve rational expressions. Suppose that a rectangle has area 48 square units and length 24 units. Then its width is 2 units. However, if the area stays the same (48 square units) and the length decreases to 12 units, the width increases to 4 units. See Figure 2 on the next page.

As length decreases, width increases. This is an example of **inverse variation,** since $W = \dfrac{A}{L}$.

Width
increases

12   4   2

24

Length decreases
$24 \cdot 2 = 12 \cdot 4 = 48$ (same area)

Figure 2

This and other types of variation are defined below, where k represents a constant.

Types of Variation

m varies inversely as p $m = \dfrac{k}{p}$

y varies directly as the square of x $y = kx^2$

r varies inversely as the square of s $r = \dfrac{k}{s^2}$

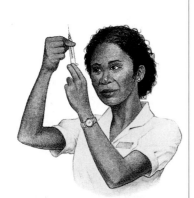

EXAMPLE 5 Using Inverse Variation

In the manufacturing of a certain medical syringe, the cost of producing the syringe varies inversely as the number produced. If 10,000 syringes are produced, the cost is $2 per unit. Find the cost per unit to produce 25,000 syringes.

Let $x =$ the number of syringes produced and

$c =$ the cost per unit.

Since c varies inversely as x, there is a constant k such that

$$c = \frac{k}{x}.$$

Find k by replacing c with 2 and x with 10,000.

$$2 = \frac{k}{10,000}$$

$$20,000 = k \qquad \text{Multiply by 10,000.}$$

Since $c = \dfrac{k}{x}$,

$$c = \frac{20,000}{25,000} = .80. \qquad \text{Let } k = 20,000 \text{ and } x = 25,000.$$

The cost per unit to make 25,000 syringes is $.80.

EXAMPLE 6 Using Inverse Variation

In electric current flow, it is found that the resistance offered by a fixed length of wire of a given material (measured in units called *ohms*) varies inversely as the square of the diameter of the wire. If a wire .01 centimeter in diameter has a resistance of .40 ohm, what is the resistance of a wire of the same length and material but .03 centimeter in diameter?

If R is the resistance and d the diameter, then

$$R = \frac{k}{d^2}$$

for some constant k. (The value of the constant k depends on the material and the temperature.) Since a wire .01 centimeter in diameter has a resistance of .40 ohm,

$$.40 = \frac{k}{(.01)^2} \qquad \text{Let } R = .40 \text{ and } d = .01.$$

$$k = .00004. \qquad \text{Multiply by } (.01)^2 = .0001.$$

Thus, for this particular length of wire,

$$R = \frac{.00004}{d^2}. \qquad \text{Let } k = .00004.$$

For the same length wire with $d = .03$,

$$R = \frac{.00004}{(.03)^2} = \frac{2}{45} \text{ ohm.}$$

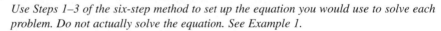

7.7 EXERCISES

Use Steps 1–3 of the six-step method to set up the equation you would use to solve each problem. Do not actually solve the equation. See Example 1.

1. The numerator of the fraction $\frac{5}{6}$ is increased by an amount so that the value of the resulting fraction is $\frac{13}{3}$. By what amount was the numerator increased?

   (a) Let $x = $ _____. (Step 1)

   (b) Write an expression for "the numerator of the fraction $\frac{5}{6}$ is increased by an amount." (Step 2)

   (c) Set up an equation to solve the problem. (Step 3)

2. If the same number is added to the numerator and subtracted from the denominator of $\frac{23}{12}$, the resulting fraction is equal to $\frac{3}{2}$. What is the number?

   (a) Let $x = $ _____. (Step 1)

   (b) Write an expression for "a number is added to the numerator of $\frac{23}{12}$." Then write an expression for "the same number is subtracted from the denominator of $\frac{23}{12}$." (Step 2)

   (c) Set up an equation to solve the problem. (Step 3)

Use the six-step method to solve each problem. See Example 1.

3. In a certain fraction, the denominator is 6 more than the numerator. If 3 is added to both the numerator and the denominator, the resulting fraction is equivalent to $\frac{5}{7}$. What was the original fraction?

4. The numerator of a certain fraction is 4 times the denominator. If 6 is added to both the numerator and the denominator, the resulting fraction is equivalent to 2. What was the original fraction?

5. A quantity, its $\frac{2}{3}$, its $\frac{1}{2}$, and its $\frac{1}{7}$, added together, become 33. What is the quantity? (From the *Rhind Mathematical Papyrus*.)

6. A quantity, its $\frac{3}{4}$, its $\frac{1}{2}$, and its $\frac{1}{3}$, added together, become 93. What is the quantity?

Solve each problem using the six-step method. See Example 2.

7. (All enrollments are in thousands.) The Medicare enrollment in 1990 was 624 more than in 1989. The ratio of these enrollments was $\frac{877}{861}$. What was the enrollment in each of these years? (*Source:* U.S. Health Care Financing Administration.)

8. In 1985, during the early years of the appearance of AIDS, there were about 3760 more cases reported than in 1984. The ratio of these numbers of cases was $\frac{205}{111}$. How many cases were reported in each of these years? (*Source:* Centers for Disease Control.)

9. In 1995, the number of female physicians in the United States was approximately $\frac{1}{4}$ the number of male physicians. The total number of physicians was approximately 720,000. How many female and how many male physicians were there? (*Source:* American Medical Association.)

10. Two of the largest national managed care firms (as ranked by total HMO enrollments) in 1996 were United Health Care Corporation and Kaiser Foundation Health Plans, Inc. The number of plans provided by United was $\frac{10}{3}$ that of Kaiser. Together, the number of plans was 52. How many plans were provided by each company? (*Source:* InterStudy Publications.)

Solve each problem. See Example 6 in Section 2.6. (*Source:* Sports Illustrated 1998 Sports Almanac.)

11. In the 1994 Olympics, Vladimir Smirnov of Kazakhstan won the 50-kilometer nordic skiing event in 2.07 hours. What was his average rate?

12. Bonnie Blair won the women's 500-meter speed skating event in the 1994 Olympics. Her average rate was 12.74 meters per second. What was her time?

13. The winner of the 1997 Indianapolis 500 (mile) race was Arie Luyendyk, with an average rate of 145.827 miles per hour. What was his time in hours?

14. In 1997, Jeff Gordon drove his Chevrolet to victory in the Daytona 500 (mile) stock car race. His average rate was 148.295 miles per hour. What was his time in hours?

15. The NCAA Division I women's champion for the 400-meter dash in 1997 was LaTarsha Stroman of Louisiana State University. Her winning time was 50.60 seconds. What was her rate in meters per second?

16. The NCAA Division I women's champion for the 200-yard freestyle event in 1996–1997 was Martina Moracova of Southern Methodist University. Her time was 48.18 seconds. What was her rate in yards per second?

Set up the equation you would use to solve each problem. Do not actually solve. See Example 3.

17. Julio flew his airplane 500 miles against the wind in the same time it took him to fly it 600 miles with the wind. If the speed of the wind was 10 miles per hour, what was the average speed of his plane? (Let x = speed of the plane in still air.)

	d	r	t
Against the wind	500	$x - 10$	
With the wind	600	$x + 10$	

18. Luvenia can row 4 miles per hour in still water. It takes as long to row 8 miles upstream as 24 miles downstream. How fast is the current? (Let x = speed of the current.)

	d	r	t
Upstream	8	$4 - x$	
Downstream	24	$4 + x$	

Solve each problem. See Example 3.

19. Suppose Stephanie walks D miles at R miles per hour in the same time that Wally walks d miles at r miles per hour. Give an equation relating D, R, d, and r.

20. If a migrating hawk travels m miles per hour in still air, what is its rate when it flies into a steady headwind of 5 miles per hour? What is its rate with a tailwind of 5 miles per hour?

21. A boat can go 20 miles against a current in the same time that it can go 60 miles with the current. The current is 4 miles per hour. Find the speed of the boat in still water.

22. A plane flies 350 miles with the wind in the same time that it can fly 310 miles against the wind. The plane has a still-air speed of 165 miles per hour. Find the speed of the wind.

23. The distance from Seattle, Washington, to Victoria, British Columbia, is about 148 miles by ferry. It takes about 4 hours less to travel by the same ferry from Victoria to Vancouver, British Columbia, a distance of about 74 miles. What is the average speed of the ferry?

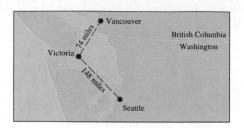

24. Sandi Goldstein flew from Dallas to Indianapolis at 180 miles per hour and then flew back at 150 miles per hour. The trip at the slower speed took 1 hour longer than the trip at the higher speed. Find the distance between the two cities.

 25. Cosmas N'Deti of Kenya won the 1995 Boston Marathon (a 26-mile race) in (about) .8 of an hour less time than the winner of the first Boston Marathon in 1897, John H. McDermott of New York. Find each runner's speed, if McDermott's speed was (about) .73 times N'Deti's speed. (*Hint:* Don't round until the end of the calculations.)

 26. Women first ran in the Boston Marathon in 1972, when Nina Kuscsik of New York won the race. In 1995, the winner was Uta Pippig of Germany, whose time was .8 of an hour less than Kuscsik's in 1972. If Pippig ran $\frac{4}{3}$ as fast as Kuscsik, find each runner's speed. (See Exercise 25 for distance.)

27. If it takes Elayn 10 hours to do a job, what is her rate?

28. If it takes Clay 12 hours to do a job, how much of the job does he do in 8 hours?

In Exercises 29 and 30, set up the equation you would use to solve each problem. Do not actually solve. See Example 4.

29. Working alone, Jorge can paint a room in 8 hours. Caterina can paint the same room working alone in 6 hours. How long will it take them if they work together? (Let x represent the time working together.)

30. Edwin Bedford can tune up his Chevy in 2 hours working alone. Beau can do the job in 3 hours working alone. How long would it take them if they worked together? (Let t represent the time working together.)

Solve each problem. See Example 4.

31. Geraldo and Luisa Hernandez operate a small laundry. Luisa, working alone, can clean a day's laundry in 9 hours. Geraldo can clean a day's laundry in 8 hours. How long would it take them if they work together?

32. Lea can groom the horses in her boarding stable in 5 hours, while Tran needs 4 hours. How long will it take them to groom the horses if they work together?

33. A pump can pump the water out of a flooded basement in 10 hours. A smaller pump takes 12 hours. How long would it take to pump the water from the basement using both pumps?

34. Doug Todd's copier can do a printing job in 7 hours. Scott's copier can do the same job in 12 hours. How long would it take to do the job using both copiers?

35. An experienced employee can enter tax data into a computer twice as fast as a new employee. Working together, it takes the employees 2 hours. How long would it take the experienced employee working alone?

36. One roofer can put a new roof on a house three times faster than another. Working together they can roof a house in 4 days. How long would it take the faster roofer working alone?

37. One pipe can fill a swimming pool in 6 hours, and another pipe can do it in 9 hours. How long will it take the two pipes working together to fill the pool $\frac{3}{4}$ full?

38. An inlet pipe can fill a swimming pool in 9 hours, and an outlet pipe can empty the pool in 12 hours. Through an error, both pipes are left open. How long will it take to fill the pool?

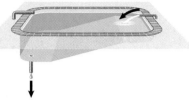

39. A cold water faucet can fill a sink in 12 minutes, and a hot water faucet can fill it in 15. The drain can empty the sink in 25 minutes. If both faucets are on and the drain is open, how long will it take to fill the sink?

40. Refer to Exercise 38. Assume the error was discovered after both pipes had been running for 3 hours, and the outlet pipe was then closed. How much more time would then be required to fill the pool? (*Hint:* How much of the job had been done when the error was discovered?)

Use personal experience or intuition to determine whether the variation between the indicated quantities is direct or inverse.*

41. The number of different lottery tickets you buy and your probability of winning that lottery

42. The rate and the distance traveled by a pickup truck in 3 hours

43. The amount of pressure put on the accelerator of a car and the speed of the car

44. The number of days from now until December 25 and the magnitude of the frenzy of Christmas shopping

45. The surface area of a balloon and its diameter

46. Your age and the probability that you believe in Santa Claus

47. The number of days until the end of the baseball season and the number of home runs that Sammy Sosa has

48. The amount of gasoline you pump and the amount you will pay

Solve each problem involving variation. See Example 6 in Section 2.6 and Examples 5 and 6 in this section.

49. If x varies directly as y, and $x = 27$ when $y = 6$, find x when $y = 2$.

50. If z varies directly as x, and $z = 30$ when $x = 8$, find z when $x = 4$.

51. If m varies inversely as p^2, and $m = 20$ when $p = 2$, find m when p is 5.

52. If a varies inversely as b^2, and $a = 48$ when $b = 4$, find a when $b = 7$.

53. If p varies inversely as q^2, and $p = 4$ when $q = \frac{1}{2}$, find p when $q = \frac{3}{2}$.

54. If z varies inversely as x^2, and $z = 9$ when $x = \frac{2}{3}$, find z when $x = \frac{5}{4}$.

55. If the constant of variation is positive and y varies directly as x, then as x increases,
   y _____.
   (increases/decreases)

56. If the constant of variation is positive and y varies inversely as x, then as x increases,
   y _____.
   (increases/decreases)

---

*The authors thank Linda Kodama of Kapiolani Community College for suggesting the inclusion of exercises of this type.

57. Over a specified distance, speed varies inversely with time. If a Dodge Viper on a test track goes a certain distance in one-half minute at 160 miles per hour, what speed is needed to go the same distance in three-fourths of a minute?

58. The pressure exerted by water at a given point varies directly with the depth of the point beneath the surface of the water. Water exerts 4.34 pounds per square inch for every 10 feet traveled below the water's surface. What is the pressure exerted on a scuba diver at 20 feet?

59. In the inversion of raw sugar, the rate of change of the amount of raw sugar varies directly as the amount of raw sugar remaining. The rate is 200 kilograms per hour when there are 800 kilograms left. What is the rate of change per hour when only 100 kilograms are left?

60. The current in a simple electrical circuit varies inversely as the resistance. If the current is 20 amps when the resistance is 5 ohms, find the current when the resistance is 8 ohms.

61. If the temperature is constant, the pressure of a gas in a container varies inversely as the volume of the container. If the pressure is 10 pounds per square foot in a container with 3 cubic feet, what is the pressure in a container with 1.5 cubic feet?

62. If the volume is constant, the pressure of gas in a container varies directly as the temperature. If the pressure is 5 pounds per square inch at a temperature of 200 Kelvin, what is the pressure at a temperature of 300 Kelvin?

63. For a constant area, the length of a rectangle varies inversely as the width. The length of a rectangle is 27 feet when the width is 10 feet. Find the width of a rectangle with the same area if the length is 18 feet.

64. Hooke's law for an elastic spring states that the distance a spring stretches varies directly with the force applied. If a force of 75 pounds stretches a certain spring 16 inches, how much will a force of 200 pounds stretch the spring?

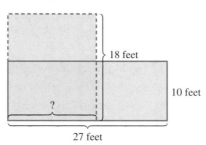

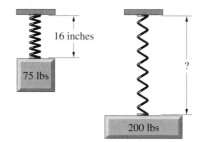

65. For a body falling freely from rest (disregarding air resistance), the distance the body falls varies directly as the square of the time. If an object is dropped from the top of a tower 400 feet high and hits the ground in 5 seconds, how far did it fall in the first 3 seconds?

66. The illumination produced by a light source varies inversely as the square of the distance from the source. If the illumination produced 4 feet from a light source is 75 foot-candles, find the illumination produced 9 feet from the same source.

TECHNOLOGY INSIGHTS (EXERCISES 67–70)

In each table, Y_1 either varies directly or varies inversely as x. Tell which type of variation is illustrated.

67.

X	Y1
0	0
1	2.5
2	5
3	7.5
4	10
5	12.5
6	15

X=0

68.

X	Y1
0	0
1	3.2
2	6.4
3	9.6
4	12.8
5	16
6	19.2

X=0

TECHNOLOGY INSIGHTS (EXERCISES 67-70) (CONTINUED)

69.

X	Y₁
1	24
2	12
3	8
4	6
5	4.8
6	4
7	3.4286

X=1

70.

X	Y₁
1	48
2	24
3	16
4	12
5	9.6
6	8
7	6.8571

X=1

Refer to the discussion in the chapter opener pertaining to heart transplants. Use the strategy discussed there to solve each of these problems. (Sources: U.S. Department of Health and Human Services, Public Health Service, Division of Organ Transplantation, and United Network for Organ Sharing.)

71. In 1995, the number of heart transplants continued to show the increase it had exhibited during the decade preceding it. In the years 1985 and 1995, the total number of heart transplants was 3080, and the ratio of the number in 1995 to the number in 1985 was approximately 33 to 10. How many heart transplants were performed in each of these two years?

72. In the years 1985 and 1995, the total number of liver transplants was 4526. The ratio of the number in 1995 to the number in 1985 was approximately 65 to 10. How many liver transplants were performed in each of these two years?

73. Students often wonder how teachers and textbook authors make up problems for them to solve. One way to do this is to start with the answer and work backwards. For example, suppose that we start with the fraction $\frac{7}{3}$ and add 3 to both the numerator and the denominator. We get $\frac{10}{6}$, which simplifies to $\frac{5}{3}$. Based on this observation, we can write the following problem.

   If a number is added to both the numerator and the denominator of $\frac{7}{3}$, the resulting fraction is equal to $\frac{5}{3}$. What is the number?

   Because of how we constructed the problem, the answer must be 3.
   Make up your own problem similar to this one, and then solve it using an equation involving rational expressions.

74. Refer to the chart in Exercise 18. Suppose that a student made the error of interchanging the positions of the expressions $4 - x$ and $4 + x$, but used the correct method of setting up the equation, applying the formula $t = \frac{d}{r}$. Solve the equation the student used, and explain how the student should immediately know that there is something wrong in the setup.

CHAPTER 7 GROUP ACTIVITY

▦ How Much Fat Do You Need?

Objective: Use ratios and proportions to calculate total fat grams needed in an individual's daily diet.

Increased risks for some leading causes of death such as heart disease, diabetes, high blood pressure, and some cancers are associated with diets high in fat. Studies show that diets low in fat are linked to the lowest rates of heart disease and some forms of cancer. How do you decide how much fat is necessary for each individual? This activity will show how to calculate total fat grams necessary to maintain or reach ideal body weight.

To calculate daily caloric need, follow these steps.

Step 1 Determine activity level factor.

A sedentary person's activity level factor = 12.

An active person's activity level factor = 15.

A very active person's activity level factor = 18.

Step 2 Determine daily caloric need.

Multiply the activity level from Step 1 by your weight in pounds. (You can use your actual weight or your ideal weight.)

A. As a group, use the above information and the following ratio to write a proportion that can be used to calculate maximum total fat grams needed daily to maintain a person's weight or to reach his/her ideal weight.

$$\frac{400 \text{ calories}}{11 \text{ total fat grams}}$$

B. Use the previous information to complete the chart below. Round answers to the nearest tenth.

Total Fat Grams Needed Daily

Person	Weight	Sedentary	Active	Very Active
A	134			
B	200			
C	153			

C. Now complete the chart for the members in your group. Complete the three activity levels. (Actual weight or ideal weight can be used.)

Total Fat Grams Needed Daily

Person	Weight	Sedentary	Active	Very Active

CHAPTER 7 SUMMARY

KEY TERMS

| 7.1 rational expression lowest terms | 7.3 least common denominator (LCD) | 7.5 complex fraction 7.7 constant of variation | inverse variation |

TEST YOUR WORD POWER

See how well you have learned the vocabulary in this chapter. Answers, with examples, are given at the bottom of the page.

1. A **rational expression** is
(a) an algebraic expression made up of a term or the sum of a finite number of terms with real coefficients and whole number exponents
(b) a polynomial equation of degree two
(c) an expression with one or more fractions in the numerator, denominator, or both

(d) the quotient of two polynomials with denominator not zero.

2. A **complex fraction** is
(a) an algebraic expression made up of a term or the sum of a finite number of terms with real coefficients and whole number exponents

(b) a polynomial equation of degree two
(c) a rational expression with one or more fractions in the numerator, denominator, or both
(d) the quotient of two polynomials with denominator not zero.

QUICK REVIEW

CONCEPTS	EXAMPLES

7.1 THE FUNDAMENTAL PROPERTY OF RATIONAL EXPRESSIONS

To find the values for which a rational expression is not defined, set the denominator equal to zero and solve the equation.

Find the values for which the expression

$$\frac{x - 4}{x^2 - 16}$$

is not defined.

$$x^2 - 16 = 0$$
$$(x - 4)(x + 4) = 0$$
$$x - 4 = 0 \quad \text{or} \quad x + 4 = 0$$
$$x = 4 \quad \text{or} \quad x = -4$$

The rational expression is not defined for 4 or -4.

To write a rational expression in lowest terms, (1) factor, and (2) use the fundamental property to remove common factors from the numerator and denominator.

Write $\dfrac{x^2 - 1}{(x - 1)^2}$ in lowest terms.

$$\frac{x^2 - 1}{(x - 1)^2} = \frac{(x - 1)(x + 1)}{(x - 1)(x - 1)}$$

$$= \frac{x + 1}{x - 1}$$

(continued)

CONCEPTS	EXAMPLES
There are often several different equivalent forms of a rational expression.	Give two equivalent forms of $-\dfrac{x-1}{x+2}$. Distribute the $-$ sign in the numerator to get $\dfrac{-x+1}{x+2}$; do so in the denominator to get $\dfrac{x-1}{-x-2}$. (There are other forms as well.)

7.2 MULTIPLICATION AND DIVISION OF RATIONAL EXPRESSIONS

Multiplication 1. Factor. 2. Multiply numerators and multiply denominators. 3. Write in lowest terms.	Multiply. $\dfrac{3x+9}{x-5}\cdot\dfrac{x^2-3x-10}{x^2-9}$ $=\dfrac{3(x+3)}{x-5}\cdot\dfrac{(x-5)(x+2)}{(x+3)(x-3)}$ $=\dfrac{3(x+3)\mathbf{(x-5)}(x+2)}{\mathbf{(x-5)}(x+3)(x-3)}$ $=\dfrac{3(x+2)}{x-3}$
Division 1. Factor. 2. Multiply the first rational expression by the reciprocal of the second. 3. Write in lowest terms.	Divide. $\dfrac{2x+1}{x+5}\div\dfrac{6x^2-x-2}{x^2-25}$ $=\dfrac{2x+1}{x+5}\div\dfrac{(2x+1)(3x-2)}{(x+5)(x-5)}$ $=\dfrac{2x+1}{\mathbf{x+5}}\cdot\dfrac{\mathbf{(x+5)}(x-5)}{(2x+1)(3x-2)}$ $=\dfrac{x-5}{3x-2}$

7.3 THE LEAST COMMON DENOMINATOR

Finding the LCD 1. Factor each denominator into prime factors. 2. List each different factor the greatest number of times it appears. 3. Multiply the factors from Step 2 to get the LCD.	Find the LCD for $\dfrac{3}{k^2-8k+16}$ and $\dfrac{1}{4k^2-16k}$. $k^2-8k+16=(k-4)^2$ $4k^2-16k=4k(k-4)$ $\mathrm{LCD}=(k-4)^2\cdot 4\cdot k$ $=4k(k-4)^2$
Writing a Rational Expression with the LCD as Denominator 1. Factor both denominators. 2. Decide what factors the denominator must be multiplied by to equal the LCD. 3. Multiply the rational expression by that factor divided by itself (multiply by 1).	Find the numerator: $\dfrac{5}{2z^2-6z}=\dfrac{}{4z^3-12z^2}$. $\dfrac{5}{2z(z-3)}=\dfrac{1}{4z^2(z-3)}$ $2z(z-3)$ must be multiplied by $2z$. $\dfrac{5}{2z(z-3)}\cdot\dfrac{2z}{2z}=\dfrac{10z}{4z^2(z-3)}=\dfrac{10z}{4z^3-12z^2}$

7.4 ADDITION AND SUBTRACTION OF RATIONAL EXPRESSIONS

Adding Rational Expressions

1. Find the LCD.

Add. $\dfrac{2}{3m + 6} + \dfrac{m}{m^2 - 4}$

$$3m + 6 = 3(m + 2)$$
$$m^2 - 4 = (m + 2)(m - 2)$$

The LCD is $3(m + 2)(m - 2)$.

2. Rewrite each rational expression with the LCD as denominator.

3. Add the numerators to get the numerator of the sum. The common denominator is the denominator of the sum.

$$= \frac{2(m - 2)}{3(m + 2)(m - 2)} + \frac{3m}{3(m + 2)(m - 2)}$$

$$= \frac{2m - 4 + 3m}{3(m + 2)(m - 2)}$$

4. Write in lowest terms.

$$= \frac{5m - 4}{3(m + 2)(m - 2)}$$

Subtracting Rational Expressions

Follow the same steps as for addition, but subtract in Step 3.

Subtract. $\dfrac{6}{k + 4} - \dfrac{2}{k}$

The LCD is $k(k + 4)$.

$$\frac{6k}{(k + 4)k} - \frac{2(k + 4)}{k(k + 4)} = \frac{6k - 2(k + 4)}{k(k + 4)}$$

$$= \frac{6k - 2k - 8}{k(k + 4)} = \frac{4k - 8}{k(k + 4)}$$

7.5 COMPLEX FRACTIONS

Simplifying Complex Fractions

Method 1 Simplify the numerator and denominator separately. Then divide the simplified numerator by the simplified denominator.

Simplify.

$$(1) \quad \frac{\dfrac{1}{a} - a}{1 - a} = \frac{\dfrac{1}{a} - \dfrac{a^2}{a}}{1 - a} = \frac{\dfrac{1 - u^2}{a}}{1 - a}$$

$$= \frac{1 - a^2}{a} \cdot \frac{1}{1 - a}$$

$$= \frac{(1 - a)(1 + a)}{a(1 - a)} = \frac{1 + a}{a}$$

Method 2 Multiply numerator and denominator of the complex fraction by the LCD of all the denominators in the complex fraction. Write in lowest terms.

$$(2) \quad \frac{\dfrac{1}{a} - a}{1 - a} = \frac{\dfrac{1}{a} - a}{1 - a} \cdot \frac{a}{a} = \frac{\dfrac{a}{a} - a^2}{(1 - a)a}$$

$$= \frac{1 - a^2}{(1 - a)a} = \frac{(1 + a)(1 - a)}{(1 - a)a}$$

$$= \frac{1 + a}{a}$$

7.6 SOLVING EQUATIONS INVOLVING RATIONAL EXPRESSIONS

Solving Equations with Rational Expressions

1. Find the LCD of all denominators in the equation.

2. Multiply each side of the equation by the LCD.

Solve $\dfrac{2}{x - 1} + \dfrac{3}{4} = \dfrac{5}{x - 1}$.

The LCD is $4(x - 1)$. Note that 1 cannot be a solution.

$$4(x - 1)\left(\frac{2}{x - 1} + \frac{3}{4}\right) = 4(x - 1)\left(\frac{5}{x - 1}\right)$$

$$4(x - 1)\left(\frac{2}{x - 1}\right) + 4(x - 1)\left(\frac{3}{4}\right) = 4(x - 1)\left(\frac{5}{x - 1}\right)$$

(continued)

CONCEPTS	EXAMPLES

3. Solve the resulting equation, which should have no fractions.

$$8 + 3(x - 1) = 20$$
$$8 + 3x - 3 = 20$$
$$3x = 15$$
$$x = 5$$

4. Check each proposed solution.

The proposed solution, 5, checks. The solution set is $\{5\}$.

7.7 APPLICATIONS OF RATIONAL EXPRESSIONS

Solving Problems about Distance

Use the six-step method.

Step 1 State what the variable represents.

On a trip from Sacramento to Monterey, Marge traveled at an average speed of 60 miles per hour. The return trip, at an average speed of 64 miles per hour, took $\frac{1}{4}$ hour less. How far did she travel between the two cities?
Let $x =$ the unknown distance.

Step 2 Use a chart to identify distance, rate, and time. Solve $d = rt$ for the unknown quantity in the chart.

	d	r	$t = \dfrac{d}{r}$
Going	x	60	$\dfrac{x}{60}$
Returning	x	64	$\dfrac{x}{64}$

Step 3 From the wording in the problem, decide the relationship between the quantities. Use those expressions to write an equation.

Since the time for the return trip was $\frac{1}{4}$ hour less, the time going equals the time returning plus $\frac{1}{4}$.

$$\frac{x}{60} = \frac{x}{64} + \frac{1}{4}$$

Step 4 Solve the equation.

$$16x = 15x + 240 \qquad \text{Multiply by 960.}$$
$$x = 240 \qquad \text{Subtract } 15x.$$

Steps 5 and 6 Answer the question; check the solution.

She traveled 240 miles.

The trip there took $\dfrac{240}{60} = 4$ hours while the return trip took $\dfrac{240}{64} = 3.75$ hours, which is $\frac{1}{4}$ hour less time. The solution checks.

Solving Problems about Work

Use the six-step method.

Step 1 State what the variable represents.

It takes the regular mail carrier 6 hours to cover her route. A substitute takes 8 hours to cover the same route. How long would it take them to cover the route together?
Let $x =$ the number of hours it would take them working together.

Step 2 Put the information from the problem in a chart. If a job is done in t units of time, the rate is $\frac{1}{t}$.

The rate of the regular carrier is $\frac{1}{6}$ job per hour; the rate of the substitute is $\frac{1}{8}$ job per hour. Multiply rate times time to get the fractional part of the job done.

	Rate	Time	Part of the Job Done
Regular	$\dfrac{1}{6}$	x	$\dfrac{1}{6}x$
Substitute	$\dfrac{1}{8}$	x	$\dfrac{1}{8}x$

CONCEPTS	EXAMPLES
Step 3 Write the equation. The sum of the fractional parts should equal 1 (whole job).	The equation is $$\frac{1}{6}x + \frac{1}{8}x = 1.$$
Step 4 Solve the equation.	The solution of the equation is $\frac{24}{7}$.
Steps 5 and 6 Answer the question; check the solution.	It would take them $\frac{24}{7}$ or $3\frac{3}{7}$ hours to cover the route together.

Solving Inverse Variation Problems

1. Write the variation equation $y = \frac{k}{x}$.

2. Find k by substituting the given values of x and y into the equation.

3. Write the equation with the value of k from Step 2 and the given value of x or y. Solve for the remaining variable.

If a varies inversely as b, and $a = 4$ when $b = 4$, find a when $b = 6$. The equation for inverse variation is $a = \frac{k}{b}$.

Substitute $a = 4$ and $b = 4$.

$$4 = \frac{k}{4}$$

$$k = 16$$

Let $k = 16$ and $b = 6$ in the variation equation.

$$a = \frac{16}{6} = \frac{8}{3}$$

CHAPTER 7 REVIEW EXERCISES

[7.1] *Find any values of the variable for which the rational expression is undefined.*

1. $\dfrac{4}{x - 3}$
    2. $\dfrac{y + 3}{2y}$
    3. $\dfrac{2k + 1}{3k^2 + 17k + 10}$

4. How would you determine the values of the variable for which a rational expression is undefined?

*Find the numerical value of each rational expression when (**a**) $x = -2$ and (**b**) $x = 4$.*

5. $\dfrac{4x - 3}{5x + 2}$
    6. $\dfrac{3x}{x^2 - 4}$

Write each rational expression in lowest terms.

7. $\dfrac{5a^3 b^3}{15a^4 b^2}$
    8. $\dfrac{m - 4}{4 - m}$

9. $\dfrac{4x^2 - 9}{6 - 4x}$
    10. $\dfrac{4p^2 + 8pq - 5q^2}{10p^2 - 3pq - q^2}$

Write four equivalent expressions for each of the following.

11. $-\dfrac{4x - 9}{2x + 3}$
    12. $\dfrac{8 - 3x}{3 + 6x}$

[7.2] *Find each product or quotient and write the answer in lowest terms.*

13. $\dfrac{18p^3}{6} \cdot \dfrac{24}{p^4}$
    14. $\dfrac{8x^2}{12x^5} \cdot \dfrac{6x^4}{2x}$

15. $\dfrac{x - 3}{4} \cdot \dfrac{5}{2x - 6}$
    16. $\dfrac{2r + 3}{r - 4} \cdot \dfrac{r^2 - 16}{6r + 9}$

17. $\dfrac{6a^2 + 7a - 3}{2a^2 - a - 6} \div \dfrac{a + 5}{a - 2}$
    18. $\dfrac{y^2 - 6y + 8}{y^2 + 3y - 18} \div \dfrac{y - 4}{y + 6}$

19. $\dfrac{2p^2 + 13p + 20}{p^2 + p - 12} \cdot \dfrac{p^2 + 2p - 15}{2p^2 + 7p + 5}$

20. $\dfrac{3z^2 + 5z - 2}{9z^2 - 1} \cdot \dfrac{9z^2 + 6z + 1}{z^2 + 5z + 6}$

[7.3] *Find the least common denominator for each list of fractions.*

21. $\dfrac{4}{9y}, \dfrac{7}{12y^2}, \dfrac{5}{27y^4}$

22. $\dfrac{3}{x^2 + 4x + 3}, \dfrac{5}{x^2 + 5x + 4}$

Rewrite each rational expression with the given denominator.

23. $\dfrac{3}{2a^3} = \dfrac{}{10a^4}$

24. $\dfrac{9}{x - 3} = \dfrac{}{18 - 6x}$

25. $\dfrac{-3y}{2y - 10} = \dfrac{}{50 - 10y}$

26. $\dfrac{4b}{b^2 + 2b - 3} = \dfrac{}{(b + 3)(b - 1)(b + 2)}$

[7.4] *Add or subtract and write each answer in lowest terms.*

27. $\dfrac{10}{x} + \dfrac{5}{x}$

28. $\dfrac{6}{3p} - \dfrac{12}{3p}$

29. $\dfrac{9}{k} - \dfrac{5}{k - 5}$

30. $\dfrac{4}{y} + \dfrac{7}{7 + y}$

31. $\dfrac{m}{3} - \dfrac{2 + 5m}{6}$

32. $\dfrac{12}{x^2} - \dfrac{3}{4x}$

33. $\dfrac{5}{a - 2b} + \dfrac{2}{a + 2b}$

34. $\dfrac{4}{k^2 - 9} - \dfrac{k + 3}{3k - 9}$

35. $\dfrac{8}{z^2 + 6z} - \dfrac{3}{z^2 + 4z - 12}$

36. $\dfrac{11}{2p - p^2} - \dfrac{2}{p^2 - 5p + 6}$

[7.5]

37. Simplify the complex fraction $\dfrac{\dfrac{a^4}{b^2}}{\dfrac{a^3}{b}}$ by

(a) Method 1 as described in Section 7.5.
(b) Method 2 as described in Section 7.5.
(c) Explain which method you prefer, and why.

Simplify each complex fraction.

38. $\dfrac{\dfrac{2}{3} - \dfrac{1}{6}}{\dfrac{1}{4} + \dfrac{2}{5}}$

39. $\dfrac{\dfrac{y - 3}{y}}{\dfrac{y + 3}{4y}}$

40. $\dfrac{\dfrac{1}{p} - \dfrac{1}{q}}{\dfrac{1}{q - p}}$

41. $\dfrac{x + \dfrac{1}{w}}{x - \dfrac{1}{w}}$

42. $\dfrac{\dfrac{1}{r + t} - 1}{\dfrac{1}{r + t} + 1}$

[7.6]

43. Before even beginning the solution process, how do you know that 2 cannot be a solution to the equation found in Exercise 46 below?

Solve each equation and check your solutions.

44. $\dfrac{4 - z}{z} + \dfrac{3}{2} = \dfrac{-4}{z}$

45. $\dfrac{3y - 1}{y - 2} = \dfrac{5}{y - 2} + 1$

46. $\dfrac{3}{m - 2} + \dfrac{1}{m - 1} = \dfrac{7}{m^2 - 3m + 2}$

Solve each formula for the specified variable.

47. $m = \dfrac{Ry}{t}$ for t

48. $x = \dfrac{3y - 5}{4}$ for y

49. $p^2 = \dfrac{4}{3m - q}$ for m

[7.7] *Solve each problem. Use the six-step method.*

50. In a certain fraction, the denominator is 4 less than the numerator. If 3 is added to both the numerator and the denominator, the resulting fraction is equal to $\dfrac{3}{2}$. Find the original fraction.

51. The denominator of a certain fraction is 3 times the numerator. If 2 is added to the numerator and subtracted from the denominator, the resulting fraction is equal to 1. Find the original fraction.

 52. On August 18, 1996, Scott Sharp won the True Value 200-mile Indy race driving a Ford with an average speed of 130.934 miles per hour. What was his time? (*Source: Sports Illustrated 1998 Sports Almanac.*)

53. A man can plant his garden in 5 hours, working alone. His daughter can do the same job in 8 hours. How long would it take them if they worked together?

54. The head gardener can mow the lawns in the city park twice as fast as his assistant. Working together, they can complete the job in $1\dfrac{1}{3}$ hours. How long would it take the head gardener working alone?

55. The longer the term of your subscription to *Monitoring Times,* the less you will have to pay per year. Is this an example of direct or inverse variation?

 56. In 1994, 38,505 American deaths were due to firearms. Of this total, the approximate ratio of male deaths to female deaths was 63 to 10. How many males and how many females died as a result of using firearms? (*Source:* National Center for Health Statistics, Monthly Vital Statistics Reports.)

57. If a parallelogram has a fixed area, the height varies inversely as the base. A parallelogram has a height of 8 centimeters and a base of 12 centimeters. Find the height if the base is changed to 24 centimeters.

58. At a given hour, two steamboats leave a city in the same direction on a straight canal. One travels at 18 miles per hour, and the other travels at 25 miles per hour. In how many hours will the boats be 35 miles apart?

RELATING CONCEPTS (EXERCISES 59–68)

In these exercises, we summarize the various concepts involving rational expressions we have covered.

Work Exercises 59–68 in order.

Let *P, Q,* and *R* be rational expressions defined as follows.

$$P = \frac{6}{x + 3} \qquad Q = \frac{5}{x + 1} \qquad R = \frac{4x}{x^2 + 4x + 3}$$

59. Find the value or values for which the expression is undefined.
   (a) *P*   **(b)** *Q*   **(c)** *R*

RELATING CONCEPTS (EXERCISES 59-68) (CONTINUED)

60. Find and express in lowest terms: $(P \cdot Q) \div R$.

61. Why is $(P \cdot Q) \div R$ not defined if $x = 0$?

62. Find the LCD for P, Q, and R.

63. Perform the operations and express in lowest terms: $P + Q - R$.

64. Simplify the complex fraction $\dfrac{P + Q}{R}$.

65. Solve the equation $P + Q = R$.

66. How does your answer to Exercise 59 help you work Exercise 65?

67. Suppose that a car travels 6 miles in $x + 3$ minutes. Explain why P represents the rate of the car (in miles per minute).

68. For what value or values of x is $R = \dfrac{40}{77}$?

Did you make the connections between the various operations with rational expressions?

MIXED REVIEW EXERCISES

Perform the indicated operations.

69. $\dfrac{\dfrac{5}{x - y} + 2}{3 - \dfrac{2}{x + y}}$

70. $\dfrac{4}{m - 1} - \dfrac{3}{m + 1}$

71. $\dfrac{8p^5}{5} \div \dfrac{2p^3}{10}$

72. $\dfrac{r - 3}{8} \div \dfrac{3r - 9}{4}$

73. $\dfrac{\dfrac{5}{x} - 1}{\dfrac{5 - x}{3x}}$

74. $\dfrac{4}{z^2 - 2z + 1} - \dfrac{3}{z^2 - 1}$

Solve.

75. $\dfrac{1}{k} + \dfrac{3}{r} = \dfrac{5}{z}$ for r

76. $\dfrac{5 + m}{m} + \dfrac{3}{4} = \dfrac{-2}{m}$

77. In 1996, the total number of people waiting for heart and liver transplants was 11,165. The ratio of those waiting for heart transplants to those waiting for liver transplants was approximately 2 to 1. How many of each type of candidate were there? (*Sources:* U.S. Department of Health and Human Services, Public Health Service, Division of Organ Transplantation, and United Network for Organ Sharing.)

78. Emily Falzon flew her plane 400 kilometers with the wind in the same time it took her to go 200 kilometers against the wind. The speed of the wind is 50 kilometers per hour. Find the speed of the plane in still air.

79. If x varies directly as y, and $x = 12$ when $y = 5$, find x when $y = 3$.

80. When Mario and Luigi work together on a job, they can do it in $3\dfrac{3}{7}$ days. Mario can do the job working alone in 8 days. How long would it take Luigi working alone?

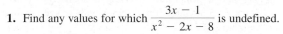

1. Find any values for which $\dfrac{3x-1}{x^2-2x-8}$ is undefined.

2. Find the numerical value of $\dfrac{6r+1}{2r^2-3r-20}$ when
   (a) $r=-2$ and **(b)** $r=4$.

3. Write four rational expressions equivalent to $-\dfrac{6x-5}{2x+3}$.

Write each rational expression in lowest terms.

4. $\dfrac{-15x^6y^4}{5x^4y}$

5. $\dfrac{6a^2+a-2}{2a^2-3a+1}$

Multiply or divide. Write each answer in lowest terms.

6. $\dfrac{5(d-2)}{9} \div \dfrac{3(d-2)}{5}$

7. $\dfrac{6k^2-k-2}{8k^2+10k+3} \cdot \dfrac{4k^2+7k+3}{3k^2+5k+2}$

8. $\dfrac{4a^2+9a+2}{3a^2+11a+10} \div \dfrac{4a^2+17a+4}{3a^2+2a-5}$

Find the least common denominator for each list of fractions.

9. $\dfrac{-3}{10p^2}, \dfrac{21}{25p^3},$ and $\dfrac{-7}{30p^5}$

10. $\dfrac{r+1}{2r^2+7r+6}$ and $\dfrac{-2r+1}{2r^2-7r-15}$

Rewrite each rational expression with the given denominator.

11. $\dfrac{15}{4p} = \dfrac{}{64p^3}$

12. $\dfrac{3}{6m-12} = \dfrac{}{42m-84}$

Add or subtract. Write each answer in lowest terms.

13. $\dfrac{4x+2}{x+5} + \dfrac{-2x+8}{x+5}$

14. $\dfrac{-4}{y+2} + \dfrac{6}{5y+10}$

15. $\dfrac{x+1}{3-x} + \dfrac{x^2}{x-3}$

16. $\dfrac{3}{2m^2-9m-5} - \dfrac{m+1}{2m^2-m-1}$

Simplify each complex fraction.

17. $\dfrac{\dfrac{2p}{k^2}}{\dfrac{3p^2}{k^3}}$

18. $\dfrac{\dfrac{1}{x+3}-1}{1+\dfrac{1}{x+3}}$

19. What are the two numbers that would have to be rejected as possible solutions for the equation $\dfrac{2}{x+1} - \dfrac{3}{x-4} = 6$?

20. Solve the equation $\dfrac{2x}{x-3} + \dfrac{1}{x+3} = \dfrac{-6}{x^2-9}$. Be sure to check your solution(s).

21. Solve the formula $F = \dfrac{k}{d-D}$ for D.

Solve each problem.

22. A boat goes 7 miles per hour in still water. It takes as long to go 20 miles upstream as 50 miles downstream. Find the speed of the current.

23. A man can paint a room in his house, working alone, in 5 hours. His wife can do the job in 4 hours. How long will it take them to paint the room if they work together?

24. Under certain conditions, the length of time that it takes for fruit to ripen during the growing season varies inversely as the average maximum temperature during the season. If it takes 25 days for fruit to ripen with an average maximum temperature of 80°, find the number of days it would take at 75°. Round your answer to the nearest whole number.

CUMULATIVE REVIEW EXERCISES CHAPTERS 1–7

1. Use the order of operations to evaluate $3 + 4 \left(\dfrac{1}{2} - \dfrac{3}{4} \right)$.

Solve.

2. $3(2y - 5) = 2 + 5y$ **3.** $A = \dfrac{1}{2}bh$ for b **4.** $\dfrac{2 + m}{2 - m} = \dfrac{3}{4}$

5. $5y \le 6y + 8$ **6.** $5m - 9 > 2m + 3$

7. For the graph of $4x + 3y = -12$,
   (a) what is the x-intercept and   **(b)** what is the y-intercept?

Sketch each graph.

8. $y = -3x + 2$ **9.** $y = -x^2 + 1$

Simplify each expression. Write with only positive exponents.

10. $\dfrac{(2x^3)^{-1} \cdot x}{2^3 x^5}$ **11.** $\dfrac{(m^{-2})^3 m}{m^5 m^{-4}}$ **12.** $\dfrac{2p^3 q^4}{8p^5 q^3}$

Perform the indicated operations.

13. $(2k^2 + 3k) - (k^2 + k - 1)$ **14.** $8x^2 y^2 (9x^4 y^5)$

15. $(2a - b)^2$ **16.** $(y^2 + 3y + 5)(3y - 1)$

17. $\dfrac{12p^3 + 2p^2 - 12p + 4}{2p - 2}$

18. A computer can do one operation in 1.4×10^{-7} seconds. How long would it take for the computer to do one trillion (10^{12}) operations?

Factor completely.

19. $8t^2 + 10tv + 3v^2$ **20.** $8r^2 - 9rs + 12s^2$ **21.** $16x^4 - 1$

Solve each equation.

22. $r^2 = 2r + 15$

23. $(r - 5)(2r + 1)(3r - 2) = 0$

Solve each problem.

24. One number is 4 more than another. The product of the numbers is 2 less than the smaller number. Find the smaller number.

25. The length of a rectangle is 2 meters less than twice the width. The area is 60 square meters. Find the width of the rectangle.

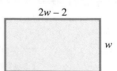

26. In 1996, the U.S. Department of Health and Human Services and the Department of Agriculture set guidelines for adult diets. A maximum percent of one's diet was suggested for calories from fat. It is a number such that when it is divided by 60, the quotient is the same as when it is divided into 15. Find this positive number to discover the maximum suggested percent from fat.

27. One of the following is equal to 1 for *all* real numbers. Which one is it?

(a) $\dfrac{k^2 + 2}{k^2 + 2}$ (b) $\dfrac{4 - m}{4 - m}$ (c) $\dfrac{2x + 9}{2x + 9}$ (d) $\dfrac{x^2 - 1}{x^2 - 1}$

28. Which one of the following rational expressions is *not* equivalent to $\dfrac{4 - 3x}{7}$?

(a) $-\dfrac{-4 + 3x}{7}$ (b) $-\dfrac{4 - 3x}{-7}$ (c) $\dfrac{-4 + 3x}{-7}$ (d) $\dfrac{-(3x + 4)}{7}$

Perform each operation and write the answer in lowest terms.

29. $\dfrac{5}{q} - \dfrac{1}{q}$

30. $\dfrac{3}{7} + \dfrac{4}{r}$

31. $\dfrac{4}{5q - 20} - \dfrac{1}{3q - 12}$

32. $\dfrac{2}{k^2 + k} - \dfrac{3}{k^2 - k}$

33. $\dfrac{7z^2 + 49z + 70}{16z^2 + 72z - 40} \div \dfrac{3z + 6}{4z^2 - 1}$

34. Simplify the complex fraction $\dfrac{\dfrac{4}{a} + \dfrac{5}{2a}}{\dfrac{7}{6a} - \dfrac{1}{5a}}$.

35. What values of x cannot possibly be solutions of the equation $\dfrac{1}{x - 4} = \dfrac{3}{2x}$?

Solve each equation. Check your solutions.

36. $\dfrac{r + 2}{5} = \dfrac{r - 3}{3}$

37. $\dfrac{1}{x} = \dfrac{1}{x + 1} + \dfrac{1}{2}$

Solve each problem.

38. On a business trip, Arlene traveled to her destination at an average speed of 60 miles per hour. Coming home, her average speed was 50 miles per hour, and the trip took $\dfrac{1}{2}$ hour longer. How far did she travel each way?

39. Juanita can weed the yard in 3 hours. Benito can weed the yard in 2 hours. How long would it take them if they worked together?

40. The force required to compress a spring varies directly as the change in the length of the spring. If a force of 12 pounds is required to compress a certain spring 3 inches, how much force is required to compress the spring 5 inches?

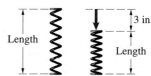

8 Equations of Lines, Inequalities, and Functions

Energy

Graphs are widely used in the media because they present a lot of information in an easy-to-understand form. As the saying goes, "A picture is worth a thousand words." The pie graph shows 1996 U.S. energy production by source. Which source produced the most energy in 1996? The line graph shows U.S. energy consumption in the years 1991–1996. By how many quadrillion BTUs did U.S. energy consumption increase from 1991–1996?* We return to these graphs later in the chapter.

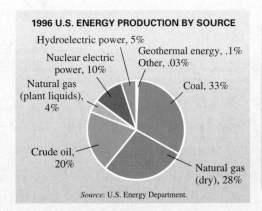

1996 U.S. ENERGY PRODUCTION BY SOURCE

Hydroelectric power, 5%
Nuclear electric power, 10%
Natural gas (plant liquids), 4%
Geothermal energy, .1%
Other, .03%
Coal, 33%
Crude oil, 20%
Natural gas (dry), 28%

Source: U.S. Energy Department.

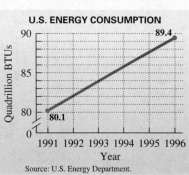

U.S. ENERGY CONSUMPTION

89.4

80.1

Source: U.S. Energy Department.

These graphs each depict some aspect of energy in the United States. Many of the applications in this chapter are based on data about energy, the theme of this chapter. We will use this theme as we work with linear equations in two variables, whose graphs are straight lines.

---

*BTU stands for British Thermal Unit and is a measure of energy consumption.

8.1 Review of Graphs and Slopes of Lines

OBJECTIVES

1 Plot ordered pairs.

2 Graph lines and find intercepts.

3 Recognize equations of vertical or horizontal lines.

4 Find the slope of a line.

5 Graph a line given its slope and a point on the line.

6 Use slope to decide whether two lines are parallel or perpendicular.

7 Solve problems involving average rate of change.

FOR EXTRA HELP

SSG Sec. 8.1
SSM Sec. 8.1

Pass the Test Software

InterAct Math Tutorial Software

Video 13

This section reviews some of the main topics of linear equations in two variables, first introduced in Chapter 4.

OBJECTIVE 1 Plot ordered pairs. Each of the pairs of numbers $(1, 2)$, $(-1, 5)$, and $(3, 7)$ is an example of an **ordered pair;** that is, a pair of numbers written within parentheses in which the order of the numbers is important. We graph an ordered pair using two perpendicular number lines that intersect at their zero points, as shown in Figure 1. Recall that the common zero point is called the **origin.** The position of any point in this plane is determined by referring to the horizontal number line, the ***x*-axis,** and the vertical number line, the ***y*-axis.** The first number in the ordered pair indicates the position relative to the *x*-axis, and the second number indicates the position relative to the *y*-axis. The *x*-axis and the *y*-axis make up a **rectangular** (or **Cartesian,** for Descartes) **coordinate system.**

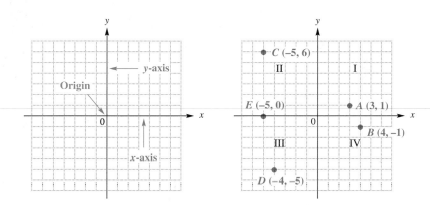

Figure 1   Figure 2

To locate, or **plot,** the point on the graph that corresponds to the ordered pair $(3, 1)$, move three units from zero to the right along the *x*-axis, and then one unit up parallel to the *y*-axis. The point corresponding to the ordered pair $(3, 1)$ is labeled A in Figure 2. The phrase "the point corresponding to the ordered pair $(3, 1)$" is often abbreviated "the point $(3, 1)$." The numbers in an ordered pair are called the **coordinates** of the corresponding point.

The four regions of the graph shown in Figure 2 are called **quadrants I, II, III,** and **IV,** reading counterclockwise from the upper right quadrant. The points of the *x*-axis and *y*-axis themselves do not belong to any quadrant. For example, point E in Figure 2 belongs to no quadrant.

OBJECTIVE 2 Graph lines and find intercepts. Each solution to an equation with two variables includes two numbers, one for each variable. To keep track of which number goes with which variable we write the solutions as ordered pairs, with the *x*-value given first. For example, we can show that $(6, -2)$ is a solution of $2x + 3y = 6$ by substitution.

$$2x + 3y = 6$$
$$2(6) + 3(-2) = 6 \quad ? \quad \text{Let } x = 6, y = -2.$$
$$12 - 6 = 6 \quad ?$$
$$6 = 6 \quad \text{True}$$

Since the pair of numbers $(6, -2)$ makes the equation true, it is a solution. On the other hand, since

$$2(5) + 3(1) = 10 + 3 = 13 \neq 6,$$

$(5, 1)$ is not a solution of the equation $2x + 3y = 6$.

To find ordered pairs that satisfy an equation, select any number for one of the variables, substitute it into the equation for that variable, and then solve for the other variable. There are infinitely many ordered pairs that will satisfy the equation. The table below shows several such ordered pairs for the equation $2x + 3y = 6$. This equation is called a **first-degree equation,** because it has no term with one variable to a power greater than one. The graph of any first-degree equation in two variables is a straight line. The graph of $2x + 3y = 6$ is the line shown in Figure 3.

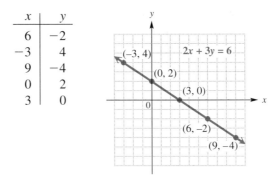

x	y
6	-2
-3	4
9	-4
0	2
3	0

Figure 3

The equation $2x + 3y = 6$ is in *standard form.*

Standard Form of a Linear Equation in Two Variables

A **linear equation in two variables** can be written in the form

$$Ax + By = C \quad (A \text{ and } B \text{ not both } 0).$$

This form is called **standard form.**

A straight line is determined if any two different points on the line are known, so finding two different points is enough to graph the line. Two useful points for graphing are the x- and y-intercepts. The **x-intercept** is the point (if any) where the line intersects the x-axis; likewise, the **y-intercept** is the point (if any) where the line intersects the y-axis. In Figure 3, the y-value of the point where the line intersects the x-axis is 0. Similarly, the x-value of the point where the line intersects the y-axis is 0. This suggests a method for finding the x- and y-intercepts.

Intercepts

Let $y = 0$ to find the x-intercept; let $x = 0$ to find the y-intercept.

EXAMPLE 1   Finding Intercepts

Find the *x*- and *y*-intercepts of $4x - y = -3$, and graph the equation.

Find the *x*-intercept by letting $y = 0$.

$$4x - \mathbf{0} = -3 \qquad \text{Let } y = 0.$$
$$4x = -3$$
$$x = -\frac{3}{4} \qquad x\text{-intercept is } \left(-\frac{3}{4}, 0\right).$$

For the *y*-intercept, let $x = 0$.

$$4(\mathbf{0}) - y = -3 \qquad \text{Let } x = 0.$$
$$-y = -3$$
$$y = 3 \qquad y\text{-intercept is } (0, 3).$$

The intercepts are the two points $\left(-\frac{3}{4}, 0\right)$ and $(0, 3)$. These ordered pairs are shown in the table with Figure 4. Use these points to draw the graph, as shown in Figure 4.

x	y
$-\frac{3}{4}$	0
0	3

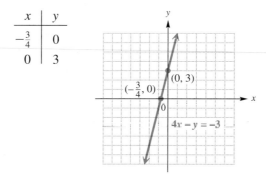

Figure 4

OBJECTIVE **3**   Recognize equations of vertical or horizontal lines. Two special cases of straight-line graphs are vertical and horizontal lines.

EXAMPLE 2   Graphing a Horizontal Line

Graph $y = 2$.

Writing $y = 2$ as $0x + 1y = 2$ shows that any value of *x*, including $x = 0$, gives $y = 2$, making the *y*-intercept $(0, 2)$. Since *y* is always 2, there is no value of *x* corresponding to $y = 0$, and so the graph has no *x*-intercept. The graph, shown with a table of ordered pairs in Figure 5 on the next page, is a horizontal line.

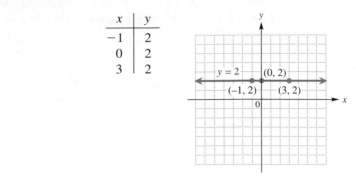

x	y
-1	2
0	2
3	2

Figure 5

EXAMPLE 3   Graphing a Vertical Line

Graph $x = -1$.

The form $1x + 0y = -1$ shows that every value of y leads to $x = -1$, and so no value of y makes $x = 0$. The graph, therefore, has no y-intercept. The only way a straight line can have no y-intercept is to be vertical, as shown in Figure 6.

x	y
-1	3
-1	0
-1	-2

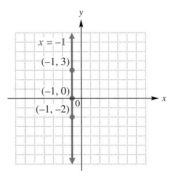

Figure 6

CONNECTIONS

When graphing with a graphing calculator, we first tell the calculator how to "set up" a rectangular coordinate system. This involves choosing the minimum and maximum x- and y-values that will determine the viewing screen. In the screen shown in Figure 7, we have chosen minimum x- and y-values of -10 and maximum x- and y-values of 10. The *scale* on each axis determines the distance between the tick marks; in the view shown, the scale is 1 for both axes. We will refer to this as the *standard window*.

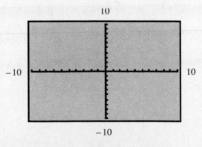

Figure 7

CONNECTIONS (CONTINUED)

For example, to graph $4x - y = 3$ with a graphing calculator, we can use the intercepts to determine an appropriate window. Here, the x-intercept is $(.75, 0)$ and the y-intercept is $(0, -3)$. We want to be able to see the intercepts on the screen. Although many window choices are possible, we choose the standard window. We must solve the equation for y to enter it into the calculator.

$$4x - y = 3$$
$$-y = -4x + 3 \qquad \text{Subtract } 4x \text{ on each side.}$$
$$y = 4x - 3 \qquad \text{Multiply both sides by } -1.$$

The graph is shown in Figures 8 and 9, which also give the intercepts at the bottoms of the screens.

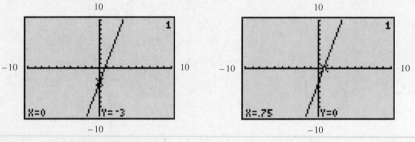

Figure 8   Figure 9

Some calculators have the capability of locating the x-intercept (called "Root" or "Zero") with a special function. Consult your owner's manual.

OBJECTIVE **4** **Find the slope of a line.** Slope (steepness) is used in many practical ways. The slope of a hill (sometimes called the *grade*) is often given as a percent. For example, a 10% $\left(\text{or } \frac{10}{100} = \frac{1}{10}\right)$ slope means the hill rises 1 unit for every 10 horizontal units. Stairs and roofs have slopes too, as shown in Figure 10.

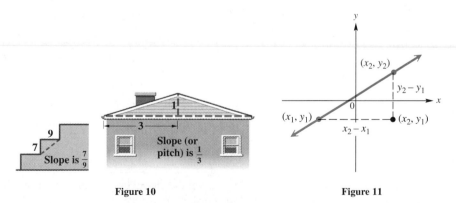

Figure 10   Figure 11

In each example that we mentioned, slope is the ratio of vertical change, or **rise,** to horizontal change, or **run.** A simple way to remember this is to think "slope is rise over run."

To get a formal definition of the slope of a line, it is convenient to use *subscripted variables* to designate two different points on the line. To differentiate between the points, we write them as (x_1, y_1) and (x_2, y_2). See Figure 11 on the previous page. (The small numbers 1 and 2 in these ordered pairs are called subscripts. Read (x_1, y_1) as "x-sub-one, y-sub-one.")

As we move along the line in Figure 11 from (x_1, y_1) to (x_2, y_2), the y-value changes from y_1 to y_2, an amount equal to $y_2 - y_1$. As y changes from y_1 to y_2, thé value of x changes from x_1 to x_2 by the amount $x_2 - x_1$. The ratio of the change in y to the change in x (the rise over the run) is called the **slope** of the line, with the letter m traditionally used for the slope.

Slope

The slope of the line through the distinct points (x_1, y_1) and (x_2, y_2) is

$$m = \frac{\text{change in } y}{\text{change in } x} = \frac{y_2 - y_1}{x_2 - x_1} \quad (x_1 \neq x_2).$$

E X A M P L E   4   Using the Definition of Slope

Find the slope of the line through the points $(2, -1)$ and $(-5, 3)$.

If $(2, -1) = (x_1, y_1)$ and $(-5, 3) = (x_2, y_2)$, then

$$m = \frac{y_2 - y_1}{x_2 - x_1} = \frac{3 - (-1)}{-5 - 2} = \frac{4}{-7} = -\frac{4}{7}.$$

See Figure 12. On the other hand, if $(2, -1) = (x_2, y_2)$ and $(-5, 3) = (x_1, y_1)$, the slope would be

$$m = \frac{-1 - 3}{2 - (-5)} = \frac{-4}{7} = -\frac{4}{7},$$

the same answer.

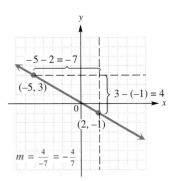

Figure 12

In calculating the slope, be careful to subtract the y-values and the x-values in the *same* order.

	Correct			Incorrect	
$\dfrac{y_2 - y_1}{x_2 - x_1}$	or	$\dfrac{y_1 - y_2}{x_1 - x_2}$	$\dfrac{y_2 - y_1}{x_1 - x_2}$	or	$\dfrac{y_1 - y_2}{x_2 - x_1}$

Also, remember that the change in y is the *numerator* and the change in x is the *denominator.*

When the equation of a line is given, one way to find the slope is to use the definition of slope by first finding two different points on the line.

EXAMPLE 5 Finding the Slope of a Line from Its Equation

Find the slope of the line $4x - y = 8$.

The intercepts can be used as the two different points needed to find the slope. Let $y = 0$ to find that the x-intercept is $(2, 0)$. Then let $x = 0$ to find that the y-intercept is $(0, -8)$. Use these two points in the slope formula. The slope is

$$m = \frac{-8 - 0}{0 - 2} = \frac{-8}{-2} = 4.$$

We review the following special cases of slope.

Slopes of Vertical and Horizontal Lines

The slope of a vertical line is undefined; the slope of a horizontal line is 0.

The slope of a line also can be found directly from its equation. Look again at the equation $4x - y = 8$ in Example 5. Solve this equation for y.

$$4x - y = 8 \qquad \text{Equation from Example 5}$$
$$-y = -4x + 8 \qquad \text{Subtract } 4x \text{ from both sides.}$$
$$y = 4x - 8 \qquad \text{Multiply both sides by } -1.$$

Notice that the slope, 4, we found using the slope formula in Example 5 is the same number as the coefficient of x in the equation $y = 4x$ 8. This always happens, *as long as the equation is solved for y.*

EXAMPLE 6 Finding the Slope of a Line from Its Equation

Find the slope of the graph of $3x - 5y = 8$.

Solve the equation for y.

$$3x - 5y = 8$$
$$-5y = -3x + 8 \qquad \text{Subtract } 3x \text{ from both sides.}$$
$$y = \frac{3}{5}x - \frac{8}{5} \qquad \text{Divide both sides by } -5.$$

The slope is given by the coefficient of x, so the slope is $\frac{3}{5}$.

OBJECTIVE **5** Graph a line given its slope and a point on the line.

EXAMPLE 7 Using the Slope and a Point to Graph a Line

(a) Graph the line that has slope $\frac{2}{3}$ and goes through the point $(-1, 4)$.

First locate the point $(-1, 4)$ on a graph as shown in Figure 13 on the next page. Then, from the definition of slope,

$$m = \frac{\text{change in } y}{\text{change in } x} = \frac{2}{3}.$$

Move *up* 2 units in the *y*-direction and then 3 units to the *right* in the *x*-direction to locate another point on the graph (labeled *P*). The line through (−1, 4) and *P* is the required graph.

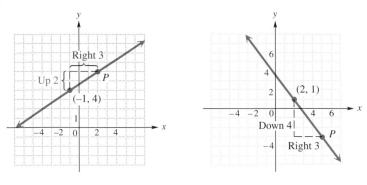

Figure 13   Figure 14

(b) Graph the line through (2, 1) that has slope $-\frac{4}{3}$.

Start by locating the point (2, 1) on the graph. Find a second point on the line by using the definition of slope.

$$\text{slope} = \frac{\text{change in } y}{\text{change in } x} = \frac{-4}{3}$$

Move *down* 4 units from (2, 1) and then 3 units to the *right*. Draw a line through this second point *P* and (2, 1), as shown in Figure 14.

The slope also could be written as

$$\frac{\text{change in } y}{\text{change in } x} = \frac{4}{-3}.$$

In this case the second point is located *up* 4 units and 3 units to the *left*. Verify that this approach produces the same line.

In Example 7(a), the slope of the line is the *positive* number $\frac{2}{3}$. The graph of the line in Figure 13 goes up from left to right. The line in Example 7(b) has a *negative* slope, $-\frac{4}{3}$. As Figure 14 shows, its graph goes down from left to right. These facts suggest the following generalization.

> A positive slope indicates that the line goes up from left to right; a negative slope indicates that the line goes down from left to right.

Figure 15 shows lines of positive, zero, negative, and undefined slopes.

OBJECTIVE 6 Use slope to decide whether two lines are parallel or perpendicular. The slopes of a pair of parallel or perpendicular lines are related in a special way.

Slopes of Parallel Lines

Two nonvertical lines with the same slope are parallel; two nonvertical parallel lines have the same slope.

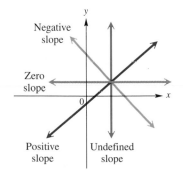

Figure 15

> **Slopes of Perpendicular Lines**
>
> If neither is vertical, perpendicular lines have slopes that are negative reciprocals; that is, their product is -1. Also, lines with slopes that are negative reciprocals are perpendicular.

EXAMPLE 8   Determining Whether Two Lines Are Parallel, Perpendicular, or Neither

Determine whether the two lines described are parallel, perpendicular, or neither.

(a) The line L_1, through $(-2, 1)$ and $(4, 5)$, and the line L_2, through $(3, 0)$ and $(0, -2)$

The slope of L_1 is

$$m_1 = \frac{5 - 1}{4 - (-2)} = \frac{4}{6} = \frac{2}{3}.$$

The slope of L_2 is

$$m_2 = \frac{-2 - 0}{0 - 3} = \frac{-2}{-3} = \frac{2}{3}.$$

Since the slopes are equal, the lines are parallel.

(b) The lines with equations $y = \frac{2}{5}x + 3$ and $y = -\frac{2}{5}x - 4$

The slopes of the lines are $\frac{2}{5}$ and $-\frac{2}{5}$. These are neither equal nor negative reciprocals, so the lines are neither parallel nor perpendicular.

(c) The lines with equations $2y = 3x - 6$ and $2x + 3y = -6$

Find the slope of each line by first solving each equation for y.

$$2y = 3x - 6 \qquad\qquad 2x + 3y = -6$$

$$y = \frac{3}{2}x - 3 \qquad\qquad 3y = -2x - 6$$

$$y = -\frac{2}{3}x - 2$$

$$\text{Slope: } \frac{3}{2} \qquad\qquad \text{Slope: } -\frac{2}{3}$$

Since the product of the slopes of the two lines is $\frac{3}{2}\left(-\frac{2}{3}\right) = -1$, the lines are perpendicular.

OBJECTIVE 7   Solve problems involving average rate of change. We know that the slope of a line is the ratio of the change in y (vertical change) to the change in x (horizontal change). This idea can be applied to real-life situations. The slope gives the average rate of change in y per unit of change in x, where the value of y depends on the value of x. The next example further illustrates this idea of average rate of change. We assume a linear relationship between x and y.

EXAMPLE 9   Interpreting Slope as Average Rate of Change

The graph in Figure 16 on the next page approximates the U.S. production of crude oil in the years 1991–1995. Find the average rate of change in production in quadrillion BTUs per year.

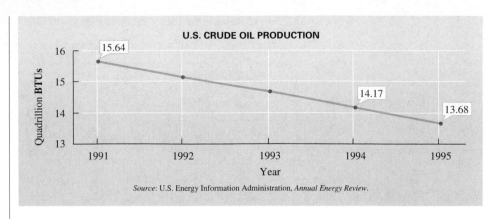

U.S. CRUDE OIL PRODUCTION

Source: U.S. Energy Information Administration, *Annual Energy Review*.

Figure 16

To use the slope formula, we need two pairs of data. From the graph, if $x = 1991$, then $y = 15.64$ and if $x = 1995$, $y = 13.68$, so we have the ordered pairs (1991, 15.64) and (1995, 13.68). By the slope formula,

$$\text{Average rate of change} = \frac{y_2 - y_1}{x_2 - x_1} = \frac{13.68 - 15.64}{1995 - 1991} = -.49.$$

This means that production of crude oil *decreased* by .49 quadrillion BTUs each year in the period from 1991–1995.

8.1 EXERCISES

Use the graph to answer the questions.

1. The graph indicates the total area of solar collectors used for heating pools from 1991–1995.

   (a) In which consecutive years did the numbers increase?

   (b) Between which two years was the decrease the greatest?

   (c) In what year was the number about 6000?

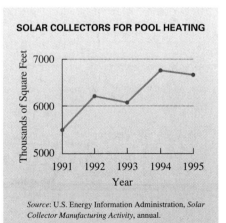

SOLAR COLLECTORS FOR POOL HEATING

Source: U.S. Energy Information Administration, *Solar Collector Manufacturing Activity*, annual.

2. The graph shows the number of solar collectors used to heat water from 1991–1995.

    (a) In what year was the number of solar collectors used to heat water the greatest?

    (b) In what one-year period was the decrease the greatest?

    (c) In which year was the number greater than in the previous year?

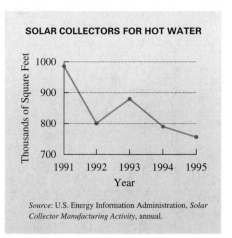

SOLAR COLLECTORS FOR HOT WATER

Source: U.S. Energy Information Administration, *Solar Collector Manufacturing Activity*, annual.

Name the quadrant, if any, in which each point is located.

3. **(a)** $(1, 6)$
    (b) $(-4, -2)$
    (c) $(-3, 6)$
    (d) $(7, -5)$
    (e) $(-3, 0)$

4. **(a)** $(-2, -10)$
    (b) $(4, 8)$
    (c) $(-9, 12)$
    (d) $(3, -9)$
    (e) $(0, -8)$

Locate the following points on a rectangular coordinate system.

5. $(2, 3)$ **6.** $(-1, 2)$

7. $(-3, -2)$ **8.** $(1, -4)$

9. $(0, 5)$ **10.** $(-2, -4)$

11. $(-2, 4)$ **12.** $(3, 0)$

13. $(-2, 0)$ **14.** $(3, -3)$

Complete the given table for each equation, and then graph the equation.

15. $x - y = 3$

x	y
0	
	0
5	
2	

16. $x + 3y = -5$

x	y
0	
	0
1	
	-1

17. Explain why the graph of $x + y = k$ cannot pass through quadrant III if $k > 0$.

18. Explain how to determine the intercepts of and graph the linear equation $4x - 3y = 12$.

Find the x-intercept and the y-intercept. Then graph each equation. See Examples 1–3.

19. $2x + 3y = 12$ **20.** $5x + 2y - 10$ **21.** $x - 3y = 6$

22. $x - 2y = -4$ **23.** $\frac{2}{3}x - 3y = 7$ **24.** $\frac{5}{7}x + \frac{6}{7}y = -2$

25. $y = 5$ **26.** $x = -3$ **27.** $x + 4 = 0$

28. $y + 2 = 0$ **29.** $x + 5y = 0$ **30.** $x - 3y = 0$

31. The screen shows the graph of one of the equations below, along with the coordinates of a point on the graph. Which one of the equations is it?
(a) $x + 2y = 4$
(b) $-3x + 5y = 15$
(c) $y = 4x - 2$
(d) $y = -2$

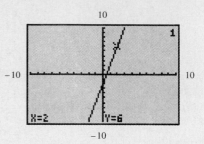

32. The screens show the graph of one of the equations below. Two views of the graph are given, along with the intercepts. Which one of the equations is it?
(a) $3x + 2y = 6$ (b) $-3x + 2y = 6$
(c) $-3x - 2y = 6$ (d) $3x - 2y = 6$

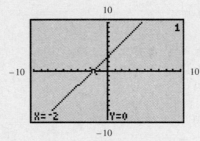 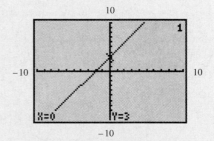

33. The table of ordered pairs shown was generated by a graphing calculator with a *table* feature.

(a) What is the x-intercept?
(b) What is the y-intercept?
(c) Which one of the equations below corresponds to this table of values?
    A. $Y_1 = 2X - 3$
    B. $Y_1 = -2X - 3$
    C. $Y_1 = 2X + 3$
    D. $Y_1 = -2X + 3$

34. The screens each show the graph of $x + y = 15$ (which was entered as $y = -x + 15$). However, different viewing windows are used. Which one of the two windows do you think would be more useful for this graph? Why?

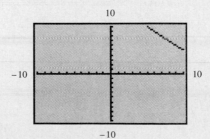 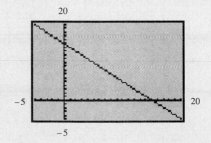

RELATING CONCEPTS (EXERCISES 35-40)

If the endpoints of a line segment are known, the coordinates of the midpoint of the segment can be found. The figure shows the coordinates of the points P and Q. Let $\overline{PQ}$ represent the line segment with endpoints at P and Q.

To derive a formula for the midpoint of $\overline{PQ}$, **work Exercises 35–40 in order.**

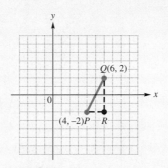

35. In the figure, R is the point with the same x-coordinate as Q and the same y-coordinate as P. Write the ordered pair that corresponds to R.

36. From the graph, determine the coordinates of the midpoint of $\overline{PR}$.

37. From the graph, determine the coordinates of the midpoint of $\overline{QR}$.

38. The x-coordinate of the midpoint M of $\overline{PQ}$ is the x-coordinate of the midpoint of $\overline{PR}$ and the y-coordinate is the y-coordinate of the midpoint of $\overline{QR}$. Write the ordered pair that corresponds to M.

39. The average of two numbers is found by dividing their sum by 2. Find the average of the x-coordinates of points P and Q. Find the average of the y-coordinates of points P and Q.

40. Comparing your answers to Exercises 36 and 37, what connection is there between the coordinates of P and Q and the coordinates of M?

Did you make the connection that the coordinates of M are the averages of the x- and y-coordinates of P and Q?

41. A ski slope drops 30 feet for every horizontal 100 feet.

Which of the following express its slope? (There are several correct choices.)

(a) $-.3$ **(b)** $-\dfrac{3}{10}$ **(c)** $-3\dfrac{1}{3}$ **(d)** $-\dfrac{30}{100}$ **(e)** $-\dfrac{10}{3}$

42. A hill has a slope of $-.05$. How many feet in the vertical direction correspond to a run of 50 feet?

Use the given figure to determine the slope of the line segment described, by counting the number of units of "rise," the number of units of "run," and then finding the quotient.

43. *AB*   **44.** *BC*   **45.** *CD*   **46.** *DE*

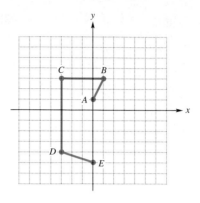

Find the slope of the line through the pair of points by using the slope formula. See Example 4.

47. $(-2, -3)$ and $(-1, 5)$ **48.** $(-4, 3)$ and $(-3, 4)$ **49.** $(-4, 1)$ and $(2, 6)$

50. $(-3, -3)$ and $(5, 6)$ **51.** $(2, 4)$ and $(-4, 4)$ **52.** $(-6, 3)$ and $(2, 3)$

Based on the figure shown here, determine which line satisfies the given description.

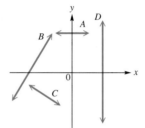

53. The line has positive slope.

54. The line has negative slope.

55. The line has slope 0.

56. The line has undefined slope.

Find the slope of each line. See Examples 5 and 6.

57. $x + 2y = 4$ **58.** $x + 3y = -6$ **59.** $-x + y = 4$

60. $3x + 4y = 12$ **61.** $y = 4x$ **62.** $y = -3x$

63. $x - 3 = 0$ **64.** $y + 5 = 0$

65. A vertical line has equation _____ = c for some constant c; a horizontal line has equation _____ = d for some constant d.

66. Explain the meaning of *slope*. Give examples.

67. Explain the procedure for graphing a straight line using its slope and a point on the line.

Use the methods shown in Example 7 to graph the line described.

68. Through $(-4, 2)$; $m = \dfrac{1}{2}$

69. Through $(-2, -3)$; $m = \dfrac{5}{4}$

70. Through $(0, -2)$; $m = -\dfrac{2}{3}$

71. Through $(0, -4)$; $m = -\dfrac{3}{2}$

72. Through $(-1, -2)$; $m = 3$

73. Through $(-2, -4)$; $m = 4$

74. If a line has slope $-\dfrac{4}{9}$, then any line parallel to it has slope _____, and any line perpendicular to it has slope _____ .

Decide whether the two lines are parallel, perpendicular, or neither. See Example 8.

75. $3x = y$ and $2y - 6x = 5$ **76.** $y = 4x$ and $4x - y = 9$

77. $4x - 3y = 8$ and $4y + 3x = 12$ **78.** $2x = y + 3$ and $2y + x = 3$

79. $4x - 3y = 5$ and $3x - 4y = 2$ **80.** $5x + y = 7$ and $5x = 3 + y$

81. The line through $(4, 6)$ and $(-8, 7)$ and the line through $(7, 4)$ and $(-5, 5)$

82. The line through $(9, 15)$ and $(-7, 12)$ and the line through $(-4, 8)$ and $(-20, 5)$

 Refer to Example 9 to solve the following problems.

83. The table gives data for the graph in Example 9, Figure 16. Use these data to find the average rate of change in crude oil production from 1991 to 1992; 1992 to 1994; 1993 to 1995. What do your answers suggest about the average rate of change?

U.S. Crude Oil Production

Year	Quadrillion BTUs
1991	15.64
1992	15.15
1993	14.66
1994	14.17
1995	13.68

Source: U.S. Energy Information Administration, *Annual Energy Review.*

84. The table gives book publishers' net dollar sales from 1995–2000. (Sales are estimated for the years 1998–2000.)

Book Publishers' Sales

Year	Sales (in millions)
1995	19,000
1996	20,000
1997	21,000
1998	22,000
1999	23,000
2000	24,000

Source: Book Industry Study Group.

(a) Find the average rate of change from 1995–1996; 1995–1999; 1998–2000. What do you notice about your answers? What does this tell you?

(b) Calculate the rates of change in part (a) as percents. What do you notice?

85. The graph shows how estimated undeveloped capacity of hydroelectric power (in millions of kilowatts) in the U.S. Pacific Division decreased from 1991–1996.

Undeveloped Water Power Capacity

Year	Kilowatts (in millions)
1991	26.2
1992	26.2
1993	26.2
1994	26.1
1995	24.0
1996	22.9

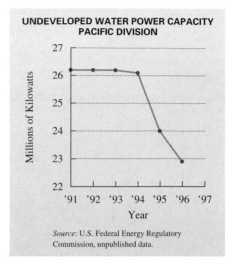

(a) Use the information given in the table of values for 1991 and 1996 to determine the average rate of change in capacity per year from 1991 through 1996.

(b) Use the information given for 1993 and 1996 to determine the average rate of change in capacity per year over that period.

(c) What do you notice about the answers for parts (a) and (b)? Explain.

86. The long-distance telephone market has changed rapidly and continuously since it was opened up to competition. The fourth-quarter market shares of AT&T for even-numbered years since 1984 are shown in the table.

AT&T Fourth-Quarter Market Shares

Year	Market Share (%)
1984	87.7
1986	81.5
1988	73.9
1990	66.5
1992	60.6
1994	57.6
1996	53.8

Source: Federal Communications Commission.

(a) Find the average annual rate of change of market share for each pair of successive years in the table, 1984 and 1986, 1986 and 1988, and so on.

(b) Plot these points on a grid using the years as *x*-values and the market share as *y*-values. Do the points appear to lie approximately on a line?

(c) How do the average rates you found in part (a) compare with your answer to part (b)?

8.2 Equations of Lines

OBJECTIVES

1 Write the equation of a line given its slope and a point on the line.

2 Write the equation of a line given two points on the line.

3 Write the equation of a line given its slope and y-intercept.

4 Find the slope and y-intercept of a line given its equation.

5 Write the equation of a line parallel or perpendicular to a given line.

6 Apply concepts of linear equations in two variables to real data.

7 Use a graphing calculator to solve linear equations in one variable.

FOR EXTRA HELP

 SSG Sec. 8.2
SSM Sec. 8.2

 Pass the Test Software

 InterAct Math Tutorial Software

 Video 13

Many real-world situations can be described by straight-line graphs. This section shows how to write a linear equation that satisfies given conditions.

OBJECTIVE **1** Write the equation of a line given its slope and a point on the line. A straight line is a set of points in the plane such that the slope between any two points is the same. In Figure 17, point P is on the line through P_1 and P_2 if the slope of the line through points P_1 and P equals the slope of the line through points P and P_2.

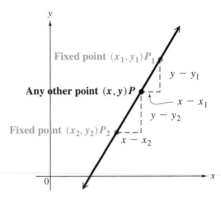

Figure 17

Setting these slopes equal to m gives

$$\frac{y - y_1}{x - x_1} = \frac{y - y_2}{x - x_2} = m,$$

$$\frac{y - y_1}{x - x_1} = m$$

$$y - y_1 = m(x - x_1). \qquad \text{Multiply both sides by } x - x_1.$$

This last equation is the *point-slope form* of the equation of the line. To use this form to write the equation of a line, we need to know the coordinates of a point (x_1, y_1) and the slope of the line, m.

Point-Slope Form

The **point-slope form** of the equation of a line is

Slope
↓
$$y - y_1 = m(x - x_1).$$
↑   ↑
Given point

EXAMPLE 1 Using the Point-Slope Form

Find the equation of the line with slope $\frac{1}{3}$, going through the point $(-2, 5)$.

Use the point-slope form of the equation of a line, with $(x_1, y_1) = (-2, 5)$ and $m = \frac{1}{3}$.

$$y - y_1 = m(x - x_1)$$

$$y - 5 = \frac{1}{3}[x - (-2)] \qquad \text{Let } y_1 = 5, m = \frac{1}{3}, x_1 = -2.$$

$$y - 5 = \frac{1}{3}(x + 2)$$

$$3y - 15 = x + 2 \qquad\qquad \text{Multiply by 3.}$$

or $\qquad\qquad x - 3y = -17 \qquad\qquad \text{Combine terms.}$

In Section 8.1, we defined *standard form* for a linear equation as

$$Ax + By = C.$$

In addition, from now on, let us agree that A, B, and C will be integers with no common factor (except 1) and $A \geq 0$. For example, the final equation found in Example 1, $x - 3y = -17$, is written in standard form.

The definition of "standard form" is not standard from one text to another. Any linear equation can be written in many different (all equally correct) forms. For example, the equation $2x + 3y = 8$ can be written as $2x = 8 - 3y$, $3y = 8 - 2x$, $x + \frac{3}{2}y = 4$, $4x + 6y = 16$, and so on. In addition to writing it in the form $Ax + By = C$ with $A \geq 0$, let us agree that the form $2x + 3y = 8$ is preferred over any multiples of both sides, such as $4x + 6y = 16$. (To write $4x + 6y = 16$ in standard form, divide both sides by 2.)

OBJECTIVE **2** **Write the equation of a line given two points on the line.** To find an equation of a line when two points on a line are known, first use the slope formula to find the slope of the line. Then use the slope with either of the given points in the point-slope form.

EXAMPLE 2 Finding an Equation of a Line When Two Points Are Known

Find an equation of the line through the points $(-4, 3)$ and $(5, -7)$.

First find the slope, using the definition.

$$m = \frac{-7 - 3}{5 - (-4)} = -\frac{10}{9}$$

Use either $(-4, 3)$ or $(5, -7)$ as (x_1, y_1) in the point-slope form of the equation of a line. If $(-4, 3)$ is used, then $-4 = x_1$ and $3 - y_1$.

$$y - y_1 = m(x - x_1) \qquad \text{Point-slope form}$$

$$y - 3 = -\frac{10}{9}[x - (-4)] \qquad \text{Let } y_1 = 3, m = -\frac{10}{9}, x_1 = -4.$$

$$y - 3 = -\frac{10}{9}(x + 4)$$

$$y + 7 = \frac{-10}{9}(x - 5)$$

$$y = \frac{-10 \, x}{9}$$

$$9(y - 3) = -10(x + 4) \qquad \text{Multiply by 9.}$$
$$9y - 27 = -10x - 40 \qquad \text{Distributive property}$$
$$10x + 9y = -13 \qquad \text{Standard form}$$

Verify that if $(5, -7)$ were used, the same equation would be found.

CONNECTIONS

Earlier examples and exercises gave equations that described real data. Now we are able to show how such equations can be found. The process of writing an equation whose graph approximates a set of data is called *data-fitting*. For example, the bar graph shows the number of multimedia personal computers (PCs), in millions, in U.S. homes.

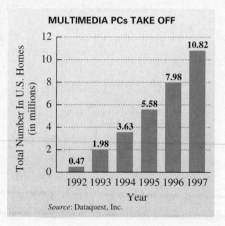

MULTIMEDIA PCs TAKE OFF

Source: Dataquest, Inc.

The data shown in the bar graph increase linearly, that is, a straight line could be drawn through the tops of any two bars that is very close to the top of each bar. This indicates that we can write a linear equation to approximate the data, allowing us to estimate the number of PCs for years other than those shown in the graph. If we let $x = 0$ represent 1992, then $x = 1$ represents 1993, $x = 2$ represents 1994, and so on. The ordered pair for 1993 is $(1, 1.98)$ and for 1996 is $(4, 7.98)$. The slope of the line through these two points is

$$\frac{y_2 - y_1}{x_2 - x_1} = \frac{7.98 - 1.98}{4 - 1} = 2.$$

Using $(1, 1.98)$ and $m = 2$, we get the following equation of the line.

$$y - y_1 = m(x - x_1)$$
$$y - 1.98 = 2(x - 1) \qquad \text{Let } m = 2, x_1 = 1, y_1 = 1.98.$$
$$y - 1.98 = 2x - 2 \qquad \text{Distributive property}$$
$$y = 2x - .02 \qquad \text{Add 1.98.}$$

The equation tells us that the number of multimedia PCs y (in millions) in year x is given by $y = 2x - .02$.

(continued)

CONNECTIONS (CONTINUED)

FOR DISCUSSION OR WRITING
Use the equation found above to predict the number of multimedia PCs in 1998. (Recall that 1992 is represented by $x = 0$.) According to the equation, in what year were there no multimedia PCs? What does the $-.02$ in the equation indicate about the number of multimedia PCs? What might cause this equation to become very inaccurate in future years?

Notice that the point-slope form does not apply to a vertical line, since the slope of a vertical line is undefined. A vertical line through the point (c, d) has equation $x = c$.

A horizontal line has slope 0. From the point-slope form, the equation of a horizontal line through the point (c, d) is

$$y - y_1 = m(x - x_1)$$
$$y - d = 0(x - c) \qquad y_1 = d, x_1 = c$$
$$y - d = 0$$
$$y = d.$$

In summary, horizontal and vertical lines have the following special equations.

Equations of Vertical and Horizontal Lines

The vertical line through (c, d) has equation $\boldsymbol{x = c,}$ and the horizontal line through (c, d) has equation $\boldsymbol{y = d.}$

OBJECTIVE **3** Write the equation of a line given its slope and y-intercept. Suppose a line has slope m and y-intercept $(0, b)$. Using the point-slope form, the equation of the line is

$$y - y_1 = m(x - x_1)$$
$$y - b = m(x - 0) \qquad x_1 = 0, y_1 = b$$
$$y = mx + b. \qquad \text{Add } b.$$

When the equation is solved for y, the coefficient of x is the slope, m, and the constant b is the y-value of the y-intercept. Because this form of the equation shows the slope and the y-intercept, it is called the *slope-intercept form.*

Slope-Intercept Form

The equation of a line with slope m and y-intercept $(0, b)$ is written in **slope-intercept form** as

$$y = mx + b.$$

$\uparrow \qquad \uparrow$
Slope y-intercept is $(0, b)$.

┌ **E X A M P L E   3**   Using the Slope-Intercept Form

Find an equation of the line with slope $-\frac{4}{5}$ and y-intercept $(0, -2)$.

Here $m = -\frac{4}{5}$ and $b = -2$. Substitute these values into the slope-intercept form.

$$y = mx + b$$

$$y = -\frac{4}{5}x - 2$$

O B J E C T I V E   **4**   Find the slope and y-intercept of a line given its equation.   If the equation of a line is written in slope-intercept form, the coefficient of x is the slope and the constant leads to the y-intercept.

┌ **E X A M P L E   4**   Finding the Slope and y-intercept from the Equation

Find the slope and y-intercept of the graph of $3y + 2x = 9$.

Write the equation in slope-intercept form by solving for y.

$$3y + 2x = 9$$

$$3y = -2x + 9$$

$$y = -\frac{2}{3}x + 3 \qquad \text{Slope-intercept form}$$

Slope ———↑   ↑——— y-intercept is $(0, 3)$.

From the slope-intercept form, the slope is $-\frac{2}{3}$ and the y-intercept is $(0, 3)$.

 The slope-intercept form of a linear equation is the most useful for several reasons. Every linear equation (of a nonvertical line) has a *unique* (one and only one) slope-intercept form. In Section 8.4 we will study *linear functions,* which are defined by the slope-intercept form. Also, this is the form we must use when graphing a line with a graphing calculator.

O B J E C T I V E   **5**   Write the equation of a line parallel or perpendicular to a given line.   As mentioned in the previous section, parallel lines have the same slope and perpendicular lines have slopes with product -1.

┌ **E X A M P L E   5**   Finding Equations of Parallel or Perpendicular Lines

Find the equation in slope-intercept form of the line passing through the point $(-4, 5)$, and **(a)** parallel to the line $2x + 3y = 6$; **(b)** perpendicular to the line $2x + 3y = 6$.

(a)   The slope of the graph of $2x + 3y = 6$ can be found by solving for y.

$$2x + 3y = 6$$

$$3y = -2x + 6 \qquad \text{Subtract } 2x \text{ on both sides.}$$

$$y = -\frac{2}{3}x + 2 \qquad \text{Divide both sides by 3.}$$

↑——— Slope

The slope is given by the coefficient of x, so $m = -\frac{2}{3}$. This means that the required equation of the line through $(-4, 5)$ and parallel to $2x + 3y = 6$ also has slope $-\frac{2}{3}$. Now use the point-slope form, with $(x_1, y_1) = (-4, 5)$ and $m = -\frac{2}{3}$.

$$y - 5 = -\frac{2}{3}[x - (-4)] \qquad y_1 = 5, m = -\frac{2}{3}, x_1 = -4$$

$$y - 5 = -\frac{2}{3}(x + 4)$$

$$y - 5 = -\frac{2}{3}x - \frac{8}{3} \qquad \text{Distributive property}$$

$$y = -\frac{2}{3}x - \frac{8}{3} + \frac{15}{3} \qquad \text{Add } 5 = \frac{15}{3} \text{ to both sides.}$$

$$y = -\frac{2}{3}x + \frac{7}{3} \qquad \text{Combine like terms.}$$

We did not clear fractions after the substitution step here because we want the equation in slope-intercept form—that is, solved for y.

(b) To be perpendicular to the line $2x + 3y = 6$, a line must have a slope that is the negative reciprocal of $-\frac{2}{3}$, which is $\frac{3}{2}$. Use the point $(-4, 5)$ and slope $\frac{3}{2}$ in the point-slope form.

$$y - 5 = \frac{3}{2}[x - (-4)] \qquad y_1 = 5, m = \frac{3}{2}, x_1 = -4$$

$$y - 5 = \frac{3}{2}(x + 4)$$

$$y - 5 = \frac{3}{2}x + 6 \qquad \text{Distributive property}$$

$$y = \frac{3}{2}x + 11 \qquad \text{Add 5 to both sides.}$$

A summary of the various forms of linear equations follows.

Summary of Forms of Linear Equations

$y - y_1 = m(x - x_1)$ **Point-slope form**
Slope is m.
Line passes through (x_1, y_1).

$y = mx + b$ **Slope-intercept form**
Slope is m.
y-intercept is $(0, b)$.

Summary of Forms of Linear Equations (continued)

$Ax + By = C$ **Standard form**
$(A \geq 0)$

Slope is $-\dfrac{A}{B}$. $(B \neq 0)$

x-intercept is $\left(\dfrac{C}{A}, 0\right)$. $(A \neq 0)$

y-intercept is $\left(0, \dfrac{C}{B}\right)$. $(B \neq 0)$

$x = c$ **Vertical line**
Undefined slope
x-intercept is $(c, 0)$.

$y = d$ **Horizontal line**
Slope is 0.
y-intercept is $(0, d)$.

O B J E C T I V E 6   Apply concepts of linear equations in two variables to real data.

E X A M P L E 6   Determining a Linear Equation to Describe Real Data

Suppose that it is time to fill up your car with gasoline. You drive into your local station and notice that 89-octane gas is selling for $1.20 per gallon. Experience has taught you that the final price you pay can be determined by the number of gallons you buy multiplied by the price per gallon (in this case, $1.20). As you pump the gas you observe two sets of numbers spinning by: one is the number of gallons you have pumped, and the other is the price you pay for that number of gallons.

The table below uses ordered pairs to illustrate this situation.

Number of Gallons Pumped	Price for This Number of Gallons
0	$0.00 = 0($1.20)
1	$1.20 = 1($1.20)
2	$2.40 = 2($1.20)
3	$3.60 = 3($1.20)
4	$4.80 = 4($1.20)

If we let x denote the number of gallons pumped, then the price y that you pay can be found by the linear equation $y = 1.20x$, where y is in dollars. This is a simple, realistic application of linear equations. Theoretically, there are infinitely many ordered pairs (x, y) that satisfy this equation, but in this application we are limited to nonnegative values for x, since we cannot have a negative number of gallons. There is also a practical maximum value for x in this situation, which varies from one car to another. What do you think determines this maximum value?

In Example 6, the ordered pair (0, 0) satisfied the equation, so the equation has the form $y = mx$, where $b = 0$. If a realistic situation involves an initial charge plus a charge per unit, the equation will have the form $y = mx + b$, where $b \neq 0$.

E X A M P L E   7   Determining an Equation and Interpreting Ordered Pairs That Satisfy It

Suppose that you can get a car wash at the gas station in Example 6 if you pay an additional $3.00.

(a) Write an equation that defines the price you will pay.

Since an additional $3.00 will be charged, you will pay $1.20x + 3.00$ dollars for x gallons of gas and a car wash. Thus, if y represents the price, the equation is $y = 1.2x + 3$. (We deleted the unnecessary zeros.)

(b) Interpret the ordered pairs $(5, 9)$ and $(10, 15)$ in relation to the equation from part (a).

The ordered pair $(5, 9)$ indicates that the price of 5 gallons of gas and a car wash is $9.00. Similarly, $(10, 15)$ indicates that the price of 10 gallons of gas and a car wash is $15.00.

O B J E C T I V E   7   Use a graphing calculator to solve linear equations in one variable.   In the Connections box in Section 8.1, we saw how a graphing calculator is used to graph a linear equation in two variables. Figure 18 shows the graph of $y = -4x + 7$. From the values at the bottom of the screen, we see that when $x = 1.75$, $y = 0$. This means that $x = 1.75$ satisfies the equation $-4x + 7 = 0$, a linear equation in one variable. Therefore, the solution set of $-4x + 7 = 0$ is $\{1.75\}$. This can be verified using the algebraic method, shown in Section 2.1. (The word "Zero" indicates that the x-intercept has been located.)

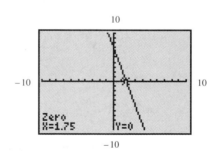

Figure 18

E X A M P L E   8   Solving an Equation with a Graphing Calculator

Use a graphing calculator to solve $-2x - 4(2 - x) = 3x + 4$.

Begin by writing the equation as an equivalent equation with 0 on one side. Do this by subtracting $3x$ and 4 from both sides.

$$-2x - 4(2 - x) - 3x - 4 = 0$$

Then graph $y = -2x - 4(2 - x) - 3x - 4$ and find the x-intercept. Notice that the viewing window must be altered from the one shown in Figure 18 because the x-intercept does not lie in the interval $[-10, 10]$. As seen in Figure 19, the x-intercept of the line is $(-12, 0)$, and thus the solution or zero of the equation is -12. The solution set is $\{-12\}$.

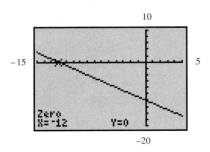

Figure 19

8.2 EXERCISES

1. The following equations all represent the same line. Which one is in standard form as defined in the text?

   (a) $3x - 2y = 5$

   (b) $2y = 3x - 5$

   (c) $\dfrac{3}{5}x - \dfrac{2}{5}y = 1$

   (d) $3x = 2y + 5$

2. Which one of the following equations is in point-slope form?

   (a) $y = 6x + 2$

   (b) $4x + y = 9$

   (c) $y - 3 = 2(x - 1)$

   (d) $2y = 3x - 7$

3. Which of the equations in Exercise 2 is in slope-intercept form?

4. Write the equation $y + 2 = -3(x - 4)$ in slope-intercept form.

5. Write the equation from Exercise 4 in standard form.

6. Write the equation $10x - 7y = 70$ in slope-intercept form.

Match each equation with the graph that it most closely resembles. (Hint: Determining the signs of m and b will help you make your decision.)

7. $y = 2x + 3$

8. $y = -2x + 3$

A.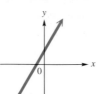

B.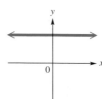

9. $y = -2x - 3$

10. $y = 2x - 3$

C.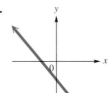

D.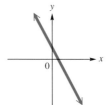

11. $y = 2x$

12. $y = -2x$

E.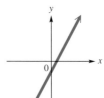

F.

13. $y = 3$

14. $y = -3$

G.

H.

Write the equation in slope-intercept form of the line satisfying the given conditions. See Example 1.

15. Through $(-2, 4)$;   slope $-\dfrac{3}{4}$

16. Through $(-1, 6)$;   slope $-\dfrac{5}{6}$

17. Through $(5, 8)$;   slope -2

18. Through $(12, 10)$;   slope 1

19. Through $(-5, 4)$;   slope $\dfrac{1}{2}$

20. Through $(7, -2)$;   slope $\dfrac{1}{4}$

21. x-intercept $(3, 0)$;   slope 4

22. x-intercept $(-2, 0)$;   slope -5

23. In your own words, list all the forms of linear equations in two variables, and describe when each form should be used.

24. Explain why the point-slope form of an equation cannot be used to find the equation of a vertical line.

Write an equation that satisfies the given conditions.

25. Through $(9, 5)$;   slope 0

26. Through $(-4, -2)$;   slope 0

27. Through $(9, 10)$;   undefined slope

28. Through $(-2, 8)$;   undefined slope

29. Through $(.5, .2)$;   vertical

30. Through $\left(\dfrac{5}{8}, \dfrac{2}{9}\right)$;   vertical

31. Through $(-7, 8)$;   horizontal

32. Through $(2, 7)$;   horizontal

Write the slope-intercept form (if possible) of the equation of the line passing through the two points. See Example 2.

33. (3, 4) and (5, 8)

34. (5, −2) and (−3, 14)

35. (6, 1) and (−2, 5)

36. (−2, 5) and (−8, 1)

37. $\left(-\dfrac{2}{5}, \dfrac{2}{5}\right)$ and $\left(\dfrac{4}{3}, \dfrac{2}{3}\right)$

38. $\left(\dfrac{3}{4}, \dfrac{8}{3}\right)$ and $\left(\dfrac{2}{5}, \dfrac{2}{3}\right)$

39. (2, 5) and (1, 5)

40. (−2, 2) and (4, 2)

41. (7, 6) and (7, −8)

42. (13, 5) and (13, −1)

43. (1, −3) and (−1, −3)

44. (−4, 6) and (5, 6)

Find the equation in slope-intercept form of the line satisfying the given conditions. See Example 3.

45. $m = 5;$ $b = 15$

46. $m = -2;$ $b = 12$

47. $m = -\dfrac{2}{3};$ $b = \dfrac{4}{5}$

48. $m = -\dfrac{5}{8};$ $b = -\dfrac{1}{3}$

49. Slope $\dfrac{2}{5}$; y-intercept (0, 5)

50. Slope $-\dfrac{3}{4}$; y-intercept (0, 7)

Write an equation in slope-intercept form of the line shown in the graph. (Hint: Use the indicated points to find the slope.)

51.

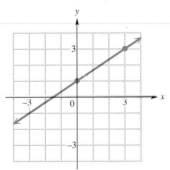

52.

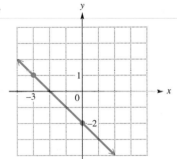

*For each equation **(a)** write in slope-intercept form, **(b)** give the slope of the line, and **(c)** give the y-intercept. See Example 4.*

53. $x + y = 12$

54. $x - y = 14$

55. $5x + 2y = 20$

56. $6x + 5y = 40$

57. $2x - 3y = 10$

58. $4x - 3y = 10$

Write an equation in slope-intercept form of the line satisfying the given conditions. See Example 5.

59. Through (7, 2); parallel to $3x - y = 8$

60. Through (4, 1); parallel to $2x + 5y = 10$

61. Through (−2, −2); parallel to $-x + 2y = 10$

62. Through (1, 3); parallel to $x + 3y = 12$

63. Through (8, 5); perpendicular to $2x - y = 7$

64. Through (2, −7); perpendicular to $5x + 2y = 18$

65. Through (−2, 7); perpendicular to $x = 9$

66. Through (8, 4); perpendicular to $x = -3$

Write an equation in the form y = mx + b for each of the following situations. Then give three ordered pairs that satisfy the equation with x-values of 1, 5, *and* 10. *See Examples 6 and 7.*

67. It costs a $15 flat fee to rent a chain saw, plus $3 per day starting with the first day. Let x represent the number of days rented, so y represents the charge to the user (in dollars).

68. It costs a borrower $.05 per day for an overdue book, plus a flat $.50 charge for all books borrowed. Let x represent the number of days the book is overdue, so y represents the total fine to the tardy user.

🖩 *Many real-world situations can be modeled by straight-line graphs. One way to find the equation of such a line is to use two typical data points from the information provided, with the point-slope form of the equation of a line. Because of the usefulness of the slope-intercept form, such equations are often given in the form y = mx + b. In Exercises 69–72, assume that the situation described can be modeled by a straight-line graph, and use the information to find the slope-intercept form of the equation of the line.*

69. The line graph at the beginning of this chapter shows a straight line that approximates U.S. energy consumption in quadrillion BTUs for the years 1991–1996. Here $x = 1$ represents 1991, $x = 2$ represents 1992, and so on. Use the ordered pairs (1, 80.1) and (6, 89.4) from the graph to write an equation of the line. (*Source:* U.S. Energy Department.)

70. The table gives nuclear power production in quadrillion BTUs per year for three years.

Nuclear Power

Year	Quadrillion BTUs
1993	6.52
1994	6.84
1995	7.19

Source: U.S. Energy Information Administration, *Annual Energy Outlook 1996.*

(a) Use the values for 1993 and 1995, with $x = 3$ representing 1993 and $x = 5$ representing 1995, to write an equation of a line through these points.

(b) Use the equation to estimate production in 1994. How does your result compare to the value for 1994 in the table?

71. Median household income of African Americans increased in recent years, as shown in the bar graph.

(a) Use the information given for the years 1993 and 1997, letting $x = 3$ represent 1993, $x = 7$ represent 1997, and y represent the median income, to write an equation that models median household income.

(b) Use the equation to approximate the median income for 1995. How does your result compare to the actual value, $23,583?

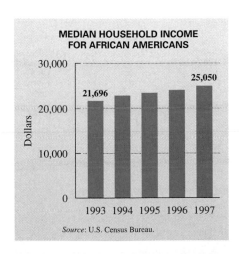

MEDIAN HOUSEHOLD INCOME FOR AFRICAN AMERICANS

Source: U.S. Census Bureau.

72. The bar graph shows median household income for Hispanics.

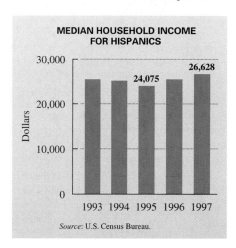

Source: U.S. Census Bureau.

(a) Use the information for the years 1995 and 1997 to write an equation. Let $x = 5$ represent 1995, $x = 7$ represent 1997, and y represent the median income.

(b) Looking at the graph, would you expect the equation from part (a) to give good approximations for 1993 and 1994 income? Would the equation give a reasonable approximation for 1998? Explain.

TECHNOLOGY INSIGHTS (EXERCISES 73–78)

In Exercises 73–76, do the following.

(a) Simplify and rewrite the equation so that the right side is 0. Then replace 0 with y.

(b) The graph of the equation for y is shown with each exercise. Use the graph to determine the solution of the given equation. See Example 8.

(c) Solve the equation using the methods of Chapter 2.

73. $2x + 7 - x = 4x - 2$

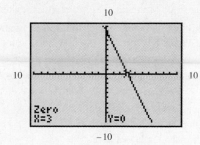

74. $7x - 2x + 4 - 5 = 3x + 1$

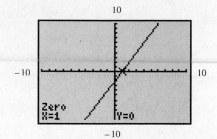

75. $3(2x + 1) - 2(x - 2) = 5$

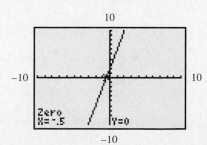

76. $4x - 3(4 - 2x) = 2(x - 3) + 6x + 2$

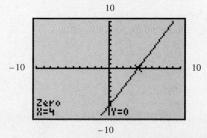

(continued)

TECHNOLOGY INSIGHTS (EXERCISES 73-78) (CONTINUED)

77. The graph of Y_1 is shown in the *standard viewing window.* Which is the only choice that could possibly be the solution of the equation $Y_1 = 0$?
  (a) -15 **(b)** 0
  (c) 5 **(d)** 15

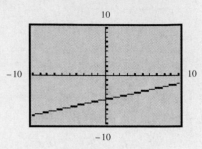

78. (a) Solve the equation $-2(x - 5) = -x - 2$ using the methods of Chapter 2.
  (b) Explain why the standard viewing window of a graphing calculator cannot provide graphical support of the solution found in part (a). What minimum and maximum x-values would make it possible for the solution to be seen?

RELATING CONCEPTS (EXERCISES 79-84)

In Section 2.4 we learned how formulas can be applied to problem solving. In Exercises 79–84, we will see how the formula that relates the Celsius and Fahrenheit temperatures is derived.

Work Exercises 79–84 in order.

79. There is a linear relationship between Celsius and Fahrenheit temperatures. When $C = 0°$, $F = $ _____°, and when $C = 100°$, $F = $ _____°.

80. Think of ordered pairs of temperatures (C, F), where C and F represent corresponding Celsius and Fahrenheit temperatures. The equation that relates the two scales has a straight-line graph that contains the two points determined in Exercise 79. What are these two points?

81. Find the slope of the line described in Exercise 80.

82. Now think of the point-slope form of the equation in terms of C and F, where C replaces x and F replaces y. Use the slope you found in Exercise 81 and one of the two points determined earlier, and find the equation that gives F in terms of C.

83. To obtain another form of the formula, use the equation you found in Exercise 82 and solve for C in terms of F.

84. The equation found in Exercise 82 is graphed on the graphing calculator screen shown here. Observe the display at the bottom, and interpret it in the context of this group of exercises.

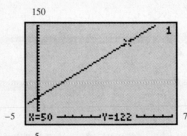

Did you make the connection between Celsius and Fahrenheit temperatures?

8.3 Linear Inequalities in Two Variables

OBJECTIVES

 1 Graph linear inequalities in two variables.

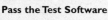 **2** Graph the intersection of two linear inequalities.

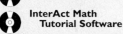 **3** Use a graphing calculator to solve linear inequalities.

FOR EXTRA HELP

📖 **SSG** Sec. 8.3
    SSM Sec. 8.3

💿 **Pass the Test Software**

💿 **InterAct Math**
    Tutorial Software

📼 **Video** 13

OBJECTIVE 1   Graph linear inequalities in two variables.   Linear inequalities in one variable were graphed on the number line in Chapter 3. In this section linear inequalities in two variables are graphed on a rectangular coordinate system.

Linear Inequality

An inequality that can be written as

$$Ax + By < C \qquad \text{or} \qquad Ax + By > C,$$

where A, B, and C are real numbers and A and B are not both 0, is called a **linear inequality in two variables.**

Also, $\leq$ and $\geq$ may replace $<$ and $>$ in the definition.

A line divides the plane into three regions: the line itself and the two half-planes on either side of the line. Recall that the graphs of linear inequalities in one variable are intervals on the number line that sometimes include an endpoint. The graphs of linear inequalities in two variables are *regions* in the real number plane and may include a *boundary line*. The **boundary line** for the inequality $Ax + By < C$ or $Ax + By > C$ is the graph of the *equation $Ax + By = C$*. To graph a linear inequality, follow these steps.

Graphing a Linear Inequality

Step 1   **Draw the boundary.**   Draw the graph of the straight line that is the boundary. Make the line solid if the inequality involves $\leq$ or $\geq$; make the line dashed if the inequality involves $<$ or $>$.

Step 2   **Choose a test point.**   Choose any point not on the line as a test point.

Step 3   **Shade the appropriate region.**   Shade the region that includes the test point if it satisfies the original inequality; otherwise, shade the region on the other side of the boundary line.

EXAMPLE 1   Graphing a Linear Inequality

Graph $3x + 2y \geq 6$.

First graph the straight line $3x + 2y = 6$. The graph of this line, the boundary of the graph of the inequality, is shown in Figure 20 on the next page. The graph of the inequality $3x + 2y \geq 6$ includes the points of the line $3x + 2y = 6$, and either the points *above* the line $3x + 2y = 6$ or the points *below* that line. To decide which, select any point not on the line $3x + 2y = 6$ as a test point. The origin, (0, 0), is often a good choice. Substitute the values from the test point (0, 0) for x and y in the inequality $3x + 2y > 6$.

$$3(0) + 2(0) > 6 \qquad ?$$
$$0 > 6 \qquad \qquad \text{False}$$

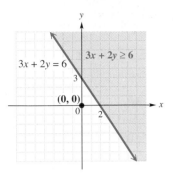

Figure 20

Since the result is false (0, 0) does not satisfy the inequality, and so the solution set includes all points on the other side of the line. This region is shaded in Figure 20. ▪

If the inequality is written in the form $y > mx + b$ or $y < mx + b$, the inequality symbol indicates which half-plane to shade.

If $y > mx + b$, shade *above* the boundary line;

if $y < mx + b$, shade *below* the boundary line.

┌ **E X A M P L E 2**   Graphing a Linear Inequality

Graph $x - 3y > 4$.

First graph the boundary line, shown in Figure 21. The points of the boundary line do not belong to the inequality $x - 3y > 4$ (since the inequality symbol is $>$, not $\geq$). For this reason, the line is dashed. Now solve the inequality for y.

$$x - 3y > 4$$
$$-3y > -x + 4$$
$$y < \frac{x}{3} - \frac{4}{3} \qquad \text{Multiply by } -\tfrac{1}{3}; \text{ change } > \text{ to } <.$$

Because of the *less than* symbol, we should shade *below* the line. As a check, we can choose a test point not on the line, say (1, 2), and substitute for x and y in the original inequality.

$$\mathbf{1} - 3(\mathbf{2}) > 4 \qquad ?$$
$$-5 > 4 \qquad\qquad \text{False}$$

This result agrees with our decision to shade below the line. The solution set, graphed in Figure 21, includes only those points in the shaded half-plane (not those on the line).

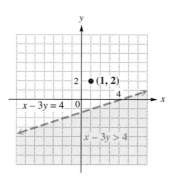

Figure 21

OBJECTIVE 2 Graph the intersection of two linear inequalities. In Section 3.2 we discussed how the words "and" and "or" are used with compound inequalities. In that section, the inequalities had one variable. Those ideas can be extended to include inequalities in two variables. A pair of inequalities joined with the word "and" is interpreted as the intersection of the solutions of the inequalities. The graph of the intersection of two or more inequalities is the region of the plane where all points satisfy all of the inequalities at the same time.

┌ E X A M P L E   3   Graphing the Intersection of Two Inequalities

Graph $2x + 4y \geq 5$ and $x \geq 1$.

To begin, we graph each of the two inequalities $2x + 4y \geq 5$ and $x \geq 1$ separately. The graph of $2x + 4y \geq 5$ is shown in Figure 22(a), and the graph of $x \geq 1$ is shown in Figure 22(b).

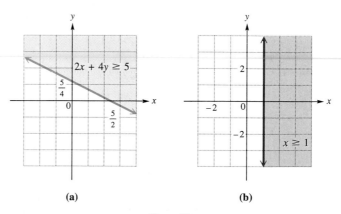

(a)   (b)

Figure 22

In practice, the two graphs in Figure 22 are graphed on the same axes. Then we use heavy shading to identify the intersection of the graphs, as shown in Figure 23 on the next page.

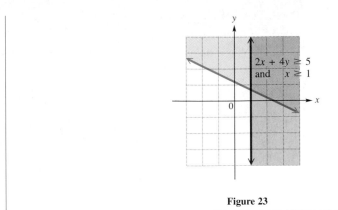

Figure 23

■ **CONNECTIONS**

Many realistic problems involve inequalities. For example, suppose a factory can have *no more than* 200 workers on a shift, but must have *at least* 100 and must manufacture *at least* 3000 units at minimum cost. The managers need to know how many workers should be on a shift in order to produce the required units at minimal cost. *Linear programming* is a method for finding the optimal (best possible) solution that meets all the conditions for such problems. The first step in solving linear programming problems with two variables is to express the conditions (constraints) as inequalities, graph the system of inequalities, and identify the region that satisfies all the inequalities at once.

FOR DISCUSSION OR WRITING

Let *x* represent the number of workers and *y* represent the number of units manufactured.

1. Write three inequalities expressing the conditions in the problem given above.

2. Graph the inequalities from Item 1 and shade the intersection.

3. The cost per worker is $50 per day and the cost to manufacture 1 unit is $100. Write an expression representing the total daily cost, *C*.

4. Find values of *x* and *y* for several points in or on the boundary of the shaded region. Include any "corner points."

5. Of the values of *x* and *y* that you chose in Item 4, which gives the least cost when substituted in the cost equation from Item 3? What does your answer mean in terms of the given problem? Is your answer reasonable? Explain why.

OBJECTIVE **3** Use a graphing calculator to solve linear inequalities. In Section 8.2 we saw that the *x*-intercept of the graph of the line $y = mx + b$ indicates the solution of the equation $mx + b = 0$. We can extend this observation to find solutions of the associated inequalities $mx + b > 0$ and $mx + b < 0$. The solution set of $mx + b > 0$ is the set of all *x*-values for which the graph of $y = mx + b$ is *above* the *x*-axis. (We consider points above because the symbol is $>$.) On the other hand, the solution set of $mx + b < 0$ is the set of all *x*-values for which the graph of $y = mx + b$ is *below* the *x*-axis. (We consider points below because the symbol is $<$.) Therefore, once we know the solution set of the equation and have the graph of the line, we can determine the solution sets of the corresponding inequalities.

In Figure 24, the x-intercept of $y = 3x - 9$ is $(3, 0)$. Therefore, as shown in Section 8.2,

$$\text{the solution set of } 3x - 9 = 0 \text{ is } \{3\}.$$

Because the graph of y lies above the x-axis for x-values greater than 3,

$$\text{the solution set of } 3x - 9 > 0 \text{ is } (3, \infty).$$

Because the graph lies below the x-axis for x-values less than 3,

$$\text{the solution set of } 3x - 9 < 0 \text{ is } (-\infty, 3).$$

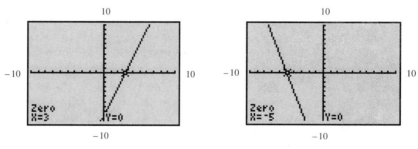

Figure 24   **Figure 25**

Suppose that we wish to solve the equation $-2(3x + 1) = -2x + 18$, and the associated inequalities $-2(3x + 1) > -2x + 18$ and $-2(3x + 1) < -2x + 18$. We begin by rewriting the equation so that the right side is equal to 0:

$$-2(3x + 1) + 2x - 18 = 0.$$

Graphing

$$y = -2(3x + 1) + 2x - 18$$

yields the x-intercept $(-5, 0)$ as shown in Figure 25. The first inequality listed above is equivalent to $y > 0$. Because the line lies *above* the x-axis for x-values less than -5,

$$\text{the solution set of } -2(3x + 1) > -2x + 18 \text{ is } (-\infty, -5).$$

Because the line lies *below* the x-axis for x-values greater than -5,

$$\text{the solution set of } -2(3x + 1) < -2x + 18 \text{ is } (-5, \infty).$$

8.3 EXERCISES

In Exercises 1–4, fill in the first blank with either solid *or* dashed. *Fill in the second blank with* above *or* below.

1. The boundary of the graph of $y \leq -x + 2$ will be a _____ line, and the shading will be _____ the line.

2. The boundary of the graph of $y < -x + 2$ will be a _____ line, and the shading will be _____ the line.

3. The boundary of the graph of $y > -x + 2$ will be a _____ line, and the shading will be _____ the line.

4. The boundary of the graph of $y \geq -x + 2$ will be a _____ line, and the shading will be _____ the line.

5. In your own words, describe the steps used to graph a linear inequality. Use examples.

6. Compare and contrast the steps for graphing a linear inequality in two variables with those used to graph a linear inequality in one variable.

Graph each linear inequality in two variables. See Examples 1 and 2.

7. $x + y \leq 2$ **8.** $x + y \leq -3$

9. $4x - y < 4$ **10.** $3x - y < 3$

11. $x + 3y \geq -2$ **12.** $x + 4y \geq -3$

13. $x + y > 0$ **14.** $x + 2y > 0$

15. $x - 3y \leq 0$ **16.** $x - 5y \leq 0$

17. $y < x$ **18.** $y \leq 4x$

Graph each compound inequality. See Example 3.

19. $x + y \leq 1$ and $x \geq 1$

20. $x - y \geq 2$ and $x \geq 3$

21. $2x - y \geq 2$ and $y < 4$

22. $3x - y \geq 3$ and $y < 3$

23. $x + y > -5$ and $y < -2$

24. $6x - 4y < 10$ and $y > 2$

Use the method described in Section 3.3 to write the absolute value inequality as an "and" statement. Then solve each compound inequality and graph its solution set in the rectangular coordinate plane.

25. $|x| < 3$ **26.** $|y| < 5$

27. $|x + 1| < 2$ **28.** $|y - 3| < 2$

When a compound inequality involves the word or, *the graph is found by graphing each individual inequality and then taking the* union *of the two. For example, the graph of*

$$2x + 4y \geq 5 \qquad or \qquad x \geq 1$$

is shown here. Use this idea to graph each compound inequality.

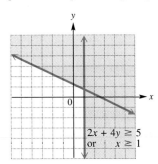

$2x + 4y \geq 5$
or $x \geq 1$

29. $x - y \geq 1$ or $y \geq 2$

30. $x + y \leq 2$ or $y \geq 3$

31. $x - 2 > y$ or $x < 1$

32. $x + 3 < y$ or $x > 3$

33. $3x + 2y < 6$ or $x - 2y > 2$

34. $x - y \geq 1$ or $x + y \leq 4$

TECHNOLOGY INSIGHTS (EXERCISES 35-42)

The graph of a linear equation $y = mx + b$ is shown on a graphing calculator screen, along with the x-value of the x-intercept of the line.

Use the screen to solve (a) $y = 0$, (b) $y < 0$, and (c) $y > 0$. See Objective 3.

35.

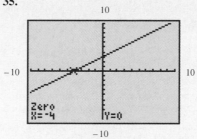

36.

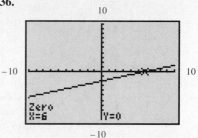

37.

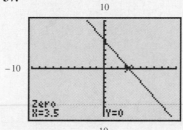

38.

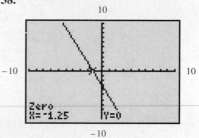

Match each inequality with its calculator-generated graph. (Hint: Use the slope, y-intercept, and inequality symbol in making your choice.)

39. $y \le 3x - 6$

40. $y \ge 3x - 6$

41. $y \le -3x - 6$

42. $y \ge -3x - 6$

A.

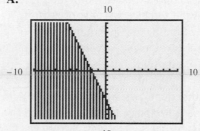

B.

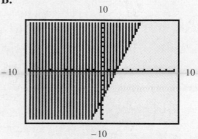

C.

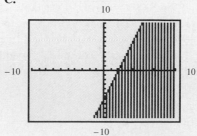

D.

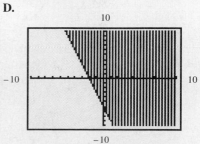

Solve the equation in part (a) and the associated inequalities in parts (b) and (c) using the methods of Chapter 2. Then graph the left side as y in the standard viewing window of a graphing calculator, and explain how the graph supports your answers in parts (a), (b), and (c).

43. (a) $5x + 3 = 0$
   (b) $5x + 3 > 0$
   (c) $5x + 3 < 0$

44. (a) $6x + 3 = 0$
   (b) $6x + 3 > 0$
   (c) $6x + 3 < 0$

45. (a) $-8x - (2x + 12) = 0$
   (b) $-8x - (2x + 12) \geq 0$
   (c) $-8x - (2x + 12) \leq 0$

46. (a) $-4x - (2x + 18) = 0$
   (b) $-4x - (2x + 18) \geq 0$
   (c) $-4x - (2x + 18) \leq 0$

 Work each problem.

47. In Section 8.2 Exercise 69, we found that the equation $y = 1.86x + 78.24$ was a reasonable approximation of U.S. energy consumption y (in quadrillion BTUs) for the years 1991–1996. The years are coded so that $x = 1$ represents 1991 and so on. According to the equation, in what years was U.S. energy consumption greater than or equal to 85 quadrillion BTUs? (*Hint:* y represents U.S. energy consumption. Substitute for y and solve for x.)

48. Refer to Section 8.2 Exercise 70. The equation $y = .335x + 5.515$ approximates nuclear power production (in quadrillion BTUs) in the years 1993–1995. The years are coded so that $x = 3$ represents 1993 and so on. Use an inequality to find the years between 1992 and 1996 when nuclear power production was less than or equal to 7 quadrillion BTUs.

8.4 Functions

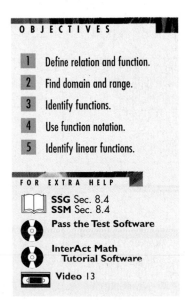

It is often useful to describe one quantity in terms of another; for example, the growth of a plant is related to the amount of light it receives, the demand for a product is related to the price of the product, the cost of a trip is related to the distance traveled, and so on. To represent these corresponding quantities, it is helpful to use ordered pairs.

For example, we can indicate the relationship between the demand for a product and its price by writing ordered pairs in which the first number represents the price and the second number represents the demand. The ordered pair (5, 1000) then could indicate a demand for 1000 items when the price of the item is $5. Since the demand depends on the price charged, we place the price first and the demand second. The ordered pair is an abbreviation for the sentence "If the price is 5 (dollars), then the demand is for 1000 (items)." Similarly, the ordered pairs (3, 5000) and (10, 250) show that a price of $3 produces a demand for 5000 items, and a price of $10 produces a demand for 250 items.

In this example, the demand depends on the price of the item. For this reason, demand is called the *dependent variable*, and price is called the *independent variable*. Generalizing, if the value of the variable y depends on the value of the variable x, then y is the **dependent variable** and x the **independent variable.**

┌── Dependent variable
(x, y)

Independent variable ──┘

OBJECTIVE ▮ Define relation and function. Since related quantities can be written using ordered pairs, the concept of *relation* can be defined as follows.

Relation

A **relation** is a set of ordered pairs.

For example, the sets

$$F = \{(1, 2), (-2, 5), (3, -1)\} \qquad \text{and} \qquad G = \{(-4, 1), (-2, 1), (-2, 0)\}$$

are both relations. In Section 4.2, we introduced a special kind of relation, called a *function*. Functions are very important in mathematics and its applications.

Function

A **function** is a relation in which, for each value of the first component of the ordered pairs, there is *exactly one value* of the second component.

 In Section 4.2, we called the first component the *input* and the second component the *output,* because the second component was determined by the first.

Of the two examples of a relation given above, only set F is a function, because for each x-value, there is exactly one y-value. In set $G,$ the last two ordered pairs have the same x-value paired with two different y-values, so G is a relation, but not a function.

In a function, there is *exactly one* value of the dependent variable, the second component, for each value of the independent variable, the first component. This is what makes functions so important in applications. It would not be as useful, for example, to know a price/demand relationship that gave more than one demand for a given price.

Another way to think of a functional relationship is to think of the independent variable as an input and the dependent variable as an output. A calculator is an input-output machine, for example. To find 8^2, we input 8, touch the squaring key, and see that the output is 64. Inputs and outputs can also be determined from a graph or a table.

A third way to describe a function is to give a rule that tells how to determine the dependent variable for a specific value of the independent variable. The rule may be given in words: the dependent variable is twice the independent variable. Usually the rule is an equation:

$$y = 2x.$$

$$\uparrow \qquad \uparrow$$

Dependent   Independent
variable   variable

This is the most efficient way to define a function.

EXAMPLE 1   Determining Independent and Dependent Variables of a Function

Determine the independent and dependent variables for each of the following functions. Give an example of an ordered pair belonging to the function.

(a) The 1996 Summer Olympics medal winners in men's basketball are {(gold, United States), (silver, Yugoslavia), (bronze, Lithuania)} (*Source:* United States Olympic Committee).

The independent variable (the first component in each ordered pair) is the type of medal; the dependent variable (the second component) is the recipient. Any of the ordered pairs could be given as an example.

(b) An input-output machine that produces square roots

The independent variables (the inputs) are nonnegative real numbers, since the square root of a negative number is not a real number. The dependent variables are their nonnegative square roots. For example, (81, 9) belongs to this function.

(c) The graph in Figure 26, which shows the relationship between the number of gallons of water in a small swimming pool and time in hours

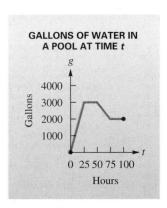

Figure 26

The independent variables are hours, and the dependent variables are the gallons of water in the pool. One ordered pair is (25, 3000).

(d) Petroleum imports in millions of barrels per day for selected years

U.S. Petroleum Imports

Year	Imports
1992	7.89
1993	8.62
1994	9.00
1995	8.83
1996	9.40

Source: U.S. Energy Department.

The independent variables are the years; the dependent variables are the imports. An example of an ordered pair is (1994, 9.00).

(e) $y = 3x + 4$

The independent variable is x, and the dependent variable is y. One ordered pair is $\left(\frac{1}{3}, 5\right)$.

OBJECTIVE `2` Find domain and range.

Domain and Range

In a relation, the set of all values of the independent variable (x) is the **domain;** the set of all values of the dependent variable (y) is the **range.**

E X A M P L E 2   Determining the Domain and Range of a Relation

Give the domain and range of each function in Example 1.

(a) The domain is the type of medal, {gold, silver, bronze}, and the range is the set of winning countries, {United States, Yugoslavia, Lithuania}.

(b) Here, the domain is restricted to nonnegative numbers: $[0, \infty)$. The range is also $[0, \infty)$.

(c) The domain is all possible values of t, the time in hours, which is the interval $[0, 100]$. The range is the number of gallons at time t, the interval $[0, 3000]$.

(d) The domain is the set of years, {1992, 1993, 1994, 1995, 1996}; the range is the set of imports (in millions of barrels per day), {7.89, 8.62, 9.00, 8.83, 9.40}.

(e) In the defining equation (or rule), $y = 3x + 4$, x can be any real number, so the domain is $\{x \mid x \text{ is a real number}\}$ or $(-\infty, \infty)$. (If the equation were $y = \dfrac{1}{x - 1}$, however, we would need to restrict the domain to all real numbers except 1, which would cause a 0 denominator.) Since every real number y can be produced by some value of x, the range is also the set $\{y \mid y \text{ is a real number}\}$ or $(-\infty, \infty)$.

The **graph of a relation** is the graph of its ordered pairs. The graph gives a picture of the relation, which can be used to determine its domain and range.

E X A M P L E 3   Finding Domains and Ranges from Graphs

Give the domain and range of each relation.

(a)

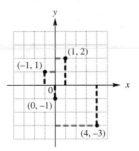

(b)

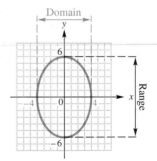

The domain is the set of x-values, $\{-1, 0, 1, 4\}$. The range is the set of y-values, $\{-3, -1, 1, 2\}$.

The x-values of the points on the graph include all numbers between -4 and 4, inclusive. The y-values include all numbers between -6 and 6, inclusive. Using interval notation,

the domain is $[-4, 4]$;
the range is $[-6, 6]$.

(c)

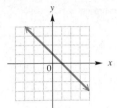

The arrowheads indicate that the line extends indefinitely left and right, as well as up and down. Therefore, both the domain and the range are the set of all real numbers, written $(-\infty, \infty)$.

(d)

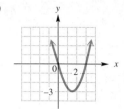

The arrowheads indicate that the graph extends indefinitely left and right, as well as upward. The domain is $(-\infty, \infty)$. Because there is a least y-value, -3, the range includes all numbers greater than or equal to -3, written $[-3, \infty)$.

Relations are often defined by equations, such as $y = 2x + 3$ and $y^2 = x$. It is sometimes necessary to determine the domain of a relation from its equation. In this book, the following agreement on the domain of a relation is assumed.

Agreement on Domain

The domain of a relation is assumed to be all real numbers that produce real numbers when substituted for the independent variable.

To illustrate this agreement, since any real number can be used as a replacement for x in $y = 2x + 3$, the domain of this function is the set of real numbers. As another example, the function defined by $y = \frac{1}{x}$ has all real numbers except 0 as domain, since y is undefined if $x = 0$. In general, the domain of a function defined by an algebraic expression is all real numbers, except those numbers that lead to division by 0 or an even root of a negative number.

OBJECTIVE **3** Identify functions. Most of the relations we have seen in the examples are functions—that is, each x-value corresponds to exactly one y-value. Now we look at ways to determine whether a given relation, defined algebraically, is a function.

In a function each value of x leads to only one value of y, so any vertical line drawn through the graph of a function must intersect the graph in at most one point. This is the **vertical line test for a function.**

Vertical Line Test

If a vertical line intersects the graph of a relation in more than one point, then the relation does not represent a function.

For example, the graph shown in Figure 27(a) is not the graph of a function, since a vertical line can intersect the graph in more than one point, while the graph in Figure 27(b) does represent a function.

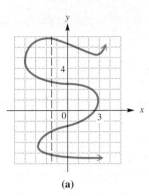

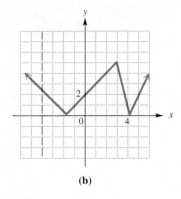

(a)   (b)

Figure 27

The vertical line test is a simple method for identifying a function defined by a graph. It is more difficult to decide whether a relation defined by an equation is a function. The next example gives some hints that may help.

EXAMPLE 4 Identifying Functions

Decide whether each of the following defines a function and give the domain.

(a) $y = \sqrt{2x - 1}$

Here, for any choice of x in the domain, there is exactly one corresponding value for y (the radical is a nonnegative number), so this equation defines a function. Since the quantity under the radical sign cannot be negative,

$$2x - 1 \geq 0$$
$$2x \geq 1$$
$$x \geq \frac{1}{2}.$$

The domain is $\left[\frac{1}{2}, \infty\right)$.

(b) $y^2 = x$

The ordered pairs $(16, 4)$ and $(16, -4)$ both satisfy this equation. Since one value of x, 16, corresponds to two values of y, 4 and -4, this equation does not define a function. Because x is equal to the square of y, the values of x must always be nonnegative. The domain of the relation is $[0, \infty)$.

(c) $y \leq x - 1$

By definition, y is a function of x if a value of x leads to exactly one value of y. In this example, a particular value of x, say 1, corresponds to many values of y. The ordered pairs $(1, 0)$, $(1, -1)$, $(1, -2)$, $(1, -3)$, and so on, all satisfy the inequality. For this reason, an inequality does not define a function. Any number can be used for x, so the domain is the set of real numbers $(-\infty, \infty)$.

(d) $y = \dfrac{5}{x - 1}$

Given any value of x in the domain, we find y by subtracting 1, then dividing the result into 5. This process produces exactly one value of y for each value in the domain, so this equation defines a function. The domain includes all real numbers except those that make the denominator 0. We find these numbers by setting the denominator equal to 0 and solving for x.

$$x - 1 = 0$$
$$x = 1$$

Thus, the domain includes all real numbers except 1. In interval notation this is written as

$$(-\infty, 1) \cup (1, \infty).$$

In summary, three variations of the definition of function are given here.

> **Variations of the Definition of Function**
>
> 1. A **function** is a relation in which, for each value of the first component of the ordered pairs, there is exactly one value of the second component.
> 2. A **function** is a set of ordered pairs in which no first component is repeated.
> 3. A **function** is a rule or correspondence that assigns exactly one range value to each domain value.

OBJECTIVE **4** Use function notation. When a function f is defined with a rule or an equation using x and y for the independent and dependent variables, we say "y is a function of x" to emphasize that y *depends on* x. We use the notation

$$y = f(x)$$

to express this. (In this special notation the parentheses do not indicate multiplication.) The letter f stands for *function*. For example, if $y = 2x - 7$, we write

$$f(x) = 2x - 7.$$

When you see the notation $f(x)$, remember that it is just another name for the dependent variable y. This **function notation** is useful for simplifying certain statements. For example, if $y = 9x - 5$, then replacing x with 2 gives

$$y = 9 \cdot 2 - 5$$
$$= 18 - 5$$
$$= \mathbf{13}.$$

The statement "if $x = 2$, then $y = \mathbf{13}$" is abbreviated with function notation as

$$f(2) = \mathbf{13}.$$

Read $f(2)$ as "f of 2" or "f at 2." Also, $f(0) = 9 \cdot 0 - 5 = -5$, and $f(-3) = 9(-3) - 5 = -32$.

These ideas and the symbols used to represent them can be explained as follows.

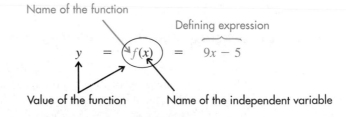

The symbol $f(x)$ *does not* indicate "f times x," but represents the y-value for the indicated x-value. As shown above, $f(2)$ is the y-value that corresponds to the x-value 2.

EXAMPLE 5 Using Function Notation

Let $f(x) = -x^2 + 5x - 3$. Find the following.

(a) $f(2)$

Replace x with 2.

$$f(2) = -2^2 + 5 \cdot 2 - 3 = -4 + 10 - 3 = 3$$

(b) $f(-1)$

$$f(-1) = -(-1)^2 + 5(-1) - 3 = -1 - 5 - 3 = -9$$

(c) $f(q)$

Replace x with q.

$$f(q) = -q^2 + 5q - 3$$

The replacement of one variable with another is important in later courses.

Sometimes letters other than f, such as g, h, or capital letters F, G, and H are used to name functions.

EXAMPLE 6 Using Function Notation

Let $g(x) = 2x + 3$. Find and simplify the following.

(a) $g(a + 1)$

Replace x with $a + 1$.

$$g(a + 1) = 2(a + 1) + 3 = 2a + 2 + 3 = 2a + 5$$

(b) $G\left(\dfrac{1}{b + 4}\right)$

$$G\left(\frac{1}{b + 4}\right) = 2\left(\frac{1}{b + 4}\right) + 3 = \left(\frac{2}{b + 4}\right) + 3$$

Replacing the variable x with an algebraic expression, like $a + 1$ in Example 6(a), is called **composition of functions.** The result is called a **composite function.**

EXAMPLE 7 Forming a Composite Function

(a) Write $f(x) = 4x - 1$ and $g(x) = x^2$ as the composite function $g[f(x)]$.

First replace $f(x)$ with $4x - 1$. Then use the fact that $g(x) = x^2$.

$$g[f(x)] = g(4x - 1) = (4x - 1)^2$$

(b) Find $f[g(x)]$ for the functions in part (a).

Work from the inside out. Replace $g(x)$ with x^2. Then use function f.

$$f[g(x)] = f(x^2) = 4x^2 - 1$$

Notice that $g[f(x)] \neq f[g(x)]$. This is generally true.

If a function is defined by an equation with x and y, not with function notation, use the following steps to find $f(x)$.

Finding an Expression for $f(x)$

If an equation that defines a function is given with x and y, to find $f(x)$:

1. solve the equation for y;

2. replace y with $f(x)$.

EXAMPLE 8   Writing Equations Using Function Notation

Rewrite each equation using function notation; then find $f(-2)$ and $f(a)$.

(a) $y = x^2 + 1$

This equation is already solved for y. Since $y = f(x)$,

$$f(x) = x^2 + 1.$$

To find $f(-2)$, we let $x = -2$:

$$f(-2) = (-2)^2 + 1$$
$$= 4 + 1$$
$$= 5.$$

We find $f(a)$ by letting $x = a$: $f(a) = a^2 + 1$.

(b) $x - 4y = 5$

First solve $x - 4y = 5$ for y. Then replace y with $f(x)$.

$$x - 4y = 5$$
$$x - 5 = 4y$$
$$y = \frac{x-5}{4} \quad \text{so} \quad f(x) = \frac{1}{4}x - \frac{5}{4}$$

Now find $f(-2)$ and $f(a)$

$$f(-2) = \frac{1}{4}(-2) - \frac{5}{4} = -\frac{7}{4}$$

and

$$f(a) = \frac{1}{4}a - \frac{5}{4}.$$

OBJECTIVE 5 Identify linear functions. Our first two-dimensional graphing was of straight lines. Linear equations (except for $x = c$) define *linear functions*.

Linear Function

A function that can be written in the form

$$f(x) = mx + b$$

for real numbers m and b is a **linear function.**

Recall from Section 8.2 that m is the slope of the line and $(0, b)$ is the y-intercept. A linear function of the form $f(x) = d$ is sometimes called a **constant function.** The domain of any linear function is $(-\infty, \infty)$. The range of a nonconstant linear function is $(-\infty, \infty)$, while the range of the constant function $f(x) = d$ is $\{d\}$.

In later chapters of this book, we will learn about several other types of functions.

8.4 EXERCISES

To work the following exercises, refer to Examples 1 and 2.

1. In your own words, define a function and give an example.

2. In your own words, define the domain of a function and give an example.

3. In an ordered pair of a relation, is the first element the independent or the dependent variable?

For each of the following relations, decide whether or not it is a function, and give the domain and range. See Examples 1–4. Use the vertical line test in Exercises 8 and 11–14.

4. $\{(1, 1), (1, -1), (2, 4), (2, -4), (3, 9), (3, -9)\}$

 5. $\{(2, 5), (3, 7), (4, 9), (5, 11)\}$

6. The set containing the top five producers of hydroelectric power and the amount each produces, in billion kilowatt-hours, is {(Canada, 319.1), (United States, 268.2), (Brazil, 225.0), (Russia, 173.4), (China, 143.1)}. (*Source:* U.S. Department of Energy, *International Energy Annual,* 1993.)

7. An input-output machine accepts positive real numbers as input, and outputs both their positive and negative roots.

8.
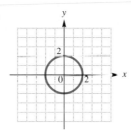

9. **Crude Oil Prices and Cost per Gallon of Gas at the Pump (1996)**

Unleaded regular	$1.22
Unleaded premium	$1.44
Crude oil	$ .21

Source: U.S. Energy Department.

10. $x = |y|$

11.
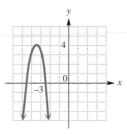

12.
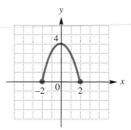

13.
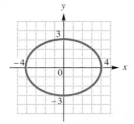

14.
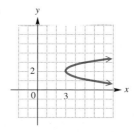

Decide whether the given relation defines y as a function of x. Give the domain. See Example 4.

15. $y = x^2$ **16.** $y = x^3$ **17.** $x = y^2$ **18.** $x = y^4$

19. $x + y < 4$ **20.** $x - y < 3$ **21.** $y = \sqrt{x}$ **22.** $y = -\sqrt{x}$

23. $xy = 1$ **24.** $xy = -3$ **25.** $y = \sqrt{4x + 2}$ **26.** $y = \sqrt{9 - 2x}$

27. $y = \dfrac{2}{x - 9}$ **28.** $y = \dfrac{-7}{x - 16}$

29. Refer to Example 1, Figure 26, to answer the questions.
   (a) What numbers are possible values of the dependent variable?
   (b) For how long is the water level increasing? Decreasing?
   (c) How many gallons are in the pool after 90 hours?
   (d) Call this function f. What is $f(0)$? What does it mean in this example?

30. The graph shows the daily megawatts of electricity used on a record-breaking summer day in Sacramento, California.

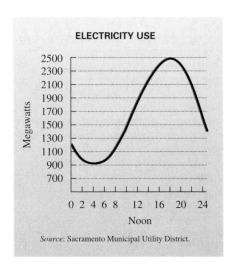

(a) Is this the graph of a function?
(b) What is the domain?
(c) Estimate the number of megawatts at 8 A.M.
(d) At what time was the most electricity used? The least electricity?

31. Give an example of a function from everyday life. (*Hint:* Fill in the blanks: _____ depends on _____, so _____ is a function of _____.)

32. Choose the correct response. The notation $f(3)$ means
   (a) the variable f times 3 or $3f$.
   (b) the value of the dependent variable when the independent variable is 3.
   (c) the value of the independent variable when the dependent variable is 3.
   (d) f equals 3.

Let $f(x) = -3x + 4$ and $g(x) = -x^2 + 4x + 1$. Find the following. See Examples 5–7.

33. $f(0)$ **34.** $f(-3)$ **35.** $f(-x)$ **36.** $f(x + 2)$

37. $g(10)$ **38.** $g(-1.5)$ **39.** $g\left(\frac{1}{2}\right)$ **40.** $g(k)$

41. $g(2p)$ **42.** $f[g(1)]$ **43.** $g[f(1)]$ **44.** $f[g(x)]$

45. Compare the answers to Exercises 42 and 43. Do you think that $f[g(x)]$ is, in general, equal to $g[f(x)]$?

46. Make up two linear functions f and g such that $f[g(2)] = 4$. (There are many ways to do this.)

An equation that defines y as a function of x is given. (a) Solve for y in terms of x, and replace y with the function notation f(x). (b) Find f(3). See Example 8.

47. $x + 3y = 12$ **48.** $x - 4y = 8$ **49.** $y + 2x^2 = 3$

50. $y - 3x^2 = 2$ **51.** $4x - 3y = 8$ **52.** $-2x + 5y = 9$

53. Fill in the blanks with the correct responses.

The equation $2x + y = 4$ has a straight _____ as its graph. One point that lies on the line is (3, _____). If we solve the equation for y and use function notation, we have a linear function $f(x) = $ _____. For this function, $f(3) = $ _____, meaning that the point (_____, _____) lies on the graph of the function.

54. Which one of the following defines a linear function?

(a) $y = \dfrac{x - 5}{4}$ (b) $y = \dfrac{1}{x}$ (c) $y = x^2$ (d) $y = \sqrt{x}$

Sketch the graph of each linear function. Give the domain.

55. $f(x) = -2x + 5$ **56.** $g(x) = 4x - 1$ **57.** $h(x) = \dfrac{1}{2}x + 2$

58. $F(x) = -\dfrac{1}{4}x + 1$ **59.** $G(x) = 2x$ **60.** $H(x) = -3x$

61. $g(x) = -4$ **62.** $f(x) = 5$

63. **(a)** Suppose that a taxicab driver charges $1.50 per mile. Fill in the chart with the correct response for the price $f(x)$ she charges for a trip of x miles.

x	$f(x)$
0	
1	
2	
3	

(b) The linear function that gives a rule for the amount charged is $f(x) = $ _____.
(c) Graph this function for the domain {0, 1, 2, 3}.

64. Suppose that a package weighing x pounds costs $f(x)$ dollars to mail to a given location, where

$$f(x) = 2.75x.$$

(a) What is the value of $f(3)$?
(b) In your own words, describe what 3 and the value $f(3)$ mean in part (a), using the terminology *independent variable* and *dependent variable*.

Forensic scientists use the lengths of certain bones to calculate the height of a person. Two bones often used are the tibia (t), the bone from the ankle to the knee, and the femur (r), the bone from the knee to the hip socket. A person's height (h) is determined from the lengths of these bones using functions defined by the following formulas. All measurements are in centimeters.

For men: $h(r) = 69.09 + 2.24r$ or $h(t) = 81.69 + 2.39t$

For women: $h(r) = 61.41 + 2.32r$ or $h(t) = 72.57 + 2.53t$

65. Find the height of a man with a femur measuring 56 centimeters.

66. Find the height of a man with a tibia measuring 40 centimeters.

67. Find the height of a woman with a femur measuring 50 centimeters.

68. Find the height of a woman with a tibia measuring 36 centimeters.

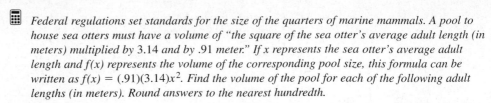

Federal regulations set standards for the size of the quarters of marine mammals. A pool to house sea otters must have a volume of "the square of the sea otter's average adult length (in meters) multiplied by 3.14 and by .91 meter." If x represents the sea otter's average adult length and f(x) represents the volume of the corresponding pool size, this formula can be written as $f(x) = (.91)(3.14)x^2$. Find the volume of the pool for each of the following adult lengths (in meters). Round answers to the nearest hundredth.

69. .8   **70.** 1.0   **71.** 1.2   **72.** 1.5

73. The linear function $f(x) = -183x + 40,034$ is a model for the number of U.S. post offices from 1990–1995, where $x = 0$ corresponds to 1990, $x = 1$ corresponds to 1991, and so on. Use this model to give the approximate number of post offices during the following years. (*Source:* U.S. Postal Service, *Annual Report of the Postmaster General and Comprehensive Statement on Postal Operations.*)
  (a) 1991
  (b) 1993
  (c) 1995
  (d) The graphing calculator screen shows a portion of the graph of $y = f(x)$ with the coordinates of a point on the graph displayed at the bottom of the screen. Interpret the meaning of the display in the context of this application.

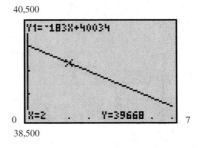

74. The linear function $f(x) = -6324x + 305,294$ is a model for U.S. defense budgets in millions of dollars from 1992–1996, where $x = 0$ corresponds to 1990, $x = 2$ corresponds to 1992, and so on. Use this model to approximate the defense budget for the following years. (*Source:* U.S. Office of Management and Budget.)
  (a) 1993
  (b) 1995
  (c) 1996
  (d) A portion of the graph of $y = f(x)$ is shown in the graphing calculator screen, with the coordinates of a point displayed at the bottom of the screen. Interpret the meaning of the display in the context of this application.

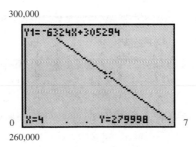

TECHNOLOGY INSIGHTS (EXERCISES 75-78)

75. The graphing calculator screen shows the graph of a linear function $y = f(x)$, along with the display of coordinates of a point on the graph. Use function notation to write what the display indicates.

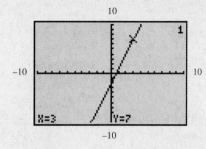

76. The table was generated by a graphing calculator for a linear function $Y_1 = f(x)$. Use the table to answer.
  (a) What is $f(2)$?
  (b) If $f(x) = -3.7$, what is the value of x?
  (c) What is the slope of the line?
  (d) What is the y-intercept of the line?
  (e) Find the expression for $f(x)$.

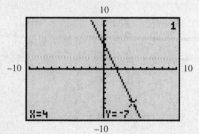

77. The two screens show the graph of the same linear function $y = f(x)$. Find the expression for $f(x)$.

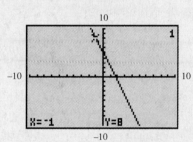

78. The formula for converting Celsius to Fahrenheit is $F = 1.8C + 32$. If we graph this formula as $y = f(x) = 1.8x + 32$ with a graphing calculator, we obtain the accompanying screen. The point $(-40, -40)$ lies on the graph, as indicated by the display. Interpret the meaning of this in the context of this exercise.

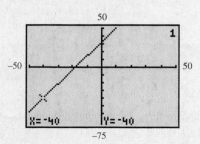

CHAPTER 8 GROUP ACTIVITY

Choosing an Energy Source (or How to Get Your Water Hot)

Objective: Write and graph linear functions that model given data.

There are many different ways to heat water. In this activity you will look at three different energy sources that may be used to provide heat for a 40-gallon home water tank.

Have each student in your group choose one of the three types of water heaters listed in the table below.

Type of Hot Water Heater	Size in Gallons	Price	Operating Cost per Month (manufacturer's estimate)	Hot Water Temperature
Kenmore Economizer 6 —Electric	40	$139.99	$35.00	120°–130°
Kenmore Economizer 6 —Natural Gas	40	$139.99	$13.25	120°–130°
Sunbather Water Heater —Solar	40	$950.00	$0.00	*

Source: Jade Mountain 1999.

---

*No temperature listed but the ad says "Best for warm climates, preheating water, summer only in cold places, or when hot water needed only in afternoons and evenings."

A. Using data for the water heater you selected, write a linear equation that represents total cost *y* of heating water with respect to time. Let *x* represent number of months.

B. Graph your equation using domain [0, 60] and range [0, 1000].

C. As a group, compare the graphs of your equations.
   1. What are the *y*-intercepts?
   2. How do the slopes compare? Which is the steepest? Which has a zero slope?

D. Discuss other factors to consider when choosing each type of water heater. Which of these water heaters would you choose to heat your home?

CHAPTER 8 SUMMARY

KEY TERMS

8.1 ordered pair
origin
x-axis
y-axis
rectangular
    (Cartesian)
    coordinate system

plot
coordinate
quadrant
first-degree equation
linear equation in two
    variables
standard form

x-intercept
y-intercept
rise
run
slope
8.2 point-slope form
slope-intercept form

8.3 linear inequality in
    two variables
boundary line
8.4 dependent variable
independent variable
relation
function

KEY TERMS

domain	vertical line test for a	composition of	linear function
range	function	functions	constant function
graph of a relation	function notation	composite function	

NEW SYMBOLS

$f(x)$ function of x (read "f of x") $f[g(x)]$ composite function

TEST YOUR WORD POWER

See how well you have learned the vocabulary in this chapter. Answers, with examples, are given at the bottom of the page.

1. A **linear equation in two variables** is an equation that can be written in the form
(a) $Ax + By < C$
(b) $ax = b$
(c) $y = x^2$
(d) $Ax + By = C$.

2. The **slope** of a line is
(a) the measure of the run over the rise of the line
(b) the distance between two points on the line
(c) the ratio of the change in y to the change in x along the line
(d) the horizontal change compared to the vertical change of two points on the line.

3. In a relationship between two variables x and y, the **independent variable** is
(a) x, if x depends on y
(b) x, if y depends on x

(c) either x or y
(d) the larger of x and y.

4. In a relationship between two variables x and y, the **dependent variable** is
(a) y, if y depends on x
(b) y, if x depends on y
(c) either x or y
(d) the smaller of x and y.

5. A **relation** is
(a) a set of ordered pairs
(b) the ratio of the change in y to the change in x along a line
(c) the set of all possible values of the independent variable
(d) all the second elements of a set of ordered pairs.

6. A **function** is
(a) the numbers in an ordered pair
(b) a set of ordered pairs in which each x-value corresponds to exactly one y-value

(c) a pair of numbers written between parentheses in which order matters
(d) the set of all ordered pairs that satisfy an equation.

7. The **domain** of a function is
(a) the set of all possible values of the dependent variable y
(b) a set of ordered pairs
(c) the difference between the x-values
(d) the set of all possible values of the independent variable x.

8. The **range** of a function is
(a) the set of all possible values of the dependent variable y
(b) a set of ordered pairs
(c) the difference between the y-values
(d) the set of all possible values of the independent variable x.

Answers to Test Your Word Power

1. (d) *Examples:* $3x + 2y = 6$, $x = y - 7$, $4x = y$ 2. (c) *Example:* The line through $(3, 6)$ and $(5, 4)$ has slope $\dfrac{4-6}{5-3} = \dfrac{-2}{2} = -1$.
3. (b) *Example:* See Item 4, which follows.   4. (a) *Example:* When borrowing money, the amount you borrow (independent variable) determines the size of your payments (dependent variable).   5. (a) *Example:* The set $\{(2, 0), (4, 3), (6, 6), (8, 9)\}$ defines a relation.   6. (b) The relation given in Item 5 is a function since the x-value of each ordered pair corresponds to exactly one y-value.   7. (d) *Example:* In the function in Item 5 above, the domain is the set of x-values, $\{0, 3, 6, 9\}$.   8. (a) *Example:* In the function in Item 5 above, the range is the set of y-values, $\{2, 4, 6, 8\}$.

QUICK REVIEW

CONCEPTS	EXAMPLES

8.1 REVIEW OF GRAPHS AND SLOPES OF LINES

Finding Intercepts

To find the x-intercept, let $y = 0$.
To find the y-intercept, let $x = 0$.

Finding Slope
If $x_2 \neq x_1$, then

$$m = \frac{\text{change in } y}{\text{change in } x} = \frac{y_2 - y_1}{x_2 - x_1}.$$

A vertical line has undefined slope.

A horizontal line has 0 slope.

Parallel lines have equal slopes.

The slopes of perpendicular lines are negative reciprocals with a product of -1.

The graph of $2x + 3y = 12$ has

$$\begin{array}{ll} x\text{-intercept} & (6, 0) \\ \text{and} \qquad y\text{-intercept} & (0, 4). \end{array}$$

For $2x + 3y = 12$,

$$m = \frac{4 - 0}{0 - 6} = -\frac{2}{3}.$$

$x = 3$ has undefined slope.

$y = -5$ has $m = 0$.

$$\begin{array}{ll} y = 2x + 3 & 4x - 2y = 6 \\ m = 2 & -2y = -4x + 6 \\ & y = 2x - 3 \\ & m = 2 \end{array}$$

These lines are **parallel**.

$$\begin{array}{ll} y = 3x - 1 & x + 3y = 4 \\ m = 3 & 3y = -x + 4 \\ & y = -\frac{1}{3}x + \frac{4}{3} \\ & m = -\frac{1}{3} \end{array}$$

These lines are **perpendicular.**

8.2 EQUATIONS OF LINES

Point-Slope Form
$y - y_1 = m(x - x_1)$

Slope-Intercept Form
$y = mx + b$

Vertical Line
$x = c$

Horizontal Line
$y = d$

Standard Form
$Ax + By = C$

$y - 3 = 4(x - 5)$ (5, 3) is on the line, $m = 4$.

$y = 2x + 3$ $m = 2$, y-intercept is (0, 3).

$x = -1$

$y = 4$

$2x - 5y = 8$

8.3 LINEAR INEQUALITIES IN TWO VARIABLES

Graphing a Linear Inequality
Step 1   Draw the graph of the line that is the boundary. Make the line solid if the inequality involves $\leq$ or $\geq$; make the line dashed if the inequality involves $<$ or $>$.

Step 2   Choose any point not on the line as a test point.

Graph $2x - 3y \leq 6$.
Draw the graph of $2x - 3y = 6$. Use a solid line because of $\leq$.

Choose (1, 2).

$$2(1) - 3(2) = 2 - 6 \leq 6 \qquad \text{True}$$

CONCEPTS	EXAMPLES
Step 3 Shade the region that includes the test point if the test point satisfies the original inequality; otherwise, shade the region on the other side of the boundary line.	Shade the side of the line that includes (1, 2). 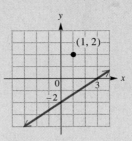

8.4 FUNCTIONS

A function is a set of ordered pairs such that for each first component there is one and only one second component. The set of first components is called the domain, and the set of second components is called the range.	$y = f(x) = x^2$ defines a function *f*, with domain $(-\infty, \infty)$ and range $[0, \infty)$.
To evaluate a function using function notation (that is, $f(x)$ notation) for a given value of *x*, substitute the value wherever *x* appears.	If $f(x) = x^2 - 7x + 12$, then $$f(1) = 1^2 - 7(1) + 12 = 6.$$
To write the equation that defines a function in function notation, solve the equation for *y*.	Given: $2x + 3y = 12$. $$3y = -2x + 12$$ $$y = -\frac{2}{3}x + 4$$
Then replace *y* with $f(x)$.	$$f(x) = -\frac{2}{3}x + 4$$

CHAPTER 8 REVIEW EXERCISES

[8.1]   *Complete the table of ordered pairs for each equation. Then graph the equation.*

1. $3x + 2y - 10$

x	y
0	
	0
2	
	-2

2. $x - y = 8$

x	y
2	
	-3
3	
	-2

Find the x- and y-intercepts and then graph each equation.

3. $4x - 3y = 12$

4. $5x + 7y = 28$

5. $2x + 5y = 20$

6. $x - 4y = 8$

7. Explain how the signs of the *x*- and *y*-coordinates of a point determine the quadrant in which the point lies.

Find the slope of each line.

8. Through $(-1, 2)$ and $(4, -5)$ **9.** Through $(0, 3)$ and $(-2, 4)$

10. $y = 2x + 3$ **11.** $3x - 4y = 5$

12. $x = 5$ **13.** Parallel to $3y = 2x + 5$

14. Perpendicular to $3x - y = 4$ **15.** Through $(-1, 5)$ and $(-1, -4)$

16. The line containing the points shown in this table

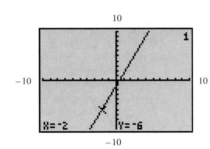

17. The line pictured in the two screens

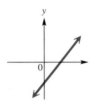 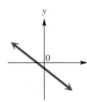 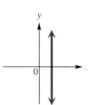

Tell whether each line has positive, negative, zero, or undefined slope.

18.   **19.**   **20.**   **21.**

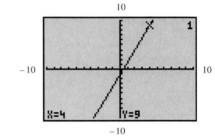

22. If a walkway rises 2 feet for every 10 feet on the horizontal, which of the following express its slope (or grade)? (There are several correct choices.)

(a) .2   **(b)** $\dfrac{2}{10}$ **(c)** $\dfrac{1}{5}$ **(d)** 20%

(e) 5   **(f)** $\dfrac{20}{100}$ **(g)** 500%   **(h)** $\dfrac{10}{2}$

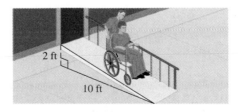

23. If the pitch of a roof is $\frac{1}{4}$, how many feet in the horizontal direction correspond to a rise of 3 feet?

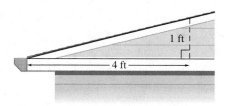

24. Family income in the United States has steadily increased for many years (primarily due to inflation). In 1970 the median family income was about $10,000 a year. In 1995 it was about $41,000 a year. Find the average rate of change of median family income over that period. (*Source:* Bureau of the Census.)

[8.2] *Write an equation in slope-intercept form for the line, if possible.*

25. Slope $-\frac{1}{3}$, y-intercept $(0, -1)$

26. Slope 0, y-intercept $(0, -2)$

27. Slope $-\frac{4}{3}$,   through $(2, 7)$

28. Slope 3,   through $(-1, 4)$

29. Vertical,   through $(2, 5)$

30. Through $(2, -5)$ and $(1, 4)$

31. Through $(-3, -1)$ and $(2, 6)$

32. Parallel to $4x - y = 3$ and through $(7, -1)$

33. Perpendicular to $2x - 5y = 7$ and through $(4, 3)$

34. The line containing the points in the table accompanying Exercise 16

35. The line pictured in Exercise 17

The equation $y = 2.1x + 230$ is a model that was recently used to predict the U.S. population, where $x = 0$ represents the year 1980 and y is the population in millions.

36. According to this equation, what was the population in 1996?

37. According to this equation, in what year would the population have reached 247 million?

[8.3] *Graph the solution set of each inequality or compound inequality.*

38. $3x - 2y \le 12$

39. $5x - y > 6$

40. $x \ge 2$

41. $2x + y \le 1$ and $x \ge 2y$

[8.4] *In Exercises 42–44, give the domain and range of each relation. Identify any functions.*

42. $\{(-4, 2), (-4, -2), (1, 5), (1, -5)\}$

43.

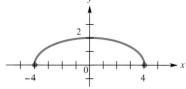

44. The number of small offices/home offices in 1996 for the top five states were $\{$(California, 71,266), (New York, 50,101), (Texas, 48,010), (Pennsylvania, 42,142), (Washington, 38,240)$\}$. (*Source:* Dun & Bradstreet's Cottage Industry File.)

Work the following problems.

45. The top 6 countries in commercial nuclear power generation in 1995 are shown in the table.

Country	U.S.	France	Japan	Germany	Canada	Russia
Billion kilowatt-hours	706	377	286	154	100	99

Source: U.S. Bureau of the Census.

(a) Suppose, in a later year, the number of billion kilowatt-hours generated in Canada and Russia was the same, 101, while the other countries continued to generate the amounts shown in the table. For the relationship between the country and the amount of power generated to be a function, which must be the independent variable: the country or the number of kilowatt-hours? Explain why.

(b) Give the domain and range of the function of part (a).

46. Refer to the line graph in the chapter introduction. Is it the graph of a function? What type of function? Give two ordered pairs that are on the line. What are the domain and range?

Determine whether each equation or inequality defines y as a function of x. Give the domain in each case. Identify any linear functions.

47. $y = 3x - 3$ **48.** $y < x + 2$ **49.** $y = |x - 4|$

50. $y = \sqrt{4x + 7}$ **51.** $x = y^2$ **52.** $y = \dfrac{7}{x - 6}$

53. Explain the test that allows us to determine whether a graph is that of a function.

Given $f(x) = -2x^2 + 3x - 6$, find each function value or expression.

54. $f(0)$ **55.** $f(2.1)$ **56.** $f\left(-\dfrac{1}{2}\right)$

57. $f(k)$ **58.** $f[f(0)]$ **59.** $f(2p)$

60. The equation $2x^2 - y = 0$ defines y as a function of x. Rewrite it using $f(x)$ notation, and find $f(3)$.

61. Suppose that $2x - 5y = 7$ defines a function. If $y = f(x)$, which one of the following defines the same function?

(a) $f(x) = \dfrac{7 - 2x}{5}$ **(b)** $f(x) = \dfrac{-7 - 2x}{5}$

(c) $f(x) = \dfrac{-7 + 2x}{5}$ **(d)** $f(x) = \dfrac{7 + 2x}{5}$

62. Can the graph of a linear function have undefined slope? Explain.

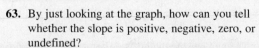 **RELATING CONCEPTS (EXERCISES 63-74)**

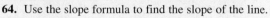

Refer to the straight-line graph shown and work Exercises 63–74 in order.

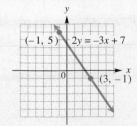

63. By just looking at the graph, how can you tell whether the slope is positive, negative, zero, or undefined?

64. Use the slope formula to find the slope of the line.

65. Find the x-intercept of the graph.

66. Find the y-intercept of the graph.

67. Use function notation to write the equation of the line. Use f to designate the function.

RELATING CONCEPTS (EXERCISES 63–74) (CONTINUED)

68. Find $f(8)$.

69. If $f(x) = -8$, what is the value of x?

70. Graph the solution set of $f(x) \geq 0$.

71. What is the solution set of $f(x) = 0$?

72. What is the solution set of $f(x) < 0$? (Use the graph and the result of Exercise 71.)

73. What is the solution set of $f(x) > 0$? (Use the graph and the result of Exercise 71.)

74. What is the slope of any line perpendicular to the line shown?

Did you make the connection between the two-variable linear functions in this chapter and the one-variable linear equations and inequalities in Chapters 2 and 3?

CHAPTER 8 TEST

1. Complete the table of ordered pairs for the equation $2x - 3y = 12$.

x	y
1	
3	
	-4

2. Find the slope of the line through the points $(6, 4)$ and $(-4, -1)$.

Find the x- and y-intercepts, and graph each equation.

3. $3x - 2y = 20$ **4.** $y = 5$ **5.** $x = 2$

6. Describe how the graph of a line with undefined slope is situated in a rectangular coordinate system.

Determine whether each pair of lines is parallel, perpendicular, or neither.

7. $5x - y = 8$ and $5y = -x + 3$ **8.** $2y = 3x + 12$ and $3y = 2x - 5$

Find the equation of each line, and write it in slope-intercept form.

9. Through $(4, -1)$; $m = -5$

10. Through $(-3, 14)$; horizontal

11. Through $(-7, 2)$ and parallel to $3x + 5y = 6$

12. Through $(-7, 2)$ and perpendicular to $y = 2x$

13. The line shown in the figures (Look at the displays at the bottom.)

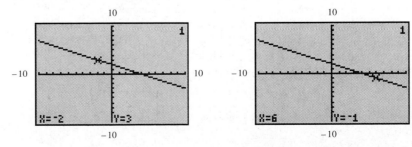

14. Which one of the following has positive slope and negative y-coordinate for its y-intercept?

(a) 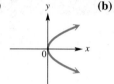   **(b)**   **(c)**   **(d)**

15. The linear equation $y = 1410x + 12{,}520$ provides a model for the number of cases served by the Child Support Enforcement program from 1990 to 1995, where $x = 0$ corresponds to 1990, $x = 1$ corresponds to 1991, and so on. Use this model to approximate the number of cases served during 1994. (*Source:* Office of Child Support Enforcement.)

16. What does the number 1410 in the equation in Exercise 15 refer to
(a) with respect to the graph?   **(b)** in the context of the problem?

Graph each inequality or compound inequality.

17. $3x - 2y > 6$ **18.** $y < 2x - 1$ and $x - y < 3$

19. Which one of the following is the graph of a function?

(a)   **(b)**   **(c)**   **(d)**

20. Which of the following does not define a function?
(a) $\{(0, 1), (-2, 3), (4, 8)\}$ **(b)** $y = 2x - 6$ **(c)** $y = \sqrt{x + 2}$
(d)

Input	Output
A	1
A	2
B	2
C	3

21. If $f(x) = -x^2 + 2x - 1$, find $f(1)$.

22. Graph the linear function $f(x) = \dfrac{2}{3}x - 1$. What is its domain? What is its range?

23. The deaths per 1000 population from 1990–1995 are shown in the table.

Death rate	8.6	8.6	8.5	8.8	8.8	8.8
Year	1990	1991	1992	1993	1994	1995

Source: U.S. National Center for Health Statistics.

Which of the following sets of ordered pairs from the table define(s) a function? Explain.
(a) {(Death rate, Year)}   **(b)** {(Year, Death rate)}

Solve each equation.

1. $-5(8 - 2z) + 4(7 - z) = 7(8 + z) - 3$

2. $A = p + prt$ for t

3. $7x^2 + 8x + 1 = 0$

4. $|2k - 7| + 4 = 11$

5. $\dfrac{2}{x - 1} = \dfrac{5}{x - 1} - \dfrac{3}{4}$

Solve each inequality and graph the solution set.

6. $-4 < 3 - 2k < 9$

7. $-.3x + 2.1(x - 4) \le -6.6$

8. $-5x + 1 \ge 11$ or $3x + 5 > 26$

Write with only positive exponents. Assume that all variables represent positive real numbers.

9. $(x^2 y^{-3})(x^{-4} y^2)$

10. $\dfrac{x^{-6} y^3 z^{-1}}{x^7 y^{-4} z}$

11. $(2m^{-2} n^3)^{-3}$

Perform the indicated operations.

12. $2(3x^2 - 8x + 1) - 4(x^2 - 3x - 9)$

13. $(3x + 2y)(5x - y)$

14. $(x + 2y)(x^2 - 2xy + 4y^2)$

15. $\dfrac{m^3 - 3m^2 + 5m - 3}{m - 1}$

Factor each polynomial completely.

16. $y^2 + 4yk - 12k^2$

17. $9x^4 - 25y^2$

18. $125x^4 - 400x^3 y + 195x^2 y^2$

19. $f^2 + 20f + 100$

20. $100x^2 + 49$

Perform each indicated operation. Express the answer in lowest terms.

21. $\dfrac{3}{2x + 6} + \dfrac{2x + 3}{2x + 6}$

22. $\dfrac{8}{x + 1} - \dfrac{2}{x + 3}$

23. $\dfrac{x^2 - 25}{3x + 6} \cdot \dfrac{4x + 8}{x^2 + 10x + 25}$

24. $\dfrac{x^2 + 2x - 3}{x^2 - 5x + 4} \cdot \dfrac{x^2 - 3x - 4}{x^2 + 3x}$

25. $\dfrac{x^2 + 5x + 6}{3x} \div \dfrac{x^2 - 4}{x^2 + x - 6}$

26. $\dfrac{6x^4 y^3 z^2}{8xyz^4} \div \dfrac{3x^2}{16y^7}$

27. Simplify the complex fraction.

$$\dfrac{\dfrac{12}{x + 6}}{\dfrac{4}{2x + 12}}$$

28. Find the x- and y-intercepts of the line with equation $3x + 5y = 12$ and graph the line.

29. Consider the points $A(-2, 1)$ and $B(3, -5)$.
    (a) Find the slope of the line AB.
    (b) Find the slope of a line perpendicular to line AB.

Give the equation of each line described in the form $y = mx + b$.

30. Through $(4, -1)$, $m = -4$

31. Through $(0, 0)$ and $(1, 4)$

32. Graph the inequality $-2x + y < -6$.

33. For the function $f(x) = -3x + 6$, **(a)** what is the domain? **(b)** what is $f(-6)$?

34. The table shows the percent of possible sunshine for the ten sunniest cities.

Which set is a function: {(Percent, City)} or {(City, Percent)}? Describe the elements of the domain of the function.

Sunniest U.S. Cities

Percent	City
90	Yuma, AZ
88	Redding, CA
86	Phoenix, AZ
85	Tucson, AZ
85	Las Vegas, NV
84	El Paso, TX
79	Fresno, CA
79	Reno, NV
78	Flagstaff, AZ
78	Sacramento, CA

Source: National Oceanic and Atmospheric Administration.

35. Does $2x - 7y = 14$ define a function? If so, write it using the function notation $f(x)$.

Solve each problem.

36. In 1996, the U.S. government raised $484.6 billion more from individual income taxes than from corporate income taxes. The total amount raised from these two sources was $828.2 billion. How much was raised from each source? (*Source:* Office of Management and Budget.)

37. Find the measure of each angle of the triangle.

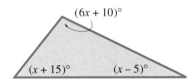

38. The length of the shorter leg of a right triangle is tripled and 4 inches is added to the result, giving the length of the hypotenuse. The longer leg is 10 inches longer than twice the shorter leg. Find the length of the shorter leg of the triangle.

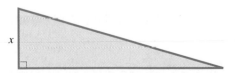

39. The cost of a pizza varies directly as the square of its radius. If a pizza with a 7-inch radius costs $6.00, how much should a pizza with a 9-inch radius cost?

40. If a man can mow his lawn in 3 hours and his wife can do the same job in 1.5 hours, how long will it take them to do the job together?

Linear Systems

Consumers are spending more money than ever on entertainment and sports. The average American household spent $1612 in 1995 on leisure pursuits including books, TV, movies, theater, and sporting events—almost as much as for health care, according to a

survey by the U.S. Labor Department. This voracious appetite for entertainment can make even leisure time exhausting. Some accommodate this plethora of entertainment choices by "multitasking"; for example, watching TV or listening to the radio as they read, channel-surfing to watch several programs at once, or using the picture-in-picture feature available in many television models to follow two sporting events at the same time.

In recent years, movies and television have been the big winners in the entertainment marketplace. From 1991 to 1996, movie box-office gross increased from $4803.2 million to $5911.5 million, or about $1.1 billion, as shown in the graph.

 Entertainment/Sports

9.1 Solving Systems of Linear Equations by Graphing

9.2 Solving Systems of Linear Equations by Substitution

9.3 Solving Systems of Linear Equations by Elimination

9.4 Linear Systems of Equations in Three Variables

9.5 Applications of Linear Systems of Equations

9.6 Solving Linear Systems of Equations by Matrix Methods

9.7 Determinants and Cramer's Rule

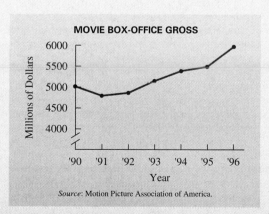

MOVIE BOX-OFFICE GROSS

Source: Motion Picture Association of America.

Although the box-office gross dipped from 1990 to 1991, it has steadily increased since then. For what year does the graph show the greatest increase? Exercises 43–48 of Section 9.3 present more information on the increase in movie attendance. Linear equations are also applied to other areas of the entertainment and sports industry in this chapter.

9.1 Solving Systems of Linear Equations by Graphing

OBJECTIVES

1 Decide whether a given ordered pair is a solution of a system.

2 Solve linear systems by graphing.

3 Solve special systems by graphing.

4 Identify special systems without graphing.

5 Recognize how a graphing calculator is used to solve a linear system.

FOR EXTRA HELP

SSG Sec. 9.1
SSM Sec. 9.1

Pass the Test Software

InterAct Math
Tutorial Software

Video 14

A **system of linear equations** consists of two or more linear equations with the same variables. Examples of systems of linear equations include

$$2x + 3y = 4 \qquad\qquad x + 3y = 1 \qquad\qquad x - y = 1$$
$$3x - y = -5 \qquad\qquad -y = 4 - 2x \quad \text{and} \qquad y = 3.$$

In the system on the right, think of $y = 3$ as an equation in two variables by writing it as $0x + y = 3$.

OBJECTIVE 1 Decide whether a given ordered pair is a solution of a system. A **solution of a system** of linear equations is an ordered pair (x, y) that makes both equations true at the same time.

EXAMPLE 1 Determining Whether an Ordered Pair Is a Solution

Is $(4, -3)$ a solution of the following systems?

(a) $x + 4y = -8$
$3x + 2y = 6$

To decide whether $(4, -3)$ is a solution of the system, substitute 4 for x and -3 for y in each equation.

$x + 4y = -8$		$3x + 2y = 6$	
$4 + 4(-3) = -8$	?	$3(4) + 2(-3) = 6$	?
$4 + (-12) = -8$	?	$12 + (-6) = 6$	?
$-8 = -8$	True	$6 = 6$	True

Since $(4, -3)$ satisfies both equations, it is a solution.

(b) $2x + 5y = -7$
$3x + 4y = 2$

Again, substitute 4 for x and -3 for y in both equations.

$2x + 5y = -7$		$3x + 4y = 2$	
$2(4) + 5(-3) = -7$	?	$3(4) + 4(-3) = 2$	?
$8 + (-15) = -7$	?	$12 + (-12) = 2$	?
$-7 = -7$	True	$0 = 2$	False

Here $(4, -3)$ is not a solution since it does not satisfy the second equation.

We discuss several methods of solving a system of two linear equations with two variables in this chapter.

OBJECTIVE 2 Solve linear systems by graphing. The set of all ordered pairs that are solutions of a system is its **solution set.** One way to find the solution set of a system of two linear equations is to graph both equations on the same axes. The graph of each line shows points whose coordinates satisfy the equation of that line. The coordinates of any point where the lines intersect give a solution of the system. Since two *different* straight lines can intersect at no more than one point, there can never be more than one solution for such a system.

EXAMPLE 2   Solving a System by Graphing

Solve the following system of equations by graphing both equations on the same axes.

$$2x + 3y = 4$$
$$3x - y = -5$$

As shown in Chapter 4, we graph these two equations by plotting several points for each line. Some ordered pairs that satisfy each equation are shown below.

$$2x + 3y = 4 \qquad 3x - y = -5$$

x	y
0	$\frac{4}{3}$
2	0
−2	$\frac{8}{3}$

x	y
0	5
$-\frac{5}{3}$	0
−2	−1

The lines in Figure 1 suggest that the graphs intersect at the point $(-1, 2)$. Check this by substituting -1 for x and 2 for y in both equations. Since $(-1, 2)$ satisfies both equations, the solution set of this system is $\{(-1, 2)\}$.

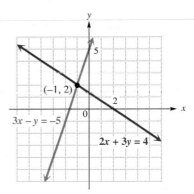

Figure 1

 A difficulty with the graphing method is that it may not be possible to determine from the graph the exact coordinates of the point that represents the solution, particularly if the solution does not involve integers. For this reason, we give algebraic methods of solution later in this chapter. The graphing method does, however, show geometrically how solutions are found.

OBJECTIVE 3   Solve special systems by graphing.   Sometimes the graphs of the two equations in a system either do not intersect at all or are the same line, as in the systems of Example 3.

E X A M P L E   3   Solving Special Systems

Solve each system by graphing.

(a) $2x + y = 2$
      $2x + y = 8$

The graphs of these lines are shown in Figure 2. The two lines are parallel and have no points in common. For such a system, there is no solution; we write the solution set as ∅.

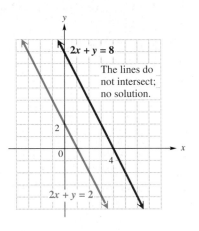

Figure 2   Figure 3

(b) $2x + 5y = 1$
      $6x + 15y = 3$

The graphs of these two equations are the same line. See Figure 3. The second equation can be obtained by multiplying both sides of the first equation by 3. In this case, every point on the line is a solution of the system, and the solution set contains an infinite number of ordered pairs. We write the solution set as $\{(x, y) \mid 2x + 5y = 1\}$. (Since they have the same graph, either equation in the system could be given in the solution set.)

The system in Example 2 has exactly one solution. A system with a solution is called a **consistent system.** A system with no solution, such as the one in Example 3(a), is called an **inconsistent system.** The equations in Example 2 are **independent equations,** equations that have different graphs. The equations of the system in Example 3(b) have the same graph. Because they are different forms of the same equation, these equations are called **dependent equations.** Examples 2 and 3 show the three cases that may occur when solving a system of equations with two variables.

Possible Types of Solutions

1. The graphs intersect at exactly one point, which gives the (single) ordered pair solution of the system. The **system is consistent** and the **equations are independent.**

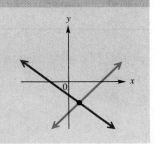

Possible Types of Solutions (continued)

2. The graphs are parallel lines, so there is no solution and the solution set is ∅. The **system is inconsistent.**

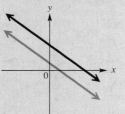

3. The graphs are the same line. There are an infinite number of solutions. The **equations are dependent.**

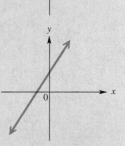

OBJECTIVE **4** Identify special systems without graphing. Example 3 showed that the graphs of an inconsistent system are parallel lines and the graphs of a system of dependent equations are the same line. We can recognize these special kinds of systems without graphing by using slopes.

┌ **EXAMPLE 4** Identifying the Three Cases Using Slopes

Describe each system without graphing.

(a) $3x + 2y = 6$
$-2y = 3x - 5$

Write each equation in slope-intercept form by solving for y.

$$3x + 2y = 6 \qquad\qquad -2y = 3x - 5$$

$$2y = -3x + 6 \qquad\qquad y = -\frac{3}{2}x + \frac{5}{2}$$

$$y = -\frac{3}{2}x + 3$$

Both equations have slope $-\frac{3}{2}$ but they have different y-intercepts, 3 and $\frac{5}{2}$. In Chapters 4 and 8 we found that lines with the same slope are parallel, so these equations have graphs that are parallel lines. The system has no solution.

(b) $2x - y = 4$
$x = \dfrac{y}{2} + 2$

Again, write the equations in slope-intercept form.

$$2x - y = 4 \qquad\qquad x = \frac{y}{2} + 2$$

$$-y = -2x + 4 \qquad\qquad \frac{y}{2} + 2 = x$$

$$y = 2x - 4 \qquad\qquad \frac{y}{2} = x - 2$$

$$y = 2x - 4$$

The equations are exactly the same; their graphs are the same line. The system has an infinite number of solutions.

(c) $x - 3y = 5$
   $2x + y = 8$
   In slope-intercept form, the equations are as follows.

$$x - 3y = 5 \qquad\qquad 2x + y = 8$$
$$-3y = -x + 5 \qquad\qquad y = -2x + 8$$
$$y = \frac{1}{3}x - \frac{5}{3}$$

The graphs of these equations are neither parallel lines nor the same line since the slopes are different. This system has exactly one solution.

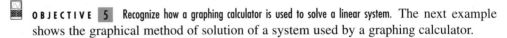

OBJECTIVE 5 Recognize how a graphing calculator is used to solve a linear system. The next example shows the graphical method of solution of a system used by a graphing calculator.

EXAMPLE 5 Illustrating the Solution of a System with a Graphing Calculator

Figures 4(a) and (b) illustrate how a graphing calculator can represent the solution of the system in Example 1(a),

$$x + 4y = -8$$
$$3x + 2y = 6.$$

In order to enter the equations, first solve each one for y.

$$x + 4y = -8 \qquad\qquad 3x + 2y = 6$$
$$4y = -x - 8 \qquad\qquad 2y = -3x + 6$$
$$y = -\frac{1}{4}x - 2 \qquad\qquad y = -\frac{3}{2}x + 3$$

The calculator allows us to enter several equations to be graphed at the same time. Designate the first one Y_1 and the second one Y_2. Graph the two equations using a standard window, and then use the capability of the calculator to find the coordinates of the point of intersection of the graphs. The display at the bottom of Figure 4(b) indicates that the solution set is $\{(4, -3)\}$.

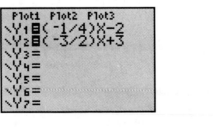

(a)

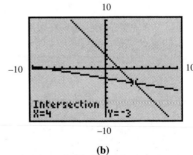

(b)

Figure 4

9.1 EXERCISES

1. Each ordered pair in Column I is a solution of one of the systems graphed in Column II. Because of the location of the point of intersection, you should be able to determine the correct system for each solution. Match each system from Column II with its solution from Column I.

I   II

(a) $(3, 4)$ **A.** 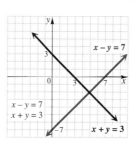   **B.**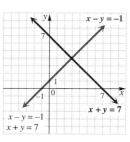

(b) $(-2, 3)$

(c) $(-4, -1)$ **C.** 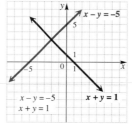   **D.**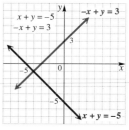

(d) $(5, -2)$

2. When a student was asked to determine whether the ordered pair $(1, -2)$ is a solution of the system

$$x + y = -1$$
$$2x + y = 4,$$

   he answered "yes." His reasoning was that the ordered pair satisfies the equation $x + y = -1$: $1 + (-2) = -1$ is true. Why is the student's answer wrong?

Decide whether the given ordered pair is a solution of the given system. See Example 1.

3. $(6, 2)$
$3x + y = 20$
$2x + 3y = 18$

4. $(3, 4)$
$2x + y = 10$
$3x + 2y = 17$

5. $(2, -3)$
$x + y = -1$
$2x + 5y = 19$

6. $(4, 3)$
$x + 2y = 10$
$3x + 5y = 3$

7. $(-1, -3)$
$3x + 5y = -18$
$4x + 2y = -10$

8. $(-9, -2)$
$2x - 5y = -8$
$3x + 6y = -39$

9. $(7, -2)$
$4x = 26 - y$
$3x = 29 + 4y$

10. $(9, 1)$
$2x = 23 - 5y$
$3x = 24 + 3y$

11. $(6, -8)$
$-2y = x + 10$
$3y = 2x + 30$

12. Write an example of a system with solution $(4, -5)$. Then explain how you found the equations in your system.

13. Which one of the ordered pairs below could not possibly be a solution of the system graphed? Why is it the only valid choice?

   (a) $(-4, -4)$ **(b)** $(-2, 2)$
   (c) $(-4, 4)$ **(d)** $(-3, 3)$

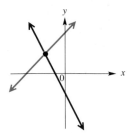

14. Which one of the ordered pairs below could possibly be a solution of the system graphed? Why is it the only valid choice?

   (a) $(2, 0)$ **(b)** $(0, 2)$
   (c) $(-2, 0)$ **(d)** $(0, -2)$

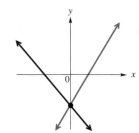

Solve each system of equations by graphing both equations on the same axes. If the system is inconsistent or the equations are dependent, say so. See Examples 2–4.

15. $x - y = 2$
    $x + y = 6$

16. $x - y = 3$
    $x + y = -1$

17. $x + y = 4$
    $y - x = 4$

18. $x + y = -5$
    $x - y = 5$

19. $x - 2y = 6$
    $x + 2y = 2$

20. $2x - y = 4$
    $4x + y = 2$

21. $3x - 2y = -3$
    $-3x - y = -6$

22. $2x - y = 4$
    $2x + 3y = 12$

23. $2x - 3y = -6$
    $y = -3x + 2$

24. $x + 2y = 4$
    $2x + 4y = 12$

25. $2x - y = 6$
    $4x - 2y = 8$

26. $2x - y = 4$
    $4x = 2y + 8$

27. $3x = 5 - y$
    $6x + 2y = 10$

28. $-3x + y = -3$
    $y = x - 3$

29. $3x - 4y = 24$
    $y = -\dfrac{3}{2}x + 3$

30. How can you tell, without graphing, that the system

$$x + y = 2$$
$$x + y = 4$$

has no solution?

31. Explain why a system of two linear equations cannot have exactly two solutions.

32. Explain the three situations that may occur (regarding the number of solutions) when solving a system of two linear equations in two variables by graphing.

33. Is it possible for a system of *three* linear equations in two variables to have a single solution? If so, give an example.

 Listening to recorded music has been an extremely popular form of entertainment for many years. During the early years, the most common format for recorded music was the vinyl phonograph record, available in various speeds (78, 45, and $33\frac{1}{3}$ revolutions per minute). The long-playing record (LP) and single (45) were the mainstays in the 1950s and remained so until the early 1980s. Another popular format among audiophiles was the reel-to-reel tape. The 1960s and 1970s saw the emergence of the now defunct four- and eight-track tape formats, and during those years the familiar audiocassette (or simply "cassette") was in its infancy. Music lovers today rely on recordings that are almost exclusively found in the cassette and compact disc (CD) formats.

THE SOUNDS OF MUSIC

Production (in millions)

500
400
300
200
100
0

Cassettes

LPs

CDs

'83 '84 '85 '86 '87 '88 '89 '90 '91 '92 '93
Year

Source: Recording Industry of America.

The graph above shows how the production of vinyl LPs, cassettes, and CDs changed over the years from 1983 to 1993. Use the graph to respond to Exercises 34–40.

34. In what year did cassette production and CD production reach equal levels?

35. (See Exercise 34.) What were those production levels?

36. Express as an ordered pair in the form (year, production level) the point of intersection of the graphs of LP production and CD production.

37. For 1987, what was the *total* production for LPs and CDs?

38. For which years in this period was the production of CDs less than the production of LPs?

39. Between which two nonconsecutive years was the production of cassettes approximately constant?

40. If a straight line were used to model the graph for CD production from 1984 to 1993, would its slope be positive, negative, or zero?

Without graphing, answer the following questions for each linear system. See Example 4.
(a) Is the system inconsistent, are the equations dependent, or neither?
(b) Is the graph a pair of intersecting lines, a pair of parallel lines, or one line?
(c) Does the system have one solution, no solution, or an infinite number of solutions?

41. $y - x = -5$
   $x + y = 1$

42. $2x + y = 6$
   $x - 3y = -4$

43. $x + 2y = 0$
   $4y = -2x$

44. $y = 3x$
   $y + 3 = 3x$

45. $5x + 4y = 7$
   $10x + 8y = 4$

46. $2x + 3y = 12$
   $2x - y = 4$

An application of mathematics in economics deals with **supply and demand.** Typically, as the price of an item increases, the demand for the item decreases, while the supply increases. (There are exceptions to this, however.) If supply and demand can be described by straight-line equations, the point at which the lines intersect determines the **equilibrium supply** and **equilibrium demand.** Suppose that an economist has studied the supply and demand for aluminum siding and has concluded that the price per unit, p, and the demand, x, are related by the demand equation

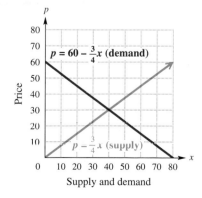

p

80
70
60
50
40
30
20
10

Price

$p = 60 - \frac{3}{4}x$ (demand)

$p = \frac{3}{4}x$ (supply)

0   10 20 30 40 50 60 70 80 x

Supply and demand

$p = 60 - \dfrac{3}{4}x$, while the supply is given by the equation $p = \dfrac{3}{4}x$. The graphs of these two equations are shown here.

Use the graph on the previous page to answer each question.

47. At what value of x does supply equal demand?

48. At what value of p does supply equal demand?

49. What are the coordinates of the point of intersection of the two lines?

50. When $x > 40$, does demand exceed supply or does supply exceed demand?

TECHNOLOGY INSIGHTS (EXERCISES 51-54)

Match the graphing calculator screens from choices A–D with the appropriate system in Exercises 51–54. See Example 5.

A.

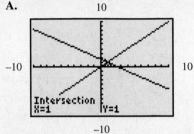

B.

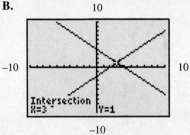

C.

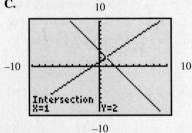

D.

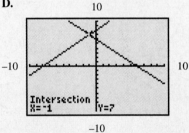

51. $x + y = 4$
$x - y = 2$

52. $x + y = 6$
$x - y = -8$

53. $2x + 3y = 5$
$x - y = 0$

54. $3x + 2y = 7$
$-x + y = 1$

 Use a graphing calculator to solve each system. See Example 5.

55. $3x + y = 2$
$2x - y = -7$

56. $x + 2y = -2$
$2x - y = 11$

RELATING CONCEPTS (EXERCISES 57-62)

In these exercises we relate solving linear equations in one variable and solving a linear system of two equations in two variables.

Work Exercises 57–62 in order.

57. Solve the linear equation $\frac{1}{2}x + 4 = 3x - 1$ using the methods described in Chapter 2.

58. Check your solution for the equation in Exercise 57 by substituting back into the original equation. What is the value that you get for both the left and right sides after you make this substitution?

RELATING CONCEPTS (EXERCISES 57–62) (CONTINUED)

59. Graph the linear system and find the solution set.

$$y = \frac{1}{2}x + 4$$
$$y = 3x - 1$$

60. How does the x-coordinate of the solution of the system in Exercise 59 compare to the solution of the linear equation in Exercise 57?

61. How does the y-coordinate of the solution of the system in Exercise 59 compare to the value you obtained in the check in Exercise 58?

62. Based on your observations in Exercises 57–61, fill in the blanks with the correct responses.

The solution of the linear equation $\frac{2}{3}x + 3 = -x + 8$ is 3. When we

substitute 3 back into the equation, we get the value _____ on both sides, verifying that 3 is indeed the solution. Now if we graph the system

$$y = \frac{2}{3}x + 3$$
$$y = -x + 8,$$

we find that the solution set of the system is {(_____, _____)}.

Did you make the connection between solving equations in one variable and solving systems in two variables?

9.2 Solving Systems of Linear Equations by Substitution

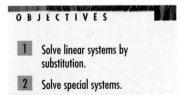

OBJECTIVES

1 Solve linear systems by substitution.

2 Solve special systems.

3 Solve linear systems with fractions.

FOR EXTRA HELP

SSG Sec. 9.2
SSM Sec. 9.2

Pass the Test Software

InterAct Math Tutorial Software

Video 14

Graphing to solve a system of equations has a serious drawback. It is difficult to estimate a solution such as $\left(\frac{1}{3}, -\frac{5}{6}\right)$ accurately from a graph (unless a graphing calculator is used).

OBJECTIVE 1 Solve linear systems by substitution. An algebraic method involving substitution is particularly useful for solving systems where either equation is solved, or can be solved easily, for one of the variables.

EXAMPLE 1 Solving a System by Substitution

Solve the system

$$3x + 5y = 26$$
$$y = 2x.$$

The second equation is solved for y. Substituting $2x$ for y in the first equation gives

$$3x + 5y = 26$$
$$3x + 5(2x) = 26 \qquad \text{Let } y = 2x.$$
$$3x + 10x = 26$$
$$13x = 26$$
$$x = 2.$$

Since $y = 2x$ and $x = 2$, $y = 2(2) = 4$. Check that the solution set of the given system is {(2, 4)} by substituting 2 for x and 4 for y in *both* equations.

To solve a system by substitution, follow these steps.

Solving Linear Systems by Substitution

Step 1 **Solve for a variable.** Solve one of the equations for either variable. (If one of the variables has coefficient 1 or -1, choose it, since the substitution method is usually easier this way.)

Step 2 **Substitute.** Substitute for that variable in the other equation. The result should be an equation with just one variable.

Step 3 **Solve.** Solve the equation from Step 2.

Step 4 **Substitute.** Substitute the result from Step 3 into the equation from Step 1 to find the value of the other variable.

Step 5 **Find the solution set.** Check the solution in both of the given equations. Then write the solution set.

E X A M P L E   2 Solving a System by Substitution

Use substitution to solve the system

$$2x = 8 - 3y$$
$$4x + 3y = 4.$$

Step 1 The substitution method requires that an equation be solved for one of the variables. Choose the first equation of the system and solve it for x.

$$2x = 8 - 3y$$
$$x = \frac{8 - 3y}{2} \qquad \text{Divide both sides by 2.}$$

Step 2 Now we substitute this value for x in the second equation of the system.

$$4x + 3y = 4$$
$$4\left(\frac{8 - 3y}{2}\right) + 3y = 4 \qquad \text{Let } x = \frac{8 - 3y}{2}.$$

Step 3 Solve this equation.

$$2(8 - 3y) + 3y = 4 \qquad \text{Divide 4 by 2.}$$
$$16 - 6y + 3y = 4 \qquad \text{Distributive property}$$
$$-3y = -12 \qquad \text{Combine terms; subtract 16.}$$
$$y = 4 \qquad \text{Divide by } -3.$$

Step 4 Find x by letting $y = 4$ in $x = \dfrac{8 - 3y}{2}$.

$$x = \frac{8 - 3 \cdot 4}{2} = \frac{8 - 12}{2} = \frac{-4}{2} = -2$$

Step 5 The solution set of the given system is $\{(-2, 4)\}$. Check the solution in both equations.

OBJECTIVE **2** Solve special systems. In the previous section we solved inconsistent systems with graphs that are parallel lines and systems of dependent equations with graphs that are the same line. We can also solve these special systems with the substitution method.

E X A M P L E   3   Solving an Inconsistent System by Substitution

Use substitution to solve the system

$$x = 5 - 2y$$
$$2x + 4y = 6.$$

Substitute $5 - 2y$ for x in the second equation.

$$2x + 4y = 6$$
$$2(5 - 2y) + 4y = 6 \qquad \text{Let } x = 5 - 2y.$$
$$10 - 4y + 4y = 6 \qquad \text{Distributive property}$$
$$10 = 6 \qquad \text{False}$$

This false result means that the system is inconsistent and its solution set is $\emptyset$. The equations of the system have graphs that are parallel lines. See Figure 5.

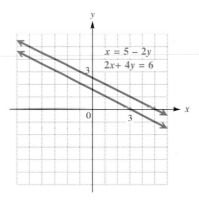

Figure 5

 It is a common error to give "false" as the answer to an inconsistent system. The correct response is $\emptyset$.

E X A M P L E   4   Solving a System with Dependent Equations by Substitution

Solve the following system by the substitution method.

$$3x - y = 4 \qquad \qquad \textbf{(1)}$$
$$-9x + 3y = -12 \qquad \textbf{(2)}$$

Begin by solving the first equation for y to get $y = 3x - 4$. Substitute $3x - 4$ for y in equation (2) and solve the resulting equation.

$$-9x + 3(3x - 4) = -12$$
$$-9x + 9x - 12 = -12 \qquad \text{Distributive property}$$
$$0 = 0 \qquad \text{Add 12; combine terms.}$$

This true result means that every solution of one equation is also a solution of the other, so the system has an infinite number of solutions: all the ordered pairs corresponding to points that lie on the common graph. The solution set is $\{(x, y) \mid 3x - y = 4\}$. A graph of the equations of this system is shown in Figure 6.

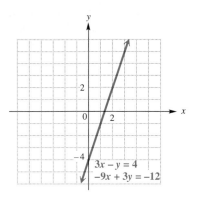

Figure 6

 It is a common error to give "true" as the answer to a dependent system. Instead, remember that a dependent system has an infinite number of solutions.

OBJECTIVE **3**   Solve linear systems with fractions.   When a system includes equations with fractions as coefficients, eliminate the fractions by multiplying both sides by a common denominator. Then solve the resulting system.

E X A M P L E   5   Solving by Substitution with Fractions as Coefficients

Solve the system

$$4x + \frac{1}{3}y = \frac{8}{3} \tag{1}$$

$$\frac{1}{2}x + \frac{3}{4}y = -\frac{5}{2}. \tag{2}$$

Begin by eliminating fractions. Clear equation (1) of fractions by multiplying both sides by 3.

$$3\left(4x + \frac{1}{3}y\right) = 3 \cdot \frac{8}{3} \qquad \text{Multiply by 3.}$$

$$3(4x) + 3\left(\frac{1}{3}y\right) = 3 \cdot \frac{8}{3} \qquad \text{Distributive property}$$

$$12x + y = 8 \tag{3}$$

Now clear equation (2) of fractions by multiplying both sides by the common denominator 4.

$$4\left(\frac{1}{2}x + \frac{3}{4}y\right) = 4\left(-\frac{5}{2}\right) \qquad \text{Multiply by 4.}$$

$$4\left(\frac{1}{2}x\right) + 4\left(\frac{3}{4}y\right) = 4\left(-\frac{5}{2}\right) \qquad \text{Distributive property}$$

$$2x + 3y = -10 \tag{4}$$

The given system of equations has been simplified to

$$12x + y = 8 \tag{3}$$

$$2x + 3y = -10. \tag{4}$$

Solve the system by the substitution method. Equation (3) can be solved for y by subtracting $12x$ on each side.

$$12x + y = 8 \tag{3}$$

$$y = -12x + 8$$

Now substitute the result for y in equation (4).

$2x + 3(-12x + 8) = -10$	Let $y = -12x + 8$.
$2x - 36x + 24 = -10$	Distributive property
$-34x = -34$	Combine terms; subtract 24.
$x = 1$	Divide by -34.

Substitute $x = 1$ in $y = -12x + 8$ to get $y = -4$. Check by substituting 1 for x and -4 for y in the original equations. The solution set is $\{(1, -4)\}$.

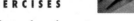

9.2 EXERCISES

1. A student solves the system

$$5x - y = 15$$

$$7x + y = 21$$

and finds that $x = 3$, which is the correct value for x. The student gives the solution set as $\{3\}$. Is this correct? Explain.

2. A student solves the system

$$x + y = 4$$

$$2x + 2y = 8$$

and obtains the equation $0 = 0$. The student gives the solution set as $\{(0, 0)\}$. Is this correct? Explain.

Solve each system by the substitution method. Check your solution. See Examples 1–4.

3. $x + y = 12$
$y = 3x$

4. $x + 3y = -28$
$y = -5x$

5. $3x + 2y = 27$
$x = y + 4$

6. $4x + 3y = -5$
$x = y - 3$

7. $3x + 5y = 25$
$x - 2y = -10$

8. $5x + 2y = -15$
$2x - y = -6$

9. $3x + 4 = -y$
$2x + y = 0$

10. $2x - 5 = -y$
$x + 3y = 0$

11. $7x + 4y = 13$
$x + y = 1$

12. $3x - 2y = 19$
$x + y = 8$

13. $3x - y = 5$
$y = 3x - 5$

14. $4x - y = -3$
$y = 4x + 3$

15. $6x - 8y = 6$
$-3x + 2y = -2$

16. $3x + 2y = 6$
$-6x + 4y = -8$

17. $2x + 8y = 3$
$x = 8 - 4y$

18. $2x + 10y = 3$
$x = 1 - 5y$

19. $12x - 16y = 8$
$3x = 4y + 2$

20. $6x + 9y = 6$
$2x = 2 - 3y$

21. Professor Brandsma gave the following item on a test in introductory algebra:
Solve the following system by the substitution method.

$$3x - y = 13$$

$$2x + 5y = 20$$

One student worked the problem by solving first for y in the first equation. Another student worked it by solving first for x in the second equation. Both students got the correct solution, (5, 2). Which student, do you think, had less work to do? Explain.

22. When you use the substitution method, how can you tell that a system has
   (a) no solution?
   (b) an infinite number of solutions?

In each system given, begin by clearing fractions and then solve by substitution. See Example 5.

23. $\dfrac{5}{3}x + 2y = \dfrac{1}{3} + y$

$2x - 3 + \dfrac{y}{3} = -2 + x$

24. $\dfrac{x}{6} + \dfrac{y}{6} = 1$

$-\dfrac{1}{2}x - \dfrac{1}{3}y = -5$

25. $\dfrac{x}{2} - \dfrac{y}{3} = \dfrac{5}{6}$

$\dfrac{x}{5} - \dfrac{y}{4} = \dfrac{1}{10}$

26. $\dfrac{x}{3} - \dfrac{3y}{4} = -\dfrac{1}{2}$

$\dfrac{2x}{3} + \dfrac{y}{2} = 3$

27. $\dfrac{x}{5} + 2y = \dfrac{8}{5}$

$\dfrac{3x}{5} + \dfrac{y}{2} = -\dfrac{7}{10}$

28. $\dfrac{x}{2} + \dfrac{y}{3} = \dfrac{7}{6}$

$\dfrac{x}{4} - \dfrac{3y}{2} = \dfrac{9}{4}$

Attending professional sporting events ranks among the most popular forms of entertainment. Use systems of equations to solve the sports entertainment problems in Exercises 29 and 30.

29. During the period from 1991 to 1996, average ticket prices rose in the National Football League from $25.21 to $35.74. If we let $x = 1$ represent 1991, $x = 2$ represent 1992, and so on, the linear equation $y = 2.1x + 22.8$ gives a good approximation for this average price, where y is in dollars. To determine the year in which the average ticket price was $28.68, solve the system

$$y = 2.1x + 22.8$$
$$y = 28.68.$$

The rounded x-value will then give us the year. Solve this system by substitution to determine the year. (*Hint:* After finding the value of x, you must then determine which year it represents.) (*Source:* Team Marketing Report, Chicago.)

30. During the period from the 1991–1992 season through the 1996–1997 season, the average price of a ticket to a National Basketball Association game rose from $23.24 to $34.08. If we let $x = 1$ represent the 1991–1992 season, $x = 2$ represent 1992–1993, and so on, the linear equation $y = 2.14x + 20.81$ gives a good approximation for this average price, where y is in dollars. To determine the period in which the average ticket price was $28.63, solve the system

$$y = 2.14x + 20.81$$
$$y = 28.63.$$

The rounded x-value will then give us the first year of the period. Solve this system by substitution to determine the period. See the hint in Exercise 29. (*Source:* Team Marketing Report, Chicago.)

RELATING CONCEPTS (EXERCISES 31–34)

A system of linear equations can be used to model the cost and the revenue of a business.

Work Exercises 31–34 in order.

31. Suppose that you start a business manufacturing and selling bicycles, and it costs you $5000 to get started. You determine that each bicycle will cost $400 to manufacture. Explain why the linear equation $y_1 = 400x + 5000$ gives your *total* cost to manufacture x bicycles (y_1 in dollars).

32. You decide to sell each bike for $600. What expression in x represents the revenue you will take in if you sell x bikes? Write an equation using y_2 to express your revenue when you sell x bikes (y_2 in dollars).

33. Form a system from the two equations in Exercises 31 and 32 and then solve the system.

34. The value of x from Exercise 33 is the number of bikes it takes to *break even*. Fill in the blanks: When _____ bikes are sold, the break-even point is reached. At that point, you have spent _____ dollars and taken in _____ dollars.

Did you make the connections between x and the number of bikes, and y and the amount taken in?

 Solve each system by substitution. Then graph both lines in the same viewing window of a graphing calculator and use the intersection feature to support your answer. See Example 5 in Section 9.1. (In Exercises 39 and 40 you will need to solve each equation for y first before graphing.)

35. $y = 6 - x$
$y = 2x$

36. $y = 4x - 4$
$y = -3x - 11$

37. $y = -\dfrac{4}{3}x + \dfrac{19}{3}$
$y = \dfrac{15}{2}x - \dfrac{5}{2}$

38. $y = -\dfrac{15}{2}x + 10$
$y = \dfrac{25}{3}x - \dfrac{65}{3}$

39. $4x + 5y = 5$
$2x + 3y = 1$

40. $6x + 5y = 13$
$3x + 3y = 4$

 41. If the point of intersection does not appear on your screen when solving a linear system using a graphing calculator, how can you find the point of intersection?

 42. Suppose that you were asked to solve the system

$$y = 1.73x + 5.28$$
$$y = -2.94x - 3.85.$$

Why would it probably be easier to solve this system using a graphing calculator than using the substitution method?

9.3 Solving Systems of Linear Equations by Elimination

OBJECTIVES

1. Solve linear systems by elimination.

2. Multiply when using the elimination method.

OBJECTIVE 1 Solve linear systems by elimination. An algebraic method that depends on the addition property of equality can also be used to solve systems. As mentioned earlier, adding the same quantity to each side of an equation results in equal sums.

$$\text{If} \quad A = B, \quad \text{then} \quad A + C = B + C.$$

This addition can be taken a step further. Adding *equal* quantities, rather than the *same* quantity, to both sides of an equation also results in equal sums.

$$\text{If} \quad A = B \quad \text{and} \quad C = D, \quad \text{then} \quad A + C = B + D.$$

3 Use an alternative method to find the second value in a solution.

4 Use the elimination method to solve special systems.

FOR EXTRA HELP

 SSG Sec. 9.3
SSM Sec. 9.3

Pass the Test Software

InterAct Math
  Tutorial Software

Video 14

Using the addition property to solve systems is called the **elimination method.** When using this method, the idea is to *eliminate* one of the variables. To do this, one of the variables in the two equations must have coefficients that are opposites. For most systems, this method is more efficient than graphing.

EXAMPLE 1 Using the Elimination Method

Use the elimination method to solve the system

$$x + y = 5$$
$$x - y = 3.$$

Each equation in this system is a statement of equality, so, as we discussed above, the sum of the right sides equals the sum of the left sides. Adding in this way gives

$$(x + y) + (x - y) = 5 + 3.$$

Combine terms and simplify to get

$$2x = 8$$
$$x = 4. \qquad \text{Divide by 2.}$$

Notice that y has been eliminated. The result, $x = 4$, gives the x-value of the solution of the given system. To find the y-value of the solution, substitute 4 for x in either of the two equations of the system. Choosing the first equation, $x + y = 5$, gives

$$x + y = 5$$
$$4 + y = 5 \qquad \text{Let } x = 4.$$
$$y = 1.$$

The solution, (4, 1), can be checked by substituting 4 for x and 1 for y in both equations of the given system.

$$
\begin{array}{llll}
x + y = 5 & & x - y = 3 & \\
4 + 1 = 5 & ? & 4 - 1 = 3 & ? \\
5 = 5 & \text{True} & 3 = 3 & \text{True}
\end{array}
$$

Since both results are true, the solution set of the given system is $\{(4, 1)\}$.

 A system is not completely solved until values for *both* x and y are found. Do not stop after finding the value of only one variable. Remember to write the solution set as a set containing an ordered pair.

In general, we use the following steps to solve a linear system of equations by the elimination method.

Solving Linear Systems by Elimination

Step 1 **Write in standard form.** Write both equations of the system in standard form $Ax + By = C$.

Step 2 **Multiply.** Multiply one or both equations by appropriate numbers (if necessary) so that the coefficients of x (or y) are opposites of each other.

Step 3 **Add.** Add the two equations to get an equation with only one variable.

> **Solving Linear Systems by Elimination (continued)**
>
> *Step 4*   **Solve.**   Solve the equation from Step 3.
>
> *Step 5*   **Substitute.**   Substitute the solution from Step 4 into either of the original equations to find the value of the remaining variable.
>
> *Step 6*   **Find the solution set.**   Check the solution in both of the original equations. Then write the solution set.

It does not matter which variable is eliminated first. Usually we choose the one that is more convenient to work with.

E X A M P L E   2   **Using the Elimination Method**

Solve the system.

$$y + 11 = 2x$$
$$4 + 5x + y = 2y + 30$$

Step 1   Rewrite both equations in the form $Ax + By = C$, getting the system

$$-2x + y = -11 \qquad \text{Subtract } 2x \text{ and } 11.$$
$$5x - y = 26. \qquad \text{Subtract 4 and } 2y.$$

Step 2   Because the coefficients of y are 1 and -1, adding will eliminate y. It is not necessary to multiply either equation by a number.

Step 3   Add the two equations. This time we use vertical addition.

$$\begin{array}{r} 2x + y = -11 \\ 5x - y = 26 \\ \hline 3x \quad\;\; = 15 \end{array} \qquad \text{Add in columns.}$$

Step 4   Solve the equation.

$$3x = 15$$
$$x = 5 \qquad \text{Divide by 3.}$$

Step 5   Find the value of y by substituting 5 for x in either of the original equations. Choosing the first gives

$$y + 11 = 2x$$
$$y + 11 = 2(5) \qquad \text{Let } x = 5.$$
$$y = 10 - 11 \qquad \text{Subtract 11.}$$
$$y = -1.$$

Step 6   The solution set is $\{(5, -1)\}$. Check the solution by substitution into both of the original equations. Let $x = 5$ and $y = -1$.

$$y + 11 = 2x \qquad\qquad\qquad 4 + 5x + y = 2y + 30$$
$$(-1) + 11 = 2(5) \quad ? \qquad\qquad 4 + 5(5) + (-1) = 2(-1) + 30 \quad ?$$
$$10 = 10 \qquad \text{True} \qquad\qquad 28 = 28 \qquad \text{True}$$

Since $(5, -1)$ is a solution of *both* equations, our solution set is correct.

OBJECTIVE **2** Multiply when using the elimination method. In both examples above, a variable was eliminated by adding the equations. Sometimes we need to multiply both sides of one or both equations in a system by some number before adding will eliminate a variable.

EXAMPLE 3 Multiplying Both Equations When Using the Elimination Method

Solve the system.

$$2x + 3y = -15 \qquad (1)$$
$$5x + 2y = 1 \qquad (2)$$

Adding the two equations gives $7x + 5y = -14$, which does not eliminate either variable. However, we can multiply each equation by a suitable number so that the coefficients of one of the two variables are opposites. For example, to eliminate x, multiply both sides of equation (1) by 5, and both sides of equation (2) by -2.

$$\begin{array}{ll} 10x + 15y = -75 & \text{Multiply equation (1) by 5.} \\ \underline{-10x - 4y = -2} & \text{Multiply equation (2) by } -2. \\ 11y = -77 & \text{Add.} \\ y = -7 & \end{array}$$

Substituting -7 for y in either equation (1) or (2) gives $x = 3$. Check that the solution set of the system is $\{(3, -7)\}$.

OBJECTIVE **3** Use an alternative method to find the second value in a solution. Sometimes it is easier to find the value of the second variable in a solution by using the elimination method twice. The next example shows this approach.

EXAMPLE 4 Finding the Second Value Using an Alternative Method

Solve the system.

$$4x = 9 - 3y \qquad (1)$$
$$5x - 2y = 8 \qquad (2)$$

Rearrange the terms in equation (1) so that like terms are aligned in columns. Add $3y$ to both sides to get the system

$$4x + 3y = 9 \qquad (3)$$
$$5x - 2y = 8.$$

One way to proceed is to eliminate y by multiplying both sides of equation (3) by 2 and both sides of equation (2) by 3, and then adding.

$$\begin{array}{l} 8x + 6y = 18 \\ \underline{15x - 6y = 24} \\ 23x = 42 \\ x = \dfrac{42}{23} \end{array}$$

Substituting $\frac{42}{23}$ for x in one of the given equations would give y, but the arithmetic involved would be messy. Instead, solve for y by starting again with the original equations and eliminating x. Multiply both sides of equation (3) by 5 and both sides of equation (2) by -4, and then add.

$$
\begin{array}{rcrl}
20x + 15y & = & 45 \\
-20x + 8y & = & -32 \\
\hline
23y & = & 13 \\
y & = & \dfrac{13}{23}
\end{array}
$$

Check that the solution set is $\left\{\left(\frac{42}{23}, \frac{13}{23}\right)\right\}$.

When the value of the first variable is a fraction, the method used in Example 4 helps avoid arithmetic errors. Of course, this method could be used to solve any system of equations.

OBJECTIVE 4 Use the elimination method to solve special systems. The next example shows the elimination method when a system is inconsistent or the equations of the system are dependent. To contrast the elimination method with the substitution method, in part (b) we use the same system solved in Example 4 of the previous section.

EXAMPLE 5 Using the Elimination Method for an Inconsistent System or Dependent Equations

Solve each system by the elimination method.

(a) $2x + 4y = 5$
$4x + 8y = -9$

Multiply both sides of $2x + 4y = 5$ by -2; then add to $4x + 8y = -9$.

$$
\begin{array}{rcrl}
-4x - 8y & = & -10 \\
4x + 8y & = & -9 \\
\hline
0 & = & -19 & \quad \text{False}
\end{array}
$$

The false statement $0 = -19$ shows that the given system has solution set $\emptyset$.

(b) $3x - y = 4$
$-9x + 3y = -12$

Multiply both sides of the first equation by 3; then add the two equations.

$$
\begin{array}{rcrl}
9x - 3y & = & 12 \\
-9x + 3y & = & -12 \\
\hline
0 & = & 0 & \quad \text{True}
\end{array}
$$

As before, this result indicates that every solution of one equation is also a solution of the other; there are an infinite number of solutions. The solution set is $\{(x, y) \mid 3x - y = 4\}$.

Summary of Situations That May Occur

One of three situations may occur when any of the methods described so far is used to solve a linear system of equations.

1. The result is a statement such as $x = 2$ or $y = -3$. The solution set will have exactly one ordered pair. The graphs of the equations of the system will intersect at exactly one point.

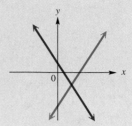

2. The final result is a false statement, such as $0 = 4$. In this case, the graphs are parallel lines and the solution set is $\emptyset$.

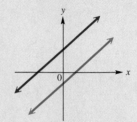

3. The final result is a true statement, such as $0 = 0$. The graphs of the equations of the system are the same line, and an infinite number of ordered pairs are solutions.

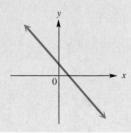

When no method of solution of a system is specified and a choice of substitution or elimination is allowed, use the following guidelines.

1. If one of the equations of the system is already solved for one of the variables, such as

$$3x + 4y = 9 \qquad \text{or} \qquad -5x + 3y = 9$$
$$y = 2x - 6 \qquad\qquad x = 3y - 7,$$

the substitution method is the better choice.

2. If both equations are in standard $Ax + By = C$ form, such as

$$4x - 11y = 3$$
$$-2x + 3y = 4,$$

and none of the variables has coefficient -1 or 1, the elimination method is the better choice.

3. If one or both of the equations are in standard form and the coefficient of one of the variables is -1 or 1, such as

$$3x + y = -2 \qquad \text{or} \qquad -x + 3y = -4$$
$$-5x + 2y = 4 \qquad\qquad 3x - 2y = 8,$$

use the elimination method, or solve for the variable with coefficient -1 or 1 and then use the substitution method.

9.3 EXERCISES

Answer true or false for each of the following statements. If false, tell why.

1. The ordered pair $(0, 0)$ *must* be a solution of a system of the form

$$Ax + By = 0$$
$$Cx + Dy = 0.$$

2. To eliminate the y-terms in the system

$$2x + 12y = 7$$
$$3x + 4y = 1,$$

we should multiply the bottom equation by 3 and then add.

3. The system

$$x + y = 1$$
$$x + y = 2$$

has $\emptyset$ as its solution set.

4. The ordered pair $(4, -5)$ cannot be a solution of a system that contains the equation $5x - 4y = 0$.

Solve each system by the elimination method. Check your answer. See Examples 1, 3, and 4.*

5. $x - y = -2$
   $x + y = 10$

6. $x + y = 10$
   $x - y = -6$

7. $2x + y = -5$
   $x - y = 2$

8. $2x + y = -15$
   $-x - y = 10$

9. $3x + 2y = 0$
   $-3x - y = 3$

10. $5x - y = 5$
    $-5x + 2y = 0$

11. $6x - 2y = -21$
    $-6x + 8y = 72$

12. $6x - 2y = -21$
    $6x + 8y = 72$

13. $2x - y = 12$
    $3x + 2y = -3$

14. $x + y - 3$
    $-3x + 2y = -19$

15. $3x + 3y = 33$
    $5x - 2y = 27$

16. $4x - 3y = -19$
    $3x + 2y = 24$

17. $5x + 4y = 12$
    $3x + 5y = 15$

18. $2x + 3y = 21$
    $5x - 2y = -14$

19. $5x - 4y = 15$
    $-3x + 6y = -9$

20. $4x + 5y = -16$
    $5x - 6y = -20$

21. $3x - 7y = 1$
    $-5x + 4y = 4$

22. $-4x + 3y = 2$
    $5x - 2y = -3$

23. Explain the steps you would use to solve the system $\begin{matrix} 2x + y = 5 \\ 5x + 3y = 11 \end{matrix}$ by substitution.

24. Explain the steps you would use to solve the system $\begin{matrix} 4x - 3y = -7 \\ 6x + 5y = 18 \end{matrix}$ by elimination.

Solve each system by elimination. See Examples 2 and 5.

25. $5x - 4y - 8x - 2 = 6x + 3y - 3$
    $4x - y = -2y - 8$

26. $2x - 8y + 3y + 2 = 5y + 16$
    $8x - 2y = 4x + 28$

27. $-2x + 3y = 12 + 2y$
    $2x - 5y + 4 = -8 - 4y$

28. $2x + 5y = 7 + 4y - x$
    $5x + 3y + 8 = 22 - x + y$

29. $7x - 9 + 2y - 8 = -3y + 4x + 13$
    $4y - 8x = -8 + 9x + 32$

30. $y + 9 = 3x - 2y + 6$
    $5 - 3x + 24 = -2x + 4y + 3$

31. $6x + 4y - 8 = x + 2y - 5$
    $-3x + 4y - 1 = -8x + 2y + 8$

32. $6x - 4y + 3 = 9x - 12y + 4$
    $5x + y + 2 = 8x - 7y + 6$

_____

*The authors thank Mitchel Levy of Broward Community College for his suggestions for this group of exercises.

In Exercises 33 and 34 (a) solve the system by the elimination method, (b) solve the system by the substitution method, and (c) tell which method you prefer for that particular system, and why.

33. $4x - 3y = -8$
    $x + 3y = 13$

34. $2x + 5y = 0$
    $x = -3y + 1$

Exercises 35 and 36 refer to the system

$$\frac{1}{3}x - \frac{1}{2}y = 7$$

$$\frac{1}{6}x + \frac{1}{3}y = 0.$$

35. One student solved the system by multiplying both equations by 6 to clear fractions, and another student multiplied by 12. Assuming they do all other work correctly, should they both get the same answer? Explain.

36. One student solved the system and wrote as his answer $\{12\}$ while another solved it and wrote $\{-6\}$ as her answer. Who, if either, was correct? Why?

Solve each system by any method. First clear all fractions.

37. $x + \frac{1}{3}y = y - 2$
    $\frac{1}{4}x + y = x + y$

38. $\frac{5}{3}x + 2y = \frac{1}{3} + y$
    $3x - 3 + \frac{y}{3} = -2 + 2x$

39. $\frac{x}{6} + \frac{y}{6} = 2$
    $-\frac{1}{2}x - \frac{1}{3}y = -8$

40. $\frac{x}{2} - \frac{y}{3} = 9$
    $\frac{x}{5} - \frac{y}{4} = 5$

41. $\frac{x}{3} - \frac{3y}{4} = -\frac{1}{2}$
    $\frac{x}{6} + \frac{y}{8} = \frac{3}{4}$

42. $\frac{x}{5} + 2y = \frac{16}{5}$
    $\frac{3x}{5} + \frac{y}{2} = -\frac{7}{5}$

▇ **RELATING CONCEPTS (EXERCISES 43–48)**

Attending the movies is one of America's favorite forms of entertainment. The graph shows how attendance gradually increased from 1991 to 1996. In 1991, attendance was 1141 million, as represented by the point $P(1991, 1141)$. In 1996, attendance was 1339 million, as represented by the point $Q(1996, 1339)$. We can find an equation of line segment PQ using a system of equations, and then use the equation to approximate the attendance in any of the years between 1991 and 1996.

Work Exercises 43–48 in order.

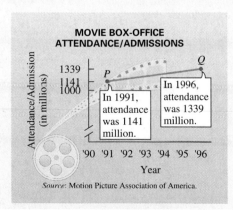

MOVIE BOX-OFFICE
ATTENDANCE/ADMISSIONS

Source: Motion Picture Association of America.

RELATING CONCEPTS (EXERCISES 43-48) (CONTINUED)

43. The line segment has an equation that can be written in the form $y = ax + b$. Using the coordinates of point P and letting $x = 1991$ and $y = 1141$, write an equation in the variables a and b.

44. Using the coordinates of point Q and letting $x = 1996$ and $y = 1339$, write a second equation in the variables a and b.

45. Write the system of equations formed from the two equations in Exercises 43 and 44, and solve the system using the elimination method.

46. What is the equation of the segment PQ?

47. Let $x = 1993$ in the equation of Exercise 46, and solve for y. How does the result compare with the actual figure of 1244 million?

48. The data for the years 1991 through 1996 do not lie in a perfectly straight line. Explain the pitfalls of relying too heavily on using the equation in Exercise 46 to predict attendance.

Did you make the connection that the equation of a line can be used to interpret and model real-life data?

9.4 Linear Systems of Equations in Three Variables

OBJECTIVES

1 Understand the geometry of systems of three equations in three unknowns.

2 Solve linear systems (with three equations and three unknowns) by elimination.

3 Solve linear systems (with three equations and three unknowns) where some of the equations have missing terms.

4 Solve special systems (with three equations and three unknowns).

FOR EXTRA HELP

 SSG Sec. 9.4
SSM Sec. 9.4

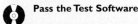 **Pass the Test Software**

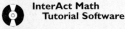 **InterAct Math**
        Tutorial Software

Video 14

A solution of an equation in three variables, such as $2x + 3y - z = 4$, is called an **ordered triple** and is written (x, y, z). For example, the ordered triples $(1, 1, 1)$ and $(10, -3, 7)$ are each solutions of $2x + 3y - z = 4$, since the numbers in these ordered triples satisfy the equation when used as replacements for x, y, and z, respectively.

In the rest of this chapter, the term *linear equation* is extended to first-degree equations of the form $Ax + By + Cz + \cdots + Dw = K$. For example, $2x + 3y - 5z = 7$ and $x - 2y - z + 3w - 2v = 8$ are linear equations, the first having three variables, and the second having five variables.

OBJECTIVE **1** Understand the geometry of systems of three equations in three unknowns. In this section we discuss the solution of a system of linear equations in three variables such as

$$4x + 8y + z = 2$$
$$x + 7y - 3z = -14$$
$$2x - 3y + 2z = 3.$$

Theoretically, a system of this type can be solved by graphing. However, the graph of a linear equation with three variables is a *plane* and not a line. Since the graph of each equation of the system is a plane, which requires three-dimensional graphing, this method is not practical. However, it does illustrate the number of solutions possible for such systems, as Figure 7 on the next page shows.

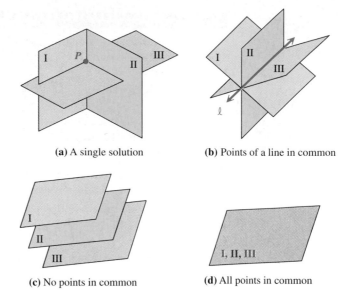

(a) A single solution   **(b)** Points of a line in common

(c) No points in common   **(d)** All points in common

Figure 7

Graphs of Linear Systems in Three Variables

1. The three planes may meet at a single, common point that is the solution of the system. See Figure 7(a).

2. The three planes may have the points of a line in common so that the set of points that satisfy the equation of the line is the solution of the system. See Figure 7(b).

3. The planes may have no points common to all three so that there is no solution for the system. See Figure 7(c).

4. The three planes may coincide so that the solution of the system is the set of all points on a plane. See Figure 7(d).

OBJECTIVE **2** Solve linear systems (with three equations and three unknowns) by elimination. Since graphing to find the solution set of a system of three equations in three variables is impractical, these systems are solved with an extension of the elimination method, summarized as follows.

Solving Linear Systems in Three Variables by Elimination

Step 1 **Eliminate a variable.** Use the elimination method to eliminate any variable from any two of the given equations. The result is an equation in two variables.

Step 2 **Eliminate the same variable again.** Eliminate the *same* variable from any *other* two equations. The result is an equation in the same two variables as in Step 1.

Step 3 **Eliminate a different variable and solve.** Use the elimination method to eliminate a second variable from the two equations in two variables that result from Steps 1 and 2. The result is an equation in one variable that gives the value of that variable.

> **Solving Linear Systems in Three Variables by Elimination (continued)**
>
> *Step 4* **Find a second value.** Substitute the value of the variable found in Step 3 into either of the equations in two variables to find the value of the second variable.
>
> *Step 5* **Find a third value.** Use the values of the two variables from Steps 3 and 4 to find the value of the third variable by substituting into any of the original equations.
>
> *Step 6* **Find the solution set.** Check the solution in all of the original equations. Then write the solution set.

E X A M P L E   1   Solving a System in Three Variables

Solve the system.

$$4x + 8y + z = 2 \tag{1}$$
$$x + 7y - 3z = -14 \tag{2}$$
$$2x - 3y + 2z = 3 \tag{3}$$

Step 1   As before, the elimination method involves eliminating a variable from the sum of two equations. The choice of which variable to eliminate is arbitrary. Suppose we decide to begin by eliminating z. To do this, multiply both sides of equation (1) by 3 and then add the result to equation (2).

$$
\begin{array}{ll}
12x + 24y + 3z = 6 & \text{Multiply both sides of (1) by 3.} \\
\underline{x + 7y - 3z = -14} & \text{(2)} \\
13x + 31y = -8 & \text{Add.} \qquad\qquad\qquad\qquad\qquad \textbf{(4)}
\end{array}
$$

Step 2   Equation (4) has only two variables. To get another equation without z, multiply both sides of equation (1) by -2 and add the result to equation (3). It is essential at this point to *eliminate the same variable, z.*

$$
\begin{array}{ll}
-8x - 16y - 2z = -4 & \text{Multiply both sides of (1) by } -2. \\
\underline{2x - 3y + 2z = 3} & \text{(3)} \\
-6x - 19y = -1 & \text{Add.} \qquad\qquad\qquad\qquad \textbf{(5)}
\end{array}
$$

Step 3   Now solve the system of equations (4) and (5) for x and y. This step is possible only if the *same* variable is eliminated in the first two steps.

$$
\begin{array}{ll}
78x + 186y = -48 & \text{Multiply both sides of (4) by 6.} \\
\underline{-78x - 247y = -13} & \text{Multiply both sides of (5) by 13.} \\
 -61y = -61 & \text{Add.} \\
y = 1 &
\end{array}
$$

Step 4   Substitute 1 for y in either equation (4) or (5). Choosing (5) gives

$$
\begin{array}{ll}
-6x - 19y = -1 & \text{(5)} \\
-6x - 19(\mathbf{1}) = -1 & \text{Let } y = 1. \\
-6x - 19 = -1 & \\
-6x = 18 & \\
x = -3. &
\end{array}
$$

Step 5 Substitute -3 for x and 1 for y in any one of the three given equations to find z. Choosing (1) gives

$$4x + 8y + z = 2 \qquad (1)$$
$$4(-3) + 8(1) + z = 2 \qquad \text{Let } x = -3 \text{ and } y = 1.$$
$$z = 6.$$

Step 6 It appears that the ordered triple $(-3, 1, 6)$ is the only solution of the system. Check that the solution satisfies all three equations of the system. We show the check here only for equation (1). The checks for equations (2) and (3) are requested in Exercise 2.

$$4x + 8y + z = 2 \qquad (1)$$
$$4(-3) + 8(1) + 6 = 2 \qquad ?$$
$$-12 + 8 + 6 = 2 \qquad ?$$
$$2 = 2 \qquad \text{True}$$

Because $(-3, 1, 6)$ also satisfies equations (2) and (3), the solution set is $\{(-3, 1, 6)\}$.

OBJECTIVE 3 Solve linear systems (with three equations and three unknowns) where some of the equations have missing terms. When this happens, one elimination step can be omitted.

EXAMPLE 2 Solving a System of Equations with Missing Terms

Solve the system.

$$6x - 12y = -5 \qquad \textbf{(6)}$$
$$8y + z = 0 \qquad \textbf{(7)}$$
$$9x - z = 12 \qquad \textbf{(8)}$$

Since equation (8) is missing the variable y, one way to begin the solution is to eliminate y again with equations (6) and (7).

$$12x - 24y = -10 \qquad \text{Multiply both sides of (6) by 2.}$$
$$\underline{24y + 3z = 0} \qquad \text{Multiply both sides of (7) by 3.}$$
$$12x + 3z = -10 \qquad \text{Add.} \qquad \textbf{(9)}$$

Use this result, together with equation (8), to eliminate z. Multiply both sides of equation (8) by 3. This gives

$$27x - 3z = 36 \qquad \text{Multiply both sides of (8) by 3.}$$
$$\underline{12x + 3z = -10} \qquad (9)$$
$$39x = 26 \qquad \text{Add.}$$
$$x = \frac{26}{39} = \frac{2}{3}.$$

Substitution into equation (8) gives

$$9x - z = 12 \qquad (8)$$

$$9\left(\frac{2}{3}\right) - z = 12 \qquad \text{Let } x = \tfrac{2}{3}.$$

$$6 - z = 12$$

$$z = \mathbf{-6}.$$

Substitution of -6 for z in equation (7) gives

$$8y + z = 0 \qquad (7)$$

$$8y - \mathbf{6} = 0 \qquad \text{Let } z = -6.$$

$$8y = 6$$

$$y = \frac{3}{4}.$$

Check in each of the original equations of the system to verify that the solution set of the system is $\left\{\left(\frac{2}{3}, \frac{3}{4}, -6\right)\right\}$.

OBJECTIVE **4** Solve special systems (with three equations and three unknowns). Linear systems with three variables may be inconsistent or may include dependent equations. The next examples illustrate these cases.

EXAMPLE 3 Solving an Inconsistent System with Three Variables

Solve the system.

$$2x - 4y + 6z = 5 \qquad \mathbf{(10)}$$

$$-x + 3y - 2z = -1 \qquad \mathbf{(11)}$$

$$x - 2y + 3z = 1 \qquad \mathbf{(12)}$$

Eliminate x by adding equations (11) and (12) to get the equation

$$y + z = 0.$$

Now, *eliminate x again,* using equations (10) and (12).

$$
\begin{array}{rl}
-2x + 4y - 6z = -2 & \text{Multiply both sides of (12) by } -2. \\
\underline{2x - 4y + 6z = 5} & (10) \\
0 = 3 & \text{False}
\end{array}
$$

The resulting false statement indicates that equations (10) and (12) have no common solution. Thus, the system is inconsistent and the solution set is $\emptyset$. The graph of this system would show at least two of the planes parallel to one another. (See Figure 7.)

NOTE If you get a false statement when adding as in Example 3, you do not need to go any further with the solution. Since two of the three planes are parallel, it is not possible for the three planes to have any common points.

┌───

E X A M P L E   4   Solving a System of Dependent Equations with Three Variables

Solve the system.

$$2x - 3y + 4z = 8 \tag{13}$$

$$-x + \frac{3}{2}y - 2z = -4 \tag{14}$$

$$6x - 9y + 12z = 24 \tag{15}$$

Multiplying both sides of equation (13) by 3 gives equation (15). Multiplying both sides of equation (14) by -6 also gives equation (15). Because of this, the equations are dependent. All three equations have the same graph, as illustrated in Figure 7(d). The solution set is written $\{(x, y, z) \mid 2x - 3y + 4z = 8\}$. Although any one of the three equations could be used to write the solution set, we prefer to use the equation with coefficients that are integers with no common factor (except 1). This is similar to our choice of a standard form for a linear equation earlier.

───┘

The method discussed in this section can be extended to solve larger systems. For example, to solve a system of four equations in four variables, eliminate a variable from three pairs of equations to get a system of three equations in three unknowns. Then proceed as shown above.

9.4 EXERCISES

1. The two equations

$$x + y + z = 6$$
$$2x - y + z = 3$$

have a common solution of $(1, 2, 3)$. Which one of the following equations would complete a system of three linear equations in three variables having solution set $\{(1, 2, 3)\}$?
(a) $3x + 2y - z = 1$ **(b)** $3x + 2y - z = 4$
(c) $3x + 2y - z = 5$ **(d)** $3x + 2y - z = 6$

2. Check that $(-3, 1, 6)$ is a solution for equations (2) and (3) in Example 1.

Solve each system of equations. See Example 1.

3. $3x + 2y + z = 8$
$2x - 3y + 2z = -16$
$x + 4y - z = 20$

4. $-3x + y - z = -10$
$-4x + 2y + 3z = -1$
$2x + 3y - 2z = -5$

5. $2x + 5y + 2z = 0$
$4x - 7y - 3z = 1$
$3x 8y - 2z = -6$

6. $5x - 2y + 3z = -9$
$4x + 3y + 5z = 4$
$2x + 4y - 2z = 14$

7. $x + y - z = -2$
$2x - y + z = -5$
$-x + 2y - 3z = -4$

8. $x + 2y + 3z = 1$
$-x - y + 3z = 2$
$-6x + y + z = -2$

Solve each system of equations. See Example 2.

9. $2x - 3y + 2z = -1$
$x + 2y + z = 17$
$2y - z = 7$

10. $2x - y + 3z = 6$
$x + 2y - z = 8$
$2y + z = 1$

11. $4x + 2y - 3z = 6$
$x - 4y + z = -4$
$-x + 2z = 2$

12. $2x + 3y - 4z = 4$
$x - 6y + z = -16$
$-x + 3z = 8$

13. $2x + y = 6$
$3y - 2z = -4$
$3x - 5z = -7$

14. $4x - 8y = -7$
$4y + z = 7$
$-8x + z = -4$

15. Using your immediate surroundings, give an example of three planes that
(a) intersect in a single point (b) do not intersect
(c) intersect in infinitely many points.

16. Explain how you can determine algebraically that a system of three linear equations in three variables has no solution. Then do the same for infinitely many solutions.

Solve each system of equations. See Examples 1, 3, and 4.

17. $2x + 2y - 6z = 5$
$-3x + y - z = -2$
$-x - y + 3z = 4$

18. $-2x + 5y + z = -3$
$5x + 14y - z = -11$
$7x + 9y - 2z = -5$

19. $-5x + 5y - 20z = -40$
$x - y + 4z = 8$
$3x - 3y + 12z = 24$

20. $x + 4y - z = 3$
$-2x - 8y + 2z = -6$
$3x + 12y - 3z = 9$

21. $2x + y - z = 6$
$4x + 2y - 2z = 12$
$-x - \frac{1}{2}y + \frac{1}{2}z = -3$

22. $2x - 8y + 2z = -10$
$-x + 4y - z = 5$
$\frac{1}{8}x - \frac{1}{2}y + \frac{1}{8}z = -\frac{5}{8}$

23. $x + y - 2z = 0$
$3x - y + z = 0$
$4x + 2y - z = 0$

24. $2x + 3y - z = 0$
$x - 4y + 2z = 0$
$3x - 5y - z = 0$

Extend the method of this section to solve each system.

25. $x + y + z - w = 5$
$2x + y - z + w = 3$
$x - 2y + 3z + w = 18$
$-x - y + z + 2w = 8$

26. $3x + y - z + 2w = 9$
$x + y + 2z - w = 10$
$x - y - z + 3w = -2$
$-x + y - z + w = -6$

■ **RELATING CONCEPTS (EXERCISES 27-36)**

Suppose that on a distant planet a function of the form

$$f(x) = ax^2 + bx + c \quad (a \neq 0)$$

describes the height in feet of a projectile x seconds after it has been projected upward.

Work Exercises 27–36 in order to see how this can be related to a system of three equations in three variables a, b, and c.

27. After 1 second, the height of a certain projectile is 128 feet. Thus, $f(1) = 128$. Use this information to find one equation in the variables a, b, and c. (*Hint:* Substitute 1 for x and 128 for $f(x)$.)

28. After 1.5 seconds, the height is 140 feet. Find a second equation in a, b, and c.

29. After 3 seconds, the height is 80 feet. Find a third equation in a, b, and c.

30. Write a system of three equations in a, b, and c, based on your answers in Exercises 27–29. Solve the system.

31. What is the function f for this particular projectile?

(continued)

RELATING CONCEPTS (EXERCISES 27-36) (CONTINUED)

32. What was the initial height of the projectile? (*Hint:* Find $f(0)$.)

33. The projectile reaches its maximum height in 1.625 seconds. Find its maximum height.

34. Verify that $f(3.25)$ also equals 0. What does this tell us about the projectile?

35. In Chapter 11 we discuss graphs of functions of the form $f(x) = ax^2 + bx + c$ ($a \neq 0$). Use a system of equations to find the values of a, b, and c for the function of this form that satisfies $f(1) = 2$, $f(-1) = 0$, and $f(-2) = 8$. Then write the expression for $f(x)$.

36. The accompanying table was generated by a graphing calculator for a function $Y_1 = ax^2 + bx + c$. Use any three points shown to find the values of a, b, and c. Then write the expression for Y_1.

X	Y1
1	8
2	15
3	24
4	35
5	48
6	63
7	80

X=1

Did you make the connection that a function can model the height of a propelled object? The height is a *function* of the time.

37. Discuss why it is necessary to eliminate the same variable in the first two steps of the elimination method with three equations and three variables.

38. In Step 3 of the elimination method for solving systems in three variables, does it matter which variable is eliminated? Explain.

9.5 Applications of Linear Systems of Equations

OBJECTIVES

1 Solve geometry problems using two variables.

2 Solve money problems using two variables.

3 Solve mixture problems using two variables.

4 Solve distance-rate-time problems using two variables.

5 Solve problems with three unknowns using a system of three equations.

Many applied problems involve more than one unknown quantity. Although most problems with two unknowns can be solved using just one variable, it often is easier to use two variables. To solve a problem with two unknowns, we must write two equations that relate the unknown quantities. The system formed by the pair of equations then can be solved using the methods of Sections 9.2 and 9.3.

The following steps, based on the six-step problem-solving method first introduced in Chapter 2, give a strategy for solving problems using more than one variable.

Solving an Applied Problem by Writing a System of Equations

Step 1 **Determine what you are to find.** Assign a variable for each unknown and *write down* what it represents.

Step 2 **Write down other information.** If appropriate, draw a figure or a diagram and label it using the variables from Step 1. Make a chart if necessary to summarize the information.

Step 3 **Write a system of equations.** Write as many equations as there are unknowns.

Step 4 **Solve the system.** Use elimination or substitution to solve the system.

FOR EXTRA HELP

SSG Sec. 9.5
SSM Sec. 9.5

Pass the Test Software

InterAct Math
    Tutorial Software

Video 15

Solving an Applied Problem by Writing a System of Equations (continued)

Step 5   **Answer the question(s).**   Be sure you have answered all questions posed.

Step 6   **Check.**   Check your solution(s) in the original problem. Be sure your answer makes sense.

OBJECTIVE ▮ Solve geometry problems using two variables. Problems about the perimeter of a geometric figure often involve two unknowns and can be solved using a system of equations.

EXAMPLE 1   Finding Dimensions of a Soccer Field

Unlike football, where the dimensions of a playing field cannot vary, a rectangular soccer field may have a width between 50 and 100 yards and a length between 50 and 100 yards. Suppose that one particular field has a perimeter of 320 yards. Its length measures 40 yards more than its width. What are the dimensions of this field? (*Source: Microsoft Encarta, Soccer.*)

Step 1   We are asked to find the dimensions of the field. Let L = the length and let W = the width.

Step 2   Figure 8 shows a soccer field with the length labeled L and the width labeled W.

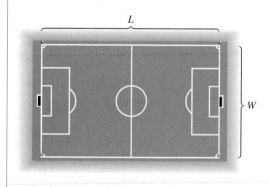

Figure 8

Step 3   Because the perimeter is 320 yards, one equation is found by using the perimeter formula:

$$2L + 2W = 320.$$

Because the length is 40 yards more than the width, we have

$$L = W + 40.$$

The system is, therefore,

$$2L + 2W = 320$$
$$L = W + 40.$$

Step 4   The second equation in the system is solved for *L,* so we can use substitution. Substitute $W + 40$ for L in the first equation, and solve for W.

$$2(W + 40) + 2W = 320 \qquad L = W + 40$$
$$2W + 80 + 2W = 320 \qquad \text{Distributive property}$$
$$4W + 80 = 320 \qquad \text{Add.}$$
$$4W = 240 \qquad \text{Subtract 80.}$$
$$\mathbf{W = 60} \qquad \text{Divide by 4.}$$

Let $W = \mathbf{60}$ in the equation $L = W + 40$.

$$L = 60 + 40 = \mathbf{100}.$$

Step 5   The length is **100** yards and the width is **60** yards.

Step 6   The perimeter of this soccer field is $2(\mathbf{100}) + 2(\mathbf{60}) = 320$ yards, and the length, 100 yards, is indeed 40 yards more than the width, since $\mathbf{100} - 40 = \mathbf{60}$. The solution is correct. ∎

OBJECTIVE **2** **Solve money problems using two variables.** Professional sport ticket prices increase annually. Average per-ticket prices in three of the four major sports (football, basketball, and hockey) now exceed $30.00.

 EXAMPLE 2   Solving a Problem about Ticket Prices

During the 1996–1997 National Hockey League and National Basketball Association seasons, 2 hockey tickets and 1 basketball ticket purchased at their average prices would have cost $110.40. One hockey ticket and 2 basketball tickets would have cost $106.32. What were the average ticket prices for the two sports? (*Source:* Team Marketing Report, Chicago.)

Let h represent the average price for a hockey ticket, and let b represent the average price for a basketball ticket. Because 2 hockey tickets and 1 basketball ticket cost a total of $110.40, one equation for the system is

$$2h + b = 110.40.$$

By similar reasoning, the second equation is

$$h + 2b = 106.32.$$

We must solve the following system:

$$2h + b = 110.40$$
$$h + 2b = 106.32.$$

To eliminate h, multiply the second equation by -2 and add.

$$2h + b = 110.40$$
$$\underline{-2h - 4b = -212.64}$$
$$-3b = -102.24 \qquad \text{Add.}$$
$$b = 34.08 \qquad \text{Divide by } -3.$$

To find the value of h, we can let $b = 34.08$ in the second equation.

$$h + 2(34.08) = 106.32 \qquad \text{Let } b = 34.08.$$
$$h + 68.16 = 106.32 \qquad \text{Multiply.}$$
$$h = 38.16 \qquad \text{Subtract 68.16.}$$

Thus, one basketball ticket costs \$34.08 and one hockey ticket costs \$38.16. A check indicates that these values satisfy both equations of the system.

CONNECTIONS

Problems that can be solved by writing a system of equations have been of interest historically. The following problem appeared in a Hindu work that dates back to about 850 A.D.

> The mixed price of 9 citrons and 7 fragrant wood apples is 107; again, the mixed price of 7 citrons and 9 fragrant wood apples is 101. O you arithmetician, tell me quickly the price of a citron and the price of a wood apple here, having distinctly separated those prices well.

(*Answer:* 8 for a citron and 5 for a wood apple.)

FOR DISCUSSION OR WRITING

What do you think is meant by "the mixed price" in the problem quoted above? Use the method discussed in this section to write a system of equations for this problem. Solve the system and compare your answer with the one given above.

OBJECTIVE 3   Solve mixture problems using two variables. We solved mixture problems earlier using one variable. For many mixture problems it seems more natural to use more than one variable and a system of equations.

EXAMPLE 3   Solving a Mixture Problem

How many ounces of 5% hydrochloric acid and of 20% hydrochloric acid must be combined to get 10 ounces of solution that is 12.5% hydrochloric acid?

Let x represent the number of ounces of 5% solution and y represent the number of ounces of 20% solution. A chart summarizes the given information.

Kind of Solution	Ounces of Solution	Ounces of Pure Acid
5%	x	$.05x$
20%	y	$.20y$
12.5%	10	$(.125)10$

Figure 9 also illustrates what is happening in the problem.

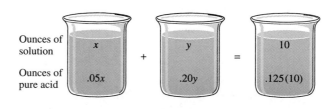

Figure 9

When x ounces of 5% solution and y ounces of 20% solution are combined, the total number of ounces is 10, so

$$x + y = 10. \tag{1}$$

The ounces of acid in the 5% solution, $.05x$, plus the ounces of acid in the 20% solution, $.20y$, should equal the total number of ounces of acid in the mixture, which is $(.125)10$, or 1.25. That is,

$$.05x + .20y = 1.25. \tag{2}$$

Eliminate x by first multiplying both sides of equation (2) by 100 to clear it of decimals, and then multiplying both sides of equation (1) by -5. Then add the results.

$$
\begin{array}{rl}
5x + 20y = 125 & \text{Multiply both sides of (2) by 100.} \\
\underline{-5x - 5y = -50} & \text{Multiply both sides of (1) by } -5. \\
15y = 75 & \text{Add.} \\
y = 5 &
\end{array}
$$

Since $y = 5$ and $x + y = 10$, x is also 5. Therefore, 5 ounces each of the 5% and the 20% solutions are required.

OBJECTIVE **4** Solve distance-rate-time problems using two variables. Motion applications require the distance formula, $d = rt$, where d is distance, r is rate (or speed), and t is time. These applications often lead to a system of equations, as in the next example.

EXAMPLE 4 Solving a Motion Problem

A car travels 250 kilometers in the same time that a truck travels 225 kilometers. If the speed of the car is 8 kilometers per hour faster than the speed of the truck, find both speeds.

A chart can be used to organize the information in problems about distance, rate, and time. Fill in the given information for each vehicle (in this case, distance) and use variables for the unknown speeds (rates) as follows.

	d	r	t
Car	250	x	
Truck	225	y	

The problem states that the car travels 8 kilometers per hour faster than the truck. Since the two speeds are x and y,

$$x = y + 8.$$

The chart shows nothing about time. To get an expression for time, solve the distance formula, $d = rt$, for t to get

$$\frac{d}{r} = t.$$

The two times can be written as $\frac{250}{x}$ and $\frac{225}{y}$. Since both vehicles travel for the same time,

$$\frac{250}{x} = \frac{225}{y}.$$

This is not a linear equation. However, multiplying both sides by xy gives

$$250y = 225x,$$

which is linear. Now solve the system.

$$x = y + 8 \tag{3}$$
$$250y = 225x \tag{4}$$

The substitution method can be used. Replace x with $y + 8$ in equation (4).

$$250y = 225(y + 8) \qquad \text{Let } x = y + 8.$$
$$250y = 225y + 1800 \qquad \text{Distributive property}$$
$$25y = 1800$$
$$y = 72$$

Since $x = y + 8$, the value of x is $72 + 8 = 80$. Check the solution in the original problem since one of the equations had variable denominators. Checking verifies that the speeds are 80 kilometers per hour for the car and 72 kilometers per hour for the truck.

OBJECTIVE **5** Solve problems with three unknowns using a system of three equations.

PROBLEM SOLVING

To solve applied problems with three or more unknowns, we extend the method given earlier for two unknowns. When three variables are used, three equations are necessary to find a solution.

EXAMPLE 5 Solving a Mixture Problem

A plant food is to be made from three chemicals. The mix must include 60% of the first two chemicals. The other two chemicals must be in a ratio of 4 to 3 by weight. How much of each chemical is needed to make 750 kilograms of the plant food?

First, choose variables to represent the three unknowns.

Let x = the number of kilograms of the first chemical;

y = the number of kilograms of the second chemical;

z = the number of kilograms of the third chemical.

Next, use the information in the problem to write three equations. To make 750 kilograms of the mix will require 60% of 750 kilograms of the first two chemicals, so

$$x + y = .60(750) = 450.$$

Since the ratio of the second and third chemicals is to be 4 to 3,

$$\frac{y}{z} = \frac{4}{3}.$$

Finally, the total amount of mix is to be 750 kilograms, so

$$x + y + z = 750.$$

Now, we must solve the system

$$x + y = 450$$
$$\frac{y}{z} = \frac{4}{3}$$
$$x + y + z = 750.$$

Use the method shown earlier to find the solution (50, 400, 300). The plant food should contain 50 kilograms of the first chemical, 400 kilograms of the second chemical, and 300 kilograms of the third chemical.

E X A M P L E   6   Solving a Business Production Problem

A company produces three color television sets, models X, Y, and Z. Each model X set requires 2 hours of electronics time, 2 hours of assembly time, and 1 hour of finishing time. Each model Y requires 1, 3, and 1 hours of electronics, assembly, and finishing time, respectively. Each model Z requires 3, 2, and 2 hours of the same time, respectively. There are 100 hours available for electronics, 100 hours available for assembly, and 65 hours available for finishing per week. How many of each model should be produced each week if all available time must be used?

Let x = the number of model X produced per week;

y = the number of model Y produced per week;

z = the number of model Z produced per week.

Organize the information in a chart.

	Each Model X	Each Model Y	Each Model Z	Totals
Hours of electronics time	2	1	3	100
Hours of assembly time	**2**	**3**	**2**	**100**
Hours of finishing time	1	1	2	65

The x model X sets require $2x$ hours of electronics, the y model Y sets require $1y$ (or y) hours of electronics, and the z model Z sets require $3z$ hours of electronics. Since 100 hours are available for electronics,

$$2x + y + 3z = 100.$$

Similarly, from the fact that 100 hours are available for assembly,

$$2x + 3y + 2z = 100,$$

and the fact that 65 hours are available for finishing leads to the equation

$$x + y + 2z = 65.$$

Solve the system

$$2x + y + 3z = 100$$
$$2x + 3y + 2z = 100$$
$$x + y + 2z = 65$$

to find $x = 15$, $y = 10$, and $z = 20$. The company should produce 15 model X, 10 model Y, and 20 model Z sets per week.

Notice the advantage of setting up the chart as in Example 6. By reading across, we can easily determine the coefficients and the constants in the system.

9.5 EXERCISES

For each application in this exercise set, select variables to represent the unknown quantities, write equations using the variables, and solve the resulting systems. Use the six-step method discussed in this section.

Use the techniques of this section to solve these sports-related problems.

1. During the 1996–1997 National Basketball Association regular season, the Chicago Bulls played 82 games. They won 56 more games than they lost. What was their win-loss record that year?

2. During the 1996–1997 National Basketball Association season, the Boston Celtics played 82 games. They lost 52 more games than they won. What was their win-loss record that year?

**1996 NBA FINAL STANDINGS
EASTERN CONFERENCE**

ATLANTIC DIVISION

Team	W	L
Miami	61	21
New York	57	25
Orlando	45	37
Washington	44	38
New Jersey	26	56
Philadelphia	22	60
Boston	—	—

CENTRAL DIVISION

Team	W	L
Chicago	—	—
Atlanta	56	26
Detroit	54	28
Charlotte	54	28
Cleveland	42	40
Indiana	39	43
Milwaukee	33	49
Toronto	30	52

Source: Sports Illustrated 1998 Sports Almanac.

3. During the 1996–1997 National Hockey League regular season, the Dallas Stars played 82 games. Together, their wins and losses totaled 74. They tied 18 fewer games than they lost. How many wins, losses, and ties did they have that year?

4. During the 1996–1997 National Hockey League season, the Boston Bruins played 82 games. Their losses and ties totaled 56, and they had 21 fewer wins than losses. How many wins, losses, and ties did they have that year?

**1996-97 NHL FINAL STANDINGS
WESTERN CONFERENCE
CENTRAL DIVISION**

	GP	W	L	T	GF	GA	Pts
Dallas	82	—	—	—	252	198	104
Detroit	82	38	26	18	253	197	94
Phoenix	82	38	37	7	240	243	83
St. Louis	82	36	35	11	236	239	83
Chicago	82	34	35	13	223	210	81
Toronto	82	30	44	8	230	273	68

**EASTERN CONFERENCE
NORTHEAST DIVISION**

	GP	W	L	T	GF	GA	Pts
Buffalo	82	40	30	12	237	208	92
Pittsburgh	82	38	36	8	285	280	84
Ottawa	82	31	36	15	226	234	77
Montreal	82	31	36	15	249	276	77
Hartford	82	32	39	11	226	256	75
Boston	82	—	—	—	234	300	61

Source: Sports Illustrated 1998 Sports Almanac.

The applications in Exercises 5–40 require solving systems with two variables, while those that follow require solving systems with three variables.

Solve each problem. See Example 1.

5. Pete and Venus measured the perimeter of a tennis court and found that it was 42 feet longer than it was wide, and had a perimeter of 228 feet. What were the length and the width of the tennis court?

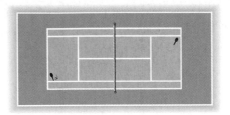

6. Scottie and Jamal found that the width of their basketball court was 44 feet less than the length. If the perimeter was 288 feet, what were the length and the width of their court?

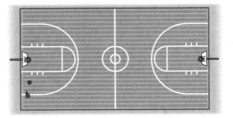

7. The length of a rectangle is 7 feet more than the width. If the length were decreased by 3 feet and the width were increased by 2 feet, the perimeter would be 32 feet. Find the length and width of the original rectangle.

8. The side of a square is 4 centimeters longer than the side of an equilateral triangle. The perimeter of the square is 24 centimeters more than the perimeter of the triangle. Find the lengths of a side of the square and a side of the triangle.

9. Find the measures of the angles marked x and y.

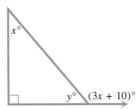

10. Find the measures of the angles marked x and y.

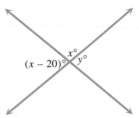

11. During 1995, two of the top-grossing concert tours were by Boyz II Men and Bruce Springsteen & the E St. Band. Together the two tours visited 174 cities. Boyz II Men visited 94 cities more than Bruce Springsteen. How many cities did each group visit? (*Source:* Pollstar.)

12. Among the ten top-selling videos in 1996, only three were not Disney films. Two of those were *Independence Day* (Fox) and *Twister* (Warner). Together, these two titles sold 31.5 million copies. *Independence Day* sold 12.5 million more copies than *Twister.* How many copies of each title were sold? (*Source:* Paul Kagan Associates Inc.)

The Fan Cost Index (FCI) represents the cost of four average-price tickets, four small soft drinks, two small beers, four hot dogs, parking for one car, two game programs, and two souvenir caps to a sporting event. For example, in 1997, the FCI for Major League Baseball was $105.63. This was by far the least for the four major professional sports. (*Source:* Team Marketing Report, Chicago.)

Solve each problem. See Example 2. Use the concept of FCI in Exercises 13 and 14.

13. For the 1996–1997 season, the FCI prices for the National Hockey League and the National Basketball Association totaled $423.12. The hockey FCI was $16.36 more than that of basketball. What were the FCIs for these sports?

14. For the 1996 season, the FCI prices for Major League Baseball and the National Football League totaled $311.03. The football FCI was $105.87 more than that of baseball. What were the FCIs for these sports?

15. Houston Community College has decided to supply its mathematics labs with color monitors. A trip to the local electronics outlet leads to the following information: 4 CGA monitors and 6 VGA monitors can be purchased for $4600, while 6 CGA monitors and 4 VGA monitors will cost $4400. What are the prices of a single CGA monitor and a single VGA monitor?

16. For his art class, Theodis bought 2 kilograms of dark clay and 3 kilograms of light clay, paying $22 for the clay. He later needed 1 kilogram of dark clay and 2 kilograms of light clay, costing $13 altogether. How much did he pay for each type of clay?

17. A factory makes use of two basic machines, *A* and *B,* which turn out two different products, yarn and thread. Each unit of yarn requires 1 hour on machine *A* and 2 hours on machine *B*, while each unit of thread requires 1 hour on *A* and 1 hour on *B*. Machine *A* runs 8 hours per day, while machine *B* runs 14 hours per day. How many units each of yarn and thread should the factory make to keep its machines running at capacity?

18. A biologist wants to grow two types of algae, green and brown. She has 15 kilograms of nutrient X and 26 kilograms of nutrient Y. A vat of green algae needs 2 kilograms of nutrient X and 3 kilograms of nutrient Y, while a vat of brown algae needs 1 kilogram of nutrient X and 2 kilograms of nutrient Y. How many vats of each type of algae should the biologist grow in order to use all the nutrients?

The formulas p = br (percentage = base × rate) and I = prt (simple interest = principal × rate × time) are used in the applications found in Exercises 23–34. To prepare for the use of these formulas, answer the questions in Exercises 19 and 20.

19. If a container of liquid contains 60 ounces of solution, what is the number of ounces of pure acid if the given solution contains the following acid concentrations?
   (a) 10%   **(b)** 25%   **(c)** 40%   **(d)** 50%

20. If $5000 is invested in an account paying simple annual interest, how much interest will be earned during the first year at the following rates?
   (a) 2%   **(b)** 3%   **(c)** 4%   **(d)** 3.5%

21. If a pound of turkey costs $.58, how much will x pounds cost?

22. If a ticket to the Kevin Costner movie *For Love of the Game* costs $7.00, and y tickets are sold, how much is collected from the sale?

Solve each problem. See Example 3.

23. How many gallons each of 25% alcohol and 35% alcohol should be mixed to get 20 gallons of 32% alcohol?

Kind of Solution	Gallons of Solution	Amount of Pure Alcohol
.25	x	.25x
.35	y	.35y
.32	20	.32(20)

24. How many liters each of 15% acid and 33% acid should be mixed to get 40 liters of 21% acid?

Kind of Solution	Liters of Solution	Amount of Pure Acid
.15	x	
.33	y	
.21	40	

25. Pure acid is to be added to a 10% acid solution to obtain 27 liters of a 20% acid solution. What amounts of each should be used?

26. A truck radiator holds 18 liters of fluid. How much pure antifreeze must be added to a mixture that is 4% antifreeze in order to fill the radiator with a mixture that is 20% antifreeze?

27. Joycelyn Lowe plans to mix pecan clusters that sell for $3.60 per pound with chocolate truffles that sell for $7.20 per pound to get a mixture that she can sell in Valentine boxes for $4.95 per pound. How much of the $3.60 clusters and the $7.20 truffles should she use to create 80 pounds of the mix?

	Number of Pounds	Price per Pound	Value of Candy
Pecan Clusters	x	3.60	3.60x
Chocolate Truffles	y	7.20	7.20y
Valentine Mixture	80	4.95	4.95(80)

28. A popular fruit drink is made by mixing fruit juices. Such a mixture with 50% juice is to be mixed with another mixture that is 30% juice to get 200 liters of a mixture that is 45% juice. How much of each should be used?

Kind of Juice	Number of Liters	Amount of Pure Fruit
.50	x	.50x
.30	y	
.45		

29. Tickets to a production of *Othello* at Nicholls State University cost $2.50 for general admission or $2.00 with a student identification. If 184 people paid to see a performance and $406 was collected, how many of each type of admission were sold?

30. A grocer plans to mix candy that sells for $1.20 a pound with candy that sells for $2.40 a pound to get a mixture that he plans to sell for $1.65 a pound. How much of the $1.20 and $2.40 candy should he use if he wants 80 pounds of the mix?

31. Jane Ann Lindstedt has been saving dimes and quarters. She has 94 coins in all. If the total value is $19.30, how many dimes and how many quarters does she have?

32. A teller at the Hibernia National Bank received a checking account deposit in twenty-dollar bills and fifty-dollar bills. She received a total of 70 bills, and the amount of the deposit was $3200. How many of each denomination were deposited?

33. A total of $3000 is invested, part at 2% simple interest and part at 4%. If the total annual return from the two investments is $100, how much is invested at each rate?

Principal	Rate	Interest
x	.02	.02x
y	.04	.04y
3000		100

34. An investor must invest a total of $15,000 in two accounts, one paying 4% annual simple interest, and the other 3%. If he wants to earn $550 annual interest, how much should he invest at each rate?

Principal	Rate	Interest
x	.04	
y	.03	
15,000		

The formula d = rt (distance = rate × time) is used in the applications found in Exercises 37–40. To prepare for the use of this formula, answer the questions in Exercises 35 and 36.

35. If the speed of a killer whale is 25 miles per hour, and the whale swims for y hours, how many miles does the whale travel?

36. If the speed of a boat in still water is 10 miles per hour, and the speed of the current of a river is x miles per hour, what is the speed of the boat
   (a) going upstream (that is, against the current) and
   (b) going downstream (that is, with the current)?

Downstream (with the current)

Upstream (against the current)

Solve each problem. See Example 4.

37. A freight train and an express train leave towns 390 kilometers apart, traveling toward one another. The freight train travels 30 kilometers per hour slower than the express train. They pass one another 3 hours later. What are their speeds?

38. A train travels 150 kilometers in the same time that a plane covers 400 kilometers. If the speed of the plane is 20 kilometers per hour less than 3 times the speed of the train, find both speeds.

39. Braving blizzard conditions on the planet Hoth, Luke Skywalker sets out at top speed in his snow speeder for a rebel base 3600 miles away. He travels into a steady headwind, and makes the trip in 2 hours. Returning, he finds that the trip back, still at top speed but now with a tailwind, takes only 1.5 hours. Find the top speed of Luke's snow speeder and the speed of the wind.

40. In his motorboat, Nguyen travels upstream at top speed to his favorite fishing spot, a distance of 36 miles, in two hours. Returning, he finds that the trip downstream, still at top speed, takes only 1.5 hours. Find the speed of Nguyen's boat and the speed of the current.

Solve each problem involving three unknowns. See Examples 5 and 6. (In Exercises 43–46, remember that the sum of the measures of the angles of a triangle is 180°.)

41. Gil Troutman has a collection of tropical fish. For each fish, he paid either $20, $40, or $65. The number of $40 fish is one less than twice the number of $20 fish. If there are 29 fish in all worth $1150, how many of each kind of fish are in the collection?

42. A motorcycle manufacturer produces three different models: the Avalon, the Durango, and the Roadripper. Production restrictions require it to make, on a monthly basis, 10 more Roadrippers than the total of the other models, and twice as many Durangos as Avalons. The shop must produce a total of 490 cycles per month. How many cycles of each type should be made per month?

43. In the figure shown, $z = x + 10$ and $x + y = 100$. Determine a third equation involving x, y, and z, and then find the measures of the three angles.

44. In the figure shown, x is 10 less than y and 20 less than z. Write a system of equations and find the measures of the three angles.

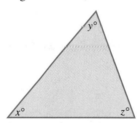

45. In a certain triangle, the measure of the second angle is $10°$ more than three times the first. The third angle measure is equal to the sum of the measures of the other two. Find the measures of the three angles.

46. The measure of the largest angle of a triangle is $12°$ less than the sum of the measures of the other two. The smallest angle measures $58°$ less than the largest. Find the measures of the angles.

47. The perimeter of a triangle is 70 centimeters. The longest side is 4 centimeters less than the sum of the other two sides. Twice the shortest side is 9 centimeters less than the longest side. Find the length of each side of the triangle.

48. The perimeter of a triangle is 56 inches. The longest side measures 4 inches less than the sum of the other two sides. Three times the shortest side is 4 inches more than the longest side. Find the lengths of the three sides.

49. A Mardi Gras trinket manufacturer supplies three wholesalers, A, B, and C. The output from a day's production is 320 cases of trinkets. She must send wholesaler A three times as many cases as she sends B, and she must send wholesaler C 160 cases less than she provides A and B together. How many cases should she send to each wholesaler to distribute the entire day's production to them?

50. A hardware supplier manufactures three kinds of clamps, types A, B, and C. Production restrictions require it to make 10 units more type C clamps than the total of the other types and twice as many type B clamps as type A. The shop must produce a total of 490 units of clamps per day. How many units of each type can be made per day?

51. The manager of a candy store wants to feature a special Easter candy mixture of jelly beans, small chocolate eggs, and marshmallow chicks. She plans to make 15 pounds of mix to sell at $1 a pound. Jelly beans sell for $.80 a pound, chocolate eggs for $2 a pound, and marshmallow chicks for $1 a pound. She will use twice as many pounds of jelly beans as eggs and chicks combined and fives times as many pounds of jelly beans as chocolate eggs. How many pounds of each candy should she use?

52. Three kinds of tickets are available for a Green Day concert: "up close," "in the middle," and "far out." "Up close" tickets cost $10 more than "in the middle" tickets, while "in the middle" tickets cost $10 more than "far out" tickets. Twice the cost of an "up close" ticket is $20 more than 3 times the cost of a "far out" seat. Find the price of each kind of ticket.

RELATING CONCEPTS (EXERCISES 53–56)

In a later chapter we will see that an equation of the form

$$x^2 + y^2 + ax + by + c = 0$$

may have a circle as its graph. It is a fact from geometry that given three noncollinear points (that is, points that do not all lie on the same straight line), there will be a circle that contains them. For example, the points $(4, 2)$, $(-5, -2)$, and $(0, 3)$ lie on the circle whose equation is

$$x^2 + y^2 - \frac{7}{5}x + \frac{27}{5}y - \frac{126}{5} = 0.$$

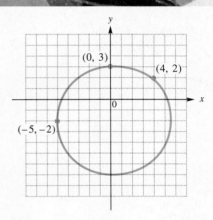

The circle is shown in the figure.

Work Exercises 53–56 in order, so that the equation of the circle passing through the points $(2, 1)$, $(-1, 0)$, *and* $(3, 3)$ *can be found.*

53. Let $x = 2$ and $y = 1$ in the equation $x^2 + y^2 + ax + by + c = 0$ to find an equation in a, b, and c.

54. Let $x = -1$ and $y = 0$ to find a second equation in a, b, and c.

55. Let $x = 3$ and $y = 3$ to find a third equation in a, b, and c.

56. Solve the system of equations formed by your answers in Exercises 53–55 to find the values of a, b, and c. What is the equation of the circle?

Did you make the connection between solving a system and finding an equation of a circle?

57. Observe the graph of the circle in the preceding Relating Concepts exercises. Explain why the graph is not that of a function.

58. Make up a problem similar to the one in Exercise 41 or Exercise 42, and solve it. (*Hint:* Start with the answer, and write the problem to fit the answer.)

9.6 Solving Linear Systems of Equations by Matrix Methods

OBJECTIVE 1 Define a matrix. An ordered array of numbers such as

$$\begin{bmatrix} 2 & 3 & 5 \\ 7 & 1 & 2 \end{bmatrix}$$

is called a **matrix.** The numbers are called **elements** of the matrix. Matrices (the plural of *matrix*) are named according to the number of **rows** and **columns** they contain. The rows are read horizontally and the columns are read vertically. For example, the first row in the matrix above is 2 3 5 and the first column is $\frac{2}{7}$. The matrix above is a 2 × 3 (read "two by three") matrix because it has 2 rows and 3 columns. The number of rows is given first, and then the number of columns.

The matrix

$$\begin{bmatrix} -1 & 0 \\ 1 & -2 \end{bmatrix}$$

is a 2 × 2 matrix, and the matrix

$$\begin{bmatrix} 8 & -1 & -3 \\ 2 & 1 & 6 \\ 0 & 5 & -3 \\ 5 & 9 & 7 \end{bmatrix}$$

is a 4 × 3 matrix. A **square matrix** is one that has the same number of rows as columns. The 2 × 2 matrix above is a square matrix.

Figure 10 shows how a graphing calculator displays the two matrices above. Work with matrices discussed in this section is made much easier by using technology when available.

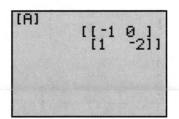
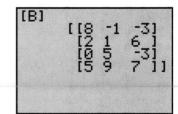

Figure 10

In this section we discuss a method of solving linear systems that uses matrices. This method is really just a very structured way of using the elimination method to solve a linear system. The advantage of this new method is that it can be done by a graphing calculator or a computer, allowing large systems of equations to be solved more easily.

OBJECTIVE 2 Write the augmented matrix for a system. To begin, we write an *augmented matrix* for the system. An **augmented matrix** has a vertical bar that separates the columns of the matrix into two groups. For example, to solve the system

$$x - 3y = 1$$
$$2x + y = -5,$$

start with the augmented matrix

$$\left[\begin{array}{rr|r} 1 & -3 & 1 \\ 2 & 1 & -5 \end{array}\right].$$

Place the coefficients of the variables to the left of the bar, and the constants to the right. The bar separates the coefficients from the constants. The matrix is just a shorthand way of writing the system of equations, so the rows of the augmented matrix can be treated the same as the equations of a system of equations.

We know that exchanging the position of two equations in a system does not change the system. Also, multiplying any equation in a system by a nonzero number does not change the system. Comparable changes to the augmented matrix of a system of equations produce new matrices that correspond to systems with the same solutions as the original system.

The following **row operations** produce new matrices that lead to systems having the same solutions as the original system.

Matrix Row Operations

1. Any two rows of the matrix may be interchanged.
2. The numbers in any row may be multiplied by any nonzero real number.
3. Any row may be changed by adding to the numbers of the row the product of a real number and the corresponding numbers of another row.

Examples of these row operations follow.

Using row operation 1,

$$\left[\begin{array}{rrr} 2 & 3 & 9 \\ 4 & 8 & -3 \\ 1 & 0 & 7 \end{array}\right] \quad \text{becomes} \quad \left[\begin{array}{rrr} 1 & 0 & 7 \\ 4 & 8 & -3 \\ 2 & 3 & 9 \end{array}\right]$$

by interchanging row 1 and row 3.

Using row operation 2,

$$\left[\begin{array}{rrr} 2 & 3 & 9 \\ 4 & 8 & -3 \\ 1 & 0 & 7 \end{array}\right] \quad \text{becomes} \quad \left[\begin{array}{rrr} 6 & 9 & 27 \\ 4 & 8 & -3 \\ 1 & 0 & 7 \end{array}\right]$$

by multiplying the numbers in row 1 by 3.

Using row operation 3,

$$\left[\begin{array}{rrr} 2 & 3 & 9 \\ 4 & 8 & -3 \\ 1 & 0 & 7 \end{array}\right] \quad \text{becomes} \quad \left[\begin{array}{rrr} 0 & 3 & -5 \\ 4 & 8 & -3 \\ 1 & 0 & 7 \end{array}\right]$$

by multiplying the numbers in row 3 by -2 and adding them to the corresponding numbers in row 1.

The third row operation corresponds to the way we eliminated a variable from a pair of equations in the previous sections.

OBJECTIVE **3** Use row operations to solve a system with two equations. Row operations can be used to rewrite a matrix until it is the matrix of a system where the solution is easy to find. The goal is a matrix in the form

$$\begin{bmatrix} 1 & a & b \\ 0 & 1 & c \end{bmatrix} \quad \text{or} \quad \begin{bmatrix} 1 & a & b & c \\ 0 & 1 & d & e \\ 0 & 0 & 1 & f \end{bmatrix}$$

for systems with two or three equations, respectively. Notice that there are 1s down the diagonal from upper left to lower right and 0s below the 1s. A matrix written this way is said to be in **row echelon form.** When these matrices are rewritten as systems of equations, the value of one variable is known, and the rest can be found by substitution. The following examples illustrate the method.

EXAMPLE 1 Using Row Operations to Solve a System with Two Variables

Use row operations to solve the system.

$$x - 3y = 1$$
$$2x + y = -5$$

We start by writing the augmented matrix of the system.

$$\begin{bmatrix} 1 & -3 & 1 \\ 2 & 1 & -5 \end{bmatrix}$$

Now we use the various row operations to change this matrix into one that leads to a system that is easier to solve.

It is best to work by columns. We start with the first column and make sure that there is a 1 in the first row, first column position. There already is a 1 in this position. Next, we get 0s in every position below the first. To get a 0 in row two, column one, we use the third row operation and add to the numbers in row two the result of multiplying each number in row one by -2. (We abbreviate this as $-2R_1 + R_2$.) Row one remains unchanged.

$$\begin{bmatrix} 1 & -3 & 1 \\ 2 + 1(-2) & 1 + -3(-2) & -5 + 1(-2) \end{bmatrix}$$

$$\uparrow \qquad \uparrow$$
Original number -2 times number
from row two   from row one

$$\begin{bmatrix} 1 & -3 & 1 \\ 0 & 7 & -7 \end{bmatrix} \qquad -2R_1 + R_2$$

The matrix now has a 1 in the first position of column one, with 0s in every position below the first.

Now we go to column two. A 1 is needed in row two, column two. We get this 1 by using the second row operation, multiplying each number of row two by $\frac{1}{7}$.

$$\begin{bmatrix} 1 & -3 & 1 \\ 0 & 1 & -1 \end{bmatrix} \qquad \tfrac{1}{7}R_2$$

This augmented matrix leads to the system of equations

$$\begin{array}{ll} 1x - 3y = 1 & \\ 0x + 1y = -1 & \end{array} \quad \text{or} \quad \begin{array}{l} x - 3y = 1 \\ y = -1. \end{array}$$

From the second equation, $y = -1$. We substitute -1 for y in the first equation to get

$$x - 3y = 1$$
$$x - 3(-1) = 1$$
$$x + 3 = 1$$
$$x = -2.$$

The solution set of the system is $\{(-2, -1)\}$. Check this solution by substitution in both equations.

If the augmented matrix of the system in Example 1 is entered as matrix A in a graphing calculator (Figure 11(a)) and the row echelon form of the matrix is found (Figure 11(b)), the system becomes

$$x + \frac{1}{2}y = -\frac{5}{2}$$
$$y = -1.$$

While this system looks different from the one we obtained in Example 1, it is equivalent, since its solution set is also $\{(-2, -1)\}$.

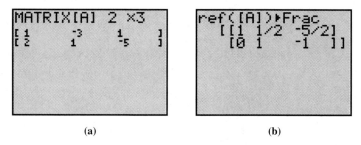

(a)   (b)

Figure 11

CONNECTIONS

One of the beautiful aspects of mathematics is that there are often many different ways to solve the same problem. It is likely that, given a particular system of equations, different students will use a variety of methods to solve the system. Regardless of the method used, they should all get the same answer.

FOR DISCUSSION OR WRITING

1. Show that the system represented by the matrix in Figure 11(b) has the same solution set as the system in Example 1.

2. Make up another system, having $y = -1$ as one of the equations, that has the same solution set as the system in Example 1.

OBJECTIVE **4** Use row operations to solve a system with three equations. A linear system with three equations is solved in a similar way. We use row operations to get 1s down the diagonal from left to right and all 0s below each 1.

EXAMPLE 2   **Using Row Operations to Solve a System with Three Variables**

Use matrix methods to solve the system.

$$x - y + 5z = -6$$
$$3x + 3y - z = 10$$
$$x + 3y + 2z = 5$$

Start by writing the augmented matrix of the system.

$$\begin{bmatrix} 1 & -1 & 5 & \bigm| & -6 \\ 3 & 3 & -1 & \bigm| & 10 \\ 1 & 3 & 2 & \bigm| & 5 \end{bmatrix}$$

This matrix already has 1 in row one, column one. Next get 0s in the rest of column one. First, add to row two the results of multiplying each number of row one by -3. This gives the matrix

$$\begin{bmatrix} 1 & -1 & 5 & \bigm| & -6 \\ 0 & 6 & -16 & \bigm| & 28 \\ 1 & 3 & 2 & \bigm| & 5 \end{bmatrix}. \qquad -3R_1 + R_2$$

Now add to the numbers in row three the results of multiplying each number of row one by -1.

$$\begin{bmatrix} 1 & -1 & 5 & \bigm| & -6 \\ 0 & 6 & -16 & \bigm| & 28 \\ 0 & 4 & -3 & \bigm| & 11 \end{bmatrix} \qquad -1R_1 + R_3$$

We get 1 in row two, column two by multiplying each number in row two by $\frac{1}{6}$.

$$\begin{bmatrix} 1 & -1 & 5 & \bigm| & -6 \\ 0 & 1 & -\dfrac{8}{3} & \bigm| & \dfrac{14}{3} \\ 0 & 4 & -3 & \bigm| & 11 \end{bmatrix} \qquad \frac{1}{6}R_2$$

Get 0 in row three, column two by adding to row three the results of multiplying each number in row two by -4.

$$\begin{bmatrix} 1 & -1 & 5 & \bigm| & -6 \\ 0 & 1 & -\dfrac{8}{3} & \bigm| & \dfrac{14}{3} \\ 0 & 0 & \dfrac{23}{3} & \bigm| & -\dfrac{23}{3} \end{bmatrix} \qquad -4R_2 + R_3$$

Finally, get 1 in row three, column three by multiplying each number in row three by $\frac{3}{23}$.

$$\begin{bmatrix} 1 & -1 & 5 & \bigm| & -6 \\ 0 & 1 & -\dfrac{8}{3} & \bigm| & \dfrac{14}{3} \\ 0 & 0 & 1 & \bigm| & -1 \end{bmatrix} \qquad \frac{3}{23}R_3$$

This final matrix gives the system of equations

$$x - y + 5z = -6$$

$$y - \frac{8}{3}z = \frac{14}{3}$$

$$z = -1.$$

Substitute -1 for z in the second equation, to get

$$y - \frac{8}{3}z = \frac{14}{3}$$

$$y - \frac{8}{3}(-1) = \frac{14}{3}$$

$$y + \frac{8}{3} = \frac{14}{3}$$

$$y = 2.$$

Finally, substitute 2 for y and -1 for z in the first equation.

$$x - y + 5z = -6$$

$$x - 2 + 5(-1) = -6$$

$$x - 2 - 5 = -6$$

$$x = 1$$

The solution set of the original system is $\{(1, 2, -1)\}$. This solution should be checked by substitution in the system.

OBJECTIVE **5** Use row operations to solve special systems. In the final example we show how to recognize inconsistent systems or systems with dependent equations when solving these systems with row operations.

EXAMPLE 3 Recognizing Inconsistent Systems or Dependent Equations

Use row operations to solve each system.

(a) $\quad 2x - 3y = 8$
$\quad -6x + 9y = 4$
Write the augmented matrix.

$$\begin{bmatrix} 2 & -3 & | & 8 \\ -6 & 9 & | & 4 \end{bmatrix}$$

Multiply the first row by $\frac{1}{2}$ to get 1 in row one, column one.

$$\begin{bmatrix} 1 & -\dfrac{3}{2} & \Big| & 4 \\ -6 & 9 & \Big| & 4 \end{bmatrix} \qquad \tfrac{1}{2}R_1$$

Multiply row one by 6 and add the results to row two.

$$\begin{bmatrix} 1 & -\dfrac{3}{2} & \Big| & 4 \\ 0 & 0 & \Big| & 28 \end{bmatrix} \qquad 6R_1 + R_2$$

The corresponding system of equations is

$$x - \frac{3}{2}y = 4$$

$$0 = 28,$$

which has no solution and so is inconsistent. The solution set is $\emptyset$.

(b) $-10x + 12y = 30$
   $5x - 6y = -15$

The augmented matrix is

$$\begin{bmatrix} -10 & 12 & \Big| & 30 \\ 5 & -6 & \Big| & -15 \end{bmatrix}$$

Multiply the first row by $-\frac{1}{10}$.

$$\begin{bmatrix} 1 & -\dfrac{6}{5} & \Big| & -3 \\ 5 & -6 & \Big| & -15 \end{bmatrix} \qquad -\tfrac{1}{10}R_1$$

Multiply the first row by -5 and add the products to row two to get

$$\begin{bmatrix} 1 & -\dfrac{6}{5} & \Big| & -3 \\ 0 & 0 & \Big| & 0 \end{bmatrix}. \qquad -5R_1 + R_2$$

The corresponding system is

$$x - \frac{6}{5}y = -3$$

$$0 = 0,$$

which has dependent equations. Clearing fractions in the first equation, we write the solution set as $\{(x, y) \mid 5x - 6y = -15\}$.

CONNECTIONS

An extension of the matrix method described in this section involves transforming an augmented matrix into **reduced row echelon form.** This form has 1s down the main diagonal and 0s above and below this diagonal. For example, the matrix for the system in Example 2 could be transformed into the following:

$$\left[\begin{array}{ccc|c} 1 & 0 & 0 & 1 \\ 0 & 1 & 0 & 2 \\ 0 & 0 & 1 & -1 \end{array}\right].$$

This would indicate that an equivalent system is

$$x = 1$$
$$y = 2$$
$$z = -1.$$

The graphing calculator screens in Figures 12(a) and (b) indicate how easily this transformation can be obtained using technology.

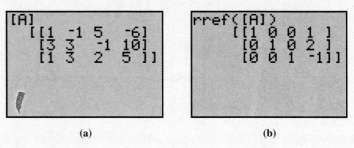

(a)   (b)

Figure 12

FOR DISCUSSION OR WRITING

1. Write the reduced row echelon form for the matrix of the system in Example 1.

2. If transforming to reduced row echelon form leads to all 0s in the final row, what kind of system is represented?

9.6 EXERCISES

1. Consider the matrix $\left[\begin{array}{ccc} -2 & 3 & 1 \\ 0 & 5 & -3 \\ 1 & 4 & 8 \end{array}\right]$ and answer the following.

   (a) What are the elements of the second row?
   (b) What are the elements of the third column?
   (c) Is this a square matrix? Explain why or why not.
   (d) Give the matrix obtained by interchanging the first and third rows.
   (e) Give the matrix obtained by multiplying the first row by $-\dfrac{1}{2}$.
   (f) Give the matrix obtained by multiplying the third row by 3 and adding to the first row.

2. Give the dimensions of each of the following matrices.

(a) $\begin{bmatrix} 3 & -7 \\ 4 & 5 \\ -1 & 0 \end{bmatrix}$
    (b) $\begin{bmatrix} 4 & 9 & 0 \\ -1 & 2 & -4 \end{bmatrix}$

(c)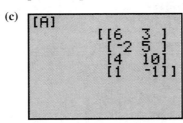
    (d)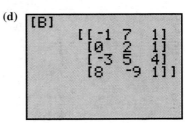

Complete the steps in the matrix solution of each system by filling in the boxes. Give the final system and the solution set. See Example 1.

3. $4x + 8y = 44$
    $2x - y = -3$

$\begin{bmatrix} 4 & 8 & | & 44 \\ 2 & -1 & | & -3 \end{bmatrix}$

$\begin{bmatrix} 1 & \blacksquare & | & \blacksquare \\ 2 & -1 & | & -3 \end{bmatrix} \quad \frac{1}{4}R_1$

$\begin{bmatrix} 1 & 2 & | & 11 \\ 0 & \blacksquare & | & \blacksquare \end{bmatrix} \quad -2R_1 + R_2$

$\begin{bmatrix} 1 & 2 & | & 11 \\ 0 & 1 & | & \blacksquare \end{bmatrix} \quad -\frac{1}{5}R_2$

4. $2x - 5y = -1$
    $3x + y = 7$

$\begin{bmatrix} 2 & -5 & | & -1 \\ 3 & 1 & | & 7 \end{bmatrix}$

$\begin{bmatrix} 1 & -\dfrac{5}{2} & | & \blacksquare \\ 3 & 1 & | & 7 \end{bmatrix} \quad \frac{1}{2}R_1$

$\begin{bmatrix} 1 & -\dfrac{5}{2} & | & -\dfrac{1}{2} \\ 0 & \blacksquare & | & \blacksquare \end{bmatrix} \quad -3R_1 + R_2$

$\begin{bmatrix} 1 & -\dfrac{5}{2} & | & -\dfrac{1}{2} \\ 0 & 1 & | & \blacksquare \end{bmatrix} \quad \frac{2}{17}R_2$

Use row operations to solve each system. See Examples 1 and 3.

5. $x + y = 5$
    $x - y = 3$

6. $x + 2y = 7$
    $x - y = -2$

7. $2x + 4y = 6$
    $3x - y = 2$

8. $4x + 5y = -7$
    $x - y = 5$

9. $3x + 4y = 13$
    $2x - 3y = -14$

10. $5x + 2y = 8$
    $3x - y = 7$

11. $-4x + 12y = 36$
    $x - 3y = 9$

12. $2x - 4y = 8$
    $-3x + 6y = 5$

13. Write a short explanation of each of the following. Include examples.
   (a) matrix
   (b) row of a matrix
   (c) column of a matrix
   (d) square matrix
   (e) augmented matrix
   (f) row operations on a matrix

14. Compare the use of the third row operation on a matrix and the elimination method of solving a system of linear equations. Give examples.

Complete the steps in the matrix solution of each system by filling in the boxes. Give the final system and the solution set. See Example 2.

15. $x + y - z = -3$
$2x + y + z = 4$
$5x - y + 2z = 23$

$$\begin{bmatrix} 1 & 1 & -1 & -3 \\ 2 & 1 & 1 & 4 \\ 5 & -1 & 2 & 23 \end{bmatrix}$$

$$\begin{bmatrix} 1 & 1 & -1 & -3 \\ 0 & ■ & ■ & ■ \\ 0 & ■ & ■ & ■ \end{bmatrix} \begin{array}{l} -2R_1 + R_2 \\ -5R_1 + R_3 \end{array}$$

$$\begin{bmatrix} 1 & 1 & -1 & -3 \\ 0 & 1 & ■ & ■ \\ 0 & -6 & 7 & 38 \end{bmatrix} -1R_2$$

$$\begin{bmatrix} 1 & 1 & -1 & -3 \\ 0 & 1 & -3 & -10 \\ 0 & 0 & ■ & ■ \end{bmatrix} 6R_2 + R_3$$

$$\begin{bmatrix} 1 & 1 & -1 & -3 \\ 0 & 1 & -3 & -10 \\ 0 & 0 & 1 & ■ \end{bmatrix} -\frac{1}{11}R_3$$

16. $2x + y + 2z = 11$
$2x - y - z = -3$
$3x + 2y + z = 9$

$$\begin{bmatrix} 2 & 1 & 2 & 11 \\ 2 & -1 & -1 & -3 \\ 3 & 2 & 1 & 9 \end{bmatrix}$$

$$\begin{bmatrix} 1 & ■ & ■ & ■ \\ 2 & -1 & -1 & -3 \\ 3 & 2 & 1 & 9 \end{bmatrix} \frac{1}{2}R_1$$

$$\begin{bmatrix} 1 & \frac{1}{2} & 1 & \frac{11}{2} \\ 0 & ■ & ■ & ■ \\ 0 & ■ & ■ & ■ \end{bmatrix} \begin{array}{l} -2R_1 + R_2 \\ -3R_1 + R_3 \end{array}$$

$$\begin{bmatrix} 1 & \frac{1}{2} & 1 & \frac{11}{2} \\ 0 & 1 & ■ & ■ \\ 0 & \frac{1}{2} & -2 & -\frac{15}{2} \end{bmatrix} -\frac{1}{2}R_2$$

$$\begin{bmatrix} 1 & \frac{1}{2} & 1 & \frac{11}{2} \\ 0 & 1 & \frac{3}{2} & 7 \\ 0 & 0 & ■ & ■ \end{bmatrix} -\frac{1}{2}R_2 + R_3$$

$$\begin{bmatrix} 1 & \frac{1}{2} & 1 & \frac{11}{2} \\ 0 & 1 & \frac{3}{2} & 7 \\ 0 & 0 & 1 & ■ \end{bmatrix} -\frac{4}{11}R_3$$

Use row operations to solve each system. See Examples 2 and 3.

17. $x + y - 3z = 1$
$2x - y + z = 9$
$3x + y - 4z = 8$

18. $2x + 4y - 3z = -18$
$3x + y - z = -5$
$x - 2y + 4z = 14$

19. $x + y - z = 6$
$2x - y + z = -9$
$x - 2y + 3z = 1$

20. $x + 3y - 6z = 7$
$2x - y + 2z = 0$
$x + y + 2z = -1$

21. $x - y = 1$
$y - z = 6$
$x + z = -1$

22. $x + y = 1$
$2x - z = 0$
$y + 2z = -2$

23. $x - 2y + z = 4$
$3x - 6y + 3z = 12$
$-2x + 4y - 2z = -8$

24. $4x + 8y + 4z = 9$
$x + 3y + 4z = 10$
$5x + 10y + 5z = 12$

The augmented matrix for the system in Exercise 3 is shown in the graphing calculator screen on the top left as matrix A. The screen on the top right shows the row echelon form for A. Compare it to the matrix shown in the answer section for Exercise 3. The screen at the bottom shows the reduced row echelon form, and from this it can be determined by inspection that the solution set of the system is $\{(1, 5)\}$.

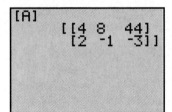
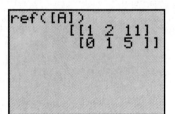
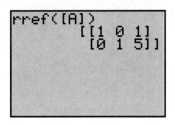

Use a graphing calculator and either one of the two matrix methods illustrated to solve each system.

25. $4x + y = 5$
$2x + y = 3$

26. $5x + 3y = 7$
$7x - 3y = -19$

27. $5x + y - 3z = -6$
$2x + 3y + z = 5$
$-3x - 2y + 4z = 3$

28. $x + y + z = 3$
$3x \quad 3y \quad 4z = 1$
$x + y + 3z = 11$

29. $x + z = -3$
$y + z = 3$
$x + y = 8$

30. $x - y = -1$
$-y + z = -2$
$x + z = -2$

9.7 Determinants and Cramer's Rule

Four methods for solving linear systems have now been presented: graphing, substitution, elimination, and a matrix method. A method of solving linear systems by using *determinants* is introduced later in this section.

Associated with every *square* matrix is a real number called the **determinant** of the matrix. A determinant is symbolized by the entries of the matrix placed between two vertical lines, such as

$$\begin{vmatrix} 2 & 3 \\ 7 & 1 \end{vmatrix} \quad \text{or} \quad \begin{vmatrix} 7 & 4 & 3 \\ 0 & 1 & 5 \\ 6 & 0 & 1 \end{vmatrix}.$$

Like matrices, determinants are named according to the number of rows and columns they contain. For example, the first determinant shown is a 2 × 2 (read "two by two") determinant. The second is a 3 × 3 determinant.

OBJECTIVE **1** Evaluate 2 × 2 determinants. The value of the 2 × 2 determinant

$$\begin{vmatrix} a & b \\ c & d \end{vmatrix}$$

is defined as follows.

Value of a 2 × 2 Determinant

$$\begin{vmatrix} a & b \\ c & d \end{vmatrix} = ad - bc$$

EXAMPLE 1 Evaluating a 2 × 2 Determinant

Evaluate the determinant.

$$\begin{vmatrix} -1 & -3 \\ 4 & -2 \end{vmatrix}$$

Here $a = -1$, $b = -3$, $c = 4$, and $d = -2$, and

$$\begin{vmatrix} -1 & -3 \\ 4 & -2 \end{vmatrix} = (-1)(-2) - (-3)(4) = 2 + 12 = 14.$$

A 3 × 3 determinant can be evaluated in a similar way.

Value of a 3 × 3 Determinant

$$\begin{vmatrix} a_1 & b_1 & c_1 \\ a_2 & b_2 & c_2 \\ a_3 & b_3 & c_3 \end{vmatrix} = (a_1 b_2 c_3 + b_1 c_2 a_3 + c_1 a_2 b_3) - (a_3 b_2 c_1 + b_3 c_2 a_1 + c_3 a_2 b_1)$$

This rule for evaluating a 3 × 3 determinant is hard to remember. A method for calculating a 3 × 3 determinant that is easier to use is based on the definition above. Rearranging terms and using the distributive property gives

$$\begin{vmatrix} a_1 & b_1 & c_1 \\ a_2 & b_2 & c_2 \\ a_3 & b_3 & c_3 \end{vmatrix} = a_1(b_2 c_3 - b_3 c_2) - a_2(b_1 c_3 - b_3 c_1) + a_3(b_1 c_2 - b_2 c_1). \qquad (1)$$

Each of the quantities in parentheses represents a 2 × 2 determinant, which is that part of the 3 × 3 determinant remaining when the row and column of the multiplier are eliminated, as shown below.

$$a_1(b_2 c_3 - b_3 c_2) \quad \begin{vmatrix} a_1 & b_1 & c_1 \\ a_2 & b_2 & c_2 \\ a_3 & b_3 & c_3 \end{vmatrix}$$

$$a_2(b_1 c_3 - b_3 c_1) \quad \begin{vmatrix} a_1 & b_1 & c_1 \\ a_2 & b_2 & c_2 \\ a_3 & b_3 & c_3 \end{vmatrix}$$

$$a_3(b_1 c_2 - b_2 c_1) \quad \begin{vmatrix} a_1 & b_1 & c_1 \\ a_2 & b_2 & c_2 \\ a_3 & b_3 & c_3 \end{vmatrix}$$

These 2×2 determinants are called **minors** of the elements in the 3×3 determinant. In the determinant above, the minors of a_1, a_2, and a_3 are, respectively,

$$\begin{vmatrix} b_2 & c_2 \\ b_3 & c_3 \end{vmatrix}, \quad \begin{vmatrix} b_1 & c_1 \\ b_3 & c_3 \end{vmatrix}, \quad \begin{vmatrix} b_1 & c_1 \\ b_2 & c_2 \end{vmatrix}.$$

OBJECTIVE 2   Use expansion by minors to evaluate 3×3 determinants.   A 3×3 determinant can be evaluated by multiplying each element in the first column by its minor and combining the products as shown in equation (1). This is called **expansion of the determinant by minors** about the first column.

E X A M P L E   2   Evaluating a 3×3 Determinant

Evaluate the determinant by expanding by minors about the first column.

$$\begin{vmatrix} 1 & 3 & -2 \\ -1 & -2 & -3 \\ 1 & 1 & 2 \end{vmatrix}$$

In this determinant, $a_1 = 1$, $a_2 = -1$, and $a_3 = 1$. Multiply each of these numbers by its minor and combine the three terms using the definition. Notice that the second term in the definition is *subtracted*.

$$\begin{vmatrix} 1 & 3 & -2 \\ -1 & -2 & -3 \\ 1 & 1 & 2 \end{vmatrix} = 1\begin{vmatrix} -2 & -3 \\ 1 & 2 \end{vmatrix} - (-1)\begin{vmatrix} 3 & -2 \\ 1 & 2 \end{vmatrix} + 1\begin{vmatrix} 3 & -2 \\ -2 & -3 \end{vmatrix}$$

$$= 1[(-2)(2) - (-3)(1)] + 1[(3)(2) - (-2)(1)]$$
$$\quad + 1[(3)(-3) - (-2)(-2)]$$
$$= 1(-1) + 1(8) + 1(-13)$$
$$= -1 + 8 - 13$$
$$= -6$$

To get equation (1) we could have rearranged terms in the definition of the determinant and factored out the three elements of the second or third columns or of any of the three rows. Therefore, expanding by minors about any row or any column results in the same value for a 3×3 determinant. To determine the correct signs for the terms of other expansions, the following **array of signs** is helpful.

Array of Signs for a 3×3 Determinant

$$\begin{matrix} + & - & + \\ - & + & - \\ + & - & + \end{matrix}$$

The signs alternate for each row and column beginning with $+$ in the first row, first column position. For example, if the expansion is to be about the second column, the first term would have a minus sign associated with it, the second term a plus sign, and the third term a minus sign.

E X A M P L E 3 Evaluating a 3 × 3 Determinant by a Different Expansion

Evaluate the determinant of Example 2 by expansion by minors about the second column.

$$\begin{vmatrix} 1 & 3 & -2 \\ -1 & -2 & -3 \\ 1 & 1 & 2 \end{vmatrix} = -3 \begin{vmatrix} -1 & -3 \\ 1 & 2 \end{vmatrix} + (-2) \begin{vmatrix} 1 & -2 \\ 1 & 2 \end{vmatrix} - 1 \begin{vmatrix} 1 & -2 \\ -1 & -3 \end{vmatrix}$$

$$= -3(1) - 2(4) - 1(-5)$$
$$= -3 - 8 + 5$$
$$= -6$$

As expected, the result is the same as in Example 2.

OBJECTIVE 3 Use a graphing calculator to evaluate determinants. The graphing calculator function det(A) assigns to each square matrix A one and only one real number, the determinant of A.

E X A M P L E 4 Evaluating Determinants Using a Graphing Calculator

Evaluate the determinants in Examples 1 and 2 using a graphing calculator.

Figure 13 shows how a graphing calculator displays the correct value for the determinant in Example 1. Similarly, Figure 14 supports the result of Example 2.

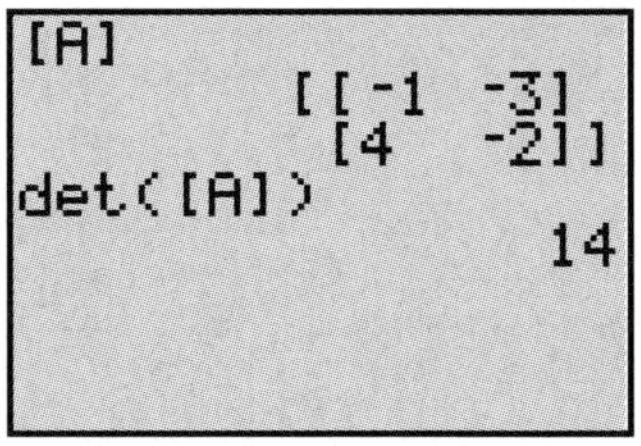

Figure 13 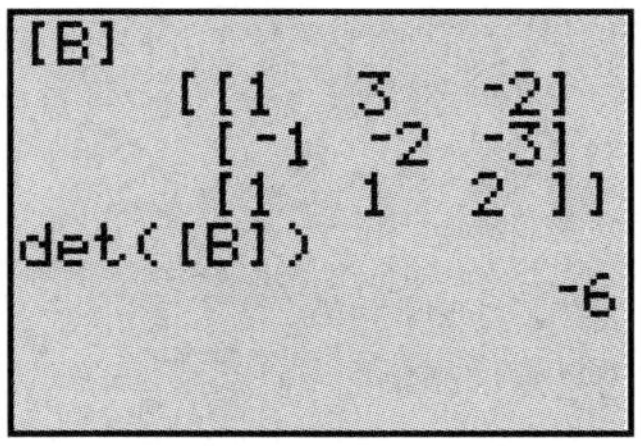   Figure 14

CONNECTIONS

Determinants of larger dimensions (such as 4 × 4) can be evaluated by extending the concepts presented thus far. However, because of the tedious calculations and chance for error, they are usually evaluated by computer or graphing calculator. For example, the determinant

$$\begin{vmatrix} -1 & -2 & 3 & 2 \\ 0 & 1 & 4 & -2 \\ 3 & -1 & 4 & 0 \\ 2 & 1 & 0 & 3 \end{vmatrix}$$

is equal to −185, as shown in the graphing calculator screen in Figure 15.

CONNECTIONS (CONTINUED)

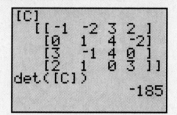

Figure 15

FOR DISCUSSION OR WRITING

Using the array of signs

$$
\begin{array}{cccc}
+ & - & + & - \\
- & + & - & + \\
+ & - & + & - \\
- & + & - & +
\end{array}
$$

evaluate the determinant above by hand, expanding about the fourth row.

OBJECTIVE 4 Understand the derivation of Cramer's rule. Determinants can be used to solve a system of the form

$$a_1x + b_1y = c_1 \qquad\qquad (1)$$
$$a_2x + b_2y = c_2. \qquad\qquad (2)$$

The result will be a formula that can be used for any system of two equations with two unknowns. To get this general solution, we eliminate y and solve for x by first multiplying both sides of equation (1) by b_2 and both sides of equation (2) by $-b_1$. Then we add these results and solve for x.

$$
\begin{array}{ll}
a_1b_2x + b_1b_2y = c_1b_2 & b_2 \text{ times both sides of equation (1)} \\
\underline{-a_2b_1x - b_1b_2y = -c_2b_1} & -b_1 \text{ times both sides of equation (2)} \\
(a_1b_2 - a_2b_1)x = c_1b_2 - c_2b_1 &
\end{array}
$$

$$x = \dfrac{c_1b_2 - c_2b_1}{a_1b_2 - a_2b_1} \quad (\text{if } a_1b_2 - a_2b_1 \neq 0)$$

To solve for y, we multiply both sides of equation (1) by $-a_2$ and both sides of equation (2) by a_1 and add.

$$
\begin{array}{ll}
-a_1a_2x - a_2b_1y = -a_2c_1 & -a_2 \text{ times both sides of (1)} \\
\underline{a_1a_2x + a_1b_2y = a_1c_2} & a_1 \text{ times both sides of (2)} \\
(a_1b_2 - a_2b_1)y = a_1c_2 - a_2c_1 &
\end{array}
$$

$$y = \dfrac{a_1c_2 - a_2c_1}{a_1b_2 - a_2b_1}$$

Both numerators and the common denominator of these values for x and y can be written as determinants, since

$$a_1c_2 - a_2c_1 = \begin{vmatrix} a_1 & c_1 \\ a_2 & c_2 \end{vmatrix},$$

$$c_1b_2 - c_2b_1 = \begin{vmatrix} c_1 & b_1 \\ c_2 & b_2 \end{vmatrix},$$

and

$$a_1b_2 - a_2b_1 = \begin{vmatrix} a_1 & b_1 \\ a_2 & b_2 \end{vmatrix}.$$

Using these results, the solutions for x and y become

$$x = \frac{\begin{vmatrix} c_1 & b_1 \\ c_2 & b_2 \end{vmatrix}}{\begin{vmatrix} a_1 & b_1 \\ a_2 & b_2 \end{vmatrix}} \quad \text{and} \quad y = \frac{\begin{vmatrix} a_1 & c_1 \\ a_2 & c_2 \end{vmatrix}}{\begin{vmatrix} a_1 & b_1 \\ a_2 & b_2 \end{vmatrix}}, \quad \text{if } \begin{vmatrix} a_1 & b_1 \\ a_2 & b_2 \end{vmatrix} \neq 0.$$

For convenience, we denote the three determinants in the solution as

$$\begin{vmatrix} a_1 & b_1 \\ a_2 & b_2 \end{vmatrix} = D, \quad \begin{vmatrix} c_1 & b_1 \\ c_2 & b_2 \end{vmatrix} = D_x, \quad \begin{vmatrix} a_1 & c_1 \\ a_2 & c_2 \end{vmatrix} = D_y.$$

Note that the elements of D are the four coefficients of the variables in the given system; the elements of D_x are obtained by replacing the coefficients of x by the respective constants; the elements of D_y are obtained by replacing the coefficients of y by the respective constants.

These results are summarized as **Cramer's rule.**

Cramer's Rule for 2 × 2 Systems

Given the system

$$a_1x + b_1y = c_1$$
$$a_2x + b_2y = c_2,$$

with

$$a_1b_2 - a_2b_1 \neq 0,$$

then

$$x = \frac{\begin{vmatrix} c_1 & b_1 \\ c_2 & b_2 \end{vmatrix}}{\begin{vmatrix} a_1 & b_1 \\ a_2 & b_2 \end{vmatrix}} = \frac{D_x}{D} \quad \text{and} \quad y = \frac{\begin{vmatrix} a_1 & c_1 \\ a_2 & c_2 \end{vmatrix}}{\begin{vmatrix} a_1 & b_1 \\ a_2 & b_2 \end{vmatrix}} = \frac{D_y}{D}$$

OBJECTIVE 5 Apply Cramer's rule to solve linear systems. To use Cramer's rule to solve a system of linear equations, find the three determinants, D, D_x, and D_y, and then write the necessary quotients for x and y.

CAUTION As indicated above, Cramer's rule does not apply if $D = a_1b_2 - a_2b_1$ is 0. When $D = 0$, the system is inconsistent or has dependent equations. For this reason, it is a good idea to evaluate D first.

EXAMPLE 5   Using Cramer's Rule for a 2 × 2 System
Use Cramer's rule to solve the system.

$$5x + 7y = -1$$
$$6x + 8y = 1$$

By Cramer's rule, $x = D_x/D$ and $y = D_y/D$. We will find D first, since if $D = 0$, Cramer's rule does not apply. If $D \neq 0$, then we will find D_x and D_y.

$$D = \begin{vmatrix} 5 & 7 \\ 6 & 8 \end{vmatrix} = 5(8) - 6(7) = -2$$

$$D_x = \begin{vmatrix} -1 & 7 \\ 1 & 8 \end{vmatrix} = (-1)8 - 7(1) = -15$$

$$D_y = \begin{vmatrix} 5 & -1 \\ 6 & 1 \end{vmatrix} = 5(1) - (-1)6 = 11$$

From Cramer's rule,

$$x = \frac{D_x}{D} = \frac{-15}{-2} = 7.5$$

and

$$y = \frac{D_y}{D} = \frac{11}{-2} = -5.5.$$

The solution set is $\{(7.5, -5.5)\}$, as we can verify by checking in the given system. ∎

Because graphing calculators can evaluate determinants, Cramer's rule can be applied using them. Figure 16 shows how the work of Example 5 is accomplished, with D the determinant of matrix A, D_x the determinant of matrix B, and D_y the determinant of matrix C.

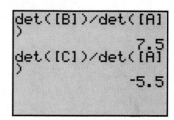

Figure 16

In a similar manner, Cramer's rule can be applied to systems of three linear equations with three variables.

Cramer's Rule for 3 × 3 Systems

Given the system

$$a_1x + b_1y + c_1z = d_1$$
$$a_2x + b_2y + c_2z = d_2$$
$$a_3x + b_3y + c_3z = d_3,$$

with

$$D = \begin{vmatrix} a_1 & b_1 & c_1 \\ a_2 & b_2 & c_2 \\ a_3 & b_3 & c_3 \end{vmatrix} \neq 0, \qquad D_x = \begin{vmatrix} d_1 & b_1 & c_1 \\ d_2 & b_2 & c_2 \\ d_3 & b_3 & c_3 \end{vmatrix},$$

$$D_y = \begin{vmatrix} a_1 & d_1 & c_1 \\ a_2 & d_2 & c_2 \\ a_3 & d_3 & c_3 \end{vmatrix}, \qquad D_z = \begin{vmatrix} a_1 & b_1 & d_1 \\ a_2 & b_2 & d_2 \\ a_3 & b_3 & d_3 \end{vmatrix},$$

then

$$x = \frac{D_x}{D}, \qquad y = \frac{D_y}{D}, \qquad z = \frac{D_z}{D}.$$

EXAMPLE 6 **Using Cramer's Rule for a 3 × 3 System**

Use Cramer's rule to solve the system.

$$x + y - z + 2 = 0$$
$$2x - y + z + 5 = 0$$
$$x - 2y + 3z - 4 = 0$$

To use Cramer's rule, we must rewrite the system in the form

$$x + y - z = -2$$
$$2x - y + z = -5$$
$$x - 2y + 3z = 4.$$

We expand by minors about row 1 to find D.

$$D = \begin{vmatrix} 1 & 1 & -1 \\ 2 & -1 & 1 \\ 1 & -2 & 3 \end{vmatrix} = 1\begin{vmatrix} -1 & 1 \\ -2 & 3 \end{vmatrix} - 1\begin{vmatrix} 2 & 1 \\ 1 & 3 \end{vmatrix} + (-1)\begin{vmatrix} 2 & -1 \\ 1 & -2 \end{vmatrix}$$

$$= 1(-1) - 1(5) - 1(-3) = -3$$

Expanding D_x by minors about row 1 gives

$$D_x = \begin{vmatrix} -2 & 1 & -1 \\ -5 & -1 & 1 \\ 4 & -2 & 3 \end{vmatrix} = -2\begin{vmatrix} -1 & 1 \\ -2 & 3 \end{vmatrix} - 1\begin{vmatrix} -5 & 1 \\ 4 & 3 \end{vmatrix} + (-1)\begin{vmatrix} -5 & -1 \\ 4 & -2 \end{vmatrix}$$

$$= -2(-1) - 1(-19) - 1(14) = 7.$$

In the same way, $D_y = -22$ and $D_z = -21$, so

$$x = \frac{D_x}{D} = \frac{7}{-3} = -\frac{7}{3}, \qquad y = \frac{D_y}{D} = \frac{-22}{-3} = \frac{22}{3}, \qquad z = \frac{D_z}{D} = \frac{-21}{-3} = 7.$$

The solution set is $\left\{\left(-\frac{7}{3}, \frac{22}{3}, 7\right)\right\}$.

As mentioned earlier, Cramer's rule does not apply when $D = 0$. The next example illustrates this case.

E X A M P L E   7   **Determining When Cramer's Rule Does Not Apply**

Show that Cramer's rule does not apply to the following system.

$$2x - 3y + 4z = 8$$
$$6x - 9y + 12z = 24$$
$$x + 2y - 3z = 5$$

We need to show that $D = 0$. Here, expanding about column 1 gives

$$D = \begin{vmatrix} 2 & -3 & 4 \\ 6 & -9 & 12 \\ 1 & 2 & -3 \end{vmatrix} = 2 \begin{vmatrix} -9 & 12 \\ 2 & -3 \end{vmatrix} - 6 \begin{vmatrix} -3 & 4 \\ 2 & -3 \end{vmatrix} + 1 \begin{vmatrix} -3 & 4 \\ -9 & 12 \end{vmatrix}$$

$$= 2(3) - 6(1) + 1(0)$$
$$= 0.$$

Since $D = 0$ here, Cramer's rule does not apply and we must use another method to solve the system.

Cramer's rule can be extended to 4×4 or larger systems. See a standard college algebra text for details.

9.7 EXERCISES

1. Which one of the following is the expression for the determinant $\begin{vmatrix} -2 & -3 \\ 4 & -6 \end{vmatrix}$?

   (a) $-2(-6) + (-3)(4)$ **(b)** $-2(-6) - 3(4)$
   (c) $-3(4) - (-2)(-6)$ **(d)** $-2(-6) - (-3)(4)$

2. Evaluate $\begin{vmatrix} 0 & 0 \\ 3 & -4 \end{vmatrix}$ and $\begin{vmatrix} 0 & 1 & 2 \\ 0 & -3 & 4 \\ 0 & 2 & 6 \end{vmatrix}$ and make a conjecture (educated guess) about the value of a determinant that has all 0s in a row or a column.

Evaluate each determinant. See Example 1.

3. $\begin{vmatrix} -2 & 5 \\ -1 & 4 \end{vmatrix}$ 4. $\begin{vmatrix} 3 & -6 \\ 2 & -2 \end{vmatrix}$ 5. $\begin{vmatrix} 1 & -2 \\ 7 & 0 \end{vmatrix}$

6. $\begin{vmatrix} -5 & -1 \\ 1 & 0 \end{vmatrix}$ 7. $\begin{vmatrix} 0 & 4 \\ 0 & 4 \end{vmatrix}$ 8. $\begin{vmatrix} 8 & -3 \\ 0 & 0 \end{vmatrix}$

Evaluate each determinant by expansion by minors about the first column. See Example 2.

9. $\begin{vmatrix} -1 & 2 & 4 \\ -3 & -2 & -3 \\ 2 & -1 & 5 \end{vmatrix}$ 10. $\begin{vmatrix} 2 & -3 & -5 \\ 1 & 2 & 2 \\ 5 & 3 & -1 \end{vmatrix}$

11. $\begin{vmatrix} 1 & 0 & -2 \\ 0 & 2 & 3 \\ 1 & 0 & 5 \end{vmatrix}$ 12. $\begin{vmatrix} 2 & -1 & 0 \\ 0 & -1 & 1 \\ 1 & 2 & 0 \end{vmatrix}$

13. Explain in your own words how to evaluate a 2×2 determinant. Illustrate with an example.

14. Explain in your own words the method of evaluating a 3 × 3 determinant. Illustrate with an example.

Evaluate each determinant by expansion by minors about any row or column. (Hint: The work is easier if you choose a row or a column with zeros.) See Example 3.

15. $\begin{vmatrix} 4 & 4 & 2 \\ 1 & -1 & -2 \\ 1 & 0 & 2 \end{vmatrix}$
16. $\begin{vmatrix} 3 & -1 & 2 \\ 1 & 5 & -2 \\ 0 & 2 & 0 \end{vmatrix}$
17. $\begin{vmatrix} 3 & 5 & -2 \\ 1 & -4 & 1 \\ 3 & 1 & -2 \end{vmatrix}$

18. $\begin{vmatrix} 0 & 0 & 3 \\ 4 & 0 & -2 \\ 2 & -1 & 3 \end{vmatrix}$
19. $\begin{vmatrix} 3 & 0 & -2 \\ 1 & -4 & 1 \\ 3 & 1 & -2 \end{vmatrix}$
20. $\begin{vmatrix} 1 & 1 & 2 \\ 5 & 5 & 7 \\ 3 & 3 & 1 \end{vmatrix}$

21. Explain why a determinant with a row or column of zeros has a value of zero.

RELATING CONCEPTS (EXERCISES 22–25)

Recall the formula for slope and the point-slope form of the equation of a line from Chapter 8.

*Use these formulas to **work Exercises 22–25 in order** and see how a determinant can be used in writing the equation of a line.*

22. Write the expression for the slope of a line passing through the points (x_1, y_1) and (x_2, y_2).

23. Using the expression from Exercise 22 as m, and the point (x_1, y_1), write the point-slope form of the equation of the line.

24. Using the equation obtained in Exercise 23, multiply both sides by $x_2 - x_1$, and write the equation so that 0 is on the right side.

25. Consider the *determinant equation*

$$\begin{vmatrix} x & y & 1 \\ x_1 & y_1 & 1 \\ x_2 & y_2 & 1 \end{vmatrix} = 0.$$

Expand by minors on the left and show that this determinant equation yields the same result that you obtained in Exercise 24.

Did you make the connection that a determinant with variables can represent one side of an equation?

TECHNOLOGY INSIGHTS (EXERCISES 26–29)

Predict the display the calculator will give for each determinant.

26.
```
[A]
        [[-6 8 ]
         [1  -3]]
det([A])
```

27.
```
[B]
        [[-1 7 1]
         [0  2 1]
         [-3 5 4]]
det([B])
```

TECHNOLOGY INSIGHTS (EXERCISES 26-29) (CONTINUED)

28.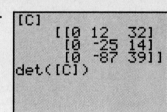

29.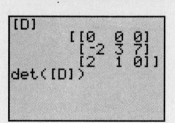

Use a graphing calculator with matrix capabilities to find each determinant.

30. $\begin{vmatrix} 1.5 & 2.6 & 9.3 \\ 5.2 & -1.4 & 8.6 \\ 0 & .7 & 1.2 \end{vmatrix}$

31. $\begin{vmatrix} \sqrt{5} & \sqrt{2} & -\sqrt{3} \\ \sqrt{7} & -\sqrt{6} & \sqrt{10} \\ -\sqrt{5} & -\sqrt{2} & \sqrt{17} \end{vmatrix}$ (To as many places as the calculator shows)

There is another method for evaluating a 3 × 3 determinant. Refer to Example 2, and carry over the first two columns to the right of the original determinant to get

$$\begin{vmatrix} 1 & 3 & -2 \\ -1 & -2 & -3 \\ 1 & 1 & 2 \end{vmatrix} \begin{matrix} 1 & 3 \\ -1 & -2 \\ 1 & 1 \end{matrix}.$$

Multiply along the diagonals as shown below, placing the product at the end of the arrow.

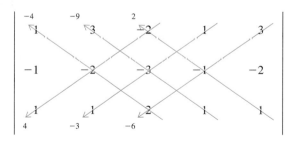

Add the top numbers: $-4 - 9 + 2 = -11.$

Add the bottom numbers: $4 - 3 - 6 = -5.$

Find the *difference* between these sums to obtain the final answer:

$$-11 - (-5) = -6.$$

Use this method to find each determinant in the indicated exercise.

32. Exercise 15   **33.** Exercise 16   **34.** Exercise 17

35. Exercise 18   **36.** Exercise 19   **37.** Exercise 20

Solve each equation by finding an expression for the determinant on the left, and then solving using the methods of Chapter 2.

38. $\begin{vmatrix} 4 & x \\ 2 & 3 \end{vmatrix} = 8$ **39.** $\begin{vmatrix} 5 & 3 \\ x & x \end{vmatrix} = 20$ **40.** $\begin{vmatrix} x & 4 \\ x & -3 \end{vmatrix} = 0$

41. Consider the system

$$4x + 3y - 2z = 1$$
$$7x - 4y + 3z = 2$$
$$-2x + y - 8z = 0.$$

Match each determinant in (a)–(d) with its correct representation from choices I, II, III, and IV.

(a) D

(b) D_x

(c) D_y

(d) D_z

I. $\begin{vmatrix} 1 & 3 & -2 \\ 2 & -4 & 3 \\ 0 & 1 & -8 \end{vmatrix}$

II. $\begin{vmatrix} 4 & 3 & 1 \\ 7 & -4 & 2 \\ -2 & 1 & 0 \end{vmatrix}$

III. $\begin{vmatrix} 4 & 1 & -2 \\ 7 & 2 & 3 \\ -2 & 0 & -8 \end{vmatrix}$

IV. $\begin{vmatrix} 4 & 3 & -2 \\ 7 & -4 & 3 \\ -2 & 1 & -8 \end{vmatrix}$

42. For the system

$$x + 3y - 6z = 7$$
$$2x - y + z = 1$$
$$x + 2y + 2z = -1,$$

$D = -43$, $D_x = -43$, $D_y = 0$, and $D_z = 43$. What is the solution set of the system?

Use Cramer's rule to solve each linear system in two variables. See Example 5.

43. $3x + 5y = -5$
$-2x + 3y = 16$

44. $5x + 2y = -3$
$4x - 3y = -30$

45. $8x + 3y = 1$
$6x - 5y = 2$

46. $3x - y = 9$
$2x + 5y = 8$

47. $2x + 3y = 4$
$5x + 6y = 7$

48. $4x + 5y = 6$
$7x + 8y = 9$

49. Look at the coefficients and constants in the systems in Exercises 47 and 48. Notice that in both cases, the six numbers are consecutive integers. Make up a system having this same pattern for its coefficients and constants, and solve it using Cramer's rule. Compare the solutions in Exercises 47, 48, and here. What do you notice?

50. Use Cramer's rule to prove that the system

$$ax + (a + 1)y = a + 2$$
$$(a + 3)x + (a + 4)y = a + 5, \quad \text{where } D \neq 0,$$

has solution set $\{(-1, 2)\}$.

Use Cramer's rule where applicable to solve each linear system in three variables. See Examples 6 and 7.

51. $2x + 3y + 2z = 15$
$x - y + 2z = 5$
$x + 2y - 6z = -26$

52. $x - y + 6z = 19$
$3x + 3y - z = 1$
$x + 9y + 2z = -19$

53. $2x - 3y + 4z = 8$
$6x - 9y + 12z = 24$
$-4x + 6y - 8z = -16$

54. $7x + y - z = 4$
$2x - 3y + z = 2$
$-6x + 9y - 3z = -6$

55. $3x + 5z = 0$
$2x + 3y = 1$
$-y + 2z = -11$

56. $-x + 2y = 4$
$3x + y = -5$
$2x + z = -1$

57. $x - 3y = 13$
$2y + z = 5$
$-x + z = -7$

58. $-5x - y = -10$
$3x + 2y + z = -3$
$-y - 2z = -13$

RELATING CONCEPTS (EXERCISES 59–62)

In this section we have seen how determinants can be used to solve systems of equations. In Exercises 22–25, the earlier Relating Concepts group, we saw how a determinant can be used to write the equation of a line given two points on the line. Here, we show how a determinant can be used to find the area of a triangle if we know the coordinates of its vertices.

RELATING CONCEPTS (EXERCISES 59-62) (CONTINUED)

Suppose that $A(x_1, y_1)$, $B(x_2, y_2)$, and $C(x_3, y_3)$ are the coordinates of the vertices of triangle ABC in the coordinate plane. Then it can be shown that the area of the triangle is given by the absolute value of

$$\frac{1}{2}\begin{vmatrix} x_1 & y_1 & 1 \\ x_2 & y_2 & 1 \\ x_3 & y_3 & 1 \end{vmatrix}.$$

Work Exercises 59–62 in order.

59. Sketch triangle ABC in the coordinate plane, given that the coordinates of A are $(0, 0)$, of B are $(-3, -4)$, and of C are $(2, -2)$.

60. Write the determinant expression described above that gives the area of triangle ABC described in Exercise 59.

61. Evaluate the absolute value of the determinant expression in Exercise 60 to find the area.

62. Use the determinant expression described above to find the area of the triangle with vertices at $(3, 8)$, $(-1, 4)$, and $(0, 1)$.

Did you make the connection that a geometric formula can be expressed using a determinant?

 Use a graphing calculator and the approach described with Figure 16 to solve each of the following systems using Cramer's rule.

63.
$$x + 2y + z = 10$$
$$2x - y - 3z = -20$$
$$-x + 4y + z = 18$$

64.
$$2x + y + 3z = 1$$
$$x - 2y + z = -3$$
$$-3x + y - 2z = -4$$

65.
$$-8w + 4x - 2y + z = -28$$
$$-w + x - y + z = -10$$
$$w + x + y + z = -4$$
$$27w + 9x + 3y + z = 2$$

66.
$$5w + 2x - 3y + z = 4.7$$
$$-2w + x + 2y - z = -3.2$$
$$w + 3x - y + 2z = 2.1$$
$$2w + x - 5y + 3z = 3.4$$

CHAPTER 9 GROUP ACTIVITY

Olympic Track and Field Results

Objective: Solve systems of equations to determine trends in Olympic track and field events.

In this activity you will compare trends for men and women in three Olympic track and field events over a span of thirty-six years. Use the data for winning distances (in meters) given in the following tables.

Gold Medal Results: Shot Put

Year	1960	1964	1968	1972	1976	1980	1984	1988	1992	1996
Men	19.68	20.33	20.54	21.18	21.05	21.35	21.26	22.47	21.70	21.62
Women	17.30	18.10	19.60	21.00	21.10	22.40	20.40	22.20	21.00	20.56

(continued)

Gold Medal Results: Javelin

Year	1960	1964	1968	1972	1976	1980	1984	1988	1992	1996
Men	84.64	82.66	90.10	90.48	94.58	91.20	86.76	84.28	89.66	88.16
Women	55.90	60.50	60.30	63.80	65.90	68.40	69.50	74.60	68.30	67.94

Gold Medal Results: Discus

Year	1960	1964	1968	1972	1976	1980	1984	1988	1992	1996
Men	59.18	61.00	63.78	64.40	67.50	66.64	66.60	68.82	65.12	69.40
Women	55.10	57.20	58.20	66.60	69.00	69.90	65.30	72.30	70.00	69.66

Source: Centre for Innovation in Mathematics Teaching; www.ex.ac.uk/cimt.

A. Have each person in the group select a different Olympic track and field event. Use the data for your selected event to do the following.

    1. Plot the data from the table for your event on a graph, using the vertical axis for distances, and the horizontal axis for the years, where $x = 0$ represents 1960, and $4x$ equals the number of years after 1960. Use different colors to differentiate between the data for men and women.

    2. Draw a straight line that would best fit the data.

    3. Estimate the intersection point of the two lines; that is, the year when the distances for men and women will be the same. Record your estimate.

B. The equations below model the data for each table.

    Shot put: $y = .25x + 19.68$ and $y = .46x + 17.3$
    Javelin: $y = .39x + 84.64$ and $y = 1.55x + 55.9$
    Discus: $y = .74x + 59.18$ and $y = 1.86x + 55.1$

Have each person solve the system of equations for the selected event.

C. As a group, write a paragraph that discusses your findings. Include answers to the following questions in your paragraph.

    1. How closely did your estimates match the results from solving the systems of equations?

    2. In which events might it be possible for women's distances to exceed men's distances?

    3. Which method for solving the systems of equations—graphing, substitution, elimination, matrix row operations, or Cramer's rule—was easiest? Explain.

CHAPTER 9 SUMMARY

KEY TERMS

9.1 system of linear equations	independent equations	column	**9.7** determinant
solution of a system	dependent equations	square matrix	minor
solution set of a system	**9.4** ordered triple	augmented matrix	expansion by minors
consistent system	**9.6** matrix	row operations	array of signs
inconsistent system	element of a matrix	row echelon form	Cramer's rule
	row		

NEW SYMBOLS

(x, y, z)	ordered triple
$\begin{bmatrix} a & b & c \\ d & e & f \end{bmatrix}$	matrix with two rows, three columns

$\begin{vmatrix} a & b \\ c & d \end{vmatrix}$ 2×2 determinant

$\begin{vmatrix} a & b & c \\ d & e & f \\ g & h & i \end{vmatrix}$ 3×3 determinant

TEST YOUR WORD POWER

See how well you have learned the vocabulary in this chapter. Answers, with examples, are given at the bottom of the page.

1. A **system of linear equations** consists of
(a) at least two linear equations with different variables
(b) two or more linear equations that have an infinite number of solutions
(c) two or more linear equations that are to be solved at the same time
(d) two or more linear inequalities that are to be solved.

2. The **solution set of a system** is
(a) all ordered pairs that satisfy one equation of the system
(b) all ordered pairs that satisfy all the equations of the system at the same time
(c) any ordered pair that satisfies one or more equations of the system
(d) the set of values that make all the equations of the system false.

3. A **consistent system** is a system of equations
(a) with one solution
(b) with no solutions
(c) with an infinite number of solutions
(d) that have the same graph.

4. An **inconsistent system** is a system of equations
(a) with one solution
(b) with no solutions
(c) with an infinite number of solutions
(d) that have the same graph.

5. **Dependent equations**
(a) have different graphs
(b) have no solutions
(c) have one solution
(d) are different forms of the same equation.

6. A **matrix** is
(a) an ordered pair of numbers
(b) an array of numbers with the same number of rows and columns
(c) a pair of numbers written between brackets
(d) a rectangular array of numbers.

7. A matrix written in **row echelon form** has
(a) elements that are all 0
(b) elements that are all 1
(c) upper left to lower right diagonal elements of 1 with 0s below the 1s
(d) upper left to lower right diagonal elements of 0 with 1s below the 0s.

8. A **determinant** is
(a) a rectangular array of numbers
(b) a real number associated with a square matrix
(c) a matrix with the same number of rows and columns
(d) an ordered pair of numbers.

9. **Expansion by minors** is
(a) a method of evaluating a 3×3 or larger determinant
(b) a way to use row operations to produce new matrices
(c) a method of evaluating a 2×2 determinant
(d) a method of evaluating any determinant.

Answers to Test Your Word Power

1. (c) *Example:* $3x - y = 3$
$$2x + y = 7$$
2. (b) *Example:* The ordered pair $(2, 3)$ satisfies both equations of the system given above, so $\{(2, 3)\}$ is the solution set of the system. **3.** (a) *Example:* The above system is consistent. The graphs of the equations intersect at exactly one point, in this case the solution $(2, 3)$. **4.** (b) *Example:* The equations of two parallel lines make up an inconsistent system; their graphs never intersect, so there is no solution to the system. **5.** (d) *Example:* The equations $4x - y = 8$ and $8x - 2y = 16$ are dependent because their graphs are the same line.

6. (d) *Examples:* $\begin{bmatrix} 3 & -1 & 2 \\ 0 & 1 & 2 \\ 4 & 3 \end{bmatrix}$, $\begin{bmatrix} 4 & 2 & 1 \\ 1 & 3 \end{bmatrix}$ **7.** (c) *Example:* $\begin{bmatrix} 1 & -3 & \frac{1}{2} \\ 0 & 1 & 4 \\ 0 & 0 & 1 \end{bmatrix}$ $\begin{bmatrix} 1 \\ -1 \\ 2 \end{bmatrix}$ $[2 \ 0 \ -5 \ 4]$

8. (b) *Examples:* $\begin{vmatrix} 2 & 1 \\ 4 & 3 \end{vmatrix}$, $\begin{vmatrix} 1 & 4 & 0 \\ 3 & -2 & -1 \\ 0 & 5 & 0 \end{vmatrix}$ **9.** (a) *Example:* See Section 9.7, Example 2.

CONCEPTS	EXAMPLES

9.1 SOLVING SYSTEMS OF LINEAR EQUATIONS BY GRAPHING

An ordered pair is a solution of a system if it makes all equations of the system true at the same time.

Is $(4, -1)$ a solution of the following system?

$$x + y = 3$$
$$2x - y = 9$$

Yes, because $4 + (-1) = 3$, and $2(4) - (-1) = 9$ are both true.

If the graphs of the equations of a system are both sketched on the same axes, the points of intersection, if any, form the solution set of the system.

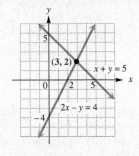

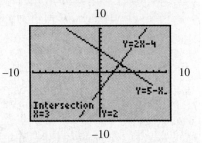

A graphing calculator can find the solution of a system by locating the point of intersection of the graphs.

$\{(3, 2)\}$ is the solution set of the system

$$x + y = 5$$
$$2x - y = 4.$$

9.2 SOLVING SYSTEMS OF LINEAR EQUATIONS BY SUBSTITUTION

Step 1 Solve one equation for one variable.

Solve by substitution.

$$x + 2y = -5 \qquad (1)$$
$$y = -2x - 1 \qquad (2)$$

Equation (2) is already solved for y.

Step 2 Substitute for that variable in the other equation to get an equation in one variable.

Substitute $-2x - 1$ for y in equation (1).

$$x + 2(-2x - 1) = -5$$

Step 3 Solve the equation from Step 2.

Solve to get $x = 1$.

Step 4 Substitute the result into the equation from Step 1 to get the value of the other variable.

To find y, let $x = 1$ in equation (2).

$$y = -2(1) - 1 = -3$$

Step 5 Check. Write the solution set.

The solution, $(1, -3)$, checks so $\{(1, -3)\}$ is the solution set.

9.3 SOLVING SYSTEMS OF LINEAR EQUATIONS BY ELIMINATION

Step 1 Write both equations in the form

$$Ax + By = C.$$

Solve by elimination.

$$x + 3y = 7$$
$$3x - y = 1$$

Step 2 Multiply one or both equations by appropriate numbers (if necessary) so that the coefficients of x (or y) are negatives of each other.

Multiply the first equation by -3 to eliminate the x-terms by addition.

Step 3 Add the equations to get an equation with only one variable.

$$\begin{array}{r} -3x - 9y = -21 \\ \underline{3x - y = 1} \\ -10y = -20 \end{array} \quad \text{Add.}$$

CONCEPTS	EXAMPLES

Step 4　Solve the equation from Step 3.

$y = 2$　　Divide by -10.

Step 5　Substitute the solution from Step 4 into either of the original equations to find the value of the remaining variable.

Substitute to get the value of x.

$$x + 3y = 7$$
$$x + 3(2) = 7 \quad \text{Let } y = 2.$$
$$x + 6 = 7 \quad \text{Multiply.}$$
$$x = 1 \quad \text{Subtract 6.}$$

Step 6　Write the solution set containing an ordered pair, and check the answer.

The solution, $(1, 2)$, checks. The solution set is $\{(1, 2)\}$.

If the result of the addition step is a false statement, such as $0 = 4$, the graphs are parallel lines and there is *no solution* for the system.

$$\begin{array}{r} x - 2y = 6 \\ -x + 2y = -2 \\ \hline 0 = 4 \end{array} \quad \text{Solution set: } \emptyset$$

If the result is a true statement, such as $0 = 0$, the graphs are the same line, and an *infinite number of ordered pairs are solutions.*

$$\begin{array}{r} x - 2y = 6 \\ -x + 2y = -6 \\ \hline 0 = 0 \end{array} \quad \text{Solution set: } \{(x, y) \mid x - 2y = 6\}$$

9.4　LINEAR SYSTEMS OF EQUATIONS IN THREE VARIABLES

Solving Linear Systems in Three Variables

Solve the system

$$x + 2y - z = 6$$
$$x + y + z = 6$$
$$2x + y - z = 7.$$

Step 1　Use the elimination method to eliminate any variable from any two of the given equations. The result is an equation in two variables.

Add the first and second equations; z is eliminated and the result is $2x + 3y = 12$.

Step 2　Eliminate the *same* variable from any *other* two equations. The result is an equation in the same two variables as in Step 1.

Eliminate z again by adding the second and third equations to get $3x + 2y = 13$. Now solve the system

$$2x + 3y = 12 \qquad (*)$$
$$3x + 2y = 13.$$

Step 3　Use the elimination method to eliminate a second variable from the two equations in two variables that result from Steps 1 and 2. The result is an equation in one variable that gives the value of that variable.

To eliminate x, multiply the top equation by -3 and the bottom equation by 2.

$$\begin{array}{r} -6x - 9y = -36 \\ 6x + 4y = 26 \\ \hline -5y = -10 \\ y = 2 \end{array}$$

Step 4　Substitute the value of the variable found in Step 3 into either of the equations in two variables to find the value of the second variable.

Let $y = 2$ in equation $(*)$.

$$2x + 3(2) = 12$$
$$2x = 6$$
$$\mathbf{x = 3}$$

Step 5　Use the values of the two variables from Steps 3 and 4 to find the value of the third variable by substituting into any of the original equations.

Let $y = 2$ and $x = 3$ in any of the original equations to find $z = 1$.

Step 6　Check in all of the original equations. Write the solution set.

Verify that when $x = 3$, $y = 2$, and $z = 1$, all three equations are satisfied. The solution set is $\{(3, 2, 1)\}$.

CONCEPTS	EXAMPLES

9.5 APPLICATIONS OF LINEAR SYSTEMS OF EQUATIONS

Use the six-step problem solving method.	The perimeter of a rectangle is 18 feet. The length is 3 feet more than twice the width. What are the dimensions of the rectangle?
Step 1 Choose a variable to represent each unknown value.	Let x represent the length and y represent the width.
Step 2 Make a figure, diagram, or chart if it will help.	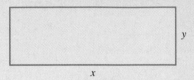
Step 3 Write two (three) equations that relate the unknowns.	From the perimeter formula, one equation is $2x + 2y = 18$. From the problem, another equation is $x = 3 + 2y$. Now solve the system $$2x + 2y = 18$$ $$x = 3 + 2y.$$
Step 4 Solve the system.	Substitute $3 + 2y$ for x in the top equation. $$2(3 + 2y) + 2y = 18$$ Solve to get $y = 2$. If $y = 2$, then $x = 3 + 2(2) = 7$.
Step 5 Answer the question(s) asked.	The length is 7 feet and the width is 2 feet.
Step 6 Check your solution in the words of the problem.	The perimeter is $2(7) + 2(2) = 18$, and $3 + 2(2) = 7$, so the solution checks.

9.6 SOLVING LINEAR SYSTEMS OF EQUATIONS BY MATRIX METHODS

Matrix Row Operations

1. Any two rows of the matrix may be interchanged.

$$\begin{bmatrix} 1 & 5 & 7 \\ 3 & 9 & -2 \\ 0 & 6 & 4 \end{bmatrix} \text{ becomes } \begin{bmatrix} 3 & 9 & -2 \\ 1 & 5 & 7 \\ 0 & 6 & 4 \end{bmatrix} \quad \begin{array}{l} \text{Interchange} \\ R_1 \text{ and } R_2. \end{array}$$

2. The numbers in any row may be multiplied by any nonzero real number.

$$\begin{bmatrix} 1 & 5 & 7 \\ 3 & 9 & -2 \\ 0 & 6 & 4 \end{bmatrix} \text{ becomes } \begin{bmatrix} 1 & 5 & 7 \\ 1 & 3 & -\dfrac{2}{3} \\ 0 & 6 & 4 \end{bmatrix} \quad \tfrac{1}{3}R_2$$

3. Any row may be changed by adding to the numbers of the row the product of a real number and the numbers of another row.

$$\begin{bmatrix} 1 & 5 & 7 \\ 3 & 9 & -2 \\ 0 & 6 & 4 \end{bmatrix} \text{ becomes } \begin{bmatrix} 1 & 5 & 7 \\ 0 & -6 & -23 \\ 0 & 6 & 4 \end{bmatrix} \quad -3R_1 + R_2$$

CONCEPTS	EXAMPLES

A system can be solved by matrix methods. Write the augmented matrix, and use row operations to obtain a matrix in row echelon form.

Solve.

$$x + 3y = 7$$
$$2x + y = 4$$

$$\begin{bmatrix} 1 & 3 & 7 \\ 2 & 1 & 4 \end{bmatrix}$$

$$\begin{bmatrix} 1 & 3 & 7 \\ 0 & -5 & -10 \end{bmatrix} \quad -2R_1 + R_2$$

$$\begin{bmatrix} 1 & 3 & 7 \\ 0 & 1 & 2 \end{bmatrix} \quad -\tfrac{1}{5}R_2$$

$$x + 3y = 7$$
$$\mathbf{y = 2}$$

When $y = 2$, $x + 3(2) = 7$, so $x = 1$. The solution set is $\{(1, 2)\}$.

9.7 DETERMINANTS AND CRAMER'S RULE

Value of a 2 × 2 Determinant

$$\begin{vmatrix} a & b \\ c & d \end{vmatrix} = ad - bc$$

Evaluate.

$$\begin{vmatrix} 3 & 4 \\ -2 & 6 \end{vmatrix} = (3)(6) - (4)(-2) = 26$$

Determinants larger than 2 × 2 are evaluated by expansion by minors about a column or row.

Array of Signs for a 3 × 3 Determinant

$$\begin{matrix} + & - & + \\ - & + & - \\ + & - & + \end{matrix}$$

Evaluate

$$\begin{vmatrix} 2 & -3 & -2 \\ -1 & -4 & -3 \\ -1 & 0 & 2 \end{vmatrix}$$

by expanding about the second column.

$$\begin{vmatrix} 2 & -3 & -2 \\ -1 & -4 & -3 \\ -1 & 0 & 2 \end{vmatrix} = -(-3)(-5) + (-4)(2) - (0)(-8)$$

$$= -15 - 8 + 0$$
$$= -23$$

Cramer's Rule for 2 × 2 Systems

Given the system

$$a_1x + b_1y = c_1$$
$$a_2x + b_2y = c_2$$

with $a_1b_2 - a_2b_1 = D \neq 0$,

then $x = \dfrac{\begin{vmatrix} c_1 & b_1 \\ c_2 & b_2 \end{vmatrix}}{\begin{vmatrix} a_1 & b_1 \\ a_2 & b_2 \end{vmatrix}} = \dfrac{D_x}{D}$

and $y = \dfrac{\begin{vmatrix} a_1 & c_1 \\ a_2 & c_2 \end{vmatrix}}{\begin{vmatrix} a_1 & b_1 \\ a_2 & b_2 \end{vmatrix}} = \dfrac{D_y}{D}$.

Solve using Cramer's rule.

$$x - 2y = -1$$
$$2x + 5y = 16$$

$$x = \dfrac{\begin{vmatrix} -1 & -2 \\ 16 & 5 \end{vmatrix}}{\begin{vmatrix} 1 & -2 \\ 2 & 5 \end{vmatrix}} = \dfrac{-5 + 32}{5 + 4} = \dfrac{27}{9} = 3$$

$$y = \dfrac{\begin{vmatrix} 1 & -1 \\ 2 & 16 \end{vmatrix}}{\begin{vmatrix} 1 & -2 \\ 2 & 5 \end{vmatrix}} = \dfrac{16 + 2}{5 + 4} = \dfrac{18}{9} = 2$$

The solution set is $\{(3, 2)\}$.

(continued)

CONCEPTS	EXAMPLES
Cramer's Rule for 3 × 3 Systems Given the system $$a_1x + b_1y + c_1z = d_1$$ $$a_2x + b_2y + c_2z = d_2$$ $$a_3x + b_3y + c_3z = d_3$$ with $$D = \begin{vmatrix} a_1 & b_1 & c_1 \\ a_2 & b_2 & c_2 \\ a_3 & b_3 & c_3 \end{vmatrix} \neq 0, \quad D_x = \begin{vmatrix} d_1 & b_1 & c_1 \\ d_2 & b_2 & c_2 \\ d_3 & b_3 & c_3 \end{vmatrix},$$ $$D_y = \begin{vmatrix} a_1 & d_1 & c_1 \\ a_2 & d_2 & c_2 \\ a_3 & d_3 & c_3 \end{vmatrix}, \quad D_z = \begin{vmatrix} a_1 & b_1 & d_1 \\ a_2 & b_2 & d_2 \\ a_3 & b_3 & d_3 \end{vmatrix},$$ then $\quad x = \dfrac{D_x}{D}, \quad y = \dfrac{D_y}{D}, \quad z = \dfrac{D_z}{D}.$	Solve using Cramer's rule. $$3x + 2y + z = -5$$ $$x - y + 3z = -5$$ $$2x + 3y + z = 0$$ Using the methods of expansion of minors, it can be shown that $D_x = 45$, $D_y = -30$, $D_z = 0$, and $D = -15$. Therefore, $$x = \frac{D_x}{D} = \frac{45}{-15} = -3,$$ $$y = \frac{D_y}{D} = \frac{-30}{-15} = 2,$$ $$z = \frac{D_z}{D} = \frac{0}{-15} = 0.$$ The solution set is $\{(-3, 2, 0)\}$.

CHAPTER 9 REVIEW EXERCISES

[9.1] *Various tools are used by business management to improve performance, but they go in and out of fashion, according to an article in the September 7, 1998, issue of* Fortune. *Three of these tools are strategic alliances, core competencies, and total quality management. Use the graph to answer the following.*

MANAGEMENT TOOLS

1. During what year was the usage rate of core competencies and total quality management the same? What was this rate?

2. On how many occasions did strategic alliances and core competencies reach the same usage level? Between these times, which one of these tools was more popular?

3. The figure shows graphs that represent the supply and demand for a certain brand of low-fat frozen yogurt at various prices per half-gallon.

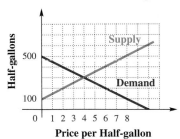

The Fortunes of Frozen Yogurt

(a) At what price does supply equal demand?
(b) For how many half-gallons does supply equal demand?
(c) What are the supply and the demand at a price of $2 per half-gallon?

4. State whether the graphed system is inconsistent or has dependent equations.

(a)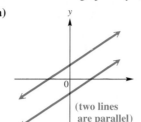

(two lines
are parallel)

(b)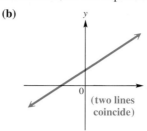

(two lines
coincide)

Decide whether the given ordered pair is a solution of the given system.

5. $(3, 4)$
$$4x - 2y = 4$$
$$5x + y = 19$$

6. $(-5, 2)$
$$x - 4y = -13$$
$$2x + 3y = 4$$

Solve each system by graphing.

7. $x + y = 4$
$$2x - y = 5$$

8. $2x + 4 = 2y$
$$y - x = -3$$

[9.2]

9. A student solves the system $\begin{array}{c}2x + y = 6 \\ -2x - y = 4\end{array}$ and gets the equation $0 = 10$. The student gives the solution set as $\{(0, 10)\}$. Is this correct? Explain.

10. Can a system of two linear equations in two unknowns have exactly three solutions? Explain.

Solve each system by the substitution method.

11. $3x + y = 7$
$$x = 2y$$

12. $2x - 5y = -19$
$$y = x + 2$$

13. $5x + 15y = 30$
$$x + 3y = 6$$

14. $\dfrac{1}{3}x + \dfrac{1}{7}y = \dfrac{52}{21}$
$$\dfrac{1}{2}x - \dfrac{1}{3}y = \dfrac{19}{6}$$

15. After solving a system of linear equations by the substitution method, a student obtained the equation "$0 = 0$." He gave the solution set of the system as $\{(0, 0)\}$. Was his answer correct? Why or why not?

16. Suppose that you were asked to solve the system $\begin{array}{c}5x - 3y = 7 \\ -x + 2y = 4\end{array}$ by substitution. Which variable in which equation would be easiest to solve for in your first step? Why?

[9.3]

17. Only one of the following systems does not require that we multiply one or both equations by a constant in order to solve the system by the elimination method. Which one is it?
(a) $-4x + 3y = 7$
$ 3x - 4y = 4$
(b) $5x + 8y = 13$
$ 12x + 24y = 36$
(c) $2x + 3y = 5$
$ x - 3y = 12$
(d) $x + 2y = 9$
$ 3x - y = 6$

18. For the system
$$2x + 12y = 7$$
$$3x + 4y = 1,$$
if we were to multiply the first equation by -3, by what number would we have to multiply the second equation in order to
(a) eliminate the x-terms when solving by the elimination method?
(b) eliminate the y-terms when solving by the elimination method?

Solve each system by the elimination method.

19. $2x - y = 13$
$x + y = 8$

20. $-4x + 3y = 25$
$6x - 5y = -39$

21. $3x - 4y = 9$
$6x - 8y = 18$

22. $2x + y = 3$
$-4x - 2y = 6$

[9.1–9.3] *Solve each system by any method.*

23. $2x + 3y = -5$
$3x + 4y = -8$

24. $6x - 9y = 0$
$2x - 3y = 0$

25. $2x + y - x = 3y + 5$
$y + 2 = x - 5$

26. $\dfrac{x}{2} + \dfrac{y}{3} = 7$

$\dfrac{x}{4} + \dfrac{2y}{3} = 8$

27. What are the five methods of solving systems discussed in this chapter? Choose one and discuss its drawbacks and advantages.

28. Why would it be easier to solve system B by the substitution method than system A?

$$\textbf{A: } -5x + 6y = 7 \qquad\qquad \textbf{B: } 2x + 9y = 13$$
$$2x + 5y = -5 \qquad\qquad\qquad y = 3x - 2$$

[9.4] *Solve each system.*

29. $2x + 3y - z = -16$
$x + 2y + 2z = -3$
$-3x + y + z = -5$

30. $4x - y = 2$
$3y + z = 9$
$x + 2z = 7$

31. $3x - y - z = -8$
$4x + 2y + 3z = 15$
$-6x + 2y + 2z = 10$

[9.5] *Solve each problem by writing a system of equations and then solving the system. Use the six-step problem solving method.*

32. The 1997 box office smash *Titanic* earned more in Europe than in the United States. This may be because average movie prices in Europe exceed those of the United States (*Source: Parade Magazine,* September 13, 1998, p. 25). For example, while the average movie ticket (to the nearest dollar) costs $5 in the United States, it costs an equivalent of $11 in London. Suppose that a group of 41 Americans and Londoners who paid these average prices spent a total of $307 for tickets. How many from each country were in the group?

33. During the 1996 National Football League season, John Elway of the Denver Broncos threw 40 passes that resulted in either a touchdown or an interception. His number of touchdowns was 2 less than twice the number of interceptions. How many passes of each type did he throw? (*Source: The Wall Street Journal Almanac 1998.*)

34. A plane flies 560 miles in 1.75 hours traveling with the wind. The return trip later against the same wind takes the plane 2 hours. Find the speed of the plane and the speed of the wind. (*Hint:* Let x represent the speed of the plane and let y represent the speed of the wind. Use the chart.)

	r	t	d
With wind	$x + y$	1.75	
Against wind	$x - y$	2	

35. Sweet's Candy Store is offering a special mix for Valentine's Day. Ms. Sweet will mix some $2-a-pound nuts with some $1-a-pound chocolate candy to get 100 pounds of mix which she will sell at $1.30 a pound. How many pounds of each should she use?

	Pounds	Price per Pound (in Dollars)	Value
Nuts	x	2	
Chocolate	y	1	
Total	100	1.30	

36. The sum of the measures of the angles of a triangle is 180°. The largest angle measures 10° less than the sum of the other two. The measure of the middle-sized angle is the average of the other two. Find the measures of the three angles.

37. Keshon Grant sells real estate. On three recent sales, he made 10% commission, 6% commission, and 5% commission. His total commissions on these sales were $17,000, and he sold property worth $280,000. If the 5% sale amounted to the sum of the other two, what were the three sales prices?

38. How many liters each of 8%, 10%, and 20% hydrogen peroxide should be mixed together to get 8 liters of 12.5% solution, if the amount of 8% solution used must be 2 liters more than the amount of 20% solution used?

 39. In the great baseball year of 1961, Yankee teammates Mickey Mantle, Roger Maris, and John Blanchard combined for 136 home runs. Mantle hit 7 fewer than Maris. Maris hit 40 more than Blanchard. What were the home run totals for each player? (*Source:* Neft, David S. and Richard M. Cohen, *The Sports Encyclopedia: Baseball 1997.*)

[9.6]   *Solve each system of equations by matrix methods.*

40. $2x + 5y = -4$
$4x - y = 14$

41. $6x + 3y = 9$
$-7x + 2y = 17$

42. $x + 2y - z = 1$
$3x + 4y + 2z = -2$
$-2x - y + z = -1$

43. $x + 3y = 7$
$3x + z = 2$
$y - 2z = 4$

[9.7]

44. Which one of the following determinants is equal to 0?

(a) $\begin{vmatrix} 3 & 2 \\ 2 & 3 \end{vmatrix}$ **(b)** $\begin{vmatrix} 4 & 2 \\ -3 & 2 \end{vmatrix}$ **(c)** $\begin{vmatrix} -1 & 1 \\ 8 & 8 \end{vmatrix}$ **(d)** $\begin{vmatrix} 1 & 2 \\ 6 & 12 \end{vmatrix}$

Evaluate each determinant.

45. $\begin{vmatrix} 2 & -9 \\ 8 & 4 \end{vmatrix}$

46. $\begin{vmatrix} 7 & 0 \\ 5 & -3 \end{vmatrix}$

47. $\begin{vmatrix} 2 & 10 & 4 \\ 0 & 1 & 3 \\ 0 & 6 & -1 \end{vmatrix}$

48. $\begin{vmatrix} -1 & 7 & 2 \\ 3 & 0 & 5 \\ -1 & 2 & 6 \end{vmatrix}$

49. Under what conditions can a system *not* be solved using Cramer's rule?

50. Why can't the system $\begin{array}{l} 3x + 2y + z = 0 \\ -x + y - 3z = 1 \end{array}$ be solved using Cramer's rule?

Use Cramer's rule to solve each system of equations.

51. $3x - 4y = -32$
$2x + y = -3$

52. $-4x + 3y = -12$
$2x + 6y = 15$

53. $4x + y + z = 11$
$x - y - z = 4$
$y + 2z = 0$

54. $-x + 3y - 4z = 4$
$2x + 4y + z = -14$
$3x - y + 2z = -8$

RELATING CONCEPTS (EXERCISES 55-60)

Consider the system of equations

$$2x + y + 3z = 1$$
$$x - 2y + z = -3$$
$$-3x + y - 2z = -4.$$

In Exercises 55–60 we examine several different ways of solving this system. **Work the exercises in order.**

55. Eliminate x from the first and second equations, and eliminate x from the second and third equations. Write the resulting system of two equations in y and z.

56. Solve the system of two equations found in Exercise 55 using elimination.

57. Complete the solution of the original system using the results of Exercise 56.

58. Solve the original system using matrix row operations (Section 9.6). Verify that your solution is the same as the one found in Exercise 57.

59. Evaluate these determinants.

(a) $D = \begin{vmatrix} 2 & 1 & 3 \\ 1 & -2 & 1 \\ -3 & 1 & -2 \end{vmatrix}$ (b) $D_x = \begin{vmatrix} 1 & 1 & 3 \\ -3 & -2 & 1 \\ -4 & 1 & -2 \end{vmatrix}$

(c) $D_y = \begin{vmatrix} 2 & 1 & 3 \\ 1 & -3 & 1 \\ -3 & -4 & -2 \end{vmatrix}$ (d) $D_z = \begin{vmatrix} 2 & 1 & 1 \\ 1 & -2 & -3 \\ -3 & 1 & -4 \end{vmatrix}$

60. Use the results of Exercise 59 and Cramer's rule to solve the original system. Verify that your solution is the same as the one found in Exercises 57 and 58.

Did you make the connection that a system can be solved by various methods, all resulting in the same solution?

MIXED REVIEW EXERCISES

Solve by any method.

61. $\dfrac{2}{3}x + \dfrac{1}{6}y = \dfrac{19}{2}$
$\dfrac{1}{3}x - \dfrac{2}{9}y = 2$

62. $2x + 5y - z = 12$
$-x + y - 4z = -10$
$-8x - 20y + 4z = 31$

63. $x = 7y + 10$
$2x + 3y = 3$

64. $x + 4y = 17$
$-3x + 2y = -9$

65. $-7x + 3y = 12$
$5x + 2y = 8$

66. $2x - 5y = 8$
$3x + 4y = 10$

67. On November 1, 1998, K-Mart advertised videocassettes and compact discs in its Sunday newspaper insert. Cheryl Joslyn went shopping and bought each of her 7 nephews a gift, either a copy of the movie *Godzilla* or the latest Aerosmith compact disc. The movie cost $14.95 and the compact disc cost $16.88, and she spent a total of $114.30. How many movies and how many compact discs did she buy? (*Source: Times Picayune.*)

68. During the Summer Olympics in Atlanta in 1996, Canada won a total of 22 medals, consisting of gold, silver, and bronze. There were 5 fewer gold medals than bronze, and 3 fewer bronze than silver. How many medals of each kind did Canada win? (*Source: The Universal Almanac 1997.*)

CHAPTER 9 TEST

 Hank Aaron and Babe Ruth are the all-time home run leaders in the major leagues. If Mark McGwire continues at the pace he is going, he will hit more home runs than either of them. Use the graphs to answer the following.

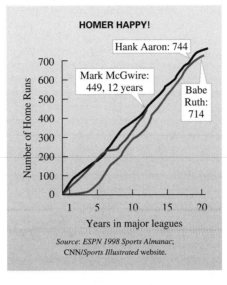

HOMER HAPPY!

Hank Aaron: 744

Mark McGwire: 449, 12 years

Babe Ruth: 714

Number of Home Runs

700
600
500
400
300
200
100
0

1   5   10   15   20

Years in major leagues

*Source: ESPN 1998 Sports Almanac;
CNN/Sports Illustrated website.*

1. Was there any year in Babe Ruth's career that he had more home runs than Hank Aaron in the same year of Aaron's career? Explain.

2. After eight years, which player had the most home runs? Who had the fewest?

3. Use a graph to solve the system.

$$x + y = 7$$
$$x - y = 5$$

Solve each system by substitution.

4. $2x - 3y = 24$
   $y = -\dfrac{2}{3}x$

5. $12x - 5y = 8$
   $3x = \dfrac{5}{4}y + 2$

Solve each system by elimination.

6. $3x + y = 12$
   $2x - y = 3$

7. $-5x + 2y = -4$
   $6x + 3y = -6$

8. $3x + 4y = 8$
   $8y = 7 - 6x$

9. $3x + 5y + 3z = 2$
   $6x + 5y + z = 0$
   $3x + 10y - 2z = 6$

Solve each problem by writing a system of equations.

10. Two cars start from points 420 miles apart and travel toward each other. They meet after 3.5 hours. Find the average speed of each car if one travels 30 miles per hour slower than the other.

11. A chemist needs 12 liters of a 40% alcohol solution. She must mix a 20% solution and a 50% solution. How many liters of each will be required to obtain what she needs?

12. In 1996, the two most popular amusement parks in the United States were Disneyland and the Magic Kingdom at Walt Disney World. Disneyland had 1.2 million more visitors than the Magic Kingdom, and together they had 28.8 million visitors. How many visitors did each park have? (*Source: The Wall Street Journal Almanac*, 1998.)

13. For an art project Kay bought 8 sheets of colored paper and 3 marker pens for $6.50. She later needed 2 more sheets of colored paper and 2 different colored pens. These items cost $3.00. Find the cost of 1 marker pen and 1 sheet of colored paper.

Solve each system by matrix methods.

14. $3x + 2y = 4$
$5x + 5y = 9$

15. $x + 3y + 2z = 11$
$3x + 7y + 4z = 23$
$5x + 3y - 5z = -14$

16. Use any method described in this chapter to solve the system.
$4x - 2y = -8$
$3y - 5z = 14$
$2x + z = -10$

Evaluate each determinant.

17. $\begin{vmatrix} 6 & -3 \\ 5 & -2 \end{vmatrix}$

18. $\begin{vmatrix} 4 & 1 & 0 \\ -2 & 7 & 3 \\ 0 & 5 & 2 \end{vmatrix}$

Solve each system by Cramer's rule.

19. $3x - y = -8$
$2x + 6y = 3$

20. $x + y + z = 6$
$2x - 2y + z = 5$
$-x + 3y + z = 0$

CUMULATIVE REVIEW EXERCISES CHAPTERS 1–9

1. List all integer factors of 40.

2. Find the value of the expression if $x = 1$ and $y = 5$.

$$\frac{3x^2 + 2y^2}{10y + 3}$$

3. Which property of real numbers justifies the statement $5 + (3 \cdot 6) = 5 + (6 \cdot 3)$?

Solve each equation.

4. $7(2x + 3) - 4(2x + 1) = 2(x + 1)$

5. $|6x - 8| = 4$

6. $ax + by = cx + d$ for x

7. $.04x + .06(x - 1) = 1.04$

Solve each inequality.

8. $\dfrac{2}{3}y + \dfrac{5}{12}y \le 20$

9. $|3x + 2| \le 4$

10. $|12t + 7| \ge 0$

Solve each problem.

11. On February 12, 1999, the U.S. Senate voted to acquit William Jefferson Clinton on both counts of impeachment (perjury and obstruction of justice). Of the 200 votes cast that day, there were 10 more "not guilty" votes than "guilty" votes. How many of each vote were there? (*Source:* MSNBC website, February 13, 1999.)

12. No baseball fan should be without a copy of *The Sports Encyclopedia: Baseball 1997* by David S. Neft and Richard M. Cohen. Now in its 17th edition, it provides exhaustive statistics for professional baseball dating back to 1876. This book has a perimeter of 38 inches, and its width measures 2.5 inches less than its length. What are the dimensions of the book?

13. Two angles of a triangle have the same measure. The measure of the third angle is 4° less than twice the measure of each of the equal angles. Find the measures of the three angles.

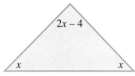

Measures are in degrees.

Perform the indicated operations.

14. $-3(-5x^2 + 3x - 10) - (x^2 - 4x + 7)$

15. $(3x - 7)(2y + 4)$

16. $\dfrac{3k^3 + 17k^2 - 27k + 7}{k + 7}$

17. Write in scientific notation: 36,500,000,000.

18. Simplify and write the answer using only positive exponents: $\left(\dfrac{x^{-4}y^3}{x^2y^4}\right)^{-1}$.

Factor completely.

19. $10m^2 + 7mp - 12p^2$

20. $64t^2 - 48t + 9$

Solve each quadratic equation.

21. $6x^2 - 7x - 3 = 0$

22. $r^2 - 121 = 0$

Perform each operation and express each answer in lowest terms.

23. $\dfrac{-3x + 6}{2x + 4} - \dfrac{-3x - 8}{2x + 4}$

24. $\dfrac{16k^2 - 9}{8k + 6} \div \dfrac{16k^2 - 24k + 9}{6}$

25. Solve the equation $\dfrac{4}{x + 1} + \dfrac{3}{x - 2} = 4$.

In Exercises 26–30, point A has coordinates $(-2, 6)$ and point B has coordinates $(4, -2)$.

26. What is the equation of the horizontal line through A?

27. What is the equation of the vertical line through B?

28. What is the slope of line AB?

29. What is the slope of a line perpendicular to line AB?

30. What is the standard form of the equation of line AB?

31. Graph the line having slope $\frac{2}{3}$ and passing through the point $(-1, -3)$.

32. Refer to the accompanying graph. If 1992 is represented by $x = 0$ and 1996 is represented by $x = 4$, what is the equation of the line joining the tops of the bars for these two years? How is the slope of this line interpreted in the context of the data represented?

TOTAL HEALTH BENEFIT COST PER EMPLOYEE FOR ACTIVE AND RETIRED WORKERS

Source: Foster Higgins.

33. Graph the inequality $-3x - 2y \le 6$.

34. Given that $f(x) = x^2 + 3x - 6$, find each of the following.
(a) $f(-3)$ (b) $f(a)$

Solve by any method.

35. $-2x + 3y = -15$
       $4x - y = 15$

36. $x + y + z = 10$
       $x - y - z = 0$
      $-x + y - z = -4$

In Exercises 37–39, solve each problem using a system of equations. Use the method of your choice: elimination, substitution, matrix row operations, or Cramer's rule.

37. Admission prices at a football game were $6 for adults and $2 for children. The total value of the tickets sold was $2528, and 454 tickets were sold. How many adults and how many children attended the game?

Kind of Ticket	Number Sold	Cost of Each (in dollars)	Total Value (in dollars)
Adult	x	6	6x
Child	y		
Total	454		

38. Two of the best-selling toys of 1996 were Tickle Me Elmo and Snacktime Kid. Based on their average retail prices, Elmo cost $8.63 less than Kid, and together they cost $63.89. What was the average retail price for each toy? (*Source:* NPD Group, Inc.)

39. At the Chalmette Nut Shop, 6 pounds of peanuts and 12 pounds of cashews cost $60, while 3 pounds of peanuts and 4 pounds of cashews cost $22. Find the cost of each type of nut.

40. The graph shows a company's costs to produce computer parts and the revenue from the sale of computer parts.

  (a) At what production level does the cost equal the revenue? What is the revenue at that point?

  (b) Profit is revenue less cost. Estimate the profit on the sale of 1100 parts.

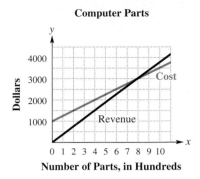

Computer Parts

Number of Parts, in Hundreds

10 Roots and Radicals

Electronics

The electronics industry encompasses producing, selling, and servicing the computers, TVs, VCRs, CD players, and cell phones that pervade our lives. The industry boomed for many years, but recently industry sales have leveled off. The graph, which gives the percent change in orders from year to year, shows the turbulence in the electronics industry from 1983 to 1998.*

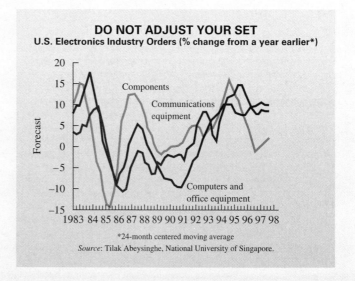

DO NOT ADJUST YOUR SET
U.S. Electronics Industry Orders (% change from a year earlier*)

*24-month centered moving average
Source: Tilak Abeysinghe, National University of Singapore.

A major reason for the downturn in electronics business is the increase in the number of companies, particularly Asian companies, that have entered the industry. This large increase in supply has caused a glut of chips, with resulting price declines. This, in turn, has contributed to the collapse of export growth throughout Eastern Asia. In the exercises for Section 10.3, we examine data on U.S. exports and imports of electronics.

*"The silicon tigers' electric shocker," *The Economist,* November 9, 1996.

Visit our Web site at www.LialAlgebra.com

10.1 Evaluating Roots

OBJECTIVES

1 Find square roots.

2 Decide whether a given root is rational, irrational, or not a real number.

3 Find decimal approximations for irrational square roots.

4 Use the Pythagorean formula.

5 Use the distance formula.

6 Find higher roots.

FOR EXTRA HELP

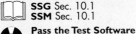 **SSG** Sec. 10.1
SSM Sec. 10.1

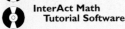 **Pass the Test Software**

**InterAct Math
Tutorial Software**

Video 16

In Section 1.2 we discussed the idea of the *square* of a number. Recall that squaring a number means multiplying the number by itself.

$$\text{If } a = 7, \text{ then } a^2 = 7 \cdot 7 = 49.$$
$$\text{If } a = -5, \text{ then } a^2 = (-5) \cdot (-5) = 25.$$
$$\text{If } a = -\frac{1}{2}, \text{ then } a^2 = \left(-\frac{1}{2}\right) \cdot \left(-\frac{1}{2}\right) = \frac{1}{4}.$$

In this chapter the opposite problem is considered.

$$\text{If } a^2 = 49, \text{ then } a = ?$$
$$\text{If } a^2 = 25, \text{ then } a = ?$$
$$\text{If } a^2 = \frac{1}{4}, \text{ then } a = ?$$

OBJECTIVE **1** **Find square roots.** To find a in the three statements above, we must find a number that when multiplied by itself results in the given number. The number a is called a **square root** of the number a^2.

EXAMPLE 1 Finding All Square Roots of a Number

Find all square roots of 49.

To find a square root of 49, think of a number that when multiplied by itself gives 49. One square root is 7, since $7 \cdot 7 = 49$. Another square root of 49 is -7, since $(-7)(-7) = 49$. The number 49 has two square roots, 7 and -7; one is positive and one is negative. ∎

The positive square root of a number is written with the symbol $\sqrt{}$. For example, the positive square root of 49 is 7, written

$$\sqrt{49} = 7.$$

The symbol $-\sqrt{}$ is used for the negative square root of a number. For example, the negative square root of 49 is -7, written

$$-\sqrt{49} = -7.$$

Most calculators have a square root key, usually labeled $\boxed{\sqrt{x}}$, that allows us to find the square root of a number. For example, if we enter 49 and use the square root key, the display will show 7.

The symbol $\sqrt{}$ is called a **radical sign** and always represents the nonnegative square root. The number inside the radical sign is called the **radicand** and the entire expression, radical sign and radicand, is called a **radical.** An algebraic expression containing a radical is called a **radical expression.**

Square Roots of a

If a is a positive real number,

$\sqrt{a}$ is the positive square root of a,

$-\sqrt{a}$ is the negative square root of a.

For nonnegative a,

$$\sqrt{a} \cdot \sqrt{a} = (\sqrt{a})^2 = a \qquad \text{and} \qquad -\sqrt{a} \cdot -\sqrt{a} = (-\sqrt{a})^2 = a.$$

Also, $\sqrt{0} = 0$.

EXAMPLE 2 Finding Square Roots

Find each square root.

(a) $\sqrt{144}$

The radical $\sqrt{144}$ represents the positive square root of 144. Think of a positive number whose square is 144.

$$12^2 = 144, \qquad \text{so} \qquad \sqrt{144} = 12.$$

(b) $-\sqrt{1024}$

This symbol represents the negative square root of 1024. A calculator with a square root key can be used to find $\sqrt{1024} = 32$. Then, $-\sqrt{1024} = -32$.

(c) $\sqrt{\dfrac{4}{9}} = \dfrac{2}{3}$
 (d) $-\sqrt{\dfrac{16}{49}} = -\dfrac{4}{7}$

As shown in the definition above, when the square root of a positive real number is squared, the result is that positive real number. (Also, $(\sqrt{0})^2 = 0$.)

EXAMPLE 3 Squaring Radical Expressions

Find the *square* of each radical expression.

(a) $\sqrt{13}$

$(\sqrt{13})^2 = 13$
   Definition of square root

(b) $-\sqrt{29}$

$(-\sqrt{29})^2 = 29$
   The square of a *negative* number is positive.

(c) $\sqrt{p^2 + 1}$

$(\sqrt{p^2 + 1})^2 = p^2 + 1$

OBJECTIVE 2 Decide whether a given root is rational, irrational, or not a real number. All numbers with square roots that are rational are called **perfect squares.** For example, 100 and $\frac{25}{4}$ are perfect squares. A number that is not a perfect square has a square root that is not a rational number. For example, $\sqrt{5}$ is not a rational number because it cannot be written as the ratio of two integers. Its decimal neither terminates nor repeats. However, $\sqrt{5}$ is a real number and corresponds to a point on the number line. As mentioned in Chapter 1, a real number that is not rational is called an *irrational number.* The number $\sqrt{5}$ is irrational. Many square roots of integers are irrational.

If a is a positive number that is not a perfect square, then $\sqrt{a}$ is irrational.

Not every number has a *real number* square root. For example, there is no real number that can be squared to get -36. (The square of a real number can never be negative.) Because of this $\sqrt{-36}$ is not a real number. A calculator may show an error message in a case like this.

If a is a negative number, then $\sqrt{a}$ is not a real number.

EXAMPLE 4   Identifying Types of Square Roots

Tell whether each square root is rational, irrational, or not a real number.

(a) $\sqrt{17}$

Since 17 is a not a perfect square, $\sqrt{17}$ is irrational.

(b) $\sqrt{64}$

The number 64 is a perfect square, 8^2, so $\sqrt{64} = 8$, a rational number.

(c) $\sqrt{85}$ is irrational.

(d) $\sqrt{81}$ is rational ($\sqrt{81} = 9$).

(e) $\sqrt{-25}$

There is no real number whose square is -25. Therefore $\sqrt{-25}$ is not a real number.

Not all irrational numbers are square roots of integers. For example, π (approximately 3.14159) is an irrational number that is not a square root of any integer.

OBJECTIVE 3   Find decimal approximations for irrational square roots. Even if a number is irrational, a decimal that approximates the number can be found by using a calculator. For example, a calculator shows that $\sqrt{10}$ is 3.16227766, although this is only a rational number *approximation* of $\sqrt{10}$.

EXAMPLE 5   Approximating Irrational Square Roots

Find a decimal approximation for each square root. Round answers to the nearest thousandth.

(a) $\sqrt{11}$

Using the square root key of a calculator gives $\sqrt{11} \approx 3.31662479 \approx 3.317$ rounded to the nearest thousandth, where $\approx$ means "is approximately equal to."

(b) $\sqrt{39} \approx 6.245$

(c) $-\sqrt{745} \approx -27.295$

(d) $\sqrt{-180}$ is not a real number.

OBJECTIVE **4** Use the Pythagorean formula. One application of square roots uses the Pythagorean formula. Recall from Section 6.6 that by this formula, if c is the length of the hypotenuse of a right triangle, and a and b are the lengths of the two legs, then

$$c^2 = a^2 + b^2.$$

See Figure 1.

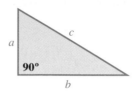

Figure 1

EXAMPLE 6 Using the Pythagorean Formula

Find the unknown length of the third side of each right triangle with sides a, b, and c, where c is the hypotenuse.

(a) $a = 3, b = 4$

Use the formula to find c^2 first.

$$
\begin{aligned}
c^2 &= a^2 + b^2 \\
&= 3^2 + 4^2 \qquad &&\text{Let } a = 3 \text{ and } b = 4. \\
&= 9 + 16 = 25 \qquad &&\text{Square and add.}
\end{aligned}
$$

Now find the positive square root of 25 to get c.

$$c = \sqrt{25} = 5$$

(Although -5 is also a square root of 25, the length of a side of a triangle must be a positive number.)

(b) $c = 9, b = 5$

Substitute the given values in the formula $c^2 = a^2 + b^2$. Then solve for a^2.

$$
\begin{aligned}
9^2 &= a^2 + 5^2 \qquad &&\text{Let } c = 9 \text{ and } b = 5. \\
81 &= a^2 + 25 \qquad &&\text{Square.} \\
56 &= a^2 \qquad &&\text{Subtract 25.}
\end{aligned}
$$

Again, we want only the positive root $a = \sqrt{56} \approx 7.483$.

Be careful not to make the common mistake of thinking that $\sqrt{a^2 + b^2}$ equals $a + b$. As Example 6(a) shows,

$$\sqrt{9 + 16} = \sqrt{25} = 5 \neq \sqrt{9} + \sqrt{16} = 3 + 4,$$

so that, in general,

$$\sqrt{a^2 + b^2} \neq a + b.$$

CONNECTIONS

Pythagoras did not actually discover the Pythagorean formula. While he may have written the first proof, there is evidence that the Babylonians knew the concept quite well. The figure on the left illustrates the formula by using a tile pattern. In the figure, the side of the square along the hypotenuse measures 5 units, while the sides along the legs measure 3 and 4 units. If we let $a = 3$, $b = 4$, and $c = 5$, the equation of the Pythagorean formula is satisfied.

$$a^2 + b^2 = c^2$$
$$3^2 + 4^2 = 5^2 \qquad ?$$
$$25 = 25 \qquad \text{True}$$

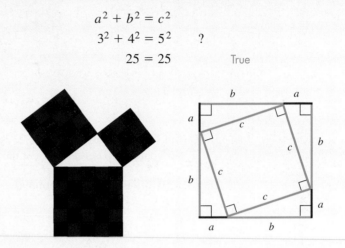

FOR DISCUSSION OR WRITING

The diagram on the right can be used to verify the Pythagorean formula. To do so, express the area of the figure in two ways: first, as the area of the large square, and then as the sum of the areas of the smaller square and the four right triangles. Finally, set the areas equal and simplify the equation.

PROBLEM SOLVING

The Pythagorean formula can be used to solve applied problems that involve right triangles. A good way to begin the solution is to sketch the triangle and label the three sides appropriately, using a variable as needed. Then use the Pythagorean formula to write an equation. This procedure is simply Steps 1–3 of the six-step problem-solving method given in Chapter 2, and used throughout the book. In Steps 4–6, we solve the equation, answer the question(s), and check the solution(s).

EXAMPLE 7   Solving an Application

A ladder 10 feet long leans against a wall. The foot of the ladder is 6 feet from the base of the wall. How high up the wall does the top of the ladder rest?

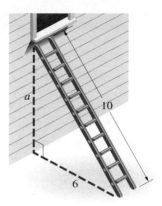

Figure 2

Steps 1 and 2   As shown in Figure 2, a right triangle is formed with the ladder as the hypotenuse. Let a represent the height of the top of the ladder.

Step 3   By the Pythagorean formula,

$$c^2 = a^2 + b^2.$$
$$10^2 = a^2 + 6^2 \qquad \text{Let } c = 10 \text{ and } b = 6.$$
$$100 = a^2 + 36 \qquad \text{Square.}$$

Step 4
$$64 = a^2 \qquad \text{Subtract 36.}$$
$$\sqrt{64} = a$$
$$a = 8 \qquad \sqrt{64} = 8$$

Step 5   Choose the positive square root of 64 since a represents a length.

Step 6   From Figure 2, we see that we must have

$$8^2 + 6^2 = 10^2 \qquad ?$$
$$64 + 36 = 100. \qquad \text{True}$$

The check shows that the top of the ladder rests 8 feet up the wall.

OBJECTIVE 5   Use the distance formula.   The Pythagorean formula is used to develop another useful formula for finding the distance between two points on the plane. Figure 3 shows the two different points (x_1, y_1) and (x_2, y_2), as well as the point (x_2, y_1). The distance between (x_2, y_2) and (x_2, y_1) is $a = y_2 - y_1$, and the distance between (x_1, y_1) and (x_2, y_1) is $b = x_2 - x_1$. From the Pythagorean formula,

$$d^2 = (x_2 - x_1)^2 + (y_2 - y_1)^2.$$

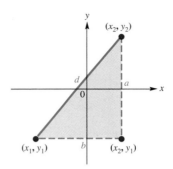

Figure 3

Taking the square root of each side, we get the **distance formula.**

Distance Formula

The distance between the points (x_1, y_1) and (x_2, y_2) is

$$d = \sqrt{(x_2 - x_1)^2 + (y_2 - y_1)^2}.$$

EXAMPLE 8   Using the Distance Formula

Find the distance between $(-3, 4)$ and $(2, 5)$.

Use the distance formula. Choose $(x_1, y_1) = (-3, 4)$ and $(x_2, y_2) = (2, 5)$.

$$d - \sqrt{(x_2 - x_1)^2 + (y_2 - y_1)^2}$$
$$= \sqrt{(2 - (-3))^2 + (5 - 4)^2} \qquad \text{Substitute.}$$
$$= \sqrt{5^2 + 1^2}$$
$$= \sqrt{26}$$

OBJECTIVE 6   **Find higher roots.** Finding the square root of a number is the inverse (reverse) of squaring a number. In a similar way, there are inverses to finding the cube of a number, or finding the fourth power of a number. These inverses are the **cube root,** written $\sqrt[3]{a}$, and the **fourth root,** written $\sqrt[4]{a}$. Similar symbols are used for higher roots. In general we have the following.

The nth root of a is written $\sqrt[n]{a}$.

In $\sqrt[n]{a}$, the number n is the **index** or **order** of the radical. It would be possible to write $\sqrt[2]{a}$ instead of $\sqrt{a}$, but the simpler symbol $\sqrt{a}$ is customary since the square root is the most commonly used root. A calculator that has a key marked $\boxed{\sqrt[x]{y}}$ or $\boxed{x^y}$ can be used to find these roots. When working with cube roots or fourth roots, it is helpful to memorize the first few *perfect cubes* ($2^3 = 8$, $3^3 = 27$, and so on) and the first few perfect fourth powers.

┌ E X A M P L E   9   **Finding Cube Roots**

Find each cube root.

(a) $\sqrt[3]{8}$

Look for a number that can be cubed to give 8. Since $2^3 = 8$, then $\sqrt[3]{8} = 2$.

(b) $\sqrt[3]{-8}$

$\sqrt[3]{-8} = -2$ because $(-2)^3 = -8$.

As Example 9 suggests, the cube root of a positive number is positive, and the cube root of a negative number is negative. *There is only one real number cube root for each real number.*

When the index of the radical is even (square root, fourth root, and so on), the radicand must be nonnegative to get a real number root. Also, for even indexes the symbols $\sqrt{}$, $\sqrt[4]{}$, $\sqrt[6]{}$, and so on are used for the *nonnegative* roots, which are called **principal roots.** The symbols $-\sqrt{}$, $-\sqrt[4]{}$, $-\sqrt[6]{}$, and so on are used for the negative roots.

┌ E X A M P L E   1 0   **Finding Higher Roots**

Find each root.

(a) $\sqrt[4]{16}$

$\sqrt[4]{16} = 2$ because 2 is positive and $2^4 = 16$.

(b) $-\sqrt[4]{16}$

From part (a), $\sqrt[4]{16} = 2$, so the negative root $-\sqrt[4]{16} = -2$.

(c) $\sqrt[4]{-16}$

To find the fourth root, the radicand must be nonnegative. There is no real number that equals $\sqrt[4]{-16}$.

(d) $\sqrt[3]{64} = 4$ since $4^3 = 64$.

(e) $-\sqrt[5]{32}$

First find $\sqrt[5]{32}$. The prime factorization of 32 as 2^5 shows that $\sqrt[5]{32} = 2$. If $\sqrt[5]{32} = 2$, then $-\sqrt[5]{32} = -2$.

1 0 . 1   E X E R C I S E S

Decide whether each statement is true or false. If false, tell why.

1. Every nonnegative number has two square roots.

2. A negative number has negative square roots.

3. Every positive number has two real square roots.

4. Every positive number has three real cube roots.

5. The cube root of every real number has the same sign as the number itself.

6. The positive square root of a positive number is its principal square root.

Find all square roots of each number. See Example 1.

7. 16   **8.** 9   **9.** 144   **10.** 225

11. $\dfrac{25}{196}$ **12.** $\dfrac{81}{400}$ **13.** 900   **14.** 1600

Find each square root that is a real number. See Examples 2 and 4(e).

15. $\sqrt{49}$ **16.** $\sqrt{81}$ **17.** $-\sqrt{121}$ **18.** $\sqrt{196}$

19. $-\sqrt{\dfrac{144}{121}}$ **20.** $-\sqrt{\dfrac{49}{36}}$ **21.** $\sqrt{-121}$ **22.** $\sqrt{-49}$

Find the square of each radical expression. See Example 3.

23. $\sqrt{100}$ **24.** $\sqrt{36}$ **25.** $-\sqrt{19}$

26. $-\sqrt{99}$ **27.** $\sqrt{3x^2 + 4}$ **28.** $\sqrt{9y^2 + 3}$

What must be true about the variable a for each statement to be true?

29. $\sqrt{a}$ represents a positive number.   **30.** $-\sqrt{a}$ represents a negative number.

31. $\sqrt{a}$ is not a real number.   **32.** $-\sqrt{a}$ is not a real number.

 Write rational, irrational, *or* not a real number *for each number. If the number is rational, give its exact value. If the number is irrational, give a decimal approximation to the nearest thousandth. Use a calculator as necessary. See Examples 4 and 5.*

33. $\sqrt{25}$ **34.** $\sqrt{169}$ **35.** $\sqrt{29}$ **36.** $\sqrt{33}$

37. $-\sqrt{64}$ **38.** $-\sqrt{500}$ **39.** $\sqrt{-29}$ **40.** $\sqrt{-47}$

41. Explain why the answers to Exercises 17 and 21 are different.

42. Explain why $\sqrt[3]{-8}$ and $-\sqrt[3]{8}$ represent the same number.

 Use a calculator with a square root key to find each root. Round to the nearest thousandth. See Example 5.

43. $\sqrt{571}$ **44.** $\sqrt{693}$ **45.** $\sqrt{798}$

46. $\sqrt{453}$ **47.** $\sqrt{3.94}$ **48.** $\sqrt{.00895}$

 Find each square root. Use a calculator and round to the nearest thousandth, if necessary. (Hint: First simplify the radicand to a single number.)

49. $\sqrt{3^2 + 4^2}$ **50.** $\sqrt{6^2 + 8^2}$ **51.** $\sqrt{8^2 + 15^2}$

52. $\sqrt{5^2 + 12^2}$ **53.** $\sqrt{2^2 + 3^2}$ **54.** $\sqrt{(-1)^2 + 5^2}$

 Use a calculator with a cube root key to find each root. Round to the nearest thousandth. (In Exercises 59 and 60, you may have to use the fact that if $a > 0$, $\sqrt[3]{-a} = -\sqrt[3]{a}$.)

55. $\sqrt[3]{12}$ **56.** $\sqrt[3]{74}$ **57.** $\sqrt[3]{130.6}$

58. $\sqrt[3]{251.8}$ **59.** $\sqrt[3]{-87}$ **60.** $\sqrt[3]{-95}$

Find the length of the unknown side of each right triangle with legs a and b and hypotenuse c. In Exercises 65 and 66, use a calculator and round to the nearest thousandth. See Example 6.

61. $a = 8$, $b = 15$ **62.** $a = 24$, $b = 10$ **63.** $a = 6$, $c = 10$

64. $b = 12$, $c = 13$ **65.** $a = 11$, $b = 4$ **66.** $a = 13$, $b = 9$

Use the Pythagorean formula to solve each problem. See Example 7.

67. The diagonal of a rectangle measures 25 centimeters. The width of the rectangle is 7 centimeters. Find the length of the rectangle.

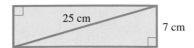

25 cm

7 cm

68. The length of a rectangle is 40 meters, and the width is 9 meters. Find the measure of the diagonal of the rectangle.

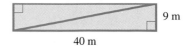

9 m

40 m

In Exercises 69–72, use these steps to solve each problem.

Step 1   *Let x represent the unknown length.*

Step 2   *Use the information in the problem to label the remaining sides of the triangle in the figure.*

Step 3   *Write an equation relating the three sides of the triangle. (Hint: Use the Pythagorean formula.)*

Step 4   *Solve the equation.*

Steps 5 and 6   *Give the answer; check it.*

69. Margaret is flying a kite on 100 feet of string. How high is it above her hand (vertically) if the horizontal distance between Margaret and the kite is 60 feet?

Not to scale

70. A guy wire is attached to the mast of a short-wave transmitting antenna. It is attached 96 feet above ground level. If the wire is staked to the ground 72 feet from the base of the mast, how long is the wire?

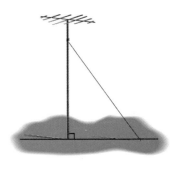

71. Two cars leave Tomball, Texas, at the same time. One travels north at 25 miles per hour and the other travels west at 60 miles per hour. How far apart are they after 3 hours? (*Hint:* Use $d = rt$ to find the distance traveled by each car.)

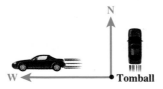

N

W

Tomball

72. A boat is being pulled toward a dock with a rope attached at water level. When the boat is 24 feet from the dock, 30 feet of rope is extended. What is the height of the dock above the water?

In Exercises 73 and 74, round to the nearest thousandth.

73. What is the value of *x* in the figure?

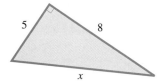

74. What is the value of *y* in the figure?

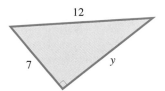

75. Use specific values for *a* and *b* different from those given in the "Caution" following Example 6 to show that $\sqrt{a^2 + b^2} \neq a + b$. Why would the values $a = 0$ and $b = 1$ *not* be satisfactory choices?

76. Explain how finding the square root of a number and squaring a number are related. Give examples.

Find the distance between each pair of points. See Example 8.

77. $(5, -3)$ and $(7, 2)$

78. $(2, 9)$ and $(-3, -3)$

79. $\left(-\frac{1}{4}, \frac{2}{3}\right)$ and $\left(\frac{3}{4}, -\frac{1}{3}\right)$

80. $\left(\frac{2}{5}, \frac{3}{2}\right)$ and $\left(-\frac{8}{5}, \frac{1}{2}\right)$

Find each root that is a real number. See Examples 9 and 10.

81. $\sqrt[3]{1000}$

82. $\sqrt[3]{8}$

83. $\sqrt[3]{-27}$

84. $\sqrt[3]{-64}$

85. $\sqrt[4]{625}$

86. $\sqrt[4]{10,000}$

87. $\sqrt[4]{-1}$

88. $\sqrt[4]{-625}$

TECHNOLOGY INSIGHTS (EXERCISES 89-92)

Exercises 89–92 illustrate the way one graphing calculator shows nth roots. Give each indicated root. Exercises 91 and 92 indicate 5th roots.

89.
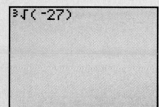

90.
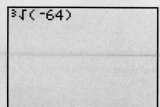

91.

92.
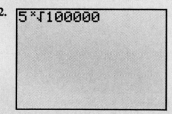

RELATING CONCEPTS (EXERCISES 93-98)

One of the many proofs of the Pythagorean formula was given in the Connections box in this section. Here is another one, attributed to the Hindu mathematician Bhāskara.

Refer to the figures and **work Exercises 93–98 in order.**

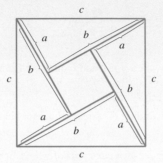

93. What is the area of the square on the left in terms of c?

94. What is the area of the small square in the middle of the figure on the left, in terms of $(b - a)$?

95. What is the sum of the areas of the two rectangles made up of triangles in the figure on the right?

96. What is the area of the small square in the figure on the right in terms of a and b?

97. The figure on the left is made up of the same square and triangles as the figure on the right. Write an equation setting the answer to Exercise 93 equal to the sum of the answers in Exercises 95 and 96.

98. Simplify the expressions you obtained in Exercise 97. What is your final result?

Did you make the connection that geometric figures can be used to prove algebraic formulas?

10.2 Multiplication and Division of Radicals

OBJECTIVES

1. Multiply radicals.
2. Simplify radicals using the product rule.
3. Simplify radical quotients using the quotient rule.
4. Use the product and quotient rules to simplify higher roots.

FOR EXTRA HELP

SSG Sec. 10.2
SSM Sec. 10.2

Pass the Test Software

InterAct Math
  Tutorial Software

Video 16

CONNECTIONS

The sixteenth century German radical symbol $\sqrt{}$ we use today is probably derived from the letter R. The radical symbol on the left below comes from the Latin word for root, *radix*. It was first used by Leonardo da Pisa (Fibonnaci) in 1220.

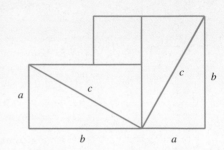

The cube root symbol shown on the right above was used by the German mathematician Christoff Rudolff in 1525. The symbol used today originated in the seventeenth century in France.

CONNECTIONS (CONTINUED)

FOR DISCUSSION OR WRITING

1. In the radical sign shown on the left, the R referred to above is clear. What other letter do you think is part of the symbol? What would the equivalent be in our modern notation?

2. How is the cube root symbol on the right related to our modern radical sign?

OBJECTIVE 1 **Multiply radicals.** Several useful rules for finding products and quotients of radicals are developed in this section. To illustrate the rule for products, notice that

$$\sqrt{4} \cdot \sqrt{9} = 2 \cdot 3 = 6 \qquad \text{and} \qquad \sqrt{4 \cdot 9} = \sqrt{36} = 6,$$

showing that

$$\sqrt{4} \cdot \sqrt{9} = \sqrt{4 \cdot 9}.$$

This result is a particular case of the more general product rule for radicals.

Product Rule for Radicals

For nonnegative real numbers x and y,

$$\sqrt{x} \cdot \sqrt{y} = \sqrt{x \cdot y} \qquad \text{and} \qquad \sqrt{x \cdot y} = \sqrt{x} \cdot \sqrt{y}.$$

That is, the product of two radicals is the radical of the product.

In general, $\sqrt{x + y} \neq \sqrt{x} + \sqrt{y}$. To see why this is so, let $x = 16$ and $y = 9$.

$$\sqrt{16 + 9} = \sqrt{25} = 5$$

but

$$\sqrt{16} + \sqrt{9} = 4 + 3 = 7.$$

EXAMPLE 1   Using the Product Rule to Multiply Radicals

Use the product rule for radicals to find each product.

(a) $\sqrt{2} \cdot \sqrt{3} = \sqrt{2 \cdot 3} = \sqrt{6}$

(b) $\sqrt{7} \cdot \sqrt{5} = \sqrt{35}$

(c) $\sqrt{11} \cdot \sqrt{a} = \sqrt{11a}$ Assume $a \geq 0$.

OBJECTIVE 2 **Simplify radicals using the product rule.** A square root radical is **simplified** when no perfect square factor remains under the radical sign. This is accomplished by using the product rule as shown in Example 2.

┌─ **E X A M P L E   2**   **Using the Product Rule to Simplify Radicals**

Simplify each radical.

(a) $\sqrt{20}$

Since 20 has a perfect square factor of 4,

$$\sqrt{20} = \sqrt{4 \cdot 5} \qquad \text{4 is a perfect square.}$$
$$= \sqrt{4} \cdot \sqrt{5} \qquad \text{Product rule}$$
$$= 2\sqrt{5}. \qquad \sqrt{4} = 2$$

Thus, $\sqrt{20} = 2\sqrt{5}$. Since 5 has no perfect square factor other than 1, $2\sqrt{5}$ is called the **simplified form** of $\sqrt{20}$. Note that $2\sqrt{5}$ represents a product, where the factors are 2 and $\sqrt{5}$; also, $2\sqrt{5} \neq \sqrt{10}$.

(b) $\sqrt{72}$

Begin by looking for the largest perfect square that is a factor of 72. This number is 36, so

$$\sqrt{72} = \sqrt{36 \cdot 2} \qquad \text{36 is a perfect square.}$$
$$= \sqrt{36} \cdot \sqrt{2} \qquad \text{Product rule}$$
$$= 6\sqrt{2}. \qquad \sqrt{36} = 6$$

We could also factor 72 into its prime factors and look for pairs of like factors. Each pair of like factors produces a factor outside the radical in the simplified form.

$$\sqrt{72} = \sqrt{2 \cdot 2 \cdot 2 \cdot 3 \cdot 3} = 2 \cdot 3 \cdot \sqrt{2} = 6\sqrt{2}$$

In either case, we obtain $6\sqrt{2}$ as the simplified form of $\sqrt{72}$; our work is simpler, however, if we begin with the largest perfect square factor.

(c) $\sqrt{300} = \sqrt{100 \cdot 3} \qquad \text{100 is a perfect square.}$
$$= \sqrt{100} \cdot \sqrt{3} \qquad \text{Product rule}$$
$$= 10\sqrt{3} \qquad \sqrt{100} = 10$$

(d) $\sqrt{15}$

The number 15 has no perfect square factor (except 1), so $\sqrt{15}$ cannot be simplified further. ∎

Sometimes the product rule can be used to simplify a product, as Example 3 shows.

┌─ **E X A M P L E   3**   **Multiplying and Simplifying Radicals**

Find each product and simplify.

(a) $\sqrt{9} \cdot \sqrt{75} = 3\sqrt{75} \qquad \sqrt{9} = 3$
$$= 3\sqrt{25 \cdot 3} \qquad \text{25 is a perfect square.}$$
$$= 3\sqrt{25} \cdot \sqrt{3} \qquad \text{Product rule}$$
$$= 3 \cdot 5\sqrt{3} \qquad \sqrt{25} = 5$$
$$= 15\sqrt{3} \qquad \text{Multiply.}$$

(b) $\sqrt{8} \cdot \sqrt{12} = \sqrt{8 \cdot 12}$ Product rule

$\quad\quad\quad\quad\quad\; = \sqrt{4 \cdot 2 \cdot 4 \cdot 3}$ Factor; 4 is a perfect square.

$\quad\quad\quad\quad\quad\; = \sqrt{4} \cdot \sqrt{4} \cdot \sqrt{2 \cdot 3}$ Product rule

$\quad\quad\quad\quad\quad\; = 2 \cdot 2 \cdot \sqrt{6}$ $\sqrt{4} = 2$

$\quad\quad\quad\quad\quad\; = 4\sqrt{6}$ Multiply.

Part (b) of Example 3 also could be simplified by using the product rule first to get

$$\sqrt{8} \cdot \sqrt{12} = \sqrt{4 \cdot 2} \cdot \sqrt{4 \cdot 3}$$
$$= 2\sqrt{2} \cdot 2\sqrt{3}$$
$$= 2 \cdot 2 \cdot \sqrt{2} \cdot \sqrt{3}$$
$$= 4\sqrt{6}.$$

Both approaches are correct.

OBJECTIVE 3 Simplify radical quotients using the quotient rule. The *quotient rule for radicals* is very similar to the product rule. It, too, can be used either way.

Quotient Rule for Radicals

If x and y are nonnegative real numbers and $y \neq 0$,

$$\frac{\sqrt{x}}{\sqrt{y}} = \sqrt{\frac{x}{y}} \quad \text{and} \quad \sqrt{\frac{x}{y}} = \frac{\sqrt{x}}{\sqrt{y}}.$$

The quotient of the radicals is the radical of the quotient.

EXAMPLE 4 Using the Quotient Rule to Simplify Radicals

Simplify each radical.

(a) $\sqrt{\dfrac{25}{9}} = \dfrac{\sqrt{25}}{\sqrt{9}} = \dfrac{5}{3}$ Quotient rule

(b) $\dfrac{\sqrt{288}}{\sqrt{2}} = \sqrt{\dfrac{288}{2}} = \sqrt{144} = 12$ Quotient rule

(c) $\sqrt{\dfrac{3}{4}} = \dfrac{\sqrt{3}}{\sqrt{4}} = \dfrac{\sqrt{3}}{2}$ Quotient rule

EXAMPLE 5 Using the Quotient Rule to Divide Radicals

Divide $27\sqrt{15}$ by $9\sqrt{3}$.

We use the quotient rule as follows.

$$\frac{27\sqrt{15}}{9\sqrt{3}} = \frac{27}{9} \cdot \frac{\sqrt{15}}{\sqrt{3}} = 3\sqrt{\frac{15}{3}} = 3\sqrt{5}$$

Some problems require both the product and quotient rules, as Example 6 shows.

⌐ E X A M P L E 6   Using Both the Product and Quotient Rules

Simplify $\sqrt{\dfrac{3}{5}} \cdot \sqrt{\dfrac{4}{5}}$.

$$\sqrt{\frac{3}{5}} \cdot \sqrt{\frac{4}{5}} = \frac{\sqrt{3}}{\sqrt{5}} \cdot \frac{\sqrt{4}}{\sqrt{5}} \qquad \text{Quotient rule}$$

$$= \frac{\sqrt{3} \cdot \sqrt{4}}{\sqrt{5} \cdot \sqrt{5}} \qquad \text{Multiply fractions.}$$

$$= \frac{\sqrt{3} \cdot 2}{\sqrt{25}} \qquad \text{Product rule; } \sqrt{4} = 2$$

$$= \frac{2\sqrt{3}}{5} \qquad \sqrt{25} = 5$$

The product and quotient rules also apply when variables appear under the radical sign, as long as all the variables represent only nonnegative numbers. For example, $\sqrt{5^2} = 5$, but $\sqrt{(-5)^2} \neq -5$.

> For a real number a, $\sqrt{a^2} = a$ only if a is nonnegative.

⌐ E X A M P L E 7   Simplifying Radicals Involving Variables

Simplify each radical. Assume all variables represent positive real numbers.

(a) $\sqrt{25m^4} = \sqrt{25} \cdot \sqrt{m^4}$ Product rule

     $= 5m^2$

(b) $\sqrt{64p^{10}} = 8p^5$ Product rule

(c) $\sqrt{r^9} = \sqrt{r^8 \cdot r}$

     $= \sqrt{r^8} \cdot \sqrt{r} = r^4\sqrt{r}$ Product rule

(d) $\sqrt{\dfrac{5}{x^2}} = \dfrac{\sqrt{5}}{\sqrt{x^2}} = \dfrac{\sqrt{5}}{x}$ Quotient rule

O B J E C T I V E **4**   Use the product and quotient rules to simplify higher roots. The product and quotient rules for radicals also work for other roots, as shown in Example 8. To simplify cube roots, look for factors that are *perfect cubes*. A **perfect cube** is a number with a rational cube root. For example, $\sqrt[3]{64} = 4$, and since 4 is a rational number, 64 is a perfect cube. Higher roots are handled in a similar manner.

Properties of Radicals

For all real numbers where the indicated roots exist,

$$\sqrt[n]{x} \cdot \sqrt[n]{y} = \sqrt[n]{xy} \qquad \text{and} \qquad \frac{\sqrt[n]{x}}{\sqrt[n]{y}} = \sqrt[n]{\frac{x}{y}}, y \neq 0.$$

EXAMPLE 8 Simplifying Higher Roots

Simplify each radical.

(a) $\sqrt[3]{32} = \sqrt[3]{8 \cdot 4}$ 8 is a perfect cube.

$$= \sqrt[3]{8} \cdot \sqrt[3]{4} = 2\sqrt[3]{4}$$

(b) $\sqrt[3]{108} = \sqrt[3]{27 \cdot 4}$ 27 is a perfect cube.

$$= \sqrt[3]{27} \cdot \sqrt[3]{4} = 3\sqrt[3]{4}$$

(c) $\sqrt[4]{32} = \sqrt[4]{16} \cdot \sqrt[4]{2} = 2\sqrt[4]{2}$ 16 is a perfect fourth power.

(d) $\sqrt[3]{\dfrac{8}{125}} = \dfrac{\sqrt[3]{8}}{\sqrt[3]{125}} = \dfrac{2}{5}$

(e) $\sqrt[4]{\dfrac{16}{625}} = \dfrac{\sqrt[4]{16}}{\sqrt[4]{625}} = \dfrac{2}{5}$

(f) $\sqrt[3]{7} \cdot \sqrt[3]{49} = \sqrt[3]{7 \cdot 49} = \sqrt[3]{7 \cdot 7^2} = \sqrt[3]{7^3} = 7$

10.2 EXERCISES

Decide whether each statement is true or false for real numbers. If false, tell why.

1. $\sqrt{4} = \pm 2$ **2.** $\sqrt{(-6)^2} = -6$ **3.** $\sqrt{-6} \cdot \sqrt{6} = -6$

4. $\sqrt[3]{(-6)^3} = -6$ **5.** $\sqrt[3]{3} \cdot \sqrt[3]{2} = \sqrt[3]{6}$ **6.** $\sqrt[3]{4} \cdot \sqrt[3]{4} = 4$

7. $\sqrt{4} \cdot \sqrt{9} = \sqrt{4 \cdot 9}$ **8.** $\sqrt{4} + \sqrt{9} = \sqrt{4 + 9}$

Use the product rule for radicals to find each product. See Example 1.

9. $\sqrt{3} \cdot \sqrt{27}$ **10.** $\sqrt{2} \cdot \sqrt{8}$ **11.** $\sqrt{6} \cdot \sqrt{15}$

12. $\sqrt{10} \cdot \sqrt{15}$ **13.** $\sqrt{13} \cdot \sqrt{13}$ **14.** $\sqrt{17} \cdot \sqrt{17}$

15. $\sqrt{13} \cdot \sqrt{r},\, r \geq 0$ **16.** $\sqrt{19} \cdot \sqrt{k},\, k \geq 0$

17. Which one of the following radicals is simplified according to the guidelines of Objective 2?

    (a) $\sqrt{47}$ **(b)** $\sqrt{45}$ **(c)** $\sqrt{48}$ **(d)** $\sqrt{44}$

18. If p is a prime number, is $\sqrt{p}$ in simplified form? Explain your answer.

Simplify each radical according to the method described in Objective 2. See Example 2.

19. $\sqrt{45}$ **20.** $\sqrt{56}$ **21.** $\sqrt{75}$ **22.** $\sqrt{18}$

23. $\sqrt{125}$ **24.** $\sqrt{80}$ **25.** $-\sqrt{700}$ **26.** $-\sqrt{600}$

27. $3\sqrt{27}$ **28.** $9\sqrt{8}$

Find each product and simplify. See Example 3.

29. $\sqrt{3} \cdot \sqrt{18}$ **30.** $\sqrt{3} \cdot \sqrt{21}$ **31.** $\sqrt{12} \cdot \sqrt{48}$

32. $\sqrt{50} \cdot \sqrt{72}$ **33.** $\sqrt{12} \cdot \sqrt{30}$ **34.** $\sqrt{30} \cdot \sqrt{24}$

35. In your own words, describe the product and quotient rules.

36. Simplify the radical $\sqrt{288}$ in two ways. First, factor 288 as $144 \cdot 2$ and then simplify completely. Second, factor 288 as $48 \cdot 6$ and then simplify completely. How do the answers compare? Make a conjecture concerning the quickest way to simplify such a radical.

Use the quotient rule and the product rule, as necessary, to simplify each radical expression. See Examples 4–6.

37. $\sqrt{\dfrac{16}{225}}$ **38.** $\sqrt{\dfrac{9}{100}}$ **39.** $\sqrt{\dfrac{7}{16}}$ **40.** $\sqrt{\dfrac{13}{25}}$

41. $\sqrt{\dfrac{5}{7}} \cdot \sqrt{35}$

42. $\sqrt{\dfrac{10}{13}} \cdot \sqrt{130}$

43. $\sqrt{\dfrac{5}{2}} \cdot \sqrt{\dfrac{125}{8}}$

44. $\sqrt{\dfrac{8}{3}} \cdot \sqrt{\dfrac{512}{27}}$

45. $\dfrac{30\sqrt{10}}{5\sqrt{2}}$

46. $\dfrac{50\sqrt{20}}{2\sqrt{10}}$

Simplify each radical. Assume that all variables represent nonnegative real numbers. See Example 7.

47. $\sqrt{m^2}$ **48.** $\sqrt{k^2}$ **49.** $\sqrt{y^4}$ **50.** $\sqrt{s^4}$

51. $\sqrt{36z^2}$ **52.** $\sqrt{49n^2}$ **53.** $\sqrt{400x^6}$ **54.** $\sqrt{900y^8}$

55. $\sqrt{z^5}$ **56.** $\sqrt{a^{13}}$ **57.** $\sqrt{x^6 y^{12}}$ **58.** $\sqrt{a^8 b^{10}}$

Simplify each radical. See Example 8.

59. $\sqrt[3]{40}$ **60.** $\sqrt[3]{48}$ **61.** $\sqrt[3]{54}$ **62.** $\sqrt[3]{192}$

63. $\sqrt[4]{80}$ **64.** $\sqrt[4]{243}$ **65.** $\sqrt[3]{\dfrac{8}{27}}$ **66.** $\sqrt[3]{\dfrac{64}{125}}$

67. $\sqrt[3]{-\dfrac{216}{125}}$ **68.** $\sqrt[3]{-\dfrac{1}{64}}$ **69.** $\sqrt[3]{5} \cdot \sqrt[3]{25}$ **70.** $\sqrt[3]{4} \cdot \sqrt[3]{16}$

71. $\sqrt[4]{4} \cdot \sqrt[4]{3}$ **72.** $\sqrt[4]{7} \cdot \sqrt[4]{4}$ **73.** $\sqrt[3]{4x} \cdot \sqrt[3]{8x^2}$ **74.** $\sqrt[3]{25p} \cdot \sqrt[3]{125p^3}$

75. In Example 2(a) we showed *algebraically* that $\sqrt{20}$ is equal to $2\sqrt{5}$. To give *numerical support* to this result, use a calculator to do the following:
  (a) Find a decimal approximation for $\sqrt{20}$ using your calculator. Record as many digits as the calculator shows.
  (b) Find a decimal approximation for $\sqrt{5}$ using your calculator, and then multiply the result by 2. Record as many digits as the calculator shows.
  (c) Your results in parts (a) and (b) should be the same. A mathematician would not accept this numerical exercise as *proof* that $\sqrt{20}$ is equal to $2\sqrt{5}$. Explain why.

76. On your calculator, multiply the approximations for $\sqrt{3}$ and $\sqrt{5}$. Now, predict what your calculator will show when you find an approximation for $\sqrt{15}$. What rule stated in this section justifies your answer?

The volume of a cube is found with the formula $V = s^3$, where s is the length of an edge of the cube. Use this information in Exercises 77 and 78.

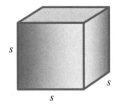

77. A container in the shape of a cube has a volume of 216 cubic centimeters. What is the depth of the container?

78. A cube-shaped box must be constructed to contain 128 cubic feet. What should the dimensions (height, width, and length) of the box be?

The volume of a sphere is found with the formula $V = \dfrac{4}{3}\pi r^3$, where r is the length of the radius of the sphere. Use this information in Exercises 79 and 80.

79. A ball in the shape of a sphere has a volume of 288π cubic inches. What is the radius of the ball?

80. Suppose that the volume of the ball described in Exercise 79 is multiplied by 8. How is the radius affected?

81. When we multiply two radicals with variables under the radical sign, such as $\sqrt{a} \cdot \sqrt{b} = \sqrt{ab}$, why is it important to know that both a and b represent nonnegative numbers?

82. Is it necessary to restrict k to a nonnegative number to say that $\sqrt[3]{k} \cdot \sqrt[3]{k} \cdot \sqrt[3]{k} = k$? Why?

█ RELATING CONCEPTS (EXERCISES 83-86)

An interesting way to represent the lengths corresponding to $\sqrt{2}, \sqrt{3}, \sqrt{4}, \sqrt{5}$, and so on is shown in the figure.

Work the following exercises in order.

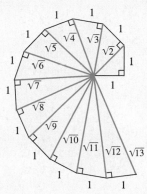

83. Use the Pythagorean formula to verify the lengths in the figure.

84. Which of the lengths indicated as radicals equal whole numbers? If the figure is continued in the same way, what would the next two whole number lengths be?

85. Find the consecutive differences between the radicands in Exercise 84. (*Hint:* The first difference is $9 - 4 = 5$.)

86. Look for a pattern that determines where these whole number lengths occur. Use this pattern to predict where the next whole number length and the one after that will occur.

Did you make the connection between the lengths of the sides of the triangles and the Pythagorean formula?

10.3 Addition and Subtraction of Radicals

OBJECTIVE 1 Add and subtract radicals. We add or subtract radicals by using the distributive property. For example,

$$8\sqrt{3} + 6\sqrt{3} = (8 + 6)\sqrt{3} \qquad \text{Distributive property}$$
$$= 14\sqrt{3}.$$

Also,

$$2\sqrt{11} - 7\sqrt{11} = -5\sqrt{11}.$$

Like radicals are terms that have multiples of the *same root* of the *same number*. Only like radicals can be combined using the distributive property. In the example above, the like radicals are $2\sqrt{11}$ and $-7\sqrt{11}$. On the other hand, examples of *unlike radicals* are

$$2\sqrt{5} \qquad \text{and} \qquad 2\sqrt{3}, \qquad \text{Different radicands}$$

as well as

$$2\sqrt{3} \qquad \text{and} \qquad 2\sqrt[3]{3}. \qquad \text{Different indexes}$$

┌─ **E X A M P L E 1**   Adding and Subtracting Like Radicals

Add or subtract, as indicated.

(a) $3\sqrt{6} + 5\sqrt{6} = (3 + 5)\sqrt{6} = 8\sqrt{6}$ Distributive property

(b) $5\sqrt{10} - 7\sqrt{10} = (5 - 7)\sqrt{10} = -2\sqrt{10}$

(c) $\sqrt[3]{5} + \sqrt[3]{5} = 1\sqrt[3]{5} + 1\sqrt[3]{5} = (1 + 1)\sqrt[3]{5} = 2\sqrt[3]{5}$

(d) $\sqrt[4]{7} + 2\sqrt[4]{7} = 1\sqrt[4]{7} + 2\sqrt[4]{7} = 3\sqrt[4]{7}$

(e) $\sqrt{3} + \sqrt{13}$ cannot be added using the distributive property.

In general, $\sqrt{x} + \sqrt{y} \neq \sqrt{x + y}$. In Example 1(e), it would be **incorrect** to try to simplify $\sqrt{3} + \sqrt{13}$ as $\sqrt{3} + \sqrt{13} = \sqrt{16} = 4$. Only *like radicals* can be combined.

O B J E C T I V E 2   Simplify radical sums and differences. Sometimes we must simplify one or more radicals in a sum or difference. Doing this may result in like radicals, which we can then add or subtract.

┌─ **E X A M P L E 2**   Adding and Subtracting Radicals That Require Simplification

Simplify as much as possible.

(a) $3\sqrt{2} + \sqrt{8} = 3\sqrt{2} + \sqrt{4 \cdot 2}$ Factor.

$\qquad\qquad\quad = 3\sqrt{2} + \sqrt{4} \cdot \sqrt{2}$ Product rule

$\qquad\qquad\quad = 3\sqrt{2} + 2\sqrt{2}$ $\sqrt{4} = 2$

$\qquad\qquad\quad = 5\sqrt{2}$ Add like radicals.

(b) $\sqrt{18} - \sqrt{27} = \sqrt{9 \cdot 2} - \sqrt{9 \cdot 3}$ Factor.

$\qquad\qquad\quad = \sqrt{9} \cdot \sqrt{2} - \sqrt{9} \cdot \sqrt{3}$ Product rule

$\qquad\qquad\quad = 3\sqrt{2} - 3\sqrt{3}$ $\sqrt{9} = 3$

Since $\sqrt{2}$ and $\sqrt{3}$ are unlike radicals, this difference cannot be simplified further.

(c) $2\sqrt{12} + 3\sqrt{75} = 2(\sqrt{4} \cdot \sqrt{3}) + 3(\sqrt{25} \cdot \sqrt{3})$ Product rule

$\qquad\qquad\quad = 2(2\sqrt{3}) + 3(5\sqrt{3})$ $\sqrt{4} = 2$ and $\sqrt{25} = 5$

$\qquad\qquad\quad = 4\sqrt{3} + 15\sqrt{3}$ Multiply.

$\qquad\qquad\quad = 19\sqrt{3}$ Add like radicals.

(d) $3\sqrt[3]{16} + 5\sqrt[3]{2} = 3(\sqrt[3]{8} \cdot \sqrt[3]{2}) + 5\sqrt[3]{2}$ Product rule

$\qquad\qquad\quad = 3(2\sqrt[3]{2}) + 5\sqrt[3]{2}$ $\sqrt[3]{8} = 2$

$\qquad\qquad\quad = 6\sqrt[3]{2} + 5\sqrt[3]{2}$ Multiply.

$\qquad\qquad\quad = 11\sqrt[3]{2}$ Add like radicals.

O B J E C T I V E 3   Simplify radical sums involving multiplication. Some radical expressions require both multiplication and addition (or subtraction). The order of operations presented in Chapter 1 still applies.

EXAMPLE 3 Simplifying Radical Sums Involving Multiplication

Simplify each radical expression. Assume that all variables represent nonnegative real numbers.

(a) $\sqrt{5} \cdot \sqrt{15} + 4\sqrt{3} = \sqrt{5 \cdot 15} + 4\sqrt{3}$ Product rule

$\qquad\qquad\qquad\quad = \sqrt{75} + 4\sqrt{3}$ Multiply.

$\qquad\qquad\qquad\quad = \sqrt{25 \cdot 3} + 4\sqrt{3}$ 25 is a perfect square.

$\qquad\qquad\qquad\quad = \sqrt{25} \cdot \sqrt{3} + 4\sqrt{3}$ Product rule

$\qquad\qquad\qquad\quad = 5\sqrt{3} + 4\sqrt{3}$ $\sqrt{25} = 5$

$\qquad\qquad\qquad\quad = 9\sqrt{3}$ Add like radicals.

(b) $\sqrt{2} \cdot \sqrt{6k} + \sqrt{27k} = \sqrt{12k} + \sqrt{27k}$ Product rule

$\qquad\qquad\qquad\quad = \sqrt{4 \cdot 3k} + \sqrt{9 \cdot 3k}$ Factor.

$\qquad\qquad\qquad\quad = \sqrt{4} \cdot \sqrt{3k} + \sqrt{9} \cdot \sqrt{3k}$ Product rule

$\qquad\qquad\qquad\quad = 2\sqrt{3k} + 3\sqrt{3k}$ $\sqrt{4} = 2$ and $\sqrt{9} = 3$

$\qquad\qquad\qquad\quad = 5\sqrt{3k}$ Add like radicals.

(c) $\sqrt[3]{2} \cdot \sqrt[3]{16m^3} - \sqrt[3]{108m^3} = \sqrt[3]{32m^3} - \sqrt[3]{108m^3}$ Product rule

$\qquad\qquad\qquad\quad = \sqrt[3]{(8m^3)4} - \sqrt[3]{(27m^3)4}$ Factor.

$\qquad\qquad\qquad\quad = 2m\sqrt[3]{4} - 3m\sqrt[3]{4}$ $\sqrt[3]{8m^3} = 2m$ and $\sqrt[3]{27m^3} = 3m$

$\qquad\qquad\qquad\quad = -m\sqrt[3]{4}$ Subtract like radicals.

 Remember that a sum or difference of radicals can be simplified only if the radicals are *like radicals*. For example, $2\sqrt{3} + 5\sqrt[3]{3}$ cannot be simplified further.

10.3 EXERCISES

Fill in each blank with the correct response.

1. $5\sqrt{2} + 6\sqrt{2} = (5 + 6)\sqrt{2} = 11\sqrt{2}$ is an example of the _____ property.

2. Like radicals have the same _____ of the same _____.

3. $\sqrt{2} + 2\sqrt{3}$ cannot be simplified because the _____ are different.

4. $2\sqrt{3} + 4\sqrt[3]{3}$ cannot be simplified because the _____ are different.

Simplify and add or subtract wherever possible. See Examples 1 and 2.

5. $14\sqrt{7} - 19\sqrt{7}$ **6.** $16\sqrt{2} - 18\sqrt{2}$ **7.** $\sqrt{17} + 4\sqrt{17}$

8. $5\sqrt{19} + \sqrt{19}$ **9.** $6\sqrt{7} - \sqrt{7}$ **10.** $11\sqrt{14} - \sqrt{14}$

11. $\sqrt{45} + 4\sqrt{20}$ **12.** $\sqrt{24} + 6\sqrt{54}$ **13.** $5\sqrt{72} - 3\sqrt{50}$

14. $6\sqrt{18} - 5\sqrt{32}$ **15.** $-5\sqrt{32} + 2\sqrt{45}$ **16.** $-4\sqrt{75} + 3\sqrt{24}$

17. $5\sqrt{7} - 3\sqrt{28} + 6\sqrt{63}$

18. $3\sqrt{11} + 5\sqrt{44} - 8\sqrt{99}$

19. $2\sqrt{8} - 5\sqrt{32} - 2\sqrt{48}$

20. $5\sqrt{72} - 3\sqrt{48} + 4\sqrt{128}$

21. $4\sqrt{50} + 3\sqrt{12} - 5\sqrt{45}$

22. $6\sqrt{18} + 2\sqrt{48} + 6\sqrt{28}$

23. $\frac{1}{4}\sqrt{288} + \frac{1}{6}\sqrt{72}$

24. $\frac{2}{3}\sqrt{27} + \frac{3}{4}\sqrt{48}$

25. The distributive property, which says $a(b + c) = ab + ac$ and $ba + ca = (b + c)a$, provides the justification for adding and subtracting like radicals. While we usually skip the step that indicates this property, we could not make the statement $2\sqrt{3} + 4\sqrt{3} = 6\sqrt{3}$ without it. Write an equation showing how the distributive property is actually used in this statement.

26. Refer to Example 1(e), and explain why $\sqrt{3} + \sqrt{7}$ cannot be further simplified. Confirm, by using calculator approximations, that $\sqrt{3} + \sqrt{7}$ is *not* equal to $\sqrt{10}$.

Perform the indicated operations. Assume that all variables represent nonnegative real numbers. See Example 3.

27. $\sqrt{6} \cdot \sqrt{2} + 9\sqrt{3}$

28. $4\sqrt{15} \cdot \sqrt{3} + 4\sqrt{5}$

29. $\sqrt{9x} + \sqrt{49x} - \sqrt{25x}$

30. $\sqrt{4a} - \sqrt{16a} + \sqrt{100a}$

31. $\sqrt{6x^2} + x\sqrt{24}$

32. $\sqrt{75x^2} + x\sqrt{108}$

33. $3\sqrt{8x^2} - 4x\sqrt{2} - x\sqrt{8}$

34. $\sqrt{2b^2} + 3b\sqrt{18} - b\sqrt{200}$

35. $-8\sqrt{32k} + 6\sqrt{8k}$

36. $4\sqrt{12x} + 2\sqrt{27x}$

37. $2\sqrt{125x^2z} + 8x\sqrt{80z}$

38. $\sqrt{48x^2y} + 5x\sqrt{27y}$

39. $4\sqrt[3]{16} - 3\sqrt[3]{54}$

40. $5\sqrt[3]{128} + 3\sqrt[3]{250}$

41. $6\sqrt[3]{8p^2} - 2\sqrt[3]{27p^2}$

42. $8k\sqrt[3]{54k} + 6\sqrt[3]{16k^4}$

43. $5\sqrt[4]{m^3} + 8\sqrt[4]{16m^3}$

44. $5\sqrt[4]{m^5} + 3\sqrt[4]{81m^5}$

45. Describe in your own words how to add and subtract radicals.

46. In the directions for Exercises 27–44, we made the assumption that all variables represent nonnegative real numbers. However, in Exercises 41 and 42, variables actually *may* represent negative numbers. Explain why this is so.

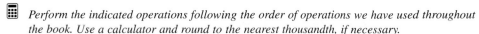

RELATING CONCEPTS (EXERCISES 47–50)

Adding and subtracting like radicals is no different than adding and subtracting other like terms.

Work Exercises 47–50 in order.

47. Combine like terms: $5x^2y + 3x^2y - 14x^2y$.

48. Combine like terms: $5(p - 2q)^2(a + b) + 3(p - 2q)^2(a + b) - 14(p - 2q)^2(a + b)$.

49. Combine like radicals: $5a^2\sqrt{xy} + 3a^2\sqrt{xy} - 14a^2\sqrt{xy}$.

50. Compare your answers in Exercises 47–49. How are they alike? How are they different?

Did you make the connection between adding and subtracting like radicals and adding and subtracting like terms?

Perform the indicated operations following the order of operations we have used throughout the book. Use a calculator and round to the nearest thousandth, if necessary.

51. $\sqrt{(-3 - 6)^2 + (2 - 4)^2}$

52. $\sqrt{(-9 - 3)^2 + (3 - 8)^2}$

53. $\sqrt{(2 - (-2))^2 + (-1 - 2)^2}$

54. $\sqrt{(3 - 1)^2 + (2 - (-1))^2}$

Find the perimeter of each figure.

55.

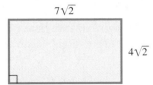

$7\sqrt{2}$

$4\sqrt{2}$

56.

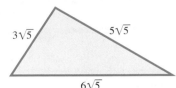

$3\sqrt{5}$ $5\sqrt{5}$

$6\sqrt{5}$

Work each problem.

57. The table shows the year in which U.S. exports of electronics were x billion dollars.

U.S. Exports

Billions of Dollars	Year
7.5	1990
19.6	1993
25.8	1994
36.4	1996

Source: U.S. Bureau of the Census; *U.S. Merchandise Trade,* series FT 900, December issue; and unpublished data.

The function defined by $y = 1.4\sqrt{x - 2.5} + 87.5$ gives a good approximation of the year y when exports were x billion dollars. Here $y = 90$ represents 1990, $y = 93$ represents 1993, and so on. Use the equation to find the years in which exports were
(a) $10 billion
(b) $50 billion.

58. U.S. imports of electronics for several years are shown in the table.

U.S. Imports

Billions of Dollars	Year
11.0	1990
19.4	1993
25.9	1994
36.8	1996

Source: U.S. Bureau of the Census; *U.S. Merchandise Trade,* series FT 900, December issue; and unpublished data.

The year when U.S. imports were x billion dollars is closely approximated by the function $y = 1.6\sqrt{x - 6} + 87$, where $y = 90$ represents 1990, and so on, as in Exercise 57.
(a) Use the equation to find the years when U.S. imports were $7 billion; $30 billion.
(b) Using the tables in Exercises 57 and 58, find the differences between U.S. exports and U.S. imports in the years listed in both tables. How do they compare?

10.4 Rationalizing the Denominator

OBJECTIVES

1 Rationalize denominators with square roots.

2 Write radicals in simplified form.

3 Rationalize denominators with cube roots.

FOR EXTRA HELP

SSG Sec. 10.4
SSM Sec. 10.4

Pass the Test Software

InterAct Math Tutorial Software

Video 16

OBJECTIVE 1 **Rationalize denominators with square roots.** Fractions are simplified by rewriting them without any radical expressions in the denominator. For example, the radical in the denominator of

$$\frac{\sqrt{3}}{\sqrt{2}}$$

can be eliminated by multiplying the numerator and the denominator by $\sqrt{2}$.

$$\frac{\sqrt{3}}{\sqrt{2}} = \frac{\sqrt{3} \cdot \sqrt{2}}{\sqrt{2} \cdot \sqrt{2}} = \frac{\sqrt{6}}{2} \qquad \text{Since } \sqrt{2} \cdot \sqrt{2} = 2$$

This process of changing the denominator from a radical (irrational number) to a rational number is called **rationalizing the denominator.** The value of the number is not changed; only the form of the number is changed, because the expression has been multiplied by 1 in the form $\dfrac{\sqrt{2}}{\sqrt{2}}$.

EXAMPLE 1   **Rationalizing Denominators**

Rationalize each denominator.

(a) $\dfrac{9}{\sqrt{6}}$

Multiply both numerator and denominator by $\sqrt{6}$.

$$\frac{9}{\sqrt{6}} = \frac{9 \cdot \sqrt{6}}{\sqrt{6} \cdot \sqrt{6}}$$

$$= \frac{9\sqrt{6}}{6} \qquad \sqrt{6} \cdot \sqrt{6} = 6$$

$$= \frac{3\sqrt{6}}{2} \qquad \text{Lowest terms}$$

(b) $\dfrac{12}{\sqrt{8}}$

The denominator here could be rationalized by multiplying by $\sqrt{8}$. However, the result can be found more directly by first simplifying the denominator.

$$\sqrt{8} = \sqrt{4} \cdot \sqrt{2} = 2\sqrt{2}$$

Then multiply numerator and denominator by $\sqrt{2}$.

$$\frac{12}{\sqrt{8}} = \frac{12}{2\sqrt{2}}$$

$$= \frac{12 \cdot \sqrt{2}}{2\sqrt{2} \cdot \sqrt{2}} \qquad \text{Multiply by } \tfrac{\sqrt{2}}{\sqrt{2}}.$$

$$= \frac{12\sqrt{2}}{2 \cdot 2} \qquad \sqrt{2} \cdot \sqrt{2} = 2$$

$$= \frac{12\sqrt{2}}{4} \qquad \text{Multiply.}$$

$$= 3\sqrt{2} \qquad \text{Lowest terms}$$

OBJECTIVE **2** Write radicals in simplified form. A radical is considered to be in simplified form if the following three conditions are met.

Simplified Form of a Radical

1. All *n*th power factors of the radicand of $\sqrt[n]{\ }$ are removed. (For example, $\sqrt{3^2} = 3$, $\sqrt[3]{5^3} = 5$, and so on.)

2. The radicand has no fractions.

3. No denominator contains a radical.

In the following examples, radicals are simplified according to these conditions.

E X A M P L E 2 Simplifying a Radical with a Fraction

Simplify $\sqrt{\dfrac{27}{5}}$ by rationalizing the denominator.

First use the quotient rule for radicals.

$$\sqrt{\frac{27}{5}} = \frac{\sqrt{27}}{\sqrt{5}}$$

Now multiply both numerator and denominator by $\sqrt{5}$.

$$\frac{\sqrt{27}}{\sqrt{5}} = \frac{\sqrt{27} \cdot \sqrt{5}}{\sqrt{5} \cdot \sqrt{5}} \qquad \text{Rationalize the denominator.}$$

$$= \frac{\sqrt{9 \cdot 3} \cdot \sqrt{5}}{5} \qquad \sqrt{5} \cdot \sqrt{5} = 5$$

$$= \frac{\sqrt{9} \cdot \sqrt{3} \cdot \sqrt{5}}{5} \qquad \text{Product rule}$$

$$= \frac{3 \cdot \sqrt{3 \cdot 5}}{5} = \frac{3\sqrt{15}}{5} \qquad \text{Product rule}$$

E X A M P L E 3 Simplifying a Product of Radicals

Simplify $\sqrt{\dfrac{5}{8}} \cdot \sqrt{\dfrac{1}{6}}$.

Use both the product and quotient rules.

$$\sqrt{\frac{5}{8}} \cdot \sqrt{\frac{1}{6}} = \sqrt{\frac{5}{8} \cdot \frac{1}{6}} \qquad \text{Product rule}$$

$$= \sqrt{\frac{5}{48}} \qquad \text{Multiply.}$$

$$= \frac{\sqrt{5}}{\sqrt{48}} \qquad \text{Quotient rule}$$

To rationalize the denominator, first simplify, then multiply the numerator and denominator by $\sqrt{3}$ as follows.

$$\frac{\sqrt{5}}{\sqrt{48}} = \frac{\sqrt{5}}{\sqrt{16} \cdot \sqrt{3}} \qquad \text{Product rule}$$

$$= \frac{\sqrt{5}}{4\sqrt{3}} \qquad \sqrt{16} = 4$$

$$= \frac{\sqrt{5} \cdot \sqrt{3}}{4\sqrt{3} \cdot \sqrt{3}} \qquad \text{Rationalize the denominator.}$$

$$= \frac{\sqrt{15}}{4 \cdot 3} \qquad \text{Product rule}$$

$$= \frac{\sqrt{15}}{12} \qquad \text{Multiply.}$$

E X A M P L E 4   Simplifying a Quotient of Radicals

Rationalize the denominator of $\dfrac{\sqrt{4x}}{\sqrt{y}}$. Assume that x and y represent positive real numbers.

Multiply numerator and denominator by $\sqrt{y}$.

$$\frac{\sqrt{4x}}{\sqrt{y}} = \frac{\sqrt{4x} \cdot \sqrt{y}}{\sqrt{y} \cdot \sqrt{y}} = \frac{\sqrt{4xy}}{y} = \frac{2\sqrt{xy}}{y}$$

OBJECTIVE 3   Rationalize denominators with cube roots. To rationalize a denominator with a cube root, change the radicand in the denominator to a perfect cube, as shown in the next example.

E X A M P L E 5   Rationalizing a Denominator with a Cube Root

Rationalize each denominator.

(a) $\sqrt[3]{\dfrac{3}{2}}$

Begin by writing the expression as a quotient of radicals. Then, multiply the numerator and the denominator by enough factors of 2 to make the denominator a perfect cube. This will eliminate the radical in the denominator. Here, multiply by $\sqrt[3]{2^2}$.

$$\sqrt[3]{\frac{3}{2}} = \frac{\sqrt[3]{3}}{\sqrt[3]{2}} = \frac{\sqrt[3]{3} \cdot \sqrt[3]{2^2}}{\sqrt[3]{2} \cdot \sqrt[3]{2^2}} = \frac{\sqrt[3]{3 \cdot 2^2}}{\sqrt[3]{2^3}} = \frac{\sqrt[3]{12}}{2} \qquad \text{Since } \sqrt[3]{2^3} = \sqrt[3]{8} = 2$$

(b) $\dfrac{\sqrt[3]{3}}{\sqrt[3]{4}}$

Since $4 \cdot 2 = 2^2 \cdot 2 = 2^3$, multiply numerator and denominator by $\sqrt[3]{2}$.

$$\frac{\sqrt[3]{3}}{\sqrt[3]{4}} = \frac{\sqrt[3]{3} \cdot \sqrt[3]{2}}{\sqrt[3]{2 \cdot 2} \cdot \sqrt[3]{2}} = \frac{\sqrt[3]{6}}{\sqrt[3]{2^3}} = \frac{\sqrt[3]{6}}{2}$$

CAUTION A common error in a problem like the one in Example 5(a) is to multiply by $\sqrt[3]{2}$ instead of $\sqrt[3]{2^2}$. Doing this would give a denominator of $\sqrt[3]{2} \cdot \sqrt[3]{2} = \sqrt[3]{4}$. Since 4 is not a perfect cube, the denominator is still not rationalized.

10.4 EXERCISES

Fill in each blank with the correct response.

1. Rationalizing the denominator means to change the denominator from a(n) _____ to a rational number.

2. To simplify $\sqrt{x^2 y}$, where $x \geq 0$, we must remove _____ from the radicand.

3. The expression $\sqrt{\dfrac{3m}{2}}$, where $m \geq 0$, is not simplified because the radical contains a(n) _____ .

4. The expression $\dfrac{2x}{\sqrt{7}}$ is not simplified because the denominator is a(n) _____ .

Rationalize each denominator. See Examples 1 and 2.

5. $\dfrac{8}{\sqrt{2}}$ 6. $\dfrac{12}{\sqrt{3}}$ 7. $\dfrac{-\sqrt{11}}{\sqrt{3}}$ 8. $\dfrac{-\sqrt{13}}{\sqrt{5}}$ 9. $\dfrac{7\sqrt{3}}{\sqrt{5}}$

10. $\dfrac{4\sqrt{6}}{\sqrt{5}}$ 11. $\dfrac{24\sqrt{10}}{16\sqrt{3}}$ 12. $\dfrac{18\sqrt{15}}{12\sqrt{2}}$ 13. $\dfrac{16}{\sqrt{27}}$ 14. $\dfrac{24}{\sqrt{18}}$

15. $\dfrac{-3}{\sqrt{50}}$ 16. $\dfrac{-5}{\sqrt{75}}$ 17. $\dfrac{63}{\sqrt{45}}$ 18. $\dfrac{27}{\sqrt{32}}$ 19. $\dfrac{\sqrt{24}}{\sqrt{8}}$

20. $\dfrac{\sqrt{36}}{\sqrt{18}}$ 21. $\sqrt{\dfrac{1}{2}}$ 22. $\sqrt{\dfrac{1}{3}}$ 23. $\sqrt{\dfrac{13}{5}}$ 24. $\sqrt{\dfrac{17}{11}}$

25. To rationalize the denominator of an expression such as $\dfrac{4}{\sqrt{3}}$, we multiply both the numerator and the denominator by $\sqrt{3}$. By what number are we actually multiplying the given expression, and what property of real numbers justifies the fact that our result is equal to the given expression?

 26. In Example 1(a), we show algebraically that $\dfrac{9}{\sqrt{6}}$ is equal to $\dfrac{3\sqrt{6}}{2}$. Support this result numerically by finding the decimal approximation of $\dfrac{9}{\sqrt{6}}$ on your calculator, and then finding the decimal approximation of $\dfrac{3\sqrt{6}}{2}$. What do you notice?

Multiply and simplify each result. See Example 3.

27. $\sqrt{\dfrac{7}{13}} \cdot \sqrt{\dfrac{13}{3}}$ 28. $\sqrt{\dfrac{19}{20}} \cdot \sqrt{\dfrac{20}{3}}$ 29. $\sqrt{\dfrac{21}{7}} \cdot \sqrt{\dfrac{21}{8}}$ 30. $\sqrt{\dfrac{5}{8}} \cdot \sqrt{\dfrac{5}{6}}$

31. $\sqrt{\dfrac{1}{12}} \cdot \sqrt{\dfrac{1}{3}}$ 32. $\sqrt{\dfrac{1}{8}} \cdot \sqrt{\dfrac{1}{2}}$ 33. $\sqrt{\dfrac{2}{9}} \cdot \sqrt{\dfrac{9}{2}}$ 34. $\sqrt{\dfrac{4}{3}} \cdot \sqrt{\dfrac{3}{4}}$

Simplify each radical. Assume that all variables represent positive real numbers. See Example 4.

35. $\sqrt{\dfrac{7}{x}}$ 36. $\sqrt{\dfrac{19}{y}}$ 37. $\sqrt{\dfrac{4x^3}{y}}$ 38. $\sqrt{\dfrac{9t^3}{s}}$

39. $\sqrt{\dfrac{18x^3}{6y}}$ 40. $\sqrt{\dfrac{24t^3}{8p}}$ 41. $\sqrt{\dfrac{9a^2r^5}{7t}}$ 42. $\sqrt{\dfrac{16x^3y^2}{13z}}$

43. Which one of the following would be an appropriate choice for multiplying the numerator and the denominator of $\dfrac{\sqrt[3]{2}}{\sqrt[3]{5}}$ in order to rationalize the denominator?

(a) $\sqrt[3]{5}$ (b) $\sqrt[3]{25}$ (c) $\sqrt[3]{2}$ (d) $\sqrt[3]{3}$

44. In Example 5(b), we multiply the numerator and denominator of $\dfrac{\sqrt[3]{3}}{\sqrt[3]{4}}$ by $\sqrt[3]{2}$ to rationalize the denominator. Suppose we had chosen to multiply by $\sqrt[3]{16}$ instead. Would we have obtained the correct answer after all simplifications were done?

Simplify. Rationalize each denominator. Assume that any variables in a denominator are nonzero. See Example 5.

45. $\sqrt[3]{\dfrac{3}{2}}$ **46.** $\sqrt[3]{\dfrac{2}{5}}$ **47.** $\dfrac{\sqrt[3]{4}}{\sqrt[3]{7}}$ **48.** $\dfrac{\sqrt[3]{5}}{\sqrt[3]{10}}$

49. $\sqrt[3]{\dfrac{3}{4y^2}}$ **50.** $\sqrt[3]{\dfrac{3}{25x^2}}$ **51.** $\dfrac{\sqrt[3]{7m}}{\sqrt[3]{36n}}$ **52.** $\dfrac{\sqrt[3]{11p}}{\sqrt[3]{49q}}$

*In Exercises 53 and 54, **(a)** give the answer as a simplified radical and **(b)** use a calculator to give the answer correct to the nearest thousandth.*

53. The period p of a pendulum is the time it takes for it to swing from one extreme to the other and back again. The value of p in seconds is given by

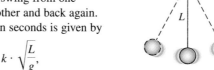

$$p = k \cdot \sqrt{\dfrac{L}{g}},$$

where L is the length of the pendulum, g is the acceleration due to gravity, and k is a constant. Find the period when $k = 6$, $L = 9$ feet, and $g = 32$ feet per second squared.

54. The velocity v of a meteorite approaching the earth is given by

$$v = \dfrac{k}{\sqrt{d}}$$

kilometers per second, where d is its distance from the center of the earth and k is a constant. What is the velocity of a meteorite that is 6000 kilometers away from the center of the earth, if $k = 450$?

10.5 Simplifying Radical Expressions

OBJECTIVES

1. Simplify products of radical expressions.
2. Simplify quotients of radical expressions.
3. Write radical expressions with quotients in lowest terms.

The conditions for which a radical is in simplest form were listed in the previous section. Below is a set of guidelines to follow when you are simplifying radical expressions. Although the guidelines are illustrated with square roots, they apply to higher roots as well.

Simplifying Radical Expressions

1. If a radical represents a rational number, use that rational number in place of the radical.

   Examples: $\sqrt{49}$ is simplified by writing 7; $\sqrt{\dfrac{169}{9}}$ by writing $\dfrac{13}{3}$.

Simplifying Radical Expressions (continued)

2. If a radical expression contains products of radicals, use the product rule for radicals, $\sqrt{x} \cdot \sqrt{y} = \sqrt{xy}$, to get a single radical.

> *Examples:* $\sqrt{3} \cdot \sqrt{2}$ is simplified to $\sqrt{6}$; $\sqrt{5} \cdot \sqrt{x}$ to $\sqrt{5x}$.

3. If a radicand has a factor that is a perfect square, express the radical as the product of the positive square root of the perfect square and the remaining radical factor.

> *Examples:* $\sqrt{20}$ is simplified to $\sqrt{20} = \sqrt{4 \cdot 5} = \sqrt{4} \cdot \sqrt{5} = 2\sqrt{5}$;
> $\sqrt[3]{16} = \sqrt[3]{8 \cdot 2} = \sqrt[3]{8} \cdot \sqrt[3]{2} = 2\sqrt[3]{2}$.

4. If a radical expression contains sums or differences of radicals, use the distributive property to combine like radicals.

> *Examples:* $3\sqrt{2} + 4\sqrt{2} = 7\sqrt{2}$, but $3\sqrt{2} + 4\sqrt{3}$ cannot be further simplified.

5. Rationalize any denominator containing a radical.

> *Examples:* $\dfrac{5}{\sqrt{3}}$ is rationalized as $\dfrac{5}{\sqrt{3}} = \dfrac{5\sqrt{3}}{\sqrt{3} \cdot \sqrt{3}} = \dfrac{5\sqrt{3}}{3}$;
>
> $\sqrt{\dfrac{3}{2}} = \dfrac{\sqrt{3}}{\sqrt{2}} = \dfrac{\sqrt{3} \cdot \sqrt{2}}{\sqrt{2} \cdot \sqrt{2}} = \dfrac{\sqrt{6}}{2}$.

OBJECTIVE 1 Simplify products of radical expressions. Use the above guidelines to do this.

EXAMPLE 1 Multiplying Radical Expressions

Find each product and simplify the answers.

(a) $\sqrt{5}(\sqrt{8} - \sqrt{32})$

Simplify inside the parentheses.

$$\sqrt{5}(\sqrt{8} - \sqrt{32}) = \sqrt{5}(2\sqrt{2} - 4\sqrt{2})$$
$$= \sqrt{5}(-2\sqrt{2}) \qquad \text{Subtract like radicals.}$$
$$= -2\sqrt{5} \cdot 2 \qquad \text{Product rule; commutative property}$$
$$= -2\sqrt{10} \qquad \text{Multiply.}$$

(b) $(\sqrt{3} + 2\sqrt{5})(\sqrt{3} - 4\sqrt{5})$

The product of these sums of radicals can be found in the same way that we found the product of binomials in Chapter 5 using the FOIL method.

$$(\sqrt{3} + 2\sqrt{5})(\sqrt{3} - 4\sqrt{5})$$

$$\qquad\quad \text{F} \qquad\qquad \text{O} \qquad\qquad \text{I} \qquad\qquad \text{L}$$
$$= \sqrt{3} \cdot \sqrt{3} + \sqrt{3}(-4\sqrt{5}) + 2\sqrt{5} \cdot \sqrt{3} + 2\sqrt{5}(-4\sqrt{5})$$
$$= 3 - 4\sqrt{15} + 2\sqrt{15} - 8 \cdot 5 \qquad\qquad \text{Product rule}$$
$$= 3 - 2\sqrt{15} - 40 \qquad\qquad\qquad\quad \text{Add like radicals.}$$
$$= -37 - 2\sqrt{15} \qquad\qquad\qquad\quad\; \text{Combine terms.}$$

(c) $(\sqrt{3} + \sqrt{21})(\sqrt{3} - \sqrt{7})$

$$(\sqrt{3} + \sqrt{21})(\sqrt{3} - \sqrt{7})$$

$$= \sqrt{3}(\sqrt{3}) + \sqrt{3}(-\sqrt{7}) + \sqrt{21}(\sqrt{3}) + \sqrt{21}(-\sqrt{7}) \qquad \text{FOIL}$$

$$= 3 - \sqrt{21} + \sqrt{63} - \sqrt{147} \qquad \text{Product rule}$$

$$= 3 - \sqrt{21} + \sqrt{9} \cdot \sqrt{7} - \sqrt{49} \cdot \sqrt{3} \qquad \text{Simplify radicals.}$$

$$= 3 - \sqrt{21} + 3\sqrt{7} - 7\sqrt{3} \qquad \sqrt{9} = 3 \text{ and } \sqrt{49} = 7$$

Since there are no like radicals, no terms may be combined.

Since radicals represent real numbers, the special products of binomials discussed in Chapter 5 can be used to find products of radicals. Example 2 uses the rule for the product that gives the difference of two squares,

$$(a + b)(a - b) = a^2 - b^2.$$

EXAMPLE 2 Using a Special Product with Radicals

Find each product.

(a) $(4 + \sqrt{3})(4 - \sqrt{3})$

Follow the pattern given above. Let $a = 4$ and $b = \sqrt{3}$.

$$(4 + \sqrt{3})(4 - \sqrt{3}) = 4^2 - (\sqrt{3})^2$$

$$= 16 - 3 = 13 \qquad 4^2 = 16 \text{ and } (\sqrt{3})^2 = 3$$

(b) $(\sqrt{12} - \sqrt{6})(\sqrt{12} + \sqrt{6}) = (\sqrt{12})^2 - (\sqrt{6})^2$

$$= 12 - 6 \qquad (\sqrt{12})^2 = 12 \text{ and } (\sqrt{6})^2 = 6$$

$$= 6$$

Both products in Example 2 resulted in rational numbers. The pairs of expressions in those products, $4 + \sqrt{3}$ and $4 - \sqrt{3}$, and $\sqrt{12} - \sqrt{6}$ and $\sqrt{12} + \sqrt{6}$, are called **conjugates** of each other.

OBJECTIVE 2 Simplify quotients of radical expressions. Products of radicals similar to those in Example 2 can be used to rationalize the denominators in quotients with binomial denominators, such as

$$\frac{2}{4 - \sqrt{3}}.$$

By Example 2(a), if this denominator, $4 - \sqrt{3}$, is multiplied by $4 + \sqrt{3}$, then the product $(4 - \sqrt{3})(4 + \sqrt{3})$ is the rational number 13. Multiplying numerator and denominator by $4 + \sqrt{3}$ gives

$$\frac{2}{4 - \sqrt{3}} = \frac{2(4 + \sqrt{3})}{(4 - \sqrt{3})(4 + \sqrt{3})}$$

$$= \frac{2(4 + \sqrt{3})}{13}.$$

The denominator now has been rationalized; it contains no radical signs.

Using Conjugates to Simplify a Radical Expression

To simplify a radical expression with two terms in the denominator, where at least one of those terms is a square root radical, multiply both the numerator and the denominator by the conjugate of the denominator.

EXAMPLE 3 Using Conjugates to Rationalize a Denominator

Simplify by rationalizing the denominator.

(a) $\dfrac{7}{3 + \sqrt{5}}$

We can eliminate the radical in the denominator by multiplying both numerator and denominator by $3 - \sqrt{5}$, the conjugate of $3 + \sqrt{5}$.

$$\frac{7}{3 + \sqrt{5}} = \frac{7(3 - \sqrt{5})}{(3 + \sqrt{5})(3 - \sqrt{5})} \qquad \text{Multiply by the conjugate.}$$

$$= \frac{7(3 - \sqrt{5})}{3^2 - (\sqrt{5})^2} \qquad (a + b)(a - b) = a^2 - b^2$$

$$= \frac{7(3 - \sqrt{5})}{9 - 5} \qquad 3^2 = 9 \text{ and } (\sqrt{5})^2 = 5$$

$$= \frac{7(3 - \sqrt{5})}{4} \qquad \text{Subtract.}$$

(b) $\dfrac{6 + \sqrt{2}}{\sqrt{2} - 5}$

Multiply numerator and denominator by $\sqrt{2} + 5$.

$$\frac{6 + \sqrt{2}}{\sqrt{2} - 5} = \frac{(6 + \sqrt{2})(\sqrt{2} + 5)}{(\sqrt{2} - 5)(\sqrt{2} + 5)}$$

$$= \frac{6\sqrt{2} + 30 + 2 + 5\sqrt{2}}{2 - 25} \qquad \text{FOIL}$$

$$= \frac{11\sqrt{2} + 32}{-23} \qquad \text{Combine terms.}$$

$$= -\frac{11\sqrt{2} + 32}{23} \qquad \frac{a}{-b} = -\frac{a}{b}$$

Rationalizing the denominator in the two expressions

$$\frac{7}{\sqrt{x + 5}} \qquad \text{and} \qquad \frac{7}{\sqrt{x} + \sqrt{5}}$$

involves different procedures. In the first denominator, which has one term, multiply both the numerator and the denominator by $\sqrt{x + 5}$. In the second denominator, which has two terms, use $\sqrt{x} - \sqrt{5}$.

O B J E C T I V E 3 Write radical expressions with quotients in lowest terms. The final example shows this.

┌─
E X A M P L E 4   Writing a Radical Quotient in Lowest Terms

Write $\dfrac{3\sqrt{3} + 15}{12}$ in lowest terms.

  Factor the numerator and denominator, and then divide numerator and denominator by any common factors.

$$\frac{3\sqrt{3} + 15}{12} = \frac{3(\sqrt{3} + 5)}{3 \cdot 4} = \frac{\sqrt{3} + 5}{4}$$

This technique is used in Chapter 11.
└─

A common error is to reduce an expression like the one in Example 4 incorrectly to lowest terms before factoring. For example,

$$\frac{4 + 8\sqrt{5}}{4} \neq 1 + 8\sqrt{5}.$$

The correct simplification is $1 + 2\sqrt{5}$. Why?

10.5 EXERCISES

Based on the work so far, many simple operations involving radicals should now be performed mentally. In Exercises 1–8, perform the operations mentally, and write the answer without doing intermediate steps.

1. $\sqrt{49} + \sqrt{36}$ **2.** $\sqrt{100} - \sqrt{81}$ **3.** $\sqrt{2} \cdot \sqrt{8}$

4. $\sqrt{8} \cdot \sqrt{8}$ **5.** $\sqrt{2}(\sqrt{32} - \sqrt{8})$ **6.** $\sqrt{3}(\sqrt{27} - \sqrt{3})$

7. $\sqrt[3]{8} + \sqrt[3]{27}$ **8.** $\sqrt{4} - \sqrt[3]{64} + \sqrt[4]{16}$

Simplify each expression. Use the five guidelines given in the text. See Examples 1 and 2.

9. $3\sqrt{5} + 2\sqrt{45}$ **10.** $2\sqrt{2} + 4\sqrt{18}$

11. $8\sqrt{50} - 4\sqrt{72}$ **12.** $4\sqrt{80} - 5\sqrt{45}$

13. $\sqrt{5}(\sqrt{3} - \sqrt{7})$ **14.** $\sqrt{7}(\sqrt{10} + \sqrt{3})$

15. $2\sqrt{5}(\sqrt{2} + 3\sqrt{5})$ **16.** $3\sqrt{7}(2\sqrt{7} + 4\sqrt{5})$

17. $3\sqrt{14} \cdot \sqrt{2} - \sqrt{28}$ **18.** $7\sqrt{6} \cdot \sqrt{3} - 2\sqrt{18}$

19. $(2\sqrt{6} + 3)(3\sqrt{6} + 7)$ **20.** $(4\sqrt{5} - 2)(2\sqrt{5} - 4)$

21. $(5\sqrt{7} - 2\sqrt{3})(3\sqrt{7} + 4\sqrt{3})$ **22.** $(2\sqrt{10} + 5\sqrt{2})(3\sqrt{10} - 3\sqrt{2})$

23. $(2\sqrt{7} + 3)^2$ **24.** $(4\sqrt{5} + 5)^2$

25. $(5 - \sqrt{2})(5 + \sqrt{2})$ **26.** $(3 - \sqrt{5})(3 + \sqrt{5})$

27. $(\sqrt{8} - \sqrt{7})(\sqrt{8} + \sqrt{7})$ **28.** $(\sqrt{12} - \sqrt{11})(\sqrt{12} + \sqrt{11})$

29. $(\sqrt{2} + \sqrt{3})(\sqrt{6} \quad \sqrt{2})$ **30.** $(\sqrt{3} + \sqrt{5})(\sqrt{15} - \sqrt{5})$

31. $(\sqrt{10} - \sqrt{5})(\sqrt{5} + \sqrt{20})$ **32.** $(\sqrt{6} - \sqrt{3})(\sqrt{3} + \sqrt{18})$

33. $(5\sqrt{7} - 2\sqrt{3})(3\sqrt{7} + 3\sqrt{3})$ **34.** $(2\sqrt{10} + 5\sqrt{2})(3\sqrt{10} - 4\sqrt{2})$

35. In Example 1(b), the original expression simplifies to $-37 - 2\sqrt{15}$. Students often try to simplify expressions like this by combining the -37 and the -2 to get $-39\sqrt{15}$, which is incorrect. Explain why.

36. If you try to rationalize the denominator of $\dfrac{2}{4 + \sqrt{3}}$ by multiplying the numerator and denominator by $4 + \sqrt{3}$, what problem arises? What should you multiply by?

Rationalize each denominator. See Example 3.

37. $\dfrac{1}{3 + \sqrt{2}}$ **38.** $\dfrac{1}{4 - \sqrt{3}}$ **39.** $\dfrac{14}{2 - \sqrt{11}}$ **40.** $\dfrac{19}{5 - \sqrt{6}}$

41. $\dfrac{\sqrt{2}}{2 - \sqrt{2}}$ **42.** $\dfrac{\sqrt{7}}{7 - \sqrt{7}}$ **43.** $\dfrac{\sqrt{5}}{\sqrt{2} + \sqrt{3}}$ **44.** $\dfrac{\sqrt{3}}{\sqrt{2} + \sqrt{3}}$

45. $\dfrac{\sqrt{12}}{\sqrt{3} + 1}$ **46.** $\dfrac{\sqrt{18}}{\sqrt{2} - 1}$ **47.** $\dfrac{\sqrt{5} + 2}{2 - \sqrt{3}}$ **48.** $\dfrac{\sqrt{7} + 3}{4 - \sqrt{5}}$

Write each quotient in lowest terms. See Example 4.

49. $\dfrac{6\sqrt{11} - 12}{6}$ **50.** $\dfrac{12\sqrt{5} - 24}{12}$ **51.** $\dfrac{2\sqrt{3} + 10}{16}$

52. $\dfrac{4\sqrt{6} + 24}{20}$ **53.** $\dfrac{12 - \sqrt{40}}{4}$ **54.** $\dfrac{9 - \sqrt{72}}{12}$

Simplify each radical expression. Assume all variables represent nonnegative real numbers.

55. $(\sqrt{5x} + \sqrt{30})(\sqrt{6x} + \sqrt{3})$ **56.** $(\sqrt{10y} - \sqrt{20})(\sqrt{2y} - \sqrt{5})$

57. $(3\sqrt{t} + \sqrt{7})(2\sqrt{t} - \sqrt{14})$ **58.** $(2\sqrt{z} - \sqrt{3})(\sqrt{z} - \sqrt{5})$

59. $(\sqrt{3m} + \sqrt{2n})(\sqrt{5m} - \sqrt{5n})$ **60.** $(\sqrt{4p} - \sqrt{3k})(\sqrt{2p} + \sqrt{9k})$

61. $\sqrt[3]{4}(\sqrt[3]{2} - 3)$ **62.** $\sqrt[3]{5}(4\sqrt[3]{5} - \sqrt[3]{25})$

63. $2\sqrt[4]{2}(3\sqrt[4]{8} + 5\sqrt[4]{4})$ **64.** $6\sqrt[4]{9}(2\sqrt[4]{9} - \sqrt[4]{27})$

65. $(\sqrt[3]{2} - 1)(\sqrt[3]{4} + 3)$ **66.** $(\sqrt[3]{9} + 5)(\sqrt[3]{3} - 4)$

67. $(\sqrt[3]{5} - \sqrt[3]{4})(\sqrt[3]{25} + \sqrt[3]{20} + \sqrt[3]{16})$ **68.** $(\sqrt[3]{4} + \sqrt[3]{2})(\sqrt[3]{16} - \sqrt[3]{8} + \sqrt[3]{4})$

RELATING CONCEPTS (EXERCISES 69–74)

Work Exercises 69–74 in order. They are designed to help you see why a common student error is indeed an error.

69. Use the distributive property to write $6(5 + 3x)$ as a sum.

70. Your answer in Exercise 69 should be $30 + 18x$. Why can we not combine these two terms to get $48x$?

71. Repeat Exercise 22 from earlier in this exercise set.

72. Your answer in Exercise 71 should be $30 + 18\sqrt{5}$. Many students will, in error, try to combine these terms to get $48\sqrt{5}$. Why is this wrong?

73. Write the expression similar to $30 + 18x$ that simplifies to $48x$. Then write the expression similar to $30 + 18\sqrt{5}$ that simplifies to $48\sqrt{5}$.

74. Write a short explanation of the similarities between combining like terms and combining like radicals.

Did you make the connection that the procedure used in combining radical terms is the same as that used in combining variable terms?

Solve each problem.

75. The radius of the circular top or bottom of a tin can with a surface area S and a height h is given by

$$r = \frac{-h + \sqrt{h^2 + .64S}}{2}.$$

What radius should be used to make a can with a height of 12 inches and a surface area of 400 square inches?

76. If an investment of P dollars grows to A dollars in two years, the annual rate of return on the investment is given by

$$r = \frac{\sqrt{A} - \sqrt{P}}{\sqrt{P}}.$$

First rationalize the denominator and then find the annual rate of return (as a percent) if $50,000 increases to $58,320.

10.6 Solving Equations with Radicals

OBJECTIVES

1. Solve equations with radicals.

2. Identify equations with no solutions.

3. Solve equations that require squaring a binomial.

FOR EXTRA HELP

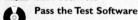
SSG Sec. 10.6
SSM Sec. 10.6

Pass the Test Software

InterAct Math Tutorial Software

Video 17

CONNECTIONS

The most common formula for the area of a triangle is $A = \frac{1}{2}bh$, where b is the length of the base and h is the height. What if the height is not known? What if we know only the lengths of the sides? Another formula, known as *Heron's formula,* allows us to calculate the area of a triangle if we know the lengths of the sides a, b, and c. First let s equal the *semiperimeter,* which is one-half the perimeter.

$$s = \frac{1}{2}(a + b + c)$$

The area A is given by the formula

$$A = \sqrt{s(s - a)(s - b)(s - c)}.$$

For example, the familiar 3–4–5 right triangle has area

$$A = \frac{1}{2}(3)(4) = 6 \text{ square units,}$$

using the familiar formula. Using Heron's formula,

$$s = \frac{1}{2}(3 + 4 + 5) = 6.$$

Therefore,

$$A = \sqrt{6(6 - 3)(6 - 4)(6 - 5)}$$
$$= \sqrt{36}$$
$$= 6$$

So $A = 6$ square units, as expected.

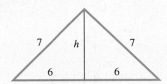

CONNECTIONS (CONTINUED)

FOR DISCUSSION OR WRITING

1. Use Heron's formula to find the area of a triangle with sides 7, 7, and 12.

2. The area of this triangle can be found with the formula $A = \frac{1}{2}bh$ as follows. Divide the triangle into two equal triangles as shown in the figure. Use the Pythagorean formula to find h using one of the small triangles. Note that h is the altitude of the original triangle. Now find the area using the formula $A = \frac{1}{2}bh$. Which way do you prefer?

OBJECTIVE **1** Solve equations with radicals. The addition and multiplication properties of equality are not enough to solve an equation with radicals such as

$$\sqrt{x + 1} = 3.$$

Solving equations that have square roots requires a new property, the **squaring property.**

Squaring Property of Equality

If both sides of a given equation are squared, all solutions of the original equation are *among* the solutions of the squared equation.

Be very careful with the squaring property. Using this property can give a new equation with *more* solutions than the original equation. For example, starting with the equation $y = 4$ and squaring each side gives

$$y^2 = 4^2, \quad \text{or} \quad y^2 = 16.$$

This last equation, $y^2 = 16$, has *two* solutions, 4 or -4, while the original equation, $y = 4$, has only *one* solution, 4. Because of this possibility, checking is more than just a guard against algebraic errors when solving an equation with radicals. It is an essential part of the solution process. *All potential solutions from the squared equation must be checked in the original equation.*

EXAMPLE 1 **Using the Squaring Property of Equality**

Solve the equation $\sqrt{p + 1} = 3$.

Use the squaring property of equality to square both sides of the equation and then solve this new equation.

$$(\sqrt{p + 1})^2 = 3^2$$
$$p + 1 = 9 \qquad (\sqrt{p + 1})^2 = p + 1$$
$$p = 8 \qquad \text{Subtract 1.}$$

Now check this answer in the original equation.

$$\sqrt{p + 1} = 3$$
$$\sqrt{8 + 1} = 3 \qquad ? \qquad \text{Let } p = 8.$$
$$\sqrt{9} = 3 \qquad ?$$
$$3 = 3 \qquad \text{True}$$

Since this statement is true, the solution set of $\sqrt{p + 1} = 3$ is $\{8\}$. In this case the squared equation had just one solution, which also satisfied the original equation.

EXAMPLE 2   Using the Squaring Property with Radicals on Each Side

Solve $3\sqrt{x} = \sqrt{x + 8}$.

Squaring both sides gives

$$(3\sqrt{x})^2 = (\sqrt{x + 8})^2$$
$$3^2(\sqrt{x})^2 = (\sqrt{x + 8})^2 \qquad (ab)^2 = a^2 b^2$$
$$9x = x + 8 \qquad (\sqrt{x})^2 = x; (\sqrt{x + 8})^2 = x + 8$$
$$8x = 8 \qquad \text{Subtract } x.$$
$$x = 1. \qquad \text{Divide by 8.}$$

Check this potential solution.

$$3\sqrt{x} = \sqrt{x + 8}$$
$$3\sqrt{1} = \sqrt{1 + 8} \qquad ? \qquad \text{Let } x = 1.$$
$$3(1) = \sqrt{9} \qquad ? \qquad \sqrt{1} = 1$$
$$3 = 3 \qquad \text{True}$$

The check shows that the solution set of the given equation is $\{1\}$.

OBJECTIVE ☐2   **Identify equations with no solutions.**   Not all equations with radicals have a solution, as shown in Examples 3 and 4.

EXAMPLE 3   Using the Squaring Property When One Side Is Negative

Solve the equation $\sqrt{y} = -3$.

Square both sides.

$$(\sqrt{y})^2 = (-3)^2$$
$$y = 9$$

Check this proposed answer in the original equation.

$$\sqrt{y} = -3$$
$$\sqrt{9} = -3 \qquad ? \qquad \text{Let } y = 9.$$
$$3 = -3 \qquad \text{False}$$

Since the statement $3 = -3$ is false, the number 9 is not a solution of the given equation and is said to be *extraneous*. In fact, $\sqrt{y} = -3$ has no solution. Since $\sqrt{y}$ represents the *nonnegative* square root of y, we might have seen immediately that there is no solution. The solution set is $\emptyset$.

Use the following steps when solving an equation with radicals.

Solving an Equation with Radicals

Step 1   **Isolate the radical.**   Arrange the terms so that the radical is alone on one side of the equation.

Step 2   **Square both sides.**

Step 3   **Combine like terms.**

Step 4   **Repeat Steps 1–3, if necessary.**   If there is still a term with a radical, repeat Steps 1–3.

Step 5   **Solve the equation.**   Find all potential solutions.

Step 6   **Check.**   All potential solutions must be checked in the original equation.

E X A M P L E   4   Using the Squaring Property with a Quadratic Expression

Solve $a = \sqrt{a^2 + 5a + 10}$.

Step 1   The radical is already alone on the right side of the equation.

Step 2   Square both sides.

$$a^2 = (\sqrt{a^2 + 5a + 10})^2$$
$$a^2 = a^2 + 5a + 10 \qquad (\sqrt{a^2 + 5a + 10})^2 = a^2 + 5a + 10$$

Step 3 $\qquad\quad 0 = 5a + 10 \qquad$ Subtract a^2.

Step 4   This step is not needed.

Step 5 $\qquad\quad a = -2 \qquad$ Subtract 10; divide by 5.

Step 6   Check this proposed solution in the original equation.

$$a = \sqrt{a^2 + 5a + 10}$$
$$-2 = \sqrt{(-2)^2 + 5(-2) + 10} \qquad ? \quad \text{Let } a = -2.$$
$$-2 = \sqrt{4 - 10 + 10} \qquad\qquad ? \quad \text{Multiply.}$$
$$-2 = 2 \qquad\qquad\qquad\qquad\quad \text{False}$$

Since $a = -2$ leads to a false result, the equation has no solution, and the solution set is $\emptyset$.

O B J E C T I V E   3   Solve equations that require squaring a binomial.   The next examples use the following facts from Section 5.4.

$$(a + b)^2 = a^2 + 2ab + b^2 \qquad \text{and} \qquad (a - b)^2 = a^2 - 2ab + b^2.$$

By the second pattern, for example,

$$(y - 3)^2 = y^2 - 2(y)(3) + (3)^2$$
$$= y^2 - 6y + 9.$$

EXAMPLE 5   Using the Squaring Property When One Side Has Two Terms

Solve the equation $\sqrt{2y - 3} = y - 3$.

Square each side, using the result above on the right side of the equation.

$$(\sqrt{2y - 3})^2 = (y - 3)^2$$
$$2y - 3 = y^2 - 6y + 9$$

This equation is quadratic because of the y^2-term. To solve it, as shown in Section 6.5, one side of the equation must be equal to 0. Subtract $2y$ and add 3, getting

$$0 = y^2 - 8y + 12.$$
$$0 = (y - 6)(y - 2) \qquad \text{Factor.}$$
$$y - 6 = 0 \quad \text{or} \quad y - 2 = 0 \qquad \text{Zero-factor property}$$
$$y = 6 \quad \text{or} \quad y = 2$$

Check both of these potential solutions in the original equation.

If $y = 6$,		If $y = 2$,	
$\sqrt{2y - 3} = y - 3$		$\sqrt{2y - 3} = y - 3$	
$\sqrt{2(6) - 3} = 6 - 3$	?	$\sqrt{2(2) - 3} = 2 - 3$	?
$\sqrt{12 - 3} = 3$	?	$\sqrt{4 - 3} = -1$	?
$\sqrt{9} = 3$	?	$\sqrt{1} = -1$	?
$3 = 3.$	True	$1 = -1.$	False

Only 6 is a solution of the equation, so the solution set is {6}.

Remember to use Step 1 and isolate the radical before squaring both sides. For example, suppose we want to solve $3\sqrt{x} - 1 = 2x$. If we skip Step 1 and square both sides, we get

$$(3\sqrt{x} - 1)^2 = (2x)^2$$
$$9x - 6\sqrt{x} + 1 = 4x^2,$$

a more complicated equation that still contains a radical. This is why we should begin by isolating the radical on one side of the equation. Example 6 shows how this is done.

EXAMPLE 6   Rewriting an Equation Before Using the Squaring Property

Solve the equation $3\sqrt{x} - 1 = 2x$.

Add 1 to both sides of the original equation.

$$3\sqrt{x} = 2x + 1$$
$$(3\sqrt{x})^2 = (2x + 1)^2 \qquad \text{Square both sides.}$$
$$9x = 4x^2 + 4x + 1$$
$$0 = 4x^2 - 5x + 1 \qquad \text{Subtract } 9x.$$
$$0 = (4x - 1)(x - 1) \qquad \text{Factor.}$$
$$4x - 1 = 0 \quad \text{or} \quad x - 1 = 0 \qquad \text{Zero-factor property}$$
$$x = \frac{1}{4} \quad \text{or} \quad x = 1$$

Check both of these potential solutions in the original equation.

If $x = \dfrac{1}{4}$,	If $x = 1$,
$3\sqrt{x} - 1 = 2x$	$3\sqrt{x} - 1 = 2x$
$3\sqrt{\dfrac{1}{4}} - 1 = 2\left(\dfrac{1}{4}\right)$?	$3\sqrt{1} - 1 = 2(1)$?
$\dfrac{3}{2} - 1 = \dfrac{1}{2}$. True	$3 - 1 = 2$. True

The solution set of the original equation is $\left\{\dfrac{1}{4}, 1\right\}$.

Errors often occur when each side of an equation is squared. For instance, in Example 6, when both sides of

$$3\sqrt{x} = 2x + 1$$

are squared, the *entire* binomial $2x + 1$ must be squared to get $4x^2 + 4x + 1$. It would be *incorrect* to square the $2x$ and the 1 separately to get $4x^2 + 1$.

Some equations with radicals require squaring twice, as in the next example.

EXAMPLE 7   Using the Squaring Property Twice

Solve $\sqrt{21 + x} = 3 + \sqrt{x}$.

   Square both sides.

$$(\sqrt{21 + x})^2 = (3 + \sqrt{x})^2$$
$$21 + x = 9 + 6\sqrt{x} + x$$
$$12 = 6\sqrt{x} \qquad \text{Combine terms; simplify.}$$
$$2 = \sqrt{x}$$
$$4 = x. \qquad \text{Square both sides again.}$$

Check the potential solution.

   If $x = 4$,
$$\sqrt{21 + x} = 3 + \sqrt{x} \qquad \text{Original equation}$$
$$\sqrt{21 + 4} = 3 + \sqrt{4} \qquad ?$$
$$5 = 5. \qquad \text{True}$$

The solution set is $\{4\}$.

10.6 EXERCISES

Solve each equation. See Examples 1–4.

1. $\sqrt{x} = 7$

2. $\sqrt{k} = 10$

3. $\sqrt{y + 2} = 3$

4. $\sqrt{x + 7} = 5$

5. $\sqrt{r - 4} = 9$

6. $\sqrt{k - 12} = 3$

7. $\sqrt{4 - t} = 7$

8. $\sqrt{9 - s} = 5$

9. $\sqrt{2t + 3} = 0$

10. $\sqrt{5x - 4} = 0$

11. $\sqrt{3x-8}=-2$ 12. $\sqrt{6y+4}=-3$
13. $\sqrt{m}-4=7$ 14. $\sqrt{t}+3=10$
15. $\sqrt{10x-8}=3\sqrt{x}$ 16. $\sqrt{17t-4}=4\sqrt{t}$
17. $5\sqrt{x}=\sqrt{10x+15}$ 18. $4\sqrt{y}=\sqrt{20y-16}$
19. $\sqrt{3x-5}=\sqrt{2x+1}$ 20. $\sqrt{5y+2}=\sqrt{3y+8}$
21. $k=\sqrt{k^2-5k-15}$ 22. $s=\sqrt{s^2-2s-6}$
23. $7x=\sqrt{49x^2+2x-10}$ 24. $6m=\sqrt{36m^2+5m-5}$

25. Explain how you can tell that the equation $\sqrt{x}=-8$ has no real number solution without performing any algebraic steps.

26. Explain why the equation $x^2=36$ has two real number solutions, while the equation $\sqrt{x}=6$ has only one real number solution.

27. In your own words, explain how to solve a radical equation with a variable radical expression.

28. The first step in solving the equation $\sqrt{2x+1}=x-7$ is to square both sides of the equation. Errors often occur when the right side is squared incorrectly. Why is the square of the right side *not* equal to x^2+49? What is the square of the right side?

Solve each equation. Remember that $(a+b)^2=a^2+2ab+b^2$. (Hint: Be sure to get the radical term alone on one side of the equation before squaring.) See Examples 5 and 6.

29. $\sqrt{2x+1}=x-7$ 30. $\sqrt{3x+3}=x-5$
31. $\sqrt{3k+10}+5=2k$ 32. $\sqrt{4t+13}+1=2t$
33. $\sqrt{5x+1}-1=x$ 34. $\sqrt{x+1}-x=1$
35. $\sqrt{6t+7}+3=t+5$ 36. $\sqrt{10x+24}=x+4$
37. $x-4-\sqrt{2x}=0$ 38. $x-3-\sqrt{4x}=0$
39. $\sqrt{x+6}=2x$ 40. $\sqrt{k+12}=k$

Solve each equation. You will need to square both sides twice. See Example 7.

41. $\sqrt{x+1}-\sqrt{x-4}=1$ 42. $\sqrt{2x+3}+\sqrt{x+1}=1$
43. $\sqrt{x}=\sqrt{x-5}+1$ 44. $\sqrt{2x}=\sqrt{x+7}-1$

45. What is wrong with the following "solution"?

$$\sqrt{3x-6}+\sqrt{x+2}=12$$
$$3x-6+x+2=144 \qquad \text{Square both sides.}$$
$$4x-4=144 \qquad \text{Combine terms.}$$
$$4x=148 \qquad \text{Add 4 on both sides.}$$
$$x=37 \qquad \text{Divide by 4.}$$

46. What is wrong with the following "solution"?

$$-\sqrt{x-1}=-4$$
$$-(x-1)=16 \qquad \text{Square both sides.}$$
$$-x+1-16 \qquad \text{Distributive property}$$
$$-x=15 \qquad \text{Subtract 1 on both sides.}$$
$$x=-15 \qquad \text{Multiply each side by } -1.$$

Solve each problem. Give answers to the nearest tenth.

47. A surveyor wants to find the height of a building. At a point 110.0 feet from the base of the building he sights to the top of the building and finds the distance to be 193.0 feet. See the figure. How high is the building?

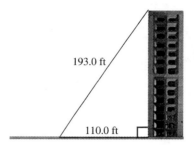

48. Two towns are separated by dense woods. To go from Town B to Town A, it is necessary to travel due west for 19.0 miles, then turn due north and travel for 14.0 miles. See the figure. How far apart are the towns?

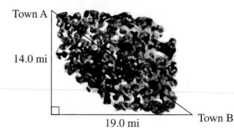

49. Police sometimes use the following procedure to estimate the speed at which a car was traveling at the time of an accident. A police officer drives the car involved in the accident under conditions similar to those during which the accident took place and then skids to a stop. If the car is driven at 30 miles per hour, then the speed at the time of the accident is given by

$$s = 30\sqrt{\frac{a}{p}},$$

where *a* is the length of the skid marks left at the time of the accident and *p* is the length of the skid marks in the police test. Find *s* for the following values of *a* and *p*.

(a) *a* = 862 feet; *p* = 156 feet
(b) *a* = 382 feet; *p* = 96 feet
(c) *a* = 84 feet; *p* = 26 feet

50. A formula for calculating the distance, *d,* one can see from an airplane to the horizon on a clear day is

$$d = 1.22\sqrt{x},$$

where *x* is the altitude of the plane in feet and *d* is given in miles. How far can one see to the horizon in a plane flying at the following altitudes? Round to the nearest tenth.

(a) 15,000 feet
(b) 18,000 feet
(c) 24,000 feet

Solve each problem.

51. In California, the number of 911 emergency calls has risen rapidly to 2.7 million, about a 9000% increase since 1985. The line graph shows the relationship between the number of calls in millions and the years from 1985 to 1997. This relationship is closely approximated by the function defined by the radical equation

$$y = \frac{1.76 + \sqrt{.0176 + .044x}}{.022}.$$

Here, y represents the year when x million calls were made. The years are coded so that 85 represents 1985, 95 represents 1995, and so on.

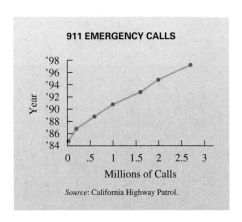

Source: California Highway Patrol.

(a) Find the year when the number of calls reached 1 million.
(b) If $y = f(x)$, find $f(3)$. Interpret your answer in the context of this problem.
(c) Use the equation to determine how many calls were made in 1995.

52. The five top-grossing films from 1995 ticket sales are listed here. The gross receipts y (in millions of dollars) and rank x are closely approximated by

$$x = \frac{9 + \sqrt{81 + 20(187 - y)}}{10}.$$

Note that the domain of x here must be restricted to whole numbers, since it includes possible ranks. Assuming the equation is still a good approximation, what rank would correspond to gross receipts of $60 million?

TOP-GROSSING FILMS

Rank/Film	Gross (millions)
1. *Toy Story*	$191.6
2. *Batman Forever*	184.0
3. *Apollo 13*	172.0
4. *Pocahontas*	141.6
5. *Ace Ventura: When Nature Calls*	108.3

Source: Exhibitor Relations Co., Inc.

TECHNOLOGY INSIGHTS (EXERCISES 53–57)

In earlier chapters we showed how to find the solution of an equation with a graphing calculator by writing the equation with one side equal to 0, replacing 0 with y, then graphing the equation $y = f(x)$. Now we can use this method to solve equations with radicals. For example, to solve $\sqrt{3x + 10} = 4$, rewrite it as $\sqrt{3x + 10} - 4 = 0$, and then as $y = \sqrt{3x + 10} - 4$. The graph of y in the graphing calculator screen on the left shows that $y = 0$ when $x = 2$, so the solution is 2.

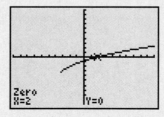 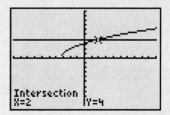

Another method, shown in Chapter 9, is to graph each side of the original equation separately, and look for the point of intersection. In the screen on the right, we have graphed $y_1 = \sqrt{3x + 10}$ and $y_2 = 4$. The value of x at the bottom of the screen again indicates that the solution is 2.

Use the information in each graph to determine the solution of the equation. Check each solution in the equation.

53. $\sqrt{5x - 4} = 4$

54. $\sqrt{2x + 6} = 2\sqrt{x}$

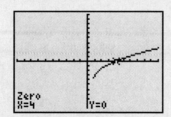 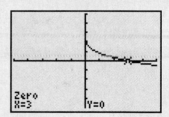

55. $\sqrt{1 - 3x} = \sqrt{5 - x}$

56. $\sqrt{2x^2 + 4x + 3} = -x$

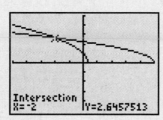 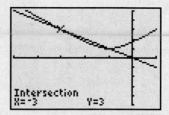

57. There are two solutions indicated in the screen for Exercise 56. They are both integers. Use the graph to determine the one not shown at the bottom of the screen.

$\overline{10.7}$ Fractional Exponents

OBJECTIVES

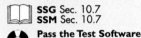

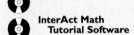

1 Define and use $a^{1/n}$.

2 Define and use $a^{m/n}$.

3 Use rules for exponents with fractional exponents.

4 Use fractional exponents to simplify radicals.

FOR EXTRA HELP

📖 **SSG** Sec. 10.7
   SSM Sec. 10.7

💿 **Pass the Test Software**

💿 **InterAct Math**
    Tutorial Software

📼 **Video** 17

OBJECTIVE 1 Define and use $a^{1/n}$. How should $5^{1/2}$ be defined? We want to define $5^{1/2}$ so that all the rules for exponents developed earlier in this book still hold. Then we should define $5^{1/2}$ so that

$$5^{1/2} \cdot 5^{1/2} = 5^{1/2+1/2} = 5^1 = 5.$$

This agrees with the product rule for exponents from Section 5.2. By definition,

$$(\sqrt{5})(\sqrt{5}) = 5.$$

Since both $5^{1/2} \cdot 5^{1/2}$ and $\sqrt{5} \cdot \sqrt{5}$ equal 5,

$$5^{1/2} \text{ should equal } \sqrt{5}.$$

Similarly,

$$5^{1/3} \cdot 5^{1/3} \cdot 5^{1/3} = 5^{1/3+1/3+1/3} = 5^{3/3} = 5,$$

and

$$\sqrt[3]{5} \cdot \sqrt[3]{5} \cdot \sqrt[3]{5} = \sqrt[3]{5^3} = 5,$$

so

$$5^{1/3} \text{ should equal } \sqrt[3]{5}.$$

These examples suggest the following definition.

$a^{1/n}$

If a is a nonnegative number and n is a positive integer,

$$a^{1/n} = \sqrt[n]{a}.$$

EXAMPLE 1 Using the Definition of $a^{1/n}$

Simplify each expression by first writing it in radical form.

(a) $16^{1/2}$

By the definition above,

$$16^{1/2} = \sqrt{16} = 4.$$

(b) $27^{1/3} = \sqrt[3]{27} = 3$ **(c)** $64^{1/3} = \sqrt[3]{64} = 4$ **(d)** $64^{1/6} = \sqrt[6]{64} = 2$

OBJECTIVE 2 Define and use $a^{m/n}$. Now a more general exponential expression like $16^{3/4}$ can be defined. By the power rule, $(a^m)^n = a^{mn}$, so that

$$16^{3/4} = (16^{1/4})^3 = (\sqrt[4]{16})^3 = 2^3 = 8.$$

However, $16^{3/4}$ could also be written as

$$16^{3/4} = (16^3)^{1/4} = (4096)^{1/4} = \sqrt[4]{4096} = 8.$$

The expression can be evaluated either way to get the same answer. As the example suggests, taking the root first involves smaller numbers and is often easier. This example suggests the following definition for $a^{m/n}$.

$a^{m/n}$

If a is a nonnegative number and m and n are integers, with $n > 0$,

$$a^{m/n} = (a^{1/n})^m = (\sqrt[n]{a})^m.$$

EXAMPLE 2 Using the Definition of $a^{m/n}$
Evaluate each expression.

(a) $9^{3/2}$

Use the definition to write

$$9^{3/2} = (9^{1/2})^3 = 3^3 = 27.$$

(b) $64^{2/3} = (64^{1/3})^2 = 4^2 = 16$

(c) $-32^{4/5} = -(32^{1/5})^4 = -2^4 = -16$

Earlier, a^{-n} was defined as

$$a^{-n} = \frac{1}{a^n}$$

for nonzero numbers a and integers n. This same result applies for negative fractional exponents.

$a^{-m/n}$

If a is a positive number and m and n are integers, with $n > 0$,

$$a^{-m/n} = \frac{1}{a^{m/n}}.$$

EXAMPLE 3 Using the Definition of $a^{-m/n}$
Write each expression with a positive exponent and then evaluate.

(a) $32^{-3/5} = \frac{1}{32^{3/5}} = \frac{1}{(32^{1/5})^3} = \frac{1}{2^3} = \frac{1}{8}$

(b) $27^{-4/3} = \frac{1}{27^{4/3}} = \frac{1}{(27^{1/3})^4} = \frac{1}{3^4} = \frac{1}{81}$

In Example 3(b), a common mistake is to write $27^{-4/3} = -27^{3/4}$. **This is incorrect.** The negative exponent does not indicate a negative number. Also, the negative exponent indicates the reciprocal of the *base*, not the reciprocal of the *exponent*.

OBJECTIVE 3 Use rules for exponents with fractional exponents. All the rules for exponents given earlier still hold when the exponents are fractions. The next examples use these rules to simplify expressions with fractional exponents.

E X A M P L E   4   Using the Rules for Exponents with Fractional Exponents

Simplify each expression. Write each answer in exponential form with only positive exponents.

(a) $3^{2/3} \cdot 3^{5/3} = 3^{2/3+5/3} = 3^{7/3}$

(b) $\dfrac{5^{1/4}}{5^{3/4}} = 5^{1/4-3/4} = 5^{-2/4} = 5^{-1/2} = \dfrac{1}{5^{1/2}}$

(c) $(9^{1/4})^2 = 9^{2(1/4)} = 9^{2/4} = 9^{1/2} = \sqrt{9} = 3$

(d) $\dfrac{2^{1/2} \cdot 2^{-1}}{2^{-3/2}} = \dfrac{2^{1/2+(-1)}}{2^{-3/2}} = \dfrac{2^{-1/2}}{2^{-3/2}} = 2^{-1/2-(-3/2)} = 2^{2/2} = 2^1 = 2$

(e) $\left(\dfrac{9}{4}\right)^{5/2} = \dfrac{9^{5/2}}{4^{5/2}} = \dfrac{(9^{1/2})^5}{(4^{1/2})^5} = \dfrac{(\sqrt{9})^5}{(\sqrt{4})^5} = \dfrac{3^5}{2^5}$

E X A M P L E   5   Using Fractional Exponents with Variables

Simplify each expression. Write each answer in exponential form with only positive exponents. Assume that all variables represent positive numbers.

(a) $m^{1/5} \cdot m^{3/5} = m^{1/5+3/5} = m^{4/5}$

(b) $\dfrac{p^{5/3}}{p^{4/3}} = p^{5/3-4/3} = p^{1/3}$

(c) $(x^2 y^{1/2})^4 = (x^2)^4 (y^{1/2})^4 = x^8 y^2$

(d) $\left(\dfrac{z^{1/4}}{w^{1/3}}\right)^5 = \dfrac{(z^{1/4})^5}{(w^{1/3})^5} = \dfrac{z^{5/4}}{w^{5/3}}$

(e) $\dfrac{k^{2/3} \cdot k^{-1/3}}{k^{5/3}} = k^{2/3+(-1/3)-5/3} = k^{-4/3} = \dfrac{1}{k^{4/3}}$

 Errors often occur in problems like those in Examples 4 and 5 because students try to convert the expressions to radicals. Remember that the *rules of exponents* apply here.

O B J E C T I V E   4   Use fractional exponents to simplify radicals.   In fact, sometimes it is easier to simplify a radical by first writing it in exponential form.

E X A M P L E   6   Simplifying Radicals by Using Rational Exponents

Simplify each radical by first writing it in exponential form.

(a) $\sqrt[6]{9^3} = (9^3)^{1/6} = 9^{3/6} = 9^{1/2} = \sqrt{9} = 3$

(b) $(\sqrt[4]{m})^2 = (m^{1/4})^2 = m^{2/4} = m^{1/2} = \sqrt{m}$

Here it is assumed that $m \geq 0$.

CONNECTIONS

Earlier in this chapter we used the rules for adding and multiplying polynomials with radical expressions. In this section we have treated radicals as exponential expressions and applied the rules for exponents. This should not be surprising since radicals are yet another way to represent real numbers, and all the rules developed in this book apply to any real number (with some exceptions). As we mentioned at the beginning of Chapter 5, polynomials are basic algebraic expressions. Our work with radicals just extends the kind of numbers that we use in polynomials.

CONNECTIONS (CONTINUED)

FOR DISCUSSION OR WRITING

Use the methods of Chapter 6 to factor the following expressions.

1. $x - 2\sqrt{x} - 3$

2. $x^2 - 10$

3. $x + 4\sqrt{x} + 4$

4. Divide the polynomial $x^2 + 3x + 5$ by $\sqrt{x}$, and simplify the result.

10.7 EXERCISES

Decide which one of the four choices is not *equal to the given expression.*

1. $49^{1/2}$ **(a)** -7 **(b)** 7 **(c)** $\sqrt{49}$ **(d)** $49^{.5}$

2. $81^{1/2}$ **(a)** 9 **(b)** $\sqrt{81}$ **(c)** $81^{.5}$ **(d)** $\dfrac{81}{2}$

3. $-64^{1/3}$ **(a)** $-\sqrt{16}$ **(b)** -4 **(c)** 4 **(d)** $-\sqrt[3]{64}$

4. $-125^{1/3}$ **(a)** $-\sqrt{25}$ **(b)** -5 **(c)** $-\sqrt[3]{125}$ **(d)** 5

Simplify each expression by first writing it in radical form. See Examples 1–3.

5. $25^{1/2}$ **6.** $121^{1/2}$ **7.** $64^{1/3}$ **8.** $125^{1/3}$ **9.** $16^{1/4}$

10. $81^{1/4}$ **11.** $32^{1/5}$ **12.** $243^{1/5}$ **13.** $4^{3/2}$ **14.** $9^{5/2}$

15. $27^{2/3}$ **16.** $8^{5/3}$ **17.** $16^{3/4}$ **18.** $64^{5/3}$ **19.** $32^{2/5}$

20. $144^{3/2}$ **21.** $-8^{2/3}$ **22.** $-27^{5/3}$ **23.** $-64^{1/3}$ **24.** $-125^{5/3}$

25. $49^{-3/2}$ **26.** $9^{-5/2}$ **27.** $216^{-2/3}$ **28.** $27^{-4/3}$ **29.** $-16^{-5/4}$

30. $-81^{-3/4}$

Simplify each expression. Write answers in exponential form with only positive exponents. Assume that all variables represent positive numbers. See Examples 3–5.

31. $2^{1/2} \cdot 2^{5/2}$ **32.** $5^{2/3} \cdot 5^{4/3}$ **33.** $6^{1/4} \cdot 6^{-3/4}$ **34.** $12^{2/5} \cdot 12^{-1/5}$

35. $\dfrac{15^{3/4}}{15^{5/4}}$ **36.** $\dfrac{7^{3/5}}{7^{-1/5}}$ **37.** $\dfrac{11^{-2/7}}{11^{-3/7}}$ **38.** $\dfrac{4^{-2/3}}{4^{1/3}}$

39. $(8^{3/2})^2$ **40.** $(5^{2/5})^{10}$ **41.** $(6^{1/3})^{3/2}$ **42.** $(7^{2/5})^{5/3}$

43. $\left(\dfrac{25}{4}\right)^{3/2}$ **44.** $\left(\dfrac{8}{27}\right)^{2/3}$ **45.** $\dfrac{2^{2/5} \cdot 2^{-3/5}}{2^{7/5}}$ **46.** $\dfrac{3^{-3/4} \cdot 3^{5/4}}{3^{-1/4}}$

47. $\dfrac{6^{-2/9}}{6^{1/9} \cdot 6^{-5/9}}$ **48.** $\dfrac{8^{6/7}}{8^{2/7} \cdot 8^{-1/7}}$ **49.** $\dfrac{z^{2/3}}{z^{-1/3}}$ **50.** $\dfrac{r^{5/4}}{r^{3/4}}$

51. $(m^3 n^{1/4})^{2/3}$ **52.** $(p^4 q^{1/2})^{4/3}$ **53.** $\left(\dfrac{a^{1/2}}{b^{1/3}}\right)^{4/3}$ **54.** $\left(\dfrac{m^{2/3}}{n^{3/4}}\right)^{1/2}$

Simplify each radical by first writing it in exponential form. Give the answer as an integer or a radical in simplest form. Assume that all variables represent nonnegative numbers. See Example 6.

55. $\sqrt[6]{4^3}$ **56.** $\sqrt[9]{8^3}$ **57.** $\sqrt[8]{16^2}$ **58.** $\sqrt[9]{27^3}$

59. $\sqrt[4]{a^2}$ **60.** $\sqrt[9]{b^3}$ **61.** $\sqrt[6]{k^4}$ **62.** $\sqrt[8]{m^4}$

The rules for multiplying and dividing radicals presented earlier in this chapter were stated for radicals having the same index. For example, we only multiplied or divided square roots, or multiplied or divided cube roots, and so on. Since we know how to write radicals with fractional exponents and from past work know how to add and subtract fractions with different denominators, we can now multiply and divide radicals having different indexes.

Work Exercises 63–68 in order, to see how to multiply $\sqrt{2}$ by $\sqrt[3]{2}$.

63. Write $\sqrt{2}$ and $\sqrt[3]{2}$ using fractional exponents.

64. Write the product $\sqrt{2} \cdot \sqrt[3]{2}$ using the expressions you found in Exercise 63.

65. What is the least common denominator of the fractional exponents in Exercise 64?

66. Repeat Exercise 64, but write the fractional exponents with the common denominator from Exercise 65.

67. Use the rule for multiplying exponential expressions with like bases to simplify the product in Exercise 66. (*Hint:* The base remains the same; do not multiply the two bases.)

68. Write the answer you obtained in Exercise 67 as a radical.

Did you make the connection that when radicals are written in exponential form, the rules for exponents may be used to multiply (or divide) unlike radical terms?

 Use the exponential key on your calculator to find the following roots. For example, to find $\sqrt[5]{32}$, enter 32 and then raise to the 1/5 power. (The exponent 1/5 may be entered as .2 if you wish.) If the root is irrational, round it to the nearest thousandth.

69. $\sqrt[6]{64}$ **70.** $\sqrt[5]{243}$ **71.** $\sqrt[7]{84}$ **72.** $\sqrt[9]{16}$

73. $\sqrt[5]{987}$ **74.** $\sqrt[6]{249}$ **75.** $\sqrt[4]{19^3}$ **76.** $\sqrt[5]{27^4}$

Each calculator screen gives the simplified real number value (or an approximation) of an exponential expression.

Use a scientific calculator to verify that the radical expression equals the exponential expression.

77.
```
6^(1/2)
            2.449489743
√(6)
```

78.
```
(-8)^(2/3)
                        4
3√((-8)²)
```

79.
```
16^(-3/4)
                    .125
1/((4×√16)^3)
```

80.
```
25^(-3/2)
                    .008
1/(√(25)^3)
```

🔢 *Solve each problem.*

81. In Section 10.6, Exercise 50, we gave the formula $d = 1.22\sqrt{x}$ for calculating the distance in miles one can see from an airplane to the horizon on a clear day. Here, x is in feet.
  (a) Write the formula using a fractional exponent.
  (b) Find d to the nearest hundredth if the altitude x is 30,000 feet.

82. A biologist has shown that the number of different plant species S on a Galápagos Island is related to the area of the island, A (in miles), by

$$S = 28.6A^{1/3}.$$

How many plant species would exist on such an island with the following areas?
  (a) 8 square miles   **(b)** 27,000 square miles

83. Explain in your own words why $7^{1/2}$ is defined as $\sqrt{7}$.

84. Explain in your own words why $7^{1/3}$ is defined as $\sqrt[3]{7}$.

CHAPTER 10 GROUP ACTIVITY

🔢 Comparing Television Sizes

Objective: Use the Pythagorean formula to find the dimensions of different television sets.

Television sets are identified by the diagonal measurement of the viewing screen. For example, a 19-inch TV measures 19 inches from one corner of the viewing screen diagonally to the other corner.

A. The chart below gives some common TV sizes as well as their corresponding widths. Use the Pythagorean formula to find the heights of the viewing screens. Round up to the next whole number. (*Hint:* The TV size is the hypotenuse.)

TV Set	TV Size	Width in Inches	Height in Inches
A	13-inch	10	
B	19-inch	15	
C	25-inch	21	
D	27-inch	22	
E	32-inch	25	
F	36-inch	30	

Source: Data based on information from a Sears catalog.

(continued)

B. The dimensions in the table above correspond only to television viewing screens. Each television set also has an outer border of $1\frac{1}{2}$ inches as well as an additional 4 inches for a control panel at the bottom of the set. Also, it is recommended that a minimum of 2 inches of space be allowed above the set for ventilation.

Consider the following entertainment centers, with dimensions for television space as given below. Find the largest television set from the chart in part A that will fit in each entertainment center.

1. 27 inches by 29 inches
2. 28 inches by 25 inches
3. 20 inches by 20 inches

CHAPTER 10 SUMMARY

KEY TERMS

10.1 square root	**fourth root**	**10.3** like radicals	**10.5** conjugate
radicand	index (order)		
radical	principal root	**10.4** rationalizing the	**10.6** squaring property
radical expression		denominator	
perfect square	**10.2** perfect cube		
cube root			

NEW SYMBOLS

$\sqrt{}$	radical sign	$\sqrt[3]{a}$	cube root of a	$a^{1/m}$	mth root of a
$\approx$	is approximately equal to	$\sqrt[n]{a}$	nth root of a		

TEST YOUR WORD POWER

See how well you have learned the vocabulary in this chapter. Answers, with examples, are given at the bottom of the next page.

1. The **square root** of a number is
(a) the number raised to the second power
(b) the number under a radical sign
(c) a number that when multiplied by itself gives the original number
(d) the inverse of the number.

2. A **radical** is
(a) a symbol that indicates the nth root
(b) an algebraic expression containing a square root

(c) the positive nth root of a number
(d) a radical sign and the number or expression under it.

3. The **principal root** of a positive number with even index n is
(a) the positive nth root of the number
(b) the negative nth root of the number
(c) the square root of the number
(d) the cube root of the number.

4. **Like radicals** are
(a) radicals in simplest form
(b) algebraic expressions containing radicals
(c) multiples of the same root of the same number
(d) radicals with the same index.

TEST YOUR WORD POWER

5. Rationalizing the denominator is the process of
(a) eliminating fractions from a radical expression
(b) changing the denominator of a fraction from a

radical to a rational number
(c) clearing a radical expression of radicals
(d) multiplying radical expressions.

6. The **conjugate** of $a + b$ is
(a) $a - b$
(b) $a \cdot b$
(c) $a \div b$
(d) $(a + b)^2$.

QUICK REVIEW

CONCEPTS	EXAMPLES

10.1 EVALUATING ROOTS

If a is a positive real number, $\sqrt{a}$ is the positive square root of a;

$-\sqrt{a}$ is the negative square root of a; $\sqrt{0} = 0$.

If a is a negative real number, $\sqrt{a}$ is not a real number.

$$\sqrt{49} = 7$$

$$-\sqrt{81} = -9$$

$$\sqrt{-25} \text{ is not a real number.}$$

If a is a positive rational number, $\sqrt{a}$ is rational if a is a perfect square. $\sqrt{a}$ is irrational if a is not a perfect square.

$$\sqrt{\frac{4}{9}}, \ \sqrt{16} \text{ are rational.}$$

$$\sqrt{\frac{2}{3}}, \ \sqrt{21} \text{ are irrational.}$$

Each real number has exactly one real cube root.

$$\sqrt[3]{27} = 3; \ \sqrt[3]{-8} = -2$$

10.2 MULTIPLICATION AND DIVISION OF RADICALS

Product Rule for Radicals
For nonnegative real numbers x and y,
$$\sqrt{x} \cdot \sqrt{y} = \sqrt{xy}$$
and
$$\sqrt{xy} = \sqrt{x} \cdot \sqrt{y}.$$

$$\sqrt{5} \cdot \sqrt{7} = \sqrt{35}$$
$$\sqrt{8} \cdot \sqrt{2} = \sqrt{16} = 4$$
$$\sqrt{48} = \sqrt{16} \cdot \sqrt{3} = 4\sqrt{3}$$

Quotient Rule for Radicals
If x and y are nonnegative real numbers and y is not 0,

$$\frac{\sqrt{x}}{\sqrt{y}} = \sqrt{\frac{x}{y}} \quad \text{and} \quad \sqrt{\frac{x}{y}} = \frac{\sqrt{x}}{\sqrt{y}}.$$

$$\sqrt{\frac{25}{64}} = \frac{\sqrt{25}}{\sqrt{64}} = \frac{5}{8}$$

$$\frac{\sqrt{8}}{\sqrt{2}} = \sqrt{\frac{8}{2}} = \sqrt{4} = 2$$

If all indicated roots are real,
$$\sqrt[n]{x} \cdot \sqrt[n]{y} = \sqrt[n]{xy}$$
and
$$\frac{\sqrt[n]{x}}{\sqrt[n]{y}} = \sqrt[n]{\frac{x}{y}}, \ y \neq 0.$$

$$\sqrt[3]{5} \cdot \sqrt[3]{3} = \sqrt[3]{15}$$

$$\frac{\sqrt[4]{12}}{\sqrt[4]{4}} = \sqrt[4]{\frac{12}{4}} = \sqrt[4]{3}$$

CONCEPTS	EXAMPLES

10.3 ADDITION AND SUBTRACTION OF RADICALS

Add and subtract like radicals by using the distributive property. Only like radicals can be combined in this way.

$$2\sqrt{5} + 4\sqrt{5} = (2 + 4)\sqrt{5}$$
$$= 6\sqrt{5}$$
$$\sqrt{8} + \sqrt{32} = 2\sqrt{2} + 4\sqrt{2}$$
$$= 6\sqrt{2}$$

10.4 RATIONALIZING THE DENOMINATOR

The denominator of a radical can be rationalized by multiplying both the numerator and denominator by the same number.

$$\frac{2}{\sqrt{3}} = \frac{2 \cdot \sqrt{3}}{\sqrt{3} \cdot \sqrt{3}} = \frac{2\sqrt{3}}{3}$$

$$\sqrt[3]{\frac{5}{6}} = \frac{\sqrt[3]{5}}{\sqrt[3]{6}} \cdot \frac{\sqrt[3]{6^2}}{\sqrt[3]{6^2}} = \frac{\sqrt[3]{180}}{6}$$

10.5 SIMPLIFYING RADICAL EXPRESSIONS

When appropriate, use the rules for adding and multiplying polynomials to simplify radical expressions.

Any denominators with radicals should be rationalized.

$$\sqrt{6}(\sqrt{5} - \sqrt{7}) = \sqrt{30} - \sqrt{42}$$
$$(\sqrt{5} - \sqrt{3})(\sqrt{5} + \sqrt{3}) = 5 - 3 = 2$$

$$\frac{3}{\sqrt{6}} = \frac{3\sqrt{6}}{6} = \frac{\sqrt{6}}{2}$$

If a radical expression contains two terms in the denominator and at least one of those terms is a radical, multiply both the numerator and the denominator by the conjugate of the denominator.

$$\frac{6}{\sqrt{7} - \sqrt{2}} = \frac{6}{\sqrt{7} - \sqrt{2}} \cdot \frac{\sqrt{7} + \sqrt{2}}{\sqrt{7} + \sqrt{2}}$$
$$= \frac{6(\sqrt{7} + \sqrt{2})}{7 - 2}$$ Multiply fractions.
$$= \frac{6(\sqrt{7} + \sqrt{2})}{5}$$ Simplify.

10.6 SOLVING EQUATIONS WITH RADICALS

Solving an Equation with Radicals

Solve $\sqrt{2x - 3} + x = 3$.

Step 1 Arrange the terms so that a radical is alone on one side of the equation.

$$\sqrt{2x - 3} = 3 - x$$ Isolate radical.

Step 2 Square each side. (By the squaring property of equality, all solutions of the original equation are *among* the solutions of the squared equation.)

$$(\sqrt{2x - 3})^2 = (3 - x)^2$$ Square.
$$2x - 3 = 9 - 6x + x^2$$

Step 3 Combine like terms.

$$0 = x^2 - 8x + 12$$ Get one side equal to 0.

Step 4 If there is still a term with a radical, repeat Steps 1–3.

$$0 = (x - 2)(x - 6)$$ Factor.
$$x - 2 = 0 \quad \text{or} \quad x - 6 = 0$$ Set each factor equal to 0.

Step 5 Solve the equation for potential solutions.

$$x = 2 \quad \text{or} \quad x = 6$$ Solve.

Step 6 Check all potential solutions from Step 5 in the original equation.

Verify that 2 is the only solution (6 is extraneous).
Solution set: $\{2\}$

10.7 FRACTIONAL EXPONENTS

Assume $a \geq 0$, m and n are integers, $n > 0$.

$$a^{1/n} = \sqrt[n]{a}$$
$$a^{m/n} = \sqrt[n]{a^m} = (\sqrt[n]{a})^m$$
$$a^{-m/n} = \frac{1}{a^{m/n}} \quad (a \neq 0)$$

$$8^{1/3} = \sqrt[3]{8} = 2$$
$$(81)^{3/4} = \sqrt[4]{81^3} = (\sqrt[4]{81})^3 = 3^3 = 27$$
$$36^{-3/2} = \frac{1}{36^{3/2}} = \frac{1}{(36^{1/2})^3} = \frac{1}{6^3} = \frac{1}{216}$$

CHAPTER 10 REVIEW EXERCISES

[10.1]   *Find all square roots of each number.*

1. 49   **2.** 81   **3.** 196   **4.** 121   **5.** 225   **6.** 729

Find each indicated root. If the root is not a real number, say so.

7. $\sqrt{16}$ **8.** $-\sqrt{36}$ **9.** $\sqrt[3]{1000}$ **10.** $\sqrt[4]{81}$

11. $\sqrt{-8100}$ **12.** $-\sqrt{4225}$ **13.** $\sqrt{\dfrac{49}{36}}$ **14.** $\sqrt{\dfrac{100}{81}}$

15. If $\sqrt{a}$ is not a real number, then what kind of number must a be?

16. Find the value of x in the figure.

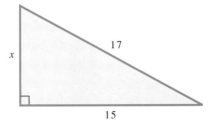

Determine whether each number is rational, irrational, *or* not a real number. *If the number is rational, give its exact value. If the number is irrational, give a decimal approximation rounded to the nearest thousandth.*

17. $\sqrt{23}$ **18.** $\sqrt{169}$ **19.** $-\sqrt{25}$ **20.** $\sqrt{-4}$

[10.2]   *Use the product rule to simplify each expression.*

21. $\sqrt{5} \cdot \sqrt{15}$ **22.** $-\sqrt{27}$ **23.** $\sqrt{160}$ **24.** $\sqrt[3]{-125}$

25. $\sqrt[3]{1728}$ **26.** $\sqrt{12} \cdot \sqrt{27}$ **27.** $\sqrt{32} \cdot \sqrt{48}$ **28.** $\sqrt{50} \cdot \sqrt{125}$

Use the product rule, the quotient rule, or both to simplify each expression.

29. $-\sqrt{\dfrac{121}{400}}$ **30.** $\sqrt{\dfrac{3}{49}}$ **31.** $\sqrt{\dfrac{7}{169}}$ **32.** $\sqrt{\dfrac{1}{6}} \cdot \sqrt{\dfrac{5}{6}}$

33. $\sqrt{\dfrac{2}{5}} \cdot \sqrt{\dfrac{2}{45}}$ **34.** $\dfrac{3\sqrt{10}}{\sqrt{5}}$ **35.** $\dfrac{24\sqrt{12}}{6\sqrt{3}}$ **36.** $\dfrac{8\sqrt{150}}{4\sqrt{75}}$

Simplify each expression. Assume that all variables represent nonnegative real numbers.

37. $\sqrt{p} \cdot \sqrt{p}$ **38.** $\sqrt{k} \cdot \sqrt{m}$ **39.** $\sqrt{r^{18}}$

40. $\sqrt{x^{10}y^{16}}$ **41.** $\sqrt{a^{15}b^{21}}$ **42.** $\sqrt{121x^{6}y^{10}}$

43. Use a calculator to find approximations for $\sqrt{.5}$ and $\dfrac{\sqrt{2}}{2}$. Based on your results, do you think that these two expressions represent the same number? If so, verify it *algebraically.*

[10.3]   *Simplify and combine terms where possible.*

44. $3\sqrt{2} + 6\sqrt{2}$ **45.** $3\sqrt{75} + 2\sqrt{27}$ **46.** $4\sqrt{12} + \sqrt{48}$

47. $4\sqrt{24} - 3\sqrt{54} + \sqrt{6}$ **48.** $2\sqrt{7} - 4\sqrt{28} + 3\sqrt{63}$ **49.** $\dfrac{2}{5}\sqrt{75} + \dfrac{3}{4}\sqrt{160}$

50. $\dfrac{1}{3}\sqrt{18} + \dfrac{1}{4}\sqrt{32}$ **51.** $\sqrt{15} \cdot \sqrt{2} + 5\sqrt{30}$

Simplify each expression. Assume that all variables represent nonnegative real numbers.

52. $\sqrt{4x} + \sqrt{36x} - \sqrt{9x}$ **53.** $\sqrt{16p} + 3\sqrt{p} - \sqrt{49p}$

54. $\sqrt{20m^{2}} - m\sqrt{45}$ **55.** $3k\sqrt{8k^{2}n} + 5k^{2}\sqrt{2n}$

[10.4] *Perform the indicated operations and write answers in simplest form. Assume that all variables represent positive real numbers.*

56. $\dfrac{8\sqrt{2}}{\sqrt{5}}$ **57.** $\dfrac{5}{\sqrt{5}}$ **58.** $\dfrac{12}{\sqrt{24}}$ **59.** $\dfrac{\sqrt{2}}{\sqrt{15}}$ **60.** $\sqrt{\dfrac{2}{5}}$

61. $\sqrt{\dfrac{5}{14}} \cdot \sqrt{28}$ **62.** $\sqrt{\dfrac{2}{7}} \cdot \sqrt{\dfrac{1}{3}}$ **63.** $\sqrt{\dfrac{r^2}{16x}}$ **64.** $\sqrt[3]{\dfrac{1}{3}}$ **65.** $\sqrt[3]{\dfrac{2}{7}}$

66. Explain how you would show, without using a calculator, that $\dfrac{\sqrt{6}}{4}$ and $\sqrt{\dfrac{48}{128}}$ represent the exact same number. Then actually perform the necessary steps.

[10.5] *Simplify each expression.*

67. $-\sqrt{3}(\sqrt{5} + \sqrt{27})$ **68.** $3\sqrt{2}(\sqrt{3} + 2\sqrt{2})$

69. $(2\sqrt{3} - 4)(5\sqrt{3} + 2)$ **70.** $(5\sqrt{7} + 2)^2$

71. $(\sqrt{5} - \sqrt{7})(\sqrt{5} + \sqrt{7})$ **72.** $(2\sqrt{3} + 5)(2\sqrt{3} - 5)$

Rationalize each denominator.

73. $\dfrac{1}{2 + \sqrt{5}}$ **74.** $\dfrac{2}{\sqrt{2} - 3}$ **75.** $\dfrac{\sqrt{8}}{\sqrt{2} + 6}$

76. $\dfrac{\sqrt{3}}{1 + \sqrt{3}}$ **77.** $\dfrac{\sqrt{5} - 1}{\sqrt{2} + 3}$ **78.** $\dfrac{2 + \sqrt{6}}{\sqrt{3} - 1}$

Write each quotient in lowest terms.

79. $\dfrac{15 + 10\sqrt{6}}{15}$ **80.** $\dfrac{3 + 9\sqrt{7}}{12}$ **81.** $\dfrac{6 + \sqrt{192}}{2}$

[10.6] *Solve each equation.*

82. $\sqrt{m} - 5 = 0$ **83.** $\sqrt{p} + 4 = 0$ **84.** $\sqrt{k + 1} = 7$

85. $\sqrt{5m + 4} = 3\sqrt{m}$ **86.** $\sqrt{2p + 3} = \sqrt{5p - 3}$ **87.** $\sqrt{4y + 1} = y - 1$

88. $\sqrt{-2k - 4} = k + 2$ **89.** $\sqrt{2 - x} + 3 = x + 7$ **90.** $\sqrt{x} - x + 2 = 0$

 Work each problem.

 91. Refer to Section 10.3, Exercise 57.

    (a) Describe the elements in the domain.

    (b) Describe the elements in the range.

    (c) According to the equation there, $y = 1.4\sqrt{x - 2.5} + 87.5$, in what year were exports about \$13 billion?

    (d) Does your answer to part (c) seem reasonable, compared to the actual data repeated here? Explain.

U.S. Exports

Billions of Dollars (x)	Year (y)
7.5	1990
19.6	1993
25.8	1994
36.4	1996

    (e) Use the equation from part (c) to find the exports in 1995. (*Hint:* Let $y = 95$ and solve for x.) Does your answer seem reasonable compared to the actual data in the table?

92. Refer to Section 10.6, Exercise 52, where $y = f(x)$ is equivalent to the equation

$$x = \frac{9 + \sqrt{81 + 20(187 - y)}}{10}.$$

Find $f(x)$. (*Hint:* Solve the equation for y, and replace y with $f(x)$.)

[10.7] *Simplify each expression. Assume that all variables represent positive real numbers.*

93. $81^{1/2}$ **94.** $-125^{1/3}$ **95.** $7^{2/3} \cdot 7^{7/3}$

96. $\dfrac{13^{4/5}}{13^{-3/5}}$ **97.** $\dfrac{x^{1/4} \cdot x^{5/4}}{x^{3/4}}$ **98.** $\sqrt[8]{49^4}$

MIXED REVIEW EXERCISES

Simplify each expression if possible. Assume all variables represent positive real numbers.

99. $64^{2/3}$ **100.** $2\sqrt{27} + 3\sqrt{75} - \sqrt{300}$

101. $\dfrac{1}{5 + \sqrt{2}}$ **102.** $\sqrt{\dfrac{1}{3}} \cdot \sqrt{\dfrac{24}{5}}$

103. $\sqrt{50y^2}$ **104.** $\sqrt[3]{-125}$

105. $-\sqrt{5}(\sqrt{2} + \sqrt{75})$ **106.** $\sqrt{\dfrac{16r^3}{3s}}$

107. $\dfrac{12 + 6\sqrt{13}}{12}$ **108.** $-\sqrt{162} + \sqrt{8}$

109. $(\sqrt{5} - \sqrt{2})^2$ **110.** $(6\sqrt{7} + 2)(4\sqrt{7} - 1)$

111. $-\sqrt{121}$ **112.** $\dfrac{x^{8/3}}{x^{2/3}}$

Solve.

113. $\sqrt{x + 2} = x - 4$ **114.** $\sqrt{k} + 3 = 0$ **115.** $\sqrt{1 + 3t} - t = -3$

RELATING CONCEPTS (EXERCISES 116–120)

In Chapter 4 we plotted points in the rectangular coordinate plane. In all cases our points had coordinates that were rational numbers. However, ordered pairs may have irrational coordinates as well.

*Use your knowledge of the material in Chapters 4 and 10 to **work Exercises 116–120 in order.***

Consider the points $A(2\sqrt{14}, 5\sqrt{7})$ and $B(-3\sqrt{14}, 10\sqrt{7})$.

116. Write an expression that represents the slope of the line containing points A and B. Do not simplify yet.

117. Simplify the numerator and the denominator in the expression from Exercise 116 by combining like radicals.

118. Write the fraction from Exercise 117 as the square root of a fraction in lowest terms.

119. Rationalize the denominator of the expression found in Exercise 118.

120. Based on your answer in Exercise 119, does line AB rise or fall from left to right?

Did you make the connection that the real number properties discussed up to this point apply to irrational numbers as well as rational numbers?

CHAPTER 10 Roots and Radicals

CHAPTER 10 TEST

On this test assume that all variables represent positive real numbers.

1. Find all square roots of 196.
2. Consider $\sqrt{142}$.
   (a) Determine whether it is rational or irrational.
   (b) Find a decimal approximation to the nearest thousandth.
3. Simplify $\sqrt[3]{216}$.

Simplify where possible.

4. $-\sqrt{27}$

5. $\sqrt{\dfrac{128}{25}}$

6. $\sqrt[3]{32}$

7. $\dfrac{20\sqrt{18}}{5\sqrt{3}}$

8. $3\sqrt{28} + \sqrt{63}$

9. $3\sqrt{27x} - 4\sqrt{48x} + 2\sqrt{3x}$

10. $\sqrt[3]{32x^2y^3}$

11. $(6 - \sqrt{5})(6 + \sqrt{5})$

12. $(2 - \sqrt{7})(3\sqrt{2} + 1)$

13. $(\sqrt{5} + \sqrt{6})^2$

14. The hypotenuse of a right triangle measures 9 inches, and one leg measures 3 inches. Find the measure of the other leg.
    (a) Give its length in simplified radical form.
    (b) Round the answer to the nearest thousandth.

Rationalize each denominator.

15. $\dfrac{5\sqrt{2}}{\sqrt{7}}$

16. $\sqrt{\dfrac{2}{3x}}$

17. $\dfrac{-2}{\sqrt[3]{4}}$

18. $\dfrac{-3}{4 - \sqrt{3}}$

Solve each equation.

19. $\sqrt{x + 1} = 5 - x$

20. $3\sqrt{x} - 1 = 2x$

Simplify each expression.

21. $8^{4/3}$

22. $-125^{2/3}$

23. $5^{3/4} \cdot 5^{1/4}$

24. $\dfrac{(3^{1/4})^3}{3^{7/4}}$

25. What is wrong with the following "solution"?

$$\sqrt{2x + 1} + 5 = 0$$
$$\sqrt{2x + 1} = -5 \qquad \text{Subtract 5.}$$
$$2x + 1 = 25 \qquad \text{Square both sides.}$$
$$2x = 24 \qquad \text{Subtract 1.}$$
$$x = 12 \qquad \text{Divide by 2.}$$

The solution set is $\{12\}$.

CUMULATIVE REVIEW EXERCISES CHAPTERS 1-10

Simplify each expression.

1. $3(6 + 7) + 6 \cdot 4 - 3^2$

2. $\dfrac{3(6 + 7) + 3}{2(4) - 1}$

3. $|-6| - |-3|$

4. $-9 + 14 + 11 + (-3 + 5)$

5. $13 - [-4 - (-2)]$

6. $-2.523 + 8.674 - 1.928$

Solve each equation or inequality.

7. $5(k - 4) - k = k - 11$

8. $-\dfrac{3}{4}y \le 12$

9. $5z + 3 - 4 > 2z + 9 + z$

Solve the problem.

10. A Macintosh Powerbook 1400C computer has dimensions 9 inches by 11.5 inches by 2 inches. Use the formula $V = LWH$ to find its volume (in cubic inches).

Graph each equation in the rectangular coordinate plane.

11. $-4x + 5y = -20$

12. $x = 2$

13. Find the slope of the line through the points $(9, -2)$ and $(-3, 8)$.

Simplify and write each expression without negative exponents. Assume that variables represent positive real numbers.

14. $(3x^6)(2x^2y)^2$

15. $\left(\dfrac{3^2y^{-2}}{2^{-1}y^3}\right)^{-3}$

16. Subtract $7x^3 - 8x^2 + 4$ from $10x^3 + 3x^2 - 9$.

17. Divide: $\dfrac{8t^3 - 4t^2 - 14t + 15}{2t + 3}$.

Factor each polynomial completely.

18. $m^2 + 12m + 32$

19. $25t^4 - 36$

20. $12a^2 + 4ab - 5b^2$

21. $81z^2 + 72z + 16$

Solve each quadratic equation.

22. $x^2 - 7x = -12$

23. $(x + 4)(x - 1) = -6$

24. For what real number(s) is the expression $\dfrac{3}{x^2 + 5x - 14}$ undefined?

Perform the indicated operations. Express each answer in lowest terms.

25. $\dfrac{x^2 - 3x - 4}{x^2 + 3x} \cdot \dfrac{x^2 + 2x - 3}{x^2 - 5x + 4}$

26. $\dfrac{t^2 + 4t - 5}{t + 5} \div \dfrac{t - 1}{t^2 + 8t + 15}$

27. $\dfrac{y}{y^2 - 1} + \dfrac{y}{y + 1}$

28. $\dfrac{2}{x + 3} - \dfrac{4}{x - 1}$

29. Simplify the complex fraction: $\dfrac{\dfrac{2}{3} + \dfrac{1}{2}}{\dfrac{1}{9} - \dfrac{1}{6}}$.

30. Graph $2x - 5y > 10$.

Solve each system of equations.

31. $4x - y = 19$
$3x + 2y = -5$

32. $2x - y = 6$
$3y = 6x - 18$

Solve each problem.

33. Des Moines and Chicago are 321 miles apart. Two cars start from these cities traveling toward each other. They meet after 3 hours. The car from Chicago has an average speed 7 miles per hour faster than the other car. Find the average speed of each car.

Des Moines   Chicago

34. The top two cable television networks in 1995 were ESPN and CNN. ESPN had .1 million more subscribers than CNN. Together, the two networks had 135.7 million subscribers. How many subscribers did each network have?

Simplify each expression if possible. Assume all variables represent nonnegative real numbers.

35. $\sqrt{27} - 2\sqrt{12} + 6\sqrt{75}$

36. $\dfrac{2}{\sqrt{3} + \sqrt{5}}$

37. $\sqrt{200x^2y^5}$

38. $16^{5/4}$

39. $(3\sqrt{2} + 1)(4\sqrt{2} - 3)$

40. Solve the equation $\sqrt{x} + 2 = x - 10$.

Quadratic Equations, Inequalities, and Graphs

11

Astronomy

Quadratic equations and quadratic functions are important in the field of astronomy. When an object moves under the influence of a constant force, its path is *parabolic* (has the shape of a parabola), and its position is described by a *quadratic equation* $y = ax^2 + bx$. Here x is in seconds, y is in feet, and a and b are real numbers. This equation varies on the surface of different planets and moons. The table compares the heights of a toy rocket propelled upward at 60 miles per hour (88 feet per second) at an angle of 60° on Earth, our moon, and Mars (as x varies from 1 to 5 seconds).

Seconds (x)	Height (y, in feet)		
	Earth	Moon	Mars
1	1.724	1.726	1.724
2	3.431	3.449	3.442
3	5.121	5.170	5.154
4	6.795	6.888	6.860
5	8.453	8.604	8.560

Source: M. Zeilik, S. Gregory, and E. Smith, *Introductory Astronomy and Astrophysics,* Saunders College Publishers, 1992.

The familiar force referred to earlier is gravity. Gravitational force is different on Earth, our moon, and Mars. It is less powerful on our moon than on Earth, and therefore, with all other conditions being equal, the rocket will rise faster on the moon than on our home planet. How does the table support this statement? Look at the figures and surmise how gravity on Mars compares to gravity on Earth and on our moon. We examine these ideas in the exercises for Section 11.1.

11.1 Solving Quadratic Equations by the Square Root Property

OBJECTIVES

1 Solve equations of the form $x^2 = $ a number.

2 Solve equations of the form $(ax + b)^2 = $ a number.

FOR EXTRA HELP

📖 SSG Sec. 11.1
   SSM Sec. 11.1

💿 Pass the Test Software

💿 InterAct Math
   Tutorial Software

📼 Video 18

In Chapter 6 we solved quadratic equations by factoring. However, since not all quadratic equations can easily be solved by factoring, it is necessary to develop other methods. In this chapter we will do just that.

Recall that a *quadratic equation* is an equation that can be written in the form

$$ax^2 + bx + c = 0$$

for real numbers a, b, and c, with $a \neq 0$. To solve $x^2 + 4x + 3 = 0$ by the zero-factor property, begin by factoring the left side and then set each factor equal to zero.

$$x^2 + 4x + 3 = 0$$

$$(x + 3)(x + 1) = 0 \qquad \text{Factor.}$$

$$x + 3 = 0 \quad \text{or} \quad x + 1 = 0 \qquad \text{Zero-factor property}$$

$$x = -3 \quad \text{or} \quad x = -1$$

The solution set is $\{-3, -1\}$.

OBJECTIVE 1 Solve equations of the form $x^2 = $ a number. To solve equations such as $x^2 = 9$, factor as follows.

$$x^2 = 9$$

$$x^2 - 9 = 0 \qquad \text{Subtract 9.}$$

$$(x + 3)(x - 3) = 0 \qquad \text{Factor.}$$

$$x + 3 = 0 \quad \text{or} \quad x - 3 = 0 \qquad \text{Zero-factor property}$$

$$x = -3 \quad \text{or} \quad x = 3$$

The solution set is $\{-3, 3\}$.

This result is generalized as the **square root property of equations.**

Square Root Property of Equations

If k is a positive number and if $a^2 = k$, then

$$a = \sqrt{k} \qquad \text{or} \qquad a = -\sqrt{k},$$

and the solution set is $\{-\sqrt{k}, \sqrt{k}\}$.

   When we solve an equation, we want to find *all* values of the variable that satisfy the equation. Therefore, we want both the positive and the negative square roots of k.

EXAMPLE 1   Solving a Quadratic Equation of the Form $x^2 = k$

Solve each equation. Write radicals in simplified form.
(a) $x^2 = 16$

By the square root property, since $x^2 = 16$,

$$x = \sqrt{16} = 4 \qquad \text{or} \qquad x = -\sqrt{16} = -4.$$

An abbreviation for "$x = 4$ or $x = -4$" is written $x = \pm 4$ and is read "x equals positive or negative 4." The solution set is $\{-4, 4\}$. Check each solution by substituting back into the original equation.

(b) $z^2 = 5$

The solutions are $z = \sqrt{5}$ or $z = -\sqrt{5}$, and the solution set may be written $\{\pm\sqrt{5}\}$.

(c) $m^2 = 8$

$$m^2 = 8$$

$$m = \sqrt{8} \qquad \text{or} \qquad m = -\sqrt{8} \qquad \text{Square root property}$$

$$m = 2\sqrt{2} \qquad \text{or} \qquad m = -2\sqrt{2} \qquad \text{Simplify } \sqrt{8}.$$

The solution set is $\{\pm 2\sqrt{2}\}$.

(d) $y^2 = -4$

Since -4 is a negative number and since the square of a real number cannot be negative, there is no real number solution for this equation. The solution set is $\emptyset$.

(e) $3x^2 + 5 = 11$

First solve the equation for x^2.

$$3x^2 + 5 = 11$$

$$3x^2 = 6 \qquad \text{Subtract 5.}$$

$$x^2 = 2 \qquad \text{Divide by 3.}$$

Now use the square root property to get the solution set $\{\pm\sqrt{2}\}$.

OBJECTIVE **2** Solve equations of the form $(ax + b)^2 = $ a number. In each of the equations in Example 1, the exponent 2 appeared with a single variable as its base. The square root property of equations can be extended to solve equations where the base is a binomial, as shown in the next example.

EXAMPLE 2 Solving a Quadratic Equation of the Form $(x + b)^2 = k$

Solve each equation.

(a) $(x - 3)^2 = 16$

Apply the square root property, using $x - 3$ as the base.

$$(x - 3)^2 = 16$$

$$x - 3 = \sqrt{16} \qquad \text{or} \qquad x - 3 = -\sqrt{16}$$

$$x - 3 = 4 \qquad \text{or} \qquad x - 3 = -4 \qquad \sqrt{16} = 4$$

$$x = 7 \qquad \text{or} \qquad x = -1 \qquad \text{Add 3.}$$

Check both answers in the original equation.

$$(x - 3)^2 = 16 \qquad\qquad\qquad (x - 3)^2 = 16$$

$$(7 - 3)^2 = 16 \quad ? \quad \text{Let } x = 7. \qquad (-1 - 3)^2 = 16 \quad ? \quad \text{Let } x = -1.$$

$$4^2 = 16 \quad ? \qquad\qquad\qquad (-4)^2 = 16 \quad ?$$

$$16 = 16 \qquad \text{True} \qquad\qquad\qquad 16 = 16 \qquad \text{True}$$

The solution set is $\{7, -1\}$.

(b) $(x - 1)^2 = 6$

By the square root property,

$$x - 1 = \sqrt{6} \qquad\qquad \text{or} \qquad x - 1 = -\sqrt{6}$$

$$x = 1 + \sqrt{6} \qquad \text{or} \qquad x = 1 - \sqrt{6}.$$

Check: $\qquad (1 + \sqrt{6} - 1)^2 = (\sqrt{6})^2 = 6;$

$\qquad\qquad\qquad\qquad (1 - \sqrt{6} - 1)^2 = (-\sqrt{6})^2 = 6.$

The solution set is $\{1 + \sqrt{6}, 1 - \sqrt{6}\}$.

The solutions in Example 2(b) may be written in abbreviated form as

$$1 \pm \sqrt{6}.$$

If they are written this way, keep in mind that *two* solutions are indicated, one with the $+$ sign and the other with the $-$ sign.

E X A M P L E   3   Solving a Quadratic Equation of the Form $(ax + b)^2 = k$

Solve $(3r - 2)^2 = 27$.

$3r - 2 = \sqrt{27}$	or	$3r - 2 = -\sqrt{27}$	Square root property
$3r - 2 = 3\sqrt{3}$	or	$3r - 2 = -3\sqrt{3}$	$\sqrt{27} = \sqrt{9 \cdot 3} = 3\sqrt{3}$
$3r = 2 + 3\sqrt{3}$	or	$3r = 2 - 3\sqrt{3}$	Add 2.
$r = \dfrac{2 + 3\sqrt{3}}{3}$	or	$r = \dfrac{2 - 3\sqrt{3}}{3}$	Divide by 3.

The solution set is $\left\{\dfrac{2 \pm 3\sqrt{3}}{3}\right\}$.

 The solutions in Example 3 are fractions that cannot be reduced, since 3 is *not* a common factor in the numerator.

E X A M P L E   4   Recognizing a Quadratic Equation with No Real Solution

Solve $(x + 3)^2 = -9$.

The square root of -9 is not a real number. The solution set is $\emptyset$.

11.1 EXERCISES

Match each equation with the correct description of its solutions.

1. $x^2 = 0$ **A.** No real number solutions

2. $x^2 = 10$ **B.** Two integer solutions

3. $x^2 = -4$ **C.** One real solution

4. $x^2 = \dfrac{9}{16}$ **D.** Two irrational solutions

5. $x^2 = 9$ **E.** Two rational solutions that are not integers

Decide whether each statement is true or false. If it is false, tell why.

6. If k is a prime number, then $x^2 = k$ has two irrational solutions.

7. If k is a positive perfect square, then $x^2 = k$ has two rational solutions.

8. If k is a positive integer, then $x^2 = k$ must have two rational solutions.

9. If $-10 < k < 0$, then $x^2 = k$ has no real solutions.

10. If $-10 < k < 10$, then $x^2 = k$ has no real solutions.

11. When a student was asked to solve $x^2 = 81$, she wrote {9} as her answer. Her teacher did not give her full credit. The student argued that because $9^2 = 81$, her answer had to be correct. Why was her answer not completely correct?

12. Explain the square root property for solving equations, and illustrate with an example.

Solve each equation by using the square root property. Express all radicals in simplest form. See Example 1.

13. $x^2 = 81$ **14.** $y^2 = 121$ **15.** $k^2 = 14$ **16.** $m^2 = 22$

17. $t^2 = 48$ **18.** $x^2 = 54$ **19.** $y^2 = -100$ **20.** $m^2 = -64$

21. $z^2 = 2.25$ **22.** $w^2 = 56.25$ **23.** $3x^2 - 8 = 64$ **24.** $2t^2 + 7 = 61$

Solve each equation by using the square root property. Express all radicals in simplest form. See Examples 2–4.

25. $(x - 3)^2 = 25$ **26.** $(y - 7)^2 = 16$ **27.** $(z + 5)^2 = -13$

28. $(m + 2)^2 = -17$ **29.** $(x - 8)^2 = 27$ **30.** $(y - 5)^2 = 40$

31. $(3k + 2)^2 = 49$ **32.** $(5t + 3)^2 = 36$ **33.** $(4x - 3)^2 = 9$

34. $(7y - 5)^2 = 25$ **35.** $(5 - 2x)^2 = 30$ **36.** $(3 - 2a)^2 = 70$

37. $(3k + 1)^2 = 18$ **38.** $(5z + 6)^2 = 75$ **39.** $\left(\frac{1}{2}x + 5\right)^2 = 12$

40. $\left(\frac{1}{3}y + 4\right)^2 = 27$ **41.** $(4k - 1)^2 - 48 = 0$ **42.** $(2s - 5)^2 - 180 = 0$

43. Johnny solved the equation in Exercise 35 and wrote his answer as $\left\{\dfrac{5 + \sqrt{30}}{2}, \dfrac{5 - \sqrt{30}}{2}\right\}$.

Linda solved the same equation and wrote her answer as $\left\{\dfrac{-5 + \sqrt{30}}{-2}, \dfrac{-5 - \sqrt{30}}{-2}\right\}$. The teacher gave them both full credit. Explain why both students were correct, although their answers look different.

44. In the solutions found in Example 3 of this section, why is it not valid to reduce the answers by dividing out the threes in the numerator and denominator?

 Use a calculator with a square root key to solve each equation. Round your answers to the nearest hundredth.

45. $(k + 2.14)^2 = 5.46$ **46.** $(r - 3.91)^2 = 9.28$

47. $(2.11p + 3.42)^2 = 9.58$ **48.** $(1.71m - 6.20)^2 = 5.41$

RELATING CONCEPTS (EXERCISES 49–54)

In Section 6.4 we saw how certain trinomials can be factored as squares of binomials.

*Use this idea and **work Exercises 49–54 in order**, considering the equation*

$$x^2 + 6x + 9 = 100.$$

49. Factor the left side of the equation as the square of a binomial, and write the resulting equation.

50. Write the equation from Exercise 49 as a compound statement using the word "or."

51. Solve the two individual equations from Exercise 50.

52. What is the solution set of the original equation?

53. Solve the equation $x^2 + 4x + 4 = 25$ using the method described in Exercises 49–52.

54. Solve the equation $4k^2 - 12k + 9 = 81$ using the method described in Exercises 49–52.

Did you make the connection that when the trinomial factors are the square of a binomial, the square root property of equations can be extended in scope?

When an object moves under the influence of a constant force, its path is parabolic, and its position can be described by a quadratic equation $y = ax^2 + bx$, where x is in seconds, y is in feet, and a and b are real numbers. This equation varies on the surface of different celestial objects. Use this fact to work Exercises 55–58. (Source: M. Zeilik, S. Gregory, and E. Smith, *Introductory Astronomy and Astrophysics, Saunders College Publishers, 1992.*)

55. If an object is propelled upward on Earth at an angle of 45° with an initial velocity of 30 miles per hour, then the equation is $y = -.017x^2 + x$. Find the height of the object in feet after 1 second. (*Hint:* Let $x = 1$ and solve for y.)

56. If an object is propelled upward on Mars at an angle of 45° with an initial velocity of 30 miles per hour, then the equation is $y = -.007x^2 + x$. Find the height of the object in feet after 1 second.

57. If an object is propelled upward on Mars at an angle of 60° with an initial velocity of 60 miles per hour, then the equation is $y = -.003x^2 + 1.727x$. Find the height of the object in feet after 2 seconds.

58. If an object is propelled upward on the moon at an angle of 60° with an initial velocity of 60 miles per hour, then the equation is $y = -.0013x^2 + 1.7273x$. Find the height of the object in feet after 2 seconds.

Solve each problem.

59. One expert at marksmanship can hold a silver dollar at forehead level, drop it, draw his gun, and shoot the coin as it passes waist level. The distance traveled by a falling object is given by

$$d = 16t^2,$$

where d is the distance (in feet) the object falls in t seconds. If the coin falls about 4 feet, use the formula to estimate the time that elapses between the dropping of the coin and the shot.

60. The illumination produced by a light source depends on the distance from the source. For a particular light source, this relationship can be expressed as

$$d^2 = \frac{4050}{I},$$

where d is the distance from the source (in feet) and I is the amount of illumination in foot-candles. How far from the source is the illumination equal to 50 foot-candles?

61. The area A of a circle with radius r is given by the formula

$$A = \pi r^2.$$

If a circle has area 81π square inches, what is its radius?

62. The surface area S of a sphere with radius r is given by the formula

$$S = 4\pi r^2.$$

If a sphere has surface area 36π square feet, what is its radius?

The amount A that P dollars invested at an annual rate of interest r will grow to in 2 years is

$$A = P(1 + r)^2.$$

63. At what interest rate will $100 grow to $110.25 in two years?

64. At what interest rate will $500 grow to $572.45 in two years?

11.2 Solving Quadratic Equations by Completing the Square

OBJECTIVES

1. Solve quadratic equations by completing the square when the coefficient of the squared term is 1.

2. Solve quadratic equations by completing the square when the coefficient of the squared term is not 1.

3. Simplify an equation before solving.

FOR EXTRA HELP

SSG Sec. 11.2
SSM Sec. 11.2

Pass the Test Software

InterAct Math Tutorial Software

Video 18

OBJECTIVE 1 Solve quadratic equations by completing the square when the coefficient of the squared term is 1. The methods studied so far are not enough to solve the equation

$$x^2 + 6x + 7 = 0.$$

If we could write the equation in the form $(x + b)^2 = k$, we could solve it with the square root property discussed in the previous section. To do that, we need to have a perfect square trinomial on one side of the equation. The next example shows how this is done.

EXAMPLE 1 Rewriting an Equation to Use the Square Root Property

Solve $x^2 + 6x + 7 = 0$.

Start by subtracting 7 from both sides of the equation to get $x^2 + 6x = -7$. To write $x^2 + 6x = -7$ in the form $(x + b)^2 = k$, the quantity on the left-hand side of $x^2 + 6x = -7$ must be made into a perfect square trinomial. The expression $x^2 + 6x + 9$ is a perfect square, since

$$x^2 + 6x + 9 = (x + 3)^2.$$

Therefore, if we add 9 to both sides of $x^2 + 6x = -7$, the equation will have a perfect square trinomial on one side, as needed.

$$x^2 + 6x + 9 = -7 + 9 \qquad \text{Add 9 to both sides.}$$
$$(x + 3)^2 = 2 \qquad \text{Factor; combine terms.}$$

Now use the square root property to complete the solution.

$$x + 3 = \sqrt{2} \qquad \text{or} \qquad x + 3 = -\sqrt{2}$$
$$x = -3 + \sqrt{2} \qquad \text{or} \qquad x = -3 - \sqrt{2}$$

The solution set is $\{-3 \pm \sqrt{2}\}$. Check this by substituting $-3 + \sqrt{2}$ and $-3 - \sqrt{2}$ for x in the original equation.

The process of changing the form of the equation in Example 1 from

$$x^2 + 6x + 7 = 0 \qquad \text{to} \qquad (x + 3)^2 = 2$$

is called **completing the square.** Completing the square changes only the form of the equation. To see this, multiply out the left side of $(x + 3)^2 = 2$ and combine terms. Then subtract 2 from both sides to see that the result is $x^2 + 6x + 7 = 0$.

Look again at the original equation

$$x^2 + 6x + 7 = 0$$

in Example 1. Note that 9 is the square of half the coefficient of x, 6.

$$\underset{\underset{\text{Coefficient of } x}{\uparrow}}{\frac{1}{2} \cdot 6} = 3 \qquad \text{and} \qquad 3^2 = 9$$

So to complete the square in Example 1, 9 was added to each side.

EXAMPLE 2 Completing the Square to Solve a Quadratic Equation

Complete the square to solve $x^2 - 8x = 5$.

To complete the square on $x^2 - 8x$, take half the coefficient of x and square it.

$$\underset{\underset{\text{Coefficient of } x}{\uparrow}}{\frac{1}{2}(-8)} = -4 \qquad \text{and} \qquad (-4)^2 = 16$$

Add the result, 16, to both sides of the equation.

$$\begin{aligned} x^2 - 8x &= 5 & \text{Given equation} \\ x^2 - 8x + \mathbf{16} &= 5 + \mathbf{16} & \text{Add 16 to both sides.} \\ (x - 4)^2 &= 21 & \text{Factor the left side as the square of a binomial.} \end{aligned}$$

Now apply the square root property.

$$\begin{aligned} x - 4 &= \pm\sqrt{21} & \text{Square root property} \\ x &= 4 \pm \sqrt{21} & \text{Add 4 to both sides.} \end{aligned}$$

A check indicates that the solution set is $\{4 \pm \sqrt{21}\}$.

To solve a quadratic equation by completing the square, follow the steps given below.

Solving a Quadratic Equation by Completing the Square

Step 1 **Be sure the squared term has coefficient 1.** If the coefficient of the squared term is 1, proceed to Step 2. If the coefficient of the squared term is not 1 but some other nonzero number a, divide both sides of the equation by a. This gives an equation that has 1 as coefficient of the squared term.

Step 2 **Put in correct form.** Make sure that all terms with variables are on one side of the equals sign and that all constants are on the other side.

Step 3 **Complete the square.** Take half the coefficient of the first-degree term, and square the result. Add the square to both sides of the equation. The side containing the variables now can be factored as a perfect square.

Step 4 **Solve.** Apply the square root property to solve the equation.

OBJECTIVE 2 Solve quadratic equations by completing the square when the coefficient of the squared term is not 1. To complete the square, the coefficient of the squared term must be 1 (Step 1). The next example shows what to do when this coefficient is not 1.

┌ **E X A M P L E   3**   Solving a Quadratic Equation by Completing the Square

Solve $4y^2 + 24y - 13 = 0$.

Step 1   Divide each side by 4 so that the coefficient of y^2 is 1.

$$4y^2 + 24y - 13 = 0$$

$$y^2 + 6y - \frac{13}{4} = 0$$

Step 2   Add $\frac{13}{4}$ to each side to get the variable terms on the left and the constant on the right.

$$y^2 + 6y = \frac{13}{4}$$

Step 3   To complete the square, take half the coefficient of y, $\frac{1}{2}(6) = 3$, and square the result: $3^2 = 9$. Add 9 to both sides of the equation, add on the right-hand side, and factor on the left.

$$y^2 + 6y + 9 = \frac{13}{4} + 9 \qquad \text{Add 9.}$$

$$y^2 + 6y + 9 = \frac{49}{4} \qquad \text{Add on the right.}$$

$$(y + 3)^2 = \frac{49}{4} \qquad \text{Factor on the left.}$$

Step 4   Use the square root property and solve for y.

$$y + 3 = \frac{7}{2} \qquad \text{or} \qquad y + 3 = -\frac{7}{2} \qquad \text{Square root property}$$

$$y = -3 + \frac{7}{2} \qquad \text{or} \qquad y = -3 - \frac{7}{2} \qquad \text{Add } -3.$$

$$y = \frac{1}{2} \qquad \text{or} \qquad y = -\frac{13}{2}$$

The solution set is $\left\{\frac{1}{2}, -\frac{13}{2}\right\}$. Check by substitution into the original equation.

┌ **E X A M P L E   4**   Solving a Quadratic Equation by Completing the Square

Solve $4p^2 + 8p + 5 = 0$.

First divide both sides by 4 to get a coefficient of 1 for the p^2-term (Step 1).

$$p^2 + 2p + \frac{5}{4} = 0$$

Add $-\frac{5}{4}$ to both sides (Step 2).

$$p^2 + 2p = -\frac{5}{4}$$

The coefficient of p is 2. Take half of 2, square the result, and add it to both sides. The left-hand side can then be factored as a perfect square (Step 3).

$$p^2 + 2p + 1 = -\frac{5}{4} + 1 \qquad [\tfrac{1}{2}(2)]^2 = 1$$

$$(p + 1)^2 = -\frac{1}{4} \qquad \text{Factor; combine terms.}$$

The square root of $-\frac{1}{4}$ is not a real number, so the solution set is $\emptyset$.

OBJECTIVE **3** Simplify an equation before solving. Sometimes an equation must be simplified before completing the square. The next example illustrates this.

EXAMPLE 5 Simplifying an Equation Before Completing the Square
Solve $(x + 3)(x - 1) = 2$.

$$(x + 3)(x - 1) = 2$$
$$x^2 + 2x - 3 = 2 \qquad \text{Use FOIL.}$$
$$x^2 + 2x = 5 \qquad \text{Add 3 to both sides.}$$
$$x^2 + 2x + 1 = 5 + 1 \qquad \text{Add 1 to get a perfect square on the left.}$$
$$(x + 1)^2 = 6 \qquad \text{Factor on the left; add on the right.}$$
$$x + 1 = \sqrt{6} \qquad \text{or} \qquad x + 1 = -\sqrt{6} \qquad \text{Square root property}$$
$$x = -1 + \sqrt{6} \qquad \text{or} \qquad x = -1 - \sqrt{6} \qquad \text{Add } -1.$$

The solution set is $\{-1 \pm \sqrt{6}\}$.

> The solutions given in Example 5 are *exact*. In applications, decimal solutions are more appropriate. Using the square root key of a calculator, $\sqrt{6} \approx 2.449$. Evaluating the two solutions gives
>
> $$x \approx 1.449 \qquad \text{and} \qquad x \approx -3.449.$$

CONNECTIONS

"Completing the square" has so far been applied only in an algebraic sense. We add an appropriate constant to a binomial to get a perfect square trinomial and then factor so that we can use the square root property. However, this procedure can literally be applied to a geometric figure so that it becomes a square. This gives a beautiful connection between algebra and geometry.

For example, to complete the square for $x^2 + 8x$, begin with a square having a side of length x. Add four rectangles of width 1 to the right side and to the bottom. To "complete the square" fill in the bottom right corner with 16 squares of area 1.

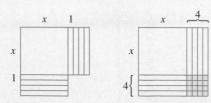

CONNECTIONS (CONTINUED)

FOR DISCUSSION OR WRITING

1. What is the area of the original square?
2. What is the area of the figure after the 8 rectangles are added?
3. What is the area of the figure after the 16 small squares are added?
4. At what point did we "complete the square"?

11.2 EXERCISES

Fill in each blank with the correct response.

1. To solve the equation $x^2 - 8x = 4$ by completing the square, the first step is to add _____ to both sides of the equation.

2. To solve the equation $3x^2 - 8x - 1 = 0$ by completing the square, the first step is to divide both sides of the equation by _____.

3. To solve $(t + 2)(t - 5) = 18$ by completing the square, we should start by _____.

4. It is not possible to solve $x^3 - x - 1 = 0$ by completing the square because _____.

5. Which one of the following steps is an appropriate way to begin solving the quadratic equation

$$2x^2 - 4x = 9$$

by completing the square?
   (a) Add 4 to both sides of the equation.   (b) Factor the left side as $2x(x - 2)$.
   (c) Factor the left side as $x(2x - 4)$.   (d) Divide both sides by 2.

6. In Example 3 of Section 6.5, we solved the quadratic equation

$$4p^2 - 26p + 40 = 0$$

by factoring. If we were to solve by completing the square, would we get the same solution set, $\left\{ \dfrac{5}{2}, 4 \right\}$?

Find the constant that should be added to each expression to make it a perfect square. See Examples 1 and 2.

7. $y^2 + 14y$ 8. $z^2 + 18z$ 9. $k^2 - 5k$

10. $m^2 - 9m$ 11. $r^2 + \dfrac{1}{2}r$ 12. $s^2 - \dfrac{1}{3}s$

Solve each equation by completing the square. See Examples 1 and 2.

13. $x^2 - 4x = -3$ 14. $y^2 - 2y = 8$ 15. $x^2 + 2x - 5 = 0$
16. $r^2 + 4r + 1 = 0$ 17. $z^2 + 6z + 9 = 0$ 18. $k^2 - 8k + 16 = 0$

Solve each equation by completing the square. See Examples 3–5.

19. $4y^2 + 4y = 3$ 20. $9x^2 + 3x = 2$ 21. $2p^2 - 2p + 3 = 0$
22. $3q^2 - 3q + 4 = 0$ 23. $3k^2 + 7k = 4$ 24. $2k^2 + 5k = 1$
25. $(x + 3)(x - 1) = 5$ 26. $(y - 8)(y + 2) = 24$ 27. $-x^2 + 2x = -5$
28. $-x^2 + 4x = 1$

Solve each equation by completing the square. Then **(a)** give the exact solutions and **(b)** give the solutions rounded to the nearest thousandth.

29. $3r^2 - 2 = 6r + 3$

30. $4p + 3 = 2p^2 + 2p$

31. $(x + 1)(x + 3) = 2$

32. $(x - 3)(x + 1) = 1$

33. In using the method of completing the square to solve $2x^2 - 10x = -8$, a student began by adding the square of half the coefficient of x (that is, $\left[\frac{1}{2}(-10)\right]^2 = 25$) to both sides of the equation. He then encountered difficulty in his later steps. What was his error? Explain the steps needed to solve the problem, and give the solution set.

34. The equation $x^4 - 2x^2 = 8$ can be solved by completing the square, even though it is not a quadratic equation. What number should be added to both sides of the equation so that the left side can be factored as the square of a binomial? Solve the equation for its real solutions.

Solve each problem.

35. A farmer has a rectangular cattle pen with perimeter 350 feet and area 7500 square feet. What are the dimensions of the pen? (*Hint:* Use the figure to set up the equation.)

36. The base of a triangle measures 1 meter more than three times the height of the triangle. Its area is 15 square meters. Find the lengths of the base and the height.

$175 - x$

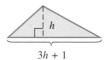

$3h + 1$

37. If an object is propelled upward from ground level on Earth with an initial velocity of 96 feet per second, its height, s (in feet), in t seconds is given by the formula $s = -16t^2 + 96t$. How long will it take for the object to be at a height of 80 feet? (*Hint:* Let $s = 80$.)

38. How long will it take the object described in Exercise 37 to be at a height of 100 feet? Round your answers to the nearest tenth.

39. If an object is propelled upward on the surface of Mars from ground level with an initial velocity of 104 feet per second, its height s (in feet) in t seconds is given by the formula $s = -13t^2 + 104t$. How long will it take for the object to be at a height of 195 feet?

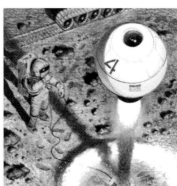

40. How long will it take the object in Exercise 39 to return to the surface? (*Hint:* When it returns to the surface, $s = 0$.)

41. Two cars travel at right angles to each other from an intersection until they are 17 miles apart. At that point one car has gone 7 miles farther than the other. How far did the slower car travel? (*Hint:* Use the Pythagorean formula.)

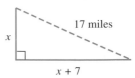

17 miles

x

$x + 7$

42. Two painters are painting a house in a development of new homes. One of the painters takes 2 hours longer to paint a house working alone than the other painter. When they do the job together, they can complete it in 4.8 hours. How long would it take the faster painter alone to paint the house? (Give your answer to the nearest tenth.)

11.3 Solving Quadratic Equations by the Quadratic Formula

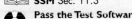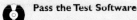

OBJECTIVES

1. Identify the values of a, b, and c in a quadratic equation.

2. Use the quadratic formula to solve quadratic equations.

3. Solve quadratic equations with only one solution.

4. Solve quadratic equations with fractions.

5. Solve an applied problem that leads to a quadratic equation.

6. Use the discriminant to determine the number and type of solutions.

FOR EXTRA HELP

📖 **SSG** Sec. 11.3
SSM Sec. 11.3

💿 **Pass the Test Software**

💿 **InterAct Math Tutorial Software**

📼 **Video** 18

Completing the square can be used to solve any quadratic equation, but the method is tedious. In this section we complete the square on the general quadratic equation $ax^2 + bx + c = 0$ to get the *quadratic formula*, a formula that gives the solution for any quadratic equation. (Note that $a \neq 0$, or we would have a linear, not a quadratic, equation.)

OBJECTIVE 1 Identify the values of a, b, and c in a quadratic equation. The first step in solving a quadratic equation by this new method is to identify the values of a, b, and c in the standard form of the quadratic equation.

EXAMPLE 1 Determining Values of a, b, and c in a Quadratic Equation

For each of the following quadratic equations, write the equation in standard form if necessary, and then identify the values of a, b, and c.

(a) $2x^2 + 3x - 5 = 0$

This equation is already in standard form. The values of a, b, and c are

$$a = 2, \qquad b = 3, \qquad \text{and} \qquad c = -5.$$

(b) $-x^2 + 2 = 6x$

First rewrite the equation with 0 on the right side to match the standard form $ax^2 + bx + c = 0$.

$$-x^2 + 2 = 6x$$
$$-x^2 - 6x + 2 = 0$$

Now identify $a = -1$, $b = -6$, and $c = 2$. (Notice that the coefficient of x^2 is understood to be -1.)

(c) $(2x - 7)(x + 4) = -23$

$$(2x - 7)(x + 4) = -23$$
$$2x^2 + x - 28 = -23 \qquad \text{Use FOIL on the left.}$$
$$2x^2 + x - 5 = 0 \qquad \text{Add 23 on each side.}$$

Now, identify the values: $a = 2$, $b = 1$, $c = -5$.

OBJECTIVE 2 Use the quadratic formula to solve quadratic equations. To develop the quadratic formula, we follow the steps for completing the square on $ax^2 + bx + c = 0$ given in the previous section. For comparison, we also show the corresponding steps for solving $2x^2 + x - 5 = 0$ (from Example 1(c)).

Step 1 Make the coefficient of the squared term equal to 1.

$$2x^2 + x - 5 = 0 \qquad\qquad\qquad ax^2 + bx + c = 0$$

$$x^2 + \frac{1}{2}x - \frac{5}{2} = 0 \quad \text{Divide by 2.} \qquad\qquad x^2 + \frac{b}{a}x + \frac{c}{a} = 0 \quad \text{Divide by } a.$$

Step 2 Get the variable terms alone on the left side.

$$x^2 + \frac{1}{2}x = \frac{5}{2} \quad \text{Add } \tfrac{5}{2}. \qquad\qquad x^2 + \frac{b}{a}x = -\frac{c}{a} \quad \text{Subtract } \tfrac{c}{a}.$$

Step 3 Add the square of half the coefficient of x to both sides, factor the left side, and combine terms on the right.

$$x^2 + \frac{1}{2}x + \frac{1}{16} = \frac{5}{2} + \frac{1}{16} \quad \text{Add } \tfrac{1}{16}. \qquad x^2 + \frac{b}{a}x + \frac{b^2}{4a^2} = -\frac{c}{a} + \frac{b^2}{4a^2} \quad \text{Add } \tfrac{b^2}{4a^2}.$$

$$\left(x + \frac{1}{4}\right)^2 = \frac{41}{16} \quad \substack{\text{Factor; add}\\\text{on right.}} \qquad \left(x + \frac{b}{2a}\right)^2 = \frac{b^2 - 4ac}{4a^2} \quad \substack{\text{Factor; add}\\\text{on right.}}$$

Step 4 Use the square root property to complete the solution.

$$x + \frac{1}{4} = \pm\sqrt{\frac{41}{16}} \qquad\qquad x + \frac{b}{2a} = \pm\sqrt{\frac{b^2 - 4ac}{4a^2}}$$

$$x + \frac{1}{4} = \pm\frac{\sqrt{41}}{4} \qquad\qquad x = -\frac{b}{2a} \pm \frac{\sqrt{b^2 - 4ac}}{2a}$$

$$x = -\frac{1}{4} \pm \frac{\sqrt{41}}{4} \qquad\qquad x = \frac{-b \pm \sqrt{b^2 - 4ac}}{2a}$$

$$x = \frac{-1 \pm \sqrt{41}}{4}$$

The final result in the column on the right is called the quadratic formula. *It is a key result that should be memorized.* Notice that there are two values, one for the $+$ sign and one for the $-$ sign.

Quadratic Formula

The solutions of the quadratic equation $ax^2 + bx + c = 0$, $a \neq 0$, are

$$x = \frac{-b + \sqrt{b^2 - 4ac}}{2a} \quad \text{and} \quad x = \frac{-b - \sqrt{b^2 - 4ac}}{2a}$$

or, in compact form,

$$x = \frac{-b \pm \sqrt{b^2 - 4ac}}{2a}.$$

 Notice that the fraction bar is under $-b$ as well as the radical. When using this formula, be sure to find the values of $-b \pm \sqrt{b^2 - 4ac}$ first, then divide those results by the value of $2a$.

E X A M P L E   2   Solving a Quadratic Equation by the Quadratic Formula

Solve $2x^2 + x - 5 = 0$ by the quadratic formula.

As found in Example 1(c), the values of a, b, and c are $a = 2$, $b = 1$, and $c = -5$. Substitute these numbers into the quadratic formula and simplify the result.

$$x = \frac{-b \pm \sqrt{b^2 - 4ac}}{2a}$$

$$x = \frac{-1 \pm \sqrt{(1)^2 - 4(2)(-5)}}{2(2)} \qquad \text{Let } a = 2, b = 1, c = -5.$$

$$x = \frac{-1 \pm \sqrt{1 + 40}}{4} \qquad \begin{array}{l}\text{Perform the operations under the} \\ \text{radical and in the denominator.}\end{array}$$

$$x = \frac{-1 \pm \sqrt{41}}{4}$$

Notice that this result agrees with the result obtained in the left column when the quadratic formula was derived above. The solution set is $\left\{ \dfrac{-1 \pm \sqrt{41}}{4} \right\}$.

E X A M P L E   3   Rewriting an Equation Before Using the Quadratic Formula

Solve $x^2 = 2x + 1$.

One side of the equation must be 0 before a, b, and c can be found. Subtract $2x$ and 1 from both sides of the equation to get

$$x^2 - 2x - 1 = 0.$$

Then $a = 1$, $b = -2$, and $c = -1$. Substitute these values into the quadratic formula.

$$x = \frac{-b \pm \sqrt{b^2 - 4ac}}{2a}$$

$$x = \frac{-(-2) \pm \sqrt{(-2)^2 - 4(1)(-1)}}{2(1)} \qquad \text{Let } a = 1, b = -2, c = -1.$$

$$x = \frac{2 \pm \sqrt{4 + 4}}{2} = \frac{2 \pm \sqrt{8}}{2}$$

$$x = \frac{2 \pm 2\sqrt{2}}{2} \qquad \sqrt{8} = \sqrt{4} \cdot \sqrt{2} = 2\sqrt{2}$$

Write the solutions in lowest terms by factoring $2 \pm 2\sqrt{2}$ as $2(1 \pm \sqrt{2})$ to get

$$x = \frac{2(1 \pm \sqrt{2})}{2} = 1 \pm \sqrt{2}.$$

The solution set is $\{1 \pm \sqrt{2}\}$.

O B J E C T I V E   3   **Solve quadratic equations with only one solution.**   The expression under the radical in the quadratic formula, $b^2 - 4ac$, is called the **discriminant.** The discriminant provides information about the number of real solutions of a quadratic equation. If the discriminant is positive, as in Examples 2 and 3, the equation has two real solutions. If it is zero, it has only one real solution (sometimes called a *double solution*). A negative discriminant indicates that there are no real solutions. The next example illustrates a quadratic equation with discriminant 0.

E X A M P L E   4   Using the Quadratic Formula When There Is One Solution

Solve $4x^2 + 25 = 20x$.

Write the equation as $4x^2 - 20x + 25 = 0$. Here, $a = 4$, $b = -20$, and $c = 25$. By the quadratic formula,

$$x = \frac{-(-20) \pm \sqrt{(-20)^2 - 400}}{8} = \frac{20 \pm 0}{8} = \frac{5}{2}.$$

The solution set is $\left\{ \frac{5}{2} \right\}$.

If the discriminant is 0, as in Example 4, then the quadratic trinomial is a perfect square, and the equation could have been solved by factoring. In Example 4, the expression $4x^2 - 20x + 25$ would be factored as $(2x - 5)^2$. Setting this equal to zero leads to the double solution $\frac{5}{2}$.

O B J E C T I V E   4   Solve quadratic equations with fractions. It is usually easier to clear quadratic equations of fractions before solving them, as shown in the next example.

E X A M P L E   5   Solving a Quadratic Equation with Fractions

Solve the equation $\dfrac{1}{10}t^2 = \dfrac{2}{5}t - \dfrac{1}{2}$.

Eliminate the denominators by multiplying both sides of the equation by the least common denominator, 10.

$$10\left(\frac{1}{10}t^2\right) = 10\left(\frac{2}{5}t - \frac{1}{2}\right)$$

$$t^2 = 4t - 5 \qquad \text{Distributive property}$$

$$t^2 - 4t + 5 = 0 \qquad \text{Standard form}$$

From this form identify $a = 1$, $b = -4$, and $c = 5$. Use the quadratic formula to complete the solution.

$$t = \frac{-(-4) \pm \sqrt{(-4)^2 - 4(1)(5)}}{2(1)} \qquad \text{Substitute into the formula.}$$

$$t = \frac{4 \pm \sqrt{16 - 20}}{2} \qquad \text{Perform the operations.}$$

$$t = \frac{4 \pm \sqrt{-4}}{2}$$

The discriminant -4 is less than 0. Because $\sqrt{-4}$ does not represent a real number, the solution set is $\emptyset$.

OBJECTIVE 5 Solve an applied problem that leads to a quadratic equation. The German astronomer Johannes Kepler (1571–1630) is responsible for discovering three laws of planetary motion. In 1619 he published his third law, which we examine in the next example. To understand this law, several terms must be defined. The *sidereal period* of a planet is the orbital period of a planet relative to the stars, or the time from one orbital point back to the same point. The *semimajor axis* of the elliptical orbit of a planet is half the length of the major (longer) axis of an ellipse. (See Figure 1.) One *astronomical unit* (AU) is the average distance between Earth and the sun.

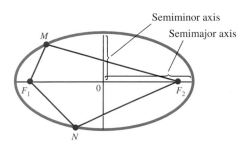

Figure 1

The orbit of Earth, as well as other planets in our solar system, takes the shape of an ellipse. In any ellipse, the sum of the distances of any point on the curve to each *focus* (labeled F_1 and F_2 in the figure) is constant. Thus $MF_1 + MF_2 = NF_1 + NF_2$, and so on for any point on the ellipse. The orbit of Earth has the sun at one focus.

EXAMPLE 6 Determining the Orbital Period *P* of Jupiter

Kepler's third law of planetary motion states that the square of the sidereal period of a planet about the sun is proportional to the cube of its orbital semimajor axis. Relative to Earth's orbit, this harmonic law is usually expressed by the equation

$$P^2 = a^3,$$

where P is the planet's sidereal period in years and a is its orbital semimajor axis in AU. For Jupiter, $a = 5.20$. Find P. (*Source:* James B. Kaler, *Astronomy! A Brief Edition*, Addison Wesley, 1997.)

Substitute 5.20 for a in the formula to get

$$P^2 = (5.20)^3$$
$$P^2 = 140.608. \qquad \text{Use a calculator.}$$

This is a quadratic equation in P. Apply the square root property. Using a calculator to find the positive square root of P to the nearest hundredth, we get

$$P = 11.86.$$

Thus, the sidereal period P for Jupiter is 11.86 years. (The negative square root, -11.86, is rejected since the period P must be a positive number.)

When we use a quadratic equation to solve an applied problem, it is important to check that the solution satisfies the physical requirements for the unknown. For example, it is impossible to have a negative length or a fractional number of people.

EXAMPLE 7 Solving an Applied Problem Using the Quadratic Formula

If an object is thrown upward on Earth from a height of 50 feet, with an initial velocity of 32 feet per second, then its height after t seconds is given by

$$h = -16t^2 + 32t + 50, \quad \text{where } h \text{ is in feet.}$$

After how many seconds will it reach a height of 30 feet?

We must find the value of t for which $h = 30$.

$$30 = -16t^2 + 32t + 50 \qquad \text{Let } h = 30.$$
$$16t^2 - 32t - 20 = 0 \qquad \text{Rewrite in standard form.}$$
$$4t^2 - 8t - 5 = 0 \qquad \text{Divide by 4.}$$

Use the quadratic formula, with $a = 4$, $b = -8$, and $c = -5$.

$$t = \frac{-b \pm \sqrt{b^2 - 4ac}}{2a}$$

$$t = \frac{-(-8) \pm \sqrt{(-8)^2 - 4(4)(-5)}}{2(4)}$$

$$t = \frac{8 \pm \sqrt{64 + 80}}{8}$$

$$t = \frac{8 \pm \sqrt{144}}{8} = \frac{8 \pm 12}{8}$$

The two solutions of the equation are

$$\frac{8 + 12}{8} = \frac{20}{8} = \frac{5}{2} \quad \text{and} \quad \frac{8 - 12}{8} = -\frac{4}{8} = -\frac{1}{2}.$$

Since t represents time, the solution $-\frac{1}{2}$ must be rejected, as time cannot be negative here. The object will reach a height of 30 feet after $\frac{5}{2}$, or 2.5, seconds.

OBJECTIVE 6 Use the discriminant to determine the number and type of solutions. In Example 5, the equation had no real solution because the quadratic formula led to an expression with $\sqrt{-4}$. Nonreal numbers of this type are called *imaginary numbers*. We will discuss them further in the next section.

The solutions of the quadratic equation $ax^2 + bx + c = 0$ are

$$x = \frac{-b \pm \sqrt{b^2 - 4ac}}{2a}.$$

If a, b, and c are integers, the type of solutions of a quadratic equation (that is, rational, irrational, or imaginary) can be determined using the discriminant. By calculating the discriminant before solving a quadratic equation, we can predict whether the solutions will be rational numbers, irrational numbers, or imaginary numbers. This can be useful in an applied problem, for example, where irrational or imaginary number solutions are not acceptable. Also, if the discriminant is a perfect square (including 0, as in Example 4), the equation can be solved by factoring. Otherwise, the quadratic formula should be used.

Discriminant

The discriminant of $ax^2 + bx + c = 0$ is given by $b^2 - 4ac$. If a, b, and c are integers, then the type of solution is determined as follows.

Discriminant	Type of Solution
Positive, and the square of an integer	Two different rational solutions
Positive, but not the square of an integer	Two different irrational solutions
Zero	One rational solution
Negative	Two different imaginary solutions

EXAMPLE 8   Using the Discriminant

Predict the number and type of solutions for the following equations.

(a) $6x^2 - x - 15 = 0$

We find the discriminant by evaluating $b^2 - 4ac$.

$$b^2 - 4ac = (-1)^2 - 4(6)(-15) \qquad a = 6, b = -1, c = -15$$
$$= 1 + 360 = 361$$

A calculator shows that $361 = 19^2$, a perfect square. Since a, b, and c are integers, the solutions will be two different rational numbers, and the equation can be solved by factoring.

(b) $3m^2 - 4m = 5$

Rewrite the equation as $3m^2 - 4m - 5 = 0$ to find $a = 3$, $b = -4$, $c = -5$. The discriminant is

$$b^2 - 4ac = (-4)^2 - 4(3)(-5) = 16 + 60 = 76.$$

Because 76 is not the square of an integer, $\sqrt{76}$ is irrational. From this and from the fact that a, b, and c are integers, the equation will have two different irrational solutions, one using $\sqrt{76}$ and one using $-\sqrt{76}$.

(c) $4x^2 + x + 1 = 0$

Since $a = 4$, $b = 1$, and $c = 1$, the discriminant is

$$1^2 - 4(4)(1) = -15.$$

Since the discriminant is negative and a, b, and c are integers, this quadratic equation will have two imaginary number solutions.

11.3 EXERCISES

Fill in each blank with the correct response.

1. For the quadratic equation $4x^2 + 5x - 9 = 0$, the values of a, b, and c are respectively _____, _____, and _____.

2. To solve the equation $3x^2 - 5x = -2$ by the quadratic formula, the first step is to add _____ to both sides of the equation.

3. When using the quadratic formula, if the discriminant $b^2 - 4ac$ is positive, the equation has _____ real solution(s).

4. The discriminant for the equation $x^2 - 6x + 9 = 0$ is _____, and therefore the equation has _____ real solution(s).

Write each equation in the standard form $ax^2 + bx + c = 0$. Then identify the values of a, b, and c. Do not actually solve the equation. See Example 1.

5. $3x^2 = 4x + 2$

6. $5x^2 = 3x - 6$

7. $3x^2 = -7x$

8. $9x^2 = 8x$

9. $(x - 3)(x + 4) = 0$

10. $(x + 6)^2 = 3$

11. $9(x - 1)(x + 2) = 8$

12. $(3x - 1)(2x + 5) = x(x - 1)$

13. Why is the restriction $a \neq 0$ necessary in the definition of a quadratic equation?

14. To solve the quadratic equation $-2x^2 - 4x + 3 = 0$, we might choose to use $a = -2$, $b = -4$, and $c = 3$. On the other hand, we might decide to first multiply both sides by -1, obtaining the equation $2x^2 + 4x - 3 = 0$, and then use $a = 2$, $b = 4$, and $c = -3$. Show that in either case, we obtain the same solution set.

15. A student writes the quadratic formula as $x = -b \pm \dfrac{\sqrt{b^2 - 4ac}}{2a}$. Is this correct? If not, explain the error, and give the correct formula.

16. Another student writes the quadratic formula as $x = -b \pm \sqrt{\dfrac{b^2 - 4ac}{2a}}$. Is this correct? If not, explain the error, and give the correct formula.

Use the quadratic formula to solve each equation. Write all radicals in simplified form, and write all answers in lowest terms. See Examples 2–4.

17. $k^2 = -12k + 13$

18. $r^2 = 8r + 9$

19. $p^2 - 4p + 4 = 0$

20. $9x^2 + 6x + 1 = 0$

21. $2x^2 + 12x = -5$

22. $5m^2 + m = 1$

23. $2y^2 = 5 + 3y$

24. $2z^2 = 30 + 7z$

25. $6x^2 + 6x = 0$

26. $4n^2 - 12n = 0$

27. $7x^2 = 12x$

28. $9r^2 = 11r$

29. $x^2 - 24 = 0$

30. $z^2 - 96 = 0$

31. $25x^2 - 4 = 0$

32. $16x^2 - 9 = 0$

33. $3x^2 - 2x + 5 = 10x + 1$

34. $4x^2 - x + 4 = x + 7$

35. $-2x^2 = -3x + 2$

36. $-x^2 = -5x + 20$

37. $2x^2 + x + 5 = 0$

38. $3x^2 + 2x + 8 = 0$

39. $(x + 3)(x + 2) = 15$

40. $(2x + 1)(x + 1) = 7$

*Use the quadratic formula to solve each equation. **(a)** Give the solutions in exact form and **(b)** use a calculator to give the solutions correct to the nearest thousandth.*

41. $2x^2 + 2x = 5$

42. $5x^2 = 3 - x$

43. $x^2 = 1 + x$

44. $x^2 = 2 + 4x$

Use the quadratic formula to solve each equation. See Example 5.

45. $\dfrac{3}{2}k^2 - k - \dfrac{4}{3} = 0$

46. $\dfrac{2}{5}x^2 - \dfrac{3}{5}x - 1 = 0$

47. $\dfrac{1}{2}x^2 + \dfrac{1}{6}x = 1$

48. $\dfrac{2}{3}y^2 - \dfrac{4}{9}y = \dfrac{1}{3}$

49. $.5x^2 = x + .5$

50. $.25x^2 = -1.5x - 1$

51. $\dfrac{3}{8}x^2 - x + \dfrac{17}{24} = 0$

52. $\dfrac{1}{3}y^2 + \dfrac{8}{9}y + \dfrac{7}{9} = 0$

53. If an applied problem leads to a quadratic equation, what must you be aware of after you have solved the equation?

54. Suppose that a problem asks you to find the length of a rectangle, and the problem leads to a quadratic equation. Which one of the following solutions to the equation cannot be an answer to the problem if L represents the length of the rectangle? Why?

    (a) $L = 9$ **(b)** $L = 5\frac{1}{4}$ **(c)** $L = \dfrac{1 + \sqrt{5}}{2}$ **(d)** $L = \dfrac{1 - \sqrt{5}}{2}$

55. Solve the formula $S = 2\pi rh + \pi r^2$ for r by first writing it in the form $ar^2 + br + c = 0$, and then using the quadratic formula.

56. Solve the formula $V = \pi r^2 h + \pi R^2 h$ for r, using the method described in Exercise 55.

In each problem, find P, the planet's sidereal period in years, using Kepler's third law of planetary motion. The given value a is its orbital semimajor axis in AU. Refer to the formula in Example 6. (Source: James B. Kaler, Astronomy! A Brief Edition, Addison Wesley, 1997.)

57. Mercury, $a = .387$

58. Venus, $a = .723$

59. Mars, $a = 1.524$

60. Ceres (the largest asteroid), $a = 2.77$

61. Uranus, $a = 19.2$

62. Neptune, $a = 30.1$

63. For Earth, $a = 1$ by definition and $P = 1$ year. Substitute these into the formula to show that Kepler's third law is valid for Earth.

Solve each problem. See Example 7.

64. The time t in seconds under certain conditions for a ball to be 48 feet in the air (on Earth) is given (approximately) by

$$48 = 64t - 16t^2.$$

Solve this equation for t. Are both answers reasonable?

65. A certain projectile is located $d = 2t^2 - 5t + 2$ feet from the ground after t seconds have elapsed. How many seconds will it take the projectile to be 14 feet from the ground?

66. A frog is sitting on a stump 3 feet above the ground. He hops off the stump and lands on the ground 4 feet away. During his leap, his height h is given by the equation $h = -.5x^2 + 1.25x + 3$, where x is the distance in feet from the base of the stump, and h is in feet. How far was the frog from the base of the stump when he was 1.25 feet above the ground?

(0, 3)

(4, 0)

67. An astronaut on the moon throws a baseball upward. The height h of the ball, in feet, x seconds after he throws it, is given by the equation

$$h = -2.7x^2 + 30x + 6.5.$$

After how many seconds is the ball 12 feet above the moon's surface?

68. A rule for estimating the number of board feet of lumber that can be cut from a log depends on the diameter of the log. To find the diameter d required to get 9 board feet of lumber, we use the equation

$$\left(\frac{d - 4}{4}\right)^2 = 9.$$

Solve this equation for d. Are both answers reasonable?

69. An old Babylonian problem asks for the length of the side of a square, given that the area of the square minus the length of a side is 870. Find the length of the side. (*Source:* Howard Eves, *An Introduction to the History of Mathematics,* 6th Edition, Saunders College Publishing, 1990.)

Use the discriminant to determine whether each equation has solutions that are (a) two different rational numbers, (b) exactly one rational number, (c) two different irrational numbers, or (d) two different imaginary numbers. See Example 8.

70. $2x^2 - x + 1 = 0$ **71.** $4x^2 - 4x + 3 = 0$ **72.** $6m^2 + 7m - 3 = 0$

73. $7x^2 - 32x - 15 = 0$ **74.** $x^2 + 4x = -4$ **75.** $4y^2 + 36y = -81$

76. $9t^2 = 30t - 15$ **77.** $25k^2 = -20k + 2$

▇ RELATING CONCEPTS (EXERCISES 78-84)

In Chapter 6 we presented methods for factoring trinomials. Some trinomials cannot be factored using integer coefficients, however. There is a way to determine beforehand whether a trinomial of the form $ax^2 + bx + c$ can be factored using the discriminant.

Work Exercises 78–84 in order.

78. For the trinomial $ax^2 + bx + c$, the expression $b^2 - 4ac$ is called the discriminant. Where have you seen the discriminant before?

79. Each of the following trinomials is factorable. Find the discriminant for each one.
   (a) $18x^2 - 9x - 2$ **(b)** $5x^2 + 7x - 6$
   (c) $48x^2 + 14x + 1$ **(d)** $x^2 - 5x - 24$

80. What do you notice about the discriminants you found in Exercise 79.

81. Factor each of the trinomials in Exercise 79.

82. Each of the following trinomials is not factorable using the methods of Chapter 6. Find the discriminant for each one.
   (a) $2x^2 + x - 5$ **(b)** $2x^2 + x + 5$ **(c)** $x^2 + 6x + 6$ **(d)** $3x^2 + 2x - 9$

83. Are any of the discriminants you found in Exercise 82 perfect squares?

84. Make a conjecture (an educated guess) concerning when a trinomial of the form $ax^2 + bx + c$ is factorable. Then use your conjecture to determine whether each trinomial is factorable. (Do not actually factor.)
   (a) $42x^2 + 117x + 66$ **(b)** $99x^2 + 186x - 24$ **(c)** $58x^2 + 184x + 27$

Did you make the connection between the nature of the discriminant and whether a trinomial with that discriminant can be factored?

11.4 Complex Numbers

OBJECTIVES

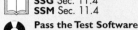

1. Write complex numbers as multiples of i.

2. Add and subtract complex numbers.

3. Multiply complex numbers.

4. Write complex number quotients in standard form.

5. Solve quadratic equations with complex number solutions.

FOR EXTRA HELP

 SSG Sec. 11.4
SSM Sec. 11.4

 Pass the Test Software

InterAct Math
Tutorial Software

 Video 19

As shown earlier in this chapter, some quadratic equations have no real number solutions. For example, the number

$$\frac{4 \pm \sqrt{-4}}{2},$$

which occurred in the solution of Example 5 in the previous section, is not a real number, because -4 appears as the radicand. For every quadratic equation to have a solution, we need a new set of numbers that includes the real numbers. This new set of numbers is defined using a new number i having the properties given below.

The Number i

$$i = \sqrt{-1} \quad \text{and} \quad i^2 = -1$$

OBJECTIVE 1 Write complex numbers as multiples of i. We can now write numbers like $\sqrt{-4}$, $\sqrt{-5}$, and $\sqrt{-8}$ as multiples of i, using a generalization of the product rule for radicals, as in the next example.

EXAMPLE 1 Simplifying Square Roots of Negative Numbers
Write each number as a multiple of i.

(a) $\sqrt{-4} = \sqrt{-1 \cdot 4} = \sqrt{-1} \cdot \sqrt{4} = i\sqrt{4} = i \cdot 2 = 2i$

(b) $\sqrt{-5} = \sqrt{-1 \cdot 5} = \sqrt{-1} \cdot \sqrt{5} = i\sqrt{5}$

(c) $\sqrt{-8} = i\sqrt{8} = i \cdot 2 \cdot \sqrt{2} = 2i\sqrt{2}$

CAUTION It is easy to mistake $\sqrt{2}\ i$ for $\sqrt{2i}$, with the i under the radical. For this reason, it is customary to write the i factor first when it is multiplied by a radical. For example, we usually write $i\sqrt{2}$ rather than $\sqrt{2}\ i$.

Numbers that are nonzero multiples of i are *imaginary numbers*. The *complex numbers* include all real numbers and all imaginary numbers.

Complex Number

A **complex number** is a number of the form $a + bi$, where a and b are real numbers. If $b \neq 0$, $a + bi$ is also an **imaginary number.**

For example, the real number 2 is a complex number since it can be written as $2 + 0i$. Also, the imaginary number $3i = 0 + 3i$ is a complex number. Other complex numbers are

$$3 - 2i, \quad 1 + i\sqrt{2}, \quad \text{and} \quad -5 + 4i.$$

In the complex number $a + bi$, a is called the **real part** and b (*not bi*) is called the **imaginary part.***

———————
*Some texts *do* refer to bi as the imaginary part.

A complex number written in the form $a + bi$ (or $a + ib$) is in **standard form.** Figure 2 shows the relationships among the various types of numbers discussed in this book. (Compare this figure to Figure 8 in Chapter 1.)

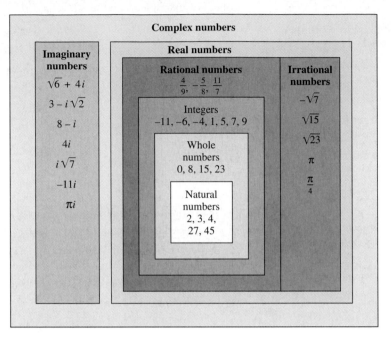

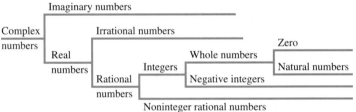

Figure 2

OBJECTIVE ⃞**2** Add and subtract complex numbers. Adding and subtracting complex numbers is similar to adding and subtracting binomials.

Addition and Subtraction of Complex Numbers

1. To add complex numbers, add their real parts and add their imaginary parts.

2. To subtract complex numbers, change the number following the subtraction sign to its negative, and then add.

The properties of Section 1.7 (commutative, associative, etc.) also hold for operations with complex numbers.

EXAMPLE 2 Adding and Subtracting Complex Numbers

Add or subtract.

(a) $(2 - 6i) + (7 + 4i) = (2 + 7) + (-6 + 4)i = 9 - 2i$

(b) $3i + (-2 - i) = -2 + (3 - 1)i = -2 + 2i$

(c) $(2 + 6i) - (-4 + i)$

Change $-4 + i$ to its negative, and then add.

$$(2 + 6i) - (-4 + i) = (2 + 6i) + (4 - i) \qquad -(-4 + i) = 4 - i$$
$$= (2 + 4) + (6 - 1)i \qquad \text{Commutative, associative, and distributive properties}$$
$$= 6 + 5i$$

(d) $(-1 + 2i) - 4 = (-1 - 4) + 2i = -5 + 2i$

OBJECTIVE 3 Multiply complex numbers. We multiply complex numbers as we do polynomials. Since $i^2 = -1$ by definition, whenever i^2 appears, we replace it with -1.

EXAMPLE 3   Multiplying Complex Numbers

Find the following products.

(a) $3i(2 - 5i) = 6i - 15i^2 \qquad$ Distributive property

$\qquad\qquad\quad = 6i - 15(-1) \qquad i^2 = -1$

$\qquad\qquad\quad = 6i + 15$

$\qquad\qquad\quad = 15 + 6i \qquad$ Commutative property

The last step gives the result in standard form.

(b) $(4 - 3i)(2 + 5i)$

Use FOIL.

$$(4 - 3i)(2 + 5i) = 4(2) + 4(5i) + (-3i)(2) + (-3i)(5i)$$
$$= 8 + 20i - 6i - 15i^2$$
$$= 8 + 14i - 15(-1)$$
$$= 8 + 14i + 15$$
$$= 23 + 14i$$

(c) $(1 + 2i)(1 - 2i) = 1 - 2i + 2i - 4i^2$

$\qquad\qquad\qquad\quad = 1 - 4(-1)$

$\qquad\qquad\qquad\quad = 1 + 4$

$\qquad\qquad\qquad\quad = 5$

OBJECTIVE 4 Write complex number quotients in standard form. The quotient of two complex numbers is expressed in standard form by changing the denominator into a real number. For example, to write

$$\frac{8 + i}{1 + 2i}$$

in standard form, the denominator must be a real number. As seen in Example 3(c), the product $(1 + 2i)(1 - 2i)$ is 5, a real number. This suggests multiplying the numerator and denominator of the given quotient by $1 - 2i$ as follows.

$$\frac{8 + i}{1 + 2i} = \frac{8 + i}{1 + 2i} \cdot \frac{1 - 2i}{1 - 2i}$$

$$= \frac{8 - 16i + i - 2i^2}{1 - 4i^2} \qquad \text{Multiply.}$$

$$= \frac{8 - 16i + i - 2(-1)}{1 - 4(-1)} \qquad i^2 = -1$$

$$= \frac{10 - 15i}{5} \qquad \text{Combine terms.}$$

$$= \frac{5(2 - 3i)}{5} = 2 - 3i \qquad \begin{array}{l}\text{Factor and write} \\ \text{in standard form.}\end{array}$$

Recall that this is the method used to rationalize some radical expressions in Chapter 10. The complex numbers $1 + 2i$ and $1 - 2i$ are *conjugates*. That is, the **conjugate** of the complex number $a + bi$ is $a - bi$. Multiplying the complex number $a + bi$ by its conjugate $a - bi$ gives the real number $a^2 + b^2$.

Product of Conjugates

$$(a + bi)(a - bi) = a^2 + b^2$$

The product of a complex number and its conjugate is the sum of the squares of the real and imaginary parts.

E X A M P L E   4   Dividing Complex Numbers

Write the following quotients in standard form.

(a) $\dfrac{-4 + i}{2 - i}$

Multiply numerator and denominator by $2 + i$, the conjugate of the denominator.

$$\frac{-4 + i}{2 - i} \cdot \frac{2 + i}{2 + i} = \frac{-8 - 4i + 2i + i^2}{4 - i^2}$$

$$= \frac{-8 - 2i - 1}{4 - (-1)} \qquad i^2 = -1$$

$$= \frac{-9 - 2i}{5}$$

$$= -\frac{9}{5} - \frac{2}{5}i \qquad \text{Standard form}$$

(b) $\dfrac{3 + i}{-i}$

Here, the conjugate of $0 - i$ is $0 + i$, or i.

$$\frac{3 + i}{-i} \cdot \frac{i}{i} = \frac{3i + i^2}{-i^2}$$

$$= \frac{-1 + 3i}{-(-1)} \qquad i^2 = -1; \text{ commutative property}$$

$$= -1 + 3i$$

CONNECTIONS

The complex number $a + bi$ is also written with the notation $\langle a, b \rangle$. (Note the similarity to an ordered pair.) This notation suggests a way to graph complex numbers on a plane in a manner similar to the way we graph ordered pairs. For graphing complex numbers, the x-axis is called the *real axis* and the y-axis is called the *imaginary axis*. For example, we graph the complex number $2 + 3i$ or $\langle 2, 3 \rangle$ just as we would the ordered pair $(2, 3)$, as shown in the figure. The figure also shows the graphs of the complex numbers $-1 - 4i$, $2i$, and -5.

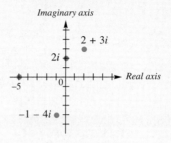

FOR DISCUSSION OR WRITING

1. Give the alternative notation for the last three complex numbers graphed above.

2. What is the real part of the complex number $-1 + 2i$? What is its imaginary part? Explain why we call the axes the real axis and the imaginary axis.

O B J E C T I V E **5** Solve quadratic equations with complex number solutions. Quadratic equations that have no real solutions do have complex solutions, as shown in the next examples.

E X A M P L E 5 Solving a Quadratic Equation with Complex Solutions (Square Root Method)
Solve $(x + 3)^2 = -25$ for complex solutions.

Use the square root property.

$$(x + 3)^2 = -25$$
$$x + 3 = \sqrt{-25} \quad \text{or} \quad x + 3 = -\sqrt{-25}$$

Since $\sqrt{-25} = 5i$,

$$x + 3 = 5i \quad \text{or} \quad x + 3 = -5i$$
$$x = -3 + 5i \quad \text{or} \quad x = -3 - 5i.$$

The solution set is $\{-3 \pm 5i\}$.

E X A M P L E 6 Solving a Quadratic Equation with Complex Solutions (Quadratic Formula)
Solve $2p^2 = 4p - 5$ for complex solutions.

Write the equation as $2p^2 - 4p + 5 = 0$. Then $a = 2$, $b = -4$, and $c = 5$. The solutions are

$$p = \frac{-(-4) \pm \sqrt{(-4)^2 - 4(2)(5)}}{2(2)}$$
$$= \frac{4 \pm \sqrt{16 - 40}}{4}$$
$$= \frac{4 \pm \sqrt{-24}}{4}.$$

Since $\sqrt{-24} = i\sqrt{24} = i \cdot \sqrt{4} \cdot \sqrt{6} = i \cdot 2 \cdot \sqrt{6} = 2i\sqrt{6}$,

$$p = \frac{4 \pm 2i\sqrt{6}}{4}$$

$$p = \frac{2(2 \pm i\sqrt{6})}{4} \qquad \text{Factor out a 2.}$$

$$p = \frac{2 \pm i\sqrt{6}}{2} \qquad \text{Lowest terms}$$

$$p = \frac{2}{2} \pm \frac{i\sqrt{6}}{2} \qquad \text{Separate into real and imaginary parts.}$$

$$p = 1 \pm \frac{\sqrt{6}}{2}i. \qquad \text{Standard form}$$

The solution set is $\left\{1 \pm \frac{\sqrt{6}}{2}i\right\}$.

11.4 EXERCISES

Write each number as a multiple of i. See Example 1.

1. $\sqrt{-9}$ **2.** $\sqrt{-36}$ **3.** $\sqrt{-20}$ **4.** $\sqrt{-27}$

5. $\sqrt{-18}$ **6.** $\sqrt{-50}$ **7.** $\sqrt{-125}$ **8.** $\sqrt{-98}$

Add or subtract as indicated. See Example 2.

9. $(2 + 8i) + (3 - 5i)$ **10.** $(4 + 5i) + (7 - 2i)$

11. $(8 - 3i) - (2 + 6i)$ **12.** $(1 + i) - (3 - 2i)$

13. $(3 - 4i) + (6 - i) - (3 + 2i)$ **14.** $(5 + 8i) - (4 + 2i) + (3 - i)$

Find each product. See Example 3.

15. $(3 + 2i)(4 - i)$ **16.** $(9 - 2i)(3 + i)$ **17.** $(5 - 4i)(3 - 2i)$

18. $(10 + 6i)(8 - 4i)$ **19.** $(3 + 6i)(3 - 6i)$ **20.** $(11 - 2i)(11 + 2i)$

TECHNOLOGY INSIGHTS (EXERCISES 21–28)

Modern graphing calculators are capable of performing operations with complex numbers. The top screen shows how the TI-83 calculator can be set for complex mode $(a + bi)$. The lower left screen shows how the square root of a negative number returns the product of a real number and i, and how addition and subtraction of complex numbers is accomplished. The lower right screen shows how multiplication is performed, how the calculator returns the real part of a complex number, and how the conjugate is given.

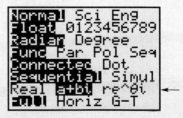

Predict the display the calculator will give for each of the following screens.

21.
```
√(-169)+√(-25)
```

22.
```
√(-36)+√(-100)
```

23.
```
(5-i)+(-2+3i)-(7
+2i)
```

24.
```
3(1+2i)-4(3-i)
```

(*Hint:* Use the order of operations.)

25.
```
conj(6-4i)
```

26.
```
imag(5+10i)
```

27.
```
(15-5i)/(3+i)
```

28.
```
(2+3i)²
```

Write each quotient in standard form. See Example 4.

29. $\dfrac{17 + i}{5 + 2i}$

30. $\dfrac{21 + i}{4 + i}$

31. $\dfrac{40}{2 + 6i}$

32. $\dfrac{13}{3 + 2i}$

33. $\dfrac{i}{4 - 3i}$

34. $\dfrac{-i}{1 + 2i}$

RELATING CONCEPTS (EXERCISES 35–40)

When you first divided whole numbers, you probably learned to check your work by multiplying your answer (the quotient) by the number doing the dividing (the divisor). For example,

$$\frac{2744}{28} = 98 \text{ is true, because } 98 \times 28 = 2744.$$

This same procedure works for real numbers other than whole numbers. Does it work for complex numbers?

(continued)

CHAPTER 11   Quadratic Equations, Inequalities, and Graphs

RELATING CONCEPTS (EXERCISES 35-40) (CONTINUED)

Work Exercises 35–40 in order to see whether it does or not.

35. Find the standard form of the quotient $\dfrac{-29 - 3i}{2 + 9i}$.

36. Multiply your answer from Exercise 35 by the divisor, $2 + 9i$. What is your answer? Is it equal to the original dividend (the numerator), $-29 - 3i$?

37. Find the standard form of the quotient $\dfrac{4 - 3i}{i}$.

38. Multiply your answer from Exercise 37 by the divisor, i. What is your answer? Is it equal to the original dividend?

39. Use the pattern established in Exercises 35–38 to determine whether the following is true: $\dfrac{14 - 5i}{4 + i} = 3 - 2i$.

40. State a rule, based on your observations, that tells whether the answer to a division problem involving complex numbers is correct.

Did you make the connection that checking complex number division is similar to checking real number division?

Solve each quadratic equation for complex solutions by the square root property. Write solutions in standard form. See Example 5.

41. $(a + 1)^2 = -4$

42. $(p - 5)^2 = -36$

43. $(k - 3)^2 = -5$

44. $(y + 6)^2 = -12$

45. $(3x + 2)^2 = -18$

46. $(4z - 1)^2 = -20$

Solve each quadratic equation for complex solutions by the quadratic formula. Write solutions in standard form. See Example 6.

47. $m^2 - 2m + 2 = 0$

48. $b^2 + b + 3 = 0$

49. $2r^2 + 3r + 5 = 0$

50. $3q^2 = 2q - 3$

51. $p^2 - 3p + 4 = 0$

52. $2a^2 = -a - 3$

53. $5x^2 + 3 = 2x$

54. $6y^2 + 2y + 1 = 0$

55. $2m^2 + 7 = -2m$

56. $4z^2 + 2z + 3 = 0$

57. $r^2 + 3 = r$

58. $4q^2 - 2q + 3 = 0$

Exercises 59–60 deal with quadratic equations having real number coefficients.

59. Suppose you are solving a quadratic equation by the quadratic formula. How can you tell, before completing the solution, whether the equation will have solutions that are not real numbers?

60. Refer to the solutions in Examples 5 and 6, and complete the following statement: If a quadratic equation has imaginary solutions, they are _____ of each other.

Answer true or false to each of the following. If false, say why.

61. Every real number is a complex number.

62. Every imaginary number is a complex number.

63. Every complex number is a real number.

64. Some complex numbers are imaginary.

65. Write a paragraph explaining how to add, subtract, multiply, and divide complex numbers. Give examples.

11.5 Equations Quadratic in Form

OBJECTIVES

1 Solve an equation with fractions by writing it in quadratic form.

2 Use quadratic equations to solve applied problems.

3 Solve an equation with radicals by writing it in quadratic form.

4 Solve an equation that is quadratic in form by substitution.

FOR EXTRA HELP

SSG Sec. 11.5
SSM Sec. 11.5

Pass the Test Software

InterAct Math
Tutorial Software

Video 19

We have introduced four methods for solving quadratic equations written in standard form $ax^2 + bx + c = 0$. The following chart gives some advantages and disadvantages of each method.

Methods for Solving Quadratic Equations

Method	Advantages	Disadvantages
Factoring	Usually the fastest method	Not all polynomials are factorable; some factorable polynomials are hard to factor.
Square root property	Simplest method for solving equations of the form $(x + a)^2 = b$	Few equations are given in this form.
Completing the square	Can always be used, although most people prefer the quadratic formula (This procedure is useful in other areas of mathematics.)	It requires more steps than other methods.
Quadratic formula	Can always be used	It is more difficult than factoring because of the square root.

OBJECTIVE **1** Solve an equation with fractions by writing it in quadratic form. A variety of nonquadratic equations can be written in the form of a quadratic equation and solved by using one of the methods in the chart. As you solve the equations in this section try to decide which is the best method for each equation.

EXAMPLE 1 Writing an Equation with Fractions in Quadratic Form

Solve $\dfrac{1}{x} + \dfrac{1}{x-1} = \dfrac{7}{12}$.

Clear fractions by multiplying each term by the common denominator, $12x(x-1)$. (Note that the domain must be restricted to $x \neq 0$ and $x \neq 1$.)

$$12x(x-1)\frac{1}{x} + 12x(x-1)\frac{1}{x-1} = 12x(x-1)\frac{7}{12}$$

$$12(x-1) + 12x = 7x(x-1)$$

$$12x - 12 + 12x = 7x^2 - 7x \qquad \text{Distributive property}$$

$$24x - 12 = 7x^2 - 7x \qquad \text{Combine terms.}$$

A quadratic equation must be in standard form $ax^2 + bx + c = 0$ before it can be solved by factoring or the quadratic formula. Combine and rearrange terms so that one side is 0. Then solve the resulting equation by factoring.

$$0 = 7x^2 - 31x + 12 \qquad \text{Standard form}$$

$$0 = (7x - 3)(x - 4) \qquad \text{Factor.}$$

Using the zero-factor property gives the solutions $\frac{3}{7}$ and 4. Check by substituting these solutions in the original equation. The solution set is $\left\{\frac{3}{7}, 4\right\}$.

OBJECTIVE **2** Use quadratic equations to solve applied problems. Earlier we solved distance-rate-time (or motion) problems that led to linear equations or rational equations. Now we extend that work to motion problems that lead to quadratic equations.

⌐ **E X A M P L E 2**   Solving a Motion Problem

A riverboat for tourists averages 12 miles per hour in still water. It takes the boat 1 hour, 4 minutes to go 6 miles upstream and return. Find the speed of the current. See Figure 3.

Figure 3

For a problem about rate (or speed), we use the distance formula, $d = rt$.

Let x = the speed of the current;

$12 - x$ = the rate upstream;

$12 + x$ = the rate downstream.

Rate upstream is the difference of the speed of the boat in still water and the speed of the current, or $12 - x$. Rate downstream is, in the same way, $12 + x$. To find time, rewrite $d = rt$ as

$$t = \frac{d}{r}.$$

This information was used to complete the following chart.

	d	r	t
Upstream	6	$12 - x$	$\dfrac{6}{12 - x}$
Downstream	6	$12 + x$	$\dfrac{6}{12 + x}$

Times in hours

Write total time, 1 hour and 4 minutes, as

$$1 + \frac{4}{60} = 1 + \frac{1}{15} = \frac{16}{15} \text{ hours.}$$

Since time upstream plus time downstream equals $\frac{16}{15}$ hours,

$$\frac{6}{12 - x} + \frac{6}{12 + x} = \frac{16}{15}.$$

Now multiply both sides of the equation by the common denominator $15(12 - x)(12 + x)$ and solve the resulting quadratic equation.

$$15(12 + x)6 + 15(12 - x)6 = 16(12 - x)(12 + x)$$

$$90(12 + x) + 90(12 - x) = 16(144 - x^2)$$

$$1080 + 90x + 1080 - 90x = 2304 - 16x^2 \qquad \text{Distributive property}$$

$$2160 = 2304 - 16x^2 \qquad \text{Combine terms.}$$

$$16x^2 = 144$$

$$x^2 = 9$$

$$x = 3 \quad \text{or} \quad x = -3 \qquad \text{Square root property}$$

The speed of the current cannot be -3, so the solution is $x = 3$ miles per hour.

OBJECTIVE 3 Solve an equation with radicals by writing it in quadratic form.

EXAMPLE 3   Solving a Radical Equation That Leads to a Quadratic Equation

Solve $x + \sqrt{x} = 6$.

This equation is not quadratic. However, squaring both sides of the equation gives a quadratic equation that can be solved by factoring.

$$\sqrt{x} = 6 - x \qquad \text{Isolate the radical on one side.}$$

$$x = 36 - 12x + x^2 \qquad \text{Square both sides.}$$

$$0 = x^2 - 13x + 36 \qquad \text{Get 0 on one side.}$$

$$0 = (x - 4)(x - 9) \qquad \text{Factor.}$$

$$x - 4 = 0 \quad \text{or} \quad x - 9 = 0 \qquad \text{Set each factor equal to 0.}$$

$$x = 4 \quad \text{or} \quad x = 9$$

Check both potential solutions in the *original* equation.

If $x = 4$,   If $x = 9$,

$4 + \sqrt{4} = 6$? $9 + \sqrt{9} = 6$?

$6 = 6$.   True $12 = 6$.   False

The solution set is $\{4\}$.

OBJECTIVE 4 Solve an equation that is quadratic in form by substitution. In Chapter 8, we discussed composition of functions of the form $g[f(x)]$, where g and f are functions. An equation that is written in composite form $a[f(x)]^2 + b[f(x)] + c = 0$, for $a \neq 0$, is called **quadratic in form.** For example, suppose

$$g(m) = 2(4m - 3)^2 + 7(4m - 3) + 5 = 0$$

and we let $f(m) = 4m - 3$. Then

$$g[f(m)] = 2[f(m)]^2 + 7[f(m)] + 5.$$

To simplify further, let $u = f(m)$; then

$$g(u) = 2u^2 + 7u + 5.$$

Solving this quadratic equation and substituting the solutions for u in $u = f(m) = 4m - 3$ will give the solutions for m.

EXAMPLE 4   Solving Equations That Are Quadratic in Form

Solve each equation.

(a) $2(4m - 3)^2 + 7(4m - 3) + 5 = 0$

Because of the repeated quantity $4m - 3$, this equation is quadratic in form with $f(m) = 4m - 3$. Let $f(m) = u$ and write

$$2(4m - 3)^2 + 7(4m - 3) + 5 = 0$$

as

$$2u^2 + 7u + 5 = 0. \qquad \text{Let } 4m - 3 = u.$$

$$(2u + 5)(u + 1) = 0 \qquad \text{Factor.}$$

By the zero-factor property, the solutions of $2u^2 + 7u + 5 = 0$ are

$$u = -\frac{5}{2} \qquad \text{or} \qquad u = -1.$$

To find m, substitute $4m - 3$ for u.

$$4m - 3 = -\frac{5}{2} \qquad \text{or} \qquad 4m - 3 = -1$$

$$4m = \frac{1}{2} \qquad \text{or} \qquad 4m = 2$$

$$m = \frac{1}{8} \qquad \text{or} \qquad m = \frac{1}{2}$$

The solution set of the original equation is $\left\{\frac{1}{8}, \frac{1}{2}\right\}$.

(b) $y^4 = 6y^2 - 3$

First write the equation as

$$y^4 - 6y^2 + 3 = 0 \qquad \text{or} \qquad (y^2)^2 - 6(y^2) + 3 = 0,$$

which is quadratic in form with $f(y) = y^2$. Substitute u for y^2 and u^2 for y^4 to get

$$u^2 - 6u + 3 = 0.$$

By the quadratic formula,

$$u = \frac{6 \pm \sqrt{36 - 12}}{2} \qquad a = 1, b = -6, c = 3$$

$$= \frac{6 \pm \sqrt{24}}{2}$$

$$= \frac{6 \pm 2\sqrt{6}}{2}$$

$$u = 3 \pm \sqrt{6}.$$

Find y by using the square root property as follows.

$$y^2 = 3 + \sqrt{6} \qquad \text{or} \qquad y^2 = 3 - \sqrt{6}$$

$$y = \pm\sqrt{3 + \sqrt{6}} \qquad \text{or} \qquad y = \pm\sqrt{3 - \sqrt{6}}$$

The solution set contains four numbers:

$$\left\{\sqrt{3 + \sqrt{6}}, -\sqrt{3 + \sqrt{6}}, \sqrt{3 - \sqrt{6}}, -\sqrt{3 - \sqrt{6}}\right\}.$$

┌─
EXAMPLE 5 Solving Equations That Are Quadratic in Form

Solve each equation.

(a) $4x^6 + 1 = 5x^3$

This equation is quadratic in form with $f(x) = x^3$. Let $x^3 = u$. Then $x^6 = u^2$. Substitute into the given equation.

$$4x^6 + 1 = 5x^3$$

$$4(x^3)^2 + 1 = 5x^3$$

$$4u^2 + 1 = 5u \qquad \text{Let } x^3 = u.$$

$$4u^2 - 5u + 1 = 0 \qquad \text{Get 0 on one side.}$$

$$(4u - 1)(u - 1) = 0 \qquad \text{Factor.}$$

$$u = \frac{1}{4} \qquad \text{or} \qquad u = 1$$

$$x^3 = \frac{1}{4} \qquad \text{or} \qquad x^3 = 1 \qquad u = x^3$$

From these equations,

$$x = \sqrt[3]{\frac{1}{4}} = \frac{\sqrt[3]{1}}{\sqrt[3]{4}} = \frac{1}{\sqrt[3]{4}} \cdot \frac{\sqrt[3]{2}}{\sqrt[3]{2}} = \frac{\sqrt[3]{2}}{2} \qquad \text{or} \qquad x = \sqrt[3]{1} = 1.$$

This method of substitution gives only real number solutions for equations with polynomials of degree $2n$ where n is odd. However, it gives all complex number solutions for equations with polynomials of degree $2n$ where n is even. The real number solution set of $4x^6 + 1 = 5x^3$ is $\left\{\frac{\sqrt[3]{2}}{2}, 1\right\}$.

(b) $2a^{2/3} - 11a^{1/3} + 12 = 0$

Let $a^{1/3} = u$; then $a^{2/3} = u^2$. Substitute into the given equation.

$$2(a^{1/3})^2 - 11a^{1/3} + 12 = 0$$

$$2u^2 - 11u + 12 = 0 \qquad \text{Let } a^{1/3} = u.$$

$$(2u - 3)(u - 4) = 0 \qquad \text{Factor.}$$

$$2u - 3 = 0 \qquad \text{or} \qquad u - 4 = 0$$

$$u = \frac{3}{2} \qquad \text{or} \qquad u = 4$$

$$a^{1/3} = \frac{3}{2} \qquad \text{or} \qquad a^{1/3} = 4 \qquad u = a^{1/3}$$

$$a = \left(\frac{3}{2}\right)^3 = \frac{27}{8} \qquad \text{or} \qquad a = 4^3 = 64 \qquad \text{Cube each side.}$$

Check that the solution set is $\left\{\frac{27}{8}, 64\right\}$.
─┘

Some people prefer to solve equations like the one in Example 5(a) by directly factoring using the given equation. For example, the equation could be solved as follows.

$$4x^6 + 1 = 5x^3 \qquad \text{Given equation}$$
$$4x^6 - 5x^3 + 1 = 0 \qquad \text{Write in standard form.}$$
$$(4x^3 - 1)(x^3 - 1) = 0 \qquad \text{Factor.}$$
$$4x^3 - 1 = 0 \quad \text{or} \quad x^3 - 1 = 0 \qquad \text{Zero-factor property}$$
$$x^3 = \frac{1}{4} \quad \text{or} \quad x^3 = 1 \qquad \text{Solve each equation.}$$

Complete the solution as in Example 5(a).

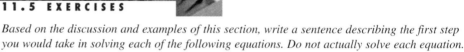

11.5 EXERCISES

Based on the discussion and examples of this section, write a sentence describing the first step you would take in solving each of the following equations. Do not actually solve each equation.

1. $\dfrac{14}{x} = x - 5$

2. $\sqrt{1 + x} + x = 5$

3. $(r^2 + r)^2 - 8(r^2 + r) + 12 = 0$

4. $3t = \sqrt{16 - 10t}$

5. Of the four methods for solving quadratic equations, it is often said that the quadratic formula is the most efficient method. Do you agree or disagree with this statement? Explain.

6. What is the relationship between the method of completing the square in solving quadratic equations and the quadratic formula?

Solve by first clearing each equation of fractions. Check your answers. See Example 1.

7. $\dfrac{14}{x} = x - 5$

8. $\dfrac{-12}{x} = x + 8$

9. $1 - \dfrac{3}{x} - \dfrac{28}{x^2} = 0$

10. $4 - \dfrac{7}{r} - \dfrac{2}{r^2} = 0$

11. $\dfrac{1}{x} + \dfrac{2}{x + 2} = \dfrac{17}{35}$

12. $\dfrac{2}{m} + \dfrac{3}{m + 9} = \dfrac{11}{4}$

13. $\dfrac{2}{x + 1} + \dfrac{3}{x + 2} = \dfrac{7}{2}$

14. $\dfrac{4}{3 - y} + \dfrac{2}{5 - y} = \dfrac{26}{15}$

15. $\dfrac{3}{2x} - \dfrac{1}{2(x + 2)} = 1$

16. $\dfrac{4}{3x} - \dfrac{1}{2(x + 1)} = 1$

17. If it takes m hours to grade a set of papers, what is the grader's rate (in job per hour)?

18. If a boat goes 20 miles per hour in still water, and the rate of the current is t miles per hour, what is the rate of the boat when it travels upstream? Downstream?

Solve each problem by writing an equation with fractions and solving it. See Example 2.

19. On a windy day Yoshiaki found that he could go 16 miles downstream and then 4 miles back upstream at top speed in a total of 48 minutes. What was the top speed of Yoshiaki's boat if the speed of the current was 15 miles per hour?

20. Lekesha flew her plane for 6 hours at a constant speed. She traveled 810 miles with the wind, then turned around and traveled 720 miles against the wind. The wind speed was a constant 15 miles per hour. Find the speed of the plane.

21. In Canada, Medicine Hat and Cranbrook are 300 kilometers apart. Harry rides his Honda 20 kilometers per hour faster than Karen rides her Yamaha. Find Harry's average speed if he travels from Cranbrook to Medicine Hat in $1\frac{1}{4}$ hours less time than Karen.

22. The distance from Jackson to Lodi is about 40 miles, as is the distance from Lodi to Manteca. Rico drove from Jackson to Lodi during rush hour, stopped in Lodi for a root beer, and then drove on to Manteca at 10 miles per hour faster. Driving time for the entire trip was 88 minutes. Find his speed from Jackson to Lodi.

23. A washing machine can be filled in 6 minutes if both the hot and cold water taps are fully opened. To fill the washer with hot water alone takes 9 minutes longer than filling with cold water alone. How long does it take to fill the tank with cold water?

24. Two pipes together can fill a large tank in 2 hours. One of the pipes, used alone, takes 3 hours longer than the other to fill the tank. How long would each pipe take to fill the tank alone?

Find all solutions in Exercises 25–32 by first squaring. Check your answers. See Example 3.

25. $2x = \sqrt{11x + 3}$

26. $4x = \sqrt{6x + 1}$

27. $3y = (16 - 10y)^{1/2}$

28. $4t = (8t + 3)^{1/2}$

29. $p - 2\sqrt{p} = 8$

30. $k + \sqrt{k} = 12$

31. $m = \sqrt{\dfrac{6 - 13m}{5}}$

32. $r = \sqrt{\dfrac{20 - 19r}{6}}$

Solve each problem. Use 3.14 as an approximation for π.

33. Artificial gravity can be created in a space station by rotating all or part of the station. To achieve gravity equal to one-half the gravity of Earth with .063 rotation per second, the distance r in meters of a point in the station from the center of rotation is given by the equation $.126\pi r = \sqrt{4.7r}$. Find the value of r (*Source:* Bernice Kastner, *Space Mathematics,* NASA: pp. 48–49.)

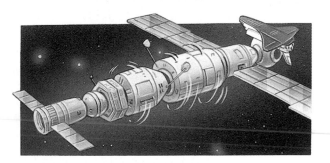

34. In Exercise 33, if Earth surface gravity is desired with a rotation rate of .04, the equation for r in meters is $.08\pi r = \sqrt{9.8r}$. Find r.

Find all solutions to each equation. Check your answers. See Examples 4 and 5.

35. $t^4 - 18t^2 + 81 = 0$

36. $y^4 - 8y^2 + 16 = 0$

37. $4k^4 - 13k^2 + 9 = 0$

38. $9x^4 - 25x^2 + 16 = 0$

39. $(x + 3)^2 + 5(x + 3) + 6 = 0$

40. $(k - 4)^2 + (k - 4) - 20 = 0$

41. $(t + 5)^2 + 6 = 7(t + 5)$

42. $3(m + 4)^2 - 8 = 2(m + 4)$

43. $2 + \dfrac{5}{3k - 1} = \dfrac{-2}{(3k - 1)^2}$

44. $3 - \dfrac{7}{2p + 2} = \dfrac{6}{(2p + 2)^2}$

45. $2 - 6(m - 1)^{-2} = (m - 1)^{-1}$

46. $3 - 2(x - 1)^{-1} = (x - 1)^{-2}$

Use substitution to solve each equation. Check your answers. See Example 5.

47. $4k^{4/3} - 13k^{2/3} + 9 = 0$

48. $9y^{2/5} = 16 - 10y^{1/5}$

49. $x^{2/3} + x^{1/3} - 2 = 0$

50. $3x^{2/3} - x^{1/3} - 24 = 0$

51. $2(1 + \sqrt{y})^2 = 13(1 + \sqrt{y}) - 6$

52. $(k^2 + k)^2 + 12 = 8(k^2 + k)$

53. $2x^4 + x^2 - 3 = 0$

54. $4k^4 + 5k^2 + 1 = 0$

55. What is wrong with the following "solution"?

$$2(m - 1)^2 - 3(m - 1) + 1 = 0$$
$$2u^2 - 3u + 1 = 0 \qquad \text{Let } u = m - 1.$$
$$(2u - 1)(u - 1) = 0$$
$$2u - 1 = 0 \qquad \text{or} \qquad u - 1 = 0$$
$$u = \frac{1}{2} \qquad\qquad u = 1$$

Solution set: $\left\{\dfrac{1}{2}, 1\right\}$

56. Explain how to solve the equation

$$y + \sqrt{y} - 6 = 0$$

using both the method of Example 3 and the method of Example 4.

The equations in Exercises 57–64 are not grouped by type. Decide which method of solution applies, and then solve each equation. Give only real solutions. See Examples 1, 3, 4, and 5.

57. $x^4 - 16x^2 + 48 = 0$

58. $\left(x - \dfrac{1}{2}\right)^2 + 5\left(x - \dfrac{1}{2}\right) - 4 = 0$

59. $\sqrt{2x + 3} = 2 + \sqrt{x - 2}$

60. $\sqrt{m + 1} = -1 + \sqrt{2m}$

61. $2m^6 + 11m^3 + 5 = 0$

62. $8x^6 + 513x^3 + 64 = 0$

63. $2 - (y - 1)^{-1} = 6(y - 1)^{-2}$

64. $3 = 2(p - 1)^{-1} + (p - 1)^{-2}$

■ **RELATING CONCEPTS (EXERCISES 65–70)**

Consider the following equation, which contains variable expressions in the denominators.

$$\frac{x^2}{(x - 3)^2} + \frac{3x}{x - 3} - 4 = 0.$$

Work Exercises 65–70 in order. *They all pertain to this equation.*

65. Why can 3 not possibly be a solution for this equation?

66. Multiply both sides of the equation by the LCD, $(x - 3)^2$, and solve. There is only one solution—what is it?

67. Write the equation in a different manner so that it is quadratic in form, with the rational expression $\dfrac{x}{x - 3}$ as the expression that you will substitute another variable for.

68. In your own words, explain why the expression $\dfrac{x}{x - 3}$ cannot equal 1.

69. Solve the equation from Exercise 67 by making the substitution $t = \dfrac{x}{x - 3}$.

You should get two values for t. Why is one of them impossible for this equation?

70. Solve the equation $x^2(x - 3)^{-2} + 3x(x - 3)^{-1} - 4 = 0$ by letting $s = (x - 3)^{-1}$. You should get two values for s. Why is this impossible for this equation?

Did you make the connection between the three solution methods and understand why one of the answers does not satisfy the equation?

11.6 Formulas and Applications

OBJECTIVES

1. Solve formulas for variables involving squares and square roots.

2. Solve applied problems about motion along a straight line.

3. Solve applied problems using the Pythagorean formula.

4. Solve applied problems using formulas for area.

FOR EXTRA HELP

SSG Sec. 11.6
SSM Sec. 11.6

Pass the Test Software

InterAct Math Tutorial Software

Video 19

OBJECTIVE 1 Solve formulas for variables involving squares and square roots. The methods presented earlier in this chapter and the previous one can be used to solve such formulas.

EXAMPLE 1 Solving for a Variable Involving a Square or a Square Root

(a) Solve $w = \dfrac{kFr}{v^2}$ for v.

Multiply each side by v^2 to clear the equation of fractions. Then solve for v, using the square root property.

$$w = \frac{kFr}{v^2}$$

$$v^2 w = kFr \qquad \text{Multiply each side by } v^2.$$

$$v^2 = \frac{kFr}{w} \qquad \text{Divide each side by } w.$$

$$v = \pm\sqrt{\frac{kFr}{w}} \qquad \text{Square root property}$$

$$v = \frac{\pm\sqrt{kFr}}{\sqrt{w}} \cdot \frac{\sqrt{w}}{\sqrt{w}} = \frac{\pm\sqrt{kFrw}}{w} \qquad \text{Rationalize the denominator.}$$

(b) Solve $d = \sqrt{\dfrac{4A}{\pi}}$ for A.

Square both sides to eliminate the radical.

$$d = \sqrt{\frac{4A}{\pi}}$$

$$d^2 = \frac{4A}{\pi} \qquad \text{Square both sides.}$$

$$\pi d^2 = 4A \qquad \text{Multiply both sides by } \pi.$$

$$\frac{\pi d^2}{4} = A \qquad \text{Divide both sides by 4.}$$

Check the solution in the original equation.

┌───

EXAMPLE 2 Solving for a Squared Variable

Solve $s = 2t^2 + kt$ for t.

Since the equation has terms with t^2 and t, write it in standard form $ax^2 + bx + c = 0$, with t as the variable instead of x.

$$s = 2t^2 + kt$$
$$0 = 2t^2 + kt - s$$

Now use the quadratic formula with $a = 2$, $b = k$, and $c = -s$.

$$t = \frac{-k \pm \sqrt{k^2 - 4(2)(-s)}}{2(2)} \qquad a = 2, b = k, c = -s$$

$$= \frac{-k \pm \sqrt{k^2 + 8s}}{4}$$

The solutions are $t = \dfrac{-k + \sqrt{k^2 + 8s}}{4}$ and $t = \dfrac{-k - \sqrt{k^2 + 8s}}{4}$.

───

OBJECTIVE 2 Solve applied problems about motion along a straight line. In the next example we use a quadratic equation to describe the distance traveled by a propelled object.

EXAMPLE 3 Solving a Straight-Line Motion Problem

If a rock on Earth is thrown upward from a 144-foot building with an initial velocity of 112 feet per second, its position (in feet above the ground) is given by $s = -16t^2 + 112t + 144$, where t is time in seconds after it was thrown. When does it hit the ground?

When the rock hits the ground, its distance above the ground is 0. Find t when s is 0 by solving the equation

$$0 = -16t^2 + 112t + 144. \qquad \text{Let } s = 0.$$
$$0 = t^2 - 7t - 9 \qquad \text{Divide both sides by } -16.$$
$$t = \frac{7 \pm \sqrt{49 + 36}}{2} \qquad \text{Quadratic formula}$$
$$t = \frac{7 \pm \sqrt{85}}{2}$$
$$t \approx 8.1 \quad \text{or} \quad t \approx -1.1 \qquad \text{Use a calculator.}$$

Discard the negative solution. The rock will hit the ground about 8.1 seconds after it is thrown.

───

CAUTION Remember to check all proposed solutions against the information of the original problem. Some solutions of the equation may not satisfy the physical conditions of the problem.

When an object moves under the influence of gravity (without air resistance) near the surface of a planet, its path has the shape of a parabola. Suppose an object is propelled upward at a 45° angle with an initial velocity of 30 miles per hour. Then the coordinates x and y of a point on the parabola in feet y at time x in seconds are given by

$$y = x - \frac{g}{1922}x^2,$$

where g is the force of gravity.

E X A M P L E 4 Solving a Problem about the Motion of a Propelled Object

Suppose a rocket is propelled from the surface of Mars at a 45° angle with an initial velocity of 30 miles per hour. See Figure 4. (*Source:* Bernice Kastner, *Space Mathematics,* NASA.)

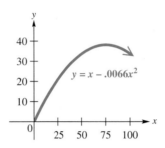

Figure 4

(a) For Mars, $g = 12.6$. Find the height of the rocket after 50 seconds.
Replacing g in the equation gives

$$y = x - \frac{12.6}{1922}x^2 \approx x - .0066x^2.$$

Now substitute 50 for x to get

$$y = 50 - .0066(50)^2 = 33.5 \text{ feet.}$$

The rocket reaches a height of 33.5 feet at 50 seconds.

(b) How long will it take for the rocket to reach 100 feet?
Let $y = 100$ and solve the equation

$$100 = x - .0066x^2.$$

Write the equation in standard form, and use the quadratic formula.

$$.0066x^2 - x + 100 = 0 \qquad \text{Get 0 on one side.}$$

$$x = \frac{1 \pm \sqrt{1 - 4(.0066)(100)}}{2(.0066)} \qquad \text{Let } a = .0066, b = -1, c = 100.$$

$$x = \frac{1 \pm \sqrt{1 - 2.64}}{.0132}$$

The quantity under the radical is negative. This indicates that the rocket never reaches a height of 100 feet.

O B J E C T I V E **3** Solve applied problems using the Pythagorean formula. The Pythagorean formula, which was introduced in Chapter 6, is used in applications involving right triangles. Such problems often require solving a quadratic equation.

E X A M P L E 5 Using the Pythagorean Formula

Two cars left an intersection at the same time, one heading due north, and the other due west. Some time later, they were exactly 100 miles apart. The car headed north had gone 20 miles farther than the car headed west. How far had each car traveled?

Let x be the distance traveled by the car headed west. Then $x + 20$ is the distance traveled by the car headed north. These distances are shown in Figure 5.

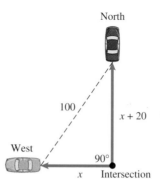

Figure 5

The cars are 100 miles apart, so the hypotenuse of the right triangle equals 100 and the two legs are equal to x and $x + 20$. Use the Pythagorean formula.

$$c^2 = a^2 + b^2$$
$$100^2 = x^2 + (x + 20)^2$$
$$10{,}000 = x^2 + x^2 + 40x + 400 \qquad \text{Square the binomial.}$$
$$0 = 2x^2 + 40x - 9600 \qquad \text{Get 0 on one side.}$$
$$0 = 2(x^2 + 20x - 4800) \qquad \text{Factor out the common factor.}$$
$$0 = x^2 + 20x - 4800 \qquad \text{Divide both sides by 2.}$$

Use the quadratic formula to find x.

$$x = \frac{-20 \pm \sqrt{400 - 4(1)(-4800)}}{2} \qquad a = 1, b = 20, c = -4800$$
$$= \frac{-20 \pm \sqrt{19{,}600}}{2}$$
$$x = 60 \quad \text{or} \quad x = -80 \qquad \text{Use a calculator.}$$

Discard the negative solution. The required distances are 60 miles and $60 + 20 = 80$ miles.

OBJECTIVE 4 Solve applied problems using formulas for area.

EXAMPLE 6 Solving an Area Problem

A rectangular reflecting pool in a park is 20 feet wide and 30 feet long. The park gardener wants to plant a strip of grass of uniform width around the edge of the pool. She has enough seed to cover 336 square feet. How wide will the strip be?

The pool is shown in Figure 6. If x represents the unknown width of the grass strip, the width of the large rectangle is given by $20 + 2x$ (the width of the pool plus two grass strips), and the length is given by $30 + 2x$.

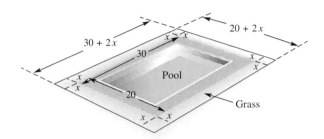

Figure 6

The area of the large rectangle is given by the product of its length and width, $(20 + 2x)(30 + 2x)$. The area of the pool is $20 \cdot 30 = 600$ square feet. The area of the large rectangle, minus the area of the pool, should equal the area of the grass strip. Since the area of the grass strip is to be 336 square feet, the equation is

$$\underset{\substack{\text{Area of} \\ \text{rectangle} \\ \downarrow}}{(20 + 2x)(30 + 2x)} - \underset{\substack{\text{Area of} \\ \text{pool} \\ \downarrow}}{600} = \underset{\substack{\text{Area of grass} \\ \downarrow}}{336}$$

$$600 + 100x + 4x^2 - 600 = 336 \qquad \text{Multiply.}$$
$$4x^2 + 100x - 336 = 0 \qquad \text{Combine terms.}$$
$$x^2 + 25x - 84 = 0 \qquad \text{Divide by 4.}$$
$$(x + 28)(x - 3) = 0 \qquad \text{Factor.}$$
$$x = -28 \quad \text{or} \quad x = 3.$$

The width cannot be -28 feet, so the grass strip should be 3 feet wide.

11.6 EXERCISES

1. What is the first step in solving a formula that has the specified variable in the denominator?

2. What is the first step in solving a formula like $gw^2 = 2r$ for w?

3. What is the first step in solving a formula like $gw^2 - kw + 24$ for w?

4. Why is it particularly important to check all proposed solutions to an applied problem against the information in the original problem?

5. In a problem like the one in Example 3, is it possible to get two correct nonzero answers?

6. In Example 4(b), if the height is changed so that the quantity under the radical becomes .472, there are two correct nonzero answers. Why is this possible?

In Exercises 7 and 8, solve for m in terms of the other variables. Remember that m must be positive.

7.

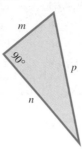

8.

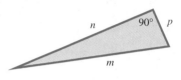

Solve each equation for the indicated variable. (While in practice we would often reject a negative value due to the physical nature of the quantity represented by the variable, leave ± in your answers here.) See Examples 1 and 2.

9. $d = kt^2$; for t

10. $s = kwd^2$; for d

11. $F = \dfrac{kA}{v^2}$; for v

12. $L = \dfrac{kd^4}{h^2}$; for h

13. $V = \dfrac{1}{3}\pi r^2 h$; for r

14. $V = \pi(r^2 + R^2)h$; for r

15. $At^2 + Bt = -C$; for t

16. $S = 2\pi rh + \pi r^2$; for r

17. $D = \sqrt{kh}$; for h

18. $F = \dfrac{k}{\sqrt{d}}$; for d

19. $p = \sqrt{\dfrac{k\ell}{g}}$; for ℓ

20. $p = \sqrt{\dfrac{k\ell}{g}}$; for g

21. If g is a positive number in the formula of Exercise 19, explain why k and ℓ must have the same sign in order for p to be a real number.

22. Refer to Example 2 of this section. Suppose that k and s both represent positive numbers.
   (a) Which one of the two solutions given is positive?
   (b) Which one is negative?
   (c) How can you tell?

Solve each problem by using a quadratic equation. Use a calculator as necessary, and round the answer to the nearest tenth. See Examples 3 and 4.

23. The Mart Hotel in Dallas, Texas, is 400 feet high. Suppose that a ball is projected upward from the top of the Mart, and its position s in feet above the ground is given by the equation $s = -16t^2 + 45t + 400$, where t is the number of seconds elapsed. How long will it take for the ball to reach a height of 200 feet above the ground? (*Source: The World Almanac and Book of Facts.*)

24. The Toronto Dominion Center in Winnipeg, Manitoba, is 407 feet high. Suppose that a ball is projected upward from the top of the Center, and its position s in feet above the ground is given by the equation $s = -16t^2 + 75t + 407$, where t is the number of seconds elapsed. How long will it take for the ball to reach a height of 450 feet above the ground? (*Source: The World Almanac and Book of Facts.*)

25. Refer to the equations in Exercises 23 and 24. Suppose that the first sentence in each problem did not give the height of the building. How could you use the equation to determine the height of the building?

26. A search light moves horizontally back and forth along a wall with the distance of the light from a starting point at t minutes given by $s = 100t^2 - 300t$. How long will it take before the light returns to the starting point?

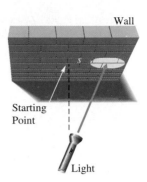

27. An object is projected directly upward from the ground. After t seconds its distance in feet above the ground is $s = 144t - 16t^2$.
 (a) After how many seconds will the object be 128 feet above the ground? (*Hint:* Look for a common factor before solving the equation.)
 (b) When does the object strike the ground?

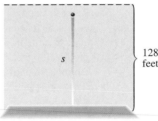

Ground level

28. The formula $D = 100t - 13t^2$ gives the distance in feet a car going approximately 68 miles per hour will skid in t seconds. Find the time it would take for the car to skid 190 feet. (*Hint:* Your answer must be less than the time it takes the car to stop, which is 3.8 seconds.)

Refer to Example 4 for Exercises 29 and 30.

29. Suppose a space vehicle leaves the moon under the conditions in Example 4. The force of gravity on the moon is approximately 3320 feet per second squared. Answer the following using four significant digits.
 (a) Write an equation for the height y of the space vehicle at x seconds.
 (b) Find the height after .25 second and after .5 second.
 (c) At what time(s) is the vehicle on the surface of the moon (that is, when is $y = 0$)?
 (d) Explain why there are two answers for part (c).

30. Refer to Exercise 29. If a rock on the surface of the moon is propelled at an angle of 60° with an initial velocity of 60 miles an hour, the equation is $y = 1.727x - .0013x^2$ feet, where x is time in seconds.
 (a) Find the height of the rock after 1 second and after 2 seconds.
 (b) How long will it take the rock to return to the surface?

Use the Pythagorean formula to solve each problem. Use a calculator as necessary, and when appropriate, round answers to the nearest tenth. See Example 5.

31. Find the lengths of the sides of the triangle.

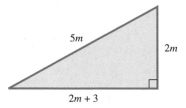

32. Find the lengths of the sides of the triangle.

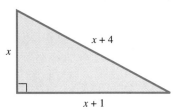

33. Refer to Exercise 23. Suppose that a wire is attached to the top of the Mart, and pulled tight. It is attached to the ground 100 feet from the base of the building, as shown in the figure. How long is the wire?

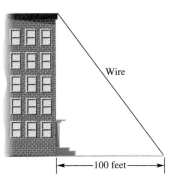

34. Refer to Exercise 24. Suppose that a wire is attached to the top of the Center, and pulled tight. The length of the wire is twice the distance between the base of the Center and the point on the ground where the wire is attached. How long is the wire?

35. Two ships leave port at the same time, one heading due south and the other heading due east. Several hours later, they are 170 miles apart. If the ship traveling south had traveled 70 miles farther than the other, how many miles had they each traveled?

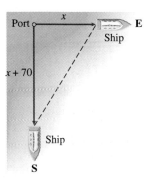

36. Kim Hobbs is flying a kite that is 30 feet farther above her hand than its horizontal distance from her. The string from her hand to the kite is 150 feet long. How high is the kite?

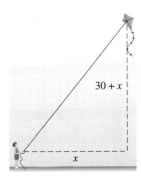

37. A toy manufacturer needs a piece of plastic in the shape of a right triangle with the longer leg 2 centimeters more than twice as long as the shorter leg, and the hypotenuse 1 centimeter more than the longer leg. How long should the three sides of the triangular piece be?

38. Michael Fuentes, a developer, owns a piece of land enclosed on three sides by streets, giving it the shape of a right triangle. The hypotenuse is 8 meters longer than the longer leg, and the shorter leg is 9 meters shorter than the hypotenuse. Find the lengths of the three sides of the property.

39. Two pieces of a large wooden puzzle fit together to form a rectangle with a length 1 centimeter less than twice the width. The diagonal, where the two pieces meet, is 2.5 centimeters in length. Find the length and width of the rectangle.

40. A 13-foot ladder is leaning against a house. The distance from the bottom of the ladder to the house is 7 feet less than the distance from the top of the ladder to the ground. How far is the bottom of the ladder from the house?

Use a quadratic equation to solve each problem. See Example 6.

41. Catarina and José want to buy a rug for a room that is 15 by 20 feet. They want to leave an even strip of flooring uncovered around the edges of the room. How wide a strip will they have if they buy a rug with an area of 234 square feet?

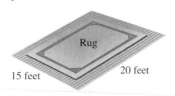

42. A club swimming pool is 30 feet wide and 40 feet long. The club members want an exposed aggregate border in a strip of uniform width around the pool. They have enough material for 296 square feet. How wide can the strip be?

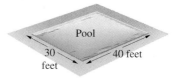

43. Arif's backyard is 20 by 30 meters. He wants to put a flower garden in the middle of the backyard, leaving a strip of grass of uniform width around the flower garden. To be happy, Arif must have 184 square meters of grass. Under these conditions what will the length and width of the garden be?

44. A rectangular piece of sheet metal has a length that is 4 inches less than twice the width. A square piece 2 inches on a side is cut from each corner. The sides are then turned up to form an uncovered box of volume 256 cubic inches. Find the length and width of the original piece of metal.

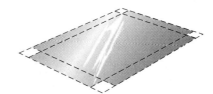

Solve each problem using a quadratic equation.

45. The formula $A = P(1 + r)^2$ gives the amount A in dollars that P dollars will grow to in 2 years at interest rate r (where r is given as a decimal), using compound interest. What interest rate will cause $2000 to grow to $2142.25 in 2 years?

46. If a square piece of cardboard has 3-inch squares cut from its corners and then has the flaps folded up to form an open-top box, the volume of the box is given by the formula $V = 3(x - 6)^2$, where x is the length of each side of the original piece of cardboard in inches. What original length would yield a box with a volume of 432 cubic inches?

47. A certain bakery has found that the daily demand for bran muffins is $\dfrac{3200}{p}$, where p is the price of a muffin in cents. The daily supply is $3p - 200$. Find the price at which supply and demand are equal.

48. In one area the demand for compact discs is $\dfrac{700}{P}$ per day, where P is the price in dollars per disc. The supply is $5P - 1$ per day. At what price does supply equal demand?

William Froude was a 19th century naval architect who used the expression $\dfrac{v^2}{g\ell}$ in shipbuilding. This expression, known as the Froude number, was also used by R. McNeill Alexander in his research on dinosaurs. (See "How Dinosaurs Ran," in Scientific American, *April 1991, pp. 130–136.) In Exercises 49 and 50, ℓ is given, as well as the value of the Froude number. Find the value of v (in meters per second). It is known that $g = 9.8$ meters per second squared.*

49. Rhinoceros: $\ell = 1.2$; Froude number = 2.57

50. Triceratops: $\ell = 2.8$; Froude number = .16

Recall that the corresponding sides of similar triangles are proportional. Use this fact to find the lengths of the indicated sides of each pair of similar triangles. Check all possible solutions in both triangles. Sides of a triangle cannot be negative.

51. Side AC

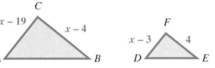

52. Side RQ

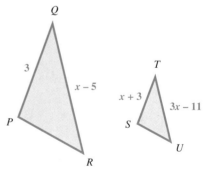

Solve each equation for the indicated variable.

53. $p = \dfrac{E^2R}{(r + R)^2}$; for R $(E > 0)$

54. $S(6S - t) = t^2$; for S

55. $10p^2c^2 + 7pcr = 12r^2$; for r

56. $S = vt + \dfrac{1}{2}gt^2$; for t

57. $LI^2 + RI + \dfrac{1}{c} = 0$; for I

58. $P = EI - RI^2$; for I

11.7 Graphs of Quadratic Functions

In Chapter 5, we graphed a few simple second-degree polynomial equations by point-plotting. In Figure 7, we repeat a table of ordered pairs for the simplest quadratic equation, $y = x^2$, and the resulting graph.

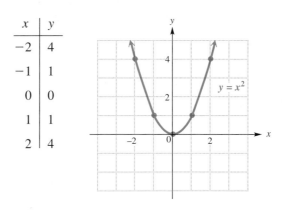

x	y
-2	4
-1	1
0	0
1	1
2	4

Figure 7

As mentioned in Chapter 5, this graph is called a **parabola.** The point (0, 0), the lowest point on the curve, is the **vertex** of this parabola. The vertical line through the vertex is the **axis** of this parabola. The parabola is **symmetric about its axis;** that is, if the graph were folded along the axis, the two portions of the curve would coincide. As Figure 7 suggests, the graph is that of a function. The domain of the function defined by $y = x^2$ is $(-\infty, \infty)$, and, since y is always nonnegative, the range is $[0, \infty)$.

OBJECTIVE 1 Graph a quadratic function. We now consider the graphs of more general *quadratic functions* as defined here.

Quadratic Function

A function that can be written in the form

$$f(x) = ax^2 + bx + c$$

for real numbers a, b, and c, with $a \neq 0$, is a **quadratic function.**

The graph of any quadratic function is a parabola with a vertical axis. For the rest of this section and the next, we use the symbols y and $f(x)$ interchangeably when discussing parabolas.

OBJECTIVE 2 Find the vertex of a parabola. Parabolas need not have their vertices at the origin, as does $f(x) = x^2$. For example, to graph a parabola of the form $f(x) = x^2 + k$, start by selecting sample values of x that were used to graph $f(x) = x^2$. The corresponding values of $f(x)$ in $f(x) = x^2 + k$ differ by k from those of $f(x) = x^2$. For this reason, the graph of $f(x) = x^2 + k$ is *shifted,* or *translated,* k units vertically compared with that of $f(x) = x^2$.

E X A M P L E   1   Graphing a Parabola with a Vertical Shift

Graph $f(x) = x^2 - 2$.

This graph has the same shape as $f(x) = x^2$, but since k here is -2, the graph is shifted 2 units downward, with vertex at $(0, -2)$. Every function value is 2 less than the corresponding function value of $f(x) = x^2$. Plotting points gives the graph in Figure 8. Note that x can be any real number, so the domain is still $(-\infty, \infty)$; since the value of y (or $f(x)$) is always greater than or equal to -2, the range is $[-2, \infty)$. The graph of $f(x) = x^2$ is shown for comparison.

x	$f(x) = x^2$	$f(x) = x^2 - 2$
-2	4	2
-1	1	-1
0	0	-2
1	1	-1
2	4	2

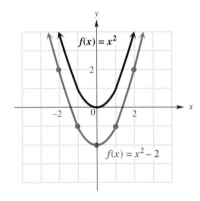

Figure 8

Vertical Shifts

The graph of $f(x) = x^2 + k$ is a parabola with the same shape as the graph of $f(x) = x^2$. The parabola is shifted k units upward if $k > 0$, and $|k|$ units downward if $k < 0$. The vertex is $(0, k)$.

The graph of $f(x) = (x - h)^2$ is also a parabola with the same shape as $f(x) = x^2$. The vertex of the parabola $f(x) = (x - h)^2$ is the lowest point on the parabola. The lowest point occurs here when $f(x)$ is 0. To get $f(x)$ equal to 0, we let $x = h$, so the vertex of $f(x) = (x - h)^2$ is at $(h, 0)$. Based on this, the graph of $f(x) = (x - h)^2$ is shifted, or translated, h units horizontally compared with that of $f(x) = x^2$.

E X A M P L E   2   Graphing a Parabola with a Horizontal Shift

Graph $f(x) = (x - 2)^2$.

When $x = 2$, then $f(x) = 0$, giving the vertex $(2, 0)$. The parabola $f(x) = (x - 2)^2$ has the same shape as $f(x) = x^2$ but is shifted 2 units to the right, as shown in Figure 9 on the next page. Again, the domain is $(-\infty, \infty)$; the range is $[0, \infty)$. As before, we show the graph of $f(x) = x^2$ for comparison.

x	$f(x) = (x-2)^2$
0	4
1	1
2	0
3	1
4	4

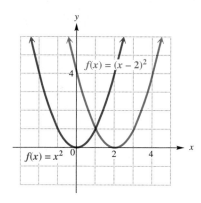

Figure 9

Horizontal Shifts

The graph of $f(x) = (x - h)^2$ is a parabola with the same shape as the graph of $f(x) = x^2$. The parabola is shifted h units horizontally: h units to the right if $h > 0$, and $|h|$ units to the left if $h < 0$. The vertex is $(h, 0)$.

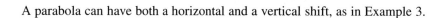

Errors frequently occur when horizontal shifts are involved. To determine the direction and magnitude of horizontal shifts, find the value that would cause the expression $x - h$ to equal 0. For example, the graph of $f(x) = (x - 5)^2$ would be shifted 5 units to the *right,* because $+5$ would cause $x - 5$ to equal 0. On the other hand, the graph of $f(x) = (x + 5)^2$ would be shifted 5 units to the *left,* because -5 would cause $x + 5$ to equal 0.

A parabola can have both a horizontal and a vertical shift, as in Example 3.

EXAMPLE 3   Graphing a Parabola with Horizontal and Vertical Shifts

Graph $f(x) = (x + 3)^2 - 2$.

This graph has the same shape as $f(x) = x^2$, but is shifted 3 units to the left (since $x + 3 = 0$ if $x = -3$), and 2 units downward (because of the -2). As shown in Figure 10, the vertex is at $(-3, -2)$. This function has domain $(-\infty, \infty)$ and range $[-2, \infty)$.

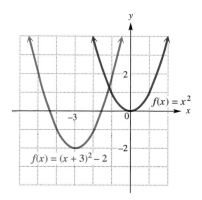

Figure 10

The characteristics of the graph of a parabola of the form $f(x) = (x - h)^2 + k$ are summarized as follows.

Vertical and Horizontal Shifts

The graph of $f(x) = (x - h)^2 + k$ is a parabola with the same shape as $f(x) = x^2$ and with vertex at (h, k). The axis is the vertical line $x = h$.

The vertical and horizontal shifts discussed in this section occur with other types of graphs as well. For example, the graph of $f(x) = \sqrt{x - h} + k$, with $h > 0$ and $k > 0$, would have the same shape as the graph of $f(x) = \sqrt{x}$ shifted h units to the right and k units up.

OBJECTIVE ‖ 3 ‖ Predict the shape and direction of the graph of a parabola from the coefficient of x^2. Not all parabolas open upward, and not all parabolas have the same shape as $f(x) = x^2$.

E X A M P L E   4   Graphing a Parabola That Opens Downward

Graph $f(x) = -\dfrac{1}{2}x^2$.

This parabola is shown in Figure 11. The coefficient $-\frac{1}{2}$ affects the shape of the graph; the $\frac{1}{2}$ makes the parabola wider (since the values of $\frac{1}{2}x^2$ increase more slowly than in $f(x) = x^2$), and the negative sign makes the parabola open downward. The graph is not shifted in any direction; the vertex is still $(0, 0)$. Here, the vertex has the *largest* function value of any point on the graph. The domain is $(-\infty, \infty)$; the range is $(-\infty, 0]$.

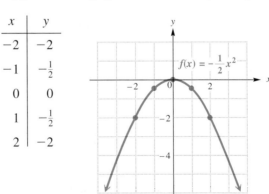

x	y
-2	-2
-1	$-\frac{1}{2}$
0	0
1	$-\frac{1}{2}$
2	-2

Figure 11

Some general principles concerning the graph of $f(x) = a(x - h)^2 + k$ are summarized as follows.

General Principles

1. The graph of the quadratic function

$$f(x) = a(x - h)^2 + k, \quad a \neq 0$$

   is a parabola with vertex at (h, k) and the vertical line $x = h$ as axis.

2. The graph opens upward if a is positive and downward if a is negative.

3. The graph is wider than $f(x) = x^2$ if $0 < |a| < 1$. The graph is narrower than $f(x) = x^2$ if $|a| > 1$.

E X A M P L E 5   Using the General Principles to Graph a Parabola

Graph $f(x) = -2(x + 3)^2 + 4$.

   The parabola opens downward (because $a < 0$), and is narrower than the graph of $f(x) = x^2$, since $|-2| = 2 > 1$, causing values of $f(x)$ to decrease more quickly than in $f(x) = -x^2$. This parabola has vertex $(-3, 4)$ as shown in Figure 12. To complete the graph, we plotted the ordered pairs $(-4, 2)$ and $(-2, 2)$. Notice that these two points are symmetric about the axis of the parabola, $x = -3$. Symmetry can be used to find additional ordered pairs that satisfy the equation.

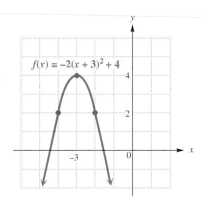

$f(x) = -2(x + 3)^2 + 4$

Figure 12

O B J E C T I V E 4   Use a quadratic function to model data.

E X A M P L E 6   Using a Quadratic Function as a Model for the Rise in Multiple Births

The number of higher-order multiple births in the United States is rising. Let x represent the number of years since 1970 and y represent the rate of increase of higher-order multiples born per 100,000 births since 1971. The data are shown in the table on the next page.

U.S. Higher-order Multiple Births

Year	x	y
1971	1	29.1
1976	6	35.0
1981	11	40.0
1986	16	47.0
1991	21	100.0
1996	26	152.6

Source: National Center for Health Statistics.

A graph of the ordered pairs (x, y) is shown in Figure 13(a). The graphs in this section suggest a parabola should approximate these points. Looking at the general shape suggested by the points in Figure 13(a), we see that the coefficient of x^2 should be positive, because the graph opens upward and should be less than 1, because the graph appears wider than the graph of $y = x^2$. Figure 13(b) shows the graph of $f(x) = .30x^2 - 3.52x + 37.3$, found by a method called *quadratic regression*. This quadratic function fits the set of data reasonably well.

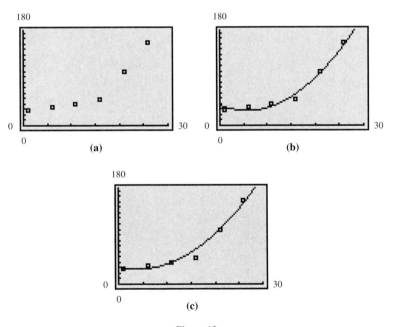

Figure 13

Another method used to find a quadratic function that fits these data is to choose three of the ordered pairs and use them to write a system of three equations. We want to find a, b, and c in the quadratic form $y = ax^2 + bx + c$. Choose any three points from the given data. We choose $(1, 29.1)$, $(11, 40)$, and $(21, 100)$. Substitute the values from these ordered pairs into the quadratic form $ax^2 + bx + c = y$ to get three equations.

$$a(1)^2 + b(1) \ + c = 29.1 \qquad \text{or} \qquad a + \ \ b + c = 29.1$$
$$a(11)^2 + b(11) + c = 40 \qquad \text{or} \qquad 121a + 11b + c = 40$$
$$a(21)^2 + b(21) + c = 100 \qquad \text{or} \qquad 441a + 21b + c = 100$$

Using one of the methods of Chapter 9 to solve this system of three equations in three variables, we find $a = .2455$, $b = -1.856$, and $c = 30.7105$, so the function we want is defined by $y = .2455x^2 - 1.856x + 30.7105$. This function is graphed with the data in Figure 13(c) and also appears to give a reasonably good approximation of the data.

11.7 EXERCISES

In Exercises 1–6, match each equation with the figure that most closely resembles its graph.

1. $g(x) = x^2 - 5$

2. $h(x) = -x^2 + 4$

3. $F(x) = (x - 1)^2$

4. $G(x) = (x + 1)^2$

5. $H(x) = (x - 1)^2 + 1$

6. $K(x) = (x + 1)^2 + 1$

A.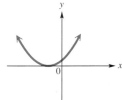

B.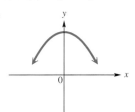

C.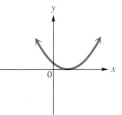

D.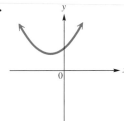

E.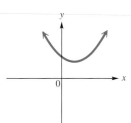

F.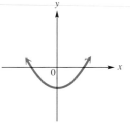

7. Explain in your own words the meaning of each term.
   (a) vertex of a parabola   **(b)** axis of a parabola

8. Explain why the axis of the graph of a quadratic function cannot be a horizontal line.

Identify the vertex of the graph of each quadratic function. See Examples 1–3.

9. $f(x) = -3x^2$

10. $f(x) = -.5x^2$

11. $f(x) = x^2 + 4$

12. $f(x) = x^2 - 4$

13. $f(x) = (x - 1)^2$

14. $f(x) = (x + 3)^2$

15. $f(x) = (x + 3)^2 - 4$

16. $f(x) = (x - 5)^2 - 8$

17. Describe how the graph of each parabola in Exercises 15 and 16 is shifted compared to the graph of $y = x^2$.

For each quadratic function, tell whether the graph opens upward or downward, and tell whether the graph is wider, narrower, or the same as the graph of $f(x) = x^2$. See Example 4.

18. $f(x) = -2x^2$

19. $f(x) = -3x^2 + 1$

20. $f(x) = .5x^2$

21. $f(x) = \frac{2}{3}x^2 - 4$

22. What does the value of a in $f(x) = a(x - h)^2 + k$ tell you about the graph of the function compared to the graph of $y = x^2$?

23. For $f(x) = a(x - h)^2 + k$, in what quadrant is the vertex if:
(a) $h > 0, k > 0$;   (b) $h > 0, k < 0$;   (c) $h < 0, k > 0$;   (d) $h < 0, k < 0$?

24. (a) What is the value of h if the graph of $f(x) = a(x - h)^2 + k$ has vertex on the y-axis?
(b) What is the value of k if the graph of $f(x) = a(x - h)^2 + k$ has vertex on the x-axis?

Sketch the graph of each parabola. Plot at least two points in addition to the vertex. Give the domain and range in Exercises 35–40. See Examples 1–5.

25. $f(x) = -2x^2$

26. $f(x) = \dfrac{1}{3}x^2$

27. $f(x) = x^2 - 1$

28. $f(x) = x^2 + 3$

29. $f(x) = -x^2 + 2$

30. $f(x) = 2x^2 - 2$

31. $f(x) = .5(x - 4)^2$

32. $f(x) = -2(x + 1)^2$

33. $f(x) = (x + 2)^2 - 1$

34. $f(x) = (x - 1)^2 + 2$

35. $f(x) = 2(x - 2)^2 - 4$

36. $f(x) = -2(x + 3)^2 + 4$

37. $f(x) = -.5(x + 1)^2 + 2$

38. $f(x) = -\dfrac{2}{3}(x + 2)^2 + 1$

39. $f(x) = 2(x - 2)^2 - 3$

40. $f(x) = \dfrac{4}{3}(x - 3)^2 - 2$

RELATING CONCEPTS (EXERCISES 41–46)

The procedures described in this section that allow the graph of $y = x^2$ to be shifted vertically and horizontally are applicable to other types of functions as well. In Section 8.4 we introduced linear functions (functions of the form $f(x) = ax + b$).

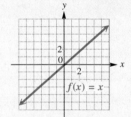

Consider the graph of the simplest linear function, $f(x) = x$, shown here, and then **work through Exercises 41–46 in order.**

41. Based on the concepts of this section, how does the graph of $y = x^2 + 6$ compare to the graph of $y = x^2$ if a *vertical* shift is considered?

42. Graph the linear function $y = x + 6$.

43. Based on the concepts of Chapter 8, how does the graph of $y = x + 6$ compare to the graph of $y = x$ if a vertical shift is considered? (*Hint:* Look at the y-intercept.)

44. Based on the concepts of this section, how does the graph of $y = (x - 6)^2$ compare to the graph of $y = x^2$ if a *horizontal* shift is considered?

45. Graph the linear function $y = x - 6$.

46. Based on the concepts of Chapter 8, how does the graph of $y = x - 6$ compare to the graph of $y = x$ if a horizontal shift is considered? (*Hint:* Look at the x-intercept.)

Did you make the connection that horizontal and vertical shifts of the graphs of equations other than $y = x^2$ occur in a similar way? Horizontal shifts of $y = f(x)$ are indicated by $y = f(x - k)$ and vertical shifts are indicated by $y = f(x) + k$.

In the following exercise, the distance formula is used to develop the equation of a parabola.

47. A parabola can be defined as the set of all points in a plane equally distant from a given point and a given line not containing the point. (The point is called the *focus* and the line is called the *directrix.*) See the figure.

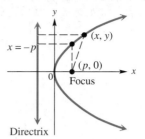

   (a) Suppose (x, y) is to be on the parabola. Suppose the directrix has equation $x = -p$. Find the distance between (x, y) and the directrix. (The distance from a point to a line is the length of the perpendicular from the point to the line.)

   (b) If $x = -p$ is the equation of the directrix, why should the focus have coordinates $(p, 0)$? (*Hint:* See the figure.)

   (c) Find an expression for the distance from (x, y) to $(p, 0)$.

   (d) Find an equation for the parabola of the figure. (*Hint:* Use the results of parts (a) and (c) and the fact that (x, y) is equally distant from the focus and the directrix.)

48. Use the equation derived in Exercise 47 to find an equation for a parabola with focus $(3, 0)$ and directrix with equation $x = -3$.

In Exercises 49 and 50, refer to Example 6.

49. The numbers of new AIDS patients who survived the first year for the years from 1991–1997 are shown in the table. In the year column, 1 represents 1991, 2 represents 1992, and so on.

AIDS Patients Who Survived the First Year

Year	Number of Patients
1	55
2	130
3	155
4	160
5	155
6	150
7	115

Source: HIV Health Services Planning Council.

   (a) Plot the ordered pairs (year, number of patients).

   (b) Use the graph to decide whether the coefficient a of x^2 in a quadratic model should be positive or negative.

   (c) Determine a quadratic function that models these data by using a system of equations. Use the ordered pairs $(2, 130)$, $(3, 155)$, and $(7, 115)$.

   (d) What would be an appropriate domain for $f(x)$? $\{1, 2, 3, 4, 5, 6, 7\}$? $(-\infty, \infty)$? Another choice?

50. CD-ROM revenues (in millions of dollars) for the years 1991–1994 are shown in the table. The years are coded so that 1 represents 1991, and so on.

CD-ROM Revenues

Year	Millions of Dollars
1	11.9
2	81.5
3	324.5
4	454.5

Source: Dataquest, Multimedia Market Trends, 1994.

Repeat parts (a)–(d) of Exercise 49 for these data. In part (c) use the data points $(1, 11.9)$, $(2, 81.5)$, and $(4, 454.5)$. In part (d) consider the set $\{1, 2, 3, 4\}$ instead of $\{1, 2, 3, 4, 5, 6, 7\}$.

TECHNOLOGY INSIGHTS (EXERCISES 51–56)

In Chapter 8 we saw that the x-value of the x-intercept of the graph of the line $y = mx + b$ is the solution of the linear equation $mx + b = 0$. In the same way, the x-values of the x-intercepts of the graph of the parabola $y = ax^2 + bx + c$ are the real solutions of the quadratic equation $ax^2 + bx + c = 0$. In Exercises 51–54, the calculator-generated graphs show the x-values of the x-intercepts of the graph of the polynomial in the equation.

Use the graphs to solve each equation.

51. $x^2 - x - 20 = 0$

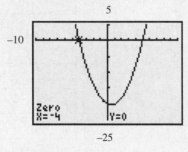 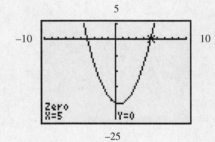

52. $x^2 + 9x + 14 = 0$

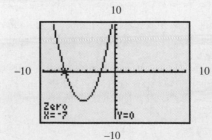 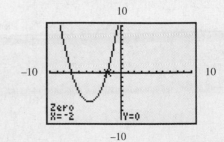

53. $-2x^2 + 5x + 3 = 0$

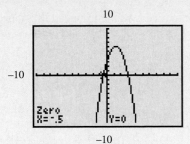 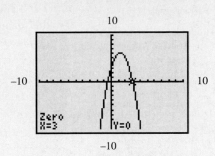

54. $-8x^2 + 6x + 5 = 0$

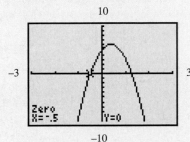 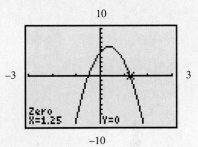

The graph of a quadratic function with $y = f(x)$ is shown in the standard viewing window, without x-axis tick marks.

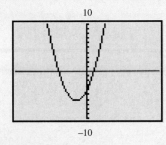

Refer to it to answer Exercises 55 and 56.

55. Which one of the following choices would be the only possible solution set for the equation $f(x) = 0$?
 (a) $\{-4, 1\}$ **(b)** $\{1, 4\}$ **(c)** $\{-1, -4\}$ **(d)** $\{4, -1\}$

56. Explain why only one choice in Exercise 55 is possible.

11.8 More About Parabolas; Applications

OBJECTIVES

1. Find the vertex of a vertical parabola.

2. Graph a quadratic function.

OBJECTIVE 1 Find the vertex of a vertical parabola. When the equation of a parabola is given in the form $f(x) = ax^2 + bx + c$, we need to locate the vertex in order to sketch an accurate graph. There are two ways to do this: complete the square as shown in Examples 1 and 2, or use a formula derived by completing the square.

3 Use the discriminant to find the number of x-intercepts of a vertical parabola.

4 Use quadratic functions to solve problems involving maximum or minimum value.

5 Graph horizontal parabolas.

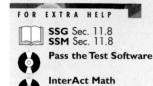
E X A M P L E   1   Completing the Square to Find the Vertex

Find the vertex of the graph of $f(x) = x^2 - 4x + 5$.

To find the vertex, express $x^2 - 4x + 5$ in the form $(x - h)^2 + k$ by completing the square. (This process was introduced in Section 11.2.) To simplify the notation, we replace $f(x)$ with y.

$$y = x^2 - 4x + 5$$

$$y - 5 = x^2 - 4x \qquad \text{Get the constant term on the left.}$$

$$y - 5 + 4 = x^2 - 4x + 4 \qquad \text{Half of } -4 \text{ is } -2; (-2)^2 = 4. \text{ Add 4 to both sides.}$$

$$y - 1 = (x - 2)^2 \qquad \text{Combine terms on the left and factor on the right.}$$

$$y = (x - 2)^2 + 1 \qquad \text{Add 1 to both sides.}$$

This form shows that the vertex of the parabola is (2, 1).

E X A M P L E   2   Completing the Square to Find the Vertex When $a \neq 1$

Find the vertex of the graph of $y = -3x^2 + 6x - 1$.

Complete the square on $-3x^2 + 6x$. Because the x^2 term has a coefficient other than 1, divide both sides by this coefficient, and then proceed as in Example 1.

$$y = -3x^2 + 6x - 1$$

$$\frac{y}{-3} = x^2 - 2x + \frac{1}{3} \qquad \text{Divide both sides by } -3.$$

$$\frac{y}{-3} - \frac{1}{3} = x^2 - 2x \qquad \text{Get the constant term on the left.}$$

$$\frac{y}{-3} - \frac{1}{3} + 1 = x^2 - 2x + 1 \qquad \begin{array}{l}\text{Half of } -2 \text{ is } -1; (-1)^2 = 1. \\ \text{Add 1 to both sides.}\end{array}$$

$$\frac{y}{-3} + \frac{2}{3} = (x - 1)^2 \qquad \begin{array}{l}\text{Combine terms on the left and} \\ \text{factor on the right.}\end{array}$$

$$\frac{y}{-3} = (x - 1)^2 - \frac{2}{3} \qquad \text{Subtract } \tfrac{2}{3} \text{ from both sides.}$$

$$y = -3(x - 1)^2 + 2 \qquad \text{Multiply by } -3 \text{ to get desired form.}$$

The vertex is (1, 2).

To derive a formula for the vertex of the graph of the quadratic function $y = ax^2 + bx + c$, complete the square on the standard form of the equation. Going through the same steps as in Example 2 gives the equation

$$y = a\left[x - \underbrace{\left(\frac{-b}{2a}\right)}_{h}\right]^2 + \underbrace{\frac{4ac - b^2}{4a}}_{k}.$$

This equation shows that the vertex (h, k) can be expressed in terms of a, b, and c. However, it is not necessary to remember this expression for k, since it can be found by replacing x with $\frac{-b}{2a}$. Using function notation, if $y = f(x)$, the y-value of the vertex is $f\left(\frac{-b}{2a}\right)$.

Vertex Formula

The graph of the quadratic function with $f(x) = ax^2 + bx + c$ has vertex

$$\left(\frac{-b}{2a}, f\left(\frac{-b}{2a} \right) \right),$$

and the axis of the parabola is the line $x = \frac{-b}{2a}$.

E X A M P L E 3   Using the Formula to Find the Vertex

Use the vertex formula to find the vertex of the graph of the function

$$f(x) = x^2 - x - 6.$$

For this function, $a = 1$, $b = -1$, and $c = -6$. The x-coordinate of the vertex of the parabola is

$$\frac{-b}{2a} = \frac{-(-1)}{2(1)} = \frac{1}{2}.$$

The y-coordinate is $f\left(\dfrac{-b}{2a} \right) = f\left(\dfrac{1}{2} \right)$.

$$f\left(\frac{1}{2} \right) = \left(\frac{1}{2} \right)^2 - \frac{1}{2} - 6 = \frac{1}{4} - \frac{1}{2} - 6 = -\frac{25}{4}$$

Finally, the vertex is $\left(\frac{1}{2}, -\frac{25}{4} \right)$.

OBJECTIVE 2   **Graph a quadratic function.** We give a general approach for graphing any quadratic function here.

Graphing a Quadratic Function f

Step 1   **Find the y-intercept.**   Find the y-intercept by evaluating $f(0)$.

Step 2   **Find the x-intercepts.**   Find the x-intercepts, if any, by solving $f(x) = 0$.

Step 3   **Find the vertex.**   Find the vertex either by using the formula or by completing the square.

Step 4   **Complete the graph.**   Find and plot additional points as needed, using symmetry about the axis.

Verify that the graph opens upward (if $a > 0$) or opens downward (if $a < 0$).

E X A M P L E 4   Using the Steps for Graphing a Quadratic Function

Graph the quadratic function $f(x) = x^2 - x - 6$.

Step 1   Find the y-intercept.

$$f(x) = x^2 - x - 6$$
$$f(0) = 0^2 - 0 - 6 \qquad \text{Find } f(0).$$
$$f(0) = -6$$

The y-intercept is $(0, -6)$.

Step 2 Find any *x*-intercepts.

$$f(x) = x^2 - x - 6$$
$$0 = x^2 - x - 6 \qquad \text{Let } f(x) = 0.$$
$$0 = (x - 3)(x + 2) \qquad \text{Factor.}$$
$$x - 3 = 0 \quad \text{or} \quad x + 2 = 0 \qquad \text{Set each factor equal to 0 and solve.}$$
$$x = 3 \quad \text{or} \quad x = -2$$

The *x*-intercepts are $(3, 0)$ and $(-2, 0)$.

Step 3 The vertex, found in Example 3, is $\left(\frac{1}{2}, -\frac{25}{4}\right)$.

Step 4 Plot the points found so far and additional points as needed using symmetry. The graph is shown in Figure 14.

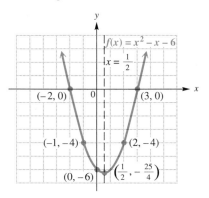

Figure 14

O B J E C T I V E **3** Use the discriminant to find the number of *x*-intercepts of a vertical parabola. The graph of a quadratic function may have two *x*-intercepts, one *x*-intercept, or no *x*-intercepts, as shown in Figure 15. Recall from Section 11.3 that the value of $b^2 - 4ac$ is called the *discriminant* of the quadratic equation $ax^2 + bx + c = 0$. We can use it to determine the number of real solutions of a quadratic equation. In a similar way, we can use the discriminant of a quadratic *function* to determine the number of *x*-intercepts of its graph. If the discriminant is positive, the parabola will have two *x*-intercepts. If the discriminant is 0, there will be only one *x*-intercept, and it will be the vertex of the parabola. If the discriminant is negative, the graph will have no *x*-intercepts.

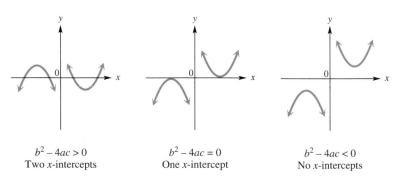

Figure 15

┌ **E X A M P L E 5**   Using the Discriminant to Determine the Number of *x*-intercepts

Determine the number of *x*-intercepts of the graph of each quadratic function. Use the discriminant.

(a) $f(x) = 2x^2 + 3x - 5$
    The discriminant is $b^2 - 4ac$. Here $a = 2$, $b = 3$, and $c = -5$, so

$$b^2 - 4ac = 9 - 4(2)(-5) = 49.$$

Since the discriminant is positive, the parabola has two *x*-intercepts.

(b) $f(x) = -3x^2 - 1$
    In this equation, $a = -3$, $b = 0$, and $c = -1$. The discriminant is

$$b^2 - 4ac = 0 - 4(-3)(-1) = -12.$$

The discriminant is negative and so the graph has no *x*-intercepts.

(c) $f(x) = 9x^2 + 6x + 1$
    Here, $a = 9$, $b = 6$, and $c = 1$. The discriminant is

$$b^2 - 4ac = 36 - 4(9)(1) = 0.$$

The parabola has only one *x*-intercept (its vertex), since the value of the discriminant is 0.

O B J E C T I V E 4   Use quadratic functions to solve problems involving maximum or minimum value. The vertex of a parabola is either the highest or the lowest point on the parabola. The *y*-value of the vertex gives the maximum or minimum value of *y*, while the *x*-value tells where that maximum or minimum occurs.

PROBLEM SOLVING

In many applied problems we must find the largest or smallest value of some quantity. When we can express that quantity as a quadratic function, the value of *k* in the vertex gives that optimum value.

┌ **E X A M P L E 6**   Finding the Maximum Area of a Rectangular Region

A farmer has 120 feet of fencing. He wants to put a fence around a rectangular plot of land next to a river. Find the maximum area he can enclose.

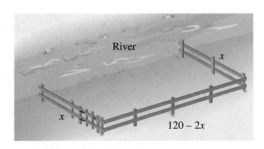

Figure 16

Figure 16 on the previous page shows the plot. Let x represent the width of the plot. Since there are 120 feet of fencing,

$$x + x + \text{length} = 120 \qquad \text{Sum of the sides is 120 feet.}$$
$$2x + \text{length} = 120 \qquad \text{Combine terms.}$$
$$\text{length} = 120 - 2x. \qquad \text{Subtract } 2x.$$

The area is given by the product of the width and length, so

$$A = x(120 - 2x) = 120x - 2x^2.$$

To determine the maximum area, find the vertex of the parabola $A = 120x - 2x^2$ using the vertex formula. Writing the equation in standard form as $A = -2x^2 + 120x$ shows that $a = -2$, $b = 120$, and $c = 0$, so

$$h = \frac{-b}{2a} = \frac{-120}{2(-2)} = \frac{-120}{-4} = 30$$

$$f(30) = -2(30)^2 + 120(30) = -2(900) + 3600 = 1800.$$

The graph is a parabola that opens downward, and its vertex is (30, 1800). Thus, the maximum area will be 1800 square feet. This area will occur if x, the width of the plot, is 30 feet.

Be careful when interpreting the meanings of the coordinates of the vertex. The first coordinate, x, gives the value for which the *function value* is a maximum or a minimum. Be sure to read the problem carefully to determine whether you are asked to find the value of the independent variable, the function value, or both.

Earlier in this chapter, we discussed the motion of a propelled object. Now we can find the maximum height of the object.

 EXAMPLE 7 Finding the Maximum Height Attained by a Rocket

In Example 4, Section 11.6, we found the height of a rocket propelled from the surface of Mars to be given by $f(x) = x - .0066x^2$, where x is time in seconds for the rocket to reach y feet. Find the maximum height the rocket will reach. (*Source:* Bernice Kastner, *Space Mathematics,* NASA.)

Use the formula for the vertex, with $a = -.0066$, $b = 1$, and $c = 0$.

$$h = \frac{-b}{2a} = \frac{-1}{2(-.0066)} \approx 75.8$$

This tells us that the maximum height is attained at 75.8 seconds. To find that height, calculate $f(75.8)$.

$$f(75.8) - 75.8 - .0066(75.8)^2$$
$$\approx 37.9 \qquad \text{Use a calculator.}$$

The rocket will attain a maximum height of approximately 37.9 feet.

OBJECTIVE 5 Graph horizontal parabolas. If x and y are exchanged in the equation $y = ax^2 + bx + c$, the equation becomes $x = ay^2 + by + c$. Because of the interchange of the roles of x and y, these parabolas are horizontal (with horizontal lines as axes), compared with the vertical ones graphed previously.

Graph of a Horizontal Parabola

The graph of $x = ay^2 + by + c$ or $x = a(y - k)^2 + h$ is a parabola with vertex at (h, k) and the horizontal line $y = k$ as axis. The graph opens to the right if a is positive and to the left if a is negative.

E X A M P L E 8   Graphing a Horizontal Parabola

Graph $x = (y - 2)^2 - 3$.

This graph has its vertex at $(-3, 2)$, since the roles of x and y are reversed. It opens to the right, the positive x-direction, and has the same shape as $y = x^2$. Plotting a few additional points gives the graph shown in Figure 17. Note that the graph is symmetric about its axis, $y = 2$.

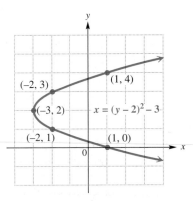

Figure 17

When a quadratic equation is given in the form $x = ay^2 + by + c$, completing the square on y changes the equation into a form in which the vertex can be more easily identified.

E X A M P L E 9   Completing the Square to Graph a Horizontal Parabola

Graph $x = -2y^2 + 4y - 3$.

Complete the square on the right to express the equation in the form $x = a(y - k)^2 + h$.

$$\frac{x}{-2} = y^2 - 2y + \frac{3}{2} \qquad \text{Divide by } -2.$$

$$\frac{x}{-2} - \frac{3}{2} = y^2 - 2y \qquad \text{Subtract } \tfrac{3}{2}.$$

$$\frac{x}{-2} - \frac{3}{2} + 1 = y^2 - 2y + 1 \qquad \text{Add 1.}$$

$$\frac{x}{-2} - \frac{1}{2} = (y - 1)^2 \qquad \text{Factor on the right; add on the left.}$$

$$\frac{x}{-2} = (y - 1)^2 + \frac{1}{2} \qquad \text{Add } \tfrac{1}{2}.$$

$$x = -2(y - 1)^2 - 1 \qquad \text{Multiply by } -2.$$

The coefficient -2 indicates the graph opens to the left (the negative x-direction) and is narrower than $y = x^2$. As shown in Figure 18, the vertex is $(-1, 1)$.

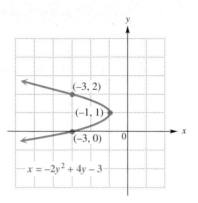

$x = -2y^2 + 4y - 3$

Figure 18

 Only quadratic equations solved for y (whose graphs are vertical parabolas) are examples of functions. The horizontal parabolas in Examples 8 and 9 are *not* graphs of functions, because they do not satisfy the vertical line test. Furthermore, the vertex formula does not apply to parabolas with horizontal axes.

In summary, the graphs of parabolas studied in this section and the previous one fall into the following categories.

Graphs of Parabolas

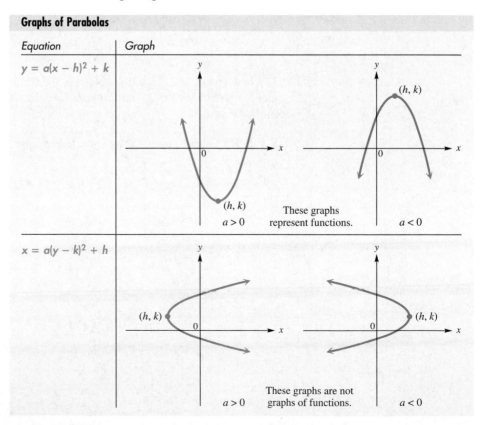

11.8 EXERCISES

1. How can you determine just by looking at the equation of a parabola whether it has a vertical or a horizontal axis?

2. Why can't the graph of a quadratic function be a parabola that opens to the left or to the right?

3. (a) How can you determine the number of x-intercepts of the graph of a quadratic function without graphing the function?
   (b) If the vertex of the graph of a quadratic function is $(1, -3)$, and the graph opens downward, how many x-intercepts does the graph have?

4. Explain how to graph a quadratic function given in the form $f(x) = ax^2 + bx + c$. Use $f(x) = x^2 - 3x - 4$ as an example.

Find the vertex of each parabola. Decide whether the graph opens upward, downward, to the left, or to the right, and state whether it is wider, narrower, or the same shape as the graph of $y = x^2$. If it is a vertical parabola, use the discriminant to determine the number of x-intercepts. See Examples 1–3, 5, 8, and 9.

5. $y = 2x^2 + 4x + 5$ 6. $y = 3x^2 - 6x + 4$ 7. $y = -x^2 + 5x + 3$

8. $x = -y^2 + 7y + 2$ 9. $x = \frac{1}{3}y^2 + 6y + 24$ 10. $x = .5y^2 + 10y - 5$

Graph the parabola using the techniques described in this section. If the graph represents a function, give its domain and range. See Examples 3, 4, 8, and 9.

11. $f(x) = x^2 + 8x + 10$ 12. $f(x) = x^2 + 10x + 23$

13. $y = -2x^2 + 4x - 5$ 14. $y = -3x^2 + 12x - 8$

15. $x = -\frac{1}{5}y^2 + 2y - 4$ 16. $x = -.5y^2 - 4y - 6$

17. $x = 3y^2 + 12y + 5$ 18. $x = 4y^2 + 16y + 11$

Use the concepts of this section to match the equation with its graph in Exercises 19–24.

A. 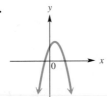   B. 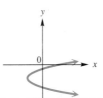   C.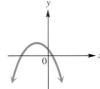

D. 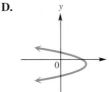   E. 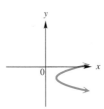   F.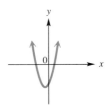

19. $y = 2x^2 + 4x - 3$ 20. $y = -x^2 + 3x + 5$ 21. $y = -\frac{1}{2}x^2 - x + 1$

22. $x = y^2 + 6y + 3$ 23. $x = -y^2 - 2y + 4$ 24. $x = 3y^2 + 6y + 5$

Solve each problem. See Examples 6 and 7.

25. Keisha Hughes has 100 meters of fencing material to enclose a rectangular exercise run for her dog. What width will give the enclosure the maximum area?

26. Morgan's Department Store wants to construct a rectangular parking lot on land bordered on one side by a highway. It has 280 feet of fencing that is to be used to fence off the other three sides. What should be the dimensions of the lot if the enclosed area is to be a maximum? What is the maximum area?

27. If an object on Earth is propelled upward with an initial velocity of 32 feet per second, then its height after t seconds is given by

$$h = 32t - 16t^2.$$

Find the maximum height attained by the object and the number of seconds it takes to hit the ground.

28. A projectile on Earth is fired straight upward so that its distance (in feet) above the ground t seconds after firing is given by

$$s(t) = -16t^2 + 400t.$$

Find the maximum height it reaches and the number of seconds it takes to reach that height.

29. If air resistance is neglected, a projectile on Earth shot straight upward with an initial velocity of 40 meters per second will be at a height s in meters given by the function

$$s(t) = -4.9t^2 + 40t,$$

where t is the number of seconds elapsed after projection. After how many seconds will it reach its maximum height, and what is this maximum height? Round your answers to the nearest tenth.

30. A space robot is propelled from the moon with its distance in feet given by $f(x) = 1.727x - .0013x^2$ feet, where x is time in seconds. (See Section 11.6, Exercise 30.) Find the maximum height the robot can reach, and the time to get there. Give answers with four significant digits.

31. The bar graph shows how Social Security assets are expected to change as the number of retirees receiving benefits increases.

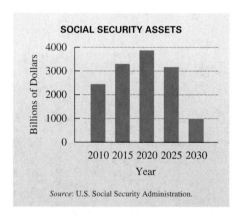

SOCIAL SECURITY ASSETS

Source: U.S. Social Security Administration.

The graph suggests that a quadratic function would be a good fit to the data. Using a statistical technique, we determined that the data are approximated by the function with $f(x) = -20.57x^2 + 758.9x - 3140$. In the equation, $x = 10$ represents 2010, $x = 15$ represents 2015, and so on, and $f(x)$ is in billions of dollars.

(a) Explain why the coefficient of x^2 in the equation is negative, based on the graph.

(b) Algebraically determine the vertex of the graph, with coordinates to four significant digits.

(c) Interpret the answer to part (b) as it applies to the application.

32. The graph shows the performance of investment portfolios with different mixtures of U.S. and foreign investments for the period January 1, 1971, to December 31, 1996.

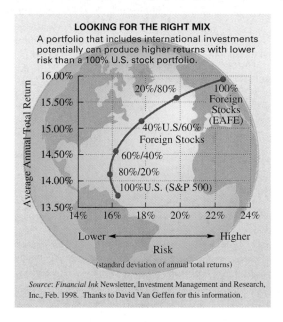

LOOKING FOR THE RIGHT MIX
A portfolio that includes international investments potentially can produce higher returns with lower risk than a 100% U.S. stock portfolio.

Source: *Financial Ink* Newsletter, Investment Management and Research, Inc., Feb. 1998. Thanks to David Van Geffen for this information.

Use the graph to answer the following questions.
  (a) Is this the graph of a function? Explain.
  (b) What investment mixture shown on the graph appears to represent the vertex? What relative amount of risk does this point represent? What return on investment does it provide?
  (c) Which point on the graph represents the riskiest investment mixture? What return on investment does it provide?

33. A charter flight charges a fare of $200 per person, plus $4 per person for each unsold seat on the plane. If the plane holds 100 passengers and if x represents the number of unsold seats, find the following.
  (a) An expression for the total revenue, R, received for the flight (*Hint:* Multiply the number of people flying, $100 - x$, by the price per ticket.)
  (b) The graph of the function from part (a)
  (c) The number of unsold seats that will produce the maximum revenue
  (d) The maximum revenue

34. For a trip to a resort, a charter bus company charges a fare of $48 per person, plus $2 per person for each unsold seat on the bus. If the bus has 42 seats and x represents the number of unsold seats, find the following.
  (a) An expression that defines the total revenue, R, from the trip (*Hint:* Multiply the total number riding, $42 - x$, by the price per ticket, $48 + 2x$)
  (b) The graph of the function from part (a)
  (c) The number of unsold seats that produces the maximum revenue
  (d) The maximum revenue

TECHNOLOGY INSIGHTS (EXERCISES 35-38)

Graphing calculators are capable of determining the coordinates of "peaks" and "valleys" of graphs. In the case of quadratic functions, these peaks and valleys are the vertices, and are called maximum and minimum points. For example, the vertex of the graph of $f(x) = -x^2 - 6x - 13$ is $(-3, -4)$, as indicated in the display at the bottom of the screen. In this case, the vertex is a maximum point.

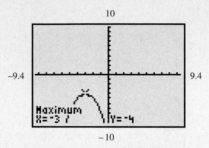

In Exercises 35–38, match the function with its calculator-generated graph by determining the vertex and using the display at the bottom of the screen.

A.

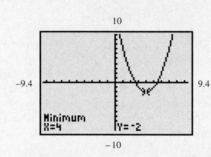

B.

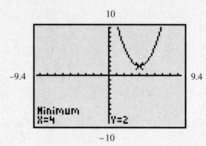

C.

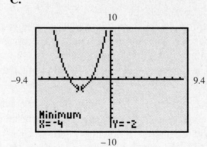

D.

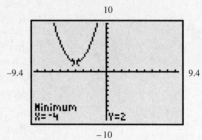

35. $f(x) = x^2 - 8x + 18$

37. $f(x) = x^2 - 8x + 14$

36. $f(x) = x^2 + 8x + 18$

38. $f(x) = x^2 + 8x + 14$

11.9 Nonlinear Inequalities

OBJECTIVES

1 Solve quadratic inequalities.

2 Solve polynomial inequalities of degree **3** or more.

3 Solve rational inequalities.

FOR EXTRA HELP

📖 **SSG** Sec. 11.9
SSM Sec. 11.9

💿 **Pass the Test Software**

💿 **InterAct Math Tutorial Software**

📼 **Video** 20

We discussed methods of solving linear inequalities in Chapter 3 and methods of solving quadratic equations in this chapter. Now we combine these ideas to solve *quadratic inequalities*.

Quadratic Inequality

A **quadratic inequality** can be written in the form

$$ax^2 + bx + c < 0 \qquad \text{or} \qquad ax^2 + bx + c > 0,$$

where a, b, and c are real numbers, with $a \neq 0$.

As before, $<$ and $>$ may be replaced with $\leq$ and $\geq$ as necessary.

OBJECTIVE 1 Solve quadratic inequalities.

┌ **EXAMPLE 1** Solving a Quadratic Inequality

Solve $x^2 - x - 12 > 0$.

First solve the quadratic *equation*

$$x^2 - x - 12 = 0.$$
$$(x - 4)(x + 3) = 0 \qquad \text{Factor.}$$
$$x - 4 = 0 \quad \text{or} \quad x + 3 = 0 \qquad \text{Zero-factor property}$$
$$x = 4 \quad \text{or} \quad x = -3$$

The numbers 4 and -3 divide the number line into three regions, as shown in Figure 19. (Be careful to graph the smaller number on the left.)

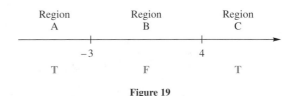

Figure 19

The numbers 4 and -3 are the only numbers that make the expression $x^2 - x - 12$ equal to 0. All other numbers make the expression either positive or negative, and the value of the expression can change from positive to negative or from negative to positive only on either side of a number that makes it 0. Therefore, if one number in a region satisfies the inequality, then all the numbers in that region will satisfy the inequality. Choose any number from Region A in Figure 19 (any number less than -3). Substitute this number for x in the original inequality. If the result is true, then all numbers in Region A satisfy the original inequality.

We choose -5 from Region A. Substitute -5 for x in the inequality.

$$x^2 - x - 12 > 0 \qquad \text{Original inequality}$$
$$(-5)^2 - (-5) - 12 > 0 \qquad ?$$
$$25 + 5 - 12 > 0 \qquad ?$$
$$18 > 0 \qquad \text{True}$$

Since -5 from Region A satisfies the inequality, all numbers from Region A are solutions.

We try 0 from Region B. If $x = 0$, then

$$0^2 - 0 - 12 > 0 \qquad ?$$
$$-12 > 0. \qquad \text{False}$$

The numbers in Region B are *not* solutions. Verify that the number 5 satisfies the inequality, so the numbers in Region C are also solutions to the inequality.

Based on these results (shown by the colored letters in Figure 19), the solution set includes the numbers in Regions A and C, as shown on the graph in Figure 20. The solution set is written

$$(-\infty, -3) \cup (4, \infty).$$

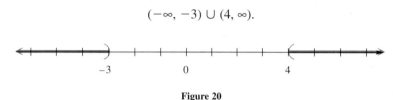

Figure 20

In summary, we solve a quadratic inequality by following these steps.

Solving a Quadratic Inequality

Step 1 **Change the inequality to an equation and solve it.**

Step 2 **Use the solutions from Step 1 to determine regions.** Graph the numbers found in Step 1 on a number line. These numbers divide the number line into regions.

Step 3 **Find the intervals that satisfy the inequality.** Substitute a number from each region into the inequality to determine the regions that satisfy the inequality. All numbers in those regions are in the solution set. A graph of the solution set will usually look like one of these.

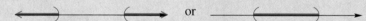

Step 4 **Consider the endpoints separately.** The numbers from Step 1 are included in the solution set if the inequality is $\leq$ or $\geq$; they are not included if it is $<$ or $>$.

Special cases of quadratic inequalities may occur, as in the next example.

EXAMPLE 2 Solving Special Cases

Solve $(2y - 3)^2 > -1$.

Since $(2y - 3)^2$ is never negative, it is always greater than -1. Thus, the solution is the set of all real numbers $(-\infty, \infty)$. In the same way, there is no solution for $(2y - 3)^2 < -1$ and the solution set is $\emptyset$.

OBJECTIVE 2 **Solve polynomial inequalities of degree 3 or more.** Higher-degree polynomial inequalities that can be factored are solved in the same way as quadratic inequalities.

EXAMPLE 3 Solving a Third-Degree Polynomial Inequality

Solve $(x - 1)(x + 2)(x - 4) \leq 0$.

This is a *cubic* (third-degree) inequality rather than a quadratic inequality, but it can be solved by the method shown above and by extending the zero-factor property to more than two factors. Begin by setting the factored polynomial *equal* to 0 and solving the equation. (Step 1)

$$(x - 1)(x + 2)(x - 4) = 0$$

$$x - 1 = 0 \quad \text{or} \quad x + 2 = 0 \quad \text{or} \quad x - 4 = 0$$

$$x = 1 \quad \text{or} \quad x = -2 \quad \text{or} \quad x = 4$$

Locate the numbers -2, 1, and 4 on a number line as in Figure 21 to determine the regions A, B, C, and D. (Step 2)

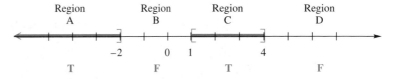

Figure 21

Substitute a number from each region into the original inequality to determine which regions satisfy the inequality. (Step 3) These results are shown below the number line in Figure 21. For example, in Region A, using $x = -3$ gives

$$(-3 - 1)(-3 + 2)(-3 - 4) \leq 0 \qquad ?$$

$$(-4)(-1)(-7) \leq 0 \qquad ?$$

$$-28 \leq 0. \qquad \text{True}$$

The numbers in Region A are in the solution set. Verify that the numbers in Region C are also in the solution set, which is written

$$(-\infty, -2] \cup [1, 4].$$

Notice that the three endpoints are included in the solution set (Step 4), graphed in Figure 21.

OBJECTIVE **3** Solve rational inequalities. Inequalities that involve fractions, called **rational inequalities,** can change sign only at values where the corresponding equation is 0 or where the denominator is 0. Steps 1, 2, and 5 in the following list consider these two cases.

Solving Rational Inequalities
Step 1 **Change the inequality to an equation and solve it.**
Step 2 **Set the denominator equal to 0 and solve the equation.**
Step 3 **Divide the number line into regions.** Use the solutions from Steps 1 and 2 to divide the number line into regions.

(continued)

Solving Rational Inequalities (continued)

Step 4 **Find the intervals that satisfy the inequality.** Test a number from each region by substituting it into the inequality to determine the intervals that satisfy the inequality.

Step 5 **Consider the endpoints.** Exclude any values that make the denominator 0.

 Remember Steps 2 and 5. Any number that makes the denominator 0 *must* be excluded from the solution set, since there will be no point on the number line at that value.

E X A M P L E 4   Solving a Rational Inequality

Solve the inequality $\dfrac{-1}{p-3} \geq 1$.

Write the corresponding equation and solve it. (Step 1)

$$\frac{-1}{p-3} = 1$$

$$-1 = p - 3 \qquad \text{Multiply by the common denominator.}$$

$$2 = p$$

Find the number that makes the denominator 0. (Step 2)

$$p - 3 = 0$$

$$p = 3$$

These two numbers, 2 and 3, divide a number line into three regions. (Step 3) (See Figure 22.)

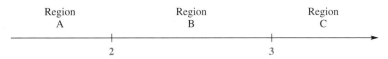

Figure 22

Testing one number from each region in the given inequality shows that the solution set is the interval [2, 3). (Step 4) Note that this interval does not include the number 3, because 3 makes the denominator of the original inequality equal to 0. (Step 5) A graph of the solution set is given in Figure 23.

Figure 23

CONNECTIONS

 As the examples in this section show, there is a close connection between equality and inequality. In Section 11.6, Example 4 discussed the motion of a rocket propelled from the surface of Mars. Under certain conditions, $y = x - .0066x^2$ gives the height of the rocket, in feet, x seconds after it leaves the surface. If we know the number of seconds it takes the rocket to reach 50 feet and return to the ground, we can determine the time intervals when it will be less than 50 feet high and more than 50 feet high. (*Source:* Bernice Kastner, *Space Mathematics,* NASA.)

FOR DISCUSSION OR WRITING

Use the quadratic formula and the first objective in this section to answer the following questions to the nearest tenth of a second.

1. At what times will the rocket be 20 feet above the surface? (*Hint:* Let $y = 20$ and solve the quadratic equation.)

2. In what time interval(s) will the rocket be more than 20 feet above the surface?

3. At what times will the rocket be at the surface? (*Hint:* Let $y = 0$ and solve the equation.)

4. In what time interval(s) will the rocket be less than 20 feet above the surface?

11.9 EXERCISES

Follow the steps in Exercises 1–4 to solve the inequality $2x^2 + x - 3 \leq 0$.

1. Change the inequality to an equation and solve it.

2. Indicate the regions determined by the solutions from Step 1 on a number line.

3. Determine the interval(s) that satisfy the inequality.

4. Determine whether the endpoints are included or not, and give the solution set in interval form.

Solve each inequality and graph the solution set. See Example 1. (Hint: In Exercises 17 and 18, use the quadratic formula.)

5. $(x + 1)(x - 5) > 0$	**6.** $(m + 6)(m - 2) > 0$	**7.** $(r + 4)(r - 6) < 0$
8. $(y + 4)(y - 8) < 0$	**9.** $x^2 - 4x + 3 \geq 0$	**10.** $m^2 - 3m - 10 \geq 0$
11. $10a^2 + 9a \geq 9$	**12.** $3r^2 + 10r \geq 8$	**13.** $9p^2 + 3p < 2$
14. $2y^2 + y < 15$	**15.** $6x^2 + x \geq 1$	**16.** $4y^2 + 7y \geq -3$
17. $y^2 - 6y + 6 \geq 0$	**18.** $3k^2 - 6k + 2 \leq 0$	

19. Explain how you determine whether to include or exclude endpoints when solving a quadratic or higher-degree inequality.

20. The solution set of the inequality $x^2 + x - 12 < 0$ is the interval $(-4, 3)$. Without actually performing any work, give the solution set of the inequality $x^2 + x - 12 \geq 0$.

Solve each inequality. See Example 2.

21. $(4 - 3x)^2 \geq -2$ **22.** $(6y + 7)^2 \geq -1$ **23.** $(3x + 5)^2 \leq -4$ **24.** $(8t + 5)^2 \leq -5$

25. Explain how you would solve and graph the solution set of a quadratic inequality. Use an example.

Solve each inequality and graph the solution set. See Example 3.

26. $(p - 1)(p - 2)(p - 4) < 0$

27. $(2r + 1)(3r - 2)(4r + 7) < 0$

28. $(a - 4)(2a + 3)(3a - 1) \geq 0$

29. $(z + 2)(4z - 3)(2z + 7) \geq 0$

30. Without actually performing any work, give the solution set of the rational inequality
$\dfrac{3}{x^2 + 1} > 0$. (*Hint:* Determine the sign of the numerator. Determine what the sign of the denominator *must* be. Then consider the inequality symbol.)

Solve each inequality and graph the solution set. See Example 4.

31. $\dfrac{x - 1}{x - 4} > 0$

32. $\dfrac{x + 1}{x - 5} > 0$

33. $\dfrac{2y + 3}{y - 5} \leq 0$

34. $\dfrac{3t + 7}{t - 3} \leq 0$

35. $\dfrac{8}{x - 2} \geq 2$

36. $\dfrac{20}{y - 1} \geq 1$

37. $\dfrac{3}{2t - 1} < 2$

38. $\dfrac{6}{m - 1} < 1$

39. $\dfrac{a}{a + 2} \geq 2$

40. $\dfrac{m}{m + 5} \geq 2$

41. $\dfrac{x}{x - 4} < 3$

42. $\dfrac{2y}{y + 1} > 4$

43. $\dfrac{4k}{2k - 1} < k$

44. $\dfrac{r}{r + 2} < 2r$

45. $\dfrac{2x - 3}{x^2 + 1} \geq 0$

46. $\dfrac{9x - 8}{4x^2 + 25} < 0$

47. $\dfrac{(3x - 5)^2}{x + 2} > 0$

48. $\dfrac{(5x - 3)^2}{2x + 1} \leq 0$

■ RELATING CONCEPTS (EXERCISES 49–52)

An alternative method for solving rational inequalities requires first writing the inequality with 0 on one side. The other side is then expressed as a single fraction.

Work Exercises 49–52 in order *to use the alternative method to solve the rational inequality in Example 4,*

$$\frac{-1}{p - 3} \geq 1.$$

49. Subtract 1 from both sides of the inequality to make one side equal to 0.

50. Rewrite the left side to get a common denominator for the two terms on the left.

51. To write the left side as a single fraction, perform the indicated subtraction and combine terms as needed.

52. Find the number that makes the numerator 0 and the number that makes the denominator 0. Complete the solution as in Example 4.

Did you make the connection between the two methods for solving rational inequalities? Which method do you prefer?

Work each problem. See the Connections box for this section.

53. The equation $y = x - 1.727x^2$, found in Section 11.6, Exercise 29, gives the height y of a space vehicle x seconds after leaving the moon. At what times will the height of the vehicle be greater than .140 foot?

54. In Section 11.6, Exercise 30, the height of a rock propelled from the surface of the moon is given as $y = 1.727x - .0013x^2$ feet, where x is time in seconds. The rock goes up and returns to the surface in approximately 1328 seconds. At what times will the height of the rock be less than 500 feet?

RELATING CONCEPTS (EXERCISES 55-58)

We know that the *x*-intercepts of the graph of a quadratic equation give the real solutions of the equation. We can also use these intercepts to solve the corresponding inequalities. The *x*-values of points on the graph that are *above* the *x*-axis give the solution set of the > inequality. Similarly, the solution set of the < inequality is found by locating the points on the graph that lie *below* the *x*-axis. For example, from the graph of $f(x) = x^2 - x - 12$ in the figure, the solution set of $x^2 - x - 12 > 0$ is $(-\infty, -3) \cup (4, \infty)$, and the solution set of $x^2 - x - 12 < 0$ is $(-3, 4)$.

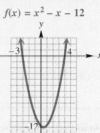

$f(x) = x^2 - x - 12$

The graph is *above* the *x*-axis for $(-\infty, -3) \cup (4, \infty)$.

In Exercises 55–58, the graph of a quadratic function f is given. Use only the graph to find the solution set of the equation or inequality. **Work through parts (a)–(c) in order each time.**

55. $f(x) = x^2 - 4x + 3$

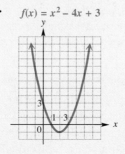

    (a) $x^2 - 4x + 3 = 0$
    (b) $x^2 - 4x + 3 > 0$
    (c) $x^2 - 4x + 3 < 0$

56. $f(x) = 3x^2 + 10x - 8$

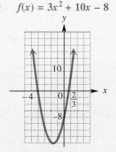

    (a) $3x^2 + 10x - 8 = 0$
    (b) $3x^2 + 10x - 8 \geq 0$
    (c) $3x^2 + 10x - 8 < 0$

RELATING CONCEPTS (EXERCISES 55-58) (CONTINUED)

57. $f(x) = -x^2 + 3x + 10$

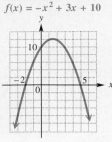

(a) $-x^2 + 3x + 10 = 0$
(b) $-x^2 + 3x + 10 \geq 0$
(c) $-x^2 + 3x + 10 \leq 0$

58. $f(x) = -2x^2 - x + 15$

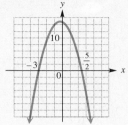

(a) $-2x^2 - x + 15 = 0$
(b) $-2x^2 - x + 15 \geq 0$
(c) $-2x^2 - x + 15 \leq 0$

Did you make the connection that the x-intercepts of the graph of a function where $y = f(x)$ give the endpoints of the solution sets of the inequalities $f(x) > 0$ and $f(x) < 0$?

Work each problem.

59. In Example 7 in Section 11.8, the height of a rocket, in feet, at time x, in seconds, is given as $f(x) = x - .0066x^2$. At what time(s) will the height of the rocket be less than 10 feet?

60. In Section 11.8, Exercise 30, the distance from the moon, in feet, of a space robot is given as $f(x) = 1.727x - .0013x^2$, where x is in seconds. At what time(s) is the robot more than 500 feet away?

CHAPTER 11 GROUP ACTIVITY

Finding the Path of a Comet

Objective: Find and graph an equation of a parabola with a given focus.

The orbit that a comet takes as it approaches the sun depends upon its velocity (as well as other factors). If the comet has enough velocity to escape the sun's pull, it takes either a parabolic orbit or a hyperbolic orbit. If it doesn't have enough velocity to escape the sun's pull, then it takes an elliptical orbit. Of course, a comet that has a parabolic or hyperbolic orbit will not come back around the sun again. Comets with elliptical orbits are the only ones we see again. For this activity we consider a comet with a parabolic orbit.

If the velocity of a comet equals escape velocity (that is, it is going just fast enough to get away from the sun), then its orbit will be parabolic. The sun is at the focus of the parabola. The vertex is the point where the comet is closest to the sun.

A. Make a sketch of the comet and the sun.

  1. Place the vertex of the parabola at the origin with the focus on the y-axis and a horizontal directrix.
  2. Assume that the comet is .75 astronomical unit from the sun at its closest point. (See Section 11.3, Objective 5 for a definition of an astronomical unit.)

B. What are the coordinates of the focus?

C. What is the equation of the directrix?

D. Using the information from Section 11.7, Exercise 47, find an equation of the parabola.

E. Graph the parabola (as a vertical parabola). Include the focus and directrix on your graph.

CHAPTER 11 SUMMARY

KEY TERMS

11.3 quadratic formula discriminant **11.4** complex number imaginary number real part	imaginary part standard form (of a complex number) conjugate (of a complex number)	**11.5** quadratic in form **11.7** parabola vertex axis	symmetric (about its axis) quadratic function **11.9** quadratic inequality rational inequality

TEST YOUR WORD POWER

See how well you have learned the vocabulary in this chapter. Answers, with examples, are given at the bottom of the next page.

1. The **quadratic formula** is
(a) a formula to find the number of solutions of a quadratic equation
(b) a formula to find the type of solutions of a quadratic equation
(c) the standard form of a quadratic equation
(d) a general formula for solving any quadratic equation.

2. The **discriminant** is
(a) the quantity under the radical in the quadratic formula
(b) the quantity in the denominator in the quadratic formula
(c) the solution set of a quadratic equation
(d) the result of using the quadratic formula.

3. A **complex number** is defined as
(a) a real number that includes a complex fraction
(b) a nonzero multiple of i
(c) a number of the form $a + bi$, where a and b are real numbers
(d) the square root of -1.

4. An **imaginary number** is
(a) a complex number $a + bi$ where $b \neq 0$
(b) a number that does not exist
(c) a complex number $a + bi$ where $b = 0$
(d) any real number.

5. A **quadratic function** is a function that can be written in the form
(a) $f(x) = mx + b$ for real numbers m and b

(b) $f(x) - \frac{P(x)}{Q(x)}$ where $Q(x) \neq 0$
(c) $f(x) = ax^2 + bx + c$ for real numbers a, b, c ($a \neq 0$)
(d) $f(x) = \sqrt{x}$ for $x \geq 0$.

6. A **parabola** is the graph of
(a) any equation in two variables
(b) a linear equation
(c) an equation of degree three
(d) a quadratic equation.

7. The **vertex** of a parabola is
(a) the point where the graph intersects the y-axis
(b) the points where the graph intersects the x-axis
(c) the lowest point on a parabola that opens up or the highest point on a parabola that opens down
(d) the origin.

8. The **axis** of a parabola is
(a) either the x-axis or the y-axis
(b) the vertical line (of a vertical parabola) or the horizontal line (of a horizontal parabola) through the vertex

(c) the lowest or highest point on the graph of a parabola
(d) a line through the origin.

9. A parabola is **symmetric about its axis** since

(a) its graph is near the axis
(b) its graph is identical when reflected across the axis
(c) its graph looks different on each side of the axis
(d) its graph intersects the axis.

QUICK REVIEW

CONCEPTS	EXAMPLES

11.1 SOLVING QUADRATIC EQUATIONS BY THE SQUARE ROOT PROPERTY

Square Root Property of Equations
If k is positive, and if $a^2 = k$, then $a = \sqrt{k}$ or $a = -\sqrt{k}$.

Solve $(2x + 1)^2 = 5$.

$$2x + 1 = \pm\sqrt{5}$$
$$2x = -1 \pm \sqrt{5}$$
$$x = \frac{-1 \pm \sqrt{5}}{2}$$

Solution set: $\left\{ \dfrac{-1 \pm \sqrt{5}}{2} \right\}$

11.2 SOLVING QUADRATIC EQUATIONS BY COMPLETING THE SQUARE

Completing the Square

1. If the coefficient of the squared term is 1, go to Step 2. If it is not 1, divide each side of the equation by this coefficient.

2. Make sure that all variable terms are on one side of the equation and all constant terms are on the other.

3. Take half the coefficient of x, square it, and add the square to each side of the equation. Factor the variable side and combine terms on the other.

4. Use the square root property to solve the equation.

Solve $2x^2 + 4x - 1 = 0$.

$$x^2 + 2x - \frac{1}{2} = 0$$

$$x^2 + 2x = \frac{1}{2}$$

$$x^2 + 2x + 1 = \frac{1}{2} + 1$$

$$(x + 1)^2 = \frac{3}{2}$$

$$x + 1 = \pm\sqrt{\frac{3}{2}} = \pm\frac{\sqrt{6}}{2}$$

$$x = -1 \pm \frac{\sqrt{6}}{2}$$

$$x = \frac{-2 \pm \sqrt{6}}{2}$$

Solution set: $\left\{ \dfrac{-2 \pm \sqrt{6}}{2} \right\}$

| CONCEPTS | EXAMPLES |

11.3 SOLVING QUADRATIC EQUATIONS BY THE QUADRATIC FORMULA

Quadratic Formula

The solutions of $ax^2 + bx + c = 0$, $(a \neq 0)$, are

$$x = \frac{-b \pm \sqrt{b^2 - 4ac}}{2a}.$$

Solve $3x^2 - 4x - 2 = 0$.

$$x = \frac{-(-4) \pm \sqrt{(-4)^2 - 4(3)(-2)}}{2(3)}$$

$a = 3, b = -4,$
$c = -2$

$$x = \frac{4 \pm \sqrt{16 + 24}}{6}$$

$$x = \frac{4 \pm \sqrt{40}}{6} = \frac{4 \pm 2\sqrt{10}}{6}$$

Discriminant: 40
There are two
real solutions.

$$x = \frac{2(2 \pm \sqrt{10})}{6} = \frac{2 \pm \sqrt{10}}{3}$$

Solution set: $\left\{ \dfrac{2 \pm \sqrt{10}}{3} \right\}$

The Discriminant

If a, b, and c are integers, then the discriminant, $b^2 - 4ac$, of $ax^2 + bx + c = 0$ determines the type of solutions as follows.

Discriminant	Solutions
Positive square of an integer	2 rational solutions
Positive, not square of an integer	2 irrational solutions
Zero	1 rational solution
Negative	2 imaginary solutions

For $x^2 + 3x - 10 = 0$, the discriminant is

$$3^2 - 4(1)(-10) = 49.$$

There are **2 rational** solutions.
For $2x^2 + 5x + 1 = 0$, the discriminant is

$$5^2 - 4(2)(1) = 17.$$

There are **2 irrational** solutions.
For $9x^2 - 6x + 1 = 0$, the discriminant is

$$(-6)^2 - 4(9)(1) = 0.$$

There is **1 rational** solution.
For $4x^2 + x + 1 = 0$, the discriminant is

$$1^2 - 4(4)(1) = -15.$$

There are **2 imaginary** solutions.

11.4 COMPLEX NUMBERS

The number $i = \sqrt{-1}$ and $i^2 = -1$.
For the positive number b,

$$\sqrt{-b} = i\sqrt{b}.$$

Simplify: $\sqrt{-19}$.

$$\sqrt{-19} = \sqrt{-1 \cdot 19} = i\sqrt{19}$$

Addition

Add complex numbers by adding the real parts and adding the imaginary parts.

Add: $(3 + 6i) + (-9 + 2i)$.

$$(3 + 6i) + (-9 + 2i) = (3 - 9) + (6 + 2)i$$
$$= -6 + 8i$$

Subtraction

To subtract complex numbers, change the number following the subtraction sign to its negative and add.

Subtract: $(5 + 4i) - (2 - 4i)$.

$$(5 + 4i) - (2 - 4i) = (5 + 4i) + (-2 + 4i)$$
$$= (5 - 2) + (4 + 4)i$$
$$= 3 + 8i$$

(continued)

CONCEPTS	EXAMPLES

Multiplication

Multiply complex numbers in the same way polynomials are multiplied. Replace i^2 with -1.

Multiply: $(7 + i)(3 - 4i)$.

$$(7 + i)(3 - 4i)$$
$$= 7(3) + 7(-4i) + i(3) + i(-4i) \quad \text{FOIL}$$
$$= 21 - 28i + 3i - 4i^2$$
$$= 21 - 25i - 4(-1) \quad\quad\quad i^2 = -1$$
$$= 21 - 25i + 4$$
$$= 25 - 25i$$

Division

Divide complex numbers by multiplying the numerator and the denominator by the conjugate of the denominator.

Divide: $\dfrac{2}{6 + i}$.

$$\frac{2}{6 + i} = \frac{2}{6 + i} \cdot \frac{6 - i}{6 - i}$$

$$= \frac{2(6 - i)}{36 - i^2}$$

$$= \frac{12 - 2i}{36 + 1}$$

$$= \frac{12 - 2i}{37}$$

$$= \frac{12}{37} - \frac{2}{37}i \quad \text{Standard form}$$

Complex Solutions

A quadratic equation may have nonreal, complex solutions. This occurs when the discriminant is negative. The quadratic formula will give complex solutions in such cases.

Solve for all complex solutions of

$$x^2 + x + 1 = 0.$$

Here, $a = 1$, $b = 1$, and $c = 1$.

$$x = \frac{-1 \pm \sqrt{1^2 - 4(1)(1)}}{2(1)}$$

$$x = \frac{-1 \pm \sqrt{1 - 4}}{2}$$

$$x = \frac{-1 \pm \sqrt{-3}}{2} \quad \text{Discriminant: } -3$$

$$x = \frac{-1 \pm i\sqrt{3}}{2} \quad \text{Two nonreal solutions}$$

Solution set: $\left\{ -\dfrac{1}{2} \pm \dfrac{\sqrt{3}}{2}i \right\}$

CONCEPTS	EXAMPLES

11.6 FORMULAS AND APPLICATIONS

To solve a formula for a squared variable, proceed as follows.

(a) The variable appears only to the second degree. Isolate the squared variable on one side of the equation, then use the square root property.

Solve $A = \dfrac{2mp}{r^2}$ for r.

$$r^2 A = 2mp \qquad \text{Multiply by } r^2.$$

$$r^2 = \frac{2mp}{A} \qquad \text{Divide by } A.$$

$$r = \pm\sqrt{\frac{2mp}{A}} \qquad \text{Take square roots.}$$

$$r = \pm\frac{\sqrt{2mpA}}{A} \qquad \text{Rationalize.}$$

(b) The variable appears to the first and second degrees. Write the equation in standard quadratic form, then use the quadratic formula.

Solve $m^2 + rm = t$ for m.

$$m^2 + rm - t = 0 \qquad \text{Standard form}$$

$$m = \frac{-r \pm \sqrt{r^2 - 4(1)(-t)}}{2(1)} \qquad a = 1, b = r, c = -t$$

$$m = \frac{-r \pm \sqrt{r^2 + 4t}}{2}$$

11.7 GRAPHS OF QUADRATIC FUNCTIONS

1. The graph of the quadratic function with $f(x) = a(x - h)^2 + k$, $a \neq 0$, is a parabola with vertex at (h, k) and the vertical line $x = h$ as axis.

2. The graph opens upward if a is positive and downward if a is negative.

3. The graph is wider than $f(x) = x^2$ if $0 < |a| < 1$ and narrower if $|a| > 1$.

Graph $f(x) = -(x + 3)^2 + 1$.

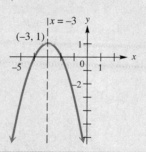

11.8 MORE ABOUT PARABOLAS; APPLICATIONS

The vertex of the graph of $f(x) = ax^2 + bx + c$, $a \neq 0$, may be found by completing the square. The vertex has coordinates

$$\left(\frac{-b}{2a}, f\left(\frac{-b}{2a}\right)\right).$$

To graph a quadratic function:
Find the y-intercept by evaluating $f(0)$. Find any x-intercepts by solving $f(x) = 0$. Find the vertex either by using the vertex formula or by completing the square. Find and plot any additional points as needed, using symmetry about the axis. Verify that the graph opens upward (if $a > 0$) or opens downward (if $a < 0$).

Graph $f(x) = x^2 + 4x + 3$.
The vertex is $(-2, -1)$.
Since $f(0) = 3$, the y-intercept is $(0, 3)$. The solutions of $x^2 + 4x + 3 = 0$ are -1 and -3, so the x-intercepts are $(-1, 0)$ and $(-3, 0)$.

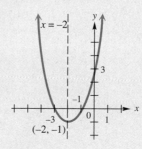

(continued)

CONCEPTS	EXAMPLES

If the discriminant, $b^2 - 4ac$, is positive, the graph of $f(x) = ax^2 + bx + c$ has two x-intercepts; if zero, one x-intercept; if negative, no x-intercepts.

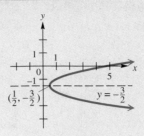

The graph of $x = ay^2 + by + c$ is a horizontal parabola, opening to the right if $a > 0$, or to the left if $a < 0$. Horizontal parabolas do not represent functions.

Graph $x = 2y^2 + 6y + 5$.
  The graph is shown above.

11.9 NONLINEAR INEQUALITIES

Solving a Quadratic (or Higher Degree Polynomial) Inequality

Step 1 Write the inequality as an equation and solve.

Step 2 Graph the numbers found in Step 1 on a number line. These numbers divide the line into regions.

Step 3 Substitute a number from each region into the inequality to determine the intervals that belong in the solution set—those intervals containing numbers that make the inequality true.

Step 4 Consider the endpoints separately.

Solve $2x^2 + 5x + 2 < 0$.
$$2x^2 + 5x + 2 = 0$$
$$x = -\frac{1}{2}, \qquad x = -2$$

$x = -3$ makes it false; $x = -1$ makes it true; $x = 0$ makes it false.

Solution set: $\left(-2, -\frac{1}{2}\right)$

Solving a Rational Inequality

Step 1 Write the inequality as an equation and solve the equation.

Step 2 Set the denominator equal to 0 and solve the equation.

Step 3 Use the solutions from Steps 1 and 2 to divide a number line into regions.

Step 4 Test a number from each region in the inequality to determine the regions that satisfy the inequality.

Step 5 Exclude any values that make the denominator 0.

Solve $\dfrac{x}{x+2} \geq 4$.

$$\frac{x}{x+2} = 4 \text{ leads to } x = -\frac{8}{3}.$$
$$x + 2 = 0$$
$$x = -2$$

-4 makes it false; $-\frac{7}{3}$ makes it true; 0 makes it false.

The solution set is $\left[-\frac{8}{3}, -2\right)$, since -2 makes the denominator 0.

CHAPTER 11 REVIEW EXERCISES

[11.1]   *Solve each equation by using the square root property. Give only real number solutions. Express all radicals in simplest form.*

1. $y^2 = 144$ **2.** $x^2 = 37$ **3.** $m^2 = 128$ **4.** $(k + 2)^2 = 25$

5. $(r - 3)^2 = 10$ **6.** $(2p + 1)^2 = 14$ **7.** $(3k + 2)^2 = -3$

8. Which one of the following equations has two real solutions?
   (a) $x^2 = 0$ **(b)** $x^2 = -4$ **(c)** $(x + 5)^2 = -16$ **(d)** $(x + 6)^2 = 25$

[11.2]   *Solve each equation by completing the square. Give only real number solutions.*

9. $m^2 + 6m + 5 = 0$ **10.** $p^2 + 4p = 7$

11. $-x^2 + 5 = 2x$ **12.** $2y^2 - 3 = -8y$

13. $5k^2 - 3k - 2 = 0$ **14.** $(4a + 1)(a - 1) = -7$

Solve each problem.

15. If an object is thrown upward on Earth from a height of 50 feet, with an initial velocity of 32 feet per second, then its height after t seconds is given by $h = -16t^2 + 32t + 50$, where h is in feet. After how many seconds will it reach a height of 30 feet?

16. Find the lengths of the three sides of the right triangle shown.

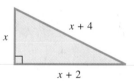

17. What must be added to $x^2 + kx$ to make it a perfect square?

[11.3]

18. Consider the equation $x^2 - 9 = 0$.
   (a) Solve the equation by factoring.
   (b) Solve the equation by the square root property.
   (c) Solve the equation by the quadratic formula.
   (d) Compare your answers. If a quadratic equation can be solved by both the factoring and the quadratic formula methods, should you always get the same results? Explain.

Solve each equation by using the quadratic formula. Give only real number solutions.

19. $x^2 - 2x - 4 = 0$ **20.** $3k^2 + 2k = -3$ **21.** $2p^2 + 8 = 4p + 11$

22. $-4x^2 + 7 = 2x$ **23.** $\dfrac{1}{4}p^2 = 2 - \dfrac{3}{4}p$

24. Use the discriminant to predict the number and type of solutions for each equation.
   (a) $a^2 + 5a + 2 = 0$ **(b)** $4x^2 = 3 - 4x$
   (c) $4x^2 = 6x - 8$ **(d)** $9z^2 + 30z + 25 = 0$

[11.4]

25. Write an explanation of the method used to divide complex numbers.

26. Use the fact that $i^2 = -1$ to complete each of the following. Do them in order.
   (a) Since $i^3 = i^2 \cdot i$, $i^3 =$ _____ .
   (b) Since $i^4 = i^3 \cdot i$, $i^4 =$ _____ .
   (c) Since $i^{48} = (i^4)^{12}$, $i^{48} =$ _____ .

Perform the indicated operations.

27. $(3 + 5i) + (2 - 6i)$ **28.** $(-2 - 8i) - (4 - 3i)$ **29.** $(6 - 2i)(3 + i)$

30. $(2 + 3i)(2 - 3i)$ **31.** $\dfrac{1 + i}{1 - i}$ **32.** $\dfrac{5 + 6i}{2 + 3i}$ **33.** $\dfrac{1}{7 - i}$

34. What is the conjugate of the real number a?

35. Is it possible to multiply a complex number by its conjugate and get an imaginary product? Explain.

Find the complex solutions of each quadratic equation.

36. $(m + 2)^2 = -3$ **37.** $(3p - 2)^2 = -8$ **38.** $3k^2 = 2k - 1$

39. $h^2 + 3h = -8$ **40.** $4q^2 + 2 = 3q$ **41.** $9z^2 + 2z + 1 = 0$

[11.5] *Solve each equation.*

42. $\dfrac{1}{y} + \dfrac{2}{y + 1} = 2$ **43.** $8(3x + 5)^2 + 2(3x + 5) - 1 = 0$

44. $-2r = \sqrt{\dfrac{48 - 20r}{2}}$ **45.** $2x^{2/3} - x^{1/3} - 28 = 0$

*46.** $5x^4 - 2x^2 - 3 = 0$

Solve each problem.

47. Lisa Wunderle drove 8 miles to pick up her friend Laurie, and then drove 11 miles to a mall at a speed 15 miles per hour faster. If Lisa's total travel time was 24 minutes, what was her speed on the trip to pick up Laurie?

48. It takes Laketa De Mol 2 hours longer to complete a project for her boss than it takes Ed Moura. Working together, it would take them 3 hours. How long would it take each one to do the job alone? Round your answers to the nearest tenth of an hour.

49. Why can't the equation $x = \sqrt{2x + 4}$ have a negative solution?

[11.6] *Solve each formula for the indicated variable. (Give answers with $\pm$.)*

50. $S = \dfrac{Id^2}{k}$;   for d **51.** $k = \dfrac{rF}{wv^2}$;   for v

52. $2\pi R^2 + 2\pi RH - S = 0$;   for R

Solve each problem.

53. The equation $s = 16t^2 + 15t$ gives the distance s in feet an object dropped off a building has fallen in t seconds. Find the time t when the object has fallen 25 feet. Round the answer to the nearest hundredth.

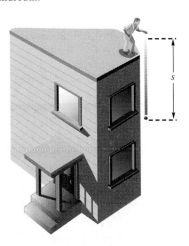

---
*Exercises identified with asterisks have imaginary number solutions.

54. A large machine requires a part in the shape of a right triangle with a hypotenuse 9 feet less than twice the length of the longer leg. The shorter leg must be $\frac{3}{4}$ the length of the of the longer leg. Find the lengths of the three sides of the part.

55. A rectangle has a length 2 meters more than its width. If one meter is cut from the length and one meter is added to the width, the resulting figure is a square with an area of 121 square meters. Find the dimensions of the original rectangle.

56. Nancy Mendoza wants to buy a mat for a photograph that measures 14 inches by 20 inches. She wants to have an even border around the picture when it is mounted on the mat. If the area of the mat she chooses is 352 square inches, how wide will the border be?

57. A lot is in the shape of a right triangle. The shortest side measures 50 meters. The longest side is 110 meters less than twice the middle side. How long is the middle side?

[11.7–11.8] *Identify the vertex of each parabola.*

58. $y = 3x^2 - 2$ **59.** $y = 6 - 2x^2$ **60.** $f(x) = -(x - 1)^2$

61. $f(x) = (x + 2)^2$ **62.** $y = (x - 3)^2 + 7$ **63.** $y = -3x^2 + 4x - 2$

Graph each parabola. Give the domain and range in Exercise 64.

64. $y = 4x^2 + 4x - 2$ **65.** $x = 2y^2 + 8y + 3$

66. If the discriminant of a quadratic function is negative, what do you know about the graph of the function?

67. Which one of the following would most closely resemble the graph of $f(x) = a(x - h)^2 + k$ if $a < 0$, $h > 0$, and $k < 0$?

(a)

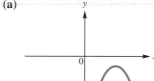

(b)

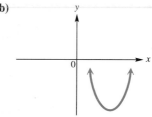

(c)

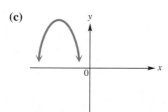

(d)

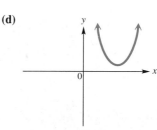

Work each problem.

68. If a missile is propelled upward on Earth at an angle of 45° with an initial velocity of 30 miles per hour, the equation $f(x) = -.017x^2 + x$ gives the height of the missile in feet at x seconds. In how many seconds does the missile reach its maximum height? What is the maximum height?

69. Consumer spending for home video games in dollars per person per year is given in the table.

Consumer Spending for Home Video Games

Year	Dollars
1990	12.39
1992	13.08
1994	15.78
1996	19.43
1997	22.71

Source: Statistical Abstract of the United States, 1997.

Let $x = 0$ represent 1990, $x = 2$ represent 1992, and so on.
(a) Use the data for 1990, 1994, and 1997 in the quadratic form $ax^2 + bx + c = y$ to write a system of three equations.
(b) Solve the system from part (a) to get a quadratic function f that models these data.

70. Find the two numbers whose sum is 40 and whose product is a maximum.

[11.9]

71. The function with $f(x) = 2x^2 - 7x - 4$ was graphed in the standard viewing window of a graphing calculator and the two x-intercepts were located. See the figures.

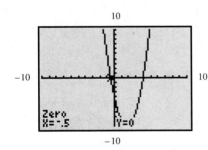 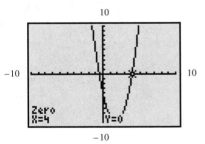

What is the solution set of
(a) $2x^2 - 7x - 4 = 0$?   **(b)** $2x^2 - 7x - 4 > 0$?   **(c)** $2x^2 - 7x - 4 \leq 0$?

Solve each inequality and graph the solution set.

72. $(x - 4)(2x + 3) > 0$

73. $x^2 + x \leq 12$

74. $2k^2 > 5k + 3$

75. $(4m + 3)^2 \leq -4$

76. $\dfrac{6}{2z - 1} < 2$

77. $\dfrac{3y + 4}{y - 2} \leq 1$

MIXED REVIEW EXERCISES

Solve.

78. $V = r^2 + R^2h$;   for R

***79.** $3t^2 - 6t = -4$

***80.** $x^4 - 1 = 0$

81. $(b^2 - 2b)^2 = 11(b^2 - 2b) - 24$

82. $(r - 1)(2r + 3)(r + 6) < 0$

83. $\dfrac{2}{x - 4} + \dfrac{1}{x} = \dfrac{11}{5}$

84. $(3k + 11)^2 = 7$

85. $p = \sqrt{\dfrac{yz}{6}}$;   for y

86. $(8k - 7)^2 \geq -1$

87. $-5x^2 = -8x + 3$

88. $6 + \dfrac{15}{s^2} = -\dfrac{19}{s}$

89. $\dfrac{-2}{x + 5} \leq -5$

Find the vertex and sketch the graph.

90. $y = \dfrac{4}{3}(x - 2)^2 + 1$

91. $x = 2(y + 3)^2 - 4$

92. $f(x) = -2x^2 + 8x - 5$

93. $x = -\dfrac{1}{2}y^2 + 6y - 14$

94. A student gave the following "solution" to the equation $b^2 = 12$.

$$b^2 = 12$$
$$b = \sqrt{12}$$
$$b = 2\sqrt{3}$$

What is wrong with this solution?

95. Explain how you would go about writing a quadratic equation whose solutions are -5 and 6. What is the standard form of one such equation?

96. The height (in feet) of a projectile t seconds after being fired from Earth into the air is given by $s(t) = -16t^2 + 160t$.
(a) Find the number of seconds required for the projectile to reach its maximum height.
(b) What is the maximum height?

97. Find the length and the width of a rectangle having a perimeter of 600 meters if the area is to be a maximum.

98. If you were to use the quadratic formula to solve $x^4 - 5x^2 + 6 = 0$, with $a = 1$, $b = -5$, and $c = 6$, what would you have to remember after you applied the formula?

RELATING CONCEPTS (EXERCISES 99–106)

Work Exercises 99–106 in order, to see the connections between equations and inequalities.

99. Use the methods of Chapters 2 and 3 to solve the equation or inequality, and graph the solution set.
(a) $3x - (4x + 2) = 0$ **(b)** $3x - (4x + 2) > 0$ **(c)** $3x - (4x + 2) < 0$

100. Use the methods of this chapter to solve the equation or inequality, and graph the solution set.
(a) $x^2 - 6x + 5 = 0$ **(b)** $x^2 - 6x + 5 > 0$ **(c)** $x^2 - 6x + 5 < 0$

101. Use the methods of Section 7.6 and Section 11.9 to solve the equation or inequality, and graph the solution set.
(a) $\dfrac{-5x + 20}{x - 2} = 0$ **(b)** $\dfrac{-5x + 20}{x - 2} > 0$ **(c)** $\dfrac{-5x + 20}{x - 2} < 0$

(continued)

RELATING CONCEPTS (EXERCISES 99-106) (CONTINUED)

Review the definition of the union of two sets in Section 3.2, and then answer the questions in Exercises 102–106.

102. Form the union of the solution sets in Exercise 99. What is their union?

103. Repeat Exercise 102 for the solution sets in Exercise 100.

104. For the equation and inequalities in Exercise 101, what value of x cannot possibly be part of any of the solution sets? What is the union of the solution sets of the equation and inequalities in Exercise 101?

105. Fill in the blanks in the following statement: If we solve a linear, quadratic, or rational equation and the two inequalities associated with it, the union of the three solution sets will be _____ ; the only exception will be in the case of the rational equation and inequalities, where the number or numbers that cause the _____ to be zero will be deleted.

106. Suppose that the solution set of a quadratic equation is $\{-5, 3\}$ and the solution set of one of the associated inequalities is $(-\infty, -5) \cup (3, \infty)$. What is the solution set of the other associated inequality?

Did you make the connection that solutions of an inequality depend on the solutions of the corresponding equation?

CHAPTER 11 TEST

*Items marked * require knowledge of complex numbers.*

Solve by using the square root property.

1. $x^2 = 39$

2. $(y + 3)^2 = 64$

Solve by completing the square.

3. $x^2 - 4x = 6$

4. $2x^2 + 12x - 3 = 0$

Solve by the quadratic formula.

5. $3w^2 + 2 = 6w$

***6.** $4x^2 + 8x + 11 = 0$

Solve by the method of your choice.

7. $p^2 - 2p - 1 = 0$

8. $(2x + 1)^2 = 18$

9. $(x - 5)(2x - 1) = 1$

10. $t^2 + 25 = 10t$

 Solve each problem.

 11. If an object is propelled into the air from ground level on Earth with an initial velocity of 64 feet per second, its height s (in feet) after t seconds is given by the formula $s = -16t^2 + 64t$. After how many seconds will the object reach a height of 64 feet?

12. Use the formula from Kepler's third law of planetary motion, $P^2 = a^3$, to determine P, in years, for Pluto, if $a = 39.4$ AU.

Perform the indicated operations.

***13.** $(3 + i) + (-2 + 3i) - (6 - i)$

***14.** $(6 + 5i)(-2 + i)$

***15.** $(3 - 8i)(3 + 8i)$

***16.** $\dfrac{15 - 5i}{7 + i}$

17. Use the discriminant to predict the number and type of solutions of $2x^2 - 8x - 3 = 0$. Do not solve.

Solve by any method.

18. $9x^4 + 4 = 37x^2$

19. $12 = (2d + 1)^2 + (2d + 1)$

20. Solve for r: $S = 4\pi r^2$. (Leave $\pm$ in your answer.)

Solve each problem by writing a quadratic equation. Use any method to solve the equation.

21. Adam Bryer has a pool 24 feet long and 10 feet wide. He wants to construct a concrete walk around the pool. If he plans for the walk to be of uniform width and cover 152 square feet, what will be the width of the walk?

22. At a point 30 meters from the base of a tower, the distance to the top of the tower is 2 meters more than twice the height of the tower. Find the height of the tower.

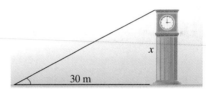

Graph each parabola. Identify the vertex. Identify the domain and range in Exercise 23.

23. $f(x) = \dfrac{1}{2}x^2 - 2$

24. $f(x) = -x^2 + 4x - 1$

25. $x = -(y - 2)^2 + 2$

26. The value (in millions) of the U.S. domestic salmon catch in the years 1990–1995 is closely approximated by the quadratic function with

$$f(x) = 22.56x^2 - 129.8x + 611.8.$$

Here, $x = 0$ represents 1990, $x = 1$ represents 1991, and so on. (*Source:* National Marine Fisheries Service.)

(a) Based on this model, how many salmon were caught in 1992?

(b) In what year did the catch reach a minimum? To the nearest million, from the model, how many salmon were caught that year? (Round your answer.)

Solve and graph each solution set.

27. $2x^2 + 7x > 15$

28. $\dfrac{5}{t - 4} \leq 1$

1. Let $S = \left\{ -\dfrac{7}{3}, -2, -\sqrt{3}, 0, .7, \sqrt{12}, \sqrt{-8}, 7, \dfrac{32}{3} \right\}$. List the elements of S that are elements of each set.
   (a) Integers   **(b)** Rational numbers

2. Simplify $2(-3)^2 + (-8)(-5) + (-17)$.

Solve each equation.

3. $7 - (4 + 3t) + 2t = -6(t - 2) - 5$

4. $|6x - 9| = |-4x + 2|$

5. $2x = \sqrt{\dfrac{5x + 2}{3}}$

6. $\dfrac{3}{x - 3} - \dfrac{2}{x - 2} = \dfrac{3}{x^2 - 5x + 6}$

7. $(r - 5)(2r + 3) = 1$

8. $b^4 - 5b^2 + 4 = 0$

Solve each inequality.

9. $-2x + 4 \le -x + 3$

10. $|3y - 7| \le 1$

11. $x^2 - 4x + 3 < 0$

12. $\dfrac{3}{y + 2} > 1$

13. Find the slope and intercepts of the line with equation $-2x + 7y = 16$.

Write an equation in slope-intercept form for the specified line in Exercises 14 and 15.

14. Through $(2, -3)$ and parallel to the line with equation $5x + 2y = 6$

15. Through $(-4, 1)$ and perpendicular to the line with equation $5x + 2y = 6$

Perform the indicated operations.

16. $(7x + 4)(2x - 3)$

17. $\left(\dfrac{2}{3}t + 9 \right)^2$

18. $(3t^3 + 5t^2 - 8t + 7) - (6t^3 + 4t - 8)$

19. Divide $4x^3 + 2x^2 - x + 26$ by $x + 2$.

Factor completely.

20. $16x - x^3$

21. $(3x + 2)^2 - 4(3x + 2) - 5$

22. $8x^3 + 27y^3$

23. $9x^2 - 30xy + 25y^2$

Perform the operations and express each answer in lowest terms. Assume denominators are nonzero.

24. $\dfrac{x^2 - 3x - 10}{x^2 + 3x + 2} \cdot \dfrac{x^2 - 2x - 3}{x^2 + 2x - 15}$

25. $\dfrac{3}{2 - k} - \dfrac{5}{k} + \dfrac{6}{k^2 - 2k}$

26. $\dfrac{\dfrac{r}{s} - \dfrac{s}{r}}{\dfrac{r}{s} + 1}$

Graph each relation. Tell whether or not each is a function, and if it is, give its domain and range.

27. $4x - 5y = 15$

28. $4x - 5y < 15$

29. $f(x) = -2(x - 1)^2 + 3$

30. The record track-qualifying speeds at North Carolina Motor Speedway since Richard Petty captured the first pole in 1965 are given in the table.

Qualifying Records

Year	Speed (in mph)
1965	116.26
1975	132.02
1985	141.85
1995	155.38
1998	156.36

Source: NASCAR.

Let $x = 0$ represent 1965, $x = 10$ represent 1975, and so on.
(a) Plot the ordered pairs (Year, Speed).
(b) Is this set of ordered pairs a function?
(c) Use the ordered pairs (0, 116.26) and (20, 141.85) to write a linear equation that models these data.
(d) Use your model to approximate the record speed for 1998 to the nearest hundredth. How does it compare to the actual value from the table?

31. Does the relation $x = 5$ define a function? Explain why or why not.

32. For the function with $f(x) = 2(x - 1)^2 - 5$, find
(a) $f(-2)$;   **(b)** the domain and the range.

Solve each of the following.

33. $2x - 4y = 10$
   $9x + 3y = 3$

34. $x + \ y + 2z = 3$
   $-x + \ y + \ z = -5$
   $2x + 3y - \ z = -8$

35. An excursion boat traveled 20 miles upriver (against the current) in 1 hour. The return trip downriver took .5 hour. Solve a system of equations to find the speed of the current and the speed of the boat in still water.

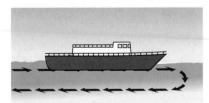

Simplify each radical expression.

36. $\sqrt[3]{\dfrac{27}{16}}$

37. $\dfrac{2}{\sqrt{7} - \sqrt{5}}$

Solve each problem.

38. The perimeter of a rectangle is 20 inches and its area is 21 square inches. What are the dimensions of the rectangle?

39. Tri rode his bicycle for 12 miles and then walked an additional 8 miles. The total time for the trip was 5 hours. If his rate while walking was 10 miles per hour less than his rate while riding, what was each rate?

40. Two cars left an intersection at the same time, one heading due south and the other due east. Later they were exactly 95 miles apart. The car heading east had gone 38 miles less than twice as far as the car heading south. How far had each car traveled?

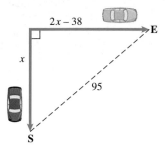

Inverse, Exponential, and Logarithmic Functions

12

One of the most important issues of our time is that of saving the environment. Industrial pollution of our air and water supplies, emission of "greenhouse gases," and climate changes as a result of global warming are concerns that must be faced now to save our planet for future generations.

Recycling is a way that we as individuals can help to protect the environment. One of the most common recyclables is plastic. The accompanying bar graph shows the increase in gross waste generated by plastics from 1960–1990. (The good news, however, is that the percentage of recovery also is increasing.) In which two 5-year periods did the waste more than double?

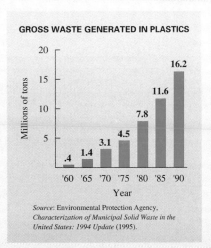

GROSS WASTE GENERATED IN PLASTICS

Source: Environmental Protection Agency, *Characterization of Municipal Solid Waste in the United States: 1994 Update* (1995).

A new type of function introduced in this chapter, the *exponential function*, can be used to model data like that shown in the graph. (See Exercise 15 in the Chapter 12 Review Exercises for an exponential function related to this graph.) Exponential functions are used throughout this chapter to analyze data involving climate and the environment, the theme of this chapter.

12.1 Inverse Functions

In this chapter we study two important types of functions, *exponential* and *logarithmic* functions. These functions are related in a special way. They are *inverses* of one another.

OBJECTIVE **1** Decide whether a function is one-to-one and, if it is, find its inverse. Suppose that G is the function $\{(-2, 2), (-1, 1), (0, 0), (1, 3), (2, 5)\}$. Another set of ordered pairs can be formed from G by interchanging the x- and y-values of each pair in G. Call this set F, with

$$F = \{(2, -2), (1, -1), (0, 0), (3, 1), (5, 2)\}.$$

To show that these two sets are related, F is called the *inverse* of G. For a function f to have an inverse, f must be *one-to-one*. In a **one-to-one function** each x-value corresponds to only one y-value and each y-value corresponds to just one x-value.

The function shown in Figure 1(a) is not one-to-one because the y-value 7 corresponds to *two* x-values, 2 and 3. That is, the ordered pairs $(2, 7)$ and $(3, 7)$ both appear in the function. The function in Figure 1(b) is one-to-one.

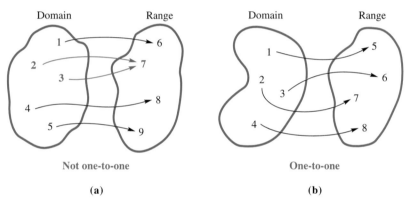

Not one-to-one

(a)

One-to-one

(b)

Figure 1

The *inverse* of any one-to-one function f is found by interchanging the components of the ordered pairs of f. The inverse of f is written f^{-1}. Read f^{-1} as "the inverse of f" or "f-inverse."

 CAUTION The symbol $f^{-1}(x)$ does not represent $\dfrac{1}{f(x)}$.

The definition of the inverse of a function follows.

Inverse of a Function

The **inverse** of a one-to-one function, f, written f^{-1}, is the set of all ordered pairs of the form (y, x) where (x, y) belongs to f. Since the inverse is formed by interchanging x and y, the domain of f becomes the range of f^{-1} and the range of f becomes the domain of f^{-1}.

For inverses f and f^{-1}, it follows that $f(f^{-1}(x)) = x$ and $f^{-1}(f(x)) = x$.

EXAMPLE 1 Deciding Whether a Function Is One-to-One

Decide whether each function is one-to-one. If it is, find the inverse function.

(a) $F = \{(-2, 1), (-1, 0), (0, 1), (1, 2), (2, 2)\}$

Each x-value in F corresponds to just one y-value. However, the y-value 2 corresponds to two x-values, 1 and 2. Also, the y-value 1 corresponds to both -2 and 0. Because some y-values correspond to more than one x-value, F is not one-to-one.

(b) $G = \{(3, 1), (0, 2), (2, 3), (4, 0)\}$

Every x-value in G corresponds to only one y-value, and every y-value corresponds to only one x-value, so G is a one-to-one function. The inverse function is found by interchanging the numbers in each ordered pair.

$$G^{-1} = \{(1, 3), (2, 0), (3, 2), (0, 4)\}$$

Notice how the domain and range of G become the range and domain, respectively, of G^{-1}.

(c) The Environmental Protection Agency has developed an indicator of air quality called the Pollutant Standard Index (PSI). If the PSI exceeds 100 on a particular day, that day is classified as unhealthy. The chart shows the number of unhealthy days in Chicago for the years 1988–1993.

Year	Number of Unhealthy Days
1988	21
1989	3
1990	3
1991	8
1992	6
1993	1

Source: Environmental Protection Agency.

Let f be the function defined in the table, with the years forming the domain and the numbers of unhealthy days forming the range. Then f is not one-to-one, because in two different years (1989 and 1990), the number of unhealthy days was the same, 3.

OBJECTIVE 2 Use the horizontal line test to determine whether a function is one-to-one. It may be difficult to decide whether a function is one-to-one just by looking at the equation that defines the function. However, by graphing the function and observing the graph, we can use the following *horizontal line test* to tell whether it is one-to-one.

Horizontal Line Test

A function is one-to-one if every horizontal line intersects the graph of the function at most once.

The horizontal line test follows from the definition of a one-to-one function. Any two points that lie on the same horizontal line have the same y-coordinate. No two ordered pairs that belong to a one-to-one function may have the same y-coordinate, and therefore no horizontal line will intersect the graph of a one-to-one function more than once.

E X A M P L E 2 Using the Horizontal Line Test

Use the horizontal line test to determine whether the graphs in Figures 2 and 3 are graphs of one-to-one functions.

(a)

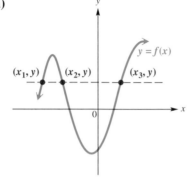

Figure 2

(b)

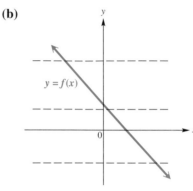

Figure 3

Because the horizontal line shown in Figure 2 intersects the graph in more than one point (actually three points in this case), the function is not one-to-one.

Every horizontal line will intersect the graph in Figure 3 in exactly one point. This function is one-to-one.

O B J E C T I V E 3 Find the equation of the inverse of a function. By definition, the inverse of a function is found by interchanging the x- and y-values of each of its ordered pairs. The equation of the inverse of a function defined by $y = f(x)$ is found in the same way.

Finding the Equation of the Inverse of $y = f(x)$

For a one-to-one function f defined by an equation $y = f(x)$, find the defining equation of the inverse as follows.

Step 1 Interchange x and y.

Step 2 Solve for y.

Step 3 Replace y with $f^{-1}(x)$.

E X A M P L E   3   Finding the Equation of the Inverse

Decide whether each of the following defines a one-to-one function. If so, find the equation of the inverse.

(a) $f(x) = 2x + 5$

By definition, this is a one-to-one function. To find the inverse, let $y = f(x)$ so that

$$y = 2x + 5$$
$$x = 2y + 5 \qquad \text{Interchange } x \text{ and } y. \text{ (Step 1)}$$
$$2y = x - 5 \qquad \text{Solve for } y. \text{ (Step 2)}$$
$$y = \frac{x - 5}{2}.$$

From the last equation,

$$f^{-1}(x) = \frac{x - 5}{2}, \qquad \text{(Step 3)}$$

which is a linear function. In the function $y = 2x + 5$, the value of y is found by starting with a value of x, multiplying by 2, and adding 5. The equation for the inverse has us *subtract* 5, and then *divide* by 2. This shows how an inverse is used to "undo" what a function does to the variable x.

(b) $y = x^2 + 2$

Both $x = 3$ and $x = -3$ correspond to $y = 11$. Because of the x^2-term, there are many pairs of x-values that each correspond to the same y-value. This means that the function defined by $y = x^2 + 2$ is not one-to-one and does not have an inverse.

If this were not noticed, following the steps given above for finding the equation of an inverse leads to

$$y = x^2 + 2$$
$$x = y^2 + 2 \qquad \text{Interchange } x \text{ and } y.$$
$$x - 2 = y^2$$
$$\pm\sqrt{x - 2} = y. \qquad \text{Square root property}$$

The last step shows that there are two y-values for each choice of $x > 2$, so the given function is not one-to-one and cannot have an inverse.

(c) $f(x) = (x - 2)^3$

Because of the cube, this is a one-to-one function. Find the inverse by replacing $f(x)$ with y and then interchanging x and y.

$$y = (x - 2)^3$$
$$x = (y - 2)^3$$

Take the cube root on each side to solve for y.

$$\sqrt[3]{x} = \sqrt[3]{(y - 2)^3}$$
$$\sqrt[3]{x} = y - 2$$
$$\sqrt[3]{x} + 2 = y$$
$$f^{-1}(x) = \sqrt[3]{x} + 2 \qquad \text{Replace } y \text{ with } f^{-1}(x).$$

OBJECTIVE **4** Graph the inverse f^{-1} from the graph of f. Suppose the point (a, b) shown in Figure 4 belongs to a one-to-one function f. Then the point (b, a) would belong to f^{-1}. The line segment connecting (a, b) and (b, a) is perpendicular to and cut in half by the line $y = x$. The points (a, b) and (b, a) are "mirror images" of each other with respect to $y = x$. For this reason the graph of f^{-1} can be found from the graph of f by locating the mirror image of each point of f with respect to $y = x$.

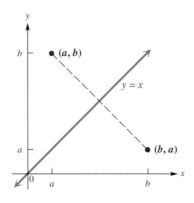

Figure 4

E X A M P L E 4 Graphing Inverses

Graph the inverses of the functions shown in Figure 5.

In Figure 5 the graphs of two functions are shown in blue and their inverses are shown in red. In each case, the graph of f^{-1} is symmetric to the graph of f with respect to the line $y = x$.

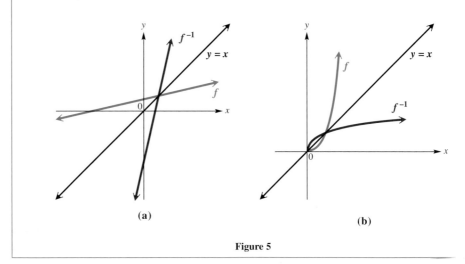

(a)   (b)

Figure 5

OBJECTIVE **5** Use a graphing calculator to graph inverse functions. We have described how inverses of one-to-one functions may be determined algebraically. We also explained how the graph of a one-to-one function f compares to the graph of its inverse f^{-1}: it is a reflection of the graph of f^{-1} across the line $y = x$. In Example 3 we showed that the inverse of the one-to-one function with $f(x) = 2x + 5$ is given by $f^{-1}(x) = \dfrac{x - 5}{2}$. If we use a square viewing window of a graphing calculator and graph $y_1 = f(x) = 2x + 5$,

$y_2 = f^{-1}(x) = \dfrac{x - 5}{2}$, and $y_3 = x$, we can see how this reflection appears on the screen. See Figure 6.

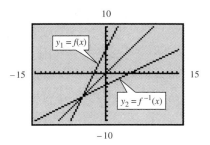

Figure 6

Some graphing calculators have the capability to "draw" the inverse of a function. Figure 7 shows the graphs of $f(x) = x^3 + 2$ and its inverse in a square viewing window.

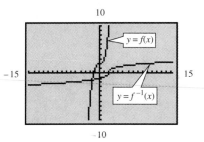

Figure 7

12.1 EXERCISES

1. The chart shows the number of uncontrolled hazardous waste sites that require further investigation to determine whether remedies are needed under the Superfund program. The ten states listed are the highest ranked in the United States.

  If this correspondence is considered to be a function that pairs each state with its number of uncontrolled waste sites, is it one-to-one? If not, explain why.

State	Number
New Jersey	108
Pennsylvania	101
California	94
New York	79
Michigan	75
Florida	53
Washington	50
Illinois	40
Wisconsin	40
Ohio	38

Source: U.S. Environmental Protection Agency.

2. The chart shows emissions of a major air pollutant, carbon monoxide, in the United States for the years 1990–1995.

Year	Amount of Emissions in Thousand Short Tons
1990	100,650
1991	97,376
1992	94,043
1993	94,133
1994	98,017
1995	92,099

Source: U.S. Environmental Protection Agency.

If this correspondence is considered to be a function that pairs each year with its emissions amount, is it one-to-one? If not, explain why.

3. The road mileage between Denver, Colorado, and several selected U.S. cities is shown in the table below.

City	Distance to Denver in Miles
Atlanta	1398
Dallas	781
Indianapolis	1058
Kansas City, MO	600
Los Angeles	1059
San Francisco	1235

If we consider this as a function that pairs each city with a distance, is it a one-to-one function? Why or why not? How could we change the answer to this question by adding 1 mile to one of the distances shown?

4. Suppose that you consider the set of ordered pairs (x, y) such that x represents a person in your mathematics class and y represents that person's mother. Explain how this function might not be a one-to-one function.

Choose the correct response from the given list.

5. If a function is made up of ordered pairs in such a way that the same y-value appears in a correspondence with two different x-values, then
(a) the function is one-to-one.
(b) the function is not one-to-one.
(c) its graph does not pass the vertical line test.
(d) it has an inverse function associated with it.

6. Which one of the following is a one-to-one function? Explain why the others are not, using specific examples.
(a) $f(x) = x$
(b) $f(x) = x^2$
(c) $f(x) = |x|$
(d) $f(x) = -x^2 + 2x - 1$

7. Only one of the graphs illustrates a one-to-one function. Which one is it?

(a)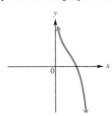

(b)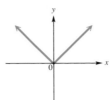

(c)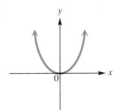

(d)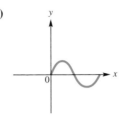

8. If a function f is one-to-one and the point (p, q) lies on the graph of f, then which one of the following *must* lie on the graph of f^{-1}?

(a) $(-p, q)$ (b) $(-q, -p)$ (c) $(p, -q)$ (d) (q, p)

If the function is one-to-one, find its inverse. See Examples 1–3.

9. $\{(3, 6), (2, 10), (5, 12)\}$

10. $\left\{(-1, 3), (0, 5), (5, 0), \left(7, -\dfrac{1}{2}\right)\right\}$

11. $\{(-1, 3), (2, 7), (4, 3), (5, 8)\}$

12. $\{(-8, 6), (-4, 3), (0, 6), (5, 10)\}$

13. $f(x) = 2x + 4$

14. $f(x) = 3x + 1$

15. $g(x) = \sqrt{x - 3}, \quad x \geq 3$

16. $g(x) = \sqrt{x + 2}, \quad x \geq -2$

17. $f(x) = 3x^2 + 2$

18. $f(x) = -4x^2 - 1$

19. $f(x) = x^3 - 4$

20. $f(x) = x^3 - 3$

Let $f(x) = 2^x$. We will see in the next section that this function is one-to-one. Find each of the following, always working part (a) before part (b).

21. (a) $f(3)$ (b) $f^{-1}(8)$

22. (a) $f(4)$ (b) $f^{-1}(16)$

23. (a) $f(0)$ (b) $f^{-1}(1)$

24. (a) $f(-2)$ (b) $f^{-1}\left(\dfrac{1}{4}\right)$

The graphs of some functions are given in Exercises 25–30. (a) Use the horizontal line test to determine whether the function is one-to-one. (b) If the function is one-to-one, graph the inverse of the function. (Remember that if f is one-to-one and (a, b) is on the graph of f, then (b, a) is on the graph of f^{-1}.) See Example 4.

25.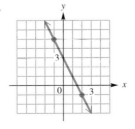

26.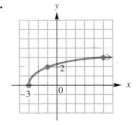

27.

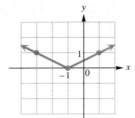

28.

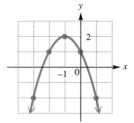

29.

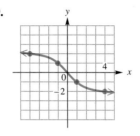

30.

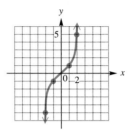

Each function defined in Exercises 31–38 is a one-to-one function. Graph the function as a solid line (or curve) and then graph its inverse on the same set of axes as a dashed line (or curve). In Exercises 35–38 you are given a table to complete so that graphing the function will be a bit easier. See Example 4.

31. $f(x) = 2x - 1$ **32.** $f(x) = 2x + 3$

33. $g(x) = -4x$ **34.** $g(x) = -2x$

35. $f(x) = \sqrt{x}, \quad x \geq 0$ **36.** $f(x) = -\sqrt{x}, \quad x \geq 0$

x	$f(x)$
0	
1	
4	

x	$f(x)$
0	
1	
4	

37. $f(x) = x^3 - 2$ **38.** $f(x) = x^3 + 3$

x	$f(x)$
-1	
0	
1	
2	

x	$f(x)$
-2	
-1	
0	
1	

■ RELATING CONCEPTS (EXERCISES 39–42)

Inverse functions are used by government agencies and other businesses to send and receive coded information. The functions they use are usually very complicated. A simple example might use the function defined by $f(x) = 2x + 5$. (Note that it is one-to-one.) Suppose that each letter of the alphabet is assigned a numerical value according to its position, as follows:

A	1	G	7	L	12	Q	17	V	22
B	2	H	8	M	13	R	18	W	23
C	3	I	9	N	14	S	19	X	24
D	4	J	10	O	15	T	20	Y	25
E	5	K	11	P	16	U	21	Z	26
F	6								

RELATING CONCEPTS (EXERCISES 39-42) (CONTINUED)

Using the function, the word ALGEBRA would be encoded as

$$7 \quad 29 \quad 19 \quad 15 \quad 9 \quad 41 \quad 7,$$

because $f(A) = f(1) = 2(1) + 5 = 7$, $f(L) = f(12) = 2(12) + 5 = 29$, and so on. The message would then be decoded by using the inverse of f, defined by $f^{-1}(x) = \dfrac{x-5}{2}$.

For example, $f^{-1}(7) = \dfrac{7-5}{2} = 1 = A$, $f^{-1}(29) = \dfrac{29-5}{2} = 12 = L$, and so on.

Work Exercises 39–42 in order.

39. Suppose that you are an agent for a detective agency and you know that today's function for your code is defined by $f(x) = 4x - 5$. Find the rule for f^{-1} algebraically.

40. You receive the following coded message today.

47 95 23 67 −1 59 27 31 51 23   7 −1 43   7 79 43 −1 75 55 67

31 71 75 27 15 23 67 15 −1 75 15 71 75 75 27 31 51

23 71 31 51   7 15 71 43 31   7 15 11   3 67 15 −1 11

Use the letter/number assignment described earlier to decode the message.

41. Why is a one-to-one function essential in this encoding/decoding process?

42. Use $f(x) = x^3 + 4$ to encode your name, using the letter/number assignment described earlier.

Did you make the connection that inverse functions can be applied to the encoding/decoding process?

 Each function defined below is one-to-one. Find the inverse algebraically, and then graph both the function and its inverse on the same square graphing calculator screen. See Objective 5.

43. $f(x) = 2x - 7$ **44.** $f(x) = -3x + 2$

45. $f(x) = x^3 + 5$ **46.** $f(x) = \sqrt[3]{x + 2}$

Some graphing calculators have the capability to draw the "inverse" of a function even if the function is not one-to-one; therefore, the inverse is not technically a function, but is a relation. For example, the graphs of $y = x^2$ and $x = y^2$ are shown in the accompanying square window.

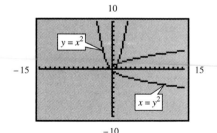

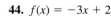 *Read your instruction manual to see if your model has this capability. Draw both y_1 and its inverse in the same square window.*

47. $y_1 = x^2 + 3x + 4$ **48.** $y_1 = x^3 - 9x$

49. Explain why the "inverse" of the function in Exercise 47 does not actually satisfy the definition of inverse as given in this section.

50. At what points do the graphs of $y = x^2$ and $x = y^2$ intersect? (See the graph above.) Verify this using algebraic methods.

12.2 Exponential Functions

O B J E C T I V E S

1 Identify exponential functions.

2 Graph exponential functions.

3 Solve exponential equations of the form $a^x = a^k$ for x.

4 Use exponential functions in applications.

F O R   E X T R A   H E L P

SSG Sec. 12.2
SSM Sec. 12.2

Pass the Test Software

InterAct Math
   Tutorial Software

Video 21

O B J E C T I V E **1** Identify exponential functions. In Section 10.7, we showed how to evaluate 2^x for rational values of x. For example,

$$2^3 = 8, \qquad 2^{-1} = \frac{1}{2}, \qquad 2^{1/2} = \sqrt{2}, \qquad 2^{3/4} = \sqrt[4]{2^3} = \sqrt[4]{8}.$$

In more advanced courses it is shown that 2^x exists for all real number values of x, both rational and irrational. (Later in the chapter, methods are given for approximating the value of 2^x for irrational x.) The following definition of an exponential function assumes that a^x exists for all real numbers x.

Exponential Function

For $a > 0$ and $a \neq 1$, and all real numbers x,

$$F(x) = a^x$$

defines an **exponential function.**

 The two restrictions on a in the definition of exponential function are important. The restriction that a must be positive is necessary so that the function can be defined for all real numbers x. For example, letting a be negative ($a = -2$, for instance) and letting $x = \frac{1}{2}$ would give the expression $(-2)^{1/2}$, which is not real. The other restriction, $a \neq 1$, is necessary because 1 raised to any power is equal to 1, and the function would then be the linear function with $F(x) = 1$.

O B J E C T I V E **2** Graph exponential functions. We can graph exponential functions by finding several ordered pairs that belong to the function. Plotting these points and connecting them with a smooth curve gives the graph.

 Be sure to plot enough points to see how rapidly the graph rises.

E X A M P L E   1   Graphing an Exponential Function with $a > 1$

Graph the exponential function with $f(x) = 2^x$.

   Choose some values of x and find the corresponding values of $f(x)$.

x	-3	-2	-1	0	1	2	3	4
$f(x) = 2^x$	$\frac{1}{8}$	$\frac{1}{4}$	$\frac{1}{2}$	1	2	4	8	16

Plotting these points and drawing a smooth curve through them gives the graph shown in Figure 8.

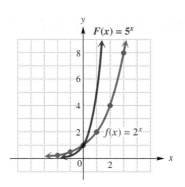

Figure 8

The graph in Figure 8 is typical of the graphs of exponential functions of the form $F(x) = a^x$, where $a > 1$. The larger value of a, faster the graph rises. To see this, compare the graph of $F(x) = 5^x$ with the graph of $f(x) = 2^x$ in Figure 8.

By the vertical line test, the graphs in Figure 8 represent functions. As these graphs suggest, the domain of an exponential function includes all real numbers. Since y is always positive, the range is $(0, \infty)$.

E X A M P L E   2   Graphing an Exponential Function with $a < 1$

Graph $g(x) = \left(\dfrac{1}{2}\right)^x$.

Again, find some points on the graph.

x	-3	-2	-1	0	1	2	3	4
$g(x) = \left(\frac{1}{2}\right)^x$	8	4	2	1	$\frac{1}{2}$	$\frac{1}{4}$	$\frac{1}{8}$	$\frac{1}{16}$

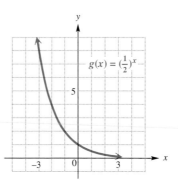

Figure 9

The graph, shown in Figure 9, is very similar to that of $f(x) = 2^x$, shown in Figure 8, except that here as x gets larger, y *decreases*. This graph is typical of the graph of a function of the form $F(x) = a^x$, where $0 < a < 1$.

The graph of $f(x) = 2^x$ and an accompanying table are shown in the graphing calculator screen in Figure 10(a) on the next page. Compare this to the results of Example 1. Similarly, the graph of $g(x) = \left(\frac{1}{2}\right)^x$ and a table are shown in Figure 10(b); compare to Example 2.

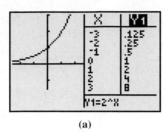

(a)

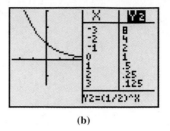

(b)

Figure 10

Based on Examples 1 and 2, we make the following generalizations about the graphs of exponential functions of the form $F(x) = a^x$.

Graph of $F(x) = a^x$

1. The graph will always contain the point $(0, 1)$.

2. When $a > 1$, the graph will *rise* from left to right. When $0 < a < 1$, the graph will *fall* from left to right. In both cases, the graph goes from the second quadrant to the first.

3. The graph will approach the x-axis, but never touch it. (Such a line is called an **asymptote.**)

4. The domain is $(-\infty, \infty)$ and the range is $(0, \infty)$.

E X A M P L E 3 Graphing a More Complicated Exponential Function

Graph $y = 3^{2x-4}$.

Find some ordered pairs. For example, if $x = 0$,

$$y = 3^{2(0)-4} = 3^{-4} = \frac{1}{81}.$$

Also, for $x = 2$,

$$y = 3^{2(2)-4} = 3^0 = 1.$$

These ordered pairs, $\left(0, \frac{1}{81}\right)$ and $(2, 1)$, along with the ordered pairs $\left(1, \frac{1}{9}\right)$ and $(3, 9)$, lead to the graph shown in Figure 11. The graph is similar to the graph of $f(x) = 2^x$ except that it is shifted to the right and rises more rapidly.

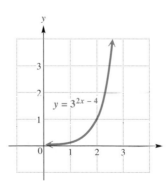

Figure 11

OBJECTIVE **3** Solve exponential equations of the form $a^x = a^k$ for x. Until now in this book, all equations that we have solved have had the variable as a base; all exponents have been constants. An **exponential equation** is an equation that has a variable in an exponent, such as

$$9^x = 27.$$

Because the exponential function defined by $F(x) = a^x$ is a one-to-one function, the following property can be used to solve many exponential equations.

Property for Solving Exponential Equations

For $a > 0$ and $a \neq 1$, if $a^x = a^y$ then $x = y$.

This property would not necessarily be true if $a = 1$.

To solve an exponential equation using this property, follow these steps.

Solving Exponential Equations

Step 1 **Each side must have the same base.** If the two sides of the equation do not have the same base, express each as a power of the same base.

Step 2 **Simplify exponents.** If necessary, use the rules of exponents to simplify the exponents.

Step 3 **Set exponents equal.** Use the above property to set the exponents equal to each other.

Step 4 **Solve.** Solve the equation obtained in Step 3.

The steps above cannot be applied to an exponential equation like

$$3^x = 12,$$

since Step 1 cannot easily be done. A method for solving such equations is given in Section 12.6.

EXAMPLE 4 Solving an Exponential Equation

Solve the equation $9^x = 27$.

We can use the property given above if both sides are changed to the same base. Since $9 = 3^2$ and $27 = 3^3$, the equation $9^x = 27$ is solved as follows.

$$(3^2)^x = 3^3 \qquad \text{Substitute. (Step 1)}$$
$$3^{2x} = 3^3 \qquad \text{Power rule for exponents (Step 2)}$$
$$2x = 3 \qquad \text{If } a^x = a^y, \text{ then } x = y. \text{ (Step 3)}$$
$$x = \frac{3}{2} \qquad \text{(Step 4)}$$

Check that the solution set is $\left\{\frac{3}{2}\right\}$ by substituting $\frac{3}{2}$ for x in the given equation.

OBJECTIVE **4** Use exponential functions in applications. Exponential functions frequently occur in applications describing growth or decay of some quantity.

E X A M P L E   5   Solving an Application of an Exponential Function

One result of the rapidly increasing world population is an increase of carbon dioxide in the air, which scientists believe may be contributing to global warming. Both population and carbon dioxide in the air are increasing exponentially. This means that the growth rate is continually increasing. The graph in Figure 12 shows the concentration of carbon dioxide (in parts per million) in the air.

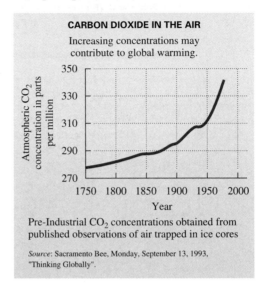

Figure 12

The data are approximated by the function with

$$f(x) = 278(1.00084)^x,$$

where x is the number of years since 1750. Use this function and a calculator to approximate the concentration of carbon dioxide in parts per million for each year.

(a) 1900

Since x represents the number of years since 1750, in this case we have $x = 1900 - 1750 = 150$. Thus, evaluate $f(150)$.

$$f(\mathbf{150}) = 278(1.00084)^{150} \qquad \text{Let } x = 150.$$
$$\approx 315 \text{ parts per million} \qquad \text{Use a calculator.}$$

(b) 1950

Use $x = 1950 - 1750 = 200$: $f(200) \approx 329$ parts per million.

E X A M P L E   6   Applying an Exponential Function

The atmospheric pressure (in millibars) at a given altitude x, in meters, can be approximated by the function defined by

$$f(x) = 1038(1.000134)^{-x},$$

for values of x between 0 and 10,000. Because the base is greater than 1 and the coefficient of x in the exponent is negative, the function values decrease as x increases. This means that as the altitude increases, the atmospheric pressure decreases. (*Source:* A. Miller and J. Thompson, *Elements of Meteorology,* Charles E. Merrill Publishing Company, 1975.)

(a) According to this function, what is the pressure at ground level?
At ground level, $x = 0$, so

$$f(0) = 1038(1.000134)^{-0} = 1038(1) = 1038.$$

The pressure is 1038 millibars.

(b) What is the pressure at 5000 meters?
Use a calculator to find $f(5000)$.

$$f(5000) = 1038(1.000134)^{-5000} \approx 531$$

The pressure is approximately 531 millibars.

12.2 EXERCISES

Choose the correct response in Exercises 1–4.

1. Which one of the following points lies on the graph of $f(x) = 2^x$?

    (a) $(1, 0)$ (b) $(2, 1)$ (c) $(0, 1)$ (d) $\left(\sqrt{2}, \frac{1}{2}\right)$

2. Which one of the following statements is true?
    (a) The y-intercept of the graph of $f(x) = 10^x$ is $(0, 10)$.
    (b) For any $a > 1$, the graph of $f(x) = a^x$ falls from left to right.
    (c) The point $\left(\frac{1}{2}, \sqrt{5}\right)$ lies on the graph of $f(x) = 5^x$.
    (d) The graph of $y = 4^x$ rises at a faster rate than the graph of $y = 10^x$.

3. The asymptote of the graph of $f(x) = a^x$
    (a) is the x-axis.   (b) is the y-axis.
    (c) has equation $x = 1$.   (d) has equation $y = 1$.

4. Which one of the following equations
   is that of the graph shown here?

    (a) $y = 1000\left(\frac{1}{2}\right)^{.3x}$

    (b) $y = 1000\left(\frac{1}{2}\right)^{x}$

    (c) $y = 1000(2)^{.3x}$
    (d) $y = 1000^x$

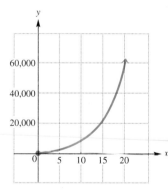

Graph each exponential function. See Examples 1–3.

5. $f(x) = 3^x$ 6. $f(x) = 5^x$ 7. $g(x) = \left(\frac{1}{3}\right)^x$ 8. $g(x) = \left(\frac{1}{5}\right)^x$

9. $y = 4^{-x}$ 10. $y = 6^{-x}$ 11. $y = 2^{2x-2}$ 12. $y = 2^{2x+1}$

13. (a) For an exponential function defined by $f(x) = a^x$, if $a > 1$, the graph _____ from left to right. If $0 < a < 1$, the graph _____ from left to right. (rises/falls)

    (b) Based on your answers in part (a), make a conjecture (an educated guess) concerning whether an exponential function defined by $f(x) = a^x$ is one-to-one. Then decide whether it has an inverse based on the concepts of Section 12.1.

14. In your own words, describe the characteristics of the graph of an exponential function. Use the exponential function defined by $f(x) = 3^x$ (Exercise 5) and the words asymptote, domain, and range in your explanation.

Solve each equation. See Example 4.

15. $100^x = 1000$ **16.** $8^x = 4$ **17.** $16^{2x+1} = 64^{x+3}$ **18.** $9^{2x-8} = 27^{x-4}$

19. $5^x = \dfrac{1}{125}$ **20.** $3^x = \dfrac{1}{81}$ **21.** $5^x = .2$ **22.** $10^x = .1$

23. $\left(\dfrac{3}{2}\right)^x = \dfrac{8}{27}$ **24.** $\left(\dfrac{4}{3}\right)^x = \dfrac{27}{64}$

 Use the exponential key of a calculator to find an approximation to the nearest thousandth.

25. $12^{2.6}$ **26.** $13^{1.8}$ **27.** $.5^{3.921}$ **28.** $.6^{4.917}$ **29.** $2.718^{2.5}$ **30.** $2.718^{-3.1}$

 31. Try to evaluate $(-2)^4$ on a scientific calculator. You may get an error message, since the exponential function key on many calculators does not allow negative bases. Discuss the concept introduced in this section that is closely related to this "peculiarity" of many scientific calculators.

32. Explain why the exponential equation $4^x = 6$ cannot be solved using the method explained in this section. Change 6 to another number that *will* allow the method of this section to be used, and then solve the equation.

The figure shown here accompanied the article "Is Our World Warming?" which appeared in the October 1990 issue of National Geographic. *It shows projected temperature increases using two graphs: one an exponential-type curve, and the other linear. From the figure, approximate the increase* **(a)** *for the exponential curve, and* **(b)** *for the linear graph for each of the following years.*

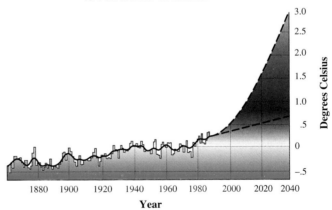

IS OUR WORLD WARMING?

Graph, "Zero Equals Average Global Temperature for the Period 1950–1979."
Dale D. Glasgow, © National Geographic Society. Reprinted by permission.

33. 2000   **34.** 2010   **35.** 2020   **36.** 2040

 Solve each problem. See Examples 5 and 6.

 37. Based on figures from 1950–1985, the number of worldwide carbon dioxide emissions in millions of short tons is approximated by the exponential function with

$$f(x) = 7147(1.0366)^x,$$

where $x = 0$ corresponds to 1950, $x = 5$ corresponds to 1955, and so on. (*Source: Carbon Dioxide Information Analysis Center,* 1994.)

(a) Use this model to approximate the number of emissions in 1950.

(b) Use this model to approximate the number of emissions in 1985.

(c) In 1990, the actual number of emissions was 25,010 million short tons. How does this compare to the number that the model provides?

38. Based on figures from 1960–1990, the gross waste generated by paper and paperboard products in millions of tons can be approximated by the exponential function with

$$f(x) = 31.28(1.028)^x,$$

where $x = 0$ corresponds to 1960, $x = 5$ corresponds to 1965, and so on. (*Source:* Environmental Protection Agency, *Characterization of Municipal Solid Waste in the United States: 1994 Update,* 1995.)

(a) Use this model to approximate the number of tons of this waste in 1960.

(b) Use this model to approximate the number of tons of this waste in 1985.

(c) In 1993, the actual number of millions of tons of this waste was 77.8. How does this compare to the number that the model provides?

39. The amount of radioactive material in an ore sample is given by the function defined by

$$A(t) = 100(3.2)^{-.5t},$$

where $A(t)$ is the amount present, in grams, of the sample t months after the initial measurement.

(a) How much was present at the initial measurement? (*Hint: $t = 0$.*)

(b) How much was present 2 months later?

(c) How much was present 10 months later?

(d) Graph the function.

40. A small business estimates that the value $V(t)$ of a copy machine is decreasing according to the function defined by

$$V(t) = 5000(2)^{-.15t},$$

where t is the number of years that have elapsed since the machine was purchased, and $V(t)$ is in dollars.

(a) What was the original value of the machine?

(b) What is the value of the machine 5 years after purchase? Give your answer to the nearest dollar.

(c) What is the value of the machine 10 years after purchase? Give your answer to the nearest dollar.

(d) Graph the function.

41. Refer to the function in Exercise 40. When will the value of the machine be $2500? (*Hint:* Let $V(t) = 2500$, divide both sides by 5000, and use the method of Example 4.)

42. Refer to the function in Exercise 40. When will the value of the machine be $1250?

The bar graph shows the average annual major league baseball player's salary for each year since free agency began. Using a technique from statistics, it was determined that the function with

$$S(x) = 74{,}741(1.17)^x$$

approximates the salary, where x = 0 corresponds to 1976, and so on, up to x = 18 representing 1994. (Salary is in dollars.)

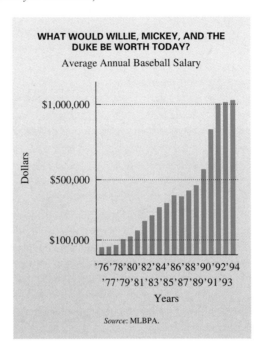

WHAT WOULD WILLIE, MICKEY, AND THE DUKE BE WORTH TODAY?

Average Annual Baseball Salary

Source: MLBPA.

43. Based on this model, what was the average salary in 1986?

44. Based on the graph, in what year did the average salary first exceed $1,000,000?

RELATING CONCEPTS (EXERCISES 45-50)

In these exercises we examine several methods of simplifying the expression $16^{3/4}$.

Work Exercises 45–50 in order.

45. Write $16^{3/4}$ as a radical expression with the exponent outside the radical. Then simplify the expression.

46. Write $16^{3/4}$ as a radical expression with the exponent under the radical. Then simplify the expression.

47. Use a calculator to find the square root of 16^3. Now find the square root of that result.

48. Explain why the result in Exercise 47 is equal to $16^{3/4}$.

49. Predict the result a calculator will give when 16 is raised to the .75 power. Then check your answer by actually performing the operation on your calculator.

50. Write $\sqrt[100]{16^{75}}$ as an exponential expression. Then write the exponent in lowest terms and evaluate this radical expression.

Did you make the connection that the expression $16^{3/4}$ can be evaluated in several different ways?

TECHNOLOGY INSIGHTS (EXERCISES 51-52)

 The number of tons, in millions, of solid waste generated in the United States during the period 1960–1990 is summarized in the following table.

Year	Millions of Tons
1960	87.8
1965	103.4
1970	121.9
1975	128.1
1980	151.5
1985	164.4
1990	195.7

Source: Environmental Protection Agency.

If $x = 0$ represents 1960, $x = 5$ represents 1965, and so on, the data can be plotted as points (0, 87.8), (5, 103.4), . . . , (30, 195.7) using a graphing calculator with statistical capability. See the graph on the left. Then, using techniques from statistics, an exponential curve can be determined to "fit" the points, as shown in the graph on the right.

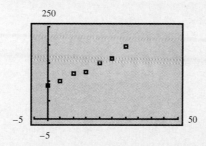 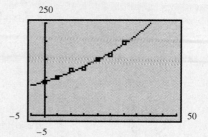

This curve is modeled by the equation

$$Y_1 = 90.11(1.0257)^x.$$

Refer to the screens in Exercises 51 and 52 and discuss the displays at the bottoms. How does the model differ from the actual figure in the table above?

51. 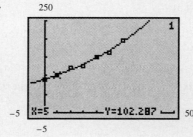   **52.**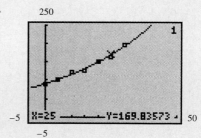

12.3 Logarithmic Functions

OBJECTIVES

1 Define a logarithm.

2 Convert between exponential and logarithmic forms.

3 Solve logarithmic equations of the form $\log_a b = k$ for a, b, or k.

4 Define and graph logarithmic functions.

5 Use logarithmic functions in applications.

FOR EXTRA HELP

SSG Sec. 12.3
SSM Sec. 12.3

Pass the Test Software

InterAct Math
Tutorial Software

Video 21

The graph of $y = 2^x$ is the curve shown in **blue** in Figure 13. Since $y = 2^x$ is a one-to-one function, it has an inverse. Interchanging x and y gives $x = 2^y$, the inverse of $y = 2^x$. As we saw in Section 12.1, the graph of the inverse is found by reflecting the graph of $y = 2^x$ about the line $y = x$. The graph of $x = 2^y$ is shown as a **red** curve in Figure 13.

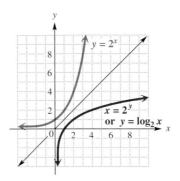

Figure 13

OBJECTIVE 1 **Define a logarithm.** We cannot solve the equation $x = 2^y$ for the dependent variable y with the methods presented up to now. The following definition is used to solve $x = 2^y$ for y.

Definition of Logarithm

For all positive numbers a, where $a \neq 1$, and all positive numbers x,

$$y = \log_a x \qquad \text{means the same as} \qquad x = a^y.$$

This key statement should be memorized. The abbreviation **log** is used for **logarithm.** Read $\log_a x$ as "the logarithm of x to the base a." To remember the location of the base and the exponent in each form, refer to the diagram that follows.

$$\text{Logarithmic form: } y = \overset{\text{Exponent}}{\underset{\text{Base}}{\log_a x}}$$

$$\text{Exponential form: } x = \overset{\text{Exponent}}{\underset{\text{Base}}{a^y}}$$

In working with logarithmic form and exponential form, remember the following.

Meaning of $\log_a x$

A **logarithm** is an exponent; $\log_a x$ is the exponent on the base a that yields the number x.

OBJECTIVE **2** Convert between exponential and logarithmic forms. We can use the definition of logarithm to write exponential statements in logarithmic form and logarithmic statements in exponential form. The chart below shows several pairs of equivalent statements.

Exponential Form	Logarithmic Form
$3^2 = 9$	$\log_3 9 = 2$
$\left(\dfrac{1}{5}\right)^{-2} = 25$	$\log_{1/5} 25 = -2$
$10^5 = 100,000$	$\log_{10} 100,000 = 5$
$4^{-3} = \dfrac{1}{64}$	$\log_4 \dfrac{1}{64} = -3$

OBJECTIVE **3** Solve logarithmic equations of the form $\log_a b = k$ for a, b, or k. A **logarithmic equation** is an equation with a logarithm in at least one term. We solve logarithmic equations of the form $\log_a b = k$ for any of the three variables by first writing the equation in exponential form.

EXAMPLE 1 Solving Logarithmic Equations

Solve the following equations.

(a) $\log_4 x = -2$

By the definition of logarithm, $\log_4 x = -2$ is equivalent to $4^{-2} = x$. Then

$$x - 4^{-2} = \frac{1}{4^2} = \frac{1}{16}$$

The solution set is $\left\{\frac{1}{16}\right\}$.

(b) $\log_{1/2} 16 = y$

First write the statement in exponential form.

$$\log_{1/2} 16 = y$$

$$\left(\frac{1}{2}\right)^y = 16 \qquad \text{Convert to exponential form.}$$

$$(2^{-1})^y = 2^4 \qquad \text{Write with the same base.}$$

$$2^{-y} = 2^4 \qquad \text{Property of exponents}$$

$$-y = 4 \qquad \text{Set exponents equal.}$$

$$y = -4 \qquad \text{Multiply by } -1.$$

The solution set is $\{-4\}$.

For any positive real number b, we know that $b^1 = b$ and $b^0 = 1$. Writing these two statements in logarithmic form gives the following two properties of logarithms.

For any positive real number b, $b \neq 1$,

$$\log_b b = 1 \qquad \text{and} \qquad \log_b 1 = 0.$$

EXAMPLE 2   Using Properties of Logarithms

Use the two properties of logarithms above to simplify each logarithm.

(a) $\log_7 7 = 1$

(b) $\log_{\sqrt{2}} \sqrt{2} = 1$

(c) $\log_9 1 = 0$

(d) $\log_{.2} 1 = 0$

OBJECTIVE **4** Define and graph logarithmic functions. Now we define the logarithmic function with base a.

Logarithmic Function

If a and x are positive numbers, with $a \neq 1$, then

$$f(x) = \log_a x$$

defines the **logarithmic function with base a.**

To graph a logarithmic function, it is helpful to write it in exponential form first. Then plot selected ordered pairs to determine the graph.

EXAMPLE 3   Graphing a Logarithmic Function

Graph $y = \log_{1/2} x$.

By writing $y = \log_{1/2} x$ in its exponential form as $x = \left(\frac{1}{2}\right)^y$, we can identify ordered pairs that satisfy the equation. Here it is easier to choose values for y and find the corresponding values of x. See the table of ordered pairs.

x	y
$\frac{1}{4}$	2
$\frac{1}{2}$	1
1	0
2	-1
4	-2
8	-3

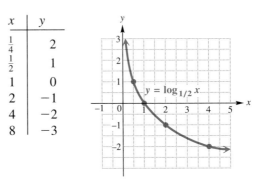

Figure 14

Plotting these points (be careful to get them in the right order) and connecting them with a smooth curve gives the graph in Figure 14. This graph is typical of logarithmic functions with $0 < a < 1$. The graph of $x = 2^y$ in Figure 13, which is equivalent to $y = \log_2 x$, is typical of graphs of logarithmic functions with base $a > 1$.

Based on the graphs of the functions $y = \log_2 x$ in Figure 13 and $y = \log_{1/2} x$ in Figure 14, we make the following generalizations about the graphs of logarithmic functions of the form $G(x) = \log_a x$.

> **Graph of $G(x) = \log_a x$**
>
> 1. The graph will always contain the point (1, 0).
>
> 2. When $a > 1$, the graph will *rise* from left to right, from the fourth quadrant to the first. When $0 < a < 1$, the graph will *fall* from left to right, from the first quadrant to the fourth.
>
> 3. The graph will approach the y-axis, but never touch it. (It is an asymptote.)
>
> 4. The domain is (0, ∞) and the range is (−∞, ∞).

Compare these generalizations to the similar ones for exponential functions in Section 12.2.

OBJECTIVE **5** Use logarithmic functions in applications. Logarithmic functions, like exponential functions, can be applied to real-world phenomena.

 EXAMPLE 4 Solving an Application of a Logarithmic Function

The function defined by

$$f(x) = 27 + 1.105 \log_{10}(x + 1)$$

approximates the barometric pressure in inches of mercury at a distance of x miles from the eye of a typical hurricane. (*Source:* A. Miller and R. Anthes, *Meteorology,* 5th edition, Charles E. Merrill Publishing Company, 1985.)

(a) Approximate the pressure 9 miles from the eye of the hurricane.

We let $x = 9$, and find $f(9)$.

$$\begin{aligned} f(9) &= 27 + 1.105 \log_{10}(9 + 1) && \text{Let } x = 9. \\ &= 27 + 1.105 \log_{10} 10 && \text{Add in parentheses.} \\ &= 27 + 1.105(1) && \log_{10} 10 = 1 \\ &= 28.105 && \text{Add.} \end{aligned}$$

The pressure 9 miles from the eye of the hurricane is 28.105 inches.

(b) Approximate the pressure 99 miles from the hurricane.

$$\begin{aligned} f(99) &= 27 + 1.105 \log_{10}(99 + 1) && \text{Let } x = 99. \\ &= 27 + 1.105 \log_{10} 100 && \text{Add in parentheses.} \\ &= 27 + 1.105(2) && \log_{10} 100 = 2 \\ &= 29.21 \end{aligned}$$

The pressure 99 miles from the eye of the hurricane is 29.21 inches.

EXAMPLE 5 Solving an Application of a Logarithmic Function

Sales (in thousands of units) of a new product are approximated by the function with

$$S(t) = 100 + 30 \log_3(2t + 1),$$

where t is the number of years after the product is introduced.

(a) What were the sales after 1 year?
Find $S(1)$.

$$\begin{aligned}
S(1) &= 100 + 30 \log_3(2 \cdot \mathbf{1} + 1) && \text{Let } t = 1.\\
&= 100 + 30 \log_3 3\\
&= 100 + 30(\mathbf{1}) && \log_3 3 = 1\\
&= 130
\end{aligned}$$

Sales were 130 thousand units after 1 year.

(b) Find the sales after 13 years.
Evaluate $S(13)$.

$$\begin{aligned}
S(13) &= 100 + 30 \log_3(2 \cdot \mathbf{13} + 1) && \text{Let } t = 13.\\
&= 100 + 30 \log_3 27\\
&= 100 + 30(\mathbf{3}) && \log_3 27 = 3\\
&= 190
\end{aligned}$$

After 13 years, sales had increased to 190 thousand units.

(c) Graph $y = S(t)$.

Use the two ordered pairs (1, 130) and (13, 190) found above. Check that (0, 100) and (40, 220) also satisfy the equation. Use these ordered pairs and knowledge of the general shape of the graph of a logarithmic function to get the graph in Figure 15.

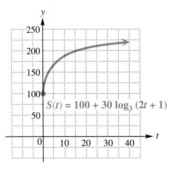

Figure 15

CONNECTIONS

In the United States, the intensity of an earthquake is rated using the *Richter scale*. The Richter scale rating of an earthquake of intensity x is given by

$$R = \log_{10} \frac{x}{x_0},$$

where x_0 is the intensity of an earthquake of a certain (small) size. Figure 16 shows Richter scale ratings for major Southern California earthquakes since 1920. As the figure indicates, earthquakes "come in bunches," and the 1990s have been an especially busy time.

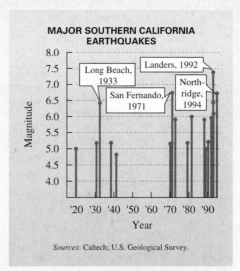

MAJOR SOUTHERN CALIFORNIA EARTHQUAKES

Long Beach, 1933

Landers, 1992

San Fernando, 1971

Northridge, 1994

Sources: Caltech; U.S. Geological Survey.

Figure 16

FOR DISCUSSION OR WRITING

Writing the logarithmic equation given above in exponential form, we get

$$10^R = \frac{x}{x_0} \quad \text{or} \quad x = 10^R x_0.$$

The 1994 Northridge earthquake had a Richter scale rating of 6.7; the Landers earthquake had a rating of 7.3. How much more powerful was the Landers earthquake than the Northridge earthquake? Compare the smallest rated earthquake in the figure (at 4.8) with the Landers quake. How much more powerful was the Landers quake?

12.3 EXERCISES

1. By definition, $\log_a x$ is the exponent to which the base a must be raised in order to obtain x. Use this definition to match the logarithm in Column I with its value from Column II. (*Example:* $\log_3 9$ is equal to 2, because 2 is the exponent to which 3 must be raised in order to obtain 9.)

I	II
(a) $\log_4 16$	**A.** -2
(b) $\log_3 81$	**B.** -1
(c) $\log_3\left(\frac{1}{3}\right)$	**C.** 2
(d) $\log_{10} .01$	**D.** 0
(e) $\log_5 \sqrt{5}$	**E.** $\frac{1}{2}$
(f) $\log_{13} 1$	**F.** 4

2. Match the logarithmic equation in Column I with the corresponding exponential equation from Column II.

I	**II**
(a) $\log_{1/3} 3 = -1$	**A.** $8^{1/3} = \sqrt[3]{8}$
(b) $\log_5 1 = 0$	**B.** $\left(\dfrac{1}{3}\right)^{-1} = 3$
(c) $\log_2 \sqrt{2} = \dfrac{1}{2}$	**C.** $4^1 = 4$
(d) $\log_{10} 1000 = 3$	**D.** $2^{1/2} = \sqrt{2}$
(e) $\log_8 \sqrt[3]{8} = \dfrac{1}{3}$	**E.** $5^0 = 1$
(f) $\log_4 4 = 1$	**F.** $10^3 = 1000$

Write in logarithmic form. See the table in Objective 2.

3. $4^5 = 1024$ **4.** $3^6 = 729$ **5.** $\left(\dfrac{1}{2}\right)^{-3} = 8$

6. $\left(\dfrac{1}{6}\right)^{-3} = 216$ **7.** $10^{-3} = .001$ **8.** $36^{1/2} = 6$

Write in exponential form. See the table in Objective 2.

9. $\log_4 64 = 3$ **10.** $\log_2 512 = 9$ **11.** $\log_{10} \dfrac{1}{10,000} = -4$

12. $\log_{100} 100 = 1$ **13.** $\log_6 1 = 0$ **14.** $\log_\pi 1 = 0$

15. When a student asked his teacher to explain how to evaluate $\log_9 3$ without showing any work, his teacher told him, "Think radically." Explain what the teacher meant by this hint.

16. A student told her teacher, "I know that $\log_2 1$ is the exponent to which 2 must be raised in order to obtain 1, but I can't think of any such number." How would you explain to the student that the value of $\log_2 1$ is 0?

Solve each equation for x. See Examples 1 and 2.

17. $x = \log_{27} 3$ **18.** $x = \log_{125} 5$ **19.** $\log_x 9 = \dfrac{1}{2}$ **20.** $\log_x 5 = \dfrac{1}{2}$

21. $\log_x 125 = -3$ **22.** $\log_x 64 = -6$ **23.** $\log_{12} x = 0$ **24.** $\log_4 x = 0$

25. $\log_x x = 1$ **26.** $\log_x 1 = 0$ **27.** $\log_x \dfrac{1}{25} = -2$ **28.** $\log_x \dfrac{1}{10} = -1$

29. $\log_8 32 = x$ **30.** $\log_{81} 27 = x$ **31.** $\log_\pi \pi^4 = x$ **32.** $\log_{\sqrt{2}} \sqrt{2}^9 = x$

33. $\log_6 \sqrt{216} = x$ **34.** $\log_4 \sqrt{64} = x$

If the point (p, q) is on the graph of $f(x) = a^x$ (for $a > 0$ and $a \neq 1$), then the point (q, p) is on the graph of $f^{-1}(x) = \log_a x$. Use this fact, and refer to the graphs required in Exercises 5–8 in Section 12.2 to graph each logarithmic function. See Example 3.

35. $y = \log_3 x$ **36.** $y = \log_5 x$ **37.** $y = \log_{1/3} x$ **38.** $y = \log_{1/5} x$

39. Explain why 1 is not allowed as a base for a logarithmic function.

40. Compare the summary of facts about the graph of $F(x) = a^x$ in Section 12.2 with the similar summary of facts about the graph of $G(x) = \log_a x$ in this section. Make a list of the facts that reinforce the concept that F and G are inverse functions.

41. The domain of $F(x) = a^x$ is $(-\infty, \infty)$, while the range is $(0, \infty)$. Therefore, since $G(x) = \log_a x$ defines the inverse of F, the domain of G is _____, while the range of G is _____.

42. The graphs of both $F(x) = 3^x$ and $G(x) = \log_3 x$ rise from left to right. Which one rises at a faster rate?

Use the graph at the right to predict the value of f(t) for the given value of t.

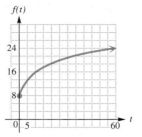

f(t)

43. $t = 0$

44. $t = 10$

45. $t = 60$

46. Show that the points determined in Exercises 43–45 lie on the graph of $f(t) = 8 \log_5(2t + 5)$.

Solve each application of a logarithmic function. See Examples 4 and 5.

47. According to selected figures from 1981–1995, the number of Superfund hazardous waste sites in the United States can be approximated by the function with

$$f(x) = 11.34 + 317.01 \log_2 x,$$

where $x = 1$ corresponds to 1981, $x = 2$ to 1982, and so on. (*Source:* Environmental Protection Agency.)

(a) Use the function to approximate the number of sites in 1984.

(b) Use the function to approximate the number of sites in 1988.

48. According to selected figures from 1980–1993, the number of trillion cubic feet of dry natural gas consumed worldwide can be approximated by the function with

$$f(x) = 51.47 + 6.044 \log_2 x,$$

where $x = 1$ corresponds to 1980, $x = 2$ to 1981, and so on. (*Source:* Energy Information Administration.)

(a) Use the function to approximate consumption in 1980.

(b) Use the function to approximate consumption in 1987.

49. A study showed that the number of mice in an old abandoned house was approximated by the function with

$$M(t) = 6 \log_4(2t + 4),$$

where t is measured in months and $t = 0$ corresponds to January 1998. Find the number of mice in the house in

(a) January 1998

(b) July 1998

(c) July 2000.

(d) Graph the function.

50. A supply of hybrid striped bass were introduced into a lake in January 1990. Biologists researching the bass population over the next decade found that the number of bass in the lake was approximated by the function with

$$B(t) = 500 \log_3(2t + 3),$$

where $t = 0$ corresponds to January 1990, $t = 1$ to January 1991, $t = 2$ to January 1992, and so on. Use this function to find the bass population in

(a) January 1990

(b) January 1993

(c) January 2002.

(d) Graph the function for $0 \le t \le 12$.

As mentioned in Section 12.1, some graphing calculators have the capability of drawing the inverse of a function. For example, the two screens that follow show the graphs of $f(x) = 2^x$ and $g(x) = \log_2 x$. The graph of g was obtained by drawing the graph of f^{-1}, since $g(x) = f^{-1}(x)$. (Compare to Figure 13 in this section.)

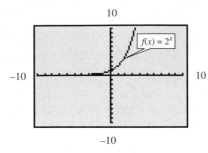

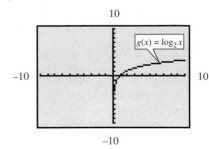

 Use a graphing calculator with the capability of drawing the inverse of a function to draw the graph of each logarithmic function. Use the standard viewing window.

51. $g(x) = \log_3 x$ (Compare to Exercise 35.)

52. $g(x) = \log_5 x$ (Compare to Exercise 36.)

53. $g(x) = \log_{1/3} x$ (Compare to Exercise 37.)

54. $g(x) = \log_{1/5} x$ (Compare to Exercise 38.)

12.4 Properties of Logarithms

OBJECTIVES

1. Use the product rule for logarithms.

2. Use the quotient rule for logarithms.

3. Use the power rule for logarithms.

4. Use properties to write alternative forms of logarithmic expressions.

FOR EXTRA HELP

 **SSG** Sec. 12.4
SSM Sec. 12.4

Pass the Test Software

InterAct Math
Tutorial Software

Video 22

Logarithms have been used as an aid to numerical calculation for several hundred years. Today the widespread use of calculators has made the use of logarithms for calculation obsolete. However, logarithms are still very important in applications and in further work in mathematics.

OBJECTIVE **1** Use the product rule for logarithms. One way in which logarithms simplify problems is by changing a problem of multiplication into one of addition. This is done with the product rule for logarithms.

Product Rule for Logarithms

If x, y, and b are positive numbers, where $b \neq 1$, then

$$\log_b xy = \log_b x + \log_b y.$$

(The logarithm of a product is the sum of the logarithms of the factors.)

 The word statement of the product rule can also be stated by replacing "logarithm" with "exponent," and the rule then becomes the familiar rule for multiplying exponential expressions: The *exponent* of a product is equal to the sum of the *exponents* of the factors.

To prove this rule, let $m = \log_b x$ and $n = \log_b y$, and recall that

$$\log_b x = m \qquad \text{means} \qquad b^m = x,$$
$$\log_b y = n \qquad \text{means} \qquad b^n = y.$$

Now consider the product xy.

$$xy = b^m \cdot b^n \qquad \text{Substitution}$$
$$xy = b^{m+n} \qquad \text{Product rule for exponents}$$
$$\log_b xy = m + n \qquad \text{Convert to logarithmic form.}$$
$$\log_b xy = \log_b x + \log_b y \qquad \text{Substitution}$$

The last statement is the result we wished to prove.

E X A M P L E 1 Using the Product Rule

Use the product rule for logarithms to rewrite the following. Assume $x > 0$.

(a) $\log_5 6 \cdot 9$

By the product rule, $\log_5 6 \cdot 9 = \log_5 6 + \log_5 9$.

(b) $\log_7 8 + \log_7 12$

$$\log_7 8 + \log_7 12 = \log_7 8 \cdot 12 = \log_7 96$$

(c) $\log_3 3x$

$$\log_3 3x = \log_3 3 + \log_3 x$$
$$\log_3 3x = 1 + \log_3 x \qquad \log_3 3 = 1$$

(d) $\log_4 x^3$

Since $x^3 = x \cdot x \cdot x$,

$$\log_4 x^3 = \log_4(x \cdot x \cdot x)$$
$$= \log_4 x + \log_4 x + \log_4 x$$
$$= 3 \log_4 x.$$

O B J E C T I V E 2 Use the quotient rule for logarithms. The rule for division is similar to the rule for multiplication.

Quotient Rule for Logarithms

If x, y, and b are positive numbers, where $b \neq 1$, then

$$\log_b \frac{x}{y} = \log_b x - \log_b y.$$

(The logarithm of a quotient is the difference between the logarithm of the numerator and the logarithm of the denominator.)

The proof of this rule is similar to the proof of the product rule.

E X A M P L E 2 Using the Quotient Rule

Use the quotient rule for logarithms to rewrite the following.

(a) $\log_4 \frac{7}{9} = \log_4 7 - \log_4 9$

(b) $\log_5 \frac{6}{x} = \log_5 6 - \log_5 x$, for $x > 0$.

(c) $\log_3 \dfrac{27}{5} = \log_3 27 - \log_3 5$

$\qquad\qquad = 3 - \log_3 5 \qquad\qquad \log_3 27 = 3$

OBJECTIVE **3** **Use the power rule for logarithms.** There is also a rule for finding the logarithm of the power of a number.

> **Power Rule for Logarithms**
>
> If x and b are positive real numbers, where $b \neq 1$, and if r is any real number, then
>
> $$\log_b x^r = r(\log_b x).$$
>
> (The logarithm of a number to a power equals the exponent times the logarithm of the number.)

As examples of this rule,

$$\log_b m^5 = 5 \log_b m \qquad \text{and} \qquad \log_3 5^{3/4} = \frac{3}{4} \log_3 5.$$

 NOTE To see an earlier illustration of this rule, refer to Example 1(d).

To prove the power rule, let

$$\log_b x = m$$
$$b^m = x \qquad\qquad \text{Convert to exponential form.}$$
$$(b^m)^r = x^r \qquad\qquad \text{Raise to the power } r.$$
$$b^{mr} = x^r \qquad\qquad \text{Power rule for exponents}$$
$$\log_b x^r = mr \qquad\qquad \text{Convert to logarithmic form.}$$
$$\log_b x^r = rm$$
$$\log_b x^r = r \log_b x. \qquad m = \log_b x$$

This is the statement to be proved.

As a special case of the rule above, let $r = \frac{1}{p}$, so that

$$\log_b \sqrt[p]{x} = \log_b x^{1/p} = \frac{1}{p} \log_b x.$$

For example, using this result, with $x > 0$,

$$\log_b \sqrt[5]{x} = \log_b x^{1/5} = \frac{1}{5} \log_b x \qquad \text{and} \qquad \log_b \sqrt[3]{x^4} = \log_b x^{4/3} = \frac{4}{3} \log_b x.$$

Another special case is

$$\log_b \frac{1}{x} = -\log_b x$$

since $\dfrac{1}{x} = x^{-1}$.

EXAMPLE 3 Using the Power Rule

Use the power rule to rewrite each of the following. Assume $a > 0$, $b > 0$, $x > 0$, $a \neq 1$, and $b \neq 1$.

(a) $\log_3 5^2 = 2 \log_3 5$ $\qquad\qquad$ (b) $\log_a x^4 = 4 \log_a x$

(c) $\log_b \sqrt{7}$

When using the power rule with logarithms of expressions involving radicals, begin by rewriting the radical expression with a rational exponent, as shown in Section 10.7.

$$\log_b \sqrt{7} = \log_b 7^{1/2} \qquad \sqrt{x} = x^{1/2}$$
$$= \frac{1}{2} \log_b 7 \qquad \text{Power rule}$$

(d) $\log_2 \sqrt[5]{x^2} = \log_2 x^{2/5} = \frac{2}{5} \log_2 x$

(e) $\log_3 \frac{1}{7} = -\log_3 7$

Two special properties involving both exponential and logarithmic expressions come directly from the fact that logarithmic and exponential functions are inverses of each other.

Special Properties

If $b > 0$ and $b \neq 1$, then

$$b^{\log_b x} = x \quad (x > 0) \qquad \text{and} \qquad \log_b b^x = x.$$

To prove the first statement, let

$$y = \log_b x.$$
$$b^y = x \qquad\qquad \text{Convert to exponential form.}$$
$$b^{\log_b x} = x \qquad\qquad \text{Replace } y \text{ with } \log_b x.$$

The proof of the second statement is similar.

EXAMPLE 4 Using the Special Properties

Find the value of the following logarithmic expressions.

(a) $\log_5 5^4$
Since $\log_b b^x = x$,

$$\log_5 5^4 = 4.$$

(b) $\log_3 9$
Since $9 = 3^2$,

$$\log_3 9 = \log_3 3^2 = 2.$$

The property $\log_b b^x = x$ was used in the last step.

(c) $4^{\log_4 10} = 10$

OBJECTIVE 4 Use properties to write alternative forms of logarithmic expressions. Doing so is important in solving equations with logarithms and in calculus.

EXAMPLE 5 Writing Logarithms in Alternative Forms

Use the properties of logarithms to rewrite each expression. Assume all variables represent positive real numbers.

(a) $\log_4 4x^3 = \log_4 4 + \log_4 x^3$ Product rule

$= 1 + 3 \log_4 x$ $\log_4 4 = 1$; power rule

(b) $\log_7 \sqrt{\dfrac{p}{q}} = \log_7 \left(\dfrac{p}{q}\right)^{1/2}$

$= \dfrac{1}{2} \log_7 \dfrac{p}{q}$ Power rule

$= \dfrac{1}{2} (\log_7 p - \log_7 q)$ Quotient rule

(c) $\log_5 \dfrac{a}{bc} = \log_5 a - \log_5 bc$ Quotient rule

$= \log_5 a - (\log_5 b + \log_5 c)$ Product rule

$= \log_5 a - \log_5 b - \log_5 c$

Notice the careful use of parentheses in the second step. Since we are subtracting the logarithm of a product, and it is being rewritten as a sum of two terms, parentheses *must* be placed around the sum.

(d) $3 \log_b x + \dfrac{1}{2} \log_b y = \log_b x^3 + \log_b y^{1/2}$ Power rule

$= \log_b x^3 \sqrt{y}$ Product rule; $y^{1/2} = \sqrt{y}$

(e) $\log_8(2p + 3r)$ cannot be rewritten by the properties of logarithms. ■

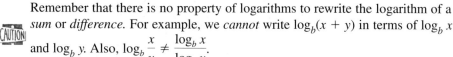

Remember that there is no property of logarithms to rewrite the logarithm of a *sum* or *difference*. For example, we *cannot* write $\log_b(x + y)$ in terms of $\log_b x$ and $\log_b y$. Also, $\log_b \dfrac{x}{y} \neq \dfrac{\log_b x}{\log_b y}$.

In the next example, we use numerical values for $\log_2 5$ and $\log_2 3$. While we use the equals sign to give these values, they are actually just approximations, since most logarithms of this type are irrational numbers. While it would be more correct to use the symbol $\approx$, we will simply use $=$ with the understanding that the values are correct to four decimal places.

EXAMPLE 6 Using the Properties of Logarithms with Numerical Values

Given that $\log_2 5 = 2.3219$ and $\log_2 3 = 1.5850$, evaluate the following.

(a) $\log_2 15$

$$\begin{aligned}\log_2 15 &= \log_2 3 \cdot 5 \\ &= \log_2 3 + \log_2 5 \qquad \text{Product rule}\\ &= 1.5850 + 2.3219 \\ &= 3.9069\end{aligned}$$

(b) $\log_2 .6$

$$\begin{aligned}\log_2 .6 &= \log_2 \frac{3}{5} \qquad .6 = \tfrac{6}{10} = \tfrac{3}{5}\\ &= \log_2 3 - \log_2 5 \qquad \text{Quotient rule}\\ &= 1.5850 - 2.3219 \\ &= -.7369\end{aligned}$$

(c) $\log_2 27$

$$\begin{aligned}\log_2 27 &= \log_2 3^3 \\ &= 3\log_2 3 \qquad \text{Power rule}\\ &= 3(1.5850) \\ &= 4.7550\end{aligned}$$

EXAMPLE 7 Deciding Whether Statements about Logarithms Are True

Decide whether each of the following statements is true or false.

(a) $\log_2 8 - \log_2 4 = \log_2 4$

Evaluate both sides.

$$\log_2 8 - \log_2 4 = \log_2 2^3 - \log_2 2^2 = 3 - 2 = 1$$
$$\log_2 4 = \log_2 2^2 = 2$$

The statement is false because $2 \neq 1$.

(b) $\log_3(\log_2 8) = \dfrac{\log_7 49}{\log_8 64}$

Evaluate both sides.

$$\log_3(\log_2 8) = \log_3(3) = 1$$
$$\frac{\log_7 49}{\log_8 64} = \frac{\log_7 7^2}{\log_8 8^2} = \frac{2}{2} = 1$$

The statement is true.

CONNECTIONS

 Long before the days of calculators and computers, the search for making calculations easier was an ongoing process. Machines built by Charles Babbage and Blaise Pascal, a system of "rods" used by John Napier, and slide rules were the forerunners of today's electronic marvels. The invention of logarithms by John Napier in the sixteenth century was a great breakthrough in the search for easier methods of calculation.

Since logarithms are exponents, their properties allowed users of tables of common logarithms to multiply by adding, divide by subtracting, raise to powers by multiplying, and take roots by dividing. Although logarithms are no longer used for computations, they play an important part in higher mathematics.

FOR DISCUSSION OR WRITING

To multiply 458.3 by 294.6 using logarithms, we add $\log_{10} 458.3$ and $\log_{10} 294.6$, then find 10 to the sum. Perform this multiplication using the log* key and the 10^x key on your calculator. Check your answer by multiplying directly with your calculator. Try division, raising to a power, and taking a root by this method.

---

*In this text, the notation $\log x$ is used to mean $\log_{10} x$. This is also the meaning of the log key on calculators.

12.4 EXERCISES

Use the indicated rule of logarithms to complete each equation in Exercises 1–5.

1. $\log_{10}(3 \cdot 4) =$ _____   (product rule)

2. $\log_{10}\left(\dfrac{3}{4}\right) =$ _____   (quotient rule)

3. $\log_{10} 3^4 =$ _____   (power rule)

4. $3^{\log_3 4} =$ _____   (special property)

5. $\log_3 3^4 =$ _____   (special property)

6. Evaluate $\log_2(8 + 8)$. Then evaluate $\log_2 8 + \log_2 8$. Are the results the same? How could you change the operation in the first expression to make the two expressions equal?

Use the properties of logarithms introduced in this section to express each logarithm as a sum or difference of logarithms, or as a single number if possible. Assume that all variables represent positive real numbers. See Examples 1–5.

7. $\log_7 \dfrac{4}{5}$ **8.** $\log_8 \dfrac{9}{11}$ **9.** $\log_2 8^{1/4}$

10. $\log_3 9^{3/4}$ **11.** $\log_4 \dfrac{3\sqrt{x}}{y}$ **12.** $\log_5 \dfrac{6\sqrt{z}}{w}$

13. $\log_3 \dfrac{\sqrt[3]{4}}{x^2 y}$ **14.** $\log_7 \dfrac{\sqrt[3]{13}}{pq^2}$ **15.** $\log_3 \sqrt{\dfrac{xy}{5}}$

16. $\log_6 \sqrt{\dfrac{pq}{7}}$ **17.** $\log_2 \dfrac{\sqrt[3]{x} \cdot \sqrt[5]{y}}{r^2}$ **18.** $\log_4 \dfrac{\sqrt[4]{z} \cdot \sqrt[5]{w}}{s^2}$

19. A student erroneously wrote $\log_a(x + y) = \log_a x + \log_a y$. When his teacher explained that this was indeed wrong, the student claimed that he had used the distributive property. Write a few sentences explaining why the distributive property does not apply in this case.

20. Write a few sentences explaining how the rules for multiplying and dividing powers of the same base are similar to the rules for finding logarithms of products and quotients.

Use the properties of logarithms introduced in this section to express each of the following as a single logarithm. Assume that all variables are defined in such a way that the variable expressions are positive, and bases are positive numbers not equal to 1. See Examples 1–5.

21. $\log_b x + \log_b y$

22. $\log_b 2 + \log_b z$

23. $3 \log_a m - \log_a n$

24. $5 \log_b x - \log_b y$

25. $(\log_a r - \log_a s) + 3 \log_a t$

26. $(\log_a p - \log_a q) + 2 \log_a r$

27. $3 \log_a 5 - 4 \log_a 3$

28. $3 \log_a 5 + \frac{1}{2} \log_a 9$

29. $\log_{10}(x + 3) + \log_{10}(x - 3)$

30. $\log_{10}(y + 4) + \log_{10}(y - 4)$

31. $3 \log_p x + \frac{1}{2} \log_p y - \frac{3}{2} \log_p z - 3 \log_p a$

32. $\frac{1}{3} \log_b x + \frac{2}{3} \log_b y - \frac{3}{4} \log_b s - \frac{2}{3} \log_b t$

To four decimal places, the values of $\log_{10} 2$ and $\log_{10} 9$ are

$$\log_{10} 2 = .3010 \qquad \log_{10} 9 = .9542.$$

Evaluate each logarithm by applying the appropriate rule or rules from this section. **DO NOT USE A CALCULATOR FOR THESE EXERCISES.** *See Example 6.*

33. $\log_{10} 18$

34. $\log_{10} \frac{9}{2}$

35. $\log_{10} \frac{2}{9}$

36. $\log_{10} 4$

37. $\log_{10} 36$

38. $\log_{10} 162$

39. $\log_{10} 3$

40. $\log_{10} \sqrt[5]{2}$

Decide whether each statement is true or false. See Example 7.

41. $\log_6 60 - \log_6 10 = 1$

42. $\log_3 7 + \log_3 \frac{1}{7} = 0$

43. $\frac{\log_{10} 7}{\log_{10} 14} = \frac{1}{2}$

44. $\frac{\log_{10} 10}{\log_{10} 100} = \frac{1}{10}$

45. Refer to the "NOTE" following the word statement of the product rule for logarithms in this section. Now, state the quotient rule in words, replacing "logarithm" with "exponent."

46. Explain why the statement for the power rule for logarithms requires that x be a positive real number.

47. Refer to Example 7(a). Change the left side of the equation using the quotient rule so that the statement becomes true, and simplify.

48. What is wrong with the following "proof" that $\log_2 16$ does not exist?

$$\log_2 16 = \log_2(-4)(-4)$$
$$= \log_2(-4) + \log_2(-4)$$

Since the logarithm of a negative number is not defined, the final step cannot be evaluated, and so $\log_2 16$ does not exist.

■ RELATING CONCEPTS (EXERCISES 49–54)

Work Exercises 49–54 in order.

49. Evaluate $\log_3 81$.

50. Write the *meaning* of the expression $\log_3 81$.

(continued)

RELATING CONCEPTS (EXERCISES 49-54) (CONTINUED)

51. Evaluate $3^{\log_3 81}$.

52. Write the *meaning* of the expression $\log_2 19$.

53. Evaluate $2^{\log_2 19}$.

54. Keeping in mind that a logarithm is an exponent, and using the results from Exercises 49–53, what is the simplest form of the expression $k^{\log_k m}$?

Did you make the connection that a logarithm is an exponent?

12.5 Evaluating Logarithms

OBJECTIVES

1 Evaluate common logarithms by using a calculator.

2 Use common logarithms in an application.

3 Evaluate natural logarithms using a calculator.

4 Use exponential functions with base e and natural logarithms in applications.

5 Use the change-of-base rule.

FOR EXTRA HELP

SSG Sec. 12.5
SSM Sec. 12.5

Pass the Test Software

InterAct Math
   Tutorial Software

Video 22

As mentioned earlier, logarithms are important in many applications of mathematics to everyday problems, particularly in biology, engineering, economics, and social science. In this section we find numerical approximations for logarithms. Traditionally, base 10 logarithms are used most often since our number system is base 10. Logarithms to base 10 are called **common logarithms** and $\log_{10} x$ is abbreviated as simply $\log x$, where the base is understood to be 10.

OBJECTIVE 1 Evaluate common logarithms by using a calculator. In the next example we give the results of evaluating some common logarithms using a calculator with a log key. (This may be a second function key on some calculators.) For simple scientific calculators, just enter the number, then press the log key. For graphing calculators, these steps are reversed. We will give all logarithms to four decimal places.

EXAMPLE 1 Evaluating Common Logarithms

Evaluate each logarithm using a calculator.

(a) log 327.1   **(b)** log 437,000   **(c)** log .0615

Figure 17 shows how a graphing calculator evaluates these logarithms. The calculator is set to give four decimal places.

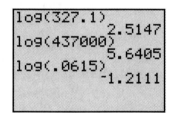

```
log(327.1)
            2.5147
log(437000)
            5.6405
log(.0615)
           -1.2111
```

Figure 17

Notice that log .0615 is found to be -1.2111, a negative result. The common logarithm of a number between 0 and 1 is always negative because the logarithm is the exponent on 10 that produces the number. For example,

$$10^{-1.2111} = .0615.$$

If the exponent (the logarithm) were positive, the result would be greater than 1, since $10^0 = 1$.

OBJECTIVE **2** Use common logarithms in an application. In chemistry, the **pH** of a solution is defined as follows.

Definition of pH

$$pH = -\log[H_3O^+],$$

where $[H_3O^+]$ is the hydronium ion concentration in moles per liter.

The pH is a measure of the acidity or alkalinity of a solution, with water, for example, having a pH of 7. In general, acids have pH numbers less than 7, and alkaline solutions have pH values greater than 7. It is customary to round pH values to the nearest tenth.

EXAMPLE 2   Using pH in an Application

Wetlands are classified as *bogs*, *fens*, *marshes*, and *swamps*. These classifications are based on pH values. A pH value between 6.0 and 7.5, such as that of Summerby Swamp in Michigan's Hiawatha National Forest, indicates that the wetland is a "rich fen." When the pH is between 4.0 and 6.0, it is a "poor fen," and if the pH falls to 3.0 or less, the wetland is a "bog." (*Source:* R. Mohlenbrock, "Summerby Swamp, Michigan," *Natural History*, March 1994.)

Suppose that the hydronium ion concentration of a sample of water from a wetland is 6.3×10^{-3}. How would this wetland be classified?

Use the definition of pH.

$$
\begin{aligned}
pH &= -\log(6.3 \times 10^{-3}) \\
&= -(\log 6.3 + \log 10^{-3}) \qquad \text{Product rule} \\
&= -[.7993 - 3(1)] \\
&= -.7993 + 3 \approx 2.2
\end{aligned}
$$

Since the pH is less than 3.0, the wetland is a bog.

EXAMPLE 3   Finding Hydronium Ion Concentration

Find the hydronium ion concentration of drinking water with a pH of 6.5.

$$
\begin{aligned}
pH = 6.5 &= -\log[H_3O^+] \\
\log[H_3O^+] &= -6.5 \qquad \text{Multiply by } -1.
\end{aligned}
$$

Solve for $[H_3O^+]$ by writing the equation in exponential form, remembering that the base is 10.

$$[H_3O^+] = 10^{-6.5} = 3.2 \times 10^{-7}$$

OBJECTIVE **3** Evaluate natural logarithms using a calculator. The most important logarithms used in applications are **natural logarithms,** which have as base the number e. The number e is irrational, like π: $e \approx 2.7182818$. Logarithms to base e are called natural logarithms because they occur in biology and the social sciences in natural situations that involve growth or decay. The base e logarithm of x is written $\ln x$ (read "el en x"). A graph of $y = \ln x$, the natural logarithmic function, is given in Figure 18.

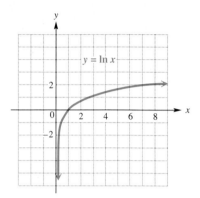

Figure 18

CONNECTIONS

The number e is a fundamental number in our universe. For this reason, e, like π, is called a *universal constant.* If there are intelligent beings elsewhere, they too will have to use e to do higher mathematics.

The letter e is used to honor Leonhard Euler, who published extensive results on the number in 1748. The first few digits of the decimal value of e are 2.7182818. Since it is an irrational number, its decimal expansion never terminates and never repeats.

The properties of e are used extensively in calculus and in higher mathematics. In Section 12.6 we see how it applies to growth and decay in the physical world.

FOR DISCUSSION OR WRITING
The value of e can be expressed as

$$e = 1 + \frac{1}{1} + \frac{1}{1 \cdot 2} + \frac{1}{1 \cdot 2 \cdot 3} + \frac{1}{1 \cdot 2 \cdot 3 \cdot 4} + \cdots.$$

Approximate e using 2 terms of this expression, then 3 terms, 4 terms, 5 terms, and 6 terms. How close is the approximation to the value of e given above with 6 terms? Does this infinite sum approach the value of e very quickly?

A calculator key labeled $\ln x$ is used to evaluate natural logarithms. If your calculator has an e^x key, but not a key labeled $\ln x$, find natural logarithms by entering the number, pressing the INV key, and then pressing the e^x key. This works because $y = e^x$ is the inverse function of $y = \ln x$ (or $y = \log_e x$).

EXAMPLE 4   Finding Natural Logarithms

Find each of the following logarithms to four decimal places.

(a) ln 192.7   **(b)** ln 10.84   **(c)** ln .5841

Figure 19 shows how a graphing calculator evaluates these natural logarithms. Like common logarithms, a number between 0 and 1 has a negative natural logarithm, as in the case of ln .5841.

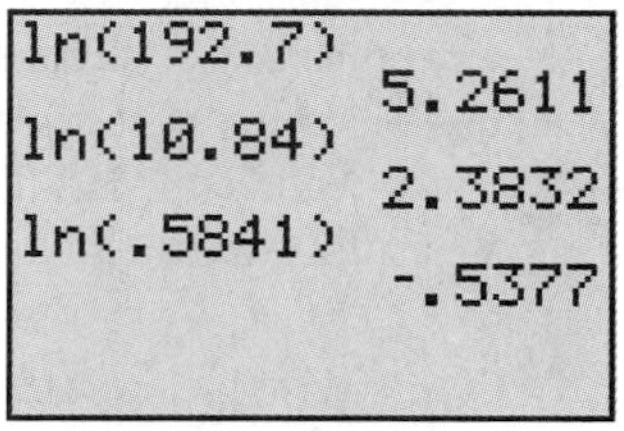

```
ln(192.7)
           5.2611
ln(10.84)
           2.3832
ln(.5841)
          -.5377
```

Figure 19

OBJECTIVE 4 Use exponential functions with base e and natural logarithms in applications. One of the most common applications of exponential functions depends on the fact that in many situations involving growth or decay of a population, the amount or number of some quantity present at time t can be closely approximated by

$$y = y_0 e^{kt},$$

where y_0 is the amount or number present at time $t = 0$, k is a constant, and e is the base of natural logarithms.

EXAMPLE 5   Applying an Exponential Function

The *greenhouse effect* refers to the phenomenon whereby emissions of gases such as carbon dioxide, methane, and chlorofluorocarbons (CFCs) have the potential to alter the climate of the earth and destroy the ozone layer. Concentrations of CFC-12, used in refrigeration technology, in parts per billion (ppb) can be modeled by the exponential function with

$$f(x) = .48 e^{.04x},$$

where $x = 0$ represents 1990. Use this function to approximate the concentration in 1998.

Since $x = 0$ represents 1990, $x = 8$ represents 1998. Evaluate $f(8)$ using a calculator.

$$f(8) = .48 e^{.04(8)} = .48 e^{.32} \approx .66$$

In 1998, the concentration of CFC-12 was about .66 ppb.

EXAMPLE 6   Applying Natural Logarithms

The number of years, $N(r)$, since two independently evolving languages split off from a common ancestral language is approximated by

$$N(r) = -5000 \ln r,$$

where r is the percent of words from the ancestral language common to both languages now. Find N if $r = 70\%$.

Write 70% as .7 and find $N(.7)$.

$$N(.7) = -5000 \ln .7$$
$$\approx -5000(-.35667)$$
$$\approx 1783$$

Approximately 1800 years have passed since the two languages separated.

O B J E C T I V E 5 Use the change-of-base rule. A calculator can be used to approximate the values of common logarithms (base 10) or natural logarithms (base e). However, sometimes we need to use logarithms to other bases. The following rule is used to convert logarithms from one base to another.

Change-of-Base Rule

If $a > 0$, $a \neq 1$, $b > 0$, $b \neq 1$, and $x > 0$, then

$$\log_a x = \frac{\log_b x}{\log_b a}.$$

To help remember the change-of-base rule, notice that x is "above" a on both sides of the equation.

Any positive number other than 1 can be used for base b in the change-of-base rule, but usually the only practical bases are e and 10, since calculators give logarithms only for these two bases.

To prove the formula for change of base, let $\log_a x = m$.

$$\log_a x = m$$
$$a^m = x \qquad \text{Change to exponential form.}$$

Since logarithmic functions are one-to-one, if all variables are positive and if $x = y$, then $\log_b x = \log_b y$.

$$\log_b(a^m) = \log_b x \qquad \text{Take logarithms on both sides.}$$
$$m \log_b a = \log_b x \qquad \text{Use the power rule.}$$
$$(\log_a x)(\log_b a) = \log_b x \qquad \text{Substitute for } m.$$
$$\log_a x = \frac{\log_b x}{\log_b a} \qquad \text{Divide both sides by } \log_b a.$$

E X A M P L E 7 Using the Change-of-Base Rule

Find each logarithm using a calculator.

(a) $\log_5 12$

Use common logarithms and the rule for change of base.

$$\log_5 12 = \frac{\log 12}{\log 5} \approx 1.5440$$

(b) $\log_2 134$

Use natural logarithms and the change-of-base rule.

$$\log_2 134 = \frac{\ln 134}{\ln 2} \approx 7.0661$$

In Example 7, the final answers were obtained *without* rounding off the intermediate values. In general, it is best to wait until the final step to round off the answer; otherwise, a build-up of round-off error may cause the final answer to have an incorrect final decimal place digit.

12.5 EXERCISES

Choose the correct response in Exercises 1–4.

1. What is the base in the expression log x?
   (a) e **(b)** 1   **(c)** 10   **(d)** x

2. What is the base in the expression ln x?
   (a) e **(b)** 1   **(c)** 10   **(d)** x

3. Since $10^0 = 1$ and $10^1 = 10$, between what two consecutive integers is the value of log 5.6?
   (a) 5 and 6   **(b)** 10 and 11   **(c)** 0 and 1   **(d)** -1 and 0

4. Since $e^1 \approx 2.718$ and $e^2 \approx 7.389$, between what two consecutive integers is the value of ln 5.6?
   (a) 5 and 6   **(b)** 2 and 3   **(c)** 1 and 2   **(d)** 0 and 1

5. Without using a calculator, give the value of $\log 10^{19.2}$.

6. Without using a calculator, give the value of $\ln e^{\sqrt{2}}$.

 You may need a calculator for the remaining exercises in this set.

Find each logarithm. Give an approximation to four decimal places. See Examples 1 and 4.

7. log 43	**8.** log 98	**9.** log 328.4
10. log 457.2	**11.** log .0326	**12.** log .1741
13. $\log(4.76 \times 10^9)$	**14.** $\log(2.13 \times 10^4)$	**15.** ln 7.84
16. ln 8.32	**17.** ln .0556	**18.** ln .0217
19. ln 388.1	**20.** ln 942.6	**21.** $\ln(8.59 \times e^2)$
22. $\ln(7.46 \times e^3)$	**23.** ln 10	**24.** log e

25. Use your calculator to find approximations of the following logarithms:
   (a) log 356.8   **(b)** log 35.68   **(c)** log 3.568.
   (d) Observe your answers and make a conjecture concerning the decimal values of the common logarithms of numbers greater than 1 that have the same digits.

26. Let k represent the number of letters in your last name.
   (a) Use your calculator to find log k.
   (b) Raise 10 to the power indicated by the number you found in part (a). What is your result?
   (c) Use the concepts of Section 12.1 to explain why you obtained the answer you found in part (b). Would it matter what number you used for k to observe the same result?

27. Try to find $\log(-1)$ using a calculator. (If you have a graphing calculator, it should be in real number mode.) What happens? Explain why this happens.

Refer to Example 2. In Exercises 28–30, suppose that water from a wetland area is sampled and found to have the given hydronium ion concentration. Determine whether the wetland is a rich fen, a poor fen, or a bog.

28. 2.5×10^{-5} **29.** 2.5×10^{-2} **30.** 2.5×10^{-7}

Find the pH of the substance with the given hydronium ion concentration. See Example 2.

31. Ammonia, 2.5×10^{-12} **32.** Sodium bicarbonate, 4.0×10^{-9}

33. Grapes, 5.0×10^{-5} **34.** Tuna, 1.3×10^{-6}

Use the formula for pH to find the hydronium ion concentration of the substance with the given pH. See Example 3.

35. Human blood plasma, 7.4   **36.** Human gastric contents, 2.0

37. Spinach, 5.4   **38.** Bananas, 4.6

 Solve each problem. See Examples 5 and 6.

39. The total expenditures in millions of current dollars for pollution abatement and control during the period from 1985 to 1993 can be approximated by the function with

$$P(x) = 70{,}967e^{.0526x},$$

where $x = 0$ corresponds to 1985, $x = 1$ to 1986, and so on. Approximate the expenditures for each of the following years. (*Source:* U.S. Bureau of Economic Analysis, *Survey of Current Business, May 1995.*)
(a) 1987   **(b)** 1990   **(c)** 1993
(d) What were the approximate expenditures for 1985?

40. The emission of the greenhouse gas nitrous oxide increased yearly during the first half of the 1990s. Based on figures during the period from 1990 to 1994, the emissions in thousands of metric tons can be modeled by the function with

$$N(x) = 446.5e^{.0118x},$$

where $x = 0$ corresponds to 1990, $x = 1$ to 1991, and so on. Approximate the emissions for each of the following years. (*Source:* U.S. Energy Information Administration, *Emission of Greenhouse Gases in the United States, annual.*)
(a) 1991   **(b)** 1992   **(c)** 1994
(d) What were the approximate emissions in 1990?

41. Based on selected figures obtained during the 1980s and 1990s, consumer expenditures on all types of books in the United States can be modeled by the function with

$$B(x) = 8768e^{.072x},$$

where $x = 0$ represents 1980, $x = 1$ represents 1981, and so on, and $B(x)$ is in millions of dollars. Approximate consumer expenditures for 1998. (*Source:* Book Industry Study Group.)

42. Based on selected figures obtained during the 1970s, 1980s, and 1990s, the total number of bachelor's degrees earned in the United States can be modeled by the function with

$$D(x) = 815{,}427e^{.0137x},$$

where $x = 1$ corresponds to 1971, $x = 10$ corresponds to 1980, and so on. Approximate the number of bachelor's degrees earned in 1994. (*Source:* U.S. National Center for Education Statistics.)

43. Suppose that the amount, in grams, of plutonium-241 present in a given sample is determined by the function with

$$A(t) = 2.00e^{-.053t},$$

where t is measured in years. Find the amount present in the sample after the given number of years.

(a) 4   **(b)** 10   **(c)** 20

(d) What was the initial amount present?

44. Suppose that the amount, in grams, of radium-226 present in a given sample is determined by the function with

$$A(t) = 3.25e^{-.00043t},$$

where t is measured in years. Find the amount present in the sample after the given number of years.

(a) 20   **(b)** 100   **(c)** 500

(d) What was the initial amount present?

For Exercises 45–48, refer to the function in Example 6.

45. Find $N(.9)$.   **46.** Find $N(.3)$.   **47.** Find $N(.5)$.

48. How many years have elapsed since the split if 80% of the words of the ancestral language are common to both languages today?

Use the change-of-base rule (with either common or natural logarithms) to find each logarithm to four decimal places. See Example 7.

49. $\log_6 12$ **50.** $\log_8 13$ **51.** $\log_{12} 6$

52. $\log_{13} 8$ **53.** $\log_{\sqrt{2}} \pi$ **54.** $\log_{\pi} \sqrt{2}$

55. Let m be the number of letters in your first name, and let n be the number of letters in your last name.

(a) In your own words, explain what $\log_m n$ means.

(b) Use your calculator to find $\log_m n$.

(c) Raise m to the power indicated by the number you found in part (b). What is your result?

56. The equation $5^x = 7$ cannot be solved using the methods described in Section 12.2. However, in solving this equation, we must find the exponent to which 5 must be raised in order to obtain 7: this is $\log_5 7$.

(a) Use the change-of-base rule and your calculator to find $\log_5 7$.

(b) Raise 5 to the number you found in part (a). What is your result?

(c) Using as many decimal places as your calculator gives, write the solution set of $5^x = 7$. (Equations of this type will be studied in more detail in Section 12.6.)

TECHNOLOGY INSIGHTS (EXERCISES 57–60)

57. The function defined by $P(x) = 70{,}967e^{.0526x}$, described in Exercise 39, is graphed in a graphing calculator-generated screen in the accompanying figure. Interpret the meanings of x and y in the display at the bottom of the screen in the context of Exercise 39.

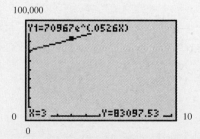

(continued)

TECHNOLOGY INSIGHTS (EXERCISES 57–60) (CONTINUED)

58. The function defined by $A(x) = 3.25e^{-.00043x}$, with $x = t$, described in Exercise 44, is graphed in a graphing calculator-generated screen in the accompanying figure. Interpret the meanings of x and y in the display at the bottom of the screen in the context of Exercise 44.

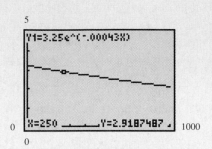

59. The screen shows a table of selected values for the function defined by

$$Y_1 = \left(1 + \frac{1}{x}\right)^x.$$

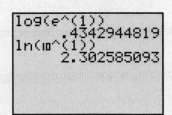

(a) Why is there an error message for $x = 0$?

(b) What number does the function value seem to approach as x takes on larger and larger values?

(c) Use a calculator to evaluate this function for $x = 1,000,000$. What value do you get? Now evaluate $e = e^1$. How close are these two values?

(d) Make a conjecture: As the values of x approach infinity, the value of $\left(1 + \frac{1}{x}\right)^x$

approaches _____.

60. Here is another property of logarithms: For $b > 0$, $x > 0$, $b \neq 1$, $x \neq 1$,

$$\log_b x = \frac{1}{\log_x b}.$$

Now observe the accompanying calculator screen.

```
log(e^(1))
            .4342944819
ln(10^(1))
            2.302585093
```

TECHNOLOGY INSIGHTS (EXERCISES 57-60) (CONTINUED)

(a) Without using a calculator, give a decimal representation for $\dfrac{1}{.4342944819}$.

Then support your answer using the reciprocal key of your calculator.

(b) Without using a calculator, give a decimal representation for $\dfrac{1}{2.302585093}$.

Then support your answer using the reciprocal key of your calculator.

Because graphing calculators are equipped with log x and ln x keys, it is possible to graph the functions defined by $f(x) = \log x$ and $g(x) = \ln x$ directly, as shown in the figures that follow.

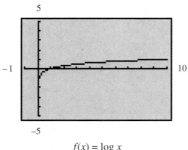

$f(x) = \log x$

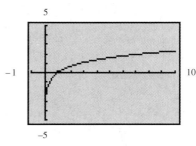

$g(x) = \ln x$

In order to graph functions defined by logarithms to bases other than 10 or e, however, we must use the change-of-base rule. For example, to graph $y = \log_2 x$, we may enter y_1 as $\dfrac{\log x}{\log 2}$ or $\dfrac{\ln x}{\ln 2}$.

This is shown in the figure that follows. (Compare it to the figure in the exercises of Section 12.3, where it was drawn using the fact that $y = \log_2 x$ is the inverse of $y = 2^x$.)

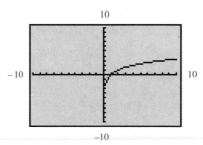

$y_1 = \log_2 x$

 Use the change-of-base rule to graph each logarithmic function with a graphing calculator. Use a viewing window with $\text{Xmin} = -1$, $\text{Xmax} = 10$, $\text{Ymin} = -5$, and $\text{Ymax} = 5$.

61. $g(x) = \log_3 x$ **62.** $g(x) = \log_5 x$

63. $g(x) = \log_{1/3} x$ **64.** $g(x) = \log_{1/5} x$

12.6 Exponential and Logarithmic Equations; Applications

OBJECTIVES

 1 Solve equations involving variables in the exponents.

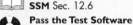 **2** Solve equations involving logarithms.

 3 Solve applications involving compound interest.

 4 Solve applications involving exponential growth and decay.

5 Use a graphing calculator to solve exponential and logarithmic equations.

FOR EXTRA HELP

SSG Sec. 12.6
SSM Sec. 12.6

Pass the Test Software

InterAct Math Tutorial Software

Video 22

As mentioned earlier, exponential and logarithmic functions are important in many applications of mathematics. Using these functions in applications requires solving exponential and logarithmic equations. Some simple equations were solved in Sections 12.2 and 12.3. More general methods for solving these equations depend on the following properties.

Properties for Equation Solving

For all real numbers $b > 0$, $b \neq 1$, and any real numbers x and y:
1. If $x = y$, then $b^x = b^y$.
2. If $b^x = b^y$, then $x = y$.
3. If $x = y$, and $x > 0$, $y > 0$, then $\log_b x = \log_b y$.
4. If $x > 0$, $y > 0$, and $\log_b x = \log_b y$, then $x = y$.

We used property 2 to solve exponential equations in Section 12.2, and property 3 was used in the proof of the change-of-base rule in the previous section. We will refer to these properties by number throughout this section.

OBJECTIVE **1** Solve equations involving variables in the exponents. The first examples illustrate a general method for solving exponential equations using property 3.

EXAMPLE 1 Solving an Exponential Equation

Solve the equation $3^x = 12$. Give the answer in decimal form.

$$3^x = 12$$
$$\log 3^x = \log 12 \qquad \text{Property 3}$$
$$x \log 3 = \log 12 \qquad \text{Power rule for logarithms}$$
$$x = \frac{\log 12}{\log 3}$$

This quotient is the exact solution. To get a decimal approximation for the solution, we use a calculator. Correct to three decimal places, a calculator gives

$$x = 2.262,$$

and the solution set is {2.262}.

CAUTION Be careful: $\frac{\log 12}{\log 3}$ is *not* equal to log 4, since log 4 = .6021, but

$$\frac{\log 12}{\log 3} = 2.262.$$

When an exponential equation has e as the base, it is easiest to use base e logarithms.

EXAMPLE 2 Solving an Exponential Equation with Base e

Solve $e^{.003x} = 40$.

Take base e logarithms on both sides.

$$\ln e^{.003x} = \ln 40$$

$$.003x = \ln 40 \qquad \ln e^k = k$$

$$x = \frac{\ln 40}{.003} \qquad \text{Divide by .003.}$$

$$x \approx 1230 \qquad \text{Use a calculator.}$$

The solution set is $\{1230\}$. Check that $e^{.003(1230)} \approx 40$.

In summary, exponential equations can be solved by one of the following methods. (The method used depends on the form of the equation.)

Solving an Exponential Equation

Method 1 **Use property 3.** Take logarithms to the same base on each side and then use the power rule of logarithms or the special property $\log_b b^x = x$. (See Examples 1 and 2 above.)

Method 2 **Use property 2.** Write both sides as exponentials with the same base and then set the exponents equal. (See Section 12.2.)

OBJECTIVE 2 Solve equations involving logarithms. The properties of logarithms from Section 12.4 are useful here, as is using the definition of a logarithm to change it to exponential form.

EXAMPLE 3 Solving a Logarithmic Equation

Solve $\log_2 (x + 5)^3 = 4$.

$$(x + 5)^3 = 2^4 \qquad \text{Convert to exponential form.}$$

$$(x + 5)^3 = 16$$

$$x + 5 = \sqrt[3]{16} \qquad \text{Take cube roots on both sides.}$$

$$x = -5 + \sqrt[3]{16}$$

$$x = -5 + 2\sqrt[3]{2} \qquad \text{Simplify the radical.}$$

Verify that the solution satisfies the equation, so the solution set is $\left\{-5 + 2\sqrt[3]{2}\right\}$.

 Recall that the domain of $y = \log_b x$ is $(0, \infty)$. For this reason, it is always necessary to check that the solution of an equation with logarithms yields only logarithms of positive numbers in the original equation.

E X A M P L E 4 Solving a Logarithmic Equation

Solve $\log_2(x + 1) - \log_2 x = \log_2 8$.

$$\log_2(x + 1) - \log_2 x = \log_2 8$$

$$\log_2 \frac{x + 1}{x} = \log_2 8 \qquad \text{Quotient rule}$$

$$\frac{x + 1}{x} = 8 \qquad \text{Property 4}$$

$$8x = x + 1 \qquad \text{Multiply by } x.$$

$$x = \frac{1}{7} \qquad \text{Subtract } x, \text{ divide by 7.}$$

Check this solution by substitution in the original equation. Here, both $x + 1$ and x must be positive. If $x = \frac{1}{7}$, this condition is satisfied, and the solution set is $\left\{\frac{1}{7}\right\}$.

E X A M P L E 5 Solving a Logarithmic Equation

Solve $\log x + \log(x - 21) = 2$.

For this equation, write the left side as a single logarithm. Then write in exponential form and solve the equation.

$$\log x + \log(x - 21) = 2$$

$$\log x(x - 21) = 2 \qquad \text{Product rule}$$

$$x(x - 21) = 10^2 \qquad \begin{array}{l} \log x = \log_{10} x; \text{ write in} \\ \text{exponential form.} \end{array}$$

$$x^2 - 21x = 100$$

$$x^2 - 21x - 100 = 0 \qquad \text{Standard form}$$

$$(x - 25)(x + 4) = 0 \qquad \text{Factor.}$$

$$x - 25 = 0 \quad \text{or} \quad x + 4 = 0 \qquad \text{Set each factor equal to } 0.$$

$$x = 25 \quad \text{or} \quad x = -4$$

The value -4 must be rejected as a solution, since it leads to the logarithm of a negative number in the original equation:

$$\log(-4) + \log(-4 - 21) = 2. \qquad \text{The left side is not defined.}$$

The only solution, therefore, is 25, and the solution set is $\{25\}$.

 Do not reject a potential solution just because it is nonpositive. Reject any value that *leads to* the logarithm of a nonpositive number.

In summary, use the following steps to solve a logarithmic equation.

Solving a Logarithmic Equation

Step 1 **Get a single logarithm on one side.** Use the product rule or quotient rule of logarithms to do this.

Step 2 **(a) Use property 4.** If $\log_b x = \log_b y$, then $x = y$. (See Example 4.)

   (b) Write the equation in exponential form. If $\log_b x = k$, then $x = b^k$. (See Examples 3 and 5.)

OBJECTIVE **3** Solve applications involving compound interest. So far in this book, problems involving applications of interest have been limited to the use of the simple interest formula, $I = prt$. In most cases, banks pay compound interest (interest paid on both principal and interest). The formula for compound interest is an important application of exponential functions.

Compound Interest Formula (for a Finite Number of Periods)

If P dollars is deposited in an account paying an annual rate of interest r compounded (paid) n times per year, the account will contain

$$A = P\left(1 + \frac{r}{n}\right)^{nt}$$

dollars after t years.

In the formula above, r is usually expressed as a decimal.

EXAMPLE 6 Solving a Compound Interest Problem

How much money will there be in an account at the end of 5 years if $1000 is deposited at 6% compounded quarterly? (Assume no withdrawals are made.)

Since interest is compounded quarterly, $n = 4$. The other values given in the problem are $P = 1000$, $r = .06$ (since 6% = .06), and $t = 5$. Substitute into the compound interest formula to get the value of A.

$$A = 1000\left(1 + \frac{.06}{4}\right)^{4 \cdot 5} \qquad \text{Substitute.}$$

$$A = 1000(1.015)^{20}$$

Now use the y^x key on a calculator, and round the answer to the nearest cent.

$$A = 1346.86$$

The account will contain $1346.86. (The actual amount of interest earned is $1346.86 - $1000 = $346.86. Why?)

Interest can be compounded annually, semiannually, quarterly, daily, and so on. The number of compounding periods can get larger and larger. If the value of n is allowed to approach infinity, we have an example of *continuous compounding*. However, the compound interest formula above cannot be used for continuous compounding, since there is no finite value for n. The formula for continuous compounding is an example of exponential growth involving the number e.

Continuous Compound Interest Formula

If a principal of P dollars is deposited at an annual rate of interest r compounded continuously for t years, the final amount on deposit is

$$A = Pe^{rt}.$$

E X A M P L E   7   Solving a Continuous Compound Interest Problem

In Example 6 we found that $1000 invested for 5 years at 6% interest compounded quarterly would grow to $1346.86.

(a) How much would this same investment grow to if interest were compounded continuously?

Use the formula for continuous compounding with $P = 1000$, $r = .06$, and $t = 5$.

$$A = Pe^{rt} \qquad \text{Formula}$$
$$= 1000e^{.06(5)} \qquad \text{Substitute.}$$
$$= 1000e^{.30}$$
$$= 1349.86 \qquad \text{Use a calculator and round to the nearest cent.}$$

Continuous compounding would cause the investment to grow to $1349.86. Notice that this is $3.00 more than the amount in Example 6, when interest was compounded quarterly.

(b) How long would it take for the initial investment to double its original amount? (This is called the **doubling time.**)

We must find the value of t that will cause A to be $2(\$1000) = \$2000$.

$$A = Pe^{rt}$$
$$2000 = 1000e^{.06t} \qquad \text{Let } A = 2P = 2000.$$
$$2 = e^{.06t} \qquad \text{Divide by 1000.}$$
$$\ln 2 = .06t \qquad \text{Take natural logarithms; } \ln e^k = k.$$
$$t = \frac{\ln 2}{.06} \qquad \text{Divide by .06.}$$
$$t \approx 11.55 \qquad \text{Use a calculator.}$$

It would take about 11.55 years for the original investment to double.

CONNECTIONS

Work Example 7(b) using just P as the original principal and $2P$ as the final amount A.

FOR DISCUSSION OR WRITING
Comment on the statement: The original amount of the investment does not affect the doubling time.

OBJECTIVE **4** Solve applications involving exponential growth and decay. We saw some applications involving exponential growth and decay in Sections 12.2 and 12.5. In many cases, quantities grow or decay according to a function defined by an exponential expression with base e. (See Section 12.5, Example 5.) You have probably heard of the carbon-14 dating process used to determine the age of fossils. The method used is based on the exponential decay function.

E X A M P L E   8   Solving an Exponential Decay Problem

Carbon-14 is a radioactive form of carbon that is found in all living plants and animals. After a plant or animal dies, the radioactive carbon-14 disintegrates according to the function

$$y = y_0 e^{-.000121t},$$

where t is time in years, and y is the amount of the sample at time t.

(a) If an initial sample contains $y_0 = 10$ grams of carbon-14, how many grams will be present after 3000 years?

Let $y_0 = 10$ and $t = 3000$ in the formula, and use a calculator.

$$y = 10e^{-.000121(3000)} \approx 6.96 \text{ grams}$$

(b) How long would it take for the initial sample to decay to half of its original amount? (This is called the **half-life.**)

Let $y = \frac{1}{2}(10) = 5$, and solve for t.

$$5 = 10e^{-.000121t} \qquad \text{Substitute.}$$

$$\frac{1}{2} = e^{-.000121t} \qquad \text{Divide by 10.}$$

$$\ln \frac{1}{2} = -.000121t \qquad \text{Take natural logarithms; } \ln e^k = k.$$

$$t = \frac{\ln \frac{1}{2}}{-.000121} \qquad \text{Divide by } -.000121.$$

$$t \approx 5728 \qquad \text{Use a calculator.}$$

The half-life is just over 5700 years.

OBJECTIVE **5** Use a graphing calculator to solve exponential and logarithmic equations. Earlier we saw that the x-intercepts of the graph of a function f correspond to the real solutions of the equation $f(x) = 0$. This idea was applied to linear and quadratic equations and can be extended to exponential and logarithmic equations as well. In Example 1, we solved the equation $3^x = 12$ algebraically using rules for logarithms and found the solution set to be $\{2.262\}$. This can be supported graphically by showing that the x-intercept of the graph of the function defined by $y = 3^x - 12$ corresponds to this solution. See Figure 20.

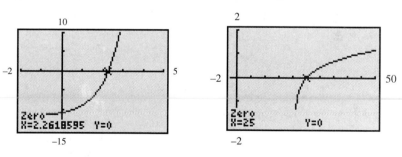

Figure 20   Figure 21

In Example 5, we solved $\log x + \log(x - 21) = 2$ and found the solution set to be $\{25\}$. Figure 21 shows that the x-intercept of the graph of the function defined by $y = \log x + \log(x - 21) - 2$ supports this result.

12.6 EXERCISES

RELATING CONCEPTS (EXERCISES 1–4)

In Section 12.2 we solved an equation such as

$$5^x = 125$$

by writing both sides as a power of the same base, setting exponents equal, and then solving the resulting equation. The equation above is solved as follows.

$5^x = 125$	Given equation
$5^x = 5^3$	$125 = 5^3$
$x = 3$	Set exponents equal.

Solution set: $\{3\}$

The method described in this section can also be used to solve this equation.

Work Exercises 1–4 in order, *to see how this is done.*

1. Take common logarithms of both sides, and write this equation.
2. Apply the power rule for logarithms on the left.
3. Get x alone on the left.
4. Use a calculator to find the decimal form of the solution. What is the solution set?

Did you make the connection that the method of solving exponential equations explained in this section can be used to solve the types studied in Section 12.2?

 Solve each equation. Give solutions to three decimal places. See Example 1.

5. $7^x = 5$
6. $4^x = 3$
7. $9^{-x+2} = 13$
8. $6^{-t+1} = 22$
9. $2^{y+3} = 5^y$
10. $6^{m+3} = 4^m$

 Use natural logarithms to solve each equation. Give solutions to three decimal places. See Example 2.

11. $e^{.006x} = 30$
12. $e^{.012x} = 23$
13. $e^{-.103x} = 7$
14. $e^{-.205x} = 9$
15. $100e^{.045x} = 300$
16. $500e^{-.003x} = 250$

17. Solve one of the equations in Exercises 11–14 using common logarithms rather than natural logarithms. (You should get the same solution.) Explain why using natural logarithms is a better choice.

18. If you were asked to solve $10^{.0025x} = 75$, would natural or common logarithms be a better choice? Explain your answer.

Solve each equation. Give exact solutions. See Example 3.

19. $\log_3(6x + 5) = 2$
20. $\log_5(12x - 8) = 3$
21. $\log_7(x + 1)^3 = 2$
22. $\log_4(y - 3)^3 = 4$

23. Suppose that in solving a logarithmic equation having the term $\log_4(x - 3)$, you obtain an apparent solution of 2. All algebraic work is correct. Explain why 2 must be rejected as a solution of the equation.

24. Suppose that in solving a logarithmic equation having the term $\log_7(3 - x)$, you obtain an apparent solution of -4. All algebraic work is correct. Should you reject -4 as a solution of the equation? Explain why or why not.

Solve each equation. Give exact solutions. See Examples 4 and 5.

25. $\log(6x + 1) = \log 3$

26. $\log(7 - x) = \log 12$

27. $\log_5(3t + 2) - \log_5 t = \log_5 4$

28. $\log_2(x + 5) - \log_2(x - 1) = \log_2 3$

29. $\log 4x - \log(x - 3) = \log 2$

30. $\log(-x) + \log 3 = \log(2x - 15)$

31. $\log_2 x + \log_2(x - 7) = 3$

32. $\log(2x - 1) + \log 10x = \log 10$

33. $\log 5x - \log(2x - 1) = \log 4$

34. $\log_3 x + \log_3(2x + 5) = 1$

35. $\log_2 x + \log_2(x - 6) = 4$

36. $\log_2 x + \log_2(x + 4) = 5$

 Solve each problem. See Examples 6–8.

37. How much money will there be in an account at the end of 6 years if \$2000 is deposited at 4% compounded quarterly? (Assume no withdrawals are made.)

38. How much money will there be in an account at the end of 7 years if \$3000 is deposited at 3.5% compounded quarterly? (Assume no withdrawals are made.)

39. A sample of 400 grams of lead-210 decays to polonium-210 according to the function defined by

$$A(t) = 400e^{-.032t},$$

where t is time in years. How much lead will be left in the sample after 25 years?

40. How long will it take the initial sample of lead in Exercise 39 to decay to half of its original amount?

41. Find the amount of money in an account after 12 years if \$5000 is deposited at 7% annual interest compounded as follows.
(a) annually **(b)** semiannually **(c)** quarterly
(d) daily (Use $n = 365$.) **(e)** continuously

42. How much money will be in an account at the end of 8 years if \$4500 is deposited at 6% annual interest compounded as follows?
(a) annually **(b)** semiannually **(c)** quarterly
(d) daily (Use $n = 365$.) **(e)** continuously

43. How much money must be deposited today to become \$1850 in 40 years at 6.5% compounded continuously?

44. How much money must be deposited today to amount to \$1000 in 10 years at 5% compounded continuously?

 45. Refer to Exercise 39 in Section 12.5. Assuming that the function continues to apply past 1993, in what year can we expect total expenditures to be 133,500 million dollars? (*Source:* U.S. Bureau of Economic Analysis, *Survey of Current Business, May 1995.*)

 46. Refer to Exercise 40 in Section 12.5. Assuming that the function continues to apply past 1994, in what year can we expect nitrous oxide emissions to be 485 thousand metric tons? (*Source:* U.S. Energy Information Administration, *Emission of Greenhouse Gases in the United States, annual.*)

47. The concentration y of a drug in a person's system decreases according to the relationship

$$y = 2e^{-.125t},$$

where y is in appropriate units, and t is in hours. Find the amount of time that it will take for the concentration to be half of its original value.

48. The number y of ants in an anthill grows according to the function defined by

$$y = 300e^{.4t},$$

where t is time measured in days. Find the time it will take for the number of ants to double.

49. Radioactive strontium decays according to the function defined by

$$y = y_0 e^{-.0239t},$$

where t is time in years.

(a) If an initial sample contains $y_0 = 5$ grams of radioactive strontium, how many grams will be present after 20 years?

(b) How many grams of the initial 5-gram sample will be present after 60 years?

(c) What is the half-life of radioactive strontium?

50. Plutonium-241 decays according to the function defined by

$$y = y_0 e^{-.053t},$$

where t is time in years.

(a) If an initial sample contains $y_0 = 200$ grams of plutonium-241, how many grams will be present after 100 years?

(b) How many grams of the initial 200-gram sample will be present after 200 years?

(c) What is the half-life of plutonium-241?

 Use a graphing calculator and the method described in Objective 5 to solve each of the following equations. Note that they were solved using algebraic methods earlier in this section.

51. $7^x = 5$ (Exercise 5)

52. $4^x = 3$ (Exercise 6)

53. $\log(6x + 1) = \log 3$ (Exercise 25)

54. $\log(7 - x) = \log 12$ (Exercise 26)

CHAPTER 12 GROUP ACTIVITY

How Much Space Do We Need?

Objective: Use natural logarithms and exponential equations to calculate how long it will take to fully populate Earth with people.

Applications of exponential growth and decay were introduced in this chapter. The formula for exponential growth is:

$$P(t) = P_0 e^{kt},$$

where $P(t)$ is population after t years, P_0 is initial population, k is annual growth rate, and t is number of years elapsed.

A. If Earth's population will double in 30 years at the current growth rate, what is this growth rate? (Express your answer as a percent.)

B. Earth has a total surface area of approximately 5.1×10^{14} square meters. Seventy percent of this surface area is rock, ice, sand, and open ocean. Another 8% of the total surface area, made up of tundra, lakes and streams, continental shelves, algae beds and reefs, and estuaries, is unfit for living space. The remaining area is suitable for growing food and for living space.

(continued)

1. Determine the surface area available for growing food.

2. Determine the surface area available for living space. Notice that the surface area available for living space is also considered available for growing food.

C. Suppose that each person needs 100 square meters of Earth's surface for living space. Earth's current population is approximately 5.5×10^9.

1. If none of the surface area available for living space is used for food, how long will it take for the livable surface of Earth to be covered with people? (Use the growth rate you found in part A and the surface area you found in Exercise B2.)

2. How much surface area would be left to grow food?

D. Measure a space that is one square meter in area. Discuss with your partner whether or not you would want to be packed this closely together on Earth. Take into account that many people live in high-rise apartment buildings and how that translates into surface area used per person.

E. Now suppose that for each person 100 square meters of Earth's surface is needed for living space and growing food.

1. Using the same population and growth rate as in part C, determine how long it will take to fill Earth with people. (Use the surface area from Exercise B2.)

2. Does 100 square meters per person for living space and growing food seem reasonable? Consider the following questions in your discussion.

How much space do you think it takes to raise animals for food? To grow grains, nuts, fruits, and vegetables?

Would food grow as well in desert areas, mountainous areas, or jungle areas?

Would there be any space left for wild animals or natural plant life?

Would there be any space left for shopping malls, movie theaters, concert halls, factories, office buildings, or parking lots?

3. Write a paragraph summarizing your results and your discussion.

CHAPTER 12 SUMMARY

KEY TERMS

12.1 one-to-one function
inverse of a function

12.2 exponential function
asymptote
exponential equation

12.3 logarithm
logarithmic equation
logarithmic function
with base a

12.5 common logarithm
natural logarithm

12.6 doubling time
half-life

NEW SYMBOLS

$f^{-1}(x)$ the inverse of $f(x)$

$\log_a x$ the logarithm of x to the base a

$\log x$ common (base 10) logarithm of x

$\ln x$ natural (base e) logarithm of x

e a constant, approximately 2.7182818

See how well you have learned the vocabulary in this chapter. Answers, with examples, are given at the bottom of the page.

1. In a **one-to-one function**
(a) each x-value corresponds to only one y-value
(b) each x-value corresponds to one or more y-values
(c) each x-value is the same as each y-value
(d) each x-value corresponds to only one y-value and each y-value corresponds to only one x-value.

2. If f is a one-to-one function, then the **inverse** of f is
(a) the set of all solutions of f
(b) the set of all ordered pairs formed by reversing the coordinates of the ordered pairs of f
(c) an equation involving an exponential expression
(d) the set of all ordered pairs that are the opposite (negative) of the coordinates of the ordered pairs of f.

3. An **exponential function** is a function defined by an expression of the form
(a) $f(x) = ax^2 + bx + c$ for real numbers a, b, c ($a \neq 0$)
(b) $f(x) = \log_a x$ for a and x positive numbers ($a \neq 1$)
(c) $f(x) = a^x$ for all real numbers x ($a > 0, a \neq 1$)
(d) $f(x) = \sqrt{x}$ for $x \geq 0$.

4. An **asymptote** is
(a) a line that a graph intersects just once
(b) a line that the graph of a function more and more closely approaches as the graph gets farther away from the origin
(c) the x-axis or y-axis
(d) a line about which a graph is symmetric.

5. A **logarithm** is
(a) an exponent
(b) a base
(c) an equation
(d) a term.

6. A **logarithmic function** is a function that is defined by an expression of the form
(a) $f(x) = ax^2 + bx + c$ for real numbers a, b, c ($a \neq 0$)
(b) $f(x) = \log_a x$ for a and x positive numbers ($a \neq 1$)
(c) $f(x) = a^x$ for all real numbers x ($a > 0, a \neq 1$)
(d) $f(x) = \sqrt{x}$ for $x \geq 0$.

QUICK REVIEW

CONCEPTS	EXAMPLES

12.1 INVERSE FUNCTIONS

Horizontal Line Test
If a horizontal line intersects the graph of a function in no more than one point, then the function is one-to-one.

Inverse Functions
For a one-to-one function f defined by an equation $y = f(x)$, the defining equation of the inverse function f^{-1} is found by interchanging x and y, solving for y, and replacing y with $f^{-1}(x)$.

Find f^{-1} if $f(x) = 2x - 3$.
The graph of f is a straight line, so f is one-to-one by the horizontal line test.

Interchange x and y in the equation $y = 2x - 3$.
$$x = 2y - 3$$
Solve for y to get $\qquad y = \frac{1}{2}x + \frac{3}{2}$.

Therefore, $\qquad f^{-1}(x) = \frac{1}{2}x + \frac{3}{2}$.

CONCEPTS	EXAMPLES

The graph of f^{-1} is a mirror image of the graph of f with respect to the line $y = x$.

The graphs of a function f and its inverse f^{-1} are given here.

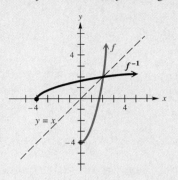

12.2 EXPONENTIAL FUNCTIONS

For $a > 0$, $a \neq 1$, $f(x) = a^x$ defines an exponential function with base a.

Graph of $F(x) = a^x$
The graph contains the point $(0, 1)$. When $a > 1$, the graph rises from left to right. When $0 < a < 1$, the graph falls from left to right. The x-axis is an asymptote. The domain is $(-\infty, \infty)$; the range is $(0, \infty)$.

$f(x) = 3^x$ defines an exponential function with base 3. Its graph is shown here.

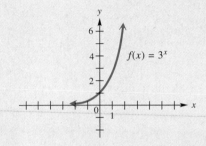

12.3 LOGARITHMIC FUNCTIONS

$y = \log_a x$ has the same meaning as $a^y = x$.

For $b > 0$, $b \neq 1$, $\log_b b = 1$ and $\log_b 1 = 0$.

For $a > 0$, $a \neq 1$, $x > 0$, $g(x) = \log_a x$ defines the logarithmic function with base a.

Graph of $G(x) = \log_a x$
The graph contains the point $(1, 0)$. When $a > 1$, the graph rises from left to right. When $0 < a < 1$, the graph falls from left to right. The y-axis is an asymptote. The domain is $(0, \infty)$; the range is $(-\infty, \infty)$.

$y = \log_2 x$ means $x = 2^y$.

$$\log_3 3 = 1, \qquad \log_5 1 = 0$$

$g(x) = \log_3 x$ defines the logarithmic function with base 3. Its graph is shown here.

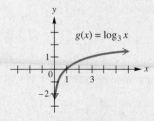

12.4 PROPERTIES OF LOGARITHMS

Product Rule

$$\log_a xy = \log_a x + \log_a y$$

Quotient Rule

$$\log_a \frac{x}{y} = \log_a x - \log_a y$$

$$\log_2 3m = \log_2 3 + \log_2 m \quad (m > 0)$$

$$\log_5 \frac{9}{4} = \log_5 9 - \log_5 4$$

(continued)

CONCEPTS	EXAMPLES
Power Rule $$\log_a x^r = r \log_a x$$	$$\log_{10} 2^3 = 3 \log_{10} 2$$
Special Properties $$b^{\log_b x} = x \quad \text{and} \quad \log_b b^x = x$$	$$6^{\log_6 10} = 10 \quad \text{and} \quad \log_3 3^4 = 4$$

12.5 EVALUATING LOGARITHMS

Change-of-Base Rule If $a > 0$, $a \neq 1$, $b > 0$, $b \neq 1$, $x > 0$, then $$\log_a x = \frac{\log_b x}{\log_b a}.$$	Approximate $\log_3 17$ to four decimal places. $$\log_3 17 = \frac{\ln 17}{\ln 3} = \frac{\log 17}{\log 3} \approx 2.5789$$

12.6 EXPONENTIAL AND LOGARITHMIC EQUATIONS; APPLICATIONS

To solve exponential equations, use these properties ($b > 0$, $b \neq 1$). **1.** If $b^x = b^y$, then $x = y$.	Solve. $$2^{3x} = 2^5$$ $$3x = 5$$ $$x = \frac{5}{3}$$ The solution set is $\left\{ \dfrac{5}{3} \right\}$.
2. If $x = y$ ($x > 0$, $y > 0$), then $\log_b x = \log_b y$.	Solve. $$5^m = 8$$ $$\log 5^m = \log 8$$ $$m \log 5 = \log 8$$ $$m = \frac{\log 8}{\log 5} \approx 1.2920$$ The solution set is $\{1.2920\}$.
To solve logarithmic equations, use these properties, where $b > 0$, $b \neq 1$, $x > 0$, $y > 0$. First use the properties of Section 12.4, if necessary, to get the equation in the proper form. **1.** If $\log_b x = \log_b y$, then $x = y$.	Solve. $$\log_3 2x = \log_3(x + 1)$$ $$2x = x + 1$$ $$x = 1$$ The solution set is $\{1\}$.
2. If $\log_b x = y$, then $b^y = x$.	Solve. $$\log_2(3a - 1) = 4$$ $$3a - 1 = 2^4 = 16$$ $$3a = 17$$ $$a = \frac{17}{3}$$ The solution set is $\left\{ \dfrac{17}{3} \right\}$.

CHAPTER 12 REVIEW EXERCISES

[12.1]   *Determine whether each graph is the graph of a one-to-one function.*

1.

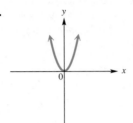

2.

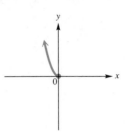

3. The chart lists the top five metropolitan areas with highest average levels of regulated particle pollution during the period from 1990–1994. (The measure is the concentration of particles 10 microns or smaller in diameter, down to 2.5 microns, measured in micrograms per cubic meter of air.) If the set of areas is the domain of the function and the set of particle pollution levels is the range, is this function one-to-one? Why or why not?

Metropolitan Area	Average Particle Pollution
1. Visalia-Tulare-Porterville, CA	60.4
2. Bakersfield, CA	54.8
3. Fresno, CA	51.7
4. Riverside-San Bernardino, CA	48.1
5. Stockton, CA	44.8

Sources: The New York Times, Environmental Protection Administration; Natural Resources Defense Council.

Determine whether each function is one-to-one. If it is, find its inverse.

4. $f(x) = -3x + 7$ **5.** $f(x) = \sqrt[3]{6x - 4}$ **6.** $f(x) = -x^2 + 3$

Each function graphed is one-to-one. Graph its inverse.

7.

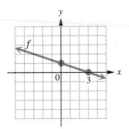

8.

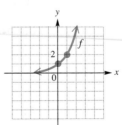

[12.2]   *Graph each function.*

9. $f(x) = 3^x$ **10.** $f(x) = \left(\dfrac{1}{3}\right)^x$ **11.** $y = 3^{x+1}$ **12.** $y = 2^{2x+3}$

Solve each equation.

13. $4^{3x} = 8^{x+4}$ **14.** $\left(\dfrac{1}{27}\right)^{x-1} = 9^{2x}$

15. The gross wastes generated in plastics, in millions of tons, from 1960–1990 can be approximated by the exponential function with

$$W(x) = .67(1.123)^x,$$

where $x = 0$ corresponds to 1960, $x = 5$ to 1965, and so on. Use this function to approximate the plastic waste amounts for the following years. Compare your answers to the values from the bar graph in the chapter introduction. (*Source:* Environmental Protection Agency, *Characterization of Municipal Solid Waste in the United States: 1994 Update,* 1995.)
(a) 1965   **(b)** 1975   **(c)** 1990

[12.3]   *Graph each function.*

16. $g(x) = \log_3 x$ (*Hint:* See Exercise 9.)   **17.** $g(x) = \log_{1/3} x$ (*Hint:* See Exercise 10.)

Solve each equation.

18. $\log_8 64 = x$ **19.** $\log_2 \sqrt{8} = x$ **20.** $\log_7\left(\dfrac{1}{49}\right) = x$

21. $\log_4 x = \dfrac{3}{2}$ **22.** $\log_k 4 = 1$ **23.** $\log_b b^2 = 2$

24. In your own words, explain the meaning of $\log_b a$.

25. Based on the meaning of $\log_b a$, what is the simplest form of $b^{\log_b a}$?

26. A company has found that total sales, in thousands of dollars, are given by the function with

$$S(x) = 100 \log_2(x + 2),$$

where x is the number of weeks after a major advertising campaign was introduced. What were the total sales 6 weeks after the campaign was introduced? Graph the function.

[12.4]   *Apply the properties of logarithms introduced in Section 12.4 to express each logarithm as a sum or difference of logarithms, or as a single number if possible. Assume that all variables represent positive real numbers.*

27. $\log_2 3xy^2$ **28.** $\log_4 \dfrac{\sqrt{x} \cdot w^2}{z}$

Use the properties of logarithms introduced in Section 12.4 to write each expression as a single logarithm. Assume that all variables represent positive real numbers, $b \neq 1$.

29. $\log_b 3 + \log_b x - 2 \log_b y$ **30.** $\log_3(x + 7) - \log_3(4x + 6)$

[12.5]   *Evaluate each logarithm. Give approximations to four decimal places.*

31. $\log 28.9$ **32.** $\log .257$ **33.** $\ln 28.9$ **34.** $\ln .257$

Use the change-of-base rule (with either common or natural logarithms) to find each logarithm. Give approximations to four decimal places.

35. $\log_{16} 13$ **36.** $\log_4 12$

Use the formula $pH = -\log[H_3O^+]$ to find the pH of each substance with the given hydronium ion concentration.

37. Milk, 4.0×10^{-7} **38.** Crackers, 3.8×10^{-9}

39. If orange juice has a pH of 4.6, what is its hydronium ion concentration?

40. Suppose the quantity, measured in grams, of a radioactive substance present at time t is given by

$$Q(t) = 500e^{-.05t},$$

where t is measured in days. Find the quantity present at the following times.
(a) $t = 0$ **(b)** $t = 4$

[12.6]   *Solve each equation. Give solutions to three decimal places.*

41. $3^x = 9.42$

42. $e^{.06x} = 3$

43. Which one of the following is *not* a solution of $7^x = 23$?

(a) $\dfrac{\log 23}{\log 7}$ **(b)** $\dfrac{\ln 23}{\ln 7}$ **(c)** $\log_7 23$ **(d)** $\log_{23} 7$

Solve each equation. Give exact solutions.

44. $\log_3(9x + 8) = 2$

45. $\log_3(p + 2) - \log_3 p = \log_3 2$

46. $\log(2x + 3) = \log 3x + 2$

47. $\log_4 x + \log_4(8 - x) = 2$

48. $\log_2 x + \log_2(x + 15) = 4$

49. Consider the logarithmic equation

$$\log(2x + 3) = \log x + 1.$$

(a) Solve the equation using properties of logarithms.

(b) If $Y_1 = \log(2x + 3)$ and $Y_2 = \log x + 1$, then the graph of $Y_1 - Y_2$ looks like this. Explain how the display at the bottom of the screen confirms the solution set found in part (a).

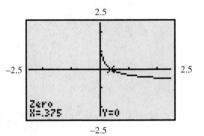

50. Explain the error in the following "solution" of the equation $\log x^2 = 2$.

$\log x^2 = 2$	Original equation
$2 \log x = 2$	Power rule for logarithms
$\log x = 1$	Divide both sides by 2.
$x = 10^1$	Write in exponential form.
$x = 10$	$10^1 = 10$

Solution set: $\{10\}$

 Solve each problem. Use a calculator as necessary.

51. If \$20,000 is deposited at 7% annual interest compounded quarterly, how much will be in the account after 5 years, assuming no withdrawals are made?

52. How much will \$10,000 compounded continuously at 6% annual interest amount to in three years?

53. Which is a better plan?

Plan A: Invest \$1000 at 4% compounded quarterly for 3 years

Plan B: Invest \$1000 at 3.9% compounded monthly for 3 years

 54. What is the half-life of the radioactive substance described in Exercise 40?

55. Based on selected figures from 1970–1995, the fractional part of the generation of municipal solid waste recovered can be approximated by the function with

$$R(x) = .0597e^{.0553x},$$

where $x = 0$ corresponds to 1970, $x = 10$ to 1980, and so on. Based on this model, what *percent* of municipal solid waste was recovered in 1990? (*Source:* Franklin Associates, Ltd., Prairie Village, KS, *Characterization of Municipal Solid Waste in the United States: 1995.*)

56. Recall from Example 6 in Section 12.5 that the number of years, $N(r)$, since two independently evolving languages split off from a common ancestral language is approximated by

$$N(r) = -5000 \ln r,$$

where r is the percent of words from the ancestral language common to both languages now. Find r if the split occurred 2000 years ago.

 A machine purchased for business use depreciates, or loses value, over a period of years. The value of the machine at the end of its useful life is called its scrap value. By one method of depreciation (where it is assumed a constant percentage of the value depreciates annually), the scrap value, S, is given by

$$S = C(1 - r)^n,$$

where C is the original cost, n is the useful life in years, and r is the constant percent of depreciation.

57. Find the scrap value of a machine costing $30,000, having a useful life of 12 years and a constant annual rate of depreciation of 15%.

58. A machine has a "half-life" of 6 years. Find the constant annual rate of depreciation.

RELATING CONCEPTS (EXERCISES 59–70)

Work Exercises 59–70 in order, *to see some of the relationships between exponential and logarithmic properties and functions.*

59. Complete the table, and graph the function $f(x) = 2^x$.

x	$f(x)$
-2	
-1	
0	
1	
2	
3	

60. Complete the table, and graph the function $g(x) = \log_2 x$.

x	$g(x)$
$\frac{1}{4}$	
$\frac{1}{2}$	
1	
2	
4	
8	

61. What do you notice about the ordered pairs found in Exercises 59 and 60? What do we call the functions f and g in relationship to each other?

62. Fill in the blanks with the word *vertical* or *horizontal:* The graph of f in Exercise 59 has a _____ asymptote, while the graph of g in Exercise 60 has a _____ asymptote.

63. Using properties of exponents, $2^2 \cdot 2^3 = 2^?$, because ___?___ + ___?___ = ___?___.

64. It is a fact that $32 = 4 \cdot 8$. Therefore, using properties of logarithms, $\log_2 32 = \log_2 \underline{\quad} + \log_2 \underline{\quad}$.

65. Use the change-of-base rule to find an approximation for $\log_2 13$. Give as many digits as your calculator displays, and store this approximation in memory.

66. In your own words, explain what $\log_2 13$ means.

67. Simplify without using a calculator: $2^{\log_2 13}$.

68. Use the exponential key of your calculator to raise 2 to the power obtained in Exercise 65. What is the result? Why is this so?

69. Based on your result in Exercise 65, the point $(13, \underline{\quad})$ lies on the graph of $g(x) = \log_2 x$.

70. Use the method of Section 12.2 to solve the equation $2^{x+1} = 8^{2x+3}$.

Did you make the connections between logarithms and exponents?

MIXED REVIEW EXERCISES

Solve.

71. $\log_3(x + 9) = 4$　　　　**72.** $\log_2 32 = x$　　　　**73.** $\log_x \dfrac{1}{81} = 2$

74. $27^x = 81$　　　　　　　　**75.** $2^{2x-3} = 8$

76. $\log_3(x + 1) - \log_3 x = 2$　　**77.** $\log(3x - 1) = \log 10$

 78. A small business estimates that the value of a copy machine is decreasing according to the function defined by

$$f(t) = 5000(2)^{-.15t},$$

where t is the number of years that have elapsed since the machine was purchased, and $f(t)$ is in dollars.

(a) What was the original value of the machine? (*Hint:* Find $f(0)$.)

(b) What is the value of the machine 5 years after purchase? Give your answer to the nearest dollar.

(c) What is the value of the machine 10 years after purchase? Give your answer to the nearest dollar.

CHAPTER 12 TEST

1. Decide whether each function is one-to-one.

(a) $f(x) = x^2 + 9$　　　(b)

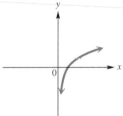

2. Find $f^{-1}(x)$ for the one-to-one function $f(x) = \sqrt[3]{x + 7}$.

3. Graph the inverse of f, given the graph of f at the right.

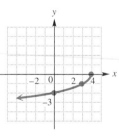

Graph each function.

4. $f(x) = 6^x$　　　　　　　　**5.** $g(x) = \log_6 x$

6. Explain how the graph of the function in Exercise 5 can be obtained from the graph of the function in Exercise 4.

Solve each equation. Give the exact solution.

7. $5^x = \dfrac{1}{625}$　　　　　　　**8.** $2^{3x-7} = 8^{2x+2}$

9. A *toxic release* is an on-site discharge of a chemical toxic to the environment. Based on figures from 1990–1993, the toxic release inventory, in millions of pounds, can be approximated by the function with

$$R(x) = 2821(.9195)^x,$$

where $x = 0$ corresponds to 1990, $x = 1$ to 1991, and so on. (*Source:* U.S. Environmental Protection Agency, *1994 Toxics Release Inventory,* June 1996.)

(a) Use this model to approximate the toxic release inventory in 1990.

(b) Use this model to approximate the toxic release inventory in 1992.

(c) In 1993, the actual toxic release inventory was 2157.4 million pounds. How does this compare to the number that the model provides?

10. Write in logarithmic form: $4^{-2} = .0625$.

11. Write in exponential form: $\log_7 49 = 2$.

Solve each equation.

12. $\log_{1/2} x = -5$ **13.** $x = \log_9 3$ **14.** $\log_x 16 = 4$

15. Fill in the blanks with the correct responses: The value of $\log_2 32$ is _____. This means that if we raise _____ to the _____ power, the result is _____.

Use properties of logarithms to write each expression as a sum or difference of logarithms. Assume variables represent positive numbers.

16. $\log_3 x^2 y$ **17.** $\log_5\left(\dfrac{\sqrt{x}}{yz}\right)$

Use properties of logarithms to write each expression as a single logarithm. Assume variables represent positive real numbers, $b \neq 1$.

18. $3 \log_b s - \log_b t$ **19.** $\dfrac{1}{4} \log_b r + 2 \log_b s - \dfrac{2}{3} \log_b t$

20. Use a calculator to approximate each logarithm to four decimal places.

(a) $\log 23.1$ **(b)** $\ln .82$

21. Use the change-of-base rule to express $\log_3 19$

(a) in terms of common logarithms.

(b) in terms of natural logarithms.

(c) correct to four decimal places.

22. Solve, giving the correct solution to four decimal places.

$$3^x = 78$$

23. Solve $\log_8(x + 5) + \log_8(x - 2) = 1$.

24. Suppose that $10,000 is invested at 4.5% annual interest, compounded quarterly. How much will be in the account in 5 years if no money is withdrawn?

25. Suppose that $15,000 is invested at 5% annual interest, compounded continuously.

(a) How much will be in the account in 5 years if no money is withdrawn?

(b) How long will it take for the initial principal to double?

CUMULATIVE REVIEW EXERCISES CHAPTERS 1–12

Let $S = \left\{ -\dfrac{9}{4}, -2, -\sqrt{2}, 0, .6, \sqrt{11}, \sqrt{-8}, 6, \dfrac{30}{3} \right\}$. *List the elements of S that are members of the set.*

1. Integers

2. Rational numbers

3. Irrational numbers

4. Real numbers

Simplify each expression.

5. $|-8| + 6 - |-2| - (-6 + 2)$

6. $-12 - |-3| - 7 - |-5|$

7. $2(-5) + (-8)(4) - (-3)$

Solve each equation or inequality.

8. $7 - (3 + 4a) + 2a = -5(a - 1) - 3$

9. $2m + 2 \le 5m - 1$

10. $|2x - 5| = 9$

11. $|3p| - 4 = 12$

12. $|3k - 8| \le 1$

13. $|4m + 2| > 10$

Graph.

14. $5x + 2y = 10$

15. $-4x + y \le 5$

Perform the indicated operations.

16. $(2p + 3)(3p - 1)$

17. $(4k - 3)^2$

18. $(3m^3 + 2m^2 - 5m) - (8m^3 + 2m - 4)$

19. Divide $6t^4 + 17t^3 - 4t^2 + 9t + 4$ by $3t + 1$.

Factor.

20. $8x + x^3$

21. $24y^2 - 7y - 6$

22. $5z^3 - 19z^2 - 4z$

23. $16a^2 - 25b^4$

24. $8c^3 + d^3$

25. $16r^2 + 56rq + 49q^2$

Perform the indicated operations.

26. $\dfrac{(5p^3)^4(-3p^7)}{2p^2(4p^4)}$

27. $\dfrac{x^2 - 9}{x^2 + 7x + 12} \div \dfrac{x - 3}{x + 5}$

28. $\dfrac{2}{k + 3} - \dfrac{5}{k - 2}$

29. $\dfrac{3}{p^2 - 4p} - \dfrac{4}{p^2 + 2p}$

30. The graph indicates that the long-term debt of the Port of New Orleans has dropped from $70,000,000 in 1986 to $25,300,000 in 1994. What is the slope of the line graphed?

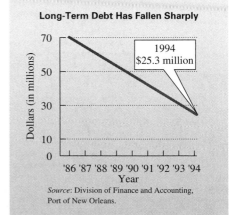

Long-Term Debt Has Fallen Sharply

1994
$25.3 million

Dollars (in millions)

'86 '87 '88 '89 '90 '91 '92 '93 '94
Year

Source: Division of Finance and Accounting, Port of New Orleans.

31. Find the standard form of the equation of the line through $(5, -1)$ and parallel to the line with equation $3x - 4y = 12$.

Solve each system.

32. $5x - 3y = 14$
    $2x + 5y = 18$

33. $x + 2y + 3z = 11$
    $3x - y + z = 8$
    $2x + 2y - 3z = -12$

34. Evaluate the determinant $\begin{vmatrix} -2 & -1 \\ 5 & 3 \end{vmatrix}$.

35. Candy worth $1.00 per pound is to be mixed with 10 pounds of candy worth $1.96 per pound to get a mixture that will be sold for $1.60 per pound. How many pounds of the $1.00 candy should be used?

Simplify.

36. $\sqrt{288}$

37. $2\sqrt{32} - 5\sqrt{98}$

38. Solve $\sqrt{2x + 1} - \sqrt{x} = 1$.

39. Multiply $(5 + 4i)(5 - 4i)$.

Solve each equation or inequality.

40. $3x^2 - x - 1 = 0$

41. $k^2 + 2k - 8 > 0$

42. $x^4 - 5x^2 + 4 = 0$

43. Find two numbers whose sum is 300 and whose product is a maximum.

44. Graph $f(x) = \dfrac{1}{3}(x - 1)^2 + 2$.

45. Graph $f(x) = 2^x$.

46. Solve $5^{x+3} = \left(\dfrac{1}{25}\right)^{3x+2}$.

47. Graph $f(x) = \log_3 x$.

48. Given that $\log_2 9 = 3.1699$, what is the value of $\log_2 81$?

49. Rewrite the following using the product, quotient, and power rules for logarithms:

$$\log \frac{x^3\sqrt{y}}{z}.$$

50. Let the number of bacteria present in a certain culture be given by

$$B(t) = 25{,}000e^{.2t},$$

where t is time measured in hours, and $t = 0$ corresponds to noon. Find, to the nearest hundred, the number of bacteria present at:
(a) noon; **(b)** 1 P.M.; **(c)** 2 P.M.; **(d)** 5 P.M.

Conic Sections

13

When a plane intersects an infinite cone at different angles, the figures formed by the intersections are called **conic sections.** See the figure. What is the difference in the angle of the plane that produces an ellipse from that of a circle? We investigated parabolas in Chapter 11. We will study the equations of the other conic sections in this chapter.

Aerospace

13.1 The Circle and the Ellipse

13.2 The Hyperbola and Square Root Functions

13.3 Nonlinear Systems of Equations

13.4 Second-Degree Inequalities; Systems of Inequalities

Johann Kepler (1571–1630) established the importance of the *ellipse* in 1609 when he discovered that orbits of the planets around the sun were elliptical, not circular. Comets and asteroids also have elliptical orbits around the sun. The orbits of the planets are nearly circular, while Halley's comet, for example, has an elliptical orbit that is long and narrow.

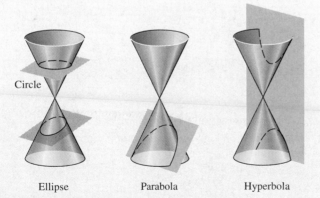

Circle

Ellipse   Parabola   Hyperbola

The theme of this chapter, aerospace, is closely related to the theme of Chapter 11, astronomy. Aerospace studies the design and navigation of space aircraft. In this chapter we will also present other applications of conic sections, such as elliptical gears, microwave antenna systems, and a location-finding system that uses the equation of a hyperbola.

13.1 The Circle and the Ellipse

OBJECTIVES

 1 Find the equation of a circle given the center and radius.

2 Determine the center and radius of a circle given its equation.

3 Recognize the equation of an ellipse.

4 Graph ellipses.

5 Graph circles and ellipses using a graphing calculator.

FOR EXTRA HELP

📖 **SSG** Sec. 13.1
SSM Sec. 13.1

💿 **Pass the Test Software**

💿 **InterAct Math Tutorial Software**

📼 **Video 23**

OBJECTIVE **1** Find the equation of a circle given the center and radius. A **circle** is the set of all points in a plane that lie a fixed distance from a fixed point. The fixed point is called the **center,** and the fixed distance is called the **radius.** We use the distance formula derived earlier to find an equation of a circle.

EXAMPLE 1 Finding the Equation of a Circle and Graphing It

Find an equation of the circle with radius 3 and center at $(0, 0)$, and graph it.

If the point (x, y) is on the circle, the distance from (x, y) to the center $(0, 0)$ is 3. By the distance formula,

$$\sqrt{(x_2 - x_1)^2 + (y_2 - y_1)^2} = d$$
$$\sqrt{(x - 0)^2 + (y - 0)^2} = 3$$
$$x^2 + y^2 = 9. \qquad \text{Square both sides.}$$

An equation of this circle is $x^2 + y^2 = 9$. The graph is shown in Figure 1.

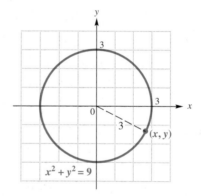

Figure 1

EXAMPLE 2 Finding the Equation of a Circle and Graphing It

Find an equation of the circle with center at $(4, -3)$ and radius 5, and graph it.

Use the distance formula again.

$$\sqrt{(x - 4)^2 + (y + 3)^2} = 5$$
$$(x - 4)^2 + (y + 3)^2 = 25$$

To graph the circle, plot the center $(4, -3)$, then move 5 units right, left, up, and down from the center. Draw a smooth curve through these four points, sketching one quarter of the circle at a time. The graph of this circle is shown in Figure 2.

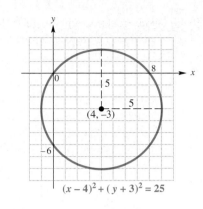

$(x-4)^2 + (y+3)^2 = 25$

Figure 2

Examples 1 and 2 suggest the form of an equation of a circle with radius r and center at (h, k). If (x, y) is a point on the circle, the distance from the center (h, k) to the point (x, y) is r. Then by the distance formula,

$$\sqrt{(x-h)^2 + (y-k)^2} = r.$$

Squaring both sides gives us the following **center-radius form** of the equation of a circle.

Equation of a Circle (Center-Radius Form)

$$(x-h)^2 + (y-k)^2 = r^2$$

is an equation of a circle of radius r with center at (h, k).

EXAMPLE 3 Using the Center-Radius Form of the Equation of a Circle
Find an equation of the circle with center at $(-1, 2)$ and radius 4.
Use the center-radius form, with $h = -1$, $k = 2$, and $r = 4$.

$$(x-h)^2 + (y-k)^2 = r^2$$
$$[x - (-1)]^2 + (y-2)^2 = 4^2$$
$$(x+1)^2 + (y-2)^2 = 16$$

OBJECTIVE 2 Determine the center and radius of a circle given its equation. In the equation found in Example 2, multiplying out $(x-4)^2$ and $(y+3)^2$ and then combining like terms gives

$$(x-4)^2 + (y+3)^2 = 25$$
$$x^2 - 8x + 16 + y^2 + 6y + 9 = 25$$
$$x^2 + y^2 - 8x + 6y = 0.$$

This general form suggests that an equation with both x^2 and y^2 terms may represent a circle. The next example shows how to tell, by completing the square.

⌐ **E X A M P L E 4** Completing the Square to Find the Center and Radius
Graph $x^2 + y^2 + 2x + 6y - 15 = 0$.

Since the equation has x^2 and y^2 terms with equal coefficients, its graph might be that of a circle. To find the center and radius, complete the square on x and y.

$$x^2 + y^2 + 2x + 6y = 15$$ Get the constant on the right.

$$(x^2 + 2x \quad) + (y^2 + 6y \quad) = 15$$ Rewrite in anticipation of completing the square.

$$\left[\frac{1}{2}(2)\right]^2 = 1 \qquad \left[\frac{1}{2}(6)\right]^2 = 9$$ Square half the coefficient of each middle term.

$$(x^2 + 2x + 1) + (y^2 + 6y + 9) = 15 + 1 + 9$$ Complete the square on both x and y.

$$(x + 1)^2 + (y + 3)^2 = 25$$ Factor on the left and add on the right.

The last equation shows that the graph is a circle with center at $(-1, -3)$ and radius 5. The graph is shown in Figure 3.

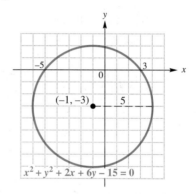

Figure 3

∎

 If the procedure of Example 4 leads to an equation of the form $(x - h)^2 + (y - k)^2 = 0$, the graph is the single point (h, k). If the constant on the right side is negative, the equation has no graph.

O B J E C T I V E 3 Recognize the equation of an ellipse. An **ellipse** is the set of all points in a plane the sum of whose distances from two fixed points is constant. These fixed points are called **foci** (singular: *focus*). Figure 4 shows an ellipse whose foci are $(c, 0)$ and $(-c, 0)$, with x-intercepts $(a, 0)$ and $(-a, 0)$ and y-intercepts $(0, b)$ and $(0, -b)$. The origin is the **center** of the ellipse.

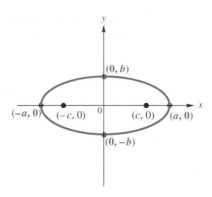

Figure 4

From the definition above, it can be shown by the distance formula that an ellipse has the following equation.

Equation of an Ellipse

The ellipse whose x-intercepts are $(a, 0)$ and $(-a, 0)$ and whose y-intercepts are $(0, b)$ and $(0, -b)$ has an equation of the form

$$\frac{x^2}{a^2} + \frac{y^2}{b^2} = 1.$$

The proof of this is outlined in the exercises. (See Exercise 51.) Note that a circle is a special case of an ellipse, where $a^2 = b^2$.

OBJECTIVE **4** Graph ellipses. To graph an ellipse, we plot the four intercepts and then sketch the ellipse through those points.

EXAMPLE 5 Graphing an Ellipse

Graph $\dfrac{x^2}{49} + \dfrac{y^2}{36} = 1$.

Here, $a^2 = 49$, so $a = \pm 7$, and the x-intercepts for this ellipse are $(7, 0)$ and $(-7, 0)$. Similarly, $b^2 = 36$, so $b = \pm 6$, and the y-intercepts are $(0, 6)$ and $(0, -6)$. Plotting the intercepts and sketching the ellipse through them gives the graph in Figure 5.

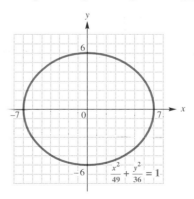

Figure 5

As with the graphs of parabolas and circles, the graph of an ellipse may be shifted horizontally and vertically, as in the next example.

E X A M P L E   6   Graphing an Ellipse Shifted Horizontally and Vertically

Graph $\dfrac{(x-2)^2}{25} + \dfrac{(y+3)^2}{49} = 1.$

Just as $(x-2)^2$ and $(y+3)^2$ would indicate that the center of a circle would be $(2, -3)$, so it is with this ellipse. Figure 6 shows that the graph goes through the four points $(2, 4)$, $(7, -3)$, $(2, -10)$, and $(-3, -3)$. The x-values of these points are found by adding $\pm a = \pm 5$ to 2, and the y-values come from adding $\pm b = \pm 7$ to -3.

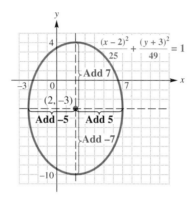

Figure 6

Notice that the graphs in this section are not graphs of functions. The only conic section whose graph is a function is the vertical parabola with equation $f(x) = ax^2 + bx + c$.

O B J E C T I V E 5 Graph circles and ellipses using a graphing calculator. A graphing calculator in function mode cannot directly graph a circle or an ellipse. We must first solve the equation for y, getting two functions y_1 and y_2. The union of these two graphs is the graph of the entire figure. To get an undistorted screen, a *square window* must be used. (See your instruction manual for details.) For example, to graph $(x+3)^2 + (y+2)^2 = 25$, begin by solving for y.

$$(x+3)^2 + (y+2)^2 = 25$$
$$(y+2)^2 = 25 - (x+3)^2 \qquad \text{Subtract } (x+3)^2.$$
$$y + 2 = \pm\sqrt{25 - (x+3)^2} \qquad \text{Take square roots.}$$
$$y = -2 \pm \sqrt{25 - (x+3)^2} \qquad \text{Subtract 2.}$$

The two functions to be graphed are

$$y_1 = -2 + \sqrt{25 - (x+3)^2} \qquad \text{and} \qquad y_2 = -2 - \sqrt{25 - (x+3)^2}.$$

See Figure 7. (*Note:* The two semicircles seem to be disconnected. This is because the graphs are nearly vertical at those points, and the calculator cannot show a true picture of the behavior there.)

$$(x + 3)^2 + (y + 2)^2 = 25$$

Square viewing window

Figure 7

CONNECTIONS

An interesting and perhaps surprising application of ellipses in our everyday life appears in gears. Elliptical bicycle gears are designed to respond to the legs' natural strengths and weaknesses. At the top and bottom of the powerstroke where the legs have the least leverage, the gear offers little resistance, but as the gear rotates, the resistance increases. This allows the legs to apply more power where it is most naturally available.

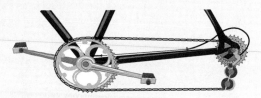

FOR DISCUSSION OR WRITING

A circle can be thought of as an ellipse in which $a = b$ in the equation. Explain how this fact distinguishes the operation of a circular gear from an elliptical gear.

13.1 EXERCISES

1. A circle has equation $x^2 + y^2 = 100$. What is the center of the circle? What is the radius?

2. The equation of an ellipse is $\dfrac{x^2}{16} + \dfrac{y^2}{7} = 1$. What is the center of the ellipse? What are the x-intercepts? The y-intercepts?

3. What is the center of the ellipse with equation $\dfrac{(y + 2)^2}{25} + \dfrac{(x + 3)^2}{49} = 1$?

4. What is the center of the circle with equation $(x - 4)^2 + (y + 1)^2 = 8$? What is the radius?

Write an equation of each circle with the given center and radius. See Examples 1–3.

5. $(0, 0)$; $r = 6$

6. $(0, 0)$; $r = 5$

7. $(-1, 3)$; $r = 4$

8. $(2, -2)$; $r = 3$

9. $(0, 4)$; $r = \sqrt{3}$

10. $(-2, 0)$; $r = \sqrt{5}$

11. Explain why a set of points that form an ellipse does not satisfy the definition of a function.

12. **(a)** How many points are there on the graph of $(x - 4)^2 + (y - 1)^2 = 0$? Explain your answer.

(b) How many points are there on the graph of $(x - 4)^2 + (y - 1)^2 = -1$? Explain your answer.

Find the center and the radius of each circle. See Example 4. (Hint: In Exercises 17 and 18 divide both sides by the greatest common factor.)

13. $x^2 + y^2 + 4x + 6y + 9 = 0$

14. $x^2 + y^2 - 8x - 12y + 3 = 0$

15. $x^2 + y^2 + 10x - 14y - 7 = 0$

16. $x^2 + y^2 - 2x + 4y - 4 = 0$

17. $3x^2 + 3y^2 - 12x - 24y + 12 = 0$

18. $2x^2 + 2y^2 + 20x + 16y + 10 = 0$

Graph each circle. Identify the center if it is not at the origin. See Examples 1, 2, and 4.

19. $x^2 + y^2 = 9$

20. $x^2 + y^2 = 4$

21. $2y^2 = 10 - 2x^2$

22. $3x^2 = 48 - 3y^2$

23. $(x + 3)^2 + (y - 2)^2 = 9$

24. $(x - 1)^2 + (y + 3)^2 = 16$

25. $x^2 + y^2 - 4x - 6y + 9 = 0$

26. $x^2 + y^2 + 8x + 2y - 8 = 0$

27. A circle can be drawn on a piece of posterboard by fastening one end of a string, pulling the string taut with a pencil, and tracing a curve as shown in the figure. Explain why this method works.

28. It is possible to sketch an ellipse on a piece of posterboard by fastening two ends of a length of string, pulling the string taut with a pencil, and tracing a curve, as shown in the drawing. Explain why this method works.

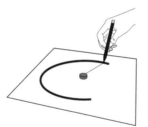

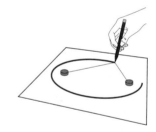

29. This figure shows the crawfish race held at the Crawfish Festival in Breaux Bridge, Louisiana. Explain why a circular "racetrack" is appropriate for such a race.

30. Discuss the similarities and differences between the equations of a circle and an ellipse.

Graph each ellipse. See Examples 5 and 6.

31. $\dfrac{x^2}{9} + \dfrac{y^2}{25} = 1$

32. $\dfrac{x^2}{9} + \dfrac{y^2}{16} = 1$

33. $\dfrac{x^2}{36} = 1 - \dfrac{y^2}{16}$

34. $\dfrac{x^2}{9} = 1 - \dfrac{y^2}{4}$

35. $\dfrac{y^2}{25} = 1 - \dfrac{x^2}{49}$

36. $\dfrac{y^2}{9} = 1 - \dfrac{x^2}{16}$

37. $\dfrac{(x + 1)^2}{64} + \dfrac{(y - 2)^2}{49} = 1$ **38.** $\dfrac{(x - 4)^2}{9} + \dfrac{(y + 2)^2}{4} = 1$

39. $\dfrac{(x - 2)^2}{16} + \dfrac{(y - 1)^2}{9} = 1$ **40.** $\dfrac{(x + 3)^2}{25} + \dfrac{(y + 2)^2}{36} = 1$

TECHNOLOGY INSIGHTS (EXERCISES 41–42)

41. The circle shown in the calculator-generated graph was created using function mode with a square viewing window. It is the graph of $(x + 2)^2 + (y - 4)^2 = 16$. What are the two functions y_1 and y_2 that were used to obtain this graph?

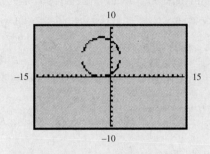

42. The ellipse shown in the calculator-generated graph was graphed using a graphing calculator in function mode, with a square viewing window. It is the graph of $\dfrac{x^2}{4} + \dfrac{y^2}{9} = 1$. What are the two functions y_1 and y_2 that were used to obtain this graph?

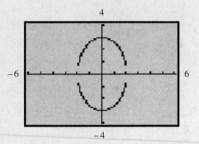

 Use a graphing calculator in function mode to graph each circle or ellipse. Use a square viewing window. See Objective 5.

43. $x^2 + y^2 = 36$ **44.** $(x - 2)^2 + y^2 = 49$

45. $\dfrac{x^2}{16} + \dfrac{y^2}{4} = 1$ **46.** $\dfrac{(x - 3)^2}{25} + \dfrac{y^2}{9} = 1$

In Exercises 47–50, see Figure 4 and use the fact that $c^2 = a^2 - b^2$. (Source for Exercises 47 and 48: James B. Kaler, *Astronomy!,* Addison Wesley, 1997.)

 47. The orbit of Mars is an ellipse with the sun at one focus. For x and y in millions of miles, the equation of the orbit is

$$\dfrac{x^2}{141.7^2} + \dfrac{y^2}{141.1^2} = 1.$$

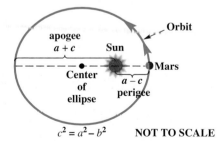

(a) Find the greatest distance (the **apogee**) from Mars to the sun.

(b) Find the smallest distance (the **perigee**) from Mars to the sun.

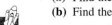 **48.** The orbit of Venus around the sun (one of the foci) is an ellipse with equation

$$\dfrac{x^2}{5013} + \dfrac{y^2}{4970} = 1,$$

where x and y are measured in millions of miles.

(a) Find the greatest distance between Venus and the sun.
(b) Find the smallest distance between Venus and the sun.

A *lithotripter* is a machine used to crush kidney stones using shock waves. The patient is placed in an elliptical tub with the kidney stone at one focus of the ellipse. A beam is projected from the other focus to the tub, so that it reflects to hit the kidney stone.

49. Suppose a lithotripter is based on the ellipse with equation

$$\frac{x^2}{36} + \frac{y^2}{9} = 1.$$

How far from the center of the ellipse must the kidney stone and the source of the beam be placed?

50. Rework Exercise 49 if the equation of the ellipse is $9x^2 + 4y^2 = 36$. (*Hint:* Write the equation in fraction form by dividing each term by 36.)

51. (a) Suppose that $(c, 0)$ and $(-c, 0)$ are the foci of an ellipse and that the sum of the distances from any point (x, y) on the ellipse to the two foci is $2a$. See the figure. Show that the equation of the resulting ellipse is

$$\frac{x^2}{a^2} + \frac{y^2}{a^2 - c^2} = 1.$$

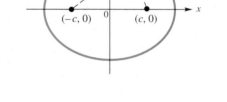

(b) Show that in the equation in part (a), the x-intercepts are $(a, 0)$ and $(-a, 0)$.

(c) Let $b^2 = a^2 - c^2$, and show that $(0, b)$ and $(0, -b)$ are the y-intercepts in the equation in part (a).

52. Use the result of Exercise 51(a) to find an equation of an ellipse with foci $(3, 0)$ and $(-3, 0)$, where the sum of the distances from any point of the ellipse to the two foci is 10.

13.2 The Hyperbola and Square Root Functions

OBJECTIVE 1 Recognize the equation of a hyperbola. A **hyperbola** is the set of all points in a plane such that the absolute value of the *difference* of the distances from two fixed points (called *foci*) is constant. Figure 8 shows a hyperbola; using the distance formula and the definition above, we can show that this hyperbola has equation

$$\frac{x^2}{16} - \frac{y^2}{12} = 1.$$

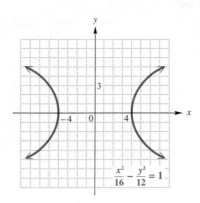

Figure 8

To graph this hyperbola, find the x-intercepts.

$$\frac{x^2}{16} - \frac{0^2}{12} = 1 \qquad \text{Let } y = 0.$$

$$\frac{x^2}{16} = 1$$

$$x^2 = 16 \qquad \text{Multiply by 16.}$$

$$x = \pm 4$$

The x-intercepts are $(4, 0)$ and $(-4, 0)$. Now, find any y-intercepts.

$$\frac{0^2}{16} - \frac{y^2}{12} = 1 \qquad \text{Let } x = 0.$$

$$\frac{-y^2}{12} = 1$$

$$y^2 = -12 \qquad \text{Multiply by } -12.$$

Because there are no *real* solutions to $y^2 = -12$, the graph has no y-intercepts.

Figure 9 gives the graph of

$$\frac{y^2}{25} - \frac{x^2}{9} = 1.$$

Here the y-intercepts are $(0, 5)$ and $(0, -5)$, and there are no x-intercepts.

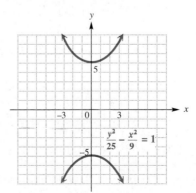

Figure 9

These examples suggest the following summary.

Equations of Hyperbolas

A hyperbola with x-intercepts $(a, 0)$ and $(-a, 0)$ has an equation of the form

$$\frac{x^2}{a^2} - \frac{y^2}{b^2} = 1,$$

and a hyperbola with y-intercepts $(0, b)$ and $(0, -b)$ has an equation of the form

$$\frac{y^2}{b^2} - \frac{x^2}{a^2} = 1.$$

OBJECTIVE **2** Graph hyperbolas by using asymptotes. The two branches of the graph of a hyperbola approach a pair of intersecting straight lines called **asymptotes.** See Figure 10. These lines are useful for sketching the graph of the hyperbola.

Asymptotes of Hyperbolas

The extended diagonals of the rectangle with corners at the points (a, b), $(-a, b)$, $(-a, -b)$, and $(a, -b)$ are the *asymptotes* of either of the hyperbolas

$$\frac{x^2}{a^2} - \frac{y^2}{b^2} = 1 \qquad \text{or} \qquad \frac{y^2}{b^2} - \frac{x^2}{a^2} = 1.$$

This rectangle is called the **fundamental rectangle.** Using the methods of Chapter 8 we could show that the equations of these asymptotes are

$$y = \frac{b}{a}x \qquad \text{and} \qquad y = -\frac{b}{a}x.$$

To graph either of the two forms of hyperbolas, $\dfrac{x^2}{a^2} - \dfrac{y^2}{b^2} = 1$ or $\dfrac{y^2}{b^2} - \dfrac{x^2}{a^2} = 1$, follow these steps.

Graphing a Hyperbola

Step 1 **Find the intercepts.** Locate the intercepts at $(a, 0)$ and $(-a, 0)$ if the x^2 term has a positive coefficient, or at $(0, b)$ and $(0, -b)$ if the y^2 term has a positive coefficient.

Step 2 **Find the fundamental rectangle.** Locate the corners of the fundamental rectangle at (a, b), $(-a, b)$, $(-a, -b)$, and $(a, -b)$.

Step 3 **Sketch the asymptotes.** The extended diagonals of the rectangle are the asymptotes of the hyperbola, and they have equations $y = \pm\dfrac{b}{a}x$.

Step 4 **Draw the graph.** Sketch each branch of the hyperbola through an intercept and approaching (but not touching) the asymptotes.

EXAMPLE 1 Graphing a Horizontal Hyperbola

Graph $\dfrac{x^2}{16} - \dfrac{y^2}{25} = 1$.

Step 1 Here $a = 4$ and $b = 5$. The x-intercepts are $(4, 0)$ and $(-4, 0)$.

Step 2 The four points $(4, 5)$, $(-4, 5)$, $(-4, -5)$, and $(4, -5)$ are the corners of the rectangle that determines the asymptotes, as shown in Figure 10.

Steps 3 and 4 The equations of the asymptotes are $y = \pm\dfrac{5}{4}x$, and the hyperbola approaches these lines as x and y get larger and larger in absolute value.

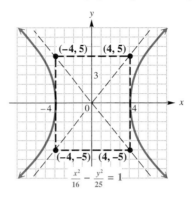

Figure 10

EXAMPLE 2 Graphing a Vertical Hyperbola

Graph $\dfrac{y^2}{49} - \dfrac{x^2}{16} = 1$.

This hyperbola has y-intercepts $(0, 7)$ and $(0, -7)$. The asymptotes are the extended diagonals of the rectangle with corners at $(4, 7)$, $(-4, 7)$, $(-4, -7)$, and $(4, -7)$. Their equations are $y = \pm\dfrac{7}{4}x$. See Figure 11.

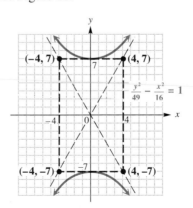

Figure 11

CAUTION When sketching the graph of a hyperbola, be sure that the branches do not touch the asymptotes.

     Hyperbolas are graphed with a graphing calculator in much the same way as circles and ellipses, by first writing the equations of two root functions that combined are equivalent to the equation of the hyperbola. A square window gives a truer shape for hyperbolas, too.

CONNECTIONS

A hyperbola and a parabola are used together in one kind of microwave antenna system. The cross sections of the system consist of a parabola and a hyperbola with the focus of the parabola coinciding with one focus of the hyperbola.

     The incoming microwaves that are parallel to the axis of the parabola are reflected from the parabola up toward the hyperbola and back to the other focus of the hyperbola, where the cone of the antenna is located to capture the signal.

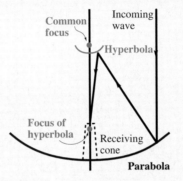

FOR DISCUSSION OR WRITING

The property of the parabola and the hyperbola that is used here is a "reflection property" of the foci. Explain why this name is appropriate.

OBJECTIVE **3**   Identify conic sections by their equations. Rewriting a second-degree equation in one of the forms given for ellipses, hyperbolas, circles, or parabolas makes it possible to determine when the graph is one of these figures. A summary of the equations and graphs of the conic sections follows.

Equation	Graph	Description	Identification
$y = a(x - h)^2 + k$	Parabola	It opens upward if $a > 0$, downward if $a < 0$. The vertex is at (h, k).	x^2 term y is not squared.
$x = a(y - k)^2 + h$	Parabola	It opens to the right if $a > 0$, to the left if $a < 0$. The vertex is at (h, k).	y^2 term x is not squared.

Equation	Graph	Description	Identification
$(x - h)^2 + (y - k)^2$ $= r^2$	Circle	The center is at (h, k), and the radius is r.	x^2 and y^2 terms have the same positive coefficient.
$\dfrac{x^2}{a^2} + \dfrac{y^2}{b^2} = 1$	Ellipse	The x-intercepts are $(a, 0)$ and $(-a, 0)$. The y-intercepts are $(0, b)$ and $(0, -b)$.	x^2 and y^2 terms have different positive coefficients.
$\dfrac{x^2}{a^2} - \dfrac{y^2}{b^2} = 1$	Hyperbola	The x-intercepts are $(a, 0)$ and $(-a, 0)$. The asymptotes are found from (a, b), $(a, -b)$, $(-a, -b)$, and $(-a, b)$.	x^2 has a positive coefficient. y^2 has a negative coefficient.
$\dfrac{y^2}{b^2} - \dfrac{x^2}{a^2} = 1$	Hyperbola	The y-intercepts are $(0, b)$ and $(0, -b)$. The asymptotes are found from (a, b), $(a, -b)$, $(-a, -b)$, and $(-a, b)$.	y^2 has a positive coefficient. x^2 has a negative coefficient.

EXAMPLE 3   Identifying the Graph of a Given Equation

Identify the graph of each equation.

(a) $9x^2 = 108 + 12y^2$

Both variables are squared, so the graph is either an ellipse or a hyperbola. (This situation also occurs for a circle, which may be considered a special case of the ellipse.) To see which one it is, rewrite the equation so that the x and y terms are on one side of the equation and 1 is on the other.

$$9x^2 = 108 + 12y^2$$
$$9x^2 - 12y^2 = 108 \qquad \text{Subtract } 12y^2.$$
$$\frac{x^2}{12} - \frac{y^2}{9} = 1 \qquad \text{Divide by 108.}$$

Because of the minus sign, the graph of this equation is a hyperbola.

(b) $x^2 = y - 3$

Only one of the two variables, x, is squared, so this is the vertical parabola $y = x^2 + 3$.

(c) $x^2 = 9 - y^2$

Get the variable terms on the same side of the equation.

$$x^2 = 9 - y^2$$

$$x^2 + y^2 = 9 \qquad \text{Add } y^2.$$

This equation represents a circle with center at the origin and radius 3.

OBJECTIVE **4** Graph certain square root functions. Recall that no vertical line will intersect the graph of a function in more than one point. Thus, horizontal parabolas and all circles, ellipses, and the hyperbolas discussed in this chapter are examples of graphs that do not satisfy the conditions of a function. However, by considering only a part of the graph of each of these we have the graph of a function, as seen in Figure 12.

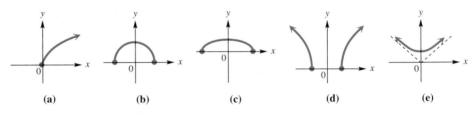

Figure 12

In parts (a), (b), (c), and (e) of Figure 12, the top portion of a conic section is shown (parabola, circle, ellipse, and hyperbola, respectively). In part (d), the top two portions of a hyperbola are shown. In each case, the graph is the graph of a function, since the graph satisfies the conditions of the vertical line test.

To get equations for the graphs shown in Figure 12, we introduce the idea of a square root function.

Square Root Function

A function of the form

$$f(x) = \sqrt{u}$$

for an algebraic expression u, with $u \geq 0$, is called a **square root function.**

EXAMPLE 4 Graphing a Square Root Function

Graph $f(x) = \sqrt{25 - x^2}$.

Replace $f(x)$ with y and square both sides to get the equation

$$y^2 = 25 - x^2, \qquad \text{or} \qquad x^2 + y^2 = 25.$$

This is the equation of a circle with center at $(0, 0)$ and radius 5. Since $f(x)$, or y, represents a principal square root in the original equation, $f(x)$ must be nonnegative. This restricts the graph to the upper half of the circle, as shown in Figure 13. Use the graph and the vertical line test to verify that it is indeed a function.

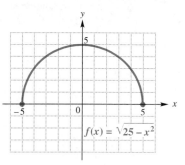

Figure 13

Refer to Figure 13. Had we wanted to graph the function

$$g(x) = -\sqrt{25 - x^2},$$

the graph would be the reflection of the graph of $f(x)$ across the x-axis. This would ensure nonpositive y-values, as indicated by the $-$ sign in the rule for $g(x)$.

Root functions, since they are functions, can be entered and graphed directly with a graphing calculator.

13.2 EXERCISES

Match each equation with the correct graph. See Examples 1 and 2, and Example 5 in Section 13.1.

1. $\dfrac{x^2}{25} + \dfrac{y^2}{9} = 1$

2. $\dfrac{x^2}{9} + \dfrac{y^2}{25} = 1$

3. $\dfrac{x^2}{9} - \dfrac{y^2}{25} = 1$

4. $\dfrac{x^2}{25} - \dfrac{y^2}{9} = 1$

A.

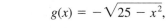

B.

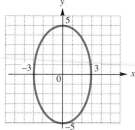

C.

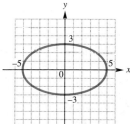

D.

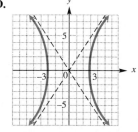

5. Write an explanation of how you can tell from the equation whether the branches of a hyperbola open up and down or open left and right.

6. Explain why the graph of a hyperbola of the type discussed in this section does not satisfy the conditions for the graph of a function.

Graph each hyperbola. See Examples 1 and 2.

7. $\dfrac{x^2}{16} - \dfrac{y^2}{9} = 1$ **8.** $\dfrac{y^2}{4} - \dfrac{x^2}{25} = 1$ **9.** $\dfrac{y^2}{9} - \dfrac{x^2}{9} = 1$

10. $\dfrac{x^2}{49} - \dfrac{y^2}{16} = 1$ **11.** $\dfrac{x^2}{25} - \dfrac{y^2}{36} = 1$ **12.** $\dfrac{y^2}{9} - \dfrac{x^2}{4} = 1$

Identify the graph of the equation as one of the four conic sections: parabolas, circles, ellipses, or hyperbolas. (It may be necessary to transform the equation into a more recognizable form.) Then sketch the graph of each equation. See Example 3.

13. $x^2 - y^2 = 16$ **14.** $x^2 + y^2 = 16$ **15.** $4x^2 + y^2 = 16$

16. $x^2 - 2y = 0$ **17.** $y^2 = 36 - x^2$ **18.** $9x^2 + 25y^2 = 225$

19. $9x^2 = 144 + 16y^2$ **20.** $x^2 + 9y^2 = 9$ **21.** $y^2 = 4 + x^2$

22. State in your own words the major difference between the definitions of *ellipse* and *hyperbola*.

Graph each square root function. See Example 4.

23. $f(x) = -\sqrt{36 - x^2}$ **24.** $f(x) = 3\sqrt{1 + \dfrac{x^2}{9}}$

25. $f(x) = \sqrt{\dfrac{x + 4}{2}}$ **26.** $f(x) = -2\sqrt{\dfrac{9 - x^2}{9}}$

Recall from Section 13.1 that the center of an ellipse may be shifted away from the origin. (See Example 6 in Section 13.1, for instance.) The same shifting process can be applied to hyperbolas. For example, the hyperbola

$$\frac{(x + 5)^2}{4} - \frac{(y - 2)^2}{9} = 1,$$

shown at the right, would have the same graph as $\dfrac{x^2}{4} - \dfrac{y^2}{9} = 1$, but centered at $(-5, 2)$. Graph each hyperbola with center shifted away from the origin.

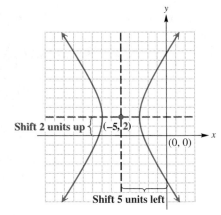

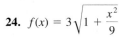

27. $\dfrac{(x - 2)^2}{4} - \dfrac{(y + 1)^2}{9} = 1$ **28.** $\dfrac{(x + 3)^2}{16} - \dfrac{(y - 2)^2}{25} = 1$

29. $\dfrac{y^2}{36} - \dfrac{(x - 2)^2}{49} = 1$ **30.** $\dfrac{(y - 5)^2}{9} - \dfrac{x^2}{25} = 1$

31. An arch has the shape of half an ellipse. The equation of the ellipse is $100x^2 + 324y^2 = 32,400$, where x and y are in meters.
   (a) How high is the center of the arch?
   (b) How wide is the arch across the bottom? (See the figure.)

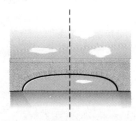

32. Two buildings in a sports complex are shaped and positioned like a portion of the branches of the hyperbola $400x^2 - 625y^2 = 250,000$, where x and y are in meters.
  (a) How far apart are the buildings at their closest point?
  (b) Find the distance d in the figure.

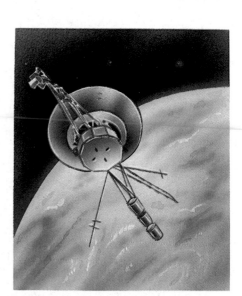

33. In rugby, after a *try* (similar to a touchdown in American football) the scoring team attempts a kick for extra points. The ball must be kicked from directly behind the point where the try was scored. The kicker can choose the distance but cannot move the ball sideways. It can be shown that the kicker's best choice is on the hyperbola with equation

$$\frac{x^2}{g^2} - \frac{y^2}{g^2} = 1,$$

where $2g$ is the distance between the goal posts. Since the hyperbola approaches its asymptotes, it is easier for the kicker to estimate points on the asymptotes instead of on the hyperbola. What are the asymptotes of this hyperbola? Why is it relatively easy to estimate them? (*Source:* Daniel C. Isaksen, "How to Kick a Field Goal," *The College Mathematics Journal,* vol. 27, no. 4, September 1996.)

34. When a satellite is launched into orbit, the shape of its trajectory is determined by its velocity. The trajectory will be hyperbolic if the velocity V, in meters per second, satisfies the inequality

$$V > \frac{2.82 \times 10^7}{\sqrt{D}},$$

where D is the distance, in meters, from the center of Earth. For what values of V will the trajectory be hyperbolic if $D = 42.5 \times 10^6$ meters? (*Source:* James B. Kaler, *Astronomy!,* Addison Wesley, 1997.)

35. The percent of women in the work force has increased steadily for many years. The line graph shows the change for the period from 1975 to 1997, where $x = 75$ represents 1975, $x = 80$ represents 1980, and so on.

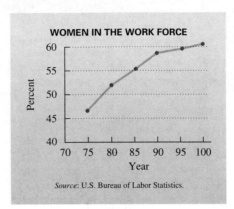

WOMEN IN THE WORK FORCE

Source: U.S. Bureau of Labor Statistics.

The graph resembles the upper branch of a horizontal hyperbola. Using statistical methods, we found the corresponding square root equation $y = .607\sqrt{383.9 + x^2}$, which closely approximates the line graph.

(a) According to the graph, what percent of women were in the work force in 1985?

(b) According to the equation, what percent of women worked in 1985? (Round to the nearest percent.)

36. Refer to Exercise 35. Use the equation to predict the year when the percent of women in the work force will be 62 percent. When will it be 65 percent?

TECHNOLOGY INSIGHTS (EXERCISES 37–38)

37. The hyperbola shown in the calculator-generated graph in the figure was graphed in function mode with a square viewing window. It is the graph of $\frac{x^2}{9} - y^2 = 1$. What are the two functions y_1 and y_2 that were used to obtain this graph?

38. Repeat Exercise 37 for the graph of $\frac{y^2}{9} - x^2 = 1$, shown in the accompanying figure.

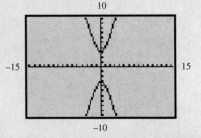

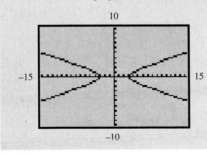

Use a graphing calculator in function mode to graph each hyperbola. Use a square viewing window.

39. $\frac{x^2}{25} - \frac{y^2}{49} = 1$ **40.** $\frac{x^2}{4} - \frac{y^2}{16} = 1$ **41.** $\frac{y^2}{9} - x^2 = 1$ **42.** $\frac{y^2}{36} - \frac{x^2}{4} = 1$

From the discussion in this section, we know that the graph of $\dfrac{x^2}{4} - y^2 = 1$ is a hyperbola. We know that the graph of this hyperbola approaches its asymptotes as x gets larger and larger.

Work Exercises 43–48 in order, *to see the relationship between the hyperbola and one of its asymptotes.*

43. Solve $\dfrac{x^2}{4} - y^2 = 1$ for y, and choose the positive square root.

44. Find the equation of the asymptote with positive slope.

45. Use a calculator to evaluate the y-coordinate of the point where $x = 50$ on the graph of the portion of the hyperbola represented by the equation obtained in Exercise 43. Round your answer to the nearest hundredth.

46. Find the y-coordinate of the point where $x = 50$ on the graph of the asymptote found in Exercise 44.

47. Compare your results in Exercises 45 and 46. How do they support the following statement? When $x = 50$, the graph of the function defined by the equation found in Exercise 43 lies *below* the graph of the asymptote found in Exercise 44.

48. What do you think will happen if we choose x-values larger than 50?

Did you make the connection between the graph of a hyperbola and the graphs of its asymptotes?

49. Suppose that a hyperbola has center at the origin, foci at $(-c, 0)$ and $(c, 0)$, and the absolute value of the difference between the distances from any point (x, y) of the hyperbola to the two foci is $2a$. See the figure. Let $b^2 = c^2 - a^2$, and show that an equation of the hyperbola is

$$\frac{x^2}{a^2} - \frac{y^2}{b^2} = 1.$$

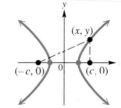

50. Use the result of Exercise 49 to find an equation of a hyperbola with center at the origin, foci at $(-2, 0)$ and $(2, 0)$, and the absolute value of the difference between the distances from any point of the hyperbola to the two foci equal to 2.

13.3 Nonlinear Systems of Equations

OBJECTIVES

1 Solve a nonlinear system by substitution.

2 Use the elimination method to solve a system with two second-degree equations.

An equation in which some terms have more than one variable or a variable of degree two or higher is called a **nonlinear equation**. A **nonlinear system of equations** includes at least one nonlinear equation.

When solving nonlinear systems, it helps to visualize the types of graphs of the equations of the system to determine the possible number of points of intersection. For

example, if a system includes two equations where the graph of one is a parabola and the graph of the other is a line, then there may be 0, 1, or 2 points of intersection. This is illustrated in Figure 14.

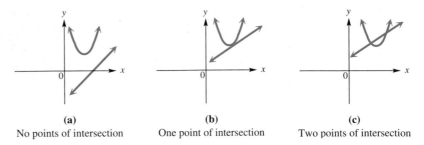

(a)	(b)	(c)
No points of intersection	One point of intersection	Two points of intersection

Figure 14

OBJECTIVE ▮ **Solve a nonlinear system by substitution.** We solve nonlinear systems by the elimination method, the substitution method, or a combination of the two. The substitution method is usually best when one of the equations is linear.

EXAMPLE 1   Using Substitution When One Equation Is Linear

Solve the system.

$$x^2 + y^2 = 9 \qquad \textbf{(1)}$$
$$2x - y = 3 \qquad \textbf{(2)}$$

The graph of (1) is a circle and the graph of (2) is a line. Visualizing the possible ways the graphs could intersect indicates that there may be 0, 1, or 2 points of intersection. It is best to solve the linear equation first for one of the two variables; then substitute the resulting expression into the nonlinear equation to obtain an equation in one variable.

$$2x - y = 3 \qquad (2)$$
$$y = 2x - 3 \qquad \textbf{(3)}$$

Substitute $2x - 3$ for y in equation (1).

$$x^2 + (2x - 3)^2 = 9$$
$$x^2 + 4x^2 - 12x + 9 = 9$$
$$5x^2 - 12x = 0$$
$$x(5x - 12) = 0 \qquad \text{Common factor is } x.$$
$$x = 0 \quad \text{or} \quad x = \frac{12}{5} \qquad \text{Zero-factor property}$$

Let $x = 0$ in equation (3) to get $y = -3$. If $x = \frac{12}{5}$, then $y = \frac{9}{5}$. The solution set of the system is $\left\{ (0, -3), \left(\frac{12}{5}, \frac{9}{5} \right) \right\}$. The graph of the system, shown in Figure 15, confirms the two points of intersection.

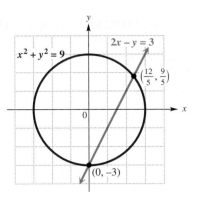

Figure 15

E X A M P L E   2   Using Substitution When One Equation Is Linear

Solve the system.

$$6x - y = 5 \qquad \textbf{(4)}$$

$$xy = 4 \qquad \textbf{(5)}$$

The graph of (4) is a line. We have not specifically mentioned equations like (5); however, it can be shown by plotting points that its graph is a hyperbola. Visualizing a line and a hyperbola indicates that there may be 0, 1, or 2 points of intersection. Since neither equation has a squared term here, solve either equation for one of the variables and then substitute the result into the other equation. Solving $xy = 4$ for x gives $x = \frac{4}{y}$. Substituting $\frac{4}{y}$ for x in equation (4) gives

$$6\left(\frac{4}{y}\right) - y = 5.$$

Clear fractions by multiplying both sides by y, noting the restriction that y cannot be 0. Then solve for y.

$$\frac{24}{y} - y = 5$$

$$24 - y^2 = 5y \qquad \text{Multiply by } y \ (y \neq 0).$$

$$0 = y^2 + 5y - 24$$

$$0 = (y - 3)(y + 8) \qquad \text{Factor.}$$

$$y = 3 \quad \text{or} \quad y = -8 \qquad \text{Zero-factor property}$$

Substitute these results into $x = \frac{4}{y}$ to obtain the corresponding values of x.

If $y = 3$, then $x = \dfrac{4}{3}$. If $y = -8$, then $x = -\dfrac{1}{2}$.

The solution set has two ordered pairs: $\left\{\left(\frac{4}{3}, 3\right), \left(-\frac{1}{2}, -8\right)\right\}$. The graph in Figure 16 on the next page shows that there are two points of intersection.

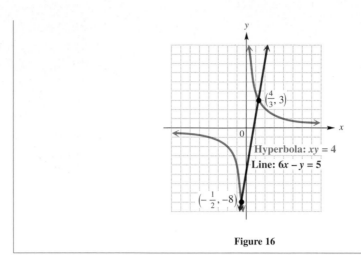

Figure 16

OBJECTIVE 2 Use the elimination method to solve a system with two second-degree equations. The elimination method is used when both equations are second degree.

EXAMPLE 3 Solving a Nonlinear System by Elimination

Solve the system.

$$x^2 + y^2 = 9 \qquad \qquad \textbf{(6)}$$
$$2x^2 - y^2 = -6 \qquad \qquad \textbf{(7)}$$

The graph of (6) is a circle, while the graph of (7) is a hyperbola. By analyzing the possibilities we conclude that there may be 0, 1, 2, 3, or 4 points of intersection. Adding the two equations will eliminate y, leaving an equation that can be solved for x.

$$
\begin{array}{rl}
x^2 + y^2 = & 9 \\
\underline{2x^2 - y^2 = } & \underline{-6} \\
3x^2 = & 3 \\
x^2 = & 1 \\
x = 1 \quad \text{or} \quad & x = -1
\end{array}
$$

Each value of x gives corresponding values for y when substituted into one of the original equations. Using equation (6) gives the following.

If $x = 1$,	**If $x = -1$,**
$1^2 + y^2 = 9$	$(-1)^2 + y^2 = 9$
$y^2 = 8$	$y^2 = 8$
$y = \sqrt{8}$ or $y = -\sqrt{8}$	$y = 2\sqrt{2}$ or $y = -2\sqrt{2}.$
$y = 2\sqrt{2}$ or $y = -2\sqrt{2}.$	

The solution set has four ordered pairs: $\{(1, 2\sqrt{2}), (1, -2\sqrt{2}), (-1, 2\sqrt{2}), (-1, -2\sqrt{2})\}$. Figure 17 shows the four points of intersection.

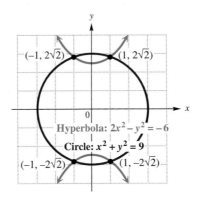

Figure 17

OBJECTIVE **3** Solve a system that requires a combination of methods. A system of second-degree equations may require a combination of methods to solve it.

EXAMPLE 4 Solving a Nonlinear System by a Combination of Methods

Solve the system.

$$x^2 + 2xy - y^2 = 7 \qquad \qquad (8)$$

$$x^2 - y^2 = 3 \qquad \qquad (9)$$

While we have not graphed equations like (8), its graph is a hyperbola. The graph of (9) is also a hyperbola. Two hyperbolas may have 0, 1, 2, 3, or 4 points of intersection. The elimination method can be used here in combination with the substitution method. Begin by eliminating the squared terms by multiplying both sides of equation (9) by -1 and then adding the result to equation (8).

$$
\begin{array}{rcr}
x^2 + 2xy - y^2 = & 7 \\
-x^2 \qquad + y^2 = & -3 \\
\hline
2xy \qquad = & 4
\end{array}
$$

Next, solve $2xy = 4$ for y. (Either variable would do.)

$$2xy = 4$$

$$y = \frac{2}{x} \qquad \qquad (10)$$

Now substitute $y = \frac{2}{x}$ into one of the original equations. It is easier to do this with equation (9).

$$x^2 - y^2 = 3 \qquad \text{(9)}$$

$$x^2 - \left(\frac{2}{x}\right)^2 = 3$$

$$x^2 - \frac{4}{x^2} = 3$$

$$x^4 - 4 = 3x^2 \qquad \text{Multiply by } x^2.$$

$$x^4 - 3x^2 - 4 = 0 \qquad \text{Subtract } 3x^2.$$

$$(x^2 - 4)(x^2 + 1) = 0 \qquad \text{Factor.}$$

$$x^2 - 4 = 0 \quad \text{or} \quad x^2 + 1 = 0$$
$$x^2 = 4 \quad \text{or} \quad x^2 = -1$$
$$x = 2 \quad \text{or} \quad x = -2 \quad\quad x = i \quad \text{or} \quad x = -i$$

Substituting the four values of x from above into equation (10) gives the corresponding values for y.

If $x = 2$, then $y = 1$.   If $x = i$, then $y = -2i$.

If $x = -2$, then $y = -1$.   If $x = -i$, then $y = 2i$.

Note that if you substitute the x-values found above into equations (8) or (9) instead of into equation (10), you get extraneous solutions. It is always wise to check all solutions in both of the given equations. There are four ordered pairs in the solution set, two with real values and two with imaginary values:

$$\{(2, 1), (-2, -1), (i, -2i), (-i, 2i)\}.$$

The graph of the system, shown in Figure 18, shows only the two real intersection points because the graph is in the real number plane. The two ordered pairs with imaginary components are solutions of the system, but do not appear on the graph.

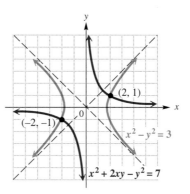

Figure 18

In the examples of this section, we analyzed the possible number of points of intersection of the graphs in each system. However, in Examples 2 and 4, we worked with equations whose graphs had not been studied. Keep in mind that it is not absolutely essential to visualize the number of points of intersection in order to solve the system. Visualizing the geometry of the graphs is only an aid to solving these systems.

OBJECTIVE **4** Use a graphing calculator to solve a nonlinear system. If the equations in a nonlinear system can be solved for y, then we can graph the equations of the system with a graphing calculator and use the capabilities of the calculator to identify all intersection points.

For instance, the two equations in Example 3 would require graphing the four separate functions

$$y_1 = \sqrt{9 - x^2}, \quad y_2 = -\sqrt{9 - x^2}, \quad y_3 = \sqrt{2x^2 + 6}, \quad \text{and} \quad y_4 = -\sqrt{2x^2 + 6}.$$

Figure 19 shows one of the points of intersection.

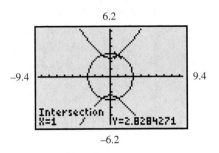

Figure 19

13.3 EXERCISES

1. Write an explanation of the steps you would use to solve the system

$$x^2 + y^2 = 25$$
$$y = x - 1$$

by the substitution method. Why would the elimination method not be appropriate for this system?

2. Write an explanation of the steps you would use to solve the system

$$x^2 + y^2 = 12$$
$$x^2 - y^2 = 13$$

by the elimination method.

Each sketch represents a system of equations. How many points are in its solution set?

3.

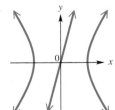

4.

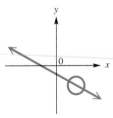

5.

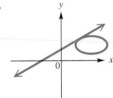

6.

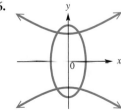

Suppose that a nonlinear system is composed of equations whose graphs are those described, and the number of points of intersection of the two graphs is as given. Make a sketch satisfying these conditions. (There may be more than one way to do this.)

7. a line and a circle;   no points

8. a line and a circle;   one point

9. a line and an ellipse;   no points

10. a line and an ellipse;   two points

11. a line and a hyperbola;   no points

12. a line and a hyperbola;   one point

13. a line and a hyperbola;   two points

14. a circle and an ellipse;   one point

15. a circle and an ellipse;   four points

16. a parabola and an ellipse;   one point

17. a parabola and an ellipse;   four points

18. a parabola and a hyperbola;   two points

Solve each system by the substitution method. See Examples 1 and 2.

19. $y = 4x^2 - x$
$y = x$

20. $y = x^2 + 6x$
$3y = 12x$

21. $y = x^2 + 6x + 9$
$x + y = 3$

22. $y = x^2 + 8x + 16$
$x - y = -4$

23. $x^2 + y^2 = 2$
$2x + y = 1$

24. $2x^2 + 4y^2 = 4$
$x = 4y$

25. $xy = 4$
$3x + 2y = -10$

26. $xy = -5$
$2x + y = 3$

27. $xy = -3$
$x + y = -2$

28. $xy = 12$
$x + y = 8$

29. $y = 3x^2 + 6x$
$y = x^2 - x - 6$

30. $y = 2x^2 + 1$
$y = 5x^2 + 2x - 7$

31. $2x^2 - y^2 = 6$
$y = x^2 - 3$

32. $x^2 + y^2 = 4$
$y = x^2 - 2$

Solve each system by the elimination method or a combination of the elimination and substitution methods. See Examples 3 and 4.

33. $3x^2 + 2y^2 = 12$
$x^2 + 2y^2 = 4$

34. $2x^2 + y^2 = 28$
$4x^2 - 5y^2 = 28$

35. $2x^2 + 3y^2 = 6$
$x^2 + 3y^2 = 3$

36. $6x^2 + y^2 = 9$
$3x^2 + 4y^2 = 36$

37. $2x^2 = 8 - 2y^2$
$3x^2 = 24 - 4y^2$

38. $5x^2 = 20 - 5y^2$
$2y^2 = 2 - x^2$

39. $x^2 + xy + y^2 = 15$
$x^2 + y^2 = 10$

40. $2x^2 + 3xy + 2y^2 = 21$
$x^2 + y^2 = 6$

41. $3x^2 + 2xy - 3y^2 = 5$
$-x^2 - 3xy + y^2 = 3$

42. $-2x^2 + 7xy - 3y^2 = 4$
$2x^2 - 3xy + 3y^2 = 4$

In Exercises 43–46, nonlinear systems are given, along with a screen showing the coordinates of one of the points of intersection of the two graphs.

Solve each system using substitution or elimination to find the coordinates of the other point of intersection.

43. $y = x^2 + 1$
    $x + y = 1$

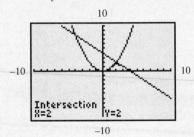

44. $y = -x^2$
    $x + y = 0$

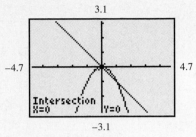

45. $y = \dfrac{1}{2}x^2$
    $x + y = 4$

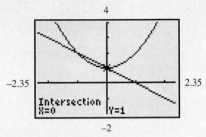

46. $y = -\dfrac{1}{3}x^2$
    $2x - y = 9$

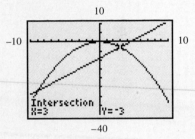

 Use a graphing calculator to solve each system. Then confirm your answer algebraically.

47. $xy = -6$
    $x + y = -1$

48. $y = 2x^2 + 4x$
    $y = -x^2 - 1$

To see how solving quadratic equations in one variable is related to solving a nonlinear system involving one quadratic equation and one linear equation, **work Exercises 49–54 in order.**

49. Solve the quadratic equation $x^2 = 3x + 10$.

50. Graph the equations $y = x^2$ and $y = 3x + 10$ on the same set of coordinate axes.

51. Solve this nonlinear system using substitution.

$$y = x^2$$
$$y = 3x + 10$$

52. How do the x-coordinates of the solutions in Exercise 51 compare to the solutions of the equation in Exercise 49?

53. Graph the quadratic function $y = x^2 - 3x - 10$.

54. How do the x-values of the x-intercepts of the quadratic function in Exercise 53 compare to the x-coordinates of the points of intersection of the graphs in Exercise 50?

Did you make the connection that the solutions of a quadratic equation $ax^2 = bx + c$ are the x-values of the solutions of the system of equations $y = ax^2$ and $y = bx + c$?

Solve each problem by using a nonlinear system.

55. The area of a rectangular rug is 84 square feet and its perimeter is 38 feet. Find the length and width of the rug.

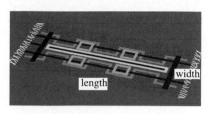

56. Find the length and width of a rectangular room whose perimeter is 50 meters and whose area is 100 square meters.

57. Historically in the United States, the number of bachelor's degrees earned by men has been greater than the number earned by women. In the 1970s, however, this began to change as the number earned by men decreased. It stayed fairly constant in the 1980s, and then in the 1990s slowly began to increase again. Meanwhile, the number of bachelor's degrees earned by women has continued to rise steadily throughout this period. Functions that model the situation are defined by the following equations, where y is the number of degrees (in thousands) granted in year x, with $x = 0$ corresponding to 1970.

$$\text{Men:} \quad y = .138x^2 + .064x + 451$$

$$\text{Women:} \quad y = 12.1x + 334$$

Solve this system of equations to find the year when the same number of bachelor's degrees was awarded to men and women. How many bachelor's degrees were awarded in that year? Give answers to the nearest whole number. (*Source:* U.S. National Center for Education Statistics, *Digest of Education Statistics,* annual.)

58. Andy Grove, chairman of chip maker Intel Corp., recently noted that decreasing prices for computers and stable prices for Internet access implied that the trend lines for these costs either have crossed or soon will. He predicted that the time is not far away when computers, like cell phones, may be given away to sell online time. To see this, assume a price of $1000 for a computer, and let x represent the number of months it will be used. (*Source:* Elizabeth Corcoran, "Can Free Computers Be that Far Away?", *Washington Post,* from *The Sacramento Bee,* February 3, 1999.)

(a) Write an equation for the monthly cost y of the computer over this period.

(b) The average monthly online cost is about $20. Assume this will remain constant and write an equation to express this cost.

(c) Solve the system of equations from parts (a) and (b). Interpret your answer in relation to the situation.

13.4 Second-Degree Inequalities; Systems of Inequalities

OBJECTIVE **1** **Graph second-degree inequalities.** The linear inequality $3x + 2y \leq 5$ is graphed by first graphing the boundary line $3x + 2y = 5$. **Second-degree inequalities** such as $x^2 + y^2 \leq 36$ are graphed in the same way. The boundary of the inequality $x^2 + y^2 \leq 36$ is the graph of the equation $x^2 + y^2 = 36$, a circle with radius 6 and center at the origin, as shown in Figure 20. The inequality $x^2 + y^2 < 36$ will include either

the points outside the circle or the points inside the circle. We decide which region to shade by substituting any point not on the circle, such as $(0, 0)$, into the original inequality. Since $0^2 + 0^2 \leq 36$ is a true statement, the original inequality includes the points inside the circle, the shaded region in Figure 20, and the boundary.

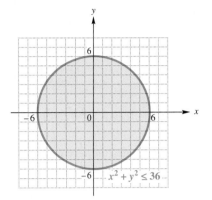

Figure 20

EXAMPLE 1   Graphing a Second-Degree Inequality

Graph $y < -2(x - 4)^2 - 3$.

The boundary, $y = -2(x - 4)^2 - 3$, is a parabola opening downward with vertex at $(4, -3)$. Using the point $(0, 0)$ as a test point gives

$$0 < -2(0 - 4)^2 - 3 \qquad ?$$
$$0 < -32 - 3 \qquad\qquad ?$$
$$0 < -35. \qquad\qquad \text{False}$$

Because the final inequality is a false statement, the points in the region containing $(0, 0)$ do not satisfy the inequality. Figure 21 shows the final graph; the parabola is drawn with a dashed line since the points of the parabola itself do not satisfy the inequality, and the region inside (or below) the parabola is shaded.

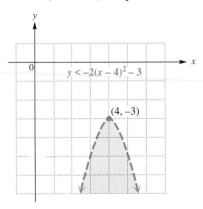

Figure 21

 The origin is the test point of choice unless the graph actually passes through $(0, 0)$.

EXAMPLE 2 Graphing a Second-Degree Inequality

Graph $16y^2 \le 144 + 9x^2$.

First rewrite the inequality as follows.

$$16y^2 - 9x^2 \le 144$$

$$\frac{y^2}{9} - \frac{x^2}{16} \le 1 \qquad \text{Divide by 144.}$$

This form of the inequality shows that the boundary is the hyperbola

$$\frac{y^2}{9} - \frac{x^2}{16} = 1.$$

The desired region will be either the region between the branches of the hyperbola or the regions above the top branch and below the bottom branch. Test the region between the branches by choosing $(0, 0)$ as a test point. Substitute into the original inequality.

$$16y^2 \le 144 + 9x^2$$
$$16(0)^2 \le 144 + 9(0)^2 \qquad ?$$
$$0 \le 144 + 0 \qquad ?$$
$$0 \le 144 \qquad\qquad \text{True}$$

Since the test point $(0, 0)$ satisfies the inequality $16y^2 \le 144 + 9x^2$, the region between the branches containing $(0, 0)$ is shaded, as shown in Figure 22.

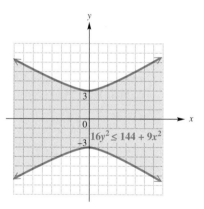

Figure 22

OBJECTIVE **2** Graph the solution set of a system of inequalities. If two or more inequalities are considered at the same time, at least one of which is nonlinear, we have a **nonlinear system of inequalities.** To find the solution set of a nonlinear system, we find the intersection of the graphs (solution sets) of the inequalities in the system. This is just an extension of the method used earlier to graph pairs of linear inequalities joined by the word *and.*

EXAMPLE 3 Graphing a Nonlinear System of Two Inequalities

Graph the solution set of

$$2x + 3y > 6$$
$$x^2 + y^2 < 16.$$

We begin by graphing the solution set of $2x + 3y > 6$. The boundary line is the graph of $2x + 3y = 6$ and is a dashed line because of the symbol $>$. The test point $(0, 0)$ leads to a false statement in the inequality $2x + 3y > 6$, so we shade the region above the line, as shown in Figure 23.

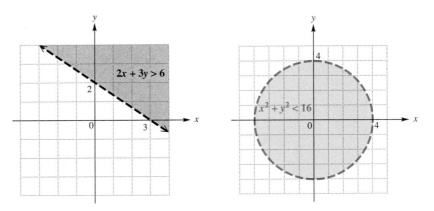

Figure 23   Figure 24

The graph of $x^2 + y^2 < 16$ is the interior of a dashed circle centered at the origin with radius 4. This is shown in Figure 24.

Finally, to get the graph of the solution set of the system, determine the intersection of the graphs of the two inequalities. The overlapping region in Figure 25 is the solution set.

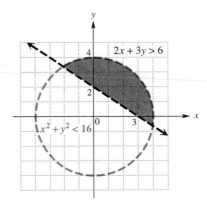

Figure 25

While the system in the following example does not contain a nonlinear inequality, it is different from those that we have solved previously.

E X A M P L E   4   Graphing a Linear System with Three Inequalities

Graph the solution set of the system

$$x + y < 1$$
$$y \le 2x + 3$$
$$y \ge -2.$$

Graph each inequality separately, on the same axes. The graph of $x + y < 1$ consists of all points below the dashed line $x + y = 1$. The graph of $y \le 2x + 3$ is the region below the solid line $y = 2x + 3$. Finally, the graph of $y \ge -2$ is the region above the solid horizontal line $y = -2$. The graph of the system, the intersection of these three graphs, is the triangular region enclosed by the three boundary lines in Figure 26, including two of its boundaries.

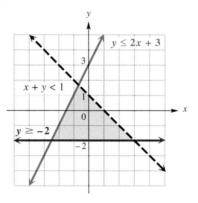

Figure 26

E X A M P L E   5   Graphing a Nonlinear System with Three Inequalities

Graph the solution set of the system

$$y \ge x^2 - 2x + 1$$
$$2x^2 + y^2 > 4$$
$$y < 4.$$

The graph of $y = x^2 - 2x + 1$ is a parabola with vertex at $(1, 0)$. Those points above (or in the interior of) the parabola satisfy the condition $y > x^2 - 2x + 1$. Thus, points on the parabola or in the interior are in the solution set of $y \ge x^2 - 2x + 1$. The graph of the equation $2x^2 + y^2 = 4$ is an ellipse. We draw it with a dashed line. To satisfy the inequality $2x^2 + y^2 > 4$, a point must lie outside the ellipse. The graph of $y < 4$ includes all points below the dashed line $y = 4$. Finally, the graph of the system is the shaded region in Figure 27 that lies outside the ellipse, inside or on the boundary of the parabola, and below the line $y = 4$.

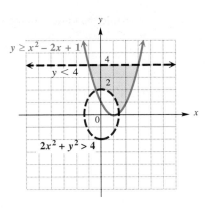

Figure 27

13.4 EXERCISES

1. Which one of the following is a description of the graph of the solution set of the system below?

$$x^2 + y^2 < 25$$

$$y > -2$$

(a) all points outside the circle $x^2 + y^2 = 25$ and above the line $y = -2$
(b) all points outside the circle $x^2 + y^2 = 25$ and below the line $y = -2$
(c) all points inside the circle $x^2 + y^2 = 25$ and above the line $y = -2$
(d) all points inside the circle $x^2 + y^2 = 25$ and below the line $y = -2$

2. Fill in each blank with the appropriate response. The graph of the system

$$y > x^2 + 1$$

$$\frac{x^2}{9} + \frac{y^2}{4} > 1$$

$$y < 5$$

consists of all points _____?_____ the parabola $y = x^2 + 1$, _____?_____ the
             (above/below)   (inside/outside)

ellipse $\frac{x^2}{9} + \frac{y^2}{4} = 1$, and _____?_____ the line $y = 5$.
                           (above/below)

3. Explain how to graph the solution set of a nonlinear inequality.

4. Explain how to graph the solution set of a system of nonlinear inequalities.

Match each nonlinear inequality with its graph.

5. $y \geq x^2 + 4$ **6.** $y \leq x^2 + 4$ **7.** $y < x^2 + 4$ **8.** $y > x^2 + 4$

A. 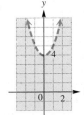   **B.** 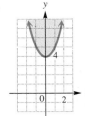   **C.** 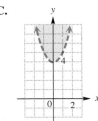   **D.**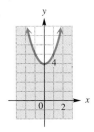

Graph each nonlinear inequality. See Examples 1 and 2.

9. $y^2 > 4 + x^2$ **10.** $y^2 \le 4 - 2x^2$ **11.** $y + 2 \ge x^2$

12. $x^2 \le 16 - y^2$ **13.** $2y^2 \ge 8 - x^2$ **14.** $x^2 \le 16 + 4y^2$

15. $y \le x^2 + 4x + 2$ **16.** $9x^2 < 16y^2 - 144$ **17.** $9x^2 > 16y^2 + 144$

18. $4y^2 \le 36 - 9x^2$ **19.** $x^2 - 4 \ge -4y^2$ **20.** $x \ge y^2 - 8y + 14$

21. $x \le -y^2 + 6y - 7$ **22.** $y^2 - 16x^2 \le 16$

Graph each system of inequalities. See Examples 3–5.

23. $2x + 5y < 10$ **24.** $3x - y > -6$ **25.** $5x - 3y \le 15$ **26.** $4x - 3y \le 0$
$\quad\;\; x - 2y < 4$ $\quad\;\; 4x + 3y > 12$ $\quad\;\; 4x + y \ge 4$ $\quad\;\; x + y \le 5$

27. $x \le 5$ **28.** $x \ge -2$ **29.** $y > x^2 - 4$ **30.** $x^2 - y^2 \ge 9$
$\quad\;\; y \le 4$ $\quad\;\; y \le 4$ $\quad\;\; y < -x^2 + 3$ $\quad\;\; \dfrac{x^2}{16} + \dfrac{y^2}{9} \le 1$

31. $y^2 - x^2 \ge 4$ **32.** $x^2 + y^2 \ge 4$ **33.** $y \le -x^2$ **34.** $y < x^2$
$\quad\; -5 \le y \le 5$ $\quad\;\; x + y \le 5$ $\quad\;\; y \ge x - 3$ $\quad\;\; y > -2$
$\quad\quad\quad\quad\quad\;\;\; x \ge 0$ $\quad\;\; y \le -1$ $\quad\;\; x + y < 3$
$\quad\quad\quad\quad\quad\;\;\; y \ge 0$ $\quad\;\; x < 1$ $\quad\;\; 3x - 2y > -6$

For each nonlinear inequality in Exercises 35–42, a restriction is placed on one or both variables. For example, the graph of

$$x^2 + y^2 \le 4, \quad x \ge 0$$

would be as shown in the figure. Only the right half of the interior of the circle and its boundary is shaded, because of the restriction that x must be nonnegative. Graph each nonlinear inequality with restrictions.

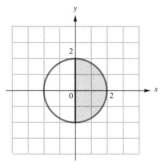

35. $x^2 + y^2 > 36, \quad x \ge 0$ **36.** $4x^2 + 25y^2 < 100, \quad y < 0$

37. $x < y^2 - 3, \quad x < 0$ **38.** $x^2 - y^2 < 4, \quad x < 0$

39. $4x^2 - y^2 > 16, \quad x < 0$ **40.** $x^2 + y^2 > 4, \quad y < 0$

41. $x^2 + 4y^2 \ge 1, \quad x \ge 0, y \ge 0$ **42.** $2x^2 - 32y^2 \le 8, \quad x \le 0, y \ge 0$

■ **TECHNOLOGY INSIGHTS (EXERCISES 43–46)**

Graphing calculators have the capability of shading above or below graphs, thus allowing them to illustrate nonlinear inequalities and systems of nonlinear inequalities. For example, the inequality discussed in Example 1 and graphed in Figure 21,

$$y < -2(x - 4)^2 - 3,$$

is shown in the accompanying screen.

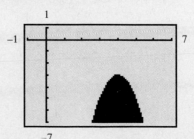

TECHNOLOGY INSIGHTS (EXERCISES 43–46) (CONTINUED)

Match each nonlinear inequality or system of nonlinear inequalities with its calculator-generated graph.

A.

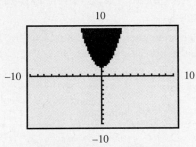

B.

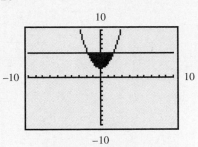

C.

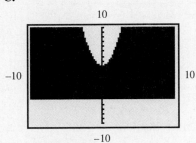

D.

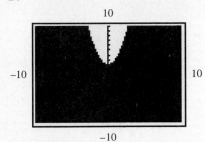

43. $y > x^2 + 2$ **44.** $y < x^2 + 2$ **45.** $y > x^2 + 2$ **46.** $y < x^2 + 2$
                                                                   $y < 5$ $y > -5$

 Use the shading feature of a graphing calculator to graph each system.

47. $y \geq x - 3$ **48.** $y \geq -x^2 + 5$
      $y \leq -x + 4$ $y \leq x^2 - 3$
49. $y < x^2 + 4x + 4$ **50.** $y > (x - 4)^2 - 3$
      $y > -3$ $y < 5$

CHAPTER 13 GROUP ACTIVITY

Finding the Paths of Natural Satellites

Objective: Write and graph equations of ellipses from given data.

(continued)

The moon, which orbits Earth, and Halley's comet, which orbits the sun, are both natural satellites. In Section 13.1, you solved problems where you were given equations of ellipses for the orbits of planets and were asked to find apogees (greatest distance from the sun) and perigees (smallest distance from the sun). This activity reverses the process; that is, given apogees and perigees you must find equations of ellipses.

A. Have each student choose a natural satellite from the chart. Predict the shape of the orbital ellipse for your satellite.

Moon or Comet	Apogee	Perigee
Moon	406.7 thousand kilometers from Earth	356.4 thousand kilometers from Earth
Halley's Comet	35 astronomical units* from the sun	.6 astronomical unit* from the sun

Source: World Book Encyclopedia 1998.
*One astronomical unit is the distance from Earth to the sun.

B. For your satellite, do the following.

1. Find values for a, b, and c. Note that apogee $= a + c$, perigee $= a - c$, and $c^2 = a^2 - b^2$.

2. Write the equation of an ellipse in the form $\dfrac{x^2}{a^2} + \dfrac{y^2}{b^2} = 1$.

3. Rewrite the equation so it can be graphed on a graphing calculator. (See Section 13.1, Objective 5.)

4. Graph your equation on a graphing calculator. Adjust the window setting in order to see the entire graph. Once the window is set correctly, get a square window to see the true shape of the ellipse.

C. Compare your graph with your partner's graph.

1. Do the graphs reflect the shapes you predicted in part A?

2. What window was used to graph each ellipse?

CHAPTER 13 SUMMARY

KEY TERMS

	conic section	**13.2** hyperbola	**13.3** nonlinear equation	nonlinear system of
13.1	circle	asymptotes of a	nonlinear system of	inequalities
	center	hyperbola	equations	
	radius	fundamental	**13.4** second-degree	
	center-radius form	rectangle	inequality	
	ellipse	square root function	system of inequalities	
	foci			

TEST YOUR WORD POWER

See how well you have learned the vocabulary in this chapter. Answers, with examples, are given at the bottom of the page.

1. Conic sections are
(a) graphs of first-degree equations
(b) the result of two or more intersecting planes
(c) graphs of first-degree inequalities
(d) figures that result from the intersection of an infinite cone with a plane.

2. A **circle** is the set of all points in a plane
(a) the difference of whose distances from two fixed points is constant
(b) that lie a fixed distance from a fixed point
(c) the sum of whose distances from two fixed points is constant
(d) that make up the graph of any second-degree equation.

3. An **ellipse** is the set of all points in a plane
(a) the difference of whose distances from two fixed points is constant
(b) that lie a fixed distance from a fixed point

(c) the sum of whose distances from two fixed points is constant
(d) that make up the graph of any second-degree equation.

4. A **hyperbola** is the set of all points in a plane
(a) the difference of whose distances from two fixed points is constant
(b) that lie a fixed distance from a fixed point
(c) the sum of whose distances from two fixed points is constant
(d) that make up the graph of any second-degree equation.

5. A **nonlinear equation** is an equation
(a) that cannot be written in the form $Ax + By = C$, for real numbers A, B, C
(b) in which the terms have only one variable
(c) of degree one
(d) of a linear function.

6. A **nonlinear system of equations** is a system
(a) with at least one linear equation
(b) with two or more inequalities
(c) with at least one nonlinear equation
(d) with at least two linear equations.

7. A **second-degree inequality** is an inequality
(a) with at least one variable of degree two and no variable of degree greater than two
(b) with any variable of degree one or greater
(c) with at least two variables of degree two
(d) with a variable of degree two or greater.

8. A **system of inequalities** is a system
(a) with at least one linear inequality
(b) with two or more inequalities
(c) with at least one nonlinear equation
(d) with at least two linear equations.

QUICK REVIEW

CONCEPTS	EXAMPLES

13.1 THE CIRCLE AND THE ELLIPSE

The circle with radius r and center at (h, k) has an equation of the form

$$(x - h)^2 + (y - k)^2 = r^2.$$

The circle $(x + 2)^2 + (y - 3)^2 = 25$ has center $(-2, 3)$ and radius 5.

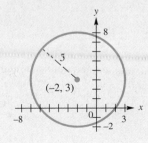

(continued)

CONCEPTS	EXAMPLES

The ellipse whose x-intercepts are $(a, 0)$ and $(-a, 0)$ and whose y-intercepts are $(0, b)$ and $(0, -b)$ has an equation of the form

$$\frac{x^2}{a^2} + \frac{y^2}{b^2} = 1.$$

Graph $\dfrac{x^2}{9} + \dfrac{y^2}{4} = 1$.

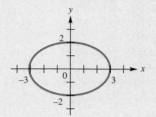

13.2 THE HYPERBOLA; MORE ON SQUARE ROOT FUNCTIONS

A hyperbola with x-intercepts $(a, 0)$ and $(-a, 0)$ has an equation of the form

$$\frac{x^2}{a^2} - \frac{y^2}{b^2} = 1,$$

and a hyperbola with y-intercepts $(0, b)$ and $(0, -b)$ has an equation of the form

$$\frac{y^2}{b^2} - \frac{x^2}{a^2} = 1.$$

Graph $\dfrac{x^2}{4} - \dfrac{y^2}{4} = 1$.

The graph has x-intercepts $(2, 0)$ and $(-2, 0)$.

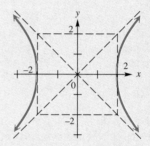

The extended diagonals of the fundamental rectangle with corners at the points (a, b), $(-a, b)$, $(-a, -b)$, and $(a, -b)$ are the asymptotes of these hyperbolas.

The fundamental rectangle has corners at $(2, 2)$, $(-2, 2)$, $(-2, -2)$, and $(2, -2)$.

To graph a square root function, square both sides so that the equation can be easily recognized. Then graph only the part indicated by the original equation.

Graph $y = -\sqrt{4 - x^2}$.

Square both sides and rearrange terms to get

$$x^2 + y^2 = 4.$$

This equation has a circle as its graph. However, graph only the lower half of the circle, since the original equation indicates that y cannot be positive.

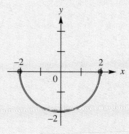

CONCEPTS	EXAMPLES

13.3 NONLINEAR SYSTEMS OF EQUATIONS

Nonlinear systems can be solved by the substitution method, the elimination method, or a combination of the two.

Solve the system

$$x^2 + 2xy - y^2 = 14$$
$$x^2 - y^2 = -16. \qquad (*)$$

Multiply equation (*) by -1 and use elimination.

$$x^2 + 2xy - y^2 = 14$$
$$\underline{-x^2 \qquad\quad + y^2 = 16}$$
$$2xy \qquad = 30$$

$$xy = 15$$

Solve for y to obtain $y = \frac{15}{x}$, and substitute into equation (*).

$$x^2 - \left(\frac{15}{x}\right)^2 = -16$$

$$x^2 - \frac{225}{x^2} = -16$$

$$x^4 + 16x^2 - 225 = 0 \qquad \text{Multiply by } x^2; \text{ add } 16x^2.$$
$$(x^2 - 9)(x^2 + 25) = 0 \qquad \text{Factor.}$$
$$x = \pm 3 \quad \text{or} \quad x = \pm 5i \qquad \text{Zero-factor property}$$

Find corresponding y-values to get the solution set

$$\{(3, 5), (-3, -5), (5i, -3i), (-5i, 3i)\}.$$

13.4 SECOND-DEGREE INEQUALITIES; SYSTEMS OF INEQUALITIES

To graph a second-degree inequality, graph the corresponding equation as a boundary and use test points to determine which region(s) form the solution set. Shade the appropriate region(s).

Graph $y \geq x^2 - 2x + 3$.

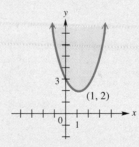

The solution set of a system of inequalities is the intersection of the solution sets of the individual inequalities.

Graph the solution set of the system

$$3x - 5y > -15$$
$$x^2 + y^2 \leq 25.$$

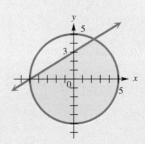

CHAPTER 13 REVIEW EXERCISES

[13.1] *Write an equation of each circle described or graphed.*

1. Center $(-2, 4)$, radius 3

2.

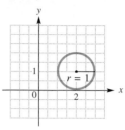

Find the center and radius of each circle.

3. $x^2 + y^2 + 6x - 4y - 3 = 0$

4. $x^2 + y^2 - 8x - 2y + 13 = 0$

5. $2x^2 + 2y^2 + 4x + 20y = -34$

6. $4x^2 + 4y^2 - 24x + 16y = 48$

[13.1–13.2] *Graph the following.*

7. $\dfrac{x^2}{16} + \dfrac{y^2}{9} = 1$

8. $\dfrac{x^2}{49} + \dfrac{y^2}{25} = 1$

9. $\dfrac{x^2}{16} - \dfrac{y^2}{25} = 1$

10. $\dfrac{y^2}{25} - \dfrac{x^2}{4} = 1$

11. $x^2 + 9y^2 = 9$

12. $f(x) = \sqrt{4 + x^2}$

13. $(x - 2)^2 + (y + 3)^2 = 16$

Identify the graph of each equation as a parabola, circle, ellipse, or hyperbola.

14. $y = 2x^2 - 3$

15. $y^2 = 2x^2 - 8$

16. $y^2 = 8 - 2x^2$

17. $x = y^2 + 4$

18. $x^2 + y^2 = 64$

19. A satellite is in an elliptical orbit around Earth with perigee altitude of 160 kilometers and apogee altitude of 16,000 kilometers. See the figure. (*Source:* Bernice Kastner, *Space Mathematics,* NASA.) Find the equation of the ellipse. (*Hint:* Refer to Section 13.1, Exercise 47.)

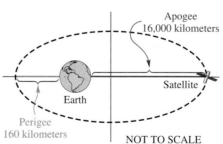

20. Ships and planes often use a location finding system called LORAN. With this system, a radio transmitter at *M* (see the figure) sends out a series of pulses. When each pulse is received at transmitter *S*, it then sends out a pulse. A ship at *P* receives pulses from both *M* and *S*. A receiver on the ship measures the difference in the arrival times of the pulses. A special map gives hyperbolas that correspond to the differences in arrival times (which give the distances d_1 and d_2 in the figure). The ship can then be located as lying on a branch of a particular hyperbola. Suppose $d_1 = 80$ miles and $d_2 = 30$ miles, and the distance between transmitters M and S is 100 miles. Use the definition to find an equation of the hyperbola the ship is located on.

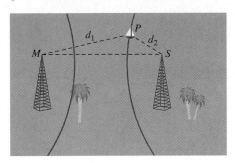

[13.3]   *Solve each system.*

21. $2y = 3x - x^2$
     $x + 2y = -12$

22. $y + 1 = x^2 + 2x$
     $y + 2x = 4$

23. $x^2 + 3y^2 = 28$
     $y - x = -2$

24. $xy = 8$
     $x - 2y = 6$

25. $x^2 + y^2 = 6$
     $x^2 - 2y^2 = -6$

26. $3x^2 - 2y^2 = 12$
     $x^2 + 4y^2 = 18$

27. How many solutions are possible for a system of two equations whose graphs are a circle and a line?

28. How many solutions are possible for a system of two equations whose graphs are a parabola and a hyperbola?

29. Walker Lake is one of two remnants of a prehistoric lake that once covered northwestern Nevada. Since farmers began diverting the Walker River to their fields in the 1860s, the lake has shrunk by half. As it shrinks, the lake becomes more salty, and fish populations begin to die. The federal government and local residents are trying to find a way to guarantee the lake more water without disrupting farm economies. The graph shows the lowering lake level and the salt concentration over the years since 1880. Interpret the point of intersection of the two curves, addressing the year, altitude, and salinity (salt concentration) at that point.

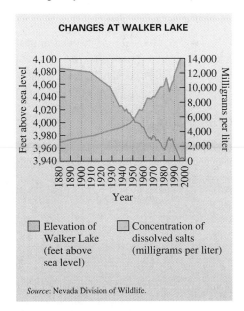

[13.4]   *Graph each nonlinear inequality.*

30. $9x^2 \geq 16y^2 + 144$

31. $4x^2 + y^2 \geq 16$

32. $y < -(x + 2)^2 + 1$

Graph each system of inequalities.

33. $2x + 5y \le 10$
$3x - y \le 6$

34. $|x| \le 2$
$|y| > 1$
$4x^2 + 9y^2 \le 36$

35. $9x^2 \le 4y^2 + 36$
$x^2 + y^2 \le 16$

MIXED REVIEW EXERCISES

Graph.

36. $\dfrac{x^2}{64} + \dfrac{y^2}{25} = 1$

37. $\dfrac{y^2}{4} - 1 = \dfrac{x^2}{9}$

38. $x^2 + y^2 = 25$

39. $y = 2(x - 2)^2 - 3$

40. $f(x) = -\sqrt{16 - x^2}$

41. $f(x) = \sqrt{4 - x}$

42. $3x + 2y \ge 0$
$y \le 4$
$x \le 4$

43. $4y > 3x - 12$
$x^2 < 16 - y^2$

RELATING CONCEPTS (EXERCISES 44-49)

In Chapter 9 we presented several methods of solving systems of linear equations in three variables. Now these methods can be used to find the equation of a circle through three points in a plane that are not on the same line. The equation of a circle can be written in the form $x^2 + y^2 + ax + by + c = 0$ for some values of a, b, and c. We will find the equation of the circle through the points $(2, 4)$, $(5, 1)$, and $(-1, 1)$.

Work Exercises 44–49 in order.

44. Determine one equation in a, b, and c by letting $x = 2$ and $y = 4$ in the general form given above. Write it with a, b, and c on the left and the constant on the right.

45. Repeat Exercise 44 for the point $(5, 1)$.

46. Repeat Exercise 44 for the point $(-1, 1)$.

47. Solve the system of equations formed by the equations found in Exercises 44–46, and give the equation of the circle that satisfies the conditions described above.

48. Use the methods of this chapter to find the center and the radius of the circle in Exercise 47.

49. Graph the circle found in Exercise 47.

Did you make the connection between solving a system of equations and writing the equation of a circle?

CHAPTER 13 TEST

1. Give an equation of the circle shown in the figure.

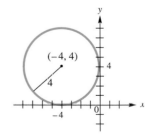

2. Find the coordinates of the center and the radius of the circle with equation $x^2 + y^2 + 8x - 2y = 8$.

Identify the graph of each equation as one of the following: parabola, hyperbola, ellipse, or circle.

3. $3x^2 + 3y^2 = 27$

4. $9x^2 + 4y^2 = 36$

5. $9x^2 = 36 + 4y^2$

6. $x = 36 - 4y^2$

Sketch the graph of each equation.

7. $x^2 + y^2 = 64$

8. $4x^2 + 9y^2 = 36$

9. $16y^2 - 4x^2 = 64$

10. $f(x) = \sqrt{16 - x^2}$

11. If a parabola and an ellipse are graphed in the same plane, what are the possible number of points of intersection of the graphs?

12. Sketch a parabola and an ellipse on the same set of axes so that there are two points of intersection of the graphs.

Solve each nonlinear system.

13. $2x - y = 9$
$xy = 5$

14. $x - 4 = 3y$
$x^2 + y^2 = 8$

15. $x^2 + y^2 = 25$
$x^2 - 2y^2 = 16$

Graph each inequality or system of inequalities.

16. $y \leq x^2 - 2$

17. $2x - 5y \geq 12$
$3x + 4y \leq 12$

18. $x^2 + 25y^2 \leq 25$
$x^2 + y^2 \leq 9$

The orbit of Mercury around the sun (a focus) is an ellipse with equation

$$\frac{x^2}{3352} + \frac{y^2}{3211} = 1,$$

where x and y are measured in million kilometers.

19. Find its apogee, its greatest distance from the sun. (*Hint:* Refer to Section 13.1, Exercise 47.)

20. Find its perigee, its smallest distance from the sun.

CUMULATIVE REVIEW EXERCISES CHAPTERS 1–13

1. Simplify $-10 + |-5| - |3| + 4$.

Solve.

2. $4 - (2x + 3) + x = 5x - 3$

3. $-4k + 7 \geq 6k + 1$

4. $|5m| - 6 = 14$

5. $|2p - 5| > 15$

6. Find the slope of the line through $(2, 5)$ and $(-4, 1)$.

Perform the indicated operations.

7. $(5y - 3)^2$

8. $(2r + 7)(6r - 1)$

9. $(8x^4 - 4x^3 + 2x^2 + 13x + 8) \div (2x + 1)$

Factor.

10. $12x^2 - 7x - 10$

11. $2y^4 + 5y^2 - 3$

12. $z^4 - 1$

13. $a^3 - 27b^3$

Simplify.

14. $\dfrac{5x - 15}{24} \cdot \dfrac{64}{3x - 9}$

15. $\dfrac{y^2 - 4}{y^2 - y - 6} \div \dfrac{y^2 - 2y}{y - 1}$

16. $\dfrac{5}{c + 5} - \dfrac{2}{c + 3}$

17. $\dfrac{p}{p^2 + p} + \dfrac{1}{p^2 + p}$

18. Kareem and Jamal want to clean an office they share. Kareem can do the job alone in 3 hours, while Jamal can do it alone in 2 hours. How long will it take them if they work together?

19. Find the equation of the line through $(-3, -2)$ and perpendicular to $2x - 3y = 7$.

Solve each system.

20. $3x - y = 12$
$2x + 3y = -3$

21. $x + y - 2z = 9$
$2x + y + z = 7$
$3x - y - z = 13$

22. $xy = -5$
$2x + y = 3$

Simplify.

23. $\left(\dfrac{4}{3}\right)^{-1}$

24. $\dfrac{(2a)^{-2}a^4}{a^{-3}}$

25. $4\sqrt[3]{16} - 2\sqrt[3]{54}$

26. $\dfrac{3\sqrt{5x}}{\sqrt{2x}}, \quad x > 0$

27. $\dfrac{5 + 3i}{2 - i}$

Solve for real values of the variable.

28. $2\sqrt{k} = \sqrt{5k + 3}$

29. $10q^2 + 13q = 3$

30. $(4x - 1)^2 = 8$

31. $\log(2x) - \log(x - 1) = \log 3$

32. $2(x^2 - 3)^2 - 5(x^2 - 3) = 12$

33. $F = \dfrac{kwv^2}{r}; \quad$ for v

Graph.

34. $3x + y = 5$

35. $f(x) = x^2 - 4x + 5$

36. $f(x) = 3^{x - 1}$

37. $\dfrac{x^2}{4} - \dfrac{y^2}{16} = 1$

38. $\dfrac{x^2}{25} + \dfrac{y^2}{16} \leq 1$

Work each problem.

39. Under certain conditions, the stopping distance of a car traveling 25 miles per hour is 61.7 feet; for a car traveling 35 miles per hour, it is 106 feet. The stopping distance in feet can be described by the equation $y = ax^2 + bx$, where x is the speed in miles per hour. (*Source: National Traffic Safety Institute Student Workbook, 1993.*)

    (a) Using the given pairs of values, write a system of equations with a and b as the variables.

    (b) Solve the system from part (a). Give answers to the nearest thousandth.

    (c) Use the result from part (b) to write a quadratic equation for the stopping distance.

    (d) Use the equation from part (c) to find the stopping distance to the nearest tenth of a foot for a car traveling 55 miles per hour.

40. The bar graph shows historic and projected annual online retail sales (in billions of dollars) over the Internet.

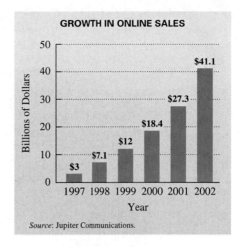

GROWTH IN ONLINE SALES

Source: Jupiter Communications.

A reasonable model for sales y in billions of dollars is the exponential function defined by $y = 1.38(1.65)^x$. The years are coded so x is the number of years since 1995.

(a) Use the model to find sales in the year 2000. (*Hint:* Let $x = 5$.)

(b) Use the model to estimate sales in the year 2003.

14 Sequences and Series

Banking/Finance

14.1 Sequences and Series

14.2 Arithmetic Sequences

14.3 Geometric Sequences

14.4 The Binomial Theorem

The popularity of mutual funds as a means of investing has shown continued growth over the past decade. The graph below indicates the number of mutual funds for each year during the period 1991–1995.

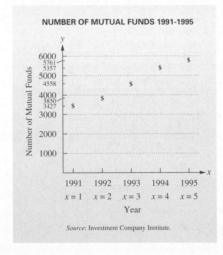

NUMBER OF MUTUAL FUNDS 1991-1995

Number of Mutual Funds

6000
5761
5357
5000
4558
4000
3850
3427
3000
2000
1000

1991   1992   1993   1994   1995
$x=1$ $x=2$ $x=3$ $x=4$ $x=5$
Year

Source: Investment Company Institute.

For the mutual fund graph, if we let $x = 1$ represent 1991, $x = 2$ represent 1992, and so on, the function

$$f(x) = -\frac{73}{6}x^4 + \frac{268}{3}x^3 - \frac{268}{3}x^2 + \frac{1489}{6}x + 3191$$

yields the exact function values (the number of funds). This function was determined using statistical methods and is based only on the positive integer inputs 1, 2, 3, 4, 5; thus, it can be used reliably only for these values. Evaluating this function for these five values yields the list, or *sequence*

$$3427, \ 3850, \ 4558, \ 5357, \ 5761.$$

Because these numbers are in a specific order, we know which year each one represents. For example, since 5357 appears in the fourth position, we know that it represents the fourth year, or 1994. What is $f(5)$? How would you interpret your result?

In this chapter we investigate sequences: functions whose domains consist of positive integers 1, 2, 3, ... The sum of the terms of a sequence is called a *series,* and we will investigate series as well. The example above, dealing with mutual funds, comes from the world of banking and finance, the theme of this chapter. We will return to this example in the exercises for Section 14.1.

Visit our Web site at www.LialAlgebra.com

14.1 Sequences and Series

OBJECTIVES

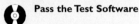

1 Find the terms of a sequence given the general term.

2 Find the general term of a sequence.

3 Use sequences to solve applied problems.

4 Use summation notation to evaluate a series.

5 Write a series using summation notation.

6 Find the arithmetic mean (average) of a group of numbers.

FOR EXTRA HELP

📖 **SSG** Sec. 14.1
   SSM Sec. 14.1

💿 **Pass the Test Software**

💿 **InterAct Math**
   Tutorial Software

📼 **Video** 24

A **sequence** is a function whose domain is the set of natural numbers. Intuitively, a sequence is a list of numbers in which the order of their appearance is important. Sequences appear in many places in daily life. For instance, the interest portions of monthly loan payments made to pay off an automobile or home loan form a sequence.

In the Palace of the Alhambra, residence of the Moorish rulers of Granada, Spain, the Sultana's quarters feature an interesting architectural pattern. There are 2 matched marble slabs inlaid in the floor, 4 walls, an octagon (8-sided) ceiling, 16 windows, 32 arches, and so on. If this pattern is continued indefinitely, the set of numbers forms an *infinite sequence.*

Infinite Sequence

An **infinite sequence** is a function with the set of positive integers as the domain.

OBJECTIVE **1** **Find the terms of a sequence given the general term.** For any positive integer n, the function value (y-value) of a sequence is written as a_n (read "a sub-n") instead of $a(n)$ or $f(n)$. The function values $a_1, a_2, a_3, \ldots$, written in order, are the **terms** of the sequence, with a_1 the first term, a_2 the second term, and so on. The expression a_n, which defines the sequence, is called the **general term** of the sequence.

In the Palace of the Alhambra example given above, the first five terms of the sequence are

$$a_1 = 2, \qquad a_2 = 4, \qquad a_3 = 8, \qquad a_4 = 16, \qquad \text{and} \qquad a_5 = 32.$$

The general term for this sequence is $a_n = 2^n$.

EXAMPLE 1   **Writing the Terms of a Sequence from the General Term**

Given an infinite sequence with $a_n = n + \dfrac{1}{n}$, find the following.

(a) The second term of the sequence
   To get a_2, the second term, replace n with 2.

$$a_2 = 2 + \frac{1}{2} = \frac{5}{2}$$

(b) The fifth term
   Replace n with 5.

$$a_5 = 5 + \frac{1}{5} = \frac{26}{5}$$

(c) $a_{10} = 10 + \dfrac{1}{10} = \dfrac{101}{10}$

(d) $a_{12} = 12 + \dfrac{1}{12} = \dfrac{145}{12}$

As mentioned earlier, a sequence is a special kind of function. Graphing calculators can be used to generate and graph sequences, as shown in Figure 1. The calculator must be in graphing dot mode, so the discrete points on the graph are not connected. Remember that the domain of a sequence consists only of natural numbers.

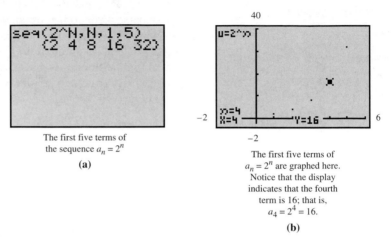

The first five terms of
the sequence $a_n = 2^n$

(a)

The first five terms of
$a_n = 2^n$ are graphed here.
Notice that the display
indicates that the fourth
term is 16; that is,
$a_4 = 2^4 = 16$.

(b)

Figure 1

OBJECTIVE **2** Find the general term of a sequence. Sometimes we need to find a general term to fit the first few terms of a given sequence. There are no rules for finding the general term of a sequence from the first few terms. In fact, it is possible to give more than one general term that produce the same first three or four terms. However, in many examples, the terms may suggest a general term.

E X A M P L E 2   Finding the General Term of a Sequence

Find an expression for the general term a_n of the sequence

$$5, \quad 10, \quad 15, \quad 20, \quad 25, \ldots.$$

The first term is 5(1), the second is 5(2), and so on. By inspection, $a_n = 5n$ will produce the given first five terms.

 One problem with using just a few terms to suggest a general term, as in Example 2, is that there may be more than one general term that gives the same first few terms.

OBJECTIVE **3** Use sequences to solve applied problems. Practical problems often involve *finite sequences*.

Finite Sequence

A **finite sequence** has a domain that includes only the first *n* positive integers.

For example, if *n* is 5, the domain is {1, 2, 3, 4, 5}, and the sequence has five terms.

EXAMPLE 3   Using a Sequence in an Application

Keshon borrows $5000 and agrees to pay $500 monthly, plus interest of 1% on the unpaid balance from the beginning of that month. Find the payments for the first four months and the remaining debt at the end of this period.

The payments and remaining balances are calculated as follows.

First month   Payment: $500 + .01(5000) = $ **550** dollars

                   Balance: $5000 - 500 = $ **4500** dollars

Second month   Payment: $500 + .01($ **4500** $) = $ **545** dollars

                   Balance: $5000 - 2 \cdot 500 = $ **4000** dollars

Third month   Payment: $500 + .01($ **4000** $) = $ **540** dollars

                   Balance: $5000 - 3 \cdot 500 = $ **3500** dollars

Fourth month   Payment: $500 + .01($ **3500** $) = $ **535** dollars

                   Balance: $5000 - 4 \cdot 500 = $ **3000** dollars

The payments for the first four months, in dollars, are

550, 545, 540, 535

and the remaining debt at the end of this period is **3000** dollars.

OBJECTIVE **4** Use summation notation to evaluate a series. By adding the terms of a sequence, we obtain a *series*.

Series

The indicated sum of the terms of a sequence is called a **series.**

For example, if we consider the sum of the payments listed in Example 3,

$$550 + 545 + 540 + 535,$$

we obtain a series that represents the total amount of payments for the first four months.

Since a sequence can be finite or infinite, there are finite or infinite series. One type of infinite series is discussed in Section 14.3, and the binomial theorem discussed in Section 14.4 defines an important finite series. In this section we discuss only finite series.

We use a compact notation, called **summation notation,** to write a series from the general term of the corresponding sequence. For example, the sum of the first six terms of the sequence with general term $a_n = 3n + 2$ is written with the Greek letter Σ (sigma) as

$$\sum_{i=1}^{6} (3i + 2).$$

We read this as "the sum from $i = 1$ to 6 of $3i + 2$." To find this sum, we replace the letter i in $3i + 2$ with 1, 2, 3, 4, 5, and 6, as follows.

$$\sum_{i=1}^{6} (3i + 2) = (3 \cdot 1 + 2) + (3 \cdot 2 + 2) + (3 \cdot 3 + 2)$$
$$+ (3 \cdot 4 + 2) + (3 \cdot 5 + 2) + (3 \cdot 6 + 2)$$
$$= 5 + 8 + 11 + 14 + 17 + 20$$
$$= 75$$

The letter i is called the **index of summation.**

[CAUTION] This use of i has no connection with the complex number i.

┌ **E X A M P L E 4**   Evaluating a Series Written in Summation Notation

Write out the terms and evaluate each of the following.

(a) $\displaystyle\sum_{i=1}^{5} (i - 4) = (1 - 4) + (2 - 4) + (3 - 4) + (4 - 4) + (5 - 4)$

$$= -3 - 2 - 1 + 0 + 1$$

$$= -5$$

(b) $\displaystyle\sum_{i=3}^{7} 3i^2 = 3(3)^2 + 3(4)^2 + 3(5)^2 + 3(6)^2 + 3(7)^2$

$$= 27 + 48 + 75 + 108 + 147$$

$$= 405$$ ∎

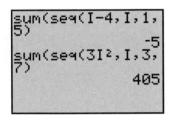

Figure 2

Figure 2 shows how a graphing calculator can be used to obtain the results found in Example 4.

O B J E C T I V E [5] Write a series using summation notation. In Example 4, we started with summation notation and wrote each series using + signs. It is possible to go in the other direction; that is, given a series, we can write it using summation notation. To do this, we observe a pattern in the terms and write the general term accordingly.

┌ **E X A M P L E 5**   Writing a Series with Summation Notation

Write the following sums with summation notation.

(a) $2 + 5 + 8 + 11$

First, find a general term a_n that will give these four terms for a_1, a_2, a_3, and a_4. Inspection (and trial and error) shows that $3i - 1$ will work for these four terms, since

$$3(1) - 1 = 2$$
$$3(2) - 1 = 5$$
$$3(3) - 1 = 8$$
$$3(4) - 1 = 11.$$

(Remember, there may be other expressions that also work. These four terms may be the first terms of more than one sequence.) Since i ranges from 1 to 4, write the sum as

$$2 + 5 + 8 + 11 = \sum_{i-1}^{4} (3i - 1).$$

(b) $8 + 27 + 64 + 125 + 216$

Since these numbers are the cubes of 2, 3, 4, 5, and 6,

$$8 + 27 + 64 + 125 + 216 = \sum_{i=2}^{6} i^3.$$ ∎

OBJECTIVE 6 Find the arithmetic mean (average) of a group of numbers. The **arithmetic mean,** or **average,** of a group of numbers is defined as the sum of all the numbers divided by the number of numbers.

Arithmetic Mean or Average

The arithmetic mean, or average, of a group of numbers is symbolized $\bar{x}$ and is found by dividing the sum of the numbers by the number of numbers. That is,

$$\bar{x} = \frac{\displaystyle\sum_{i=1}^{n} x_i}{n}.$$

Here the values of x_i represent the individual numbers in the group, and n represents the number of numbers.

EXAMPLE 6 Finding the Arithmetic Mean or Average

The following table shows the number of companies listed on the New York Stock Exchange for each year during the period 1990–1996. What was the average number of listings for this seven-year period?

Year	Number of Listings
1990	1774
1991	1885
1992	2088
1993	2361
1994	2570
1995	2675
1996	2907

Source: New York Stock Exchange.

Let $x_1 = 1774$, $x_2 = 1885$, and so on. Since there are 7 numbers in the group, $n = 7$. Therefore,

$$\bar{x} = \frac{\displaystyle\sum_{i=1}^{7} x_i}{7}$$

$$= \frac{1774 + 1885 + 2088 + 2361 + 2570 + 2675 + 2907}{7}$$

$$= 2323 \quad \text{(rounded to the nearest unit).}$$

The average number of listings for this seven-year period was 2323.

CONNECTIONS

One of the most famous sequences in mathematics is the **Fibonacci sequence:**

$$1, 1, 2, 3, 5, 8, 13, 21, 34, 55, \ldots .$$

This sequence is named for the Italian mathematician Leonardo of Pisa (1170–1250), who was also known as Fibonacci. The Fibonacci sequence is found in numerous places in nature. For example, male honeybees hatch from eggs that have not been fertilized, so a male bee has only one parent, a female. On the other hand, female honeybees hatch from fertilized eggs, so a female has two parents, one male and one female. The number of ancestors in consecutive generations of bees follows the Fibonacci sequence. Successive terms in the sequence also appear in plants: in the daisy head, the pineapple, and the pine cone, for instance.

FOR DISCUSSION OR WRITING

1. See if you can discover the pattern in the Fibonacci sequence. (*Hint:* Can you explain how to find the next term, given the preceding two terms?)

2. Draw a tree showing the number of ancestors of a male bee in each generation following the description given above.

14.1 EXERCISES

1. Suppose that f is a function with domain all real numbers, where $f(x) = 2x + 4$. Suppose that an infinite sequence is defined by $a_n = 2n + 4$. Discuss the similarities and the differences between the function and the sequence. Give examples using each.

2. What is wrong with the following? For the sequence defined by $a_n = 2n + 4$, find $a_{1/2}$.

Write out the first five terms of each sequence. See Example 1.

3. $a_n = \dfrac{n + 3}{n}$ 4. $a_n = \dfrac{n + 2}{n + 1}$ 5. $a_n = 3^n$ 6. $a_n = 1^{n-1}$

7. $a_n = \dfrac{1}{n^2}$ 8. $a_n = \dfrac{n^2}{n + 1}$ 9. $a_n = (-1)^n$ 10. $a_n = (-1)^{2n-1}$

Find the indicated term for each sequence. See Example 1.

11. $a_n = -9n + 2$; a_8 12. $a_n = 3n - 7$; a_{12}

13. $a_n = \dfrac{3n + 7}{2n - 5}$; a_{14} 14. $a_n = \dfrac{5n - 9}{3n + 8}$; a_{16}

15. $a_n = (n + 1)(2n + 3)$; a_8 16. $a_n = (5n - 2)(3n + 1)$; a_{10}

Find a general term, a_n, for the given terms of each sequence. See Example 2.

17. $4, 8, 12, 16, \ldots$ 18. $-10, -20, -30, -40, \ldots$

19. $\dfrac{1}{3}, \dfrac{1}{9}, \dfrac{1}{27}, \dfrac{1}{81}, \ldots$ 20. $\dfrac{1}{2}, \dfrac{2}{3}, \dfrac{3}{4}, \dfrac{4}{5}, \ldots$

Solve each applied problem by writing the first few terms of a sequence. See Example 3.

21. Anne borrows $1000 and agrees to pay $100 plus interest of 1% on the unpaid balance each month. Find the payments for the first six months and the remaining debt at the end of this period.

22. *Larissa Perez* is offered a new modeling job with a salary of $20{,}000 + 2500n$ dollars per year at the end of the nth year. Write a sequence showing her salary at the end of each of the first five years. If she continues in this way, what will her salary be at the end of the tenth year?

23. Suppose that an automobile loses $\frac{1}{5}$ of its value each year; that is, at the end of any given year, the value is $\frac{4}{5}$ of the value at the beginning of that year. If a car cost $20,000 new, what is its value at the end of 5 years?

24. A certain car loses $\frac{1}{2}$ of its value each year. If this car cost $40,000 new, what is its value at the end of 6 years?

Write out each series and evaluate it. See Example 4.

25. $\displaystyle\sum_{i=1}^{3} (i^2 + 2)$ **26.** $\displaystyle\sum_{i=1}^{4} i(i + 3)$ **27.** $\displaystyle\sum_{i=2}^{5} \frac{1}{i}$ **28.** $\displaystyle\sum_{i=0}^{4} \frac{i}{i + 1}$

29. $\displaystyle\sum_{i=1}^{6} (-1)^i$ **30.** $\displaystyle\sum_{i=1}^{5} (-1)^i \cdot i$ **31.** $\displaystyle\sum_{i=3}^{7} (i - 3)(i + 2)$ **32.** $\displaystyle\sum_{i=2}^{6} \frac{i^2 + 1}{2}$

Write out the terms of each series.

33. $\displaystyle\sum_{i=1}^{5} 2x \cdot i$ **34.** $\displaystyle\sum_{i=1}^{6} x^i$ **35.** $\displaystyle\sum_{i=1}^{5} i \cdot x^i$ **36.** $\displaystyle\sum_{i=2}^{6} \frac{x + i}{x - i}$

Write each series using summation notation. See Example 5.

37. $3 + 4 + 5 + 6 + 7$ **38.** $1 + 4 + 9 + 16$

39. $\dfrac{1}{2} + \dfrac{1}{3} + \dfrac{1}{4} + \dfrac{1}{5} + \dfrac{1}{6}$ **40.** $-1 + 2 - 3 + 4 - 5 + 6$

41. Explain the basic difference between a sequence and a series.

42. Evaluate $\displaystyle\sum_{i=1}^{3} 5i$ and $5 \displaystyle\sum_{i=1}^{3} i$. Notice that the sums are the same. Explain how the distributive property plays a role in assuring us that the two sums are equal.

Find the arithmetic mean for each collection of numbers. See Example 6.

43. 8, 11, 14, 9, 3, 6, 8   **44.** 10, 12, 8, 19, 23   **45.** 5, 9, 8, 2, 4, 7, 3, 2   **46.** 2, 1, 4, 8, 3, 7

Solve each problem. See Example 6.

47. As mentioned in the chapter introduction, the number of mutual funds available to investors for each year during the period 1991–1995 is given in the following table.

Year	Number of Funds Available
1991	3427
1992	3850
1993	4558
1994	5357
1995	5761

Source: Investment Company Institute.

To the nearest whole number, what was the average number of funds available during this period?

48. The total assets of mutual funds, in billions of dollars, for each year during the period 1992–1996 are shown in the following table.

Year	Assets (in billions of dollars)
1992	1646.3
1993	2075.4
1994	2161.5
1995	2820.4
1996	3539.2

Source: Investment Company Institute.

To the nearest tenth (in billions of dollars), what were the average assets during this period?

RELATING CONCEPTS (EXERCISES 49–56)

The following properties of series provide useful shortcuts for evaluating series.

If $a_1, a_2, a_3, \ldots, a_n$ and $b_1, b_2, b_3, \ldots, b_n$ are two sequences, and c is a constant, then for every positive integer n,

(a) $\displaystyle\sum_{i=1}^{n} c = nc$ **(b)** $\displaystyle\sum_{i=1}^{n} ca_i = c \sum_{i=1}^{n} a_i$

(c) $\displaystyle\sum_{i=1}^{n} (a_i + b_i) = \sum_{i=1}^{n} a_i + \sum_{i=1}^{n} b_i$ **(d)** $\displaystyle\sum_{i=1}^{n} (a_i - b_i) = \sum_{i=1}^{n} a_i - \sum_{i=1}^{n} b_i.$

Work these exercises in order.

49. Use property (c) to write $\displaystyle\sum_{i=1}^{6} (i^2 + 3i + 5)$ as the sum of three summations.

50. Use property (b) to rewrite the second summation from Exercise 49.

51. Use property (a) to rewrite the third summation from Exercise 49.

52. Rewrite $1 + 2 + 3 + 4 + \cdots + n = \dfrac{n(n + 1)}{2}$ in summation notation.

53. Rewrite $1^2 + 2^2 + 3^2 + 4^2 + \cdots + n^2 = \dfrac{n(n + 1)(2n + 1)}{6}$ in summation notation.

54. Use the summations you wrote in Exercises 52 and 53 and the properties given above to evaluate the three summations from Exercises 49–51. This gives the value of $\displaystyle\sum_{i=1}^{6} (i^2 + 3i + 5)$ without writing out all six terms.

55. Use the properties and summations given above to evaluate $\displaystyle\sum_{i=1}^{12} (i^2 - i).$

56. Use the properties and summations given above to evaluate $\displaystyle\sum_{i=1}^{20} (2 + i - i^2).$

Did you make the connection that evaluating series can be made easier using shortcuts that are true in general?

14.2 Arithmetic Sequences

OBJECTIVES

1. Find the common difference for an arithmetic sequence.

2. Find the general term of an arithmetic sequence.

3. Use an arithmetic sequence in an application.

4. Find any specified term or the number of terms of an arithmetic sequence.

5. Find the sum of a specified number of terms of an arithmetic sequence.

FOR EXTRA HELP

 SSG Sec. 14.2
SSM Sec. 14.2

 Pass the Test Software

 **InterAct Math Tutorial Software**

 Video 24

OBJECTIVE **1** **Find the common difference for an arithmetic sequence.** In this section we introduce a special type of sequence that has many applications.

> **Arithmetic Sequence**
>
> A sequence in which each term after the first differs from the preceding term by a constant amount is called an **arithmetic sequence** or **arithmetic progression.**

For example, the sequence

$$6, 11, 16, 21, 26, \ldots$$

is an arithmetic sequence, since the difference between any two adjacent terms is always 5. The number 5 is called the **common difference** of the arithmetic sequence. The common difference, d, is found by subtracting any pair of terms a_n and a_{n+1}. That is,

$$d = a_{n+1} - a_n.$$

EXAMPLE 1 **Finding the Common Difference**

Find d for the arithmetic sequence

$$-11, -4, 3, 10, 17, 24, \ldots.$$

Since the sequence is arithmetic, d is the difference between any two adjacent terms. Choosing the terms 10 and 17 gives

$$d = 17 - 10$$
$$= 7.$$

The terms -11 and -4 would give $d = -4 - (-11) = 7$, the same result.

EXAMPLE 2 **Writing the Terms of a Sequence from the First Term and Common Difference**

Write the first five terms of the arithmetic sequence with first term 3 and common difference -2.

The second term is found by adding -2 to the first term 3, getting 1. For the next term, add -2 to 1, and so on. The first five terms are

$$3, 1, -1, -3, -5.$$

OBJECTIVE **2** **Find the general term of an arithmetic sequence.** Generalizing from Example 2, if we know the first term, a_1, and the common difference, d, of an arithmetic sequence, then the sequence is completely defined as

$$a_1, \quad a_2 = a_1 + d, \quad a_3 = a_1 + 2d, \quad a_4 = a_1 + 3d, \ldots.$$

Writing the terms of the sequence in this way suggests the following rule.

General Term of an Arithmetic Sequence

The general term of an arithmetic sequence with first term a_1 and common difference d is

$$a_n = a_1 + (n - 1)d.$$

Since $a_n = a_1 + (n - 1)d = dn + (a_1 - d)$ is a linear function in n, any linear expression of the form $kn + c$, where k and c are real numbers, defines an arithmetic sequence.

E X A M P L E   3   Finding the General Term of an Arithmetic Sequence

Find the general term for the arithmetic sequence

$$-9, -6, -3, 0, 3, 6, \ldots .$$

Then use the general term to find a_{20}.

Here the first term is $a_1 = -9$. To find d, subtract any two adjacent terms. For example,

$$d = -3 - (-6) = 3.$$

Now find a_n.

$$
\begin{aligned}
a_n &= a_1 + (n - 1)d && \text{Formula for } a_n \\
&= -9 + (n - 1)(\mathbf{3}) && \text{Let } a_1 = -9,\ d = 3. \\
&= -9 + 3n - 3 && \text{Distributive property} \\
a_n &= 3n - 12 && \text{Combine terms.}
\end{aligned}
$$

Thus, the general term is $a_n = 3n - 12$. To find a_{20}, let $n = 20$.

$$a_{20} = 3(\mathbf{20}) - 12 = 60 - 12 = 48$$

O B J E C T I V E **3** Use an arithmetic sequence in an application.

E X A M P L E   4   Applying an Arithmetic Sequence

Howie Sorkin's uncle decides to start a fund for Howie's education. He makes an initial contribution of \$3000 and each month deposits an additional \$500. Thus, after one month there will be \$3000 + \$500 = \$3500. How much will there be after 24 months? (Disregard any interest.)

The contributions can be described using an arithmetic sequence. After n months, the fund will contain

$$a_n = 3000 + 500n \text{ dollars.}$$

To find the amount in the fund after 24 months, find a_{24}.

$$
\begin{aligned}
a_{24} &= 3000 + 500(\mathbf{24}) && \text{Let } n = 24. \\
&= 3000 + 12{,}000 && \text{Multiply.} \\
&= 15{,}000 && \text{Add.}
\end{aligned}
$$

The account will contain \$15,000 (disregarding interest) after 24 months.

OBJECTIVE 4 Find any specified term or the number of terms of an arithmetic sequence. The formula for the general term has four variables: a_n, a_1, n, and d. If we know any three of these, the formula can be used to find the value of the fourth variable. The next example shows how to find a particular term.

E X A M P L E 5   Finding a Specified Term

Find the indicated term for each of the following arithmetic sequences.

(a) $a_1 = -6$, $d = 12$; a_{15}
Use the formula $a_n = a_1 + (n-1)d$. Since we want $a_n = a_{15}$, $n = 15$.

$$\begin{aligned} a_{15} &= a_1 + (\mathbf{15}-1)d &&\text{Let } n = 15. \\ &= \mathbf{-6} + 14(\mathbf{12}) &&\text{Let } a_1 = -6, d = 12. \\ &= 162 \end{aligned}$$

(b) $a_5 = 2$ and $a_{11} = -10$; a_{17}
Any term can be found if a_1 and d are known. Use the formula for a_n with the two given terms.

$$\begin{aligned} a_5 &= a_1 + (5-1)d & a_{11} &= a_1 + (11-1)d \\ a_5 &= a_1 + 4d & a_{11} &= a_1 + 10d \\ 2 &= a_1 + 4d \quad a_5 = 2 & -10 &= a_1 + 10d \quad a_{11} = -10 \end{aligned}$$

This gives a system of two equations with two variables, a_1 and d. Find d by adding -1 times one equation to the other to eliminate a_1.

$$\begin{aligned} -10 &= a_1 + 10d \\ \underline{-2} &= \underline{-a_1 - 4d} \qquad \text{Multiply } 2 = a_1 + 4d \text{ by } -1. \\ -12 &= 6d \qquad\quad \text{Add.} \\ -2 &= d \qquad\qquad\quad\ \text{Divide by 6.} \end{aligned}$$

Now find a_1 by substituting -2 for d into either equation.

$$\begin{aligned} -10 &= a_1 + 10(\mathbf{-2}) &&\text{Let } d = -2. \\ -10 &= a_1 - 20 \\ \mathbf{10} &= a_1 \end{aligned}$$

Use the formula for a_n to find a_{17}.

$$\begin{aligned} a_{17} &= a_1 + (\mathbf{17}-1)d &&\text{Let } n = 17. \\ &= a_1 + 16d \\ &= \mathbf{10} + 16(\mathbf{-2}) &&\text{Let } a_1 = 10, d = -2. \\ &= -22 \end{aligned}$$

Sometimes we need to find out how many terms are in a sequence as shown in the following example.

E X A M P L E 6   Finding the Number of Terms in a Sequence

Find the number of terms in the arithmetic sequence

$$-8, -2, 4, 10, \ldots, 52.$$

Let n represent the number of terms in the sequence. Since $a_n = 52$, $a_1 = -8$, and $d = -2 - (-8) = 6$, use the formula $a_n = a_1 + (n - 1)d$ to find n. Substituting the known values into the formula gives

$$a_n = a_1 + (n - 1)d$$
$$52 = -8 + (n - 1)6 \qquad \text{Let } a_n = 52, a_1 = -8, d = 6.$$
$$52 = -8 + 6n - 6 \qquad \text{Distributive property}$$
$$66 = 6n \qquad \text{Combine terms.}$$
$$n = 11. \qquad \text{Divide by 6.}$$

The sequence has 11 terms.

OBJECTIVE 5 Find the sum of a specified number of terms of an arithmetic sequence. To find a formula for the sum, S_n, of the first n terms of an arithmetic sequence, we can write out the terms as

$$S_n = a_1 + (a_1 + d) + (a_1 + 2d) + \cdots + [a_1 + (n - 1)d].$$

This same sum can be written in reverse as

$$S_n = a_n + (a_n - d) + (a_n - 2d) + \cdots + [a_n - (n - 1)d].$$

Now add the corresponding terms of these two expressions for S_n to get

$$2S_n = (a_1 + a_n) + (a_1 + a_n) + (a_1 + a_n) + \cdots + (a_1 + a_n).$$

The right-hand side of this expression contains n terms, each equal to $a_1 + a_n$, so

$$2S_n = n(a_1 + a_n)$$
$$S_n = \frac{n}{2}(a_1 + a_n).$$

EXAMPLE 7 Finding the Sum of the First n Terms

Find the sum of the first five terms of the arithmetic sequence in which $a_n = 2n - 5$.
We can use the formula $S_n = \frac{n}{2}(a_1 + a_n)$ to find the sum of the first five terms. Here $n = 5$, $a_1 = 2(1) - 5 = -3$, and $a_5 = 2(5) - 5 = 5$. From the formula,

$$S_5 = \frac{5}{2}(-3 + 5) = \frac{5}{2}(2) = 5.$$

It is sometimes useful to express the sum of an arithmetic sequence, S_n, in terms of a_1 and d, the quantities that define the sequence. We can do this as follows. Since

$$S_n = \frac{n}{2}(a_1 + a_n) \qquad \text{and} \qquad a_n = a_1 + (n - 1)d,$$

by substituting the expression for a_n into the expression for S_n, we get

$$S_n = \frac{n}{2}(a_1 + [a_1 + (n - 1)d])$$

$$S_n = \frac{n}{2}[2a_1 + (n - 1)d].$$

The following summary gives both of the alternative forms that may be used to find the sum of the first n terms of an arithmetic sequence.

Sum of the First n Terms of an Arithmetic Sequence

The sum of the first n terms of the arithmetic sequence with first term a_1, nth term a_n, and common difference d is

$$S_n = \frac{n}{2}(a_1 + a_n) \qquad \text{or} \qquad S_n = \frac{n}{2}[2a_1 + (n-1)d].$$

E X A M P L E   8   Finding the Sum of the First n Terms

Find the sum of the first 8 terms of the arithmetic sequence having first term 3 and common difference -2.

Since the known values, $a_1 = 3$, $d = -2$, and $n = 8$, appear in the second formula for S_n, we use it.

$$S_n = \frac{n}{2}[2a_1 + (n-1)d]$$

$$S_8 = \frac{8}{2}[2(3) + (8-1)(-2)] \qquad \text{Let } a_1 = 3, d = -2, n = 8.$$

$$= 4[6 - 14]$$

$$= -32$$

As mentioned above, linear expressions of the form $kn + c$, where k and c are real numbers, define an arithmetic sequence. For example, the sequences defined by $a_n = 2n + 5$ and $a_n = n - 3$ are arithmetic sequences. For this reason,

$$\sum_{i=1}^{n} (ki + c)$$

represents the sum of the first n terms of an arithmetic sequence having first term $a_1 = k(1) + c = k + c$ and general term $a_n = k(n) + c = kn + c$. We can find this sum with the first formula for S_n given above, as shown in the next example.

E X A M P L E   9   Using S_n to Evaluate a Summation

Find $\displaystyle\sum_{i=1}^{12} (2i - 1)$.

This is the sum of the first 12 terms of the arithmetic sequence having $a_n = 2n - 1$. This sum, S_{12}, is found with the formula for S_n,

$$S_n = \frac{n}{2}(a_1 + a_n).$$

Here $n = 12$, $a_1 = 2(1) - 1 = 1$, $a_{12} = 2(12) - 1 = 23$. Substitute these values into the formula to get

$$S_{12} = \frac{12}{2}(1 + 23) = 6(24) = 144.$$

Figure 3 shows how a graphing calculator supports the result of Example 9.

Figure 3

14.2 EXERCISES

1. Using several examples, explain the meaning of *arithmetic sequence*.
2. Can any two terms of an arithmetic sequence be used to find the common difference? Explain.

If the given sequence is arithmetic, find the common difference, d. If the sequence is not arithmetic, say so. See Example 1.

3. 1, 2, 3, 4, 5, . . .   4. 2, 5, 8, 11, . . .
5. 2, −4, 6, −8, 10, −12, . . .   6. −6, −10, −14, −18, . . .
7. −10, −5, 0, 5, 10, . . .   8. 1, 2, 4, 7, 11, 16, . . .
9. 3.42, 5.57, 7.72, 9.87, . . .   10. $1, \frac{3}{2}, 2, \frac{5}{2}, 3, \frac{7}{2}, \ldots$
11. $-\frac{5}{3}, -1, -\frac{1}{3}, \frac{1}{3}, \ldots$ 12. $\frac{1}{2}, \frac{1}{3}, \frac{1}{4}, \frac{1}{5}, \frac{1}{6}, \ldots$

Use the formula for a_n to find the general term for each arithmetic sequence. See Example 3.

13. $a_1 = 2, d = 5$ 14. $a_1 = 5, d = -3$
15. $3, \frac{15}{4}, \frac{9}{2}, \frac{21}{4}, \ldots$ 16. 4, 14, 24, . . .
17. −3, 0, 3, . . .   18. −10, −5, 0, 5, 10, . . .

Find the indicated term for each arithmetic sequence. See Examples 2 and 5.

19. $a_1 = 4, d = 3; \quad a_{25}$ 20. $a_1 = 1, d = -\frac{1}{2}; \quad a_{12}$
21. 2, 4, 6, . . . ; a_{24} 22. 1, 5, 9, . . .; a_{50}
23. $a_{12} = -45, a_{10} = -37; \quad a_1$ 24. $a_{10} = -2, a_{15} = -8; \quad a_3$

Find the number of terms in each arithmetic sequence. See Example 6.

25. 3, 5, 7, . . . , 33   26. $2, \frac{3}{2}, 1, \frac{1}{2}, \ldots, -5$
27. $\frac{3}{4}, 3, \frac{21}{4}, \ldots, 12$ 28. 4, 1, −2, . . . , −32

29. In the formulas for S_n, what does n represent?
30. Explain when you would use each of the two formulas for S_n.

RELATING CONCEPTS (EXERCISES 31-34)

The following exercises show how to find the sum $1 + 2 + 3 + \cdots + 99 + 100$ in an ingenious way.

Work these exercises in order.

31. Consider the following:

$$S = \quad 1 + \quad 2 + \quad 3 + \cdots + 99 + 100$$
$$S = 100 + 99 + 98 + \cdots + \quad 2 + \quad 1.$$

Add the left sides of this equation. The result is _____. Add the columns on the right side. The sum _____ appears _____ times, so by multiplication, the sum of the right sides of the equations is _____.

32. Form an equation by setting the sum of the left sides equal to the sum of the right sides.

33. Solve the equation from Exercise 32 to find that the desired sum, S, is _____.

34. Find the sum $S = 1 + 2 + 3 + \cdots + 199 + 200$ using the procedure described in Exercises 31–33.

Did you make the connection that a sum of the form $1 + 2 + \cdots + n$ can be found using a pattern?

Find S_6 for each arithmetic sequence. See Examples 7 and 8.

35. $a_1 = 6, d = 3$ **36.** $a_1 = 5, d = 4$ **37.** $a_1 = 7, d = -3$

38. $a_1 = -5, d = -4$ **39.** $a_n = 4 + 3n$ **40.** $a_n = 9 + 5n$

Use a formula for S_n to evaluate each series. See Example 9.

41. $\displaystyle\sum_{i=1}^{10} (8i - 5)$ **42.** $\displaystyle\sum_{i=1}^{17} (i - 1)$ **43.** $\displaystyle\sum_{i=1}^{20} (2i - 5)$

44. $\displaystyle\sum_{i=1}^{10} \left(\frac{1}{2}i - 1\right)$ **45.** $\displaystyle\sum_{i=1}^{250} i$ **46.** $\displaystyle\sum_{i=1}^{2000} i$

Solve each applied problem. (Hint: Determine whether you need to find a specific term of a sequence or the sum of the terms of a sequence immediately after reading the problem.) See Example 4.

47. Nancy Bondy's aunt has promised to deposit $1 to her account on the first day of her birthday month, $2 on the second day, $3 on the third day, and so on for 30 days. How much will this amount to over the entire month?

48. Repeat Exercise 47, but assume that the deposits are $2, $4, $6, and so on, and that the month is February of a leap year.

49. Suppose that Randy Morgan is offered a job at $1600 per month with a guaranteed increase of $50 every 6 months for 5 years. What will his salary be at the end of this period of time?

50. Repeat Exercise 49, but assume that the starting salary is $2000 per month, and the guaranteed increase is $100 every 4 months for 3 years.

51. A seating section in a theater-in-the-round has 20 seats in the first row, 22 in the second row, 24 in the third row, and so on for 25 rows. How many seats are there in the last row? How many seats are there in the section'?

52. José Valdevielso has started on a fitness program. He plans to jog 10 minutes per day for the first week, and then add 10 minutes per day each week until he is jogging an hour each day. In which week will this occur? What is the total number of minutes he will run during the first four weeks?

53. A child builds with blocks, placing 35 blocks in the first row, 31 in the second row, 27 in the third row, and so on. Continuing this pattern, can she end with a row containing exactly 1 block? If not, how many blocks will the last row contain? How many rows can she build this way?

54. A stack of firewood has 28 pieces on the bottom, 24 on top of those, then 20, and so on. If there are 108 pieces of wood, how many rows are there? (*Hint: n ≤ 7.*)

RELATING CONCEPTS (EXERCISES 55–58)

Let f(x) = mx + b. **Work these exercises in order.**

55. Find $f(1), f(2)$, and $f(3)$.

56. Consider the sequence $f(1), f(2), f(3), \ldots$. Is it an arithmetic sequence?

57. If the sequence is arithmetic, what is the common difference?

58. What is a_n for the sequence described in Exercise 56?

Did you make the connection between a linear function and an arithmetic sequence?

14.3 Geometric Sequences

OBJECTIVES

1 Find the common ratio of a geometric sequence.

2 Find the general term of a geometric sequence.

3 Find any specified term of a geometric sequence.

4 Find the sum of a specified number of terms of a geometric sequence.

5 Apply the formula for the future value of an ordinary annuity.

6 Find the sum of an infinite number of terms of certain geometric sequences.

FOR EXTRA HELP

 SSG Sec. 14.3
SSM Sec. 14.3

Pass the Test Software

InterAct Math Tutorial Software

Video 24

In an arithmetic sequence, each term after the first is found by *adding* a fixed number to the previous term. A *geometric sequence* is defined as follows.

Geometric Sequence

A **geometric sequence** or **geometric progression** is a sequence in which each term after the first is a constant multiple of the preceding term.

OBJECTIVE 1 Find the common ratio of a geometric sequence. We find the constant multiplier, called the **common ratio,** by dividing any term after the first by the preceding term. That is, the common ratio is

$$r = \frac{a_{n+1}}{a_n}.$$

For example,

$$2, 6, 18, 54, 162, \ldots$$

is a geometric sequence in which the first term, a_1, is 2 and the common ratio is

$$r = \frac{6}{2} = \frac{18}{6} = \frac{54}{18} = \frac{162}{54} = 3.$$

EXAMPLE 1 Finding the Common Ratio

Find r for the geometric sequence

$$15, \frac{15}{2}, \frac{15}{4}, \frac{15}{8}, \ldots.$$

To find r, choose any two adjacent terms and divide the second one by the first. Choosing the second and third terms of the sequence,

$$r = \frac{a_3}{a_2} = \frac{15}{4} \div \frac{15}{2} = \frac{1}{2}.$$

Any other two adjacent terms could have been used to find r. Additional terms of the sequence can be found by multiplying each successive term by $\frac{1}{2}$.

OBJECTIVE **2** Find the general term of a geometric sequence. The general term a_n of a geometric sequence $a_1, a_2, a_3, \ldots$ is expressed in terms of a_1 and r by writing the first few terms as

$$a_1, \quad a_2 = a_1 r, \quad a_3 = a_1 r^2, \quad a_4 = a_1 r^3, \ldots,$$

which suggests the following rule.

General Term of a Geometric Sequence

The general term of the geometric sequence with first term a_1 and common ratio r is

$$a_n = a_1 r^{n-1}.$$

 CAUTION Be careful to use the correct order of operations when finding $a_1 r^{n-1}$. The value of r^{n-1} must be found first. Then multiply the result by a_1.

EXAMPLE 2 Finding the General Term

Find the general term of the sequence in Example 1.

The first term is $a_1 = 15$ and the common ratio is $r = \frac{1}{2}$. Substituting into the formula for the general term gives

$$a_n = a_1 r^{n-1} = 15\left(\frac{1}{2}\right)^{n-1},$$

the required general term. Notice that it is not possible to simplify further, because the exponent must be applied before the multiplication can be done.

OBJECTIVE **3** Find any specified term of a geometric sequence. We can use the formula for the general term to find any particular term.

EXAMPLE 3 Finding a Specified Term

Find the indicated term for each geometric sequence.

(a) $a_1 = 4, r = -3; \quad a_6$

Let $n = 6$. From the general term $a_n = a_1 r^{n-1}$,

$$\begin{aligned}
a_6 &= a_1 \cdot r^{6-1} && \text{Let } n = 6. \\
&= 4 \cdot (-3)^5 && \text{Let } a_1 = 4, r = -3. \\
&= -972. && \text{Evaluate } (-3)^5 \text{ first.}
\end{aligned}$$

(b) $\dfrac{3}{4}, \dfrac{3}{8}, \dfrac{3}{16}, \ldots ; \quad a_7$

Here, $r = \dfrac{1}{2}$, $a_1 = \dfrac{3}{4}$, and n is 7.

$$a_7 = \frac{3}{4} \cdot \left(\frac{1}{2}\right)^6 = \frac{3}{4} \cdot \frac{1}{64} = \frac{3}{256}$$

E X A M P L E   4   Writing the Terms of a Sequence

Write the first five terms of the geometric sequence whose first term is 5 and whose common ratio is $\frac{1}{2}$.

Using the formula $a_n = a_1 r^{n-1}$,

$$a_1 = 5,$$

$$a_2 = 5\left(\frac{1}{2}\right) = \frac{5}{2},$$

$$a_3 = 5\left(\frac{1}{2}\right)^2 = \frac{5}{4},$$

$$a_4 = 5\left(\frac{1}{2}\right)^3 = \frac{5}{8},$$

$$a_5 = 5\left(\frac{1}{2}\right)^4 = \frac{5}{16}.$$

OBJECTIVE 4   Find the sum of a specified number of terms of a geometric sequence. It is convenient to have a formula for the sum of the first n terms of a geometric sequence, S_n. We can develop a formula by first writing out S_n.

$$S_n = a_1 + a_1 r + a_1 r^2 + a_1 r^3 + \cdots + a_1 r^{n-1}$$

Next, we multiply both sides by r.

$$r S_n = a_1 r + a_1 r^2 + a_1 r^3 + a_1 r^4 + \cdots + a_1 r^n$$

We subtract the first result from the second.

$$r S_n - S_n = (a_1 r - a_1) + (a_1 r^2 - a_1 r) + (a_1 r^3 - a_1 r^2)$$
$$+ (a_1 r^4 - a_1 r^3) + \cdots + (a_1 r^n - a_1 r^{n-1})$$

Using the commutative and associative properties to rearrange the terms on the right, we get

$$r S_n - S_n = (a_1 r - a_1 r) + (a_1 r^2 - a_1 r^2)$$
$$+ (a_1 r^3 - a_1 r^3) + \cdots + (a_1 r^n - a_1) \qquad \text{Distributive property}$$

so, if $r \neq 1$,

$$S_n = \frac{a_1 r^n - a_1}{r - 1} = \frac{a_1(r^n - 1)}{r - 1}. \qquad \text{Divide by } r - 1.$$

A summary of this discussion follows.

Sum of the First n Terms of a Geometric Sequence

The sum of the first n terms of the geometric sequence with first term a_1 and common ratio r is

$$S_n = \frac{a_1(r^n - 1)}{r - 1} \quad (r \neq 1).$$

If $r = 1$, $S_n = a_1 + a_1 + a_1 + \cdots + a_1 = na_1$.

Multiplying the formula for S_n by $\frac{-1}{-1}$ gives us an alternative form that is sometimes preferable.

$$S_n = \frac{a_1(r^n - 1)}{r - 1} \cdot \frac{-1}{-1} = \frac{a_1(1 - r^n)}{1 - r}$$

EXAMPLE 5 Finding the Sum of the First n Terms

Find the sum of the first six terms of the geometric sequence with first term -2 and common ratio 3.

Substitute $n = 6$, $a_1 = -2$, and $r = 3$ into the formula for S_n.

$$S_n = \frac{a_1(r^n - 1)}{r - 1}$$

$$S_6 = \frac{-2(3^6 - 1)}{3 - 1} \qquad \text{Let } n = 6, a_1 = -2, r = 3.$$

$$= \frac{-2(729 - 1)}{2} \qquad \text{Evaluate the exponential.}$$

$$= -728$$

A series of the form

$$\sum_{i=1}^{n} a \cdot b^i$$

represents the sum of the first n terms of a geometric sequence having first term $a_1 = a \cdot b^1 = ab$ and common ratio b. The next example illustrates this form.

EXAMPLE 6 Using the Formula for S_n to Find a Summation

Find $\displaystyle\sum_{i=1}^{4} 3 \cdot 2^i$.

Since the series is in the form

$$\sum_{i=1}^{n} a \cdot b^i,$$

it represents the sum of the first n terms of the geometric sequence with $a_1 = a \cdot b^1$ and $r = b$. The sum is found by using the formula

$$S_n = \frac{a_1(r^n - 1)}{r - 1}.$$

Here $n = 4$. Also, $a_1 = 6$ and $r = 2$. Now substitute into the formula for S_n.

$$S_4 = \frac{6(2^4 - 1)}{2 - 1} \qquad \text{Let } n = 4, a_1 = 6, r = 2.$$

$$= \frac{6(16 - 1)}{1} \qquad \text{Evaluate } 2^4.$$

$$= 90$$

Figure 4 shows how a graphing calculator can store the terms in a list, and then find the sum of these terms. This supports the result of Example 6.

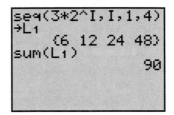

Figure 4

OBJECTIVE **5** **Apply the formula for the future value of an ordinary annuity.** A sequence of equal payments made at equal periods of time is called an **annuity.** If the payments are made at the end of the time period, and if the frequency of payments is the same as the frequency of compounding, the annuity is called an **ordinary annuity.** The time between payments is the **payment period,** and the time from the beginning of the first payment period to the end of the last period is called the **term of the annuity.** The **future value of the annuity,** the final sum on deposit, is defined as the sum of the compound amounts of all the payments, compounded to the end of the term.

For example, suppose $1500 is deposited at the end of the year for the next 6 years in an account paying 8% per year compounded annually.

To find the future value of this annuity, look separately at each of the $1500 payments. The first of these payments will produce a compound amount of

$$1500(1 + .08)^5 = 1500(1.08)^5.$$

Use 5 as the exponent instead of 6 since the money is deposited at the *end* of the first year and earns interest for only 5 years. The second payment of $1500 will produce a compound amount of $1500(1.08)^4$. Continuing in this way and finding the sum of all the terms gives

$$1500(1.08)^5 + 1500(1.08)^4 + 1500(1.08)^3 + 1500(1.08)^2 + 1500(1.08)^1 + 1500.$$

(The last payment earns no interest at all.) Reading this in reverse order, we see that it is just the sum of the first six terms of a geometric sequence with $a_1 = 1500$, $r = 1.08$, and $n = 6$. Therefore, the sum is

$$\frac{a_1(r^n - 1)}{r - 1} = \frac{1500[(1.08)^6 - 1]}{1.08 - 1} = 11{,}003.89$$

or $11,003.89.

We state the following formula without proof.

Future Value of an Ordinary Annuity

$$S = R\left[\frac{(1+i)^n - 1}{i}\right]$$

where

S is future value,
R is the payment at the end of each period,
i is the interest rate per period,
n is the number of periods.

EXAMPLE 7 Applying the Formula for the Future Value of an Annuity

(a) Rocky Rhodes is an athlete who feels that his playing career will last 7 years. To prepare for his future, he deposits $22,000 at the end of each year for 7 years in an account paying 6% compounded annually. How much will he have on deposit after 7 years?

His payments form an ordinary annuity with $R = 22,000$, $n = 7$, and $i = .06$. The future value of this annuity (by the formula above) is

$$S = 22,000\left[\frac{(1.06)^7 - 1}{.06}\right] = 184,664.43, \quad \text{Use a calculator.}$$

or $184,664.43.

(b) Experts say that the baby boom generation (born between 1946 and 1960) cannot count on a company pension or Social Security to provide a comfortable retirement, as their parents did. It is recommended that they start to save early and regularly. Judy Zahrndt, a baby boomer, has decided to deposit $200 at the end of each month in an account that pays interest of 7.2% compounded monthly for retirement in 20 years. How much will be in the account at that time?

Because the interest is compounded monthly, $i = \frac{.072}{12}$. Also, $R = 200$ and $n = 12(20)$. The future value is

$$S = 200\left[\frac{\left(1 + \frac{.072}{12}\right)^{12(20)} - 1}{\frac{.072}{12}}\right] = 106,752.47,$$

or $106,752.47.

OBJECTIVE 6 Find the sum of an infinite number of terms of certain geometric sequences. Now, consider an infinite geometric sequence such as

$$\frac{1}{3}, \frac{1}{6}, \frac{1}{12}, \frac{1}{24}, \frac{1}{48}, \ldots$$

Can the sum of the terms of such a sequence be found somehow? The sum of the first two terms is

$$S_2 = \frac{1}{3} + \frac{1}{6} = \frac{1}{2} = .5.$$

In a similar manner,

$$S_3 = S_2 + \frac{1}{12} = \frac{1}{2} + \frac{1}{12} = \frac{7}{12} \approx .583,$$

$$S_4 = S_3 + \frac{1}{24} = \frac{7}{12} + \frac{1}{24} = \frac{15}{24} = .625,$$

$$S_5 = \frac{31}{48} \approx .64583,$$

$$S_6 = \frac{21}{32} = .65625,$$

$$S_7 = \frac{127}{192} \approx .6614583.$$

Each term of the geometric sequence is smaller than the preceding one, so each additional term is contributing less and less to the sum. In decimal form (to the nearest thousandth) the first seven terms and the tenth term are given below.

Term	a_1	a_2	a_3	a_4	a_5	a_6	a_7	a_{10}
Value	.333	.167	.083	.042	.021	.010	.005	.001

As the table suggests, the value of a term gets closer and closer to zero as the number of the term increases. To express this idea, we say that as n increases without bound (written $n \to \infty$), the limit of the term a_n is zero, written

$$\lim_{n \to \infty} a_n = 0.$$

A number that can be defined as the sum of an infinite number of terms of a geometric sequence can be found by starting with the expression for the sum of a finite number of terms:

$$S_n = \frac{a_1(r^n - 1)}{r - 1}.$$

If $|r| < 1$, then as n increases without bound the value of r^n gets closer and closer to zero. For example, in the infinite sequence discussed above, $r = \frac{1}{2} = .5$. The chart below shows how $r^n = (.5)^n$, given to the nearest thousandth, gets smaller as n increases.

n	1	2	3	4	5	6	7	10
r^n	.5	.25	.125	.063	.031	.016	.008	.001

As r^n approaches 0, $r^n - 1$ approaches $0 - 1 = -1$, and S_n approaches the quotient $\frac{-a_1}{r - 1}$. Thus,

$$\lim_{r^n \to 0} S_n = \lim_{r^n \to 0} \frac{a_1(r^n - 1)}{r - 1}$$

$$= \frac{a_1(0 - 1)}{r - 1}$$

$$= \frac{-a_1}{r - 1} = \frac{a_1}{1 - r}.$$

This limit is defined to be the sum of the infinite geometric sequence:

$$a_1 + a_1r + a_1r^2 + a_1r^3 + \cdots = \frac{a_1}{1-r}, \quad \text{if } |r| < 1.$$

What happens if $|r| > 1$? For example, suppose the sequence is

$$6, 12, 24, \ldots, 3(2)^n, \ldots.$$

In this kind of sequence, as n increases, the value of r^n also increases and so does the sum S_n. Since each new term adds a larger and larger amount to the sum, there is no limit to the value of S_n, and the sum S_n does not exist. A similar situation exists if $r = 1$.

In summary, the sum of the terms of an infinite geometric sequence is as follows.

Sum of the Terms of an Infinite Geometric Sequence

The sum S of the terms of an infinite geometric sequence with first term a_1 and common ratio r, where $|r| < 1$, is

$$S = \frac{a_1}{1-r}.$$

If $|r| \geq 1$, the sum does not exist.

EXAMPLE 8 Finding the Sum of the Terms of an Infinite Geometric Sequence

Find the sum of the terms of the infinite geometric sequence with $a_1 = 3$ and $r = -\frac{1}{3}$.

From the rule above, the sum is

$$S = \frac{a_1}{1-r} = \frac{3}{1-(-1/3)}$$

$$= \frac{3}{4/3}$$

$$= \frac{9}{4}.$$

In summation notation, the sum of an infinite geometric sequence is written as

$$\sum_{i=1}^{\infty} a_i.$$

For instance, the sum in Example 8 would be written

$$\sum_{i=1}^{\infty} 3\left(-\frac{1}{3}\right)^{i-1}.$$

EXAMPLE 9 Finding the Sum of the Terms of an Infinite Geometric Series

Find $\displaystyle\sum_{i=1}^{\infty} \left(\frac{1}{2}\right)^i$.

This is the infinite geometric series

$$\frac{1}{2} + \frac{1}{4} + \frac{1}{8} + \cdots,$$

with $a_1 = \frac{1}{2}$ and $r = \frac{1}{2}$. Since $|r| < 1$, we find the sum as follows.

$$S = \frac{a_1}{1-r} = \frac{\frac{1}{2}}{1-\frac{1}{2}} = \frac{\frac{1}{2}}{\frac{1}{2}} = 1$$

14.3 EXERCISES

1. Using several examples, explain the meaning of *geometric sequence*.

2. Explain why the sequence 5, 5, 5, 5, . . . can be considered either arithmetic or geometric.

If the given sequence is geometric, find the common ratio, r. If the sequence is not geometric, say so. See Example 1.

3. 4, 8, 16, 32, . . .

4. 5, 15, 45, 135, . . .

5. $\frac{1}{3}, \frac{2}{3}, \frac{3}{3}, \frac{4}{3}, \frac{5}{3}, \ldots$

6. $\frac{1}{3}, \frac{2}{3}, \frac{4}{3}, \frac{8}{3}, \ldots$

7. 1, −3, 9, −27, 81, . . .

8. 1, −3, 7, −11, . . .

9. $1, -\frac{1}{2}, \frac{1}{4}, -\frac{1}{8}, \frac{1}{16}, \ldots$

10. $\frac{2}{3}, \frac{2}{15}, \frac{2}{75}, \frac{2}{375}, \ldots$

Find a general term for each geometric sequence. See Example 2.

11. 5, 10, . . .

12. −2, −6, . . .

13. $\frac{1}{9}, \frac{1}{3}, \ldots$

14. $-3, \frac{3}{2}, \ldots$

15. 10, −2, . . .

16. −4, 8, . . .

Find the indicated term for each geometric sequence. See Example 3.

17. 2, 10, 50, . . . ; a_{10}

18. −1, −3, −9, . . . ; a_{15}

19. $\frac{1}{2}, \frac{1}{6}, \frac{1}{18}, \ldots$; a_{12}

20. $\frac{2}{3}, -\frac{1}{3}, \frac{1}{6}, \ldots$; a_{18}

21. $a_3 = \frac{1}{2}, a_7 = \frac{1}{32}$; a_{25}

22. $a_5 = 48, a_8 = -384$; a_{10}

RELATING CONCEPTS (EXERCISES 23–26)

In Chapter 1 we learned that any repeating decimal is a rational number; that is, it can be expressed as a quotient of integers. Thus, the repeating decimal

$$.99999 \ldots,$$

an endless string of 9s, must be a rational number.

Work Exercises 23–26 in order to discover the surprising simplest form of this rational number.

23. Use long division or your previous experience to write a repeating decimal representation for $\frac{1}{3}$.

RELATING CONCEPTS (EXERCISES 23-26) (CONTINUED)

24. Use long division or your previous experience to write a repeating decimal representation for $\frac{2}{3}$.

25. Because $\frac{1}{3} + \frac{2}{3} = 1$, the sum of the decimal representations in Exercises 23 and 24 must also equal 1. Line up the decimals in the usual vertical method for addition, and obtain the repeating decimal result. The value of this decimal is exactly 1.

26. The repeating decimal .99999 . . . can be written as the sum of the terms of a geometric sequence with $a_1 = .9$ and $r = .1$:

$$.99999 \ldots = .9 + .9(.1) + .9(.1)^2 + .9(.1)^3 + .9(.1)^4 + .9(.1)^5 + \ldots$$

Since $|.1| < 1$, this sum can be found using the formula $S = \frac{a_1}{1 - r}$. Use this formula to support the result you found another way in Exercises 23–25.

Did you make the connection that, although it may not seem to be true, the value of .99999 . . . is 1?

Use the formula for S_n to find the sum for each geometric sequence. See Examples 5 and 6. In Exercises 29–34, give the answer to the nearest thousandth.

27. $\dfrac{1}{3}, \dfrac{1}{9}, \dfrac{1}{27}, \dfrac{1}{81}, \dfrac{1}{243}$

28. $\dfrac{4}{3}, \dfrac{8}{3}, \dfrac{16}{3}, \dfrac{32}{3}, \dfrac{64}{3}, \dfrac{128}{3}$

29. $-\dfrac{4}{3}, -\dfrac{4}{9}, -\dfrac{4}{27}, -\dfrac{4}{81}, -\dfrac{4}{243}, -\dfrac{4}{729}$

30. $\dfrac{5}{16}, -\dfrac{5}{32}, \dfrac{5}{64}, -\dfrac{5}{128}, \dfrac{5}{256}$

31. $\displaystyle\sum_{i=1}^{7} 4\left(\dfrac{2}{5}\right)^i$

32. $\displaystyle\sum_{i=1}^{8} 5\left(\dfrac{2}{3}\right)^i$

33. $\displaystyle\sum_{i=1}^{10} (-2)\left(\dfrac{3}{5}\right)^i$

34. $\displaystyle\sum_{i=1}^{6} (-2)\left(-\dfrac{1}{2}\right)^i$

Solve each problem involving an ordinary annuity. See Example 7.

35. A father opened a savings account for his daughter on the day she was born, depositing $1000. Each year on her birthday he deposits another $1000, making the last deposit on her twenty-first birthday. If the account pays 9.5% interest compounded annually, how much is in the account at the end of the day on the daughter's twenty-first birthday?

36. A 45-year-old man puts $1000 in a retirement account at the end of each quarter $\left(\frac{1}{4}\text{ of a year}\right)$ until he reaches the age of 60. If the account pays 11% annual interest compounded quarterly, how much will be in the account at that time?

37. At the end of each quarter a 50-year-old woman puts $1200 in a retirement account that pays 7% interest compounded quarterly. When she reaches age 60, she withdraws the entire amount and places it in a mutual fund that pays 9% interest compounded monthly. From then on she deposits $300 in the mutual fund at the end of each month. How much is in the account when she reaches age 65?

38. John Bray deposits $10,000 at the beginning of each year for 12 years in an account paying 5% compounded annually. He then puts the total amount on deposit in another account paying 6% compounded semiannually for another 9 years. Find the final amount on deposit after the entire 21-year period.

Find the sum, if it exists, of the terms of each infinite geometric sequence. See Examples 8 and 9.

39. $a_1 = 6, r = \dfrac{1}{3}$

40. $a_1 = 10, r = \dfrac{1}{5}$

41. $a_1 = 1000, r = -\dfrac{1}{10}$

42. $a_1 = 8500, r = \dfrac{3}{5}$

43. $\displaystyle\sum_{i=1}^{\infty} \dfrac{9}{8}\left(-\dfrac{2}{3}\right)^i$

44. $\displaystyle\sum_{i=1}^{\infty} \dfrac{3}{5}\left(\dfrac{5}{6}\right)^i$

45. $\displaystyle\sum_{i=1}^{\infty} \dfrac{12}{5}\left(\dfrac{5}{4}\right)^i$

46. $\displaystyle\sum_{i=1}^{\infty} \left(-\dfrac{16}{3}\right)\left(-\dfrac{9}{8}\right)^i$

Solve each application. (Hint: Determine whether you need to find a specific term of a sequence or the sum of the terms of a sequence immediately after reading the problem.)

47. A certain ball when dropped from a height rebounds $\dfrac{3}{5}$ of the original height. How high will the ball rebound after the fourth bounce if it was dropped from a height of 10 feet?

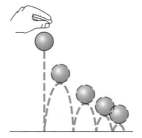

48. A fully wound yo-yo has a string 40 inches long. It is allowed to drop and on its first rebound, it returns to a height 15 inches lower than its original height. Assuming this "rebound ratio" remains constant until the yo-yo comes to rest, how far does it travel on its third trip up the string?

49. A particular substance decays in such a way that it loses half its weight each day. In how many days will 256 grams of the substance be reduced to 32 grams? How much of the substance is left after 10 days?

50. A tracer dye is injected into a system with an input and an excretion. After one hour $\dfrac{2}{3}$ of the dye is left. At the end of the second hour $\dfrac{2}{3}$ of the remaining dye is left, and so on. If one unit of the dye is injected, how much is left after 6 hours?

51. In a certain community the consumption of electricity has increased about 6% per year.
  (a) If a community uses 1.1 billion units of electricity now, how much will it use five years from now?
  (b) Find how many years it will take for the consumption to double.

52. Suppose the community in Exercise 51 reduces its increase in consumption to 2% per year.
  (a) How much will it use five years from now?
  (b) Find the number of years for the consumption to double.

53. A machine depreciates by $\frac{1}{4}$ of its value each year. If it cost $50,000 new, what is its value after 8 years?

54. Refer to Exercise 48. Theoretically, how far does the yo-yo travel before coming to rest?

RELATING CONCEPTS (EXERCISES 55-58)

Let $g(x) = ab^x$. ***Work Exercises 55–58 in order.***

55. Find $g(1)$, $g(2)$, and $g(3)$.

56. Consider the sequence $g(1), g(2), g(3), \ldots$. Is it a geometric sequence?

57. If the sequence is geometric, what is the common ratio?

58. What is a_n for the sequence described in Exercise 56?

Did you make the connection between an exponential function and a geometric sequence?

14.4 The Binomial Theorem

OBJECTIVES

1 Expand a binomial raised to a power.

2 Find any specified term of the expansion of a binomial.

FOR EXTRA HELP

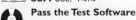
SSG Sec. 14.4
SSM Sec. 14.4

Pass the Test Software

InterAct Math
  Tutorial Software

Video 24

OBJECTIVE 1 **Expand a binomial raised to a power.** Writing out the binomial expression $(x + y)^n$ for nonnegative integer values of n gives a family of expressions that is important in the study of mathematics and its applications. For example,

$$(x + y)^0 = 1,$$
$$(x + y)^1 = x + y,$$
$$(x + y)^2 = x^2 + 2xy + y^2,$$
$$(x + y)^3 = x^3 + 3x^2y + 3xy^2 + y^3,$$
$$(x + y)^4 = x^4 + 4x^3y + 6x^2y^2 + 4xy^3 + y^4,$$
$$(x + y)^5 = x^5 + 5x^4y + 10x^3y^2 + 10x^2y^3 + 5xy^4 + y^5.$$

Inspection shows that these expansions follow a pattern. By identifying the pattern we can write a general expression for $(x + y)^n$.

First, if n is a positive integer, each expansion after $(x + y)^0$ begins with x raised to the same power to which the binomial is raised. That is, the expansion of $(x + y)^1$ has a first term of x^1, the expansion of $(x + y)^2$ has a first term of x^2, the expansion of $(x + y)^3$ has a first term of x^3, and so on. Also, the last term in each expansion is y to this same power, so the expansion of $(x + y)^n$ should begin with the term x^n and end with the term y^n.

The exponents on x decrease by one in each term after the first, while the exponents on y, beginning with y in the second term, increase by one in each succeeding term. Thus the *variables* in the expansion of $(x + y)^n$ have the following pattern.

$$x^n, \quad x^{n-1}y, \quad x^{n-2}y^2, \quad x^{n-3}y^3, \ldots, xy^{n-1}, \quad y^n$$

This pattern suggests that the sum of the exponents on x and y in each term is n. For example, in the third term above, the variable part is $x^{n-2}y^2$ and the sum of the exponents, $n - 2$ and 2, is n.

Now examine the pattern for the *coefficients* in the terms of the expansions shown above. Writing the coefficients alone in a triangular pattern gives **Pascal's triangle,** in honor of the seventeenth-century mathematician Blaise Pascal, one of the first to use it extensively.

Pascal's Triangle

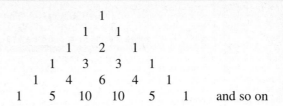

and so on

Arranging the coefficients in this way shows that each number in the triangle is the sum of the two numbers just above it (one to the right and one to the left). For example, starting with 1 as the first row, in the fifth row from the top, 1 is the sum of 1 (the only number above it), 4 is the sum of 1 and 3, while 6 is the sum of 3 and 3, and so on.

We get the coefficients for $(x + y)^6$ by attaching the seventh row to the table by adding pairs of numbers from the sixth row.

$$1 \quad 6 \quad 15 \quad 20 \quad 15 \quad 6 \quad 1$$

Use these coefficients to expand $(x + y)^6$ as

$$(x + y)^6 = x^6 + 6x^5y + 15x^4y^2 + 20x^3y^3 + 15x^2y^4 + 6xy^5 + y^6.$$

CONNECTIONS

Over the years, many interesting patterns have been discovered in Pascal's triangle. In the figure below, the triangular array is written in a different form. The indicated sums along the diagonals shown are the terms of the *Fibonacci sequence,* mentioned in Section 14.1. The presence of this sequence in the triangle apparently was not recognized by Pascal.

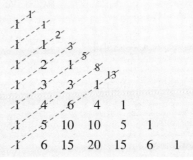

CONNECTIONS (CONTINUED)

Triangular numbers are found by counting the number of points in triangular arrangements of points. The first few triangular numbers are shown below.

Triangular numbers

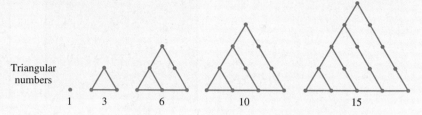

1   3   6   10   15

The number of points in these figures form the sequence 1, 3, 6, 10, . . . , a sequence that is found in Pascal's triangle, as shown in the next figure.

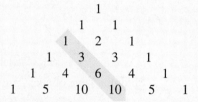

FOR DISCUSSION OR WRITING

1. Predict the next two numbers in the sequence of sums of the diagonals of Pascal's triangle.

2. Predict the next five numbers in the list of triangular numbers.

3. Describe other sequences that can be found in Pascal's triangle.

Although it is possible to use Pascal's triangle to find the coefficients of $(x + y)^n$ for any positive integer value of n, it is impractical for large values of n. A more efficient way to determine these coefficients is needed. It is helpful to use the following notational shorthand. The symbol $n!$ (read "n factorial") is defined as follows.

n Factorial ($n!$)

For any positive integer n,

$$n(n - 1)(n - 2)(n - 3) \cdots (2)(1) = n!.$$

For example,

$$3! = 3 \cdot 2 \cdot 1 = 6 \quad \text{and} \quad 5! = 5 \cdot 4 \cdot 3 \cdot 2 \cdot 1 = 120.$$

From the definition of n factorial, $n[(n - 1)!] = n!$. If $n = 1$, then $1(0!) = 1! = 1$. Because of this, 0! is defined as

$$0! = 1.$$

Scientific and graphing calculators have the capability of computing factorials. The three factorial expressions above are shown in Figure 5(a). Figure 5(b) shows some larger factorials.

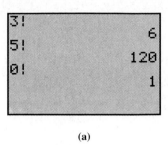

(a)

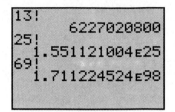

A graphing calculator with a 10-digit display will give the exact value of $n!$ for $n \le 13$ and approximate values of $n!$ for $14 \le n \le 69$.

(b)

Figure 5

EXAMPLE 1   Evaluating Expressions with $n!$

Find the value of each of the following.

(a) $\dfrac{5!}{4!\,1!} = \dfrac{5 \cdot 4 \cdot 3 \cdot 2 \cdot 1}{(4 \cdot 3 \cdot 2 \cdot 1)(1)} = 5$

(b) $\dfrac{5!}{3!\,2!} = \dfrac{5 \cdot 4 \cdot 3 \cdot 2 \cdot 1}{(3 \cdot 2 \cdot 1)(2 \cdot 1)} = \dfrac{5 \cdot 4}{2 \cdot 1} = 10$

(c) $\dfrac{6!}{3!\,3!} = \dfrac{6 \cdot 5 \cdot 4 \cdot 3 \cdot 2 \cdot 1}{(3 \cdot 2 \cdot 1)(3 \cdot 2 \cdot 1)} = \dfrac{6 \cdot 5 \cdot 4}{3 \cdot 2 \cdot 1} = 20$

(d) $\dfrac{4!}{4!\,0!} = \dfrac{4 \cdot 3 \cdot 2 \cdot 1}{(4 \cdot 3 \cdot 2 \cdot 1)(1)} = 1$

Now look again at the coefficients of the expansion

$$(x + y)^5 = x^5 + 5x^4y + 10x^3y^2 + 10x^2y^3 + 5xy^4 + y^5.$$

The coefficient of the second term is 5 and the exponents on the variables in that term are 4 and 1. From Example 1(a), $5!/(4!\,1!) = 5$. The coefficient of the third term is 10, and the exponents are 3 and 2. From Example 1(b), $5!/(3!\,2!) = 10$. Similar results hold true for the remaining terms. The first term can be written as $1x^5y^0$ and the last term can be written as $1x^0y^5$. Then the coefficient of the first term should be $5!/(5!\,0!) = 1$, and the coefficient of the last term would be $5!/(0!\,5!) = 1$. Generalizing, the coefficient for a term of $(x + y)^n$ in which the variable part is $x^r y^{n-r}$ will be

$$\frac{n!}{r!(n - r)!}.$$

 NOTE The denominator factorials in the coefficient of a term are the same as the exponents on the variables in that term.

The expression $\dfrac{n!}{r!(n - r)!}$ is often represented by the symbol $_nC_r$. This comes from the fact that if we choose *combinations* of n things taken r at a time, the result is given

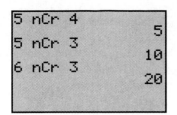

Figure 6

by that expression. A graphing calculator can evaluate this expression for particular values of n and r. Figure 6 shows how a calculator evaluates $_5C_4$, $_5C_3$, and $_6C_3$. Compare these results to parts (a), (b), and (c) of Example 1.

Summarizing this work gives the **binomial theorem,** or the **general binomial expansion.**

Binomial Theorem

For any positive integer n,

$$(x + y)^n = x^n + \frac{n!}{(n-1)!\,1!}x^{n-1}y + \frac{n!}{(n-2)!\,2!}x^{n-2}y^2$$

$$+ \frac{n!}{(n-3)!\,3!}x^{n-3}y^3 + \cdots + \frac{n!}{1!(n-1)!}xy^{n-1} + y^n.$$

The binomial theorem can be written in summation notation as

$$(x + y)^n = \sum_{i=0}^{n} \frac{n!}{(n-i)!\,i!}x^{n-i}y^i.$$

 The letter i is used here instead of r because we are using summation notation. It is not the imaginary number i.

┌ E X A M P L E   2   Using the Binomial Theorem

Expand $(2m + 3)^4$.

$$(2m + 3)^4 = (2m)^4 + \frac{4!}{3!\,1!}(2m)^3(3) + \frac{4!}{2!\,2!}(2m)^2(3)^2 + \frac{4!}{1!\,3!}(2m)(3)^3 + 3^4$$

$$= 16m^4 + 4(8m^3)(3) + 6(4m^2)(9) + 4(2m)(27) + 81$$

$$= 16m^4 + 96m^3 + 216m^2 + 216m + 81$$

┌ E X A M P L E   3   Using the Binomial Theorem

Expand $\left(a - \dfrac{b}{2}\right)^5$.

$$\left(a - \frac{b}{2}\right)^5 = a^5 + \frac{5!}{4!\,1!}a^4\left(-\frac{b}{2}\right) + \frac{5!}{3!\,2!}a^3\left(-\frac{b}{2}\right)^2 + \frac{5!}{2!\,3!}a^2\left(-\frac{b}{2}\right)^3$$

$$+ \frac{5!}{1!\,4!}a\left(-\frac{b}{2}\right)^4 + \left(-\frac{b}{2}\right)^5$$

$$= a^5 + 5a^4\left(-\frac{b}{2}\right) + 10a^3\left(\frac{b^2}{4}\right) + 10a^2\left(-\frac{b^3}{8}\right)$$

$$+ 5a\left(\frac{b^4}{16}\right) + \left(-\frac{b^5}{32}\right)$$

$$= a^5 - \frac{5}{2}a^4b + \frac{5}{2}a^3b^2 - \frac{5}{4}a^2b^3 + \frac{5}{16}ab^4 - \frac{1}{32}b^5$$

> When the binomial is the *difference* of two terms as in Example 3, the signs of the terms in the expansion will alternate. Those terms with odd exponents on the second variable expression $\left(-\frac{b}{2}\right.$ in Example 3$)$ will be negative, while those with even exponents on the second variable expression will be positive.

OBJECTIVE 2 Find any specified term of the expansion of a binomial. Any single term of a binomial expansion can be determined without writing out the whole expansion. For example, if $n \geq 10$, the tenth term of $(x + y)^n$ has y raised to the ninth power (since y has the power of 1 in the second term, the power of 2 in the third term, and so on). Since the exponents on x and y in any term must have a sum of n, the exponent on x in the tenth term is $n - 9$. These quantities, 9 and $n - 9$, determine the factorials in the denominator of the coefficient. Thus,

$$\frac{n!}{(n-9)!\,9!}\,x^{n-9}y^9$$

is the tenth term of $(x + y)^n$. A generalization of this idea follows.

---

r th Term of the Binomial Expansion

If $n \geq r - 1$, the rth term of the expansion of $(x + y)^n$ is

$$\frac{n!}{[n-(r-1)]!(r-1)!}\,x^{n-(r-1)}y^{r-1}.$$

This general expression is confusing. Remember to start with the exponent on y, which is 1 less than the term number r. Then subtract that exponent from n to get the exponent on x : $n - (r - 1)$. The two exponents are then used as the factorials in the denominator of the coefficient.

EXAMPLE 4 Finding a Single Term of a Binomial Expansion

Find the fourth term of $(a + 2b)^{10}$.

In the fourth term, $2b$ has an exponent of $4 - 1 = 3$ and a has an exponent of $10 - 3 = 7$. The fourth term is

$$\frac{10!}{7!\,3!}\,(a^7)(2b)^3 = \frac{10 \cdot 9 \cdot 8}{3 \cdot 2 \cdot 1}\,(a^7)(8b^3)$$
$$= 120a^7(8b^3)$$
$$= 960a^7b^3.$$

14.4 EXERCISES

Evaluate each of the following. See Example 1.

1. $2!$

2. $7!$

3. $\dfrac{6!}{4!\,2!}$

4. $\dfrac{7!}{3!\,4!}$

5. $_6C_2$

6. $_7C_4$

7. $\dfrac{4!}{0!\,4!}$

8. $\dfrac{5!}{5!\,0!}$

TECHNOLOGY INSIGHTS (EXERCISES 9-12)

Predict the answer that the calculator will display for the given entry.

9.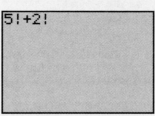

10.

11.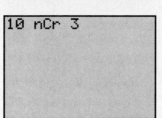

12.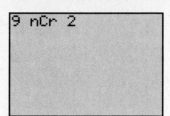

Use the binomial theorem to expand each of the following. See Examples 2 and 3.

13. $(m + n)^4$

14. $(x + r)^5$

15. $(a - b)^5$

16. $(p - q)^4$

17. $(2x + 3)^3$

18. $\left(\dfrac{x}{2} - y\right)^4$

19. $\left(\dfrac{x}{3} + 2y\right)^5$

20. $(x^2 + 1)^4$

21. $(mx - n^2)^3$

22. $(2p^2 - q^2)^3$

Write the first four terms of each binomial expansion.

23. $(r + 2s)^{12}$

24. $(m - n)^{20}$

25. $(3x - y)^{14}$

26. $(2p + 3q)^{11}$

27. $(t^2 + u^2)^{10}$

28. $(x^2 - y^2)^{15}$

Find the indicated term of each binomial expansion. See Example 4.

29. $(2m + n)^{10}$;   fourth term

30. $(a - 3b)^{12}$;   fifth term

31. $\left(x + \dfrac{y}{2}\right)^8$;   seventh term

32. $(3p - 2q)^{15}$;   eighth term

33. $(k - 1)^9$;   third term

34. $(-4 - s)^{11}$;   fourth term

35. The middle term of $(x^2 - 2y)^6$

36. The middle term of $(m^3 + 3)^8$

37. The term with x^9y^4 in $(3r^3 - 4y^2)^5$

38. The term with x^{10} in $\left(x^3 - \dfrac{2}{x}\right)^6$

CHAPTER 14 GROUP ACTIVITY

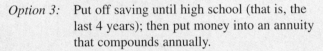

▦ Investing for the Future

Objective: Calculate compound interest; understand the effects of compounding monthly, quarterly and annually.

In this chapter you have seen many different types of financing and investing options including loans, mutual funds, savings accounts, and annuities. In this activity you will analyze annuities with different periods of compound interest.

Consider a family that has a 10-year-old child, for whom they want to save for college. The family is considering three different savings options.

Option 1: An annuity that compounds quarterly.

Option 2: An annuity that compounds monthly.

Option 3: Put off saving until high school (that is, the last 4 years); then put money into an annuity that compounds annually.

A. Have each member of your group calculate total savings for one of the three options. Use the following formula, where S is future value, R is payment amount made each *period, i* is annual interest rate divided by the number of periods per year, n is total number of compounding periods, along with the specific information given below for each option.

$$S = R\left[\frac{(1 + i)^n - 1}{i}\right]$$

Option 1: The family plans to save $300 per quarter (3 months), the interest rate is 5%, and the annuity compounds quarterly. Savings will be for 8 years.

Option 2: The family plans to save $100 per month, the interest rate is 5%, and the annuity compounds monthly. Again, savings will be for 8 years.

Option 3: The family plans to wait and save only the last 4 years. They will save $2400 a year, the interest rate is 5%, and the annuity compounds annually.

B. Compare your answers.

1. How much is being invested using each option?

2. Which option resulted in the largest amount of savings?

3. Explain why this option produced more savings.

4. What other considerations might be involved in deciding how to save?

CHAPTER 14 SUMMARY

KEY TERMS

14.1 sequence
infinite sequence
terms of a sequence
general term
finite sequence
series
summation notation
index of summation

arithmetic mean
(average)
14.2 arithmetic
sequence
(arithmetic
progression)
common difference
14.3 geometric sequence

(geometric
progression)
common ratio
annuity
ordinary annuity
payment period
future value of an
annuity

term of an annuity
14.4 Pascal's triangle
binomial theorem
(general binomial
expansion)

NEW SYMBOLS

a_n nth term of a sequence

$\displaystyle\sum_{i=1}^{n} a_i$ summation notation

S_n sum of first n terms of a
sequence

$\displaystyle\lim_{n \to \infty} a_n$ limit of a_n as n gets larger
and larger

$\displaystyle\sum_{i=1}^{\infty} a_i$ sum of an infinite number
of terms

$n!$ n factorial

$_nC_r$ binomial coefficient
(combinations of n things
taken r at a time)

TEST YOUR WORD POWER

See how well you have learned the vocabulary in this chapter. Answers, with examples, are given at the bottom of the page.

1. An **infinite sequence** is
(a) the values of a function
(b) a function whose domain is the set
of natural numbers
(c) the sum of the terms of a function
(d) the average of a group of numbers.

2. A **series** is
(a) the sum of the terms of a sequence
(b) the product of the terms of a
sequence
(c) the average of the terms of a
sequence
(d) the function values of a sequence.

3. An **arithmetic sequence** is a
sequence in which
(a) each term after the first is a constant
multiple of the preceding term
(b) the numbers are written in a
triangular array

(c) the terms are added
(d) each term after the first differs
from the preceding term by a
common amount.

4. A **geometric sequence** is a
sequence in which
(a) each term after the first is a constant
multiple of the preceding term
(b) the numbers are written in a
triangular array
(c) the terms are multiplied
(d) each term after the first differs
from the preceding term by a
common amount.

5. The **common difference** is
(a) the average of the terms in a
sequence
(b) the constant multiplier in a
geometric sequence

(c) the difference between any two
adjacent terms in an arithmetic
sequence
(d) the sum of the terms of an
arithmetic sequence.

6. The **common ratio** is
(a) the average of the terms in a
sequence
(b) the constant multiplier in a
geometric sequence
(c) the difference between any two
adjacent terms in an arithmetic
sequence
(d) the product of the terms of a
geometric sequence.

Answers to Test Your Word Power

1. (b) *Example:* The ordered list of numbers 3, 6, 9, 12, 15, . . . is an infinite sequence.   **2.** (a) *Example:* $3 + 6 + 9 + 12 + 15$, written in summation notation as $\displaystyle\sum_{i=1}^{5} 3i$, is a series.   **3.** (d) *Example:* The sequence $-3, 2, 7, 12, 17, . . .$ is arithmetic.   **4.** (a) *Example:* The sequence 1, 4, 16, 64, 256, . . . is geometric.   **5.** (c) *Example:* The common difference of the arithmetic sequence in Item 3 above is 5 since $2 - (-3) = 5$, $7 - 2 = 5$, $12 - 7 = 5$, and so on.   **6.** (b) *Example:* The common ratio of the geometric sequence in Item 4 above is 4 since $\frac{4}{1} = \frac{16}{4} = \frac{64}{16} = \frac{256}{64} = 4$.

QUICK REVIEW

CONCEPTS	EXAMPLES

14.1 SEQUENCES AND SERIES

Sequence

General Term a_n

Series

$$1, \frac{1}{2}, \frac{1}{3}, \frac{1}{4}, \ldots, \frac{1}{n}$$

has general term $\frac{1}{n}$.

The corresponding series is the *sum*

$$1 + \frac{1}{2} + \frac{1}{3} + \frac{1}{4} + \cdots + \frac{1}{n}.$$

14.2 ARITHMETIC SEQUENCES

Assume a_1 is the first term, a_n is the nth term, and d is the common difference.

Common Difference

$d = a_{n+1} - a_n$

nth Term

$a_n = a_1 + (n - 1)d$

Sum of the First n Terms

$$S_n = \frac{n}{2}(a_1 + a_n) \quad \text{or}$$

$$S_n = \frac{n}{2}[2a_1 + (n - 1)d]$$

The arithmetic sequence 2, 5, 8, 11, . . . has $a_1 = 2$.

$$d = 5 - 2 = 3$$

(Any two successive terms could have been used.)

Suppose that $n = 10$. Then the 10th term is

$$a_{10} = 2 + (10 - 1)3$$
$$= 2 + 9 \cdot 3 = 29.$$

The sum of the first 10 terms is

$$S_{10} = \frac{10}{2}(2 + a_{10})$$
$$= 5(2 + 29) = 5(31) = 155$$

or $\quad\quad S_{10} = \frac{10}{2}[2(2) + (10 - 1)3]$

$$= 5(4 + 9 \cdot 3)$$
$$= 5(4 + 27) = 5(31) = 155.$$

14.3 GEOMETRIC SEQUENCES

Assume a_1 is the first term, a_n is the nth term, and r is the common ratio.

Common Ratio

$$r = \frac{a_{n+1}}{a_n}$$

nth Term

$$a_n = a_1 r^{n-1}$$

Sum of the First n Terms

$$S_n = \frac{a_1(r^n - 1)}{r - 1} \quad (r \neq 1)$$

Future Value of an Ordinary Annuity

$$S = R\left[\frac{(1 + i)^n - 1}{i}\right],$$

The geometric sequence 1, 2, 4, 8, . . . has $a_1 = 1$.

$$r = \frac{8}{4} = 2$$

(Any two successive terms could have been used.)

Suppose that $n = 6$. Then the 6th term is

$$a_6 = (1)(2)^{6-1} = 1(2)^5 = 32.$$

The sum of the first 6 terms is

$$S_6 = \frac{1(2^6 - 1)}{2 - 1} = \frac{64 - 1}{1} = 63.$$

If \$5800 is deposited into an ordinary annuity at the end of each quarter for four years, and interest is earned at 6.4% compounded quarterly, then

$$R = \$5800, i = \frac{.064}{4} = .016, n = 4(4) = 16,$$

CONCEPTS	EXAMPLES

where S is future value, R is the payment at the end of each period, i is the interest rate per period, and n is the number of periods.

and

$$S = 5800\left[\frac{(1 + .016)^{16} - 1}{.016}\right]$$

$$= \$104,812.44.$$

Sum of the Terms of an Infinite Geometric Sequence with $|r| < 1$

$$S = \frac{a_1}{1 - r}$$

The sum S of the terms of an infinite geometric sequence with $a_1 = 1$ and $r = \frac{1}{2}$ is

$$S = \frac{1}{1 - \frac{1}{2}} = \frac{1}{\frac{1}{2}} = 2.$$

14.4 THE BINOMIAL THEOREM

For any positive integer n,

$$n(n - 1)(n - 2) \ldots (2)(1) = n!.$$

$$0! = 1$$

$$_nC_r = \frac{n!}{r!(n - r)!}$$

$$4! = 4 \cdot 3 \cdot 2 \cdot 1$$
$$= 24$$

$$_5C_3 = \frac{5!}{3!\,(5 - 3)!}$$

$$= \frac{5!}{3!\,2!}$$

$$= \frac{5 \cdot 4 \cdot 3 \cdot 2 \cdot 1}{3 \cdot 2 \cdot 1 \cdot 2 \cdot 1}$$

$$= 10$$

General Binomial Expansion

For any positive integer n,

$$(x + y)^n = x^n + \frac{n!}{(n - 1)!\,1!}x^{n-1}y$$

$$+ \frac{n!}{(n - 2)!\,2!}x^{n-2}y^2$$

$$+ \frac{n!}{(n - 3)!\,3!}x^{n-3}y^3 + \cdots$$

$$+ \frac{n!}{1!(n - 1)!}xy^{n-1} + y^n.$$

$$(2m + 3)^4 = (2m)^4 + \frac{4!}{3!\,1!}(2m)^3(3) + \frac{4!}{2!\,2!}(2m)^2(3)^2$$

$$+ \frac{4!}{1!\,3!}(2m)(3)^3 + 3^4$$

$$- 2^4m^4 + 4(2)^3m^3(3)$$

$$+ 6(2)^2m^2(9) + 4(2m)(27) + 81$$

$$= 16m^4 + 12(8)m^3 + 54(4)m^2 + 216m + 81$$

$$= 16m^4 + 96m^3 + 216m^2 + 216m + 81$$

rth Term of the Binomial Expansion of $(x + y)^n$

$$\frac{n!}{[n - (r-1)]!\,(r-1)!}x^{n-(r-1)}y^{r-1}$$

The 8th term of $(a - 2b)^{10}$ is

$$\frac{10!}{3!\,7!}a^3(-2b)^7 = \frac{10 \cdot 9 \cdot 8}{3 \cdot 2 \cdot 1}a^3(-2)^7b^7$$

$$= 120(-128)a^3b^7$$

$$= -15,360a^3b^7.$$

CHAPTER 14 REVIEW EXERCISES

[14.1] *Write out the first four terms of each sequence.*

1. $a_n = 2n - 3$ **2.** $a_n = \frac{n - 1}{n}$ **3.** $a_n = n^2$

4. $a_n = \left(\dfrac{1}{2}\right)^n$

5. $a_n = (n + 1)(n - 1)$

Write each series as a sum of terms.

6. $\displaystyle\sum_{i=1}^{5} i^2 x$

7. $\displaystyle\sum_{i=1}^{6} (i + 1)x^i$

Evaluate each series.

8. $\displaystyle\sum_{i=1}^{4} (i + 2)$

9. $\displaystyle\sum_{i=1}^{6} 2^i$

10. $\displaystyle\sum_{i=4}^{7} \dfrac{i}{i + 1}$

11. Find the arithmetic mean, or average, of the share volume of the five most active trading days on the New York Stock Exchange as of the end of 1997.

Date	Volume (in thousands)
Jan. 23, 1997	684,588
July 16, 1996	680,913
Dec. 20, 1996	654,110
June 20, 1997	652,945
July 16, 1997	652,848

Source: New York Stock Exchange.

[14.2–14.3] *Decide whether each sequence is arithmetic, geometric, or neither. If the sequence is arithmetic, find the common difference, d. If it is geometric, find the common ratio, r.*

12. $2, 5, 8, 11, \ldots$

13. $-6, -2, 2, 6, 10, \ldots$

14. $\dfrac{2}{3}, -\dfrac{1}{3}, \dfrac{1}{6}, -\dfrac{1}{12}, \ldots$

15. $-1, 1, -1, 1, -1, \ldots$

16. $64, 32, 8, \dfrac{1}{2}, \ldots$

17. $64, 32, 16, 8, \ldots$

18. $10, 8, 6, 4, \ldots$

[14.2] *Find the indicated term for each arithmetic sequence.*

19. $a_1 = -2, d = 5; \quad a_{16}$

20. $a_6 = 12, a_8 = 18; \quad a_{25}$

Find the general term for each arithmetic sequence.

21. $a_1 = -4, d = -5$

22. $6, 3, 0, -3, \ldots$

Find the number of terms in each arithmetic sequence.

23. $7, 10, 13, \ldots, 49$

24. $5, 1, -3, \ldots, -79$

Find S_8 for each arithmetic sequence.

25. $a_1 = -2, d = 6$

26. $a_n = -2 + 5n$

[14.3] *Find the general term for each geometric sequence.*

27. $-1, -4, \ldots$

28. $\dfrac{2}{3}, \dfrac{2}{15}, \ldots$

Find the indicated term for each geometric sequence.

29. $2, -6, 18, \ldots ; \quad a_{11}$

30. $a_3 = 20, a_5 = 80; \quad a_{10}$

Find each sum, if it exists.

31. $\displaystyle\sum_{i=1}^{5} \left(\frac{1}{4}\right)^i$

32. $\displaystyle\sum_{i=1}^{8} \frac{3}{4}(-1)^i$

33. $\displaystyle\sum_{i=1}^{\infty} 4\left(\frac{1}{5}\right)^i$

34. $\displaystyle\sum_{i=1}^{\infty} 2(3)^i$

[14.4] *Use the binomial theorem to expand each binomial.*

35. $(2p - q)^5$

36. $(x^2 + 3y)^4$

37. $(\sqrt{m} + \sqrt{n})^4$

38. Write the fourth term of the expansion of $(3a + 2b)^{19}$.

39. Write the twenty-third term of the expansion of $(-2k + 3)^{25}$.

MIXED REVIEW EXERCISES

Find the indicated term and S_{10} for each sequence.

40. a_{40}: arithmetic; $1, 7, 13, \ldots$

41. a_{10}: geometric; $-3, 6, -12, \ldots$

42. a_9: geometric; $a_1 = 1, r = -3$

43. a_{15}: arithmetic; $a_1 = -4, d = 3$

Find the general term for each arithmetic or geometric sequence.

44. $2, 7, 12, \ldots$

45. $2, 8, 32, \ldots$

46. $27, 9, 3, \ldots$

47. $12, 9, 6, \ldots$

Solve each problem.

48. When Mary's sled goes down the hill near her home, she covers 3 feet in the first second, then for each second after that she goes 4 feet more than in the preceding second. If the distance she covers going down is 210 feet, how long does it take her to reach the bottom?

49. An ordinary annuity is set up so that $672 is deposited at the end of each quarter for 7 years. The money earns 8% annual interest compounded quarterly. What is the future value of the annuity?

50. The school population in Pfleugerville has been dropping 3% per year. The current population is 50,000. If this trend continues, what will the population be in 6 years?

51. A pump removes $\frac{1}{2}$ of the liquid in a container with each stroke. What fraction of the liquid is left in the container after 7 strokes?

52. Consider the repeating decimal number $.55555 \ldots$.
  (a) Write it as the sum of the terms of an infinite geometric sequence.
  (b) What is r for this sequence?
  (c) Find this infinite sum, if it exists, and write it as a common fraction in lowest terms.

53. Can the sum of the terms of the infinite geometric sequence with $a_n = 5(2)^n$ be found? Explain.

54. Can any two terms of a geometric sequence be used to find the common ratio? Explain.

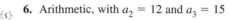

CHAPTER 14 TEST

Write the first five terms of each sequence described.

1. $a_n = (-1)^n + 1$

2. Arithmetic, with $a_1 = 4$ and $d = 2$

3. Geometric, with $a_4 = 6$ and $r = \frac{1}{2}$

Find a_4 for each sequence described.

4. Arithmetic, with $a_1 = 6$ and $d = -2$

5. Geometric, with $a_5 = 16$ and $a_7 = 9$

Find S_5 for each sequence described.

6. Arithmetic, with $a_2 = 12$ and $a_3 = 15$

7. Geometric, with $a_5 = 4$ and $a_7 = 1$

 8. The share volume (in millions) of the five most active stocks on the American Stock Exchange in 1996 is shown in the following table.

Stock	Share Volume (in millions)
Viacom (Class B)	271.9
Trans World Airlines	199.8
Echo Bay Mines	198.5
IVAX	183.5
Ampex (Class A)	179.1

Source: American Stock Exchange.

 To the nearest tenth of a million, what was the average share volume for these five stocks?

9. If $4000 is deposited in an ordinary annuity at the end of each quarter for 7 years and earns 6% interest compounded quarterly, how much will be in the account at the end of this term?

10. Under what conditions does an infinite geometric series have a sum?

Find each sum that exists.

11. $\displaystyle\sum_{i=1}^{5} (2i + 8)$

12. $\displaystyle\sum_{i=1}^{6} (3i - 5)$

13. $\displaystyle\sum_{i=1}^{500} i$

14. $\displaystyle\sum_{i=1}^{3} \frac{1}{2} (4^i)$

15. $\displaystyle\sum_{i=1}^{\infty} \left(\frac{1}{4}\right)^i$

16. $\displaystyle\sum_{i=1}^{\infty} 6\left(\frac{3}{2}\right)^i$

 17. Cheryl bought a new sewing machine for $300. She agreed to pay $20 per month for 15 months plus interest of 1% each month on the unpaid balance. Find the total cost of the machine.

 18. During the summer months, the population of a certain insect colony triples each week. If there are 20 insects in the colony at the end of the first week in July, how many are present by the end of September? (Assume exactly four weeks in a month.)

19. Write the fifth term of $\left(2x - \dfrac{y}{3}\right)^{12}$.

20. Expand $(3k - 5)^4$.

This set of exercises may be considered a final examination for the course.

Let $P = \left\{ -\dfrac{8}{3}, 10, 0, \sqrt{13}, -\sqrt{3}, \dfrac{45}{15}, \sqrt{-7}, .82, -3 \right\}$. *List the elements of P that are members of each set.*

1. Integers

2. Rational numbers

3. Irrational numbers

4. Real numbers

Simplify each expression.

5. $|-7| + 6 - |-10| - (-8 + 3)$

6. $-15 - |-4| - 10 - |-6|$

7. $4(-6) + (-8)(5) - (-9)$

Solve each equation or inequality.

8. $9 - (5 + 3a) + 5a = -4(a - 3) - 7$

9. $7m + 18 \leq 9m - 2$

10. $|4x - 3| = 21$

11. $\dfrac{x + 3}{12} - \dfrac{x - 3}{6} = 0$

12. $2x > 8$ or $-3x > 9$

13. $|2m - 5| \geq 11$

Perform the indicated operations.

14. $(4p + 2)(5p - 3)$

15. $(3k - 7)^2$

16. $(2m^3 - 3m^2 + 8m) - (7m^3 + 5m - 8)$

17. Divide $6t^4 + 5t^3 - 18t^2 + 14t - 1$ by $3t - 2$.

Factor.

18. $7x + x^3$

19. $14y^2 + 13y - 12$

20. $6z^3 + 5z^2 - 4z$

21. $49a^4 - 9b^2$

22. $c^3 + 27d^3$

23. $64r^2 + 48rq + 9q^2$

Solve each equation or inequality.

24. $2x^2 + x = 10$

25. $k^2 - k - 6 \leq 0$

Simplify.

26. $\left(\dfrac{2}{3}\right)^{-2}$

27. $\dfrac{(3p^2)^3(-2p^6)}{4p^3(5p^7)}$

28. Find any values for which the rational expression $\dfrac{2}{x^2 - 81}$ is undefined.

Simplify.

29. $\dfrac{x^2 - 16}{x^2 + 2x - 8} \div \dfrac{x - 4}{x + 7}$

30. $\dfrac{5}{p^2 + 3p} - \dfrac{2}{p^2 - 4p}$

Solve.

31. $\dfrac{4}{x - 3} - \dfrac{6}{x + 3} = \dfrac{24}{x^2 - 9}$

32. $6x^2 + 5x = 8$

33. $\sqrt{3x - 2} = x$

34. The president of InstaTune, a chain of franchised automobile tune-up shops, reports that people who buy a franchise and open a shop pay a weekly fee (in dollars) to company headquarters, according to the linear function $f(x) = .07x + 135$, where $f(x)$ is the fee and x is the total amount of money taken in during the week by the shop. Find the weekly fee if $2000 is taken in for the week. (*Source: Business Week.*)

35. Find the slope of the line through $(4, -5)$ and $(-12, -17)$.

36. Find the standard form of the equation of the line through $(-2, 10)$ and parallel to the line with equation $3x + y = 7$.

Graph.

37. $x - 3y = 6$ **38.** $4x - y < 4$

Solve each system of equations.

39. $2x + 5y = -19$ **40.** $x + 2y + z = 8$
        $-3x + 2y = -19$ $2x - y + 3z = 15$
                                            $-x + 3y - 3z = -11$

Evaluate each determinant.

41. $\begin{vmatrix} -3 & -2 \\ 6 & 9 \end{vmatrix}$ **42.** $\begin{vmatrix} 2 & 4 & 1 \\ 1 & 3 & 6 \\ 2 & 3 & -1 \end{vmatrix}$

43. Nuts worth $3 per pound are to be mixed with 8 pounds of nuts worth $4.25 per pound to get a mixture that will be sold for $4 per pound. How many pounds of the $3 nuts should be used?

44. Simplify $5\sqrt{72} - 4\sqrt{50}$.

45. Multiply $(8 + 3i)(8 - 3i)$.

46. Find $f^{-1}(x)$, if $f(x) = 9x + 5$.   **47.** Graph $g(x) = \left(\dfrac{1}{3}\right)^x$.

48. Solve $3^{2x-1} = 81$.   **49.** Graph $y = \log_{1/3} x$.

50. Solve $\log_8 x + \log_8(x + 2) = 1$.

Graph.

51. $f(x) = 2(x - 2)^2 - 3$ **52.** $\dfrac{x^2}{9} + \dfrac{y^2}{25} = 1$ **53.** $x^2 - y^2 = 9$

54. Find the equation of a circle with center at $(-5, 12)$ and radius 9.

55. Write the first five terms of the sequence defined by $a_n = 5n - 12$.

56. Find the sum of the first six terms of the arithmetic sequence with $a_1 = 8$ and $d = 2$.

57. Find the sum of the geometric series $15 - 6 + \dfrac{12}{5} - \dfrac{24}{25} + \cdots$.

58. Find the sum: $\displaystyle\sum_{i=1}^{4} 3i$.

59. Use the binomial theorem to expand $(2a - 1)^5$.

60. What is the fourth term in the expansion of $\left(3x^4 - \dfrac{1}{2}y^2\right)^5$?

APPENDIX A   REVIEW OF DECIMALS AND PERCENTS

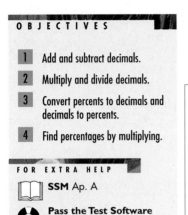

OBJECTIVES

1 Add and subtract decimals.

2 Multiply and divide decimals.

3 Convert percents to decimals and decimals to percents.

4 Find percentages by multiplying.

FOR EXTRA HELP

📖 **SSM** Ap. A

💿 **Pass the Test Software**

💿 **InterAct Math Tutorial Software**

📼 **Video** N/A

OBJECTIVE **1** Add and subtract decimals. A **decimal** is a number written with a decimal point, such as 4.2. The operations on decimals—addition, subtraction, multiplication, and division—are explained in the next examples.

EXAMPLE 1 Adding and Subtracting Decimals

Add or subtract as indicated.

(a) $6.92 + 14.8 + 3.217$

Place the numbers in a column, with decimal points lined up, then add. If you like, attach zeros to make all the numbers the same length; this is a good way to avoid errors. For example,

$$
\begin{array}{r}
\text{Decimal points} \\
\text{lined up}
\end{array}
\quad
\begin{array}{r}
6.92 \\
14.8 \\
+\;3.217 \\
\hline
24.937
\end{array}
\quad \text{or} \quad
\begin{array}{r}
6.920 \\
14.800 \\
+\;3.217 \\
\hline
24.937.
\end{array}
$$

(b) $47.6 - 32.509$

Write the numbers in a column, attaching zeros to 47.6.

$$
\begin{array}{r}
47.6 \\
-32.509 \\
\end{array}
\quad \text{becomes} \quad
\begin{array}{r}
47.600 \\
-32.509 \\
\hline
15.091
\end{array}
$$

(c) $3 - .253$

$$
\begin{array}{r}
3.000 \\
-\;.253 \\
\hline
2.747
\end{array}
$$

OBJECTIVE **2** Multiply and divide decimals. Multiplication and division of decimals are similar to the same operations with whole numbers.

EXAMPLE 2 Multiplying Decimals

Multiply.

(a) 29.3×4.52

Multiply as if the numbers were whole numbers. To find the number of decimal places in the answer, add the numbers of decimal places in the factors.

$$
\begin{array}{r}
29.3 \\
\times 4.52 \\
\hline
5\,86 \\
14\,6\,5 \\
117\,2 \\
\hline
132.4\,36
\end{array}
$$

1 decimal place
2 decimal places

3 decimal places in answer

905

(b) 7.003×55.8

$$
\begin{array}{rl}
7.003 & \text{3 decimal places} \\
\times\ 55.8 & \text{1 decimal place} \\
\hline
5\ 602\ 4 & \\
35\ 015 & \\
350\ 15 & \\
\hline
390.767\ 4 & \text{4 decimal places in answer}
\end{array}
$$

EXAMPLE 3 Dividing Decimals

Divide: $279.45 \div 24.3$.

Move the decimal point in 24.3 one place to the right, to get 243. Move the decimal point the same number of places in 279.45. By doing this, 24.3 is converted into the whole number 243.

$$243.\overline{)2794.5}$$

Bring the decimal point straight up and divide as with whole numbers.

$$
\begin{array}{r}
11.5 \\
243.\overline{)2794.5} \\
\underline{243} \\
364 \\
\underline{243} \\
121\ 5 \\
\underline{121\ 5} \\
0
\end{array}
$$

OBJECTIVE 3 Convert percents to decimals and decimals to percents. One of the main uses of decimals is in percent problems. The word **percent** means "per one hundred." Percent is written with the sign %. One percent means "one per one hundred."

Percent

$$1\% = .01 \quad \text{or} \quad 1\% = \frac{1}{100}$$

EXAMPLE 4 Converting between Decimals and Percents

Convert.

(a) 75% to a decimal
Since $1\% = .01$,

$$75\% = 75 \cdot 1\% = 75 \cdot (.01) = .75.$$

The fraction form $1\% = \frac{1}{100}$ can also be used to convert 75% to a decimal.

$$75\% = 75 \cdot 1\% = 75 \cdot \left(\frac{1}{100}\right) = .75$$

(b) 2.63 to a percent

$$2.63 = 263 \cdot (.01) = 263 \cdot 1\% = 263\%$$

OBJECTIVE **4** Find percentages by multiplying. A part of a whole is called a **percentage.** For example, since 50% represents $\frac{50}{100} = \frac{1}{2}$ of a whole, 50% of 800 is half of 800, or 400. Multiply to find percentages, as in the next example.

EXAMPLE 5 Finding Percentages

Find the percentages.

(a) 15% of 600

The word *of* indicates multiplication here. For this reason, 15% of 600 is found by multiplying.

$$15\% \cdot 600 = (.15) \cdot 600 = 90$$

(b) 125% of 80

$$125\% \cdot 80 = (1.25) \cdot 80 = 100$$

(c) What percent of 52 is 7.8?

We can translate this sentence to symbols word by word.

$$\text{What percent} \quad \text{of} \quad 52 \quad \text{is} \quad 7.8?$$
$$p \qquad\qquad \cdot \quad 52 \quad = \quad 7.8$$

Solve the equation.

$$52p = 7.8$$
$$p = .15 \qquad \text{Divide both sides by 52.}$$
$$p = 15\% \qquad \text{Change to percent.}$$

EXAMPLE 6 Using Percent in a Consumer Problem

A video movie with a regular price of $18 is on sale at 22% off. Find the amount of the discount.

The discount is 22% of $18. Change 22% to .22 and use the fact that *of* indicates multiplication.

$$22\% \cdot 18 = .22 \cdot 18 = 3.96$$

The discount is $3.96.

APPENDIX A EXERCISES

Perform the indicated operations. See Examples 1–3.

1. $14.23 + 9.81 + 74.63 + 18.715$

2. $89.416 + 21.32 + 478.91 + 298.213$

3. $19.74 - 6.53$

4. $27.96 - 8.39$

5. $219 - 68.51$

6. $283 - 12.42$

7.
$$\begin{array}{r} 48.96 \\ 37.421 \\ +\ 9.72 \\ \hline \end{array}$$

8.
$$\begin{array}{r} 9.71 \\ 4.8 \\ 3.6 \\ 5.2 \\ +8.17 \\ \hline \end{array}$$

9. 8.6
−3.751

10. 27.8
−13.582

11. 39.6×4.2 **12.** 18.7×2.3 **13.** 42.1×3.9 **14.** 19.63×4.08

15. $.042 \times 32$ **16.** 571×2.9 **17.** $24.84 \div 6$ **18.** $32.84 \div 4$

19. $7.6266 \div 3.42$ **20.** $14.9202 \div 2.43$ **21.** $2496 \div .52$ **22.** $.56984 \div .034$

Convert each percent to a decimal. See Example 4(a).

23. 53% **24.** 38% **25.** 129% **26.** 174%

27. 96% **28.** 11% **29.** .9% **30.** .1%

Convert each decimal to a percent. See Example 4(b).

31. .80 **32.** .75 **33.** .007 **34.** 1.4

35. .67 **36.** .003 **37.** .125 **38.** .983

Respond to each statement or question. Round your answer to the nearest hundredth if appropriate. See Example 5.

39. What is 14% of 780? **40.** Find 12% of 350.

41. Find 22% of 1086. **42.** What is 20% of 1500?

43. 4 is what percent of 80? **44.** 1300 is what percent of 2000?

45. What percent of 5820 is 6402? **46.** What percent of 75 is 90?

47. 121 is what percent of 484? **48.** What percent of 3200 is 64?

49. Find 118% of 125.8. **50.** Find 3% of 128.

51. What is 91.72% of 8546.95? **52.** Find 12.741% of 58.902.

53. What percent of 198.72 is 14.68? **54.** 586.3 is what percent of 765.4?

Solve each problem. See Example 6.

55. A retailer has $23,000 invested in her business. She finds that she is earning 12% per year on this investment. How much money is she earning per year?

56. Harley Dabler recently bought a duplex for $144,000. He expects to earn 16% per year on the purchase price. How many dollars per year will he earn?

57. For a recent tour of the eastern United States, a travel agent figured that the trip totaled 2300 miles, with 35% of the trip by air. How many miles of the trip were by air?

58. Capitol Savings Bank pays 3.2% interest per year. What is the annual interest on an account of $3000?

59. An ad for steel-belted radial tires promises 15% better mileage when the tires are used. Alexandria's Escort now goes 420 miles on a tank of gas. If she switched to the new tires, how many extra miles could she drive on a tank of gas?

60. A home worth $77,000 is located in an area where home prices are increasing at a rate of 12% per year. By how much would the value of this home increase in one year?

61. A family of four with a monthly income of $2000 spends 90% of its earnings and saves the rest. Find the *annual* savings of this family.

APPENDIX B SETS

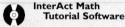
FOR EXTRA HELP

📖 **SSM** Ap. B

💿 **Pass the Test Software**

💿 **InterAct Math**
   Tutorial Software

📼 **Video** N/A

OBJECTIVE **1** List the elements of a set. A **set** is a collection of things. The objects in a set are called the **elements** of the set. A set is represented by listing its elements between **set braces,** { }. The order in which the elements of a set are listed is unimportant.

EXAMPLE 1 Listing the Elements of a Set

Represent the following sets by listing the elements.

(a) The set of states in the United States that border on the Pacific Ocean = {California, Oregon, Washington, Hawaii, Alaska}.

(b) The set of all counting numbers less than 6 = {1, 2, 3, 4, 5}.

OBJECTIVE **2** Learn the vocabulary and symbols used to discuss sets. Capital letters are used to name sets. To state that 5 is an element of

$$S = \{1, 2, 3, 4, 5\},$$

write $5 \in S$. The statement $6 \notin S$ means that 6 is not an element of S.

A set with no elements is called the **empty set,** or the **null set.** The symbols $\emptyset$ or { } are used for the empty set. If we let A be the set of all cats that fly, then A is the empty set.

$$A = \emptyset \qquad \text{or} \qquad A = \{ \ \}$$

CAUTION Do not make the common error of writing the empty set as {$\emptyset$}.

In any discussion of sets, there is some set that includes all the elements under consideration. This set is called the **universal set** for that situation. For example, if the discussion is about presidents of the United States, then the set of all presidents of the United States is the universal set. The universal set is denoted U.

OBJECTIVE **3** Decide whether a set is finite or infinite. In Example 1, there are five elements in the set in part (a), and five in part (b). If the number of elements in a set is either 0 or a counting number, then the set is **finite.** On the other hand, the set of natural numbers, for example, is an **infinite** set, because there is no final number. We can list the elements of the set of natural numbers as

$$N = \{1, 2, 3, 4 \ldots\}$$

where the three dots indicate that the set continues indefinitely. Not all infinite sets can be listed in this way. For example, there is no way to list the elements in the set of all real numbers between 1 and 2.

E X A M P L E   2   Distinguishing between Finite and Infinite Sets

List the elements of each set, if possible. Decide whether each set is finite or infinite.

(a) The set of all integers
One way to list the elements is $\{\ldots, -2, -1, 0, 1, 2, \ldots\}$. The set is infinite.

(b) The set of all natural numbers between 0 and 5
$\{1, 2, 3, 4\}$ The set is finite.

(c) The set of all irrational numbers
This is an infinite set whose elements cannot be listed.

Two sets are **equal** if they have exactly the same elements. Thus, the set of natural numbers and the set of positive integers are equal sets. Also, the sets

$$\{1, 2, 4, 7\} \quad \text{and} \quad \{4, 2, 7, 1\}$$

are equal. The order of the elements does not make a difference.

OBJECTIVE 4 Decide whether a given set is a subset of another set. If all elements of a set A are also elements of a new set B, then we say A is a **subset** of B, written $A \subseteq B$. We use the symbol $A \nsubseteq B$ to mean that A is not a subset of B.

E X A M P L E   3   Using Subset Notation

Let $A = \{1, 2, 3, 4\}$, $B = \{1, 4\}$, and $C = \{1\}$. Then $B \subseteq A$, $C \subseteq A$, and $C \subseteq B$, but $A \nsubseteq B$, $A \nsubseteq C$, and $B \nsubseteq C$.

The set $M = \{a, b\}$ has four subsets: $\{a, b\}$, $\{a\}$, $\{b\}$, and $\emptyset$. The empty set is defined to be a subset of any set. How many subsets does $N = \{a, b, c\}$ have? There is one subset with 3 elements: $\{a, b, c\}$. There are three subsets with 2 elements:

$$\{a, b\}, \quad \{a, c\}, \quad \text{and} \quad \{b, c\}.$$

There are three subsets with 1 element:

$$\{a\}, \quad \{b\}, \quad \text{and} \quad \{c\}.$$

There is one subset with 0 elements: $\emptyset$. Thus, set N has eight subsets.
The following generalization can be made.

Number of Subsets of a Set

A set with n elements has 2^n subsets.

To illustrate the relationships between sets, **Venn diagrams** are often used. A rectangle represents the universal set, U. The sets under discussion are represented by regions within the rectangle. The Venn diagram in Figure 1 shows that $B \subseteq A$.

OBJECTIVE 5 Find the complement of a set. For every set A, there is a set A', the **complement** of A, that contains all the elements of U that are not in A. The shaded region in the Venn diagram in Figure 2 represents A'.

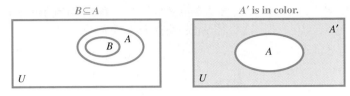

Figure 1

Figure 2

E X A M P L E 4 Determining the Complement of a Set

Given $U = \{a, b, c, d, e, f, g\}$, $A = \{a, b, c\}$, $B = \{a, d, f, g\}$, and $C = \{d, e\}$, find A', B', and C'.

(a) $A' = \{d, e, f, g\}$ **(b)** $B' = \{b, c, e\}$ **(c)** $C' = \{a, b, c, f, g\}$

O B J E C T I V E 6 Find the union and the intersection of two sets. The **union** of two sets A and B, written $A \cup B$, is the set of all elements of A together with all elements of B. Thus, for the sets in Example 4,

$$A \cup B = \{a, b, c, d, f, g\}$$

and

$$A \cup C = \{a, b, c, d, e\}.$$

In Figure 3 the shaded region is the union of sets A and B.

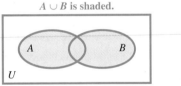

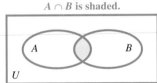

Figure 3

Figure 4

E X A M P L E 5 Finding the Union of Two Sets

If $M = \{2, 5, 7\}$ and $N = \{1, 2, 3, 4, 5\}$, then

$$M \cup N = \{1, 2, 3, 4, 5, 7\}.$$

The **intersection** of two sets A and B, written $A \cap B$, is the set of all elements that belong to both A and B. For example if,

$$A = \{\text{Jose, Ellen, Marge, Kevin}\}$$

and

$$B = \{\text{Jose, Patrick, Ellen, Sue}\},$$

then

$$A \cap B = \{\text{Jose, Ellen}\}.$$

The shaded region in Figure 4 represents the intersection of the two sets A and B.

E X A M P L E 6 Finding the Intersection of Two Sets

Suppose that $P = \{3, 9, 27\}$, $Q = \{2, 3, 10, 18, 27, 28\}$, and $R = \{2, 10, 28\}$.

(a) $P \cap Q = \{3, 27\}$ **(b)** $Q \cap R = \{2, 10, 28\} = R$ **(c)** $P \cap R = \emptyset$

Sets like P and R in Example 6 that have no elements in common are called **disjoint sets.** The Venn diagram in Figure 5 shows a pair of disjoint sets.

Disjoint sets; $A \cap B = \emptyset$

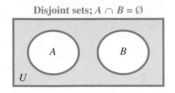

Figure 5

┌───
E X A M P L E   7   Using Set Operations

Let $U = \{2, 5, 7, 10, 14, 20\}$, $A = \{2, 10, 14, 20\}$, $B = \{5, 7\}$, and $C = \{2, 5, 7\}$. Find each of the following.

(a) $A \cup B = \{2, 5, 7, 10, 14, 20\} = U$

(b) $A \cap B = \emptyset$

(c) $B \cup C = \{2, 5, 7\} = C$

(d) $B \cap C = \{5, 7\} = B$

(e) $A' = \{5, 7\} = B$
─── ■

APPENDIX B   EXERCISES

List the elements of each set. See Examples 1 and 2.

1. The set of all natural numbers less than 8

2. The set of all integers between 4 and 10

3. The set of seasons

4. The set of months of the year

5. The set of women presidents of the United States

6. The set of all living humans who are more than 200 years old

7. The set of letters of the alphabet between K and M

8. The set of letters of the alphabet between D and H

9. The set of positive even integers

10. The set of all multiples of 5

11. Which of the sets described in Exercises 1–10 are infinite sets?

12. Which of the sets described in Exercises 1–10 are finite sets?

Tell whether each statement is true or false.

13. $5 \in \{1, 2, 5, 8\}$

14. $6 \in \{1, 2, 3, 4, 5\}$

15. $2 \in \{1, 3, 5, 7, 9\}$

16. $1 \in \{6, 2, 5, 1\}$

17. $7 \notin \{2, 4, 6, 8\}$

18. $7 \notin \{1, 3, 5, 7\}$

19. $\{2, 4, 9, 12, 13\} = \{13, 12, 9, 4, 2\}$

20. $\{7, 11, 4\} = \{7, 11, 4, 0\}$

Let

$$A = \{1, 3, 4, 5, 7, 8\}$$
$$B = \{2, 4, 6, 8\}$$
$$C = \{1, 3, 5, 7\}$$
$$D = \{1, 2, 3\}$$
$$E = \{3, 7\}$$
$$U = \{1, 2, 3, 4, 5, 6, 7, 8, 9, 10\}.$$

Tell whether each statement is true or false. See Examples 3, 5, 6, and 7.

21. $A \subseteq U$ **22.** $D \subseteq A$ **23.** $\emptyset \subseteq A$ **24.** $\{1, 2\} \subseteq D$ **25.** $C \subseteq A$

26. $A \subseteq C$ **27.** $D \subseteq B$ **28.** $E \subseteq C$ **29.** $D \nsubseteq E$ **30.** $E \nsubseteq A$

31. There are exactly 4 subsets of E.   **32.** There are exactly 8 subsets of D.

33. There are exactly 12 subsets of C.   **34.** There are exactly 16 subsets of B.

35. $\{4, 6, 8, 12\} \cap \{6, 8, 14, 17\} = \{6, 8\}$ **36.** $\{2, 5, 9\} \cap \{1, 2, 3, 4, 5\} = \{2, 5\}$

37. $\{3, 1, 0\} \cap \{0, 2, 4\} = \{0\}$ **38.** $\{4, 2, 1\} \cap \{1, 2, 3, 4\} = \{1, 2, 3\}$

39. $\{3, 9, 12\} \cap \emptyset = \{3, 9, 12\}$ **40.** $\{3, 9, 12\} \cup \emptyset = \emptyset$

41. $\{4, 9, 11, 7, 3\} \cup \{1, 2, 3, 4, 5\} = \{1, 2, 3, 4, 5, 7, 9, 11\}$

42. $\{1, 2, 3\} \cup \{1, 2, 3\} = \{1, 2, 3\}$

43. $\{3, 5, 7, 9\} \cup \{4, 6, 8\} = \emptyset$

44. $\{5, 10, 15, 20\} \cup \{5, 15, 30\} = \{5, 15\}$

Let

$$U = \{a, b, c, d, e, f, g, h\}$$
$$A = \{a, b, c, d, e, f\}$$
$$B = \{a, c, e\}$$
$$C = \{a, f\}$$
$$D = \{d\}.$$

List the elements in each set. See Examples 4–7.

45. A' **46.** B' **47.** C' **48.** D'

49. $A \cap B$ **50.** $B \cap A$ **51.** $A \cap D$ **52.** $B \cap D$

53. $B \cap C$ **54.** $A \cup B$ **55.** $B \cup D$ **56.** $B \cup C$

57. $C \cup B$ **58.** $C \cup D$ **59.** $A \cap \emptyset$ **60.** $B \cup \emptyset$

61. Name every pair of disjoint sets among A–D above.

APPENDIX C JOINT AND COMBINED VARIATION

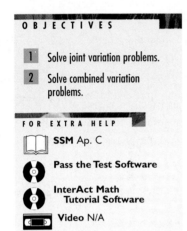

OBJECTIVES

1 Solve joint variation problems.

2 Solve combined variation problems.

FOR EXTRA HELP

📖 **SSM** Ap. C

💿 **Pass the Test Software**

💿 **InterAct Math Tutorial Software**

📼 **Video** N/A

In Sections 2.5 (Example 6) and 7.7 (Examples 5 and 6) we discussed direct and inverse variation. These types of variation are reviewed below, where k represents a constant.

Types of Variation

y varies directly as x	$y = kx$
y varies directly as the nth power of x	$y = kx^n$
y varies inversely as x	$y = \dfrac{k}{x}$
y varies inversely as the nth power of x	$y = \dfrac{k}{x^n}$

Recall that in direct variation, for $k > 0$, as the value of x increases, the value of y also increases. Similarly, as x decreases, y decreases. In inverse variation, for $k > 0$, as x increases, y decreases, and as x decreases, y increases.

We followed these steps to solve direct and inverse variation problems.

Solving a Variation Problem

Step 1 Write the variation equation.

Step 2 Substitute the initial values and solve for k.

Step 3 Rewrite the variation equation with the value of k from Step 2.

Step 4 Substitute the remaining values, solve for the unknown, and find the required answer.

As a review, the steps are labeled in the following example involving direct variation as a power.

E X A M P L E 1   Solving a Problem Involving Direct Variation as a Power

The distance a body falls from rest varies directly as the square of the time it falls (here we disregard air resistance). If an object falls 64 feet in 2 seconds, how far will it fall in 8 seconds?

Step 1   If d represents the distance the object falls and t the time it takes to fall,

$$d = kt^2$$

for some constant k.

Step 2   To find the value of k, use the fact that the object falls 64 feet in 2 seconds.

$$d = kt^2 \qquad \text{Formula}$$
$$64 = k(2)^2 \qquad \text{Let } d = 64 \text{ and } t = 2.$$
$$k = 16 \qquad \text{Find } k.$$

914

Step 3   With this result, the variation equation becomes

$$d = 16t^2.$$

Step 4   Now let $t = 8$ to find the number of feet the object will fall in 8 seconds.

$$d = 16(8)^2 \qquad \text{Let } t = 8.$$
$$= 1024$$

The object will fall 1024 feet in 8 seconds.

OBJECTIVE 1 Solve joint variation problems. It is common for one variable to depend on several others. For example, if one variable varies as the product of several other variables (perhaps raised to powers), the first variable is said to **vary jointly** as the others.

EXAMPLE 2   Solving a Joint Variation Problem

The strength of a rectangular beam varies jointly as its width and the square of its depth. If the strength of a beam 2 inches wide by 10 inches deep is 1000 pounds per square inch, what is the strength of a beam 4 inches wide and 8 inches deep?

If S represents the strength, w the width, and d the depth, then

$$S = kwd^2$$

for some constant, k. Since $S = 1000$ if $w = 2$ and $d = 10$,

$$1000 = k(2)(10)^2. \qquad \text{Let } S = 1000, w = 2, \text{ and } d = 10.$$

Solving this equation for k gives

$$1000 = k \cdot 2 \cdot 100$$
$$1000 - 200k$$
$$k - 5,$$

so

$$S = 5wd^2.$$

Find S when $w = 4$ and $d = 8$ by substitution in $S = 5wd^2$.

$$S = 5(4)(8)^2 \qquad \text{Let } w = 4 \text{ and } d = 8.$$
$$= 1280$$

The strength of the beam is 1280 pounds per square inch.

OBJECTIVE 2 Solve combined variation problems. There are many combinations of direct and inverse variation. The final example shows a typical **combined variation** problem.

EXAMPLE 3   Solving a Combined Variation Problem

The body-mass index, or BMI, is used by physicians to assess a person's level of fatness.* The BMI varies directly as an individual's weight in pounds and inversely as the

___

*Source: *Reader's Digest*, October 1993.

square of the individual's height in inches. A person who weighs 118 pounds and is 64 inches tall has BMI 20. (The BMI is rounded to the nearest whole number.) Find the BMI of a person who weighs 165 pounds with height 70 inches.

Let B represent BMI, w weight, and h height. Then

$$B = \frac{kw}{h^2}. \quad \leftarrow \text{BMI varies directly as the weight.}$$
$$\phantom{B = \frac{kw}{h^2}.} \quad \leftarrow \text{BMI varies inversely as the square of the height.}$$

To find k, let $B = 20$, $w = 118$, and $h = 64$.

$$20 = \frac{k(118)}{64^2}$$

$$k = \frac{20(64^2)}{118} \qquad \text{Multiply by } 64^2; \text{ divide by } 118.$$

$$\approx 694 \qquad \text{Use a calculator.}$$

Now find B when $k = 694$, $w = 165$, and $h = 70$.

$$B = \frac{694(165)}{70^2} \approx 23 \quad \text{(rounded)}$$

The required BMI is 23. A BMI from 20–26 is considered desirable.

APPENDIX C EXERCISES

Determine whether each equation represents direct, inverse, joint, or combined variation.

1. $y = \dfrac{3}{x}$ **2.** $y = \dfrac{8}{x}$ **3.** $y = 10x^2$ **4.** $y = 2x^3$

5. $y = 3xz^4$ **6.** $y = 6x^3z^2$ **7.** $y = \dfrac{4x}{wz}$ **8.** $y = \dfrac{6x}{st}$

Solve each problem.

9. If x varies directly as y, and $x = 9$ when $y = 3$, find x when $y = 12$.

10. If x varies directly as y, and $x = 10$ when $y = 7$, find y when $x = 50$.

11. If z varies inversely as w, and $z = 10$ when $w = .5$, find z when $w = 8$.

12. If t varies inversely as s, and $t = 3$ when $s = 5$, find s when $t = 5$.

13. p varies jointly as q and r^2, and $p = 200$ when $q = 2$ and $r = 3$. Find p when $q = 5$ and $r = 2$.

14. f varies jointly as g^2 and h, and $f = 50$ when $g = 4$ and $h = 2$. Find f when $g = 3$ and $h = 6$.

Solve each variation problem. Use a calculator as necessary. See Examples 1–3.

15. The amount of water emptied by a pipe varies directly as the square of the diameter of the pipe. For a certain constant water flow, a pipe emptying into a canal will allow 200 gallons of water to escape in an hour. The diameter of the pipe is 6 inches. How much water would a 12-inch pipe empty into the canal in an hour, assuming the same water flow?

16. The pressure exerted by a certain liquid at a given point varies directly as the depth of the point beneath the surface of the liquid. The pressure at 30 meters is 80 newtons per square centimeter. What pressure is exerted at 50 meters?

17. For a body falling freely from rest (disregarding air resistance), the distance the body falls varies directly as the square of the time. If an object is dropped from the top of a tower 576 feet high and hits the ground in 6 seconds, how far did it fall in the first 4 seconds?

18. The illumination produced by a light source varies inversely as the square of the distance from the source. If the illumination produced 1 meter from a certain light source is 768 footcandles, find the illumination produced 6 meters from the same source.

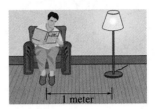

19. The frequency of a vibrating string varies inversely as its length. That is, a longer string vibrates fewer times in a second than a shorter string. Suppose a piano string 2 feet long vibrates 250 cycles per second. What frequency would a string 5 feet long have?

20. The force with which the earth attracts an object above the earth's surface varies inversely with the square of the distance of the object from the center of the earth. If an object 4000 miles from the center of the earth is attracted with a force of 160 pounds, find the force of attraction if the object were 6000 miles from the center of the earth.

21. Natural gas provides 35.8% of U.S. energy. (*Source:* U.S. Energy Department.) The volume of a gas varies inversely as the pressure and directly as the temperature. (Temperature must be measured in *degrees Kelvin* (K), a unit of measurement used in physics.) If a certain gas occupies a volume of 1.3 liters at 300 K and a pressure of 18 newtons per square centimeter, find the volume at 340 K and a pressure of 24 newtons per square centimeter.

22. One source of renewable energy is wind, although as of 1995, it provided less than 5 trillion BTUs in the United States. (*Source:* U.S. Energy Information Administration, *Annual Energy Review.*) The force of the wind blowing on a vertical surface varies jointly as the area of the surface and the square of the velocity. If a wind of 40 miles per hour exerts a force of 50 pounds on a surface of $\frac{1}{2}$ square foot, how much force will a wind of 80 miles per hour place on a surface of 2 square feet?

23. It is shown in engineering that the maximum load a cylindrical column of circular cross-section can hold varies directly as the fourth power of the diameter and inversely as the square of the height. If a column 9 feet high and 3 feet in diameter will support a load of 8 tons, how great a load will be supported by a column 12 feet high and 2 feet in diameter?

24. The force needed to keep a car from skidding on a curve varies inversely as the radius of the curve and jointly as the weight of the car and the square of the speed. If 242 pounds of force keep a 2000-pound car from skidding on a curve of radius 500 feet at 30 miles per hour, what force would keep the same car from skidding on a curve of radius 750 feet at 50 miles per hour?

25. The maximum load of a horizontal beam that is supported at both ends varies directly as the width and the square of the height and inversely as the length between the supports. A beam 6 meters long, .1 meter wide, and .06 meter high supports a load of 360 kilograms. What is the maximum load supported by a beam 16 meters long, .2 meter wide, and .08 meter high?

26. The number of long-distance phone calls between two cities in a certain time period varies directly as the populations p_1 and p_2 of the cities, and inversely as the distance between them. If 80,000 calls are made between two cities 400 miles apart, with populations of 70,000 and 100,000, how many calls are made between cities with populations of 50,000 and 75,000 that are 250 miles apart?

27. NBA basketball player Chris Webber weighs 260 pounds and is 6 feet, 10 inches tall. Use the information in Example 3 to find his body-mass index. (*Source:* Internet: "The Unofficial Chris Webber Page.")

28. A body-mass index from 27 through 29 carries a slight risk of weight-related health problems, while one of 30 or more indicates a great increase in risk. Use your own height and weight and the information in Example 3 to determine whether you are at risk.

APPENDIX D SYNTHETIC DIVISION

OBJECTIVES

1. Use synthetic division to divide by a polynomial of the form $x - k$.

2. Use the remainder theorem to evaluate a polynomial.

3. Decide whether a given number is a solution of an equation.

FOR EXTRA HELP

SSM Ap.D

Pass the Test Software

InterAct Math Tutorial Software

Video N/A

OBJECTIVE 1 Use synthetic division to divide by a polynomial of the form $x - k$. Many times when one polynomial is divided by a second, the second polynomial is of the form $x - k$, where the coefficient of the x term is 1. There is a shortcut way for doing these divisions. To see how this shortcut works, look first below left, where the division of $3x^3 - 2x + 5$ by $x - 3$ is shown. Notice that 0 was inserted for the missing x^2 term.

$$
\begin{array}{r}
3x^2 + 9x + 25 \\
x - 3\overline{\smash{)}3x^3 + 0x^2 - 2x + 5} \\
\underline{3x^3 - 9x^2} \\
9x^2 - 2x \\
\underline{9x^2 - 27x} \\
25x + 5 \\
\underline{25x - 75} \\
80
\end{array}
\qquad
\begin{array}{r}
3 \quad\ 9 \quad\ 25 \\
1 - 3\overline{\smash{)}3 \quad\ 0 \quad -2 \quad\ 5} \\
\underline{3 \quad -9} \\
9 \quad -2 \\
\underline{9 \quad -27} \\
25 \quad\ 5 \\
\underline{25 \quad -75} \\
80
\end{array}
$$

On the right, exactly the same division is shown written without the variables. This is why it is *essential* to use 0 as a placeholder in synthetic division. All the numbers in color on the right are repetitions of the numbers directly above them, so they may be omitted, as shown on the left below.

$$
\begin{array}{r}
3 \quad\ 9 \quad\ 25 \\
1 - 3\overline{\smash{)}3 \quad\ 0 \quad -2 \quad\ 5} \\
\underline{-9} \\
9 \quad -2 \\
\underline{-27} \\
25 \quad\ 5 \\
\underline{-75} \\
80
\end{array}
\qquad
\begin{array}{r}
3 \quad\ 9 \quad\ 25 \\
1 - 3\overline{\smash{)}3 \quad\ 0 \quad -2 \quad\ 5} \\
\underline{-9} \\
9 \\
\underline{-27} \\
25 \\
\underline{-75} \\
80
\end{array}
$$

The numbers in color on the left are again repetitions of the numbers directly above them; they too may be omitted, as shown on the right above.

Now the problem can be condensed. If the 3 in the dividend is brought down to the beginning of the bottom row, the top row can be omitted, since it duplicates the bottom row.

$$
\begin{array}{r}
1 - 3\overline{\smash{)}3 \quad\ 0 \quad -2 \quad\ 5} \\
\underline{-9 \quad -27 \quad -75} \\
3 \quad\ 9 \quad\ 25 \quad\ 80
\end{array}
$$

Finally, the 1 at the upper left can be omitted. Also, to simplify the arithmetic, subtraction in the second row is replaced by addition. We compensate for this by changing the -3 at the upper left to its additive inverse, 3. The result of doing all this is shown below.

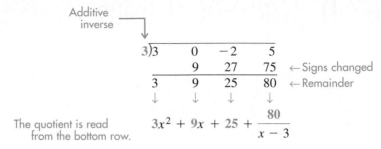

Additive inverse

$$
\begin{array}{r|rrrr}
3) & 3 & 0 & -2 & 5 \\
  &   & 9 & 27 & 75 \\ \hline
  & 3 & 9 & 25 & 80 \\
  & \downarrow & \downarrow & \downarrow & \downarrow
\end{array}
$$
$\leftarrow$ Signs changed
$\leftarrow$ Remainder

The quotient is read from the bottom row. $3x^2 + 9x + 25 + \dfrac{80}{x-3}$

The first three numbers in the bottom row are the coefficients of the quotient polynomial with degree 1 less than the degree of the dividend. The last number gives the remainder.

Synthetic Division

This shortcut procedure is called **synthetic division.** It is used only when dividing a polynomial by a binomial of the form $x - k$.

EXAMPLE 1   Using Synthetic Division

Use synthetic division to divide $5x^2 + 16x + 15$ by $x + 2$.

As mentioned above, we use synthetic division only when dividing by a polynomial of the form $x - k$. Get $x + 2$ in this form by writing it as

$$x + 2 = x - (-2),$$

where $k = -2$. Now write the coefficients of $5x^2 + 16x + 15$, placing -2 to the left.

$x + 2$ leads to -2.

$$
-2)\overline{\begin{array}{rrr} 5 & 16 & 15 \end{array}} \quad \leftarrow \text{Coefficients}
$$

Bring down the 5, and multiply: $-2 \cdot 5 = -10$.

$$
\begin{array}{r|rrr}
-2) & 5 & 16 & 15 \\
    &   & -10 & \\ \hline
    & 5 &   &
\end{array}
$$

Add 16 and -10, getting 6. Multiply 6 and -2 to get -12.

$$
\begin{array}{r|rrr}
-2) & 5 & 16 & 15 \\
    &   & -10 & -12 \\ \hline
    & 5 & 6 &
\end{array}
$$

Add 15 and -12, getting 3.

$$
\begin{array}{r|rrr}
-2) & 5 & 16 & 15 \\
    &   & -10 & -12 \\ \hline
    & 5 & 6 & 3
\end{array}
$$
$\leftarrow$ Remainder

The result is read from the bottom row.

$$\frac{5x^2 + 16x + 15}{x + 2} = 5x + 6 + \frac{3}{x + 2}$$

EXAMPLE 2   Using Synthetic Division with Missing Terms

Use synthetic division to find $(-4x^5 + x^4 + 6x^3 + 2x^2 + 50) \div (x - 2)$.

Use the steps given above, inserting a 0 for the missing x term.

$$
\begin{array}{r|rrrrrr}
2) & -4 & 1 & 6 & 2 & 0 & 50 \\
   &   & -8 & -14 & -16 & -28 & -56 \\
\hline
   & -4 & -7 & -8 & -14 & -28 & -6
\end{array}
$$

Read the result from the bottom row.

$$\frac{-4x^5 + x^4 + 6x^3 + 2x^2 + 50}{x - 2} = -4x^4 - 7x^3 - 8x^2 - 14x - 28 + \frac{-6}{x - 2}$$

OBJECTIVE 2 Use the remainder theorem to evaluate a polynomial. We can use synthetic division to evaluate polynomials. For example, in the synthetic division of Example 2, where the polynomial was divided by $x - 2$, the remainder was -6.

Replacing x in the polynomial with 2 gives

$$
\begin{aligned}
-4x^5 + x^4 + 6x^3 + 2x^2 + 50 &= -4 \cdot 2^5 + 2^4 + 6 \cdot 2^3 + 2 \cdot 2^2 + 50 \\
&= -4 \cdot 32 + 16 + 6 \cdot 8 + 2 \cdot 4 + 50 \\
&= -128 + 16 + 48 + 8 + 50 \\
&= -6,
\end{aligned}
$$

the same number as the remainder; that is, dividing by $x - 2$ produced a remainder equal to the result when x is replaced with 2. This always happens, as the following remainder theorem states.

Remainder Theorem

If the polynomial $P(x)$ is divided by $x - k$, then the remainder is equal to $P(k)$.

This result is proved in more advanced courses.

EXAMPLE 3   Using the Remainder Theorem

Let $P(x) = 2x^3 - 5x^2 - 3x + 11$. Find $P(-2)$.

Use the remainder theorem; divide $P(x)$ by $x - (-2)$.

$$
\begin{array}{r|rrrr}
-2) & 2 & -5 & -3 & 11 \\
    &   & -4 & 18 & -30 \\
\hline
    & 2 & -9 & 15 & -19
\end{array}
$$
← Remainder

By this result, $P(-2) = -19$.

OBJECTIVE 3 Decide whether a given number is a solution of an equation. The remainder theorem also can be used to show that a given number is a solution of an equation.

EXAMPLE 4 Using the Remainder Theorem

Show that -5 is a solution of the equation

$$2x^4 + 12x^3 + 6x^2 - 5x + 75 = 0.$$

One way to show that -5 is a solution is by substituting -5 for x in the equation. However, an easier way is to use synthetic division and the remainder theorem given above.

$$
\begin{array}{c}
\text{Proposed solution} \rightarrow\ -5)\overline{\begin{array}{ccccc} 2 & 12 & 6 & -5 & 75 \\ & -10 & -10 & 20 & -75 \\ \hline 2 & 2 & -4 & 15 & 0 \end{array}} \ \leftarrow \text{Remainder}
\end{array}
$$

Since the remainder is 0, the polynomial has a value of 0 when $k = -5$, and so -5 is a solution of the given equation.

The synthetic division above also shows that $x - (-5)$ divides the polynomial with 0 remainder. Thus $x - (-5) = x + 5$ is a *factor* of the polynomial and

$$2x^4 + 12x^3 + 6x^2 - 5x + 75 = (x + 5)(2x^3 + 2x^2 - 4x + 15).$$

The second factor is the quotient polynomial found in the last row of the synthetic division.

APPENDIX D EXERCISES

1. What is the purpose of synthetic division?

2. What type of polynomial divisors may be used with synthetic division?

Use synthetic division to perform the division. See Examples 1 and 2.

3. $\dfrac{x^2 - 6x + 5}{x - 1}$

4. $\dfrac{x^2 - 4x - 21}{x + 3}$

5. $\dfrac{4m^2 + 19m - 5}{m + 5}$

6. $\dfrac{3k^2 - 5k - 12}{k - 3}$

7. $\dfrac{2a^2 + 8a + 13}{a + 2}$

8. $\dfrac{4y^2 - 5y - 20}{y - 4}$

9. $(p^2 - 3p + 5) \div (p + 1)$

10. $(z^2 + 4z - 6) \div (z - 5)$

11. $\dfrac{4a^3 - 3a^2 + 2a - 3}{a - 1}$

12. $\dfrac{5p^3 - 6p^2 + 3p + 14}{p + 1}$

13. $(x^5 - 2x^3 + 3x^2 - 4x - 2) \div (x - 2)$

14. $(2y^5 - 5y^4 - 3y^2 - 6y - 23) \div (y - 3)$

15. $(-4r^6 - 3r^5 - 3r^4 + 5r^3 - 6r^2 + 3r + 3) \div (r - 1)$

16. $(2t^6 - 3t^5 + 2t^4 - 5t^3 + 6t^2 - 3t - 2) \div (t - 2)$

17. $(-3y^5 + 2y^4 - 5y^3 - 6y^2 - 1) \div (y + 2)$

18. $(m^6 + 2m^4 - 5m + 11) \div (m - 2)$

19. $\dfrac{y^3 + 1}{y - 1}$

20. $\dfrac{z^4 + 81}{z - 3}$

Use the remainder theorem to find P(k). See Example 3.

21. $P(x) = 2x^3 - 4x^2 + 5x - 3; \quad k = 2$

22. $P(y) = y^3 + 3y^2 - y + 5; \quad k = -1$

23. $P(r) = -r^3 - 5r^2 - 4r - 2; \quad k = -4$

24. $P(z) = -z^3 + 5z^2 - 3z + 4; \quad k = 3$

25. $P(y) = 2y^3 - 4y^2 + 5y - 33; \quad k = 3$ **26.** $P(x) = x^3 - 3x^2 + 4x - 4; \quad k = 2$

27. Explain why a zero remainder in synthetic division of $P(x)$ by $x - k$ indicates that k is a solution of the equation $P(x) = 0$.

28. Explain why it is important to insert zeros as placeholders for missing terms before performing synthetic division.

Use synthetic division to decide whether the given number is a solution of the equation. See Example 4.

29. $x^3 - 2x^2 - 3x + 10 = 0; \quad x = -2$

30. $x^3 - 3x^2 - x + 10 = 0; \quad x = -2$

31. $m^4 + 2m^3 - 3m^2 + 8m - 8 = 0; \quad m = -2$

32. $r^4 - r^3 - 6r^2 + 5r + 10 = 0; \quad r = -2$

33. $3a^3 + 2a^2 - 2a + 11 = 0; \quad a = -2$

34. $3z^3 + 10z^2 + 3z - 9 = 0; \quad z = -2$

35. $2x^3 - x^2 - 13x + 24 = 0; \quad x = -3$

36. $5p^3 + 22p^2 + p - 28 = 0; \quad p = -4$

APPENDIX E REVIEW OF EXPONENTS, POLYNOMIALS, AND FACTORING

(TRANSITION FROM BEGINNING TO INTERMEDIATE ALGEBRA)

OBJECTIVES

1 Review the basic rules for exponents.

2 Review addition, subtraction, and multiplication of polynomials.

3 Review factoring techniques.

FOR EXTRA HELP

📖 SSM Ap. E

💿 Pass the Test Software

💿 InterAct Math Tutorial Software

📼 Video N/A

OBJECTIVE 1 **Review the basic rules for exponents.** In Sections 5.2 and 5.5 we introduced the following definitions and rules for working with exponents.

Definitions and Rules for Exponents

If no denominators are zero, for any integers m and n:

		Examples
Product rule	$a^m \cdot a^n = a^{m+n}$	$7^4 \cdot 7^3 = 7^7$
Zero exponent	$a^0 = 1$	$(-3)^0 = 1$
Negative exponent	$a^{-n} = \dfrac{1}{a^n}$	$5^{-3} = \dfrac{1}{5^3}$
Quotient rule	$\dfrac{a^m}{a^n} = a^{m-n}$	$\dfrac{2^2}{2^5} = 2^{-3} = \dfrac{1}{2^3}$
Power rules (a)	$(a^m)^n = a^{mn}$	$(4^2)^3 = 4^6$
(b)	$(ab)^m = a^m b^m$	$(3k)^4 = 3^4 k^4$
(c)	$\left(\dfrac{a}{b}\right)^m = \dfrac{a^m}{b^m}$	$\left(\dfrac{2}{3}\right)^{10} = \dfrac{2^{10}}{3^{10}}$
(d)	$\dfrac{a^{-m}}{b^{-n}} = \dfrac{b^n}{a^m}$	$\dfrac{5^{-3}}{3^{-5}} = \dfrac{3^5}{5^3}$
(e)	$\left(\dfrac{a}{b}\right)^{-m} = \left(\dfrac{b}{a}\right)^m$	$\left(\dfrac{4}{7}\right)^{-2} = \left(\dfrac{7}{4}\right)^2$

EXAMPLE 1 Applying Definitions and Rules for Exponents

Apply the definitions and rules for exponents to simplify each of the following, expressing the final answer using only positive exponents. Assume that all variables represent nonzero real numbers.

(a) $(x^2 y^{-3})(x^{-5} y^7) = (x^{2+(-5)})(y^{-3+7})$

$$= x^{-3} y^4$$

$$= \frac{1}{x^3} y^4$$

$$= \frac{y^4}{x^3}$$

(b) $(-5)^0 + (-5^0) = 1 + (-1) = 0$ $-5^0 = -1 \cdot 5^0 = -1 \cdot 1 = -1$

(c) $\dfrac{(t^5 s^{-4})^2}{(t^{-3} s^5)^3} = \dfrac{t^{10} s^{-8}}{t^{-9} s^{15}} = \dfrac{t^{10} t^9}{s^{15} s^8} = \dfrac{t^{19}}{s^{23}}$

(d) $\left(\dfrac{-3x^{-4} y}{x^5 y^{-4}}\right)^{-2} = \left(\dfrac{x^5 y^{-4}}{-3x^{-4} y}\right)^2 = \dfrac{x^{10} y^{-8}}{9x^{-8} y^2} = \dfrac{x^{18}}{9y^{10}}$

(e) $(2x^2 y^3 z)^2 (x^4 y^2)^3 = (4x^4 y^6 z^2)(x^{12} y^6) = 4x^{16} y^{12} z^2$

O B J E C T I V E 2   Review addition, subtraction, and multiplication of polynomials. These arithmetic operations with polynomials were covered in Sections 5.1, 5.3, and 5.4. We review them here.

Adding and Subtracting Polynomials

To add polynomials, add like terms. To subtract polynomials, change all signs on the second polynomial and add the result to the first polynomial.

E X A M P L E 2   Adding and Subtracting Polynomials

Add or subtract as indicated.

(a) $(-4x^3 + 3x^2 - 8x + 2) + (5x^3 - 8x^2 + 12x - 3)$

$\quad = (-4 + 5)x^3 + (3 - 8)x^2 + (-8 + 12)x + (2 - 3)$

$\quad = x^3 - 5x^2 + 4x - 1$

(b) $-4(x^2 + 3x - 6) - (2x^2 - 3x + 7) = -4x^2 - 12x + 24 - 2x^2 + 3x - 7$

$\qquad\qquad\qquad\qquad\qquad\qquad\qquad\qquad\quad = -6x^2 - 9x + 17$

(c) Subtract.
$$\begin{array}{r} 2t^2 - 3t - 4 \\ -8t^2 + 4t - 1 \end{array}$$

Change the sign of each term in $-8t^2 + 4t - 1$, and add.

$$\begin{array}{r} 2t^2 - 3t - 4 \\ 8t^2 - 4t + 1 \\ \hline 10t^2 - 7t - 3 \quad \text{Add.} \end{array}$$

Multiplying Polynomials

To multiply two polynomials, multiply each term of the second polynomial by each term of the first polynomial and add the products. In particular, when multiplying two binomials, use the FOIL method. (See Section 5.3.)

There are also several special product rules that are useful when multiplying binomials.

Special Product Rules

$$(x + y)(x - y) = x^2 - y^2$$
$$(x + y)^2 = x^2 + 2xy + y^2$$
$$(x - y)^2 = x^2 - 2xy + y^2$$

EXAMPLE 3 Multiplying Polynomials

Find each product.

(a) $(4y - 1)(3y + 2) = 4y(3y) + 4y(2) - 1(3y) - 1(2)$ FOIL
$$= 12y^2 + 8y - 3y - 2$$
$$= 12y^2 + 5y - 2$$

(b) $(3x + 5y)(3x - 5y) = (3x)^2 - (5y)^2$
$$= 9x^2 - 25y^2$$

(c) $(2t + 3)^2 = (2t)^2 + 2(2t)(3) + 3^2$
$$= 4t^2 + 12t + 9$$

(d) $(5x - 1)^2 = (5x)^2 - 2(5x)(1) + 1^2$
$$= 25x^2 - 10x + 1$$

(e) $(3x + 2)(9x^2 - 6x + 4)$

Multiply vertically.

$$
\begin{array}{r}
9x^2 - 6x + 4 \\
3x + 2 \\
\hline
18x^2 - 12x + 8 \\
27x^3 - 18x^2 + 12x \\
\hline
27x^3 \qquad\qquad + 8
\end{array}
$$

The product is the sum of two cubes, $27x^3 + 8$.

OBJECTIVE 3 Review factoring techniques. Factoring, which involves writing a polynomial as a product, was covered in Chapter 6. Here are some general guidelines to use when factoring.

Factoring a Polynomial

Step 1 Is there a common factor?

Step 2 How many terms are in the polynomial?
Four terms Can the polynomial be factored by grouping?
Three terms Is it a perfect square trinomial? If the trinomial is not a perfect square, check to see whether the coefficient of the squared term is 1. If so, use the method of Section 6.2. If the coefficient of the squared term is not 1, use the general factoring methods of Section 6.3.
Two terms Check to see whether it is either the difference of two squares or the sum or difference of two cubes (Section 6.4).

Step 3 Can any factors be factored further?

EXAMPLE 4 Factoring Polynomials

Factor each polynomial completely.

(a) $6x^2y^3 - 12x^3y^2 = 6x^2y^2(y - 2x)$ $6x^2y^2$ is the greatest common factor.

(b) $3x^2 - x - 2 = (3x + 2)(x - 1)$

To find the factors, find two terms that multiply to give $3x^2$ ($3x$ and x) and two terms that multiply to give -2 ($+2$ and -1). Make sure that the sum of the outer and inner products in the factored form is $-x$, since $-3x + 2x = -x$. To check, multiply the factors using the FOIL method.

(c) $100t^4 - 81 = (10t^2)^2 - 9^2$ Difference of two squares

$$= (10t^2 + 9)(10t^2 - 9)$$

(d) $4x^2 + 20xy + 25y^2 = (2x + 5y)^2$ Perfect square trinomial

The terms $4x^2$ and $25y^2$ are both perfect squares, so factor as a perfect square trinomial. To check, take twice the product of the two terms in the squared binomial.

$$2(2x)(5y) = 20xy$$

Since $20xy$ is the middle term of the trinomial, the trinomial is a perfect square and can be factored as $(2x + 5y)^2$.

(e) $1000x^3 - 27 = (10x)^3 - 3^3$ Difference of two cubes

$$= (10x - 3)[(10x)^2 + 10x(3) + 3^2]$$
$$= (10x - 3)(100x^2 + 30x + 9)$$

(f) $6xy - 3x + 4y - 2$

Since there are four terms, try factoring by grouping.

$$6xy - 3x + 4y - 2 = (6xy - 3x) + (4y - 2)$$
$$= 3x(2y - 1) + 2(2y - 1)$$
$$= (2y - 1)(3x + 2)$$

In the final step, factor out the greatest common factor, the binomial $2y - 1$.

APPENDIX E EXERCISES

Apply the definitions and rules for exponents to simplify each of the following, expressing the final answer using only positive exponents. Assume that all variables represent positive real numbers. See Example 1.

1. $(a^4b^{-3})(a^{-6}b^2)$ **2.** $(t^{-3}s^{-5})(t^8s^{-2})$

3. $(5x^{-2}y)^2(2xy^4)^2$ **4.** $(7x^{-3}y^4)^3(2x^{-1}y^{-4})^2$

5. $-6^0 + (-6)^0$ **6.** $(-12)^0 - 12^0$

7. $\dfrac{(2w^{-1}x^2y^{-1})^3}{(4w^5x^{-2}y)^2}$ **8.** $\dfrac{(5p^{-3}q^2r^{-4})^2}{(10p^4q^{-1}r^5)^{-1}}$

9. $\left(\dfrac{-4a^{-2}b^4}{a^3b^{-1}}\right)^{-3}$ **10.** $\left(\dfrac{r^{-3}s^{-8}}{-6r^2s^{-4}}\right)^{-2}$

11. $(7x^{-4}y^2z^{-2})^{-2}(7x^4y^{-1}z^3)^2$ **12.** $(3m^{-5}n^2p^{-4})^3(3m^4n^{-3}p^5)^{-2}$

Add or subtract as indicated. See Example 2.

13. $(2a^4 + 3a^3 - 6a^2 + 5a - 12) + (-8a^4 + 8a^3 - 14a^2 + 21a - 3)$

14. $(-6r^4 - 3r^3 + 12r^2 - 9r + 9) + (8r^4 - 13r^3 - 14r^2 - 10r - 3)$

15. $(6x^3 - 12x^2 + 3x - 4) - (-2x^3 + 6x^2 - 3x + 12)$

16. $(10y^3 - 4y^2 + 8y + 7) - (7y^3 + 5y^2 - 2y - 13)$

17. Add.
$$5x^2y + 2xy^2 + y^3$$
$$\underline{-4x^2y - 3xy^2 + 5y^3}$$

18. Add.
$$6ab^3 - 2a^2b^2 + 3b^5$$
$$\underline{8ab^3 + 12a^2b^2 - 8b^5}$$

19. $3(5x^2 - 12x + 4) - 2(9x^2 + 13x - 10)$

20. $-4(2t^3 - 3t^2 + 4t - 1) - 3(-8t^3 + 3t^2 - 2t + 9)$

21. Subtract.
$$6x^3 - 2x^2 + 3x - 1$$
$$\underline{-4x^3 + 2x^2 - 6x + 3}$$

22. Subtract.
$$-9y^3 - 2y^2 + 3y - 8$$
$$\underline{-8y^3 + 4y^2 + 3y + 1}$$

Find each product. See Example 3.

23. $(3x + 1)(2x - 7)$

24. $(5z + 3)(2z - 3)$

25. $(4x - 1)(x - 2)$

26. $(7t - 3)(t - 4)$

27. $(4t + 3)(4t - 3)$

28. $(6x + 1)(6x - 1)$

29. $(2y^2 + 4)(2y^2 - 4)$

30. $(3b^3 + 2t)(3b^3 - 2t)$

31. $(4x - 3)^2$

32. $(9t + 2)^2$

33. $(6r + 5y)^2$

34. $(8m - 3n)^2$

35. $(c + 2d)(c^2 - 2cd + 4d^2)$

36. $(f + 3g)(f^2 - 3fg + 9g^2)$

37. $(4x - 1)(16x^2 + 4x + 1)$

38. $(5r - 2)(25r^2 + 10r + 4)$

39. $(7t + 5s)(2t^2 + 5st - s^2)$

40. $(8p + 3q)(2p^2 - 4pq + q^2)$

Factor each polynomial completely. See Example 4.

41. $8x^3y^4 + 12x^2y^3 + 36xy^4$

42. $10m^5n + 4m^2n^3 + 18m^3n^2$

43. $x^2 - 2x - 15$

44. $x^2 + x - 12$

45. $2x^2 - 9x - 18$

46. $3x^2 + 2x - 8$

47. $36t^2 - 25$

48. $49r^2 - 9$

49. $16t^2 + 24t + 9$

50. $25t^2 + 90t + 81$

51. $4m^2p - 12mnp + 9n^2p$

52. $16p^2r - 40pqr + 25q^2r$

53. $x^3 + 1$

54. $x^3 + 27$

55. $8t^3 + 125$

56. $27s^3 + 64$

57. $t^6 - 125$

58. $w^6 - 27$

59. $5xt + 15xr + 2yt + 6yr$

60. $3am + 18mb + 2an + 12nb$

61. $6ar + 12br - 5as - 10bs$

62. $7mt + 35ms - 2nt - 10ns$

63. $t^4 - 1$

64. $r^4 - 81$

65. $4x^2 + 12xy + 9y^2 - 1$

66. $81t^2 + 36ty + 4y^2 - 9$

Answers to Selected Exercises

In this section we provide the answers that we think most students will obtain when they work the exercises using the methods explained in the text. If your answer does not look exactly like the one given here, it is not necessarily wrong. In many cases there are equivalent forms of the answer that are correct. For example, if the answer section shows $\frac{3}{4}$ and your answer is .75, you have obtained the right answer but written it in a different (yet equivalent) form. Unless the directions specify otherwise, .75 is just as valid an answer as $\frac{3}{4}$.

In general, if your answer does not agree with the one given in the text, see whether it can be transformed into the other form. If it can, then it is the correct answer. If you still have doubts, talk with your instructor.

CHAPTER 1 THE REAL NUMBER SYSTEM

SECTION 1.1 (PAGE 9)

CONNECTIONS **Page 2:** Answers will vary.

EXERCISES **1.** true **3.** false; The fraction $\frac{17}{51}$ can be reduced to $\frac{1}{3}$. **5.** false; *Product* refers to multiplication, so the product of 8 and 2 is 16. **7.** prime **9.** composite; $2 \cdot 2 \cdot 2 \cdot 2 \cdot 2 \cdot 2$ **11.** composite; $2 \cdot 7 \cdot 13 \cdot 19$ **13.** neither **15.** composite; $2 \cdot 3 \cdot 5$ **17.** composite; $2 \cdot 2 \cdot 5 \cdot 5 \cdot 5$ **19.** composite; $2 \cdot 2 \cdot 31$ **21.** prime **23.** $\frac{1}{2}$ **25.** $\frac{5}{6}$ **27.** $\frac{1}{3}$ **29.** $\frac{6}{5}$ **31.** (c) **33.** $\frac{24}{35}$ **35.** $\frac{6}{25}$ **37.** $\frac{6}{5}$ or $1\frac{1}{5}$ **39.** $\frac{232}{15}$ or $15\frac{7}{15}$ **41.** $\frac{10}{3}$ or $3\frac{1}{3}$ **43.** 12 **45.** $\frac{1}{16}$ **47.** $\frac{84}{47}$ or $1\frac{37}{47}$ **51.** $\frac{2}{3}$ **53.** $\frac{8}{9}$ **55.** $\frac{27}{8}$ or $3\frac{3}{8}$ **57.** $\frac{17}{36}$ **59.** $\frac{11}{12}$ **61.** $\frac{4}{3}$ or $1\frac{1}{3}$ **63.** 6 cups **65.** 34 dollars **67.** $\frac{9}{16}$ inch **69.** $618\frac{3}{4}$ feet **71.** $5\frac{5}{24}$ inches **73.** $\frac{1}{3}$ cup **75.** 650 **77.** (a) Crum (b) Jordan (c) Jordan (d) Baldock (e) Tobin and Perry; $\frac{1}{2}$ **79.** (b)

SECTION 1.2 (PAGE 19)

EXERCISES **1.** false; $4 + 3(8 - 2) = 4 + 3 \cdot 6 = 4 + 18 = 22$. The common error leading to 42 is adding 4 to 3 and then multiplying by 6. One must follow the rules for order of operations. **3.** false; The correct interpretation is $4 = 16 - 12$.

5. 49   **7.** 144   **9.** 64   **11.** 1000   **13.** 81   **15.** 1024   **17.** $\dfrac{16}{81}$ **19.** .000064   **23.** 32   **25.** $\dfrac{49}{30}$ or $1\dfrac{19}{30}$
27. 12   **29.** 23.01   **31.** 95   **33.** 90   **35.** 14   **37.** 9   **41.** true   **43.** false   **45.** true   **47.** true
49. false   **51.** false   **53.** true   **55.** $15 = 5 + 10$ **57.** $9 > 5 - 4$ **59.** $16 \neq 19$ **61.** $2 \leq 3$ **63.** Seven is less than nineteen. True   **65.** Three is not equal to six. True   **67.** Eight is greater than or equal to eleven. False **69.** Answers will vary. One example is $5 + 3 \geq 2 \cdot 2$.   **71.** $30 > 5$ **73.** $3 \leq 12$ **75.** is younger than   **77.** The inequality symbol $\geq$ implies a true statement if 12 equals 12 *or* if 12 is greater than 12.   **79.** December 1996, January 1997, May 1997, June 1997, November 1997   **81.** (a) .7   (b) 1.3%

SECTION 1.3 (PAGE 25)

EXERCISES   **1.** 10   **3.** $12 + x$; 21   **5.** no   **7.** $2x^3 = 2 \cdot x \cdot x \cdot x$, while $2x \cdot 2x \cdot 2x = (2x)^3$.   **9.** The exponent 2 applies only to its base, which is x. (The expression $(4x)^2$ would require multiplying 4 by $x = 3$ first.)   **11.** (Answers will vary.) Two such pairs are $x = 0, y = 6$ and $x = 1, y = 4$. To determine them, choose a value for x, substitute it into the expression $2x + y$, and then subtract the value of $2x$ from 6.   **13.** (a) 13   (b) 15   **15.** (a) 20   (b) 30   **17.** (a) 64
(b) 144   **19.** (a) $\dfrac{5}{3}$ (b) $\dfrac{7}{3}$ **21.** (a) $\dfrac{7}{8}$ (b) $\dfrac{13}{12}$ **23.** (a) 52   (b) 114   **25.** (a) 25.836   (b) 38.754
27. (a) 24   (b) 28   **29.** (a) 12   (b) 33   **31.** (a) 6   (b) $\dfrac{9}{5}$ **33.** (a) $\dfrac{4}{3}$ (b) $\dfrac{13}{6}$ **35.** (a) $\dfrac{2}{7}$ (b) $\dfrac{16}{27}$
37. (a) 12   (b) 55   **39.** (a) 1   (b) $\dfrac{28}{17}$ **41.** (a) 3.684   (b) 8.841   **43.** $12x$ **45.** $x + 7$ **47.** $x - 2$
49. $7 - x$ **51.** $x - 6$ **53.** $\dfrac{12}{x}$ **55.** $6(x - 4)$ **57.** No, it is a connective word that joins the two factors: the number and 6.   **59.** yes   **61.** no   **63.** yes   **65.** yes   **67.** yes   **69.** $x + 8 = 18$; 10   **71.** $16 - \dfrac{3}{4}x = 13$; 4
73. $2x + 1 = 5$; 2   **75.** $3x = 2x + 8$; 8   **77.** expression   **79.** equation   **81.** equation   **83.** $10.50; less by $.33
85. $12.10; less by $.27

SECTION 1.4 (PAGE 34)

EXERCISES   **1.** 4   **3.** 0   **5.** One example is $\sqrt{12}$. There are others.   **7.** true   **9.** true   **11.** (a) 3, 7   (b) 0, 3, 7
(c) $-9, 0, 3, 7$ (d) $-9, -1\dfrac{1}{4}, -\dfrac{3}{5}, 0, 3, 5.9, 7$ (e) $-\sqrt{7}, \sqrt{5}$ (f) All are real numbers.   **15.** 93,000   **17.** $-30°$
19. $-31,532$ **21.** -8 **23.** [number line from -6 to 2]   **25.** [number line from -6 to 4]   **27.** $-3\dfrac{4}{5}$ $-1\dfrac{5}{8}$ $\dfrac{1}{4}$ $2\dfrac{1}{2}$ [number line from -4 to 4]
29. (a) A   (b) A   (c) B   (d) B   **31.** (a) 2   (b) 2   **33.** (a) -6 (b) 6   **35.** (a) -3 (b) 3   **37.** (a) 0   (b) 0
39. $a - b$ **41.** -12 **43.** -8 **45.** 3   **47.** $|-3|$ or 3   **49.** $-|-6|$ or -6 **51.** $|5 - 3|$ or 2   **53.** true
55. true   **57.** true   **59.** false   **61.** true   **63.** false   **65.** Softwood plywood from 1995 to 1996 represents the greatest drop.   **67.** true

Answers will vary in Exercises 69–77.   **69.** true: $a = 0$ or $b = 0$ or both $a = 0$ and $b = 0$; false: Choose any values for a and b so that neither a nor b is zero   **71.** true: Choose a to be the opposite of b ($a = -b$); false: $a \neq -b$ **73.** $\dfrac{1}{2}, \dfrac{5}{8}, 1\dfrac{3}{4}$
75. $-3\dfrac{1}{2}, -\dfrac{2}{3}, \dfrac{3}{7}$ **77.** $\sqrt{5}, \pi, -\sqrt{3}$

SECTION 1.5 (PAGE 44)

EXERCISES   **1.** negative   **3.** -3; 5   **7.** 2   **9.** -3 **11.** -10 **13.** -13 **15.** -15.9 **17.** -1 **19.** 13
21. -3 **23.** -4 **25.** -10 **27.** -16 **29.** 11   **31.** 19   **33.** -4 **35.** 5   **37.** 0   **39.** $\dfrac{3}{4}$ **41.** -8
43. $\dfrac{15}{8}$ **45.** -6.3 **47.** -24 **49.** -16 **51.** no   **53.** Answers will vary. One example is $-8 - (-2) = -6$.
55. -1 **57.** $\dfrac{17}{9}$ or $1\dfrac{8}{9}$ **59.** $-5 + 12 + 6$; 13   **61.** $[-19 + (-4)] + 14$; -9 **63.** $[-4 + (-10)] + 12$; -2

65. $[8 + (-18)] + 4; -6$ **67.** $4 - (-8); 12$ **69.** $-2 - 8; -10$ **71.** $[9 + (-4)] - 7; -2$
73. $[8 - (-5)] - 12; 1$ **75.** \$13.2 billion **77.** $-\$26.0$ billion **79.** 50,395 feet **81.** 1345 feet **83. (a)** -10
(b) 5 **(c)** -12 **(d)** -31 **85.** 45°F **87.** -41°F **89.** 14,776 feet **91.** 365 pounds **93.** \$323.83

SECTION 1.6 (PAGE 57)

EXERCISES 1. greater than 0 **3.** greater than 0 **5.** less than 0 **7.** greater than 0 **9.** equal to 0 **11.** 20
13. -28 **15.** 80 **17.** 0 **19.** $\dfrac{5}{6}$ **21.** $\dfrac{3}{2}$ **23.** $-32, -16, -8, -4, -2, -1, 1, 2, 4, 8, 16, 32$
25. $-40, -20, -10, -8, -5, -4, -2, -1, 1, 2, 4, 5, 8, 10, 20, 40$ **27.** $-31, -1, 1, 31$ **29.** -3 **31.** -2
33. 16 **35.** 0 **37.** 25.63 **39.** $\dfrac{3}{2}$ **43.** -11 **45.** -2 **47.** 35 **49.** 6 **51.** -18 **53.** 67
55. -8 **57.** 3 **59.** 7 **61.** 4 **63.** -3 **65.** First, substitute -3 for x and 4 for y to get $3(-3) + 2(4)$. Then
perform the multiplications to get $-9 + 8$. Finally, add to get -1. **67.** 47 **69.** 72 **71.** $-\dfrac{78}{25}$ **73.** 0 **75.** -23
77. 2 **79.** $9 + (-9)(2); -9$ **81.** $-4 - 2(-1)(6); 8$ **83.** $(1.5)(-3.2) - 9; -13.8$ **85.** $12[9 - (-8)]; 204$
87. $\dfrac{-12}{-5 + (-1)}; 2$ **89.** $\dfrac{15 + (-3)}{4(-3)}; -1$ **91.** $\dfrac{\left(-\dfrac{1}{2}\right)\left(\dfrac{3}{4}\right)}{-\dfrac{2}{3}}; \dfrac{9}{16}$ **93.** $6x = -42; -7$ **95.** $\dfrac{x}{3} = -3; -9$
97. $x - 6 = 4; 10$ **99.** $x + 5 = -5; -10$ **101.** $8\dfrac{2}{5}$ **103.** 2 **105.** \$12.60 (to the nearest cent)
107. 0 **109. (a)** 6 is divisible by 2. **(b)** 9 is not divisible by 2. **111. (a)** 64 is divisible by 4. **(b)** 35 is not divisible
by 4. **113. (a)** 2 is divisible by 2 and $1 + 5 + 2 + 4 + 8 + 2 + 2 = 24$ is divisible by 3. **(b)** While 0 is divisible by 2,
$2 + 8 + 7 + 3 + 5 + 9 + 0 = 34$ is not divisible by 3. **115. (a)** $4 + 1 + 1 + 4 + 1 + 0 + 7 = 18$ is divisible by 9.
(b) $2 + 2 + 8 + 7 + 3 + 2 + 1 = 25$ is not divisible by 9.

SECTION 1.7 (PAGE 67)

EXERCISES 1. B **3.** C **5.** B **7.** B **9.** G **11.** commutative property **13.** associative property
15. associative property **17.** inverse property **19.** inverse property **21.** identity property **23.** commutative
property **25.** distributive property **27.** identity property **29.** distributive property **31.** identity property
35. $7 + r$ **37.** s **39.** $-6x + (-6) \cdot 7; -6x - 42$ **41.** $w + [5 + (-3)]; w + 2$ **43.** 11 **45.** 0 **47.** -38
49. 1 **51.** Subtraction is not associative. **53.** The expression following the first equals sign should be $-3(4) - 3(-6)$.
The student forgot that 6 should be preceded by a $-$ sign. The correct work is $-3(4 - 6) = -3(4) - 3(-6) = -12 + 18 = 6$.
55. $(5 + 1)x; 6x$ **57.** $4t + 12$ **59.** $-8r - 24$ **61.** $-5y + 20$ **63.** $-16y - 20z$ **65.** $8(z + w)$
67. $7(2v + 5r)$ **69.** $24r + 32s - 40y$ **71.** $(1 + 1 + 1)q; 3q$ **73.** $(-5 + 1)x; -4x$ **75.** $-4t - 3m$
77. $5c + 4d$ **79.** $3q - 5r + 8s$ **81.** for example, "putting on your socks" and "putting on your shoes" **83.** 0
84. $-3(5) + (-3)(-5)$ **85.** -15 **86.** We must interpret $(-3)(-5)$ as 15, since it is the additive inverse of -15.

SECTION 1.8 (PAGE 73)

EXERCISES 1. false **3.** true **5.** (c) **7.** (a) **9.** $4r + 11$ **11.** $5 + 2x - 6y$ **13.** $-7 + 3p$ **15.** -12
17. 5 **19.** 1 **21.** -1 **23.** 74 **25.** Answers will vary. For example, $-3x$ and $4x$. **27.** like **29.** unlike
31. like **33.** unlike **35.** Apples and oranges are examples of unlike fruits, just like x and y are unlike terms. We cannot
add x and y to get an expression any simpler than $x + y$; we cannot add, for example, 2 apples and 3 oranges to obtain 5 fruits
that are all alike. **37.** $9k - 5$ **39.** $-\dfrac{1}{3}t - \dfrac{28}{3}$ **41.** $-4.1r + 5.6$ **43.** $-2y^2 + 3y^3$ **45.** $-19p + 16$
47. $-4y + 22$ **49.** $-16y + 63$ **51.** $4k - 7$ **53.** $-23.7y - 12.6$ **55.** $(x + 3) + 5x; 6x + 3$
57. $(13 + 6x) - (-7x); 13 + 13x$ **59.** $2(3x + 4) - (-4 + 6x); 12$ **61.** Wording will vary. One example is "the
difference between 9 times a number and the sum of the number and 2." **63.** 2, 3, 4, 5 **64.** 1 **65. (a)** 1, 2, 3, 4
(b) 3, 4, 5, 6 **(c)** 4, 5, 6, 7 **66.** The value of $x + b$ also increases by 1 unit. **67. (a)** 2, 4, 6, 8 **(b)** 2, 5, 8, 11
(c) 2, 6, 10, 14 **68.** m **69. (a)** 7, 9, 11, 13 **(b)** 5, 8, 11, 14 **(c)** 1, 5, 9, 13 In comparison, we see that while the
values themselves are different, the number of units of increase is the same as the corresponding parts of Exercise 67. **70.** m

CHAPTER 1 REVIEW EXERCISES (PAGE 79)

1. $\dfrac{3}{4}$ **3.** $\dfrac{9}{40}$ **5.** 625 **7.** .0000000032 **9.** 27 **11.** 39 **13.** true **15.** false **17.** $5 + 2 \neq 10$

19. 30 **21.** 14 **23.** $x + 6$ **25.** $6x - 9$ **27.** yes **29.** $2x - 6 = 10; 8$ **31.**

$$\begin{array}{c} -\tfrac{1}{2} \quad 2.5 \\ \xleftarrow{\hspace{0.3cm}}|+|\bullet\bullet|+|\bullet|+|\xrightarrow{\hspace{0.3cm}} \\ -4 \;\; -2 \;\;\; 0 \;\;\; 2 \;\;\; 4 \end{array}$$

33. rational numbers, real numbers **35.** -10 **37.** $-\dfrac{3}{4}$ **39.** true **41.** true **43. (a)** 9 **(b)** 9 **45. (a)** -6
(b) 6 **47.** 12 **49.** -19 **51.** -6 **53.** -17 **55.** -21.8 **57.** -10 **59.** -11 **61.** 7 **63.** 10.31
65. 2 **67.** $(-31 + 12) + 19; 0$ **69.** $-4 - (-6); 2$ **71.** -2 **73.** \$26.25 **75.** $-\$29$ **77.** It gained 4 yards.
79. 36 **81.** $\dfrac{1}{2}$ **83.** -20 **85.** -24 **87.** 4 **89.** $-\dfrac{3}{4}$ **91.** -1 **93.** 1 **95.** -18 **97.** 125

99. $-4(5) - 9; -29$ **101.** $\dfrac{12}{8 + (-4)}; 3$ **103.** $8x = -24; -3$ **105. (a)** \$61,374,559 **(b)** \$2,245,068

107. identity property **109.** inverse property **111.** associative property **113.** distributive property

115. $(7 + 1)y; 8y$ **117.** $3(2s + 5y)$ **119.** $25 - (5 - 2) = 22$ and $(25 - 5) - 2 = 18$. Because different groupings lead
to different results, we conclude that in general subtraction is not associative. **121.** $11m$ **123.** $16p^2 + 2p$

125. $-2m + 29$ **127.** 16 **129.** $\dfrac{8}{3}$ **131.** 2 **133.** $-1\dfrac{1}{2}$ **135.** $-\dfrac{28}{15}$ **137.** $8x^2 - 21y^2$ **139.** When
dividing 0 *by* a nonzero number, the quotient will be 0. However, dividing a number *by* 0 is undefined. **141.** $5(x + 7)$;
$5x + 35$ **143.** -79

CHAPTER 1 TEST (PAGE 85)

[1.1] 1. $\dfrac{7}{11}$ **2.** $\dfrac{241}{120}$ **3.** $\dfrac{19}{18}$ **4. (a)** 492 million **(b)** 861 million **[1.2] 5.** true **[1.4] 6.** $\begin{array}{c}\xleftarrow{\hspace{0.2cm}}+\bullet|+\bullet|+\bullet|+|\bullet|\xrightarrow{\hspace{0.2cm}} \\ -4 \;\; -2 \;\;\; 0 \;\;\; 2 \;\;\; 4\end{array}$

7. rational numbers, real numbers **8.** If -8 and -1 are both graphed on a number line, we see that the point for -8 is
to the *left* of the point for -1. This indicates $-8 < -1$. **9.** $\dfrac{-6}{2 + (-8)}; 1$ **[1.2] 10. (a)** Japanese, Vietnamese, Filipino,
Hawaiian, Other Asian and Pacific Islander **(b)** Asian Indian, Chinese, Korean **[1.1, 1.4–1.6] 11.** 4 **12.** $-2\dfrac{5}{6}$

13. 2 **14.** 6 **15.** 108 **16.** 3 **17.** $\dfrac{30}{7}$ **[1.3, 1.5, 1.6] 18.** 6 **19.** 4 **[1.4–1.6] 20.** -70 **21.** 3
[1.5, 1.6] 22. (a) $-\$3100$ **(b)** \$3600 **(c)** \$6800 **(d)** \$4200 **23.** 15 **24. (a)** -14 **(b)** 55 **(c)** 31
[1.7] 25. B **26.** D **27.** E **28.** A **29.** C **30.** distributive property **31. (a)** -18 **(b)** -18 **(c)** The
distributive property assures us that the answers must be the same, because $a(b + c) = ab + ac$ for all a, b, c. **[1.8] 32.** $21x$
33. $15x - 3$

CHAPTER 2 LINEAR EQUATIONS AND APPLICATIONS

SECTION 2.1 (PAGE 95)

EXERCISES **1. (a)** and **(c)** **5.** $\{6\}$ **7.** $\{6.3\}$ **9.** $\{-6\}$ **11.** $\{-2\}$ **13.** $\{4\}$ **15.** $\{0\}$ **17.** $\{-2\}$
19. $\{-7\}$ **23.** $\{13\}$ **25.** $\{-4\}$ **27.** $\{0\}$ **29.** $\left\{\dfrac{7}{15}\right\}$ **31.** $\{7\}$ **33.** $\{-4\}$ **35.** $\{13\}$ **37.** $\{29\}$
39. $\{18\}$ **41.** If both sides of an equation are multiplied by 0, the resulting equation is $0 = 0$. This is true, but does not help
to solve the equation. **43.** $\{6\}$ **45.** $\{-5\}$ **47.** $\left\{-\dfrac{18}{5}\right\}$ **49.** $\{12\}$ **51.** $\{0\}$ **53.** $\{-48\}$ **55.** $\{-12\}$

57. $\left\{\dfrac{4}{7}\right\}$ **59.** {40} **61.** {3} **63.** {7} **65.** {−35} **67.** $\left\{\dfrac{35}{2}\right\}$ **69.** $\left\{-\dfrac{27}{35}\right\}$ **71.** {−12.2}

73. Answers will vary. For example, $\dfrac{3}{2}x = -6$. **75.** $3x = 2x + 17$; {17} **77.** $5x + 3x = 7x + 9$; {9}

79. $\dfrac{x}{-5} = 2$; {−10}

SECTION 2.2 (PAGE 103)

EXERCISES **3.** {4} **5.** {−1} **7.** $\left\{-\dfrac{4}{5}\right\}$ **9.** {−6} **11.** ∅ **13.** {all real numbers} **15.** {1} **17.** ∅

19. No, it is incorrect to divide by a variable. If $-3x$ is added to both sides, the equation becomes $4x = 0$, so $x = 0$ and {0} is the correct solution set. **21.** Multiply both sides by the LCD of all fractions in the equation or by the power of 10 that

makes all decimal numbers integers. **23.** {5} **25.** {0} **27.** $\left\{-\dfrac{7}{5}\right\}$ **29.** {120} **31.** {6} **33.** {15,000}

35. 800 **36.** Yes, you will get $(100 \cdot 2) \cdot 4 = 200 \cdot 4 = 800$. This is a result of the associative property. **37.** No, because $(100a)(100b) = 10,000ab \neq 100ab$. **38.** The distributive property involves the operation of *addition* as well. **39.** Yes, the associative property of multiplication is used here. **40.** no **41.** {8} **43.** {0} **45.** {4} **47.** {20}

49. {all real numbers} **51.** ∅ **53.** $11 - q$ **55.** $x + 7$ **57.** $a + 12$; $a - 5$ **59.** $\dfrac{t}{5}$

SECTION 2.3 (PAGE 112)

CONNECTIONS **Page 105:** Polya's Step 1 corresponds to our Steps 1 and 2. Polya's Step 2 corresponds to our Step 3. Polya's Step 3 corresponds to our Steps 4 and 5. Polya's Step 4 corresponds to our Step 6. Trial and error or guessing and checking fit into Polya's Step 2, devising a plan.

EXERCISES **1.** (c) **5.** 3 **7.** −4 **9.** 53 Republicans, 46 Democrats **11.** 6562 men, 3779 women **13.** Fernandez, 285; Irwin, 286 **15.** Airborne Express: 3; Federal Express: 9; United Parcel Service: 1 **17.** 36 million

miles **19.** Smoltz. $253\dfrac{2}{3}$; Maddux: 245; Wohlers: $77\dfrac{1}{3}$ **21.** *A* and *B*: 40°; *C*: 100° **23.** exertion: 9443 calories;

regulating body temperature: 1757 calories **25.** 18 prescriptions **27.** peanuts: $22\dfrac{1}{2}$ ounces; cashews: $4\dfrac{1}{2}$ ounces

29. $k - m$ **31.** no **33.** $x - 1$ **35.** 80° **37.** 26° **39.** 55° **41.** 68, 69 **43.** 10, 12 **45.** 101, 102 **47.** 10, 11 **49.** 15, 17, 19 **51.** 1993: $2.78 billion; 1994: $3.33 billion; 1995: $3.53 billion

SECTION 2.4 (PAGE 122)

EXERCISES **1.** The perimeter of a geometric figure is the sum of the lengths of the sides. **3.** area **5.** perimeter **7.** area **9.** area **11.** $P = 20$ **13.** $P = 24$ **15.** $A = 70$ **17.** $r = 40$ **19.** $I = 875$ **21.** $A = 91$ **23.** $r = 1.3$ **25.** $A = 452.16$ **27.** 4 **29.** $V = 384$ **31.** $V = 48$ **33.** $V = 904.32$ **35.** about 154,000 square

feet **37.** 1978 feet **39.** perimeter: 172 inches; area: 1785 square inches **41.** $\dfrac{729}{32}$ or $22\dfrac{25}{32}$ cubic inches

43. 23,800.10 square feet **45.** 107°, 73° **47.** 75°, 75° **49.** 139°, 139° **51.** $r = \dfrac{d}{t}$ **53.** $p = \dfrac{I}{rt}$

55. $a = P - b - c$ **57.** $b = \dfrac{2A}{h}$ **59.** $r = \dfrac{A - p}{pt}$ **61.** $h = \dfrac{V}{\pi r^2}$ **63.** $m = \dfrac{y - b}{x}$ **65.** (a) $P - 2L = 2W$

(b) $W = \dfrac{P - 2L}{2}$ **66. (a)** $\dfrac{P}{2} = L + W$ **(b)** $\dfrac{P}{2} - L = W$ **67. (a)** multiplication identity property **(b)** An expression

divided by 1 is equal to itself. **(c)** rule for multiplication of fractions **(d)** rule for subtraction of fractions **68.** $\dfrac{5T}{4} + 1$

SECTION 2.5 (PAGE 131)

EXERCISES   **1.** $\dfrac{5}{8}$ **3.** $\dfrac{1}{4}$ **5.** $\dfrac{2}{1}$ **7.** $\dfrac{3}{1}$ **9.** (d)   **13.** true   **15.** false   **17.** true   **19.** {35}   **21.** {27}

23. {−1}   **25.** 6.875 ounces   **27.** $670.48   **29.** $9.30   **31.** 4 feet   **33.** 12,500 fish   **35.** (a) $\dfrac{26}{100} = \dfrac{x}{350}$;

$91 million   **(b)** $112 million; $11.2 million   **(c)** $119 million   **37.** 30 count   **39.** 31-ounce size   **41.** 32-ounce size
43. 4   **45.** 1   **47. (a)**   **(b)** 54 feet   **49.** $242   **51.** $255   **53.** $4850   **55.** 30

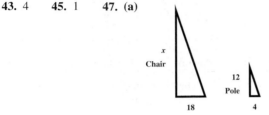

56. (a) $5x = 12$ **(b)** $\dfrac{12}{5}$ **57. (a)** $5x = 12$ **(b)** $\dfrac{12}{5}$ **58.** They are the same. Solving by cross products yields the same

solution as multiplying by the least common denominator.   **59.** $40.32   **61.** $\dfrac{80}{3}$ or $26\dfrac{2}{3}$ inches

63. approximately 2593 miles

SECTION 2.6 (PAGE 143)

EXERCISES   **1.** 35 milliliters   **3.** $350   **5.** $14.15   **7.** 3.334 hours   **9.** 7.95 meters per second   **11.** $533
13. 284%   **15. (a)** about $785 million   **(b)** about $1726 million   **(c)** about $471 million   **17.** 0 liters

19. 160 gallons   **21.** $53\dfrac{1}{3}$ kilograms   **23.** 4 liters   **25.** 25 milliliters   **27.** (d)   **29.** $5000 at 3%; $1000 at 5%

31. $40,000 at 3%; $110,000 at 4%   **33.** 25 fives   **35.** 84 fives; 42 tens   **37.** 20 pounds   **39.** Yes. In Example 2, it

would be reasonable to mix, say, $9\dfrac{1}{2}$ liters with 20 liters to get $29\dfrac{1}{2}$ liters of mixture. However, in Example 4, it is not possible

to have a fractional part of a five- or ten-dollar bill.   **40. (a)** $.05x + .10(3400 − x) = 290$ **(b)** 1000 nickels; 2400 dimes
41. (a) $.05x + .10(3400 − x) = 290$ **(b)** $1000 at 5%; $2400 at 10%   **42.** They are the same.   **43.** No, you will get a
different solution to the equation, but you *will* get the same answers to the problem.   **45.** 530 miles   **47.** No, since the

rate is given in miles per *hour*, the time must be in hours. So the distance is $45 \times \dfrac{1}{2} = \dfrac{45}{2} = 22\dfrac{1}{2}$ miles.   **49.** 10 hours

51. northbound: 60 miles per hour; southbound: 80 miles per hour   **53.** 5 hours   **55.** single man: $51.58; single woman:
$40.71   **57.** Business/Government: 6,100,000; K–12: 5,000,000; HED: 900,000; Homes: 9,800,000   **59.** 120 miles
per hour

CHAPTER 2 REVIEW EXERCISES (PAGE 154)

1. {6}   **3.** {7}   **5.** {11}   **7.** {5}   **9.** {5}   **11.** $\left\{\dfrac{64}{5}\right\}$ **13.** {all real numbers}   **15.** {all real numbers}

17. ∅   **19.** 112 nations   **21.** Hawaii: 6425 square miles; Rhode Island: 1212 square miles   **23.** 80°   **25.** $A = 28$

27. $V = 904.32$ **29.** $h = \dfrac{2A}{b + B}$ **31.** 100°; 100°   **33.** 42.2°; 92.8°   **35.** $\dfrac{3}{2}$ **37.** $\dfrac{3}{4}$ **39.** $\left\{\dfrac{7}{2}\right\}$ **41.** $\left\{\dfrac{25}{19}\right\}$

43. .2 meter or 200 millimeters   **45.** 375 kilometers   **47.** 25.5-ounce size   **49.** approximately $15.4 million

51. $5000 at 5%; $5000 at 6%   **53.** 13 hours   **55.** $\dfrac{5}{6}$ hour or 50 minutes   **57.** 28 games   **59.** Answers will vary.

61. {7}   **63.** {2}   **65.** {70}   **67.** ∅   **69.** The first step is to distribute the negative sign over *both* terms in the
parentheses, to get $3 − 8 − 4x$ on the left side of the equation. The student got $3 − 8 + 4x$ instead. The correct solution set is
{−2}.   **71.** Golden Gate Bridge: 4200 feet; Brooklyn Bridge: 1595 feet   **73.** 64-ounce size   **75.** faster train: 80 miles

per hour; slower train: 50 miles per hour   **77.** 8 centimeters   **79.** No. Only equations that have one fractional term on each side can be solved directly by cross multiplication. This equation has two terms on the left.

CHAPTER 2 TEST   (PAGE 158)

[2.1–2.2] **1.** $\{-6\}$ **2.** $\{21\}$ **3.** $\emptyset$ **4.** $\{30\}$ **5.** {all real numbers}   [2.3] **6.** Hawaii: 4021 square miles;

Maui: 728 square miles; Kauai: 551 square miles   [2.4] **7. (a)** $W = \dfrac{P - 2L}{2}$ or $W = \dfrac{P}{2} - L$ **(b)** 18   **8.** 75°, 75°

9. 50°   [2.5] **10.** $\{-29\}$ **11.** 62.8 centimeters   **12.** 2300 miles   [2.6] **13.** \$8000 at 3%; \$14,000 at 4.5%

14. 236%   **15.** 4 hours   **16.** East 132, West 120   **17.** 26 points

CUMULATIVE REVIEW EXERCISES CHAPTERS 1-2   (PAGE 159)

[1.1] **1.** $\dfrac{3}{4}$ **2.** $\dfrac{37}{60}$ **3.** $\dfrac{48}{5}$ [1.2] **4.** $\dfrac{1}{2}x - 18$ **5.** $\dfrac{6}{x + 12} = 2$ [1.4] **6.** true   [1.5–1.6] **7.** 11

8. -8 **9.** 28   [1.3] **10.** $-\dfrac{19}{3}$ [1.7] **11.** distributive property   **12.** commutative property   [1.8] **13.** $2k - 11$

[2.1–2.2] **14.** $\{-1\}$ **15.** $\{-1\}$ **16.** $\{-12\}$ **17.** $\{2\}$ [2.5] **18.** $\{26\}$ **19.** $\{-13\}$ [2.4] **20.** $y = \dfrac{24 - 3x}{4}$

21. $n = \dfrac{A - P}{iP}$ [2.3] **22.** 4 centimeters; 9 centimeters; 27 centimeters   [2.4] **23.** 1029.92 feet

[2.5] **24.** $\dfrac{25}{6}$ or $4\dfrac{1}{6}$ cups   [2.6] **25.** 40 miles per hour; 60 miles per hour

CHAPTER 3 LINEAR INEQUALITIES AND ABSOLUTE VALUE

SECTION 3.1   (PAGE 170)

EXERCISES **1.** D   **3.** B   **5.** F   **7.** Use a parenthesis when an endpoint is not included; use a bracket when it is included.   **9.** $[5, \infty)$ **11.** $(7, \infty)$ **13.** $(-4, \infty)$

15. $(-\infty, -40]$ **17.** $(-\infty, 4]$ **19.** $\left(-\infty, -\dfrac{15}{2}\right)$

21. $\left[\dfrac{1}{2}, \infty\right)$ **23.** $(3, \infty)$ **25.** $(-\infty, 4)$

27. $\left(-\infty, \dfrac{23}{6}\right]$ **29.** $\left(-\infty, \dfrac{76}{11}\right)$ **31.** $(-\infty, \infty)$

33. $\emptyset$ **35.** It is incorrect. The inequality symbol should be reversed only when multiplying or dividing by a negative number. Since 5 is positive, the inequality symbol should not be reversed.   **37.** $(1, 11)$

39. $[-14, 10]$ **41.** $[-5, 6]$ **43.** $\left[-\dfrac{14}{3}, 2\right]$

45. $\left[-\dfrac{1}{2}, \dfrac{35}{2}\right]$ **47.** $\left(-\dfrac{1}{3}, \dfrac{1}{9}\right)$ **49.** April, May, June, July

51. January, February, March, August, September, October, November, December   **53.** 1990–1991, 1991–1992, 1992–1993,

1993–1994 **55.** at least 8.3 million tickets **57.** 2 miles **59.** at least 80 **61.** 50 miles **63.** 26 tapes
65. $\{-9\}$ **66.** $(-9, \infty)$ **67.** $(-\infty, -9)$
68. the set of all real numbers **69.** $(-\infty, -3)$ **71.** There is no such number y, since $4 \not< 1$.

SECTION 3.2 (PAGE 178)

EXERCISES 1. true **3.** false; The union is $(-\infty, 8) \cup (8, \infty)$. **5.** false; The intersection is $\emptyset$. **7.** $\{1, 3, 5\}$ or B
9. $\{4\}$ or D **11.** $\emptyset$ **13.** $\{1, 2, 3, 4, 5, 6\}$ or A **15.** $\{1, 3, 5, 6\}$ **17.** $\{1, 4, 6\}$ **19.** Each is equal to $\{1\}$. This
illustrates the associative property of set intersection. **21.** Many answers are possible. One example is the intersection of
two streets is the region common to *both* streets. **23.** **25.**
27. $[5, 9]$ **29.** $(-3, -1)$ **31.** $(-\infty, 4]$
33. **35.** **37.** $(-\infty, -5) \cup (5, \infty)$
39. $(-\infty, -1) \cup (2, \infty)$ **41.** $[-4, -1]$ **43.** $[-9, -6]$ **45.** $(-\infty, 3)$ **47.** $[3, 9)$
49. intersection; $(-5, -1)$ **51.** union; $(-\infty, 4)$
53. union; $(-\infty, 0] \cup [2, \infty)$ **55.** intersection; $[4, 12]$
57. 1995 and 1996 **59.** Maria, Joe **60.** none of them **61.** none of them **62.** Luigi, Than **63.** Maria, Joe
64. all of them

SECTION 3.3 (PAGE 187)

CONNECTIONS Page 186: The filled carton may contain between 30.4 and 33.6 ounces, inclusive.

EXERCISES 1. (a) E **(b)** C **(c)** D **(d)** B **(e)** A **3. (a)** one **(b)** two **(c)** none **5.** $\{-12, 12\}$ **7.** $\{-5, 5\}$
9. $\{-6, 12\}$ **11.** $\{-4, 3\}$ **13.** $\left\{-3, \dfrac{11}{2}\right\}$ **15.** $\left\{-\dfrac{19}{2}, \dfrac{9}{2}\right\}$ **17.** $\{-10, -2\}$ **19.** $\left\{-8, \dfrac{32}{3}\right\}$ **21. (a)** use *or*
(b) use *and* **(c)** use *or* **23.** $(-\infty, -3) \cup (3, \infty)$ **25.** $(-\infty, -4] \cup [4, \infty)$
27. $(-\infty, -12) \cup (8, \infty)$ **29.** $(-\infty, -2) \cup (8, \infty)$
31. $[-3, 3]$ **33.** $(-4, 4)$ **35.** $[-12, 8]$
37. $[-2, 8]$ **39.** $(-\infty, -5) \cup (13, \infty)$
41. $(-\infty, -25) \cup (15, \infty)$ **43.** $\{-6, -1\}$ **45.** $\left[-\dfrac{10}{3}, 4\right]$
47. $\left[-\dfrac{7}{6}, -\dfrac{5}{6}\right]$ **49.** $\left(-\infty, -\dfrac{7}{3}\right] \cup [3, \infty)$ **51.** $|x - 4| = 9$
(or $|4 - x| = 9$) **53.** $\{-5, -3\}$ **55.** $(-\infty, -3) \cup (2, \infty)$ **57.** $[-10, 0]$ **59.** $\{-1, 3\}$ **61.** $\left\{-3, \dfrac{5}{3}\right\}$
63. $\left\{-\dfrac{1}{3}, -\dfrac{1}{15}\right\}$ **65.** $\left\{-\dfrac{5}{4}\right\}$ **67.** $\emptyset$ **69.** $\left\{-\dfrac{1}{4}\right\}$ **71.** $\emptyset$ **73.** $(-\infty, \infty)$ **75.** $\left\{-\dfrac{3}{7}\right\}$ **77.** $(-\infty, \infty)$
79. $\left(-\infty, -\dfrac{7}{10}\right) \cup \left(-\dfrac{7}{10}, \infty\right)$ **81.** 460.2 feet **82.** Federal Office Building, City Hall, Kansas City Power and Light,
Hyatt Regency **83.** Southwest Bell Telephone, City Center Square, Commerce Tower, Federal Office Building, City Hall,
Kansas City Power and Light, Hyatt Regency **84. (a)** $|x - 460.2| \geq 75$ **(b)** $x \geq 535.2$ or $x \leq 385.2$ **(c)** Pershing

Road Associates, AT&T Town Pavilion, One Kansas City Place   **(d)** It makes sense because it includes all buildings *not* listed earlier.   **85. (a)** $|x - 16| \le .5$ **(b)** $15.5 \le x \le 16.5$ **(c)** $x \le 15.5$ or $x \ge 16.5$

SUMMARY: EXERCISES ON SOLVING LINEAR AND ABSOLUTE VALUE EQUATIONS AND INEQUALITIES (PAGE 190)

1. $\{12\}$ **3.** $\{7\}$ **5.** $\emptyset$ **7.** $\left[-\dfrac{2}{3}, \infty\right)$ **9.** $\{-3\}$ **11.** $(-\infty, 5]$ **13.** $\{2\}$ **15.** $\emptyset$ **17.** $(-5.5, 5.5)$

19. $\left\{-\dfrac{96}{5}\right\}$ **21.** $(-\infty, -24)$ **23.** $\left\{\dfrac{7}{2}\right\}$ **25.** $(-\infty, \infty)$ **27.** $(-\infty, -4) \cup (7, \infty)$ **29.** $\left\{-\dfrac{1}{5}\right\}$

31. $\left[-\dfrac{1}{3}, 3\right]$ **33.** $\left\{-\dfrac{1}{6}, 2\right\}$ **35.** $(-\infty, -1] \cup \left[\dfrac{5}{3}, \infty\right)$ **37.** $\left\{-\dfrac{5}{2}\right\}$ **39.** $\left[-\dfrac{9}{2}, \dfrac{15}{2}\right]$ **41.** $(-\infty, \infty)$

43. $(-\infty, \infty)$ **45.** $\{-2\}$ **47.** $(-\infty, -1) \cup (2, \infty)$

CHAPTER 3 REVIEW EXERCISES (PAGE 194)

1. $(-9, \infty)$ **3.** $\left(\dfrac{3}{2}, \infty\right)$ **5.** $[3, 5)$ **7.** any grade greater than or equal to 61% **9.** 99 tickets or less   **11.** $\{a, c\}$

13. $\{a, c, e, f, g\}$ **15.** $(-\infty, 3)$ **17.** $(-\infty, -1] \cup (5, \infty)$

19. $(-3, 4)$ **21.** $(4, \infty)$ **23. (a)** managerial and professional specialty   **(b)** managerial and professional specialty, mathematical and computer scientists   **25.** $\left\{-\dfrac{1}{3}, 5\right\}$ **27.** $\left\{-\dfrac{3}{2}, \dfrac{1}{2}\right\}$ **29.** $\left\{-\dfrac{1}{2}\right\}$ **31.** $[-3, -2]$

33. $\left(-\infty, -\dfrac{8}{5}\right) \cup (2, \infty)$ **35.** $\left\{\dfrac{3}{11}\right\}$ **37.** $(-2, \infty)$ **39.** $(-\infty, \infty)$ **41.** any amount greater than or equal to $1100

43. $\left(-\infty, \dfrac{14}{17}\right)$ **45.** $(-\infty, 1]$ **47.** $(-\infty, \infty)$ **49.** $[-4, -2]$ **51.**

CHAPTER 3 TEST (PAGE 195)

[3.1]   **1.** We must reverse the direction of the inequality symbol.   **2.** $[1, \infty)$

3. $(-\infty, 28)$ **4.** $[-3, 3]$ **5. (c)**   **6.** 1992 and 1993; 1991 and 1992

7. 82% **8.** $[500, \infty)$ [3.2] **9. (a)** $\{1, 5\}$ **(b)** $\{1, 2, 5, 7, 9, 12\}$ **10. (a)** $[2, 9)$ **(b)** $(-\infty, 3) \cup [6, \infty)$

[3.3]   **11.** $\left\{-1, \dfrac{5}{2}\right\}$ **12.** $(-\infty, -1) \cup \left(\dfrac{5}{2}, \infty\right)$ **13.** $\left(-1, \dfrac{5}{2}\right)$ **14.** $\left\{-\dfrac{5}{7}, \dfrac{11}{3}\right\}$ **15.** $\left(\dfrac{1}{3}, \dfrac{7}{3}\right)$

16. (a) $\emptyset$ **(b)** $(-\infty, \infty)$ **(c)** $\emptyset$

CUMULATIVE REVIEW EXERCISES CHAPTERS 1-3 (PAGE 196)

[1.4]   **1.** $9, 6$ **2.** $0, 9, 6$ **3.** $-8, 0, 9, 6$ **4.** $-8, -\dfrac{2}{3}, 0, \dfrac{4}{5}, 9, 6$ **5.** $-\sqrt{6}$ **6.** All are real numbers.

[1.1, 1.5]   **7.** $-\dfrac{22}{21}$ [1.4, 1.5] **8.** 8 [1.2, 1.6] **9.** -243 **10.** $\dfrac{216}{343}$ [1.4] **11.** $\sqrt{-36}$ is not a real number.

[1.6]   **12.** $\dfrac{4 + 4}{4 - 4}$ is undefined.   [1.3, 1.6] **13.** -16 **14.** -34 **15.** 184 **16.** $\dfrac{27}{16}$ [1.8] **17.** $-20r + 17$

18. $13k + 42$ [1.7] **19.** commutative property   **20.** distributive property   **21.** inverse property

[1.6]   **22.** $-\dfrac{3}{2}$ [2.1, 2.2] **23.** $\{5\}$ **24.** $\{30\}$ **25.** $\{15\}$ [2.4] **26.** $b = P - a - c$

[3.1]   **27.** $[-14, \infty)$ ———[———→
                                           -14

28. $\left[\dfrac{5}{3}, 3\right)$ ←—[————)——→
                                           $\frac{5}{3}$ 3

[3.2]   **29.** $(-\infty, 0) \cup (2, \infty)$ ←—)——(—→
                                                        0 2

[3.3]   **30.** $\left(-\infty, -\dfrac{1}{7}\right] \cup [1, \infty)$ ←—]——[—→
                                                                              $-\frac{1}{7}$ 1

[2.6]   **31.** $5000   **[3.1]**   **32.** $6\dfrac{1}{3}$ grams   **33.** 74 or greater   **[2.6]**   **34.** $\dfrac{1}{8}$ hour   **35.** 2 liters

36. 9 pennies, 12 nickels, 8 quarters   **[2.4]**   **37.** 44 milligrams   **38.** 20 drops   **[2.3, 2.6]**   **39.** **(a)** 79   **(b)** 4.9%
[2.4]   **40.** 25.7

CHAPTER 4 LINEAR EQUATIONS IN TWO VARIABLES

SECTION 4.1   (PAGE 207)

EXERCISES   **1.** does; do not   **3.** II   **5.** 3   **7.** between 1990 and 1991; $21,000 and between 1992 and 1993; $122,000
9. 1993   **11.** (d)   **13.** inkjet   **15.** 1994–1995; 1993–1994   **17.** 4800; 3400; yes   **19.** yes   **21.** yes
23. no   **25.** yes   **27.** yes   **29.** no   **33.** No, for two ordered pairs to be equal, the x-values must be equal, and the
y-values must be equal. Here, $4 \neq -1$ and $-1 \neq 4$.   **35.** 11   **37.** $-\dfrac{7}{2}$ **39.** -4 **41.** -5 **43.** Substituting $\dfrac{1}{3}$ for
x in $y = 6x + 2$ gives $y = 6\left(\dfrac{1}{3}\right) + 2 = 2 + 2 = 4$. Because $6\left(\dfrac{1}{7}\right) = \dfrac{6}{7}$, calculating y requires working with fractions.
45. 4; 6; -6 **47.** 3; -5; -15 **49.** -9; -9; -9 **51.** -6; -6; -6 **53.–60.**

61. negative; negative   **63.** positive; negative   **65.** -3; 6; -2; 4

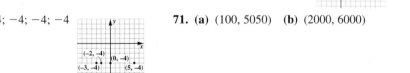

67. -3; 4; -6; $-\dfrac{4}{3}$

69. -4; -4; -4; -4

71. (a) $(100, 5050)$ **(b)** $(2000, 6000)$

73. (a) Yes, there is an approximate linear
relationship between the years since 1980 and the
minutes beyond two hours to complete a game.

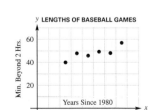

(b) Yes, a prediction could be made by
determining the linear equation that
describes the linear relationship. However,
only input values for the years included in
the given data, 1980–1996, should be used.
(c) input: years since 1980; output:
minutes beyond two hours

75. (a) $(30, 133)$, $(40, 126)$, $(60, 112)$, $(70, 105)$ **(b)** input: age; output: lower limit of the target heart rate zone
77. No. To go beyond the given data at either end assumes that the graph continues in the same way. This may not be true.

SECTION 4.2 (PAGE 220)

CONNECTIONS   **Page 217:**   **1.** $3x + 4 - 2x - 7 - 4x - 3 = 0$ **2.** $5x - 15 - 3(x - 2) = 0$

EXERCISES   **1.** (c)   **3.** (d)   **5.** (b)   **7.** 5; 5; 3   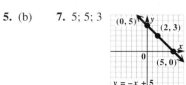   **9.** 1; 3; −1

11. −6; −2; −5   **13.** (12, 0); (0, −8)   **15.** (0, 0); (0, 0)   **17.** $y = 0$ **19.** Choose a value other

than 0 for either x or y. For example, if $x = -5$, $y = 4$.   **21.**   **23.**   **25.**

27.   **29.**   **31.**   **33.**   **35.**   **37.** 2

39. 5   **41.** The y-value of each ordered pair gives the value of the expression in x for each x-value. The solution of the equation is the x-value that corresponds to $y = 0$, which is the x-intercept of the graph.   **43. (a)** 1980: 52.2 thousand; 1985: 66.25 thousand; 1990: 80.3 thousand; 1994: 91.54 thousand   **(b)** 1980: 54 thousand; 1985: 67 thousand; 1990: 77 thousand; 1994: 93 thousand   **(c)** Yes, they are quite close.   **45. (a)** 163.2 centimeters   **(b)** 171 centimeters   **(c)** 151.5 centimeters
(d) 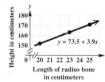   **47. (a)** $1250   **(b)** $1250   **(c)** $1250   **(d)** $6250   **49.** yes   **51.** yes   **53.** no   **55.** yes

SECTION 4.3 (PAGE 231)

CONNECTIONS   **Page 225:**   Ski slopes, "grade" on a treadmill, and slope (or "fall") of a sewer pipe are some additional examples.

EXERCISES   **1.** Rise is the vertical change between two different points on a line. Run is the horizontal change between two different points on a line.   **3.** 4   **5.** $-\dfrac{1}{2}$ **7.** 0   **9.** Yes; it doesn't matter which point you start with. Both differences will be the negatives of the differences in Exercise 8, and the quotient will be the same.

In Exercises 11–13, sketches will vary.
11. The line must rise from left to right.   **13.** The line must be horizontal.   **15.** Because he found the difference $3 - 5 = -2$ in the numerator, he should have subtracted in the same order in the denominator to get $-1 - 2 = -3$. The correct slope is $\dfrac{-2}{-3} = \dfrac{2}{3}$.   **17.** $\dfrac{5}{4}$ **19.** $\dfrac{3}{2}$ **21.** 0   **23.** undefined   **25.** $-\dfrac{1}{2}$ **27.** 5   **29.** $\dfrac{1}{4}$ **31.** $\dfrac{3}{2}$
33. undefined   **35. (a)** negative   **(b)** zero   **37. (a)** positive   **(b)** negative   **39. (a)** zero   **(b)** negative
41. (a)   **43.** $-\dfrac{2}{5}$; $-\dfrac{2}{5}$; parallel   **45.** $\dfrac{8}{9}$; $-\dfrac{4}{3}$; neither   **47.** $\dfrac{3}{2}$; $-\dfrac{2}{3}$; perpendicular   **49.** $\dfrac{3}{10}$ **51.** 232,000
52. positive; increased   **53.** 232,000 students   **54.** −2   **55.** negative; decreased   **56.** 2 students per computer

57. (a) Possible ordered pairs are (1991, 37.8), (1992, 38.6), (1993, 39.4), (1994, 40.2), (1995, 41.0), (1996, 41.8). The input numbers are 1991, 1992, 1993, 1994, 1995, 1996. The output numbers (in thousands) are 37.8, 38.6, 39.4, 40.2, 41.0, 41.8. **(b)** Yes, they are all within .4 thousand (400).   **59.** The change for each year is .1 billion (or 100,000,000) square feet, so the graph is a straight line.   **61.** 2   **63.** $\dfrac{2}{5}$ **65.** (0, 4)

CHAPTER 4 REVIEW EXERCISES (PAGE 241)

1. -1; 2; 1   **3.** 7; 7; 7   **5.** no   **7.** I

9. none

11. I or III

13. $\left(-\dfrac{5}{2}, 0\right)$; (0, 5)

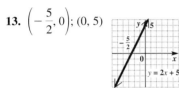

15. $(-4, 0)$; $(0, -2)$

17. function   **19.** function

21. $-\dfrac{1}{2}$ **23.** undefined   **25.** $\dfrac{2}{3}$ **27.** $-\dfrac{4}{7}$ **29.** $\dfrac{1}{3}$ **31.** parallel   **33.** neither   **35.** 0; $\dfrac{8}{3}$; -9

37. $\left(-\dfrac{3}{2}, 0\right)$; (0, 3); 2   **39.** $\dfrac{1}{3}$ **41.** yes   **43.**

45. It is wise to plot three points before drawing the line, because if the three points do not lie in a line, then an error has been made and can be corrected.

CHAPTER 4 TEST (PAGE 243)

[4.1] **1. (b)**   **2.** 16%   **3.** -6; -10; -15 **4.** -12; -12; -12 **5.** no   [4.2] **6.** To find the *x*-intercept, let $y = 0$, and to find the *y*-intercept, let $x = 0$.   **7.** (2, 0); (0, 6)   **8.** no *x*-intercept; $(0, -3)$

9. Yes, because each input determines exactly one output.   **10.** The inputs are the years from 1990 to 1996, and the outputs are the corresponding Super Bowl advertising costs in thousands of dollars.

[4.3] **11.** $-\dfrac{8}{3}$ **12.** -2 **13.** $\dfrac{5}{2}$ **14.** $\dfrac{1}{2}$ **15.** -4 **16.** $\dfrac{1}{4}$

CUMULATIVE REVIEW EXERCISES CHAPTERS 1–4 (PAGE 244)

[1.1] **1.** $\dfrac{551}{40}$ or $13\dfrac{31}{40}$ **2.** $\dfrac{6}{5}$ [1.5] **3.** 7   [1.6] **4.** $\dfrac{73}{18}$ [1.1] **5.** $\dfrac{5}{12}$ [1.4] **6.** true   [1.3] **7.** -134

[1.7] **8.** distributive property   [1.8] **9.** $-p + 2$ [2.4] **10.** $h = \dfrac{3V}{\pi r^2}$ [2.2] **11.** $\{-1\}$ **12.** $\{2\}$

[2.2, 2.5] **13.** $\{-13\}$ [3.1] **14.** $(-\infty, 2]$ **15.** $\left(-\dfrac{8}{3}, -\dfrac{4}{3}\right)$ **16.** \$300 or more   [2.4] **17.** 6 miles

[2.6] **18.** 10 liters   [4.1] **19.**

x	y
12	89.45
28	81.95
36	78.20

20. \$7000; \$10,000   [4.2] **21.** *x*-intercept: (4, 0); *y*-intercept: (0, 6)

22.   [4.3] **23.** $-\dfrac{3}{2}$ **24.** perpendicular   [1.4, 4.2] **25.** Yes; (Taiwan, -6.6), (South Korea, -40.5);

To determine the smallest and largest *change,* in the usual sense, we need to consider the absolute value of the changes.

CHAPTER 5 POLYNOMIALS AND EXPONENTS

SECTION 5.1 (PAGE 254)

CONNECTIONS **Page 250:** The polynomial gives the following: for 1996, 5730 million (greater than actual); for 1997, 5594 million (less than actual); for 1999, 5260 million (less than actual); for 2000, 5062 million (less than actual); for 2001, 4844 million (greater than actual).

EXERCISES **1.** 7; 5   **3.** 8   **5.** 26   **7.** 0   **9.** 1; 6   **11.** 1; 1   **13.** 2; $-19, -1$ **15.** 3; 1, 8, 5   **17.** $2m^5$
19. $-r^5$ **21.** cannot be simplified   **23.** $-5x^5$ **25.** $5p^9 + 4p^7$ **27.** $-2y^2$ **29.** already simplified; 4; binomial
31. $11m^4 - 7m^3 - 3m^2$; 4; trinomial   **33.** x^4; 4; monomial   **35.** 7; 0; monomial   **37.** (a) 36   (b) -12
39. (a) 14   (b) -19 **41.** -3 **43.** 4; \$5.00   **44.** 6; \$27   **45.** 2.5; 130   **46.** 459.87 billion   **47.** $5m^2 + 3m$
49. $4x^4 - 4x^2$ **51.** $\dfrac{7}{6}x^2 - \dfrac{2}{15}x + \dfrac{5}{6}$ **53.** $6m^3 + m^2 + 12m - 14$ **55.** $15m^3 - 13m^2 + 8m + 11$
57. Answers will vary.   **59.** $5m^2 - 14m + 6$ **61.** $4x^3 + 2x^2 + 5x$ **63.** $-11y^4 + 8y^2 + y$ **65.** $a^4 - a^2 + 1$
67. $5m^2 + 8m - 10$ **69.** $-6x^2 - 12x + 12$ **71.** -10 **73.** $4b - 5c$ **75.** $6x - xy - 7$
77. $-3x^2y - 15xy - 3xy^2$ **79.** $8x^2 + 8x + 6$ **81.** (a) $23y + 5t$ (b) approximately $37.26°$, $79.52°$, and $63.22°$
83. $-6x^2 + 6x - 7$ **85.** $-7x - 1$ **87.** $0, -3, -4, -3, 0$ **89.** $7, 1, -1, 1, 7$

91. $0, 3, 4, 3, 0$ **93.** $4, 1, 0, 1, 4$

SECTION 5.2 (PAGE 263)

EXERCISES **1.** false   **3.** false   **5.** w^6 **7.** $\dfrac{1}{4^4}$ **9.** $(-7x)^4$ **11.** $\left(\dfrac{1}{2}\right)^6$ **13.** In $(-3)^4$, -3 is the base, while in
-3^4, 3 is the base. $(-3)^4 = 81$, while $-3^4 = -81$.   **15.** base: 3; exponent: 5; 243   **17.** base: -3; exponent: 5; -243
19. base: $-6x$; exponent: 4   **21.** base: x; exponent: 4   **23.** $5^2 + 5^3$ is a sum, not a product. $5^2 + 5^3 = 25 + 125 = 150$.
25. 5^8 **27.** 4^{12} **29.** $(-7)^9$ **31.** t^{24} **33.** $-56r^7$ **35.** $42p^{10}$ **37.** The product rule applies only when the
bases are the same. $3^2 \cdot 4^3 = 9 \cdot 64 = 576$.   **39.** $-32; -32; 1$ **41.** $729; -729; 0$ **43.** 4^6 **45.** t^{20} **47.** 7^3r^3
49. $5^5x^5y^5$ **51.** 5^{12} **53.** -8^{15} **55.** $8q^3r^3$ **57.** $\dfrac{1}{2^3}$ **59.** $\dfrac{a^3}{b^3}$ **61.** $\dfrac{9^8}{5^8}$ **63.** $\dfrac{5^5}{2^5}$ **65.** $\dfrac{9^5}{8^3}$
67. $2^{12}x^{12}$ **69.** -6^5p^5 **71.** $6^5x^{10}y^{15}$ **73.** x^{21} **75.** $2^2w^4x^{26}y^7$ **77.** $-r^{18}s^{17}$ **79.** $\dfrac{5^3a^6b^{15}}{c^{18}}$
81. This is incorrect. Using the product rule, it is simplified as follows: $(10^2)^3 = 10^{2\cdot3} = 10^6 = 1{,}000{,}000$.
83. $12x^5$ **85.** $6p^7$ **87.** $125x^6$ **91.** \$304.16   **93.** \$1843.88

SECTION 5.3 (PAGE 270)

EXERCISES   **1.** (a) B   (b) D   (c) A \ (d) C   **3.** $40a^{14}$ **5.** $-6m^2 - 4m$ **7.** $24p - 18p^2 + 36p^4$
9. $-16z^2 - 24z^3 - 24z^4$ **11.** $14x^5y^3 + 21x^3y^2 - 28x^2y^2$ **13.** $12x^3 + 26x^2 + 10x + 1$
15. $20m^4 - m^3 - 8m^2 - 17m - 15$ **17.** $6x^6 - 3x^5 - 4x^4 + 4x^3 - 5x^2 + 8x - 3$ **19.** $5x^4 - 13x^3 + 20x^2 + 7x + 5$
21. $n^2 + n - 6$ **23.** $x^2 - 36$ **25.** $8r^2 - 10r - 3$ **27.** $9x^2 - 4$ **29.** $9q^2 + 6q + 1$ **31.** $3x^2 - 5xy - 2y^2$
33. $-3t^2 - 14t + 24$ **35.** $6y^5 - 21y^4 - 45y^3$ **37.** $36x^2 + 24x + 4$ **41.** $6p^2 - \dfrac{5}{2}pq - \dfrac{25}{12}q^2$
43. $2m^6 - 5m^3 - 12$ **45.** $2k^5 - 6k^3h^2 + k^2h^2 - 3h^4$ **47.** $6p^8 + 15p^7 + 12p^6 + 36p^5 + 15p^4$
49. $-24x^8 - 28x^7 + 32x^6 + 20x^5$ **51.** $14x + 49$ **53.** $\pi x^2 - 9$ **55.** $30x + 60$ **56.** $30x + 60 = 600$
57. 18   **58.** 10 yards by 60 yards   **59.** \$2100   **60.** 140 yards   **61.** \$1260   **62.** (a) $30kx + 60k$ (dollars)
(b) $6rx + 32r$

SECTION 5.4 (PAGE 275)

EXERCISES   **1.** (a) $4x^2$ (b) $12x$ (c) 9   (d) $4x^2 + 12x + 9$ **3.** $p^2 + 4p + 4$ **5.** $a^2 - 2ac + c^2$
7. $16x^2 - 24x + 9$ **9.** $64t^2 + 112st + 49s^2$ **11.** $25x^2 + 4xy + \dfrac{4}{25}y^2$ **13.** $4x^3 + 20x^2 + 25x$
15. $-16r^2 + 16r - 4$ **17.** (a) $49x^2$ (b) 0   (c) $-9y^2$ (d) $49x^2 - 9y^2$; Because 0 is the identity element for addition,
it is not necessary to write "+ 0."   **19.** $q^2 - 4$ **21.** $4w^2 - 25$ **23.** $100x^2 - 9y^2$ **25.** $4x^4 - 25$
27. $49x^2 - \dfrac{9}{49}$ **29.** $9p^3 - 49p$ **31.** $(a + b)^2$ **32.** a^2 **33.** $2ab$ **34.** b^2 **35.** $a^2 + 2ab + b^2$ **36.** They
both represent the area of the entire large square.   **37.** 1225   **38.** $30^2 + 2(30)(5) + 5^2$ **39.** 1225   **40.** They are
equal.   **41.** 9999   **43.** 39,999   **45.** $399\dfrac{3}{4}$ **47.** $m^3 - 15m^2 + 75m - 125$ **49.** $8a^3 + 12a^2 + 6a + 1$
51. $81r^4 - 216r^3t + 216r^2t^2 - 96rt^3 + 16t^4$ **53.** $x^2 + y^2$ is the sum of squares, while $(x + y)^2$ is the square of a sum.
$(x + y)^2 = x^2 + 2xy + y^2$, and thus there is another term, $2xy$.   **55.** $\dfrac{1}{2}m^2 - 2n^2$ **57.** $9a^2 - 4$ **59.** $\pi x^2 + 4\pi x + 4\pi$
61. $x^3 + 6x^2 + 12x + 8$

SECTION 5.5 (PAGE 283)

CONNECTIONS   **Page 283:**   (All are in millions of dollars.) 3501.09; 4792.64; 5607.39; 6560.65; 7675.96

EXERCISES   **1.** false   **3.** true   **5.** false   **7.** 1   **9.** 1   **11.** -1 **13.** 0   **15.** 0   **17.** 2   **19.** $\dfrac{1}{64}$
21. 16   **23.** $\dfrac{49}{36}$ **25.** $\dfrac{1}{81}$ **27.** $\dfrac{8}{15}$ **29.** 5^3 **31.** $\dfrac{1}{9}$ **33.** 5^2 **35.** x^{15} **37.** 6^3 **39.** $2r^4$ **41.** $\dfrac{5^2}{4^3}$
43. $\dfrac{p^5}{q^8}$ **45.** r^9 **47.** $\dfrac{x^5}{6}$ **49.** $a + b$ **51.** $(x + 2y)^2$ **53.** 1   **54.** $\dfrac{5^2}{5^2}$ **55.** 5^0 **56.** $1 = 5^0$; This supports
the definition for 0 as an exponent.   **57.** 7^3 or 343   **59.** $\dfrac{1}{x^2}$ **61.** $\dfrac{64x}{9}$ **63.** $\dfrac{x^2z^4}{y^2}$ **65.** $6x$ **67.** $\dfrac{1}{m^{10}n^5}$
69. $\dfrac{1}{xyz}$ **71.** x^3y^9 **73.** The student attempted to use the quotient rule with unequal bases. The correct way to simplify
this expression is $\dfrac{16^3}{2^2} = \dfrac{(2^4)^3}{2^2} = \dfrac{2^{12}}{2^2} = 2^{10} = 1024.$

SECTION 5.6 (PAGE 290)

CONNECTIONS   **Page 290:**   **1.** -104 **2.** -104 **3.** They are both -104.   **4.** The answers should agree.

EXERCISES   **1.** $6x^2 + 8$; 2; $3x^2 + 4$ **3.** $3x^2 + 4$; 2 (These may be reversed.); $6x^2 + 8$ **5.** The first is a polynomial
divided by a monomial, covered in Objective 1. This section does not cover dividing a monomial by a polynomial of several
terms.

7. $30x^3 - 10x + 5$ **9.** $4m^3 - 2m^2 + 1$ **11.** $4t^4 - 2t^2 + 2t$ **13.** $a^4 - a + \dfrac{2}{a}$ **15.** $4x^3 - 3x^2 + 2x$

17. $1 + 5x - 9x^2$ **19.** $\dfrac{12}{x} + 8 + 2x$ **21.** $\dfrac{4x^2}{3} + x + \dfrac{2}{3x}$ **23.** $9r^3 - 12r^2 + 2r + \dfrac{26}{3} - \dfrac{2}{3r}$

25. $-m^2 + 3m - \dfrac{4}{m}$ **27.** $4 - 3a + \dfrac{5}{a}$ **29.** $\dfrac{12}{x} - \dfrac{6}{x^2} + \dfrac{14}{x^3} - \dfrac{10}{x^4}$ **31.** $\dfrac{2}{3}x$ would not be an acceptable form,

because $\dfrac{2}{3}x = \dfrac{2}{3} \cdot \dfrac{x}{1} = \dfrac{2x}{3}$, which is *not* equivalent to $\dfrac{2}{3x}$.   **33.** $5x^3 + 4x^2 - 3x + 1$ **35.** $-63m^4 - 21m^3 - 35m^2 + 14m$

37. 1423 **38.** $(1 \times 10^3) + (4 \times 10^2) + (2 \times 10^1) + (3 \times 10^0)$ **39.** $x^3 + 4x^2 + 2x + 3$ **40.** They are similar in that the coefficients of powers of ten are equal to the coefficients of the powers of x. They are different in that one is a constant while the other is a polynomial. They are equal if $x = 10$ (the base of our decimal system).   **41.** $x + 2$ **43.** $2y - 5$

45. $p - 4 + \dfrac{44}{p + 6}$ **47.** $r - 5$ **49.** $6m - 1$ **51.** $2a - 14 + \dfrac{74}{2a + 3}$ **53.** $4x^2 - 7x + 3$ **55.** $4k^3 - k + 2$

57. $5y^3 + 2y - 3$ **59.** $3k^2 + 2k - 2 + \dfrac{6}{k - 2}$ **61.** $2p^3 - 6p^2 + 7p - 4 + \dfrac{14}{3p + 1}$ **63.** $r^2 - 1 + \dfrac{4}{r^2 - 1}$

65. $y^2 - y + 1$ **67.** $a^2 + 1$ **69.** $x^2 - 4x + 2 + \dfrac{9x - 4}{x^2 + 3}$ **71.** $x^3 + 3x^2 - x + 5$ **73.** $\dfrac{3}{2}a - 10 + \dfrac{77}{2a + 6}$

75. The process stops when the degree of the remainder is less than the degree of the divisor.   **77.** $x^2 + x - 3$ units
79. $5x^2 - 11x + 14$ hours   **81. (a)** is correct, **(b)** is incorrect.   **82. (a)** is incorrect, **(b)** is correct.
83. (a) is correct, **(b)** is incorrect.   **84.** Because any power of 1 is 1, to evaluate a polynomial for 1 we simply add the coefficients. If the divisor is $x - 1$, this method does not apply, since $1 - 1 = 0$ and division by 0 is undefined.

SECTION 5.7 (PAGE 296)

CONNECTIONS   **Page 294:**   **1.** Move the decimal point 2 places to the right and use zeros as placeholders as necessary; use the same procedure, but move 4 places to the right.   **2.** Move the decimal point 2 places to the left and use zeros as placeholders as necessary; use the same procedure, but move 4 places to the left.   **3.** In 32,000, zero is used as a placeholder for the hundreds, tens, and units place values. In .00032, zero is used as a placeholder for the tenths, hundredths, and thousandths place values.

EXERCISES   **1.** A   **3.** C   **5.** in scientific notation   **7.** not in scientific notation; 5.6×10^6 **9.** not in scientific notation; 8×10^1 **11.** not in scientific notation; 4×10^3 **15.** 5.876×10^9 **17.** 8.235×10^4 **19.** 7×10^{-6}
21. 2.03×10^{-3} **23.** $750{,}000$ **25.** $5{,}677{,}000{,}000{,}000$ **27.** 6.21 **29.** $.00078$ **31.** $.000000005134$
33. $600{,}000{,}000{,}000$ **35.** $15{,}000{,}000$ **37.** $60{,}000$ **39.** $.0003$ **41.** 40 **43.** $.000013$ **45.** $4.7\text{E}-7$
47. $2\text{E}7$ **49.** $1\text{E}1$ **51.** 1×10^{10} **53.** $2{,}000{,}000{,}000$ **55.** $\$7.326 \times 10^9$ **57.** $\$3.0262 \times 10^{10}$
59. about $15{,}300$ seconds   **61. (a)** 1995: 1.063×10^{11}; 1996: 1.124×10^{11}; 1997: 1.250×10^{11} **(b)** 9.35×10^9

CHAPTER 5 REVIEW EXERCISES (PAGE 304)

1. $22m^2$; degree 2; monomial   **3.** already in descending powers; degree 5; none of these   **5.** $7r^4 - 4r^3 + 1$; degree 4;
trinomial   **7.** $a^3 + 4a^2$ **9.** $-13k^4 - 15k^2 + 18k$ **11.**

x	-2	-1	0	1	2
y	10	1	-2	1	10

$y = 3x^2 - 2$

13. 4^{11} **15.** $-72x^7$ **17.** $19^5 x^5$ **19.** $5p^4 t^4$ **21.** $6^2 x^{16} y^4 z^{16}$ **23.** $a^3 - 2a^2 - 7a + 2$ **25.** $5p^5 - 2p^4 - 3p^3 + 25p^2 + 15p$ **27.** $6k^2 - 9k - 6$ **29.** $12k^2 - 32kq - 35q^2$ **31.** $2x^2 + x - 6$ **33.** $a^2 + 8a + 16$
35. $36m^2 - 25$ **37.** $r^3 + 6r^2 + 12r + 8$ **39. (a)** Answers will vary. For example, let $x = 1$ and $y = 2$. $(1 + 2)^2 \neq 1^2 + 2^2$, because $9 \neq 5$.   **(b)** Answers will vary. For example, let $x = 1$ and $y = 2$. $(1 + 2)^3 \neq 1^3 + 2^3$, because $27 \neq 9$.
41. In both cases, $x = 0$ and $y = 1$ lead to 1 on each side of the inequality. This would not be sufficient to show that *in general* the inequality is true. It would be necessary to choose other values of x and y.   **43.** $\dfrac{4}{3}\pi(x + 1)^3$ or $\dfrac{4}{3}\pi x^3 + 4\pi x^2 +$
$4\pi x + \dfrac{4}{3}\pi$ cubic inches   **45.** 0 **47.** $-\dfrac{1}{49}$ **49.** 5^8 **51.** $\dfrac{3}{4}$ **53.** x^2 **55.** $\dfrac{r^2}{6}$ **57.** $\dfrac{1}{a^3 b^5}$ **59.** $\dfrac{-5y^2}{3}$

61. $-2m^2n + mn + \dfrac{6n^3}{5}$ **63.** This is not correct. The friend wrote the second term of the quotient as $-12x$ rather than

$-2x$. Here is the correct method: $\dfrac{6x^2 - 12x}{6} = \dfrac{6x^2}{6} - \dfrac{12x}{6} = x^2 - 2x.$ **65.** $2a^2 + 3a - 1 + \dfrac{6}{5a - 3}$ **67.** $m^2 + 4m - 2$

69. 2.8988×10^{10} **71.** 24,000   **73.** .000000897   **75.** 4,000,000   **77.** .000002   **79.** 4.2×10^{42}
81. (a) 2.82×10^9 **(b)** 5.784×10^{11} **(c)** 1.77×10^{10} **83. (a)** 2   **(b)** 92   **(c)** $x^3 + x^2 - 10x - 6$
(d) Both results are 94.   **84. (a)** 1   **(b)** 33   **(c)** $-x^3 - x^2 + 12x$ **(d)** Both results are -32.   **85. (a)** -6
(b) 12   **(c)** $x^4 - 2x^3 - 14x^2 + 30x + 9$ **(d)** Both results are -72.   **86. (a)** 3   **(b)** 183   **(c)** $x^2 + 4x + 1$
(d) Both results are 61.   **87.** Answers will vary.   **88.** We could not choose 3 because it would make the denominator
equal to zero. It is the only invalid replacement for x.   **89.** 2   **91.** $144a^2 - 1$ **93.** $\dfrac{1}{8^{12}}$ **95.** $\dfrac{2}{3m^3}$ **97.** r^{13}
99. $-y^2 - 4y + 4$ **101.** $y^2 + 5y + 1$ **103.** $10p^2 - 3p - 5$ **105.** $49 - 28k + 4k^2$

CHAPTER 5 TEST   (PAGE 308)

[5.1] **1.** $-7x^2 + 8x$; 2; binomial   **2.** $4n^4 + 13n^3 - 10n^2$; 4; trinomial   **3.** $4, -2, -4, -2, 4$

$y = 2x^2 - 4$

4. $-2y^2 - 9y + 17$ **5.** $-21a^3b^2 + 7ab^5 - 5a^2b^2$ **6.** $-12t^2 + 5t + 8$ [5.3] **7.** $-27x^5 + 18x^4 - 6x^3 + 3x^2$
8. $t^2 - 5t - 24$ **9.** $8x^2 + 2xy - 3y^2$ [5.4] **10.** $25x^2 - 20xy + 4y^2$ **11.** $100v^2 - 9w^2$ **12.** $2r^3 + r^2 - 16r + 15$
13. $9x^2 + 54x + 81$ [5.5] **14.** $\dfrac{1}{625}$ **15.** 2   **16.** $\dfrac{7}{12}$ [5.2] **17.** $9x^3y^5$ [5.5] **18.** 8^5 **19.** x^2y^6

20. Disagree, because $3^{-4} = \dfrac{1}{3^4} = \dfrac{1}{81}$, which is positive.   [5.6] **21.** $4y^2 - 3y + 2 + \dfrac{5}{y}$ **22.** $-3xy^2 + 2x^3y^2 + 4y^2$

23. $3x^2 + 6x + 11 + \dfrac{26}{x - 2}$ [5.7] **24. (a)** 4.5×10^{10} **(b)** .0000036   **(c)** .00019   **25. (a)** 1.7×10^5; 1×10^3
(b) (more than) 1.7×10^8 pounds

CUMULATIVE REVIEW EXERCISES CHAPTERS 1-5   (PAGE 309)

[1.1] **1.** $\dfrac{7}{4}$ **2.** 5   **3.** $\dfrac{19}{24}$ **4.** $-\dfrac{1}{20}$ **5.** $31\dfrac{1}{4}$ cubic yards   [1.5] **6.** \$1836   [1.6] **7.** 1, 3, 5, 9, 15, 45

[1.5] **8.** positive   [1.6] **9.** -8 **10.** 24   **11.** $\dfrac{1}{2}$ **12.** -4 [1.7] **13.** associative property   **14.** distributive

property   [1.5] **15. (a)** -175 **(b)** 575   [1.8] **16.** $-10x^2 + 21x - 29$ [2.1, 2.2] **17.** $\left\{\dfrac{13}{4}\right\}$ **18.** $\emptyset$

[2.4] **19.** $r = \dfrac{d}{t}$ [2.5] **20.** $\{-5\}$ [2.1, 2.2] **21.** $\{-12\}$ **22.** $\{20\}$ **23.** {all real numbers} or $(-\infty, \infty)$

[2.3] **24.** mouse: 160; elephant: 10   **25.** 4   **26.** 1995 and 1996: 22; 1994: 24   [3.1] **27.** $[10, \infty)$ **28.** $\left(-\infty, -\dfrac{14}{5}\right)$

or $(-\infty, \infty)$ **29.** $[-4, 2)$ [2.4] **30.** 11 feet and 22 feet   [4.1] **31.**

x	0	-4	2	-2	4
y	2	0	3	1	4

[4.2] **32.**

$y = -3x + 6$

[5.1] **33.**

$y = (x + 4)^2$

34. $11x^3 - 14x^2 - x + 14$ [5.3] **35.** $18x^7 - 54x^6 + 60x^5$

36. $63x^2 + 57x + 12$ [5.4] **37.** $25x^2 + 80x + 64$ [5.6] **38.** $2x^2 - 3x + 1$ **39.** $y^2 - 2y + 6$

[5.5] **40.** $\dfrac{5}{4}$ **41.** 2   **42.** 1   **43.** $\dfrac{2b}{a^{10}}$ [5.7] **44.** 3.45×10^4 **45.** .000000536

CHAPTER 6 FACTORING AND APPLICATIONS

SECTION 6.1 (PAGE 318)

EXERCISES   1. no; 6   **3.** multiplication   **5.** Answers will vary. One example is 5, 10, 15.   **7.** 8   **9.** $10x^3$
11. $6m^3n^2$ **13.** xy^2 **15.** 6   **17.** factored   **19.** not factored   **21.** yes; x^3y^2 **23.** 2   **25.** x
27. $3m^2$ **29.** $2z^4$ **31.** $2mn^4$ **33.** $-7x^3y^2$ **35.** $12(y - 2)$ **37.** $10a(a - 2)$ **39.** $5y^6(13y^4 + 7)$
41. no common factor (except 1)   **43.** $8m^2n^2(n + 3)$ **45.** $13y^2(y^6 + 2y^2 - 3)$ **47.** $9qp^3(5q^3p^2 + 4p^3 + 9q)$
49. $a^3(a^2 + 2b^2 - 3a^2b^2 + 4ab^3)$ **51.** $(x + 2)(c - d)$ **53.** $(m + 2n)(m + n)$ **55.** not in factored form;
$(7t + 4)(8 + x)$ **57.** in factored form   **59.** not in factored form; $(y + 4)(18x^2 + 7)$ **61.** The quantities in
parentheses are not the same, so there is no common factor in the two terms, $12k^3(s - 3)$ and $7(s + 3)$.   **63.** $(p + 4)(p + 3)$
65. $(a - 2)(a + 5)$ **67.** $(z + 2)(7z - a)$ **69.** $(3r + 2y)(6r - x)$ **71.** $(a^2 + b^2)(3a + 2b)$ **73.** $(1 - a)(1 - b)$
75. $(4m - p^2)(4m^2 - p)$ **77.** $(5 - 2p)(m + 3)$ **79.** $(6r - y)(3r + 2y)$ **81.** $(a^5 - 3)(1 + 2b)$ **83.** commutative
property and associative property   **84.** $(2xy - 8x) + (-3y) + 12$; $2x$ **85.** $(2xy - 8x) + (-3y + 12)$; yes
86. $2x(y - 4) - 3(y - 4)$ **87.** No, because it is the difference between two terms, $2x(y - 4)$ and $3(y - 4)$.
88. $(2x - 3)(y - 4)$; yes   **89. (a)** yes   **(b)** When either one is multiplied out, the product is $1 - a + ab - b$.

SECTION 6.2 (PAGE 323)

EXERCISES   1. a and b must have different signs, one positive and one negative.   **5.** 1 and 48, -1 and -48, 2 and 24, -2
and -24, 3 and 16, -3 and -16, 4 and 12, -4 and -12, 6 and 8, -6 and -8; the pair with a sum of -19 is -3 and -16.
7. 1 and -24; -1 and 24, 2 and -12, -2 and 12, 3 and -8, -3 and 8, 4 and -6, -4 and 6; the pair with a sum of -5 is 3
and -8.   **9.** (c)   **11.** $p + 6$ **13.** $x + 11$ **15.** $x - 8$ **17.** $y - 5$ **19.** $x + 11$ **21.** $y - 9$ **23.** 1
24. -1 **25.** The sums are opposites as well.   **26.** The product is $x^2 - x - 12$. This is not correct because the middle
term is incorrect.   **27.** The product is $x^2 + x - 12$. This is correct because we obtain the exact trinomial we were given to
factor.   **28.** It is the opposite of what it should be.   **29.** reverse the signs of the two second terms of the binomials
30. $(x - 5)(x + 3)$ **31.** $(y + 8)(y + 1)$ **33.** $(b + 3)(b + 5)$ **35.** $(m + 5)(m - 4)$ **37.** $(y - 5)(y - 3)$
39. prime   **41.** $(t - 4)^2$ or $(t - 4)(t - 4)$ **43.** $(r - 6)(r + 5)$ **45.** prime   **47.** $(r + 2a)(r + a)$
49. $(t + 2z)(t - 3z)$ **51.** $(x + y)(x + 3y)$ **53.** $(v - 5w)(v - 6w)$ **55.** $4(x + 5)(x - 2)$ **57.** $2t(t + 1)(t + 3)$
59. $2x^4(x - 3)(x + 7)$ **61.** $mn(m - 6n)(m - 4n)$ **63.** The factored form $(2x + 4)(x - 3)$ is incorrect because $2x + 4$
has a common factor of 2, which must be factored out for the trinomial to be completely factored.   **65.** $a^3(a + 4b)(a - b)$
67. $yz(y + 3z)(y - 2z)$ **69.** $z^8(z - 7y)(z + 3y)$ **71.** $(a + b)(x + 4)(x - 3)$ **73.** $(2p + q)(r - 9)(r - 3)$
75. $a^2 + 13a + 36$

SECTION 6.3 (PAGE 330)

EXERCISES   1. 2; -21 **3.** $-6x$; $7x$ or $7x$; $-6x$ **5.** $x - 3$ **7.** $-4b$; $2a$ **9.** $-3x - 10$; -5; $+2$ or $-3x - 10$;
$+2$; -5 **11. (a)**   **13. (a)**   **15.** The binomial $2x - 6$ cannot be a factor, because it has a common factor of 2, but the
polynomial does not.
The order of the factors is irrelevant in Exercises 17–63.
17. $(3a + 7)(a + 1)$ **19.** $(4r - 3)(r + 1)$ **21.** $(3m - 1)(5m + 2)$ **23.** $(4m + 1)(2m - 3)$ **25.** $(4x + 3)(5x - 1)$
27. $(3m + 1)(7m + 2)$ **29.** $(4y - 1)(5y + 11)$ **31.** $(2b + 1)(3b + 2)$ **33.** $3(4x - 1)(2x - 3)$
35. $q(5m + 2)(8m - 3)$ **37.** $2m(m - 4)(m + 5)$ **39.** $3n^2(5n - 3)(n - 2)$ **41.** $3x^3(2x + 5)(3x - 5)$
43. $y^2(5x - 4)(3x + 1)$ **45.** $(3p + 4q)(4p - 3q)$ **47.** $(5a + 2b)(5a + 3b)$ **49.** $(3a - 5b)(2a + b)$
51. $m^4n(3m + 2n)(2m + n)$ **53.** $(5 - x)(1 - x)$ **55.** $(4 + 3x)(4 + x)$ **57.** $-5x(2x + 7)(x - 4)$
59. $-1(x + 7)(x - 3)$ **61.** $-1(3x + 4)(x - 1)$ **63.** $-1(a + 2b)(2a + b)$ **65.** Yes, $(x + 7)(3 - x)$ is equivalent to
$-1(x + 7)(x - 3)$, because $-1(x - 3) = -x + 3 = 3 - x$.   **67.** $(m + 1)^3(5q - 2)(5q + 1)$

69. $(r + 3)^3(5x + 2y)(3x - 8y)$ **71.** $-4, 4$ **73.** $-11, -7, 7, 11$ **75.** $5 \cdot 7$ **76.** $(-5)(-7)$ **77.** The product of $3x - 4$ and $2x - 1$ is indeed $6x^2 - 11x + 4$. **78.** The product of $4 - 3x$ and $1 - 2x$ is indeed $6x^2 - 11x + 4$.
79. The factors in Exercise 78 are the opposites of the factors in Exercise 77. **80.** $(3 - 7t)(5 - 2t)$ **81.** $-a; -b$
82. $-P; -Q$

SECTION 6.4 (PAGE 338)

EXERCISES **1.** 1; 4; 9; 16; 25; 36; 49; 64; 81; 100; 121; 144; 169; 196; 225; 256; 289; 324; 361; 400 **3.** 1; 8; 27; 64; 125;
216; 343; 512; 729; 1000 **5. (a)** both of these **(b)** a perfect cube **(c)** a perfect square **(d)** a perfect square

7. $(y + 5)(y - 5)$ **9.** $(3r + 2)(3r - 2)$ **11.** $\left(6m + \dfrac{4}{5}\right)\left(6m - \dfrac{4}{5}\right)$ **13.** $4(3x + 2)(3x - 2)$

15. $(14p + 15)(14p - 15)$ **17.** $(4r + 5a)(4r - 5a)$ **19.** prime **21.** $(p^2 + 7)(p^2 - 7)$
23. $(x^2 + 1)(x + 1)(x - 1)$ **25.** $(p^2 + 16)(p + 4)(p - 4)$ **27.** The teacher was justified, because it was not factored
completely; $x^2 - 9$ can be factored as $(x + 3)(x - 3)$. The complete factored form is $(x^2 + 9)(x + 3)(x - 3)$.

29. $(w + 1)^2$ **31.** $(x - 4)^2$ **33.** $\left(t + \dfrac{1}{2}\right)^2$ **35.** $(x - .5)^2$ **37.** $2(x + 6)^2$ **39.** $(4x - 5)^2$ **41.** $(7x - 2y)^2$

43. $(8x + 3y)^2$ **45.** $-2(5h - 2y)^2$ **47.** $(a + 1)(a^2 - a + 1)$ **49.** $(a - 1)(a^2 + a + 1)$
51. $(p + q)(p^2 - pq + q^2)$ **53.** $(3x - 1)(9x^2 + 3x + 1)$ **55.** $(2p + 9q)(4p^2 - 18pq + 81q^2)$
57. $(y - 2x)(y^2 + 2yx + 4x^2)$ **59.** $(3a - 4b)(9a^2 + 12ab + 16b^2)$ **61.** $(5t + 2s)(25t^2 - 10ts + 4s^2)$
64. $(5x - 2)(2x + 3)$ **65.** $5x - 2$ **66.** Yes. If $10x^2 + 11x - 6$ factors as $(5x - 2)(2x + 3)$, then when $10x^2 + 11x - 6$
is divided by $2x + 3$, the quotient should be $5x - 2$. **67.** $x^2 + x + 1; (x - 1)(x^2 + x + 1)$ **69.** $-2b(3a^2 + b^2)$
71. $3(r - k)(1 + r + k)$ **72.** $(x + 1)^2 - 2^2$ or $(x + 1)^2 - 4$ **73.** $[(x + 1) - 2][(x + 1) + 2]$ **74.** $(x - 1)(x + 3)$
75. $x^2 + 2x - 3$ **76.** $(x - 1)(x + 3)$ **77.** The results are the same. (Preference is a matter of individual choice.)
79. 10 **81.** 9

SUMMARY: EXERCISES ON FACTORING (PAGE 341)

EXERCISES **1.** $(a - 6)(a + 2)$ **3.** $6(y - 2)(y + 1)$ **5.** $6(a + 2b + 3c)$ **7.** $(p - 11)(p - 6)$
9. $(5z - 6)(2z + 1)$ **11.** $(m + n + 5)(m - n)$ **13.** $8a^3(a - 3)(a + 2)$ **15.** $(z - 5a)(z + 2a)$ **17.** $(x - 5)(x - 4)$
19. $(3n - 2)(2n - 5)$ **21.** $4(4x + 5)$ **23.** $(3y - 4)(2y + 1)$ **25.** $(6z + 1)(z + 5)$ **27.** $(2k - 3)^2$
29. $6(3m + 2z)(3m - 2z)$ **31.** $(3k - 2)(k + 2)$ **33.** $7k(2k + 5)(k - 2)$ **35.** $(y^2 + 4)(y + 2)(y - 2)$
37. $8m(1 - 2m)$ **39.** $(z - 2)(z^2 + 2z + 4)$ **41.** prime **43.** $8m^3(4m^6 + 2m^2 + 3)$ **45.** $(4r + 3m)^2$
47. $(5h + 7g)(3h - 2g)$ **49.** $(k - 5)(k - 6)$ **51.** $3k(k - 5)(k + 1)$ **53.** $(10p + 3)(100p^2 - 30p + 9)$
55. $(2 + m)(3 + p)$ **57.** $(4z - 1)^2$ **59.** $3(6m - 1)^2$ **61.** prime **63.** $8z(4z - 1)(z + 2)$ **65.** $(4 + m)(5 + 3n)$
67. $2(3a - 1)(a + 2)$ **69.** $(a - b)(a^2 + ab + b^2 + 2)$ **71.** $(8m - 5n)^2$ **73.** $(4k - 3h)(2k + h)$
75. $(m + 2)(m^2 + m + 1)$ **77.** $(5y - 6z)(2y + z)$ **79.** $(8a - b)(a + 3b)$ **81.** $(x^3 - 1)(x^3 + 1)$
82. $(x - 1)(x^2 + x + 1)(x + 1)(x^2 - x + 1)$ **83.** $(x^2 - 1)(x^4 + x^2 + 1)$ **84.** $(x - 1)(x + 1)(x^4 + x^2 + 1)$
85. The result in Exercise 82 is completely factored. **86.** Show that $x^4 + x^2 + 1 = (x^2 + x + 1)(x^2 - x + 1)$.
87. difference of squares **88.** $(x - 3)(x^2 + 3x + 9)(x + 3)(x^2 - 3x + 9)$

SECTION 6.5 (PAGE 347)

CONNECTIONS **Page 346:** **1.** 4 seconds **2.** 64 feet

EXERCISES **1.** $ax^2 + bx + c$ **3.** factoring **5.** $\left\{-\dfrac{8}{3}, -7\right\}$ **7.** $\{0, -4\}$ **9.** To solve $2x(3x - 4) = 0$, set each

variable factor equal to 0 to get $x = 0$ or $3x - 4 = 0$. Then solve the second equation to get $x = \dfrac{4}{3}$.

11. Because $(x - 9)^2 = (x - 9)(x - 9) = 0$ leads to two solutions of 9, we call 9 a double solution. **13.** $\{-2, -1\}$

15. $\{1, 2\}$ **17.** $\{-8, 3\}$ **19.** $\{-1, 3\}$ **21.** $\{-2, -1\}$ **23.** $\{-4\}$ **25.** $\left\{-2, \dfrac{1}{3}\right\}$ **27.** $\left\{-\dfrac{4}{3}, \dfrac{1}{2}\right\}$

29. $\left\{-\dfrac{2}{3}\right\}$ **31.** $\{-3, 3\}$ **33.** $\left\{-\dfrac{7}{4}, \dfrac{7}{4}\right\}$ **35.** $\{-11, 11\}$ **37.** Another solution is -11. **39.** $\{0, 7\}$

41. $\left\{0, \dfrac{1}{2}\right\}$ **43.** $\{2, 5\}$ **45.** $\left\{-4, \dfrac{1}{2}\right\}$ **47.** $\left\{-12, \dfrac{11}{2}\right\}$ **49.** $\{-1, 3\}$ **51.** $\left\{-\dfrac{5}{2}, \dfrac{1}{3}, 5\right\}$ **53.** $\left\{-\dfrac{7}{2}, -3, 1\right\}$

55. $\left\{-\dfrac{7}{3}, 0, \dfrac{7}{3}\right\}$ **57.** $\{-2, 0, 4\}$ **59.** $\{-5, 0, 4\}$ **61.** $\{-3, 0, 5\}$ **63.** $\left\{-\dfrac{4}{3}, -1, \dfrac{1}{2}\right\}$ **65.** $\left\{-\dfrac{2}{3}, 4\right\}$

67. To use the zero-factor property, one side of the equation must be 0. Before solving this equation, multiply the factors on the left side, and subtract 1 from both sides so the right side is 0.   **69.** $\{-.5, .1\}$ **71.** $\{-1.1, -2.5\}$

SECTION 6.6 (PAGE 354)

CONNECTIONS **Page 353:** The diagonal of the floor should be $\sqrt{208} \approx 14.4$ ft, or about 14 feet, 5 inches. The carpenter is off by 3 inches and must correct the error to avoid major construction problems.

EXERCISES   **1.** a variable; the unknown   **3.** an equation   **5. (a)** $80 = (x + 8)(x - 8)$ **(b)** 12   **(c)** length: 20 units; width: 4 units   **7. (a)** $60 = \dfrac{1}{2}(3x + 6)(x + 5)$ **(b)** 3   **(c)** base: 15 units; height: 8 units   **9.** length: 7 inches; width: 4 inches   **11.** length: 11 inches; width: 8 inches   **13.** base: 12 inches; height: 5 inches   **15.** height: 13 inches; width: 10 inches   **17.** 12 centimeters   **19.** 8 feet   **21.** 6 meters   **23.** 1 second   **25.** 3 seconds   **27.** 1 second and 3 seconds   **29.** 4 seconds   **31.** 9 centimeters   **33.** 256 feet   **35.** 5.7 seconds   **37.** 8.0 seconds   **39.** $-3, -2$ or 4, 5   **41.** 7, 9, 11   **43.** $-2, 0, 2$ or 6, 8, 10   **45. (a)** 1990   **(b)** 8.8 billion passengers; This approximation is .2 billion less than the value in the table.   **(c)** 25   **(d)** 8.0 billion passengers; This is correct to the nearest tenth billion.   **(e)** 1995   **47. (a)** approximately 17,000 injuries   **(b)** approximately .7 or 7 to 10; This indicates that in 1998, on the average, 1.4 people were injured per accident.   **49.** c^2 **50.** b^2 **51.** a^2 **52.** $a^2 + b^2 = c^2$; This is the Pythagorean formula.

CHAPTER 6 REVIEW EXERCISES (PAGE 363)

1. $7(t + 2)$ **3.** $(x - 4)(2y + 3)$ **5.** $(x + 3)(x + 2)$ **7.** $(q + 9)(q - 3)$ **9.** $(r + 8s)(r - 12s)$ **11.** $8p(p + 2)(p - 5)$ **13.** $p^5(p - 2q)(p + q)$ **15.** r and $6r$; $2r$ and $3r$ **17.** $(2k - 1)(k - 2)$ **19.** $(3r + 2)(2r - 3)$ **21.** $(v + 3)(8v - 7)$ **23.** $-3(x + 2)(2x - 5)$ **25. (b)**   **27.** $(n + 7)(n - 7)$ **29.** $(7y + 5w)(7y - 5w)$ **31.** prime   **33.** $(3t - 7)^2$ **35.** $(5k + 4x)(25k^2 - 20kx + 16x^2)$ **37.** $\{-3, -1\}$ **39.** $\{3, 5\}$ **41.** $\left\{-\dfrac{8}{9}, \dfrac{8}{9}\right\}$ **43.** $\{-1, 6\}$ **45.** $\{6\}$ **47.** length: 10 meters; width: 4 meters   **49.** length: 6 meters; height: 5 meters   **51.** 112 feet   **53.** 256 feet   **55.** 2 inches   **57. (a)** $17.4 million **(b)** $-$48.6 million   **(c)** The equation is based on the data for 1994 to 1996. To use it to predict much beyond 1996 may lead to incorrect results, since it is possible that the conditions the equation was based on will change.   **59.** $(3k + 5)(k + 2)$ **61.** $(y^2 + 25)(y + 5)(y - 5)$ **63.** $8abc(3b^2c - 7ac^2 + 9ab)$ **65.** $6xyz(2xz^2 + 2y - 5x^2yz^3)$ **67.** $(2r + 3q)(6r - 5q)$ **69.** $(7t + 4)^2$ **71.** $\{-5, 2\}$ **73. (a)** 417,000 vehicles   **(b)** The estimate may be unreliable because the conditions that prevailed in the years 1995–1997 may have changed, causing either a greater increase or a greater decrease in the numbers of alternative-fueled vehicles.   **75.** width: 10 meters; length: 17 meters   **77.** 34 miles   **79.** 25 miles

CHAPTER 6 TEST (PAGE 368)

[6.1–6.4] **1. (d)**   **2.** $m^2n(2mn + 3m - 5n)$ **3.** $(x + 3)(x - 8)$ **4.** $(2x + 3)(x - 1)$ **5.** $(5z - 1)(2z - 3)$ **6.** prime   **7.** prime   **8.** $(2 - a)(6 + b)$ **9.** $(3y + 8)(3y - 8)$ **10.** $(2x - 7y)^2$ **11.** $-2(x + 1)^2$ **12.** $3t^2(2t + 9)(t - 4)$ **13.** $(r - 5)(r^2 + 5r + 25)$ **14.** $8(k + 2)(k^2 - 2k + 4)$ **15.** The product $(p + 3)(p + 3) = p^2 + 6p + 9$, which does not equal $p^2 + 9$. The binomial $p^2 + 9$ is a prime polynomial.   [6.5] **16.** $\left\{6, \dfrac{1}{2}\right\}$ **17.** $\left\{-\dfrac{2}{5}, \dfrac{2}{5}\right\}$ **18.** $\{10\}$ **19.** $\{-3, 0, 3\}$ [6.6] **20.** $1\dfrac{1}{2}$ seconds and $4\dfrac{1}{2}$ seconds   **21. (a)** $3x - 7$ **(b)** $2x - 1$ **(c)** 17 feet   **22.** July 1, 1995: 341,000; July 1, 1996: 368,000   [6.5] **23.** Another solution is $-\dfrac{2}{3}$.

CUMULATIVE REVIEW EXERCISES CHAPTERS 1-6 (PAGE 369)

[2.2] **1.** $\{0\}$ **2.** $\{.05\}$ **3.** $\{6\}$ [2.4] **4.** $P = \dfrac{A}{1 + rt}$ [2.3] **5.** $110°$ and $70°$ **6.** exports: $741 million; imports: $426 million   [2.5] **7.** 35 pounds   [4.1] **8. (a)** negative; positive   **(b)** negative; negative

9. $(2, 27{,}000)$, $(5, 63{,}000)$ [4.2] **10.** x-intercept: $\left(-\dfrac{1}{4}, 0\right)$; y-intercept: $(0, 3)$

11.

[4.3] **12.** 209.2; A slope of 209.2 means that the number of radio stations increased on the average by about 209 per year.

[5.2, 5.5] **13.** 4   **14.** $\dfrac{16}{9}$ **15.** 256   **16.** $\dfrac{1}{p^2}$ [5.1] **17.** $-4k^2 - 4k + 8$ [5.3] **18.** $6m^8 - 15m^6 + 3m^4$

19. $3y^3 + 8y^2 + 12y - 5$ [5.4] **20.** $4p^2 - 9q^2$ [5.6] **21.** $4x^3 + 6x^2 - 3x + 10$ **22.** $6p^2 + 7p + 1 + \dfrac{7}{2p - 2}$

[6.2–6.3] **23.** $(2a - 1)(a + 4)$ **24.** $(2m + 3)(5m + 2)$ **25.** $(5x + 3y)(3x - 2y)$ [6.4] **26.** $(3x + 1)^2$

27. $-2(4t + 7z)^2$ **28.** $(5r + 9t)(5r - 9t)$ [6.1] **29.** $25(4x^2 + 1)$ [6.3] **30.** $2pq(3p + 1)(p + 1)$

[6.1] **31.** $(2x + y)(a - b)$ [6.5] **32.** $\left\{\dfrac{3}{2}, -2, 6\right\}$ **33.** $\left\{-\dfrac{2}{3}, \dfrac{1}{2}\right\}$ [6.6] **34.** $-4, -2$ or $8, 10$

35. 5 meters, 12 meters, 13 meters

CHAPTER 7 RATIONAL EXPRESSIONS

SECTION 7.1 (PAGE 378)

CONNECTIONS   **Page 378:** **1.** $3x^2 + 11x + 8$ cannot be factored, so this quotient cannot be reduced. By long division the quotient is $3x + 5 + \dfrac{-2}{x + 2}$.   **2.** The numerator factors as $(x - 2)(x^2 + 2x + 4)$, so by reducing, the quotient is $x - 2$. Long division gives the same quotient.

EXERCISES **1. (a)** $3; -5$ **(b)** $q; -1$ **5.** 0   **7.** $-\dfrac{5}{3}$ **9.** $-3, 2$ **11.** never undefined   **13. (a)** 1   **(b)** $\dfrac{17}{12}$

15. (a) 0   **(b)** $-\dfrac{10}{3}$ **17. (a)** $\dfrac{9}{5}$ **(b)** undefined   **19. (a)** $\dfrac{2}{7}$ **(b)** $\dfrac{13}{3}$ **21.** Any number divided by itself is 1,

provided the number is not 0. This expression is equal to $\dfrac{1}{x + 2}$ for all values of x except -2 and 2.   **23.** $3r^2$ **25.** $\dfrac{2}{5}$

27. $\dfrac{x - 1}{x + 1}$ **29.** $\dfrac{7}{5}$ **31.** $m - n$ **33.** $\dfrac{3(2m + 1)}{4}$ **35.** $\dfrac{3m}{5}$ **37.** $\dfrac{3r - 2s}{3}$ **39.** $\dfrac{z - 3}{z + 5}$ **41.** $k - 3$

43. $\dfrac{x + 1}{x - 1}$ **45.** -1 **47.** $-(m + 1)$ **49.** -1 **51.** already in lowest terms   **53.** $x^2 + 3$ **Answers may vary**

in Exercises 55–59.   **55.** $\dfrac{-(x + 4)}{x - 3}, \dfrac{-x - 4}{x - 3}, \dfrac{x + 4}{-(x - 3)}, \dfrac{x + 4}{-x + 3}$ **57.** $\dfrac{-(2x - 3)}{x + 3}, \dfrac{-2x + 3}{x + 3}, \dfrac{2x - 3}{-(x + 3)}, \dfrac{2x - 3}{-x - 3}$

59. $-\dfrac{3x - 1}{5x - 6}, \dfrac{-(3x - 1)}{5x - 6}, \dfrac{-3x + 1}{-5x + 6}, \dfrac{3x - 1}{-5x + 6}$ **61.** $\dfrac{m + n}{2}$ **63.** $-\dfrac{b^2 + ba + a^2}{a + b}$ **65.** $\dfrac{z + 3}{z}$ **67.** $x + 3$

69. $x + 5$ **71.** $x - 3$

SECTION 7.2 (PAGE 385)

EXERCISES **1. (a)** B **(b)** D **(c)** C **(d)** A   **3.** $\dfrac{3a}{2}$ **5.** $-\dfrac{4x^4}{3}$ **7.** $\dfrac{2}{c + d}$ **9.** 5   **11.** $-\dfrac{3}{2t^4}$ **13.** $\dfrac{1}{4}$

15. -4 causes the denominator in the first fraction to equal 0.   **16.** -5 causes the denominator in the second fraction to equal 0.   **17.** -7 causes the numerator in the divisor to equal 0, meaning that we would be dividing by 0. This is undefined.

18. We *are* allowed to divide 0 by a nonzero number.   **21.** $\dfrac{10}{9}$ **23.** $-\dfrac{3}{4}$ **25.** $-\dfrac{9}{2}$ **27.** $\dfrac{p+4}{p+2}$

29. $\dfrac{(k-1)^2}{(k+1)(2k-1)}$ **31.** $\dfrac{4k-1}{3k-2}$ **33.** $\dfrac{m+4p}{m+p}$ **35.** $\dfrac{m+6}{m+3}$ **37.** $\dfrac{y+3}{y+4}$ **39.** $\dfrac{m}{m+5}$ **41.** $\dfrac{r+6s}{r+s}$

43. $\dfrac{(q-3)^2(q+2)^2}{q+1}$ **45.** $\dfrac{x+10}{10}$ **47.** $\dfrac{3-a-b}{2a-b}$ **49.** $-\dfrac{(x+y)^2(x^2-xy+y^2)}{3y(y-x)(x-y)}$ or $\dfrac{(x+y)^2(x^2-xy+y^2)}{3y(x-y)^2}$

51. $\dfrac{5xy^2}{4q}$

SECTION 7.3   (PAGE 391)

EXERCISES   1. (c)   **3.** (c)   **5.** 60   **7.** 1800   **9.** x^5 **11.** $30p$ **13.** $180y^4$ **15.** $15a^5b^3$ **17.** $12p(p-2)$ **19.** $2^3 \cdot 3 \cdot 5$ **20.** $(t+4)^3(t-3)(t+8)$ **21.** The similarity is that 2 is replaced by $t+4$, 3 is replaced by $t-3$, and 5 is replaced by $t+8$.   **22.** The procedure used is the same. The only difference is that for algebraic fractions, the factors may contain variables, while in a common fraction, the factors are constants.   **23.** $18(r-2)$ **25.** $12p(p+5)^2$ **27.** $8(y+2)(y+1)$ **29.** $m-3$ or $3-m$ **31.** $p-q$ or $q-p$ **33.** $a(a+6)(a-3)$ **35.** $(k+3)(k-5)(k+7)(k+8)$ **37.** Yes, because $(2x-5)^2 = (5-2x)^2$.   **39.** 7   **40.** 1   **41.** identity property of multiplication   **42.** 7   **43.** 1   **44.** identity property of multiplication   **45.** $\dfrac{60m^2k^3}{32k^4}$ **47.** $\dfrac{57z}{6z-18}$

49. $\dfrac{-4a}{18a-36}$ **51.** $\dfrac{6(k+1)}{k(k-4)(k+1)}$ **53.** $\dfrac{36r(r+1)}{(r-3)(r+2)(r+1)}$ **55.** $\dfrac{ab(a+2b)}{2a^3b+a^2b^2-ab^3}$

57. $\dfrac{(t-r)(4r-t)}{t^3-r^3}$ **59.** $\dfrac{2y(z-y)(y-z)}{y^4-z^3y}$ or $\dfrac{-2y(y-z)^2}{y^4-z^3y}$

SECTION 7.4   (PAGE 399)

EXERCISES   1. $\dfrac{5}{4}$ **2.** $\dfrac{-5}{-7+3}$ **3.** 1   **4.** $\dfrac{-5}{-4} = \dfrac{5}{4}$; The two answers are equal.   **5.** Jack's answer was correct. His answer can be obtained by multiplying Jill's correct answer by $\dfrac{-1}{-1} = 1$, the identity element for multiplication.

6. Changing the sign of each term in the fraction is a way of multiplying by $\dfrac{-1}{-1}$ or 1, the identity element for multiplication.

7. Putting a negative sign in front of a fraction gives the opposite of the fraction. Changing the signs in either the numerator or the denominator also gives the opposite. Thus we have multiplied by $(-1)(-1) = 1$, the identity element for multiplication.

8. (a) not equivalent   **(b)** equivalent   **(c)** equivalent   **(d)** not equivalent   **(e)** not equivalent   **(f)** not equivalent

9. $\dfrac{11}{m}$ **11.** b **13.** x **15.** $y-6$ **19.** $\dfrac{3z+5}{15}$ **21.** $\dfrac{10-7r}{14}$ **23.** $\dfrac{-3x-2}{4x}$ **25.** $\dfrac{x+1}{2}$ **27.** $\dfrac{5x+9}{6x}$

29. $\dfrac{7-6p}{3p^2}$ **31.** $\dfrac{x+8}{x+2}$ **33.** $\dfrac{3}{t}$ **35.** $m-2$ or $2-m$ **37.** $\dfrac{-2}{x-5}$ or $\dfrac{2}{5-x}$ **39.** -4 **41.** $\dfrac{-5}{x-y^2}$ or $\dfrac{5}{y^2-x}$

43. $\dfrac{x+y}{5x-3y}$ or $\dfrac{-x-y}{3y-5x}$ **45.** $\dfrac{-6}{4p-5}$ or $\dfrac{6}{5-4p}$ **47.** $\dfrac{-(m+n)}{2(m-n)}$ **49.** $\dfrac{-x^2+6x+11}{(x+3)(x-3)(x+1)}$

51. $\dfrac{-5q^2-13q+7}{(3q-2)(q+4)(2q-3)}$ **53.** $\dfrac{9r+2}{r(r+2)(r-1)}$ **55.** $\dfrac{2x^2+6xy+8y^2}{(x+y)(x+y)(x+3y)}$ or $\dfrac{2x^2+6xy+8y^2}{(x+y)^2(x+3y)}$

57. $\dfrac{15r^2+10ry-y^2}{(3r+2y)(6r-y)(6r+y)}$ **59.** $\dfrac{2k^2-10k+6}{(k-3)(k-1)^2}$ **61.** $\dfrac{7k^2+31k+92}{(k-4)(k+4)^2}$ **63. (a)** $\dfrac{9k^2+6k+26}{5(3k+1)}$ **(b)** $\dfrac{1}{4}$

SECTION 7.5   (PAGE 406)

CONNECTIONS   Page 406: 1.413793103; 2

EXERCISES   1. (a) $6; \dfrac{1}{6}$ **(b)** $12; \dfrac{3}{4}$ **(c)** $\dfrac{1}{6} \div \dfrac{3}{4}$ **(d)** $\dfrac{2}{9}$ **3.** Choice (d) is correct, because every sign has been changed in the fraction.   **7.** -6 **9.** $\dfrac{1}{xy}$ **11.** $\dfrac{2a^2b}{3}$ **13.** $\dfrac{m(m+2)}{3(m-4)}$ **15.** $\dfrac{2}{x}$ **17.** $\dfrac{8}{x}$ **19.** $\dfrac{a^2-5}{a^2+1}$ **21.** $\dfrac{31}{50}$

23. $\dfrac{y^2 + x^2}{xy(y - x)}$ **25.** $\dfrac{40 - 12p}{85p}$ **27.** $\dfrac{5y - 2x}{3 + 4xy}$ **29.** $\dfrac{a - 2}{2a}$ **31.** $\dfrac{z - 5}{4}$ **33.** $\dfrac{-m}{m + 2}$ **35.** $\dfrac{3m(m - 3)}{(m - 1)(m - 8)}$

37. division   **39.** $\dfrac{\frac{3}{8} + \frac{5}{6}}{2}$ **40.** $\dfrac{29}{48}$ **41.** $\dfrac{29}{48}$ **42.** Answers will vary.   **43.** $\dfrac{5}{3}$ **45.** $\dfrac{13}{2}$ **47.** $\dfrac{19r}{15}$

SECTION 7.6 (PAGE 414)

EXERCISES   **1.** expression; $\dfrac{43}{40}x$ **3.** equation; $\left\{\dfrac{40}{43}\right\}$ **5.** expression; $-\dfrac{1}{10}y$ **7.** equation; $\{-10\}$ **9.** $-2, 0$

11. $-3, 4, -\dfrac{1}{2}$ **13.** $-9, 1, -2, 2$ **17.** $\left\{\dfrac{1}{4}\right\}$ **19.** $\left\{-\dfrac{3}{4}\right\}$ **21.** $\{-15\}$ **23.** $\{7\}$ **25.** $\{-15\}$ **27.** $\{-5\}$

29. $\{-6\}$ **31.** $\emptyset$ **33.** $\{5\}$ **35.** $\{4\}$ **37.** $\{1\}$ **39.** $\{4\}$ **41.** $\{5\}$ **43.** $\{-2, 12\}$ **45.** $\emptyset$ **47.** $\{3\}$

49. $\{3\}$ **51.** $\left\{-\dfrac{1}{5}, 3\right\}$ **53.** $\left\{-\dfrac{1}{2}, 5\right\}$ **55.** $\{3\}$ **57.** $\left\{-\dfrac{1}{3}, 3\right\}$ **59.** $\left\{-6, \dfrac{1}{2}\right\}$ **61.** $\{6\}$

63. Transform the equation so that the terms with k are on one side and the remaining term is on the other.   **65.** $F = \dfrac{ma}{k}$

67. $a = \dfrac{kF}{m}$ **69.** $R = \dfrac{E - Ir}{I}$ or $R = \dfrac{E}{I} - r$ **71.** $A = \dfrac{h(B + b)}{2}$ **73.** $a = \dfrac{2S - ndL}{nd}$ or $a = \dfrac{2S}{nd} - L$

75. $y = \dfrac{xz}{x + z}$ **77.** $z = \dfrac{3y}{5 - 9xy}$ or $z = \dfrac{-3y}{9xy - 5}$ **79.** The solution set is $\{-5\}$. The number 3 must be rejected.

80. The simplified form is $x + 5$.   **(a)** It is the same as the actual solution.   **(b)** It is the same as the rejected solution.

81. The solution set is $\{-3\}$. The number 1 must be rejected.   **82.** The simplified form is $\dfrac{x + 3}{2(x + 1)}$.   **(a)** It is the same as the actual solution.   **(b)** It is the same as the rejected solution.   **83.** Answers will vary.   **84.** If we transform so that 0 is on one side, then perform the operation(s) and *reduce to lowest terms* before solving, no rejected values will appear.

85. $-\dfrac{3}{10}$ **87.** An error message would occur.   **89.** An error message would occur.

SUMMARY: EXERCISES ON OPERATIONS AND EQUATIONS WITH RATIONAL EXPRESSIONS (PAGE 419)

1. operation; $\dfrac{10}{p}$ **3.** operation; $\dfrac{1}{2x^2(x + 2)}$ **5.** operation; $\dfrac{y + 2}{y - 1}$ **7.** equation; $\{39\}$ **9.** operation; $\dfrac{13}{3(p + 2)}$

11. equation; $\left\{\dfrac{1}{7}, 2\right\}$ **13.** operation; $\dfrac{7}{12z}$ **15.** operation; $\dfrac{3m + 5}{(m + 3)(m + 2)(m + 1)}$ **17.** equation; $\emptyset$

19. operation; $\dfrac{t + 2}{2(2t + 1)}$

SECTION 7.7 (PAGE 427)

CONNECTIONS   **Page 423:**   54.5 miles per hour; 57.4 miles per hour; yes; More time is spent at the slower speed, and thus the average speed is less than the average of the two speeds.

EXERCISES   **1. (a)** an amount   **(b)** $5 + x$ **(c)** $\dfrac{5 + x}{6} = \dfrac{13}{3}$ **3.** $\dfrac{12}{18}$ **5.** $\dfrac{1386}{97}$ **7.** 1989: 33,579 (thousand); 1990: 34,203 (thousand)   **9.** female: 144,000; male: 576,000   **11.** 24.15 kilometers per hour   **13.** 3.429 hours

15. 7.91 meters per second   **17.** $\dfrac{500}{x - 10} - \dfrac{600}{x + 10}$ **19.** $\dfrac{D}{R} - \dfrac{d}{r}$ **21.** 8 miles per hour   **23.** $18\dfrac{1}{2}$ miles per hour

25. N'Deti: 12.02 miles per hour; McDermott: 8.78 miles per hour   **27.** $\dfrac{1}{10}$ job per hour   **29.** $\dfrac{1}{8}x + \dfrac{1}{6}x = 1$ or $\dfrac{1}{8} + \dfrac{1}{6} = \dfrac{1}{x}$ **31.** $4\dfrac{4}{17}$ hours   **33.** $5\dfrac{5}{11}$ hours   **35.** 3 hours   **37.** $2\dfrac{7}{10}$ hours   **39.** $9\dfrac{1}{11}$ minutes   **41.** direct

43. direct   **45.** direct   **47.** inverse   **49.** 9   **51.** $\dfrac{16}{5}$ **53.** $\dfrac{4}{9}$ **55.** increases   **57.** $106\dfrac{2}{3}$ miles per hour

59. 25 kilograms per hour   **61.** 20 pounds per square foot   **63.** 15 feet   **65.** 144 feet   **67.** direct   **69.** inverse
71. approximately 2364 in 1995 and 716 in 1985 (The actual numbers were 2361 and 719.)

CHAPTER 7 REVIEW EXERCISES   (PAGE 439)

1. 3   **3.** $-5, -\dfrac{2}{3}$ **5. (a)** $\dfrac{11}{8}$ **(b)** $\dfrac{13}{22}$ **7.** $\dfrac{b}{3a}$ **9.** $\dfrac{-(2x+3)}{2}$

Answers may vary in Exercise 11.

11. $\dfrac{-(4x-9)}{2x+3}, \dfrac{-4x+9}{2x+3}, \dfrac{4x-9}{-(2x+3)}, \dfrac{4x-9}{-2x-3}$ **13.** $\dfrac{72}{p}$ **15.** $\dfrac{5}{8}$ **17.** $\dfrac{3a-1}{a+5}$ **19.** $\dfrac{p+5}{p+1}$ **21.** $108y^4$

23. $\dfrac{15a}{10a^4}$ **25.** $\dfrac{15y}{50-10y}$ **27.** $\dfrac{15}{x}$ **29.** $\dfrac{4k-45}{k(k-5)}$ **31.** $\dfrac{-2-3m}{6}$ **33.** $\dfrac{7a+6b}{(a-2b)(a+2b)}$

35. $\dfrac{5z-16}{z(z+6)(z-2)}$ **37. (a)** $\dfrac{a}{b}$ **(b)** $\dfrac{a}{b}$ **(c)** Answers will vary.   **39.** $\dfrac{4(y-3)}{y+3}$ **41.** $\dfrac{xw+1}{xw-1}$ **43.** It would cause

the first and third denominators to equal 0.   **45.** $\emptyset$ **47.** $t = \dfrac{Ry}{m}$ **49.** $m = \dfrac{4+p^2q}{3p^2}$ **51.** $\dfrac{2}{6}$ **53.** $3\dfrac{1}{13}$ hours

55. inverse   **57.** 4 centimeters   **59. (a)** -3 **(b)** -1 **(c)** $-3, -1$ **60.** $\dfrac{15}{2x}$ **61.** If $x = 0$, the divisor R is

equal to 0, and division by 0 is undefined.   **62.** $(x+3)(x+1)$ **63.** $\dfrac{7}{x+1}$ **64.** $\dfrac{11x+21}{4x}$ **65.** $\emptyset$ **66.** We

know that -3 is not allowed because P and R are undefined for $x = -3$.   **67.** Rate is equal to distance divided by time.

Here, distance is 6 miles and time is $x+3$ minutes, so rate $= \dfrac{6}{x+3}$, which is the expression for P.   **68.** $\dfrac{6}{5}, \dfrac{5}{2}$

69. $\dfrac{(5+2x-2y)(x+y)}{(3x+3y-2)(x-y)}$ **71.** $8p^2$ **73.** 3   **75.** $r = \dfrac{3kz}{5k-z}$ or $r = \dfrac{-3kz}{z-5k}$ **77.** approximately 7443 for hearts

and 3722 for livers (The actual numbers were 7467 and 3698.)   **79.** $\dfrac{36}{5}$

CHAPTER 7 TEST   (PAGE 443)

[7.1] **1.** $-2, 4$ **2. (a)** $\dfrac{11}{6}$ **(b)** undefined   **3.** (Answers may vary.) $\dfrac{-(6x-5)}{2x+3}, \dfrac{-6x+5}{2x+3}, \dfrac{6x-5}{-(2x+3)}, \dfrac{6x-5}{-2x-3}$

4. $-3x^2y^3$ **5.** $\dfrac{3a+2}{a-1}$ [7.2] **6.** $\dfrac{25}{27}$ **7.** $\dfrac{3k-2}{3k+2}$ **8.** $\dfrac{a-1}{a+4}$ [7.3] **9.** $150p^5$ **10.** $(2r+3)(r+2)(r-5)$

11. $\dfrac{240p^2}{64p^3}$ **12.** $\dfrac{21}{42m-84}$ [7.4] **13.** 2   **14.** $\dfrac{-14}{5(y+2)}$ **15.** $\dfrac{-x^2+x+1}{3-x}$ or $\dfrac{x^2-x-1}{x-3}$

16. $\dfrac{-m^2+7m+2}{(2m+1)(m-5)(m-1)}$ [7.5] **17.** $\dfrac{2k}{3p}$ **18.** $\dfrac{-2-x}{4+x}$ [7.6] **19.** $-1, 4$ **20.** $\left\{-\dfrac{1}{2}\right\}$ **21.** $D = \dfrac{dF-k}{F}$ or

$D = \dfrac{k-dF}{-F}$ [7.7] **22.** 3 miles per hour   **23.** $2\dfrac{2}{9}$ hours   **24.** 27 days

CUMULATIVE REVIEW EXERCISES CHAPTERS 1-7   (PAGE 444)

[1.2, 1.5, 1.6] **1.** 2   [2.2] **2.** $\{17\}$ [2.4] **3.** $b = \dfrac{2A}{h}$ [2.5] **4.** $\left\{-\dfrac{2}{7}\right\}$ [3.1] **5.** $[-8, \infty)$ **6.** $(4, \infty)$

[4.1] **7. (a)** $(-3, 0)$ **(b)** $(0, -4)$ [4.2] **8.**   [5.1] **9.**

[5.2, 5.5] **10.** $\dfrac{1}{2^4 x^7}$ **11.** $\dfrac{1}{m^6}$ **12.** $\dfrac{q}{4p^2}$ [5.1] **13.** $k^2 + 2k + 1$ [5.2] **14.** $72x^6 y^7$ [5.4] **15.** $4a^2 - 4ab + b^2$

[5.3] **16.** $3y^3 + 8y^2 + 12y - 5$ [5.6] **17.** $6p^2 + 7p + 1 + \dfrac{3}{p-1}$ [5.7] **18.** 1.4×10^5 seconds

[6.3] **19.** $(4t + 3v)(2t + v)$ **20.** prime [6.4] **21.** $(4x^2 + 1)(2x + 1)(2x - 1)$ [6.5] **22.** $\{-3, 5\}$

23. $\left\{ 5, -\dfrac{1}{2}, \dfrac{2}{3} \right\}$ [6.6] **24.** -2 or -1 **25.** 6 meters **26.** 30 (The maximum percent is 30%.) [7.1] **27.** (a)

28. (d) [7.4] **29.** $\dfrac{4}{q}$ **30.** $\dfrac{3r + 28}{7r}$ **31.** $\dfrac{7}{15(q - 4)}$ **32.** $\dfrac{-k - 5}{k(k + 1)(k - 1)}$ [7.2] **33.** $\dfrac{7(2z + 1)}{24}$

[7.5] **34.** $\dfrac{195}{29}$ [7.6] **35.** 4, 0 **36.** $\left\{ \dfrac{21}{2} \right\}$ **37.** $\{-2, 1\}$ [7.7] **38.** 150 miles **39.** $1\dfrac{1}{5}$ hours **40.** 20 pounds

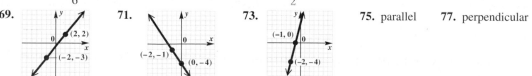

CHAPTER 8 EQUATIONS OF LINES, INEQUALITIES, AND FUNCTIONS

SECTION 8.1 (PAGE 456)

EXERCISES **1.** (a) 1991–1992 and 1993–1994 (b) 1992–1993 (c) 1993 **3.** (a) I (b) III (c) II (d) IV
(e) none **5–14.** **15.** $-3; 3; 2; -1$ **17.** In quadrant III, both coordinates of the

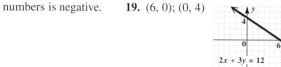

ordered pairs are negative. If $x + y = k$ and k is positive, then either x or y must be positive, because the sum of two negative numbers is negative. **19.** $(6, 0); (0, 4)$ **21.** $(6, 0); (0, -2)$

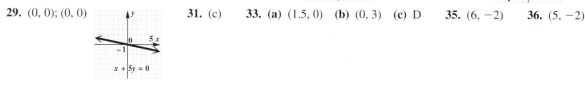

23. $\left(\dfrac{21}{2}, 0 \right); \left(0, -\dfrac{7}{3} \right)$ **25.** none; $(0, 5)$ **27.** $(-4, 0)$; none

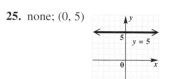
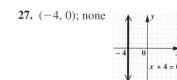

29. $(0, 0); (0, 0)$ **31.** (c) **33.** (a) $(1.5, 0)$ (b) $(0, 3)$ (c) D **35.** $(6, -2)$ **36.** $(5, -2)$

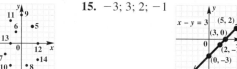

37. $(6, 0)$ **38.** $(5, 0)$ **39.** 5; 0 **40.** The x-coordinate of M is the average of the x-coordinates of P and Q. The y-coordinate of M is the average of the y-coordinates of P and Q. **41.** (a), (b), and (d) **43.** 2 **45.** undefined

47. 8 **49.** $\dfrac{5}{6}$ **51.** 0 **53.** B **55.** A **57.** $-\dfrac{1}{2}$ **59.** 1 **61.** 4 **63.** undefined **65.** $x; y$

69. **71.** **73.** **75.** parallel **77.** perpendicular

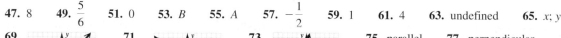

79. neither   **81.** parallel   **83.** The average rate of change is the same, no matter which two ordered pairs are selected to calculate it.   **85. (a)** −.66 million kilowatts   **(b)** −1.1 million kilowatts   **(c)** The graph is not a straight line, so the average rate of change varies for different pairs of years.

SECTION 8.2 (PAGE 471)

CONNECTIONS   PAGE 466:   approximately 12 million; 1992; Since −.02 is close to 0, the number of PCs doubles each year in the indicated years.

EXERCISES   **1.** (a)   **3.** (a)   **5.** $3x + y = 10$ **7.** A   **9.** C   **11.** H   **13.** B   **15.** $y = -\frac{3}{4}x + \frac{5}{2}$

17. $y = -2x + 18$ **19.** $y = \frac{1}{2}x + \frac{13}{2}$ **21.** $y = 4x - 12$ **25.** $y = 5$ **27.** $x = 9$ **29.** $x = .5$ **31.** $y = 8$

33. $y = 2x - 2$ **35.** $y = -\frac{1}{2}x + 4$ **37.** $y = \frac{2}{13}x + \frac{6}{13}$ **39.** $y = 5$ **41.** $x = 7$ **43.** $y = -3$

45. $y = 5x + 15$ **47.** $y = -\frac{2}{3}x + \frac{4}{5}$ **49.** $y = \frac{2}{5}x + 5$ **51.** $y = \frac{2}{3}x + 1$ **53. (a)** $y = -x + 12$ **(b)** -1

(c) $(0, 12)$ **55. (a)** $y = -\frac{5}{2}x + 10$ **(b)** $-\frac{5}{2}$ **(c)** $(0, 10)$ **57. (a)** $y = \frac{2}{3}x - \frac{10}{3}$ **(b)** $\frac{2}{3}$ **(c)** $\left(0, -\frac{10}{3}\right)$

59. $y = 3x - 19$ **61.** $y = \frac{1}{2}x - 1$ **63.** $y = -\frac{1}{2}x + 9$ **65.** $y = 7$ **67.** $y = 3x + 15$; $(1, 18), (5, 30), (10, 45)$

69. $y = 1.86x + 78.24$ **71. (a)** $y = 838.5x + 19,180.5$ **(b)** \$23,373; It is close to the actual value.

73. (a) $y = -3x + 9$ **(b)** 3   **(c)** $\{3\}$ **75. (a)** $y = 4x + 2$ **(b)** $-.5$ **(c)** $\{-.5\}$ **77.** (d)   **79.** 32; 212

80. $(0, 32)$ and $(100, 212)$ **81.** $\frac{9}{5}$ **82.** $F = \frac{9}{5}C + 32$ **83.** $C = \frac{5}{9}(F - 32)$ **84.** When the Celsius temperature

is 50°, the Fahrenheit temperature is 122°.

SECTION 8.3 (PAGE 481)

CONNECTIONS   PAGE 480:   **1.** $x \leq 200, x \geq 100, y \geq 3000$ **2.**   **3.** $C = 50x + 100y$

4. Some examples are (100, 5000), (150, 3000), and (150, 5000). The corner points are (100, 3000) and (200, 3000).
5. The least cost occurs when $x = 100$ and $y = 3000$. The company should use 100 workers and manufacture 3000 units to achieve the lowest possible cost.

EXERCISES   **1.** solid; below   **3.** dashed; above   **7.**   **9.**   **11.**

13.   **15.**   **17.**   **19.**   **21.**

23.   **25.** $-3 < x < 3$ **27.** $-2 < x + 1 < 2$

29.   **31.**   **33.**   **35. (a)** $\{-4\}$ **(b)** $(-\infty, -4)$ **(c)** $(-4, \infty)$

37. (a) $\{3.5\}$ **(b)** $(3.5, \infty)$ **(c)** $(-\infty, 3.5)$ **39.** C   **41.** A   **43. (a)** $\{-.6\}$ **(b)** $(-.6, \infty)$ **(c)** $(-\infty, -.6)$
The graph of $y_1 = 5x + 3$ has x-intercept $(-.6, 0)$, supporting the result of part (a). The graph of y_1 lies *above* the x-axis for values of x *greater than* $-.6$, supporting the result of part (b). The graph of y_1 lies *below* the x-axis for values of x *less than* $-.6$, supporting the result of part (c).

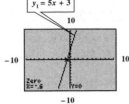

45. (a) $\{-1.2\}$ **(b)** $(-\infty, -1.2]$ **(c)** $[-1.2, \infty)$ The graph of $y_1 = -8x - (2x + 12)$ has x-intercept $(-1.2, 0)$, supporting the result of part (a). The graph of y_1 lies *above or on* the x-axis for values of x *less than or equal to* -1.2, supporting the result of part (b). The graph of y_1 lies *below or on* the x-axis for values of x *greater than or equal to* -1.2, supporting the result of part (c).   **47.** 1994–1996

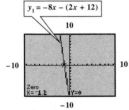

SECTION 8.4 (PAGE 493)

EXERCISES   **3.** the independent variable   **5.** function; domain: $\{2, 3, 4, 5\}$; range: $\{5, 7, 9, 11\}$ **7.** not a function; domain: $(0, \infty)$; range: $(-\infty, 0) \cup (0, \infty)$ **9.** function; domain: {unleaded regular, unleaded premium, crude oil}; range: $\{1.22, 1.44, .21\}$ **11.** function; domain: $(-\infty, \infty)$; range: $(-\infty, 4]$ **13.** not a function; domain: $[-4, 4]$; range: $[-3, 3]$ **15.** function; domain: $(-\infty, \infty)$ **17.** not a function; domain: $[0, \infty)$ **19.** not a function; domain: $(-\infty, \infty)$
21. function; domain: $[0, \infty)$ **23.** function; domain: $(-\infty, 0) \cup (0, \infty)$ **25.** function; domain: $\left[-\dfrac{1}{2}, \infty\right)$
27. function; domain: $(-\infty, 9) \cup (9, \infty)$ **29. (a)** $[0, 3000]$ **(b)** 25 hours; 25 hours **(c)** 2000 gallons **(d)** $f(0) = 0$; The pool is empty at time zero.   **31.** Here is one example. The cost of gasoline; number of gallons used; cost; number of gallons
33. 4   **35.** $3x + 4$ **37.** -59 **39.** $\dfrac{11}{4}$ **41.** $-4p^2 + 8p + 1$ **43.** 4   **45.** No—in general, $f[g(x)] \neq g[f(x)]$.
47. (a) $f(x) = \dfrac{12 - x}{3}$ **(b)** 3   **49. (a)** $f(x) = 3 - 2x^2$ **(b)** -15 **51. (a)** $f(x) = \dfrac{8 - 4x}{-3}$ **(b)** $\dfrac{4}{3}$
53. line; -2; $-2x + 4$; -2; 3; -2 **55.** domain: $(-\infty, \infty)$ **57.** domain: $(-\infty, \infty)$

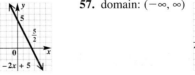

59. domain: $(-\infty, \infty)$ **61.** domain: $(-\infty, \infty)$

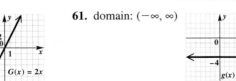

63. (a) $0; $1.50; $3.00; $4.50   **(b)** $1.50x$ **(c)**   **65.** 194.53 centimeters   **67.** 177.41 centimeters

69. 1.83 cubic meters   **71.** 4.11 cubic meters   **73. (a)** 39,851   **(b)** 39,485   **(c)** 39,119   **(d)** In 1992, there were 39,668 post offices in the U.S.   **75.** $f(3) = 7$ **77.** $f(x) = -3x + 5$

CHAPTER 8 REVIEW EXERCISES   (PAGE 501)

1.

x	y
0	5
$\dfrac{10}{3}$	0
2	2
$\dfrac{14}{3}$	-2

3. $(3, 0); (0, -4)$

5. $(10, 0); (0, 4)$

7. If both coordinates are positive, the point lies in quadrant I. If the first coordinate is negative and the second is positive, the point lies in quadrant II. To lie in quadrant III, the point must have both coordinates negative. To lie in quadrant IV, the first coordinate must be positive and the second must be negative.   **9.** $-\dfrac{1}{2}$ **11.** $\dfrac{3}{4}$ **13.** $\dfrac{2}{3}$ **15.** undefined

17. $\dfrac{5}{2}$ **19.** negative   **21.** undefined   **23.** 12 feet   **25.** $y = -\dfrac{1}{3}x - 1$ **27.** $y = -\dfrac{4}{3}x + \dfrac{29}{3}$

29. $x = 2$ (Slope-intercept form is not possible.)   **31.** $y = \dfrac{7}{5}x + \dfrac{16}{5}$ **33.** $y = -\dfrac{5}{2}x + 13$ **35.** $y = \dfrac{5}{2}x - 1$

37. 1988   **39.**   **41.**   **43.** domain: $[-4, 4]$; range: $[0, 2]$; function

45. (a) The independent variable must be the country, because, for each country, there is exactly one amount of power. On the other hand, for a specific amount of power, there would be more than one country generating that amount.   **(b)** domain: {United States, France, Japan, Germany, Canada, Russia}; range: {101, 154, 286, 377, 706}

47. function; domain: $(-\infty, \infty)$; linear function   **49.** function; domain: $(-\infty, \infty)$ **51.** not a function; domain: $[0, \infty)$

53. If no vertical line intersects the graph in more than one point, then it is the graph of a function.   **55.** -8.52

57. $-2k^2 + 3k - 6$ **59.** $-8p^2 + 6p - 6$ **61.** (c)   **63.** Because it falls from left to right, the slope is negative.

64. $-\dfrac{3}{2}$ **65.** $\left(\dfrac{7}{3}, 0\right)$ **66.** $\left(0, \dfrac{7}{2}\right)$ **67.** $f(x) = -\dfrac{3}{2}x + \dfrac{7}{2}$ **68.** $f(8) = -\dfrac{17}{2}$ **69.** $x - \dfrac{23}{3}$

70.   **71.** $\left\{\dfrac{7}{3}\right\}$ **72.** $\left(\dfrac{7}{3}, \infty\right)$ **73.** $\left(-\infty, \dfrac{7}{3}\right)$ **74.** $\dfrac{2}{3}$

CHAPTER 8 TEST   (PAGE 505)

[8.1] 1.

x	y
1	$-\dfrac{10}{3}$
3	-2
0	-4

2. $\dfrac{1}{2}$ **3.** $\left(\dfrac{20}{3}, 0\right); (0, -10)$

4. none; $(0, 5)$

5. (2, 0); none   **6.** It is a vertical line.   **7.** perpendicular   **8.** neither   [8.2] **9.** $y = -5x + 19$

10. $y = 14$ **11.** $y = -\dfrac{3}{5}x - \dfrac{11}{5}$ **12.** $y = -\dfrac{1}{2}x - \dfrac{3}{2}$ **13.** $y = -\dfrac{1}{2}x + 2$ **14.** (b)   **15.** 18,160

16. (a) It is the slope of the line.   **(b)** It is the annual increase in the number of cases served.   [8.3] **17.**

18.   [8.4] **19.** (d)   **20.** (d)   **21.** 0   **22.** domain: $(-\infty, \infty)$; range: $(-\infty, \infty)$

23. (b); The set in (a) includes, for example, the two ordered pairs (8.6, 1990) and (8.6, 1991). In a function, the independent variable (with value 8.6 here) cannot correspond to more than one dependent variable.

CUMULATIVE REVIEW EXERCISES CHAPTERS 1–8 (PAGE 507)

[2.2] **1.** $\{-65\}$ [2.4] **2.** $t = \dfrac{A - p}{pr}$ [6.5] **3.** $\left\{-1, -\dfrac{1}{7}\right\}$ [3.3] **4.** $\{0, 7\}$ [7.6] **5.** $\{5\}$

[3.1] **6.** $\left(-3, \dfrac{7}{2}\right)$ **7.** $(-\infty, 1]$

[3.2] **8.** $(-\infty, -2] \cup (7, \infty)$ [5.1–5.5] **9.** $\dfrac{1}{x^2 y}$ **10.** $\dfrac{y^7}{x^{13}z^2}$ **11.** $\dfrac{m^6}{8n^9}$ **12.** $2x^2 - 4x + 38$

13. $15x^2 + 7xy - 2y^2$ **14.** $x^3 + 8y^3$ [5.6] **15.** $m^2 - 2m + 3$ [6.1–6.4] **16.** $(y + 6k)(y - 2k)$

17. $(3x^2 - 5y)(3x^2 + 5y)$ **18.** $5x^2(5x - 13y)(5x - 3y)$ **19.** $(f + 10)^2$ **20.** prime   [7.4] **21.** 1

22. $\dfrac{6x + 22}{(x + 1)(x + 3)}$ or $\dfrac{2(3x + 11)}{(x + 1)(x + 3)}$ [7.2] **23.** $\dfrac{4(x - 5)}{3(x + 5)}$ **24.** $\dfrac{x + 1}{x}$ **25.** $\dfrac{(x + 3)^2}{3x}$ **26.** $\dfrac{4xy^4}{z^2}$ [7.5] **27.** 6

[4.2, 8.1] **28.** x-intercept: (4, 0); y-intercept: $\left(0, \dfrac{12}{5}\right)$ 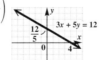   [4.3, 8.1] **29. (a)** $-\dfrac{6}{5}$ **(b)** $\dfrac{5}{6}$

[8.2] **30.** $y = -4x + 15$ **31.** $y = 4x$ [8.3] **32.**   [8.4] **33. (a)** $(-\infty, \infty)$ **(b)** 24

34. {(City, Percent)}; The elements of the domain are the names of the cities.   **35.** yes; $f(x) = \dfrac{2}{7}x - 2$

[2.3] **36.** corporate income taxes: $171.8 billion; individual income taxes: $656.4 billion   [2.6] **37.** 15°, 35°, 130°

[6.6] **38.** 7 inches   [2.5] **39.** $9.92   [7.7] **40.** 1 hour

CHAPTER 9 LINEAR SYSTEMS

SECTION 9.1 (PAGE 515)

EXERCISES   **1. (a)** B **(b)** C **(c)** D **(d)** A   **3.** yes   **5.** no   **7.** yes   **9.** yes   **11.** no   **13. (a)**; The ordered pair solution must be in quadrant II, and $(-4, -4)$ is in quadrant III.   **15.** $\{(4, 2)\}$

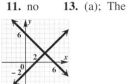

17. $\{(0, 4)\}$ **19.** $\{(4, -1)\}$

In Exercises 21–29, we do not show the graphs.
21. $\{(1, 3)\}$ **23.** $\{(0, 2)\}$ **25.** $\emptyset$ (inconsistent system)   **27.** $\{(x, y) \mid 3x = 5 - y\}$ (dependent equations)
29. $\{(4, -3)\}$ **33.** Yes, it is possible. For example, the system $x + y = 5$ has the single solution $(2, 3)$.
$$x - y = -1$$
$$2x - y = 1$$

35. about 350 million for each format   **37.** 200 million   **39.** between 1984 and 1986, and between 1988 and 1990
41. (a) neither **(b)** intersecting lines **(c)** one solution   **43. (a)** dependent **(b)** one line **(c)** infinite number of solutions   **45. (a)** inconsistent **(b)** parallel lines **(c)** no solution   **47.** 40   **49.** (40, 30)   **51.** B   **53.** A
55. $\{(-1, 5)\}$ **57.** $\{2\}$ **58.** 5   **59.** $\{(2, 5)\}$ **60.** The x-coordinate, 2, is equal to the solution
of the equation.   **61.** The y-coordinate, 5, is equal to the value we obtained on both sides when checking.   **62.** 5; 3; 5

SECTION 9.2 (PAGE 523)

EXERCISES   **1.** No, it is not correct, because the solution set is $\{(3, 0)\}$. The y-value in the ordered pair must also be determined.
3. $\{(3, 9)\}$ **5.** $\{(7, 3)\}$ **7.** $\{(0, 5)\}$ **9.** $\{(-4, 8)\}$ **11.** $\{(3, -2)\}$ **13.** $\{(x, y) \mid 3x - y = 5\}$
15. $\left\{\left(\frac{1}{3}, -\frac{1}{2}\right)\right\}$ **17.** $\emptyset$ **19.** $\{(x, y) \mid 3x = 4y + 2\}$ **21.** The first student had less work to do, because the coefficient of y in the first equation is -1. The second student had to divide by 2, introducing fractions into the expression for x.
23. $\{(2, -3)\}$ **25.** $\{(3, 2)\}$ **27.** $\{(-2, 1)\}$ **29.** 1993   **31.** To find the total cost, multiply the number of bicycles (x) by the cost per bicycle (\$400), and add the fixed cost (\$5000). Thus, $y_1 = 400x + 5000$ gives this total cost (in dollars).

32. $y_2 = 600x$ **33.** $y_1 = 400x + 5000$
$ y_2 = 600x;$ solution set: $\{(25, 15{,}000)\}$ **34.** 25; 15,000; 15,000
35. $\{(2, 4)\}$ **37.** $\{(1, 5)\}$

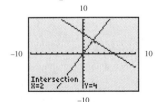

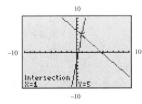

39. $\{(5, -3)\}$; The equations to input are $y_1 = \dfrac{5 - 4x}{5}$ and $y_2 = \dfrac{1 - 2x}{3}$.

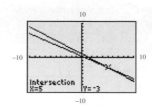

41. Adjust the viewing window so that it does appear.

SECTION 9.3 (PAGE 531)

EXERCISES **1.** true   **3.** true   **5.** $\{(4, 6)\}$ **7.** $\{(-1, -3)\}$ **9.** $\{(-2, 3)\}$ **11.** $\left\{\left(-\dfrac{2}{3}, \dfrac{17}{2}\right)\right\}$ **13.** $\{(3, -6)\}$

15. $\{(7, 4)\}$ **17.** $\{(0, 3)\}$ **19.** $\{(3, 0)\}$ **21.** $\left\{\left(-\dfrac{32}{23}, -\dfrac{17}{23}\right)\right\}$ **25.** $\{(-3, 4)\}$ **27.** $\{(x, y) \mid 2x - y = -12\}$
29. $\{(0, 6)\}$ **31.** $\emptyset$ **33. (a)** $\{(1, 4)\}$ **(b)** $\{(1, 4)\}$ **(c)** Answers will vary.   **35.** Yes, they should both get the same
answer, since both procedures are mathematically valid.   **37.** $\{(0, 3)\}$ **39.** $\{(24, -12)\}$ **41.** $\{(3, 2)\}$
43. $1141 = 1991a + b$ **44.** $1339 = 1996a + b$ **45.** $1991a + b = 1141$
$\qquad\qquad\qquad\qquad\qquad\qquad\qquad\qquad\qquad\qquad\qquad 1996a + b = 1339;$ solution set: $\{(39.6, -77{,}702.6)\}$
46. $y = 39.6x - 77{,}702.6$ **47.** 1220.2 (million); This is slightly less than the actual figure.
48. It is not realistic to expect the data to lie in a perfectly straight line; as a result, the quantity obtained from an equation
determined in this way will probably be "off" a bit. One cannot put too much faith in models such as this one, because not all
data points are linear in nature.

SECTION 9.4 (PAGE 538)

EXERCISES **1. (b)**   **3.** $\{(1, 4, -3)\}$ **5.** $\{(0, 2, -5)\}$ **7.** $\left\{\left(-\dfrac{7}{3}, \dfrac{22}{3}, 7\right)\right\}$ **9.** $\{(4, 5, 3)\}$

11. $\{(2, 2, 2)\}$ **13.** $\left\{\left(\dfrac{8}{3}, \dfrac{2}{3}, 3\right)\right\}$ **17.** $\emptyset$

The solution sets in Exercises 19–21 may be given in other equivalent forms.
19. $\{(x, y, z) \mid x - y + 4z = 8\}$ **21.** $\{(x, y, z) \mid 2x + y - z = 6\}$ **23.** $\{(0, 0, 0)\}$ **25.** $\{(2, 1, 5, 3)\}$
27. $128 = a + b + c$ **28.** $140 = 2.25a + 1.5b + c$ **29.** $80 = 9a + 3b + c$ **30.** $a + \ \ b + c = 128$
$\qquad\qquad\qquad\qquad\qquad\qquad\qquad\qquad\qquad\qquad\qquad\qquad\qquad\qquad\qquad\qquad\qquad\qquad 2.25a + 1.5b + c = 140$
$\qquad\qquad\qquad\qquad\qquad\qquad\qquad\qquad\qquad\qquad\qquad\qquad\qquad\qquad\qquad\qquad\qquad\qquad\quad 9a + \ \ 3b + c = 80$
$\qquad\qquad\qquad\qquad\qquad\qquad\qquad\qquad\qquad\qquad\qquad\qquad\qquad\qquad\qquad\qquad\qquad\qquad\quad \{(-32, 104, 56)\}$
31. $f(x) = -32x^2 + 104x + 56$ **32.** 56 feet   **33.** 140.5 feet   **34.** It tells us the projectile hits the ground 3.25 seconds after
it is projected.   **35.** $a = 3, b = 1, c = -2; f(x) = 3x^2 + x - 2$ **36.** $a = 1, b = 4, c = 3; Y_1 = x^2 + 4x + 3$ **37.** If one
were to eliminate *different* variables in the first two steps, the result would be two equations in three variables, and it would not be
possible to solve for a single variable in the next step.

SECTION 9.5 (PAGE 547)

CONNECTIONS **PAGE 543:** "Mixed price" refers to the price of a mixture of the two products. The system is $9x + 7y = 107$,
$7x + 9y = 101$, where x represents the price of a citron and y represents the price of a wood apple.

EXERCISES **1.** wins: 69; losses: 13   **3.** wins: 48; losses: 26; ties: 8   **5.** length: 78 feet; width: 36 feet
7. length: 12 feet; width: 5 feet   **9.** $x = 40, y = 50$, so the angles measure 40° and 50°.   **11.** Boyz II Men: 134; Bruce
Springsteen & the E St. Band: 40   **13.** National Hockey League: $219.74; National Basketball Association: $203.38
15. CGA monitor: $400; VGA monitor: $500   **17.** 6 units of yarn; 2 units of thread   **19. (a)** 6 ounces **(b)** 15 ounces
(c) 24 ounces **(d)** 30 ounces   **21.** $.58x$ **23.** 6 gallons of 25%; 14 gallons of 35%   **25.** 3 liters of pure acid;
24 liters of 10% acid   **27.** 50 pounds of $3.60 clusters; 30 pounds of $7.20 truffles   **29.** 76 general admission; 108 with
student identification   **31.** 28 dimes; 66 quarters   **33.** $1000 at 2%; $2000 at 4%   **35.** $25y$ miles   **37.** freight train:
50 kilometers per hour; express train: 80 kilometers per hour   **39.** top speed: 2100 miles per hour; wind speed: 300 miles per
hour   **41.** 8 fish at $20; 15 fish at $40; 6 fish at $65   **43.** $x + y + z = 180$; angle measures: 70°, 30°, 80°

45. first: 20°; second: 70°; third: 90°   **47.** shortest: 12 centimeters; middle: 25 centimeters; longest: 33 centimeters
49. A: 180 cases; B: 60 cases; C: 80 cases   **51.** 10 pounds of jelly beans; 2 pounds of chocolate eggs; 3 pounds of
marshmallow chicks   **53.** $2a + b + c = -5$ **54.** $-a + c = -1$ **55.** $3a + 3b + c = -18$ **56.** $a = 1, b = -7,$
$c = 0; x^2 + y^2 + x - 7y = 0$ **57.** It fails the vertical line test.

SECTION 9.6 (PAGE 562)

CONNECTIONS   **PAGE 558:**   **1.** The solution set is also $\{(-2, -1)\}$.   **2.** Answers will vary. One example is $3x + y = -7$
$y = -1.$

PAGE 562:   **1.** $\begin{bmatrix} 1 & 0 & | & -2 \\ 0 & 1 & | & -1 \end{bmatrix}$ **2.** a system of dependent equations

EXERCISES   **1.** (a) $0, 5, -3$ (b) $1, -3, 8$ (c) yes; The number of rows is the same as the number of columns (three).

(d) $\begin{bmatrix} 1 & 4 & 8 \\ 0 & 5 & -3 \\ -2 & 3 & 1 \end{bmatrix}$ (e) $\begin{bmatrix} 1 & -\frac{3}{2} & -\frac{1}{2} \\ 0 & 5 & -3 \\ 1 & 4 & 8 \end{bmatrix}$ (f) $\begin{bmatrix} 1 & 15 & 25 \\ 0 & 5 & -3 \\ 1 & 4 & 8 \end{bmatrix}$ **3.** $\begin{bmatrix} 1 & 2 & | & 11 \\ 2 & -1 & | & -3 \end{bmatrix}; \begin{bmatrix} 1 & 2 & | & 11 \\ 0 & -5 & | & -25 \end{bmatrix}; \begin{bmatrix} 1 & 2 & | & 11 \\ 0 & 1 & | & 5 \end{bmatrix};$

$x + 2y = 11, y = 5; \{(1, 5)\}$ **5.** $\{(4, 1)\}$ **7.** $\{(1, 1)\}$ **9.** $\{(-1, 4)\}$ **11.** $\emptyset$

15. $\begin{bmatrix} 1 & 1 & -1 & | & -3 \\ 0 & -1 & 3 & | & 10 \\ 0 & -6 & 7 & | & 38 \end{bmatrix}; \begin{bmatrix} 1 & 1 & -1 & | & -3 \\ 0 & 1 & -3 & | & -10 \\ 0 & -6 & 7 & | & 38 \end{bmatrix}; \begin{bmatrix} 1 & 1 & -1 & | & -3 \\ 0 & 1 & -3 & | & -10 \\ 0 & 0 & -11 & | & -22 \end{bmatrix}; \begin{bmatrix} 1 & 1 & -1 & | & -3 \\ 0 & 1 & -3 & | & -10 \\ 0 & 0 & 1 & | & 2 \end{bmatrix}; x + y - z = -3,$

$y - 3z = -10, z = 2; \{(3, -4, 2)\}$ **17.** $\{(4, 0, 1)\}$ **19.** $\{(-1, 23, 16)\}$ **21.** $\{(3, 2, -4)\}$
23. $\{(x, y) \mid x - 2y + z = 4\}$ **25.** $\{(1, 1)\}$ **27.** $\{(-1, 2, 1)\}$ **29.** $\{(1, 7, -4)\}$

SECTION 9.7 (PAGE 573)

CONNECTIONS   **PAGE 568:** -185

EXERCISES   **1.** (d)   **3.** -3 **5.** 14   **7.** 0   **9.** 59   **11.** 14   **15.** -22 **17.** 20   **19.** -5 **21.** By
choosing that row or column to expand about, all terms will have a factor of 0, and so the sum of all these terms will be 0.
22. $\frac{y_2 - y_1}{x_2 - x_1}$ **23.** $y - y_1 = \frac{y_2 - y_1}{x_2 - x_1}(x - x_1)$ **24.** $x_2 y - x_1 y - x_2 y_1 - xy_2 + x_1 y_2 + xy_1 = 0$ **25.** The result is the
same as in Exercise 24.   **27.** -18 **29.** 0   **31.** -22.04285452 **33.** 16   **35.** -12 **37.** 0
39. $\{10\}$ **41.** (a) IV   (b) I   (c) III   (d) II   **43.** $\{(-5, 2)\}$ **45.** $\left\{\left(\frac{11}{58}, -\frac{5}{29}\right)\right\}$ **47.** $\{(-1, 2)\}$

49. Answers will vary. One example is $\begin{array}{l} 6x + 7y = 8 \\ 9x + 10y = 11. \end{array}$ Each system has the solution set $\{(-1, 2)\}$.   **51.** $\{(-2, 3, 5)\}$

53. Cramer's rule does not apply.   **55.** $\{(20, -13, -12)\}$ **57.** $\left\{\left(\frac{62}{5}, -\frac{1}{5}, \frac{27}{5}\right)\right\}$

59.   **60.** $\frac{1}{2} \begin{vmatrix} 0 & 0 & 1 \\ -3 & -4 & 1 \\ 2 & -2 & 1 \end{vmatrix}$ **61.** 7   **62.** 8   **63.** $\{(-1, 3, 5)\}$ **65.** $\{(1, -3, 2, -4)\}$

CHAPTER 9 REVIEW EXERCISES (PAGE 584)

1. In 1995, they both reached a rate of about 72%.   **3.** (a) $4   (b) 300 half-gallons   (c) supply: 200 half-gallons;
demand: 400 half-gallons   **5.** yes   **7.** $\{(3, 1)\}$ **9.** No, this is not correct. A false statement indicates that the solution
set is $\emptyset$.   **11.** $\{(2, 1)\}$ **13.** $\{(x, y) \mid x + 3y = 6\}$ **15.** His answer was incorrect since the system has infinitely many
solutions (as indicated by the true statement $0 = 0$).   **17.** (c)   **19.** $\{(7, 1)\}$ **21.** $\{(x, y) \mid 3x - 4y = 9\}$
23. $\{(-4, 1)\}$ **25.** $\{(9, 2)\}$ **29.** $\{(1, -5, 3)\}$ **31.** $\emptyset$ **33.** touchdown passes: 26; interceptions: 14
35. 30 pounds of $2-a-pound nuts; 70 pounds of $1-a-pound candy   **37.** $40,000 at 10%; $100,000 at 6%; $140,000 at 5%
39. Mantle: 54; Maris: 61; Blanchard: 21   **41.** $\{(-1, 5)\}$ **43.** $\{(1, 2, -1)\}$ **45.** 80   **47.** -38

49. Cramer's rule does not apply if $D = 0$.   **51.** $\{(-4, 5)\}$ **53.** $\{(3, -2, 1)\}$ **55.** $\begin{aligned}5y + z &= 7 \\ -5y + z &= -13\end{aligned}$

56. $\{(2, -3)\}$ **57.** $\{(4, 2, -3)\}$ **58.** $\{(4, 2, -3)\}$ **59. (a)** -10 **(b)** -40 **(c)** -20 **(d)** 30
60. $\{(4, 2, -3)\}$ **61.** $\{(12, 9)\}$ **63.** $\{(3, -1)\}$ **65.** $\{(0, 4)\}$ **67.** 2 copies of *Godzilla*; 5 Aerosmith compact discs

CHAPTER 9 TEST   (PAGE 589)

[9.1] **1.** No; The graph for Babe Ruth lies completely below the graph for Aaron, indicating that Ruth's total was always lower than Aaron's.   **2.** Aaron had the most and Ruth had the fewest.   **3.** $\{(6, 1)\}$ [9.2] **4.** $\{(6, -4)\}$

5. $\{(x, y) \mid 12x - 5y = 8\}$ [9.3] **6.** $\{(3, 3)\}$ **7.** $\{(0, -2)\}$ **8.** $\emptyset$ [9.4] **9.** $\left\{\left(-\dfrac{2}{3}, \dfrac{4}{5}, 0\right)\right\}$ [9.5] **10.** slower car: 45 miles per hour; faster car: 75 miles per hour   **11.** 4 liters of 20%; 8 liters of 50%   **12.** Disneyland: 15.0 million; Magic Kingdom: 13.8 million   **13.** marker pen: \$1.10; colored paper sheet: \$.40   [9.6] **14.** $\left\{\left(\dfrac{2}{5}, \dfrac{7}{5}\right)\right\}$

15. $\{(-1, 2, 3)\}$ [9.1–9.4, 9.6, 9.7] **16.** $\{(-3, -2, -4)\}$ [9.7] **17.** 3   **18.** 0   **19.** $\left\{\left(-\dfrac{9}{4}, \dfrac{5}{4}\right)\right\}$ **20.** $\{(5, 2, -1)\}$

CUMULATIVE REVIEW EXERCISES CHAPTERS 1-9   (PAGE 590)

[1.6] **1.** $-1, 1, -2, 2, -4, 4, -5, 5, -8, 8, -10, 10, -20, 20, -40, 40$ [1.3] **2.** 1   [1.7] **3.** commutative property
[2.2] **4.** $\left\{-\dfrac{15}{4}\right\}$ [3.3] **5.** $\left\{\dfrac{2}{3}, 2\right\}$ [2.4] **6.** $x = \dfrac{d - by}{a - c}$ or $x = \dfrac{by - d}{c - a}$ [2.2] **7.** $\{11\}$ [3.1] **8.** $\left(-\infty, \dfrac{240}{13}\right]$
[3.3] **9.** $\left[-2, \dfrac{2}{3}\right]$ **10.** $(-\infty, \infty)$ [2.3, 2.4] **11.** not guilty: 105; guilty: 95   **12.** width: $8\dfrac{1}{4}$ inches; length: $10\dfrac{3}{4}$ inches
13. $46°, 46°, 88°$ [5.1] **14.** $14x^2 - 5x + 23$ [5.3] **15.** $6xy + 12x - 14y - 28$ [5.6] **16.** $3k^2 - 4k + 1$

[5.7] **17.** 3.65×10^{10} [5.5] **18.** $x^6 y$ [6.3] **19.** $(5m - 4p)(2m + 3p)$ [6.4] **20.** $(8t - 3)^2$ [6.5] **21.** $\left\{-\dfrac{1}{3}, \dfrac{3}{2}\right\}$

22. $\{-11, 11\}$ [7.4] **23.** $\dfrac{7}{x + 2}$ [7.2] **24.** $\dfrac{3}{4k - 3}$ [7.6] **25.** $\left\{-\dfrac{1}{4}, 3\right\}$ [8.1] **26.** $y = 6$ **27.** $x = 4$

[4.3, 8.1] **28.** $-\dfrac{4}{3}$ **29.** $\dfrac{3}{4}$ [8.2] **30.** $4x + 3y = 10$ [8.1] **31.**   [8.2] **32.** $y = 103.25x + 3502;$

The slope represents the average yearly increase in health benefit cost during the period.   [8.3] **33.**

[8.4] **34. (a)** -6 **(b)** $a^2 + 3a - 6$ [9.1–9.3, 9.6, 9.7] **35.** $\{(3, -3)\}$ [9.4, 9.6, 9.7] **36.** $\{(5, 3, 2)\}$
[9.5] **37.** 405 adults and 49 children   **38.** Tickle Me Elmo: \$27.63; Snacktime Kid: \$36.26   **39.** peanuts: \$2 per pound; cashews: \$4 per pound   [9.1] **40. (a)** $x = 8$ or 800 items; \$3000   **(b)** about \$500

CHAPTER 10 ROOTS AND RADICALS

SECTION 10.1 (PAGE 602)

CONNECTIONS **Page 599:** The area of the large square is $(a + b)^2$ or $a^2 + 2ab + b^2$. The sum of the areas of the smaller square and the four right triangles is $c^2 + 2ab$. Set these equal to each other and subtract $2ab$ from both sides to get $a^2 + b^2 = c^2$.

EXERCISES **1.** false; Zero has only one square root. **3.** true **5.** true **7.** $-4, 4$ **9.** $-12, 12$ **11.** $-\dfrac{5}{14}, \dfrac{5}{14}$

13. $-30, 30$ **15.** 7 **17.** -11 **19.** $-\dfrac{12}{11}$ **21.** not a real number **23.** 100 **25.** 19 **27.** $3x^2 + 4$

29. a must be positive. **31.** a must be negative. **33.** rational; 5 **35.** irrational; 5.385 **37.** rational; -8
39. not a real number **41.** The answer to Exercise 17 is the negative square root of a positive number. However, in Exercise 21, the square root of a negative number is not a real number. **43.** 23.896 **45.** 28.249 **47.** 1.985 **49.** 5
51. 17 **53.** 3.606 **55.** 2.289 **57.** 5.074 **59.** -4.431 **61.** $c = 17$ **63.** $b = 8$ **65.** $c = 11.705$
67. 24 centimeters **69.** 80 feet **71.** 195 miles **73.** 9.434 **75.** Answers will vary. For example, if $a = 2$ and $b = 7$, $\sqrt{a^2 + b^2} = \sqrt{53}$, while $a + b = 9$. $\sqrt{53} \neq 9$. If $a = 0$ and $b = 1$, then $\sqrt{a^2 + b^2}$ is equal to $a + b$ (which is not true in general). **77.** $\sqrt{29}$ **79.** $\sqrt{2}$ **81.** 10 **83.** -3 **85.** 5 **87.** not a real number **89.** -3 **91.** 3
93. c^2 **94.** $(b - a)^2$ **95.** $2ab$ **96.** $(b - a)^2 = a^2 - 2ab + b^2$ **97.** $c^2 = 2ab + (a^2 - 2ab + b^2)$
98. $c^2 = a^2 + b^2$

SECTION 10.2 (PAGE 611)

CONNECTIONS **Page 607: 1.** x; $\sqrt{x}$ **2.** The last part of it ($\sqrt{}$) is used as part of the radical symbol $\sqrt{}$.

EXERCISES **1.** false; $\sqrt{4}$ represents only the principal (positive) square root. **3.** false; $\sqrt{-6}$ is not a real number.
5. true **7.** true **9.** 9 **11.** $3\sqrt{10}$ **13.** 13 **15.** $\sqrt{13r}$ **17.** (a) **19.** $3\sqrt{5}$ **21.** $5\sqrt{3}$ **23.** $5\sqrt{5}$

25. $-10\sqrt{7}$ **27.** $9\sqrt{3}$ **29.** $3\sqrt{6}$ **31.** 24 **33.** $6\sqrt{10}$ **37.** $\dfrac{4}{15}$ **39.** $\dfrac{\sqrt{7}}{4}$ **41.** 5 **43.** $\dfrac{25}{4}$ **45.** $6\sqrt{5}$

47. m **49.** y^2 **51.** $6z$ **53.** $20x^3$ **55.** $z^2\sqrt{z}$ **57.** x^3y^6 **59.** $2\sqrt[5]{5}$ **61.** $3\sqrt[3]{2}$ **63.** $2\sqrt[4]{5}$ **65.** $\dfrac{2}{3}$

67. $-\dfrac{6}{5}$ **69.** 5 **71.** $\sqrt[4]{12}$ **73.** $2x\sqrt[3]{4}$

In Exercise 75, the number of displayed digits will vary among calculator models. Also, less sophisticated models may exhibit round-off error in the final decimal place.
75. (a) 4.472135955 **(b)** 4.472135955 **(c)** The numerical results are not a proof because both answers are approximations and they might differ if calculated to more decimal places.
77. 6 centimeters **79.** 6 inches **81.** The product rule for radicals requires that both a and b must be nonnegative. Otherwise $\sqrt{a}$ and $\sqrt{b}$ would not be real numbers (except when $a = b = 0$). **83.** To verify the first length, we show, in the first triangle, that $1^2 + 1^2 = 2$, so the hypotenuse has length $\sqrt{2}$. Similarly, in the second triangle, $(\sqrt{2})^2 + 1^2 = 3$, so the hypotenuse has length $\sqrt{3}$, and so on. **84.** $\sqrt{4}$, $\sqrt{9}$, and so on; $\sqrt{16} = 4$ and $\sqrt{25} = 5$ **85.** Look at the radicands: $9 - 4 = 5$, $16 - 9 = 7$, $25 - 16 = 9$, and so on. **86.** The differences between consecutive whole number lengths increase by 2 each time, so we can predict that the next one will be $\sqrt{25 + 11} = \sqrt{36}$, and the one after that will be $\sqrt{36 + 13} = \sqrt{49}$.

SECTION 10.3 (PAGE 615)

EXERCISES **1.** distributive **3.** radicands **5.** $-5\sqrt{7}$ **7.** $5\sqrt{17}$ **9.** $5\sqrt{7}$ **11.** $11\sqrt{5}$ **13.** $15\sqrt{2}$
15. $-20\sqrt{2} + 6\sqrt{5}$ **17.** $17\sqrt{7}$ **19.** $-16\sqrt{2} - 8\sqrt{3}$ **21.** $20\sqrt{2} + 6\sqrt{3} - 15\sqrt{5}$ **23.** $4\sqrt{2}$
25. $2\sqrt{3} + 4\sqrt{3} = (2 + 4)\sqrt{3} - 6\sqrt{3}$ **27.** $11\sqrt{3}$ **29.** $5\sqrt{x}$ **31.** $3x\sqrt{6}$ **33.** 0 **35.** $-20\sqrt{2k}$
37. $42x\sqrt{5z}$ **39.** $-\sqrt[3]{2}$ **41.** $6\sqrt[3]{p^2}$ **43.** $21\sqrt[4]{m^3}$ **47.** $-6x^2y$ **48.** $-6(p - 2q)^2(a + b)$ **49.** $-6a^2\sqrt{xy}$
50. The answers are alike because the numerical coefficient of the three answers is the same: -6. Also, the first variable factor is raised to the second power, and the second variable factor is raised to the first power. The answers are different because the

variables are different: x and y, then $p - 2q$ and $a + b$, and then a and $\sqrt{xy}$. **51.** 9.220 **53.** 5 **55.** $22\sqrt{2}$
57. (a) 1991 **(b)** 1997

SECTION 10.4 (PAGE 621)

EXERCISES **1.** radical **3.** fraction **5.** $4\sqrt{2}$ **7.** $\dfrac{-\sqrt{33}}{3}$ **9.** $\dfrac{7\sqrt{15}}{5}$ **11.** $\dfrac{\sqrt{30}}{2}$ **13.** $\dfrac{16\sqrt{3}}{9}$ **15.** $\dfrac{-3\sqrt{2}}{10}$

17. $\dfrac{21\sqrt{5}}{5}$ **19.** $\sqrt{3}$ **21.** $\dfrac{\sqrt{2}}{2}$ **23.** $\dfrac{\sqrt{65}}{5}$ **25.** 1; identity property for multiplication **27.** $\dfrac{\sqrt{21}}{3}$ **29.** $\dfrac{3\sqrt{14}}{4}$

31. $\dfrac{1}{6}$ **33.** 1 **35.** $\dfrac{\sqrt{7x}}{x}$ **37.** $\dfrac{2x\sqrt{xy}}{y}$ **39.** $\dfrac{x\sqrt{3xy}}{y}$ **41.** $\dfrac{3ar^2\sqrt{7rt}}{7t}$ **43. (b)** **45.** $\dfrac{\sqrt[3]{12}}{2}$ **47.** $\dfrac{\sqrt[3]{196}}{7}$

49. $\dfrac{\sqrt[3]{6y}}{2y}$ **51.** $\dfrac{\sqrt[3]{42mn^2}}{6n}$ **53. (a)** $\dfrac{9\sqrt{2}}{4}$ seconds **(b)** 3.182 seconds

SECTION 10.5 (PAGE 626)

EXERCISES **1.** 13 **3.** 4 **5.** 4 **7.** 5 **9.** $9\sqrt{5}$ **11.** $16\sqrt{2}$ **13.** $\sqrt{15} - \sqrt{35}$ **15.** $2\sqrt{10} + 30$
17. $4\sqrt{7}$ **19.** $57 + 23\sqrt{6}$ **21.** $81 + 14\sqrt{21}$ **23.** $37 + 12\sqrt{7}$ **25.** 23 **27.** 1 **29.** $2\sqrt{3} - 2 + 3\sqrt{2} - \sqrt{6}$
31. $15\sqrt{2} - 15$ **33.** $87 + 9\sqrt{21}$ **35.** Because multiplication must be performed before addition, it is incorrect to add
-37 and -2. Since $-2\sqrt{15}$ cannot be simplified, the expression cannot be written in a simpler form, and the final answer is
$-37 - 2\sqrt{15}$. **37.** $\dfrac{3 - \sqrt{2}}{7}$ **39.** $-4 - 2\sqrt{11}$ **41.** $1 + \sqrt{2}$ **43.** $-\sqrt{10} + \sqrt{15}$ **45.** $3 - \sqrt{3}$

47. $2\sqrt{5} + \sqrt{15} + 4 + 2\sqrt{3}$ **49.** $\sqrt{11} - 2$ **51.** $\dfrac{\sqrt{3} + 5}{8}$ **53.** $\dfrac{6 - \sqrt{10}}{2}$ **55.** $x\sqrt{30} + \sqrt{15x} + 6\sqrt{5x} + 3\sqrt{10}$
57. $6t - 3\sqrt{14t} + 2\sqrt{7t} - 7\sqrt{2}$ **59.** $m\sqrt{15} + \sqrt{10mn} - \sqrt{15mn} - n\sqrt{10}$ **61.** $2 - 3\sqrt[3]{4}$ **63.** $12 + 10\sqrt[4]{8}$
65. $-1 + 3\sqrt[3]{2} - \sqrt[3]{4}$ **67.** 1 **69.** $30 + 18x$ **70.** They are not like terms. **71.** $30 + 18\sqrt{5}$ **72.** They are not
like radicals. **73.** Make the first term $30x$, so that $30x + 18x = 48x$; make the first term $30\sqrt{5}$, so that $30\sqrt{5} + 18\sqrt{5} = 48\sqrt{5}$.
74. When combining like terms, we add (or subtract) the coefficients of the common factors of the terms: $2xy + 5xy = 7xy$.
When combining like radicals, we add (or subtract) the coefficients of the common radical: $2\sqrt{ab} + 5\sqrt{ab} = 7\sqrt{ab}$.
75. 4 inches

SECTION 10.6 (PAGE 633)

CONNECTIONS **Page 629:** **1.** $6\sqrt{13} \approx 21.63$ (to the nearest hundredth) **2.** $h = \sqrt{13}$; $6\sqrt{13} \approx 21.63$

EXERCISES **1.** $\{49\}$ **3.** $\{7\}$ **5.** $\{85\}$ **7.** $\{-45\}$ **9.** $\left\{-\dfrac{3}{2}\right\}$ **11.** $\emptyset$ **13.** $\{121\}$ **15.** $\{8\}$ **17.** $\{1\}$
19. $\{6\}$ **21.** $\emptyset$ **23.** $\{5\}$ **25.** Since $\sqrt{x}$ must be greater than or equal to zero for any replacement for x, it cannot
equal -8, a negative number. **29.** $\{12\}$ **31.** $\{5\}$ **33.** $\{0, 3\}$ **35.** $\{-1, 3\}$ **37.** $\{8\}$ **39.** $\{4\}$ **41.** $\{8\}$
43. $\{9\}$ **45.** We cannot square term by term. The left side must be squared as a binomial in the first step. **47.** 158.6 feet
49. (a) 70.5 miles per hour **(b)** 59.8 miles per hour **(c)** 53.9 miles per hour **51. (a)** 1991 **(b)** 1997; Here, $f(3) \approx 97.5$.
Thus, 1997 is the year when about 3 million calls were made. **(c)** about 2.1 million **53.** 4 **55.** -2 **57.** -1

SECTION 10.7 (PAGE 641)

CONNECTIONS **Page 641:** **1.** $(\sqrt{x} - 3)(\sqrt{x} + 1)$ **2.** $(x + \sqrt{10})(x - \sqrt{10})$ **3.** $(\sqrt{x} + 2)^2$
4. $x\sqrt{x} + 3\sqrt{x} + \dfrac{5\sqrt{x}}{x}$

EXERCISES **1. (a)** **3. (c)** **5.** 5 **7.** 4 **9.** 2 **11.** 2 **13.** 8 **15.** 9 **17.** 8 **19.** 4 **21.** -4
23. -4 **25.** $\dfrac{1}{343}$ **27.** $\dfrac{1}{36}$ **29.** $-\dfrac{1}{32}$ **31.** 2^3 **33.** $\dfrac{1}{6^{1/2}}$ **35.** $\dfrac{1}{15^{1/2}}$ **37.** $11^{1/7}$ **39.** 8^3 **41.** $6^{1/2}$

43. $\dfrac{5^3}{2^3}$ **45.** $\dfrac{1}{2^{8/5}}$ **47.** $6^{2/9}$ **49.** z **51.** $m^2 n^{1/6}$ **53.** $\dfrac{a^{2/3}}{b^{4/9}}$ **55.** 2 **57.** 2 **59.** $\sqrt{a}$ **61.** $\sqrt[3]{k^2}$
63. $\sqrt{2} = 2^{1/2}$ and $\sqrt[3]{2} = 2^{1/3}$ **64.** $2^{1/2} \cdot 2^{1/3}$ **65.** 6 **66.** $2^{3/6} \cdot 2^{2/6}$ **67.** $2^{5/6}$ **68.** $\sqrt[6]{2^5}$ or $\sqrt[6]{32}$ **69.** 2
71. 1.883 **73.** 3.971 **75.** 9.100 **81.** (a) $d = 1.22x^{1/2}$ (b) 211.31 miles **83.** Because $(7^{1/2})^2 = 7$ and
$(\sqrt{7})^2 = 7$, they should be equal, so we define $7^{1/2}$ to be $\sqrt{7}$.

CHAPTER 10 REVIEW EXERCISES (PAGE 647)

1. $-7, 7$ **3.** $-14, 14$ **5.** $-15, 15$ **7.** 4 **9.** 10 **11.** not a real number **13.** $\dfrac{7}{6}$ **15.** a must be negative.

17. irrational; 4.796 **19.** rational; -5 **21.** $5\sqrt{3}$ **23.** $4\sqrt{10}$ **25.** 12 **27.** $16\sqrt{6}$ **29.** $-\dfrac{11}{20}$ **31.** $\dfrac{\sqrt{7}}{13}$

33. $\dfrac{2}{15}$ **35.** 8 **37.** p **39.** r^9 **41.** $a^7 b^{10}\sqrt{ab}$ **43.** Yes, because both approximations are .7071067812.

45. $21\sqrt{3}$ **47.** 0 **49.** $2\sqrt{3} + 3\sqrt{10}$ **51.** $6\sqrt{30}$ **53.** 0 **55.** $11k^2\sqrt{2n}$ **57.** $\sqrt{5}$ **59.** $\dfrac{\sqrt{30}}{15}$

61. $\sqrt{10}$ **63.** $\dfrac{r\sqrt{x}}{4x}$ **65.** $\dfrac{\sqrt[3]{98}}{7}$ **67.** $-\sqrt{15} - 9$ **69.** $22 - 16\sqrt{3}$ **71.** -2 **73.** $-2 + \sqrt{5}$

75. $\dfrac{-2 + 6\sqrt{2}}{17}$ **77.** $\dfrac{-\sqrt{10} + 3\sqrt{5} + \sqrt{2} - 3}{7}$ **79.** $\dfrac{3 + 2\sqrt{6}}{3}$ **81.** $3 + 4\sqrt{3}$ **83.** $\varnothing$ **85.** $\{1\}$ **87.** $\{6\}$
89. $\{-2\}$ **91.** (a) billions of dollars in exports (b) years (c) 1992 (d) Yes, because \$13 billion is between \$7.5
billion and \$19.6 billion, so the year should be between 1990 and 1993. (e) \$31.2 billion; yes **93.** 9 **95.** 7^3 or 343
97. $x^{3/4}$ **99.** 16 **101.** $\dfrac{5 - \sqrt{2}}{23}$ **103.** $5y\sqrt{2}$ **105.** $-\sqrt{10} - 5\sqrt{15}$ **107.** $\dfrac{2 + \sqrt{13}}{2}$ **109.** $7 - 2\sqrt{10}$
111. -11 **113.** $\{7\}$ **115.** $\{8\}$ **117.** $-\dfrac{5\sqrt{7}}{5\sqrt{14}}$ **119.** $-\dfrac{\sqrt{2}}{2}$

CHAPTER 10 TEST (PAGE 650)

[10.1] **1.** $-14, 14$ **2.** (a) irrational (b) 11.916 [10.2–10.5] **3.** 6 **4.** $-3\sqrt{3}$ **5.** $\dfrac{8\sqrt{2}}{5}$ **6.** $2\sqrt[3]{4}$ **7.** $4\sqrt{6}$
8. $9\sqrt{7}$ **9.** $-5\sqrt{3x}$ **10.** $2y\sqrt[3]{4x^2}$ **11.** 31 **12.** $6\sqrt{2} + 2 - 3\sqrt{14} - \sqrt{7}$ **13.** $11 + 2\sqrt{30}$
[10.1] **14.** (a) $6\sqrt{2}$ inches (b) 8.485 inches [10.4] **15.** $\dfrac{5\sqrt{14}}{7}$ **16.** $\dfrac{\sqrt{6x}}{3x}$ **17.** $-\sqrt[3]{2}$ **18.** $\dfrac{-12 - 3\sqrt{3}}{13}$
[10.6] **19.** $\{3\}$ **20.** $\left\{\dfrac{1}{4}, 1\right\}$ [10.7] **21.** 16 **22.** -25 **23.** 5 **24.** $\dfrac{1}{3}$ [10.6] **25.** 12 is not a solution. A check
shows that it does not satisfy the original equation.

CUMULATIVE REVIEW EXERCISES CHAPTERS 1–10 (PAGE 650)

[1.5–1.6] **1.** 54 **2.** 6 **3.** 3 **4.** 18 **5.** 15 **6.** 4.223 [2.2] **7.** $\{3\}$ [3.1] **8.** $[-16, \infty)$ **9.** $(5, \infty)$
[2.4] **10.** 207 cubic inches [4.2, 8.1] **11.** **12.** [4.3, 8.1] **13.** $-\dfrac{5}{6}$ [5.2] **14.** $12x^{10}y^2$

[5.5] **15.** $\dfrac{y^{15}}{5832}$ [5.1] **16.** $3x^3 + 11x^2 - 13$ [5.6] **17.** $4t^2 - 8t + 5$ [6.2–6.4] **18.** $(m + 8)(m + 4)$
19. $(5t^2 + 6)(5t^2 - 6)$ **20.** $(6a + 5b)(2a - b)$ **21.** $(9z + 4)^2$ [6.5] **22.** $\{3, 4\}$ **23.** $\{-2, -1\}$ [7.1] **24.** 2, -7
[7.2] **25.** $\dfrac{x + 1}{x}$ **26.** $(t + 5)(t + 3)$ [7.4] **27.** $\dfrac{y^2}{(y + 1)(y - 1)}$ **28.** $\dfrac{-2x - 14}{(x + 3)(x - 1)}$ [7.5] **29.** -21

[8.3] **30.**   [9.1–9.3] **31.** $\{(3, -7)\}$ **32.** $\{(x, y) \mid 2x - y = 6\}$ [9.5] **33.** from Chicago: 57 mph;

from Des Moines: 50 mph   **34.** CNN: 67.8 million; ESPN: 67.9 million   [10.3–10.5] **35.** $29\sqrt{3}$ **36.** $-\sqrt{3} + \sqrt{5}$
37. $10xy^2\sqrt{2y}$ [10.7] **38.** 32   [10.5] **39.** $21 - 5\sqrt{2}$ [10.6] **40.** $\{16\}$

CHAPTER 11 QUADRATIC EQUATIONS, INEQUALITIES, AND GRAPHS

SECTION 11.1 (PAGE 656)

EXERCISES **1.** C   **3.** A   **5.** B   **7.** true   **9.** true   **11.** According to the square root property, -9 is also a
solution, so her answer was not completely correct. The solution set is $\{\pm 9\}$.   **13.** $\{\pm 9\}$ **15.** $\{\pm\sqrt{14}\}$
17. $\{\pm 4\sqrt{3}\}$ **19.** $\emptyset$ **21.** $\{\pm 1.5\}$ **23.** $\{\pm 2\sqrt{6}\}$ **25.** $\{-2, 8\}$ **27.** $\emptyset$ **29.** $\{8 \pm 3\sqrt{3}\}$
31. $\left\{-3, \dfrac{5}{3}\right\}$ **33.** $\left\{0, \dfrac{3}{2}\right\}$ **35.** $\left\{\dfrac{5 \pm \sqrt{30}}{2}\right\}$ **37.** $\left\{\dfrac{-1 \pm 3\sqrt{2}}{3}\right\}$ **39.** $\{-10 \pm 4\sqrt{3}\}$ **41.** $\left\{\dfrac{1 \pm 4\sqrt{3}}{4}\right\}$
43. Johnny's first solution, $\dfrac{5 + \sqrt{30}}{2}$, is equivalent to Linda's second solution, $\dfrac{-5 - \sqrt{30}}{-2}$. This can be verified by multiplying
$\dfrac{5 + \sqrt{30}}{2}$ by 1 in the form $\dfrac{-1}{-1}$. Similarly, Johnny's second solution is equivalent to Linda's first one.   **45.** $\{-4.48, .20\}$
47. $\{-3.09, -.15\}$ **49.** $(x + 3)^2 = 100$ **50.** $x + 3 = -10$ or $x + 3 = 10$ **51.** $\{-13\}; \{7\}$ **52.** $\{-13, 7\}$
53. $\{-7, 3\}$ **54.** $\{-3, 6\}$ **55.** .983 foot   **57.** 3.442 feet   **59.** about $\dfrac{1}{2}$ second   **61.** 9 inches   **63.** 5%

SECTION 11.2 (PAGE 663)

CONNECTIONS **Page 663:** **1.** x^2 **2.** $x^2 + 8x$ **3.** $x^2 + 8x + 16$ **4.** It occurred when we added the 16 squares.

EXERCISES **1.** 16   **3.** multiplying $(t + 2)(t - 5)$ to get $t^2 - 3t - 10$ **5.** (d)   **7.** 49   **9.** $\dfrac{25}{4}$ **11.** $\dfrac{1}{16}$

13. $\{1, 3\}$ **15.** $\{-1 \pm \sqrt{6}\}$ **17.** $\{-3\}$ **19.** $\left\{-\dfrac{3}{2}, \dfrac{1}{2}\right\}$ **21.** $\emptyset$ **23.** $\left\{\dfrac{-7 \pm \sqrt{97}}{6}\right\}$ **25.** $\{-4, 2\}$
27. $\{1 \pm \sqrt{6}\}$ **29.** (a) $\left\{\dfrac{3 \pm 2\sqrt{6}}{3}\right\}$ (b) $\{-.633, 2.633\}$ **31.** (a) $\{-2 \pm \sqrt{3}\}$ (b) $\{-3.732, -.268\}$
35. 75 feet by 100 feet   **37.** 1 second and 5 seconds   **39.** 3 seconds and 5 seconds   **41.** 8 miles

SECTION 11.3 (PAGE 671)

EXERCISES **1.** 4; 5; -9 **3.** 2   **5.** $a = 3, b = -4, c = -2$ **7.** $a = 3, b = 7, c = 0$ **9.** $a = 1, b = 1, c = -12$
11. $a = 9, b = 9, c = -26$ **13.** If a were 0, the equation would be linear, not quadratic.   **15.** No, because $2a$ should be
the denominator for $-b$ as well. The correct formula is $x = \dfrac{-b \pm \sqrt{b^2 - 4ac}}{2a}$.   **17.** $\{-13, 1\}$ **19.** $\{2\}$
21. $\left\{\dfrac{-6 \pm \sqrt{26}}{2}\right\}$ **23.** $\left\{-1, \dfrac{5}{2}\right\}$ **25.** $\{-1, 0\}$ **27.** $\left\{0, \dfrac{12}{7}\right\}$ **29.** $\{\pm 2\sqrt{6}\}$ **31.** $\left\{\pm\dfrac{2}{5}\right\}$ **33.** $\left\{\dfrac{6 \pm 2\sqrt{6}}{3}\right\}$
35. $\emptyset$ **37.** $\emptyset$ **39.** $\left\{\dfrac{-5 \pm \sqrt{61}}{2}\right\}$ **41.** (a) $\left\{\dfrac{-1 \pm \sqrt{11}}{2}\right\}$ (b) $\{-2.158, 1.158\}$ **43.** (a) $\left\{\dfrac{1 \pm \sqrt{5}}{2}\right\}$
(b) $\{-.618, 1.618\}$ **45.** $\left\{-\dfrac{2}{3}, \dfrac{4}{3}\right\}$ **47.** $\left\{\dfrac{-1 \pm \sqrt{73}}{6}\right\}$ **49.** $\{1 \pm \sqrt{2}\}$ **51.** $\emptyset$ **53.** The solution(s) must

make sense in the original problem. (For example, a length cannot be negative.)   **55.** $r = \dfrac{-\pi h \pm \sqrt{\pi^2 h^2 + \pi S}}{\pi}$

57. $P = .241$ (about 88 days)   **59.** $P = 1.881$ **61.** $P = 84.130$ **63.** $1^2 = 1^3$, or $1 = 1$. This is a true statement.
65. 4 seconds   **67.** approximately .2 second and 10.9 seconds   **69.** 30 units   **71.** (d)   **73.** (a)   **75.** (b)
77. (c)   **78.** It is the radicand in the quadratic formula.   **79.** (a) 225 (b) 169 (c) 4 (d) 121   **80.** Each is a
perfect square.   **81.** (a) $(3x - 2)(6x + 1)$ (b) $(x + 2)(5x - 3)$ (c) $(8x + 1)(6x + 1)$ (d) $(x + 3)(x - 8)$
82. (a) 41 (b) -39 (c) 12 (d) 112   **83.** no   **84.** If the discriminant is a perfect square the trinomial is factorable.
(a) yes (b) yes (c) no

SECTION 11.4 (PAGE 680)

CONNECTIONS   **Page 679:**   **1.** $\langle -1, -4 \rangle, \langle 0, 2 \rangle, \langle -5, 0 \rangle$ **2.** $-1; 2$

EXERCISES   **1.** $3i$ **3.** $2i\sqrt{5}$ **5.** $3i\sqrt{2}$ **7.** $5i\sqrt{5}$ **9.** $5 + 3i$ **11.** $6 - 9i$ **13.** $6 - 7i$ **15.** $14 + 5i$
17. $7 - 22i$ **19.** 45   **21.** $18i$ **23.** -4 **25.** $6 + 4i$ **27.** $4 - 3i$ **29.** $3 - i$ **31.** $2 - 6i$
33. $-\dfrac{3}{25} + \dfrac{4}{25}i$ **35.** $-1 + 3i$ **36.** The product of $-1 + 3i$ and $2 + 9i$ is $-29 - 3i$, which *is* the original dividend.
37. $-3 - 4i$ **38.** The product of $-3 - 4i$ and i is $4 - 3i$, which *is* the original dividend.   **39.** Because
$(3 - 2i)(4 + i) = 14 - 5i$ is true, the given statement is true.   **40.** The product of the quotient and the divisor must equal
the dividend.   **41.** $\{-1 \pm 2i\}$ **43.** $\{3 \pm i\sqrt{5}\}$ **45.** $\left\{-\dfrac{2}{3} \pm i\sqrt{2}\right\}$ **47.** $\{1 \pm i\}$ **49.** $\left\{-\dfrac{3}{4} \pm \dfrac{\sqrt{31}}{4}i\right\}$
51. $\left\{\dfrac{3}{2} \pm \dfrac{\sqrt{7}}{2}i\right\}$ **53.** $\left\{\dfrac{1}{5} \pm \dfrac{\sqrt{14}}{5}i\right\}$ **55.** $\left\{-\dfrac{1}{2} \pm \dfrac{\sqrt{13}}{2}i\right\}$ **57.** $\left\{\dfrac{1}{2} \pm \dfrac{\sqrt{11}}{2}i\right\}$ **59.** If the discriminant $b^2 - 4ac$
is negative, the equation will have solutions that are not real.   **61.** true   **63.** false; For example, $3 + 2i$ is a complex
number but it is not real.

SECTION 11.5 (PAGE 688)

EXERCISES   **1.** Multiply by the LCD, x.   **3.** Substitute for $r^2 + r$.   **7.** $\{-2, 7\}$ **9.** $\{-4, 7\}$ **11.** $\left\{-\dfrac{14}{17}, 5\right\}$
13. $\left\{-\dfrac{11}{7}, 0\right\}$ **15.** $\left\{\dfrac{-1 + \sqrt{13}}{2}, \dfrac{-1 - \sqrt{13}}{2}\right\}$ **17.** $\dfrac{1}{m}$ job per hour   **19.** 25 miles per hour
21. 80 kilometers per hour   **23.** 9 minutes   **25.** $\{3\}$ **27.** $\left\{\dfrac{8}{9}\right\}$ **29.** $\{16\}$ **31.** $\left\{\dfrac{2}{5}\right\}$ **33.** 30 meters
35. $\{-3, 3\}$ **37.** $\left\{-\dfrac{3}{2}, -1, 1, \dfrac{3}{2}\right\}$ **39.** $\{-6, -5\}$ **41.** $\{-4, 1\}$ **43.** $\left\{-\dfrac{1}{3}, \dfrac{1}{6}\right\}$ **45.** $\left\{-\dfrac{1}{2}, 3\right\}$
47. $\left\{-\dfrac{27}{8}, -1, 1, \dfrac{27}{8}\right\}$ **49.** $\{-8, 1\}$ **51.** $\{25\}$ **53.** $\left\{-1, 1, -\dfrac{\sqrt{6}}{2}i, \dfrac{\sqrt{6}}{2}i\right\}$ **55.** The solutions given are for u.
Each must be set equal to $m - 1$ and solved for m. The correct solutions are $\dfrac{3}{2}$ and 2.   **57.** $\{-2\sqrt{3}, -2, 2, 2\sqrt{3}\}$
59. $\{3, 11\}$ **61.** $\left\{-\sqrt[3]{5}, -\dfrac{\sqrt[3]{4}}{2}\right\}$ **63.** $\left\{-\dfrac{1}{2}, 3\right\}$ **65.** It would cause both denominators to be 0, and division by 0 is
undefined.   **66.** The solution is $\dfrac{12}{5}$.   **67.** $\left(\dfrac{x}{x-3}\right)^2 + 3\left(\dfrac{x}{x-3}\right) - 4 = 0$ **68.** The numerator can never equal the
denominator, since the denominator is 3 less than the numerator.   **69.** $\left\{\dfrac{12}{5}\right\}$; The values for t are -4 and 1. The value 1 is
impossible because it leads to a contradiction (since $\dfrac{x}{x-3}$ is never equal to 1).   **70.** $\left\{\dfrac{12}{5}\right\}$; The values for s are $\dfrac{1}{x}$ and $\dfrac{-4}{x}$.
The value $\dfrac{1}{x}$ is impossible, since $\dfrac{1}{x} \neq \dfrac{1}{x-3}$ for all x.

SECTION 11.6 (PAGE 695)

EXERCISES 1. Multiply both sides by the common denominator. **3.** Write it in standard form (with 0 on one side in decreasing powers of w). **5.** No. There is only one time after the rock starts moving when it hits the ground.

7. $m = \sqrt{p^2 - n^2}$ **9.** $t = \dfrac{\pm\sqrt{dk}}{k}$ **11.** $v = \dfrac{\pm\sqrt{kAF}}{F}$ **13.** $r = \dfrac{\pm\sqrt{3\pi Vh}}{\pi h}$ **15.** $t = \dfrac{-B \pm \sqrt{B^2 - 4AC}}{2A}$

17. $h = \dfrac{D^2}{k}$ **19.** $\ell = \dfrac{p^2 g}{k}$ **21.** If g is positive, the only way to have a real value for p is to have $k\ell$ positive, since the quotient of two positive numbers is positive. If k and ℓ have different signs, their product is negative, leading to a negative radicand. **23.** 5.2 seconds **25.** Find s when $t = 0$. **27. (a)** 1 second and 8 seconds **(b)** 9 seconds after it is projected **29. (a)** $y = x - 1.727x^2$ **(b)** .006825 foot; .1421 foot **(c)** 0 seconds; .5790 second **(d)** One represents the time before the vehicle leaves the surface. The other represents the time it returns to the surface. **31.** 2.3, 5.3, 5.8 **33.** 412.3 feet **35.** eastbound ship: 80 miles; southbound ship: 150 miles **37.** 5 centimeters, 12 centimeters, 13 centimeters **39.** length: 2 centimeters; width: 1.5 centimeters **41.** 1 foot **43.** length: 26 meters; width: 16 meters

45. .035 or 3.5% **47.** $.80 **49.** 5.5 meters per second **51.** 5 or 14 **53.** $R = \dfrac{E^2 - 2pr \pm E\sqrt{E^2 - 4pr}}{2p}$

55. $r = \dfrac{5pc}{4}$ or $r = -\dfrac{2pc}{3}$ **57.** $I = \dfrac{-cR \pm \sqrt{c^2R^2 - 4cL}}{2cL}$

SECTION 11.7 (PAGE 707)

EXERCISES 1. F **3.** C **5.** E **9.** $(0, 0)$ **11.** $(0, 4)$ **13.** $(1, 0)$ **15.** $(-3, -4)$ **17.** In Exercise 15, the parabola is shifted 3 units to the left and 4 units down. The parabola in Exercise 16 is shifted 5 units to the right and 8 units down. **19.** downward; narrower **21.** upward; wider **23. (a)** I **(b)** IV **(c)** II **(d)** III

25. **27.** **29.** **31.** **33.**

35. domain: $(-\infty, \infty)$; range: $[-4, \infty)$ **37.** domain: $(-\infty, \infty)$; range: $(-\infty, 2]$

39. domain: $(-\infty, \infty)$; range: $[-3, \infty)$ **41.** It is shifted 6 units upward. **42.**

43. It is shifted 6 units upward. **44.** It is shifted 6 units to the right. **45.**

46. It is shifted 6 units to the right. **47. (a)** $|x + p|$ **(b)** The distance from the focus to the origin should equal the distance from the directrix to the origin. **(c)** $\sqrt{(x - p)^2 + y^2}$ **(d)** $y^2 = 4px$

49. (a)

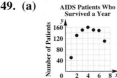

AIDS Patients Who Survived a Year

(b) negative   **(c)** $f(x) = -7x^2 + 60x + 38$ **(d)** Many answers are possible. Since the data give

one y-value for each year, $\{1, 2, 3, 4, 5, 6, 7\}$ may be most appropriate.   **51.** $\{-4, 5\}$ **53.** $\{-.5, 3\}$ **55.** (a)

SECTION 11.8 (PAGE 719)

EXERCISES   1. If x is squared, it has a vertical axis; if y is squared, it has a horizontal axis.   **3. (a)** Use the discriminant of the function. If it is positive, there are two x-intercepts. If it is zero, there is one x-intercept (at the vertex), and if it is negative, there is no x-intercept.   **(b)** none   **5.** $(-1, 3)$; upward; narrower; no x-intercepts   **7.** $\left(\frac{5}{2}, \frac{37}{4}\right)$; downward; same;

two x-intercepts   **9.** $(-3, -9)$; to the right; wider   **11.** domain: $(-\infty, \infty)$; range: $[-6, \infty)$

$f(x) = x^2 + 8x + 10$

13. domain: $(-\infty, \infty)$; range: $(-\infty, -3]$

$y = -2x^2 + 4x - 5$

15. not a function

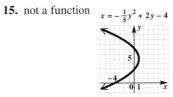

$x = -\frac{1}{5}y^2 + 2y - 4$

17. not a function

$x = 3y^2 + 12y + 5$

19. F   **21.** C   **23.** D   **25.** 25 meters   **27.** 16 feet; 2 seconds

29. 4.1 seconds; 81.6 meters   **31. (a)** The coefficient of x^2 is negative, because the parabola opens downward.
(b) $(18.45, 3860)$ **(c)** In 2018 social security assets will reach their maximum value of \$3860 billion.   **33. (a)** $R(x) = (100 - x)(200 + 4x) = 20,000 + 200x - 4x^2$ **(b)**

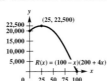

$R(x) = (100 - x)(200 + 4x)$

(c) 25   **(d)** \$22,500   **35.** B   **37.** A

SECTION 11.9 (PAGE 727)

CONNECTIONS   Page 727:   1. 31.6 seconds and 119.9 seconds   **2.** between 31.6 seconds and 119.9 seconds
3. 0 seconds (when it is first propelled) and 151.5 seconds (when it hits the ground)   **4.** between 0 and 31.6 seconds and also between 119.9 and 151.5 seconds

EXERCISES   1. $\left\{-\frac{3}{2}, 1\right\}$ **3.** $\left(-\frac{3}{2}, 1\right)$ **5.** $(-\infty, -1) \cup (5, \infty)$ **7.** $(-4, 6)$

9. $(-\infty, 1] \cup [3, \infty)$ **11.** $\left(-\infty, -\frac{3}{2}\right] \cup \left[\frac{3}{5}, \infty\right)$

13. $\left(-\frac{2}{3}, \frac{1}{3}\right)$ **15.** $\left(-\infty, -\frac{1}{2}\right] \cup \left[\frac{1}{3}, \infty\right)$

17. $(-\infty, 3 - \sqrt{3}] \cup [3 + \sqrt{3}, \infty)$ **19.** Include the endpoints if the symbol is $\geq$ or $\leq$. Exclude the

endpoints if the symbol is $>$ or $<$. **21.** $(-\infty, \infty)$ **23.** $\emptyset$ **27.** $\left(-\infty, -\dfrac{7}{4}\right) \cup \left(-\dfrac{1}{2}, \dfrac{2}{3}\right)$

29. $\left[-\dfrac{7}{2}, -2\right] \cup \left[\dfrac{3}{4}, \infty\right)$ **31.** $(-\infty, 1) \cup (4, \infty)$

33. $\left[-\dfrac{3}{2}, 5\right)$ **35.** $(2, 6]$ **37.** $\left(-\infty, \dfrac{1}{2}\right) \cup \left(\dfrac{5}{4}, \infty\right)$

39. $[-4, -2)$ **41.** $(-\infty, 4) \cup (6, \infty)$ **43.** $\left(0, \dfrac{1}{2}\right) \cup \left(\dfrac{5}{2}, \infty\right)$

45. $\left[\dfrac{3}{2}, \infty\right)$ **47.** $\left(-2, \dfrac{5}{3}\right) \cup \left(\dfrac{5}{3}, \infty\right)$ **49.** $\dfrac{-1}{p-3} - 1 \geq 0$

50. $\dfrac{-1}{p-3} - \dfrac{p-3}{p-3} \geq 0$ **51.** $\dfrac{-p+2}{p-3} \geq 0$ **52.** $2, 3; [2, 3)$ **53.** between approximately .237 and .342 second

55. (a) $\{1, 3\}$ **(b)** $(-\infty, 1) \cup (3, \infty)$ **(c)** the open interval $(1, 3)$ **56. (a)** $\left\{-4, \dfrac{2}{3}\right\}$ **(b)** $(-\infty, -4] \cup \left[\dfrac{2}{3}, \infty\right)$

(c) the open interval $\left(-4, \dfrac{2}{3}\right)$ **57. (a)** $\{-2, 5\}$ **(b)** $[-2, 5]$ **(c)** $(-\infty, -2] \cup [5, \infty)$ **58. (a)** $\left\{-3, \dfrac{5}{2}\right\}$

(b) $\left[-3, \dfrac{5}{2}\right]$ **(c)** $(-\infty, -3] \cup \left[\dfrac{5}{2}, \infty\right)$ **59.** between 0 and 10.8 seconds and between 140.8 and 151.5 seconds

CHAPTER 11 REVIEW EXERCISES (PAGE 737)

1. $\{\pm 12\}$ **3.** $\{\pm 8\sqrt{2}\}$ **5.** $\{3 \pm \sqrt{10}\}$ **7.** $\emptyset$ **9.** $\{-5, -1\}$ **11.** $\{-1 \pm \sqrt{6}\}$ **13.** $\left\{-\dfrac{2}{5}, 1\right\}$

15. 2.5 seconds **17.** $\left(\dfrac{k}{2}\right)^2$ or $\dfrac{k^2}{4}$ **19.** $\{1 \pm \sqrt{5}\}$ **21.** $\left\{\dfrac{2 \pm \sqrt{10}}{2}\right\}$ **23.** $\left\{\dfrac{-3 \pm \sqrt{41}}{2}\right\}$ **25.** Multiply both the

numerator and the denominator by the conjugate of the denominator. **27.** $5 - i$ **29.** 20 **31.** i **33.** $\dfrac{7}{50} + \dfrac{1}{50}i$

35. No, the product $(a + bi)(a - bi) = a^2 + b^2$ will always be the sum of the squares of two real numbers, which is a real

number. **37.** $\left\{\dfrac{2}{3} \pm \dfrac{2\sqrt{2}}{3}i\right\}$ **39.** $\left\{-\dfrac{3}{2} \pm \dfrac{\sqrt{23}}{2}i\right\}$ **41.** $\left\{-\dfrac{1}{9} \pm \dfrac{2\sqrt{2}}{9}i\right\}$ **43.** $\left\{-\dfrac{11}{6}, -\dfrac{19}{12}\right\}$

45. $\left\{-\dfrac{343}{8}, 64\right\}$ **47.** 40 miles per hour **49.** Because x appears on the left side alone, and because it is equal to the

nonnegative square root of $2x + 4$, it cannot be negative. **51.** $v = \dfrac{\pm\sqrt{rFkw}}{kw}$ **53.** .87 second **55.** 10 meters by

12 meters **57.** 120 meters **59.** $(0, 6)$ **61.** $(-2, 0)$ **63.** $\left(\dfrac{2}{3}, -\dfrac{2}{3}\right)$ **65.** **67. (a)**

69. (a) $c = 12.39, 16a + 4b + c = 15.78, 49a + 7b + c = 22.71$ **(b)** $f(x) = .2089x^2 + .0118x + 12.39$
71. (a) $\{-.5, 4\}$ **(b)** $(-\infty, -.5) \cup (4, \infty)$ **(c)** $[-.5, 4]$ **73.** $[-4, 3]$ **75.** $\emptyset$

77. $[-3, 2)$ **79.** $\left\{\dfrac{3 + i\sqrt{3}}{3}, \dfrac{3 - i\sqrt{3}}{3}\right\}$ **81.** $\{-2, -1, 3, 4\}$ **83.** $\left\{\dfrac{4}{11}, 5\right\}$ **85.** $y = \dfrac{6p^2}{z}$

87. $\left\{\dfrac{3}{5}, 1\right\}$ **89.** $\left(-5, -\dfrac{23}{5}\right]$ **91.** $(-4, -3)$ **93.** $(4, 6)$

95. Work backwards as follows.

$x = -5$ or $x = 6$ Given

$x + 5 = 0$ or $x - 6 = 0$

$(x + 5)(x - 6) = 0$ Zero-factor property

$x^2 - x - 30 = 0$ Multiply the factors.

The equation $x^2 - x - 30 = 0$ has -5 and 6 as solutions.

97. length: 150 meters; width: 150 meters

99. (a) $\{-2\}$ **(b)** $(-\infty, -2)$ **(c)** $(-2, \infty)$

100. (a) $\{1, 5\}$ **(b)** $(-\infty, 1) \cup (5, \infty)$ **(c)** $(1, 5)$

101. (a) $\{4\}$ **(b)** $(2, 4)$ **(c)** $(-\infty, 2) \cup (4, \infty)$

102. $(-\infty, \infty)$ **103.** $(-\infty, \infty)$ **104.** $2; (-\infty, 2) \cup (2, \infty)$ **105.** $(-\infty, \infty)$; denominator   **106.** $(-5, 3)$

CHAPTER 11 TEST (PAGE 742)

[11.1] **1.** $\{\pm\sqrt{39}\}$ **2.** $\{-11, 5\}$ [11.2] **3.** $\{2 \pm \sqrt{10}\}$ **4.** $\left\{\dfrac{-6 \pm \sqrt{42}}{2}\right\}$ [11.3] **5.** $\left\{\dfrac{3 \pm \sqrt{3}}{3}\right\}$

6. $\left\{-1 \pm \dfrac{\sqrt{7}}{2}i\right\}$ [11.1–11.3] **7.** $\{1 \pm \sqrt{2}\}$ **8.** $\left\{\dfrac{-1 \pm 3\sqrt{2}}{2}\right\}$ **9.** $\left\{\dfrac{11 \pm \sqrt{89}}{4}\right\}$ **10.** $\{5\}$ **11.** 2 seconds

12. 247.3 years   [11.4] **13.** $-5 + 5i$ **14.** $-17 - 4i$ **15.** 73   **16.** $2 - i$ [11.3] **17.** two irrational solutions

[11.1–11.3, 11.5] **18.** $\left\{-2, -\dfrac{1}{3}, \dfrac{1}{3}, 2\right\}$ **19.** $\left\{-\dfrac{5}{2}, 1\right\}$ [11.6] **20.** $r = \dfrac{\pm\sqrt{\pi S}}{2\pi}$ **21.** 2 feet

22. 16 meters   [11.7] **23.** vertex: $(0, -2)$; domain: $(-\infty, \infty)$; range: $[-2, \infty)$ [11.8] **24.** vertex: $(2, 3)$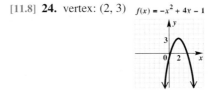

25. vertex: $(2, 2)$ **26. (a)** 442 million   **(b)** 1993; 425 million

[11.9] **27.** $(-\infty, -5) \cup \left(\dfrac{3}{2}, \infty\right)$ **28.** $(-\infty, 4) \cup [9, \infty)$

CUMULATIVE REVIEW EXERCISES CHAPTERS 1–11 (PAGE 744)

[1.4] **1. (a)** $-2, 0, 7$ **(b)** $-\dfrac{7}{3}, -2, 0, .7, 7, \dfrac{32}{3}$ [1.5, 1.6] **2.** 41   [2.2] **3.** $\left\{\dfrac{4}{5}\right\}$ [3.3] **4.** $\left\{\dfrac{11}{10}, \dfrac{7}{2}\right\}$ [10.6] **5.** $\left\{\dfrac{2}{3}\right\}$

[7.6] **6.** $\emptyset$ [11.1–11.3] **7.** $\left\{\dfrac{7 + \sqrt{177}}{4}, \dfrac{7 - \sqrt{177}}{4}\right\}$ [11.5] **8.** $\{-2, -1, 1, 2\}$ [3.1] **9.** $[1, \infty)$ [3.3] **10.** $\left[2, \dfrac{8}{3}\right]$

[11.9] **11.** $(1, 3)$ **12.** $(-2, 1)$ [4.2, 4.3] **13.** $m = \dfrac{2}{7}$; x-intercept: $(-8, 0)$; y-intercept: $\left(0, \dfrac{16}{7}\right)$

[8.2] **14.** $y = -\dfrac{5}{2}x + 2$ **15.** $y = \dfrac{2}{5}x + \dfrac{13}{5}$ [5.3] **16.** $14x^2 - 13x - 12$ [5.4] **17.** $\dfrac{4}{9}t^2 + 12t + 81$

[5.1] **18.** $-3t^3 + 5t^2 - 12t + 15$ [5.6] **19.** $4x^2 - 6x + 11 + \dfrac{4}{x + 2}$ [6.1–6.4] **20.** $x(4 + x)(4 - x)$

21. $9(x - 1)(x + 1)$ **22.** $(2x + 3y)(4x^2 - 6xy + 9y^2)$ **23.** $(3x - 5y)^2$ [7.2] **24.** $\dfrac{x - 5}{x + 5}$ [7.4] **25.** $-\dfrac{8}{k}$

[7.5] **26.** $\dfrac{r - s}{r}$ [4.2, 8.4] **27.** function; domain: $(-\infty, \infty)$; range: $(-\infty, \infty)$

[8.3, 8.4] **28.** not a function   [11.7] **29.** function; domain: $(-\infty, \infty)$; range: $(-\infty, 3]$

[8.4] **30. (a)** Qualifying Records   **(b)** yes   **(c)** $y = 1.279x + 116.26$ **(d)** 158.47; It is a little too high.

31. No, because the graph is a vertical line, which is not the graph of a function by the vertical line test.   **32. (a)** 13
(b) domain: $(-\infty, \infty)$; range: $[-5, \infty)$ [9.1–9.3] **33.** $\{(1, -2)\}$ [9.4] **34.** $\{(3, -4, 2)\}$ [9.5] **35.** current: 10 miles per
hour; boat: 30 miles per hour   [10.4] **36.** $\dfrac{3\sqrt[3]{4}}{4}$ [10.5] **37.** $\sqrt{7} + \sqrt{5}$ [2.4] **38.** 7 inches by 3 inches

[7.7] **39.** biking: 12 miles per hour; walking: 2 miles per hour   [11.6] **40.** southbound car: 57 miles; eastbound car: 76 miles

CHAPTER 12 INVERSE, EXPONENTIAL, AND LOGARITHMIC FUNCTIONS

SECTION 12.1 (PAGE 753)

EXERCISES   1. It is not one-to-one, because both Illinois and Wisconsin are paired with the same range element, 40.   **5. (b)**

7. (a)   9. $\{(6, 3), (10, 2), (12, 5)\}$ **11.** not one-to-one   **13.** $f^{-1}(x) = \dfrac{x - 4}{2}$ **15.** $g^{-1}(x) = x^2 + 3$,

$x \geq 0$ **17.** not one-to-one   **19.** $f^{-1}(x) = \sqrt[3]{x + 4}$ **21. (a)** 8   **(b)** 3   **23. (a)** 1   **(b)** 0
25. (a) one-to-one   **(b)**   **27. (a)** not one-to-one   **29. (a)** one-to-one   **(b)**

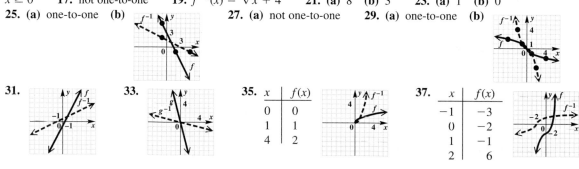

31.   **33.**   **35.**

x	$f(x)$
0	0
1	1
4	2

37.

x	$f(x)$
-1	-3
0	-2
1	-1
2	6

39. $f^{-1}(x) = \dfrac{x + 5}{4}$ **40.** My graphing calculator is the greatest thing since sliced bread.   **41.** If the function were not

one-to-one, there would be ambiguity in some of the characters, as they could represent more than one letter.
42. Answers will vary. For example, Jane Doe is   1004 5 2748 129 68 3379 129.

43. $f^{-1}(x) = \dfrac{x + 7}{2}$ 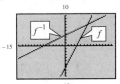   **45.** $f^{-1}(x) = \sqrt[3]{x - 5}$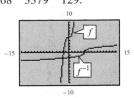

47. 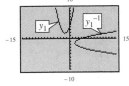   **49.** It is not a one-to-one function.

SECTION 12.2 (PAGE 763)

EXERCISES   **1.** (c)   **3.** (a)   **5.**   **7.**   **9.**   **11.**

13. **(a)** rises; falls **(b)** It is one-to-one and thus has an inverse.   **15.** $\left\{\dfrac{3}{2}\right\}$ **17.** $\{7\}$ **19.** $\{-3\}$ **21.** $\{-1\}$

23. $\{-3\}$ **25.** 639.545   **27.** .066   **29.** 12.179   **31.** In the definition of the exponential function defined by
$F(x) = a^x$, a must be a positive number. This corresponds to the peculiarity that some scientific calculators do not allow
negative bases.   **33.** **(a)** .5°C **(b)** .35°C   **35.** **(a)** 1.6°C **(b)** .5°C   **37.** **(a)** 7147 million short tons
(b) 25,149 million short tons **(c)** This is less than the 30,100 million short tons the model provides.
39. **(a)** 100 grams **(b)** 31.25 grams **(c)** .30 gram (to the nearest hundredth) **(d)**

41. 6.67 years after it was purchased   **43.** about \$360,000   **45.** $(\sqrt[4]{16})^3$; 8   **46.** $\sqrt[4]{16^3}$; 8   **47.** 64; 8
48. Because $\sqrt{\sqrt{x}} = (x^{1/2})^{1/2} = x^{1/4} = \sqrt[4]{x}$, the fourth root of 16^3 can be found by taking the square root twice.
49. 8   **50.** $16^{75/100}$; $16^{3/4} = (\sqrt[4]{16})^3 = 2^3 = 8$ **51.** The display indicates that for the year 1965, the model gives a value
of 102.287 million tons, which is slightly less than the actual value of 103.4 million tons.

SECTION 12.3 (PAGE 773)

CONNECTIONS   **Page 773:** almost 4 times as powerful; about 300 times as powerful

EXERCISES   **1.** **(a)** C **(b)** F **(c)** B **(d)** A **(e)** E **(f)** D   **3.** $\log_4 1024 = 5$ **5.** $\log_{1/2} 8 - -3$

7. $\log_{10} .001 = -3$ **9.** $4^3 = 64$ **11.** $10^{-4} = \dfrac{1}{10,000}$ **13.** $6^0 = 1$ **15.** By using the word "radically," the teacher

meant for him to consider roots. Because 3 is the square (2nd) root of 9, $\log_9 3 = \dfrac{1}{2}$.   **17.** $\left\{\dfrac{1}{3}\right\}$ **19.** $\{81\}$ **21.** $\left\{\dfrac{1}{5}\right\}$

23. $\{1\}$ **25.** $\{x \mid x > 0, x \neq 1\}$ **27.** $\{5\}$ **29.** $\left\{\dfrac{5}{3}\right\}$ **31.** $\{4\}$ **33.** $\left\{\dfrac{3}{2}\right\}$ **35.**

37.   **39.** Every power of 1 is equal to 1, and thus it cannot be used as a base.   **41.** $(0, \infty)$; $(-\infty, \infty)$

43. 8   **45.** 24   **47. (a)** 645 sites **(b)** 962 sites   **49. (a)** 6 **(b)** 12 **(c)** 18 **(d)**

51. 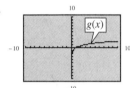   **53.**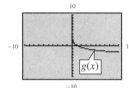

SECTION 12.4 (PAGE 782)

CONNECTIONS   **Page 782:**

$$\begin{aligned} \log_{10} 458.3 &\approx 2.661149857 \\ + \log_{10} 294.6 &\approx 2.469232743 \\ \hline &\approx 5.130382600 \end{aligned}$$

$10^{5.130382600} \approx 135,015.18$

A calculator gives $(458.3)(294.6) = 135,015.18$.

EXERCISES   **1.** $\log_{10} 3 + \log_{10} 4$ **3.** $4 \log_{10} 3$ **5.** 4   **7.** $\log_7 4 - \log_7 5$ **9.** $\dfrac{1}{4} \log_2 8$ or $\dfrac{3}{4}$

11. $\log_4 3 + \dfrac{1}{2} \log_4 x - \log_4 y$ **13.** $\dfrac{1}{3} \log_3 4 - 2 \log_3 x - \log_3 y$ **15.** $\dfrac{1}{2} \log_3 x + \dfrac{1}{2} \log_3 y - \dfrac{1}{2} \log_3 5$

17. $\dfrac{1}{3} \log_2 x + \dfrac{1}{5} \log_2 y - 2 \log_2 r$ **21.** $\log_b xy$ **23.** $\log_a \dfrac{m^3}{n}$ **25.** $\log_a \dfrac{rt^3}{s}$ **27.** $\log_a \dfrac{125}{81}$

29. $\log_{10}(x^2 - 9)$ **31.** $\log_p \dfrac{x^3 y^{1/2}}{z^{3/2} a^3}$ **33.** 1.2552   **35.** $-.6532$ **37.** 1.5562   **39.** .4771   **41.** true

43. false   **45.** The exponent of a quotient is the difference between the exponent of the numerator and the exponent of the denominator.   **47.** $\log_2 8 - \log_2 4 = \log_2 \dfrac{8}{4} = \log_2 2 = 1$ **49.** 4   **50.** It is the exponent to which 3 must be raised in order to obtain 81.   **51.** 81   **52.** It is the exponent to which 2 must be raised in order to obtain 19.   **53.** 19   **54.** m

SECTION 12.5 (PAGE 789)

CONNECTIONS   **Page 786:** $2, 2.5; 2.\overline{6}; 2.708\overline{3}; 2.716$; The difference is .0016151618. It approaches e fairly quickly.

EXERCISES   **1.** (c)   **3.** (c)   **5.** 19.2   **7.** 1.6335   **9.** 2.5164   **11.** -1.4868 **13.** 9.6776   **15.** 2.0592
17. -2.8896 **19.** 5.9613   **21.** 4.1506   **23.** 2.3026   **25. (a)** 2.552424846 **(b)** 1.552424846 **(c)** .552424846
(d) The whole number parts will vary but the decimal parts are the same.   **27.** An error message appears, because we cannot find the common logarithm of a negative number.   **29.** bog   **31.** 11.6   **33.** 4.3   **35.** 4.0×10^{-8}
37. 4.0×10^{-6} **39. (a)** 78,840 million dollars **(b)** 92,316 million dollars **(c)** 108,095 million dollars

(d) 70,967 million dollars   **41.** 32,044 million dollars   **43. (a)** 1.62 grams   **(b)** 1.18 grams   **(c)** .69 gram
(d) 2.00 grams   **45.** 527 years   **47.** 3466 years   **49.** 1.3869   **51.** .7211   **53.** 3.3030   **55.** Answers will vary. Suppose the name is Jeffery Cole, with $m = 7$ and $n = 4$.   **(a)** $\log_7 4$ is the exponent to which 7 must be raised in order to obtain 4.   **(b)** .7124143742   **(c)** 4   **57.** It means that in 1988, expenditures were approximately 83,097 million dollars.

59. (a) The expression $\dfrac{1}{x}$ in the base cannot be evaluated, since division by 0 is not defined.   **(b)** e **(c)** 2.718280469; 2.718281828; they differ in the sixth decimal place.   **(d)** e **61.**

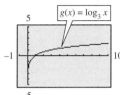

63.

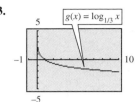

SECTION 12.6 (PAGE 800)

CONNECTIONS   Page 798: When the equation is set up, the common factor P can be divided into both sides, and thus its particular value does not affect the solution of the equation.

EXERCISES   1. $\log 5^x = \log 125$ **2.** $x \log 5 = \log 125$ **3.** $x = \dfrac{\log 125}{\log 5}$ **4.** $\dfrac{\log 125}{\log 5} = 3; \{3\}$ **5.** $\{.827\}$

7. $\{.833\}$ **9.** $\{2.269\}$ **11.** $\{566.866\}$ **13.** $\{-18.892\}$ **15.** $\{24.414\}$ **19.** $\left\{\dfrac{2}{3}\right\}$ **21.** $\{-1 + \sqrt[3]{49}\}$

23. The apparent solution, 2, causes the expression $\log_4(x - 3)$ to be $\log_4(-1)$, and we cannot take the logarithm of a negative number.   **25.** $\left\{\dfrac{1}{3}\right\}$ **27.** $\{2\}$ **29.** $\emptyset$ **31.** $\{8\}$ **33.** $\left\{\dfrac{4}{3}\right\}$ **35.** $\{8\}$ **37.** $2539.47

39. about 180 grams   **41. (a)** $11,260.96   **(b)** $11,416.64   **(c)** $11,497.99   **(d)** $11,580.90   **(e)** $11,581.83
43. $137.41   **45.** 1997   **47.** 5.5 hours   **49. (a)** 3.10 grams   **(b)** 1.19 grams   **(c)** 29 years

51. $\{.827\}$

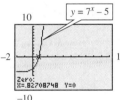

53. $\left\{\dfrac{1}{3}\right\}$ or $\{.333\}$

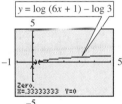

CHAPTER 12 REVIEW EXERCISES (PAGE 807)

1. not one-to-one   **3.** one-to-one; Each element of the domain corresponds to only one element of the range, and vice versa.

5. $f^{-1}(x) = \dfrac{x^3 + 4}{6}$ **7.**

9.

$f(x) = 3^x$

11.

$y = 3^{x+1}$

13. $\{4\}$

15. (a) 1.2 million tons (less than actual)   **(b)** 3.8 million tons (less than actual)   **(c)** 21.8 million tons (more than actual)

17.

$g(x) = \log_{1/3} x$

19. $\left\{\dfrac{3}{2}\right\}$ **21.** $\{8\}$ **23.** $\{b \mid b > 0, b \neq 1\}$ **25.** a **27.** $\log_2 3 + \log_2 x + 2 \log_2 y$

29. $\log_b \dfrac{3x}{y^2}$ **31.** 1.4609   **33.** 3.3638   **35.** .9251   **37.** 6.4   **39.** 2.5×10^{-5} **41.** $\{2.042\}$ **43. (d)**

45. $\{2\}$ **47.** $\{4\}$ **49. (a)** $\left\{\dfrac{3}{8}\right\}$ **(b)** The x-value of the x-intercept is .375, the decimal equivalent of $\dfrac{3}{8}$.

51. $28,295.56 **53.** Plan A is better, since it would pay $2.92 more. **55.** about 18.04% **57.** $4267

59. $\dfrac{1}{4}, \dfrac{1}{2}, 1, 2, 4, 8$ **60.** $-2, -1, 0, 1, 2, 3$ 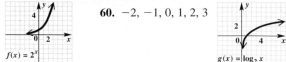 **61.** The roles of x and y are reversed. They

are inverses. **62.** horizontal; vertical **63.** 5; 2; 3; 5 **64.** 4; 8 (or 8; 4) **65.** 3.700439718 (The number of displayed digits may vary.) **66.** $\log_2 13$ is the exponent to which 2 must be raised in order to obtain 13. **67.** 13

68. 13; The number in Exercise 65 is the exponent to which 2 must be raised in order to obtain 13. **69.** 3.700439718

70. $\left\{-\dfrac{8}{5}\right\}$ **71.** $\{72\}$ **73.** $\left\{\dfrac{1}{9}\right\}$ **75.** $\{3\}$ **77.** $\left\{\dfrac{11}{3}\right\}$

CHAPTER 12 TEST (PAGE 811)

[12.1] **1. (a)** not one-to-one **(b)** one-to-one **2.** $f^{-1}(x) = x^3 - 7$ **3.**   [12.2] **4.**

[12.3] **5.**   [12.1–12.3] **6.** Once the graph of $f(x) = 6^x$ is sketched, interchange the x- and y-values of its ordered

pairs. The resulting points will be on the graph of $g(x) = \log_6 x$, since f and g are inverses. [12.2] **7.** $\{-4\}$ **8.** $\left\{-\dfrac{13}{3}\right\}$

9. (a) 2821 million pounds **(b)** 2385 million pounds **(c)** This is less than the 2193 million pounds that the model provides.

[12.3] **10.** $\log_4 .0625 = -2$ **11.** $7^2 = 49$ **12.** $\{32\}$ **13.** $\left\{\dfrac{1}{2}\right\}$ **14.** $\{2\}$ **15.** 5; 2; 5th; 32

[12.4] **16.** $2\log_3 x + \log_3 y$ **17.** $\dfrac{1}{2}\log_5 x - \log_5 y - \log_5 z$ **18.** $\log_b \dfrac{s^3}{t}$ **19.** $\log_b \dfrac{r^{1/4}s^2}{t^{2/3}}$

[12.5] **20. (a)** 1.3636 **(b)** $-.1985$ **21. (a)** $\dfrac{\log 19}{\log 3}$ **(b)** $\dfrac{\ln 19}{\ln 3}$ **(c)** 2.6801

[12.6] **22.** $\{3.9656\}$ **23.** $\{3\}$ **24.** $12,507.51 **25. (a)** $19,260.38 **(b)** approximately 13.9 years

CUMULATIVE REVIEW EXERCISES CHAPTERS 1–12 (PAGE 813)

[1.4] **1.** $-2, 0, 6, \dfrac{30}{3}$ (or 10) **2.** $-\dfrac{9}{4}, -2, 0, .6, 6, \dfrac{30}{3}$ (or 10) **3.** $-\sqrt{2}, \sqrt{11}$ **4.** $-\dfrac{9}{4}, -2, -\sqrt{2}, 0, .6, \sqrt{11}, 6,$

$\dfrac{30}{3}$ (or 10) [1.4–1.6] **5.** 16 **6.** -27 **7.** -39 [2.2] **8.** $\left\{-\dfrac{2}{3}\right\}$ [3.1] **9.** $[1, \infty)$ [3.3] **10.** $\{-2, 7\}$

11. $\left\{\pm\dfrac{16}{3}\right\}$ **12.** $\left[\dfrac{7}{3}, 3\right]$ **13.** $(-\infty, -3) \cup (2, \infty)$ [4.2] **14.**  [8.3] **15.**

[5.3] **16.** $6p^2 + 7p - 3$ [5.4] **17.** $16k^2 - 24k + 9$ [5.1] **18.** $-5m^3 + 2m^2 - 7m + 4$

[5.6] **19.** $2t^3 + 5t^2 - 3t + 4$ [6.1–6.4] **20.** $x(8 + x^2)$ **21.** $(3y - 2)(8y + 3)$ **22.** $z(5z + 1)(z - 4)$

[5.6] **23.** $(4a + 5b^2)(4a - 5b^2)$ **24.** $(2c + d)(4c^2 - 2cd + d^2)$ **25.** $(4r + 7q)^2$ [5.2, 5.5] **26.** $-\dfrac{1875p^{13}}{8}$

[7.2] **27.** $\dfrac{x + 5}{x + 4}$ [7.4] **28.** $\dfrac{-3k - 19}{(k + 3)(k - 2)}$ **29.** $\dfrac{22 - p}{p(p - 4)(p + 2)}$ [8.1] **30.** $-5{,}587{,}500$ [8.2] **31.** $3x - 4y = 19$

[9.1–9.3] **32.** $\{(4, 2)\}$ [9.4] **33.** $\{(1, -1, 4)\}$ [9.7] **34.** -1 [9.5] **35.** 6 pounds [10.2] **36.** $12\sqrt{2}$

[10.3] **37.** $-27\sqrt{2}$ [10.6] **38.** $\{0, 4\}$ [11.4] **39.** 41 [11.3] **40.** $\left\{\dfrac{1 \pm \sqrt{13}}{6}\right\}$ [11.9] **41.** $(-\infty, -4) \cup (2, \infty)$

[11.5] **42.** $\{\pm 1, \pm 2\}$ [11.6] **43.** 150 and 150 [11.7] **44.** $f(x) = \frac{1}{3}(x - 1)^2 + 2$ [12.2] **45.**

[12.6] **46.** $\{-1\}$ [12.3] **47.** [12.4] **48.** 6.3398 **49.** $3 \log x + \dfrac{1}{2} \log y - \log z$

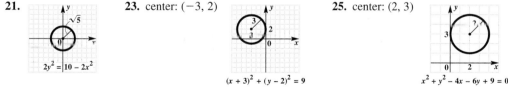

[12.6] **50.** **(a)** 25,000 **(b)** 30,500 **(c)** 37,300 **(d)** 68,000

CHAPTER 13 CONIC SECTIONS

SECTION 13.1 (PAGE 821)

CONNECTIONS **Page 821:** Answers will vary.

EXERCISES **1.** $(0, 0)$; 10 **3.** $(-3, -2)$ **5.** $x^2 + y^2 = 36$ **7.** $(x + 1)^2 + (y - 3)^2 = 16$ **9.** $x^2 + (y - 4)^2 - 3$
11. By the vertical line test the set is not a function, because a vertical line may intersect the graph of an ellipse in two points.
13. center: $(-2, -3)$; radius: 2 **15.** center: $(-5, 7)$; radius: 9 **17.** center: $(2, 4)$; radius: 4 **19.**

21. **23.** center: $(-3, 2)$ **25.** center: $(2, 3)$

27. The taut string represents the constant radius, and the point where it is fastened serves as the center. **29.** A circular racetrack is most appropriate because the crawfish can move in any direction. Distance from the center determines the winner.

31. **33.** **35.** **37.** **39.**

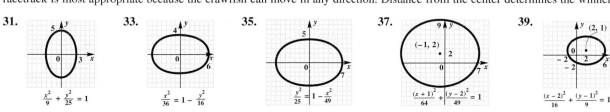

41. $y_1 = 4 + \sqrt{16 - (x + 2)^2}$, $y_2 = 4 - \sqrt{16 - (x + 2)^2}$ **43.**

45.

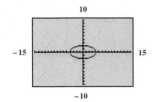

47. (a) 154.7 million miles **(b)** 128.7 million miles (Answers are rounded.)

49. $3\sqrt{3}$ units

SECTION 13.2 (PAGE 831)

CONNECTIONS **Page 828:** Answers will vary.

EXERCISES **1.** C **3.** D **5.** If the x^2 term is positive, the graph has only x-intercepts and opens left and right. If the y^2 term is positive, the graph has only y-intercepts and opens up and down. **7.** **9.**

11. 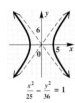 **13.** hyperbola **15.** ellipse **17.** circle

19. hyperbola **21.** hyperbola **23.** **25.**

27. $\dfrac{(x-2)^2}{4} - \dfrac{(y+1)^2}{9} = 1$ **29.** $\dfrac{y^2}{36} - \dfrac{(x-2)^2}{49} = 1$ **31. (a)** 10 meters **(b)** 36 meters **33.** $y = \pm x$; The lines $y = \pm x$ form

 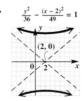

a 45° angle with the line through the goal posts. Most people can estimate a 45° angle fairly easily. **35. (a)** about 55 percent

(b) 53 percent   **37.** $y_1 = \sqrt{\dfrac{x^2}{9} - 1}$, $y_2 = -\sqrt{\dfrac{x^2}{9} - 1}$ **39.**

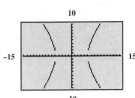

41.

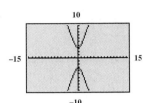

43. $y = \sqrt{\dfrac{x^2}{4} - 1}$ **44.** $y = \dfrac{1}{2}x$ **45.** $y \approx 24.98$ **46.** $y = 25$

47. Because $24.98 < 25$, the graph of $y = \sqrt{\dfrac{x^2}{4} - 1}$ lies below the graph of $y = \dfrac{1}{2}x$.   **48.** The y-values on the hyperbola will approach the y-values on the line, but will always be less.

SECTION 13.3 (PAGE 841)

EXERCISES   **3.** one   **5.** none   **7.**   **9.**   **11.**   **13.**

15.   **17.**   **19.** $\left\{(0, 0), \left(\dfrac{1}{2}, \dfrac{1}{2}\right)\right\}$ **21.** $\{(-6, 9), (-1, 4)\}$ **23.** $\left\{\left(-\dfrac{1}{5}, \dfrac{7}{5}\right), (1, -1)\right\}$

25. $\left\{(-2, -2), \left(-\dfrac{4}{3}, -3\right)\right\}$ **27.** $\{(-3, 1), (1, -3)\}$ **29.** $\left\{\left(-\dfrac{3}{2}, -\dfrac{9}{4}\right), (-2, 0)\right\}$ **31.** $\{(-\sqrt{3}, 0), (\sqrt{3}, 0),$ $(-\sqrt{5}, 2), (\sqrt{5}, 2)\}$ **33.** $\{(-2, 0), (2, 0)\}$ **35.** $\{(\sqrt{3}, 0), (-\sqrt{3}, 0)\}$ **37.** $\{(-2i\sqrt{2}, -2\sqrt{3}), (-2i\sqrt{2}, 2\sqrt{3}),$ $(2i\sqrt{2}, -2\sqrt{3}), (2i\sqrt{2}, 2\sqrt{3})\}$ **39.** $\{(-\sqrt{5}, -\sqrt{5}), (\sqrt{5}, \sqrt{5})\}$ **41.** $\{(i, 2i), (-i, -2i), (2, -1), (-2, 1)\}$
43. $(-1, 2)$ **45.** $(-4, 8)$ **47.** $\{(2, -3), (-3, 2)\}$

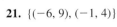
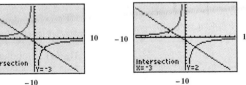

49. $\{-2, 5\}$ **50.**   **51.** $\{(-2, 4), (5, 25)\}$ **52.** They are the same.   **53.**

54. They are the same.   **55.** length: 12 feet; width: 7 feet   **57.** 1981; 467 thousand

SECTION 13.4 (PAGE 849)

EXERCISES **1.** (c) **5.** B **7.** A **9.** **11.** **13.**

15. **17.** **19.** **21.** **23.**

25. **27.** **29.** **31.** **33.**

35. 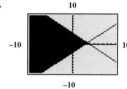 **37.** 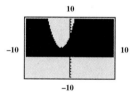 **39.** **41.** **43.** A **45.** B

47. **49.**

CHAPTER 13 REVIEW EXERCISES (PAGE 856)

1. $(x + 2)^2 + (y - 4)^2 = 9$ **3.** center: $(-3, 2)$; radius: 4 **5.** center: $(-1, -5)$; radius: 3 **7.**

9. **11.** **13.** **15.** hyperbola **17.** parabola

19. $\dfrac{x^2}{65{,}286{,}400} + \dfrac{y^2}{2{,}560{,}000} = 1$ **21.** $\{(6, -9), (-2, -5)\}$ **23.** $\{(4, 2), (-1, -3)\}$ **25.** $\{(-\sqrt{2}, 2), (-\sqrt{2}, -2),$
$(\sqrt{2}, -2), (\sqrt{2}, 2)\}$ **27.** 0, 1, or 2 **29.** The graph shows that in about 1950 the altitude was reduced to approximately 4005 feet above sea level, causing the salinity to increase to about 6000 milligrams per liter. **31.**

33.
$2x + 5y \le 10$
$3x - y \le 6$

35.
$9x^2 \le 4y^2 + 36$
$x^2 + y^2 \le 16$

37.
$\dfrac{y^2}{4} - 1 = \dfrac{x^2}{9}$

39.
$y = 2(x - 2)^2 - 3$

41.
$f(x) = \sqrt{4 - x}$

43.
$4y > 3x - 12$
$x^2 < 16 - y^2$

44. $2a + 4b + c = -20$ **45.** $5a + b + c = -26$ **46.** $-a + b + c = -2$

47. $\{(-4, -2, -4)\}; \ x^2 + y^2 - 4x - 2y - 4 = 0$ **48.** center: (2, 1); radius: 3   **49.**
$x^2 + y^2 - 4x - 2y - 4 = 0$

CHAPTER 13 TEST (PAGE 858)

[13.1] **1.** $(x + 4)^2 + (y - 4)^2 = 16$ **2.** center: $(-4, 1)$; radius: 5   [13.2] **3.** circle   **4.** ellipse   **5.** hyperbola
6. parabola   [13.1] **7.**
$x^2 + y^2 = 64$
8.
$4x^2 + 9y^2 = 36$
[13.2] **9.**
$16y^2 - 4x^2 = 64$
10.
$f(x) = \sqrt{16 - x^2}$

[13.3] **11.** 0, 1, 2, 3, or 4   **12.**   **13.** $\left\{\left(-\dfrac{1}{2}, -10\right), (5, 1)\right\}$ **14.** $\left\{(-2, -2), \left(\dfrac{14}{5}, -\dfrac{2}{5}\right)\right\}$

15. $\{(-\sqrt{22}, -\sqrt{3}), (-\sqrt{22}, \sqrt{3}), (\sqrt{22}, -\sqrt{3}), (\sqrt{22}, \sqrt{3})\}$ [13.4] **16.**
$y \le x^2 - 2$
17.
$2x - 5y \ge 12$
$3x + 4y \le 12$

18.
$x^2 + 25y^2 \le 25$
$x^2 + y^2 \le 9$
[13.1] **19.** 69.8 million kilometers   **20.** 46.0 million kilometers

CUMULATIVE REVIEW EXERCISES CHAPTERS 1–13 (PAGE 859)

[1.4, 1.5] **1.** -4 [2.2] **2.** $\left\{\dfrac{2}{3}\right\}$ [3.1] **3.** $\left(-\infty, \dfrac{3}{5}\right]$ [3.3] **4.** $\{-4, 4\}$ **5.** $(-\infty, -5) \cup (10, \infty)$ [4.3] **6.** $\dfrac{2}{3}$

[5.4] **7.** $25y^2 - 30y + 9$ [5.3] **8.** $12r^2 + 40r - 7$ [5.6] **9.** $4x^3 - 4x^2 + 3x + 5 + \dfrac{3}{2x + 1}$

[6.3] **10.** $(3x + 2)(4x - 5)$ **11.** $(2y^2 - 1)(y^2 + 3)$ [6.4] **12.** $(z^2 + 1)(z + 1)(z - 1)$ **13.** $(a - 3b)(a^2 + 3ab + 9b^2)$

[7.2] **14.** $\dfrac{40}{9}$ **15.** $\dfrac{y - 1}{y(y - 3)}$ [7.4] **16.** $\dfrac{3c + 5}{(c + 5)(c + 3)}$ **17.** $\dfrac{1}{p}$ [7.7] **18.** $\dfrac{6}{5}$ or $1\dfrac{1}{5}$ hours

[8.2] **19.** $3x + 2y = -13$ [9.1–9.3] **20.** $\{(3, -3)\}$ [9.4] **21.** $\{(4, 1, -2)\}$ [13.3] **22.** $\left\{(-1, 5), \left(\dfrac{5}{2}, -2\right)\right\}$

[5.5] **23.** $\dfrac{3}{4}$ **24.** $\dfrac{a^5}{4}$ [10.3] **25.** $2\sqrt[3]{2}$ [10.4] **26.** $\dfrac{3\sqrt{10}}{2}$ [11.4] **27.** $\dfrac{7}{5} + \dfrac{11}{5}i$ [10.6] **28.** $\emptyset$

[6.5] **29.** $\left\{\dfrac{1}{5}, -\dfrac{3}{2}\right\}$ [11.1–11.3] **30.** $\left\{\dfrac{1 - 2\sqrt{2}}{4}, \dfrac{1 + 2\sqrt{2}}{4}\right\}$ [12.6] **31.** $\{3\}$ [11.5] **32.** $\left\{-\dfrac{\sqrt{6}}{2}, \dfrac{\sqrt{6}}{2}, -\sqrt{7}, \sqrt{7}\right\}$

[11.6] **33.** $v = \dfrac{\pm\sqrt{rFkw}}{kw}$ [4.2] **34.**   [11.8] **35.**   [12.2] **36.**

[13.2] **37.**   [13.4] **38.**   [9.5] **39.** **(a)** $61.7 = 625a + 25b$, $106 = 1225a + 35b$

(b) $a = .056$, $b = 1.067$ **(c)** $y = .056x^2 + 1.067x$ **(d)** 228.1 feet   [12.6] **40.** **(a)** \$16.9 billion   **(b)** \$75.8 billion

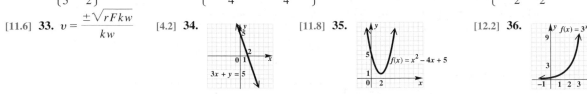

CHAPTER 14 SEQUENCES AND SERIES

SECTION 14.1 (PAGE 868)

CONNECTIONS **Page 868: 1.** After the first term, each term is the sum of the two preceding terms.   **2.** We show only four generations in this sketch.

```
M   1
|
F   1
/\
F   M   2
/\  |
F M F   3
```

EXERCISES **3.** $4, \dfrac{5}{2}, 2, \dfrac{7}{4}, \dfrac{8}{5}$ **5.** 3, 9, 27, 81, 243   **7.** $1, \dfrac{1}{4}, \dfrac{1}{9}, \dfrac{1}{16}, \dfrac{1}{25}$ **9.** $-1, 1, -1, 1, -1$ **11.** -70 **13.** $\dfrac{49}{23}$

15. 171   **17.** $4n$ **19.** $\dfrac{1}{3^n}$ **21.** \$110, \$109, \$108, \$107, \$106, \$105; \$400   **23.** \$6554   **25.** $3 + 6 + 11 = 20$

27. $\dfrac{1}{2} + \dfrac{1}{3} + \dfrac{1}{4} + \dfrac{1}{5} = \dfrac{77}{60}$ **29.** $-1 + 1 - 1 + 1 - 1 + 1 = 0$ **31.** $0 + 6 + 14 + 24 + 36 = 80$

33. $2x + 4x + 6x + 8x + 10x$ **35.** $x + 2x^2 + 3x^3 + 4x^4 + 5x^5$

Answers may vary for Exercises 37 and 39.

37. $\displaystyle\sum_{i=1}^{5}(i + 2)$ **39.** $\displaystyle\sum_{i=1}^{5}\dfrac{1}{i + 1}$ **41.** A sequence is a list of terms in a specific order, while a series is the indicated sum of

the terms of a sequence.   **43.** $\dfrac{59}{7}$ **45.** 5   **47.** 4591   **49.** $\displaystyle\sum_{i=1}^{6}i^2 + \sum_{i=1}^{6}3i + \sum_{i=1}^{6}5$ **50.** $3\displaystyle\sum_{i=1}^{6}i$ **51.** $6 \cdot 5 = 30$

52. $\displaystyle\sum_{i=1}^{n}i = \dfrac{n(n + 1)}{2}$ **53.** $\displaystyle\sum_{i=1}^{n}i^2 = \dfrac{n(n + 1)(2n + 1)}{6}$ **54.** $91 + 63 + 30 = 184$ **55.** 572   **56.** -2620

SECTION 14.2 (PAGE 876)

EXERCISES **3.** $d = 1$ **5.** not arithmetic **7.** $d = 5$ **9.** $d = 2.15$ **11.** $d = \dfrac{2}{3}$ **13.** $a_n = 5n - 3$

15. $a_n = \dfrac{3}{4}n + \dfrac{9}{4}$ **17.** $a_n = 3n - 6$ **19.** 76 **21.** 48 **23.** -1 **25.** 16 **27.** 6 **29.** n represents the number of terms. **31.** $2S$; 101; 100; 10,100 **32.** $2S = 10,100$ **33.** 5050 **34.** 20,100 **35.** 81 **37.** -3 **39.** 87 **41.** 390 **43.** 320 **45.** 31,375 **47.** \$465 **49.** \$2100 per month **51.** 68; 1100 **53.** no; 3; 9 **55.** $m + b, 2m + b, 3m + b$ **56.** yes **57.** m **58.** $a_n = mn + b$

SECTION 14.3 (PAGE 886)

EXERCISES **3.** $r = 2$ **5.** not geometric **7.** $r = -3$ **9.** $r = -\dfrac{1}{2}$ **11.** $a_n = 5(2)^{n-1}$ **13.** $a_n = \dfrac{3^{n-1}}{9}$

15. $a_n = 10\left(-\dfrac{1}{5}\right)^{n-1}$ **17.** $2(5)^9$ **19.** $\dfrac{1}{2}\left(\dfrac{1}{3}\right)^{11}$ **21.** $2\left(\dfrac{1}{2}\right)^{24} = \dfrac{1}{2^{23}}$ **23.** .33333 . . . **24.** .66666 . . .

25. .99999 . . . **26.** $\dfrac{a_1}{1 - r} = \dfrac{.9}{1 - .1} = \dfrac{.9}{.9} = 1$; Therefore, .99999 . . . = 1. **27.** $\dfrac{121}{243}$ **29.** -1.997 **31.** 2.662

33. -2.982 **35.** \$66,988.91 **37.** \$130,159.72 **39.** 9 **41.** $\dfrac{10,000}{11}$ **43.** $-\dfrac{9}{20}$ **45.** does not exist

47. $10\left(\dfrac{3}{5}\right)^4 \approx 1.3$ feet **49.** 3 days; $\dfrac{1}{4}$ gram **51. (a)** $1.1(1.06)^5 \approx 1.5$ billion units **(b)** approximately 12 years

53. $\$50,000\left(\dfrac{3}{4}\right)^8 \approx \$5000$ **55.** ab, ab^2, ab^3 **56.** yes **57.** b **58.** $a_n = ab^n$

SECTION 14.4 (PAGE 894)

CONNECTIONS **Page 891:** **1.** 21 and 34 **2.** 15, 21, 28, 36, 45 **3.** Answers will vary.

EXERCISES **1.** 2 **3.** 15 **5.** 15 **7.** 1 **9.** 122 **11.** 120 **13.** $m^4 + 4m^3n + 6m^2n^2 + 4mn^3 + n^4$

15. $a^5 - 5a^4b + 10a^3b^2 - 10a^2b^3 + 5ab^4 - b^5$ **17.** $8x^3 + 36x^2 + 54x + 27$ **19.** $\dfrac{x^5}{243} + \dfrac{10x^4y}{81} + \dfrac{40x^3y^2}{27} +$

$\dfrac{80x^2y^3}{9} + \dfrac{80xy^4}{3} + 32y^5$ **21.** $m^3x^3 - 3m^2n^2x^2 + 3mn^4x - n^6$ **23.** $r^{12} + 24r^{11}s + 264r^{10}s^2 + 1760r^9s^3$

25. $3^{14}x^{14} - 14(3^{13})x^{13}y + 91(3^{12})x^{12}y^2 - 364(3^{11})x^{11}y^3$ **27.** $t^{20} + 10t^{18}u^2 + 45t^{16}u^4 + 120t^{14}u^6$ **29.** $120(2^7)m^7n^3$

31. $\dfrac{7x^2y^6}{16}$ **33.** $36k^7$ **35.** $-160x^6y^3$ **37.** $4320x^9y^4$

CHAPTER 14 REVIEW EXERCISES (PAGE 899)

1. $-1, 1, 3, 5$ **3.** 1, 4, 9, 16 **5.** 0, 3, 8, 15 **7.** $2x + 3x^2 + 4x^3 + 5x^4 + 6x^5 + 7x^6$ **9.** 126

11. 665,081 (thousand) **13.** arithmetic; $d = 4$ **15.** geometric; $r = -1$ **17.** geometric; $r = \dfrac{1}{2}$

19. 73 **21.** $a_n = -5n + 1$ **23.** 15 **25.** 152 **27.** $a_n = -1(4)^{n-1}$ **29.** $2(-3)^{10} = 118,098$ **31.** $\dfrac{341}{1024}$

33. 1 **35.** $32p^5 - 80p^4q + 80p^3q^2 - 40p^2q^3 + 10pq^4 - q^5$ **37.** $m^2 + 4m\sqrt{mn} + 6mn + 4n\sqrt{mn} + n^2$

39. $-18,400(3)^{22}k^3$ **41.** $a_{10} = 1536; S_{10} = 1023$ **43.** $a_{15} = 38; S_{10} = 95$ **45.** $a_n = 2(4)^{n-1}$

47. $a_n = -3n + 15$ **49.** \$24,898.41 **51.** $\dfrac{1}{128}$ **53.** No, the sum cannot be found, because $r = 2$, and this value of r does not satisfy $|r| < 1$.

CHAPTER 14 TEST   (PAGE 902)

[14.1] **1.** 0, 2, 0, 2, 0   [14.2] **2.** 4, 6, 8, 10, 12   [14.3] **3.** 48, 24, 12, 6, 3   [14.2] **4.** 0   [14.3] **5.** $\dfrac{64}{3}$ or $-\dfrac{64}{3}$

[14.2] **6.** 75   [14.3] **7.** 124 or 44   [14.1] **8.** 206.6 million   [14.3] **9.** \$137,925.91   **10.** It has a sum if $|r| < 1.$

[14.2] **11.** 70   **12.** 33   **13.** 125,250   [14.3] **14.** 42   **15.** $\dfrac{1}{3}$ **16.** The sum does not exist.   [14.1] **17.** \$324

[14.3] **18.** $20(3^{11}) = 3{,}542{,}940$ [14.4] **19.** $\dfrac{14{,}080x^8y^4}{9}$ **20.** $81k^4 - 540k^3 + 1350k^2 - 1500k + 625$

CUMULATIVE REVIEW EXERCISES CHAPTERS 1–14   (PAGE 903)

[1.4] **1.** $10, 0, \dfrac{45}{15}$ (or 3), -3 **2.** $-\dfrac{8}{3}, 10, 0, \dfrac{45}{15}$ (or 3), .82, -3 **3.** $\sqrt{13}, -\sqrt{3}$ **4.** all except $\sqrt{-7}$

[1.4–1.6] **5.** 8   **6.** -35 **7.** -55 [2.2] **8.** $\left\{\dfrac{1}{6}\right\}$ [3.1] **9.** $[10, \infty)$ [3.3] **10.** $\left\{-\dfrac{9}{2}, 6\right\}$ [2.5] **11.** $\{9\}$

[3.2] **12.** $(-\infty, -3) \cup (4, \infty)$ [3.3] **13.** $(-\infty, -3] \cup [8, \infty)$ [5.3] **14.** $20p^2 - 2p - 6$ [5.4] **15.** $9k^2 - 42k + 49$

[5.1] **16.** $-5m^3 - 3m^2 + 3m + 8$ [5.6] **17.** $2t^3 + 3t^2 - 4t + 2 + \dfrac{3}{3t - 2}$ [6.1] **18.** $x(7 + x^2)$

[6.2, 6.3] **19.** $(7y - 4)(2y + 3)$ **20.** $z(3z + 4)(2z - 1)$ [6.4] **21.** $(7a^2 + 3b)(7a^2 - 3b)$

22. $(c + 3d)(c^2 - 3cd + 9d^2)$ **23.** $(8r + 3q)^2$ [6.5] **24.** $\left\{-\dfrac{5}{2}, 2\right\}$ [11.9] **25.** $[-2, 3]$ [5.2, 5.5] **26.** $\dfrac{9}{4}$

27. $-\dfrac{27p^2}{10}$ [7.1] **28.** $-9, 9$ [7.2] **29.** $\dfrac{x + 7}{x - 2}$ [7.4] **30.** $\dfrac{3p - 26}{p(p + 3)(p - 4)}$ [7.6] **31.** $\varnothing$

[11.3] **32.** $\left\{\dfrac{-5 + \sqrt{217}}{12}, \dfrac{-5 - \sqrt{217}}{12}\right\}$ [10.6] **33.** $\{1, 2\}$ [8.4] **34.** \$275   [4.3, 8.1] **35.** $\dfrac{3}{4}$

[8.2] **36.** $3x + y = 4$ [4.3, 8.1] **37.**   [8.3] **38.**   [9.1–9.3] **39.** $\{(3, -5)\}$

[9.4] **40.** $\{(2, 1, 4)\}$ [9.7] **41.** -15 **42.** 7   [9.5] **43.** 2 pounds   [10.3] **44.** $10\sqrt{2}$ [11.4] **45.** 73

[12.1] **46.** $f^{-1}(x) = \dfrac{x - 5}{9}$ [12.2] **47.**   **48.** $\left\{\dfrac{5}{2}\right\}$ [12.3] **49.**   [12.6] **50.** $\{2\}$

[11.7] **51.**   [13.1] **52.**   [13.2] **53.**

[13.1] **54.** $(x + 5)^2 + (y - 12)^2 = 81$ [14.1] **55.** $-7, -2, 3, 8, 13$ [14.2] **56.** 78   [14.3] **57.** $\dfrac{75}{7}$ [14.2] **58.** 30

[14.4] **59.** $32a^5 - 80a^4 + 80a^3 - 40a^2 + 10a - 1$ **60.** $-\dfrac{45x^8y^6}{4}$

APPENDIX A   (PAGE 907)

1. 117.385   **3.** 13.21   **5.** 150.49   **7.** 96.101   **9.** 4.849   **11.** 166.32   **13.** 164.19   **15.** 1.344
17. 4.14   **19.** 2.23   **21.** 4800   **23.** .53   **25.** 1.29   **27.** .96   **29.** .009   **31.** 80%   **33.** .7%
35. 67%   **37.** 12.5%   **39.** 109.2   **41.** 238.92   **43.** 5%   **45.** 110%   **47.** 25%   **49.** 148.44
51. 7839.26   **53.** 7.39%   **55.** $2760   **57.** 805 miles   **59.** 63 miles   **61.** $2400

APPENDIX B   (PAGE 912)

1. {1, 2, 3, 4, 5, 6, 7}   **3.** {winter, spring, summer, fall}   **5.** $\emptyset$ **7.** {L}   **9.** {2, 4, 6, 8, 10, . . . }
11. The sets in Exercises 9 and 10 are infinite sets.   **13.** true   **15.** false   **17.** true   **19.** true   **21.** true
23. true   **25.** true   **27.** false   **29.** true   **31.** true   **33.** false   **35.** true   **37.** true   **39.** false
41. true   **43.** false   **45.** {g, h}   **47.** {b, c, d, e, g, h}   **49.** {a, c, e} = B **51.** {d}   **53.** {a}
55. {a, c, d, e}   **57.** {a, c, e, f}   **59.** $\emptyset$ **61.** B and D; C and D

APPENDIX C   (PAGE 916)

1. inverse   **3.** direct   **5.** joint   **7.** combined   **9.** 36   **11.** .625   **13.** $222\frac{2}{9}$ **15.** 800 gallons

17. 256 feet   **19.** 100 cycles per second   **21.** 1.105 liters   **23.** $\frac{8}{9}$ ton   **25.** 480 kilograms   **27.** 27

APPENDIX D   (PAGE 922)

1. Synthetic division provides a quick, easy way to divide a polynomial by a binomial of the form $x - k$.
3. $x - 5$ **5.** $4m - 1$ **7.** $2a + 4 + \dfrac{5}{a + 2}$ **9.** $p - 4 + \dfrac{9}{p + 1}$ **11.** $4a^2 + a + 3$

13. $x^4 + 2x^3 + 2x^2 + 7x + 10 + \dfrac{18}{x - 2}$ **15.** $-4r^5 - 7r^4 - 10r^3 - 5r^2 - 11r - 8 + \dfrac{-5}{r - 1}$

17. $-3y^4 + 8y^3 - 21y^2 + 36y - 72 + \dfrac{143}{y + 2}$ **19.** $y^2 + y + 1 + \dfrac{2}{y - 1}$ **21.** 7   **23.** -2 **25.** 0
27. By the remainder theorem, a zero remainder means that $P(k) = 0$; that is, k is a number that makes $P(x) = 0$.
29. yes   **31.** no   **33.** no   **35.** yes

APPENDIX E   (PAGE 927)

1. $\dfrac{1}{a^2 b}$ **3.** $\dfrac{100y^{10}}{x^2}$ **5.** 0   **7.** $\dfrac{x^{10}}{2w^{13}y^5}$ **9.** $\dfrac{a^{15}}{-64b^{15}}$ **11.** $\dfrac{x^{16}z^{10}}{y^6}$ **13.** $-6a^4 + 11a^3 - 20a^2 + 26a - 15$
15. $8x^3 - 18x^2 + 6x - 16$ **17.** $x^2 y - xy^2 + 6y^3$ **19.** $-3x^2 - 62x + 32$ **21.** $10x^3 - 4x^2 + 9x - 4$
23. $6x^2 - 19x - 7$ **25.** $4x^2 - 9x + 2$ **27.** $16t^2 - 9$ **29.** $4y^4 - 16$ **31.** $16x^2 - 24x + 9$
33. $36r^2 + 60ry + 25y^2$ **35.** $c^3 + 8d^3$ **37.** $64x^3 - 1$ **39.** $14t^3 + 45st^2 + 18s^2 t - 5s^3$
41. $4xy^3(2x^2 y + 3x + 9y)$ **43.** $(x + 3)(x - 5)$ **45.** $(2x + 3)(x - 6)$ **47.** $(6t + 5)(6t - 5)$ **49.** $(4t + 3)^2$
51. $p(2m - 3n)^2$ **53.** $(x + 1)(x^2 - x + 1)$ **55.** $(2t + 5)(4t^2 - 10t + 25)$ **57.** $(t^2 - 5)(t^4 + 5t^2 + 25)$
59. $(5x + 2y)(t + 3r)$ **61.** $(6r - 5s)(a + 2b)$ **63.** $(t^2 + 1)(t + 1)(t - 1)$ **65.** $(2x + 3y - 1)(2x + 3y + 1)$

Index

of an ellipse, 819
equivalent, 89
exponential, 761, 794
extraneous solution of, 630
Fahrenheit to Celsius, 476
first-degree, 448
graph of, 448
of a horizontal line, 215, 449, 466
of a hyperbola, 826
independent, 512
of an inverse function, 751
of a line, 213
linear in one variable, 89
linear in two variables, 201, 448, 463
linear system of, 533
logarithmic, 794
with no solution, 630
nonlinear, 835
of a parabola, 701
quadratic, 253, 312, 342, 653, 654
quadratic in form, 685
with radicals, 629
with rational expressions, 409
solution of, 23, 89
solution set of, 89
square root property of, 654
of a vertical line, 216, 450, 466
from word statements, 56
Equilibrium demand, 517
Equilibrium supply, 517
Equivalent equations, 89
Equivalent inequalities, 164
Euler, Leonhard, 786
Evaluating
determinants, 566
logarithms, 784
Even consecutive integers, 110
Expansion of a determinant, 567
Exponential equations, 761, 794
applications, of, 797
calculator graphing method for
solving, 799
properties for solving, 761, 794
solving, 761, 794
Exponential expressions, 14, 258
base of, 14, 258
evaluating, 14
exponent of, 258
Exponential functions, 747, 758
applications of, 762
graph of, 758
from logarithmic functions, 769
Exponents, 14, 258
application of, 294

fractional, 638
integer, 278
negative, 278
power rules for, 260, 261
product rule for, 259
quotient rule for, 281
review of, 924
and scientific notation, 294
summary of rules for, 282
zero, 248, 277
Expressions
algebraic, 21
distinguishing from equations, 25, 57, 409
exponential, 14, 258
mathematical, 18
radical, 595
rational, 371
simplifying, 69, 72
terms of, 248
Extraneous solution, 630

F

Factorial notation, 891
Factoring, 314
difference of two cubes, 335
difference of two squares, 332
by grouping, 316
perfect square trinomials, 334
review of, 926
sum of two cubes, 337
trinomials, 320, 325
Factoring method for solving quadratic
equations, 342, 654
Factoring trinomials, 320
by grouping, 325
in two variables, 329
using FOIL, 326
Factors, 2, 313
common, 313
greatest common, 3, 313
of integers, 50, 313
of a number, 2, 313
prime, 2
Fahrenheit-Celsius relationship, 476
Fibonnaci, 606, 868
Fibonacci sequence, 868, 890
Finite sequence, 864
application of, 865
Finite set, 909
First-degree equations, 448. *See also*
Linear equations
Foci
of an ellipse, 818

of a hyperbola, 824
Focus of a parabola, 709
FOIL method, 268, 320
inner product of, 268
outer product of, 268
Fourth root, 601
Fraction bar, 2
Fractional exponents, 638
definition of, 638
rules for, 640
Fractions, 2
addition of, 5
basic principle of, 3
complex, 401
continued, 406
division of, 5
least common denominator of, 388
linear equations with, 100
lowest terms of, 3
mixed numbers, 6
multiplication of, 4
reciprocals of, 4, 51
subtraction of, 7
Froude, William, 700
Froude number, 700
Function(s), 219, 347, 485
composition of, 491
constant, 492
definitions of, 490
domain of, 487
exponential, 747, 758
identity, 488
inverse of, 748, 751
linear, 219, 492
logarithmic, 770
notation for, 490
one-to-one, 748
polynomial, 247
quadratic, 701
range of, 487
square root, 830
vertical line test for, 488
Fundamental property of rational
expressions, 374
Fundamental rectangle of a hyperbola, 826
Future value of an ordinary annuity, 882
$f(x)$ notation, 490

G

Galileo, 346
General binomial expansion, 893

Index of Applications

Producer Price Index, 36, 43
Sales tax, 132, 156
Supply, 157, 517, 584, 585, 700

EDUCATION

Bachelor's degrees in the U.S., 790, 844
Funding for Head Start programs, 115
Master's degrees in business management, 222
Population of a school, 901
Public school enrollment, 234
Students per computer, 234
Test scores, 170, 172, 194, 196, 197

ENVIRONMENT

Acres of Texas farmland, 299
Atmospheric pressure, 762
Barometric pressure, 771
Carbon dioxide emissions, 762, 764
Carbon monoxide emissions, 754
Classifying wetlands based on pH values, 785, 790
Consumption of electricity, 889
Depth of trenches, 46
Elevation, 35, 48
Global warming, 764
Greenhouse effect, 787, 790, 801
Gross waste generated by paper products, 765
Gross waste generated in plastics, 747, 808
Hazardous waste sites, 753, 775
Height of mountains, 35, 46
Height of water falls, 154
Insecticide ingredients, 114
Land area, 154, 158
Municipal solid waste recovered, 809
Natural gas consumption, 775
Particle pollution, 807
Pollutant Standard Index, 749
Richter scale rating of an earthquake, 772
Solid waste generated in the U.S., 767
Sunniest U.S. cities, 508
Temperature, 43, 47, 48, 82, 84
Tornado sightings, 171
Toxic release inventory, 812
Volcano, 245
Wind as a source of renewable energy, 917
Wind-chill factor, 35

FINANCE

Annual interest, 146, 908
Baby boomers' inheritance, 245
Borrowing money and paying it back, 865, 868
Checkbook balance, 82
Compound interest, 266, 797, 801, 809, 812, 887, 888, 896, 902
Continuous compound interest, 798, 801, 809, 812
Credit card account, 48
Debt, 82, 143

Deposits, 551, 872, 877, 887
Future value of an annuity, 882, 883, 901, 902
Interest earned, 136, 143
Interest rate, 135, 659, 700
Investment, 137, 145, 146, 147, 149, 153, 156, 159, 197, 310, 550, 551, 628, 809, 908
Monetary value, 143
Mutual funds, 862, 869, 870
Retirement account, 887, 888
Savings, 908
Simple interest rate, 551
Sponsorship spending, 241

GEOMETRY

Angle measurement, 113, 114, 119, 155, 508, 553, 587, 591
Area of a circular region, 123, 124, 155
Area of a rectangular region, 124, 293
Area of a trapezoid, 124
Area of a triangle, 135
Base and height of a triangular object, 118, 355, 367, 664
Circumference of a circle, 155, 159, 160
Circumference of a dome, 124
Complementary and supplementary angles, 110, 115, 154, 158
Diagonal of a rectangle, 604
Diameter of a circle, 155
Dimensions of a box-shaped object, 123, 355, 364, 700
Dimensions of a cube, 612
Dimensions of a rectangular object, 194, 272, 349, 350, 355, 364, 367, 432, 444, 541, 548, 664, 699, 739, 745, 844
Dimensions of a square, 365
Fencing and laying sod, 180, 272
Height of a parallelogram, 441
Height of a trapezoid, 158
Length of a side of a square, 158, 548, 674
Maximum area of a rectangular region, 715, 719, 720, 741
Maximum volume, 123
Perimeter of a rectangular region, 124, 272
Perimeter of a square, 135
Radius of a circle, 155, 156, 658
Radius of a cone, 245
Radius of a cylinder, 628
Radius of a sphere, 117, 612, 658
Right triangles, 355, 356, 363, 364, 370, 737
Sides of a triangle, 157, 311, 508, 553, 699, 739
Similar triangles, 133
Strip around a rectangular object, 365, 695, 699, 739, 743
Volume of a box, 123, 124, 359, 651
Volume of a pyramid, 357

GOVERNMENT

Coded messages, 757
Exports and imports, 246, 369, 486, 617, 648

Income taxes, 508
Long-term debt, 813
NASA's budget for space station research, 250, 256
Nuclear power, 474, 484, 504
Number of Democrats vs. Republicans, 112, 154
Number of U.S. post offices, 496
Pollution abatement and control expenditures, 790, 801
Social security assets, 720
Total defense budget, 46
U.S. crude oil production, 455, 461
U.S. defense budget, 496
U.S. energy consumption, 446, 474, 484
U.S. energy production, 446
Votes to acquit William Jefferson Clinton, 591

HEALTH/LIFE SCIENCES

AIDS cases, 428, 709
Adult diet guidelines, 445
Body-mass index, 198, 915, 918
Cases served by the Child Support Enforcement program, 506
Child's vs. adult's drug dosage, 198
Death rate, 506
Diet, 197
Drug concentration in a person's system, 801
Drug solution, 145, 156
Fat grams needed daily, 434
Fitness program, 877
Health benefits, 592
Heart transplants, 371, 433, 442
Height of a woman, 223
Higher-order multiple births, 705
Languages, 787, 791, 810
Leading causes of death in the U.S., 421
Lithotripter, 824
Liver transplants, 433, 442
Male pattern baldness, 145
Male vs. female physicians, 428
Medicare enrollment, 428
National health care firms, 428
Pharmacist's prescriptions, 114
Target heart rate zone, 211
Tracer dye injected into a system, 888
Weight of a man, 223

LABOR

Construction industry employees, 82
Hourly earnings, 27, 60
Household income, 245, 474, 475, 503
Job categories for white employees, 12
Job categories for African Americans, 9
Job duties, 113
Number of workers on a shift, 480
Salary, 83, 869, 877
Time required to complete a job alone, 430, 441, 442, 665, 738
Union members, 60
Weekly earnings of full-time workers, 194

Formulas

Figure	*Formulas*	*Examples*
Square	Perimeter: $P = 4s$ Area: $A = s^2$	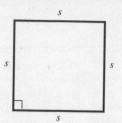
Rectangle	Perimeter: $P = 2L + 2W$ Area: $A = LW$	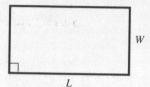
Triangle	Perimeter: $P = a + b + c$ Area: $A = \dfrac{1}{2}bh$	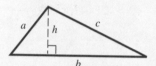
Pythagorean Formula (for right triangles)	In a right triangle with legs a and b and hypotenuse c, $c^2 = a^2 + b^2$.	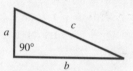
Sum of the Angles of a Triangle	$A + B + C = 180°$	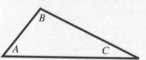
Circle	Diameter: $d = 2r$ Circumference: $C = 2\pi r$ $ C = \pi d$ Area: $A = \pi r^2$	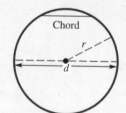
Parallelogram	Area: $A = bh$ Perimeter: $P = 2a + 2b$	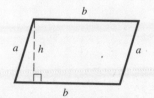
Trapezoid	Area: $A = \dfrac{1}{2}(B + b)h$ Perimeter: $P = a + b + c + B$	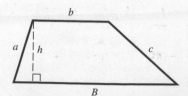